DISCOVER BIOLOGY

Sixth Edition

Anu Singh-Cundy
Western Washington University

Gary Shin
California State University, Long Beach

W. W. NORTON & COMPANY
NEW YORK • LONDON

W. W. Norton & Company has been independent since its founding in 1923, when William Warder Norton and Mary D. Herter Norton first published lectures delivered at the People's Institute, the adult education division of New York City's Cooper Union. The firm soon expanded its program beyond the Institute, publishing books by celebrated academics from America and abroad. By midcentury, the two major pillars of Norton's publishing program—trade books and college texts—were firmly established. In the 1950s, the Norton family transferred control of the company to its employees, and today—with a staff of four hundred and a comparable number of trade, college, and professional titles published each year—W. W. Norton & Company stands as the largest and oldest publishing house owned wholly by its employees.

Printed in the United States of America

Editor: Betsy Twitchell

Assistant Editor: Courtney Shaw

Project Editor: Christine D'Antonio

Developmental Editors: Jody Larson and Michael Zierler

Editorial Assistant: Katie Callahan

Manuscript Editor: Stephanie Hiebert

Managing Editor, College: Marian Johnson

Managing Editor, College Digital Media: Kim Yi

Associate Director of Production: Benjamin Reynolds

Media Editor: Robin Kimball

Associate Media Editors: Cailin Barrett-Bressack and Callinda Taylor

Media Project Editor: Kristin Sheerin

Marketing Manager: Meredith Leo

Design Director: Rubina Yeh

Photo Editor: Nelson Colón

Photo Researcher: Donna Ranieri

Permissions Manager: Megan Jackson

Permissions Clearer: Bethany Salminen

Composition: CodeMantra; Project Manager: Sofia Buono

Illustrations: Imagineering Media Services, Inc.; Project Manager: Wynne Au-Yeung

Manufacturing: LSC Comunications—Kendallville, IN

Permission to use copyrighted material is included in the credits section of this book, which begins on page CR1.

ISBN 978-0-393-93673-5

W. W. Norton & Company, Inc., 500 Fifth Avenue, New York, NY 10110-0017
wwnorton.com

W. W. Norton & Company Ltd., 15 Carlisle Street, London W1D 3BS
5 6 7 8 9 0

Preface

Using stem cells to repair or replace damaged organs. Discovering every last branch on the vast evolutionary tree of life. Seeking a cure for cancer. These are but a few of the many reasons why biology is a gripping subject for us and, we hope, for students using this book. These topics are simultaneously intensely interesting and critically important. Because the scientific understanding of fundamental biological principles is growing by leaps and bounds, this is an exciting time to write, teach, and learn about all areas of the biological sciences.

But the very things that make biology so interesting—the rapid pace of new discoveries and the many applications of these discoveries by human societies—can make it a difficult subject to teach and to learn. The problem is only made worse by the wide variation in backgrounds and interests of the students who take this course. When we set out to write the Sixth Edition of *Discover Biology*, we asked ourselves, "How can we convey the excitement, breadth, and relevance of biology to this varied group of students without burying them in an avalanche of information?" Similarly, we asked ourselves, "How can we encourage the retention of core concepts in biology by facilitating active learning inside and outside of the classroom?"

We answered these questions in several ways. In considering which topics and details to include from the vast group of possibilities, we had the following goals:

- To pique students' interest with fascinating stories about life on Earth in all its grandeur
- To take the feedback of the extensive reviews we received for every chapter to present core biological concepts at just the right level of detail for nonscience students
- To pair the textbook with instructor and student resources that enrich the learning process through applications, activities, and assessment

With this edition, we also welcome new author Gary Shin. Gary teaches hundreds of students from very diverse backgrounds each term at California State University, Long Beach, and at community colleges in southern California. He is truly a teacher in the trenches with a keen eye for which topics most engage students and which ones they struggle with. In addition, Gary's background in evolutionary genetics and wildlife ecology has resulted in a fresh perspective on the Evolution and Ecology units (Units 3 and 4).

What's New in the Sixth Edition

NEW, MORE READABLE CHAPTERS. We recognize how important it is for any non-majors biology textbook to strike a balance between including enough detail to promote a true understanding of concepts and not overwhelming students with a deluge of information. Reviewers, as well as adopters of the previous edition, praise our chapters for clearly and accessibly guiding students toward a rich appreciation for biological concepts and processes. But we are not ones to rest on our laurels. The Sixth Edition narrative is more focused, more direct, and 10 percent shorter overall.

NEW "EXTREME BIOLOGY" EXAMPLES. The inspiration to create this new feature arose out of the popularity of the "Extreme Diversity" spreads in the Fifth Edition of our book. These spreads have been retained in Chapters 17–19 in the Sixth Edition (pages 386–387, 410–411, 418, 423, and 456–457). But now, "Extreme Biology" examples throughout every chapter highlight the amazing diversity of life on Earth by providing examples of extreme traits, adaptations, and behaviors. Each "Extreme Biology" example relates directly to the concept being covered in the text. For example:

- EXTREME RUNNING provides a memorable example of the meaning of calories: The longest ultramarathon was 350 miles, lasting 80 hours and 44 minutes. The runner burned a whopping 42,858 calories! (Chapter 5: Energy, Metabolism, and Enzymes)
- EXTREME BUZZ explains how banana molecules mimic an alarm pheromone that affects gene expression in bee brains. (Chapter 11: DNA and Genes)
- EXTREME DIMORPHISM uses elephant seals as an extraordinary example of this evolutionary concept. Bulls can grow up to 20 feet long and weigh 4 tons, while cows are typically only 10 feet long and weigh 1 ton. (Chapter 14: How Evolution Works)

NEW "BIOLOGY IN THE NEWS" ARTICLES. For each edition, we select an entirely new set of recent articles from the popular press and broadcast media for this unique feature. News stories for the Sixth Edition hail from diverse sources, including the *Los Angeles Times*, the *Globe and Mail*, CBS, National Public Radio, the *Huffington*

Post, and more. The most biologically relevant passages in each article are reproduced at the end of the chapter, along with a brief analysis and discussion questions geared specifically toward nonmajors biology students. For instructors who would like to assign this feature to their students outside of class, short quizzes for each article are included in Norton Smartwork and the accompanying coursepacks.

NEW ULTIMATE GUIDE TO TEACHING BIOLOGY. Inspired by efforts across the country to bring active, inquiry-based learning into the classroom, for the Sixth Edition we partnered with Julie Harless from Lone Star College to create this one-stop resource. *The Ultimate Guide* is a curated collection of rich in-class activities from dozens of biology instructors across the country. It provides instructors with group activities designed to be used in a variety of classroom sizes and setups, think-pair-share activities, video clip recommendations with discussion questions, clicker questions, sample syllabi, and sample lecture plans to ease the transition to a new book.

NEW ADAPTIVE SOFTWARE. InQuizitive recognizes the unique background of each student to offer an individualized learning path. By providing deeper instruction to students who need it, InQuizitive addresses the biomodality of students taking nonmajors biology and ensures that all students come to class prepared. InQuizitive includes a variety of question types that emphasize visual learning, gamelike elements that fuel student curiosity, and an easy-to-use interface that works on mobile devices.

Fascinating Stories

A story on a contemporary topic, from the possibility of arsenic-based life-forms to the reality of climate change, opens each chapter. The story is then revisited in more depth at the end of the chapter, showing students how a mastery of the chapter's content makes them more scientifically literate.

25 Global Change

TOP OF THE FOOD CHAIN, TOP OF THE WORLD. As global warming melts the Arctic ice cap, receding summer ice is increasingly leaving polar bears stranded on land or ice floes. Some bears are moving south into Canada and Alaska, but global warming may also affect the bears in a surprising way—by damaging marine food chains from the bottom up.

568

Is the Cupboard Bare?

Polar bears are in trouble. The big, white bears live by hunting seals on vast stretches of sea ice covering the Arctic Ocean. Yet each summer a larger area of the ice cap melts, leaving the bears stranded on land, often with hungry cubs. Until the ice returns in winter, the giant bears fast or eat garbage at Canadian and Russian garbage dumps. As the planet heats up, the ice melts earlier each spring and freezes later each fall, extending the bears' months of fasting by weeks. Unless they abandon the Arctic and begin migrating south into Canada and Siberia to compete with grizzly bears each summer, polar bears may go extinct.

Surprisingly, global warming could also wallop the bears from a completely different direction. In the summer of 2010, ecologists reported that global populations of tiny but critically important organisms called phytoplankton had declined by 40 percent since the 1950s. Like plants, phytoplankton live by photosynthesizing—that is, making sugar molecules from carbon dioxide and water using energy from the sun. On land, the big photosynthesizers are trees and other plants. But in the world's oceans, photosynthesis is carried out by phytoplankton—a mix of bacteria, algae, diatoms, and other protists.

Floating at the surface of millions of square miles of ocean, phytoplankton perform half of all the photosynthesis on the planet and supply half of all the new oxygen in Earth's atmosphere. Photosynthetic phytoplankton are the foundation for all the great marine food webs. Every marine animal—from tiny shrimplike krill to fish, seals, and polar bears—depends on the energy and building blocks stored in phytoplankton. A 40 percent decline in phytoplankton would lead to a drastic decline in krill and marine fish. Without fish, there would be no seals, to say nothing of polar bears, poised precariously at the top of the energy pyramid.

What could cause such a decline in phytoplankton? Could a shortage of phytoplankton lead to a collapse of marine food chains?

In 2010, researchers reported that phytoplankton had declined in 8 out of 10 regions of the ocean since 1950. In this chapter we'll learn how the expanding human population is dramatically affecting the biosphere.

ace. The consequences for humans

o the observed rise in average global emperature over the past century.

The warming of Earth's climate has led to melting of polar and glacial ice, rising sea evels, acidification of the oceans, and shifts n where many species live.

Climate models predict an increase in the requency of severe weather and changes n rainfall patterns.

Many species are likely to become extinct over the next 100 years.

569

APPLYING WHAT WE LEARNED

Bye-Bye, Food Chain?

Vast ocean populations of tiny phytoplankton are responsible for about half of all the photosynthesis on Earth. These tiny organisms supply all of the energy for marine food chains and about half of all newly generated oxygen. Every organism on Earth depends on the work of phytoplankton—not just polar bears and fish, but humans too. Every time we take a breath, some of the oxygen comes from phytoplankton. Earth's atmosphere of 20 percent oxygen comes largely from the work of photosynthetic ocean bacteria that lived 2.5 billion years ago. And much of the oil that fuels our vehicles and industry comes from phytoplankton that lived millions of years ago.

So, when researchers at Dalhousie University in Nova Scotia reported in 2010 that marine phytoplankton were in sharp decline, ecologists were shocked. Without phytoplankton, life in the oceans would die and atmospheric oxygen levels would begin to decline. Were things that bad?

Oceanographers have known for a long time that rising temperatures hurt phytoplankton. As ocean waters warm up, they become more "stratified," with warmer water sitting on top of colder water. These static layers of water don't turn over, so there's no upwelling of nutrients from the seafloor. And without nutrients, phytoplankton can't grow. Data collected from satellite images suggest that warming oceans have caused a 6 percent drop in phytoplankton numbers since the early 1980s.

How does global warming hurt phytoplankton? We know that burning fossil fuels adds carbon dioxide to the atmosphere and that a lot of the extra carbon dioxide is absorbed by the oceans. Since carbon dioxide is a building block used for making sugars during photosynthesis, you might think that more carbon dioxide would be good for phytoplankton. And in some parts of the ocean, populations of phytoplankton *are* increasing. Unfortunately, carbon dioxide dissolved in water also makes water acidic, and acidic water contains less iron, an essential nutrient for phytoplankton.

Different parts of the ocean are warming up faster or slower, and some parts are becoming more acidic while other parts are not. Oceanographers aren't sure where all this is going. To find out whether the changes they had observed were short-term (and perhaps not serious) or long-term (and serious), the Nova Scotia researchers looked at half a million measurements of ocean clarity and color collected since the late 1800s. In general, cloudier water and greener water contain more phytoplankton. After analyzing the data, these researchers concluded that phytoplankton had declined by 40 percent since 1950.

Not so fast, said researcher Mark Ohman, of the Scripps Institution of Oceanography. Ohman didn't object to the data, but he claimed that the other researchers' analysis contained errors. For example, those researchers had reported that 59 percent of local ocean measurements had shown declines in phytoplankton. But Ohman pointed out that only 38 percent of those declines were statistically significant. The researchers had also used measurements from only the top 20 meters of the ocean, but in some parts of the ocean, phytoplankton live much deeper than that.

Although ocean scientists agree that phytoplankton have declined in number, and that this decline is most likely related to global warming and ocean acidification, just how extensively phytoplankton populations have declined is, for now, unsettled. Marine fisheries *are* collapsing, but mostly because of overfishing. That said, we need to halt ongoing declines in phytoplankton. Otherwise, it will be bye-bye, marine food chain and bye-bye, polar bears.

584

Art That Promotes Understanding

A consistent system of banners, labels, and bubble captions is built into every figure, helping students to better navigate the information hierarchy. First, banners and part labels help students identify and understand the big picture of the concept being illustrated. Then, extensive bubble captions guide students through the figure's most important elements, helping them to develop a more complete understanding of the concept.

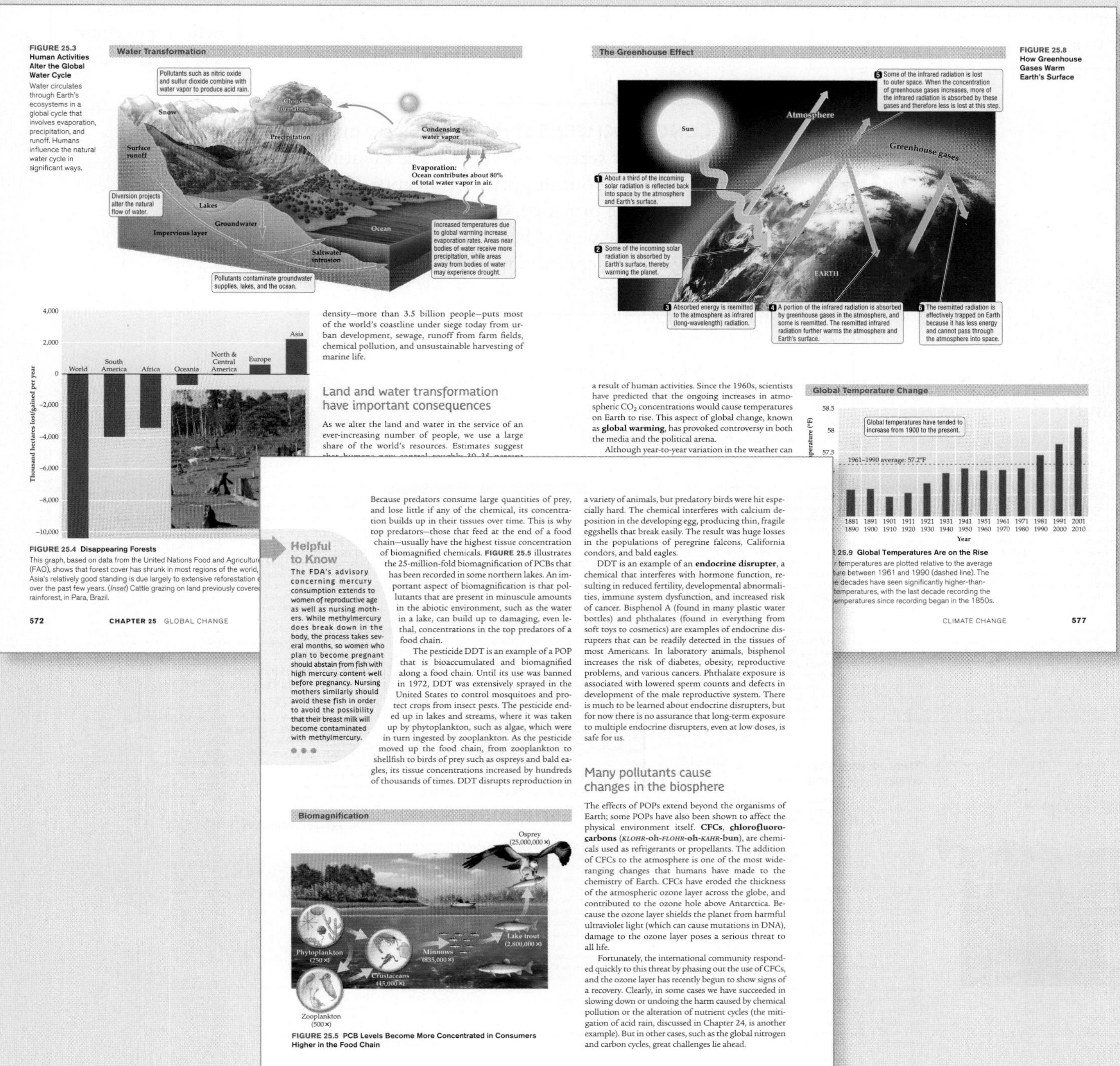

FIGURE 25.3 Human Activities Alter the Global Water Cycle Water circulates through Earth's ecosystems in a global cycle that involves evaporation, precipitation, and runoff. Humans influence the natural water cycle in significant ways.

Water Transformation

Pollutants such as nitric oxide and sulfur dioxide combine with water vapor to produce acid rain.
Cloud formation
Snow
Precipitation
Condensing water vapor
Surface runoff
Evaporation: Ocean contributes about 80% of total water vapor in air.
Diversion projects alter the natural flow of water.
Lakes
Groundwater
Impervious layer
Ocean
Saltwater intrusion
Increased temperatures due to global warming increase evaporation rates. Areas near bodies of water receive more precipitation, while areas away from bodies of water may experience drought.
Pollutants contaminate groundwater supplies, lakes, and the ocean.

Thousand hectares lost/gained per year
4,000
2,000
0
-2,000
-4,000
-6,000
-8,000
-10,000
World
South America
Africa
Oceania
North & Central America
Europe
Asia

FIGURE 25.4 Disappearing Forests This graph, based on data from the United Nations Food and Agriculture (FAO), shows that forest cover has shrunk in most regions of the world, Asia's relatively good standing is due largely to extensive reforestation over the past few years. (*Inset*) Cattle grazing on land previously covered rainforest, in Para, Brazil.

density—more than 3.5 billion people—puts most of the world's coastline under siege today from urban development, sewage, runoff from farm fields, chemical pollution, and unsustainable harvesting of marine life.

Land and water transformation have important consequences

As we alter the land and water in the service of an ever-increasing number of people, we use a large share of the world's resources. Estimates suggest

572 CHAPTER 25 GLOBAL CHANGE

The Greenhouse Effect

FIGURE 25.8 How Greenhouse Gases Warm Earth's Surface

5 Some of the infrared radiation is lost to outer space. When the concentration of greenhouse gases increases, more of the infrared radiation is absorbed by these gases and therefore less is lost at this step.
Sun
Atmosphere
Greenhouse gases
1 About a third of the incoming solar radiation is reflected back into space by the atmosphere and Earth's surface.
2 Some of the incoming solar radiation is absorbed by Earth's surface, thereby warming the planet.
EARTH
3 Absorbed energy is reemitted to the atmosphere as infrared (long-wavelength) radiation.
4 A portion of the infrared radiation is absorbed by greenhouse gases in the atmosphere, and some is reemitted. The reemitted infrared radiation further warms the atmosphere and Earth's surface.
6 The reemitted radiation is effectively trapped on Earth because it has less energy and cannot pass through the atmosphere into space.

a result of human activities. Since the 1960s, scientists have predicted that the ongoing increases in atmospheric CO_2 concentrations would cause temperatures on Earth to rise. This aspect of global change, known as **global warming**, has provoked controversy in both the media and the political arena.

Although year-to-year variation in the weather can

Global Temperature Change

58.5
58
57.5
Global temperatures have tended to increase from 1900 to the present.
1961–1990 average: 57.2°F
1881 1890
1891 1900
1901 1910
1911 1920
1921 1930
1931 1940
1941 1950
1951 1960
1961 1970
1971 1980
1981 1990
1991 2000
2001 2010
Year

25.9 Global Temperatures Are on the Rise
temperatures are plotted relative to the average ture between 1961 and 1990 (dashed line). The e decades have seen significantly higher-than- temperatures, with the last decade recording the emperatures since recording began in the 1850s.

CLIMATE CHANGE 577

Helpful to Know

The FDA's advisory concerning mercury consumption extends to women of reproductive age as well as nursing mothers. While methylmercury does break down in the body, the process takes several months, so women who plan to become pregnant should abstain from fish with high mercury content well before pregnancy. Nursing mothers similarly should avoid these fish in order to avoid the possibility that their breast milk will become contaminated with methylmercury.

Because predators consume large quantities of prey, and lose little if any of the chemical, its concentration builds up in their tissues over time. This is why top predators—those that feed at the end of a food chain—usually have the highest tissue concentration of biomagnified chemicals. FIGURE 25.5 illustrates the 25-million-fold biomagnification of PCBs that has been recorded in some northern lakes. An important aspect of biomagnification is that pollutants that are present in minuscule amounts in the abiotic environment, such as the water in a lake, can build up to damaging, even lethal, concentrations in the top predators of a food chain.

The pesticide DDT is an example of a POP that is bioaccumulated and biomagnified along a food chain. Until its use was banned in 1972, DDT was extensively sprayed in the United States to control mosquitoes and protect crops from insect pests. The pesticide ended up in lakes and streams, where it was taken up by phytoplankton, such as algae, which were in turn ingested by zooplankton. As the pesticide moved up the food chain, from zooplankton to shellfish to birds of prey such as ospreys and bald eagles, its tissue concentrations increased by hundreds of thousands of times. DDT disrupts reproduction in a variety of animals, but predatory birds were hit especially hard. The chemical interferes with calcium deposition in the developing egg, producing thin, fragile eggshells that break easily. The result was huge losses in the populations of peregrine falcons, California condors, and bald eagles.

DDT is an example of an **endocrine disrupter**, a chemical that interferes with hormone function, resulting in reduced fertility, developmental abnormalities, immune system dysfunction, and increased risk of cancer. Bisphenol A (found in many plastic water bottles) and phthalates (found in everything from soft toys to cosmetics) are examples of endocrine disrupters that can be readily detected in the tissues of most Americans. In laboratory animals, bisphenol increases the risk of diabetes, obesity, reproductive problems, and various cancers. Phthalate exposure is associated with lowered sperm counts and defects in development of the male reproductive system. There is much to be learned about endocrine disrupters, but for now there is no assurance that long-term exposure to multiple endocrine disrupters, even at low doses, is safe for us.

Biomagnification

Osprey (25,000,000×)
Lake trout (2,800,000×)
Phytoplankton (250×)
Minnows (835,000×)
Crustaceans (45,000×)
Zooplankton (500×)

FIGURE 25.5 PCB Levels Become More Concentrated in Consumers Higher in the Food Chain

Many pollutants cause changes in the biosphere

The effects of POPs extend beyond the organisms of Earth; some POPs have also been shown to affect the physical environment itself. **CFCs**, **chlorofluorocarbons** (*KLOHR*-**oh**-*FLOHR*-**oh**-*KAHR*-**bun**), are chemicals used as refrigerants or propellants. The addition of CFCs to the atmosphere is one of the most wide-ranging changes that humans have made to the chemistry of Earth. CFCs have eroded the thickness of the atmospheric ozone layer across the globe, and contributed to the ozone hole above Antarctica. Because the ozone layer shields the planet from harmful ultraviolet light (which can cause mutations in DNA), damage to the ozone layer poses a serious threat to all life.

Fortunately, the international community responded quickly to this threat by phasing out the use of CFCs, and the ozone layer has recently begun to show signs of a recovery. Clearly, in some cases we have succeeded in slowing down or undoing the harm caused by chemical pollution or the alteration of nutrient cycles (the mitigation of acid rain, discussed in Chapter 24, is another example). But in other cases, such as the global nitrogen and carbon cycles, great challenges lie ahead.

574 CHAPTER 25 GLOBAL CHANGE

Unique Pedagogy

Distinctive pedagogical features throughout each chapter promote long-term retention of key concepts and a deeper understanding of new terminology. In the Sixth Edition, every major chapter section includes a Concept Check with answers.

PRONUNCIATION GUIDES are provided alongside new and unfamiliar terms in the chapter text, with the expectation that if students can pronounce key terms, they will be more likely to speak up in class.

HELPFUL TO KNOW boxes demystify new concepts and terms, making it more likely that students will retain them for the test and beyond.

CONCEPT CHECKS at the end of every major chapter section ask students to identify and think about important concepts. The answers are provided, but upside down, striking the perfect balance between challenging students and making sure they get the answer.

Helpful to Know

The FDA's advisory concerning mercury consumption extends to women of reproductive age as well as nursing mothers. While methylmercury does break down in the body, the process takes several months, so women who plan to become pregnant should abstain from fish with high mercury content well before pregnancy. Nursing mothers similarly should avoid these fish in order to avoid the possibility that their breast milk will become contaminated with methylmercury.

Because predators consume large quantities of prey, and lose little if any of the chemical, its concentration builds up in their tissues over time. This is why top predators—those that feed at the end of a food chain—usually have the highest tissue concentration of biomagnified chemicals. **FIGURE 25.5** illustrates the 25-million-fold biomagnification of PCBs that has been recorded in some northern lakes. An important aspect of biomagnification is that pollutants that are present in minuscule amounts in the abiotic environment, such as the water in a lake, can build up to damaging, even lethal, concentrations in the top predators of a food chain.

The pesticide DDT is an example of a POP that is bioaccumulated and biomagnified along a food chain. Until its use was banned in 1972, DDT was extensively sprayed in the United States to control mosquitoes and protect crops from insect pests. The pesticide ended up in lakes and streams, where it was taken up by phytoplankton, such as algae, which were in turn ingested by zooplankton. As the pesticide moved up the food chain, from zooplankton to shellfish to birds of prey such as ospreys and bald eagles, its tissue concentrations increased by hundreds of thousands of times. DDT disrupts reproduction in a variety of animals, but predatory birds were hit especially hard. The chemical interferes with calcium deposition in the developing egg, producing thin, fragile eggshells that break easily. The result was huge losses in the populations of peregrine falcons, California condors, and bald eagles.

DDT is an example of an **endocrine disrupter**, a chemical that interferes with hormone function, resulting in reduced fertility, developmental abnormalities, immune system dysfunction, and increased risk of cancer. Bisphenol A (found in many plastic water bottles) and phthalates (found in everything from soft toys to cosmetics) are examples of endocrine disrupters that can be readily detected in the tissues of most Americans. In laboratory animals, bisphenol increases the risk of diabetes, obesity, reproductive problems, and various cancers. Phthalate exposure is associated with lowered sperm counts and defects in development of the male reproductive system. There is much to be learned about endocrine disrupters, but for now there is no assurance that long-term exposure to multiple endocrine disrupters, even at low doses, is safe for us.

Biomagnification

FIGURE 25.5 PCB Levels Become More Concentrated in Consumers Higher in the Food Chain

Many pollutants cause changes in the biosphere

The effects of POPs extend beyond the organisms of Earth; some POPs have also been shown to affect the physical environment itself. **CFCs, chlorofluorocarbons** (*KLOHR*-**oh**-*FLOHR*-**oh**-*KAHR*-**bun**), are chemicals used as refrigerants or propellants. The addition of CFCs to the atmosphere is one of the most wide-ranging changes that humans have made to the chemistry of Earth. CFCs have eroded the thickness of the atmospheric ozone layer across the globe, and contributed to the ozone hole above Antarctica. Because the ozone layer shields the planet from harmful ultraviolet light (which can cause mutations in DNA), damage to the ozone layer poses a serious threat to all life.

Fortunately, the international community responded quickly to this threat by phasing out the use of CFCs, and the ozone layer has recently begun to show signs of a recovery. Clearly, in some cases we have succeeded in slowing down or undoing the harm caused by chemical pollution or the alteration of nutrient cycles (the mitigation of acid rain, discussed in Chapter 24, is another example). But in other cases, such as the global nitrogen and carbon cycles, great challenges lie ahead.

574 CHAPTER 25 GLOBAL CHANGE

Concept Check

1. Compare and contrast bioaccumulation and biomagnification. What are some distinctive characteristics of chemicals that tend to bioaccumulate?
2. In a food chain, which organisms are most affected by biomagnification?

25.3 Human Impacts on the Global Carbon Cycle

Nearly all of us have had a hand in changing the world's nutrient cycles, at least a tiny bit. We affect nutrient cycles when we sprinkle fertilizer on our lawns and gardens, and when we send our waste to landfills, sewage plants, or septic tanks. The cheap and abundant food that people in rich countries take for granted comes for the most part from intensive farming, with its heavy input of fertilizer and energy from nonrenewable sources. We add huge amounts of carbon dioxide, nitrogen, phosphorus, and sulfur to our environment. Of particular concern in the context of climate change is our disruption of the global carbon cycle.

Atmospheric carbon dioxide levels have risen dramatically

Although CO_2 makes up less than 0.04 percent of Earth's atmosphere, it is far more important than its low concentration might suggest. As we saw in earlier chapters, CO_2 is an essential raw material for photosynthesis, on which most life depends. CO_2 is also the most important of the atmospheric gases that contribute to global warming. Therefore, scientists took notice in the early 1960s when new measurements showed that the concentration of CO_2 in the atmosphere was rising rapidly.

Scientists have been measuring the concentration of CO_2 in the atmosphere since 1958. By also measuring CO_2 concentrations in air bubbles trapped in ice, scientists have been able to estimate the concentration of CO_2 in the atmosphere over the last several hundred thousand years (**FIGURE 25.6**). Both types of measurements show that CO_2 levels have risen dramatically during the past two centuries. Overall, of the current yearly increase in atmospheric CO_2 levels, about 75 percent is due to the burning of fossil fuels. Logging and burning of forests are responsible for most of the remaining 25 percent, but industrial processes such as cement manufacturing also make a significant contribution.

The recent increase in CO_2 levels is striking for two reasons. First, the increase happened quickly: the concentration of CO_2 increased from 280 to 380 parts per million (ppm) in roughly 200 years. Measurements from ice bubbles show that this rate of increase is greater than even the most sudden increase that occurred naturally during the past 420,000 years. Second, CO_2 levels are higher than those estimated for any time during that same period. In the middle of 2013, global carbon dioxide concentrations stood at 397 ppm, with the levels increasing at the rate of about 3 ppm per year.

Increased carbon dioxide concentrations have many biological effects

An increase in the concentration of CO_2 in the air can have large effects on plants (**FIGURE 25.7**). Many plants increase their rate of photosynthesis and use water more efficiently, and therefore grow more rapidly, when more CO_2 is available. When CO_2 levels remain high, some plant species keep growing at higher rates, but others drop their growth rates over time. As CO_2 concentrations in the atmosphere rise, species that maintain rapid

Atmospheric Carbon Dioxide

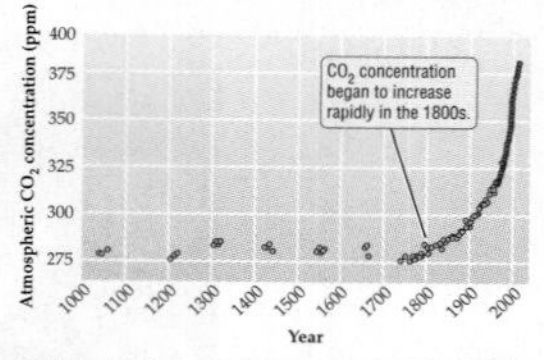

FIGURE 25.6 Atmospheric CO_2 Levels Are Rising Rapidly Atmospheric CO_2 levels (measured in parts per million, or ppm) have increased greatly in the past 200 years. The red circles are direct measurements at the Mauna Loa Observatory in Hawaii, at 11,135 feet above sea level. The green circles indicate CO_2 levels measured from bubbles of air trapped in ice that formed many hundreds of years ago.

Helpful to Know

The FDA banned the use of bisphenol A (BPA) in children's bottles and sippy cups in 2012, and from infant formula packaging in 2013. However, BPA is still found in a wide variety of products, including children's toys, canned goods, and register receipts. Its use in everyday products is so widespread that the CDC has found detectable levels of BPA in 93 percent of the Americans it tested.

Concept Check Answers

1. In bioaccumulation, a chemical accumulates within an organism's tissues at concentrations higher than in the surroundings; it takes place within a trophic level. In biomagnification, accumulated chemicals are passed on from one consumer to the next—that is, from one trophic level to the next. Chemicals that bioaccumulate are long-lived and not easily secreted, often because they bind to proteins or fats.
2. Top predators, because toxins accumulate between each trophic level.

HUMAN IMPACTS ON THE GLOBAL CARBON CYCLE 575

Extreme Biology

EXTREME BIOLOGY features help students remember key concepts by highlighting engaging examples from the amazing diversity of the living world.

EXTREME ICE CORE

Scientists are able to deduce global prehistoric climate patterns and CO_2 levels by collecting ice core samples from around the world and analyzing tiny air bubbles trapped within the ice. Cores are removed from ice sheets—some as much as 11,000 feet deep.

the concentration of CO_2 and other greenhouse gases in the atmosphere—a conclusion that has been supported by hundreds of studies published since 1995.

Some predicted consequences of climate change are now being seen

Long-term and large-scale changes in the state of Earth's climate are broadly known as climate change. Global warming is one component of climate change, and some of its effects on the biosphere are now evident (**TABLE 25.1**). Consistent with the warming trend, satellite images show that Arctic sea ice has been declining by 2.7 percent per decade since 1978 (**FIGURE 25.10**). Sea levels rose by an average of 1.8 millimeters per year between 1961 and 1993, and they have been rising by an average of 3.1 millimeters per year since then. Thermal expansion—the increase in volume as water warms up—has contributed to sea level rise, as has the melting of glaciers (**FIGURE 25.11**) and polar ice. As atmospheric carbon dioxide levels rise, more of the gas is absorbed by the oceans, leading to ocean acidification. Since the industrial revolution, the pH of the world's oceans has declined from an average value of about 8.25 to 8.14.

Helpful to Know

Ocean acidification occurs with higher CO_2 levels because CO_2 combines with water to form carbonic acid (H_2CO_3), which is why excessive consumption of carbonated beverages leads to erosion of tooth enamel. On a global scale, as CO_2 levels rise, levels of carbonic acid in the ocean increase.

Decline of the Arctic Ice Cap

(a) 1980

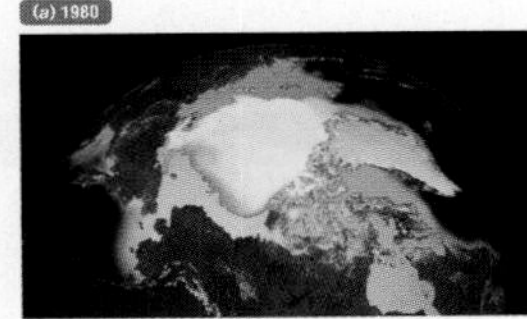

(b) 2012

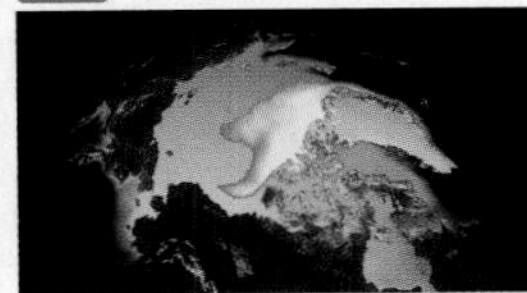

FIGURE 25.10 The Extent of Polar Sea Ice Has Declined Sharply
Summer sea ice in the Arctic has declined by almost 25 percent compared to preindustrial levels. Climate change has affected wind and ocean currents in different ways across the globe, explaining why the Antarctic ice sheet is relatively stable. The satellite-based illustrations show the extent of the polar ice cap in the Arctic and the ice sheet on Greenland in 1980 (a) and 2012 (b).

TABLE 25.1 Some Consequences of Climate Change

ABIOTIC CHANGES	SOME BIOTIC CONSEQUENCE
Increase in near-surface and ocean temperatures	Ecosystem disruption, loss of ec
Melting of glaciers	Spring floods, summer drought
Loss of summer sea ice	Species extinction, loss of cultu
Rise in sea levels (from melting ice, thermal expansion)	Loss of habitat, human habitatio
Ocean acidification	Loss of marine organisms with damage to fisheries
Increased frequency of severe weather	Habitat destruction; loss of hum
Change in rainfall pattern, drought in some regions	Ecosystem degradation; severe

humidity may decrease, since the lack of trees exposes the ground to direct sunlight and leads to an increase in evaporation rates. Such climatic changes can make it less likely that the forest will regrow even if the logging stops. In addition, as we will see shortly, the cutting and burning of forests increases the amount of carbon dioxide in the atmosphere—an aspect of global change that can alter the climate worldwide.

Concept Check

1. Give some examples of human activities that lead to land transformation.
2. Describe some causes of the degradation of coastal ecosystems.

25.2 Changes in the Chemistry of Earth

In Chapter 24 we learned that life on Earth depends on and participates in the cycling of nutrients in ecosystems. Net primary productivity often depends on the amount of nitrogen and phosphorus available to producers, for example, and an overabundance of sulfuric acid in rainfall lowers the pH of lakes and rivers, destroying fish populations. In the next two sections we look at how human activities, like manufacturing, motorized transportation, and overpopulation, have altered the natural nutrient cycles (see also Figures 24.10, 24.11, 24.13, and 24.14).

Bioaccumulation concentrates pollutants up the food chain

Humans release many synthetic chemicals and pollutants into the air, water, and soil that then cycle through ecosystems. These human-made chemicals can be ingested, inhaled, or absorbed by organisms. If a chemical binds to cells or tissues and stays there, then we say it **bioaccumulates** in an individual. As a substance bioaccumulates, its concentration within an organism ex-

classified as **persistent organic pollutants (POPs)**. Some of the most damaging POPs that are widespread in our biosphere include different types of PCBs (polychlorinated biphenyls, used in the production of electronics) and dioxins (a by-product of many industrial processes, such as the bleaching of paper pulp). Because many of these pollutants have an atmospheric cycle, they can be transported over vast distances across the globe to contaminate food chains in remote places where the chemicals have never been used.

Heavy metals such as mercury, cadmium, and lead can also bioaccumulate in a wide variety of organisms. Mercury enters the food chain when bacteria absorb it from soil or water and convert it to an organic form known as methylmercury. Methylmercury is much more toxic than inorganic forms of mercury, in part because the organic form bioaccumulates more readily, being stored in muscle tissues of shellfish, fish, and humans. Methylmercury bioaccumulated by bacteria is passed on to consumers, such as zooplankton (microscopic aquatic animals), that feed on mercury-accumulating bacteria. In this way, the methylmercury is progressively transferred to other consumers throughout the food web. The FDA has issued an advisory suggesting that pregnant women, in particular, abstain from eating mackerel, shark, swordfish, and tilefish, because these predatory fishes tend to accumulate higher levels of mercury.

The increase in the tissue concentrations of a bioaccumulated chemical at successively higher trophic levels in a food chain is known as **biomagnification**. Bioaccumulation and biomagnification might seem similar at first glance. Bioaccumulation is the accumulation of a substance in an individual within a trophic level, and biomagnification is the increase in tissue concentrations of a chemical as organic matter is passed from one trophic level to the next in a food chain.

EXTREME DEFORESTATION

The Amazon rainforest has been particularly influenced by human activities. Since 1970, 270,000 square miles of the forest has been destroyed—an area roughly the size of Texas. Deforestation in the Amazon in some years has exceeded 10,000 square miles.

Concept Check Answers

1. Destruction of habitats for resources or agriculture; alterations of habitats due to activities such as grazing.
2. Coastal ecosystems are affected by urban development, sewage dumping, excessive nutrient runoff, chemical pollution, and overfishing.

A Melting Glacier

(a) 1913

(b) 2009

FIGURE 25.11 Many Glaciers Are in Retreat
The extent of Shepard Glacier in Glacier National Park, Montana, in 1913 (a) versus 2009 (b). Most of the world's glaciers are in retreat, although some, especially in parts of South America and Central Asia, are either stable or growing slightly.

The additional heat energy that warmer temperatures generate, especially over the tropical oceans, is increasing the frequency of severe weather and lengthening the storm season. Since the middle of the twentieth century, the number of tropical storms sweeping into North America has not changed significantly, but the number of class 3 and class 4 hurricanes has nearly tripled. Rainfall patterns have changed: there is more rain in the eastern United States and northern Europe, and less in parts of the Mediterranean, northeastern and southern Africa, and parts of South Asia. Some recent climate simulations predict that global warming will worsen ozone depletion—with the highest increase in UV radiation in tropical rather than polar regions—because of alterations in wind flow patterns in the upper atmosphere.

Climate change has brought many species to the brink

Recent temperature increases have also changed the biotic (living) component of ecosystems. Many northern ecosystems are shifting their range poleward at a rate of about 0.42 kilometer (a quarter of a mile) per year, as species migrate north in an attempt to find their "comfort zone." For example, as temperatures increased in Europe during the twentieth century, dozens of bird and butterfly species shifted their geographic ranges to the north (see Figure 1.3). Similarly, the length of the growing season has increased for plants in northern latitudes as temperatures have warmed since 1980. However, some species—Arctic and alpine plants and animals, for example—have nowhere else to go (**FIGURE 25.12**). Canadian researchers have recorded a 60 percent decline in caribou and reindeer populations worldwide. There is higher calf mortality among the herds, and the animals suffer more from attacks by biting insects, whose populations have climbed.

In tropical waters, high temperatures combined with lower pH result in *coral bleaching*, caused by a loss of the algal symbiotic partner and often resulting in the death of the coral animal as well. About a third of the tropical coral reefs have been destroyed in the last few decades, succumbing to the collective onslaught of coral bleaching, pollution, and physical damage from an increase in severe storms.

Although the magnitude of warming is much larger in the northern latitudes, scientists expect a more severe impact on tropical ecosystems. Plants and animals in the moist tropics are adapted to a stable habitat and therefore live very close to the limits of their tolerance. Any change in that previously stable environment—increased temperature and reduced moisture, for example—puts them in jeopardy. In general, species with specialized habitat requirements are the most vulnerable. Experts studying species vulnerability warn that only 18–45 percent of the plants and animals native to the moist tropics are likely to survive beyond 2100. According to the International Union for Conservation of Nature

EXTREME SHORTCUT

The receding Arctic ice cap is bad news for polar bears but good news for transport companies. Climate scientists anticipate that within the next few decades, ships will be able to ferry goods between Asia and Europe by sailing through the Arctic Ocean, rather than making the long transit around Africa or through the Suez Canal.

Applied Features

BIOLOGY MATTERS boxes in nearly every chapter connect biology to real-life relevant topics that students care about: their health, society, and the environment.

BIOLOGYMATTERS

Toward a Sustainable Society

Many different lines of evidence suggest that the current human impact on the biosphere is not sustainable (see Figure 1). An action or process is *sustainable* if it can be continued indefinitely without serious damage being caused to the environment. Consider our use of fossil fuels. Although fossil fuels provide abundant energy now, our use of these fuels is not sustainable: they are not renewable, and hence supplies will run out, perhaps sooner rather than later (see Figure 2). Already, the volume of new sources of oil discovered worldwide has dropped steadily from over 200 billion barrels during the period from 1960 to 1965, to less than 30 billion barrels during 1995–2000. In 2007, the world used about 31 billion barrels of oil, but only 5 billion barrels of new oil was discovered in that year.

Actions that cause serious damage to the environment are also considered unsustainable, in part because our economies depend on clean air, clean water, and healthy soils. People currently use over 50 percent of the world's annual supply of available freshwater, and demand is expected to rise as populations increase. Many regions of the world already experience problems with either the amount of water available or its quality and safety. Declining water resources are a serious issue today, and experts are worried that matters may get much worse.

To illustrate the problem, let's look at water pumped from underground sources, or *groundwater*. How does the rate at which people use groundwater compare with the rate at which it is replenished by rainfall? The answer is that we often use water in an unsustainable way: we pump it from *aquifers* (underground bodies of water, sometimes bounded by impermeable layers of rock) much more rapidly than it is renewed.

In Texas, for example, for 100 years water has been pumped from the vast Ogallala aquifer faster than it has been replenished, causing the Texan portion of the aquifer to lose half its original volume. If that rate of use were to continue, in another 100 years the water would be gone, and many of the farms and industries that depend on it would collapse. Texas is not alone. Rapid drops in groundwater levels (about 1 meter per year) in China pose a severe threat to its recent agricultural and economic gains; and at current rates of use, large agricultural regions in India will completely run out of water in 5–10 years. In Mexico City, pumping has caused land within the city to sink by an average of 7.5 meters (more than 24 feet) since 1900, damaging buildings, destroying sewers, and causing floods.

Sustainability is one aspect of ecology where each of us has a role. We can build a more sustainable society by supporting legislation that fosters less destructive and more efficient use of natural resources; by patronizing businesses that take measures to lessen their negative impact on the planet; by supporting sustainable agriculture; and by modifying our own lifestyle to reduce our Ecological Footprint (see the "Biology Matters" box in Chapter 21, page 497). For example, we can:

- Increase our use of renewable energy and energy-efficient appliances;
- Reduce all unnecessary use of fossil fuels (for instance, by biking to work or using public transportation);
- Support organic farming; buy seafood from sustainable fisheries;
- Use "green" building materials; and reduce, reuse, and recycle waste.
- Support aid efforts that provide education, health care, and family-planning services in poor countries.

Experts estimate that more than 200 million women around the world wish to limit their family size but have no access to family planning.

FIGURE 1 The Most Inconvenient Truth: Climate Change Is Caused by Overpopulation and Overconsumption

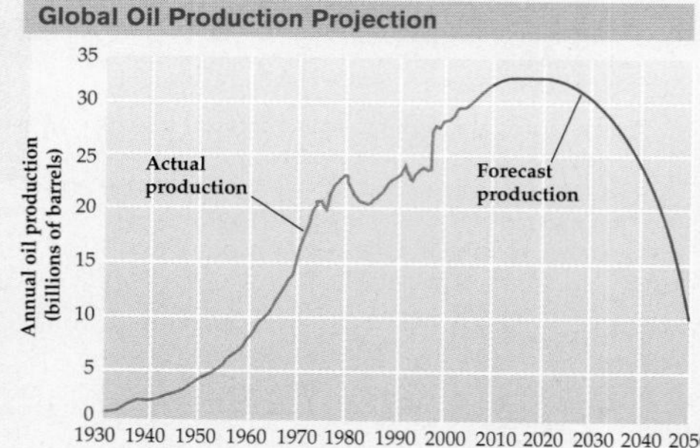

FIGURE 2 Running Out of Oil
Many experts predict that the annual global production of oil will peak, and then decline, sometime before 2020.

CLIMATE CHANGE 581

BIOLOGY IN THE NEWS features serve as a capstone to each chapter, reinforcing the chapter opening and closing story by providing students with an example of how they might encounter a related issue in their own lives—by reading about it in the news. The news article excerpts are accompanied by author analysis and discussion questions.

BIOLOGY IN THE NEWS

Blind, Starving Cheetahs: The New Symbol of Climate Change?

BY ADAM WELZ • *Guardian*, June 21, 2013

The world's fastest land animal is in trouble. The cheetah, formerly found across much of Africa, the Middle East and the Indian subcontinent, has been extirpated from at least 27 countries and is now on the Red List of threatened species.

Namibia holds by far the largest remaining population of the speedy cat. Between 3,500 and 5,000 cheetahs roam national parks, communal rangelands and private commercial ranches of this vast, arid country in south-western Africa, where they face threats like gun-toting livestock farmers and woody plants.

Yes, woody plants. Namibia is under invasion by multiplying armies of thorny trees and bushes, which are spreading across its landscape and smothering its grasslands.

So-called bush encroachment has transformed millions of hectares of Namibia's open rangeland into nearly impenetrable thicket and hammered its cattle industry . . .

Bush encroachment can also be bad news for cheetahs, which evolved to use bursts of extreme speed to run down prey in open areas. Low-slung thorns and the locked-open eyes of predators in "kill mode" are a nasty combination. Conservationists have found starving cheetahs that lost their sight after streaking through bush encroached habitats in pursuit of fleet footed food.

. . . An emerging body of science indicates that rapidly increasing atmospheric carbon dioxide may be boosting the onrushing waves of woody vegetation.

Savanna ecosystems, such as those that cover much of Africa, can be seen as battlegrounds between trees and grasses, each trying to take territory from the other. The outcomes of these battles are determined by many factors including periodic fire, an integral part of African savannas.

In simple terms, fire kills small trees and therefore helps fire-resilient grasses occupy territory. Trees have to have a long-enough break from fire to grow to a sufficient size—about four metres high—to be fireproof and establish themselves in the landscape. The faster trees grow, the more likely they are to reach four metres before the next fire.

Lab research shows that many savanna trees grow significantly faster as atmospheric CO_2 rises, and a new analysis of satellite images indicates that so-called "CO_2 fertilisation" has caused a large increase in plant growth in warm, arid areas worldwide.

. . . Increased atmospheric CO_2 seems to be upsetting many savanna ecosystems' vegetal balance of power in favour of trees and shrubs.

If increasing atmospheric carbon dioxide is causing climate change and also driving bush encroachment that results in blind cheetahs, should blind, starving cheetahs be a new symbol of climate change?

The astounding predatory behavior of cheetahs is well known among the general public. Incredible bursts that propel the cheetah to speeds of 60 miles per hour have been captured by videographers for decades. This speed, however, is becoming a liability as climate change alters the savanna landscape. Woody vegetation is encroaching on once open plains and presents dangerous obstacles to the famously fleet African cat. Trying to negotiate dense vegetation at high speed has led many cheetahs to become blinded by woody growth.

This study illustrates the practical consequences of a changing climate. The issue of climate change is largely discussed on a global scale by climatologists, but biologists are gaining more insight into the direct effects of such changes on organisms and ecosystems. For example, the melting Arctic sea ice has significantly affected populations of polar bears, which need the ice sheets for hunting grounds. Tropical frogs are experiencing increased fungal infections as temperatures warm. Coral reefs have been decimated by bleaching events, in which corals eject their photosynthetic symbionts as ocean temperatures warm.

Just as climate scientists are unable to predict the exact effects of global climate change, biologists are unable to deduce the impact of climate change on species. However, these examples of how warming temperatures are affecting biological organisms and communities illustrate that the effects of global climate change resonate throughout all levels of biological organization.

Evaluating the News

1. From your reading of this chapter, what factors do you think are influencing the distribution of woody vegetation, and therefore the cheetahs?
2. While the expansion of woody vegetation is bad for the cheetah, some species may benefit from the change in vegetative structure. How might the savanna community change in the face of global climate change?
3. Do you think studies such as the one reported in this article or the numerous studies on the impact of warming on polar bear populations may motivate people to address global climate change?

RESOURCES FOR STUDENTS AND INSTRUCTORS

The Ultimate Guide to Teaching Biology: Discover Biology

More and more, instructors want to spend class time engaging in inquiry-based activities that build students' science *skills* while teaching them the science *facts*. *The Ultimate Guide* helps instructors spend class time enriching students' appreciation for biology through resources including activities that can be carried out in a variety of classroom sizes and setups, think-pair-share activities, video clip recommendations with discussion questions, clicker questions, sample syllabi, and sample lecture plans.

NEW! InQuizitive

Norton's new formative and adaptive quizzing program, **InQuizitive**, preserves valuable lecture and lab time by personalizing quiz questions for each student and building knowledge outside of class.

- A variety of question types test student knowledge in different ways: matching, ranking, drag-and-drop, point-and-click, drag-and-fill, images, and more.
- Engaging, gamelike elements built into InQuizitive fuel student curiosity and motivate students as they learn. Students set their confidence level on each question to reflect their knowledge, track their own progress easily, earn point bonuses for high performance, and review learning objectives they might not have mastered.
- Quizzes are structured in sections that build knowledge through levels of Bloom's taxonomy, and across all the learning objectives of each chapter.
- Links to the ebook make it easy for students to reference their textbook as they work.
- The program is easy to use. InQuizitive works on mobile devices, and premade assignments are ready to go out of the box.

Norton Smartwork

Norton **Smartwork** includes high-quality questions and answer-specific feedback that help students apply, analyze, and evaluate key concepts. Norton Smartwork is device agnostic and not Flash based, so it can be used on tablets and smartphones. New author Gary Shin has selected the two most difficult concepts for students in each chapter in the textbook and authored extensive feedback for all the questions in Norton Smartwork about those concepts. This feedback is designed to direct students back to the sections in the book or ebook that they should review before trying the question again. Every question has benefited from the careful eyes of authors who are also Norton Smartwork power users; they have reviewed every question and revised many for accuracy, accessibility, and efficacy.

Ebook

An affordable and convenient alternative, the enhanced ebook can be viewed on any device—laptop, tablet, phone, even a public computer—and will stay synced between devices. Art expands for a closer look, pop-up key terms provide a quick vocabulary check, and direct links from Norton Smartwork make sure that students see the connection between their assessment and their reading. Further, it's easy to highlight and take notes, print chapters, and search the text.

Coursepacks

Free and easy-to-use coursepacks include review and study materials such as animations, flashcards, chapter quizzes with feedback, "Biology in the News" quizzes, and test banks. Coursepacks are available in Blackboard, Desire2Learn (D2L), ANGEL, Canvas, and Moodle.

RESOURCES FOR STUDENTS AND INSTRUCTORS

Presentation Tools

- **Lecture Slides.** These slides feature selected art from the text with detailed lecture outlines and links to the animations on **DiscoverBiology.com**.
- **Art Slides.** These PowerPoint sets contain all the art from the book.
- **Unlabeled Art Slides.** The figure labels are removed in these sets so that students can fill in the blanks.
- **Active Art Slides.** These slides feature complex pieces of artwork broken down into movable, editable components. This function allows instructors to customize figures within PowerPoint, choosing exactly what their students will see during a lecture.

Animations

Key figures in the book are presented as HTML5 animations, which are embedded in the PowerPoint lecture outlines for projection in class, and available for students in the ebook and coursepacks. Questions in Norton Smartwork and InQuizitive incorporate the animations either as the basis for a question or in the solution feedback.

Test Bank

The **Test Bank** for *Discover Biology* is based on an evidence-centered design that was collaboratively developed by some of the brightest minds in educational testing—including leading academic researchers and advisers with the Educational Testing Service (ETS). The result is an assessment resource that (1) defines expected student competencies, (2) evenly distributes concepts and topics, and (3) ensures specific links to topics.

Each chapter of the Test Bank is structured around the Key Concepts from the textbook and evaluates students according to the first five levels of Bloom's taxonomy of knowledge types: Remembering, Understanding, Applying, Analyzing, and Evaluating. Questions are further classified by section and difficulty, and they are provided in multiple-choice, fill-in-the-blank (completion), and true/false formats.

Acknowledgments

Reviewers of the Sixth Edition

Mari Aanenson, Western Illinois University
Holly Ahern, SUNY Adirondack
Christine Andrews, Lane Community College
Bert Atsma, Union County College
Robert E. Bailey, Central Michigan University
William David Barnes, SUNY Canton
Tiffany Bensen, The University of Mississippi
Valerie Bishop, Meridian Community College
Mark Bland, University of Central Arkansas
Claire Carpenter, Yakima Valley Community College
Maitreyee Chandra, Diablo Valley College
Rhonda Crotty, Tarrant County College
Deborah Dardis, Southeastern Louisiana University
Begona De Velasco, California State University–Dominguez Hills
Mary Dion, St. Louis Community College–Meramec
Danielle DuCharme, Waubonsee Community College
Robert G. Ewy, SUNY Potsdam
Edison R. Fowlks, Hampton University
Amanda Gilleland, St. Petersburg College
Jerrie Hanible, Southeastern Louisiana University
Chadwick Hanna, California University of Pennsylvania
Mario Hollomon, Texas Southern University
Mesha Hunte-Brown, Drexel University
Wanda Jester, Salisbury University
Suzanne Kempke, St. Johns River State College
Dubear Kroening, University of Wisconsin–Fox Valley
Rukmani Kuppuswami, Laredo Community College
Suzanne Long, Monroe Community College
Stephanie Loveless, Danville Area Community College
Leroy R. McClenaghan Jr., San Diego State University
Malinda McMurry, Morehead State University
Dana Newton, College of the Albemarle
Brigid C. O'Donnell, Plymouth State University
Sean O'Keefe, Morehead State University
Krista Peppers, University of Central Arkansas
David Peyton, Morehead State University
Claire Prouty, Sam Houston State University
Logan Randolph, Polk State College
Laura Ritt, Burlington County College
Lori Rose, Sam Houston State University
Dorothy Scholl, The University of New Orleans
Pramila Sen, Houston Community College
Brian Seymour, Edward Waters College
Marek Sliwinski, University of Northern Iowa
Ayodotun Sodipe, Texas Southern University
Ronald Tavernier, SUNY Canton
Jeff Taylor, SUNY Canton
Nicholas Tippery, University of Wisconsin–Whitewater
Sophia Ushinsky, Concordia University
Jennifer Wiatrowski, Pasco-Hernando State College
Rachel Wiechman, West Liberty University
Lance R. Williams, University of Texas at Tyler
Lawrence Williams, University of Houston
Holly Woodruff, Central Piedmont Community College

Survey Participants

David Bailey, St. Norbert College
Verona Barr, Heartland Community College
James Barron, Montana State University–Billings
Mark Belk, Brigham Young University
Steven Brumbaugh, Green River Community College
Rob Channell, Fort Hays State University
Andy Cook, Wheeling Jesuit University
Kathy Gallucci, Elon University
Tamar Liberman Goulet, The University of Mississippi
Joby Jacob, Borough of Manhattan Community College
Suzanne Kempke, St. Johns River State College
Andrea Kozol, Framingham State University
Dubear Kroening, University of Wisconsin–Fox Valley
Jennifer Landin, North Carolina State University
Gabrielle McLemore, Morgan State University
Owen Meyers, Borough of Manhattan Community College
Steve Muzos, Austin Community College
Brigid C. O'Donnell, Plymouth State University
Melinda Ostraff, Brigham Young University
Kimberly Regier, University of Colorado Denver
Brian Rehill, United States Naval Academy
Laura Ritt, Burlington County College
Lori Rose, Sam Houston State University
Georgianna Saunders, Missouri State University
Brian Seymour, Edward Waters College
Indrani Sindhuvalli, Florida State College at Jacksonville–South Campus
Paul Smith, Virginia Commonwealth University
Jennifer Snekser, Long Island University–Post
Rissa Springs, Texas Lutheran University
Tim Tripp, Sam Houston State University
Koshy Varghese, Eastfield College
Michael Wenzel, Folsom Lake College
Rachel Wiechman, West Liberty University
Carolyn Zanta, Clarkson University

Reviewers of Previous Editions

Michael Abruzzo, California State University–Chico
James Agee, University of Washington
Holly Ahern, Adirondack Community College
Mac Alford, University of Southern Mississippi
Laura Ambrose, University of Regina
Marjay Anderson, Howard University
Angelika M. Antoni, Kutztown University
Idelisa Ayala, Broward College
Caryn Babaian, Bucks County College
Neil R. Baker, Ohio State University
Marilyn Banta, Texas State University–San Marcos
Sarah Barlow, Middle Tennessee State University
Christine Barrow, Prince George's Community College
Gregory Beaulieu, University of Victoria
Craig Benkman, New Mexico State University

Elizabeth Bennett, Georgia College and State University
Stewart Berlocher, University of Illinois–Urbana
Robert Bernatzky, University of Massachusetts–Amherst
Nancy Berner, University of the South
Robert Bevins, Georgetown College
Janice M. Bonner, College of Notre Dame of Maryland
Juan Bouzat, University of Illinois–Urbana
Bryan Brendley, Gannon University
Randy Brewton, University of Tennessee–Knoxville
Peggy Brickman, University of Georgia
Sarah Bruce, Towson University
Christine Buckley, Rose-Hulman Institute of Technology
Neil Buckley, SUNY Plattsburgh
Art Buikema, Virginia Tech University
John Burk, Smith College
Kathleen Burt-Utley, The University of New Orleans
Wilbert Butler Jr., Tallahassee Community College
David Byres, Florida Community College at Jacksonville–South Campus
Naomi Cappuccino, Carleton University
Kelly Cartwright, College of Lake County
Aaron Cassill, University of Texas at San Antonio
Heather Vance Chalcraft, East Carolina University
Van Christman, Ricks College
Jerry Cook, Sam Houston State University
Keith Crandall, Brigham Young University
Helen Cronenberger, University of Texas at Austin
Chad Cryer, Austin Community College
Francie Cuffney, Meredith College
Kathleen Curran, Wesley College
Gregory Dahlem, Northern Kentucky University
Don Dailey, Austin Peay State University
Judith D'Aleo, Plymouth State University
Vern Damsteegt, Montgomery College
Paul da Silva, College of Marin
Garry Davies, University of Alaska–Anchorage
Angela Davis, Danville Area Community College
Sandra Davis, University of Louisiana–Monroe
Kathleen DeCicco-Skinner, American University
Véronique Delesalle, Gettysburg College
Pablo Delis, Hillsborough Community College
Lisa J. Delissio, Salem State College
Alan de Queiroz, University of Colorado
Jean de Saix, University of North Carolina–Chapel Hill
Joseph Dickinson, University of Utah
Gregg Dieringer, Northwest Missouri State University
Deborah Donovan, Western Washington University
Christian d'Orgeix, Virginia State University
Harold Dowse, University of Maine
John Edwards, University of Washington
Jean Engohang-Ndong, Brigham Young University–Hawaii
Susan Epperson, University of Colorado Colorado Springs
Jonathon Evans, University of the South
William Ezell, University of North Carolina–Pemberton
Deborah Fahey, Wheaton College
Susan Farmer, Abraham Baldwin Agricultural College
Richard Farrar, Idaho State University
Marion Fass, Beloit College
Tracy M. Felton, Union County College
Linda Fergusson-Kolmes, Portland Community College
Richard Finnell, Texas A&M University
Ryan Fisher, Salem State College
Susan Fisher, Ohio State University
Paul Florence, Jefferson Community & Technical College
April Ann Fong, Portland Community College–Sylvania Campus
Edison Fowlks, Hampton University
Jennifer Fritz, University of Texas at Austin
Kathy Gallucci, Elon University
Wendy Garrison, The University of Mississippi
Gail Gasparich, Towson University
Aiah A. Gbakima, Morgan State University
Dennis Gemmell, Kingsborough Community College
Alexandros Georgakilas, East Carolina University
Kajal Ghoshroy, Museum of Natural History–Las Cruces
Caitlin Gille, Pasco-Hernando Community College
Beverly Glover, Western Oklahoma State College
Jack Goldberg, University of California–Davis
Andrew Goliszek, North Carolina Agricultural and Technological State University
Glenn Gorelick, Citrus College
Tamar Goulet, The University of Mississippi
Bill Grant, North Carolina State University
Harry W. Greene, Cornell University
John Griffis, University of Southern Mississippi
Cindy Gustafson-Brown, University of California–San Diego
Ronald Gutberlet, Salisbury University
Laura Haas, New Mexico State University
Barbara Hager, Cazenovia College
Blanche Haning, University of North Carolina–Chapel Hill
Robert Harms, St. Louis Community College–Meramec
Jill Harp, Winston-Salem State University
Chris Haynes, Shelton State Community College
Thomas Hemmerly, Middle Tennessee State University
Nancy Holcroft-Benson, Johnson County Community College
Tom Horvath, SUNY Oneonta
Anne-Marie Hoskinson, Minnesota State University, Mankato
Daniel J. Howard, New Mexico State University
Laura F. Huenneke, New Mexico State University
Tonya Huff, Riverside Community College
James L. Hulbert, Rollins College
Meshagae Hunte-Brown, Drexel University
Brenda Hunzinger, Lake Land College
Karen Jackson, Jacksonville University
Karel Jacobs, Chicago State University
Sayna Jahangiri, Folsom Lake College
Jane Jefferies, Brigham Young University
Denim Jochimsen, University of Idaho
Mark Johnson, Georgetown College
Robert M. Jonas, Texas Lutheran University
Anthony Jones, Tallahassee Community College
Arnold Karpoff, University of Louisville
Paul Kasello, Virginia State University
Laura Katz, Smith College
Andrew Keth, Clarion University of Pennsylvania
Tasneem Khaleel, Montana State University
Joshua King, Central Connecticut State University
Yolanda Kirkpatrick, Pellissippi State Community College
John Knesel, University of Louisiana–Monroe
Will Kopachik, Michigan State University
Olga Kopp, Utah Valley University
Erica Kosal, North Carolina Wesleyan College
Hans Landel, North Seattle Community College
Jennifer Landin, North Carolina State University
Allen Landwer, Hardin-Simmons University
Katherine C. Larson, University of Central Arkansas
Neva Laurie-Berry, Pacific Lutheran University
Paula Lemons, University of Georgia
Shawn Lester, Montgomery College
Margaret Liberti, SUNY Cobleskill
Harvey Liftin, Broward County Community College
Lee Likins, University of Missouri–Kansas City
Cynthia Littlejohn, University of Southern Mississippi
Suzanne Long, Monroe Community College
Craig Longtine, North Hennepin Community College
Melanie Loo, California State University–Sacramento
Kenneth Lopez, New Mexico State University
David Loring, Johnson County Community College

Ann S. Lumsden, Florida State University
Monica Macklin, Northeastern State University
Blasé Maffia, University of Miami
Patricia Mancini, Bridgewater State College
Lisa Maranto, Prince George's Community College
Boriana Marintcheva, Bridgewater State College
Roy Mason, Mount San Jacinto College
Catarina Mata, Borough of Manhattan Community College
Joyce Maxwell, California State University–Northridge
Phillip McClean, North Dakota State University
Quintece Miel McCrary, University of Maryland–Eastern Shore
Amy McCune, Cornell University
Bruce McKee, University of Tennessee
Bob McMaster, Holyoke Community College
Dorian McMillan, College of Charleston
Alexie McNerthney, Portland Community College
Susan Meacham, University of Nevada, Las Vegas
Susan Meiers, Western Illinois University
Gretchen Meyer, Williams College
Steven T. Mezik, Herkimer County Community College
James Mickle, North Carolina State University
Brook Milligan, New Mexico State University
Ali Mohamed, Virginia State University
James Mone, Millersville University
Daniela Monk, Washington State University
Brenda Moore, Truman State University
Ruth S. Moseley, S. D. Bishop Community College
Elizabeth Nash, Long Beach Community College
Jon Nickles, University of Alaska–Anchorage
John Niedzwiecki, Belmont University
Zia Nisani, Antelope Valley College
Benjamin Normark, University of Massachusetts–Amherst
Ikemefuna Nwosu, Lake Land College
Douglas Oba, University of Wisconsin–Marshfield
Mary O'Connell, New Mexico State University
Jonas Okeagu, Fayetteville State University
Brady Olson, Western Washington University
Alexander E. Olvido, Longwood University
Marcy Osgood, University of Michigan
Melinda Ostraff, Brigham Young University
Jason Oyadomari, Finlandia University
Donald Padgett, Bridgewater State College
Penelope Padgett, University of North Carolina–Chapel Hill
Kevin Padian, University of California–Berkeley
Brian Palestis, Wagner College
John Palka, University of Washington
Anthony Palombella, Longwood College
Snehlata Pandey, Hampton University
Murali T. Panen, Luzerne County Community College
Robert Patterson, North Carolina State University
Nancy Pelaez, California State University–Fullerton
Pat Pendarvis, Southeastern Louisiana University
Brian Perkins, Texas A&M University
Patrick Pfaffle, Carthage College
Patricia Phelps, Austin Community College
Massimo Pigliucci, University of Tennessee
Joel Piperberg, Millersville University
Jeffrey Podos, University of Massachusetts–Amherst
Robert Pozos, San Diego State University
Ralph Preszler, New Mexico State University
Jim Price, Utah Valley University
Todd Primm, Sam Houston State University
Jerry Purcell, Alamo Community College
Ashley Rall McGee, Valdosta State University
Stuart Reichler, University of Texas at Austin
Mindy Reynolds-Walsh, Washington College
Richard Ring, University of Victoria
Michelle Rogers, Austin Peay State University
Lori Ann (Henderson) Rose, Sam Houston State University
Allison Roy, Kutztown University
Barbara Rundell, College of DuPage
Ron Ruppert, Cuesta College
Lynette Rushton, South Puget Sound Community College
Michael Rutledge, Middle Tennessee State University
Shamili Sandiford, College of DuPage
Barbara Schaal, Washington University
Jennifer Schramm, Chemeketa Community College
John Richard Schrock, Emporia State University
Kurt Schwenk, University of Connecticut
Harlan Scott, Howard Payne University
Erik Scully, Towson University
Tara A. Scully, George Washington University
David Secord, University of Washington
Brian Seymour, Sonoma State University
Marieken Shaner, University of New Mexico
Erica Sharar, Irvine Valley College
William Shear, Hampden-Sydney College
Cara Shillington, Eastern Michigan University
Barbara Shipes, Hampton University
Mark Shotwell, Slippery Rock University
Shaukat Siddiqi, Virginia State University
Jennie Skillen, College of Southern Nevada
Donald Slish, SUNY Plattsburgh
Julie Smit, University of Windsor
James Smith, Montgomery College
Philip Snider, University of Houston
Julie Snyder, Hudson High School
Mary Lou Soczek, Fitchburg State University
Michael Sovic, Ohio State University
Ruth Sporer, Rutgers–Camden
Jim Stegge, Rochester Community and Technical College
Richard Stevens, Monroe Community College
Neal Stewart, University of North Carolina–Greensboro
Tim Stewart, Longwood College
Bethany Stone, University of Missouri
Nancy Stotz, New Mexico State University
Steven Strain, Slippery Rock University
Allan Strand, College of Charleston
Marshall Sundberg, Emporia State University
Kirsten Swinstrom, State Rosa Junior College
Alana Synhoff, Florida Community College
Joyce Tamashiro, University of Puget Sound
Steve Tanner, University of Missouri
Josephine Taylor, Stephen F. Austin State University
Kristina Teagarden, West Virginia University
John Trimble, Saint Francis College
Mary Tyler, University of Maine
Doug Ure, Chemeketa Community College
Rani Vajravelu, University of Central Florida
Roy Van Driesche, University of Massachusetts–Amherst
Cheryl Vaughan, Harvard University
John Vaughan, St. Petersburg College
William Velhagen, Longwood College
Mary Vetter, Luther College
Alain Viel, Harvard Medical School
Carol Wake, South Dakota State University
Jerry Waldvogel, Clemson University
Elsbeth Walker, University of Massachusetts–Amherst
Holly Walters, Cape Fear Community College
Daniel Wang, University of Miami
Stephen Warburton, New Mexico State University
Carol Weaver, Union University
Paul Webb, University of Michigan
Teresa Weglarz-Hall, University of Wisconsin–Fox Valley
Michael Wenzel, California State University–Sacramento
Cindy White, University of Northern Colorado

Jennifer Wiatrowski, Pasco-Hernando Community College
Antonia Wijte, Irvine Valley College
Peter Wilkin, Purdue University North Central
Daniel Williams, Winston-Salem State University
Elizabeth Willott, University of Arizona
Peter Wimberger, University of Puget Sound
Allan Wolfe, Lebanon Valley College
Edwin Wong, Western Connecticut State University
David Woodruff, University of California–San Diego
Louise Wootton, Georgian Court University
Silvia Wozniak, Winthrop University
Robin Wright, University of Washington
Donald Yee, University of Southern Mississippi
Calvin Young, Fullerton College
Carolyn A. Zanta, Clarkson University

Thanks to the *Discover Biology* Team

As always, revising this textbook was a monumental task, but a fun one too. We are thankful to the many editors, researchers, and assistants at W. W. Norton who helped shepherd this book through the significant revisions in text, photos, and artwork that you see here. In particular, we'd like to thank our editor, Betsy Twitchell, for helping us plan and execute the book you are now holding in your hands. Her enthusiasm for our book and keen market sense have been invaluable. The keen attention to details and talented wordsmithing of our developmental editors, Jody Larson and Michael Zierler, had an enormously positive effect on the readability of the text. Thanks to our eagle-eyed copy editor, Stephanie Hiebert, a most superbly meticulous, perceptive, and skillful editor. Thanks also to Christine D'Antonio for seamlessly coordinating the movement and synthesis of the innumerable parts of this book. Our thanks also to Ben Reynolds for skillfully overseeing the final assembly into a tangible, beautiful book. Photo researcher Donna Ranieri and photo editor Nelson Colón also contributed enormously to the visual appeal of this beautiful revision. The herculean efforts by media editor Robin Kimball and associate editor Cailin Barrett-Bressack have resulted in the highest-quality and most robust media package this book has ever had. Media project editor Kristin Sheerin's careful attention to the digital content has ensured its high level of consistency with the printed text. With marketing manager Meredith Leo's tireless advocacy of this book in the marketplace, we're confident that it will reach as wide an audience as possible. Katie Callahan deserves thanks for making sure that the many parallel tracks of reviewing, revising, and correcting eventually converged at the right time and place. Finally, we would like to thank our families for support during the long process that is a textbook revision, especially Don, Ryan, and Erika Singh-Cundy and Susan, Thomas, and Julia Shin.

About the Authors

Anu Singh-Cundy received her PhD from Cornell University and did postdoctoral research in cell and molecular biology at Penn State. She is an associate professor at Western Washington University, where she teaches a variety of undergraduate and graduate courses, including organismal biology, cell biology, plant developmental biology, and plant biochemistry. She has taught introductory biology to nonmajors for over 15 years and is recognized for pedagogical innovations that communicate biological principles in a manner that engages the nonscience student and emphasizes the relevance of biology in everyday life. Her research focuses on cell-cell communication in plants, especially self-incompatibility and other pollen-pistil interactions. She has published over a dozen research articles and has received several awards and grants, including a grant from the National Science Foundation.

After earning a degree in English literature at UC Berkeley, **Gary Shin** turned his academic sights toward biology, first working as a field biologist and then moving into the lab to study population genetics. He earned his PhD at UCLA, studying the evolutionary genetics of viruses. Dr. Shin has worked at the Centers for Disease Control and Prevention and now teaches general biology at California State University, Long Beach, sharing his passion for biology with nonmajors.

Contents

Preface iii
Guided Tour vi
Resources for Students and Instructors xii
Acknowledgments xv
About the Authors xix

1 The Nature of Science and the Characteristics of Life 2
Earthbound Extraterrestrial? Or Just Another Microbe in the Mud? 3
1.1 The Nature of Science 4
Biology Matters: Science and the Citizen 8
1.2 The Process of Science 9
1.3 Scientific Facts and Theories 13
1.4 The Characteristics of Living Organisms 14
1.5 Biological Evolution and the Unity and Diversity of Life 17
1.6 The Biological Hierarchy 20
Researchers Wrangle over Bacteria 23
Biology in the News: Curbing the Enthusiasm on Daily Multivitamins 24

UNIT 1 Cells: The Basic Units of Life

2 The Chemistry of Life 28
How the "Cookie Monster" Tackled Trans Fats 29
2.1 Matter, Elements, and Atomic Structure 30
2.2 The Bonds That Link Atoms 33
2.3 The Special Properties of Water 36
2.4 Chemical Reactions 39
2.5 The pH Scale 40
2.6 The Chemical Building Blocks of Life 41
2.7 Carbohydrates 43
2.8 Proteins 45
2.9 Lipids 49
Biology Matters: Dietary Lipids: The Good, the Bad, and the Truly Ugly 51
2.10 Nucleotides and Nucleic Acids 53
How Bad Are Trans Fats? 55
Biology in the News: For a Better, Leaner Burger, Get to Know Your Proteins 56

3 Cell Structure and Internal Compartments 60
Wanted: Long-Term Roommate; Must Help Keep House and Have Own DNA 61
3.1 Cells: The Smallest Units of Life 63
3.2 The Plasma Membrane 66
3.3 Prokaryotic and Eukaryotic Cells 68
3.4 Internal Compartments of Eukaryotic Cells 70
Biology Matters: Organelles and Human Disease 74
3.5 The Cytoskeleton 76
The Evolution of Eukaryotes 80
Biology in the News: New Technology Can Make It So 1 Baby Has 3 Parents: And Prevent That Child from Being Born with Serious Mitochondrial Diseases 81

4 Cell Membranes, Transport, and Communication 84
Mysterious Memory Loss 85
4.1 The Plasma Membrane as Gate and Gatekeeper 86
4.2 Osmosis 89
Biology Matters: Osmosis in the Kitchen and Garden 90
4.3 Facilitated Membrane Transport 92
4.4 Exocytosis and Endocytosis 95
4.5 Cellular Connections 97
4.6 Cell Signaling 99
Cholesterol in the Brain 100
Biology in the News: Genetically Engineered Tomato Mimics Good Cholesterol 101

5 Energy, Metabolism, and Enzymes 104
Kick-Start Your Metabolic Engine! 105
5.1 The Role of Energy in Living Systems 106
5.2 Metabolism 109
5.3 Enzymes 112
Biology Matters: Enzymes in Action 113
5.4 Metabolic Pathways 115
Food, Folks, and Metabolism 118

Biology in the News: Lemurs' Long-Buried Secrets Revealed 119

6 Photosynthesis and Cellular Respiration 122
Every Breath You Take 123
6.1 Molecular Energy Carriers 124
6.2 An Overview of Photosynthesis and Cellular Respiration 126
6.3 Photosynthesis: Energy from Sunlight 128
Biology Matters: The Rainbow Colors of Plant Pigments 130
6.4 Cellular Respiration: Energy from Food 137
Waiting to Exhale 143
Biology in the News: Cassava and Mental Deficits 144

7 Cell Division 148
Olympic-Class Algal Bloom 149
7.1 Why Cells Divide 150
7.2 The Cell Cycle 153
Biology Matters: Programmed Cell Death: Going Out in Style 155
7.3 The Chromosomal Organization of Genetic Material 156
7.4 Mitosis and Cytokinesis: From One Cell to Two Identical Cells 159
7.5 Meiosis: Halving the Chromosome Set to Make Gametes 161
The Great Divide 168
Biology in the News: The Science of Making Babies Becomes Commonplace 169

8 Cancer and Human Health 172
Henrietta Lacks's Immortal Cells 173
8.1 Cancer: Good Cells Gone Bad 174
8.2 Cancer-Critical Genes 178
8.3 The Progression to Cancer 180
8.4 Treatment and Prevention 181
Biology Matters: Avoiding Cancer by Avoiding Chemical Carcinogens 185
How HeLa Cells Changed Biomedicine 186
Biology in the News: Cancer Gene Has Led Jolie and Others to Surgery 187

UNIT 2 Genetics

9 Patterns of Inheritance 190
The Lost Princess 191
9.1 Principles of Genetics: An Overview 192
9.2 Basic Patterns of Inheritance 197
9.3 Mendel's Laws of Inheritance 199
9.4 Extensions of Mendel's Laws 203
Biology Matters: Know Your Type 205
Solving the Mystery of the Lost Princess 211
Biology in the News: White Tiger Genetic Secret Unveiled: Single Mutation in Single Gene Removes Orange Color 212

10 Chromosomes and Human Genetics 216
Family Ties 217
10.1 The Role of Chromosomes in Inheritance 218
10.2 Genetic Linkage and Crossing-Over 221
10.3 Human Genetic Disorders 223
Biology Matters: Most Chronic Diseases Are Complex Traits 226
10.4 Autosomal Inheritance of Single-Gene Mutations 227
10.5 Sex-Linked Inheritance of Single-Gene Mutations 229
10.6 Inherited Chromosomal Abnormalities 230
Testing for Huntington's Disease 233
Biology in the News: Stanford University Students Study Their Own DNA 234

11 DNA and Genes 240
The Man from the Copper Age 241
11.1 An Overview of DNA and Genes 242
11.2 The Three-Dimensional Structure of DNA 245
11.3 How DNA Is Replicated 247
11.4 Repairing Replication Errors and Damaged DNA 248
11.5 Genome Organization 250
Biology Matters: Prenatal Genetic Screening 251
11.6 DNA Packing in Eukaryotes 253
11.7 Patterns of Gene Expression 254
CSI: Copper Age 256
Biology in the News: The Human Genome Project: How It Changed Biology Forever 257

12 From Gene to Protein 260
Greek Myths and One-Eyed Sheep 261
12.1 How Genes Work 262
12.2 Transcription: Information Flow from DNA to RNA 264
12.3 The Genetic Code 267
12.4 Translation: Information Flow from mRNA to Protein 269
12.5 The Effect of Mutations on Protein Synthesis 270
12.6 How Cells Control Gene Expression 272

Biology Matters: One Allele Makes You Strong, Another Helps You Endure 275
From Gene Expression to Cyclops 276
Biology in the News: BPA Could Affect Brain Development by Impacting Gene Regulation, Study Finds 277

13 DNA Technology 280
Eduardo Kac's "Plantimal" 281
13.1 The Brave New World of DNA Technology 282
13.2 DNA Fingerprinting 286
13.3 Genetic Engineering 287
Biology Matters: Have You Had Your GMO Today? 288
13.4 Reproductive Cloning of Animals 291
13.5 Stem Cells: Dedicated to Division 292
13.6 Human Gene Therapy 297
13.7 Ethical and Social Dimensions of DNA Technology 299
13.8 A Closer Look at Some Tools of DNA Technology 300
How to Make a Plantimal 304
Biology in the News: Gene Therapy Shows New Signs of Promise 305

UNIT 3 Evolution

14 How Evolution Works 308
Finches Feasting on Blood 309
14.1 Evolution and Natural Selection 310
14.2 Mechanisms of Evolutionary Change 312
14.3 Natural Selection Leads to Adaptive Evolution 316
Biology Matters: Testing Whether Evolution Is Occurring in Natural Populations 317
14.4 Adaptations 321
14.5 Sexual Selection 324
14.6 The Evidence for Biological Evolution 325
14.7 The Impact of Evolutionary Thought 331
Darwin's Finches: Evolution in Action 333
Biology in the News: Evolution via Roadkill 335

15 The Origin of Species 338
Cichlid Mysteries 339
15.1 What Are Species? 340
15.2 Speciation: Generating Biodiversity 342
Biology Matters: Islands Are Centers for Speciation—and Extinction 345
15.3 Adaptive Radiations: Increases in the Diversity of Life 348
15.4 Evolution Can Explain the Unity and Diversity of Life 349
15.5 Rates of Speciation 351
Lake Victoria: Center of Speciation 352
Biology in the News: First Love Child of Human, Neanderthal Found 353

16 The Evolutionary History of Life 356
Puzzling Fossils in a Frozen Wasteland 357
16.1 Macroevolution: Large-Scale Body Changes 358
16.2 The Fossil Record: A Guide to the Past 359
16.3 The History of Life on Earth 361
16.4 The Effects of Plate Tectonics 365
16.5 Mass Extinctions: Worldwide Losses of Species 366
16.6 Rapid Macroevolution through Differential Gene Expression 368
Biology Matters: Is a Mass Extinction Under Way? 369
16.7 Phylogenetics: Reconstructing Evolutionary Relationships 371
When Antarctica Was Green 374
Biology in the News: Fish's DNA May Explain How Fins Turned to Feet 375

17 Bacteria, Archaea, and Viruses 378
A Hitchhiker's Guide to the Human Body 379
17.1 The Diversity of Life 380
Biology Matters: The Importance of Biodiversity 382
17.2 Bacteria and Archaea: Tiny, Successful, and Abundant 383
17.3 How Prokaryotes Affect Our World 391
17.4 Viruses: Nonliving Infectious Agents 395
All of Us Together 398
Biology in the News: Gut Microbiome Largely Stable over Time, Feces Study Finds 399

18 Protista, Plantae, and Fungi 402
Did Plants Teach Rivers to Wander? 403
18.1 The Dawn of Eukarya 405
18.2 Protista: The First Eukaryotes 407
18.3 Plantae: The Green Mantle of Our World 413
18.4 Fungi: A World of Decomposers 419
Biology Matters: The Many Threats to Biodiversity 420
18.5 Lichens and Mycorrhizae: Collaborations between Kingdoms 424
The Root of the Problem: Why Rivers Meander 426
Biology in the News: Nectar That Gives Bees a Buzz Lures Them Back for More 427

19 Animalia 430
Who We Are 431
19.1 The Evolutionary Origins of Animalia 432

19.2 Characteristics of Animals 433
19.3 The First Invertebrates: Sponges, Jellyfish, and Relatives 438
19.4 The Protostomes 440
19.5 The Deuterostomes 448
19.6 Chordates like Us: The Vertebrates 450
Biology Matters: Good-bye, Catch of the Day? 454
Clues to the Evolution of Multicellularity 460
Biology in the News: Prime Number Cicadas 461

UNIT 4 The Biosphere

20 The Biosphere 464
Invasion of the Zebra Mussels 465
20.1 Ecology: Understanding the Interconnected Web 466
20.2 Climate's Large Effect on the Biosphere 468
20.3 Terrestrial Biomes 471
20.4 Aquatic Biomes 477
Biology Matters: The Great Pacific Garbage Patch 479
How Invasive Mussels Can Harm Whole Ecosystems 484
Biology in the News: Kenyan Company Turns Old Sandals into Colorful Array of Toys and Safari Animals 485

21 Growth of Populations 488
The Tragedy of Easter Island 489
21.1 What Is a Population? 490
21.2 Changes in Population Size 491
21.3 Exponential Growth 492
21.4 Logistic Growth and the Limits on Population Size 494
Biology Matters: How Big Is Your Ecological Footprint? 497
21.5 Applications of Population Ecology 498
What Does the Future Hold? 501
Biology in the News: Eating Bugs: Would You Dine on Cicadas? Crickets? Buttered Beetles? 503

22 Animal Behavior 506
The Evolution of Niceness 507
22.1 Sensing and Responding: The Nature of Behavioral Responses 508
Biology Matters: Drinking and the Dark Sides of Human Behavior 509
22.2 Fixed and Learned Behaviors in Animals 510
22.3 Social Behavior in Animals 513
22.4 Facilitating Behavioral Interactions through Communication 515
22.5 Mating Behaviors 517
The Genetics of Domestication 520
Biology in the News: Study Discovers DNA That Tells Mice How to Construct Their Homes 521

23 Ecological Communities 524
Fatal Feline Attraction 525
23.1 Species Interactions 527
Biology Matters: Introduced Species: Taking Island Communities by Stealth 532
23.2 How Species Interactions Shape Communities 534
23.3 How Communities Change over Time 536
23.4 Human Impacts on Community Structure 539
How a Parasite Can Hijack Your Brain 542
Biology in the News: Beyond the Kiss, Mistletoe Helps Feed Forests, Study Suggests 543

24 Ecosystems 546
Deepwater Horizon: Death of an Ecosystem? 547
24.1 How Ecosystems Function: An Overview 548
24.2 Energy Capture in Ecosystems 550
24.3 Energy Flow through Ecosystems 552
24.4 Biogeochemical Cycles 554
24.5 Human Actions Can Alter Ecosystem Processes 559
Biology Matter: Is There a Free Lunch? Ecosystems at Your Service 563
What Happens When the Worst Happens? 564
Biology in the News: Too Much Deer Pee Changing Northern Forests 565

25 Global Change 568
Is the Cupboard Bare? 569
25.1 Land and Water Transformation 570
25.2 Changes in the Chemistry of Earth 573
25.3 Human Impacts on the Global Carbon Cycle 575
25.4 Climate Change 576
Biology Matter: Toward a Sustainable Society 581
25.5 Timely Action Can Avert the Worst-Case Scenarios 583
Bye-Bye, Food Chain? 584
Biology in the News: Blind, Starving Cheetahs: The New Symbol of Climate Change? 585

Appendix: The Hardy-Weinberg Equilibrium APP1
Table of Metric-English Conversion T1
Periodic Table of the Elements T2
Self-Quiz Answers AS1
Analysis and Application Answers AA1
Glossary GL1
Credits CR1
Index I1

DISCOVER BIOLOGY

Sixth Edition

1 The Nature of Science and the Characteristics of Life

ARSENIC BACTERIA. Geomicrobiologist Felisa Wolfe-Simon collects mud from the bottom of Mono Lake, California. Bacteria from the lake tolerate high concentrations of salt and arsenic.

Earthbound Extraterrestrial? Or Just Another Microbe in the Mud?

A few years ago, NASA made an announcement that put the national science news media on a roller-coaster ride of excitement. The press release launched a debate that spilled from the sedate pages of scientific journals to the world of social media. NASA-funded researchers had discovered a bacterium that could grow in high concentrations of poisonous arsenic. Not only that, but this bizarre bacterium builds arsenic right into its DNA, according to the researchers. The chemical building blocks of DNA are the same in all organisms, as far as anyone knew. But NASA's announcement suggested that this one bacterium was replacing the phosphorus in DNA with arsenic. It was comparable, they said, to something you'd find on another planet.

The news media ran with the story. The *Huffington Post* wrote, "In a bombshell that upends long-held assumptions about the basic building blocks of life, scientists have discovered a whole new type of creature: a microbe that lives on arsenic." And NASA's breathless press release—titled "Get Your Biology Textbook ... and an Eraser!"—said that the discovery "begs a rewrite of biology textbooks by changing our understanding of how life is formed from its most basic elemental building blocks." But wait. Don't start erasing just yet.

Does this bizarre bacterium *really* build arsenic into its DNA? Why was NASA's announcement met with both excitement and skepticism?

We will return to this story toward the end of the chapter. We will see that back-and-forth arguments are common at the cutting edge of science. Radical scientific claims invite intense scrutiny and vigorous debate—as they should. Far from being a weakness, skeptical inquiry—and the insistence on sound evidence—is the greatest strength of science.

We begin this chapter with a focus on what science is and what it is not. As we proceed, you'll see that the process of science—also known as the scientific method—is what distinguishes science from other ways of understanding the world around us. Next, we'll turn our attention to the subject that is at the heart of the rest of the book: biology, the scientific study of the living world.

MAIN MESSAGE The scientific method is an evidence-based system for understanding our world, including living organisms. Because of their common evolutionary origin, all living organisms share certain key characteristics.

KEY CONCEPTS

- Science is a body of knowledge about the natural world and an evidence-based process for generating that knowledge. Biology is the scientific study of the living world.
- Scientific inquiry begins with observations of nature. The scientific method involves generating and testing hypotheses about those observations.
- Hypotheses can be tested with observational studies, experiments, or both.
- In an experiment, investigators manipulate one aspect of nature (independent variable) and study how that action affects another aspect of nature (dependent variable).
- A scientific fact is a direct and repeatable observation of a particular aspect of the natural world.
- A scientific theory is a major explanation about the natural world that has been repeatedly confirmed in diverse ways and is accepted as part of scientific knowledge.
- All living organisms are composed of one or more cells, reproduce using DNA, acquire energy from their environment, sense and respond to their environment, maintain their internal state, and evolve.
- Biological evolution is a change in the overall genetic characteristics of a group of organisms over successive generations.
- Life on Earth can be studied on many levels, from atom to biosphere.

EARTHBOUND EXTRATERRESTRIAL? OR JUST ANOTHER MICROBE IN THE MUD? 3

1.1 The Nature of Science 4
1.2 The Process of Science 9
1.3 Scientific Facts and Theories 13
1.4 The Characteristics of Living Organisms 14
1.5 Biological Evolution and the Unity and Diversity of Life 17
1.6 The Biological Hierarchy 20

APPLYING WHAT WE LEARNED 23
Researchers Wrangle over Bacteria

THIS BOOK IS ABOUT YOU, the rest of the living world around you, and the intricate web that connects living beings to one another and to their surroundings. As you explore the story of life, you will develop an appreciation of how science works and a deeper understanding of how life works.

Science is at the heart of many of the big issues we face as a society. A few examples are genetic testing and the confidentiality of personal genetic data, research on embryonic stem cells, and what to do about climate change. Then there are issues that are not as urgent but that also stir up controversy; for example, should we use DNA technology to bring back prehistoric beasts that became extinct many thousands of years ago (**FIGURE 1.1**)? What do you think?

FIGURE 1.1 Is De-Extinction a Good Idea?
Woolly mammoths, like these characters from the film *Ice Age: Continental Drift*, were hunted to extinction toward the end of the last ice age. The last of them survived until about 3,000 years ago on an island off Siberia. It is theoretically possible to take DNA from cells preserved in the permafrost and bring these ancient behemoths back to life. Is "resurrection" of extinct life-forms a good idea? Should public funds be spent on these projects? Come back to this figure toward the end of your biology class and see if you want to revise your answers.

Opinions on these issues are often influenced by personal values and individual concerns. Commercial and political interests also have an impact on the application of scientific knowledge. But a shared understanding of the underlying science offers the hope of rational debate and constructive social action on these complex issues.

We begin this chapter with a look at science as a way of knowing and as a body of knowledge about the natural world. Next we turn our attention to **biology**, the scientific study of life, by asking what, exactly, is meant by that powerful word: "life." As you will see, all living things, diverse though they are, are related and have certain characteristics in common. Furthermore, all living organisms are part of an interlinked pattern we call the hierarchy of life.

1.1 The Nature of Science

You probably asked a lot of questions when you were a child: What is that? How does it work? Why does it do that? We are driven by a deep-seated curiosity about the world around us, a tendency to ask questions that we seem to express most freely when we are children (**FIGURE 1.2**). Through the centuries, that spirit of inquiry has been the main driving force behind science.

Beyond the universal thirst for understanding, science offers many practical benefits. **Technology** refers to the practical application of scientific techniques and principles. Science is behind technologies like satellites and the TV receivers that use them, lifesaving medical procedures and drugs, microwave ovens, and every text, tweet, and image we send over the Internet. But beyond being a provider of technologies, science is a way of understanding the world.

The scientific way of looking at the world—let's call it *scientific thinking*—is logical, strives for objectivity, and values evidence over all other ways of discovering the truth. Scientific thinking is one of the most democratic of human endeavors because it is not owned by any group, tribe, or nation, nor is it presided over by any human authority that is elevated above ordinary humans.

People like you are contributing to the advance of science

In recent years, hundreds of Citizen Science projects have been undertaken. In these projects, people from all walks of life partner with professional researchers to advance scientific knowledge (**TABLE 1.1**). Citizen scientists have volunteered their DNA, tracked bees, recorded the blooming and fruiting of local plants, monitored invasive species, cataloged roadkills, donated their gaming skills to help predict protein structure, or simply loaned the computing capacity of their workstations to scientists over the Internet.

Crowdsourcing using the services of the online community to create content and solve problems—has made large contributions to understanding genetic variation in humans.

The Christmas Bird Count, organized by the National Audubon Society since 1900, is the longest-running Citizen Science project. Data gathered by hundreds of thousands of volunteers have contributed to over 200 technical papers and helped scientists understand how climate change is affecting bird migration (**FIGURE 1.3**).

FIGURE 1.2 Curiosity Is at the Heart of Scientific Inquiry
Do you recall any questions about nature that you asked when you were little?

Science is a body of knowledge and a process for generating that knowledge

Science takes its name from *scientia*, a Latin word for "knowledge." Science is a particular kind of knowledge, one that deals with the natural world. By "natural world" we mean the observable universe around us—that which can be seen or measured or detected in some way by humans. We can define **science** as a body of knowledge about the natural world and an evidence-based process for acquiring that knowledge (**TABLE 1.2**).

As implied in the definition, science is much more than a mountain of knowledge. Science is a particular system for generating knowledge. The processes that generate scientific knowledge have traditionally been called the **scientific method**, a label that originated with nineteenth-century philosophers of science. Although "method" is singular, the scientific method is not one single sequence of steps or a set recipe that all scientists follow in a rigid manner. Instead, the term represents the core logic of how science works. Some people prefer to speak of the "process of science," rather than the scientific method. Whatever we call it, the procedures that generate scientific knowledge can be applied in a broad range of disciplines—from social psychology to forensics.

The scientific method can be illustrated in a concept map like the one in **FIGURE 1.4**. A concept map is a diagram illustrating how the components of a particular structure, organization, or process relate to each other. Concept maps help us visualize how parts fit together and flow from one another. Keep this in mind as we explore the process of science in this chapter.

TABLE 1.1 Citizen Science Projects: Collaborations between Citizens and Professional Scientists

PROJECT NAME	ROLE OF CITIZEN SCIENTIST	PROJECT HOST
Cell Slider	Classify images of cancer cells, to help accelerate cancer research.	Cancer Research UK
uBiome	Donate samples of microbes (microscopic organisms) from your own body, to help catalog diversity of microbes that make their home on the human body. Note: Participants are asked to make a donation.	Research lab, Oxford University, UK
Tag a Tiny	Catch, measure, and release juvenile Atlantic bluefin tuna after attaching ID tags.	Large Pelagics Research Center, Gloucester, MA
Mastodon Matrix Project	Analyze samples of fossil dirt (found around mastodon bones) mailed to your home by the researchers.	Paleontological Research Institution, Ithaca, NY
Play with Your Dog	Send video of you playing with your dog, to help dissect the human-canine relationship.	Horowitz Dog Cognition Lab, NYC

NOTE: For more examples of such projects, and information on how to participate, visit scistarter.com.

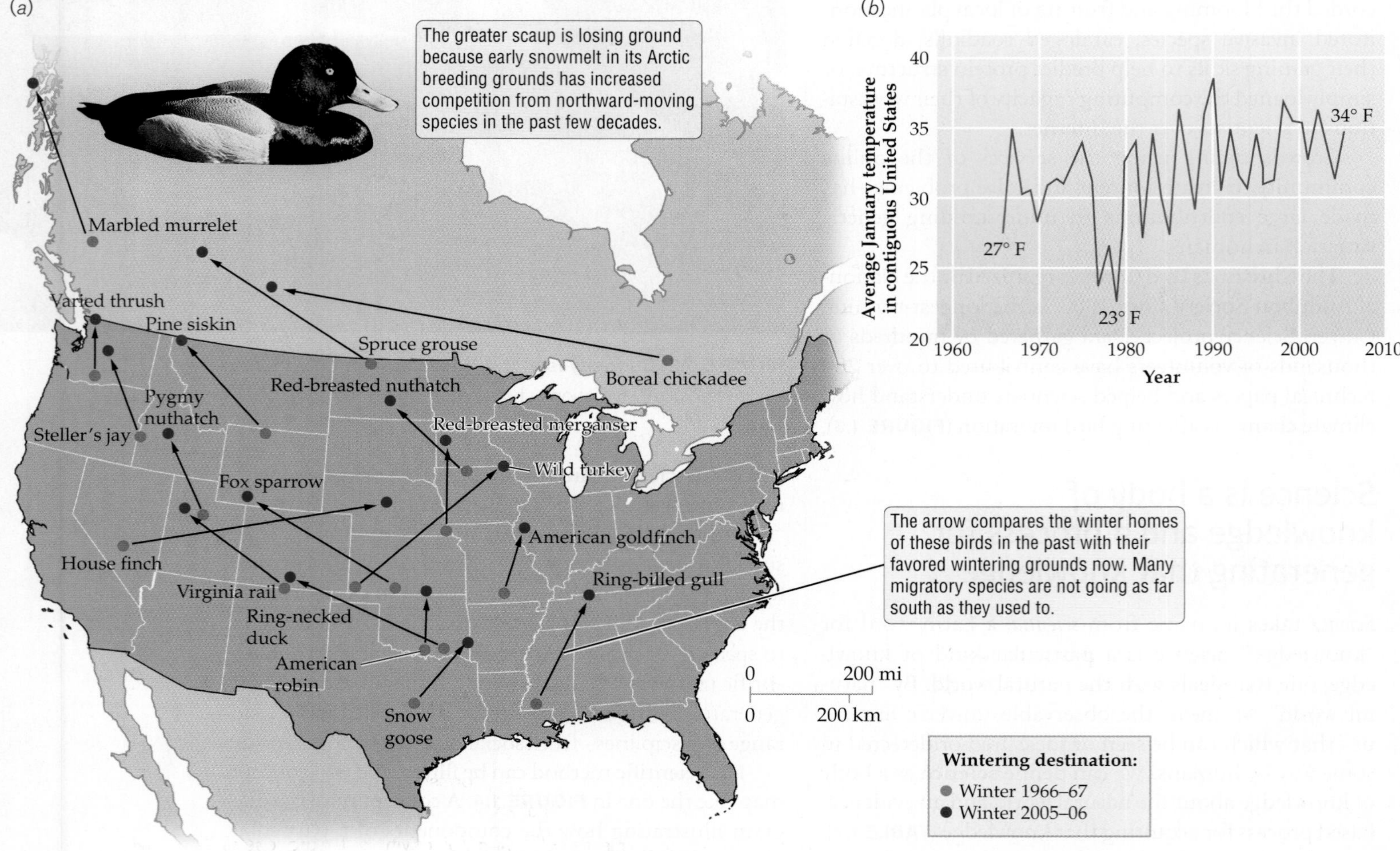

FIGURE 1.3 The Winter Range of Many Migratory Birds Has Shifted Northward
This range map (*a*) is based on data collected by volunteers participating in the Christmas Bird Count, the longest-running Citizen Science project. It shows that many migratory birds in North America are not going as far south in the nonbreeding season as they did just 40 years ago. The graph (*b*) shows the change in average January temperature in the United States over the same period. Globally, the average surface temperature of Earth has increased by about 1°C in the last 100 years—a phenomenon known as global warming.

TABLE 1.2 Characteristics of Science

SCIENCE
▪ deals with the natural world, which can be detected, observed, and measured.
▪ is based on evidence from observations and/or experiments.
▪ is subject to independent validation and peer review.
▪ is open to challenge by anyone at any time on the basis of evidence.
▪ is a self-correcting endeavor.

Scientific hypotheses must be testable

A **scientific hypothesis** (plural "hypotheses") is an educated guess that seeks to explain observations of nature. In science, a hypothesis is useless unless it is testable. The tests could be observational studies or experiments or both. Who conducts these tests? No matter how creative and plausible the hypothesis is, the burden of testing it rests on the person proposing the hypothesis.

When a hypothesis is tested and upheld, it is said to be *supported*. A supported hypothesis is one in which we can be relatively confident. If the tests show the hypothesis to be wrong, then the hypothesis has been refuted (shown to be false). Sometimes a test neither supports

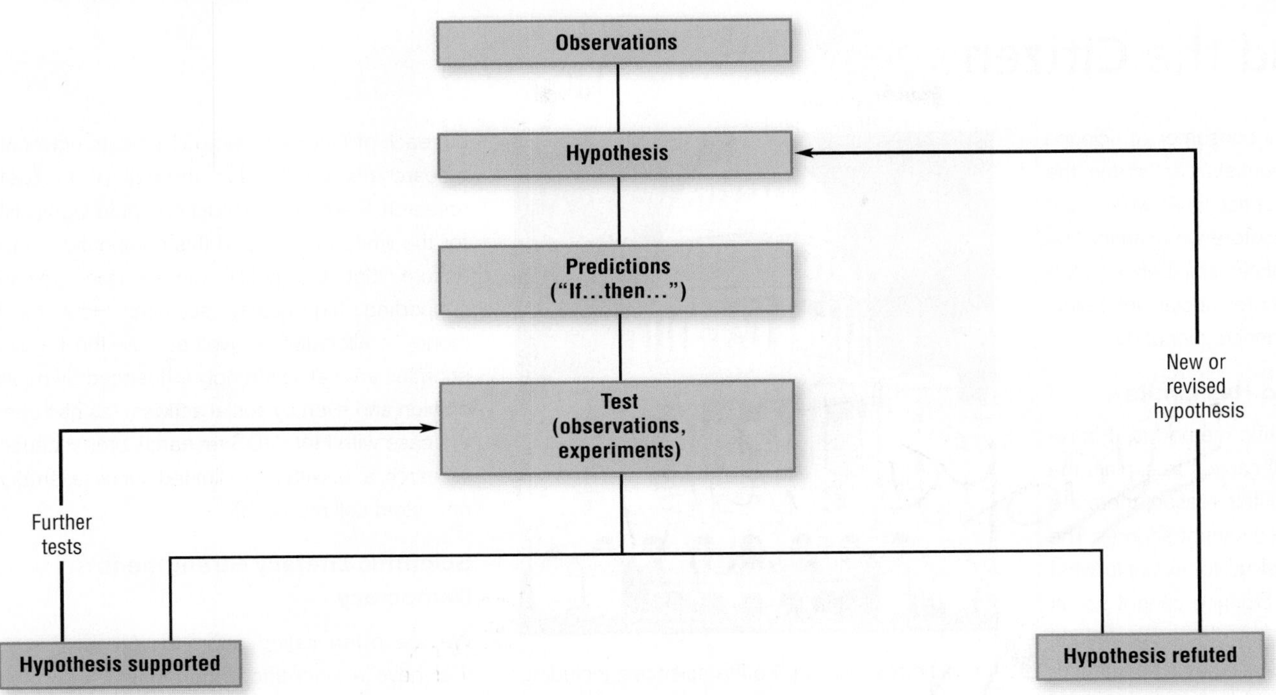

FIGURE 1.4 Scientific Hypotheses Must Be Testable

nor refutes a hypothesis, in which case the test is declared inconclusive and the investigators must find a better test.

The scientific method requires objectivity

An absolute requirement of the scientific method is that evidence must be based on observations or experiments or both. Furthermore, the observations and experiments that furnish the evidence must be subject to testing by others; independent researchers should be able to make the same observations, or obtain the same experimental results if they use the same conditions. In addition, the evidence must be collected in an objective fashion—that is, as free of bias as possible. As you might imagine, freedom from bias is more an ideal than something we can depend on. However, modern science has safeguards in place to ensure that scientific knowledge will come closer to meeting that ideal over time.

The main protection against bias, and even outright fraud, is the requirement for *peer-reviewed publication*. Claims of evidence that are confined to a scientist's notebook or the blogosphere do not meet the criterion of peer-reviewed publication. A *peer* is someone at an equal level—in this case, another scientist who is recognized as expert in the field. Peer-reviewed publications are scientific journals that publish original research only after it has passed the scrutiny of experts who have no direct involvement in the research under review (**FIGURE 1.5**).

Concept Check

1. What characteristics of the process of science set it apart from other ways of knowing?
2. What mechanisms help bring objectivity to the process of science?

Concept Check Answers

1. Scientific knowledge is acquired through evidence, and its objectivity and accuracy are policed through peer review.
2. Repeatability of observations and experiments, and the requirement for peer-reviewed publication of scientific findings.

FIGURE 1.5 Some Peer-Reviewed Science Journals The criterion of peer-reviewed publication is one means of enforcing rigor and objectivity in the application of the scientific method.

BIOLOGYMATTERS

Science and the Citizen

The public is not simply a consumer of science and its spin-offs. Nonscientists can shape the course of science and influence what, where, and how technology is used. Before we examine the many ways in which the relationship between science and the citizen promotes social well-being, let's first consider what science *cannot* do.

The Scientific Method Has Limits

As powerful as the scientific method is, it is restricted to seeking natural causes to explain the workings of our world. For this reason, there are areas of inquiry that science cannot address. The scientific method cannot tell us, for example, what is morally right or wrong. Science cannot speak to the existence of God or any other supernatural being. Nor can science tell us what is beautiful or ugly, which poems are lyrical, or which paintings most inspiring. So although science can exist comfortably alongside different belief systems—religious, political, and personal—it cannot answer all their questions.

According to a 2010 poll by the Pew Research Center, 61 percent of the American public sees no conflict between science and their own beliefs. The same poll shows that 85 percent of the American public views science as having a mostly positive effect on society.

Public-Funded Research Contributes to the Advancement of Science

In North America, the vast majority of *basic research* in science is funded by the federal government—that is, by taxpayers. Basic research is intended to expand the fundamental knowledge base of science. Many industries and businesses spend a great deal of money on *applied research*, which seeks to commercialize the knowledge gained from basic research. The new drugs, diagnostic tests, and medical technology that biomedical companies introduce each year are, ultimately, the fruit of the public investment in basic research.

In the United States, the federal government appropriates about $40 billion each year for basic research in the life sciences, including biomedicine and agriculture. These funds are disbursed mainly to four federal agencies: the National Institutes of Health (NIH), the National Science Foundation (NSF), the U.S. Department of Energy (DOE), and the U.S. Department of Agriculture (USDA). Some of these agencies have their own research institutes and laboratories, but each of them also awards funds to university researchers, who conduct the bulk of the basic research. Researchers must compete vigorously for the limited funds, and this competition helps ensure that the public money goes toward supporting high-quality science. How much money is allocated, as well as how the funding priorities are set, is strongly influenced by public opinion and even by social activism (as has been the case with HIV-AIDS research, breast cancer research, and, with more limited success, embryonic stem cell research).

Scientific Literacy Strengthens Democracy

We are often called upon to vote on issues that have a scientific underpinning. The table below lists some of the science-related ballot measures that have been put to the vote during state and local elections in the United States in recent years. Although our personal values and political leanings are likely to influence how we vote, most would agree that the underlying science should be taken into consideration.

Some Statewide Ballot Measures on Science-Related Issues

INTENT OF PROPOSED INITIATIVE/ REFERENDUM	STATE	YEAR INTRODUCED/ OUTCOME	INITIATIVE OR REFERENDUM
Proposal 2012-03	Michigan	2012/passed	To require that at least 25% of the state's energy is from renewable sources
Medical Marijuana Initiative	Massachusetts	2012/passed	To legalize the sale of limited amounts of marijuana to patients with a doctor's prescription
Proposition 37	California	2012/failed	To require labeling of foods containing parts of genetically modified organisms (GMOs)
Initiative 1107	Washington	2010/failed	To repeal the 2-cent sales tax on candy, soda pop, and bottled water, legislated initially for health and environmental reasons

NOTE: A *ballot measure* is a referendum or initiative that is put to the vote in state or local elections. A *referendum* originates with the state legislature, whereas an *initiative* is brought forward by a petition from citizens (who could be backed by special interests). Citizen initiatives are given different names in different states ("proposition," "proposal," or "measure," for example), and not all states have a system of citizen initiatives.

1.2 The Process of Science

Scientific inquiry generally begins as an attempt to explain observations about the natural world. For example, since the 1970s, nutritionists have noticed that death from heart disease is less common in communities that eat a lot of fish and other seafood. In science, just as in everyday life, observations lead to questions, and questions lead to potential explanations. Why is fish consumption linked to better heart health? Could it be that a particular nutrient in fish protects people from heart disease? In the next section we see how the scientific method can be used to solve riddles of this sort.

FIGURE 1.6 Through Observations, We Collect Descriptive Data and Find Patterns in Natural Phenomena

Observations are the wellspring of science

Science aims to explain observations about the natural world. An **observation** is a description, measurement, or record of any object or phenomenon. We can study nature in many different ways: by looking through a microscope, diving to the ocean floor, walking through a meadow, studying satellite images of forest cover, running chemical tests with sophisticated instruments, or using remote cameras to photograph a secretive animal (**FIGURE 1.6**).

To be of any use in science, *an observation must be reproducible*: independent observers should be able to see or detect the object or phenomenon at least some of the time. Sightings of Sasquatch ("Big Foot"), for example, lack credibility precisely because fans of "cryptobiology" have failed to produce samples or recorded images or sounds that stand up to scrutiny by independent observers.

Observations of nature can be purely descriptive, reporting information (**data**) about what is found in nature: where, when, how much. Mapping the types of sea creatures found in different zones on a rocky shore, listing the flowers in bloom through the growing season in an alpine meadow, counting how many birds of prey and how many perching birds are found on an island—all these are examples of descriptive studies.

Observations take an analytical turn when they identify patterns in nature and ask what *causes* those patterns. As examples, people who eat a lot of fish seem to have better heart health than those who do not each much fish (**FIGURE 1.7**). One species of small barnacles is always found above the high-tide mark,

The Scientific Method in Action

The observations

Native Alaskans and Greenland Inuits have high per capita fish consumption. These populations have some of the lowest rates of heart disease in the world.

The hypothesis

Fish oils in the diet reduce the risk of death from heart disease.

The predictions

If the hypothesis is true, *then* . . .

(1) Those who eat more fish will have a lower risk of death from heart disease.

(2) In people at risk for heart attacks, mortality will be lower in individuals who take fish oil supplements compared to those who don't.

FIGURE 1.7 The Scientific Method Begins with Observations, Hypothesis, and Predictions

but a related species is seen only low on the shoreline. Yellow-flowered plants bloom earlier than red-flowered ones in a particular alpine meadow. Perching birds always outnumber birds of prey.

Scientific hypotheses make clear-cut predictions

A scientific hypothesis is an informed, logical, and plausible explanation for observations of the natural world. Investigators know what is plausible, or possible, if they have a good understanding of what is already known; that is why a new scientific hypothesis is often called an *educated* guess. Keeping established knowledge in mind helps researchers to avoid "reinventing the wheel" and to choose among alternative, arriving at one that is most probable.

A well-constructed hypothesis should be stated clearly and should avoid vagueness or ambiguity. For example, "Fish is good for you" is too vague. What is meant by "good"? How would someone test for "goodness"? Now consider a more precisely worded hypothesis: "In a large and representative population of humans, individuals who eat fish regularly are less likely to die of heart disease than those who consume little or no fish."

A precisely stated hypothesis enables clear-cut predictions, which are essential for testing that hypothesis. Predictions made by scientific hypothesis can be cast as "If . . . then" statements. To frame these predictions, the investigator asks, "*If* my hypothesis is correct, *then* what else can I expect to happen?" For the hypothesis shown in Figure 1.7, we can make this prediction: *If* it is true that fish consumption is good for heart health, *then* in a large population of humans, individuals who consume more fish will have a lower risk of death from heart disease than will individuals who consume little or no fish.

Scientific hypotheses must be refutable but cannot be proved beyond all doubt

It should be possible—at least in principle—to show that a scientific hypothesis is false. A classic example of a refutable hypothesis is the statement "All swans are white." Finding a single black swan would show this hypothesis to be false (as it happens, there *is* a species of swan in Australia that is black).

TABLE 1.3 Criteria for Scientific Hypotheses

A SCIENTIFIC HYPOTHESIS

- is an educated guess that seeks to explain observed phenomena.
- makes clear predictions that can be arranged in "if … then" statements.
- must be testable repeatedly and independently.
- must be potentially refutable.
- can never be proved, but only supported or refuted.

Irrefutable hypotheses are common in *pseudoscience*, which is the practice of using arguments that sound vaguely scientific without actually using the scientific method. Consider this claim: space aliens are among us, but because of their advanced extraterrestrial technology, humans cannot detect them. There is no way to test this hypothesis, and there is no way to show it to be false either.

The hypothesis about the heart benefits of fish consumption is both testable and refutable. What if observational tests show that people who eat fish have *higher* odds of dying from heart disease than those who eat no fish at all, all other aspects of their lifestyle being equal? In this case, not only would the hypothesis lack supporting evidence, but we would know it to be false.

An experiment can provide strong support for a hypothesis, increasing our confidence that it is correct, but no hypothesis can be *proved* beyond all doubt (**TABLE 1.3**). One would have to know everything about everything to be certain that every test that could ever be devised would always support a certain hypothesis. Such certainty, of course, is not possible. Albert Einstein famously said, "No amount of experimentation can ever prove me right; a single experiment can prove me wrong."

What if the original experimenter failed to take into account an important factor affecting the outcome of an experiment simply because he or she was unaware of its importance? For example, what if those who eat a lot of fish also eat less saturated fat, and it

EXTREME REFUTATION

A single black swan disproves the hypothesis that all swans are white. This species is native to Australia.

is lower fat consumption that protects against fatal heart disease rather than fish consumption per se?

Good researchers try to be alert to alternative explanations, but there is no guarantee that any and all complexities have been anticipated and accounted for by the investigator. In fact, unforeseen complexities, together with the limits imposed by the technologies of the day, are the most common reason that accepted ideas in science are sometimes overturned. However, our confidence in a hypothesis grows as many researchers—perhaps testing the same hypothesis in different ways and in different contexts—add their independent support.

Hypotheses can be tested with observational studies

We can test a hypothesis through observational studies or experiments or both. In studying nature, no matter what the approach, we focus on variables. A **variable** is any aspect of nature that is capable of changing. Fish consumption is a variable, as is death (mortality) from heart disease. Observations help us identify variables that are potentially interesting or that could potentially explain the patterns we might observe in nature. We can then design observational studies or experiments to tease out the relationship between any two variables.

The graph in **FIGURE 1.8** is based on data pooled from several observational studies. The data show an association between fish consumption and the risk of dying from heart disease. But a question arises: Could this association be a coincidence? That is, could the link be a matter of chance? The researchers used the tools of **statistics**—a mathematical science that uses probability theory to estimate the reliability of data—to show it was very unlikely that the link between fish consumption and lower risk of death from heart disease was just random.

Correlation is not causation

The statistical analysis showed that fish consumption and heart disease mortality were correlated. **Correlation** means that two or more aspects of the natural world behave in an interrelated manner: if we know the value of one variable (fish consumption per week), we can predict a particular value for the other variable (risk of mortality from heart disease). Does the statistical correlation mean that fish consumption was the *cause* of better heart health among the study subjects? The answer is: no, not necessarily!

What if fish consumption and heart disease were separately linked to a *third variable*—higher income, for example? It could be that wealthier people eat more fish because they can afford to, and they are less likely to die of heart disease because they have better access to high-quality health care. In that case, higher income is the cause of both the higher fish intake and the lower risk of death from heart disease. If so, the apparent link between fish intake and heart health is termed *spurious*. In statistics, the word "spurious" means doubtful, inaccurate, or false. A classic example of a *spurious correlation* is the strong link between the crowing of a rooster and daybreak. Does that correlation mean that the rooster's crowing *caused* the dawning of the day?

Observational studies that indicate a link between two variables cannot demonstrate **causation**; that is, they cannot show that change in one variable *causes* change in the other. However, such studies

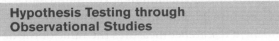

EXTREME CORRELATION

Is global warming driving young people to college? In the last 50 years, college enrollment has gone from 46% to 70% of high school grads, and Earth has warmed by 0.62°C. A third variable—modernization—is the best explanation for this correlation.

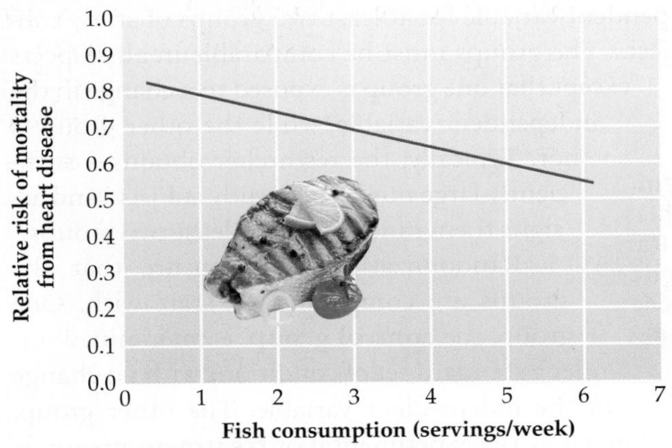

FIGURE 1.8 An Observational Test of the Hypothesis That Fish Consumption Reduces the Risk of Death from Heart Disease

The data plotted in the graph are based on aggregate studies of about 200,000 people over 20 years. The data support the hypothesis but do not demonstrate that fish consumption is the *cause* of the better heart health enjoyed by those who eat fish often.

are very useful in showing the *possibility* of a causal connection. By identifying variables that are linked, observational studies become the basis for additional research. For example, researchers can try to tease out the possibility that a third variable is the cause. In the case of fish consumption, the heart benefit was seen even when researchers compared people in the same income group. By eliminating other variables as possible causes, observational studies can build a persuasive case for causation. However, experiments are the most effective for showing causation.

Experiments are the gold standard for establishing causation

An **experiment** is a repeatable manipulation of one or more aspects of the natural world. In conducting a scientific experiment, an investigator typically manipulates a single variable, known as the **independent variable** or manipulated variable. Any variable that responds, or could potentially respond, to the changes in the independent variable is called the **dependent variable** or responding variable. If we think of the independent variable as the cause, then the dependent variable is the effect. In the simplest experimental design, a single independent variable is changed, and its effects upon a presumed dependent variable are measured.

The **controlled experiment** is a common and particularly useful experimental design. In a controlled experiment, a researcher measures the value of the dependent variable for at least two groups of study subjects. The groups must be comparable in all respects except that one group is exposed to a change in the independent variable, while the other group is not. Typically, the researcher obtains a sufficiently large number of study subjects and assigns them randomly to at least two groups.

Randomization helps ensure that the groups are comparable to start with. One group, the **control group**, is maintained under a standard set of conditions with no change in the independent variable. The other group, known as the experimental or **treatment group**, is maintained under the same standard set of conditions as the control group, but is manipulated in a way that changes the independent variable. In a well-designed experiment, the control and treatment groups are as similar to each other as possible and all variables, other than the independent variable, are held constant.

Helpful to Know

If you find independent and dependent variables too confusingly similar, remember that the *in*dependent variable is the one that the *in*vestigator manipulates.

Hypothesis Testing through Experimental Studies

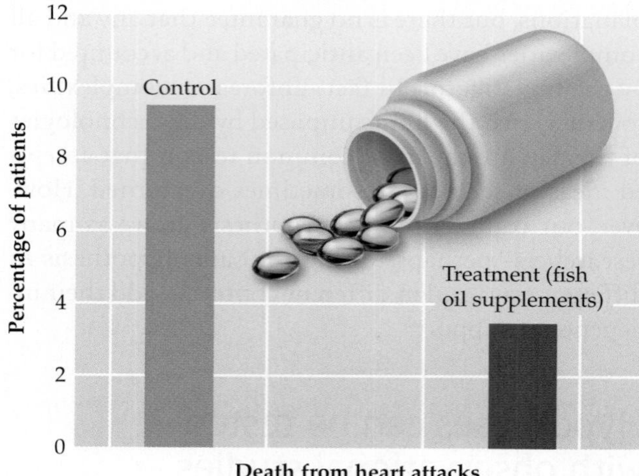

FIGURE 1.9 Effect of Fish Oil Supplements on Mortality from Heart Attacks in an At-Risk Population of Males
In an experimental study of over 2,000 men who had been diagnosed with heart disease, researchers randomly assigned about half to a control group and the other half to a treatment group. The men in the treatment group were asked to take 900 milligrams of fish oils daily, whereas the men in the control group were not directed to alter their diet in any way.

Observational studies had shown that eating cold-water fish such as salmon, trout, and sardines was even more strongly linked to heart health than was eating warm-water fish. Cold-water fish have higher levels of omega-3 oils, which have a particular chemical structure. Researchers designed randomized controlled experiments to test the hypothesis that moderate consumption of fish oils reduces heart disease mortality. They recruited about 2,000 men who had been diagnosed with heart problems, and asked roughly half of them to take 900 milligrams of fish oils daily; the rest of the men, who were not directed to alter their diet in any way, served as the control group.

At the end of the 2-year study period, there were 62 percent fewer deaths in the treatment group compared to the control group (**FIGURE 1.9**). The experimental test therefore supported the hypothesis that fish oils reduced heart disease mortality. However, a potential drawback of the experimental design was that study subjects knew whether or not they were receiving fish oil supplements.

- In a **single-blind experiment**, the study subjects do not know whether they belong to the control group or the treatment group. Instead of the treatment, the control group receives a **placebo**, which is a dummy pill or sham treatment that mimics the actual treatment. This type of experiment avoids the possibility of a **placebo effect**, the sense among study participants that they are feeling better because they have received a beneficial treatment.

- In a **double-blind experiment**, neither the study subjects nor the researchers know which participants are receiving the treatment and which are controls. A double-blind experiment avoids conscious or unconscious bias on the part of the experimenter. **FIGURE 1.10** shows the results of a double-blind experiment conducted on U.S. Navy SEAL trainees to test the hypothesis that caffeine enables sleep-deprived cadets to react more quickly in demanding situations than they could otherwise. All participants ate snack bars—with or without caffeine—that tasted the same.

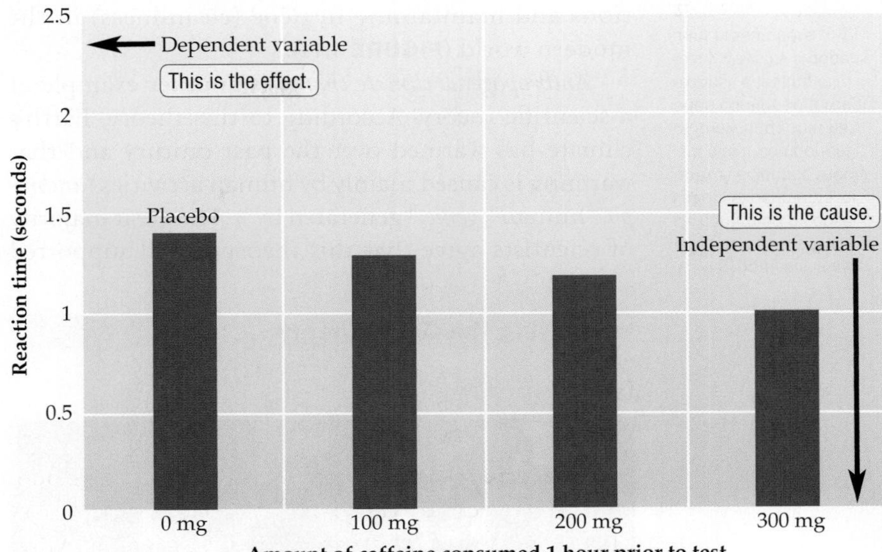

FIGURE 1.10 Effect of Caffeine Consumption on Reaction Time

Reaction time was tested in Navy SEAL trainees who had been sleep deprived for two days. Cadets in the three treatment groups ate a snack bar containing 100, 200, or 300 milligrams of caffeine. Those in the control group ate a snack bar that tasted the same but contained no caffeine. At the time of the test, neither the trainees nor the investigators knew whether a test taker belonged to the treatment group or the control group.

Concept Check

1. Which type of test would you use to investigate a causal link between two variables: an observational study or an experiment?
2. How is a control group different from the treatment (experimental) group?

1.3 Scientific Facts and Theories

What are "facts"? In casual conversation, we typically use the term to mean things that are known to be true, as opposed to things that are simply guesses. A **scientific fact** is a direct and repeatable observation of any aspect of the natural world. An example of such an observation might be that an apple, when dropped, falls to the ground—not up into the sky. Figure 1.3 noted a fact: the average temperature of Earth has increased by about 1 °C in the last 100 years—a phenomenon known as global warming. We noted another fact in Figure 1.7: in the 1970s, Greenland Inuits consumed more fish than the average North American.

What does the word "theory" mean to you? Outside of science, people often use the word to mean an unproven explanation. If something unusual occurs, someone might say, "I have a theory about how that happened." The theory could be anything from a wild guess to a well-considered explanation, but either way it is "just a theory."

Scientists use the term "theory" to mean something very different. If an idea is merely one of many explanations, it is a hypothesis. In contrast, a **scientific theory** is a major explanation about the natural world that has been confirmed through extensive testing in diverse ways by independent researchers. Furthermore, competing hypotheses are ruled out. When experts in the field recognize the validity of one hypothesis, or a set of related hypotheses, these explanations of nature become accepted as a scientific theory.

Scientific theories are not "iffy" ideas. They have such power to predict outcomes that we can base our everyday actions on them. Just one example is the *germ theory of disease*, experimentally tested and verified by Robert Koch in 1890. The germ theory of disease holds that some diseases are caused by

Concept Check Answers

1. A controlled experiment.
2. The two groups are maintained under identical conditions, except that the independent variable stays constant for the control group, while changing for the treatment group.

Concept Check Answers

1. A hypothesis is an educated guess that can be confirmed by observation, experiments, or both. A scientific theory is a major component of knowledge that has been confirmed through extensive testing in many ways by independent researchers.

2. Theory.

microbes, minute organisms visible only with a microscope. This theory is the basis for treating infections and maintaining hygiene (cleanliness) in the modern world (**FIGURE 1.11**).

Anthropogenic climate change is another example of a scientific theory. According to this theory, Earth's climate has warmed over the past century and that warming is caused mainly by human activities (*anthropo*, "human"; *genic*, "generated by"). The great majority of scientists agree that this theory is well supported by many observations and scientific experiments from fields as diverse as climatology, geology, and ecology.

Some of the predictions of this theory have been borne out. Climate models had predicted the melting of glaciers in many parts of the world, as well as a reduction in the summer sea ice in the Arctic Ocean, well before these predictions could be reliably measured. Citizen scientists participating in the Christmas Bird Count have brought to light evidence supporting another prediction: that some migratory species will shift their range northward in response to climate change (see Figure 1.3).

Applying the Germ Theory

(*a*)

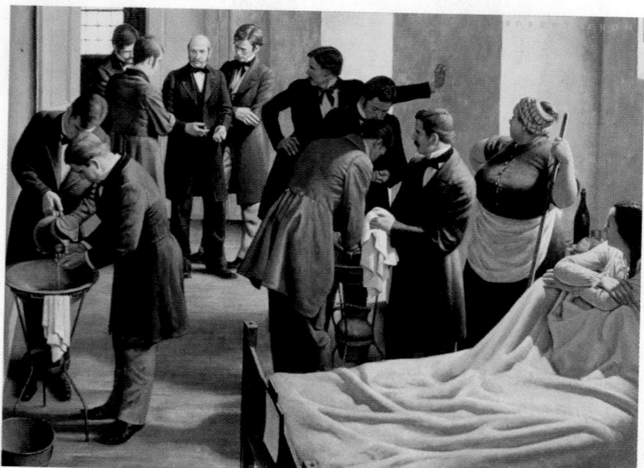

(*b*)

FIGURE 1.11 The Germ Theory of Disease Is a Scientific Theory

According to the germ theory, invisible microorganisms can cause diseases. Application of this theory, through such measures as scrupulous hand washing (*a*), cut death rates in hospital wards by half in the late nineteenth century. (*b*) Subway commuters in Mexico City attempted to protect themselves from infection during a swine flu outbreak in 2009 by wearing surgical masks.

Concept Check

1. How is a scientific hypothesis different from a scientific theory?
2. Is anthropogenic climate change an example of a scientific hypothesis, fact, or theory?

1.4 The Characteristics of Living Organisms

The science that we focus on in the rest of this book is biology, the science of life. But what is life? Although many have tried, no one has produced a simple definition of life that encompasses the great diversity of living forms—from massive redwoods to microscopic bacteria and everything in between. But according to the theory of evolution, all living organisms are descendants of a common ancestor that arose billions of years ago. The theory predicts that as a consequence of their common origin, all living things share some important characteristics (**TABLE 1.4**). We consider each of these characteristics in the sections that follow.

TABLE 1.4 Characteristics of Living Organisms

LIVING ORGANISMS
▪ are composed of one or more cells.
▪ reproduce using DNA.
▪ obtain energy from their environment to support metabolism.
▪ sense their environment and respond to it.
▪ maintain a constant internal environment (homeostasis).
▪ can evolve as groups.

Cellular Organization and DNA

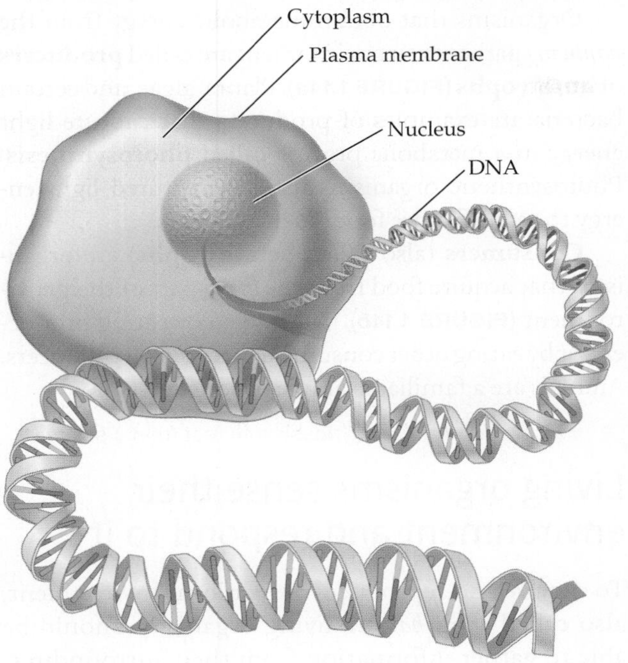

Cytoplasm
Plasma membrane
Nucleus
DNA

FIGURE 1.12 The Cell Is the Smallest Unit of Life

Each cell has a plasma membrane that acts as a barrier between its contents and the surrounding environment. Cells contain DNA, which carries genetic information from one generation to the next.

Living organisms are composed of cells

The first organisms were single cells that existed about 3.8 billion years ago. The **cell** is the smallest and most basic unit of life, the fundamental building block of all living things (**FIGURE 1.12**). Every cell is bounded by an oily layer called the **plasma membrane**. The interior of a cell—known as the **cytoplasm**—is a thick, aqueous (water-based) fluid studded with a number of structures that have specialized functions.

All organisms are made up of one or more cells. A bacterium (plural "bacteria") consists of a single cell. In contrast, large organisms, such as orangutans and oak trees, are made up of many different kinds of specialized cells and are known as **multicellular organisms**. The human body, for example, is composed of about 10 trillion cells of 220 different types, including skin cells, muscle cells, brain cells, and cells that fight disease (immune cells).

Living organisms reproduce themselves via DNA

One of the key characteristics of living organisms is **reproduction**, the ability to produce offspring. Single-celled organisms, such as bacteria, can reproduce by dividing into two cells that are virtually identical copies of the original cell.

Multicellular organisms can reproduce in a variety of ways. Most plants and animals use sexual reproduction to produce *offspring*, which is the next generation of the organism. In **sexual reproduction**, male cells (*sperm*) fuse with female cells (*eggs*) to produce a single cell (*zygote*) in the process known as **fertilization** (**FIGURE 1.13**). The zygote divides many times to produce a juvenile (young) offspring that matures into an adult—a process known as **development**.

Some organisms—including many plants and some animals—can multiply themselves through **asexual reproduction**—that is, without the involvement of specialized reproductive cells such as sperm and eggs. Sponges can bud off new individuals, and many plants can produce side shoots that can break off and develop into new individuals.

Sexual Reproduction

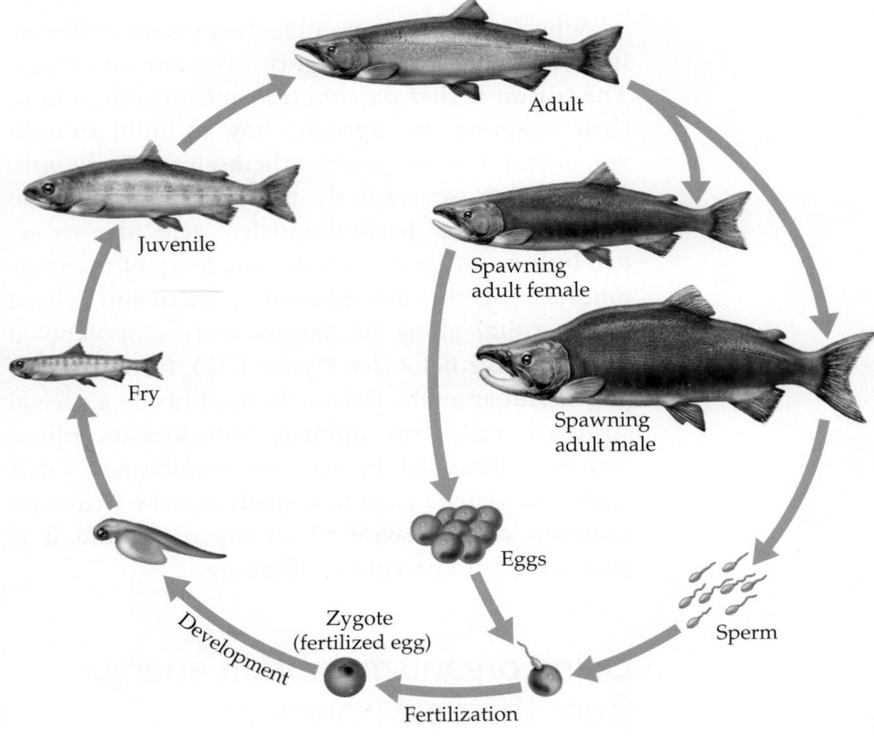

FIGURE 1.13 In Sexual Reproduction, Offspring Receive Genetic Information from Two Different Individuals

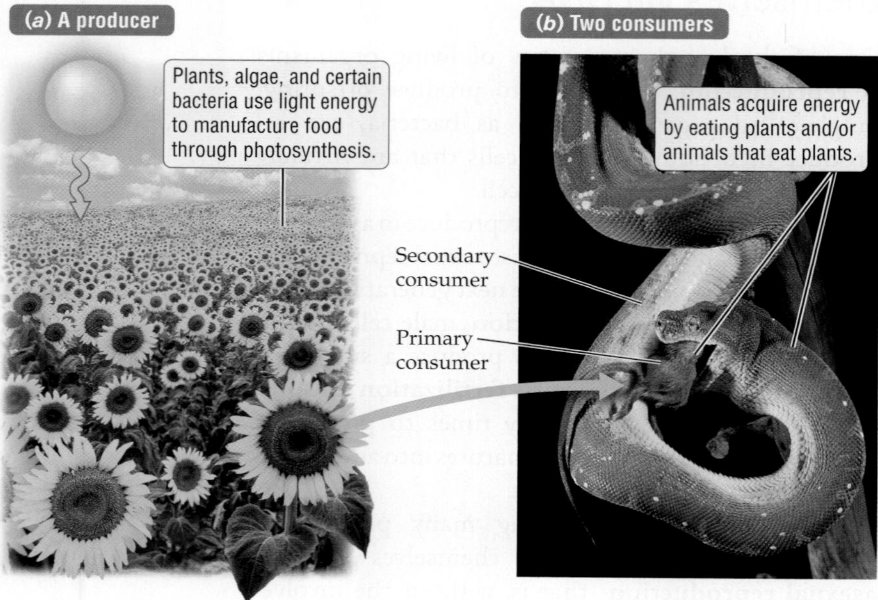
Producers and Consumers
(a) A producer
Plants, algae, and certain bacteria use light energy to manufacture food through photosynthesis.
(b) Two consumers
Animals acquire energy by eating plants and/or animals that eat plants.
Secondary consumer
Primary consumer

FIGURE 1.14 Energy Capture by Organisms
Whereas plants can capture energy from sunlight through photosynthesis (*a*), animals must get their energy by eating other organisms (*b*). The green tree python pictured in (*b*) is ingesting a source of energy (a rat) that itself derived energy from eating plants.

Why do offspring resemble their parents? Why do little acorns grow into mighty oaks, not into frogs? The reason is that parents transmit information to their offspring that specifies how to build an individual very like the parents themselves. That *genetic information* is carried in the form of a large chemical called **DNA**, or deoxyribonucleic acid (**dee-*OX*-ee-*RYE*-boh-noo-*CLAY*-ic**). DNA is made up of many atoms held together in a ladderlike pattern and twisted into a spiral along its length—a structure known as the double helix (see Figure 1.12). In the cells of plants and animals, DNA is housed inside a special structure, called the **nucleus** (***NOO*-klee-us**; plural "nuclei"), bounded by its own membranes. DNA stores the genetic code that spells out the structure, function, and behavior of an organism, and it is passed on from parents to offspring.

Living organisms obtain energy from their environment

All organisms need energy to persist. Organisms use a wide variety of methods to capture this energy from their environment. The capture, storage, and use of energy by living organisms is known as **metabolism**.

Organisms that obtain metabolic energy from the *nonliving* part of their environment are called **producers** or **autotrophs** (**FIGURE 1.14*a***). Plants, algae, and certain bacteria are examples of producers that capture light energy in a metabolic process called **photosynthesis**. Photosynthetic organisms use the captured light energy to manufacture food, such as sugars.

Consumers (also called **heterotrophs**) are organisms that acquire food from the *living* part of their environment (**FIGURE 1.14*b***). They do so by eating producers, or by eating other consumers that feed on producers. Animals are a familiar example of consumers.

Living organisms sense their environment and respond to it

To survive and reproduce in their local environment, also called their *habitat*, living organisms should be able to gather information from their surroundings and react to it in ways that are beneficial to them. Living organisms sense many aspects of their external environment, from the direction of sunlight (as the sunflowers in Figure 1.14*a* illustrate) to the presence of food and the suitability of a potential mate. In biology, sensing and responding to external cues is termed **behavior**. Behavior is most obvious—often dramatically so—in animals; however, all kinds of organisms, from bacteria to plants, exhibit behavior.

Like humans, many animals can smell, hear, taste, touch, and see the environment around them. Some organisms can sense things that humans are not good at perceiving, such as ultraviolet light, electrical fields, and ultrasonic sounds. Plants can sense gravity and ensure that their roots grow toward it, while their shoots grow away from it. Some bacteria have magnetic particles inside that act like a miniature compass; they direct the migration of the cell by signaling which way is north or south, up or down.

Living organisms actively maintain their internal conditions

Living organisms sense and respond not only to external conditions, but also to their internal state. Most cells, and many multicellular organisms as well, maintain remarkably constant internal conditions—a process known as **homeostasis**. For example, the internal temperature of the human body is held at

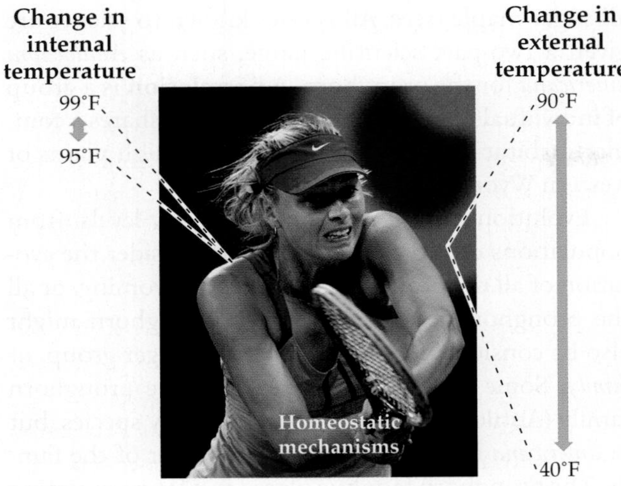

FIGURE 1.15 Homeostatic Mechanisms Enable Organisms to Maintain a Relatively Constant Internal State

Core body temperature varies slightly from person to person, and over the course of the day in a healthy individual. The average daily oral temperature for most people is 37°C or 98.6°F.

98.6°F (37°C) on average. When heat or cold threatens to alter the core body temperature, our bodies quickly respond—for example, by sweating to cool us (**FIGURE 1.15**) or by shivering to warm us.

Groups of living organisms can evolve

Biological evolution refers to a change in the overall genetic characteristics of a group of organisms from one generation to the next. Biological evolution is considered a fact because scientists can demonstrate evolutionary change through repeatable experiments—in a matter of days using bacteria growing in a lab dish, for example.

The process by which humans cause evolutionary change in other organisms is known as *artificial selection* (**FIGURE 1.16**). The dramatic changes brought about by artificial selection in domesticated plants and animals inspired nineteenth-century biologists to wonder whether nature—instead of human intervention—could cause new forms of life to evolve from ancestral forms.

Today, a great deal of evidence shows that groups of organisms evolve in their natural environment. An ancestral group can give rise to one or more descendant groups that are genetically unique. If the groups become so different from each other that they can no longer interbreed, we recognize them as different *species*. *Why* do groups of organisms evolve in their natural environment? We discuss one of the mechanisms—*natural selection*—in Section 1.5.

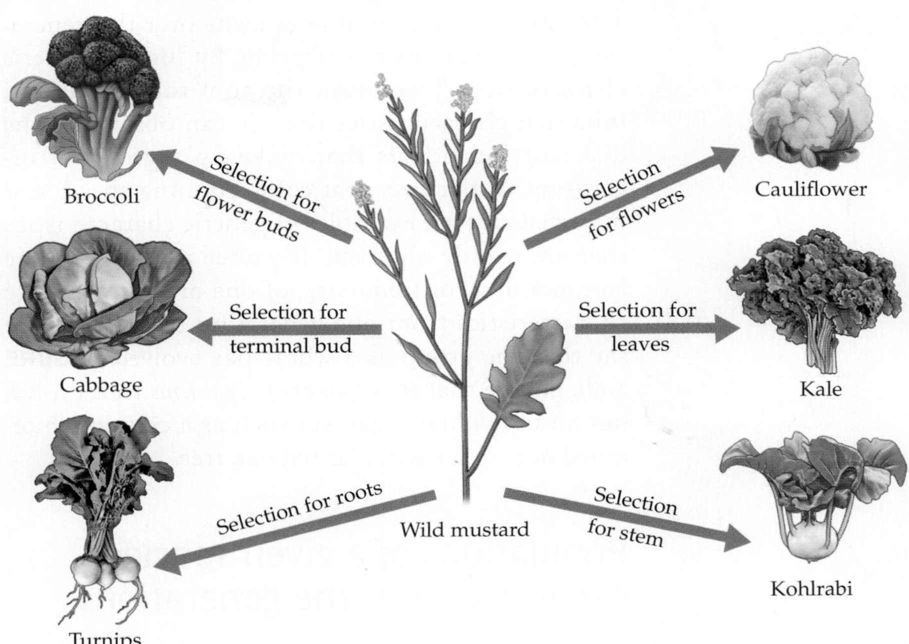

FIGURE 1.16 Evolution of Wild Mustard through Domestication

In artificial selection, humans are the selective agents. All the crops shown here were created by Mediterranean gardeners through selective breeding of genetically varied wild mustard plants over the course of about 200 years.

Concept Check

1. What is the role of DNA in reproduction?
2. Are humans producers or consumers in the food web? Why?

Concept Check Answers

1. DNA stores information that can be passed from parent to offspring, through asexual or sexual reproduction.
2. We are consumers because we must obtain energy from the living, rather than the nonliving, part of our environment.

1.5 Biological Evolution and the Unity and Diversity of Life

The fact of biological evolution, and the theories that explain the mechanisms by which it occurs, are central to our modern understanding of life. Evolution explains the *unity* of life: why all living organisms are like each other in certain key aspects (see Table 1.4). Evolution also explains the *diversity* of life: how lifeforms, from arsenic bacteria to woolly mammoths, became different from each other.

As noted in the preceding section, **biological evolution** is a change in the overall genetic characteristics of a group of organisms over the generations, from parents to offspring. By "overall genetic characteristics," we mean the sum total of all the inherited characteristics that we can observe in the different individuals that make up a group of organisms. Antler size, coat color, running speed, and maternal care are examples of genetic characteristics that are readily observed. If we see a change in the commonness, or frequency, of one or more of these characteristics from one generation to the next, we say that the group as a whole has evolved (**FIGURE 1.17**). Notice that it is *groups of organisms* that evolve, not an individual organism such as a certain white-tailed deer or a particular red oak tree.

Populations of a given species can evolve over the generations

We define a **species** as all individuals that can interbreed in their natural surroundings to produce fertile offspring. For example, all pronghorn belong to a single species, as do all monarch butterflies, and all sugar maple trees. All species known to science are given a two-part scientific name, such as *Antilocapra americana* for the pronghorn. A **population** is a group of individuals of a particular species that shares a common habitat—all the pronghorn in the high plains of western Wyoming, for example.

Evolution can be considered at many levels, from populations on up. Scientists might consider the evolution of all the pronghorn in western Wyoming, or all the pronghorn in North America. Pronghorn might also be considered in the context of a larger group, or *family*. Some 5–20 million years ago, the pronghorn family (Antilocapridae) consisted of many species, but *A. americana* is the sole surviving member of the family. The pronghorn family is placed in a larger grouping (*order*) that includes all animals with two even hooves (encompassing pronghorn, cattle, deer, and pigs). All two-hoofed animals are placed in the very large grouping (*class*) known as mammals, in which females produce milk to feed their young. All mammals are part of a yet larger group (*phylum*); phylum Chordata includes all animals with a backbone and also some less familiar relatives, such as sea squirts, that have developmental similarities. Chordates, along with all other animals, are placed in the *kingdom* Animalia.

Evolution by Means of Natural Selection
(a) Genetic variability
(b) Differential reproduction
(c) Adaptation
Original population
Next generation
Time
Some animals run faster than others.
Faster animals are more likely to survive and reproduce.
Fast runners are more common than in the original population. Therefore, this population has evolved from one generation to the next.

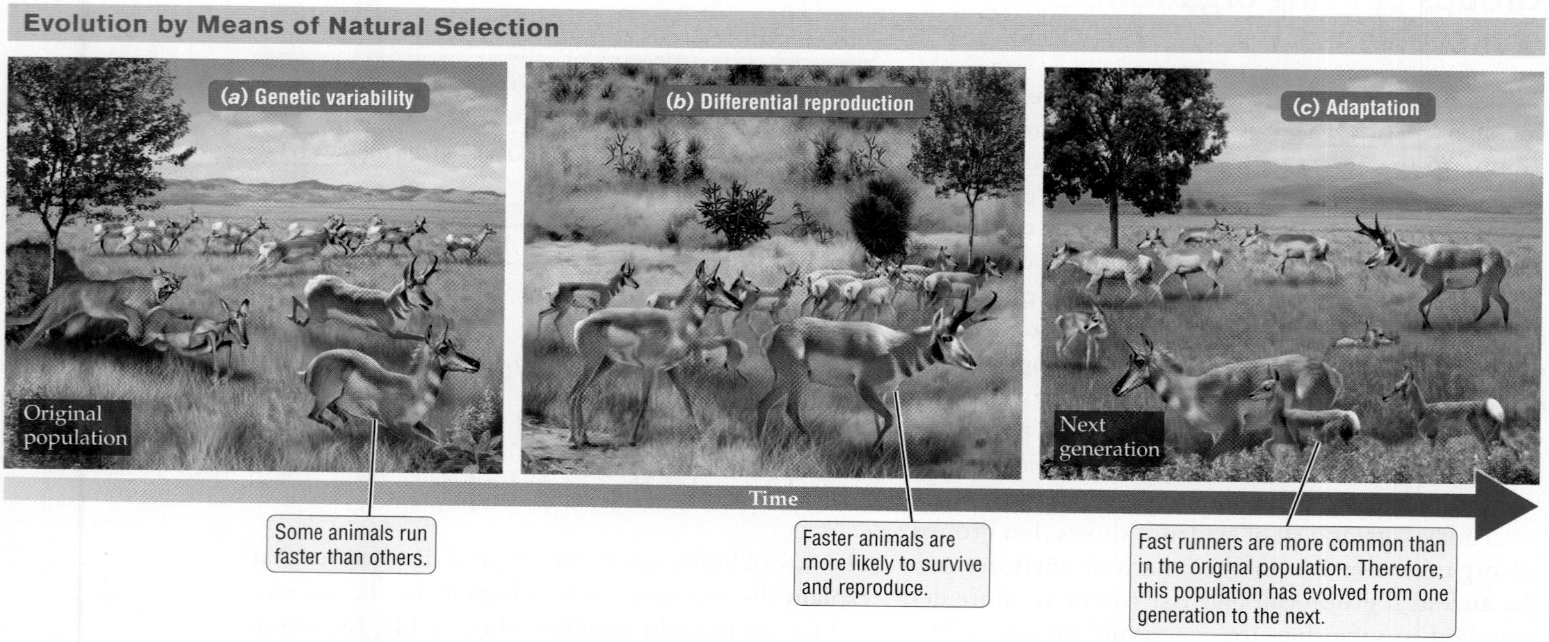

FIGURE 1.17 Groups of Living Organisms Can Evolve
Natural selection, exerted through predators looking for a meal, favors the survival and reproduction of fast pronghorn over the slower animals in the population. As a result, the next generation of pronghorn contains many more fast runners. Over the generations, the pronghorn population has evolved to exhibit a greater average running speed than the ancestral population had.

Natural selection favors individuals with adaptive traits

Natural selection is an evolutionary mechanism that changes the overall genetic composition of a population from one generation to the next by favoring the survival and reproduction of individuals best suited to their environment. Which individuals are favored by natural selection? In a genetically diverse population, individuals that have certain advantageous characteristics, known as **adaptive traits**, are favored to survive and reproduce. For a prey species such as pronghorn, being fleet of foot is an adaptive trait. In a pronghorn population, speedy animals have greater reproductive success compared to slower animals (see Figure 1.17).

Natural selection adapts a population to its habitat

Natural selection causes the population as whole to become better adapted to its habitat. **Adaptation** refers to a good match between a population and its particular habitat. By favoring individuals with adaptive traits over those with maladaptive ones, natural selection drives the evolution of the whole population so that the adaptive traits become common in the population.

Pronghorn, which live on the dry plains and grasslands of western North America, are exquisitely adapted to their environment (**TABLE 1.5**). The original population of pronghorn must have been genetically varied with respect to the traits listed in the table. The presence of predators led to differential survival and reproduction. For example, swifter individuals were more likely to outrace their predators, whereas slower individuals were more likely to be eaten. More of the swift pronghorn survived to have offspring, and the trait consequently became more common in the descendant population (see Figure 1.17).

Darwin and Wallace explained how natural selection produces new species

British naturalists Charles Darwin (1809–1882) and Alfred Wallace (1823–1913) independently recognized that if two populations of a single species come to occupy very different habitats, they could become very different from one another. The reason is that over time, natural selection would drive the two populations along very different paths of adaptation. Eventually, these populations could become so different that they would lose the ability to interbreed and produce fertile offspring. The two newly evolved populations would then be recognized as different species, descended from a common ancestor.

This insight led Darwin to the conclusion that all present-day species are descended "with modification" from ancestral species: humans and all other mammals are descended from a common ancestor that lived many millions of years ago, all animals are descended from a common ancestor that lived even further back in time, and all animals and plants have a common ancestor yet further back in the history of life.

We can categorize all known species into three domains of life

To date, scientists have described about 1.7 million species, but this is only the tip of the diversity iceberg. A vast number of species remain to be discovered and cataloged. Even so, by comparing the similarities and dissimilarities among the known species—especially in the information encoded in their DNA—biologists have succeeded in placing most of the major life-forms into an elaborate "tree

TABLE 1.5 Adaptive Traits in Pronghorn

- Camouflage coloration
- High running speed
- Large eyes set high on the head for 320-degree field of vision
- Very large lungs and heart for sustained running
- Hollow hair for insulation against winter cold
- Strong grinding teeth (molars) for breaking up tough plant food
- Four-chambered stomach for digesting plant food with the help of bacteria
- Long intestines for digesting and absorbing nutrient-poor plant material

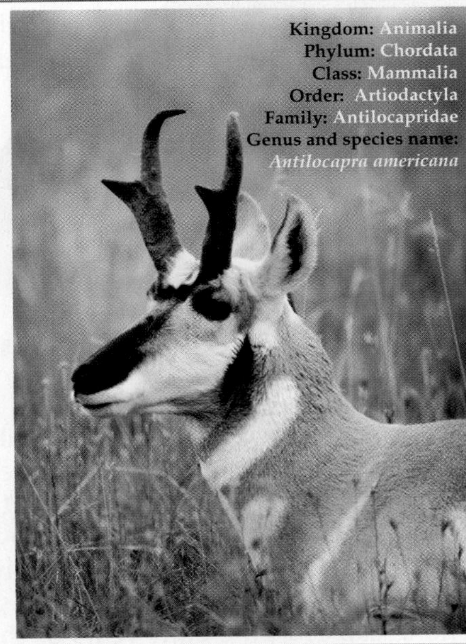

Kingdom: Animalia
Phylum: Chordata
Class: Mammalia
Order: Artiodactyla
Family: Antilocapridae
Genus and species name:
Antilocapra americana

Concept Check Answers

1. An individual deer does not evolve. A population of deer, however, may evolve larger average antler size.
2. Eukarya.

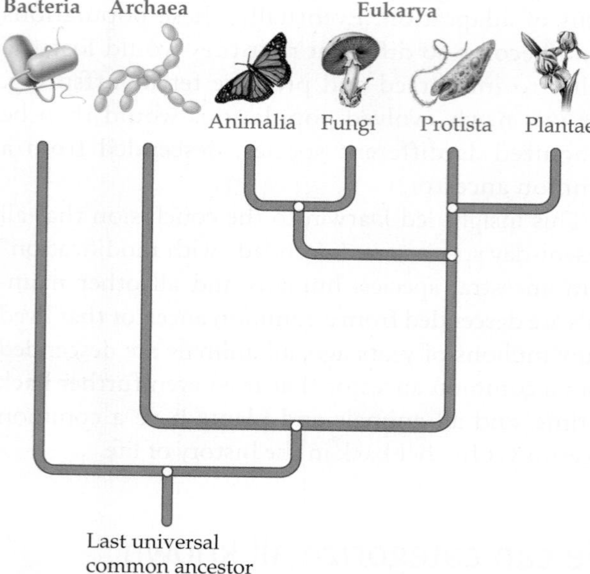

FIGURE 1.18 The Tree of Life

This evolutionary tree depicts the origin and diversification of the three domains of life (in red). The four main kingdoms of the domain Eukarya are shown in green. The evolutionary relationships of the Protista are not well understood.

of life" (**FIGURE 1.18**). The main branches of this tree are the three biggest categories of life-forms, known as domains. The three **domains** of life are Bacteria, Archaea, and Eukarya.

Bacteria and **Archaea** (**ahr-*KEE*-uh**) are composed of single-celled organisms, the great majority of which are microscopic. DNA evidence tells us that Bacteria and Archaea are substantially different because they diverged from each other billions of years ago and have been evolving separately ever since. But because they are superficially similar, bacteria and archaeans have traditionally been lumped together as **prokaryotes**, a label of convenience.

Eukarya (**yoo-*KAYR*-ee-uh**), the third domain of life, includes both single-celled and multicellular forms. This domain is divided into four kingdoms: **Protista** (algae, amoebas, and their relatives); **Plantae** (all the plants); **Fungi** (from yeasts to mushrooms); and **Animalia**, including animals with backbones (*vertebrates*) and those without (*invertebrates*).

EXTREME SENSORY PERCEPTION

Sharks are more sensitive to electrical fields than any other animal. Electroreceptors on the snout enable sharks to detect minute electrical currents generated by the metabolic processes of their prey. Shark attacks destroyed a lot of under sea phone cables in the early years of the telegraph.

Concept Check

1. Explain what is wrong with this statement: Over time, a male white-tailed deer evolves larger antlers.
2. Which domain of life do humans belong to?

1.6 The Biological Hierarchy

The **biological hierarchy** is essentially a linear concept map for visualizing the breadth and scope of life, from the smallest structures that are meaningful in biology to the broadest interactions between living and nonliving systems that we can comprehend (**FIGURE 1.19**). The biological hierarchy has many levels of organization, ranging from atoms at the lowest level up to the entire biosphere at the highest level. In scale, the hierarchy ranges from less than one ten-billionth of a meter (the approximate size of an atom) to 12 million meters (the diameter of Earth).

At its lowest level, the biological hierarchy begins with **atoms**, which are the building blocks of matter, the material of which the universe is composed. Two or more atoms held together by strong chemical bonds become a **molecule**, the next level in the hierarchy. We use the term **biomolecules** to refer to molecules that are found in living cells. Carbon atoms are prominent in biomolecules, which is why we say that life on Earth is carbon based. DNA, the genetic material that carries the code for building an organism, is an example of a biomolecule.

As noted earlier, the cell is the basic unit of life; and some organisms, such as bacteria, consist of only a single cell. Multicellular organisms also form tissues, the next level in the biological hierarchy. A **tissue** is a group of cells that performs a unique but fairly narrow set of tasks in the body. Plants and animals have many different types of tissues, each with unique functions. Nervous tissue, for example, performs the important function of transmitting electrical signals in the animal body. Muscle tissue can contract, enabling animals to move their bodies.

Plants and animals have **organs**, which are body parts composed of different types of tissues functioning in a coordinated manner. Organs perform a broader range of functions than any one tissue can carry out

The Biological Hierarchy

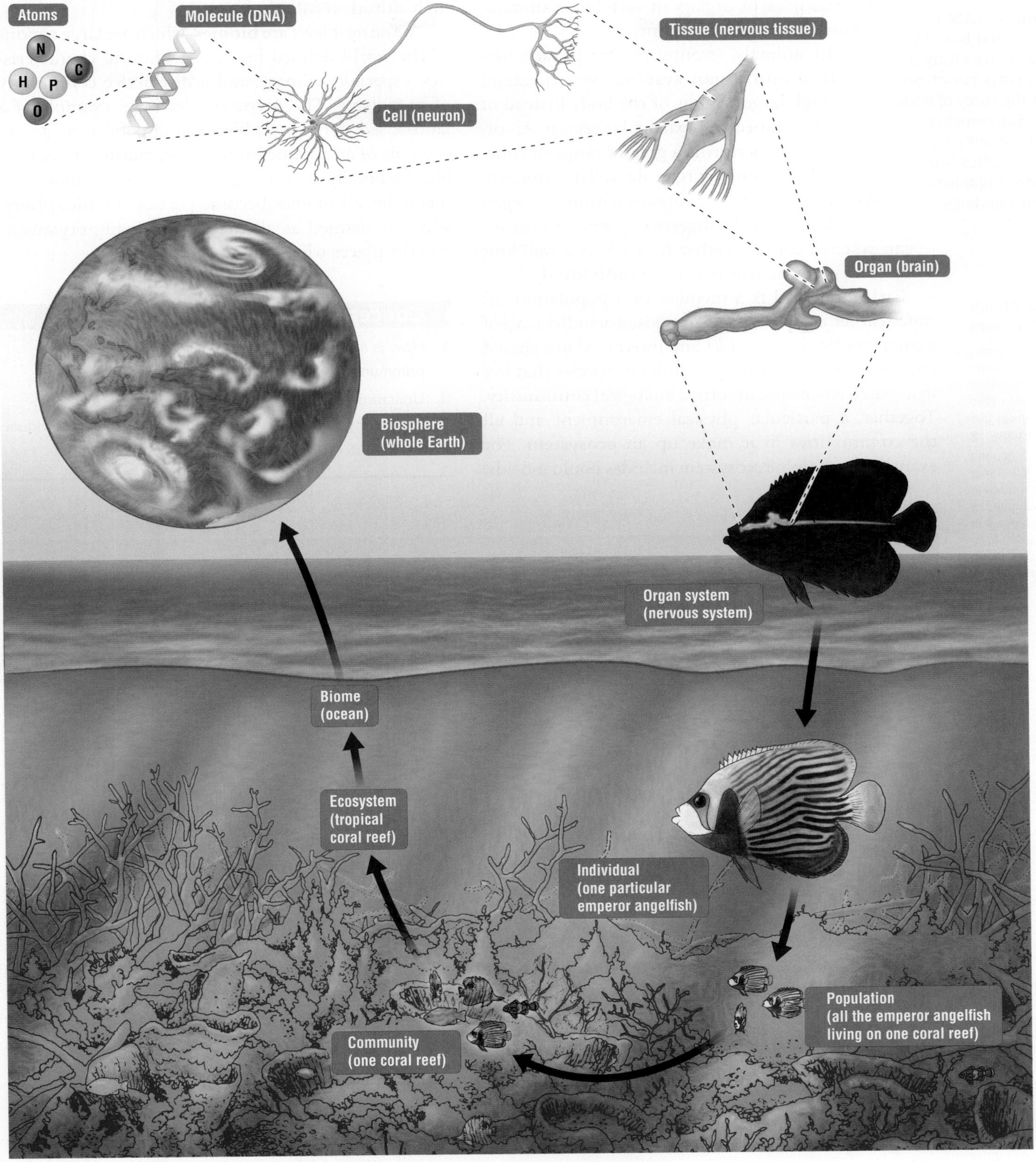

FIGURE 1.19 The Biological Hierarchy Extends from the Atom to the Biosphere
Levels of biological organization can be traced from atoms and molecules found in organisms all the way up to the biosphere, which includes all living organisms and their nonliving environment.

Helpful to Know

There are many academic disciplines within biology. *Physiology* is the study of how organisms function. *Anatomy* is the study of body structures. *Biochemistry* focuses on the chemistry of life. *Ecology* studies interactions among organisms and their surroundings.

on its own. An organ has a discrete boundary and a specific location in the body. The heart and brain are examples of organs in vertebrate animals, those that have a backbone.

In animals, groups of organs are networked into **organ systems**, which extend through large regions of the body instead of being confined to a particular region. An organ system performs a greater range of functions than a single organ does. The stomach, liver, and intestines are organs within the organ system known as the digestive system. All the organ systems come together to work as a well-knit whole that we recognize as a single **individual**.

Each individual is a member of a population. As noted earlier, a population is composed of individuals of a single species that interact and interbreed in a shared environment. Populations of different species that live in a shared environment form a biological **community**. Together, a particular physical environment and all the communities in it make up an **ecosystem**. For example, a subalpine ecosystem includes boulder fields, avalanche tracks, glacial streams, and thin air, together with a variety of hardy organisms adapted for life at high altitude (**FIGURE 1.20**).

At the next level are **biomes**, which are large regions of the world defined by shared physical characteristics, especially climate, and a distinctive community of organisms. The Arctic tundra is an example of a land-based (terrestrial) biome, and coral reefs are an example of an aquatic (in this case, marine or oceanic) biome. Finally, at the highest level of the biological hierarchy, all biomes become part of the **biosphere**, which is defined as all the world's living organisms and the places where they live.

Concept Check

1. How is a population different from a biological community?
2. Unscramble this scrambled biological hierarchy: community, organ system, ecosystem, atom, tissue, individual, biosphere, organ, cell, biome, population, molecule.

Concept Check Answers

1. A population is composed of members of the same species in a shared habitat. A community refers to all members of all species that share a common habitat.
2. Atom, molecule, cell, tissue, organ, organ system, individual, population, community, ecosystem, biome, biosphere.

FIGURE 1.20 An Ecosystem Includes All Living Organisms in a Habitat Plus the Nonliving World around Them
The photo shows a subalpine ecosystem in North Cascades National Park in Washington state. An ecosystem is composed of communities of living organisms and their environment functioning as a distinct ecological unit. Environmental characteristics—short growing season, bitter cold, avalanches, thin air, and intense UV radiation, for example—govern the communities that are found in an ecosystem. There are several distinct communities in this ecosystem, including forest clumps dominated by subalpine fir (*Abies lasiocarpa*), wet seeps dominated by sedges, and drier meadows dominated by heather. Animal life includes mountain goats and rodents such as pikas and mountain marmots.

APPLYING WHAT WE LEARNED

Researchers Wrangle over Bacteria

NASA's announcement about the arsenic-loving bacterium was the science sensation of the year. Felisa Wolfe-Simon and her colleagues at the U.S. Geological Survey had found a bacterium that could survive what for most organisms would be deadly amounts of toxic arsenic. In the lab, Wolfe-Simon grew the bacteria in increasing amounts of arsenic with almost no phosphorus. Without phosphorus, cells cannot make new DNA; and without new DNA, cells cannot divide and multiply. Yet despite being deprived of phosphorus, these particular bacterial cells continued to multiply. How were they doing it?

Arsenic has chemical properties like those of phosphorus, and Wolfe-Simon hypothesized that the bacteria were simply substituting arsenic for phosphorus. And if bacteria on Earth could pull off such a switch, there was no telling what life on other planets might do. Biologists say that all life on Earth requires six major elements: carbon, hydrogen, nitrogen, oxygen, sulfur, and phosphorus. If a bacterium could do without phosphorus, maybe none of the six were absolutely required for life. Such a conclusion meant that life might flourish on many more planets than previously thought.

It seemed like an eye-opening discovery, and the media played it up (despite the lack of actual extraterrestrials). But a number of scientists challenged Wolfe-Simon's conclusions, which were published at the same time in a peer-reviewed scientific journal. They wanted more evidence and wrote letters, tweets, and blogs that openly criticized Wolfe-Simon's work.

FIGURE 1.21 Mono Lake at Sunset
The Rim Fire lights up the horizon in this 2013 photo. The wildfire consumed more than 250,000 acres, including parts of Yosemite National Park. Despite the severity of such wildfires, some burrowing animals and the seeds and roots of many plants survive the scorching temperatures, highlighting the toughness and tenacity of life.

Critics argued that Wolfe-Simon's team had not ruled out the possibility that the bacteria still had enough phosphorus to build new DNA. For example, they pointed out that the "phosphorus-free" test tubes actually contained small amounts of phosphorus from dying bacteria and other sources. Wolfe-Simon's team needed to determine whether that was enough phosphorus for the bacteria to multiply, they said. Mono Lake, the lake where the bacteria live, contains large amounts of phosphorus (**FIGURE 1.21**). With such a rich supply of phosphorus, some scientists said, there could be no evolutionary advantage to evolving the unique ability to live without it.

In 2012, Rosie Redfield at the University of British Columbia claimed that careful analysis by her team showed there wasn't a trace of arsenic in the DNA of this strange bacterium. The arsenic monster has typical DNA, she said: it contains phosphorus, not arsenic. The dust did not settle right away, because Redfield and her colleagues chose to present their findings in an online research blog instead of a peer-reviewed journal. But studies by other microbiologists published later that year also refuted the claim that the bacterium contains arsenic, not phosphorus, in its DNA. As for growing in a *completely* phosphorus-free test tube, none of these microbiologists could coax the bacterium to do that. Repeatability of the test, as you have seen, is a crucial criterion in showing that a hypothesis is valid.

Science is a process of asking questions and trying to answer those questions through hypothesis testing. Wolfe-Simon and her colleagues took a chance on an interesting question: What do organisms actually need to live? Although the DNA chemistry of these bacteria turned out to be ordinary, their strategies for flourishing in an arsenic-rich habitat continue to boggle the mind. How do they manage to extract phosphorus from their environment to build into their DNA, when they are awash in great quantities of a very similar chemical, arsenic? Wolfe-Simon has the last word when she points out that many questions about this Mono Lake inhabitant have yet to be answered.

BIOLOGY IN THE NEWS

Curbing the Enthusiasm on Daily Multivitamins

BY RONI CARYN RABIN • *New York Times*, October 10, 2012

Can you reduce your risk of cancer by taking a multivitamin every day?

Last week, Boston researchers announced that one of the largest long-term clinical trials of multivitamins in the United States—encompassing 14,000 male physicians 50 and older, and lasting over a decade—found that taking a common combination of essential vitamins and minerals every day decreased the incidence of cancer by 8 percent, compared with a placebo pill.

Men who had already had cancer earlier in life were most likely to benefit, the study found. Cancer deaths also were lower among those who took vitamins, though that may have been a chance finding. Curiously, the vitamin regimen did not reduce the rate of prostate cancer, the most common cancer affecting men.

The researchers also looked for side effects and found that daily vitamins caused only minor problems, like occasional skin rashes.

Even though an 8 percent reduction in the overall cancer rate is fairly modest, Dr. Demetrius Albanes, senior investigator at the National Cancer Institute, said the potential public health implications were vast. "If you think of the hundreds of thousands of new cases of cancer every year, 8 percent can add up quite a bit," he said.

Yet no one is rushing out to urge more Americans to take multivitamins. Although half the population already takes some kind of supplement, previous studies have yielded decidedly mixed results. Some trials of high doses of nutrients believed to be cancer-fighters were shut down prematurely when they backfired, driving up cancer rates instead of reducing them.

Current federal dietary guidelines and American Cancer Society recommendations encourage people to eat a balanced diet rich in fruits and vegetables. Until now, the consensus has been that there is insufficient scientific evidence to justify taking a multivitamin to prevent cancer or other chronic diseases.

Here we go again, you might say in frustration. Why can't scientists make up their minds? Why do these studies say opposite things?

Of all the 1.7 million organisms known to science, humans are the most difficult to study. For ethical reasons, an ideal experimental design may not be allowed when human subjects are involved. For example, scientists cannot test a substance that is suspected of being harmful in human patients. Nonhuman animals, most commonly rats and mice, are often used as surrogates in such experiments because their metabolism is similar in many ways.

The type of large-scale double-blind controlled experiments we described as the gold standard for establishing causality are, in practice, difficult with human subjects. Some types of blinding are not possible—for example, when a treatment is impossible to mask because of its strong taste. Longitudinal studies—in which researchers track a cohort of participants over many years—run afoul of inaccurate reporting by the participants. Investigations show that participants commonly misremember, and sometimes tell outright fibs (about how many cookies they *actually* ate, for example). On top of that, large-scale studies are enormously expensive, and the logistics of sustaining them over decades are so great that repeating such studies with the benefit of hindsight is rarely feasible.

The studies that produced the conflicting results—frighteningly *higher* rates of cancer in supplement takers—used very high doses of single vitamins (vitamin E or folic acid). In contrast, the physicians clinical trial cited in the news story used standard brand-name multivitamins. This trial avoided some of the common pitfalls of long-term studies by concentrating on a well-educated group—physicians—who are likely to follow directions and give accurate reports. It was a double-blind study in which participants were randomly assigned to take either a multivitamin or a placebo.

A limitation of this study is that all the participants were older white males. Therefore, study results may not apply to younger people, women, and nonwhites. The participants also had a healthier lifestyle than the average person; most ate four servings of fruits and vegetables every day and consumed little red meat. Would the beneficial effects of multivitamin supplementation be more pronounced in the average person, who may not eat as healthy a diet? We don't know. But the study suggests that taking a standard multivitamin does no harm and may do some good, at least for people similar to the study subjects.

Evaluating the News

1. Is the physicians clinical trial an example of an observational test or an experimental test of a hypothesis? State the hypothesis that the researchers were testing. Is the hypothesis refutable? Explain.

2. Some people are inclined to say that nutritional science is so error prone that we should simply ignore all of it. Explain your viewpoint, giving reasons.

CHAPTER REVIEW

Summary

1.1 The Nature of Science

- Science is both a body of knowledge about the natural world and an evidence-based process for generating that knowledge.
- The scientific method represents the core logic of the process by which scientific knowledge is generated. The scientific method requires that we (1) make observations, (2) devise a hypothesis to explain the observations, (3) generate predictions from that hypothesis, and (4) test those predictions.

1.2 The Process of Science

- We can test hypotheses by making additional observations or by performing experiments (controlled, repeated manipulations of nature) that will either uphold the predictions or show them to be incorrect.
- A hypothesis cannot be proved true; it can only be upheld or not upheld. If the predictions of a hypothesis are not upheld, the hypothesis is rejected or modified. If the predictions are upheld, the hypothesis is supported.
- Correlation means that two or more aspects of the natural world behave in an interrelated manner: if we know a particular value for one aspect, we can predict a particular value for the other aspect. However, correlation between two variables does not necessarily mean that one is the cause of the other.
- In a scientific experiment, the independent variable is the one that is manipulated by the investigator. Any variable that can potentially respond to the changes in the independent variable is called a dependent variable.

1.3 Scientific Facts and Theories

- A scientific fact is a direct and repeatable observation of any aspect of the natural world.
- A scientific theory is a major idea that has been supported by many different observations and experiments.

1.4 The Characteristics of Living Organisms

- Because of their common evolutionary origins, all living organisms have certain key characteristics in common.
- All living organisms (1) are built of cells; (2) reproduce, using DNA to pass genetic information from parent to offspring; (3) take in energy from their environment; (4) sense and respond to their environment; (5) maintain constant internal conditions; and (6) evolve as groups.

1.5 Biological Evolution and the Unity and Diversity of Life

- Biological evolution is a change in the overall genetic characteristics of a population across generations, from parent to offspring.
- Natural selection is a key evolutionary mechanism: it causes evolution by favoring the survival and reproduction of individuals that are best suited to their environment.
- Natural selection makes a population or species better adapted to its environment.
- All life on Earth can be classified into three domains: Bacteria, Archaea, and Eukarya.

1.6 The Biological Hierarchy

- The term "biological hierarchy" refers to the many levels at which life can be studied: atom, molecule, cell, tissue, organ, organ system, individual, population, community, ecosystem, biome, biosphere.
- The individuals of a given species in a particular area constitute a population. Populations of different species in an area make up a community. Communities, along with the physical habitat they live in, constitute ecosystems.
- Ecosystems make up biomes, large regions of the world that are defined by the climate and the distinctive communities found there. All the biomes on Earth make up our one single biosphere.

Key Terms

adaptation (p. 19)
adaptive trait (p. 19)
Animalia (p. 20)
Archaea (p. 20)
asexual reproduction (p. 15)
atom (p. 20)
autotroph (p. 16)
Bacteria (p. 20)
behavior (p. 16)
biological evolution (p. 18)
biological hierarchy (p. 20)
biology (p. 4)
biome (p. 22)
biomolecule (p. 20)
biosphere (p. 22)
causation (p. 11)
cell (p. 15)
community (p. 22)
consumer (p. 16)
control group (p. 12)
controlled experiment (p. 12)
correlation (p. 11)
cytoplasm (p. 15)
data (p. 9)
dependent variable (p. 12)
development (p. 15)
DNA (p. 16)

domain (p. 20)
double-blind experiment (p. 13)
ecosystem (p. 22)
Eukarya (p. 20)
experiment (p. 12)
fertilization (p. 15)
Fungi (p. 20)
heterotroph (p. 16)
homeostasis (p. 16)
independent variable (p. 12)
individual (p. 22)
metabolism (p. 16)
microbe (p. 14)
molecule (p. 20)
multicellular organism (p. 15)
natural selection (p. 19)
nucleus (p. 16)
observation (p. 9)
organ (p. 20)
organ system (p. 22)
photosynthesis (p. 16)
placebo (p. 13)
placebo effect (p. 13)
Plantae (p. 20)
plasma membrane (p. 15)
population (p. 18)
producer (p. 16)
prokaryote (p. 20)
Protista (p. 20)
reproduction (p. 15)
science (p. 5)
scientific fact (p. 13)
scientific hypothesis (p. 6)
scientific method (p. 5)
scientific theory (p. 13)
sexual reproduction (p. 15)
single-blind experiment (p. 12)
species (p. 18)
statistics (p. 11)
technology (p. 4)
tissue (p. 20)
treatment group (p. 12)
variable (p. 11)

Self-Quiz

1. A scientific hypothesis is
   a. an educated guess explaining an observation.
   b. a prediction based on an observation.
   c. a scientific theory.
   d. an idea that can be proved beyond all doubt through experimentation.
2. A scientific theory
   a. is so well established that it is not open to challenge on the basis of new evidence.
   b. is a major explanation that has been supported by many and diverse lines of evidence.
   c. is considered a scientific fact in that it is a direct and repeatable observation of nature.
   d. is a scientific hypothesis that can be tested through experimental but not observational studies.
3. For an investigator who is following the scientific method, which one of the steps listed here would come immediately before rejecting a hypothesis?
   a. making a prediction
   b. developing a new or revised hypothesis
   c. making observations of nature
   d. conducting controlled, repeated manipulations of nature
4. Which of the following questions could *not* be used to develop a testable hypothesis?
   a. Can arsenic replace phosphorus in the DNA of a bacterium?
   b. Is there a relationship between high consumption of soda pop and obesity?
   c. Should smokers pay the same premium for medical insurance as nonsmokers?
   d. Do imported grapes have higher levels of pesticides than grapes grown in the United States?
5. Which of the following is manipulated by the experimenter in a controlled experiment?
   a. confounding variable
   b. control group
   c. dependent variable
   d. independent variable
6. Which of the following is a property shared by all life-forms on Earth?
   a. use of DNA for reproduction
   b. presence of a cell wall composed of long chains of sugars
   c. storing of genetic information in the nucleus
   d. ability to capture energy from the nonliving environment for all metabolic needs
7. Natural selection
   a. is a random process in which some individuals survive and others die off over time.
   b. is readily observed within a single generation of individuals.
   c. tends to make a whole population better adapted to its surroundings.
   d. occurs at the level of populations but not at the level of species.

8. An organ in the human body
   a. is the basic unit of life.
   b. has a discrete shape and specific location in the body.
   c. consists of a single tissue type.
   d. is composed of two or more organ systems.

9. The biome
   a. consists of two or more tissues conducting specialized functions in an integrated manner.
   b. encompasses all organisms on Earth plus their environment.
   c. extends over large regions of Earth that share similar climate and plant communities.
   d. consists of members of one species that share the same habitat.

Analysis and Application

1. Describe three characteristics of science as a way of knowing. Can science answer all types of questions that humans might raise? Explain.

2. What does the statement "Correlation is not causation" mean? Give an example of two variables for which you know the two are correlated but neither one is likely to be the cause of the other. (*Hint:* For an example, think back on what else was going on when you last came down with a cold.)

3. Describe one observation, one hypothesis, and one experiment that led researchers to the conclusion that something in cold-water fish might be beneficial for those at risk from heart disease.

4. What are the key characteristics of a *scientific* hypothesis?

5. What is a scientific fact? How does it differ from a scientific theory?

6. Consider a biological community that includes grasses, lions, sunshine, wildebeest (even-hoofed antelope), and ticks (small bloodsucking animals related to spiders). Identify the producers and consumers, and arrange them in a chain that describes who eats whom (a food chain).

7. What are the levels of the biological hierarchy? Arrange them in their proper relationship with respect to one another, from smallest to largest. Give an example for each level that you know from your own experience.

8. The manufacturers of toning shoes make a variety of fitness claims for their products, most promising that you will burn more calories and improve muscle function if you wear the shoes regularly. Study the advertising for a toning-shoe brand, and then design an experiment to test the claims made by the manufacturer.
   (a) State a hypothesis that you might be able to test for this shoe. Is your hypothesis stated in terms you can measure?
   (b) Devise a test for your hypothesis. What would be your independent variable? Your dependent variable? What controls would be appropriate? What size and composition would the groups you study have?
   (c) How would you measure the dependent variables?
   (d) State how the results you might obtain would either verify your hypothesis or refute it.

   Now compare your experimental design with one conducted by professional researchers sponsored by the American Council on Exercise (available free at www.acefitness.org) or by *Consumer Reports*.

2 The Chemistry of Life

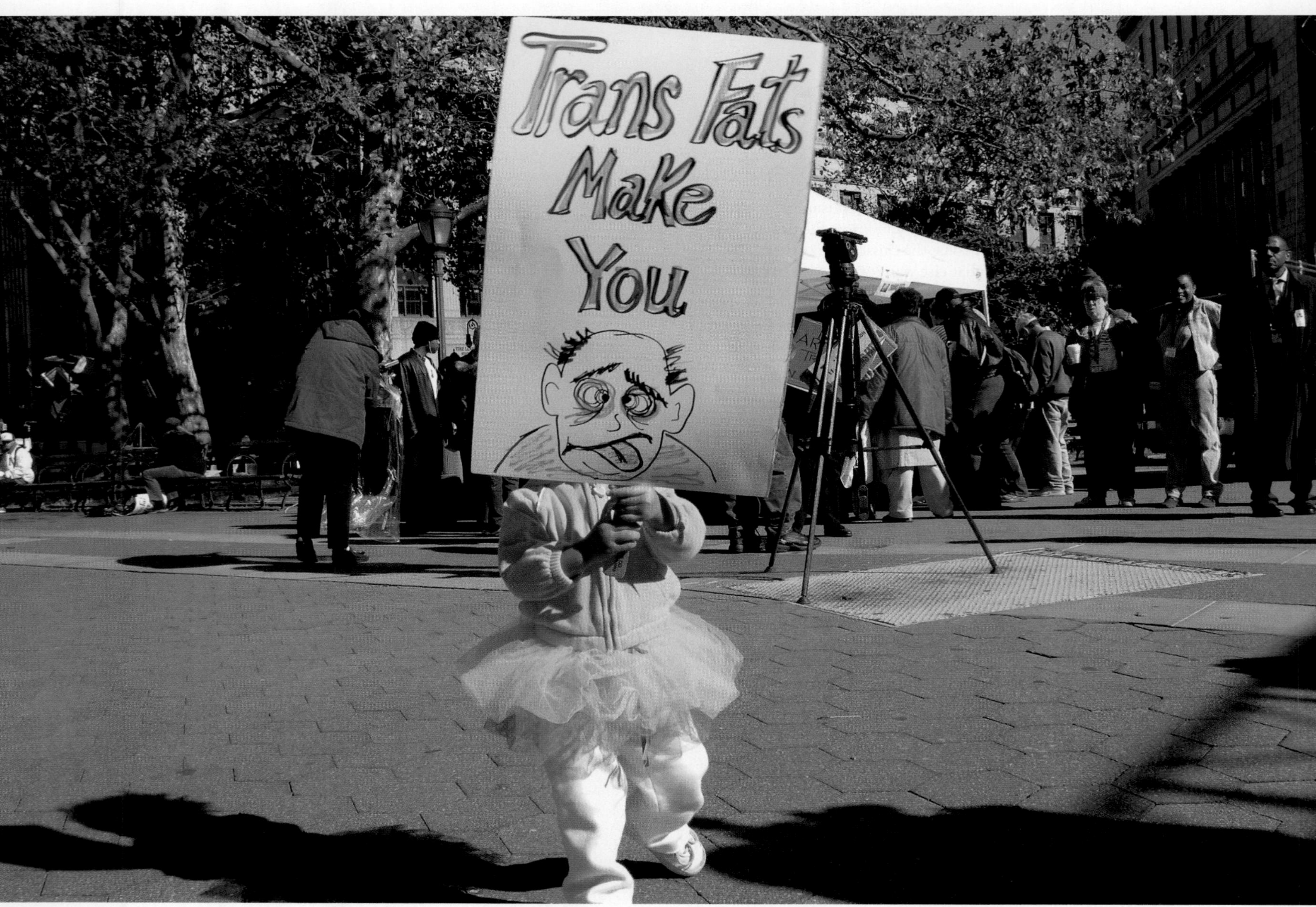

AN ANTI–TRANS FAT RALLY. A young protestor makes her views known as the New York City Board of Health holds public hearings on a plan to ban trans fats from city restaurants.

How the "Cookie Monster" Tackled Trans Fats

In 2011, California became the first state to ban trans fats in restaurants, bakeries, hospitals, and other places that prepare food. Although for years the restaurant industry had argued that such laws would keep them from making the foods that customers wanted and even put some restaurants out of business, the date the law went into effect passed with barely a ripple. Restaurants didn't go out of business, and nobody went without hot, salty fries.

Five years earlier, New York City's Board of Health had stunned and horrified restaurant owners by banning trans fats from the city's famous restaurants, giving restaurateurs just 6 months to stop frying in trans fat–laden cooking oils, and 18 months to eliminate trans fats from all food. An outraged spokesman for the National Restaurant Association argued that the city had "no business banning a product the Food and Drug Administration has already approved."

The restaurant association had a point: the momentum for banning trans fats hadn't come from the federal government or even from the prestigious American Heart Association, but from a California public interest attorney named Stephen Joseph. Joseph's father had died of heart disease, and as Joseph learned how important trans fats are in the development of heart disease, he decided to take action. In 2003 he filed a lawsuit against Nabisco, asking the company to stop selling Oreo cookies containing trans fats to children. Nabisco's parent company, Kraft Foods, told CNN, "We stand behind Oreo, a wholesome snack people have known and loved for more than 90 years."

What is a trans fat and how is it different from other fats? Are all fats bad for our health?

Joseph—dubbed the "Cookie Monster" by the *San Francisco Chronicle*—argued that Oreo cookies made with trans fats were literally dangerous. Talk show hosts called the suit "ridiculous" and suggested banning lawyers instead. Within days, Joseph withdrew his lawsuit. But, as we'll see, the Cookie Monster attorney had the last laugh.

In this chapter you'll learn about atoms and how they combine to form molecules. You'll also learn about the molecules that serve as building blocks of living things. One group of these building blocks is the lipids, which includes trans fats.

MAIN MESSAGE Life on Earth is based on carbon-containing molecules, and four types of these—carbohydrates, proteins, lipids, and nucleic acids—are common to all life-forms on our planet.

KEY CONCEPTS

- Understanding the chemistry of life helps us understand life itself.
- Four elements—oxygen, carbon, hydrogen, and nitrogen—account for about 96 percent of the weight of a living cell.
- A molecule contains two or more atoms linked through covalent bonds. In a covalent bond, two atoms share a pair of electrons.
- The gain or loss of electrons produces ions, which have either negative or positive charge. A salt is composed of ions held together by the mutual attraction between their opposite electrical charge.
- Water molecules associate with each other through weak attractions known as hydrogen bonds. The unique properties of water have a profound influence on the chemistry of life.
- A chemical reaction occurs when bonds between atoms are formed or broken. To sustain life, thousands of different chemical reactions must occur inside even the simplest cell.
- An acid releases hydrogen ions, and a base accepts them. All living things have some ability to maintain an internal acid-base balance.
- Four main classes of molecules are common to all living organisms: carbohydrates, proteins, lipids, and nucleic acids.

HOW THE "COOKIE MONSTER" TACKLED TRANS FATS 29
2.1 Matter, Elements, and Atomic Structure 30
2.2 The Bonds That Link Atoms 33
2.3 The Special Properties of Water 36
2.4 Chemical Reactions 39
2.5 The pH Scale 40
2.6 The Chemical Building Blocks of Life 41
2.7 Carbohydrates 43
2.8 Proteins 45
2.9 Lipids 49
2.10 Nucleotides and Nucleic Acids 53
APPLYING WHAT WE LEARNED
How Bad Are Trans Fats? 55

Helpful to Know

The mass of an object is a measure of all the material in it. The more mass an object has, the more difficult it is to move it. Weight measures how strongly the mass is pulled on by gravity. The mass of an object on Earth is the same as its weight.

FOR ALL ITS REMARKABLE DIVERSITY, life as we know it is built from a rather limited variety of atoms. The fact that all cells share this limited range of atomic ingredients reminds us of the common evolutionary heritage of all life on Earth.

In this chapter we begin our exploration of cellular life by identifying the chemical components shared by all cells. We use the term **biomolecules** to refer to chemical substances unique to living cells. The cell makes and uses many different types of biomolecules, small and large. Among these are carbohydrates, proteins, lipids, and nucleic acids. All biomolecules have a backbone of carbon atoms, which is why we say that life on our planet is carbon based.

We also consider why life is critically dependent on water, and why most of the chemical reactions that are vital for life occur in an aqueous (watery) environment. In short, this chapter is about how living organisms function at the chemical level. Many of the topics introduced in this chapter serve as a foundation for the deeper investigation of life that follows in later chapters. Understanding the chemistry of life helps us understand life itself.

EXTREME DIFFERENCE

How is the chemistry of this humpback whale different from that of the ice-coated rocks along the Alaskan shore? We all know that water is necessary for life. But what's so special about the water molecule?

2.1 Matter, Elements, and Atomic Structure

What is the world made of? The answer is: *matter*. **Matter** is defined as anything that has mass and occupies space. Think of it as the "stuff" the universe is composed of. All matter consists of one or more chemical elements. An **element** is a pure substance that has distinctive physical and chemical properties and cannot be broken down to other substances by ordinary chemical methods. Each element is identified by a one- or two-letter symbol; for example, oxygen is identified as O, calcium as Ca. Chemists have discovered 92 naturally occurring elements in the universe. Hydrogen (H) is the most abundant element.

The overall chemical composition of living beings is distinctly different from that of the nonliving part of Earth. Silicon (Si), for example, makes up 28 percent of Earth's crust, but less than 0.001 percent of the human body (**FIGURE 2.1**). Much of the silicon in rock and sand is combined with oxygen atoms (as silicates, SiO_3), which is why the oxygen content of Earth's crust is also high.

Just four elements—carbon (C), hydrogen (H), oxygen (O), and nitrogen (N)—account for more than 96 percent of the mass of the average cell. Water, which contains hydrogen and oxygen atoms, makes up 70 percent of the mass of the cell. Water is also abundant on the surface of our planet, 71 percent of which is covered by oceans. In contrast, carbon is scarce in Earth's crust, but it is the third most abundant element in a cell.

An **atom** is defined as the smallest unit of an element that still has the distinctive chemical properties of that element. Atoms are so small that more than a trillion of them could easily fit on the head of a pin. Because there are 92 naturally occurring elements, there are also 92 different types of atoms. The uniqueness of an element comes from the special characteristics of the atoms that compose it.

What makes the atoms of one element different from those of another? The answer lies in the specific combination of three atomic components. The first two components are electrically charged: **protons** have a positive charge (+), and **electrons** have a negative charge (−). As the name of the third component implies, **neutrons** lack any electrical charge. These three components—especially electrons—determine the physical and chemical properties of an element and how its atoms interact with other atoms.

A single atom has a dense central core, called the **nucleus** (*NOO*-**klee-us**; plural "nuclei"), that contains

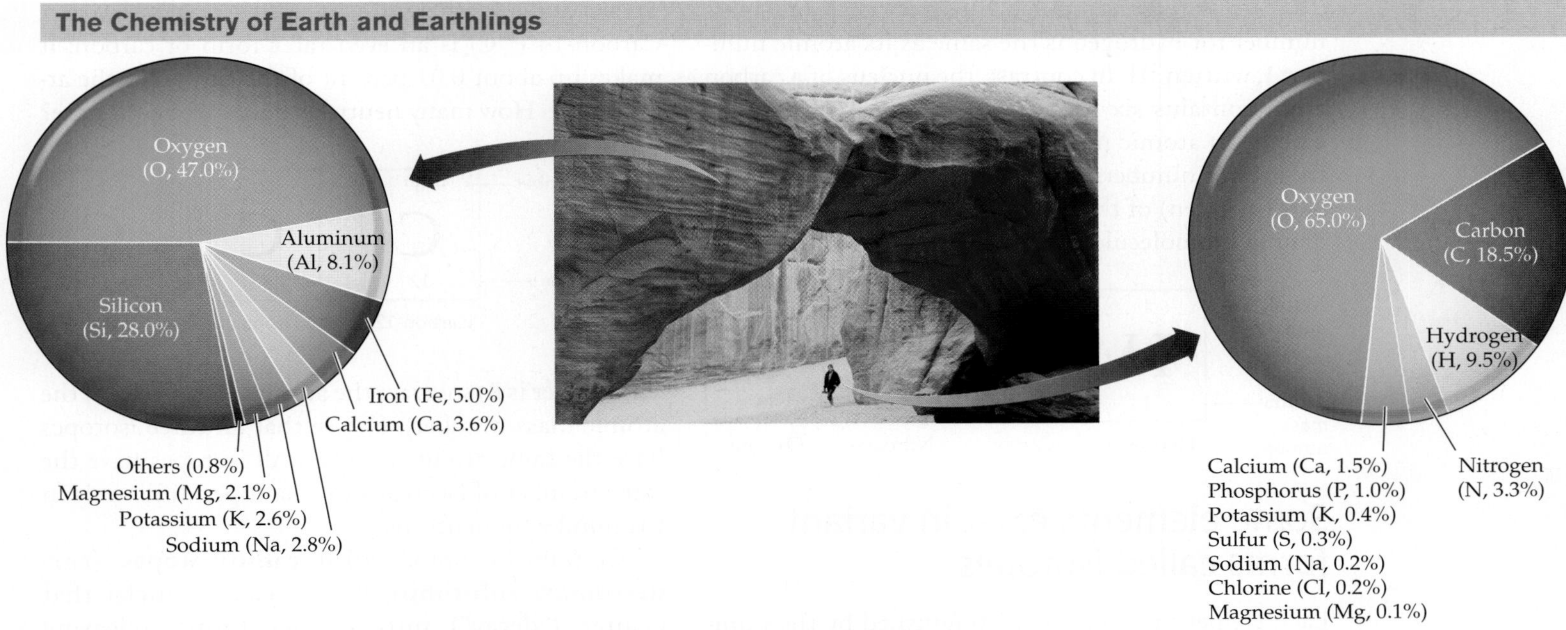

FIGURE 2.1 Chemical Composition of Earth's Crust and the Human Body

one or more protons and is therefore positively charged. One or more negatively charged electrons move around the nucleus, in defined volumes of space known as **electron shells** (**FIGURE 2.2**). And, except in the case of the most common form of hydrogen, the nucleus also contains one or more neutrons. As a whole, the positive charge on the nucleus balances the negative charges of the electrons, making atoms electrically neutral.

To get an idea of the scale of an atom, imagine that a hydrogen nucleus is the size of a marble; the electron would move around this nucleus in a space as big as the Houston Astrodome.

The size and structure of atoms can be described with numbers

The characteristics of an atom can be summarized by numbers that describe the atom's structure and mass. The number of protons in an atom's nucleus is the **atomic number** of that particular element. Hydrogen, having a single proton, has an atomic number of 1; carbon, which has six protons, has an atomic number of 6.

The sum of an atom's protons and neutrons, another distinguishing feature of each element, is its **atomic mass number**. The mass of an electron is negligible—only about 1/2,000 that of a proton or neutron. Protons and neutrons have about the same mass, so the atomic mass number of an element is based on the total number of protons and neutrons contained in the nucleus of the atom. Hydrogen has

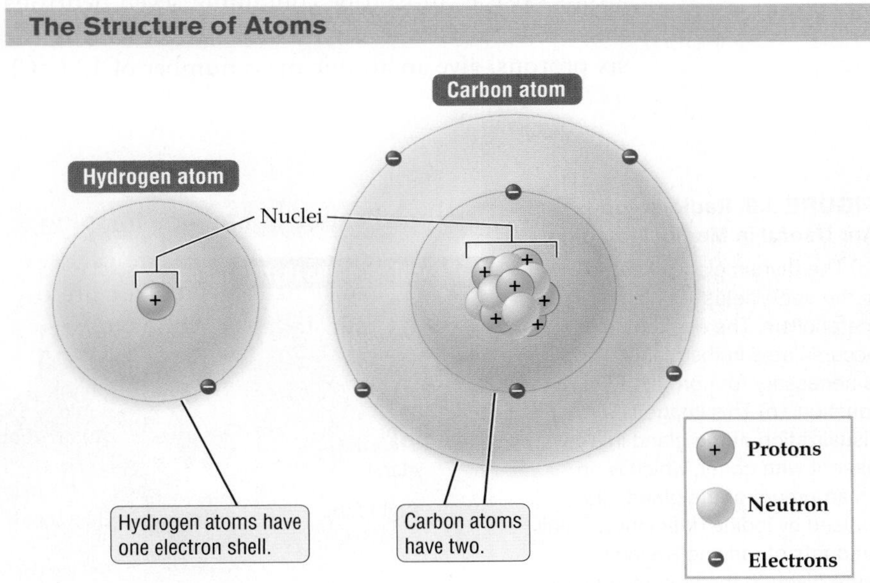

FIGURE 2.2 Atomic Structure

The electrons, protons, neutrons, and nuclei of these hydrogen and carbon atoms are shown greatly enlarged in relation to the size of the whole atom. An electron shell is a simplified way of representing the space that electrons move in as they orbit the nucleus.

a single proton and no neutrons, so the atomic mass number for hydrogen is the same as its atomic number: 1, written ^{1}H. In contrast, the nucleus of a carbon atom contains six protons and six neutrons, giving carbon an atomic mass number of 12 (^{12}C). Here are the atomic numbers (in black) and atomic mass numbers (in green) of the four atoms that are most abundant in biomolecules:

Some elements exist in variant forms called isotopes

Each element is uniquely distinguished by the number of protons in its nucleus. However, the atoms of some elements come in different forms, known as *isotopes*, that vary only in the number of neutrons in their nuclei. **Isotopes** of an element have the same atomic number, but different atomic mass numbers. For example, over 99 percent of the carbon atoms found in atmospheric carbon dioxide gas (CO_2) have an atomic mass number of 12 (^{12}C). However, a tiny fraction—slightly less than 1 percent—of those carbon atoms exist as an isotope containing seven neutrons instead of six. These seven neutrons, together with the six protons, give an atomic mass number of 13 (^{13}C).

Helpful to Know

The atoms commonly found in living cells are shown in the periodic table of the elements (see page T2).

This isotope is therefore referred to as carbon-13. Carbon-14 (^{14}C) is an even rarer form of carbon; it makes up about 0.01 percent of the carbon in the atmosphere. How many neutrons does carbon-14 have?

The answer is 8 (subtract the atomic number from the atomic mass number). Notice that all three isotopes have the same atomic number. All isotopes have the same number of electrons too. Isotopes differ only in the number of neutrons.

Certain isotopes, called **radioisotopes** (*RAY*-**dee-oh-***EYE***-soh-tohp**), have unstable nuclei that change ("decay") into simpler forms, releasing high-energy radiation in the process. For example, carbon-14 is a radioisotope. Only a fraction of known isotopes are radioactive; some of these, such as carbon-14, phosphorus-32, and the two radioisotopes of hydrogen (deuterium, ^{2}H; and tritium, ^{3}H), have important uses in both research and medicine.

The radiation given off by radioisotopes can be detected in a number of ways, ranging from simple film exposure to the use of sophisticated scanning machines. These methods allow the location and quantity of radioisotopes to be tracked fairly easily—a characteristic that makes them useful in medical diagnostics (**FIGURE 2.3**). For example, the thyroid

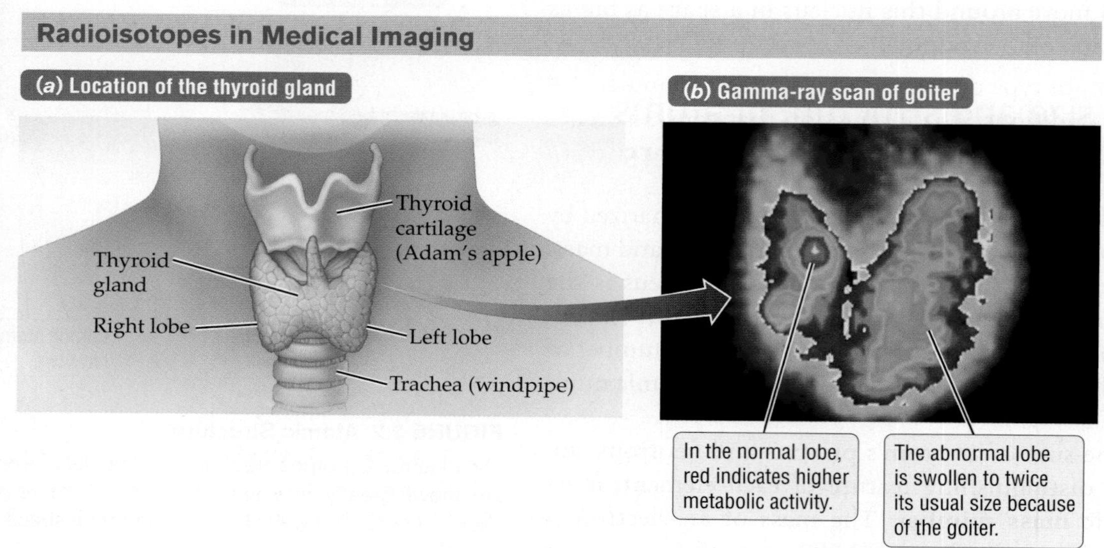

FIGURE 2.3 Radioisotopes Are Useful in Medical Imaging
(*a*) The thyroid gland, located in the neck, helps regulate metabolism. The element iodine accumulates in the gland and is necessary for normal thyroid function. (*b*) This image is a visualization of the gland in a patient with goiter, which is an enlargement of the gland, often caused by iodine deficiency. Small amounts of radioactive iodine were given to the patient, and the accumulated radioisotope was detected with a gamma-ray scan.

gland takes up iodine for producing a special type of hormone required by the body. By administering a low dose of an iodine radioisotope (iodine-131) to patients with thyroid disease and then using an imaging device, physicians can see how the thyroid takes up the radioisotope (**FIGURE 2.3*b***). If a patient is found to be suffering from cancer of the thyroid, repeated doses of iodine-131 can be administered as a therapy because the accumulation of radioactivity in the thyroid tends to kill the cancer cells.

Concept Check

1. An uncharged atom of iron (Fe) contains 26 electrons and 30 neutrons. (a) How many protons does it have? (b) How many neutral particles? (c) What is its atomic mass number?
2. True or false: All isotopes are dangerous because they release high-energy radiation.

2.2 The Bonds That Link Atoms

The number of electrons in an atom and how they are distributed around the nucleus are the main determinants of an atom's chemical behavior. Some elements are chemically *inert*: their atoms tend not to interact with other atoms by losing, gaining, or sharing electrons. Most elements, however—and all elements that are biologically important—have atoms that are far more social. Such atoms have a tendency to donate electrons, accept electrons, and even share electrons, if the right type of atom becomes available under the right conditions. The interaction that causes two atoms to associate with each other is known as a **chemical bond**. In chemistry, two main types of chemical bonds are recognized:

- **Covalent bonds**, in which atoms share electrons to form **molecules**.
- **Ionic bonds**, in which atoms with opposite electrical charge are held together by their mutual attraction. Atoms held together by ionic bonds are **salts**.

Chemists have developed a simple shorthand, known as a **chemical formula** (**FIGURE 2.4*a***), to represent the atomic composition of molecules and salts. The formulas use the letter symbol of each element and a subscript number to the right of the symbol to show how many atoms of that element are contained in the molecules or salts. For example, the chemical formula for a water molecule is H_2O (2 hydrogens plus 1 oxygen). The chemical formula for table sugar (sucrose), which has 12 carbons, 22 hydrogens, and 11 oxygens per molecule, is $C_{12}H_{22}O_{11}$.

Chemical formulas describe ionic substances (such as table salt) as well. Table salt has equal numbers of sodium (Na) and chlorine (Cl) atoms in electrically charged forms called **ions**. Sodium ion is represented as Na^+, and chlorine ion as Cl^-. The formula for table salt is NaCl.

A **chemical compound** is a substance in which atoms from two or more *different* elements are bonded together, each in a precise ratio. Salts, by definition, are compounds. As we noted above, a salt contains at least two elements with opposite charge: sodium ions (Na^+) and chloride (Cl^-) in the case of table salt, NaCl. A molecule such as table sugar is clearly a compound. But molecular oxygen (O_2), composed of just two oxygen atoms, is a molecule that is not a compound.

H, CH_2SH, C=C, H_3C, H

2-Butene-1-thiol, a stinky molecule

EXTREME STINK

Skunks produce short-chain thiols in their anal glands and spray them as a defense. Like many sulfur-containing molecules, thiols stink.

Representations of a Molecule

(*a*) Chemical formula

H_2O

(*b*) Structural formula

H–O–H

(*c*) Ball-and-stick model

O H H

(*d*) Space-filling model

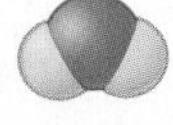

FIGURE 2.4 Covalent Bonds between Atoms Can Be Represented in Various Ways

Different ways of representing the water molecule are shown here. By convention, hydrogen is shown in white and oxygen in red.

Concept Check Answers

1. (a) 26; (b) 30; (c) 56.
2. False. Only some isotopes—those that have unstable nuclei—fall apart spontaneously, releasing high-energy radiation.

EXTREME ANTIDOTE

One remedy for a skunked pet is to wash it in a solution of baking soda and 3 percent hydrogen peroxide. Hydrogen peroxide (H_2O_2) and sodium bicarbonate ($NaHCO_3$) are both compounds. Commonly known as baking soda, $NaHCO_3$ finds hundreds of popular uses.

Covalent bonds form by electron sharing between atoms

A molecule contains at least two atoms held together by covalent bonds. A single covalent bond represents the sharing of *one pair* of electrons between two atoms. In a structural formula, which displays how atoms are connected within a molecule (**FIGURE 2.4*b***), a single covalent bond is indicated by a single straight line. The atoms joined in a covalent bond can be of the same type (**FIGURE 2.5*a***), or they can be atoms of two different elements (which is the case for all the molecules in **FIGURE 2.5*b*** except hydrogen gas and oxygen gas).

What drives atoms to share electrons with one another? To answer that question, let's consider how electrons are distributed in the space around the nucleus. The electrons of every atom can be visualized as moving around the nucleus in concentric layers called *shells* (see Figure 2.2). The maximum number of electrons in any shell is fixed. When all of an atom's shells are filled to capacity, it is in its most stable state. Atomic shells are filled starting from the innermost shell, which can hold two electrons at most. The next shell outside it can hold a maximum of eight electrons.

Atoms that have unfilled outer shells can achieve a more stable state by interacting in ways that will achieve maximum occupancy of the outermost shell (called the *valence shell*). One way for an atom to achieve greater stability is to fill its outermost shell by sharing one or more of its outer-shell electrons with a neighboring atom. Each atom in this arrangement contributes one electron to every pair of shared electrons. Each pair of electrons that is shared between the two atoms constitutes one covalent bond. When

Each hydrogen atom has a single electron contained in a shell that can hold a total of two electrons.

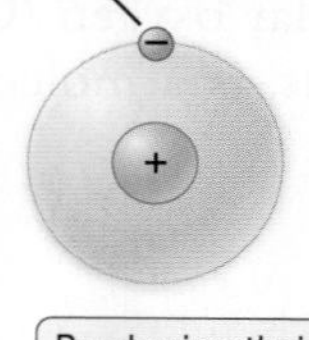

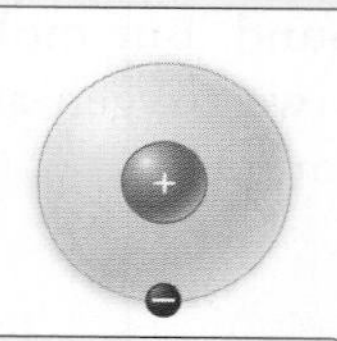

By sharing their electrons, both of these hydrogen atoms have filled their shells.

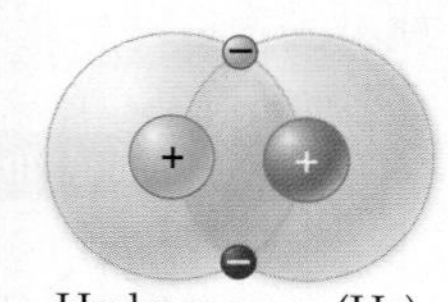

Hydrogen gas (H_2)

The outermost shell of an oxygen atom has six electrons but can hold a maximum of eight.

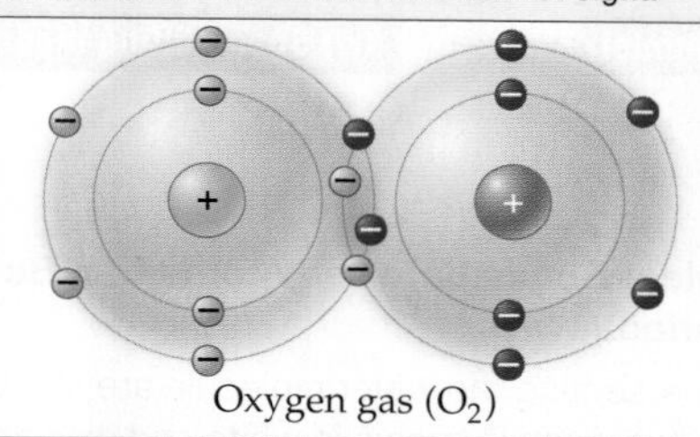

Oxygen gas (O_2)

By sharing *two* pairs of electrons, each oxygen atom fills its outermost electron shell to capacity.

(*b*) Some biologically important atoms

ATOM	SYMBOL	NUMBER OF POSSIBLE BONDS	SAMPLE MOLECULES	
Hydrogen	H	1	H—H	Hydrogen gas (H_2)
Oxygen	O	2	O=O	Oxygen gas (O_2)
Sulfur	S	2	H—S—H	Hydrogen sulfide (H_2S)
Nitrogen	N	3	H—N(—H)—H	Ammonia (NH_3)
Carbon	C	4	H—C(—H)(—H)—H	Methane (CH_4)

FIGURE 2.5 Covalent Bonds and Electron Shells

Atoms in biologically important molecules have as many as four electron shells. The innermost shell holds a maximum of two electrons, and the next shell outside it holds a maximum of eight. (*a*) Two hydrogen atoms can share one pair of electrons, forming one covalent bond. (*b*) The number of covalent bonds an atom can form depends on the number of electrons needed to fill its outermost shell.

two atoms share two pairs of electrons, the result is a *double bond*, as in oxygen gas, O_2 (see Figure 2.5*a*).

The number of covalent bonds an atom can form is equal to the number of electrons needed to fill its outermost shell. Consider the electron sharing that occurs between hydrogen and oxygen in a water molecule. Hydrogen has one electron in its single shell, so that shell is one electron short of maximum occupancy and therefore maximum stability. The inner shell of oxygen is filled, but its outer shell is not: it has six electrons, when it can hold as many as eight. This situation can be resolved by mutual borrowing between two hydrogen atoms and an oxygen atom: each hydrogen atom shares its one electron with oxygen, and the oxygen atom shares two of its electrons, one with each of the hydrogen atoms:

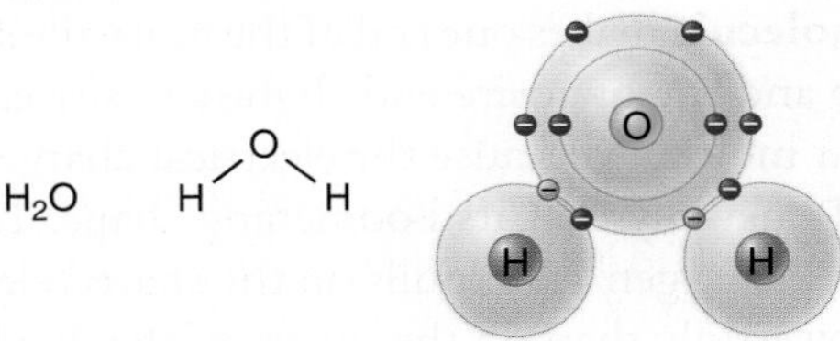

So the atoms contribute electrons in a way that makes the outer shells of all three atoms complete, at least on a shared basis. This kind of sharing requires an intimate association between the atoms, which is why covalent bonds are so strong.

A *double bond*, representing two covalent bonds, exists when *two pairs* of electrons are shared between two atoms. Some atoms may even share *three pairs* of electrons, in which case they are said to have *triple bonds*. Nitrogen gas, also known as molecular nitrogen (N_2) to distinguish it from elemental nitrogen (N), consists of two nitrogen atoms bound by triple covalent bonds (three pairs of shared electrons):

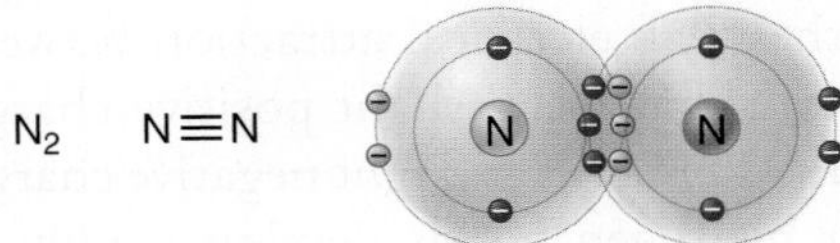

The physical and chemical properties of a molecule are often greatly influenced by the shape of that molecule in three-dimensional space. A *ball-and-stick model* shows the angles made by adjacent covalent bonds in a molecule (**FIGURE 2.4*c***). Atoms are represented by spheres ("balls"), and a single covalent bond is represented by a line or rod ("stick"). A *space-filling model* is yet another way to depict the shape of a molecule in three-dimensional space (**FIGURE 2.4*d***). Unlike a ball-and-stick representation of a molecule, a space-filling model shows the width of atoms (atomic radii) and the distance between nuclei (bond length) in accurate proportions.

Ionic bonds form between atoms of opposite charge

Certain elements have a tendency to gain or lose electrons, becoming charged ions, as mentioned earlier. Ions with opposite electrical charges can associate through ionic bonds. For example, an electron in the outer shell of a neutral sodium atom can be transferred to the outer shell of a neutral chlorine atom, as shown in **FIGURE 2.6**. The loss of an electron from the sodium atom and the gain of an electron by chlorine converts both neutral atoms into ions with charges that are equal and opposite. For maximum stability, the ions created through this electron transfer must remain closely associated.

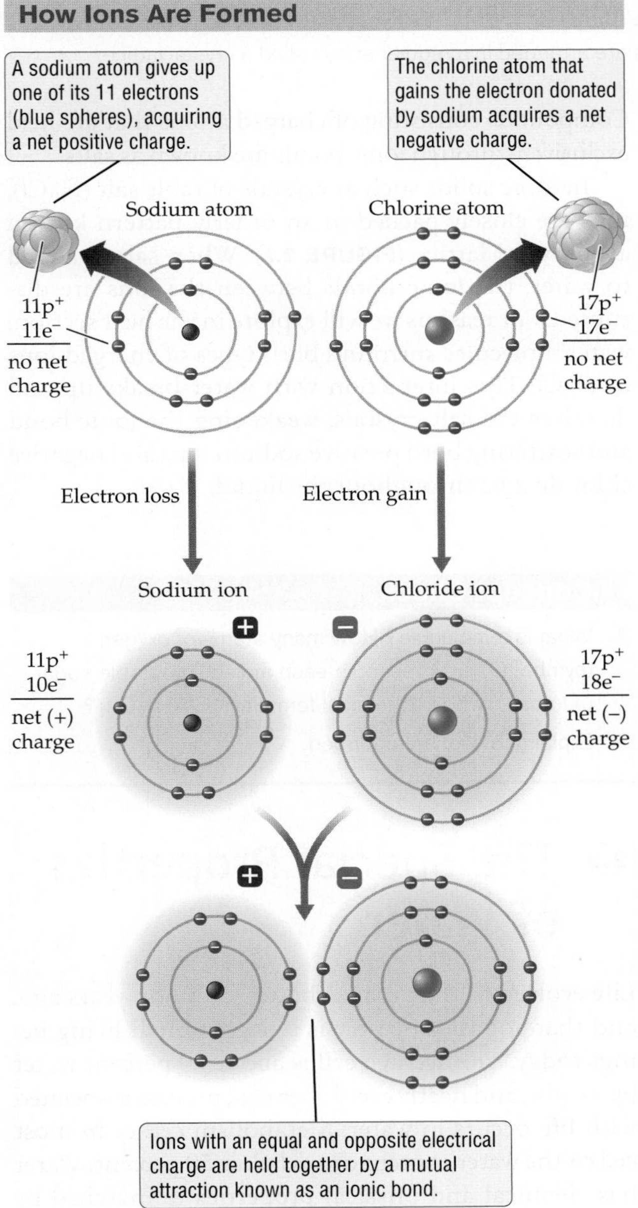

FIGURE 2.6 Ions Are Created through the Loss or Gain of Electrons Negatively charged ions are often given special names, such as "chloride" for the chlorine ion (Cl^-) and "fluoride" for the fluorine ion (F^-).

Structure of Salt Crystals

FIGURE 2.7 Salts Are Ionic Compounds
In an ionic solid, the ions are arranged in a regular array called a crystal lattice.

Compounds consisting of charged atoms that are held exclusively through ionic bonds are known as salts.

In ionic solids such as crystals of table salt (NaCl), ions are closely packed in an orderly pattern known as a crystal lattice (**FIGURE 2.7**). When salt is added to water, the ionic bonds between the ions are disrupted. For reasons we will explore in the next section, water molecules surround both types of charged ions in NaCl. This interaction with water breaks up and dissolves the salt crystals, weakening the ionic bond and scattering both positive sodium ions and negative chloride ions throughout the liquid.

Concept Check Answers

1. A molecule contains two or more atoms held together by covalent bonds. Each molecule of sucrose has 11 atoms of oxygen.

2. Ions form when the atoms of one element lose electrons and the atoms of another element gain electrons, generating positively and negatively charged ions, respectively.

Concept Check

1. What is a molecule? How many atoms of oxygen (symbol O) are present in each molecule of table sugar (sucrose), whose chemical formula is $C_{12}H_{22}O_{11}$?
2. Explain how ions are formed.

2.3 The Special Properties of Water

Life evolved in the oceans about 3.5 billion years ago, and that aquatic ancestry is reflected in all living beings today. The average cell is about 70 percent water by weight, and nearly every chemical process associated with life occurs in water. Metabolism ceases in most cells if the water content drops below 50 percent. Water has chemical and physical properties unmatched by any other known substance, and water's ability to sustain life is based on these qualities.

Water is a polar molecule

We noted earlier that each molecule of water is made up of two hydrogen atoms and one oxygen atom held together by covalent bonds. In some covalent bonds, however, electrons are not *equally* shared by the two atoms. In these molecules, one atom has greater electron-grabbing power (technically known as *electronegativity*). The more electronegative atom exerts a stronger pull on the shared electrons, resulting in a polar covalent bond. "Polar" means the two ends are the opposite of each other, like the two poles of Earth. The uneven distribution of electrical charge in a **polar molecule** makes one end of the molecule slightly negative and the opposite end slightly positive. Water is a polar molecule because the electrical charge is distributed unevenly over its boomerang shape. The nucleus of the oxygen atom pulls on the shared electrons more powerfully than do the nuclei of the two hydrogen atoms. The oxygen end of the molecule therefore has a slightly negative charge, while the two hydrogen atoms have a slightly positive charge. In the following schematic representation of the water molecule, the two polar regions are shown in green and blue:

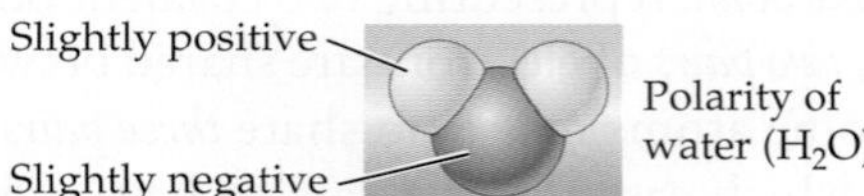

Because opposite electrical charges attract, the slightly positive hydrogen atoms of one water molecule are drawn toward the slightly negative oxygen atom of a neighboring water molecule. A **hydrogen bond** is the weak electrical attraction between a hydrogen atom that has a slight positive charge and a neighboring atom with a slight negative charge. Intermolecular hydrogen bonds develop readily between adjacent water molecules:

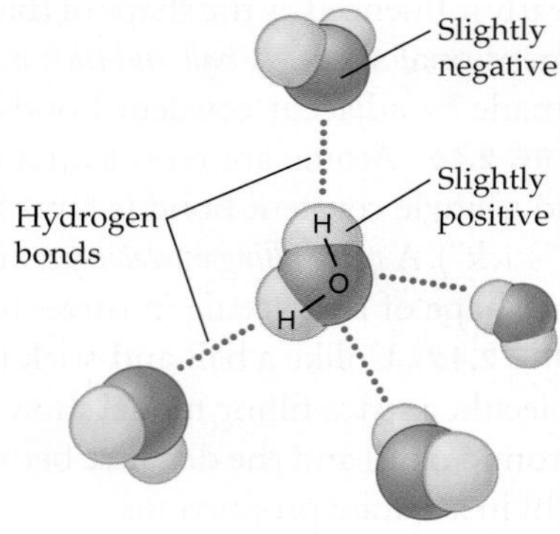

Although a single hydrogen bond is about 20 times weaker than a covalent bond, collectively the cross-linking of many water molecules through many hydrogen bonds amounts to a potent force. The polarity of water and the resultant hydrogen bonding explain nearly all of the special properties of water that we explore next.

Water is a solvent for charged or polar substances

Water molecules can form hydrogen bonds with other polar molecules, which is why polar compounds dissolve in water—that is, mix completely with the water. Ions also dissolve readily in water, because water molecules can form a network around the ions. For example, when table salt is added to water, the solid crystals dissolve. The crystals break apart as the ions in the salt crystal are surrounded by water molecules and held in solution throughout the liquid (**FIGURE 2.8**).

Because dissolved compounds are so common in and around living cells, chemists and biologists use specific terms to describe these mixtures: a **solution** is any combination of a **solute** (a dissolved substance) and a **solvent** (the fluid into which the solute has dissolved). A cup of black coffee has hundreds of different types of solutes in it, caffeine being one of them. Stirring a spoonful of table sugar (sucrose) into the cup adds yet another solute. Water, which makes up more than 98 percent of the volume of that cup of coffee, is the solvent in which the many solutes are dissolved.

How Salts Dissolve in Water

Solutes: Na^+ (sodium ions)
Cl^- (chloride ions)
Solvent: H_2O (water)
Solution: solutes + solvent

Na^+

Cl^-

The negative pole of the water molecule orients to the positive ion.

The positive pole of the water molecule orients to the negative ion.

FIGURE 2.8 Charged Substances Dissolve in Water to Form Solutions

Water is also the solvent of life: the cytoplasm is a thick solution that contains thousands of different types of solutes and is about 70 percent water in most cells. Because so many biologically important ions and molecules function as solutes, water is commonly called the universal solvent.

Because of their polar nature, water molecules do not interact with uncharged or nonpolar substances. Electron sharing is largely symmetrical in **nonpolar molecules**. When added to water, nonpolar molecules fail to go into solution and tend to cluster among their own kind instead. This is exactly what happens when olive oil, composed of nonpolar molecules, is added to vinegar, an aqueous solution containing ions and polar molecules. To make a salad dressing, we must shake the oil and vinegar vigorously to mix them; left alone, the polar and nonpolar components will tend to separate.

Oil molecules are hydrophobic. They are excluded from water and tend to clump together.
Olive oil
Vinegar
Vinegar molecules are hydrophilic. They are held in solution by water molecules.

Molecules that associate with water (such as sugar and salt) are called **hydrophilic** (*hydro*, "water"; *philic*, "loving"); molecules that are excluded from water (such as oil and wax) are called **hydrophobic** (*phobic*, "fearing"). The reason that olive oil does not dissolve easily in vinegar is that the ingredients in the oil are hydrophobic and water molecules therefore do not interact with them. Being lighter than water, the "shunned" oil molecules clump together and float on top of the vinegar.

Helpful to Know

Heat is the *total energy* in a body due to the movement of the atoms within it. *Temperature* is a measure of the *average speed* with which the atoms are moving. An object's size affects its total heat but not its temperature. A bathtub with 30°C water has more heat than a teacup with 30°C water, but because water molecules move at the same average speed in both they have the same temperature.

Hydrogen bonding accounts for the physical properties of liquid water and ice

Water can exist in all three states of matter: liquid, solid, and gas. The interplay of hydrogen bonds explains the physical properties of the liquid and solid phases of water. At moderate temperatures, hydrogen bonds among water molecules are constantly forming and breaking. This nonstop jostling is the reason that water is a liquid at room temperature: the molecules cannot be packed together tightly enough to form a solid.

EXTREME UNDER-ICE HARVEST

Inuits in the Canadian Arctic collect mussels off the ocean floor at low tide, after climbing down through holes they cut in the winter sea ice. The ice, unsupported by water, can collapse on the mussel gatherers, who must also watch for the returning tide.

At 0°C, water molecules have less energy and cannot move about as vigorously. A more stable network of hydrogen bonds emerges as water turns into ice. A particular volume of liquid water expands when it turns into ice, taking up more space. The water molecules are farther apart in ice, locked into an orderly pattern known as a crystal lattice (**FIGURE 2.9**). Because ice is less dense, it floats on liquid water. If ice were denser than water, it would sink when it froze. Lakes in cold parts of the world would freeze from the bottom up. Instead, the ice that floats on the lake surface acts like an insulating blanket, enabling aquatic organisms to survive the winter in the liquid water below.

Water moderates temperature swings

It takes a lot of heat energy to increase the temperature of water because a large proportion of the energy input goes toward breaking hydrogen bonds. The temperature of water—that is, the average speed of water molecules—can rise only after the hydrogen bonds are broken. For cells, the practical consequence is that it takes a relatively large amount of heat to increase the temperature of their watery contents.

Water is also a very effective heat reservoir. If air temperatures start to drop and water cools, some of the water's stored heat is released, warming up the environment in the process. Seaside locations experience milder winters for this very reason. Cells, too, are buffered from a sharp decrease in internal temperature when external temperatures drop.

The evaporation of water has a cooling effect

When heat energy is supplied to a liquid, some molecules become energetic enough to make the transition from the liquid to the gaseous state—a phenomenon known as **evaporation**. In the case of water, a substantial amount of heat energy must first be invested in snapping the network of hydrogen bonds before water molecules in the liquid state can move fast enough to escape as water vapor (steam).

Hydrogen Bonding

Hydrogen bonding in liquid water

Hydrogen bonds are constantly forming…

…and breaking.

Hydrogen bonding in ice

When water freezes, the hydrogen bonds become more rigid as water molecules become stacked in a three-dimensional network of hexagons, forming an ice crystal.

Water molecules are placed farther apart in an ice crystal, making ice less dense and able to float on liquid water.

FIGURE 2.9 Hydrogen Bonding between Water Molecules

A hydrogen bond is a weak attraction between a hydrogen atom with a partial positive charge and the partially negative region of any polar molecule. In vapor form, water molecules move too rapidly, and are too far apart, to form hydrogen bonds.

The evaporation of water molecules removes heat from a surface, cooling it in the process. *Evaporative cooling* is the lowering of temperature associated with the evaporation of liquids. The fastest molecules evaporate first because they alone have enough energy to break loose from the surface of the liquid. The departure of the high-speed molecules causes the average temperature of the remaining liquid to drop, because the molecules that stay behind are slower and therefore cooler. Sweaty humans and panting dogs are cooled when their body heat goes toward disrupting hydrogen bonds and increasing the speed of water molecules in human sweat or canine slobber.

Hydrogen bonding accounts for the cohesion of water molecules

Cohesion is the attractive force that holds atoms or molecules of the same kind to one another. Substances with high cohesion tend to stick together. The tendency of water molecules to cling to each other through hydrogen bonding generates strong cohesion. You can experience cohesion when you use a straw to drink water or soda: the liquid is pulled up in one continuous stream because all the water molecules are linked together by hydrogen bonds. This cohesive strength is also critical for the ascent of water into the canopy of tall trees.

Surface tension is an important property of water resulting from the cohesiveness of water molecules. This force tends to minimize the surface area of water at its boundary with air. Water has the highest surface tension of any liquid except mercury. The pull of hydrogen bonds generates surface tension, and this force holds the water surface taut, resisting stretching or breaking. Water's surface tension is strong enough to support very light objects: aquatic organisms like water spiders, or a paper clip sitting on top of water in a cup filled to the brim. In the case of the spider, its long legs distribute its weight, and hydrophobic material on its legs repels water; the legs dimple the water surface without breaking through the air-water boundary (**FIGURE 2.10**).

You do not feel the surface tension of water in a swimming pool if you dive in neatly, with your fingertips leading the way. But if you expose more of your body surface, as you do when you belly flop into the pool, you might feel the collective smack of the interlinked water molecules under tension at the air-water boundary.

Surface Tension

The weight of a water spider, with its long, water-repellent legs, dimples the water surface.
Surface tension opposes the spider's weight, preventing the spider from breaking through the sheet of hydrogen-bonded water molecules at the air-water boundary.

FIGURE 2.10 Hydrogen Bonding between Water Molecules Contributes to the High Surface Tension of Water

The cohesion of water molecules gives rise to surface tension, a force that resists the stretching of the water surface at any air-water boundary.

Concept Check

1. Why don't oil and water mix?
2. New York City and Pittsburgh are at the same latitude (about 40° north), but Pittsburgh is farther inland. Which one has a more moderate climate, with less difference between average midwinter and midsummer temperatures? Why?

Concept Check Answers

1. Oil molecules are nonpolar and therefore hydrophobic. They cannot dissolve in water because they have no electrical attraction for the polar water molecules, so they separate from water instead.
2. New York City, because of its location. Large bodies of water moderate temperatures, releasing heat in the winter and absorbing it in the summer.

2.4 Chemical Reactions

Many biological processes require atoms to break existing connections or form new ones. The process of breaking or creating chemical bonds is known as a **chemical reaction**. A **reactant** is a substance that undergoes a chemical reaction, either alone or in conjunction with other reactants. The alteration of electron-sharing patterns through a chemical reaction yields at least one chemical substance that is different from the reactants, and any newly formed substances are called the **products** of the chemical reaction.

The standard notation for chemical reactions, the *chemical equation*, displays reactants to the left of an arrow and products to the right of that arrow. Hydrogen and oxygen, for example, can combine to produce water molecules in an explosive reaction. The reaction releases so much energy that it is used in liquid-fuel rockets:

$$2\,H_2 + O_2 \longrightarrow 2\,H_2O + \text{Energy}$$

(Reactants) (Product)

The arrow in the equation indicates that the molecules on the left side of the equation are converted to the product, water. The numbers in front of the molecules define how many molecules participate in the reaction. In this case, two molecules of molecular hydrogen (H_2) combine with one molecule of molecular oxygen (O_2) to produce two molecules of water. (Note that when a single molecule of a substance is intended, the numeral 1 is generally omitted.)

Some chemical reactions release energy, but others will not occur without an input of energy. The manufacture of ammonia, a key ingredient in many synthetic (human-made) fertilizers, requires large amounts of energy, usually provided by fossil fuels such as natural gas or petroleum. Nitrogen and hydrogen gases are combined to produce ammonia (NH_3). The chemical equation for this reaction is

$$\text{Energy}\ \downarrow$$

$$3\,H_2 + N_2 \longrightarrow 2\,NH_3$$

(Reactants) (Product)

Chemical bonds in atoms are rearranged during a chemical reaction, but the process can neither create nor destroy atoms. Therefore, the reaction must begin and end with the same number of atoms of each element. In the reaction depicting ammonia manufacture above, there are six hydrogen atoms for each pair of nitrogen atoms among the reactants, and all six hydrogen atoms and both nitrogen atoms are accounted for in the two molecules of product (ammonia).

Concept Check

This reaction summarizes the burning of methane (CH_4), or "swamp gas": $CH_4 + 2\,O_2 \rightarrow CO_2 + 2\,H_2O$.

1. Name the products and the reactants in the equation above.
2. How many molecules of molecular oxygen (O_2) are needed to produce one molecule of carbon dioxide (CO_2) through this reaction?

Concept Check Answers

1. Methane and oxygen are reactants; carbon dioxide and water are products.
2. Two.

2.5 The pH Scale

In any volume of pure water, a few water molecules spontaneously split apart—or *dissociate*—to produce a hydrogen (H^+) ion and a hydroxyl (OH^-) ion. Only a tiny number of molecules are dissociated at any one time, and they balance each other electrically. The dissociation of water molecules is summarized by this equation:

$$\underset{\text{Water}}{H_2O} \rightleftharpoons \underset{\text{Hydrogen ion}}{H^+} + \underset{\text{Hydroxyl ion}}{OH^-}$$

Some compounds that mix in water change this balance by either adding more hydrogen ions to the solution or removing them from solution by "sopping them up." Compounds that release hydrogen ions are called **acids**. The following equation depicts how hydrochloric acid (HCl) releases hydrogen ions in an aqueous solution:

$$\underset{\text{Hydrochloric acid}}{HCl} \rightleftharpoons \underset{\text{Hydrogen ion}}{H^+} + \underset{\text{Chloride ion}}{Cl^-}$$

A compound that removes hydrogen ions from an aqueous solution, or adds hydroxyl ions to it, is called a **base**. The following equation depicts how a base called sodium hydroxide releases OH^- ions in an aqueous solution:

$$\underset{\text{Sodium hydroxide}}{NaOH} \rightleftharpoons \underset{\text{Sodium ion}}{Na^+} + \underset{\text{Hydroxyl ion}}{OH^-}$$

If you add a base such as NaOH to water, the solution becomes more basic. That is because the added hydroxide ions (OH^-) combine with some of the hydrogen ions (H^+) released by water molecules, leaving fewer free H^+ in the solution.

Acids and bases are common in everyday life. Mildly acidic solutions have a tangy taste, and moderately basic solutions often feel soapy. Examples of familiar acids include citric acid, which gives lemon juice its mouth-puckering taste. Familiar bases include antacids, baking soda,

EXTREME REACTION

The 250-ton space shuttle was lifted in orbit by the explosive reaction of liquid molecular oxygen (O_2) with liquid molecular hydrogen (H_2).

and ammonia cleaners. In all these substances, what makes them acids or bases is the concentration of hydrogen ions that are free in the solution.

Hydrogen ion concentration is commonly expressed on a scale from 0 to 14, where 0 represents an extremely high concentration of free hydrogen ions, and 14 represents the lowest concentration. On this scale, called the **pH scale**, each number represents a 10-fold increase or decrease in the concentration of hydrogen ions (**FIGURE 2.11**). In pure water, the concentrations of free hydrogen ions and hydroxide ions are equal, and the pH is said to be neutral, or in the middle of the scale, at pH 7.

Our stomach juices are able to break down food because they are very acidic (about pH 2). At this low pH, hydrogen bonds and ionic bonds are disrupted by the high concentration of free hydrogen ions, and even the covalent bonds of some biomolecules are broken. At the other extreme, a very basic substance, such as oven cleaner (pH 13.5), can also break or disrupt biomolecules. This is why extremes of pH can be dangerous, causing chemical burns on the skin.

Most living systems function best at an internal pH close to neutral. Because hydrogen ions can move so freely during normal life processes, organisms must have ways of preventing dramatic changes in their internal pH. Substances called **buffers** meet this need by maintaining the concentration of hydrogen ions within narrow limits. They do so by releasing hydrogen ions when the surroundings become too basic (excessive OH^- ions, high pH) and accepting hydrogen ions when the surroundings become too acidic (excessive H^+ ions, low pH).

Our blood pH is maintained between pH 7.34 and 7.45 in part by a buffer system made up of carbonic acid (H_2CO_3) and bicarbonate ions (HCO_3^-). This buffer system resists a change in pH caused by the addition of small amounts of acid or base. The blood's buffer system can be overwhelmed, however, by a large excess of hydrogen ions (termed acidosis) or a large excess of hydroxide ions (alkalosis). The lungs and kidneys assist in regulating the blood pH and buffer system. When the function of these organs is impaired, as could happen with kidney failure or severe pneumonia, acidosis or alkalosis can develop.

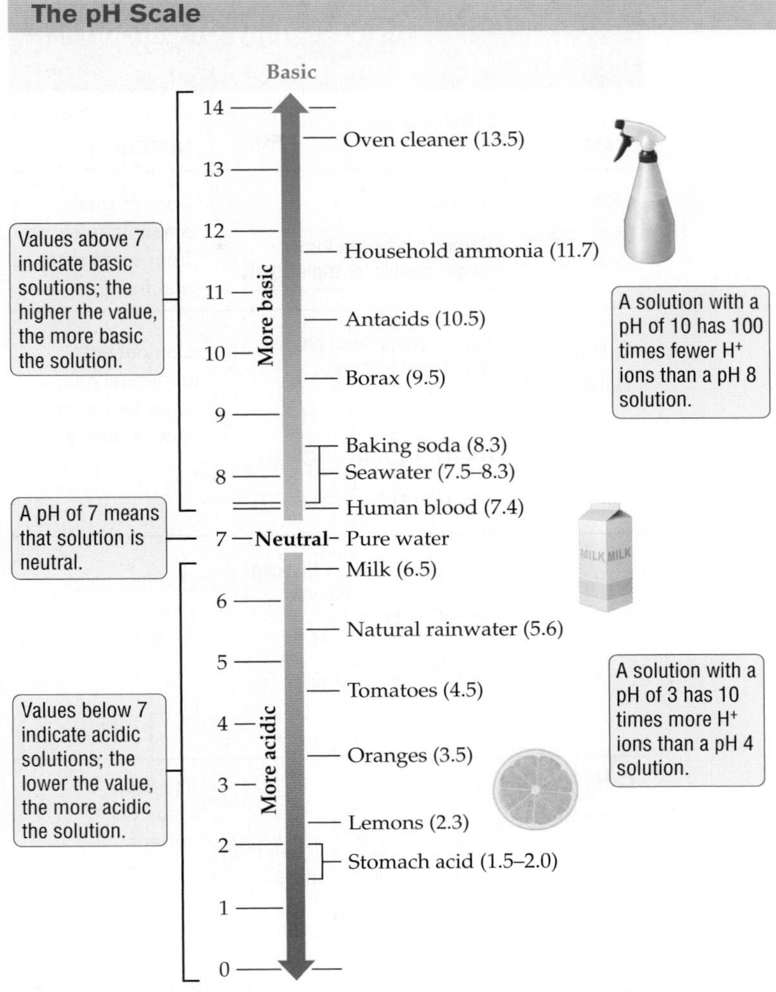

The pH Scale
Basic
14
13
12
11
10
9
8
7
6
5
4
3
2
1
0
More basic
Neutral
More acidic
Acidic
Oven cleaner (13.5)
Household ammonia (11.7)
Antacids (10.5)
Borax (9.5)
Baking soda (8.3)
Seawater (7.5–8.3)
Human blood (7.4)
Pure water
Milk (6.5)
Natural rainwater (5.6)
Tomatoes (4.5)
Oranges (3.5)
Lemons (2.3)
Stomach acid (1.5–2.0)
Values above 7 indicate basic solutions; the higher the value, the more basic the solution.
A pH of 7 means that solution is neutral.
Values below 7 indicate acidic solutions; the lower the value, the more acidic the solution.
A solution with a pH of 10 has 100 times fewer H+ ions than a pH 8 solution.
A solution with a pH of 3 has 10 times more H+ ions than a pH 4 solution.
MILK MILK

FIGURE 2.11 The pH Scale Indicates Hydrogen Ion Concentration, a Measure of Acidity versus Basicity

Concept Check

1. What is the difference between an acid and a base?
2. Which has a higher concentration of free hydrogen ions: vinegar, pH 3.0; or tomato juice, pH 6.0?

2.6 The Chemical Building Blocks of Life

If all the water in any living organism were removed, four major classes of molecules would remain, all of them critical for living cells: carbohydrates, proteins, lipids, and nucleic acids. Each of these biologically important molecules is built on a framework of covalently linked carbon atoms associated with hydrogen. Oxygen, nitrogen, phosphorus, and sulfur atoms are also found in some of these molecules.

Concept Check Answers

1. An acid releases hydrogen ions; a base absorbs them.

2. Vinegar; its H^+ concentration is 1,000 times greater.

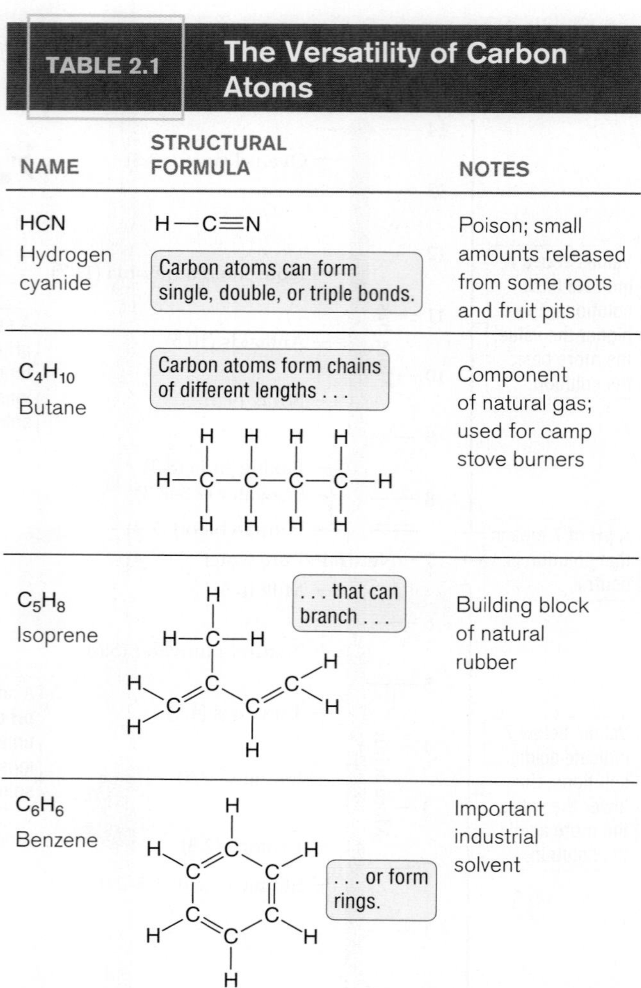

TABLE 2.1 The Versatility of Carbon Atoms

NAME	STRUCTURAL FORMULA	NOTES
HCN Hydrogen cyanide	H—C≡N (Carbon atoms can form single, double, or triple bonds.)	Poison; small amounts released from some roots and fruit pits
C_4H_{10} Butane	(Carbon atoms form chains of different lengths . . .)	Component of natural gas; used for camp stove burners
C_5H_8 Isoprene	(. . . that can branch . . .)	Building block of natural rubber
C_6H_6 Benzene	(. . . or form rings.)	Important industrial solvent

Helpful to Know

In biology, an *organic molecule* is one that contains carbon bonded to at least one hydrogen atom. Today, scientists can make thousands of organic molecules in a test tube, but in the nineteenth century, organic molecules were known only from *organisms*, the root word for "organic." You see the old meaning in every day words such as "organic food," which is raised without artificial fertilizers, pesticides, or drugs.

Carbon provides the framework for life

Carbon is the predominant element in living systems partly because it can form large molecules that contain thousands of atoms. A single carbon atom can form strong covalent bonds with up to four other atoms. Even more important, carbon can bond to carbon, forming long chains, branched molecules, or even rings (**TABLE 2.1**). The diversity of biological processes depends on the wide variety of molecular structures that can be built from a carbon-carbon framework bonded to a small handful of other types of atoms. No other element is as versatile as carbon.

Biomolecules that include at least one carbon-hydrogen bond are referred to as **organic molecules**. Sugars and amino acids are examples of small organic molecules (having up to 20 or so atoms). Small organic molecules can link up via covalent bonds to create larger assemblies of atoms, called **macromolecules** (*macro*, "large"). Starch and proteins are examples of macromolecules.

Small molecules that serve as repeating units in a macromolecule are called **monomers** (*mono*, "one"; *mer*, "part"). Macromolecules that contain monomers as building blocks are called **polymers** (**FIGURE 2.12**). A polymer (*poly*, "many") may contain hundreds or even thousands of monomers linked through covalent bonds. Polymers account for most of an organism's dry weight (its weight after all water is removed) and are essential for every structure and chemical process that we associate with life.

Functional groups have distinct chemical properties

In living organisms, fewer than 70 different biological monomers combine in an endless variety of ways

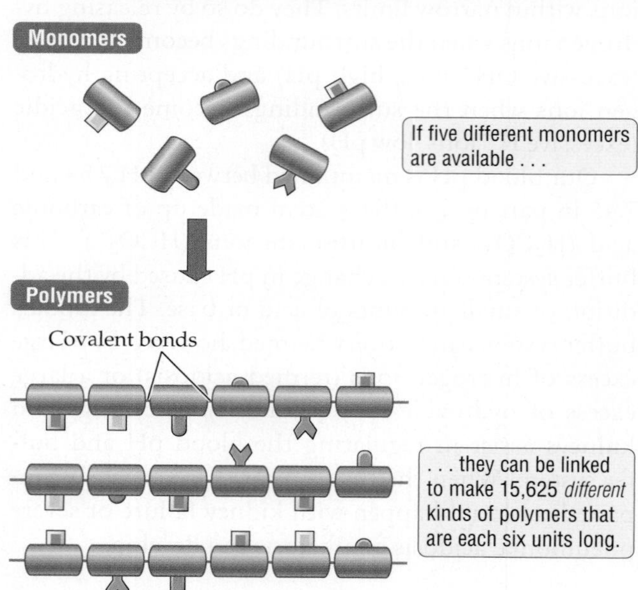
Building Polymers from Monomers
Monomers
If five different monomers are available . . .
Polymers
Covalent bonds
. . . they can be linked to make 15,625 *different* kinds of polymers that are each six units long.

FIGURE 2.12 Many Different Polymers Can Be Made by Joining Monomers in Different Ways
A handful of monomers can be assembled into a great variety of polymer chains just by being linked in a different order.

to produce polymers with many different properties. Polymers are therefore a step up from monomers in complexity, and they have chemical properties beyond those of monomers. Furthermore, the properties of organic polymers often emerge from attached clusters of atoms called *functional groups* (**TABLE 2.2**).

Functional groups are clusters of covalently bonded atoms that have the same distinctive chemical properties no matter what molecule they are found in. Some functional groups help establish covalent linkages between monomers; others have more general effects on the chemical characteristics of a polymer, such as whether it is acidic or basic, how soluble it is in water, and how easily it can form associations with other molecules.

TABLE 2.2 Important Functional Groups

FUNCTIONAL GROUP	STRUCTURAL FORMULA	BALL-AND-STICK MODEL
Amino group	$-NH_2$	Bond to carbon atom; N, H, H
Carboxyl group	$-COOH$	C, O, O, H
Hydroxyl group	$-OH$	O, H
Phosphate group	$-PO_4$	P, O, O, O^-, O^-

Concept Check

1. In biology, what is an organic molecule?
2. Offer an explanation for the observation that many important biomolecules are polymers. Why would being able to form polymers from monomers be an advantage to a cell?

Concept Check Answers

1. Organic molecules include at least one carbon-hydrogen bond.
2. Many different kinds of polymers can be made from a handful of monomers, and this versatility increases the diversity of molecules a cell can make.

2.7 Carbohydrates

Sugars are familiar as compounds that taste sweet. Although not all sugars are sweet-tasting, these small organic molecules are an important source of energy for nearly all organisms. Sugars and their polymers are *carbohydrates*. This name comes from the ratio of C, H, and O in the compounds, in which for each carbon atom (*carbo*) there are two hydrogens and one oxygen (corresponding to a molecule of water, a *hydrate*). We can define a **carbohydrate** as a molecule that contains carbon, hydrogen, and oxygen in a ratio of 1:2:1.

The simplest sugar molecules are called **monosaccharides** (*MAH*-**noh**-*SAK*-**uh-ride**; *mono*, "one"; *sacchar*, "sugar"). Monosaccharides are often referred to by the number of carbon atoms they contain. For example, a sugar with the general molecular formula $(CH_2O)_5$ is known as a five-carbon sugar. The more common way to express the molecular formula for this sugar is $C_5H_{10}O_5$.

When monosaccharides with five or more carbon atoms are dissolved in water, the sugar molecules may exist in either chain form or ring form. Here are the chain and ring forms of a five-carbon sugar called ribose:

Ribose: chain form **Ribose: ring form**

The one monosaccharide that is found in almost all cells is **glucose** ($C_6H_{12}O_6$). Glucose has a key role as an energy source within the cell, and nearly all the chemical reactions that produce energy for living organisms involve the manufacture or breakdown of this sugar. Fructose, fruit sugar, has the

Simple Carbohydrates: Sugars

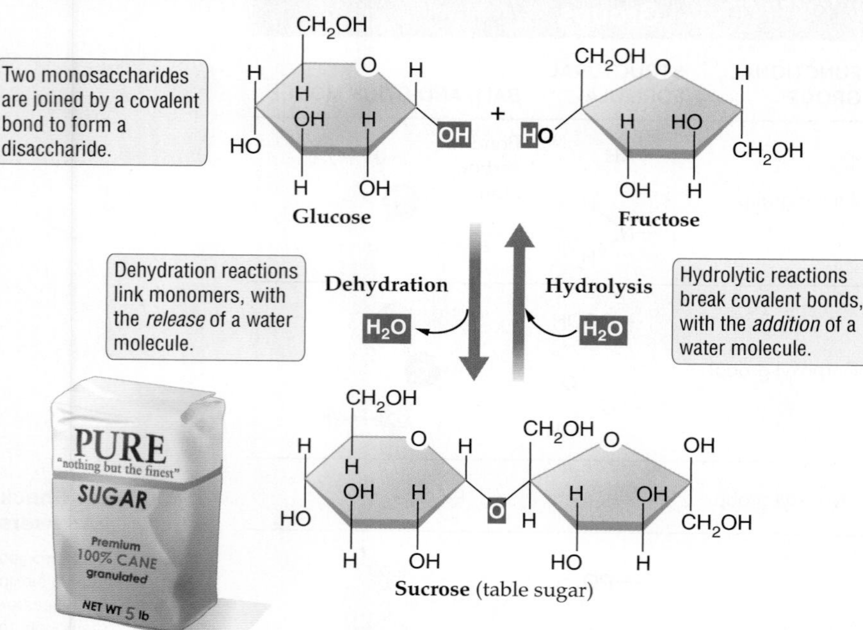

Two monosaccharides are joined by a covalent bond to form a disaccharide.
Glucose
Fructose
Dehydration reactions link monomers, with the *release* of a water molecule.
Dehydration
Hydrolysis
Hydrolytic reactions break covalent bonds, with the *addition* of a water molecule.
PURE
"nothing but the finest"
SUGAR
Premium 100% CANE granulated
NET WT 5 lb
Sucrose (table sugar)

FIGURE 2.13 Monosaccharides Can Bond Together to Form Disaccharides
Glucose and fructose are sugar monomers that, when linked by a covalent bond, form the disaccharide sucrose, or table sugar.

same molecular formula as glucose ($C_6H_{12}O_6$), but the atoms are connected in a different pattern, resulting in very different physical and chemical properties. Fructose is nearly twice as sweet as glucose, and it is widely used as a sweetener in processed foods because corn is a cheap source of the sugar.

Scientists often use a shorthand notation to represent the ring form of carbon-containing molecules: the symbols for the carbon atoms, and most of the hydrogen and oxygen atoms, are left out. Here are the structural formula and shorthand notation for glucose:

Glucose

Monosaccharides can combine to form larger, more complex molecules. Two covalently joined monosaccharides form a **disaccharide** (*di*, "two"). Our familiar table sugar, sucrose, is a disaccharide built by linking a molecule of glucose and a molecule of fructose, with the removal of a water molecule (**FIGURE 2.13**).

The chemical reaction in which a water molecule is *removed* as a covalent bond forms is known as a **dehydration reaction**. The reverse reaction, in which a water molecule is added to break a covalent bond, is called a **hydrolytic reaction**. Sucrose is broken down in our digestive system through hydrolytic reactions, and the released monomers are absorbed by the intestinal wall and eventually delivered to the bloodstream.

Polysaccharides are large polymers built by linking many monosaccharides. Polysaccharides perform a variety of functions in living organisms (**FIGURE 2.14**). **Cellulose**, for example, is a polysaccharide that is bundled into strong parallel fibers that help support the plant body (**FIGURE 2.14*a***). Cotton fabric, made from special cells on the surface of cotton seeds, is mostly cellulose. Carbohydrates are polysaccharides that provide metabolic energy, as we have already seen in the case of glucose. **Starch**—abundant in a dish of mashed potatoes or steamed rice—is a polysaccharide that serves as an energy storage molecule inside plant cells (**FIGURE 2.14*b***).

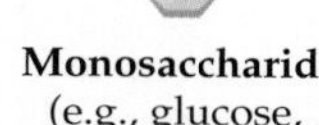

Monosaccharide (e.g., glucose, fructose)
Disaccharide (e.g., lactose, sucrose)
Polysaccharide (e.g., glycogen, starch, cellulose)

Cellulose and starch are both built from glucose, but they differ in how the monosaccharides are linked. Starch is water-soluble and easily broken down in our digestive system. Cellulose, on the other hand, is not water-soluble, which is fortunate for owners of 100 percent cotton clothes, who would literally lose their shirts in the wash otherwise. Unlike starch, cellulose cannot be broken down in the human digestive system, and only some bacteria and fungi can use it for energy.

Glycogen is the main storage polysaccharide in animal cells (**FIGURE 2.14*c***), although, as we will see later, most of the surplus energy ingested by animals is stockpiled in the form of storage lipids ("fat") rather than carbohydrate. The majority of the glycogen reserve in our bodies is stored inside liver cells and skeletal muscle cells.

Helpful to Know

Humans have only ~200–500 grams of glycogen, stored mainly in the liver and muscles. That's worth 800–2,000 Calories of stored energy. About 14 hours without food, or a couple of hours of intense exercise, will use up most of the glycogen stored in a typical adult. The fatigue that results is called "hitting the wall" in the world of long-distance runners and bikers. When glycogen runs low, the body switches to "burning" stored lipids, but energy from lipid reserves is released more slowly than from glycogen reserves.

Complex Carbohydrates: Polysaccharides

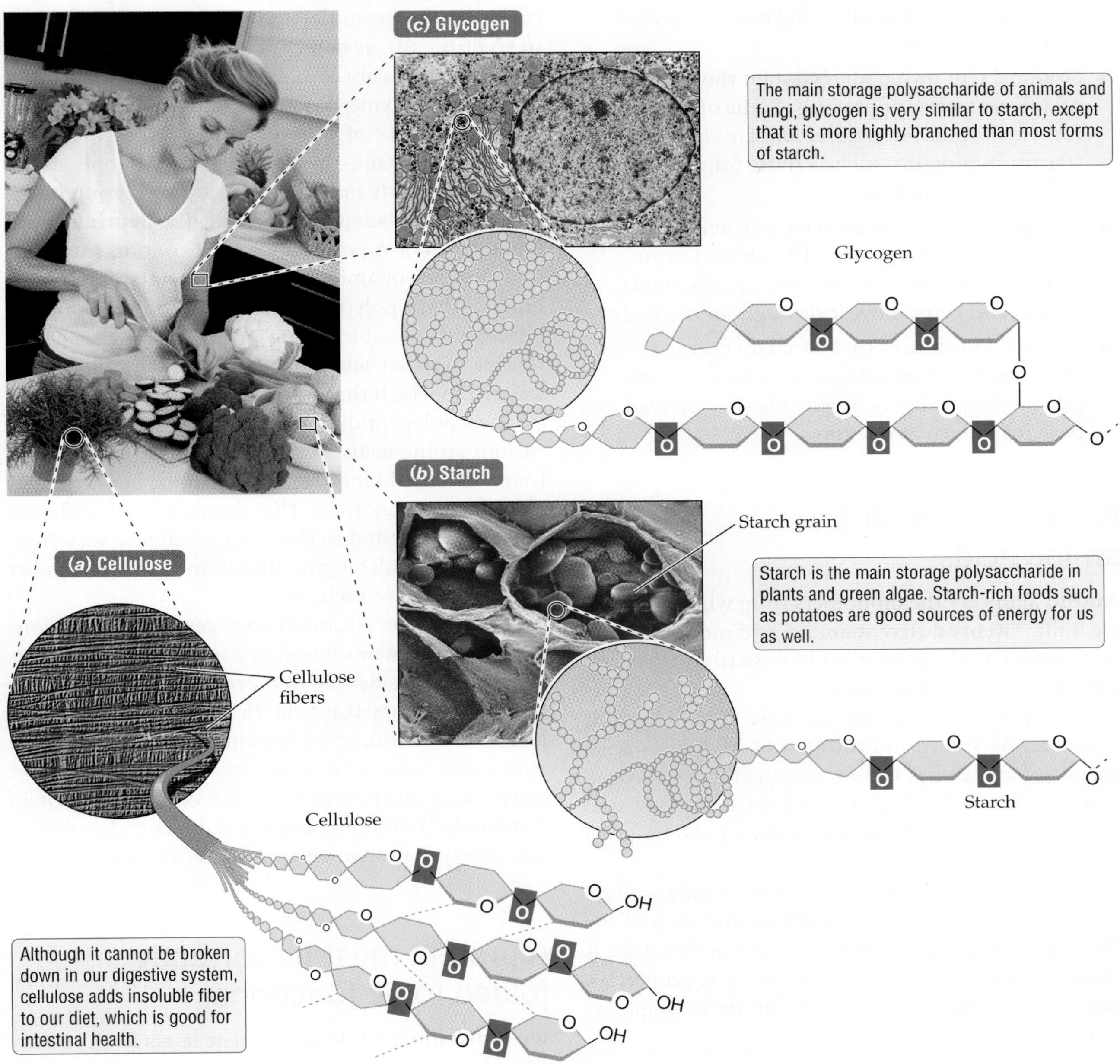

(c) Glycogen
The main storage polysaccharide of animals and fungi, glycogen is very similar to starch, except that it is more highly branched than most forms of starch.
Glycogen
(b) Starch
Starch grain
Starch is the main storage polysaccharide in plants and green algae. Starch-rich foods such as potatoes are good sources of energy for us as well.
Starch
(a) Cellulose
Cellulose fibers
Cellulose
Although it cannot be broken down in our digestive system, cellulose adds insoluble fiber to our diet, which is good for intestinal health.

FIGURE 2.14 Monosaccharides Can Bond Together to Form Polysaccharides
Cellulose, starch, and glycogen are all polymers built from glucose subunits.

Concept Check

1. Which atoms are found in all carbohydrates?
2. Which of the following is a polysaccharide and a key structural component of plant cell walls: glucose, sucrose, monosaccharide, cellulose, or glycogen?

2.8 Proteins

Proteins, which are composed of amino acid monomers, are among the most familiar biomolecules. We often hear and read about proteins in connection with diet and nutrition. We can categorize proteins by some of the main functions they perform:

Concept Check Answers

1. Hydrogen, oxygen, and carbon.
2. Cellulose.

- *Storage.* Bird eggs and plant seeds contain storage proteins, whose function is to supply the building blocks that offspring need for growth and development.
- *Structure.* Our own bodies contain thousands of different types of proteins. Some of these form anatomical structures and are classified as structural proteins, such as those found in bones, cartilage, hair, and nails.
- *Transport.* Some proteins ferry nutrients and other materials within the body. The protein hemoglobin, abundant in our red blood cells, binds oxygen and helps move it throughout the body.
- *Catalysis.* Substances that speed up chemical reactions are called *catalysts*. Almost all chemical reactions in living organisms are catalyzed by proteins known as **enzymes**.

Proteins are built from amino acids

Amino acids are the monomers from which proteins are built. Twenty different amino acid monomers can be arranged in a huge number of ways to construct an enormous variety of proteins.

The chemical "anatomy" of these 20 amino acids is similar. As shown in **FIGURE 2.15**, all amino acids consist of an "alpha" carbon attached to a hydrogen atom, a chemical side chain called the *R group*, and two functional groups: an *amino group* (—NH_2) and a *carboxyl group* (—COOH).

One amino acid differs from another only with respect to the type of R group present, and each of the 20 possible R groups is unique to one amino acid. R groups vary in terms of size, acidic or basic properties, and whether they are hydrophobic or hydrophilic.

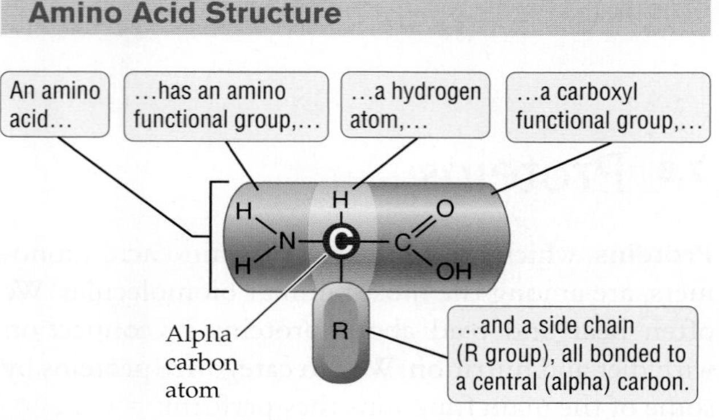
Amino Acid Structure
An amino acid...
...has an amino functional group,...
...a hydrogen atom,...
...a carboxyl functional group,...
H
H
N
C
H
C
O
OH
R
Alpha carbon atom
...and a side chain (R group), all bonded to a central (alpha) carbon.

FIGURE 2.15 The General Structure of an Amino Acid

The R groups found in amino acids range from just one hydrogen atom (in glycine), to longer chains (in lysine, for example) and ring structures (such as in tryptophan), as depicted in **FIGURE 2.16**.

Amino acids are covalently linked in a linear chain to create a polymer known as a **polypeptide**. Every protein consists of one or more polypeptides. In a polypeptide chain, the amino group of one amino acid is covalently linked to the carboxyl group of another via a covalent linkage called a **peptide bond** (**FIGURE 2.17**). A polypeptide may contain hundreds to thousands of amino acids held together by peptide bonds. These polypeptides are built from the same pool of 20 possible amino acids, so the crucial difference between one polypeptide and another is the *sequence* in which the amino acids are linked.

Two polypeptides may differ in the amounts of various amino acids found in the chain; for instance, lysine may be absent in one polypeptide but extremely abundant in another. The thousands of different polypeptides found in the average cell also vary enormously in overall length—that is, in the total number of amino acids in each.

How can just 20 amino acids generate the millions of different proteins found in nature? To see how this is possible, consider the number of different sentences that can be written using the 26 letters of the English alphabet. If we think of the protein alphabet as having 20 amino acid "letters," it is easy to see that this alphabet can produce an enormous number of different protein "sentences." Life's complexity and diversity depends on this variety in protein structure and function.

A protein must be correctly folded to be functional

We can distinguish four different levels in the structure of a protein. These levels are termed *primary*, *secondary*, *tertiary*, and *quaternary*. The structure at each of these levels affects the protein's function.

The **primary structure** of a polypeptide is its sequence of amino acids (**FIGURE 2.18*a***). A polypeptide must acquire a higher level of organization, beyond its primary structure, before it can function as a protein or part of a protein.

The **secondary structure** of a protein is created by the *local* folding of the amino acid chain into specific three-dimensional patterns (**FIGURE 2.18*b***). Alpha (α) helices and beta (β) sheets are two of the most common types of secondary structure. An *alpha helix* is a spiral pattern, like curled ribbon on a gift package. A

Amino Acid Diversity

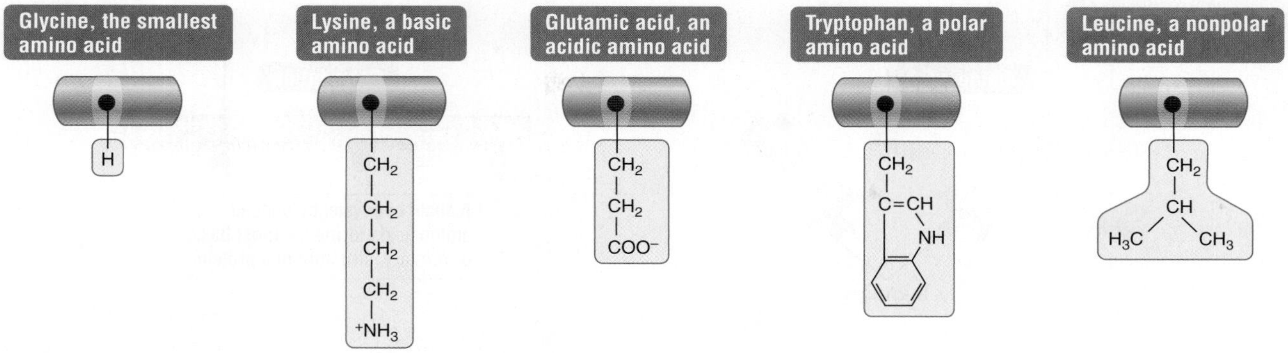

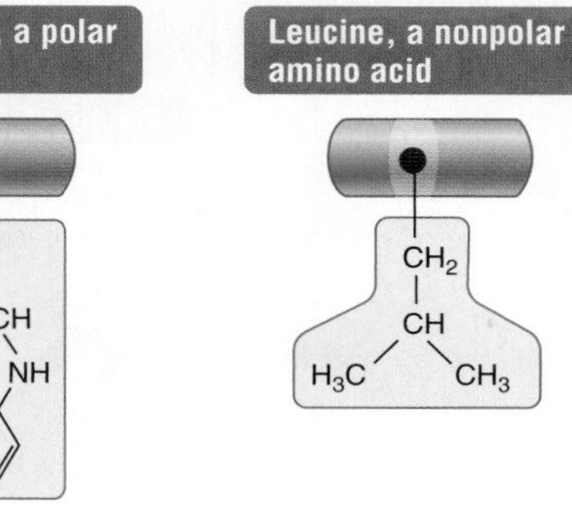

FIGURE 2.16 The Diversity of Amino Acids

Twenty different amino acids can be found in proteins, each differing from the other only in the nature of its R group. Five examples are illustrated here. The R group is responsible for the distinctive properties of each amino acid.

beta sheet is created when the polypeptide backbone is bent into "ridges" and "valleys," as in a paper fan.

The **tertiary structure** of a protein is a very specific three-dimensional shape (**FIGURE 2.18c**). This shape is attained not merely through local patterns of folding, as in the secondary structure, but through interactions between *distant* segments of the polypeptide chain. Tertiary (*TERT*-**shee-er-ee**) structure is stabilized by noncovalent associations such as ionic bonds, and often by covalent links between distant amino acids.

Some proteins are composed of more than one polypeptide, in which case they have yet another level of organization, called the **quaternary structure** (*KWAH*-**ter-nayr-ee**). In such a protein, the quaternary structure must be achieved before it can become biologically active. Hemoglobin, which transports oxygen in our blood, is an example of a protein with quaternary structure (**FIGURE 2.18*d***).

The activity of most proteins is critically dependent on their three-dimensional structure. Extremes of temperature, pH, and salt concentration can change or destroy the structure of a protein, and consequently change or destroy its activity. Why? The answer is that most properly folded proteins tend to have hydrophilic R groups exposed on the surface, and hydrophobic R groups buried deep inside the folded structure. This precise arrangement enables the protein to form hydrogen bonds with surrounding water molecules and remain dissolved. When a protein is heated beyond a certain temperature, its weak noncovalent bonds break and the protein unfolds, losing its orderly three-dimensional shape.

The destruction of a protein's three-dimensional structure, resulting in loss of protein activity, is known as **denaturation**. Some proteins are more easily denatured than others. We witness the denaturation of egg proteins when we cook eggs. At room temperature, albumin, the predominant protein in eggs, is

Formation of a Peptide Bond

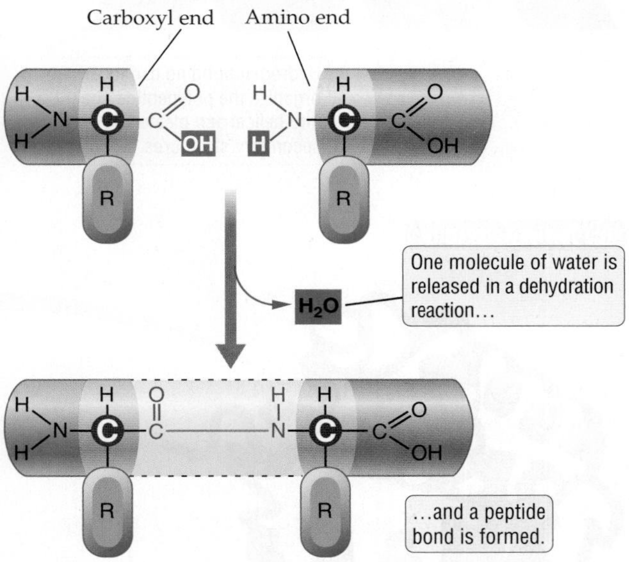

Carboxyl end

Amino end

One molecule of water is released in a dehydration reaction...

H_2O

...and a peptide bond is formed.

FIGURE 2.17 Peptide Bonds Link Amino Acids

EXTREME DENATURATION

Park officials demonstrated protein denaturation by cooking eggs on bare ground as temperatures reached 128°F in Death Valley National Park in June 2013.

Proteins

(a) Primary structure

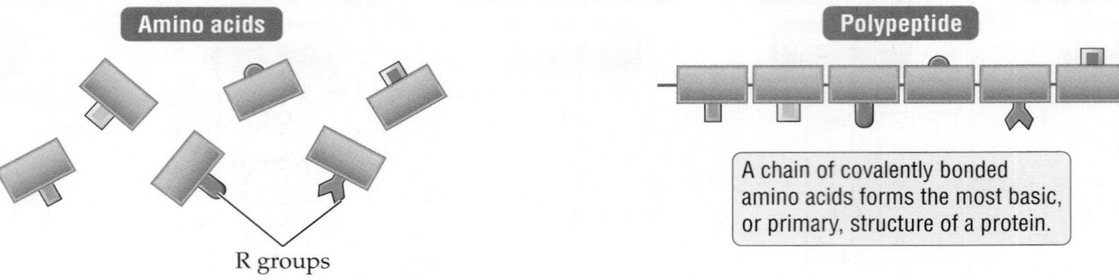

Amino acids
R groups
Polypeptide
A chain of covalently bonded amino acids forms the most basic, or primary, structure of a protein.

(b) Secondary structure

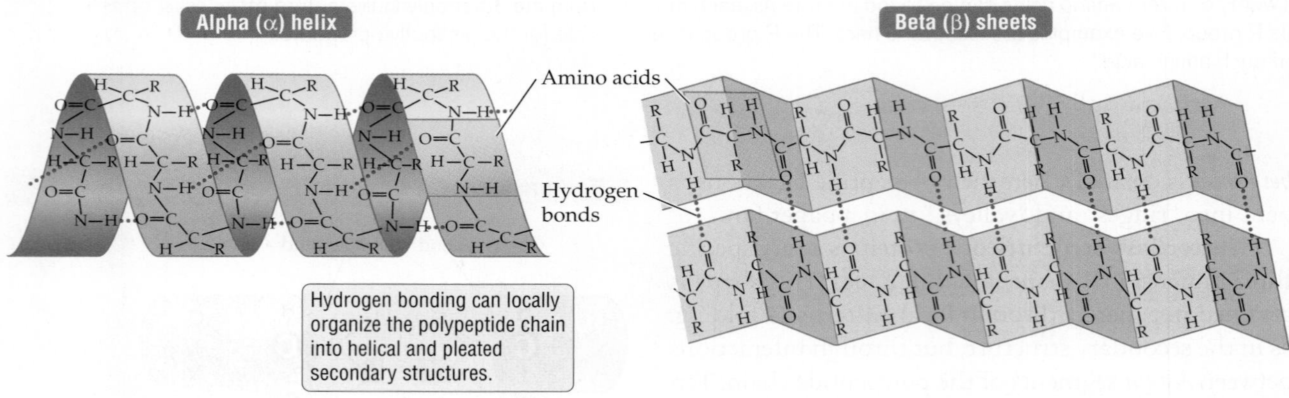

Alpha (α) helix
Beta (β) sheets
Amino acids
Hydrogen bonds
Hydrogen bonding can locally organize the polypeptide chain into helical and pleated secondary structures.

(c) Tertiary structure

(d) Quaternary structure

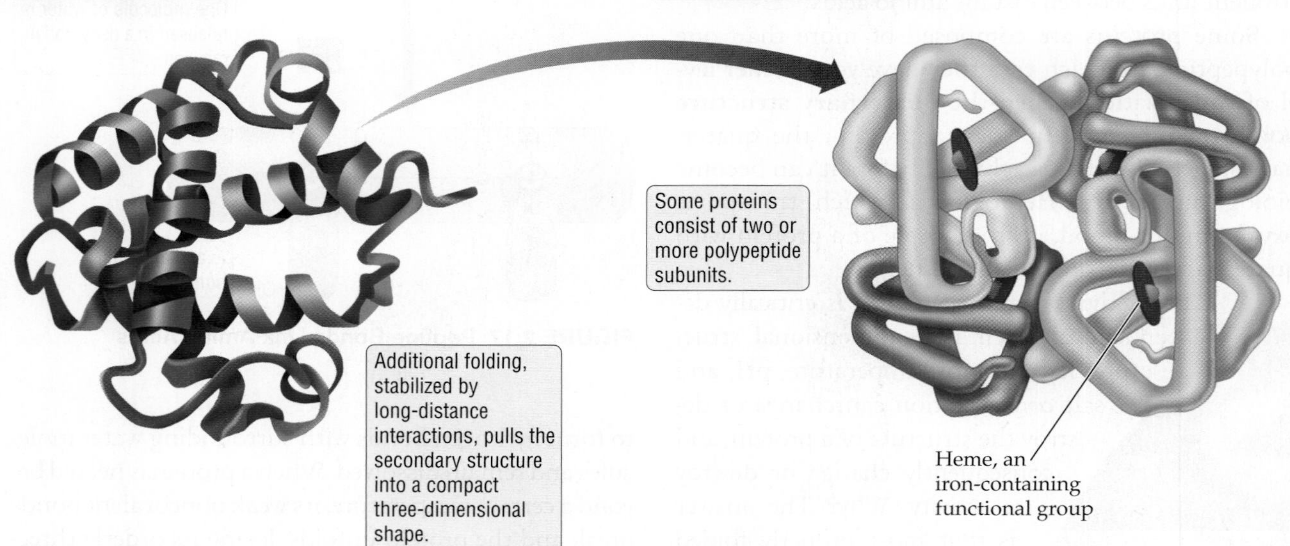

Some proteins consist of two or more polypeptide subunits.
Additional folding, stabilized by long-distance interactions, pulls the secondary structure into a compact three-dimensional shape.
Heme, an iron-containing functional group

FIGURE 2.18 The Four Levels of Protein Structure

(*a*) All proteins have a primary structure composed of a linear chain of amino acids known as a polypeptide. (*b*) Local coiling and folding of the polypeptide chain gives a protein its secondary structure. (*c*) Further folding, stabilized by long-distance interactions, gives a protein its tertiary structure. (*d*) Proteins composed of more than one polypeptide have a quaternary structure stabilized by bonds between the chains.

dissolved in the aqueous egg white. Albumin is denatured as the egg is cooked, causing the polypeptides to clump together into the white-colored solid mass in a fried egg. Most denatured proteins cannot regain their original three-dimensional structure, which is why you cannot "unscramble" an egg!

Concept Check

1. What monomers are the basic building blocks of all proteins? How many kinds are found in proteins?
2. What is protein denaturation?

2.9 Lipids

Lipids are hydrophobic biomolecules built from chains or rings of *hydrocarbon*, which, as you might guess, consists of carbon and hydrogen atoms. Fatty acids, glycerides, sterols, and waxes are examples of lipids; we will discuss these in greater detail shortly.

Most lipids are built from one or more **fatty acids**. A fatty acid has a long hydrocarbon chain that is strongly hydrophobic. At the other end of the hydrocarbon chain is a carboxyl group, a functional group that is polar and therefore hydrophilic. The hydrocarbon chains found in fatty acids contain many carbon atoms—16 to 22 in the lipids common in our foods. The carbon atoms in a fatty acid chain can be linked in different ways:

- In **saturated fatty acids** (**FIGURE 2.19*a***), all the carbon atoms are linked by single covalent bonds. Each carbon in the fatty acid chain is "saturated" because all of its available electrons are bonded to other atoms, which include at least two hydrogen atoms.

Concept Check Answers

1. Amino acids. All proteins are built from a pool of 20 different amino acids.
2. Denaturation is the destruction of the three-dimensional structure of a protein, resulting in loss of biological activity.

Fatty Acids

(*a*) A saturated fatty acid

It contains no double bonds in its fatty acid chain.

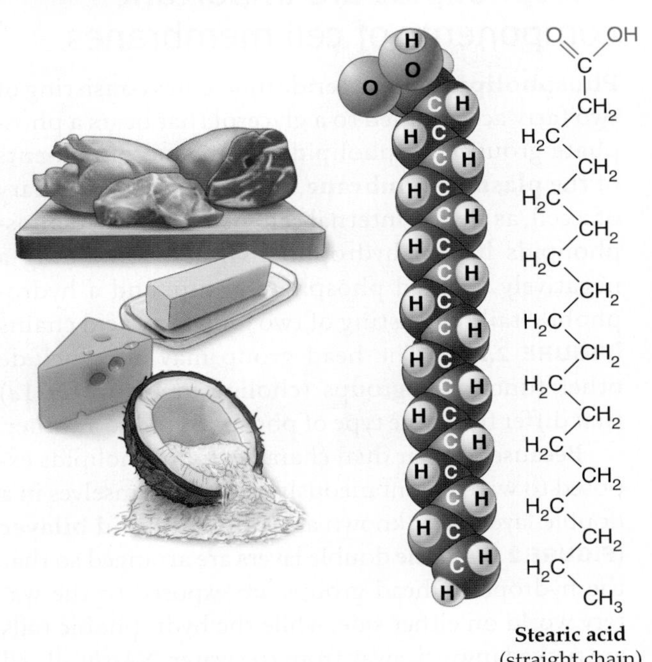

(*b*) An unsaturated fatty acid

It has one or more double bonds in its fatty acid chain.

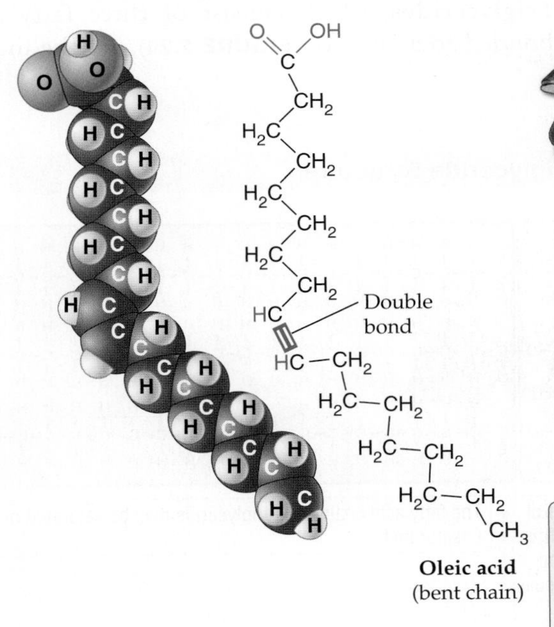

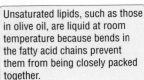

Unsaturated lipids, such as those in olive oil, are liquid at room temperature because bends in the fatty acid chains prevent them from being closely packed together.

FIGURE 2.19 Saturated and Unsaturated Are the Two Main Types of Fatty Acids
The space-filling models of (*a*) stearic acid and (*b*) oleic acid show that a saturated fatty acid is a straight molecule, whereas an unsaturated fatty acid molecule has a bend in it. Saturated fatty acids can pack tightly to form a solid at room temperature, but unsaturated fats cannot.

- In **unsaturated fatty acids** (**FIGURE 2.19*b***), one or more pairs of carbon atoms are linked by double covalent bonds. Each of these carbons is bonded to only one hydrogen atom, so the chain is not "saturated" with hydrogen.

The significance of the double bonds in unsaturated fatty acids goes beyond the difference in the number of hydrogen atoms. Saturated hydrocarbon chains tend to be straight, but the presence of double bonds in unsaturated chains can introduce kinks. The straight-chain fatty acids can pack together very tightly, forming solids or semisolids at room temperature. Unsaturated fatty acids with kinks cannot pack tightly, so these lipids tend to be liquid at room temperature.

Animals store surplus energy as triglycerides

Familiar foods such as butter and olive oil are actually complex mixtures of different types of lipids, with small amounts of other substances, such as milk protein in butter and vitamin E in most vegetable oils. The lipids in butter and olive oil include fatty acids of different types and a class of lipids called glycerides. A *glyceride* contains one to three fatty acids covalently bonded to a three-carbon molecule called glycerol.

Triglycerides, which consist of three fatty acids bonded to a glycerol (**FIGURE 2.20**), are the most common glyceride in our diet. Triglycerides built largely from saturated fatty acids tend to be solid at room temperature and are informally known as *fats*. Butter and lard are rich in triglycerides containing saturated fatty acids. In contrast, triglycerides rich in unsaturated fatty acids tend to be liquid at room temperature and are informally known as *oils*. Canola oil, olive oil, and flaxseed oil are examples, which is why all of these lipids are liquid at room temperature. Some tropical "oils," such as those derived from coconut and palm kernels, are solid at room temperature because they also contain saturated lipids—as much as in butter and lard, or even more.

A wide variety of organisms store surplus energy in the form of triglycerides, usually deposited in the cytoplasm of cells as lipid droplets. Lipids are efficient as storage reserves because they contain slightly more than twice the energy found in an equal weight of carbohydrate or protein, while occupying only one-sixth the volume. Carbohydrates and proteins take up more space inside a cell because they are hydrophilic; these macromolecules are extensively associated with water molecules, and all these extra molecules add bulk.

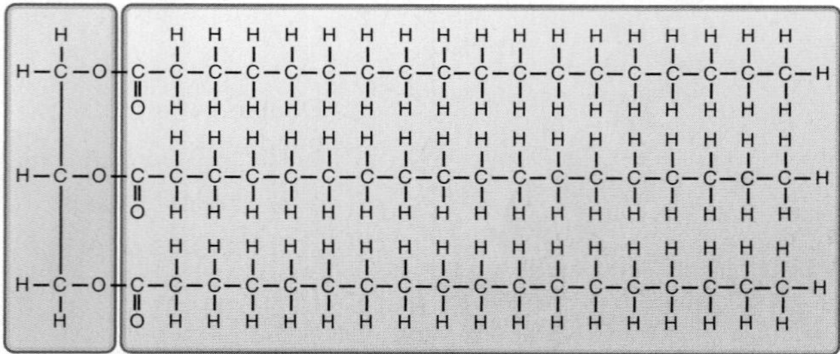
Triglyceride Structure
Glycerol is a three-carbon molecule.
The fatty acid chains in a triglyceride may be saturated or unsaturated.

FIGURE 2.20 Triglycerides Contain Three Fatty Acids Bound to a Glycerol

Glycerides consist of a three-carbon sugar alcohol called *glycerol* bound to one, two, or three fatty acids. Triglycerides have three fatty acids, one linked to each of the three carbons of glycerol. The triglyceride depicted here is glyceryl tristearate, the most common storage lipid in animal cells.

Phospholipids are important components of cell membranes

Phospholipids are glyceride molecules consisting of two fatty acids joined to a glycerol that bears a phosphate group. Phospholipids are major components of the **plasma membrane**, the outermost boundary of a cell, as well as internal cell membranes. All phospholipids have a hydrophilic "head" containing a negatively charged phosphate group, and a hydrophobic "tail" consisting of two long fatty acid chains (**FIGURE 2.21*a***). The head group may also include other functional groups (choline in Figure 2.21*a*) that differ from one type of phospholipid to another.

Because of their dual character, phospholipids exposed to water spontaneously arrange themselves in a double-layer sheet known as a **phospholipid bilayer** (**FIGURE 2.21*b***). The double layers are arranged so that the hydrophilic head groups are exposed to the watery world on either side, while the hydrophobic tails are tucked inward, away from the water. Nearly all cell membranes are organized as lipid bilayers. Cell membranes control the exchange of ions and molecules between the cells and their external environment, and also between various compartments within a cell (see Chapter 3).

BIOLOGYMATTERS

Dietary Lipids: The Good, the Bad, and the Truly Ugly

Our bodies can make nearly all the lipids we need from the organic molecules we consume as food. However, a moderate intake of lipids, especially certain types of lipids, is an important part of a healthy diet. Nutritionists recommend that we consume modest amounts of unsaturated fatty acids, such as those found in olive oil and canola oil.

One class of unsaturated fatty acids, known as *omega-3 fatty acids*, is known for its health benefits. Flaxseed and walnuts, certain algae, and cold-water fish are good sources of omega-3 fatty acids. Ample evidence shows that EPA (eicosapentaenoic acid) and DHA (docosahexaenoic acid), the main omega-3 fatty acids in fish oil, have anti-inflammatory effects in the human body and protect against heart disease (see Chapter 1).

How saturated lipids affect human health is a complex and contentious subject. Harmful effects are seen in laboratory rats fed a diet rich in saturated lipids; the effects include increased occurrence of heart disease and some types of cancer. Studies of human populations, however, present a confusing picture, in part because human subjects are difficult to study. Some populations, such as Amish farmers and South Pacific island communities, have low rates of heart disease despite a very high intake of saturated lipids. It is possible that these populations are protected from potential negative effects because their total calorie consumption is lower or because they exercise more.

A recent analysis, which pooled data from multiple studies, suggests that the *ratio* of saturated to unsaturated lipids in our diet has a significant influence on disease risk. According to this report, consuming large amounts of saturated lipids, and relatively low amounts of unsaturated lipids, is correlated with greater risk of heart disease. Despite the uncertainties, most nutrition experts say that no more than 7 percent of our total calories should come from saturated fat.

Better understood are the harmful effects of trans unsaturated lipids, known as *trans fats*. Trans fats contain unsaturated fatty acids whose hydrophobic tails are relatively straight compared to the bent shape of the more common (cis) type of unsaturated fatty acids. Because their straighter chains are readily compacted, trans fats are semisolid at room temperature. The overwhelming majority of trans fat in the American diet comes from partial *hydrogenation* of vegetable oils. The goal of this industrial process is to convert the unsaturated lipids in liquid vegetable oils to semisolid products such as margarine. As the oil is treated with hydrogen gas (H_2), some of the cis fatty acid molecules are turned into saturated fatty acids through complete hydrogenation (and therefore loss) of the carbon-carbon double bonds. In some of the cis fatty acid molecules, however, the double bonds remain but swivel in a way that straightens the fatty acid tail; these straight-chain unsaturated fatty acids are trans fats.

Trans fats have been popular in the processed-food industry because they are cheaper than the alternatives and less prone to becoming rancid. Foods prepared with trans fats last well on the shelf and do not need expensive refrigeration. Cis unsaturated fatty acids, in contrast, are very susceptible to attack by oxygen gas (O_2).

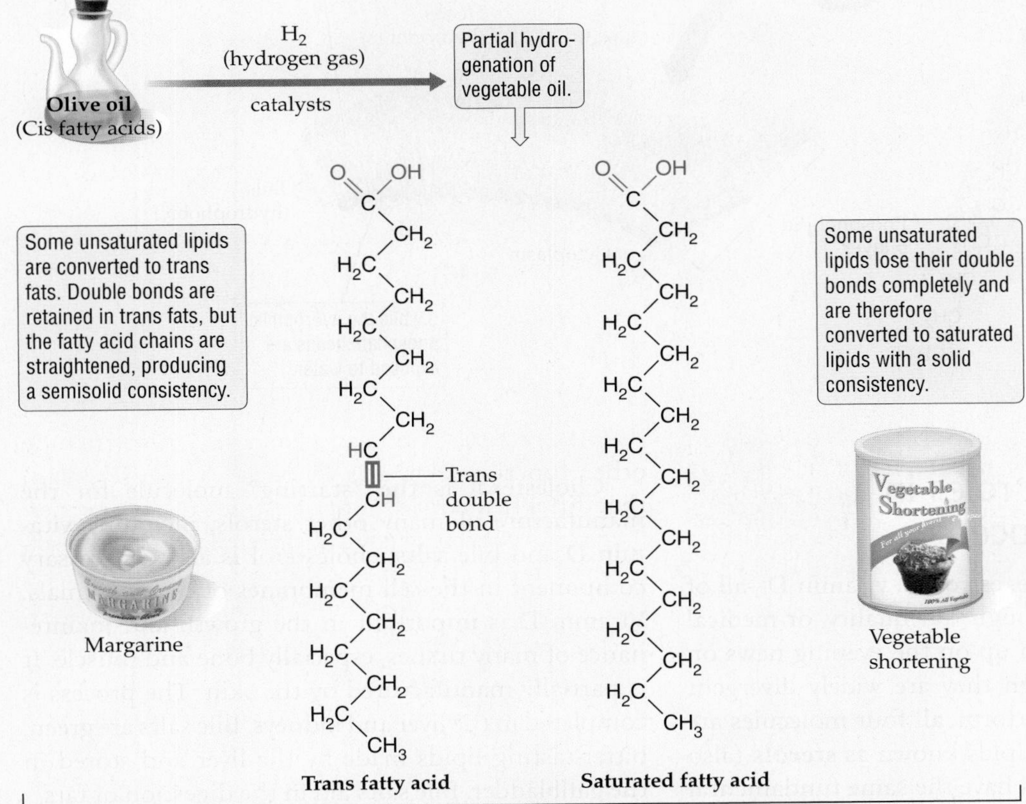

Olive oil
(Cis fatty acids)
H2
(hydrogen gas)
catalysts
Partial hydrogenation of vegetable oil.
Some unsaturated lipids are converted to trans fats. Double bonds are retained in trans fats, but the fatty acid chains are straightened, producing a semisolid consistency.
Some unsaturated lipids lose their double bonds completely and are therefore converted to saturated lipids with a solid consistency.
Trans double bond
Margarine
Vegetable Shortening
Vegetable shortening
Trans fatty acid
Saturated fatty acid
Straight-chain molecules

FIGURE 2.21 Membranes Contain Double Sheets of Phospholipids

(*a*) There are many types of phospholipids, differing in the chemistry of their head group and the length of their fatty acid tails. The one shown here is phosphatidylcholine, which has choline (blue) attached to the phosphate (red) and glycerol (green) in its head group.
(*b*) Phospholipids spontaneously orient themselves into a double-layer sheet in which their hydrophilic head groups are oriented toward the watery environments of the cell interior and exterior, while the hydrophobic fatty acid chains convene in the middle of the "sandwich."

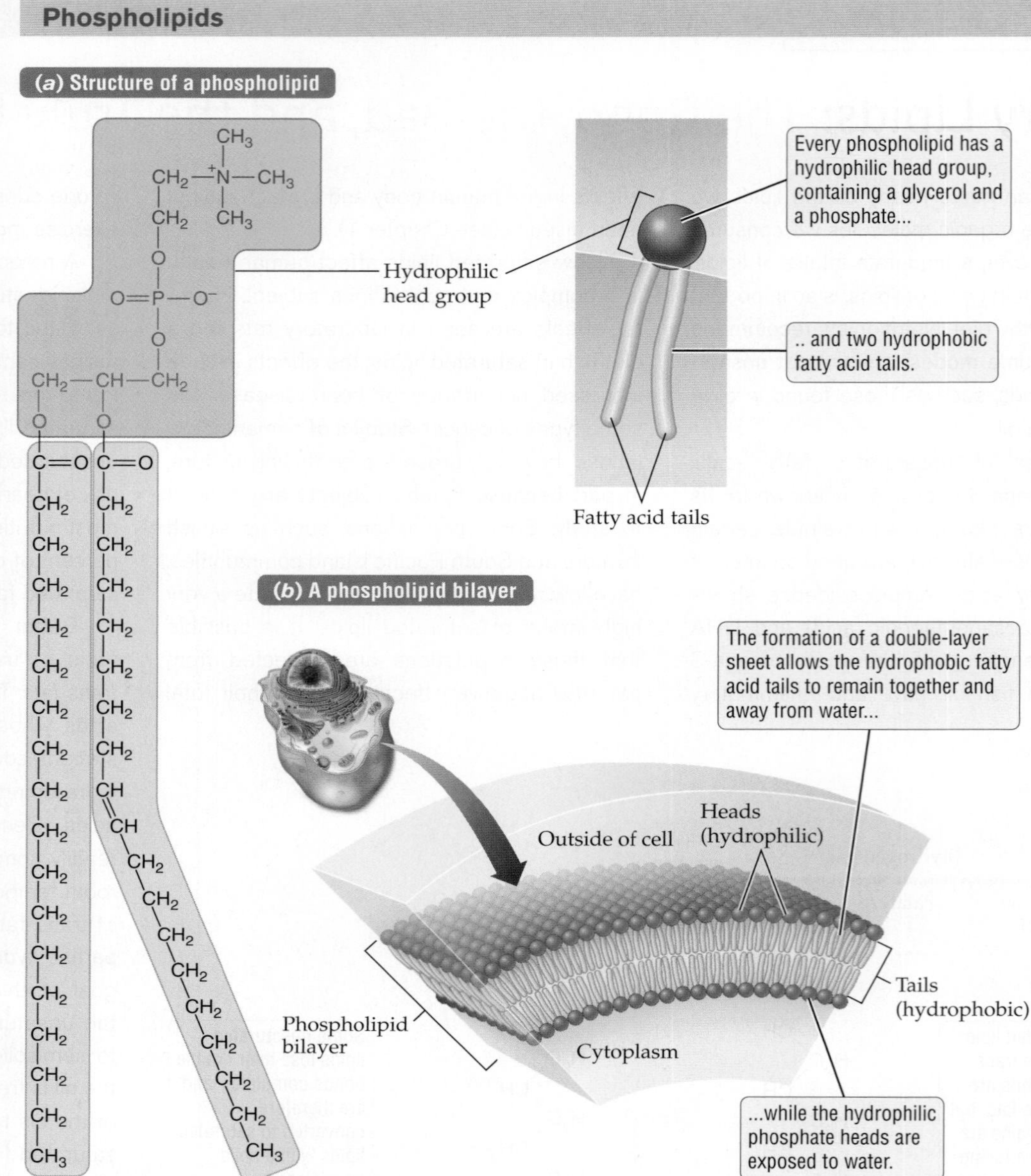

Phospholipids
(a) Structure of a phospholipid
Hydrophilic head group
Every phospholipid has a hydrophilic head group, containing a glycerol and a phosphate...
.. and two hydrophobic fatty acid tails.
Fatty acid tails
(b) A phospholipid bilayer
The formation of a double-layer sheet allows the hydrophobic fatty acid tails to remain together and away from water...
Outside of cell
Heads (hydrophilic)
Tails (hydrophobic)
Phospholipid bilayer
Cytoplasm
...while the hydrophilic phosphate heads are exposed to water.

Sterols play vital roles in a variety of life processes

Cholesterol, testosterone, estrogen, vitamin D—all of these are lipids with enough star quality, or medical notoriety, that they turn up on the evening news on a regular basis. Although they are widely divergent in the functions they perform, all four molecules are classified in a group of lipids known as **sterols** (also called *steroids*). All sterols have the same fundamental structure: four hydrocarbon rings fused to each other. They differ in the number, type, and position of functional groups, and in the carbon side chains linked to the four hydrocarbon rings (**FIGURE 2.22**).

Cholesterol is the "starting" molecule for the manufacture of many other sterols, including vitamin D and bile salts; cholesterol is also a necessary component in the cell membranes of many animals. Vitamin D is important in the growth and maintenance of many tissues, especially bone and muscle. It is partially manufactured by the skin. The process is completed in the liver and kidneys. Bile salts are green, bitter-tasting lipids made by the liver and stored in the gallbladder. Bile salts aid in the digestion of fats.

Cholesterol is also the "starting" molecule in the production of steroid *hormones*, including sex hormones such as estrogen and testosterone. **Hormones** are signaling molecules that are active in very small

Sterols

A B C D

Four fused hydrocarbon rings, basic structure of all sterols

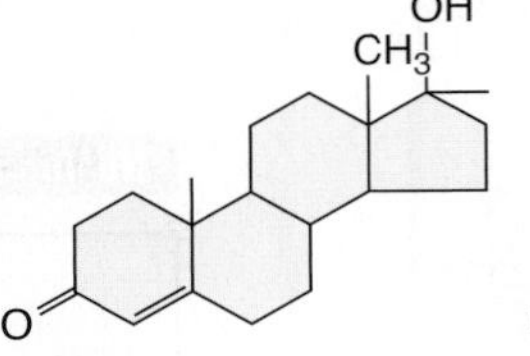

Testosterone

HO

Cholesterol

FIGURE 2.22 Sterols Are Lipids Built from Four Fused Hydrocarbon Rings
All sterols share the same basic four-ring structure but have different functional groups attached to these rings. Testosterone is a hormone that controls male sexual characteristics in many animals, including the male wood duck (photo). Cholesterol is an important constituent of the cell membranes of all birds and mammals, and of many other animals as well.

amounts and control a great variety of processes in plants and animals. The sex hormones, such as estrogen and testosterone, promote the development and maintenance of the reproductive system in animals. Testosterone, in its several natural forms and numerous synthetic forms, is an anabolic steroid (*anabolic*, "putting together"). Among its many effects is the promotion of muscle growth.

The use of anabolic steroids by competitive athletes is seen as unfair advantage, and the drugs are banned by all major sports organizations. The regular use of anabolic steroids is associated with significant health risks, including higher odds of heart attack, stroke, liver damage, and liver and kidney cancer.

Concept Check

1. What is the single biggest difference between carbohydrates and lipids?
2. How are saturated and unsaturated fatty acids different in terms of chemical structure?

2.10 Nucleotides and Nucleic Acids

Nucleotides are important monomers in all organisms because they are the building blocks of the hereditary material. A **nucleotide** is a small organic molecule with three chemical components: (1) a **nitrogenous base** (nitrogen-containing base) that is covalently bonded to (2) a five-carbon sugar, which in turn is covalently bonded to (3) a **phosphate group**, a functional group consisting of a phosphate atom and four oxygen atoms (**FIGURE 2.23**).

Five different nucleotides—each containing one of the nitrogenous bases adenine, cytosine, guanine, thymine, or uracil (see Figure 2.23)—serve as the components for a class of polymers called nucleic acids. **Nucleic acids** in living cells are of two kinds: deoxyribonucleic acid (**DNA**) and ribonucleic acid (**RNA**). DNA is distinguished from RNA both by the type of sugar in its nucleotides and by two of the nitrogenous bases that bond with that sugar. Ribose, the sugar in RNA, differs from deoxyribose, the sugar in DNA, in that it has one more oxygen atom (see Figure 2.23).

Of the five different kinds of nitrogenous bases, thymine is found only in DNA, and uracil is found only in RNA. The nucleotides in RNA and DNA are bonded through covalent linkages known as *phosphodiester bonds* between the sugar and phosphate groups of each successive nucleotide. The result is a chain of nucleotides, or *polynucleotide*. RNA consists of a single polynucleotide chain (or "strand"); DNA is composed of two polynucleotide chains (it is "double-stranded"), which are twisted in a spiral pattern to form the *DNA double helix*.

Nucleotides perform two essential functions in the cell: genetic information storage, and energy transfer. Every organism has nucleic acid "software" dictating how that organism will live, grow, reproduce, and respond to the external world around it. DNA can be copied in a way that preserves its sequence of nucleotides, and therefore the information coded in it.

Some types of nucleotides function as energy delivery molecules, or energy carriers. The most universal of these energy carriers is the nucleotide known as

Concept Check Answers

1. Lipids are nonpolar and therefore hydrophobic.
2. A saturated fatty acid contains only single bonds in its hydrocarbon tail; an unsaturated fatty acid contains one or more carbon-carbon double bonds.

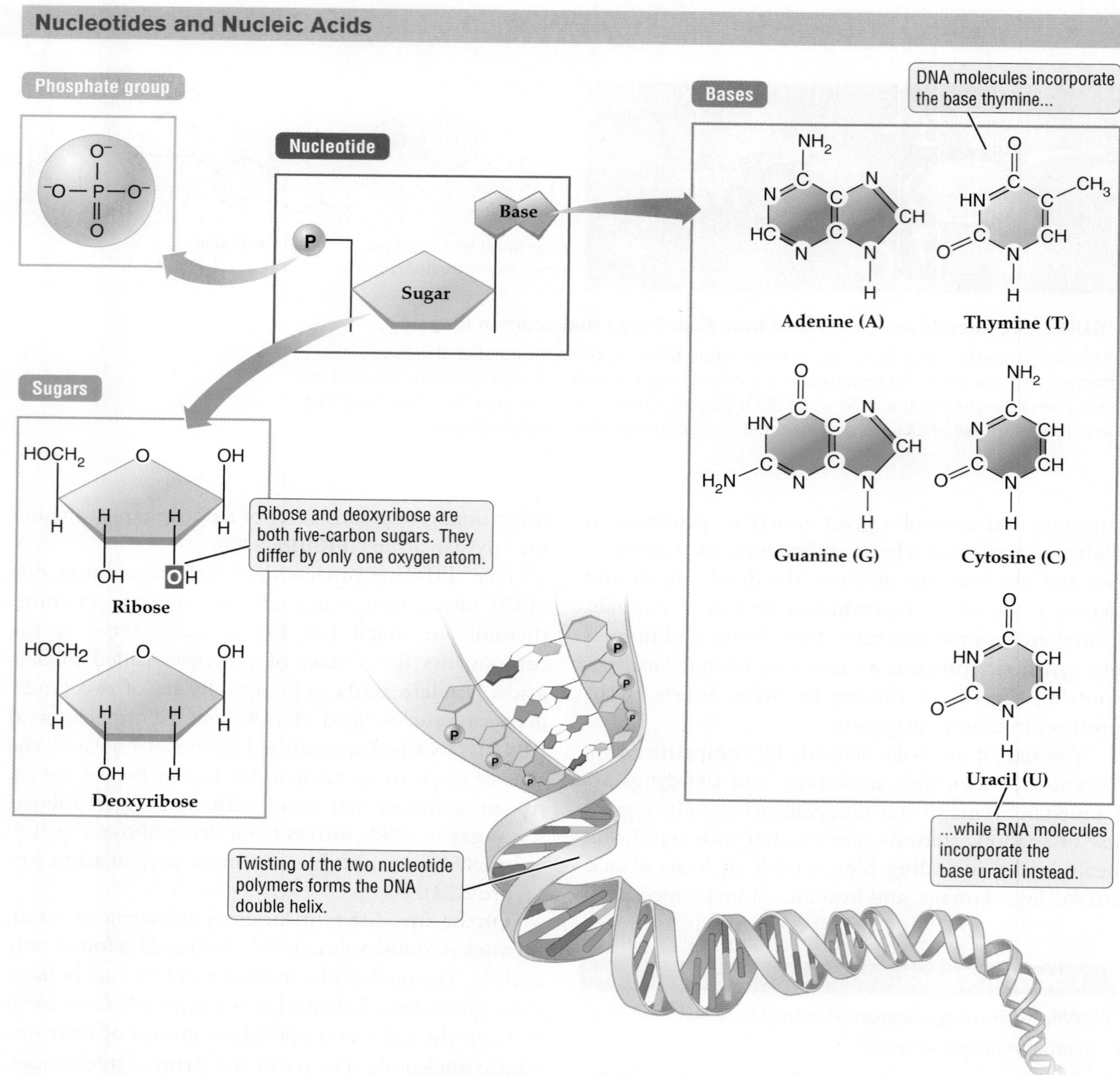

FIGURE 2.23 Nucleotides Are the Building Blocks of Nucleic Acids
Each nucleotide consists of a five-carbon sugar linked to a nitrogenous base and one or more phosphate groups. The bases adenine, guanine, cytosine, and thymine, when linked to the sugar deoxyribose, form the building blocks of DNA. The bases adenine, guanine, cytosine, and uracil, when linked to the sugar ribose, form the building blocks of RNA.

adenosine triphosphate, or **ATP** (**FIGURE 2.24**). The ATP molecule is made up of adenine (A) bonded to a ribose, plus three phosphate groups. ATP is the universal energy carrier for living organisms; many cellular chemical reactions depend on energy delivered by ATP. The energy of ATP is stored in the covalent bonds that link the three phosphate groups. The breaking of the bond between two phosphate groups releases energy that is used to power other chemical reactions.

Concept Check

1. Monomers of what type are the building blocks of all nucleic acids?
2. List two differences between DNA and RNA.

Concept Check Answers

1. Nucleotides.
2. DNA is a double-stranded molecule, whereas RNA is generally single-stranded. RNA has uracil instead of thymine.

Nucleotides as Energy Carriers: ATP

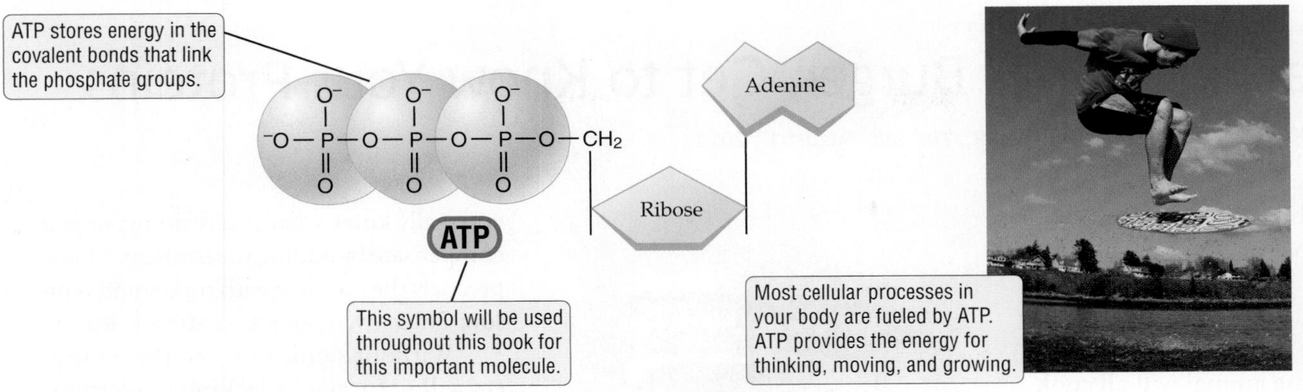

ATP stores energy in the covalent bonds that link the phosphate groups.
Adenine
Ribose
ATP
This symbol will be used throughout this book for this important molecule.
Most cellular processes in your body are fueled by ATP. ATP provides the energy for thinking, moving, and growing.

FIGURE 2.24 The Nucleotide ATP Serves as an Energy Carrier in Every Living Cell The phosphate groups in ATP are held together by energy-rich covalent bonds. Energy is released when these bonds are broken, and the released energy powers a great variety of biological processes, including the motion of the skimboarder pictured here.

APPLYING WHAT WE LEARNED

How Bad Are Trans Fats?

Back in 2003, attorney Stephen Joseph sued Kraft Foods, demanding that the food conglomerate stop selling Oreo cookies containing trans fats. At the time, most people had never heard of trans fats. But they'd heard of Oreo cookies, and Joseph's suit was an instant sensation. Joseph didn't care if everyone laughed at him. What mattered was that people wanted to know what trans fats were; and when they found out, they took a second, doubtful look at their Oreos.

Just days after filing the lawsuit, however, Joseph withdrew it. Media coverage of his lawsuit and of trans fats was so widespread, he said, that he could no longer argue in court that consumers didn't know the cookies contained dangerous trans fats. Just as important, within a day of the first major news stories, Kraft reversed itself and announced it would phase out the dangerous fats. Since the "Cookie Monster" lawyer had won his point through publicity alone, he had no reason to pursue an expensive lawsuit. In fact, by the end of 2005, just 31 months later, Kraft said it had voluntarily eliminated or reduced trans fats in Oreo cookies and all of its other products.

Because saturated animal fats were thought to contribute to cardiovascular disease, most people's grandparents and great grandparents were told that margarine and other artificial "spreads" were healthier for the heart than butter. But until recently, nearly all margarines were made from partially hydrogenated vegetable oils, which virtually always contain high levels of trans fats.

Trans fats change the ratio of two forms of cholesterol in our blood, increasing the risk of heart disease. As a result, trans fats are much worse for us than the saturated fats in butter. Experts estimate that trans fats contribute to 30,000–100,000 deaths from heart disease per year. In fact, a tiny increase in the consumption of industrial trans fats—2 percent, as measured in calories—results in a 20–30 percent increase in the risk of death from heart disease.

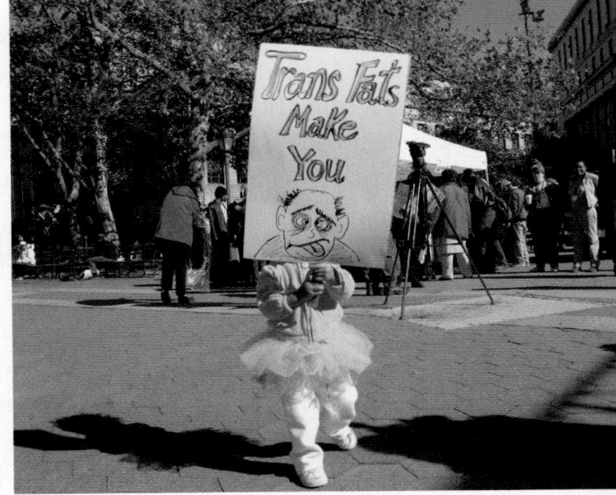

The U.S. Food and Drug Administration (FDA) recommends that Americans consume minimal amounts of trans fats—as little as possible. Animal fats—including those in butter, ice cream, milk, cheese, and yogurt—contain small amounts of naturally occurring trans fats, but researchers are not yet sure if these are as dangerous as the industrial trans fats in partially hydrogenated oils.

Today, nutrition labels in the United States must list how many grams of trans fats a "serving" contains. But don't assume that "zero" means zero. If a food contains less than 0.5 gram of trans fats per serving, the manufacturer is allowed to list that as "0 g"—even if the food contains up to 0.49 gram of trans fats per serving. If a "serving" is small enough, almost any food can be listed as having "0 grams of trans fats." As a result, it's easy to consume enough trans fats to affect your health without knowing it. A person who ate five servings of something that contained 0.49 gram of trans fat would exceed the 2-gram daily maximum recommended by the American Heart Association. In general, it's safe to assume that any food containing partially hydrogenated oils contains trans fats—even if the package says "Zero Trans Fats."

For a Better, Leaner Burger, Get to Know Your Proteins

BY AMY BLASZYK • National Public Radio (NPR), *The Salt*, August 17, 2012

We love our hamburgers ... [but] all that bliss doesn't come easy—all meat is not created equal ...

"Meat from different species, different ages in the same species, and different cuts from the same animal will all cook differently. This is mainly because of the specific proteins in the meat as well as the fat in and around the muscle group," [says food expert Michael Chu.]

According to Chu, there are a couple types of protein that matter in the kitchen: collagen, myosin and actin. Collagen is the fibrous—and most abundant—protein found in animals. It connects and supports body tissues, including tendons and ligaments. Myosin and actin work together to help cells move—contracting muscles ...

So what does this all mean when it comes to grilling up a tasty burger?

When you cook meat, the collagen begins to break down into gelatin. It breaks down more slowly at low cooking temperatures, which is why you roast more fibrous cuts from the chest (brisket) or leg (shank) longer to break down the meat properly. Unfortunately, before collagen breaks down, Chu says, it does the worst thing imaginable—it tightens up. That translates into tough meat.

Chu says with burgers, the meat—and thus the collagen—is so ground up, it's pretty much destroyed. That removes collagen from the equation, explaining why you can make a good burger out of a less tender cut, like shoulder (chuck). But now, in addition to the myosin and actin, we have something else to think about: fat. That white marbly stuff most of us spend our lives trying to lose is what makes meat in general and hamburgers in particular taste so darn good.

In cooking, fats lubricate. Using a ground meat that's 95 percent lean will produce a dry, crumbly burger—unless you really know what you're doing, or you compensate by adding moisture in. This is precisely the problem with turkey and venison: These two meats are naturally lean.

"Without significant fat, these burgers will naturally be lacking tenderness, flavor and succulence. Without the fat, the proteins will have very little to interfere with their natural tendencies while being heated—in particular, myosin, which, when denatured, will cross-link with each other to form a gel, like a meat glue. Too much myosin cross-linking will result in noticeable shrinkage and a rubbery texture."

If you're making beef burgers, Chu suggests sticking with meat that has a 20 to 25 percent fat content. With all burgers, particularly those made from leaner meats, binders like egg and breadcrumbs will do wonders.

Another suggestion from Chu if you are definitely into lean meats: browning, which binds up the proteins and will help even those leaner cuts of meat hold together. Sear your patties in a pan with a little oil before transferring them to the grill.

Cooking is a series of chemistry experiments. And creative cooking is all about clever chemistry. Take meat, poultry, and fish. Animal flesh is 70–87 percent water. The protein content of animal foods varies from 16 to 23 percent. The lipid content of muscle tissue is as low as 1 percent in lean fish to as much as 25 percent in well-marbled rib-eye steak.

The amount and distribution of proteins and lipids varies depending on the age of the animal and other physical factors, such as what it's eating and how much exercise it's getting. Wild-animal meat is relatively tough because game animals run around more than stockyard animals do. Lean cuts of meat tend to be tougher because they have proportionately more collagen. As noted in the news story, collagen shrinks when cooked. That's because high temperatures denature the protein and also drive out some of the water associated with this highly hydrophilic polymer.

Marinades with low pH—those containing lemon juice, vinegar, wine, or yogurt, for example—help tenderize the meat by breaking collagen into smaller polypeptides. Protein-degrading enzymes—such as papain from papaya, bromelain from pineapples, and actinidin from kiwifruit—are also useful in meat marinades. Pounding meat, or grinding it, also breaks up the collagen, so that's another way to tenderize it. The practice of brining—soaking in a saltwater bath for several hours—is useful for lean poultry.

Evaluating the News

1. On the basis of function, how would you classify collagen: Does it function in storage, structure, transport, or catalysis?

2. Why is pasteurized pineapple not effective in a meat marinade? (Pasteurization involves heating food to about 70°C [158°F] to kill microbes or slow their growth.) Some people are against the pasteurization of milk. Research the topic and decide whether you agree or disagree, stating your reasons.

CHAPTER REVIEW

Summary

2.1 Matter, Elements, and Atomic Structure

- The physical world is composed of matter, which is anything that has mass and occupies space. Matter consists of 92 different types of chemical elements, each with unique properties.
- An atom is the smallest unit of an element that has the chemical properties of that element. Atoms contain positively charged protons, uncharged neutrons, and negatively charged electrons. The number and arrangement of electrons in the atom of an element determines the chemical properties of that element.
- The atomic number of an element is the number of protons in its nucleus, and its atomic mass number is the sum of the number of protons and neutrons.
- Isotopes of an element have different numbers of neutrons but the same number of protons. Radioisotopes are isotopes that give off radiation.

2.2 The Bonds That Link Atoms

- The chemical interactions that cause atoms to associate with each other are known as chemical bonds.
- Covalent bonds are formed by the sharing of electrons between atoms. Atoms share electrons with other atoms to fill their outermost electron shells to capacity. The bonding properties of an atom are determined by the number of electrons in its outermost shell. A molecule contains at least two atoms that are held together by covalent bonds.
- When an atom loses or gains electrons, it becomes a positively or negatively charged ion, respectively. Ions of opposite charge are held together by ionic bonds, and atoms that are bound exclusively through such bonds are known as salts.
- Chemical compounds contain atoms from at least two different elements. All salts are compounds, but a molecule is a compound only if it contains atoms from at least two different elements.

2.3 The Special Properties of Water

- Hydrogen bonds are weak associations between two molecules such that a partially positive hydrogen atom within one molecule is attracted to a partially negative region of the other molecule. Partial electrical charges result from the unequal sharing of electrons between atoms, giving rise to polar molecules.
- Water is a polar molecule. Hydrogen bonding between water molecules accounts for the special properties of water, including its high heat capacity and high heat of vaporization, properties that enable water to moderate temperature swings.
- Water is a universal solvent for ions and polar molecules, which are hydrophilic and therefore readily dissolve in water. Nonpolar molecules cannot associate with water and are therefore hydrophobic. Nonpolar molecules are excluded by water, causing them to clump together.
- Water has the highest surface tension of any liquid except mercury. Surface tension is a force that tends to minimize the surface area of water at an air-water boundary, and it influences many biological phenomena.

2.4 Chemical Reactions

- In chemical reactions, bonds between atoms are formed or broken. Although the participants in a chemical reaction (reactants) are modified to give rise to new ions or molecules (products), atoms are neither created nor destroyed in the process.
- Some chemical reactions release energy; others cannot proceed without an input of energy.

2.5 The pH Scale

- The life-supporting chemical reactions of a cell are conducted in a watery medium. Acids donate hydrogen ions in a solution; bases accept hydrogen ions.
- The concentration of free hydrogen ions in water is expressed by the pH scale.
- Buffers help maintain a constant pH in an aqueous solution.

2.6 The Chemical Building Blocks of Life

- Carbon atoms can link with each other and with other atoms to generate a great diversity of compounds.
- The four main biomolecules are carbohydrates, proteins, lipids, and nucleic acids.

2.7 Carbohydrates

- Carbohydrates include simple sugars (monosaccharides), as well as disaccharides and more complex polymers (polysaccharides).
- Carbohydrates provide energy and physical support for living organisms.

2.8 Proteins

- Amino acids are the building blocks of proteins. A chain of amino acids linked together makes a polypeptide, which constitutes the primary structure of a protein.
- The three-dimensional shape of a protein is critical for its biological function.

2.9 Lipids

- Lipids are hydrophobic substances containing one or more rings or chains of hydrocarbons. Fatty acids, the building blocks of most lipids, are saturated or unsaturated, depending on the absence or presence, respectively, of double covalent bonds in their hydrocarbon chains.
- Triglycerides are important for energy storage.
- Phospholipids are the basic components of biological membranes.
- Sterols include cholesterol and sex hormones.

2.10 Nucleotides and Nucleic Acids

- Each nucleotide consists of a five-carbon sugar, one of five nitrogenous bases (adenine, cytosine, guanine, thymine, or uracil), and a phosphate group.
- Nucleotides are the building blocks of the nucleic acids DNA and RNA. DNA polymers, made up of four types of nucleotides, form the blueprint for life and govern the anatomical features and chemical reactions of a living organism.
- ATP is an energy-rich molecule that delivers energy for a great variety of cellular processes.

Key Terms

acid (p. 40)
amino acid (p. 46)
atom (p. 30)
atomic mass number (p. 31)
atomic number (p. 31)
ATP (adenosine triphosphate) (p. 54)
base (p. 40)
biomolecule (p. 30)
buffer (p. 41)
carbohydrate (p. 43)
cellulose (p. 44)
chemical bond (p. 33)
chemical compound (p. 33)
chemical formula (p. 33)
chemical reaction (p. 39)
cohesion (p. 39)
covalent bond (p. 33)
dehydration reaction (p. 44)
denaturation (p. 47)
disaccharide (p. 44)
DNA (p. 53)
electron (p. 30)
electron shell (p. 31)
element (p. 30)
enzyme (p. 46)
evaporation (p. 38)
fatty acid (p. 49)
functional group (p. 43)
glucose (p. 43)
glycogen (p. 44)
hormone (p. 52)
hydrogen bond (p. 36)
hydrolytic reaction (p. 44)
hydrophilic (p. 37)
hydrophobic (p. 37)
ion (p. 33)
ionic bond (p. 33)
isotope (p. 32)
lipid (p. 49)
macromolecule (p. 42)
matter (p. 30)
molecule (p. 33)
monomer (p. 42)
monosaccharide (p. 43)
neutron (p. 30)
nitrogenous base (p. 53)
nonpolar molecule (p. 37)
nucleic acid (p. 53)
nucleotide (p. 53)
nucleus (p. 30)
organic molecule (p. 42)
peptide bond (p. 46)
pH scale (p. 41)
phosphate group (p. 53)
phospholipid (p. 50)
phospholipid bilayer (p. 50)
plasma membrane (p. 50)
polar molecule (p. 36)
polymer (p. 42)
polypeptide (p. 46)
polysaccharide (p. 44)
primary structure (p. 46)
product (p. 39)
protein (p. 45)
proton (p. 30)
quaternary structure (p. 47)
radioisotope (p. 32)
reactant (p. 39)
RNA (p. 53)
salt (p. 33)
saturated fatty acid (p. 49)
secondary structure (p. 46)
solute (p. 37)
solution (p. 37)
solvent (p. 37)
starch (p. 44)
sterol (p. 52)
surface tension (p. 39)
tertiary structure (p. 47)
triglyceride (p. 50)
unsaturated fatty acid (p. 50)

Self-Quiz

1. The neutral atoms of a single element
   a. all have the same number of electrons.
   b. can form linkages only with other atoms of the same element.
   c. can have different numbers of electrons.
   d. can never be part of a chemical compound.
2. Two atoms can form a covalent bond by
   a. sharing protons.
   b. swapping nuclei.
   c. sharing electrons.
   d. sticking together because they have opposite electrical charges.
3. Which of the following statements about molecules is true?
   a. A single molecule cannot have atoms from two different elements.
   b. Atoms in a molecule are linked only via ionic bonds.
   c. Molecules are found only in living organisms.
   d. Molecules can contain as few as two atoms.
4. Which of the following statements about ionic bonds is *not* true?
   a. They cannot exist without water molecules.
   b. They are not the same as hydrogen bonds.
   c. They involve electrical attraction between atoms with opposite charge.
   d. They are known to exist in crystals of table salt, NaCl.
5. Hydrogen bonds are especially important for living organisms because
   a. they occur only inside of organisms.
   b. they are stronger than covalent bonds and maintain the physical stability of molecules.
   c. they enable polar molecules to dissolve in water, which is the universal medium for life processes.
   d. once formed, they never break.
6. Glucose is an important example of a
   a. protein.
   b. carbohydrate.
   c. lipid.
   d. nucleic acid.
7. Peptide bonds in proteins
   a. connect amino acids to sugar monomers.
   b. bind phosphate groups to adenine.
   c. connect amino acids together.
   d. connect nitrogenous bases to ribose monomers.
8. An alpha helix is an example of __________ protein structure.
   a. primary
   b. secondary
   c. tertiary
   d. quaternary
9. Sterols are classified as
   a. sugars.
   b. amino acids.
   c. nucleotides.
   d. lipids.

10. Unlike saturated fatty acids, unsaturated fatty acids
   a. are solid at room temperature.
   b. pack more tightly because they have straight chains.
   c. have one or more double bonds in their hydrocarbon chain.
   d. have the full complement (maximum number) of hydrogen atoms covalently bonded to each carbon atom in the hydrocarbon chain.

11. Which of the following statements about the nature of matter is *not* true?
   a. Anything that is matter must occupy space.
   b. Anything that is matter must have mass.
   c. About 92 different elements and 20 types of matter occur naturally in our universe.
   d. The smallest unit of matter that has all the properties of a particular element is the atom.

12. Lactase is a protein whose function is to break apart a milk sugar called lactose. RNA polymerase is a protein whose function is to join nucleotides together to create RNA molecules. The structure and function of these two proteins are different most likely because
   a. there are different amounts of monosaccharides in the backbone of the two proteins.
   b. the two proteins differ in their sensitivity to pH.
   c. one protein is more abundant in the cytoplasm than the other.
   d. the two proteins differ in their amino acid sequence.

13. Hydrogen cyanide (HCN), which has the structure H—C≡N,
   a. is an organic molecule.
   b. is a salt.
   c. contains a carbon atom that shares all the electrons in its outermost electron shell with one atom of nitrogen.
   d. contains a carbon atom that shares two pairs of electrons with a hydrogen atom.

14. The water molecule (H_2O)
   a. is nonpolar.
   b. can form a network around both negatively charged ions and positively charged ions.
   c. can form hydrogen bonds with hydrophobic as well as hydrophilic chemicals.
   d. has symmetrical sharing of electrons between the oxygen atom and each of the two hydrogen atoms.

Analysis and Application

1. What is a monomer, and what is its relationship to a polymer? Should lipids be regarded as polymers? Why or why not?
2. A sample of pure water contains no added acids or bases. Predict the pH of the water and explain your reasoning.
3. What are hydrogen bonds? Explain how the polarity of water molecules contributes to their tendency to form hydrogen bonds.
4. Describe the chemical properties of carbon atoms that make them especially suitable for forming so many different molecules of life.
5. Describe one function that is relevant to biological processes for each of the following compounds: carbohydrates, nucleic acids, proteins, lipids.
6. Take a look at the photo of the hamburger on page 56. Identify one macromolecule that is especially abundant in each of the main ingredients of the hamburger. Name the building blocks that each macromolecule is composed of, and name at least one important function it performs in the human body.
7. The photo below shows a lone boulder in Joshua Tree National Park. In what way is the chemistry of your body different from the chemistry of this rock? Name two elements that are abundant in your body but scarce in the rock. Name a small organic molecule that contains both of these elements. What type of polymer, if any, can be formed by this organic molecule?

8. Name the solvent, and some of the solutes, we might find in a cup of sweetened black coffee topped with whipped cream. Which of these substances are hydrophilic? Which are hydrophobic?

9. The photo shows an Inuit fisherman with Arctic char that he has pulled from his ice-fishing hole. Explain how fish can thrive under the ice covering this lake. Why has the lake not frozen solid from the bottom up? Compare the arrangement of water molecules in the lake ice and the liquid water below it.

3 Cell Structure and Internal Compartments

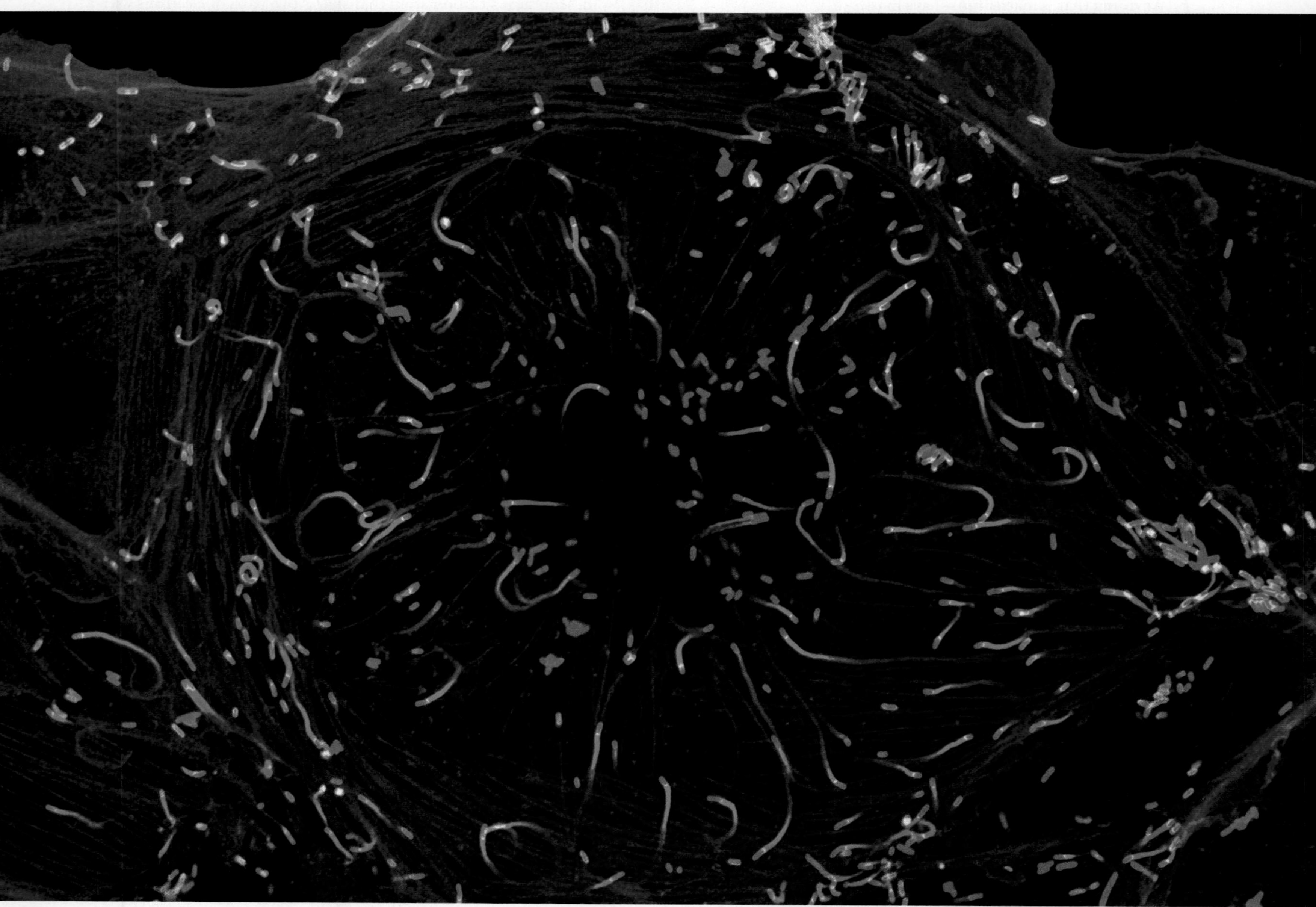

INVADERS IN INNER SPACE. Bacteria called *Listeria monocytogenes* (red) have infected this human cell. The bacteria are pushed along on cables of protein (blue and green strands) that are part of the internal architecture of eukaryotic cells. The bacteria move within the cell by hitching a ride on cables made up of a protein called actin.

Wanted: Long-Term Roommate; Must Help Keep House and Have Own DNA

Of the hundreds of trillions of cells in your body, only a fraction are actually you. The rest are mostly bacteria inhabiting different parts of your body. Although most of these fellow travelers are an important part of who we are, not all of them are benign or helpful. Parasites, which grow and multiply at our expense, include not only animals such as worms, but also disease-causing microbes. Our bodies have defensive immune cells that attack these invaders, but some parasitic microbes actually hide inside our cells and take over their internal machinery.

Consider *Listeria monocytogenes*, a bacterium that can lurk in raw milk, meat, and fish. In the United States, this bacterium causes nearly 2,000 cases of severe food poisoning each year, killing about 400 people. Inside the body, giant immune cells called macrophages attack and engulf *L. monocytogenes*. But the bacterium narrowly escapes being digested by the macrophage and then handily turns the tables on its attacker. While still in the macrophage, *L. monocytogenes* uses the hapless macrophage's own cellular skeleton to bury itself deep within the cell's cytoplasm, where it can continue to divide and increase its numbers.

Surprisingly, this business of one cell diving inside another is not only common but also ancient. Many kinds of bacteria, both nasty and nice, infect or inhabit the cells of animals, plants, insects, amoebas, and even other bacteria. Aphids—insects that commonly infest houseplants—often harbor a bacterium that's itself infected by another bacterium! And the malaria parasite *Plasmodium falciparum*, which is a eukaryote, has been filmed burglarizing human red blood cells.

The hijacking of one organism's cellular machinery and energy resources by another organism is a form of *parasitism*. And parasitism is just one form of *symbiosis*, a relationship in which organisms of different species live together in close association. The ability of cells to survive inside other cells is a clue to an amazing story of how eukaryotic cells came to be. At the end of this chapter we'll look more closely at these primal relationships. But first, let's examine the general structures of prokaryotic and eukaryotic cells.

What is the evolutionary link between prokaryotes and eukaryotes? Do prokaryotes still lurk inside our cells, a few billion years after the original home invasion?

MAIN MESSAGE Prokaryotic and eukaryotic cells are both bounded by a plasma membrane that encloses an aqueous cytoplasm, but eukaryotic cells have a greater variety of internal membrane-enclosed compartments.

KEY CONCEPTS

- All living organisms are made up of one or more basic units called cells.
- Most cells are small. Small size optimizes surface area relative to volume.
- Multicellularity makes possible a larger body size and division of labor among the different types of cells.
- The plasma membrane forms the boundary of a cell, controls the movement of materials in and out of the cell, and determines how the cell communicates.
- Prokaryotes are single-celled organisms that lack a nucleus. Eukaryotes are single-celled or multicellular organisms whose cells have a nucleus and several other internal compartments.
- The membrane-enclosed compartments of the eukaryotic cell have diverse and specialized functions.
- The cytoskeleton is a network of protein cables and cylinders that facilitates internal transport, gives shape and mechanical strength to a cell, and enables it to move.

WANTED: LONG-TERM ROOMMATE; MUST HELP KEEP HOUSE AND HAVE OWN DNA 61

3.1 Cells: The Smallest Units of Life 63
3.2 The Plasma Membrane 66
3.3 Prokaryotic and Eukaryotic Cells 68
3.4 Internal Compartments of Eukaryotic Cells 70
3.5 The Cytoskeleton 76

APPLYING WHAT WE LEARNED 80
The Evolution of Eukaryotes

CERTAIN LARGE BIOMOLECULES are common to all life-forms. By themselves, these building blocks of life, which were the focus of Chapter 2, are inanimate. Only when working together in a cell do these molecules become part of a living entity, a *cell*. A **cell** is a self-contained structure that can replicate itself and use energy to maintain its complex organization. From bacteria to blue whales, the cell is the smallest unit of life (**FIGURE 3.1**).

This chapter explores life at the level of the cell. After a broad overview of the unity and diversity of living cells, followed by a comparison of cellular organization in prokaryotes and eukaryotes, we examine the internal structures and compartments that enable a cell to function as an efficient and well-coordinated whole. We begin the tour at the outer boundary of the cell and then work our way inward.

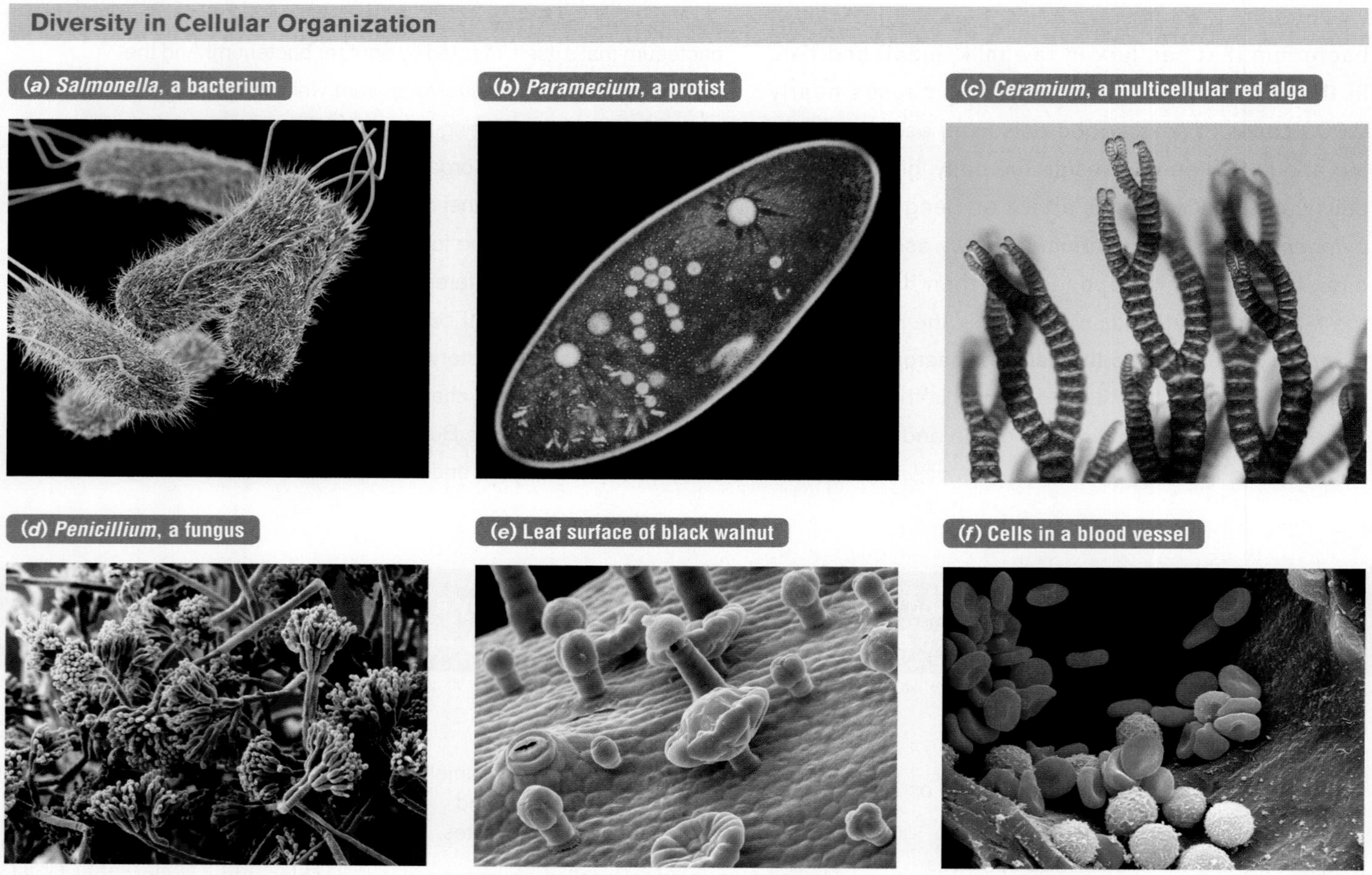

Diversity in Cellular Organization

(*a*) *Salmonella*, a bacterium

(*b*) *Paramecium*, a protist

(*c*) *Ceramium*, a multicellular red alga

(*d*) *Penicillium*, a fungus

(*e*) Leaf surface of black walnut

(*f*) Cells in a blood vessel

FIGURE 3.1 The Cell Is the Basic Unit of Life

(*a*) *Salmonella typhimurium*, a prokaryote that causes food poisoning, is single-celled. (*b*) *Paramecium caudatum* is a single-celled eukaryote that lives in freshwater. (*c*) *Ceramium pacificum*, a multicellular red alga, is a marine protist. (*d*) *Penicillium camembertii*, a multicellular fungus, produces spores (green) for asexual reproduction. (*e*) Air pores and protective hairs are visible on the surface of the black walnut (*Juglans nigra*) leaf pictured here. (*f*) Red blood cells and white blood cells are shown here inside a blood vessel. What characteristics are shared by all of these cells? At a basic level, how are your white blood cells different from the bacterial cell?

3.1 Cells: The Smallest Units of Life

Every living thing is composed of one or more cells, and all cells living today came from a preexisting cell. These two concepts are the main tenets of the **cell theory**, proposed in the middle of the nineteenth century. Today the cell theory is one of the unifying principles of biology, and we regard the cell as the smallest unit of life on Earth. If we disassemble a cell, we find that the constituent parts do not retain the distinctive characteristics of life, such as the capacity to reproduce.

Cells contain many different types of organelles

Every cell has an aqueous interior that contains many different types of biomolecules, and it is surrounded by a lipid-based boundary known as the **plasma membrane**. The contents of a cell internal to the plasma membrane are collectively referred to as the **cytoplasm**. The cytoplasm contains a thick fluid called the **cytosol** that is composed of a multitude of ions and biomolecules mixed in water.

Embedded in the cytosol, or adrift in it, are vital structures called **organelles**, which perform unique functions in the cell. **Ribosomes**, which are key components of the cell's protein-making machinery, are minute organelles that lack membranes. Many thousands of ribosomes are found in the cytosol of both prokaryotic and eukaryotic cells.

Some organelles are wrapped in one or more lipid-based membranes. The **nucleus** (plural "nuclei"), found only in eukaryotic cells, contains DNA enveloped in double membranes. The **mitochondrion** (*MYE*-**tuh**-*KAHN*-**dree-un**; plural "mitochondria"), often dubbed the powerhouse of the cell because it supplies energy, is another example of an organelle bounded by two membranes.

The cell or cells that make up an individual organism come in a wonderful diversity of shapes, sizes, life strategies, and behaviors. Prokaryotes—bacteria and archaeans—are generally regarded as single-celled organisms (**FIGURE 3.1*a***). Protists, a grab bag category of eukaryotes, may be single-celled like *Paramecium* (**FIGURE 3.1*b***) or multicellular like red seaweed (**FIGURE 3.1*c***). Fungi include single-celled species, like baker's yeast (*Saccharomyces cerevisiae*), and many multicellular forms, such as *Penicillium* (**FIGURE 3.1*d***) and the familiar mushrooms. All members of the plant and animal kingdoms are multicellular (**FIGURE 3.1e and *f***).

The microscope is a window into the life of a cell

Our awareness of cells as the basic units of life is based largely on our ability to see them. The instrument that opened the eyes of the scientific world to the existence of cells—the *light microscope*—was invented in the last quarter of the sixteenth century. The key components of early light microscopes were ground-glass lenses that bent incoming rays of light to produce magnified images of tiny specimens (**FIGURE 3.2**).

The study of cells began in the seventeenth century when Robert Hooke examined a piece of cork under a microscope and noticed that it was made up of little compartments. Hooke called the compartments "cells," after the Latin word for a monk's tiny room.

Although the light microscope has a place in the early history of biology, similar instruments are just as important in ongoing research today. The quality of lenses, however, has improved significantly since their first use: today's standard light microscopes can achieve well over a 1,000-fold magnification. This degree of magnification enables us to distinguish structures as small as 1/2,000,000 of a meter, or 0.5 micrometer (μm). Modern light microscopes reveal not just animal and plant cells (5–100 μm), but also organelles such as mitochondria, which are about 1 μm wide.

Cell biologists can also mark, or tag, membranes, organelles, or even an individual biomolecule such as a

Helpful to Know

Organelle is Latin for "little organ," as in the tiny organs of a cell. Some biologists reserve the term for membrane-enclosed compartments alone. In recent years, most cell biologists have moved toward the usage you see in this textbook: an organelle is any structure in the cytoplasm that carries out a unique function. *Structure* here means a discrete functional unit composed of several to many macromolecules.

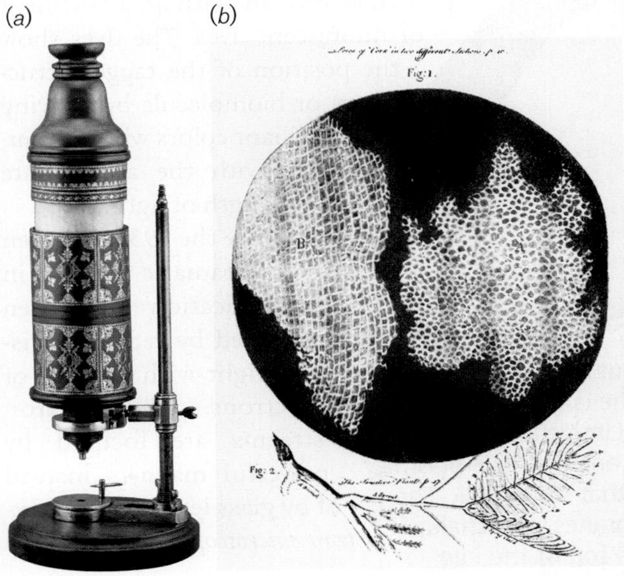

(*a*) (*b*)

FIGURE 3.2 Light Microscope Used by Robert Hooke (1635–1703)
(*a*) Hooke's microscope.
(*b*) A piece of cork examined under Hooke's microscope.

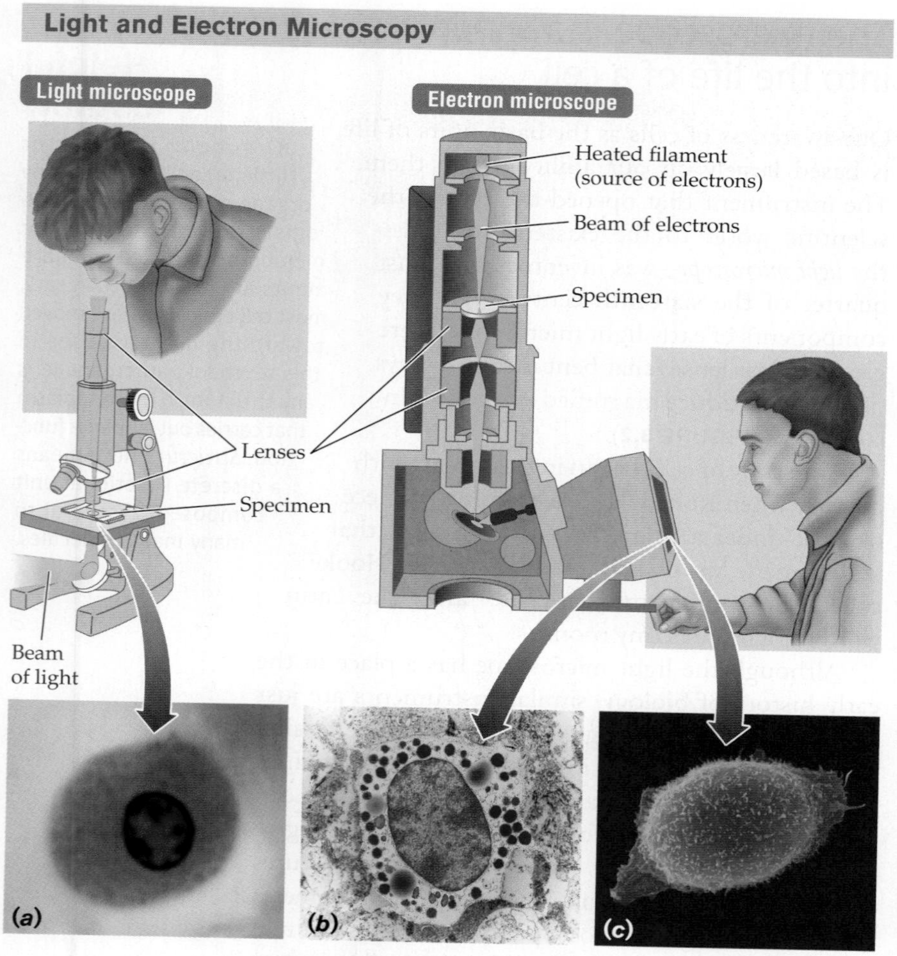

Light and Electron Microscopy

FIGURE 3.3 Microscopy Enables Us to Visualize Cells and Cell Structures

The photos show human mast cells, part of the immune system, imaged through (*a*) light microscopy, (*b*) transmission electron microscopy (TEM), and (*c*) scanning electron microscopy (SEM).

particular protein, with an assortment of fluorescent dyes. The dyes show the position of the tagged structure or biomolecule by glowing in brilliant colors when illuminated with the appropriate wavelength of light.

Since the 1930s, an even more dramatic increase in magnification has been achieved by replacing visible light with streams of electrons. The electron streams are focused by powerful magnets instead of by glass lenses. These *electron microscopes* can magnify a specimen more than 100,000 times (**FIGURE 3.3**). One version of this instrument—the *transmission electron microscope* (*TEM*)—can produce detailed views of the smallest organelles, such as an individual ribosome. Biological material must be sliced into extremely thin sections to prepare it for TEM imaging. Electron beams pass through the thin section to form a detailed image of even the tiniest structures in the specimen.

Another type of electron microscope—the *scanning electron microscope* (*SEM*)—moves an electron beam back and forth across the specimen to generate a three-dimensional view of its surface. Because SEM images have great depth of field, all visible parts of the specimen are in sharp focus. A specimen does not have to be sectioned to visualize its surfaces with an SEM.

EXTREME CELL SIZE

The ostrich egg—about 16 inches and up to 5 pounds—contains the largest single cell on the planet. The egg cell inside a bird's egg has a nucleus, cytoplasm, and a saclike plasma membrane extension that stores yolk. The egg white, shell membranes, and chalky shell are deposited on top of the egg cell after it has been fertilized.

The ratio of surface area to volume limits cell size

Some cells are giants. Your body contains some nerve cells that extend all the way from the base of the spine to your big toe, a distance of about 1 meter (3 feet) in an average adult. These exceptionally long cells are only about 10 μm at the widest point, so you cannot actually see them without a microscope. Egg cells, engorged with food for the offspring that might develop from them, are often quite large. Frog eggs, about 1 millimeter in diameter, are easily seen with the unaided eye.

Big cells are not the norm, however. The great majority of cells are microscopic—too small to be seen with the naked eye (**FIGURE 3.4**). On average, prokaryotic cells are smaller than eukaryotic cells. Most bacteria average about 1 μm in width, and the smallest of them—known as mycoplasmas—are only 0.1 μm across. Most animal cells are about 10 μm in diameter. Plant cells tend to be larger, most of them falling in the range of 20–100 μm.

Why are most cells small? The reason is that every cell must exchange materials with its environment, and that exchange becomes more of a challenge as cell size increases. The exchange of nutrients and waste products takes place across the plasma membrane. A big cell has more cytoplasm and more metabolic activity, and therefore a greater need for nutrients and a larger waste output, than a smaller cell.

As the width of a typical cell increases, the volume increases by a greater proportion than the surface area does, as illustrated in **FIGURE 3.5**. Notice that when the

Size Range of Biological Structures

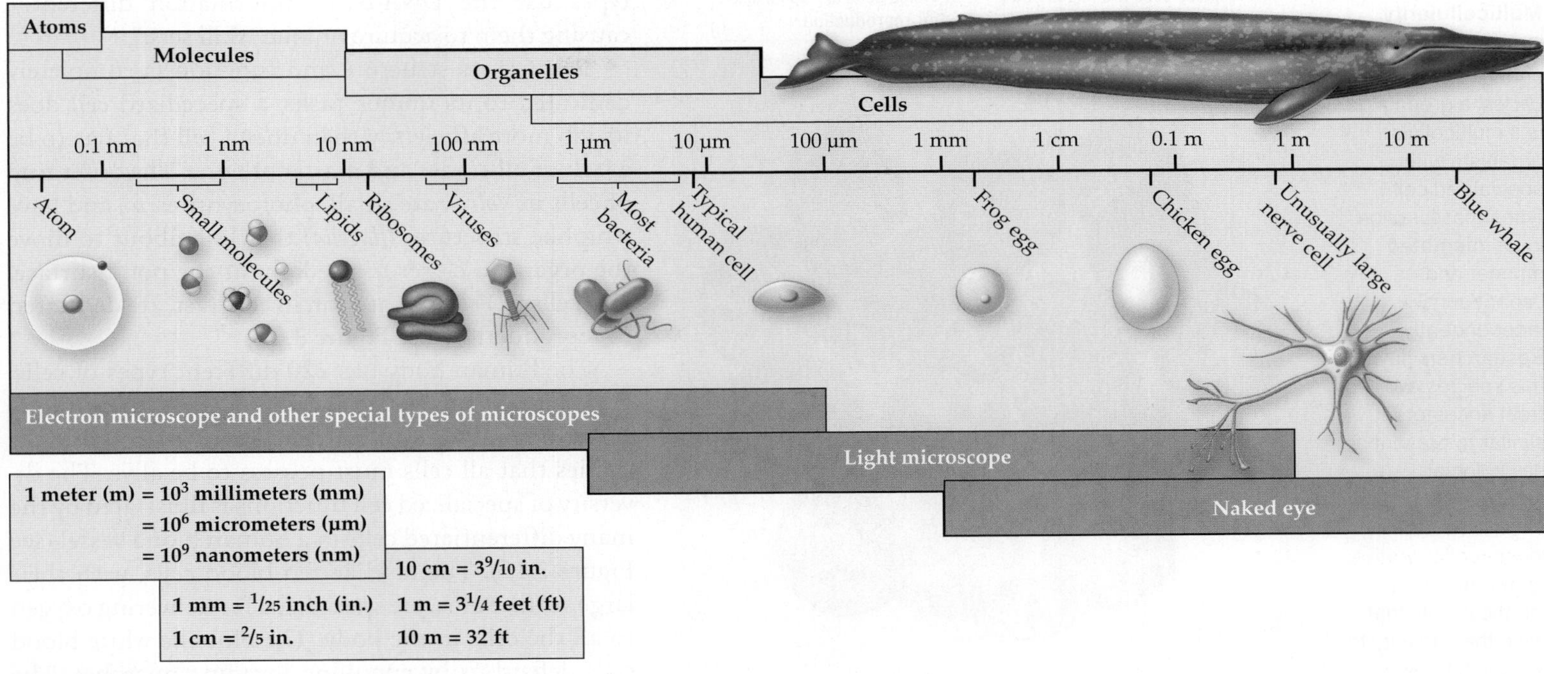

FIGURE 3.4 Most Cells Are Microscopic

length of the side of a cube-shaped cell is doubled, its volume increases eightfold but its surface area increases only fourfold. This relationship is a basic law of geometry, and it applies to spherical cells as well: as the diameter increases, the *ratio* between the surface area and volume decreases. The surface area–to-volume ratio puts an upper limit of about 100 μm on the width of most cells. Any larger, and the cell could starve for nutrients that cannot get in, and succumb to the buildup of waste products that cannot get out.

Some cells get around this size limit with unusual shapes. Because they are long and narrow, the nerve cells that run from spine to big toe have about the same cytoplasmic volume as the average animal cell, but a much larger surface area. Giant eggs, too, are not constrained by this limit; a large egg is crammed with metabolically inactive food stores, with the cytoplasm confined to a narrow band near the cell's plasma membrane, where exchange is easiest.

Multicellularity enables larger body size and efficiency through division of labor

Bacteria and other microscopic organisms demonstrate beyond a shadow of a doubt that being tiny can be an extremely successful life strategy. However, being bigger than the competition can deliver benefits, depending on how and where an organism lives. A large individual

How Surface Area Changes Relative to Volume

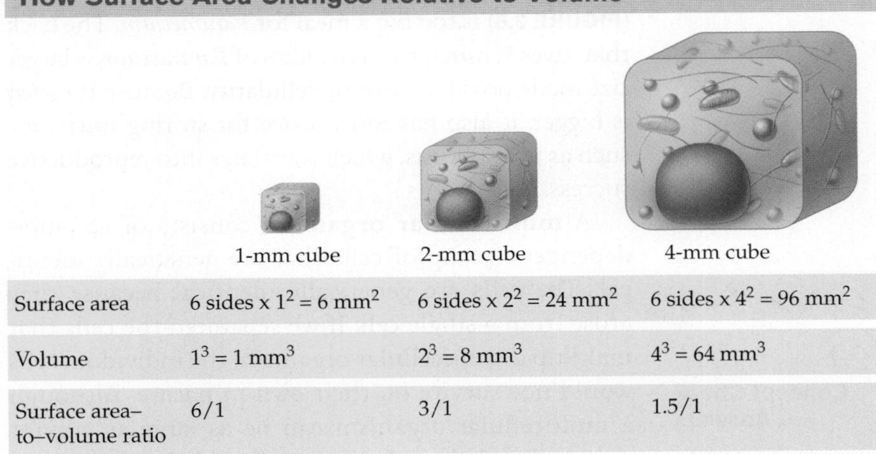

	1-mm cube	2-mm cube	4-mm cube
Surface area	6 sides x 1^2 = 6 mm^2	6 sides x 2^2 = 24 mm^2	6 sides x 4^2 = 96 mm^2
Volume	1^3 = 1 mm^3	2^3 = 8 mm^3	4^3 = 64 mm^3
Surface area–to–volume ratio	6/1	3/1	1.5/1

FIGURE 3.5 The Ratio of Surface Area to Volume Limits Cell Size

As the width of a cell increases, the volume increases more steeply than the surface area. Larger cells have higher metabolic capacity because they have a larger volume of cytoplasm, but they have less surface area in proportion to that volume. If a cell gets too large, it will not have the surface area it needs to exchange materials with its environment.

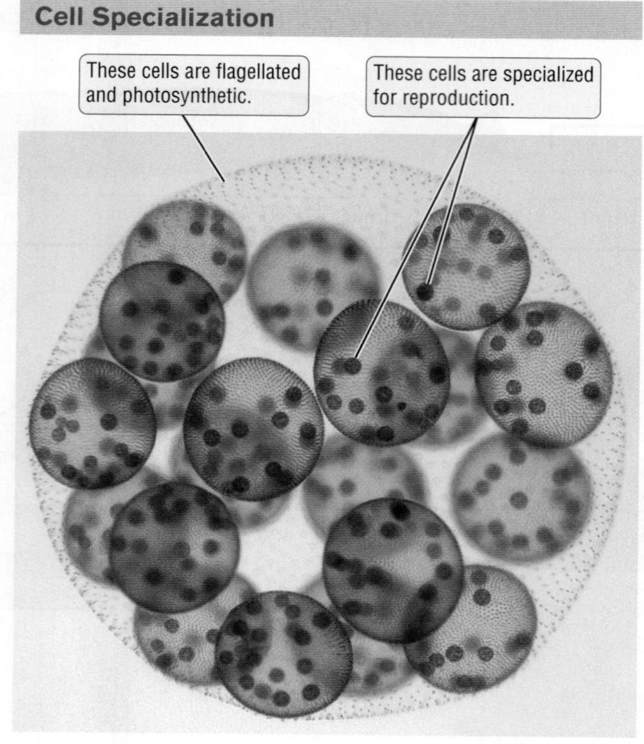

Cell Specialization
These cells are flagellated and photosynthetic.
These cells are specialized for reproduction.

FIGURE 3.6 Cell Specialization Is One Benefit of Multicellularity Recent studies show that *Volvox carteri*, a green alga, is a multicellular organism with specialized cell types that function in an integrated manner and cannot survive independently. Researchers say this species evolved from ancestors similar to present-day colonial relatives. Colonial organisms are looser collaboratives of genetically identical cells that have the capacity to live on their own.

may be better positioned to recruit resources from its environment than a smaller individual. For example, a predatory single-celled organism such as a *Paramecium* (see Figure 3.1*b*) can ingest several thousand bacteria in a day. But with a diameter of 2 millimeters, *Volvox carteri* (**FIGURE 3.6**) is too big a meal for *Paramecium*. The trick that saves *V. carteri* from the likes of *Paramecium* is larger size made possible by multicellularity. Because *V. carteri* is bigger, it also has more space for storing nutrients, such as phosphorus, which translates into reproductive success.

A **multicellular organism** consists of an interdependent group of cells that are genetically identical. The cells are genetically identical because they arose from a single cell. If we separated the cells that make up a multicellular organism, the individual cells would not survive on their own in nature. Although a multicellular organism can be as large as a giant redwood or a blue whale, the cells in it are themselves quite small, so they do not exceed the limits imposed by the ratio of surface area to volume.

The adaptive benefits of multicellularity go beyond getting food or avoiding becoming food; multicellularity makes cell specialization possible. Cell specialization enables division of labor among the cells. Multicellular organisms have different cell types, each specialized for particular jobs. All of the cells have the same DNA, but during development, different cell types use the DNA-based information differently, causing them to acquire unique "skill sets."

Because its structure and function is completely dedicated to its unique tasks, a specialized cell does its job more efficiently than does a cell that has to be a jack-of-all-trades and master of none. The outer ring of cells in *Volvox* are small, photosynthesize, and have whiplike structures (*flagella*) that lash about to move the organism closer to the light at the pond surface. The cells in *Volvox*'s interior, in contrast, are dedicated to reproduction (see Figure 3.6).

The human body has 220 different types of cells. Each *differentiated* (specialized) cell performs unique functions, in addition to conducting some basic activities that all cells must possess to be alive. The diversity of specialized cell functions is illustrated by the many differentiated cells in a human blood vessel (see Figure 3.1*f*). The disclike red blood cells, with their large surface area, are specialized for delivering oxygen to all the cells in the body. The bloblike white blood cells defend us by engulfing invading microbes. The blood vessel walls contain an even greater diversity of cell types, including endothelial cells that line the blood vessel and muscle cells that contract or relax to reduce or increase blood flow.

Concept Check

1. Why are most cells small?
2. Which has more surface area collectively: a pound of little potatoes or a pound of big potatoes?

Concept Check Answers

1. A larger cell has less surface area relative to its volume and can't exchange materials with its environment as efficiently as a smaller cell.
2. Altogether, little potatoes have more surface area than the same weight of big potatoes.

3.2 The Plasma Membrane

A key characteristic of every cell is the existence of a plasma membrane separating that cell from its surrounding environment. Most of the chemical reactions required for sustaining life take place within the cytoplasm, the main compartment formed by this boundary. The lipid boundary created by the plasma membrane has the effect of enclosing and concentrating necessary raw materials in a limited space, thereby facilitating chemical processes.

We noted in Chapter 2 that biological membranes, such as the plasma membrane, consist of a bilayer of phospholipids. In this bilayer, all the hydrophilic head groups are exposed to the aqueous environments either outside the cell or toward the cytoplasm, while the hydrophobic tails of the phospholipids congregate in the interior of the membrane (see Figure 2.21).

If the plasma membrane had no function other than to mark the boundary of the cell and to hold in its contents, a simple phospholipid bilayer would suffice. However, the plasma membrane has additional, critical functions:

- It enables the cell to capture essential molecules while shutting out unwanted ones.
- It releases waste products but prevents needed molecules from leaving the cell.
- It interacts with the outside world by receiving and sending signals as necessary.
- In cells that are held firmly in place, it provides an anchoring function.

The diverse functions of the plasma membrane are made possible chiefly by the many different types of *membrane proteins* associated with, or embedded in, the phospholipid bilayer (**FIGURE 3.7**).

The plasma membrane must be permeable to many substances, but selective about which ones. This *selective permeability* comes from the different types of proteins embedded in the phospholipid bilayer: transport proteins, receptor proteins, and adhesion proteins.

Transport proteins are membrane-spanning proteins whose function is to assist the import or export of substances. Some transport proteins form tunnels that allow the passage of selected ions and molecules. The activity of membrane transport proteins is usually tightly regulated to match the needs of the cell.

Receptor proteins act as sites for signal perception, and as such they are key components of a cell's communication system. Some receptor proteins are found in all or nearly all cell types in the human body; others are unique to a particular cell type. Each receptor protein generally binds a specific type of signaling molecule. When a signaling molecule docks on its target receptor, it triggers a change in cellular activity. One example is the insulin receptor; when insulin binds to this receptor, the cell steps up its import of glucose.

Most cells in the animal body are attached to other cells, or to a dense mat of biomolecules, called the **extracellular matrix (ECM)**, that is deposited on the outside surface of cells. Cell attachment is highly specific, with a given cell adhering only to particular cell types or to an ECM with a particular chemical composition. In many cases, *adhesion proteins* embedded in the plasma membrane are the main link between a cell and its extracellular neighborhood (see Figure 3.7).

Chains of sugars are covalently linked to the cell surface side of adhesion proteins, and these carbohydrate groups help in both the recognition and the interlinking that are necessary for cell attachment. Our skin cells have transmembrane adhesion proteins, called integrins, that tether those cells to an ECM made mostly of collagen. Collagen is the most abundant protein in the human body, and destruction of collagen over time leads to wrinkling and sagging as we age.

Membrane Proteins

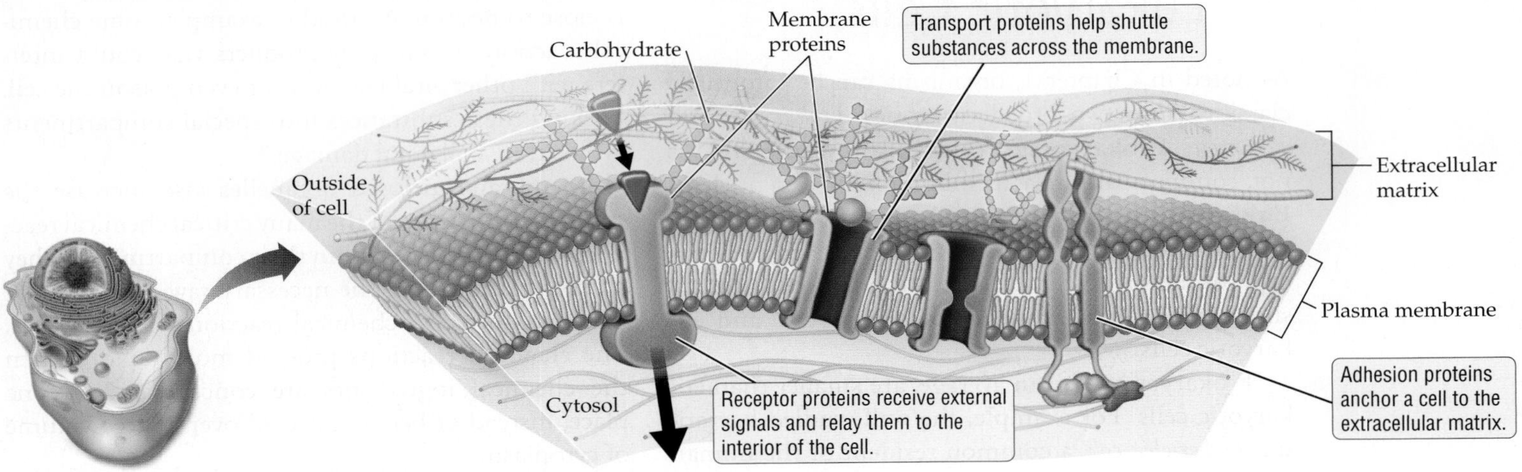

FIGURE 3.7 The Many Functions of Membrane Proteins

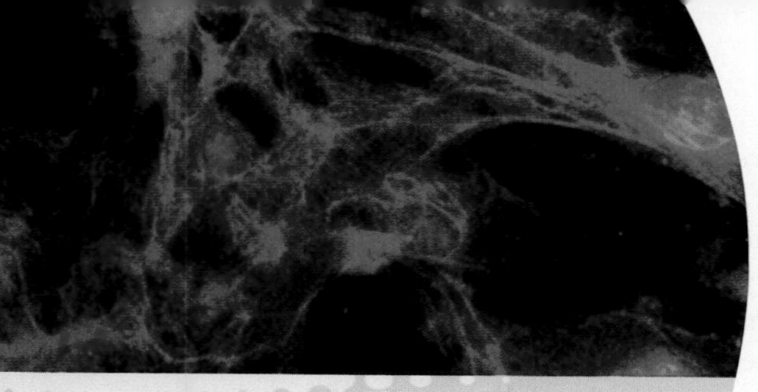

EXTREME ECM

Cells (nuclei stained blue) make up only 10% of the volume of a living bone. The extracellular matrix (ECM) of bone is rich in collagen (red) and calcium compounds, and it accounts for 90% of the weight and volume of bone.

Unless they are anchored to structures inside or outside the cell, most plasma membrane proteins are free to drift within the plane of the phospholipid bilayer. The **fluid mosaic model** describes the plasma membrane as a highly mobile mixture of phospholipids, other types of lipids, and many different types of membrane proteins. The term "mosaic" reflects this mixture of components, just as in art a mosaic is a mixture of different-colored stones or tiles. "Fluid" in this model refers to the free-flowing and flexible nature of the membrane, and this property is critical for many of its functions.

The flexibility of the plasma membrane facilitates many cellular functions—for example, whole-cell movement. Cells that move from place to place—such as white blood cells in pursuit of invaders—could not crawl about if their cell membranes were rigid and unchangeable.

Concept Check

1. Name two functions performed by plasma membrane proteins.
2. What is meant by the fluid mosaic nature of the plasma membrane?

Concept Check Answers

1. Cell communication; transport of substances in and out of the cell; cell adhesion.
2. The plasma membrane consists of many different types of lipids and proteins (it is a mosaic), and many of them can move freely within the bilayer (it is fluid).

3.3 Prokaryotic and Eukaryotic Cells

As noted in Chapter 1, organisms can be informally classified into two broad categories: *prokaryotes* and *eukaryotes*. **Prokaryotes** are organisms whose DNA is not confined within a membrane-enclosed nucleus. **Eukaryotes** are organisms whose DNA is enclosed in a nucleus. Prokaryotes are single-celled; eukaryotes can be single-celled or multicellular. **FIGURE 3.8** shows generalized drawings of prokaryotic and eukaryotic cells.

Prokaryotic cells, on average, are smaller than eukaryotic cells. For example, the well-studied bacterium *Escherichia coli*, a common resident of the human intestine, is only two-millionths of a meter, or 2 micrometers, long. About 125 *E. coli* would fit end to end across the period at the end of this sentence. The average eukaryotic cell has roughly a thousand times the volume of the average prokaryotic cell.

Most prokaryotes have a tough cell wall outside the plasma membrane (**FIGURE 3.8*a***) that helps maintain the shape and structural integrity of the organism. Some bacteria have additional protective layers, the *capsule*, made of slippery polysaccharides.

Plants, fungi, and some protists (mostly algae) have cell walls made of polysaccharide, but the polysaccharides differ among these groups of eukaryotes and differ yet again from those of prokaryotic cell walls. Animal cells lack a polysaccharide cell wall, but many of them are encased in, or attached to, the meshwork of protein and carbohydrate polymers that makes up the extracellular matrix (ECM; shown in Figure 3.7).

A striking difference between prokaryotic and eukaryotic cells is the diversity and extent of internal compartments within the cytoplasm of eukaryotes. In addition to the nucleus, several other types of membrane-enclosed organelles, each with a distinct function, are seen in eukaryotic cells (**FIGURE 3.8*b***). In contrast, membrane-enclosed organelles are lacking in most prokaryotes.

We mentioned earlier that one advantage of multicellularity is the division of labor among different cells. In a similar way, membrane-enclosed organelles enable an intracellular division of labor. As one example, unique chemical environments can be maintained within a membrane-enclosed compartment. The reactions that break apart polymers work best under highly acidic conditions, so the organelles that specialize in that task maintain a very low pH, even though the pH of the cytosol outside the organelles is close to neutral. As another example, some chemical reactions produce by-products that could interfere with other vital reactions or even poison the cell. Locking these substances into special compartments avoids this "collateral damage."

Membrane-enclosed organelles also increase the speed and efficiency of the many critical chemical reactions that take place within their compartments. They do so by stockpiling the necessary raw material—the reactants for these chemical reactions—in one place. The chemical reactions proceed more rapidly when the chemical ingredients are concentrated in one place, instead of being scattered over a large volume of cytoplasm.

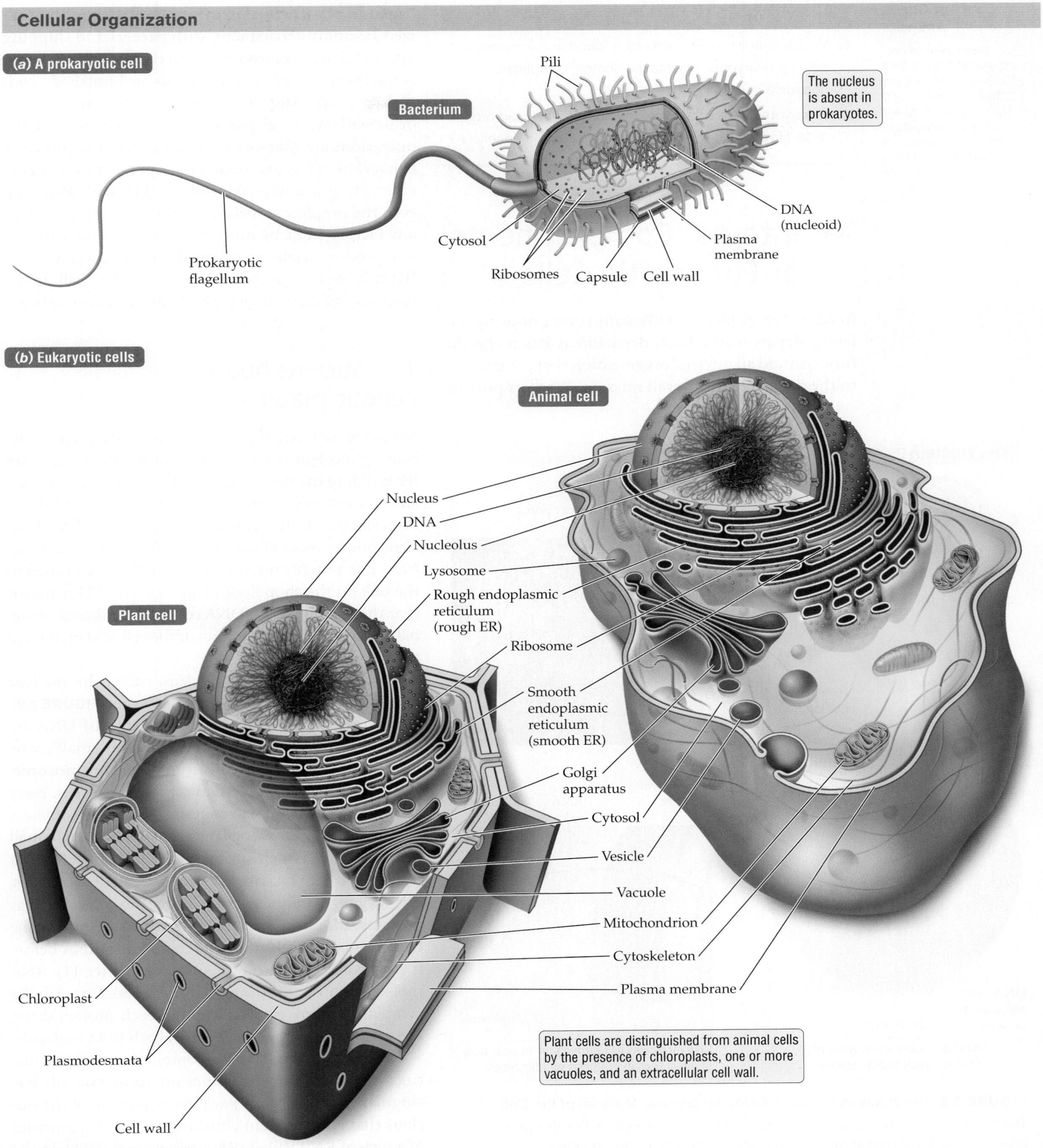

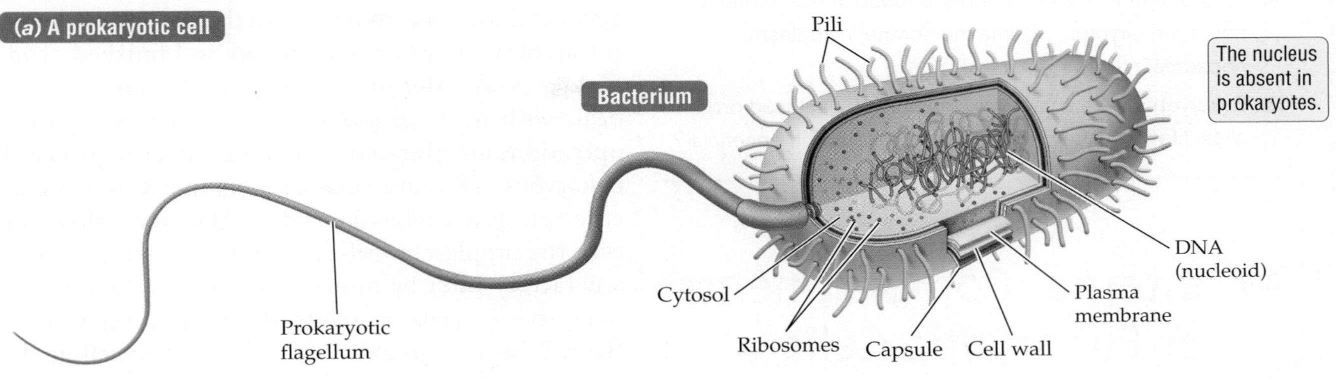
(a) A prokaryotic cell
Bacterium
Pili
The nucleus is absent in prokaryotes.
DNA (nucleoid)
Plasma membrane
Cell wall
Capsule
Ribosomes
Cytosol
Prokaryotic flagellum
(b) Eukaryotic cells
Animal cell
Plant cell
Nucleus
DNA
Nucleolus
Lysosome
Rough endoplasmic reticulum (rough ER)
Ribosome
Smooth endoplasmic reticulum (smooth ER)
Golgi apparatus
Cytosol
Vesicle
Vacuole
Mitochondrion
Cytoskeleton
Plasma membrane
Chloroplast
Plasmodesmata
Cell wall
Plant cells are distinguished from animal cells by the presence of chloroplasts, one or more vacuoles, and an extracellular cell wall.

FIGURE 3.8 Prokaryotic and Eukaryotic Cells Compared

Concept Check Answers

1. Nucleus.
2. Chemical reactions proceed faster because reactants are concentrated, special chemical environments (for example, low pH) can be created, and toxic substances can be compartmentalized.

Concept Check

1. Which of these cell structures is found in eukaryotes but not in prokaryotes: plasma membrane, cytoplasm, ribosome, nucleus?
2. Eukaryotes have a variety of membrane-enclosed organelles. How is this type of internal organization beneficial?

3.4 Internal Compartments of Eukaryotic Cells

Imagine a large factory with many rooms housing different departments. Each department has a specific function and an internal organization that contributes to the department's overall mission. Workers putting together a particular item are located in a centralized assembly department, with packers and shipping agents taking over from them in the next department. A warehouse stores raw materials and finished products, a power station supplies energy, another site deals with waste disposal and recycling, and all the operations are overseen by an administrative office. A eukaryotic cell is just such a highly structured, highly efficient, and energy-dependent "factory." However, even the simplest amoeba is vastly more complex than any factory built by humans, and it has the capacity to reproduce itself—a quality unique to living systems. We will begin our tour of the eukaryotic cell at the nucleus, the cellular factory's "administrative office."

The Nucleus

The nuclear envelope has openings called nuclear pores
Nuclear envelope
Nuclear pore
RNA
Ribosomes
DNA, with associated proteins
Nucleolus
Cytoplasm
Ribosomes and some types of RNA are made in the nucleolus.
RNA molecules exit through the nuclear pore to direct protein synthesis on ribosomes.

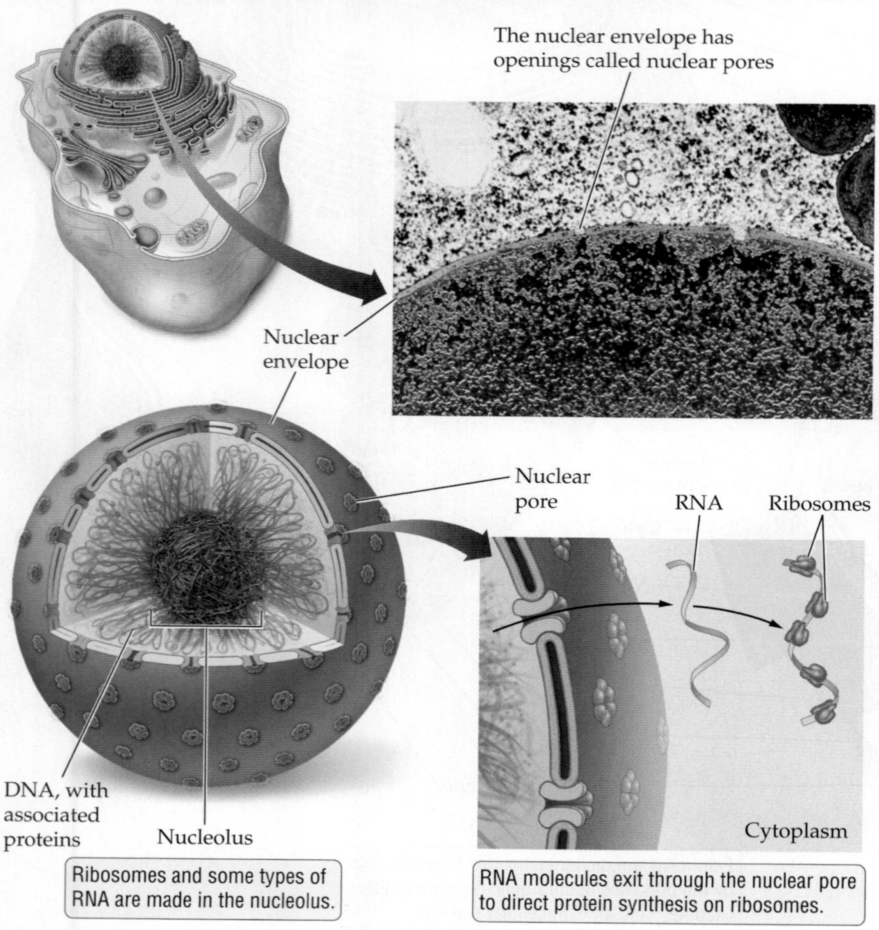

FIGURE 3.9 The Nucleus Contains DNA, the Genetic Material of the Cell
The nucleus is bounded by a double-membrane nuclear envelope. Nuclear pores provide a regulated passageway for molecules entering and exiting the nucleus.

The nucleus houses genetic material

No living cell can function without DNA—the code-bearing molecule that contains information necessary for building the cell, managing its day-to-day activities, and controlling its growth and reproduction. In eukaryotic cells, the clearly delineated membrane-enclosed nucleus houses most of the cell's DNA. The nucleus functions like a highly responsive head office, well tuned to the talk on the shop floor of the "factory." This means that the readout of the DNA code can be altered by signals received from other parts of the cell, and even from the outside environment.

The boundary of the nucleus, called the **nuclear envelope**, is made up of two lipid bilayers (**FIGURE 3.9**). Inside the nuclear envelope, long strands of DNA are packaged with proteins into a remarkably small space. Each DNA double helix constitutes one **chromosome**. With the exception of certain reproductive cells, every cell in the average human contains 46 chromosomes.

The nuclear envelope contains thousands of small openings, called **nuclear pores** (see Figure 3.9). Nuclear pores allow free passage to ions and small molecules, but they regulate the entry of larger molecules such as proteins, admitting some and shutting out others. The code carried in DNA must be copied into RNA before it can be decoded into proteins (see Chapter 11). RNA molecules must pass through nuclear pores to reach the cytoplasm, where ribosomes build each protein using the particular "recipe" delivered by each RNA molecule.

The nuclei of most cells contain one or more distinct regions, known as nucleoli (**noo-*KLEE*-uh-lye**; singular "nucleolus"). A *nucleolus* is a region of the nucleus that specializes in churning out large quantities of a type of RNA called rRNA (ribosomal RNA). In the nucleolus, rRNA is bundled with special proteins into partially assembled ribosomes, and these structures exit the nucleus through the nuclear pores to be fully assembled in the cytoplasm (see Figure 3.9). Nucleoli are most prominent in cells making large amounts of proteins, in which the demand for ribosomes is high.

The endoplasmic reticulum manufactures certain lipids and proteins

If the nucleus functions as the administrative office of the cell, the cytoplasm is the main factory floor putting together the great majority of proteins and other chemical components of the cell. However, certain types of lipids and proteins are made in the **endoplasmic reticulum** (**reh-*TIK*-yoo-lum**), an extensive and interconnected network of tubes and flattened sacs (**FIGURE 3.10**). The endoplasmic reticulum (ER) functions like a specialized department, preparing items for transport to other parts of the cell or export outside the cell.

The boundary of the endoplasmic reticulum is formed by a single membrane that is usually joined with the outer membrane of the nuclear envelope. The space inside the ER membrane is a **lumen**, a general term for the cavity inside any closed structure.

The membranes of the ER are classified into two types based on their microscopic appearance: smooth and rough (see Figure 3.10). Enzymes associated with the surface of the **smooth ER** manufacture various types of lipids destined for other cellular compartments, including the plasma membrane. In some cell types, smooth-ER membranes also break down organic compounds that could be toxic if they accumulated in the body. For example, the smooth ER in human liver cells has enzymes that break down caffeine, alcohol, and ibuprofen (the active ingredient in some painkillers).

Rough ER is so named because of its bumpy appearance, which results from the many ribosomes attached on the side facing the cytosol. Some proteins manufactured on the rough ER go to other compartments within the cytoplasm; others are shipped to the cell surface, either for incorporation into the plasma membrane or for release into the world outside the cell (the extracellular space).

Transport vesicles move materials

How do macromolecules such as lipids, proteins, and carbohydrates move from one internal compartment to another, and even to the extracellular space? First

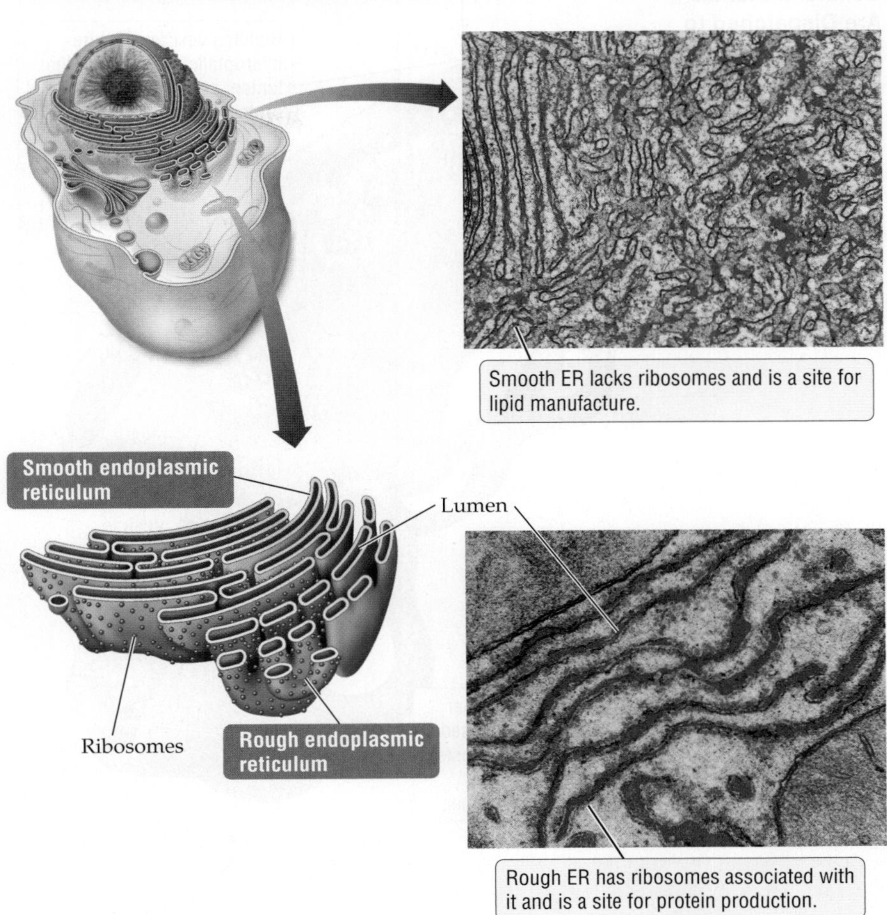
The Endoplasmic Reticulum
Smooth ER lacks ribosomes and is a site for lipid manufacture.
Smooth endoplasmic reticulum
Lumen
Ribosomes
Rough endoplasmic reticulum
Rough ER has ribosomes associated with it and is a site for protein production.

FIGURE 3.10 Some Lipids and Proteins Are Made in the Endoplasmic Reticulum

the macromolecule in question is packaged into a **transport vesicle**: a small, spherical, membrane-enclosed sac that moves material between cellular compartments (**FIGURE 3.11**). These vesicles are like the carts used to move goods between different departments in a factory. Hydrophobic cargo (such as lipids and hydrophobic proteins) is incorporated into the single membrane that forms the boundary of each transport vesicle. Hydrophilic material (mainly carbohydrates and hydrophilic proteins) is carried in the lumen of the transport vesicle.

EXTREME BAD LUCK?

Chocolate isn't good for Fido. Why? Animals that are mainly meat eaters lack many of the ER enzymes that plant eaters use to degrade plant toxins, and some plant chemicals (such as those in chocolate) can therefore be harmful to them.

FIGURE 3.11 Cellular Materials Are Dispatched to a Wide Variety of Destinations via Vesicles

Here, molecules are being shipped from the ER to the Golgi apparatus.

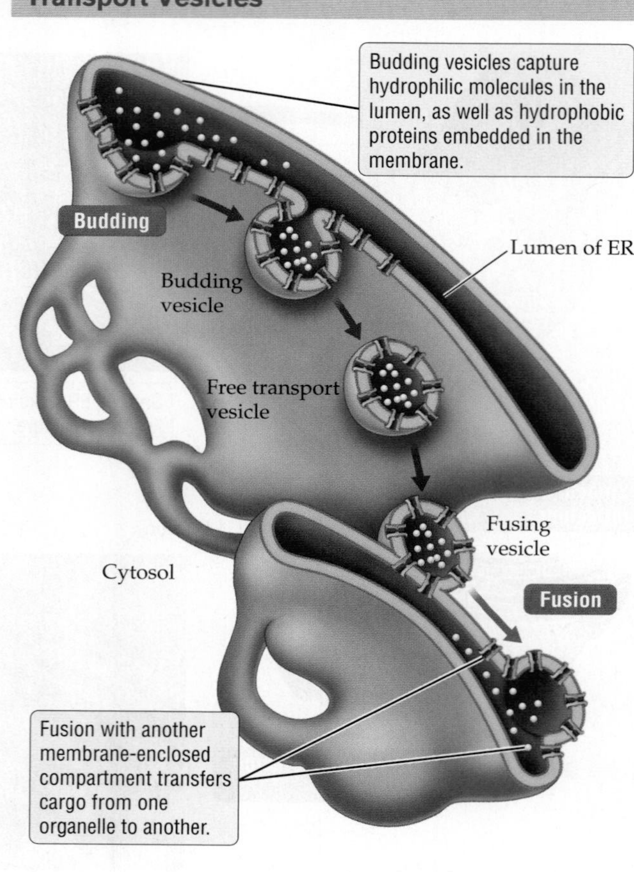

Transport Vesicles
Budding vesicles capture hydrophilic molecules in the lumen, as well as hydrophobic proteins embedded in the membrane.
Budding
Budding vesicle
Lumen of ER
Free transport vesicle
Fusing vesicle
Cytosol
Fusion
Fusion with another membrane-enclosed compartment transfers cargo from one organelle to another.

A transport vesicle buds off from a membrane, such as the ER membrane, like a soap bubble emerging from a bubble blower. The vesicle delivers the cargo to its destination simply by fusing with the membrane of the target compartment, like a small soap bubble merging with a bigger one. The contents of a transport vesicle are determined by the compartment where it originated. For example, transport vesicles that pinch off from the ER enclose a small portion of the ER lumen and are enclosed by a patch of ER membrane.

The Golgi apparatus sorts and ships macromolecules

Another membranous organelle, the **Golgi apparatus**, directs proteins and lipids produced by the ER to their final destinations, either inside or outside the cell. The Golgi apparatus functions as a sorting station, much like the shipping department in a factory. In a shipping department, goods destined for different locations get address tags that indicate where they should be sent. Similarly, in the Golgi apparatus the addition of specific chemical groups to proteins and lipids helps target them to other destinations. The chemical tags include carbohydrate molecules and phosphate groups.

Under the electron microscope, the Golgi apparatus looks like a series of flattened membrane sacs stacked together and surrounded by many small transport vesicles (**FIGURE 3.12**). The vesicles move lipids and proteins from the ER to the Golgi apparatus and carry them between the various sacs of the Golgi apparatus.

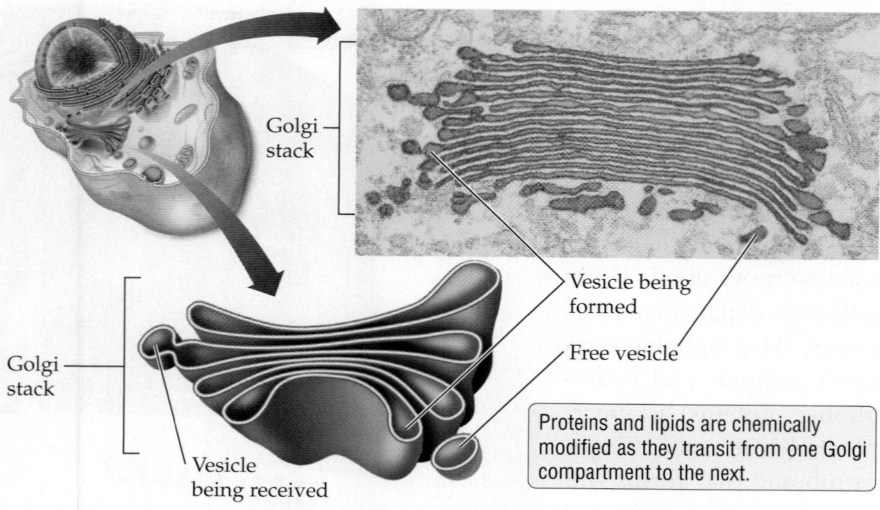

The Golgi Apparatus
Golgi stack
Golgi stack
Vesicle being formed
Free vesicle
Vesicle being received
Proteins and lipids are chemically modified as they transit from one Golgi compartment to the next.

FIGURE 3.12 The Golgi Apparatus Routes Proteins and Lipids to Their Final Destinations

Proteins and lipids are chemically modified, sorted, and shipped to their final destinations, inside or outside the cell, by the Golgi apparatus.

Lysosomes and vacuoles disassemble macromolecules

In animal cells, large molecules destined to be broken down are directed to organelles called *lysosomes*. **Lysosomes** (*lyso*, "to break"; *soma*, "body") are membranous organelles that degrade macromolecules and release the subunits into the cytoplasm. In other words, lysosomes are the junkyard and recycling center of the cell. The macromolecules destined for destruction are delivered to lysosomes by transport vesicles. Even whole organelles, such as damaged mitochondria, can fuse with lysosomes and be taken apart in the lysosomal lumen.

A single membrane forms the boundary of each lysosome (**FIGURE 3.13**). Inside it are a variety of enzymes, each specializing in degrading specific

Lysosomes

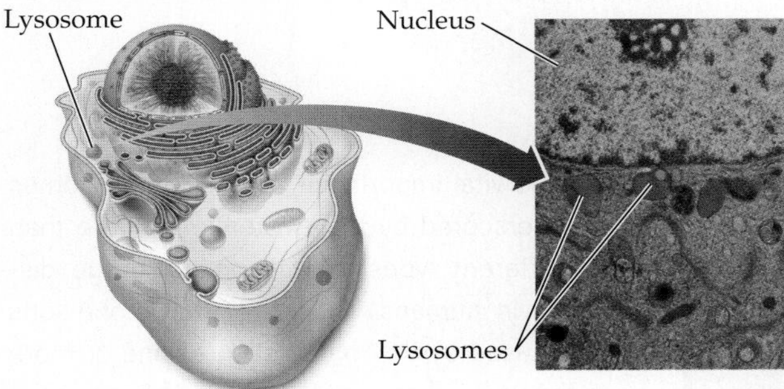

FIGURE 3.13 Lysosomes Degrade Macromolecules
Lysosomes are found in animal cells. Lysosomes help to digest molecules taken up from outside and to break down cell components whose molecules can be repurposed.

macromolecules, such as a particular class of lipid or protein. Many of the breakdown products—among them fatty acids, amino acids, and sugars—are transported across the lysosomal membrane and released into the cytoplasm for reuse. Lysosomes can adopt a variety of irregular shapes, but all are characterized by an acidic interior with a pH of about 5. (For comparison, the pH of the cytosol is close to 7.) Lysosomal enzymes work best at the lower pH.

The plant organelles called **vacuoles** perform many of the same functions as the lysosomes of animal cells, as well as some others. Most mature plant cells have a central vacuole that can occupy more than a third of a plant cell's total volume (**FIGURE 3.14**). Besides containing enzymes that break down macromolecules, plant vacuoles store a variety of ions and water-soluble molecules: calcium ions, sugars, and colorful pigments, for example. Like lysosomes, most plant vacuoles have an acidic pH because they accumulate hydrogen ions.

Some plant vacuoles stockpile noxious compounds that could discourage animals from feeding on the plant. For example, the vacuoles of tobacco leaves accumulate nicotine, a nervous system toxin that is released when the leaf cells are damaged.

Large vacuoles filled with water also contribute to the overall rigidity of the nonwoody parts of a plant. The vacuolar contents exert a physical pressure, known as *turgor pressure*, against the cytoplasm, the plasma membrane, and the cell walls. Turgor pressure keeps a plant cell plumped up the way air pressure in an inner tube keeps a car tire inflated. Loss of turgor pressure leads to the droopy appearance of houseplants that have gone too long without water.

Mitochondria power the cell

So far, we have explored the administrative offices, factory floors, and shipping department within the cellular factory. None of these offices or specialized departments could function without a source of energy to run the machines that produce the goods and services. In most eukaryotic cells, the main source of this energy is the mitochondrion, a double-membrane organelle that fuels cellular activities by extracting energy from food molecules. All kinds of eukaryotes, including plants, need mitochondria to convert the chemical energy in food molecules into a form useful for powering cellular activities.

Mitochondria are pod shaped and bounded by double membranes. The space between the two membranes is called the **intermembrane space**. The inner mitochondrial membrane is thrown into large folds, called **cristae** (*KRIS*-**tee**; singular "crista") that present a large surface area. The space interior to the cristae is called the **matrix** (plural "matrices"; **FIGURE 3.15**). Mitochondria use chemical reactions to transform the energy of food molecules into ATP (adenosine triphosphate), the universal cellular fuel. The energy

Plant Vacuoles

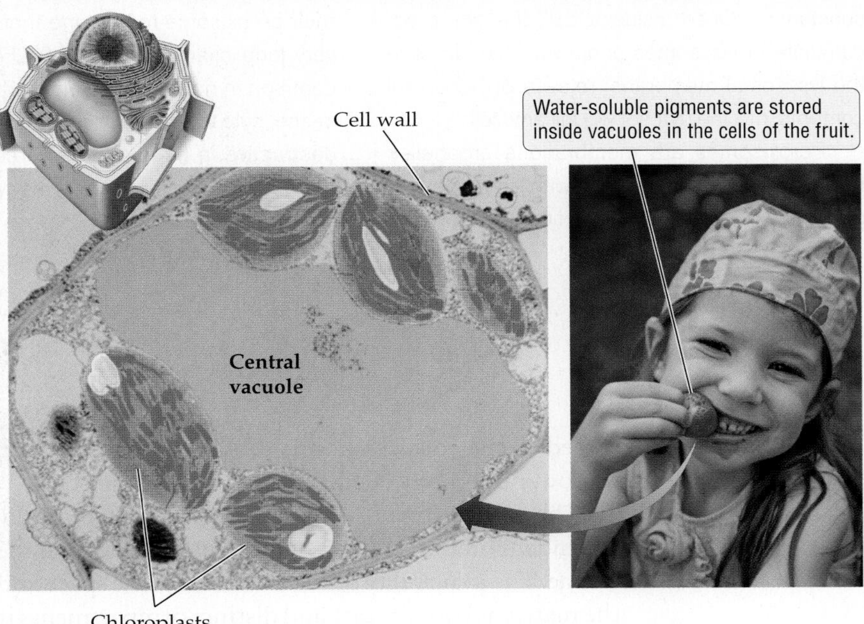

Cell wall
Water-soluble pigments are stored inside vacuoles in the cells of the fruit.
Central vacuole
Chloroplasts

FIGURE 3.14 Plant Vacuoles Store, Recycle, and Provide Turgor
Plant vacuoles contain enzymes for degrading large macromolecules. They also store water, ions, sugars, and other nutrients, and they may contain pigments that attract pollinators and/or toxins that deter herbivores. The fluid pressure that develops inside the vacuole keeps plant cells plumped up.

BIOLOGYMATTERS

Organelles and Human Disease

The normal functioning of cellular structures is crucial to our survival. The *complete* failure of any one of the organelle types in *all* cells of the body would be incompatible with life. Small errors in the workings of an organelle can make us ill, sometimes very ill. For example, mitochondrial malfunctions are implicated in a variety of *neurodegenerative diseases*, those that result in the death of brain or nerve cells. Defects in ER enzymes produce a host of disorders broadly classified as ER dysfunctions. Similarly, defects in Golgi apparatus enzymes lead to a variety of disorders, such as the carbohydrate-deficient glycoprotein (CDG) syndromes. Many of these conditions are genetic disorders, caused by a corruption of the DNA code that normally directs production of a functional protein. However, environmental substances can also damage organelles. *Asbestosis* is the destruction of lung tissue from exposure to asbestos, a mineral substance. Asbestos fibers are engulfed by immune cells, whose job is to defend the body from foreign substances. Once inside a cell, the fibers accumulate in lysosomes and, over time, damage the lysosomal membrane; release of lysosomal contents into the cytosol will kill any cell.

Peroxisomes are membranous organelles that degrade lipids. Patients with adrenoleukodystrophy (ALD) have a defective transporter in their peroxisome membrane that fails to import very long chain fatty acids (VLFAs), which are common in normal diets. The fatty acids pile up in the cytosol; the accumulation is especially destructive in brain and nerve cells and in the hormone-producing organs known as adrenal glands.

The movie *Lorenzo's Oil* tells the true story of a couple's battle to prolong the life of their son Lorenzo, who was born with the peroxisome disorder adrenoleukodystrophy (ALD). Untrained in scientific research, the couple developed a vegetable oil formula that slows the excessive buildup of lipids within peroxisomes. Lorenzo died at age 30, more than two decades later than his doctors had predicted. He died of pneumonia, not ALD.

The vital importance of healthy lysosomes is underscored by the existence of more than 40 different types of lysosomal storage disorders in humans. These inherited conditions are caused by the malfunction of one or more of the many lysosomal enzymes whose job it is to degrade specific macromolecules. When a lysosomal enzyme is absent or fails to work properly, the macromolecule it would normally degrade piles up inside the lysosomes. The consequences are devastating, and most of these disorders are fatal in childhood.

Tay-Sachs disease is one such metabolic disorder. The lysosomal enzyme responsible for breaking down a membrane lipid found in brain tissue fails to do its job, with the result that large amounts of this lipid accumulate in nerve cells, compromising the function of these cells and eventually destroying them. Tay-Sachs disease is rare in the population as a whole, but it was once unusually common in some populations that tend to marry among themselves, such as certain French-Canadian communities and Orthodox Jews from eastern Europe. Through a combination of genetic testing and genetic counseling for individuals contemplating marriage, Tay-Sachs disease has been virtually eradicated among Jewish communities in North America and Israel.

stored in the covalent bonds of ATP is used, in turn, to power the many chemical reactions of the cell.

The production of ATP by mitochondria is critically dependent on the activities of proteins embedded in the cristae. The availability of the intermembrane space and the matrix as two separate and distinct compartments is also crucial for the process. With this unique setup, mitochondria are able to trap some of the chemical energy released when food molecules are broken down; this energy is then used to synthesize ATP through a process that requires oxygen. In other words, oxygen gas (O_2) and molecules derived from food are the raw materials that fuel the mitochondrial power station. As in most human-made power stations, carbon dioxide (CO_2) and water (H_2O) are released as by-products of this energy-producing process, which is called **cellular respiration**.

Chloroplasts capture energy from sunlight

Mitochondria provide life-sustaining ATP to eukaryotic cells (both plant and animal), but the cells of plants and the protists known as algae have additional organelles,

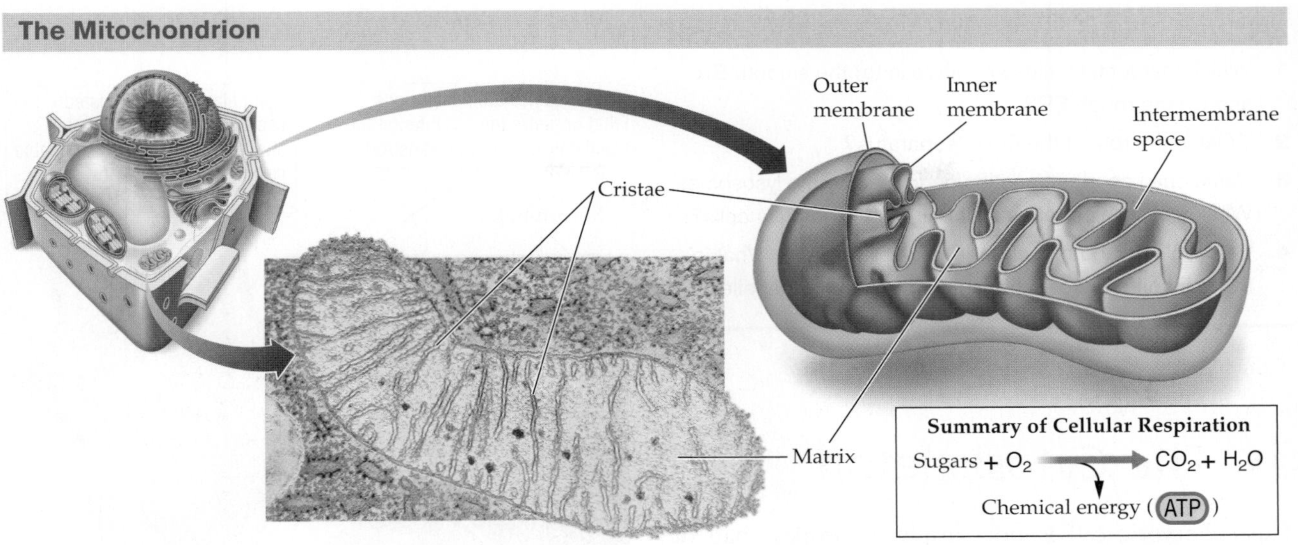

FIGURE 3.15 Mitochondria Extract Energy from Food Molecules to Make ATP

Each mitochondrion has a double membrane. The infoldings of the inner membrane (cristae) create a large surface area with room for many ATP-generating enzymes.

called **chloroplasts** (FIGURE 3.16), that capture energy from sunlight and use it to manufacture sugars—a process known as **photosynthesis**. The energy in these sugars is used directly by plant cells and indirectly by all organisms that eat plants. At this very moment, as you read this page, your brain and the muscles that move your eyes are using energy from food molecules that were originally produced in chloroplasts through photosynthesis.

During photosynthesis, water molecules are broken down, releasing oxygen gas. The oxygen produced in photosynthesis sustains life for us and many other life-forms. Mitochondria depend on a continual supply of that oxygen to produce ATP.

Chloroplasts are enclosed by two membranes that form the chloroplast envelope. Inside the envelope lies an internal network of membranes, some of which are arranged like stacked pancakes (see Figure 3.16). Each "pancake" in the stack is called a **thylakoid** (*THYE*-**luh-koyd**). Embedded in the thylakoid membranes are special light-absorbing pigments, notably **chlorophyll**, that enable chloroplasts to capture energy from sunlight. Chlorophyll absorbs red and blue wavelengths of light, but not green. Green wavelengths are reflected by it, which is why the pigment appears green.

The light absorbed by chlorophyll is used to generate energy carriers such as ATP. Enzymes present in the space surrounding the thylakoids use these energy carriers to synthesize (manufacture) sugars from water and from carbon dioxide that the plant takes in from the air around it.

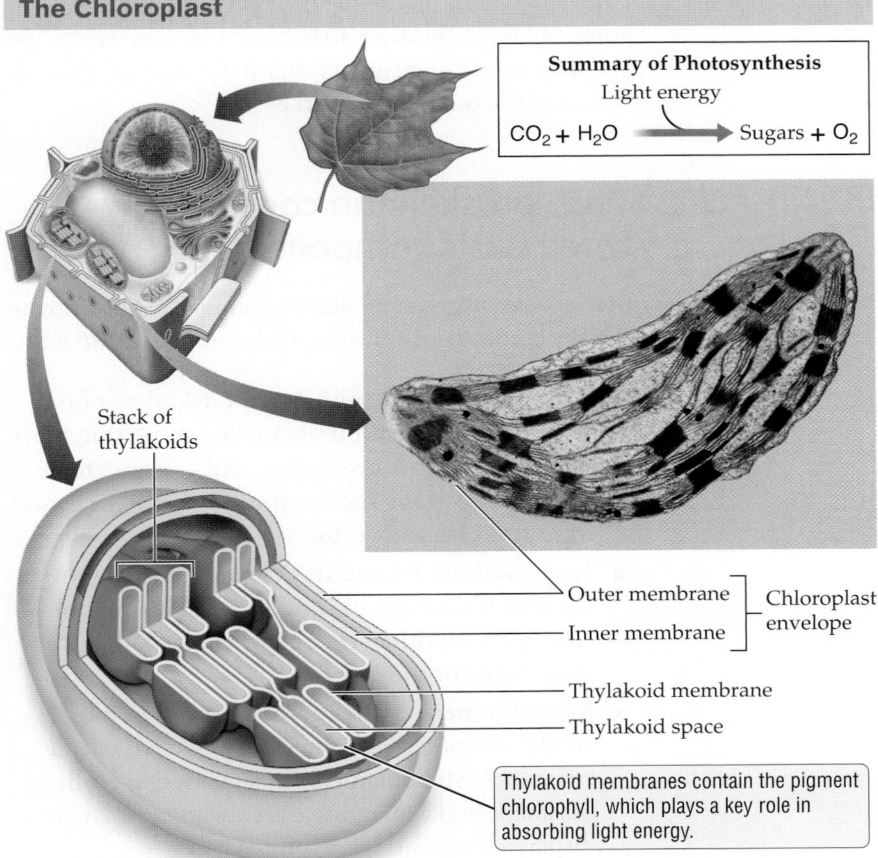

FIGURE 3.16 Chloroplasts Capture Energy from Sunlight and Use It to Make Sugars

Chloroplasts are found in green plant parts and in the protists known as algae.

Concept Check Answers

1. (a) Lipids; (b) proteins.
2. Receiving newly made lipids and proteins from the ER, adding specific chemical groups ("address labels") to them, and shipping them out to their final destination.
3. Degradation of macromolecules such as proteins and lipids. In plants, the vacuole performs this function and others.
4. Chloroplasts release O_2 as they use sunlight to fuel the conversion of CO_2 and water into sugars. Mitochondria break down sugars, consuming O_2 and releasing CO_2 and water.

Concept Check

1. Which macromolecules are made in (a) the smooth ER and (b) the rough ER?
2. What is the role of the Golgi apparatus?
3. Name one important function performed by the lysosomes. What organelle performs a similar function in plant cells?
4. Chloroplasts and mitochondria both make ATP. What is the main difference in the function of these organelles?

3.5 The Cytoskeleton

The eukaryotic cell is not simply a formless bag of membrane with cytosol and organelles sloshing around inside. A network of protein cylinders and filaments, collectively known as the **cytoskeleton**, organizes the interior of the cell, supports the intracellular movement of organelles such as transport vesicles, gives shape to wall-less cells, and even enables whole-cell movement in some cell types (**FIGURE 3.17**). In this section we discuss the main types of elements that make up the cytoskeleton and the functions that they provide.

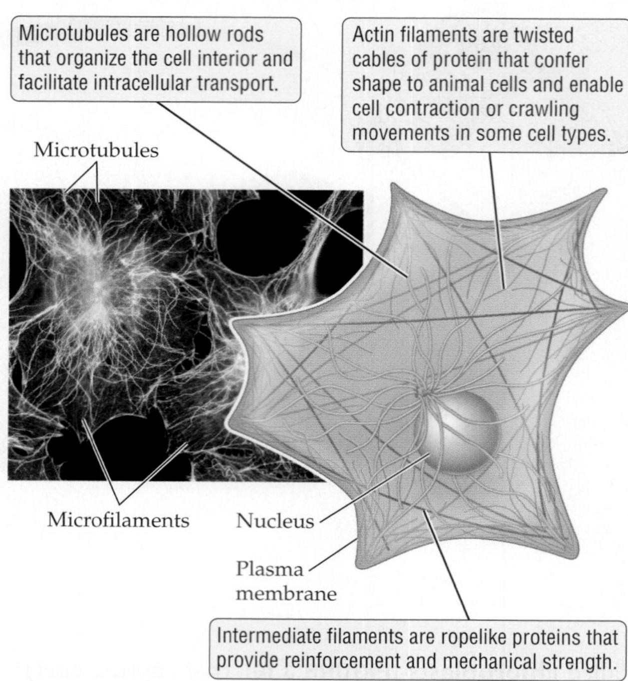

The Cytoskeleton
Microtubules are hollow rods that organize the cell interior and facilitate intracellular transport.
Actin filaments are twisted cables of protein that confer shape to animal cells and enable cell contraction or crawling movements in some cell types.
Microtubules
Microfilaments
Nucleus
Plasma membrane
Intermediate filaments are ropelike proteins that provide reinforcement and mechanical strength.

FIGURE 3.17 An Overview of the Cytoskeletal System

The cytoskeleton consists of three basic components

Microtubules, intermediate filaments, and *microfilaments* are the key components of the cytoskeleton (**FIGURE 3.18**).

- **Microtubules** are rigid, hollow cylinders of protein that help position organelles, move transport vesicles and other organelles, and generate force in cell projections such as the *cilia* or *flagella* found in some eukaryotic cells.
- **Intermediate filaments** are ropelike cables of protein, the nature of which can vary from one cell type to another, that provide mechanical reinforcement for the cell.
- **Microfilaments** (also known as *actin filaments*) are the thinnest of the three types of cytoskeletal structures. Microfilaments are involved in creating cell shape and generating the crawling movements displayed by some eukaryotic cells.

Cells like *Amoeba* that creep along on a solid surface depend on the changeable nature of cytoskeletal components (mainly microfilaments) for their movement, because cell crawling involves dramatic changes in cell shape. Changes to the cytoskeleton enable the extension and retraction of the plasma membrane at the leading and trailing edges of the cell, respectively. Organisms that can swim in a liquid, like the *Paramecium* and *Volvox* we met earlier, depend on cytoskeletal elements (microtubules) for their movement.

Microtubules support movement inside the cell

Microtubules are the thickest of the cytoskeleton filaments, with a diameter of about 25 nanometers (**FIGURE 3.18*a***). Each microtubule is a cylindrical structure made from subunits of a protein called **tubulin** (TOO-**byoo-lin**). A microtubule can grow or shrink in length by adding or losing tubulin monomers. Most animal cells have a system of microtubules that radiate from the center of the cell and terminate at the plasma membrane (see Figure 3.17). The radial pattern of microtubules serves as an internal scaffold that helps position organelles such as the ER and the Golgi apparatus.

Microtubules also define the paths along which vesicles are guided in their travels from one organelle to another, or from an organelle to the cell surface.

Main Components of the Cytoskeleton

(*a*) Microtubules

Microtubules are composed of tubulin subunits.

(*b*) Intermediate filaments

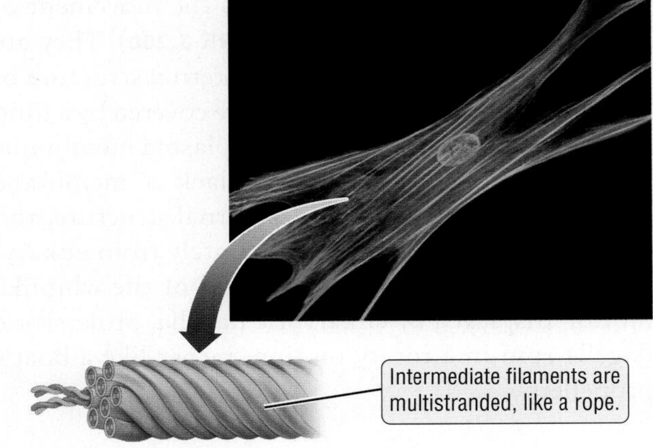

Intermediate filaments are multistranded, like a rope.

(*c*) Microfilaments

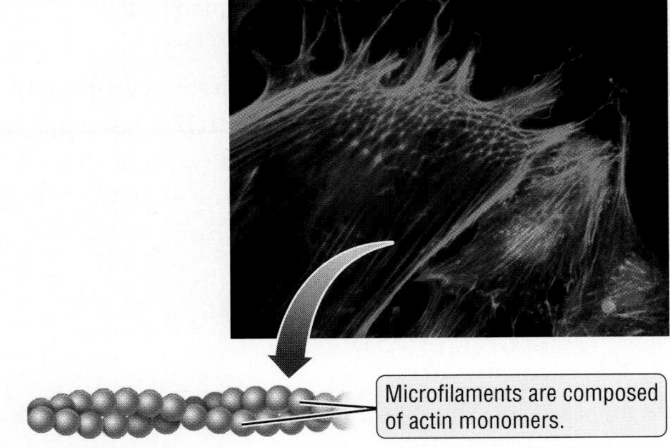

Microfilaments are composed of actin monomers.

FIGURE 3.18 The Structure of Microtubules, Intermediate Filaments, and Microfilaments

The cytoskeleton is composed of three basic units: (*a*) microtubules, (*b*) intermediate filaments, and (*c*) microfilaments.

The ability of microtubules to act as "railroad tracks" for vesicles depends on the action of *motor proteins.* A motor protein attaches to a vesicle by its "tail" end and associates with a microtubule through its "head" end. The motor protein then uses the energy of ATP to move along the microtubule, carrying the vesicle with it.

Helpful to Know

"Skeleton" may seem an odd word to apply to cell structures, since cells do not have bones as we do. However, biologists give the word a larger meaning: any structure or framework that supports and/or protects a cell or an organism. "Cytoskeleton" refers to the structures within cells (*cyto*, "cell") that give them shape and structure. "Endoskeletons" (*endo*, "inner") are found in many animals, such as fish and mammals; and "exoskeletons" (*exo*, "outer") are found in insects and crustaceans.

Intermediate filaments provide mechanical reinforcement

Intermediate filaments are a diverse class of ropelike filaments about 8–12 nanometers in diameter. They are thinner than microtubules but thicker than microfilaments (**FIGURE 3.18*b***).

Intermediate filaments serve as structural supports, in the way beams and girders support a building. For example, intermediate filaments consisting of the protein keratin strengthen the living cells in our skin. Skin cells lacking functional keratin cannot withstand even mild physical pressure, and their rupture results in severe blistering and other types of skin lesions. Intermediate filaments also provide mechanical reinforcement for internal cell membranes such as the nuclear envelope.

Microfilaments are involved in cell movement

Of the three filament types, microfilaments have the smallest diameter, about 7 nanometers (**FIGURE 3.18*c***). Each microfilament is a very flexible spirally wound cable built from monomers of the protein *actin*. Like microtubules, microfilaments are dynamic structures that can shorten or lengthen in either direction through rapid disassembly, or rapid assembly, at one or both ends. Microfilaments are responsible for the ability of cells such as white blood cells and amoebas to crawl along a surface (**FIGURE 3.19**).

Cell crawling enables protists such as amoebas and slime molds to find food and mating partners. Fibroblasts, which play an important role in the healing of skin wounds, migrate to the site of injury using the microfilament-based system of cell crawling. Cell migration is also crucial in the embryonic development of many animals. However, cell migration is a devastating step in the development of cancer because it

How Microfilaments Enable Cell Crawling

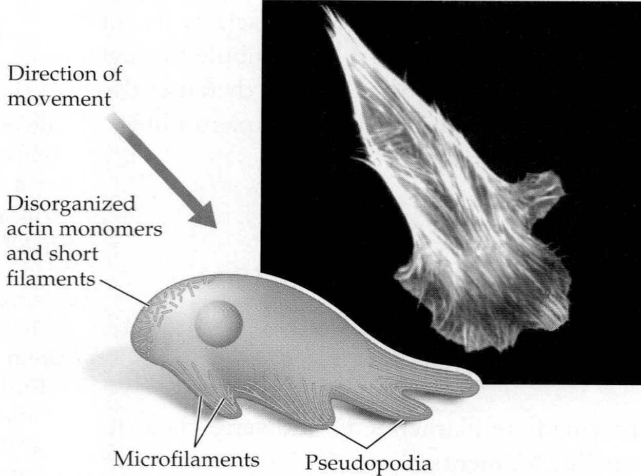

FIGURE 3.19 Microfilaments Drive Some Types of Whole-Cell Movement

At the leading edge, microfilaments lengthen in parallel arrays, pushing the plasma membrane out to form the pseudopodia (literally, "false feet"; singular "pseudopodium"). At the same time, microfilaments at the trailing end disassemble altogether. The plasma membrane at the trailing end detaches from the solid surface, while motor proteins generate forces that pull the rear of the cell forward.

enables cancer cells to invade other tissues and even spread through the body.

Cilia and flagella enable whole-cell movement

Many protists and animals have cells covered in a large number of hairlike projections, called **cilia** (singular "cilium"), that can be moved back and forth, like the oars of a rowboat, to move the whole cell through a liquid or to move a liquid over the cell surface. Inside each cilium is a flexible cytoskeletal apparatus created by bundles of microtubules. This apparatus consists of nine pairs ("doublets") of microtubules arranged in a ring around a central pair. Motor proteins interlinking the microtubules use the energy of ATP to flex the microtubules against each other, causing the whole cilium to bend.

Many aquatic protists, like the *Paramecium* shown in Figure 3.1*b*, use cilia to move about in the waters they inhabit. Other cells, such as those that line our respiratory passages, use cilia to move an overlying fluid layer (**FIGURE 3.20*a***). The cilia propel unwanted material, caught in a layer of mucus, out of the lungs and into the throat for elimination by coughing or swallowing.

Many bacteria, archaeans, and protists, as well as the sperm cells of some plants and all animals, can propel themselves through a fluid using one or more whiplike structures called **flagella** (**fluh-*JELL*-uh**; singular "flagellum"). **Eukaryotic flagella** are lashed about in a pattern that resembles the movement of a circus ringmaster's whip (**FIGURE 3.20*b***). They are much longer than cilia, but the internal structure of the two is very similar, and both are covered by a lipid bilayer that is an extension of the plasma membrane. **Prokaryotic flagella**, however, lack a membrane covering, have a very different internal structure, and are believed to have evolved separately from eukaryotic flagella (**FIGURE 3.20c**). Instead of the whiplike motion displayed by eukaryotic flagella, prokaryotic flagella spin in a rotary motion, rather like a boat's propellers.

Concept Check

1. List three important functions performed by the cytoskeleton.
2. Compare eukaryotic cilia and flagella.

Concept Check Answers

1. Positioning organelles within cells; facilitating transport of organelles within cells; enabling cells to travel by cell crawling or the movement of flagella or cilia.
2. Flagella are longer and move in a whiplike fashion, while cilia move like oars. The arrangement of microtubules within the two is similar, and both are covered by a plasma membrane.

How the Cytoskeleton Enables Movement

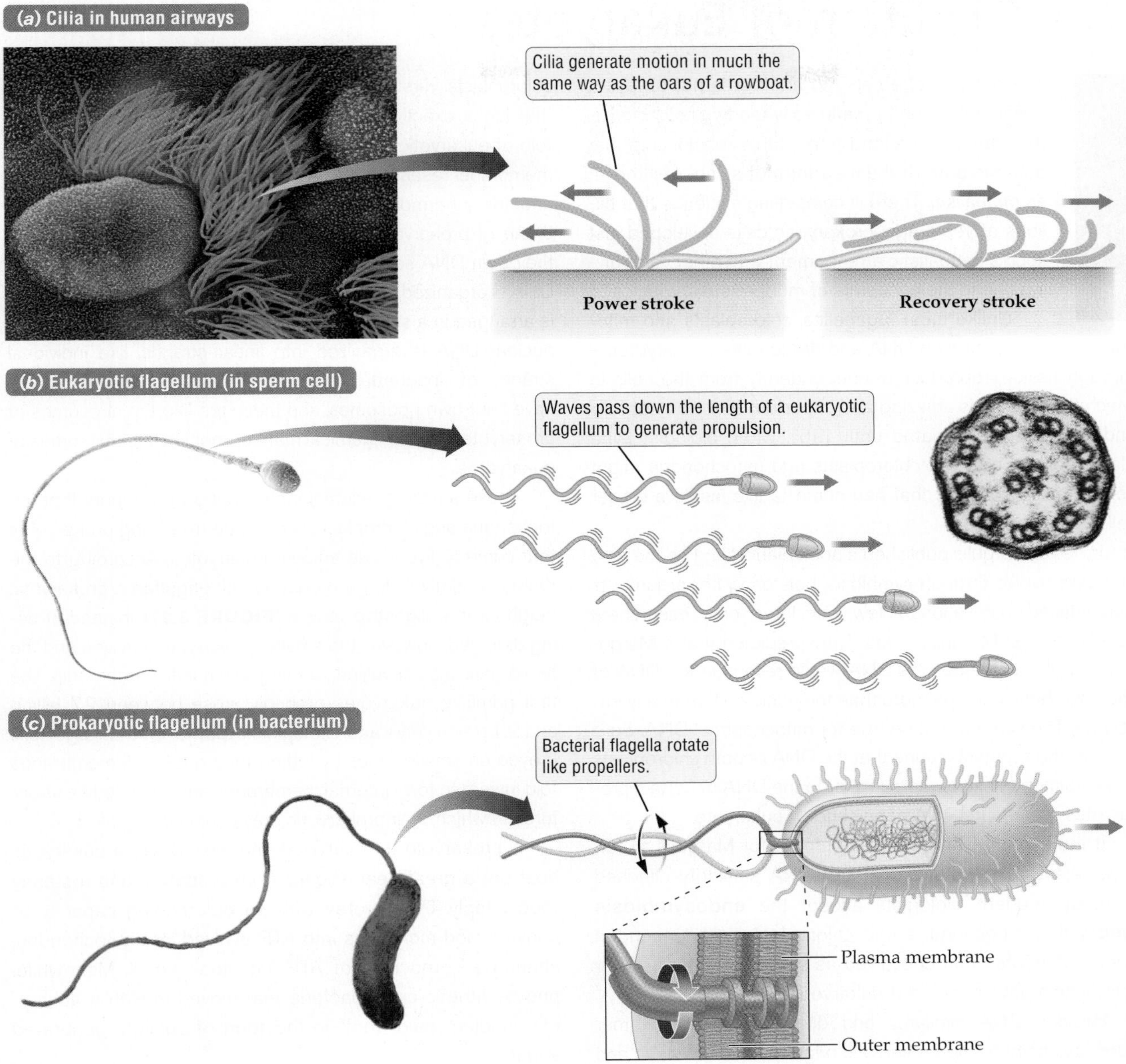

(a) Cilia in human airways

Cilia generate motion in much the same way as the oars of a rowboat.

Power stroke

Recovery stroke

(b) Eukaryotic flagellum (in sperm cell)

Waves pass down the length of a eukaryotic flagellum to generate propulsion.

(c) Prokaryotic flagellum (in bacterium)

Bacterial flagella rotate like propellers.

Plasma membrane

Outer membrane

FIGURE 3.20 Cilia and Flagella Generate Movement

Many organisms, especially single-celled organisms, use cilia or flagella to generate movement. (*a*) Tufts of cilia are present on the cells that line our breathing tubes (bronchi **[brahng-KYE]**; singular "bronchus"). (*b*) Eukaryotic flagella, such those in sperm cells, are much longer than cilia. Eukaryotic cilia and flagella contain bundles of microtubules arranged in a 9+2 pattern (*inset*) and are covered by a plasma membrane. Prokaryotic flagella have a very different structure. (*c*) A prokaryotic flagellum, such as the one on this bacterium (*Bdellovibrio bacteriovorus*), consists of ropelike proteins attached to protein complexes anchored in the cell membranes.

APPLYING WHAT WE LEARNED

The Evolution of Eukaryotes

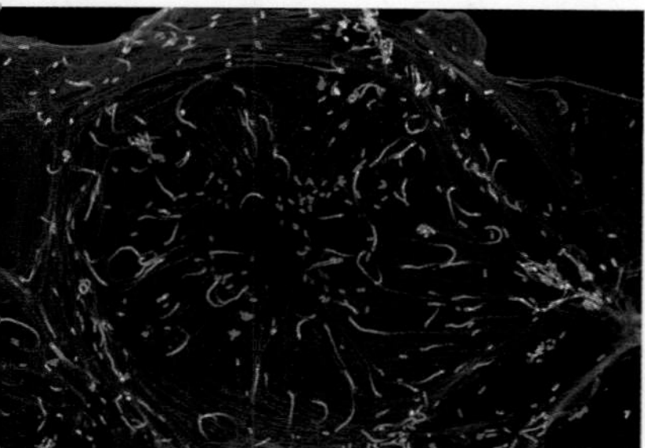

The entry of one cell into another is not always a hostile takeover. Sometimes when two cells merge, they enter into a long-term stable relationship that benefits both of them—a form of symbiosis known as *mutualism*. There is compelling evidence that billions of years ago, prokaryotic cells developed just such a mutualistic arrangement with other cells, giving rise to the ancestors of modern eukaryotic cells.

Unlike most organelles, chloroplasts and mitochondria have their own DNA and divide—like prokaryotes—through fission, reproducing independently from the cells in which they live. This amazing and important fact was not fully understood or appreciated until 1967, when biologist Lynn Margulis proposed that chloroplasts and mitochondria might be former prokaryotes that had come to live inside a bigger cell.

In 1970, Margulis published a book elaborating on the idea that cells evolve through symbiosis, bolstering her arguments with detailed observations. A few years later, some researchers decided to test Margulis's idea. They predicted that if Margulis was right, a chloroplast's DNA would resemble the DNA of photosynthetic bacteria more than the nuclear DNA in any eukaryote. The same would be true for mitochondrial DNA. Sure enough, the scientists found that the DNA of both chloroplasts and mitochondria is far more similar to the DNA of certain prokaryotes than to the DNA of any known eukaryote.

It took years of tenacious argument for Margulis to fully persuade the skeptics. But the DNA similarity clinched it. Today, modern biologists accept the **endosymbiosis theory** that mitochondria and chloroplasts are descended from prokaryotes that coevolved as symbionts inside larger cells, which gave rise to the eukaryotic lineage.

Besides DNA similarity and division by fission, other clues also point to the prokaryotic origins of mitochondria and chloroplasts. First, both have an arrangement of membranes that looks exactly as if a bacterium became enveloped in a fold of eukaryotic cell membrane: in both organelles the outer membrane resembles the plasma membrane of eukaryotes, and the innermost membrane resembles the plasma membrane of prokaryotes. Second, both organelles not only have their own DNA separate from the cell's nuclear DNA, but that DNA is organized just like a prokaryote's. Each organelle's DNA is arranged in a small, closed loop, like a necklace; in contrast, nuclear DNA is organized into linear strands, like individual strands of spaghetti. Finally, mitochondria and chloroplasts have their own ribosomes, and these are like the ribosomes of prokaryotes in their chemical makeup, not like the ribosomes of eukaryotes.

These amazing resemblances support the theory that mitochondria and chloroplasts were once free-living prokaryotes that came to live inside ancient eukaryotes. According to the theory, a relatively large predatory cell engulfed a prokaryotic neighbor in a digestive vesicle (**FIGURE 3.21**). Instead of being digested, however, the smaller prokaryote survived and the two organisms coevolved, forming a symbiotic partnership. The first primitive eukaryotes probably arose between 2.7 billion and 2.1 billion years ago. Biologists hypothesize that large cells preyed on smaller ones by letting their outer cell membranes fold inward to form internal membranes and vacuole-like structures in which their prokaryotic prey could be engulfed.

A prokaryote that survived and moved into a eukaryotic host got a great deal: shelter from predators and a steady food supply. Prokaryotes with an outstanding capacity to convert food molecules into ATP evolved into mitochondria, sharing a cornucopia of ATP with their hosts. Meanwhile, photosynthetic cyanobacteria that moved in with a eukaryotic landlord paid "rent" in the form of sunlight-generated sugar.

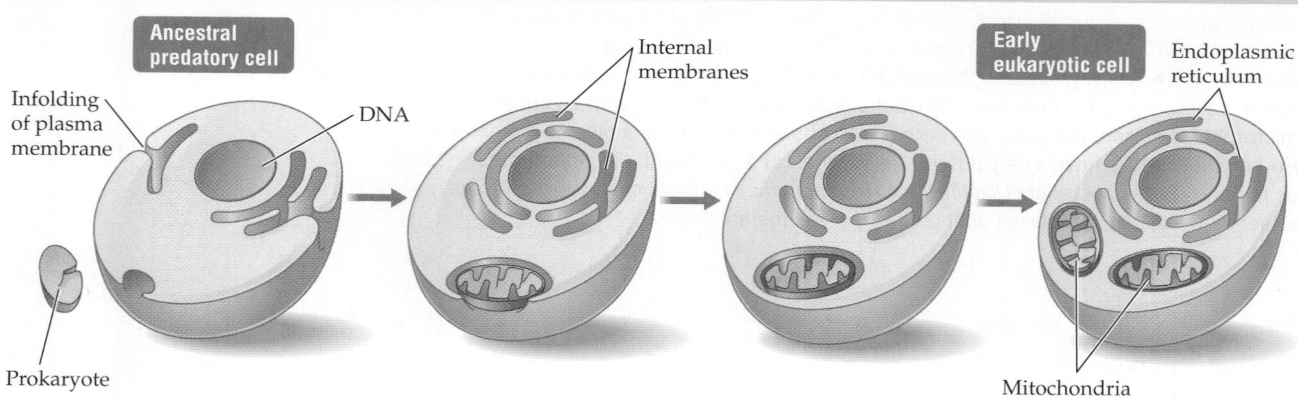

FIGURE 3.21 How Ancestral Eukaryotes Acquired Membrane-Enclosed Organelles Some organelles, such as mitochondria and chloroplasts, are likely descendants of engulfed prokaryotes. Other membrane-enclosed organelles, such as the endoplasmic reticulum, probably arose through an infolding of the plasma membrane.

BIOLOGY IN THE NEWS

New Technology Can Make It So 1 Baby Has 3 Parents: And Prevent That Child from Being Born with Serious Mitochondrial Diseases

CBS NEW YORK • March 29, 2013

New technology could bring babies into the world with three biological parents. The procedure is close to being legalized in the United Kingdom, and is aimed at preventing a genetic disorder.

But some people argue that more attention needs to be paid to the ethics behind the process.

Doctors at Pacific Fertility Center recently told CBS 2 Kristine Johnson that they are now able to use technology to screen and select embryos that are free of genetic diseases.

"The goal is always the same: identify embryos that are unaffected, and are going to produce normal children and only transfer those back," explained Dr. Rusty Herbert.

Now, in Britain, technology is about to take a huge leap, and create babies using the DNA from three different people in an effort to prevent mitochondrial disorders.

Mitochondria [are] the main power generator[s] inside of human cells. If [they are] defective, mothers could pass on devastating diseases to their children.

A three-person biological baby involves two fertilized eggs.

An egg is taken from a woman with mitochondrial problem and a sperm is taken from the father. A second egg is taken from another woman with healthy mitochondria.

DNA from the donor egg is replaced with DNA from the first egg, which is then implanted in the first woman. The baby is born with less than 1 percent of the DNA from the donor mother.

Not everyone is thrilled about the new advance.

"People have characterized this as sliding down a slippery slope. This one actually throws us off a cliff," said Marcy Darnovsky, a spokesperson for the Center for Genetics and Society.

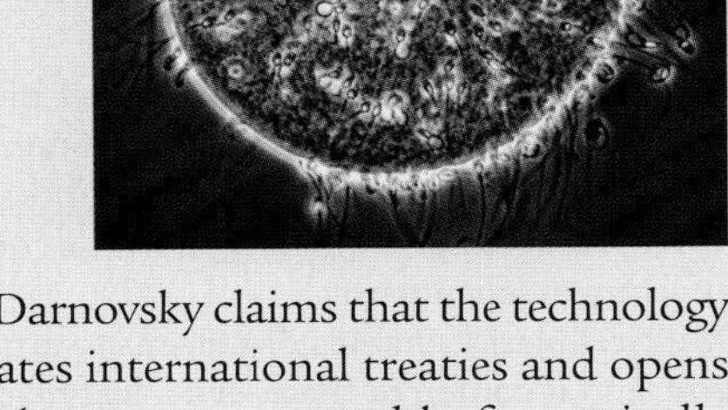

Darnovsky claims that the technology violates international treaties and opens the door to a new world of genetically enhanced "super babies."

"We may find ourselves in that kind of world before we know it and I think most us don't want to be there," she said.

However, others say "regulate," don't deny.

"For a disease situation, for mitochondrial disease, this should be strongly explored," Dr. Herbert said.

The technology has received support from the British public, according to published reports, but lawmakers have yet to make a decision on whether doctors can move forward.

The average human cell contains hundreds of mitochondria. If a few were abnormal, we'd hardly know. But roughly one in 4,000 babies in the United States is born with a large proportion of unhealthy mitochondria. The symptoms of mitochondrial disease range from relatively mild muscle weakness to hearing and vision loss and severe muscle and nerve impairment.

Some patients have mistakes in one of the nearly 3,000 nuclear genes that code for mitochondrial proteins. In about 15 percent of those afflicted with mitochondrial disease, the error lies in the mitochondrial DNA, not the nuclear DNA. Mitochondria, as you learned in this chapter, have their own DNA and make many copies of themselves. Mitochondrial DNA carries information for 37 proteins crucial for the normal function, especially the copying, of this organelle.

Although the DNA in your nucleus is a half-and-half blend of genetic information from your mother and father, all of your mitochondria are descended from those that were in the egg cell that became you. That is, children inherit their mitochondria from their mothers. A woman with defective mitochondrial DNA faces the agony that any children she bears will almost certainly inherit her disease.

If Parliament approves, the United Kingdom is likely to see the first attempts to make test-tube babies using mitochondrial replacement. Women with defective mitochondrial DNA could bear healthy babies if nuclear DNA from their eggs could be transplanted into "empty" eggs that have healthy mitochondria. In experiments with lab animals, the technique has successfully produced babies with cytoplasm of one individual and nuclear information from two other individuals, the male and female parents. Scientists at Oregon Health & Science University have used the technology to make human embryos with three parents, although these cannot be implanted into a would-be mother until new legislation is passed to allow this type of therapy.

Evaluating the News

1. The mitochondrial disorders Alpers' disease and Leigh's disease are caused by errors in nuclear DNA and mitochondrial DNA, respectively. Can mitochondrial replacement therapy be used to prevent one or both of these diseases in children born to an affected mother? Explain.

2. As noted in the article, some people are opposed to this technology, believing it to be an attempt to "create super babies." What is your stance on this issue? Give reasons. What is the best way for societies to resolve differences in opinion on such matters? Should those who weigh in be expected to understand the scientific bases of these techniques?

CHAPTER REVIEW

Summary

3.1 Cells: The Smallest Units of Life

- The cell is the basic unit of all living organisms.
- Most cells are small because the ratio of surface area to volume limits cell size. As a cell's width increases, its volume increases vastly more than its surface area, so a larger cell has proportionately less plasma membrane area to import and export substances but must support a much larger cytoplasmic volume.
- Broadly speaking, larger organisms are more effective predators, less susceptible prey, and better able to obtain and store nutrients.
- A multicellular organism is a closely integrated group of cells, with a common developmental origin, whose constituent cells are incapable of living independently.
- Multicellularity enabled organisms to attain larger size, and it conferred the added advantage of greater efficiency through division of labor among the multiple cell types.

3.2 The Plasma Membrane

- Every cell is surrounded by a plasma membrane that separates the chemical reactions of life from the surrounding environment.
- According to the fluid mosaic model, the plasma membrane is a highly mobile assemblage of lipids and proteins, many of which can move within the plane of the membrane.
- Proteins in the plasma membrane perform a variety of functions. Receptor proteins facilitate communication, transport proteins mediate the movement of substances across the membrane, and adhesion proteins help cells attach to one another.

3.3 Prokaryotic and Eukaryotic Cells

- Living organisms are classified as either prokaryotes or eukaryotes.
- Prokaryotes are single-celled organisms lacking a nucleus and complex internal compartments. Eukaryotes may be single-celled or multicellular, and their cells typically possess many membrane-enclosed compartments, such as the nucleus.
- The cytoplasm is all the cell contents enclosed by the plasma membrane. It consists of an aqueous cytosol (a thick fluid that contains many ions and molecules) and organelles (internal structures with unique functions).
- By volume, eukaryotic cells can be a thousand times larger than prokaryotic cells. They require internal compartments that concentrate and organize cellular chemical reactions for optimal function.

3.4 Internal Compartments of Eukaryotic Cells

- The nucleus contains DNA. It is bounded by the nuclear envelope, which has many pores. Information stored in DNA is conveyed by RNA molecules to the cytoplasm.
- Lipids are made in the smooth endoplasmic reticulum (ER). Some proteins are manufactured in the rough ER.
- Molecules move among organelles via vesicles that bud off one compartment to fuse with a target membrane.
- The Golgi apparatus receives proteins and lipids, sorts them, and directs them to their final destinations.
- Lysosomes break down large organic molecules such as proteins into simpler compounds that can be used by the cell. Vacuoles are similar to lysosomes but also store ions and molecules and lend physical support to plant cells.
- Mitochondria produce chemical energy for eukaryotic cells in the form of ATP.
- Chloroplasts harness the energy of sunlight to make sugars through photosynthesis.

3.5 The Cytoskeleton

- Eukaryotic cells depend on the cytoskeleton for structural support, and for the ability to move and change shape.
- The cytoskeleton consists of three types of filaments: microtubules, intermediate filaments, and microfilaments. Microtubules position organelles and can move them inside the cell. Microfilaments give shape to the cell and enable cell crawling. Intermediate filaments provide mechanical strength to cells.
- Some protists, sperm cells, archaeans, and bacteria move using cilia or flagella. Eukaryotic flagella are different in structure and action from prokaryotic flagella.

Key Terms

cell (p. 62)
cell theory (p. 63)
cellular respiration (p. 74)
chlorophyll (p. 75)
chloroplast (p. 75)
chromosome (p. 70)
cilium (p. 78)
crista (p. 73)
cytoplasm (p. 63)
cytoskeleton (p. 76)
cytosol (p. 63)
endoplasmic reticulum (ER) (p. 71)
endosymbiosis theory (p. 80)
eukaryote (p. 68)
eukaryotic flagellum (p. 78)
extracellular matrix (ECM) (p. 67)
flagellum (p. 78)
fluid mosaic model (p. 68)
Golgi apparatus (p. 72)
intermediate filament (p. 76)
intermembrane space (p. 73)
lumen (p. 71)
lysosome (p. 72)
matrix (p. 73)
microfilament (p. 76)
microtubule (p. 76)
mitochondrion (p. 63)
multicellular organism (p. 66)
nuclear envelope (p. 70)
nuclear pore (p. 70)
nucleus (p. 63)
organelle (p. 63)
photosynthesis (p. 75)
plasma membrane (p. 63)
prokaryote (p. 68)
prokaryotic flagellum (p. 78)
ribosome (p. 63)
rough ER (p. 71)
smooth ER (p. 71)
thylakoid (p. 75)
transport vesicle (p. 71)
tubulin (p. 76)
vacuole (p. 73)

Self-Quiz

1. In contrast to the average prokaryotic cells, eukaryotic cells
   a. have no nucleus.
   b. have many different types of internal compartments.
   c. have ribosomes in their plasma membranes.
   d. lack a plasma membrane.
2. Which of the following would be found in a plasma membrane?
   a. proteins
   b. DNA
   c. mitochondria
   d. endoplasmic reticulum
3. Which of the following organelles has ribosomes attached to it?
   a. Golgi apparatus
   b. smooth endoplasmic reticulum
   c. rough endoplasmic reticulum
   d. microtubule
4. Which organelle captures energy from sunlight?
   a. mitochondrion
   b. cell nucleus
   c. Golgi apparatus
   d. chloroplast
5. Which organelle uses oxygen to extract energy from sugars?
   a. chloroplast
   b. mitochondrion
   c. nucleus
   d. plasma membrane
6. Which organelle contains both thylakoids and cristae?
   a. chloroplast
   b. mitochondrion
   c. nucleus
   d. none of the above
7. The internal system of protein cables and cylinders that makes whole-cell movement possible is called the
   a. endoplasmic reticulum.
   b. cytoskeleton.
   c. lysosomal system.
   d. mitochondrial matrix.
8. Which of the following is *not* part of the cytoskeleton?
   a. pseudopodium
   b. intermediate filament
   c. microtubule
   d. microfilament
9. How is a prokaryotic flagellum different from a eukaryotic flagellum?
   a. It moves in a whiplike manner.
   b. It is not covered by plasma membrane.
   c. It evolved from eukaryotic flagella.
   d. It is composed of many cilia.
10. Which of the following organelles are thought to have arisen from primitive prokaryotes?
    a. endoplasmic reticulum and nucleus
    b. Golgi apparatus and lysosomes
    c. chloroplasts and mitochondria
    d. vacuoles and transport vesicles

Analysis and Application

1. What features are common to all cells, and what is the function of each?
2. Describe the major components of the plasma membrane, and explain why we say that the membrane has a fluid mosaic nature.
3. Compare mitochondria and chloroplasts in terms of their occurrence (in what types of cells, in what types of organisms), structure, and function.
4. Living cells in your skin make a protein called elastin and secrete it into the extracellular matrix (ECM). The protein contributes to the elasticity, or springiness, of skin. Is elastin manufactured by the smooth ER or the rough ER? Describe the journey of an elastin molecule from the ER to the outside of the cell, mentioning all the organelles it passes through.
5. What are the possible adaptive benefits of multicellularity?
6. List the advantages of membrane-enclosed internal compartments.
7. The image at the right is a transmission electron micrograph of a plant cell. Label the following structures on this image: cell wall, nucleus, vacuole, chloroplast. Explain the main function(s) of each structure.

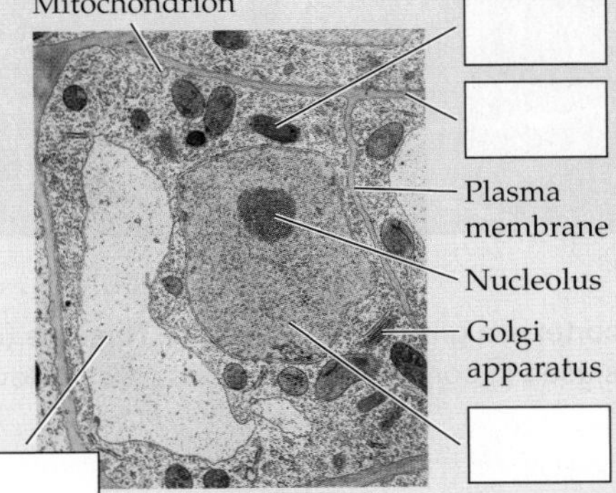

8. Complete the table below by listing the main functions of the cellular structures specified. In the column on the right, state whether the organelle is found in *most* prokaryotic, eukaryotic, plant, or animal cells.

CELLULAR STRUCTURE	MAIN FUNCTION(S)	FOUND IN: PROKARYOTES? EUKARYOTES? PLANTS? ANIMALS?
Plasma membrane		
Cytoplasm		
Nucleus		
Ribosome		
Endoplasmic reticulum		
Golgi apparatus		
Lysosome		
Mitochondrion		
Chloroplast		
Cytoskeleton		

4 Cell Membranes, Transport, and Communication

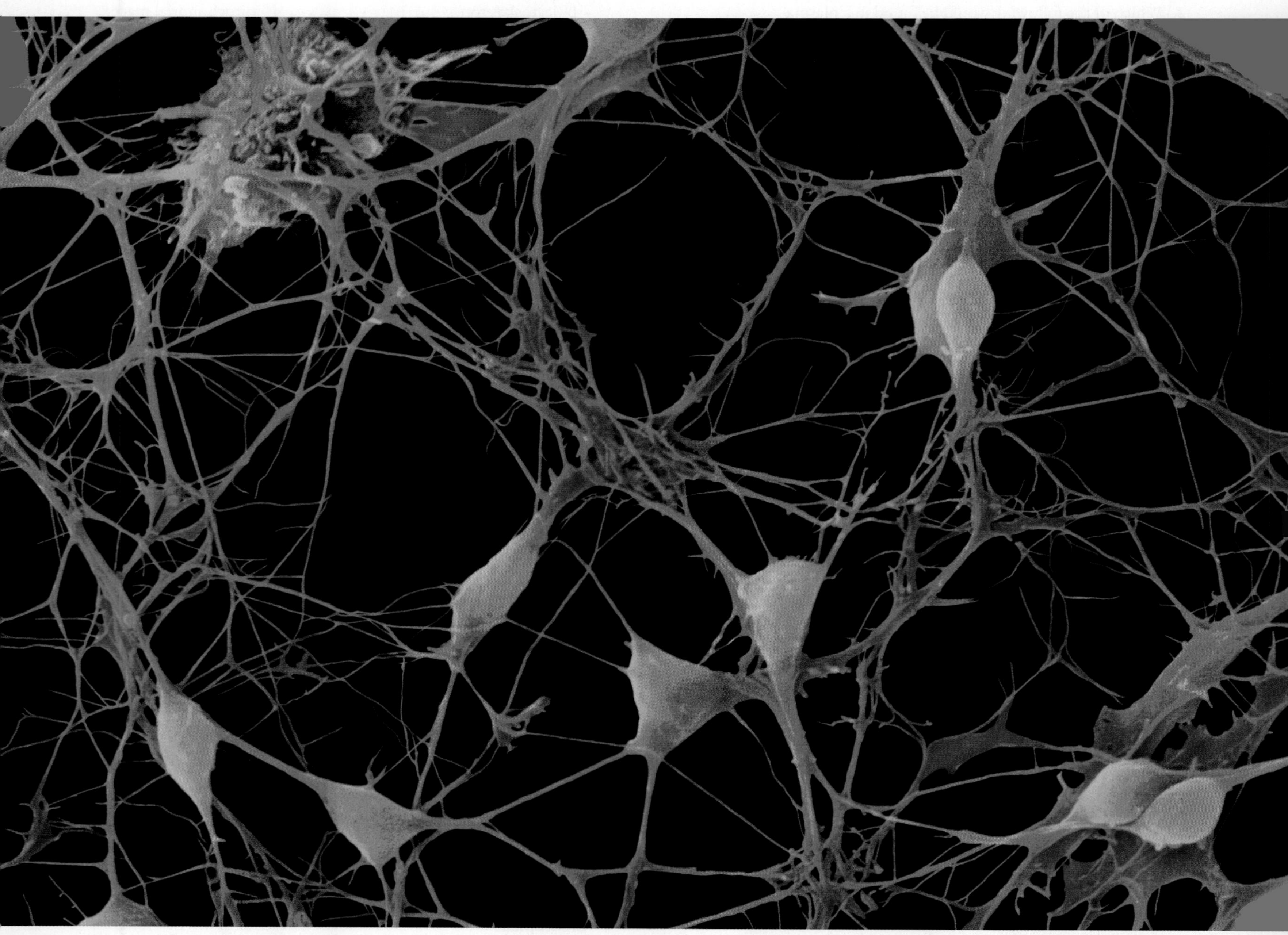

BRAIN CELLS. These cells from the brain cortex are growing in a lab dish. Neurons (orange) have large cell bodies and many projections used in communication. Glial cells (yellow) support and protect the neurons.

Mysterious Memory Loss

Duane Graveline, a retired astronaut and army flight surgeon, returned home one day from his regular morning walk but then stopped in his driveway, feeling lost. His wife soon found him wandering in front of their house, but he seemed to have no idea who she was and, suspicious of her, he refused to come inside their house or get in his own car for a trip to the doctor. Eventually his wife got him to a doctor, but it was 6 hours after she found him before he recognized her (and his doctor) and seemed back to normal. Yet Graveline had no memory of what had happened to him. The doctor diagnosed "transient global amnesia"—meaning he temporarily forgot everything. While accurate, it wasn't a very helpful diagnosis. What had caused his amnesia?

Graveline suspected it might have been a drug he was taking. A few weeks earlier, he had reported to NASA's Johnson Space Center for his annual physical. The doctor there had told him that his cholesterol levels were on the high side and prescribed a "statin" drug. Cholesterol is a normal component of our cell membranes. But high cholesterol levels in the blood are associated with an increased risk of heart disease, the most common cause of death in older people. Lowering cholesterol with statins is standard medical practice, and Graveline, 68 years old and himself a doctor and medical researcher, accepted the statin prescription without question.

Why was the astronaut losing his memory? Can *low* cholesterol be a problem? Does cholesterol do anything *good* for us, or is it all bad?

After his bout of amnesia, Graveline became suspicious and stopped taking the statin drug. A year went by with no more amnesia. When he returned for his next physical, the doctor dismissed Graveline's suspicion that the statin might have caused his amnesia and persuaded him to start taking it again, although at a lower dose. Six weeks later, Graveline had another bout of amnesia. For 12 hours he couldn't remember anything that had happened to him after high school. Why was this happening to him? After exploring the structure and function of biological membranes, we will return to this mystery at the end of the chapter.

MAIN MESSAGE The plasma membrane controls how a cell exchanges materials with its surroundings and how it communicates with other cells.

KEY CONCEPTS

- The movement of materials into and out of a cell across the plasma membrane is highly selective.
- In passive transport, substances move from one point to another without an input of energy. Diffusion is the movement of a substance from a region where it is at higher concentration to a region where its concentration is lower.
- In active transport, energy is required. Active transport may involve moving substances "uphill," against a concentration gradient.
- In osmosis, water molecules diffuse across a selectively permeable membrane.
- Some materials can move within a cell in transport vesicles, and into and out of a cell through endocytosis and exocytosis.
- Neighboring cells in a multicellular organism are often connected through cell junctions, which can be specialized for communication or cell-to-cell attachment.
- Cells communicate over both short and long distances through signaling molecules and membrane-localized signal receptors.
- The cellular response to a signaling molecule can be rapid (less than a second) or slow (over an hour).
- Hydrophilic signaling molecules bind to receptor proteins localized to the plasma membrane. Hydrophobic signaling molecules can cross cell membranes and bind to an intracellular receptor.

MYSTERIOUS MEMORY LOSS 85

4.1 The Plasma Membrane as Gate and Gatekeeper 86
4.2 Osmosis 89
4.3 Facilitated Membrane Transport 92
4.4 Exocytosis and Endocytosis 95
4.5 Cellular Connections 97
4.6 Cell Signaling 99

APPLYING WHAT WE LEARNED 100
Cholesterol in the Brain

MOST LIFE-SUSTAINING CHEMICAL REACTIONS cannot take place outside of cells. To maintain an internal chemistry that supports life, cells must carefully manage the traffic of substances across their membranes. All cells must be able to move materials into and out of themselves, as well as to control which materials can enter or leave at any given time. Dysfunction and death follow, if this precarious balance goes awry (**FIGURE 4.1**).

In this chapter we consider how cells manage their relationship with the surroundings. We begin by examining the role of the plasma membrane as both gate and gatekeeper for substances entering and leaving the cell. Then we consider the movement of water, and why water management is a matter of life or death for any cell. We show how cell membranes serve as manufacturing and packaging centers and also as "luggage" transporters. We discuss the ways cells are physically connected to one another, and how these connections facilitate communication between cells. We conclude with a look at the role of signaling molecules in cell communication.

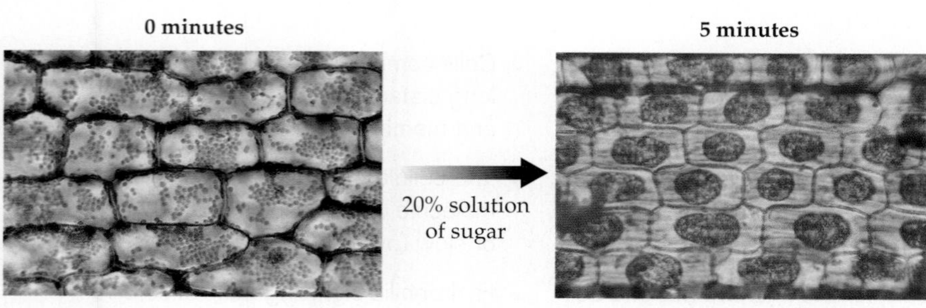

FIGURE 4.1 The Plasma Membrane Manages a Cell's Relationship with the External World

Elodea is a common aquarium plant with leaves that are just a few cells thick. If you place the leaves in an overly syrupy solution instead of the water, you will see the plasma membrane pull away from the cell walls and the cytoplasm shrink. If you leave the plants in such a solution for a good while, they will die. Why?

4.1 The Plasma Membrane as Gate and Gatekeeper

We noted in Chapter 3 that the plasma membrane separates the inside of a cell from the environment outside. The plasma membrane is as universal a feature of life on Earth as the DNA-based genetic code. Although some biologically important materials can pass directly through the plasma membrane's phospholipid bilayer, most cannot. Embedded in the phospholipid bilayer are many different types of proteins (see Figure 3.7), which collectively make up more than half the weight of the typical plasma membrane. Some of the membrane-spanning proteins, known as **transport proteins**, provide pathways by which materials can enter or leave cells.

As noted in Chapter 3, the plasma membrane must allow some molecules and ions to pass through; that is, the membrane is permeable. But biological membranes must be "choosy" about which substances can go in or out at any point in time. This **selective permeability** means that some substances can cross the membrane at any and all times, others are excluded at all times, and yet others pass through the membrane aided by transport proteins when needed by the cell. For example, small molecules like oxygen gas can move across a biological membrane at any time. Larger substances and ions can cross a membrane only with the aid of transport proteins.

Because of the plasma membrane's selectivity, the environment of the cell interior is chemically very different from the surroundings outside. Even in a multicellular organism, whose cells are continually bathed in some sort of fluid, the chemical composition of the cytoplasm is distinctly different from that of the extracellular environment (**FIGURE 4.2**).

Our cells, for example, are bathed in a fluid, such as blood, that has high concentrations of sodium and calcium ions. In contrast, the cytosol (the fluid part of cytoplasm) has very low concentrations of these ions and higher concentrations of potassium ions than are found in the extracellular fluid (as shown in Figure 4.2). A large portion of the metabolic energy spent by a cell goes toward maintaining the very special chemistry of the cell interior. The loss of selective membrane permeability is one of the surest signs of cell death.

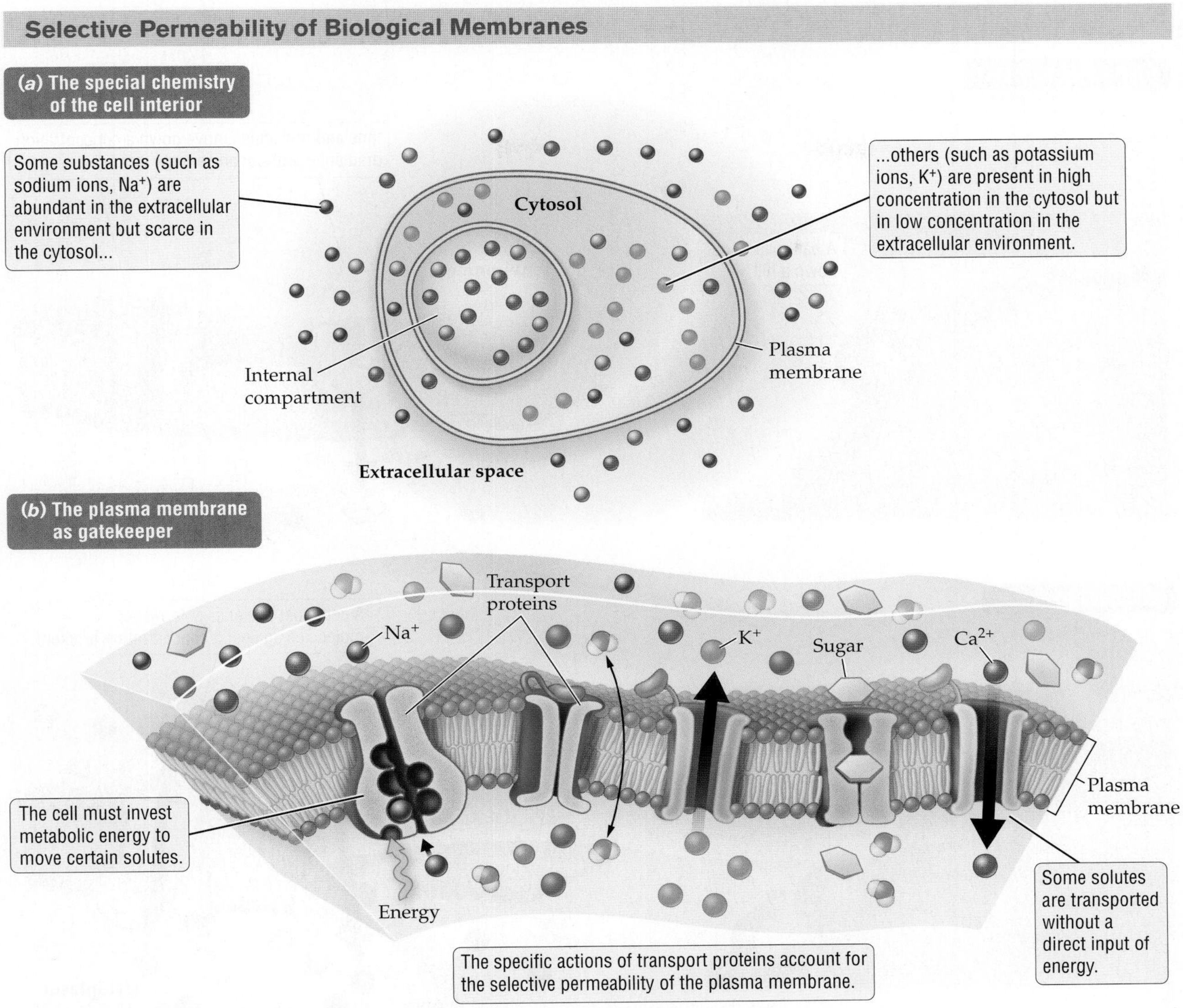

Selective Permeability of Biological Membranes
(a) The special chemistry of the cell interior
Some substances (such as sodium ions, Na+) are abundant in the extracellular environment but scarce in the cytosol...
...others (such as potassium ions, K+) are present in high concentration in the cytosol but in low concentration in the extracellular environment.
Cytosol
Internal compartment
Plasma membrane
Extracellular space
(b) The plasma membrane as gatekeeper
Transport proteins
Na+
K+
Sugar
Ca2+
Plasma membrane
The cell must invest metabolic energy to move certain solutes.
Energy
The specific actions of transport proteins account for the selective permeability of the plasma membrane.
Some solutes are transported without a direct input of energy.

FIGURE 4.2 The Plasma Membrane Is a Barrier and a Gatekeeper
(*a*) The chemistry of the cytosol is distinctly different from that of the extracellular environment, in part because the plasma membrane moves substances in a highly selective fashion.
(*b*) The selectivity of biological membranes is determined in large part by the types of membrane proteins in their phospholipid bilayer.

In diffusion, substances move passively down a concentration gradient

What drives the transport of any substance from one point to another? Two general rules can help us understand the movement of substances anywhere in our universe.

1. **Passive transport** is the spontaneous movement of a substance and can take place without any input of energy.
2. **Active transport** is the movement of a substance in response to an input of energy.

Passive and active transport are often described using the physical example of a ball moving down or up a hill, in which the ball represents a chemical substance (**FIGURE 4.3**). The ball rolls downhill on its own (passive transport), but it cannot roll uphill unless it is actively pushed, which requires energy (active transport).

Diffusion is the passive transport of a substance from a region where it is more concentrated to a region where it is less concentrated. A difference

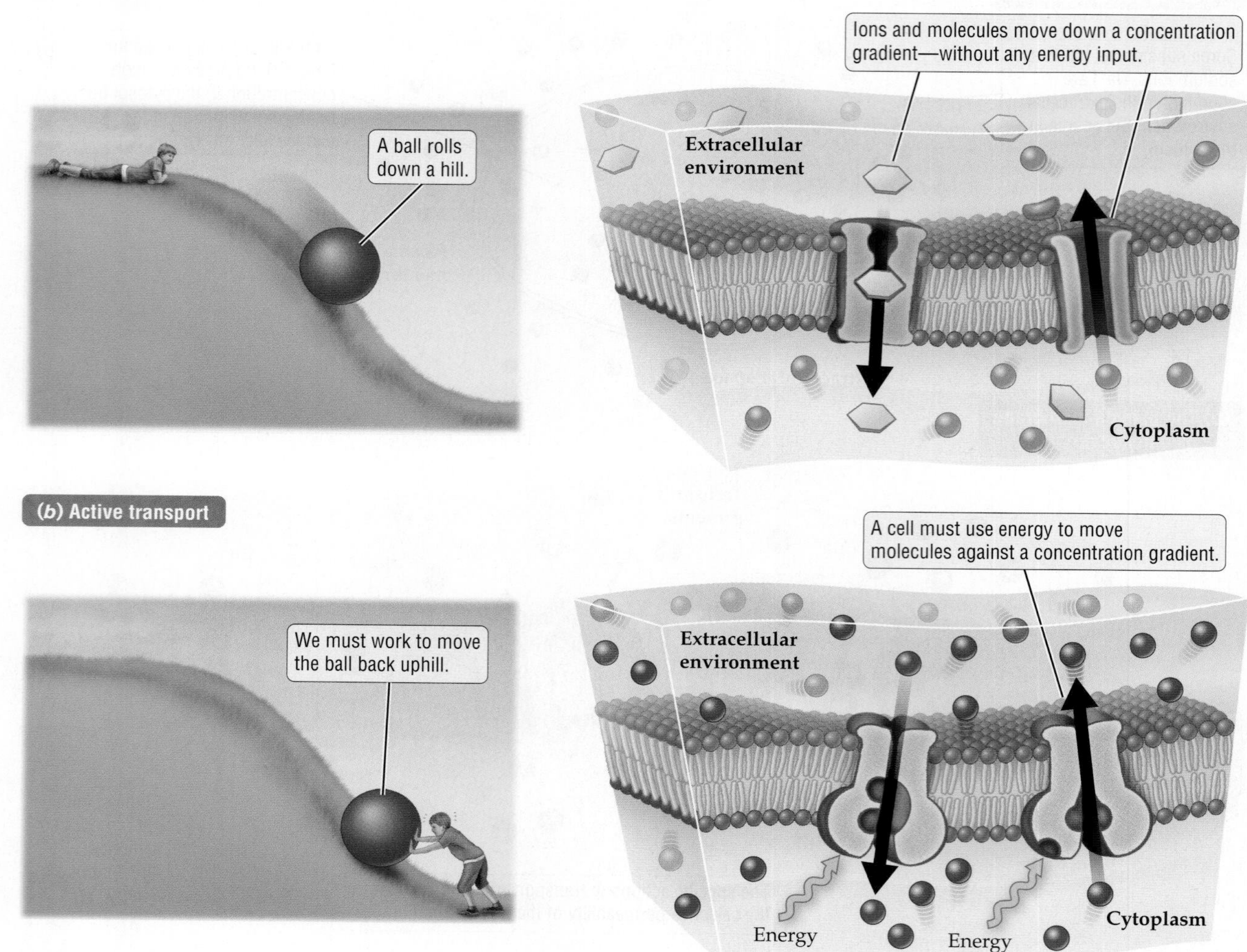
Types of Facilitated Transport
(a) Passive transport
A ball rolls down a hill.
Ions and molecules move down a concentration gradient—without any energy input.
Extracellular environment
Cytoplasm
(b) Active transport
We must work to move the ball back uphill.
A cell must use energy to move molecules against a concentration gradient.
Extracellular environment
Energy
Energy
Cytoplasm

FIGURE 4.3 Active versus Passive Movement of Substances
Materials can move into and out of organisms either passively (without an input of energy) or actively (with an input of energy). (*a*) Substances can move passively through a membrane from a region where they are present at high concentration to a region of low concentration. (*b*) Energy is required to move substances from regions of low concentration to regions of high concentration.

in concentration between two regions is termed a **concentration gradient**. Molecules distributed unevenly in a solution naturally move from higher to lower concentration, which is called moving *down* the gradient.

Imagine emptying a packet of drink mix, with food coloring, into a pitcher of water. Because the ions and molecules in any fluid are in constant motion, ingredients in the drink mix immediately begin to blend with the water, even without stirring. You can watch the food coloring move down its concentration gradient: it spreads from the area of high concentration—the spot where the mix was poured in—into the surrounding water, where the food coloring is not abundant (**FIGURE 4.4**).

Once the food coloring is distributed evenly, we say that *equilibrium* has been reached; the concentration difference that drives the diffusion has disappeared. Although the chemicals in the drink mix will continue to move about in the water, the average movement is equal in all directions.

Small substances—whether atoms or molecules or tiny particles of food dye—diffuse faster than large ones. The rate, or speed, of diffusion increases with increasing temperature because any substance has more energy, and therefore moves faster, at warm temperatures than at cooler temperatures. The food coloring in the drink mix will diffuse into warm water faster than it will diffuse into cold water, for example. The steeper the concentration gradient between two points—that is, the greater the difference in concentration between those two points—the faster the substance diffuses.

You will see throughout this chapter that the same passive process, diffusion, plays a key role in the transfer of some molecules, such as water, oxygen, and carbon dioxide, into and out of cells. However, the larger molecules that the cell needs are present in fairly low concentrations in a cell's surroundings, but are found at relatively high concentrations within the cell. The continued uptake of these substances requires energy—that is, active transport—because they must be brought in *against* a concentration gradient. Without active transport, or the energy to fuel it, no organism would survive very long.

Some small molecules can diffuse through the phospholipid bilayer

Materials that readily cross a phospholipid bilayer on their own do so through simple diffusion. **Simple diffusion** is the passive movement of a substance across a membrane *without* the assistance of any membrane components. Oxygen and carbon dioxide enter and leave cells by simple diffusion. These small, uncharged molecules slip through the hydrophobic phospholipid bilayer without hindrance.

Most hydrophobic molecules, even fairly large ones, can pass through cell membranes because they mix readily in the hydrophobic core of the phospholipid bilayer. Many early pesticides, such as DDT, were effective in killing insects precisely because they could easily get into cells this way. Unfortunately, DDT accumulates in animal fat for the same reason, and it is toxic to more than just insects.

Although some small substances can slip through, the phospholipid bilayer is generally an effective barrier to large molecules and ions. Sugars, amino acids, and nucleotides are examples of molecules that are too big to cross a membrane via simple diffusion. Substances with electrical charge—such as sodium and potassium ions—are hydrophilic and are therefore repelled by the hydrophobic interior of the phospholipid bilayer. These substances cannot cross a biological membrane without the help of transport proteins.

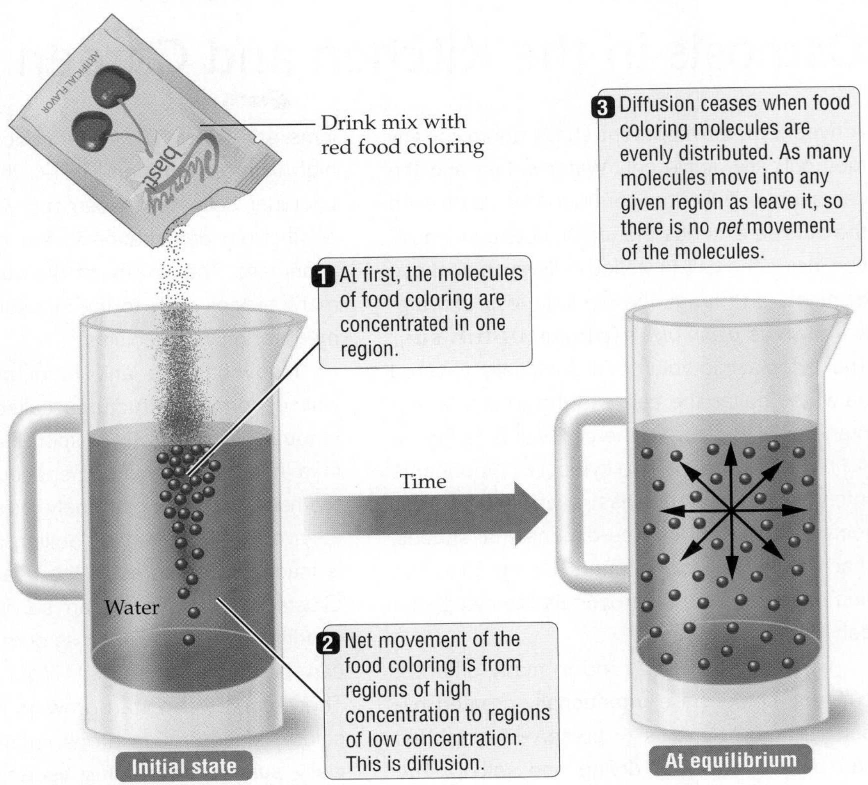

Diffusion
Drink mix with red food coloring
1 At first, the molecules of food coloring are concentrated in one region.
Water
Time
2 Net movement of the food coloring is from regions of high concentration to regions of low concentration. This is diffusion.
3 Diffusion ceases when food coloring molecules are evenly distributed. As many molecules move into any given region as leave it, so there is no *net* movement of the molecules.
Initial state
At equilibrium

FIGURE 4.4 Diffusion Is a Passive Process
Diffusion is the spontaneous movement of a substance (such as the food coloring in a drink mix) from a region of high concentration to a region of low concentration without an input of energy. Equilibrium is reached when the substance becomes uniformly distributed.

Concept Check

1. What is meant by the selective permeability of biological membranes?
2. How is passive transport different from active transport by biological membranes?

Concept Check Answers

1. Biological membranes allow some substances to enter or leave the cell, but block other substances.
2. Passive transport occurs spontaneously and moves substances down their concentration gradient without the input of energy. Active transport, which moves substances against their concentration gradient, requires the cell to expend energy.

4.2 Osmosis

Water is the medium of life: most cells are about 70 percent water, and nearly all cellular processes take place in an aqueous environment. So it will come as no surprise that maintaining a proper water balance is vital for every cell. How do cells take up water, and how do they deal with an excess or lack of water?

Water molecules are so small, compared to the phospholipids and other hydrophobic components of the plasma membrane, that some can slip across

BIOLOGYMATTERS

Osmosis in the Kitchen and Garden

A hypertonic environment spells doom for any metabolically active cell. Water is lost and the cell shrinks if the concentration of solutes on the outside exceeds the total solute concentration inside the cell. In walled cells, such as those of plants and fungi, the osmotic loss of water is known as *plasmolysis* (**plaz-*MAH*-luh-sus**). The cytoplasmic volume is drastically lowered as water leaves the cell, and the plasma membrane pulls away from the cell wall (see Figure 4.1). The crowding of the cytosolic components into a concentrated mass is often fatal. Proteins lose their vital three-dimensional shapes, macromolecular components clump together, and organelles are destroyed as the cytoplasm dehydrates.

For many millennia, and in many different cultures, cooks have intentionally created hypertonic environments to preserve food. Meat can be preserved by drying and salting. The salt hastens the drying by drawing out the water through osmosis. Salt also discourages the growth of bacteria and fungi: most bacterial cells or fungal strands that begin to grow on the salty surface will quickly wither from plasmolysis. Plasmolysis is also the culinary secret to producing many pickles, preserves, chutneys, jams, and jellies. The sugar concentration is so high in most jams and jellies that fungal and bacterial cells suffer osmotic water loss and death from dehydration in the hypertonic environment. The acidity of the vinegar used in many pickles adds to the inhospitable environment for microorganisms.

Many bacteria and fungi produce thick-walled dormant structures called *spores*. The sprouting and growth of spores are prevented in a hypertonic environment, but the spores themselves can be extremely resistant and may survive if sterilization by boiling and steaming is inadequate. Spores of the deadly bacterium *Clostridium botulinum* can survive hypertonic conditions in a food such as pure honey, which can be safely stored at room temperature. These spores can start growing if the external solute concentration drops. Infants are especially susceptible to this bacterium, which is why experts say honey should not be given to children younger than 1 year.

High concentrations of fertilizer can kill a plant through irreversible plasmolysis. Plant food—or *fertilizer*, to use the more accurate term—is usually sold as a concentrated powder or liquid that must be sufficiently diluted before it is applied to plant roots or foliage. Plants suffering from "fertilizer burn" have a wilted look because the hypertonic environment removes water from the plant by osmosis. Plants that grow in brackish water—sea grasses and mangrove trees, for example—cope with the saltiness by increasing the solute concentration of their cytosol. The road salt used for deicing pavement in winter is usually not concentrated enough to plasmolyze plant roots, but upon washing into lakes and ponds it can damage the delicately balanced osmotic equilibrium of aquatic species, especially wall-less protists and the eggs and larvae of animals.

Helpful to Know

The term "net movement" refers to the fact that molecules are always moving in all directions because of random motion. Where a gradient exists, more molecules move in one direction than in the other. For example, if 4,000 molecules are moving down a gradient, perhaps 500 molecules are moving in the opposite direction just by chance. The difference between the two, 4,000 − 500 = 3,500 molecules, is the net movement.

the plasma membrane, like a small child weaving among the adults in a crowded room. But when a large inflow of water is needed, cells can activate special tunnel-like protein complexes, called *aquaporins*, that allow rapid uptake of water.

Osmosis is the diffusion of water across a selectively permeable membrane. Because it is a type of diffusion, a passive process, no energy is expended when water enters or leaves a cell by osmosis. As in any type of diffusion, a concentration difference drives the net movement: water molecules tend to move from a region where they are more abundant to a region where they are less abundant. But osmosis is a special case of diffusion: it describes the net movement of water across a membrane that is permeable to water molecules but not to most solutes.

Pond-dwelling protists such as *Paramecium* take up water by osmosis because the concentration of water molecules is higher in pond water than it is inside the cell. The cytosol is about 70 percent water, and teeming with solute particles: many, many millions of ions and small organic molecules, such as sugars and amino acids, and many millions of larger macromolecules, such as proteins and nucleic acids. The cytosol therefore has a much higher concentration of solute particles than pond water has, which means a proportionately lower concentration of water molecules. The difference in concentration of water molecules is what drives the net flow of water molecules from the pond water into the *Paramecium* (**FIGURE 4.5**).

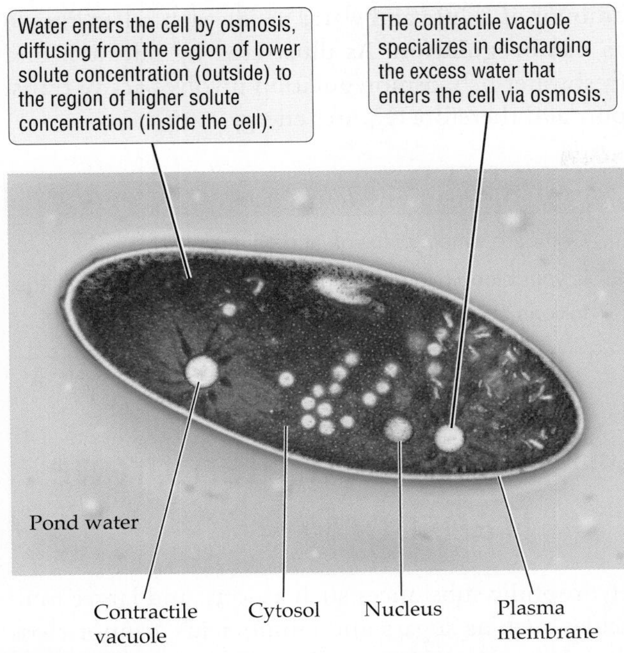

Water enters the cell by osmosis, diffusing from the region of lower solute concentration (outside) to the region of higher solute concentration (inside the cell).
The contractile vacuole specializes in discharging the excess water that enters the cell via osmosis.
Pond water
Contractile vacuole
Cytosol
Nucleus
Plasma membrane

FIGURE 4.5 Osmotic Balance in *Paramecium*
Paramecium is a wall-less, single-celled protist that ingests bacteria and other small organisms at lake bottoms.

The convention in science is to describe osmosis with reference to the concentration of *solutes*, rather than the concentration of water molecules, which are the *solvent* in biological systems. So in practice, osmosis is defined as the net movement of water molecules across a selectively permeable membrane from a region of lower solute concentration to a region of higher solute concentration.

The water content of cells is continually affected by osmosis, and too much or too little water inside a cell can be disastrous. Cells can find themselves in an external environment that has too low a solute concentration, has too high a solute concentration, or is just right (**FIGURE 4.6**).

- A **hypotonic solution** is an external medium that has a lower solute concentration than the cytosol of the cell has, so more water flows into the cell than out of it; unchecked, this movement can cause wall-less cells to burst.

Helpful to Know

The prefixes *hyper* ("more, too many, too much"), *hypo* ("less, fewer, not enough"), and *iso* ("equal") occur in a number of scientific terms. Here, for example, a "hypertonic" environment is one in which the surroundings have a higher concentration of solutes than the cytosol of the cell. Notice that these are all terms describing the relationship of one solution to another.

Osmosis

	Isotonic solution: Cells neither gain nor lose water	Hypertonic solution: Cells lose water	Hypotonic solution: Cells gain water
	Total solute concentration outside **equals** total solute concentration inside.	Total solute concentration outside **exceeds** total solute concentration inside.	Total solute concentration outside is **lower than** total solute concentration inside.
Animal cells	H_2O / H_2O	H_2O / H_2O	H_2O / H_2O
Plant cells	H_2O / H_2O	H_2O / H_2O	H_2O / H_2O / Vacuole

Wall-less cells can gain so much water in a hypotonic solution that they burst.

Walled cells, such as those of plants, are protected from bursting in a hypotonic solution because the cell wall resists expansion.

FIGURE 4.6 Water Moves into and out of Cells by Osmosis
Osmosis is the diffusion of water across a selectively permeable membrane. Cells lose water in a hypertonic solution and gain water in a hypotonic solution. Our cells are bathed in an isotonic fluid to prevent osmotic swelling and bursting.

Concept Check Answers

1. Osmosis is a special case of simple diffusion: it is the passive movement of water across a selectively permeable membrane.

2. Isotonic. Severe illness would result if our internal fluids did not match the osmotic state of our cells.

- A **hypertonic solution** is an external medium that has higher solute concentration than the cytosol has, so more water flows out of the cell than into it. This movement causes the cell to shrink.
- An **isotonic solution** is "just right" in that its solute concentration is the same as that inside the cell. In this situation, the concentration of solutes is the same on both sides of the plasma membrane, so just as much water leaves the cell as enters it, with the result that there is no net movement of water across the membrane.

Most of our cells are bathed in an isotonic solution. Human blood, for example, is isotonic with the cells in the body. The fluid part of blood—known as *plasma*—contains many ions (such as sodium and calcium), many small organic molecules (such as glucose), and a variety of plasma proteins. Wall-less protists that live in a hypotonic environment, as the pond-dwelling *Paramecium* does, must "bail out" the excess water they take in using a specialized organelle, the contractile vacuole (see Figure 4.5). Walled cells, such as plant or fungal cells, do not burst in hypotonic solutions. Like the tire around an inner tube, the cell wall resists the fluid pressure (called turgor pressure) that builds as water enters the cell. When the wall pressure equals the turgor pressure, the cell stops taking in water.

Because organisms inhabit such a wide range of different habitats, cells do not necessarily find themselves in a perfect world of isotonic solutions at all times. Some organisms that live in a hypertonic world, such as ocean-dwelling fish, have adaptations to help their cells avoid the tendency to lose water in a hypertonic environment. They actively transport salt out of their gills, while their kidneys help them retain more of the water they drink.

Freshwater fish face the opposite challenge: a tendency to absorb too much water from their hypotonic surroundings. These fish have transport proteins in the gills that enable them to actively absorb salts, while their kidneys help them excrete excess water. This constant balancing act to maintain an appropriate amount of solute and water inside each cell is known as **osmoregulation**. As illustrated by saltwater and freshwater fish, osmoregulation involves active transport and therefore requires energy.

EXTREME CONFORMITY

The salt content of the Great Salt Lake is 27 percent in many places, compared to 3.5 percent for the ocean. Yet brine shrimp have adapted to the high salinity. Instead of spending energy to osmoregulate, the tiny shrimp match their internal solute concentration to that of the lake.

Concept Check

1. Compare simple diffusion and osmosis.
2. Is your blood hypertonic, hypotonic, or isotonic for all the trillions of cells in your body?

4.3 Facilitated Membrane Transport

Hydrophilic substances such as ions, and larger molecules such as sugars and amino acids, cannot cross the plasma membrane without assistance. Even nutrients such as the simplest sugars and amino acids, which consist of no more than 30 atoms or so, are too large or too hydrophilic to diffuse through the hydrophobic core of the phospholipid bilayer. Despite being small, ions such as H^+ (hydrogen ions) or Na^+ (sodium ions) cannot get through, because they are repelled by the hydrophobic tails in the middle of the phospholipid bilayer. As a result, ions, polar molecules, and all large substances need help to cross the plasma membrane. That help is provided by membrane transport proteins.

When membrane proteins move substances across a biological membrane, the process is known as **facilitated transport**. Two types of transport proteins help substances move across the plasma membrane: *channel proteins* and *carrier proteins* (**FIGURE 4.7**). Passive transport by these membrane proteins is called **facilitated diffusion**, in contrast with simple diffusion, which is unaided by proteins. Some carrier proteins are active transporters, moving substances against a concentration gradient.

Channel proteins move substances passively

Channel proteins, or *membrane channels*, as they are also known, enable substances of the right size and charge to move passively through the plasma membrane, down a concentration gradient (**FIGURE 4.7*b***). Channel proteins form tunnels that span the thickness of the phospholipid bilayer. A channel protein has just the right width, and just the right chemistry, to enable the transit of a specific substance. A calcium channel allows the transmembrane flow of calcium ions, for example, but it does not allow passage of potassium ions. No direct input of energy is needed.

Aquaporins, which facilitate the rapid transmembrane flow of water, are an example of channel proteins. Most channel proteins, however, specialize in moving some type of ion in or out of the cell.

Cystic fibrosis is a genetic disorder in which salt accumulates on the skin and on the lining of the lungs and other organs. A channel protein called CFTR (for cystic fibrosis transmembrane conductance regulator) specializes in the facilitated diffusion of chloride ions (Cl^-) across the plasma membrane of cells that line the lungs, small intestines, sweat glands, and other organs. The CFTR channel protein is defective in people with cystic fibrosis, which affects one in 2,500 people of northern European descent. The defective chloride channel is responsible for the salt accumulation and other effects of the disease.

Transmembrane Transport

Type of transport:	Passive Transport		Active Transport
Mechanism:	**(*a*) Simple diffusion** Molecules slip between phospholipids.	**(*b*) Facilitated diffusion** Diffusion is facilitated by channel proteins or carrier proteins.	**(*c*) Energy input required** Active transport is facilitated by carrier proteins.
Outside Hydrophilic heads Hydrophobic tails Cytoplasm	Oxygen Carbon dioxide	Channel protein Carrier protein	Energy
Types of molecules that typically cross the membrane:	Small molecules such as oxygen and carbon dioxide	**Channel proteins:** ions, water **Carrier proteins:** ions, various polar or charged molecules	**Active carrier proteins:** ions, various polar or charged molecules

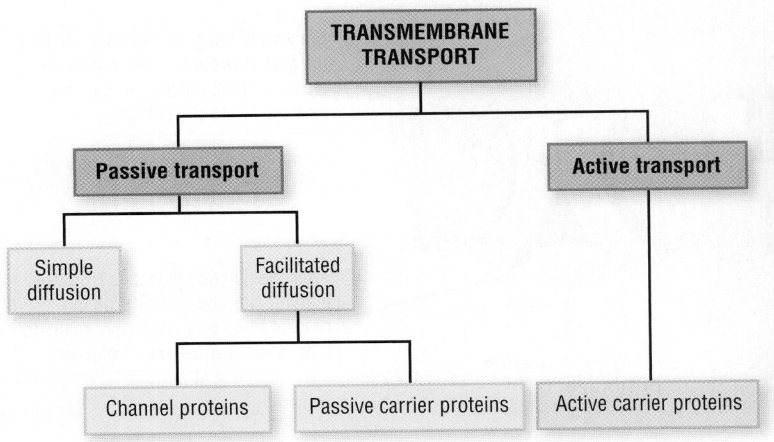

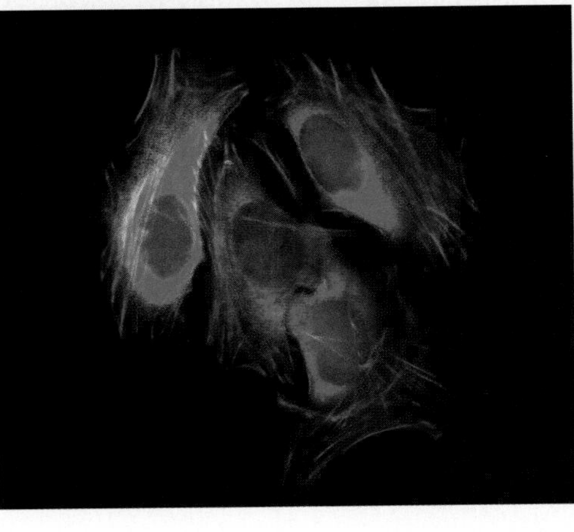

FIGURE 4.7 The Plasma Membrane Controls What Enters and Leaves the Cell

Proteins that span the plasma membrane (*b,c*) play an important role in moving materials into and out of cells. The flowchart summarizes the different types of processes that move materials across biological membranes. The photo shows the localization of glucose transporters in the plasma membrane of human cells cultured in a lab dish. These passive carrier proteins have been tagged with a green fluorescent dye. The nucleus is stained blue, and the cytoskeletal protein actin has been tagged with a red fluorescent dye.

Carrier proteins bind to molecules to help them cross the membrane

Carrier proteins function more like a revolving door than an open tunnel (**FIGURE 4.7c**). A carrier protein recognizes, binds, and transports a specific target, such as a particular ion or a certain type of sugar. The selectivity comes from the fact that only the target ion or molecule can fit into the folds on the surface of a specific carrier protein. When the cargo binds to the carrier protein, the protein changes shape in such a way that the cargo is now exposed on the other side of the membrane. The shape change also decreases the carrier protein's affinity ("clinginess") for the cargo, causing the ion or molecule to be let go. In this way, the molecule is picked up on one side of the membrane and discharged on the other side. Carrier proteins transport a great variety of substances across biological membranes—ions, amino acids, sugars, and nucleotides, for example. Carrier proteins are of two types:

- *Passive carrier proteins* move substances down a concentration gradient and therefore do not require energy.
- *Active carrier proteins* mobilize substances against a concentration gradient and cannot function without an input of energy.

Helpful to Know

Because the cell must use energy anytime molecules or ions have to be pumped "uphill," the amount of energy it spends on active transport is quite large. It may surprise you to learn that 30–40 percent of the energy used by a resting human body is spent on fueling active transport across the cells' plasma membranes!

Passive carrier proteins mediate facilitated diffusion

Passive carrier proteins assist in the diffusion of ions and molecules down their concentration gradients. Glucose carriers known as GLUT proteins are an example of passive carrier proteins (**FIGURE 4.8**). All cells in the human body need glucose for energy. The plasma membrane of every cell has glucose carrier proteins to absorb the sugar from the bloodstream. The majority of our cells can pick up glucose from the blood in a passive manner because blood normally contains about 10 times as much glucose as the cytoplasm the average cell has. Glucose can simply "roll" down its concentration gradient, aided by GLUT proteins.

Active carrier proteins move materials against a concentration gradient

As we have seen, ions and molecules can cross a plasma membrane against a concentration gradient only by active transport. **Active carrier proteins**, also known as *membrane pumps*, move molecules across the plasma membrane with the aid of ATP or another source of energy. Like passive carrier proteins, active carrier proteins bind only to certain ions or molecules: those that can fit into specific folds in the protein (**FIGURE 4.9**). In this case, however, the addition of energy brings about a shape change in the active carrier protein. This shape change forcibly releases the molecule being transferred, regardless of the concentration of that molecule near the site of release. This energy-driven mechanism enables active carrier proteins to move their cargo against the gradient—from regions of low concentration to regions of high concentration.

Passive Transport: Facilitated Diffusion of Glucose by a Carrier Protein

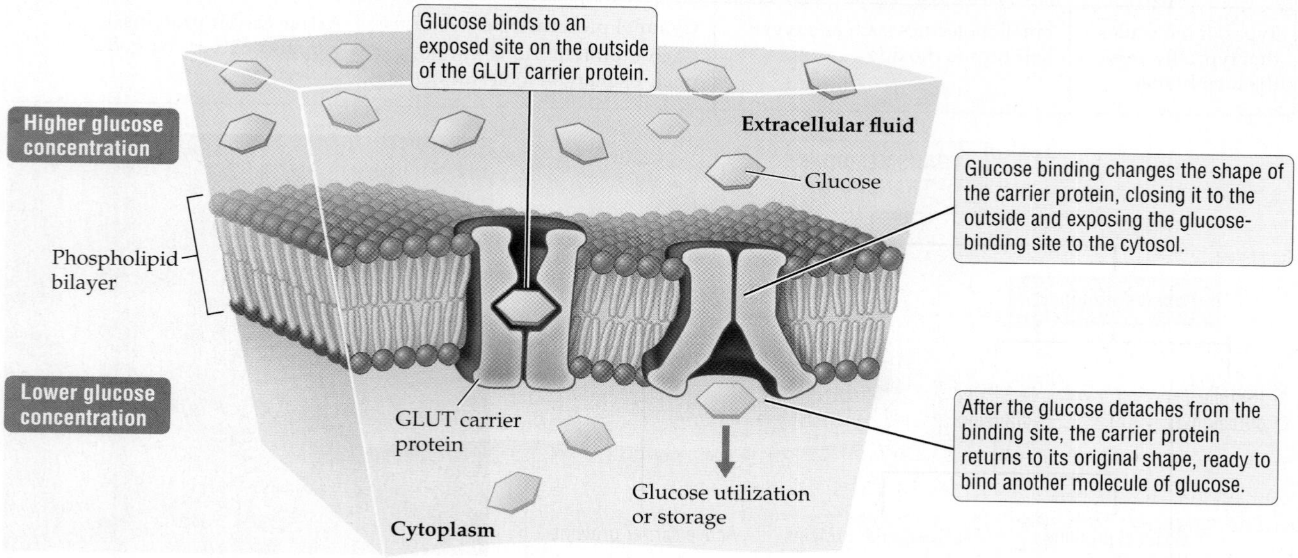

Glucose binds to an exposed site on the outside of the GLUT carrier protein.

Higher glucose concentration

Extracellular fluid

Glucose

Glucose binding changes the shape of the carrier protein, closing it to the outside and exposing the glucose-binding site to the cytosol.

Phospholipid bilayer

Lower glucose concentration

GLUT carrier protein

Glucose utilization or storage

Cytoplasm

After the glucose detaches from the binding site, the carrier protein returns to its original shape, ready to bind another molecule of glucose.

FIGURE 4.8 Passive Carrier Proteins Help Substances Diffuse across the Membrane

The facilitated diffusion of glucose into our cells is mediated by a class of passive carriers called GLUT proteins. No energy is required for the transport of glucose by carrier proteins. The sugar moves from the side of the membrane where it is at high concentration to the side where it is at lower concentration.

The *sodium-potassium pump* is one of the most important active carrier proteins in our cells. It is present in the plasma membrane of virtually all the cells in our bodies, and it is so vital that most animal cells would die quickly if the sodium-potassium pump stopped working. The sodium-potassium pump creates and maintains the large, but opposite, concentration gradients of sodium and potassium ions across the plasma membrane in most animal cells. Blood and other body fluids have high concentrations of sodium ions (Na^+) but low concentrations of potassium ions (K^+). Within our cells, the situation is reversed: Na^+ is scarce in the cytoplasm but K^+ is plentiful. The sodium-potassium pump maintains these concentration differences by exporting sodium ions from the cell while importing potassium ions. It picks up Na^+ ions from the cytoplasm and moves them "uphill" to the outside of the cell, using energy from the breakdown of ATP.

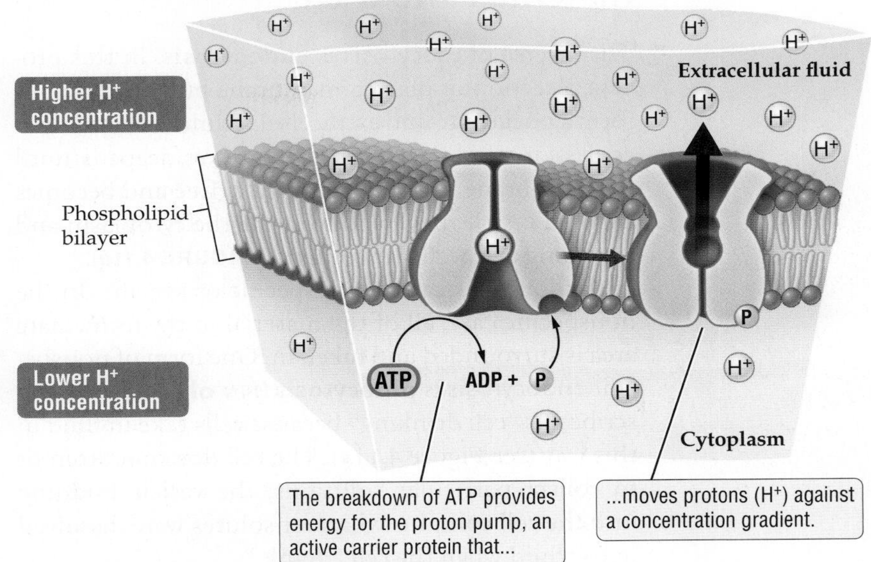

FIGURE 4.9 Active Carrier Proteins Move Substances against a Gradient

Active carrier proteins use energy to move materials from regions of low concentration to regions of high concentration. When food arrives, proton pumps secrete hydrogen ions (H^+) into your stomach. Because the cells lining the stomach have a lower concentration of H^+ than the stomach cavity has, these cells must use energy to secrete protons against the concentration gradient.

Concept Check

1. Explain why ions, such as Na^+, cannot move across a phospholipid bilayer unassisted.
2. Compare channel proteins and carrier proteins.

Concept Check Answers

1. The electrical charge on ions makes them hydrophilic, so they cannot cross the lipid bilayer, which is hydrophobic.

2. Both types of transport proteins facilitate the movement of substances across the membrane. But channel proteins allow passive movement of materials and do not need energy input. Carrier proteins transport a great variety of substances either passively or actively.

4.4 Exocytosis and Endocytosis

In Chapter 3 we saw that many molecules are transported from place to place within a cell wrapped in small membrane packages called *transport vesicles*. These "molecular ferries" can also package chemical substances for export or import across the plasma membrane.

Exocytosis moves bulk materials out of the cell

In **exocytosis** (*EX*-**oh-sye**-*TOH*-**sus**) cells release substances into their surroundings by fusing membrane-enclosed vesicles with the plasma membrane (**FIGURE 4.10**). The substance to be exported is packaged into transport vesicles by the ER-Golgi network of membranes inside the cell. As the transport vesicle approaches the plasma membrane, a portion of the vesicular membrane makes contact with the plasma membrane and fuses with it. In the process, the inside of the vesicle (the lumen) is opened to the exterior of the cell, discharging the contents.

Many of the chemical messages released into the bloodstream in humans, and many other animals, are discharged via exocytosis by the cells that produce them. For example, after we eat a sugary snack, specialized cells in the pancreas release the hormone insulin via exocytosis. Insulin moves through the bloodstream to other cells and signals them to take up the glucose released from the snack.

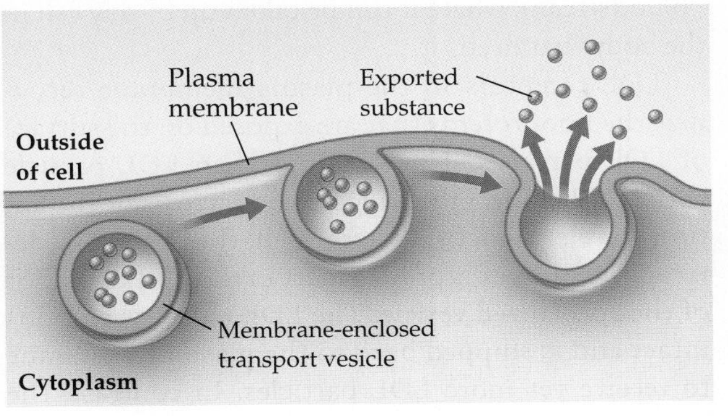

FIGURE 4.10 Cell Contents Are Exported through Exocytosis

Exocytosis exports materials from the cell. The vesicle's membrane fuses with the plasma membrane, releasing its content to the external environment.

Endocytosis brings bulk materials into the cell

The reverse of exocytosis is **endocytosis**. In this process, a section of plasma membrane bulges inward to form a pocket around extracellular fluid, selected molecules, or whole particles. The pocket deepens until the membrane it is made of breaks free and becomes a closed vesicle, now wholly inside the cytoplasm and enclosing extracellular contents (**FIGURE 4.11*a***).

Endocytosis can be nonspecific or specific. In the nonspecific case, all of the material in the immediate area is surrounded and taken in. One form of nonspecific endocytosis is **pinocytosis** (*PIN*-**oh-...**), often described as "cell drinking" because cells take in fluid in this way (see Figure 4.11*a*). The cell does not attempt to collect particular solutions; the vesicle budding into the cell contains whatever solutes were dissolved in the fluid when the cell "drank."

Endocytosis can be so specific that only one type of molecule is enveloped and imported. How does a particular section of plasma membrane "know" what to endocytose? The answer lies in the presence in the membrane of specific **receptors**, proteins that recognize and interact with particular target substances in the external environment. In **receptor-mediated endocytosis**, these receptor proteins determine which substances are incorporated into the vesicles that arise from a particular plasma membrane region (**FIGURE 4.11*b***). The receptors select the cargo by recognizing and binding to surface characteristics of the material they bring in.

Our cells use receptor-mediated endocytosis to take up cholesterol-containing packages called low-density lipoprotein (LDL) particles. The liver produces cholesterol, and because this lipid is hydrophobic, it must be packaged with proteins (called "apolipoproteins") that help it mix into an aqueous environment (**FIGURE 4.12**). Only as a part of the LDL particle can cholesterol be released into the bloodstream, where it can be taken up by any cell in the body that needs it.

LDL receptors in the plasma membrane recognize the lipoproteins that are exposed on the surface of LDL particles. The docking of an LDL particle with an LDL receptor triggers endocytosis of the entire complex. Once inside the cell, the LDL particle-receptor complex is pulled apart in the acidic interior of the specialized vesicle. The LDL receptor remains intact and is shipped back to the plasma membrane to receive yet more LDL particles. In contrast, the LDL particle is sent to the lysosome, an organelle that specializes in taking apart biomolecules. The

Helpful to Know

Exo ("out, outward") and *endo* ("in, inward") are prefixes that are common in scientific terminology. In "exocytosis" and "endocytosis" these prefixes are combined with *cyt* (the term for "cell") and the suffix *osis*, which means "action or condition." Literally, the two terms mean "outward cell action" and "inward cell action."

Endocytosis

(*a*) Mechanism of pinocytosis

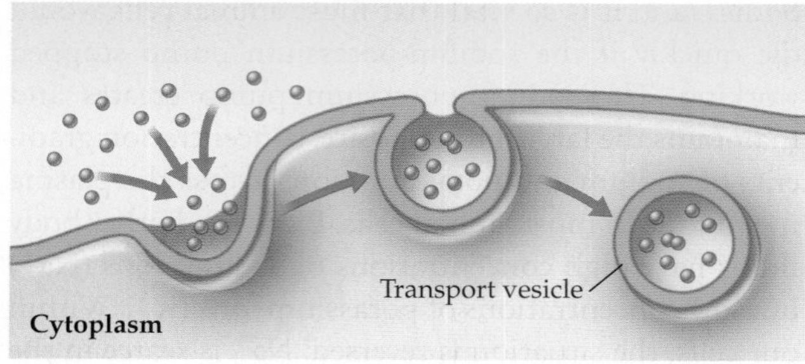

(*b*) Mechanism of receptor-mediated endocytosis

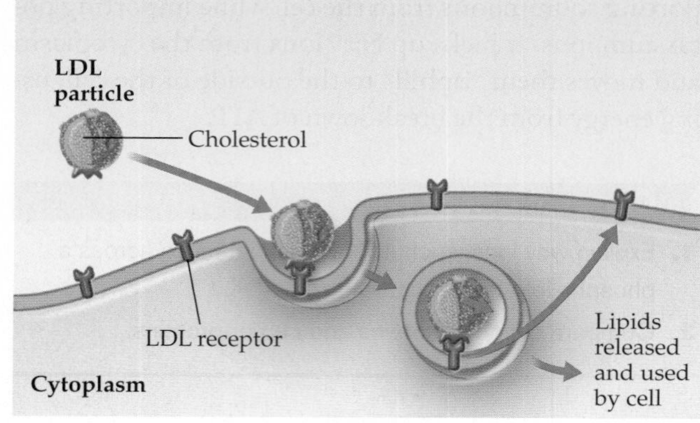

(*c*) Mechanism of phagocytosis

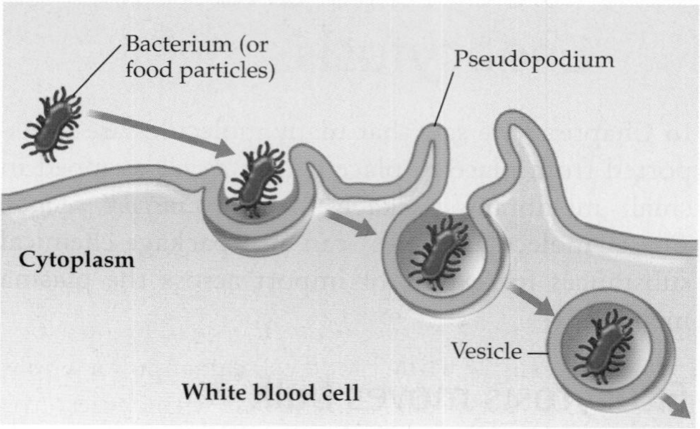

FIGURE 4.11 Extracellular Substances Are Imported through Endocytosis

Endocytosis brings material from the outside of the cell to the inside, wrapped in membrane vesicles. (*a*) Pinocytosis, a nonspecific form of endocytosis that packages external fluid into a vesicle, is illustrated here. (*b*) Receptor-mediated endocytosis is a highly selective process in which only certain extracellular molecules are recognized by, and bound to, special plasma membrane receptors. (*c*) Phagocytosis is endocytosis on a large scale.

Structure of an LDL Particle

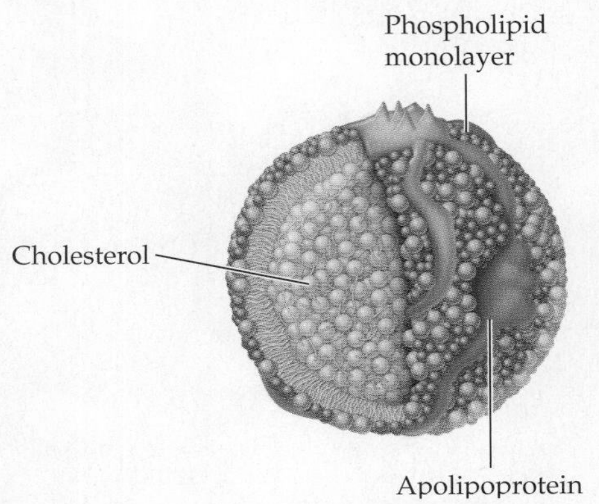

FIGURE 4.12 LDL Particles Deliver Lipids throughout the Body

An LDL particle is a complex of protein (apolipoprotein) and various lipids, including cholesterol.

cholesterol and amino acids from the apolipoprotein are released into the cytosol.

In individuals who have familial hypercholesterolemia, the LDL receptors are defective or lacking. Because the cells cannot take delivery of the LDL particles, cholesterol and other lipids build up and are deposited in the lining of the blood vessels as fatty plaques. Plaques can break loose from the blood vessel wall to form a clot that can block the flow of blood. A heart attack occurs when blood supply to the heart is blocked; a stroke results when blood flow in the brain is restricted.

Phagocytosis, or "cell eating," is a large-scale version of endocytosis in that it involves the ingestion of particles considerably larger than macromolecules, such as an entire bacterium or virus (**FIGURE 4.11c**). This remarkable process occurs in specialized cells, such as the white blood cells that defend us from infection. A single white blood cell can engulf a whole bacterium or yeast cell. As is the case with cholesterol uptake, receptors in the membrane of the white blood cell enable it to recognize and ingest harmful microorganisms.

Concept Check

1. Which process is more selective in terms of the cargo transported: pinocytosis or receptor-mediated endocytosis?
2. What is the fate of LDL particles that bind to cell surface receptors?

4.5 Cellular Connections

Multicellular organisms benefit from having a variety of cells and tissues, each of which performs a narrow range of specialized tasks. An organism that can maximize efficiency through cell specialization is better adapted to the challenges presented by its surroundings. But all the many cells in a multicellular body must be woven together properly, with appropriate means of communication among them, for the body to function efficiently.

From seaweed to walruses, multicellular organisms have at least some cells that are interconnected in particular ways. Plasma membrane structures that interconnect adjacent cells are known as **cell junctions**. Vertebrate animals possess three main types of cell junctions: *anchoring junctions*, *tight junctions*, and *gap junctions* (**FIGURE 4.13a**).

- **Anchoring junctions** (also known as *desmosomes*) are structures formed by patches of protein located for the most part on the cytoplasmic face of the plasma membrane. Extensions from each protein patch pass through the plasma membrane to "hook" to similar extensions protruding from an adjacent cell. The main function of anchoring junctions is to link cells together to brace them collectively against forces that would rupture an isolated cell. Anchoring junctions are especially abundant in tissues that experience heavy structural stress, such as heart muscle.
- **Tight junctions** are structures formed by belts of proteins that run along the plasma membrane. Adjacent cells are bound together by their belts of protein, collectively forming a leak-proof sheet of cells (**FIGURE 4.14**). Most molecules cannot pass from one side of the sheet to the other side, because tight junctions keep these substances from creeping between cells, the way grout prevents water from creeping between bathroom tiles. Tight junctions are especially common in *epithelial cells*,

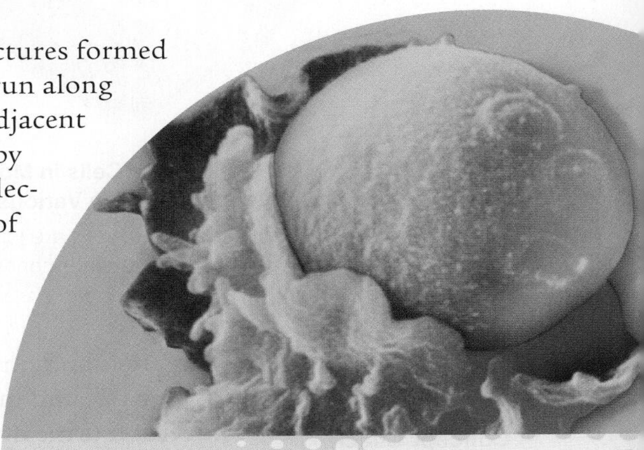

EXTREME CHUGALUGGING

A macrophage (blue) is engulfing an invading yeast cell (yellow). These specialized white blood cells defend the body from infections and foreign substances. Macrophages can engulf particles many times larger than they are—about the equivalent of a person swallowing a Thanksgiving Day turkey whole!

Concept Check Answers

1. Receptor-mediated endocytosis.
2. They are internalized via receptor-mediated endocytosis and taken apart in the lysosome.

Cellular Connections

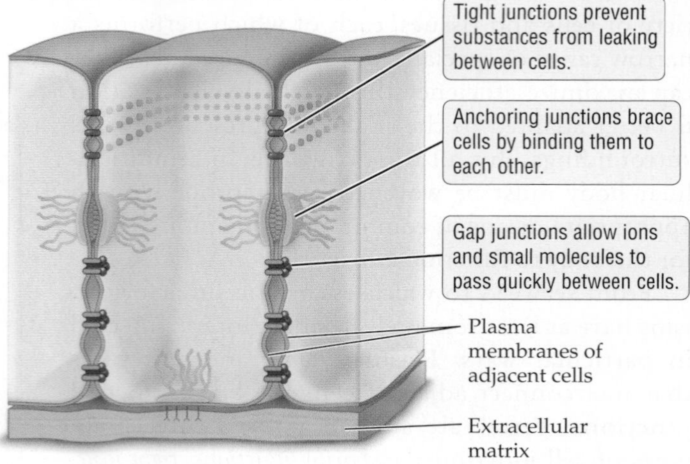

(a) Animal cells: cell junctions
Tight junctions prevent substances from leaking between cells.
Anchoring junctions brace cells by binding them to each other.
Gap junctions allow ions and small molecules to pass quickly between cells.
Plasma membranes of adjacent cells
Extracellular matrix

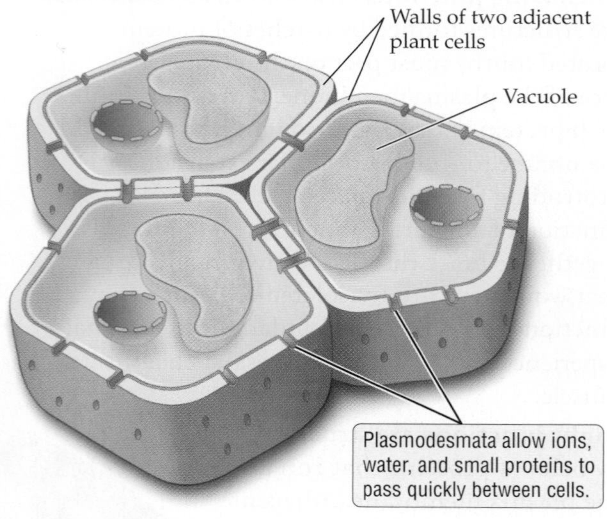

(b) Plant cells: plasmodesmata
Walls of two adjacent plant cells
Vacuole
Plasmodesmata allow ions, water, and small proteins to pass quickly between cells.

FIGURE 4.13 Cells in Multicellular Organisms Are Interconnected in Various Ways

(*a*) Many animal cells are joined by different types of junctions. (*b*) Plant cells are interconnected by plasmodesmata.

which are found on the surface of the body, in most organs, and in the lining of body cavities. The reason urine does not leak from the bladder into other body tissues is that tight junctions in the epithelial cells lining the bladder block its passage to the other side.

- **Gap junctions** are the most widespread type of cellular connection in animals. Gap junctions are direct cytoplasmic connections between two cells. They consist of protein-lined tunnels that span the small intercellular space separating adjacent cells. Gap junctions allow the rapid and direct passage of ions and small molecules, including signaling molecules. Electrical signals can be transmitted extremely quickly through gap junctions, and this speed is critical for such activities as the coordinated contraction of heart muscle and the communication between brain cells that enables us to think or feel emotions.

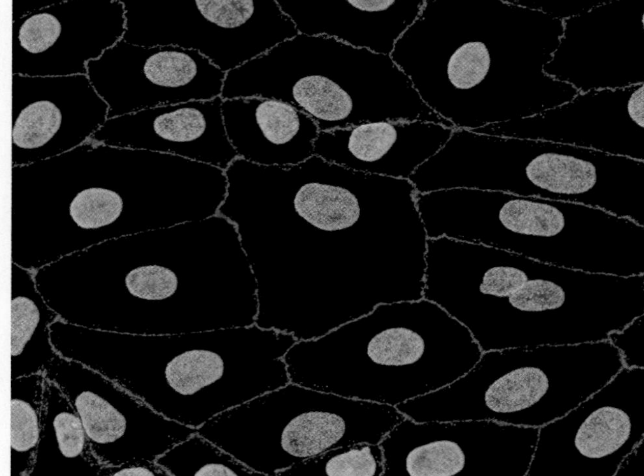

FIGURE 4.14 Cells Held Together by Tight Junctions Form Leak-Proof Sheets

Tight junction proteins (stained red using a technique called immunolocalization) are located all along the surface of these pig kidney cells. Nuclei are stained blue. Cells like these line the millions of tubules in our kidneys that form and concentrate urine.

Recall that plant cells, unlike animal cells, are enclosed in a polysaccharide cell wall that surrounds their plasma membrane. Plants use communication channels called **plasmodesmata** (**plaz-moh-DEZ-muh-tuh**; singular "plasmodesma") that are functionally similar to the gap junctions of animals. Plasmodesmata are tunnels that breach the cell walls between two cells and connect their cytoplasms (**FIGURE 4.13*b***). They are lined by the merged plasma membranes of the two cells and provide a pathway for the direct and rapid flow of ions, water, and molecules such as small proteins.

Concept Check

1. What is the main function of tight junctions?
2. How are the gap junctions of animal cells similar to the plasmodesmata found in plant cells?

Concept Check Answers

1. To create a leak-proof sheet of cells.
2. Both are cytoplasmic connections that act as tunnels to allow ions and small molecules to move between neighboring cells.

4.6 Cell Signaling

In general, communication between cells is based on the release and perception of **signaling molecules**, which could be ions, small molecules the size of amino acids, or larger molecules such as proteins. The signaling molecule is sensed by another cell, the **target cell**, usually through the means of *receptors proteins*. Signaling molecules, their target cells, and the receptor proteins in those target cells are therefore the key components of any signaling system in the living world. Most signaling molecules are short-lived, being destroyed or removed from the vicinity of the target cell within seconds. Some, however, are long-lived, lasting in the body for many days.

The specificity of signal perception and response comes from the receptor protein. Within the cytoplasm, the receptors may lie in the cytosol or inside an organelle such as the nucleus. Plasma membrane receptors bind their signaling molecules at the cell surface and must relay receipt of the signal to the cytoplasm through a series of cellular events, which are collectively known as **signal transduction pathways**.

Signaling molecules that bind to receptor proteins inside the cell—that is, in the cytoplasm—must cross the plasma membrane and may need to enter membrane-enclosed compartments such as the nucleus (**FIGURE 4.15**). Because they must diffuse across the hydrophobic phospholipid bilayer of these membranes, signaling molecules that bind to receptors inside the cell tend to be hydrophobic lipids themselves. Examples are sex hormones such as testosterone and estrogen, and other sterols (see Chapter 2).

If you have ever jumped in response to a sudden noise, you have experienced the almost instantaneous work of the fast-acting signaling molecules released by nerve cells, called *neurotransmitters*. If these signaling molecules acted slowly, a "jump" would take hours or even days to occur. Nerve signals are narrowly targeted: each nerve releases its neurotransmitters in a small space (the *synaptic cleft*) very close to the plasma membrane of its target cell, such as a muscle cell. Most neurotransmitters disappear soon after their release.

All multicellular organisms use hormones to coordinate the activities of different cells and tissues. **Hormones** are long-lasting signaling molecules that can act over long distances. In contrast to nerve cell signals, hormones are released into the bloodstream, which disseminates them throughout the body. Typically they act more slowly than neurotransmitters, but some, such as the hormone adrenaline, can act quite rapidly, triggering a response within seconds.

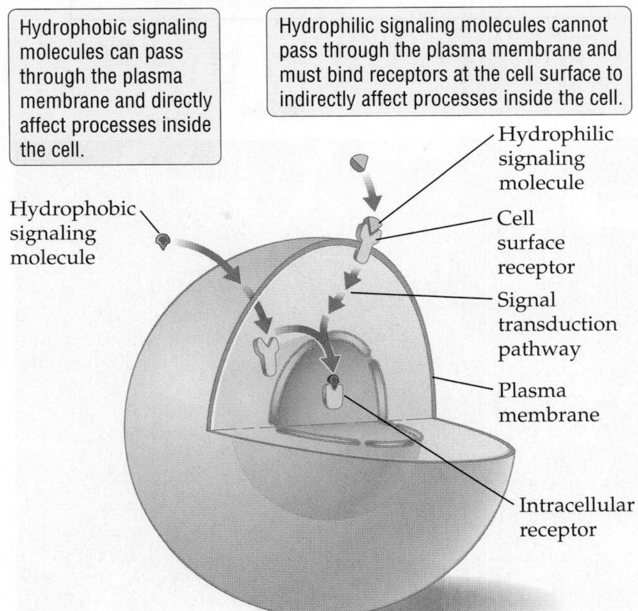

FIGURE 4.15 Receptors for Signaling Molecules
Intracellular receptors reside in the cytosol or the nucleus and bind to hydrophobic signaling molecules that can cross the plasma membrane. Cell surface receptors are embedded in the plasma membrane and bind to hydrophilic signaling molecules that cannot cross the membrane.

Human growth hormone (hGH) is an example of a slower-acting signaling molecule. This hormone is at work, stimulating the growth of bones and other tissues, as we grow in size throughout childhood. If hGH acted as fast as a neurotransmitter, the term "growth spurt" would have a whole new meaning! In vertebrate animals, most hormones are produced by cells in one part of the body and transported through the bloodstream to target cells in another part of the body, ensuring rapid and widespread distribution.

Concept Check

1. Would you expect the receptor for a small lipid signal, such as the hormone testosterone, to be located within the cytoplasm or in the plasma membrane? Would it require signal transduction? Explain.
2. Compare hormone signaling with signaling by a neurotransmitter.

Concept Check Answers

1. Within the cytoplasm. Because lipid signals are hydrophobic and can cross the plasma membrane, they do not dock with cell surface receptors. As a result, signal transduction is not required.

2. Hormones are long-lasting and widely broadcast throughout the body. Neurotransmitters are released very close to their target cells and are quickly removed.

APPLYING WHAT WE LEARNED

Cholesterol in the Brain

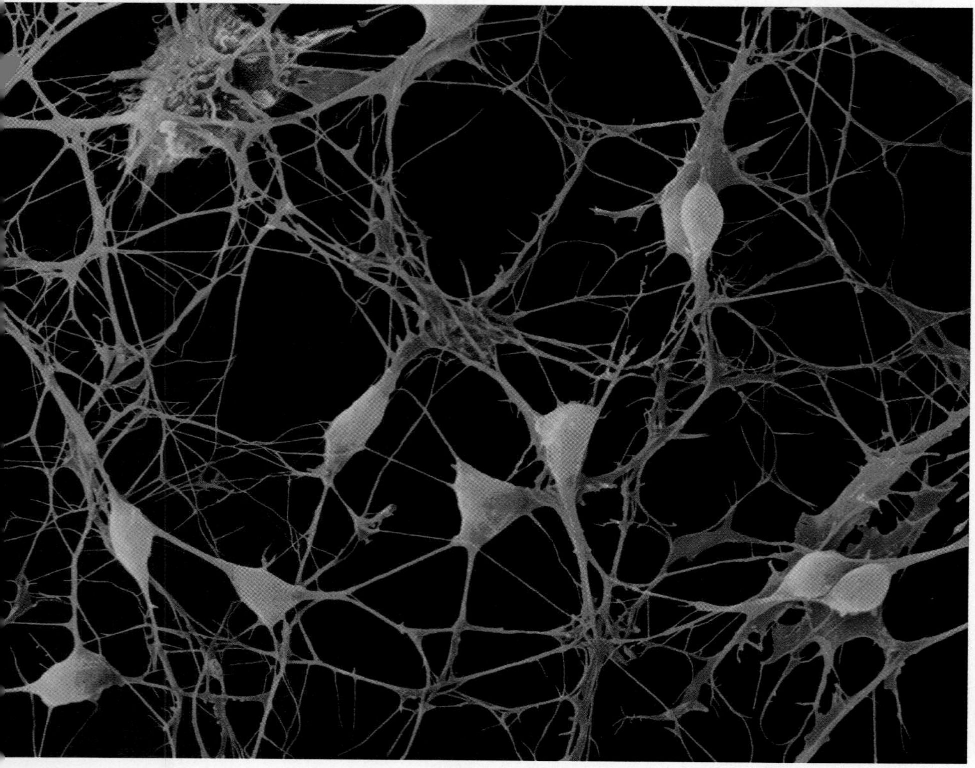

At the beginning of this chapter you learned about a retired flight surgeon and medical researcher, named Duane Graveline, who became convinced that a drug he was taking to lower his cholesterol had caused him to experience a temporary but dramatic case of amnesia—not once but twice. Was it really reasonable for him to think that cholesterol-lowering drugs might cause memory problems? Isn't cholesterol just some bad actor that clogs arteries?

Cholesterol is found in the cell membranes of all animals with a backbone (vertebrates). A sterol similar to cholesterol is found in invertebrates such as shrimp, and plants have their own version (plant sterols). These sterols keep membranes from becoming leaky by plugging the space between the phospholipid molecules that make up much of the lipid bilayer.

The membranes of brain cells—the plasma membrane as well as organellar membranes such as mitochondrial membranes—have the most cholesterol. In fact, a quarter of all the cholesterol in the body is in the brain, and all of that cholesterol is made in the brain itself.

Cholesterol appears to be essential to forming connections between one nerve cell and the next in the brain—which is the current model of how we make and store memories. Our brains contain two kinds of nerve cells: *neurons*, which connect to one another to transmit information and form memories; and *glial cells*, which care for the neurons. One of the things that glial cells do is supply the neurons with cholesterol. When glial cells can't make cholesterol, the neurons don't make as many connections.

Much of the cholesterol in the body is manufactured by the liver, and statins reduce blood cholesterol mainly by interfering with a critical liver enzyme. But can statins prevent cholesterol production by glial cells? Most regions of the brain have a *blood-brain barrier*, an anatomical arrangement that prevents most substances, except key nutrients such as oxygen and glucose, from leaving the blood vessels there. However, some types of statins could breach the blood-brain barrier. If they escaped from the blood vessels to enter glial cells, these statins could block cholesterol production in the brain.

In a 2013 study of over 100,000 statin users, 17 percent discontinued use of the drugs for reasons ranging from expense to significant side effects. Six of every 10,000 patients experienced memory loss, and just like Graveline, they recovered when they stopped taking the drugs. Other known side effects, such as irreversible muscle damage, were rarer. It is not clear why a small percentage of statin users suffer side effects, but the majority don't.

Statins have complex interactions with many substances that a person might take, perhaps unknowingly. Grapefruits must be avoided, for example, because a chemical in the fruit blocks the enzymes that break down statins in the body, resulting in an overdose. Genetic differences between people may also affect how the body deals with statins.

Today, statins remain some of the most widely prescribed drugs in developed countries. There is no question that these drugs save lives. Death from heart disease is lowered by 25–35 percent in high-risk individuals who take the drug. But doctors warn that the potential benefits must be weighed against the risks, and those who take these powerful drugs should be monitored for side effects. Thanks to concerns raised by Dr. Graveline and others, the warning label on the prescription now lists memory loss as a potential side effect of statins.

As for the former astronaut, he has not only lowered his blood cholesterol with exercise and changes to his diet, but he has also written four books advocating drug-free strategies for better heart health.

BIOLOGY IN THE NEWS

Genetically Engineered Tomato Mimics Good Cholesterol

BY MONTE MORIN • *Los Angeles Times*, March 20, 2013

Researchers at UCLA have genetically engineered tomatoes that, when fed to mice, mimic the beneficial qualities of good cholesterol, according to a new study . . .

[They] used bacteria to insert genes into the cells of tomato plants, so that they would produce a peptide that mimics the actions of HDL, or "good" cholesterol.

Later generations of those genetically engineered tomatoes were frozen, ground up and then fed to female mice who were themselves bred to be highly susceptible to LDL, or "bad" cholesterol . . .

[Mice that] received the peptide-enhanced tomatoes as part of their diet had significantly lower levels of inflammation and less plaque in their arteries, according to senior author Dr. Alan Fogelman, director of atherosclerosis research at the David Geffen School of Medicine . . .

The peptide involved in the study was 6F, which mimics the actions of apoA-1, the main protein in high-density lipoprotein, or HDL.

Despite the experiment's success, do not expect to find such genetically engineered tomatoes in the supermarket anytime soon. Researchers are still trying to determine exactly how the peptide functions in living animals.

If the levels of LDL particles in the bloodstream are high—either because too much cholesterol is absorbed from the gut, too much is produced by the liver, or not enough is removed by cells in the body—the LDL particles deposit cholesterol and other lipids in the arteries, increasing the risk of a heart attack. For this reason, cholesterol found in LDL particles is popularly called "bad" cholesterol. So what exactly is "good cholesterol," and why the angelic reputation?

Cholesterol associated with HDL particles is popularly known as "good cholesterol." HDL particles are so named because they contain high-density lipoprotein (HDL). LDL particles contain low-density lipoprotein (LDL). Think of LDL and HDL particles as two types of trucks. LDL particles are like light delivery vans that keep your local convenience store stocked with milk and corn chips. HDL particles are like big, heavy recycling trucks that pick up used bottles and newspapers from every house in the neighborhood. HDL particles collect cholesterol from body tissues and return it to the liver, ovaries, testes, and adrenal glands for recycling into other materials. They also sponge up cholesterol from arterial walls and cart it off to the liver.

Going just by its job description, you could predict that high levels of HDL particles must be good for heart health. A large number of observational studies, following thousands of people, have shown repeatedly that people who have high levels of HDL, relative to LDL, particles are much less likely to die from heart attack or stroke.

Why do some people have higher levels of HDL particles? Before age 50, women have more favorable ratios of HDL to LDL particles than men have because female sex hormones boost HDL particles. Other factors that improve the levels of HDL particles include aerobic exercise, not smoking, eating whole grains and vegetables, limiting consumption of saturated fats and especially trans fats, and drinking alcohol in *moderation* (that means the alcohol equivalent of one 4-ounce glass of wine for women, and two for men, each day).

	MEN	WOMEN
Ratio of LDL/HDL cholesterol	Average: 5 to 1 Optimal: 4 to 1 or less	Average: 4.4 to 1 Optimal: 4 to 1 or less
Ratio of ApoB/ApoA1	Average: 1.1 Optimal: 0.65 or less	Average: 1.0 Optimal: 0.62 or less

The peptide discovered by the UCLA researchers mimics apolipoprotein A-1, the type of protein found in HDL particles. Scientists already know that it's not the *cholesterol* in the particles but the protein in them that wins HDL particles their halo. In fact, the most stringent test of one's heart disease risk is to measure the ratio of apolipoprotein B (the one in LDL particles) to apolipoprotein A-1 (the protein in HDL particles). To avoid confusing the public, most physicians still discuss the ratio of total cholesterol to HDL, but increasingly, medical labs are reporting the ratio of ApoB to ApoA1. If UCLA's Apo-mimicking tomatoes do for men what they do for mice, GMO pasta sauce might be one option for getting on the right side of the Apo ratio.

Evaluating the News

1. Compare the structure and function of LDL and HDL particles.
2. What can people do to increase their HDL levels?

CHAPTER REVIEW

Summary

4.1 The Plasma Membrane as Gate and Gatekeeper

- The plasma membrane is a selectively permeable phospholipid bilayer with embedded proteins.
- In passive transport, cells carry substances across the plasma membrane without the direct expenditure of energy. Active transport by cells requires an energy input.
- Diffusion is the passive transport of a substance from a region where it is at a higher concentration to a region where it is at a lower concentration.

4.2 Osmosis

- Osmosis is the diffusion of water across a selectively permeable membrane. When placed in a hypotonic solution, a cell gains water. In a hypertonic solution, water moves out of cells. In an isotonic solution, there is no net uptake of water by the cell.
- Cells can actively balance their water content by osmoregulation.

4.3 Facilitated Membrane Transport

- Hydrophilic substances and larger molecules cannot cross the plasma membrane without the assistance of membrane-spanning transport proteins. Channel proteins move substances passively; carrier proteins moves substances either passively or actively.
- Passive carrier proteins facilitate the passive transport of molecules and ions down a concentration gradient.
- Active carrier proteins move substances into or out of the cell against a concentration gradient and require an input of energy (from an energy source such as ATP) to do so.

4.4 Exocytosis and Endocytosis

- Cells export materials by exocytosis and import materials by endocytosis.
- In receptor-mediated endocytosis, receptor proteins in the plasma membrane recognize and bind the substance to be brought into the cell.

4.5 Cellular Connections

- Cell junctions hold cell communities together.
- Three types of cellular junctions connect neighboring animal cells. Anchoring junctions attach adjacent cells and make them resistant to breaking forces. Tight junctions bind cells together to form leak-proof sheets. Gap junctions are cytoplasmic tunnels that allow the passage of small molecules.
- Plasmodesmata are cytoplasmic tunnels that connect neighboring plant cells.

4.6 Cell Signaling

- Cell signaling requires signaling molecules, the receptor proteins the signals bind to, and the target cells that the receptor protein are located in.
- Hydrophilic signaling molecules cannot enter a cell; they bind to receptor proteins located in the plasma membrane. The binding triggers signal transduction pathways that relay the message within the cytoplasm.
- Hydrophobic signaling molecules can pass through the plasma membrane; they bind to preceptor proteins located in the cytoplasm.
- Hormones are long-distance signaling molecules that are broadly distributed in the body.

Key Terms

active carrier protein (p. 94)
active transport (p. 87)
anchoring junction (p. 97)
carrier protein (p. 94)
cell junction (p. 97)
channel protein (p. 92)
concentration gradient (p. 88)
diffusion (p. 87)
endocytosis (p. 96)
exocytosis (p. 95)
facilitated diffusion (p. 92)
facilitated transport (p. 92)
gap junction (p. 98)
hormone (p. 99)
hypertonic solution (p. 92)
hypotonic solution (p. 91)
isotonic solution (p. 92)
osmoregulation (p. 92)
osmosis (p. 90)
passive carrier protein (p. 94)
passive transport (p. 87)
phagocytosis (p. 97)
pinocytosis (p. 96)
plasmodesma (p. 98)
receptor (p. 96)
receptor-mediated endocytosis (p. 96)
selective permeability (p. 86)
signal transduction pathway (p. 99)
signaling molecule (p. 99)
simple diffusion (p. 89)
target cell (p. 99)
tight junction (p. 97)
transport protein (p. 86)

Self-Quiz

1. Which of the following are *not* part of the plasma membrane?
   a. proteins
   b. phospholipids
   c. receptors
   d. genes

2. A direct input of energy is needed for
   a. diffusion.
   b. active transport.
   c. osmosis.
   d. passive transport.

3. Which of the following can move across a plasma membrane through simple diffusion?
   a. oxygen gas (O_2)
   b. hydrogen ions (H^+)
   c. aspartic acid, a charged amino acid
   d. human growth hormone, a hydrophilic protein

4. Water would move out of a cell in
   a. a hypotonic solution.
   b. an isotonic solution.
   c. a hypertonic solution.
   d. none of the above

5. Channel proteins are different from carrier proteins in that they
   a. are needed for simple diffusion but not for facilitated diffusion.
   b. help in the transmembrane transport of water, but not of ions.
   c. cannot transport substances actively.
   d. cannot function without the direct input of energy in the form of ATP.

6. Which of the following describes movement of material out of a cell?
   a. pinocytosis
   b. phagocytosis
   c. endocytosis
   d. exocytosis

7. Which of these cellular connections creates a leak-proof sheet of cells, such as is found in the cells lining the urinary bladder?
   a. anchoring junction
   b. tight junction
   c. plasmodesma
   d. gap junction

8. Animal cells can directly exchange water and other small molecules through
   a. gap junctions.
   b. microfilaments.
   c. anchoring junctions.
   d. tight junctions.

9. Cell signaling involves
   a. receptor proteins.
   b. signaling molecules.
   c. target cells.
   d. all of the above

10. A nerve signal (neurotransmitter)
    a. must travel through the bloodstream to reach target cells.
    b. acts on a target cell that is nearby.
    c. must be long-lived.
    d. must be hydrophobic in nature.

Analysis and Application

1. Imagine that you release some scented air freshener in one corner of a room. Is the spread of the scent molecules through the room an example of diffusion? When equilibrium is reached, probably many hours later, has diffusion ceased? Have the scent molecules stopped moving about? Explain.

2. *Paramecium* is a wall-less, single-celled protist that lives in freshwater ponds. Are its natural surroundings hypertonic, hypotonic, or isotonic with respect to the cell? What osmoregulatory problem does this organism face in pond water, and how does it cope with that problem?

3. A classmate has come down with strep throat caused by group A *Streptococcus* bacteria. She has a fever and sore throat, and the pus on her tonsils is a sign that her white blood cells are doing battle with the invading bacteria (the pus contains the remains of white blood cells that died after doing their share to destroy the invaders). Explain the important role of cell membranes in the mechanism by which your white blood cells destroy invading bacteria.

4. The epithelial cells lining your intestines encounter a great variety of substances, many of which could produce ill effects if they were to cross the epithelial layer to enter the bloodstream. How do cell junctions in intestinal epithelial cells help keep potential toxins out of your bloodstream?

5. Figure 4.12 shows the structure of an LDL particle. An LDL particle has a highly hydrophobic core that consists of cholesterol covalently linked to fatty acids ("esterified"). The shell-like surface is made up of phospholipids, unesterified cholesterol, and a single large protein (apolipoprotein B). What is the role of LDL particles? Which chemical component of the LDL particle is recognized by the LDL receptor? Describe how LDL particles are internalized by the many cells in the human body that import cholesterol.

5 Energy, Metabolism, and Enzymes

METABOLIC KICK. The kinetic energy of these spectacular moves comes from the chemical energy in the food the soccer players ate. It takes enzymes to unleash that energy in a usable form.

Kick-Start Your Metabolic Engine!

"Fourteen ways to boost your metabolism!" "Speed up your metabolism and shed the pounds fast!!" "Feeling sluggish? Tropical fruit revs up your internal engine." You hear the sales pitch in advertisements for energy drinks, diet foods, and herbal supplements. At the checkout stand, every other magazine seems to promise health and vitality. If only we learn how to master our metabolism.

But what *is* metabolism? It's a general term for all the chemical reactions organisms use to capture, store, and use energy. All living organisms need energy to build and sustain every part of the body—from bones to DNA. Growing, reproducing, fighting off pathogens, even lazing around—it all takes energy. The liver and the brain are the biggest energy spenders: of all the energy used by a human body at rest, the liver accounts for 27 percent, and the brain uses about 20 percent.

Your metabolic rate is a measure of how much energy you're using. If you're sitting still and you are a young woman of about 140 pounds, you're using as much energy per hour as a 75-watt lightbulb. How much energy you spend when you're at rest depends on a number of things, including your genetics, age, height, weight, muscle mass, and gender. Because of the genes they've inherited, some people do have a faster metabolism, and others metabolize more slowly than the average person. Body size and shape also affect the rate at which we use energy. Generally speaking, a taller, heavier person burns more calories at rest than does a smaller, lighter person. Children have a higher metabolism than adults, and metabolic rate declines as we get older. People who are physically active tend to have a higher metabolic rate: an athletic woman likely has a higher resting metabolic rate than does a couch-potato man.

How does exercise affect metabolism? Can what you eat affect your metabolic rate? How does caffeine affect how many calories you burn through the day? Will eating jalapeño peppers crank up the calorie burn?

Before we address these questions, let's explore why cells need energy and how they use it to create organized structures. In this chapter we will see that a living cell is a highly organized, energy-dependent chemical factory whose thousands of reactions proceed with the help of enzymes.

MAIN MESSAGE Metabolism, the capture and use of energy, is vital for life. Enzymes, which speed up the many chemical reactions within a cell, are also essential.

KEY CONCEPTS

- Living organisms obey the universal laws of energy conversion and chemical change.
- The sun is the ultimate source of energy for most living organisms. Photosynthetic organisms capture energy from the sun and use it to synthesize sugars from carbon dioxide and water. Most organisms can break down sugars to release energy.
- Metabolism refers to the capture, storage, and use of energy by a living cell. All the chemical reactions that occur inside a cell are part of that cell's metabolism.
- Metabolic pathways are sequences of enzyme-controlled chemical reactions.
- Catabolic reactions break down biomolecules to release energy, and anabolic reactions use energy to build (synthesize) biomolecules.
- Enzymes greatly increase the rate of chemical reactions. Enzymes bind to specific substrates and position these reactants so that they interact more easily to make products.
- The activity of enzymes is sensitive to temperature, pH, and salt concentration.
- A metabolic pathway is a multistep sequence of chemical reactions, with each step catalyzed by a different enzyme.

KICK-START YOUR METABOLIC ENGINE!		105
5.1	The Role of Energy in Living Systems	106
5.2	Metabolism	109
5.3	Enzymes	112
5.4	Metabolic Pathways	115
APPLYING WHAT WE LEARNED Food, Folks, and Metabolism		118

ALL LIVING CELLS REQUIRE ENERGY, which they must obtain from the living or nonliving components of their environment. Organisms use energy to manufacture the many chemical compounds that make up living cells, and for growth, reproduction, and defense. Thousands of different types of chemical reactions are required to sustain life in even the simplest cell.

In this chapter we look first at energy basics: the different kinds of energy, and the conversion of energy from one form to another. Next we examine the role played by energy in the chemical reactions that maintain living systems—that is, in the metabolism of living things. Finally, we discuss the special properties of enzymes and explain how these remarkable biomolecules speed up chemical reactions that would otherwise be too slow to sustain life.

EXTREME RUNNING

Dean Karnazes ran 350 miles without stopping. His record-setting ultramarathon, lasting 80 hours and 44 minutes, was fueled by an estimated 42,858 Calories. A Calorie (same as a kilocalorie) is a measure of the chemical energy in food molecules. How is the way that Karnazes obtains energy different from the way a corn plant acquires energy?

5.1 The Role of Energy in Living Systems

Any discussion about chemical processes in cells is at heart a discussion about the capture and use of *energy*. Every atom, molecule, particle, or object in the physical world possesses energy. We can define **energy** as the capacity of any object to do work. Work, in turn, can be defined as the capacity to bring about a change in a defined system. In the context of energy, the word "system" refers to any portion of the universe we choose to study. In speaking of energy in the living world, a system can be a biomolecule, an organelle, a single bacterial cell, a mat of algae in a pond, a community of organisms in an oak-maple woodland ... on up to the biosphere.

The energy of any system is an attribute of that system—a physical quantity associated with that system. Energy can be recognized and expressed in many different ways, depending on which aspects of the system we wish to describe.

There are different forms of energy

The many different forms of energy can be organized into two broad categories: *potential energy* and *kinetic energy*. **Potential energy** is the energy stored in any system as a consequence of its position. A rock on a hilltop, water in a dam, and Lady Gaga's headgear—all have potential energy, a capacity to do work. Their potential energy is a consequence of their position relative to their surroundings. For example, the potential energy of the water in a dam is due to the force of gravity, and this type of potential energy is known as gravitational energy.

Chemical energy is another form of potential energy; it is the energy stored in the bonds between atoms and ions. The covalent bonds that hold the atoms in a molecule, for example, store substantial amounts of chemical energy. A spoonful of table sugar (sucrose) harbors the chemical energy of many millions of carbon, hydrogen, and oxygen atoms linked via covalent bonds. A pinch of table salt contains the chemical energy of many millions of sodium and chloride ions chained together through ionic bonds.

Kinetic energy is the energy that a system possesses as a consequence of its state of motion. Consider what happens when we use an electric blender to whip strawberries and ice cream into a smoothie. Some of the *electrical energy*, which is the energy associated with the flow of electrons, is turned into the *mechanical energy* of the whirling blades that whip up the ingredients in the blender. Another form of kinetic energy, *light energy*, is the energy associated with the wavelike movement of packets of energy called *photons*. Electrical energy, mechanical energy, and light energy are all examples of kinetic energy.

Heat energy, also known as thermal energy, can be considered a type of kinetic energy. All atoms and molecules move to some degree, either vibrating in place or careening randomly from one point to another. When these particles of matter collide with other particles, they transfer some of their energy to their target, increasing its speed. Heat energy is that portion of the total energy of a particle that *can flow* from one particle to another.

Energy Transformations in Living Systems

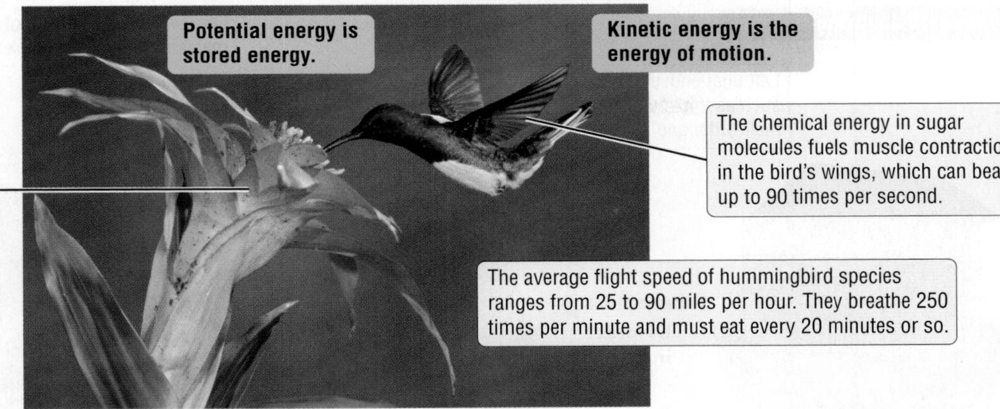

Potential energy is stored energy.
Kinetic energy is the energy of motion.
The covalent bonds in the sugar molecules in nectar possess chemical energy, a form of potential energy.
The chemical energy in sugar molecules fuels muscle contractions in the bird's wings, which can beat up to 90 times per second.
The average flight speed of hummingbird species ranges from 25 to 90 miles per hour. They breathe 250 times per minute and must eat every 20 minutes or so.

FIGURE 5.1 Potential Energy in Food Is Converted to Kinetic Energy in a Hummingbird's Body

One form of energy can often be converted into another form. Potential energy is released as kinetic energy when a rock rolls down a hill, water rushes down the spillway of a dam, or headgear comes crashing to the floor. A falling object may release *sound energy*, another type of kinetic energy.

A hovering hummingbird beats its wings 30–90 times per second, fueled by the conversion of chemical energy stored in nectar into the kinetic energy of contracting muscles (**FIGURE 5.1**). If you lift a book and place it on a high shelf, some of the chemical energy from your breakfast is converted into the kinetic energy of moving muscles. The book on the bookshelf, because you placed it up high, now has more potential energy.

The laws of thermodynamics apply to living systems

Certain universal principles—known as the *laws of thermodynamics*—specify that our universe contains a fixed amount of energy, and that although energy can be converted from one form to another, it can be neither created nor destroyed. These powerful concepts apply to living organisms just as much as they describe energy transactions in the engine of a gas-electric hybrid car or in the center of a distant galaxy.

When applied to living systems, the laws of thermodynamics explain *why* a living cell cannot exist without an input of energy. They set the ground rules for energy conversions inside a cell—which reactions can happen spontaneously, and which reactions will not take place without an energy input. As we will see shortly, the laws of thermodynamics even predict that anytime energy is released in a cell, a certain portion of that energy will be "wasted"—in that it will be unavailable for any cellular work.

The **first law of thermodynamics** states that energy can neither be created nor destroyed, but only converted from one form to another. According to the first law, also known as the *law of conservation of energy*, the total energy of any closed system remains the same over time. Energy can, however, be converted from one form into another. Some of the light energy that a plant absorbs can be turned into the chemical energy of the sugar molecules in the plant's nectar. This chemical energy can be converted into electrical energy in a hummingbird's brain cells, into mechanical energy in its muscle cells, and into heat energy in any of its cells whose metabolism is fueled by the nectar.

Helpful to Know

The term "thermodynamics" comes from physics and is derived from Greek words meaning the movement (*dynamics*) of heat (*thermo*). Heat always moves from warmer to cooler areas—a basic principle governing all biological processes.

The **second law of thermodynamics** states that the natural tendency of the universe is to become less organized, more disorderly. Any system exhibits this tendency unless energy from elsewhere in the universe is used to organize that system. If any part of the universe, whether a cell or a toolshed, displays a high level of order, we can be certain that energy has been captured from elsewhere to create and maintain that order (**FIGURE 5.2*a***). In using energy to create internal order, a living system reduces the order of its surroundings, so on the whole, the universe becomes more disordered.

The tremendous structural and functional complexity of the cell exists in the midst of a general tendency toward chaos. To counteract the natural tendency toward disorganization, the cell must capture, store, and use energy.

One of the many implications of the second law of thermodynamics is that the capture, storage, or use of energy by living cells is never 100 percent efficient, so at least some of the invested energy is lost as a disordered and unusable form of energy called *metabolic heat*.

FIGURE 5.2 An Input of Energy Is Needed to Create or Maintain a High Level of Organization
(*a*) The disorder, or *entropy*, of a system tends to increase unless that tendency is countered by an input of energy. (*b*) Living cells maintain their complex organization through continual input of energy from the environment.

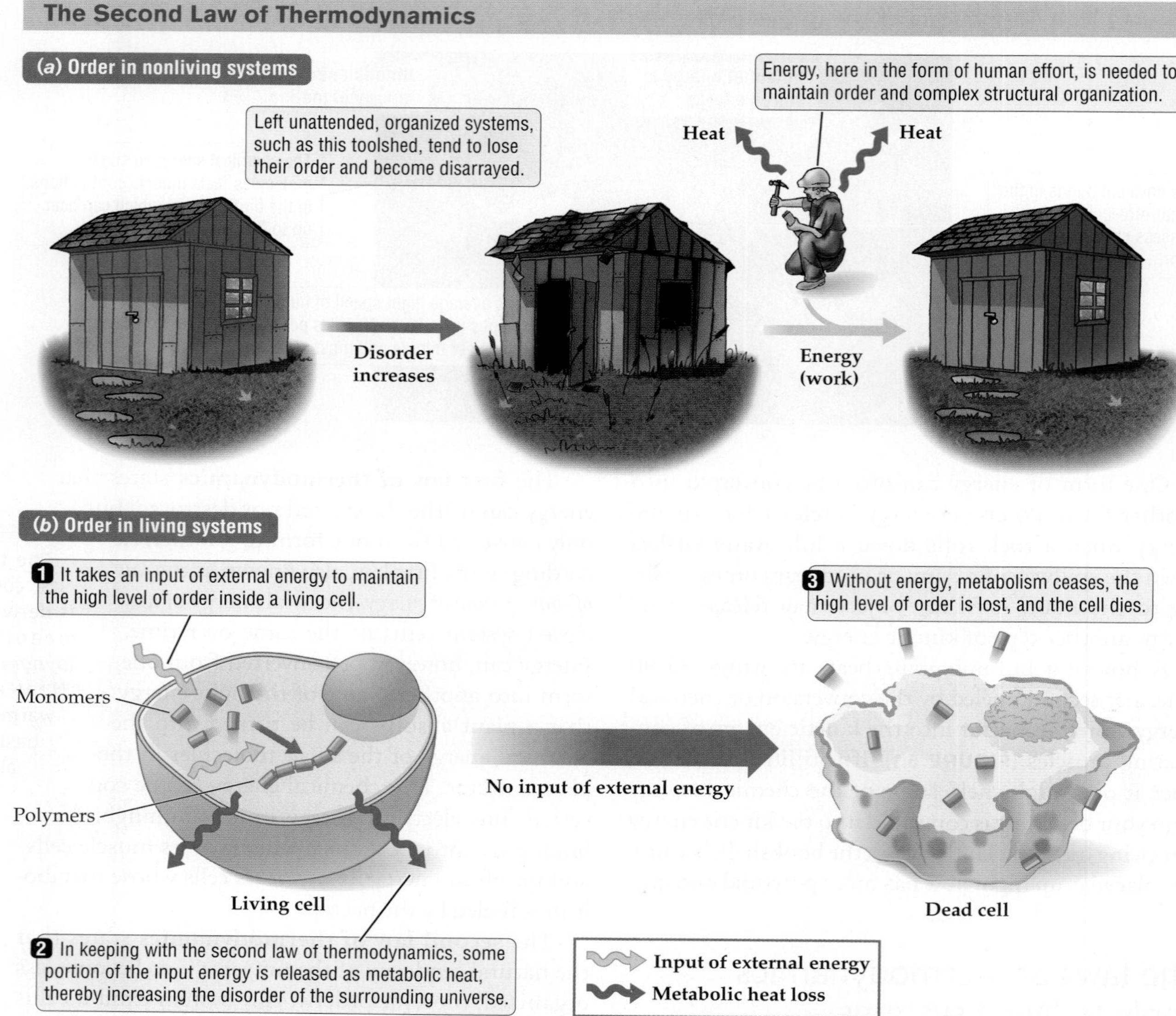

The Second Law of Thermodynamics
(a) Order in nonliving systems
Left unattended, organized systems, such as this toolshed, tend to lose their order and become disarrayed.
Energy, here in the form of human effort, is needed to maintain order and complex structural organization.
Heat
Heat
Disorder increases
Energy (work)
(b) Order in living systems
1 It takes an input of external energy to maintain the high level of order inside a living cell.
Monomers
Polymers
Living cell
No input of external energy
3 Without energy, metabolism ceases, the high level of order is lost, and the cell dies.
Dead cell
2 In keeping with the second law of thermodynamics, some portion of the input energy is released as metabolic heat, thereby increasing the disorder of the surrounding universe.
Input of external energy
Metabolic heat loss

In other words, through the very act of creating order within, living systems add to the disorder of the universe by releasing metabolic heat into their surroundings (**FIGURE 5.2*b***). Consequently, only a portion of available energy, usually a relatively small portion, is available to fuel cellular processes.

Helpful to Know

The word "respiration" can cause some confusion because in everyday use it means "breathing in and out." We use "cellular respiration" to refer to the energy-harnessing reactions in cells that consume oxygen and produce carbon dioxide as a by-product.

The flow of energy connects living things with the environment

Where does the energy used to generate order in the cell come from? We know from the first law of thermodynamics that the cell cannot create energy from nothing; the necessary energy must come from outside the cell.

As you learned in Chapter 1, *producers* (*autotrophs*) obtain energy from the nonliving part of their environment. Plants, algae, and certain bacteria make food through *photosynthesis*, a sequence of reactions that is ultimately fueled by light energy. We can define food as biomolecules rich in chemical energy. *Consumers* (*heterotrophs*) obtain food by eating other organisms or absorbing their dead remains.

Energy flows through an ecosystem in a single direction, passing from producers to consumers, with some energy escaping as metabolic heat at every step, as dictated by the second law of thermodynamics. In contrast to the one-way flow of energy, much of the matter in an ecosystem is recycled within it. Carbon atoms and other essential elements of living things pass from producers to consumers, and then back to producers after cycling through nonliving parts of the environment.

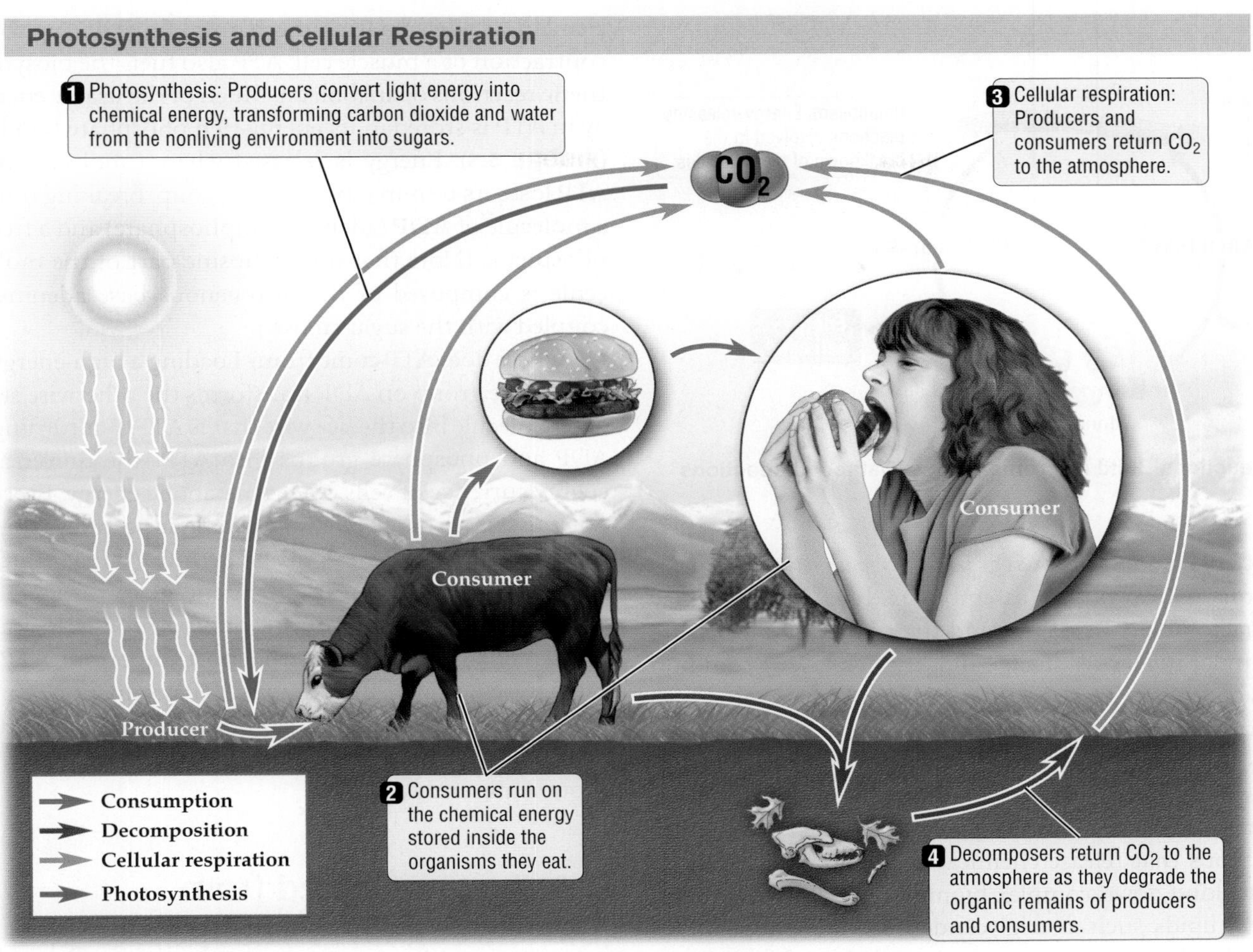

FIGURE 5.3 Photosynthesis and Cellular Respiration Are Complementary Processes

Matter, in the form of carbon atoms, cycles among producers, consumers, and the environment.

For example, carbon-containing molecules from living cells are returned to the nonliving part of the ecosystem as organisms break down food molecules through an energy-releasing process known as *cellular respiration*: the carbon-carbon bonds are broken, and each carbon atom is combined with oxygen and released into the environment as a molecule of carbon dioxide (CO_2).

Consumers, including *decomposers* such as bacteria and fungi, are not the only organisms that break down food molecules and release carbon dioxide through cellular respiration. Producers such as plants also rely on cellular respiration. Photosynthetic cells manufacture energy-rich sugar molecules, but they rely on cellular respiration to extract energy from these molecules to meet their daily energy needs.

What photosynthetic cells can do that most other cells cannot is absorb carbon dioxide from the environment and link its carbon atoms into the backbone of biological molecules. In this way, carbon atoms are continually cycled from carbon dioxide in the atmosphere to sugars and other molecules made by producers, and then back to the atmosphere as carbon dioxide released by respiring producers and consumers. **FIGURE 5.3** illustrates the relationship between photosynthesis and cellular respiration. If you pay close attention to which molecules are used and which are released by photosynthesis and cellular respiration, respectively, you will see that the two are complementary processes.

Concept Check

1. A Big Mac cheeseburger has 704 Calories. Is the energy in the Big Mac an example of potential energy or kinetic energy?
2. Is all the energy in a Big Mac at least theoretically available to you to move your body?

Concept Check Answers

1. A Big Mac contains chemical energy, a type of potential energy.

2. No. Quite apart from the Calories that are not absorbed or used up for other metabolic purposes, it follows from the second law of thermodynamics that the chemical energy in the Big Mac cannot be converted into kinetic energy in our muscles with 100 percent efficiency. Some of that chemical energy will be lost as metabolic heat during the energy conversion.

5.2 Metabolism

As noted earlier, **metabolism** refers to all the chemical reactions within a living cell that capture, store, or use energy. Most chemical reactions in a cell occur in chains

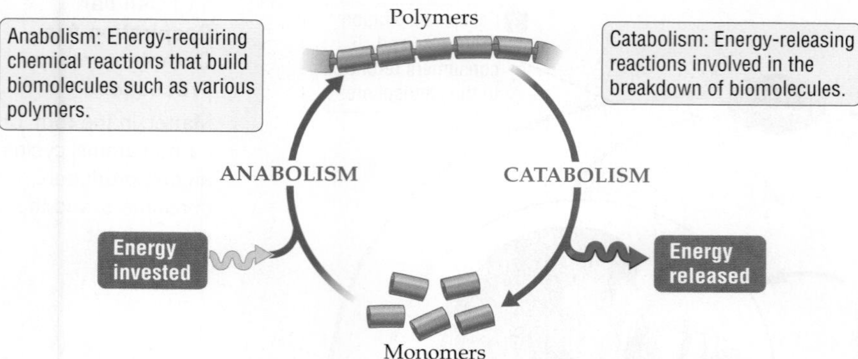

FIGURE 5.4 Anabolic Reactions Build Macromolecules; Catabolic Reactions Degrade Them

of linked events known as *metabolic pathways*. The metabolic pathways that assemble or disassemble the key macromolecules of life, and their building blocks, are similar in all organisms—another sign of our common evolutionary heritage. Nearly all of those metabolic reactions are facilitated by *enzymes*.

All living cells require two main types of metabolism: *catabolism* and *anabolism* (**FIGURE 5.4**). **Catabolism** (**kuh-*TAB*-uh-*LIH*-zum**) refers to the linked chain of reactions that release chemical energy in the process of breaking down complex biomolecules. Carbohydrates, and lipids such as triglycerides, are the complex biomolecules most commonly degraded by these energy-releasing pathways.

Anabolism (**uh-*NAB*-uh-*LIH*-zum**) refers to the linked chain of energy-requiring reactions that create complex biomolecules from smaller organic compounds. Anabolic reactions are also known as *biosynthetic pathways* (*bios*, "life"; *synthetikos*, "of composition") because complex biomolecules, such as glycogen, proteins, and triglycerides, are put together from simpler building blocks during these reactions.

Helpful to Know

It may seem odd that *gaining* electrons is called *reduction*. The terms were coined by eighteenth-century chemists for reactions involving the gain (oxidation) or loss (reduction) of oxygen atoms. But modern chemists have broadened the definitions to include electron transfer reactions. This mnemonic may help you keep things straight: OIL RIG, Oxidation Is Loss; Reduction Is Gain.

ATP delivers energy to anabolic pathways and is regenerated via catabolic pathways

As we mentioned in Chapter 2, every living cell uses the small, energy-rich organic molecule **ATP** (adenosine triphosphate) as an energy-delivery service. Energy from ATP powers a variety of activities in the cell, such as moving molecules and ions in or out of the cell, moving organelles through the cytosol on tracks formed by cytoskeletal elements, and generating mechanical force during the contraction of a muscle cell. ATP also fuels the biosynthetic reactions of anabolism. Much of the usable energy in ATP is stored in its energy-rich phosphate bonds (**FIGURE 5.5**). Energy is released when a molecule of ATP loses its terminal phosphate group, breaking into a molecule of **ADP** (adenosine diphosphate) and a free phosphate. (Note that the adenosine part of the molecule is composed of the nitrogenous base adenine, coupled with the sugar ribose.)

Where does ATP come from? Loading a high-energy phosphate group on ADP transforms the otherwise sedate molecule into the live wire that is ATP. But turning ADP and phosphate groups into ATP—the universal energy currency of cells—takes metabolic energy. Every cell must have energy-releasing catabolic pathways that can transform ADP and phosphate into ATP. Producers, such as plants, can use light energy to turn ADP and phosphate into ATP during photosynthesis. In animals, cellular respiration is the most important ATP-generating pathway. Continuous ATP production is an urgent priority for the human body; if it were halted, each cell would consume its entire supply of ATP in about a minute.

Energy is extracted from food through a series of oxidation-reduction reactions

Many metabolic pathways, such as photosynthesis and cellular respiration, consist of a series of chemical reactions in which electrons are transferred from one molecule or atom to another. **Oxidation** is the loss of electrons from a molecule, atom, or ion. **Reduction** is just the opposite: the gain of electrons by a molecule, atom, or ion (**FIGURE 5.6**). Because the two processes are complementary, oxidation and reduction go hand in hand, and the paired processes are called oxidation-reduction reactions, or simply **redox reactions**.

The gain or loss of oxygen atoms is a telltale sign of a redox reaction. Oxygen is a powerful oxidizer in that it can pull electrons from many other atoms. For example, oxygen atoms pull electrons from iron (Fe) to form the crumbly red material we know as rust (Fe_2O_3). The oxygen atoms in rust are reduced; the iron atoms, having lost electrons, are oxidized.

The gain or loss of hydrogen atoms is another way to spot a redox reaction. A hydrogen atom covalently bonded to a different atom, such as carbon or oxygen, becomes "electron-poor" because the shared electrons are pulled more strongly by the partner atom. An atom

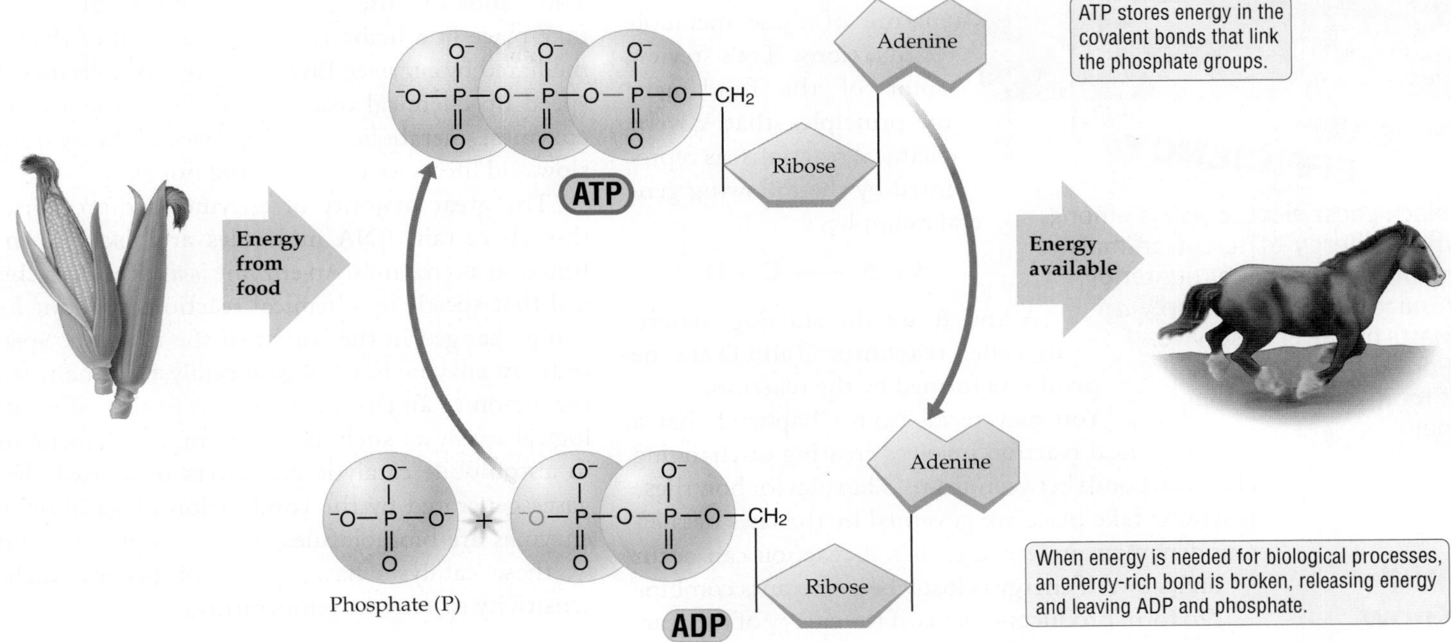

FIGURE 5.5 ATP Functions as an Energy-Storing Molecule in All Cells
When the terminal high-energy bond of ATP breaks, energy is released, and ATP is turned into ADP with the release of a phosphate group. Energy yielding metabolic pathways, such as photosynthesis and cellular respiration, are needed to manufacture ATP from ADP and phosphate.

or molecule that gains one or more hydrogen atoms becomes "richer" in electrons, and is therefore reduced.

Redox reactions involving the gain or loss of oxygen and hydrogen atoms are common in living cells. **Cellular respiration**, the oxygen-dependent catabolic reaction that your cells use to extract energy from food is an example of a redox reaction:

$$\underset{\text{Glucose}}{C_6H_{12}O_6} + \underset{\text{Oxygen gas}}{6\,O_2} \longrightarrow \underset{\text{Carbon dioxide}}{6\,CO_2} + \underset{\text{Water}}{6\,H_2O} + \underset{\text{ATP}}{\text{Energy}}$$

Each of the six carbon atoms in glucose is *oxidized* as it gains oxygen atoms and loses hydrogen atoms to become carbon dioxide. Each oxygen atom in O_2 strips two hydrogen atoms from glucose to become a molecule of water. The gain of hydrogen atoms by the oxygen atom signifies that it has been *reduced*.

In dramatic contrast, consider the summary equation for **photosynthesis**:

$$\underset{\text{Carbon dioxide}}{6\,CO_2} + \underset{\text{Water}}{6\,H_2O} + \underset{\text{Light}}{\text{Energy}} \longrightarrow \underset{\text{Glucose}}{C_6H_{12}O_6} + \underset{\text{Oxygen gas}}{6\,O_2}$$

Photosynthesis is essentially the reverse of cellular respiration. In this anabolic pathway, carbon dioxide is *reduced* as it gains electrons and hydrogen atoms to be transformed into glucose. Water molecules lose electrons and hydrogen atoms, and are therefore *oxidized* to molecular oxygen (O_2).

Chemical reactions are governed by the laws of thermodynamics

How does the cell control such a powerful event as the oxidation of a glucose molecule and break it down

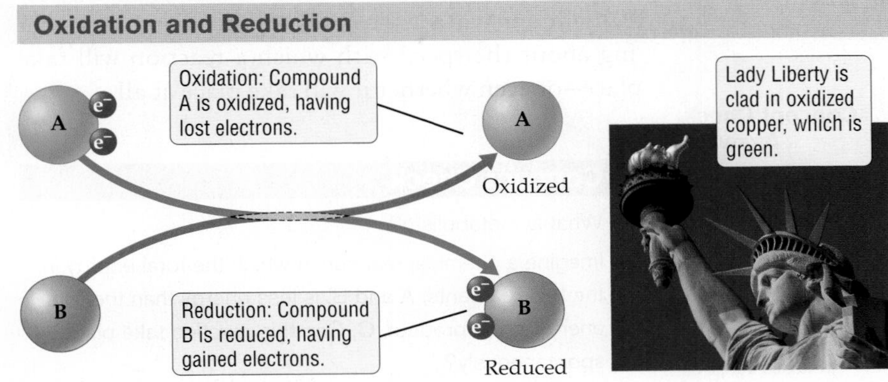

FIGURE 5.6 Oxidation and Reduction Go Together
An electron lost by an atom must be gained by another atom.

EXTREME EFFICIENCY

The black ghost electric fish is among the most energy-efficient animals on the planet. This night hunter of the Amazon River needs only 4 milliwatts of power to survive. An iPod uses a thousand times more energy in an hour.

into smaller, more manageable and useful steps? The answer lies in the stepwise nature of these metabolic reactions. Let's review some of the fundamental principles that govern chemical reactions, as represented by the following general example:

$$A + B \longrightarrow C + D$$

A and B are the starting materials, called **reactants**; C and D are the **products** formed by the reaction.

You may recall from Chapter 2 that a chemical reaction involves creating or changing chemical bonds between atoms. The rules for how these reactions take place are governed by the laws of thermodynamics presented earlier. A reaction can occur on its own if energy is lost when reactants combine to form products—that is, if the energy of the reactants is higher than that of the products formed. These "downhill" reactions can happen spontaneously. An example is the rusting of iron mentioned in the preceding section. Spontaneous reactions can occur because the reactants have a higher state of order than the products (the second law of thermodynamics).

In contrast, reactions in which reactants have a *lower* energy than the products require an input of energy to occur; they are "uphill" reactions. These reactions will not occur spontaneously; they need energy from another source.

The second law says that "downhill" reactions *can* happen spontaneously—but that does not mean such a reaction *will* occur. The second law makes the reaction possible but says nothing about whether it is *probable*; it also says nothing about the speed with which a reaction will take place—or even whether it will take place at all.

Helpful to Know

This chapter introduces several similar-sounding words beginning with *cata*. "Catabolism" refers to the breakdown reactions of metabolism. "Catalysis" (**kuh-TAL-uh-sus**), however, refers to increasing the rate of a reaction, not whether molecules are broken down or built up. Most catabolic reactions are catalytic (**CAT-uh-LIH-tik**): they require enzymes (catalysts) in order to proceed at speeds necessary for cell function. Enzymes are also required as catalysts for biosynthetic (anabolic) reactions in the cell.

Concept Check

1. What is metabolism?
2. Imagine a chemical reaction in which the total energy in the two reactants, A and B, is less energy than the total energy in the product, C. Can this reaction take place spontaneously?

Concept Check Answers

1. Metabolism refers to all the chemical reactions that capture, store, and use energy in a living cell.
2. No. The second law dictates that C cannot be made from A and B, without an input of energy.

5.3 Enzymes

Thousands of different kinds of chemical reactions take place in a living cell, and nearly all of them are mediated by *enzymes*. **Enzymes** are biomolecules that speed up chemical reactions. Without the action of enzymes, metabolic reactions would be extremely slow and life as we know it could not exist.

The great majority of enzymes are proteins, although certain RNA molecules are also known to function as enzymes. An enzyme is a **catalyst**, a chemical that speeds up chemical reactions without itself being changed in the course of the reaction. Specifically, an enzyme is a *biological* catalyst. In many ways, the action of an enzyme is similar to that of nonbiological catalysts such as platinum, an element used in automobile catalytic converters to detoxify waste gases generated by the combustion of gasoline. But enzymes are biomolecules, and as we will see shortly, these catalysts have special properties, such as sensitivity to extreme temperatures.

Enzymes remain unaltered and are reused in the course of a reaction

An important characteristic of catalysts is that, unlike reactants, they remain chemically unaltered after the reaction is over (**TABLE 5.1**). Because enzyme molecules are used over and over, relatively small amounts

TABLE 5.1 Properties of Enzymes

ENZYMES

- are usually proteins.
- increase the rate of chemical reactions, often by a million-fold or more.
- generally act on one or a few specific substrates
- remain unchanged by the reaction.
- are reused over and over, catalyzing the transformation of many substrate molecules.
- are sensitive to temperature, pH, and salt concentration.
- may need the assistance of special cofactors (specific ions or molecules).
- may be inhibited by specific ions or molecules (inhibitors).
- are usually tightly regulated within the cell or inside the body of a multicellular organism.

BIOLOGY MATTERS

Enzymes in Action

Enzymes are the workhorses of the cell. There are thousands of different types of enzymes in the human body, and if any one of them fails to function properly, some sort of illness is likely to follow. All 50 states have regulations that require newborns to be screened for several *hereditary disorders*, including *phenylketonuria*, a condition caused by the failure of an enzyme called phenylalanine hydroxylase (PAH). The enzyme catalyzes degradation of the amino acid phenylalanine.

Without PAH activity, large amounts of brain-damaging chemicals, called phenylketones, accumulate in the blood and urine. About one in 15,000 Americans is born with phenylketonuria. If the condition is identified early, managing the patient's diet in the first 16 years of life can avoid brain impairment. Affected individuals must avoid foods high in phenylalanine (meat, cheese, and legumes, as well as food containing the artificial sweetener aspartame) and take special amino acid formulations to prevent protein deficiency.

Enzymes help us digest the foods we eat, starting with the starch-degrading enzymes (*amylases*) in the mouth. The lining of the stomach produces protein-degrading enzymes

(broadly known as *proteases*) that break apart proteins under acidic conditions. The pancreas and small intestine produce a variety of proteases, as well as *lipases* that break down lipids. In the small intestines, an enzyme called *lactase* breaks apart a milk disaccharide, *lactose*, to release the two monosaccharides glucose and galactose. About 65 percent of the world's adults are *lactose intolerant* because their small intestines stop making lactase once they are past childhood.

Some fruits contain proteases—papain in papayas and bromelain in pineapples—that are useful as meat tenderizers. The enzymes break up some of the large fibrous proteins, such as collagen, that are abundant in tougher cuts of meat. The bromelain in fresh pineapples also destroys any dessert made from gelatin, which is rich in collagen. Canned pineapples are no threat to gelatin desserts because canned fruit is treated with heat to prevent microbial growth, which inactivates the bromelain also.

Enzymes are widely used in household products, pharmaceuticals, food manufacturing, and industrial processes. Many clothing and dishwashing detergents contain amylases, proteases, and lipases, to help remove stains and residues from organic material. Enzymes are also important in the paper industry and are expected to assume greater importance in the emerging biofuel industry.

Rennin, an enzyme from the gut of cows and sheep, has been used for hundreds of years to make cheese. The enzyme denatures milk proteins, causing them to separate from the whey (liquid) in clumps (curds) that are further processed to make the finished product. In the mash stage of beer brewing, a number of carbohydrate-degrading enzymes are used, and proteases are employed to remove cloudiness. *Cellulases* and *pectinases*—enzymes that break down polysaccharides—are used to clarify fruit juices.

of an enzyme are needed to catalyze a reaction. Most enzymes are highly specific in their action, catalyzing only one type of chemical reaction or, at most, a small number of very similar reactions.

The specific reactants that bind to a particular enzyme are called the **substrates** of that enzyme. Substrates bind to an enzyme in an orientation that favors the making and breaking of chemical bonds. Most enzymes have docking points—known as *active sites*—for one or more substrates. The size, geometry, and chemistry of an active site determine which substrate can bind to it, and this selectivity is the source of enzymes' substrate specificity. Enzymes are often named by the addition of *ase* to the name of the substrate they act on. Lactase, for example, is the digestive enzyme that breaks down the major milk sugar, lactose. Many more examples are listed in the "Biology Matters" box on this page.

The function of enzymes—like the function of most proteins—is crucially dependent on their three-dimensional shape. Conditions that change their shape, or that alter their binding to substrates, can make them nonfunctional.

- High temperatures denature (destroy) the three-dimensional configuration of most enzymes (see Chapter 2). Most human enzymes, for example, work best at the average core body temperature, 37°C (98.6°F), and many will lose activity at temperatures even 5°C above or below that optimum temperature.

- Extremes of pH—high acidity or high alkalinity—also disrupt the function of most enzymes, often by altering the chemistry of the active site.
- Some enzymes need particular ions or small molecules—known as *cofactors*—to reach highest activity. For example, carbonic anhydrase, an important enzyme for maintaining blood pH, needs zinc ions as cofactors.
- Some enzymes need a particular salt concentration for optimal activity, and work poorly at higher or lower salt concentrations. At very high salt concentrations, the three-dimensional structure of proteins is destroyed. The denaturing effect of salt is put to work in the manufacture of some cheese and tofu products: sea salt or calcium salts are used to curdle soy-milk proteins in making tofu, for example.

The shape of an enzyme determines its function

The binding of an enzyme to its particular substrate depends on a match between the three-dimensional shapes of both the substrate and the enzyme molecules. In the same way that a lock accepts only a key with just the right shape, each enzyme has an **active site** that fits only substrates with the correct three-dimensional shape and chemical characteristics (**FIGURE 5.7**).

Substrates
A
B
1 Substrates bind to the active site.
2 Enzyme facilitates the reaction.
Catalysis
Enzyme
Active site
Enzyme
Induced fit: As substrates enter, the active site changes shape to mold snugly around the bound substrates.
3 Product is released.
A B
Enzyme
4 The enzyme is not permanently changed by the reaction and can be recycled.

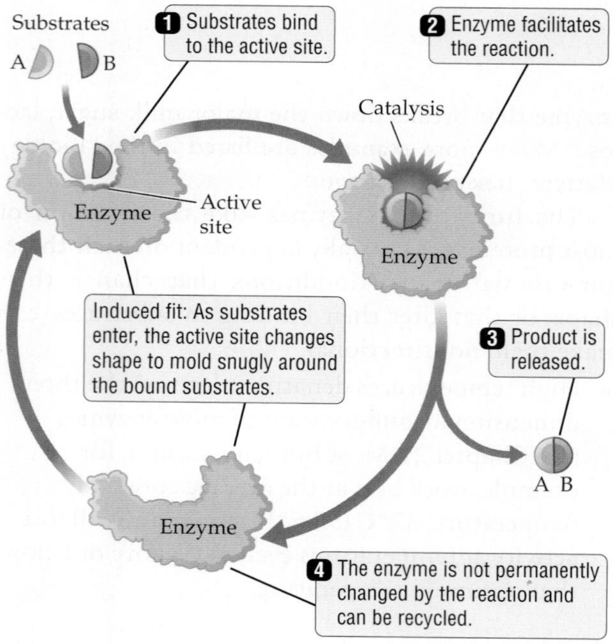

FIGURE 5.7 Enzymes as Molecular Negotiators

An enzyme brings together two reactants (A and B) to form the product AB.

The shape of an active site is somewhat flexible, and a substrate can tweak it to create an even better fit between it and the enzyme. According to the **induced fit model** of substrate-enzyme interaction, as a substrate enters the active site the parts of the enzyme shift about slightly to enable the active site to mold itself around the substrate. This is similar to the way a limp glove (enzyme) takes on the shape of your hand (substrate) as you put it on. The ability of a substrate to induce a tighter and more accurate fit for itself in the active site of an enzyme stabilizes the interaction between the two and enables catalysis to proceed.

Carbonic anhydrase is a vital blood enzyme that speeds up the removal of carbon dioxide from our tissues. It accelerates the reaction of water and carbon dioxide by a factor of nearly 10 million. In fact, a single carbonic anhydrase molecule can process more than 10,000 molecules of carbon dioxide in just one second. Without it, carbon dioxide would react with water so slowly that little of it would dissolve in the blood, and we would not be able to rid our bodies of carbon dioxide fast enough to survive. Carbonic anhydrase binds both carbon dioxide and water in its active site. By bringing these two substrates together in exactly the right positions, the active site of carbonic anhydrase promotes the reaction (**FIGURE 5.8**). Continual and rapid transfer of carbon dioxide from cells into the blood is a vital function made possible by the catalytic action of carbonic anhydrase.

Enzymes increase reaction rates by lowering the energy barrier

What does it take for atoms or molecules to react with each other? The reactants must bump into each other often enough, fast enough, and in the correct orientation to allow their chemical bonds to be rearranged. This set of conditions constitutes an energy "hump," or *energy barrier*, that must be overcome before the atoms or molecules can react. The minimum energy input that enables atoms and molecules to react is called the **activation energy**. A chemical reaction is fast if a large fraction of the potential reactants is above the activation energy threshold.

Heat is one type of energy that can push reactants over the energy barrier. Atoms and molecules move faster at higher temperatures, so raising the temperature causes more of them to collide forcefully enough to react. The reason a safety match does not ignite

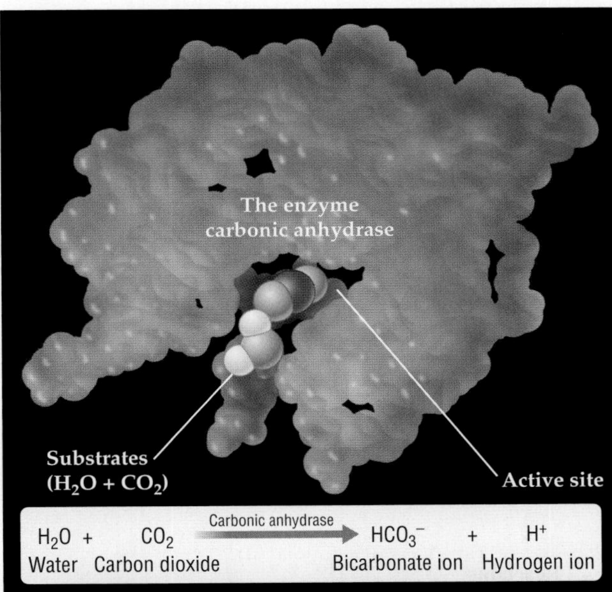

FIGURE 5.8 An Enzyme in Action: Carbonic Anhydrase

Our cell's spew out carbon dioxide as a byproduct of cellular respiration. To help get rid of the waste gas, carbonic anhydrase speeds up the reaction between carbon dioxide and water molecules. The resulting bicarbonate (HCO_3-) ions dissolve readily in blood, which whisks them away to the lungs, where bicarbonate is converted back to carbon dioxide and exhaled.

spontaneously at room temperature is that its chemical ingredients react very slowly with molecular oxygen in the air. But a small rise in temperature—from friction against a rough surface, for example—can provide the activation energy necessary for the chemicals in the match head to react with oxygen and burst into flame.

How do chemical reactions inside a cell overcome the energy barrier? Relying on heat as a source of activation energy is not a workable solution for most cellular processes, because heat acts indiscriminately and can damage complex molecules. The cell must be selective about which chemical reactions are allowed to proceed at any given time. An enzyme catalyst acts to lower the energy barrier, so that more of the reactants can cross it. As mentioned earlier, the larger the proportion of reactants that make it over the energy barrier, the faster a reaction proceeds.

The great majority of chemical reactions in the cell take place when the energy barrier for the reaction is lowered by an enzyme (**FIGURE 5.9**). An enzyme lowers the energy barrier of a reaction by bringing substrates close enough together. Like a skillful negotiator that is not above arm twisting, an enzyme prods the bound reactants by straining their chemical bonds. Bonds are broken, formed, or rearranged as a result.

Note that an enzyme simply increases the rate, or speed, of a reaction that could occur on its own—a spontaneous reaction. An enzyme does not provide energy for a reaction that is thermodynamically "uphill"—that is, one that would create products having more energy than the reactants had. An enzyme is not altered by the chemical reaction in any way. Like a good negotiator, it can be used over and over again. A small amount of enzyme can turn thousands of substrate molecules into product in a fraction of a second.

Concept Check

1. What is an enzyme? Describe two characteristics that are important to its function.
2. How does an enzyme catalyze a chemical reaction?

Concept Check Answers

1. An enzyme is a biological catalyst that speeds up chemical reactions. It acts on specific substrates and, is not used up by the reaction it catalyzes.

2. By lowering the activation energy barrier, making it possible for more reactant molecules to engage in the reaction and thereby effectively speeding up the reaction.

5.4 Metabolic Pathways

So far, we have discussed the activity of a single enzyme promoting a single chemical reaction, but enzymes working in isolation is not the most common state in the cell. Typically, enzymes are involved in catalyzing the steps in sequences of chemical reactions known as **metabolic pathways**. Metabolic pathways are responsible for the production of key biological molecules, such as amino acids and nucleotides. The enzyme-driven pathways of photosynthesis and cellular respiration are also organized in multistep sequences.

Multistep metabolic pathways can proceed rapidly and efficiently because enzymes in a particular pathway are usually located close together, and because the products of one enzyme-catalyzed step serve as substrates for the next reaction in the series. An enzyme in such a pathway does not have to "wait around" for a substrate to appear. The result is that the steps in the pathway are "pushed" toward one specific outcome.

We can represent the sequential reactions of a metabolic pathway as follows:

$$A \xrightarrow{E1} B \xrightarrow{E2} C \xrightarrow{E3} D$$

The enzyme E1 catalyzes the conversion of A to B, enzyme E2 catalyzes the conversion of B to C, and so on, ensuring that D is produced in the end.

The challenge faced by all enzyme-catalyzed reactions is the need for the enzyme and its substrates to encounter each other often enough. Enzymes in multistep

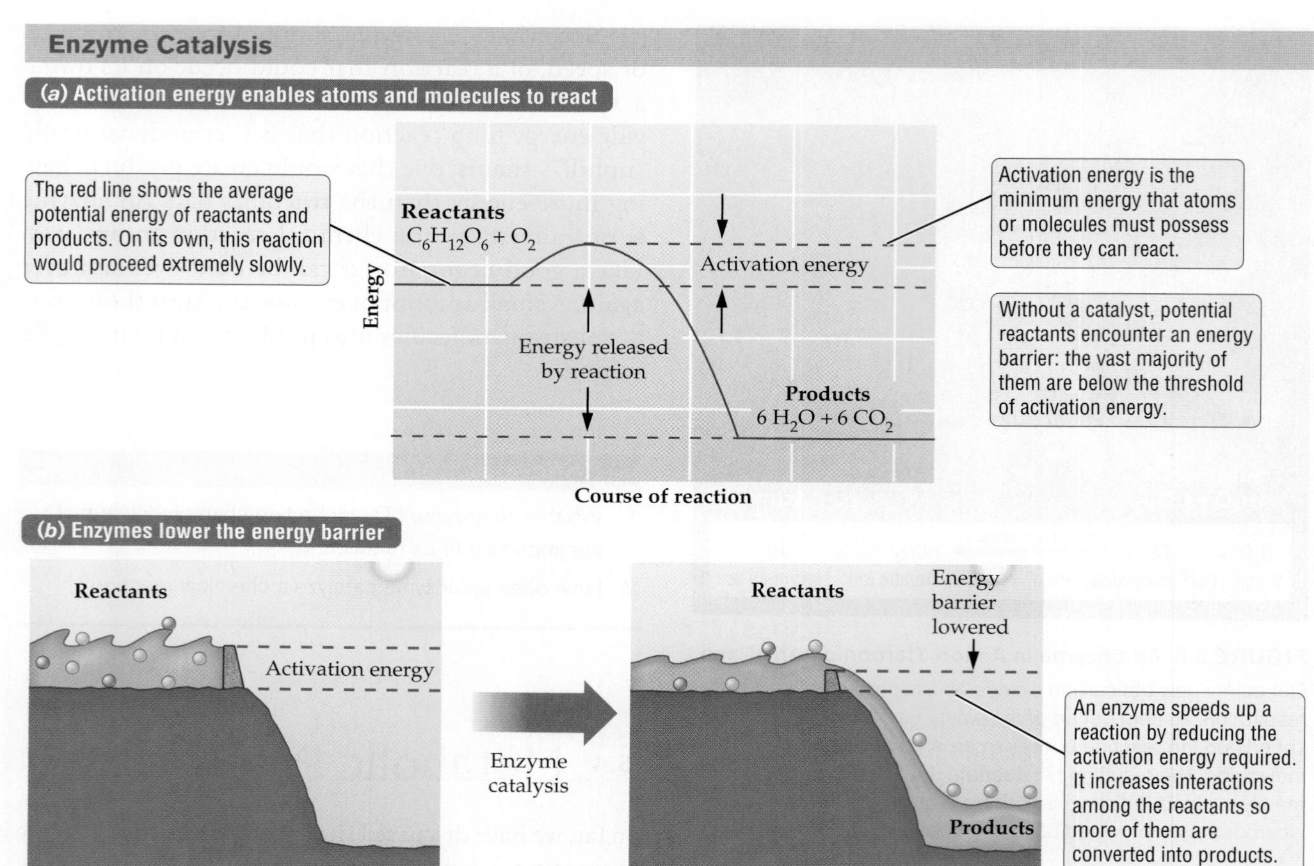
Enzyme Catalysis
(a) Activation energy enables atoms and molecules to react
The red line shows the average potential energy of reactants and products. On its own, this reaction would proceed extremely slowly.
Reactants
$C_6H_{12}O_6 + O_2$
Energy
Activation energy
Energy released by reaction
Products
$6\ H_2O + 6\ CO_2$
Course of reaction
Activation energy is the minimum energy that atoms or molecules must possess before they can react.
Without a catalyst, potential reactants encounter an energy barrier: the vast majority of them are below the threshold of activation energy.
(b) Enzymes lower the energy barrier
Reactants
Activation energy
Enzyme catalysis
Reactants
Energy barrier lowered
Products
An enzyme speeds up a reaction by reducing the activation energy required. It increases interactions among the reactants so more of them are converted into products.

FIGURE 5.9 Enzymes Reduce the Activation Energy Needed to Initiate a Reaction
(*a*) Potential reactants face an energy barrier they must overcome in order for the reaction to proceed at an appreciable rate. (*b*) In the analogy shown here, the reactants are represented by water in a reservoir, held back by a dam (energy barrier). On the left, the energy barrier is so high that most of the reactants (colored spheres) cannot spill over it. But if less activation energy is required, as shown on the right, a large proportion of the reactants can overcome the barrier and be turned into products (green).

pathways are frequently located close to one another in cells—for example, in cell membranes. Because the enzymes are close together, products of one reaction are, in effect, "aimed" at the next enzyme in the pathway.

Another way of increasing the odds of encounters between an enzyme and its substrate is to contain and concentrate the two inside a membrane-enclosed compartment, such as in a mitochondrion (**FIGURE 5.10**). As we saw in Chapter 3, membrane-bound organelles concentrate the proteins and chemical compounds required for certain biological processes. For example, the enzymes that break down organic molecules to release carbon dioxide are highly concentrated in the aqueous matrix that lies inside every mitochondrion. Other enzymes involved in the production of ATP are embedded in the inner mitochondrial membrane in a precise order. Likewise, the proteins and pigments that capture light and synthesize ATP are highly organized on the elaborate membranes of the chloroplast. We discuss these pathways in greater detail in the next chapter.

Concept Check

1. What is meant by the term "metabolic pathway"?
2. What is the adaptive value of organizing chemical reactions into metabolic pathways?

Concept Check Answers

1. A metabolic pathway is a multistep sequence of chemical reactions in which the products of one chemical reaction serve as the reactants for the next reaction in the series.

2. This type of metabolic organization delivers speed and efficiency by concentrating reactants where they are needed.

Arrangement of Metabolic Pathways

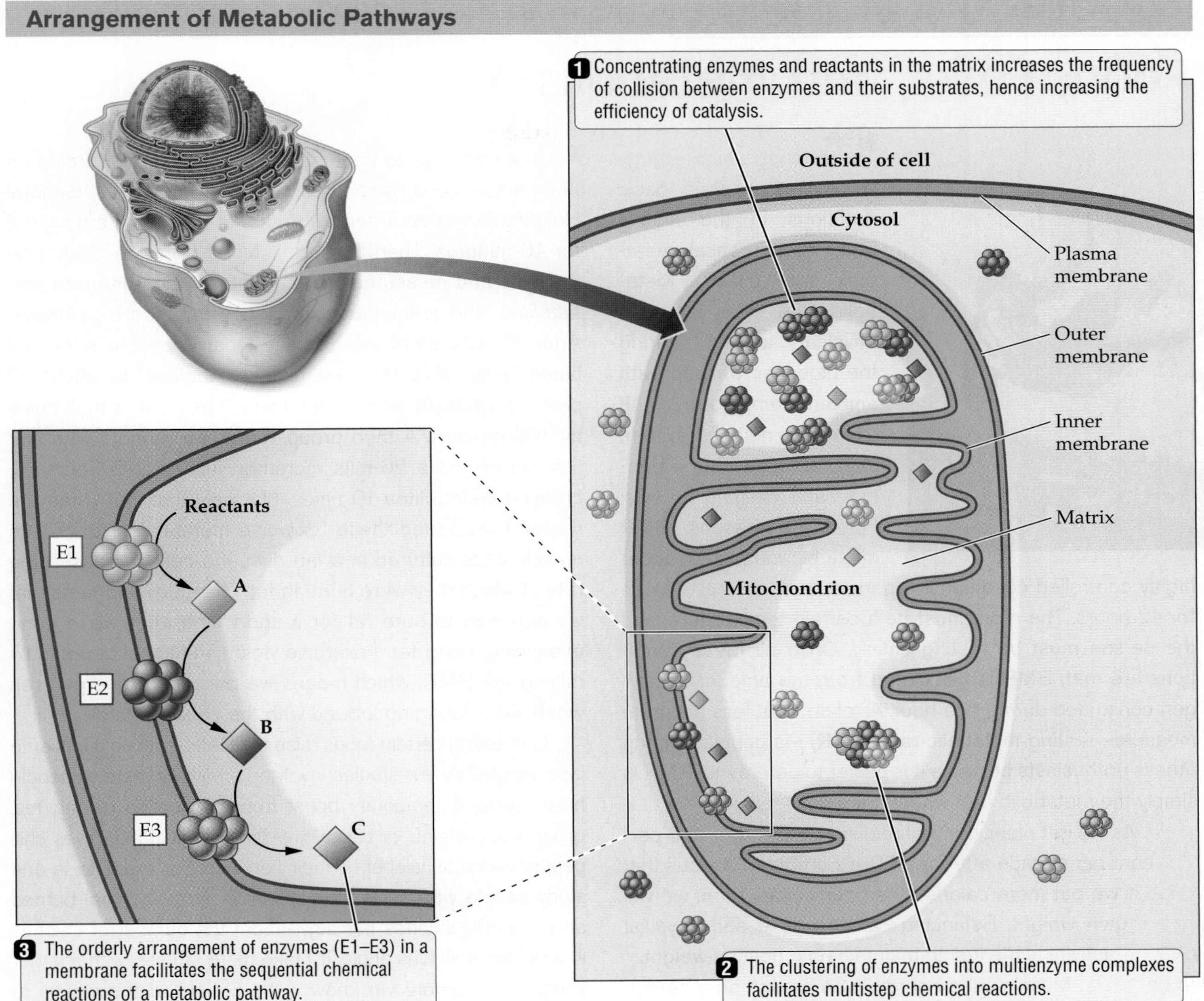

1 Concentrating enzymes and reactants in the matrix increases the frequency of collision between enzymes and their substrates, hence increasing the efficiency of catalysis.
Outside of cell
Cytosol
Plasma membrane
Outer membrane
Inner membrane
Matrix
Mitochondrion
Reactants
E1
A
E2
B
E3
C
3 The orderly arrangement of enzymes (E1–E3) in a membrane facilitates the sequential chemical reactions of a metabolic pathway.
2 The clustering of enzymes into multienzyme complexes facilitates multistep chemical reactions.

FIGURE 5.10 Metabolic Pathways Are Organized in Ways That Increase Their Efficiency

Enzymes are often arranged in the cell in ways that facilitate the orderly series of chemical reactions that make up a metabolic pathway. These arrangements include the concentration of enzymes in organelles (in this case, a mitochondrion), their localization in membranes, and their clustering in multienzyme complexes.

APPLYING WHAT WE LEARNED

Food, Folks, and Metabolism

The rate at which energy is used to sustain basic functions in the animal body is the basal metabolic rate (BMR). Metabolic rate goes up with physical activity, during the digestion of food, with ambient temperature, and even with nervous system stimulation such as watching an exciting movie. For these reasons, BMR must be measured under highly controlled conditions. A person must not have eaten for 12 hours. The room must be a certain temperature, and the person must be resting calmly. Once all these conditions are met, BMR is calculated from the amount of oxygen consumed during one hour. A related but less accurate measure—resting metabolic rate (RMR)—is popular among fitness enthusiasts because it is easier to determine; RMR is simply the metabolic rate when a person is resting.

As we get older, our metabolism slows—by 1 to 3 percent per decade after age 20. It's probably obvious that if we eat more calories than our bodies burn, we will gain weight. Balancing calorie intake and physical activity is the key to maintaining a healthy weight.

Muscle uses energy even at rest, so a person with more muscle expends more calories at rest than a person the same height and weight who has less muscle. In fact, every pound of muscle uses about 16 Calories per day simply for maintenance. So exercising burns calories directly and also builds muscle that burns more energy while you're sitting at your computer. Because most men have more muscle than women, men tend to have a higher BMR (**TABLE 5.2**). The body responds to endurance training by stepping up its energy-generating capacity—by increasing the number of mitochondria in muscle cells, for example. That's why some of the highest BMRs recorded belong to distance runners and other endurance athletes.

Helpful to Know

A calorie is one way of measuring heat energy. Chemists define 1 kilocalorie (1,000 calories) as the amount of energy needed to raise the temperature of 1 liter of water by 1°C. Somewhat confusingly, nutritionists and food manufacturers use the term "Calorie" (with a capital "C") to mean one kilocalorie. The Calories listed on Nutrition Facts labels are what chemists would call kilocalories. So, a dietitian would tell you that a stick of celery contains 9 Calories, whereas a chemist would say it has 9 kilocalories.

In an attempt to understand the effects of exercise on the human body, researchers at Massachusetts General Hospital put three different groups of people on a treadmill for 10 minutes. Then they drew blood from the study participants and measured the amounts of 200 different *metabolites*—the various end products of metabolic pathways. After 10 minutes of exercise, the least fit people increased blood levels of certain "exercise metabolites" by about 50 percent, while fit adults increased these same molecules by 100 percent. A third group, marathon runners who had just completed a 26-mile marathon lasting 3–5 hours, increased metabolites 10 times. Not only that, but when the researchers added these "exercise metabolites" to mouse muscle cells cultured in a lab dish, the cells increased the rate at which they were burning fat. The study suggests that we continue to burn fat for a short time after we're done exercising. Long-term exercise yields additional benefits by raising the BMR, which means we burn more energy even when we're lounging around with the couch potatoes.

Can eating certain foods raise the BMR, making it easier to lose weight? A few studies involving small numbers of people have shown a temporary boost from consuming oolong tea, green tea, caffeine, or capsaicin—the chemical that gives chili peppers such as jalapeños their fiery bite. For example, in one study people who consumed capsaicin at every meal burned an extra 100 calories per day—about the number of calories in a golden delicious apple. It'll take more studies, with a larger sample size, before we know if the hot-chili diet amounts to more than a hill of beans!

TABLE 5.2 Basal Metabolic Rates for Average Adults and Elite Athletes

	WEIGHT		*BMR*	
	(KG)	**(LB)**	**(WATTS/H)**	**(CALORIES/H)**
Woman	60	132	68	58
Man	70	154	87	75
Endurance athlete	70	154	193	166

BIOLOGY IN THE NEWS

Lemurs' Long-Buried Secrets Revealed

ABC SCIENCE • March 3, 2013

Scientists curious about the winter vanishing act of a Madagascar dwarf lemur were astonished to find the animals curled up asleep in underground burrows.

The discovery, published in *Nature Scientific Reports*, makes the island country's eastern dwarf lemurs the only primates in the world known to hibernate underground ...

"You don't see them, trap them or find them during the dry season (winter time) while walking the forests at night," says study co-author Marina Blanco of Germany's Hamburg University ...

And so the team fitted radio-transmitter collars on 12 lemurs from two eastern species in summer, and waited ...

Setting out in winter with signal trackers, the team fully expected to find the lemurs sleeping in tree holes.

"We were tracking the collar's signal and pointing our antenna up in the air, towards the tip of a tree. But the signal was coming from the ground, so we thought the animal had lost the collar," says Blanco.

"We looked around and didn't see anything so we started to dig up the area and found a furry ball, the dwarf lemur was curled up and cold to the touch, still wearing its collar."

The tiny bundles weigh about 250 to 350 grams, depending on which species they belong to.

They hibernate for anything from three to six months buried 10 to 40 centimetres under a spongy layer of tree roots, soil and decaying plant matter ...

During the Madagascar winter, lemurs are exposed to drastic daily temperature fluctuations of as much as 30 degrees Celsius.

In the highland rainforests, ambient temperatures can drop to between zero and five deg C in winter—cold for animals used to summer averages in the 30s.

So in retrospect, underground hibernation in the tropics is not surprising, say the researchers.

"Burrows provide more insulation than tree holes or nests given that soil temperature is generally lower than ambient temperature during the day but higher during the night," they write.

Animals that do not burn up food just to keep themselves warm and toasty are popularly described as cold-blooded. Cold-blooded animals—or *ectotherms*—must gain heat from their environment in order to warm themselves. Snakes, lizards, most insects, and other invertebrates are typically cold-blooded. Because their basal metabolic rate (BMR) is low, their food needs are low too. The downside of their metabolic penny pinching is that cold-blooded animals cannot be up and about in frosty weather.

Not many amphibians and reptiles live in places that freeze over in the winter, and the few species that do must cope with the bitter cold or migrate to balmy places. Hibernation is one coping strategy. Hibernators drop their BMR as cold weather arrives, allowing their body temperature to plummet. Even many warm-blooded animals (*endotherms*)—such as bears, squirrels, possums, skunks, and badgers—deal with cold winters by hibernating. Some hibernators—ground squirrels, for example—allow their body temperature to drop to within a degree or so of freezing. The body temperature of black bears drops by only a few degrees, but every little bit helps when food is scarce or absent.

The discovery of hibernation in the tropics took everyone by surprise. However, winter temperatures can plunge to near freezing in higher elevations of Madagascar. A dwindling food supply, combined with cold, makes hibernation an effective strategy. The fat-tailed lemur, a relative of the dwarf lemur that hibernates in tree holes rather than underground, drops its BMR and lives off the chemical energy stockpiled in its tail. The tiny dwarf lemur balances its energy budget by not only dropping its BMR but also minimizing heat loss by burrowing underground. The rarity of these game plans have both startled and delighted scientists and nature enthusiasts alike.

Evaluating the News

1. What are the adaptive advantages and disadvantages of being warm-blooded?

2. Explain why hibernation is a good adaptive strategy for the fat-tailed lemur and the dwarf lemur. Which metabolic pathway—catabolism or anabolism—predominates in the winter in the fat-storing cells in the tail of the fat-tailed lemur?

CHAPTER REVIEW

Summary

5.1 The Role of Energy in Living Systems

- Energy is the capacity to do work. Work is the capacity to bring about a change in a system.
- Potential energy is the energy stored in any system as a consequence of its position. Chemical energy is one type of potential energy.
- Kinetic energy is the energy a system possesses as a consequence of its state of motion. Heat energy, a form of kinetic energy, is that portion of the total energy of a particle of matter that can flow to another particle.
- The first law of thermodynamics states that energy can be converted from one form to another but is never created or destroyed.
- The second law of thermodynamics states that the natural tendency of the universe is to become less organized. The creation of biological order therefore requires energy. The creation of internal order in living organisms is always accompanied by the transfer of disorder to the environment, generally in the form of metabolic heat.
- The sun is the source of energy fueling most living organisms. Producers capture the sun's energy through photosynthesis. Plants, algae, and some bacteria gain energy from their environment through photosynthesis. Many producers and consumers use cellular respiration to extract usable energy from food molecules.
- Matter, in the form of chemical elements such as carbon, cycles between living organisms and their environment.

5.2 Metabolism

- All the many chemical reactions involved in the capture, storage, and use of energy by living organisms are collectively known as metabolism.
- Energy-releasing breakdown reactions are catabolism; energy-requiring synthesis reactions are anabolism.
- The energy-rich molecule ATP supplies much of the energy needed to fuel cellular activities.
- In oxidation, electrons are lost from a molecule, atom, or ion; in reduction, electrons are gained.
- The minimum energy required to initiate a chemical reaction is called the activation energy. Most chemical reactions must overcome an activation energy barrier to proceed at an appreciable rate.

5.3 Enzymes

- Enzymes are biological catalysts that speed up chemical reactions. Enzymes position bound reactant molecules in such a way that they collide more often in the orientation that favors product formation. Like all catalysts, enzymes lower the activation energy barrier of a reaction.
- The activity of enzymes is highly specific. Each enzyme binds to a specific substrate or substrates and catalyzes a specific chemical reaction. The specificity of an enzyme is based on the three-dimensional shape and chemical characteristics of its active site.
- The three-dimensional shape of an enzyme, and therefore its activity, can be affected by temperature, pH, and salt concentration. Some enzymes must work with other chemicals, called cofactors, to be maximally effective.

5.4 Metabolic Pathways

- A metabolic pathway is a multistep sequence of chemical reactions, with each step catalyzed by a different enzyme.
- Metabolic pathways proceed rapidly and efficiently because all necessary components are placed close together, at high concentrations, and in the correct order. The products of one enzyme-catalyzed step serve as substrates for the next reaction in the series.

Key Terms

activation energy (p. 114)
active site (p. 114)
ADP (p. 110)
anabolism (p. 110)
ATP (p. 110)
catabolism (p. 110)
catalyst (p. 112)
cellular respiration (p. 111)
chemical energy (p. 106)
energy (p. 106)
enzyme (p. 112)
first law of thermodynamics (p. 107)
heat energy (p. 106)
induced fit model (p. 114)
kinetic energy (p. 106)
metabolic pathway (p. 115)
metabolism (p. 109)
oxidation (p. 110)
photosynthesis (p. 111)
potential energy (p. 106)
product (p. 112)
reactant (p. 112)
redox reaction (p. 110)
reduction (p. 110)
second law of thermodynamics (p. 107)
substrate (p. 113)

Self-Quiz

1. Which of the following statements is true?
   a. Cells can produce their own energy from nothing.
   b. Cells use energy only to generate heat and move molecules around.
   c. Cells obey the same physical laws of energy as the nonliving environment.
   d. Most animals obtain energy from minerals to fuel their metabolic needs.

2. Living organisms need energy to
   a. organize chemical compounds into complex biological structures.
   b. decrease the disorder of the surrounding environment.
   c. transform metabolic heat into kinetic energy.
   d. keep themselves separate from the nonliving environment.

3. The carbon atoms contained in organic compounds such as proteins
   a. are manufactured by cells for use in the organism.
   b. are recycled from the nonliving environment.
   c. differ from those found in CO_2 gas.
   d. cannot be oxidized under any circumstances.

4. Oxidation is the
   a. removal of oxygen atoms from a molecule.
   b. gain of electrons by an atom.
   c. loss of electrons by an atom.
   d. synthesis of complex molecules.

5. Which of these molecules is in a reduced state?
   a. CO_2
   b. N_2
   c. O_2
   d. CH_4

6. The minimum input of energy that initiates a chemical reaction
   a. is called activation energy.
   b. is independent of the laws of thermodynamics.
   c. is known as the activation energy barrier.
   d. always takes the form of heat.

7. Activation energy is most like
   a. the energy released by a ball rolling down a hill.
   b. the energy required to push a ball from the bottom of a hill to the top.
   c. the energy required to get a ball over a hump and onto a downward slope.
   d. the energy that keeps a ball from moving.

8. Enzymes
   a. provide energy for anabolic but not catabolic pathways.
   b. are consumed during the reactions that they speed up.
   c. catalyze reactions that would otherwise never occur.
   d. catalyze reactions that would otherwise occur much more slowly.

9. The active site of an enzyme
   a. has the same shape for all known enzymes.
   b. binds the products of reaction, not the substrate.
   c. does not play a direct role in catalyzing a reaction.
   d. can bring molecules together in a way that promotes a reaction between them.

10. Metabolic pathways
    a. always break down large molecules into smaller units.
    b. only link smaller molecules together to create polymers.
    c. are often organized as a multistep sequence of reactions.
    d. occur only in mitochondria.

Analysis and Application

1. Think back to what you ate for breakfast. What has become of all the chemical energy locked in those food molecules? List the different types of energy transformations that have occurred in your body as you have gone about your day, starting with the chemical energy in your breakfast.
2. Describe the role of the second law of thermodynamics in living systems.
3. Compare anabolism and catabolism. Is photosynthesis an anabolic or catabolic process?
4. Explain what is wrong with this statement: An enzyme provides energy for reactions that cannot proceed without an investment of energy.
5. Explain the induced fit model of interaction between an enzyme and its substrate.
6. A Chinese herbal medicine called *ma huang* (*Ephedra sinica*) increases BMR significantly and was once widely used in weight loss pills, usually in combination with caffeine. Its use in supplements was banned in 2006, after reports of deaths among users of the herb. Research the metabolic effects of ephedrine, the active ingredient in the herb, to explain why a very high metabolism can kill.

Ephedra sinica

6 Photosynthesis and Cellular Respiration

BIG-WAVE SURFER JAY MORIARITY. Jay Moriarity—renowned for a spectacular wipeout that ran on the cover of *Surfer* magazine in May 1995—died in 2001 while on a photo shoot in the Maldive Islands, in the Indian Ocean. But Moriarity didn't die surfing. He died meditating 45 feet below the surface of the water without oxygen—a practice called "static apnea" that is common among surfers and divers.

Every Breath You Take

How long can you hold your breath? A minute? Two minutes? Doctors fear brain death in anyone who has gone without oxygen for more than 5 minutes. But German engineering student Tom Siestas held his breath underwater for 11 minutes 35 seconds and claimed the men's record for the rather intimidating sport of *static apnea* (literally, "motionless, without air"). Russian free diver Natalia Molchanova claimed the women's record with a time of 8 minutes 23 seconds.

Athletes like Siestas and Molchanova are extremely unusual—certainly in their training and possibly in their genetic makeup—which is why static apnea should be left to trained professionals. In fact, health professionals suspect that even experts may be putting their long-term health at risk with their arduous oxygen deprivation training regimens, not to mention the even more dangerous competitions. In 2001, famed big-wave surfer Jay Moriarity drowned while practicing static apnea as part of his regular training. He was last seen alive while meditating 45 feet below the surface of the Indian Ocean.

Static-apnea training includes endurance exercise (which depends heavily on high oxygen uptake) and training at high altitude (which also improves oxygen delivery). Endurance athletes such as 2013 Tour de France winner Chris Froome have a phenomenal ability to deliver oxygen to their tissues. Before a competition, the British bicyclist doubles his caloric intake to about 7,000 Calories per day, most of it from carbohydrates. In the digestive tract, carbohydrates break down into glucose, a key molecule in the metabolic pathways we explore in this chapter.

Is efficient use of oxygen and glucose the main difference between elite endurance athletes and the rest of us? What species is the static-apnea champion?

Whether elite athletes or couch potatoes, we all need oxygen and glucose to live. But *why* do our cells need these molecules? Where do these molecules come from anyway? We tackle these questions in this chapter. You'll see what becomes of your lunch after it reaches a cell in your body. And we'll give you the greatest recipe ever: how to make lunch from sunshine, air, and water.

MAIN MESSAGE Photosynthesis and cellular respiration are complementary processes; these pathways furnish chemical energy for cells.

KEY CONCEPTS

- All cells need energy carriers, such as ATP, to store and deliver usable energy.
- In plants and algae, photosynthesis takes place in special organelles called chloroplasts.
- During the light reactions of photosynthesis, sunlight and water are used to produce energy carriers, releasing oxygen gas in the process.
- In the Calvin cycle reactions of photosynthesis, the energy carriers are used to manufacture sugars from carbon dioxide.
- Photosynthesis in C_4 plants is highly efficient because Calvin cycle reactions take place in tissues that maintain a high CO_2 environment.
- In plants with CAM (crassulacean acid metabolism), photosynthesis runs on CO_2 obtained through stomata that open only at night, when the risk of dehydration is lower.
- Most eukaryotes rely on cellular respiration, which requires oxygen (O_2) to extract energy from sugars and other food molecules.
- Cellular respiration has three main stages: glycolysis, the Krebs cycle, and oxidative phosphorylation.
- Fermentation enables certain organisms and certain cell types to generate ATP without oxygen, through glycolysis alone.

EVERY BREATH YOU TAKE 123

6.1 Molecular Energy Carriers 124

6.2 An Overview of Photosynthesis and Cellular Respiration 126

6.3 Photosynthesis: Energy from Sunlight 128

6.4 Cellular Respiration: Energy from Food 137

APPLYING WHAT WE LEARNED 143
Waiting to Exhale

ENERGY IS NECESSARY for all types of cellular activity. *Metabolism* encompasses all the chemical reactions involved in the capture, storage, and utilization of energy in a cell, and it is a fundamental necessity for every living thing. A *metabolic pathway* consists of a series of chemical reactions by which energy is transformed as biomolecules are changed, step by step, into either simpler or more complex forms. Many of these reactions are critically dependent on small molecules known as energy carriers, which function as an energy delivery service inside the cell.

We begin this chapter by exploring the nature of energy carriers. With this knowledge we can better appreciate two of the most important and widespread metabolic pathways on Earth: *photosynthesis*, which captures light energy to make sugars, and *cellular respiration*, which releases energy from food molecules to fuel cellular activities. Photosynthesis gives off oxygen (O_2), a by-product that is crucial for the survival of consumers like us because it is necessary for cellular respiration. While photosynthesizers *use up* carbon dioxide (CO_2) and water (H_2O) to make sugars, producers and consumers alike *release* these molecules as they extract energy from sugars during cellular respiration (**FIGURE 6.1**).

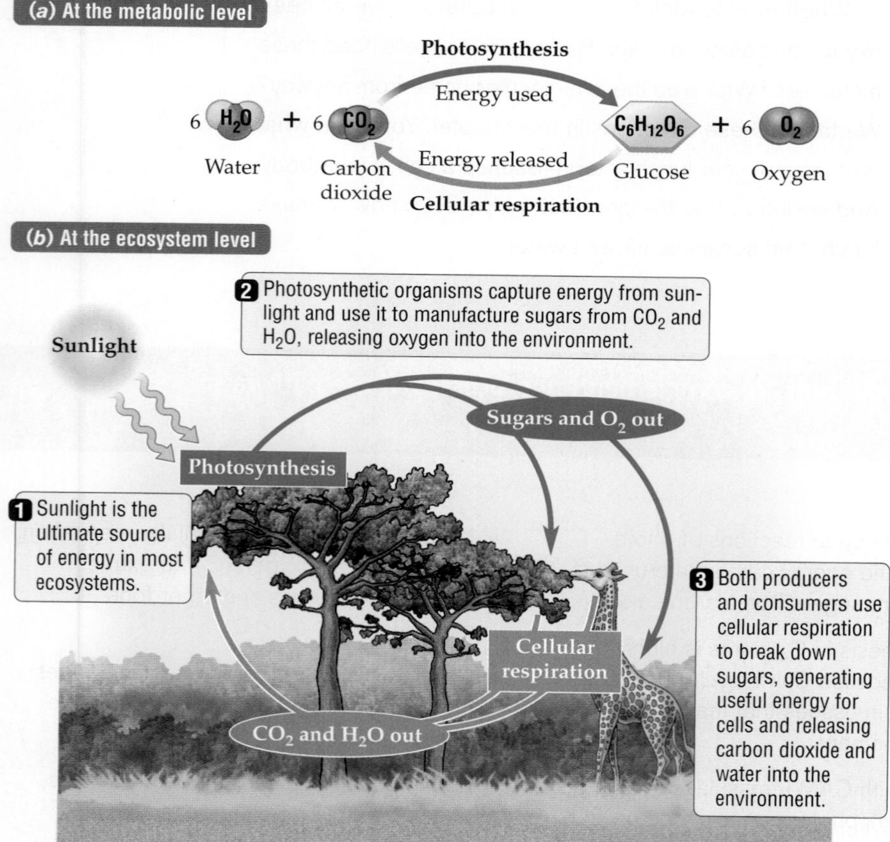

Photosynthesis and Cellular Respiration Are Complementary
(a) At the metabolic level
Photosynthesis
Energy used
6 H_2O + 6 CO_2
$C_6H_{12}O_6$ + 6 O_2
Water
Carbon dioxide
Energy released
Glucose
Oxygen
Cellular respiration
(b) At the ecosystem level
2 Photosynthetic organisms capture energy from sunlight and use it to manufacture sugars from CO_2 and H_2O, releasing oxygen into the environment.
Sunlight
Sugars and O_2 out
Photosynthesis
1 Sunlight is the ultimate source of energy in most ecosystems.
3 Both producers and consumers use cellular respiration to break down sugars, generating useful energy for cells and releasing carbon dioxide and water into the environment.
Cellular respiration
CO_2 and H_2O out

FIGURE 6.1 The Relationship between Photosynthesis and Cellular Respiration

Can animals exist without plants? Can plants exist without animals?

6.1 Molecular Energy Carriers

You may think you could run very nicely on bacon double cheeseburgers and triple fudge sundaes, but you should know that no food is of any *direct* use to the trillions of cells in your body. Cellular activities are powered not by bacon fat, beef protein, or carbohydrate from a sesame bun, but by tiny molecules that are generically known as **energy carriers**: small organic molecules specialized for receiving, storing, and delivering energy within the cell. A molecular energy carrier is rather like a rechargeable battery, and the cell is like a miniature machine that runs on energy from many millions of these batteries.

Energy carriers become "fully charged" when they receive energy from metabolic pathways that release energy, and they "lose the charge" as they deliver energy to the many thousands of chemical reactions that could not proceed without this energy (**FIGURE 6.2**). Any cell that runs out of fully charged energy carriers is a dead cell.

Of the common energy carriers in a cell, *ATP* (adenosine triphosphate) is the most versatile: it delivers energy to the largest number and greatest diversity of cellular processes. ATP releases stored energy when it loses a terminal phosphate (P) to become *ADP* (adenosine *di*phosphate). The released energy fuels cellular activities ranging from the manufacture of biomolecules to cell division (see Figure 6.2).

What becomes of the low-energy ADP molecules that are spun off every time an ATP molecule discharges

Cellular Activities Are Fueled by Energy Carriers

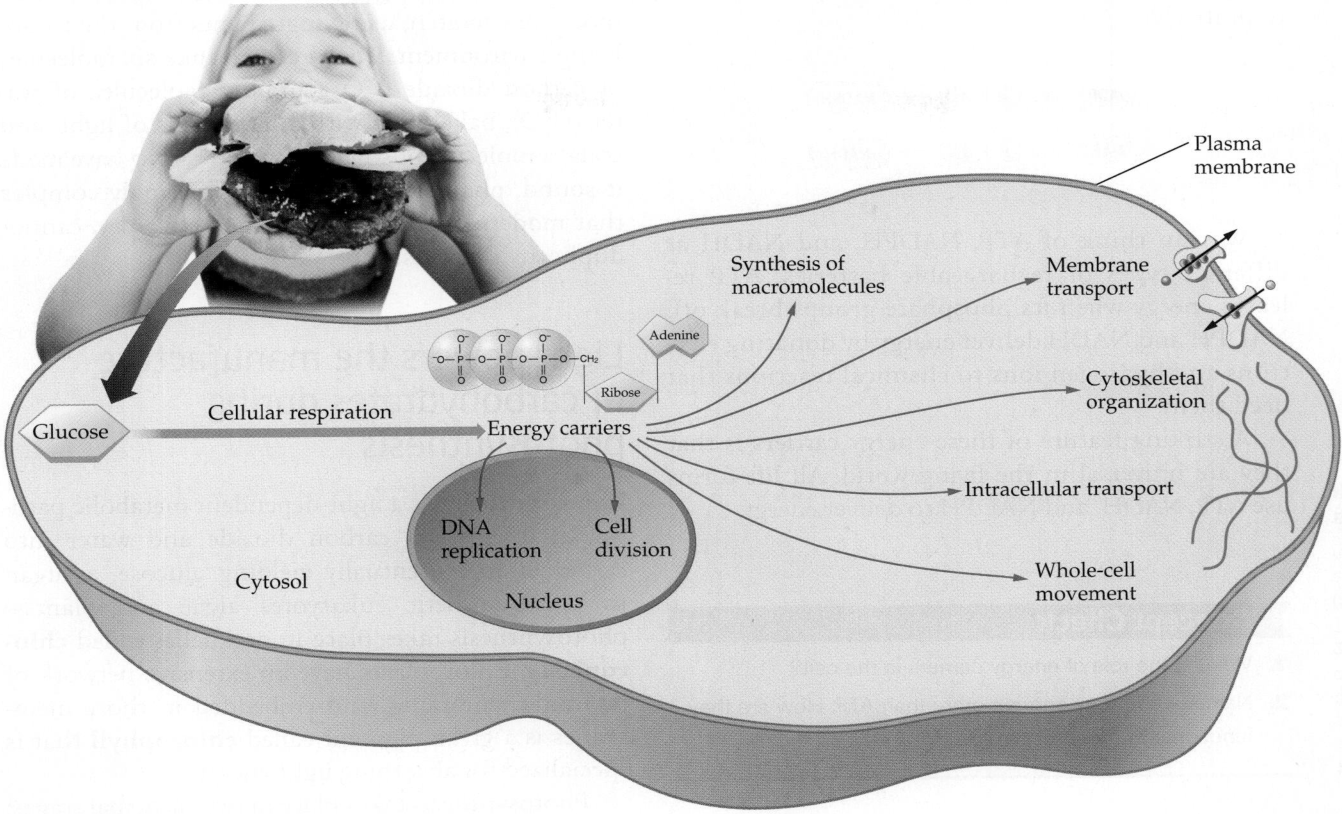

FIGURE 6.2 Lunch Must Be Turned into Energy Carriers to Do Us Any Good
Just about all cellular processes run on molecular energy carriers. Your body needs a mechanism to turn lunch into energy carriers. Cellular respiration is a pathway that generates ATP, the universal energy carrier used by all cells.

its energy? ADP can combine with phosphate and be converted back to ATP, but this is not a simple reaction. The recharging requires energy-releasing pathways, such as photosynthesis and cellular respiration. In photosynthesis, the energy for turning ADP into ATP comes from sunlight. In cellular respiration, this energy comes from the breakdown of food molecules. These two pathways are rather like battery chargers, turning low-energy ADP molecules into high-energy ATP molecules:

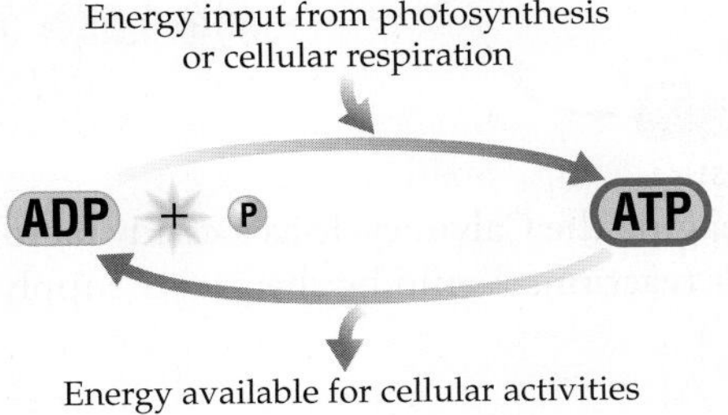

ATP is not the only energy carrier that cells rely on. **NADPH** and **NADH** are energy carriers that hold energy and matter in the form of certain loosely bound electrons and hydrogen atoms. (Recall that one hydrogen atom is the same as one electron plus one hydrogen ion, H^+.) NADPH delivers electrons and hydrogen ions to metabolic pathways that build macromolecules (*anabolic* pathways); NADH specializes in picking up the electrons and hydrogen ions released by metabolic pathways that take macromolecules apart (*catabolic* pathways).

How do NADPH and NADH acquire their high-energy, loosely bound electrons and hydrogen ions? $NADP^+$ (nicotinamide adenine dinucleotide phosphate) is the precursor to NADPH, and NAD^+ (nicotinamide adenine dinucleotide) is the precursor to NADH, in the same way that ADP is the precursor to ATP. Each of these precursors can pick up two electrons plus a hydrogen ion and be transformed into the high-energy forms NADPH and NADH, respectively:

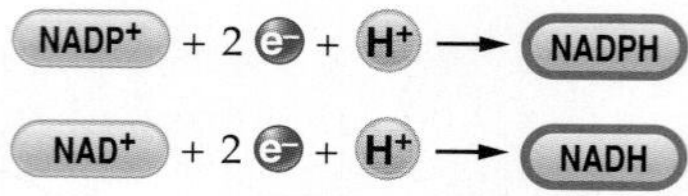

We can think of ATP, NADPH, and NADH as different types of rechargeable batteries. ATP releases energy when its phosphate groups break off. NADPH and NADH deliver energy by donating electrons and hydrogen ions to chemical reactions that need them.

A striking feature of these energy carriers is that they are universal in the living world. All life-forms use ATP, NADH, and NADPH to deliver energy.

Concept Check

1. What is the role of energy carriers in the cell?
2. Name two energy carriers other than ATP. How are they functionally different from ATP?

Concept Check Answers

1. Energy carriers are specialized for receiving, storing, and delivering the energy that powers cellular activities.
2. NADPH and NADH. They deliver atoms (hydrogen ions and electrons) to chemical reactions.

6.2 An Overview of Photosynthesis and Cellular Respiration

Photosynthesis and cellular respiration are two of the most important metabolic pathways in living organisms. In most ecosystems on Earth, the sun is the ultimate source of energy for living cells. Plants and other photosynthetic organisms, which trap energy from sunlight to manufacture their own food, are the *producers* in most ecosystems. Producers, in turn, support *consumers,* which acquire energy by eating producers or other consumers. Whether producers or consumers, all eukaryotes and many prokaryotes need cellular respiration. Cellular respiration enables organisms to harvest the chemical energy that is locked in the covalent bonds of food molecules and turn it into a directly usable form: the chemical energy of ATP.

The only photosynthetic producers on Earth are certain groups of bacteria, the protists known as algae, and plants. Photosynthetic organisms do not need to eat other organisms, because they make their food from scratch, using ingredients from their nonliving environment. Their recipe? Take six molecules of carbon dioxide (CO_2) and six molecules of water (H_2O), bake them with a few beams of light, and voilà, a molecule of sugar! As simple as we have made it sound, photosynthesis is so staggeringly complex that modern chemists, for all their wizardry, cannot duplicate it in a test tube.

Light powers the manufacture of carbohydrates during photosynthesis

Photosynthesis is a light-dependent metabolic pathway that converts carbon dioxide and water into carbohydrates, eventually yielding glucose, a sugar. In photosynthetic eukaryotes—algae and plants—photosynthesis takes place in organelles called **chloroplasts**. Chloroplasts have an extensive network of internal membranes, and embedded in those membranes is a green pigment called **chlorophyll** that is specialized for absorbing light energy.

Photosynthesis takes place in two principal stages: the *light reactions* and the *Calvin cycle* (**FIGURE 6.3**).

- During the **light reactions**, water molecules are split and oxygen gas is released when chlorophyll absorbs light energy. Electrons and protons (H^+) extracted from water molecules ultimately generate ATP and NADPH.
- In the **Calvin cycle**, a series of enzyme-catalyzed reactions converts carbon dioxide (CO_2) into sugar through the use of energy delivered by ATP and of electrons and hydrogen ions donated by NADPH.

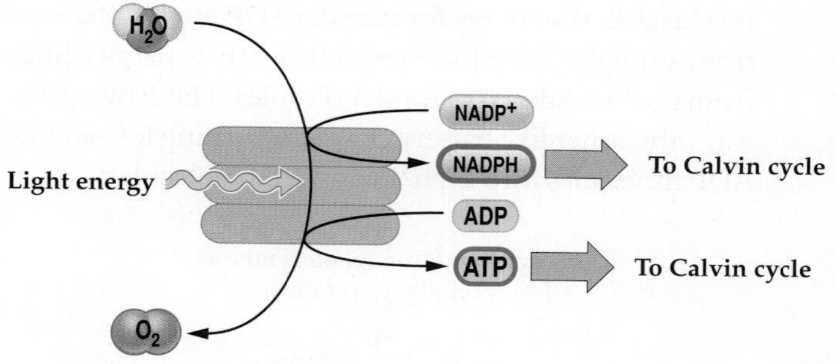

If we imagine the Calvin cycle as a candy factory, then the light reactions would be the power supply for the factory.

Energy from sugars is used to make ATP during cellular respiration

Cellular respiration is an oxygen-dependent metabolic pathway through which food molecules, such as sugars, are broken down into carbon dioxide and water, and the released energy is used to generate ATP. We refer to this process as *cellular* respiration to distinguish it from *whole-body* respiration, which, in the case of animals with lungs, is the inhaling and exhaling of air (breathing). However, breathing in animals, including humans, is directly related to cellular respiration. The air we breathe out is rich in carbon dioxide and water vapor, both by-products of cellular respiration. We must inhale air rich in oxygen because oxygen is essential to cellular respiration.

Cellular respiration begins in the cytosol and is completed in the **mitochondrion** (*MYE*-**tuh**-*KAHN*-**dree-un**; plural "mitochondria"), an organelle enclosed in double membranes. Mitochondria are especially abundant in cell types that have large energy demands, such as skeletal muscles, brain cells, and liver cells. Liver tissue, for example, contains over a thousand mitochondria per cell.

Cellular respiration takes place in three stages: *glycolysis*, the *Krebs cycle*, and *oxidative phosphorylation* (**FIGURE 6.4**).

- The first stage, **glycolysis** (**glye**-*KAH*-**luh-sus**), takes place in the cytosol, the fluid portion of the cytoplasm. During glycolysis, sugars (mainly glucose) are split to make a three-carbon compound (*pyruvate*), releasing two molecules of ATP and two molecules of NADH for each glucose molecule that is split.
- In the **Krebs cycle**, pyruvate enters the mitochondrion and is completely degraded by enzyme-driven reactions, releasing carbon dioxide. The degradation of carbon backbones by the Krebs cycle produces a large bounty of energy carriers, including ATP and NADH.
- In the last and final step of cellular respiration, **oxidative phosphorylation**, the chemical energy of NADH is converted into the chemical energy of ATP through a membrane-dependent process. Electrons and hydrogen atoms removed from NADH are handed over to molecular oxygen, creating water. In the process, a large amount of ATP is generated. Oxidative phosphorylation generates at least 15 times more ATP than does glycolysis alone.

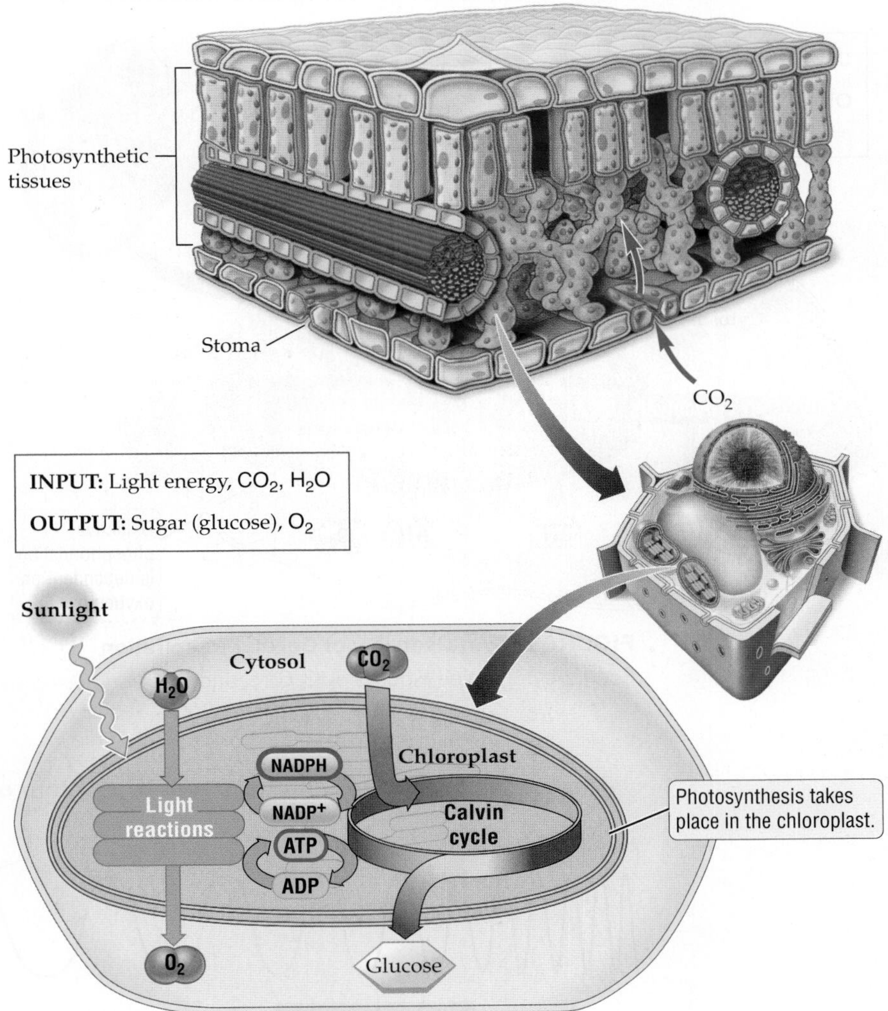

FIGURE 6.3 An Overview of Photosynthesis
For more detail, see Figures 6.8 and 6.9.

Oxidative phosphorylation needs oxygen gas. We cannot survive more than a few minutes without oxygen because, in the absence of this gas, our cells cannot make enough ATP to run the many activities that rely on energy delivered by this extraordinary energy carrier.

Concept Check

1. How are the light reactions different from the Calvin cycle reactions?
2. Explain what is wrong with this statement: Cellular respiration occurs in consumers like us, but not in plants.

Concept Check Answers

1. In the light reactions, light energy is used to split water molecules, releasing oxygen gas, and light energy is converted into the chemical energy of ATP and NADPH. In the Calvin cycle, carbon dioxide is fixed into sugars using energy from ATP and hydrogen ions and electrons delivered by NADPH.

2. Cellular respiration also takes place in producers like plants, which need a pathway to extract energy from food molecules such as glucose.

The Three Stages of Cellular Respiration

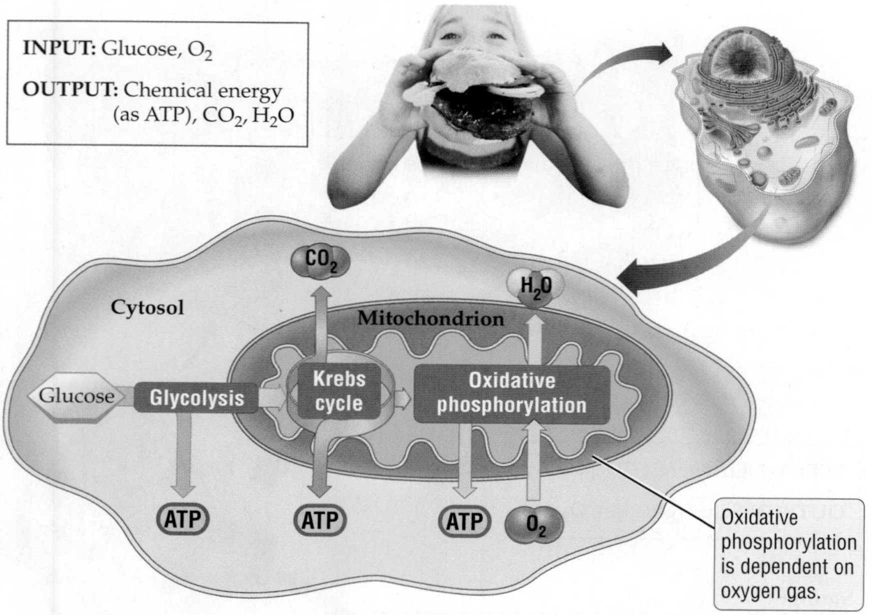

FIGURE 6.4 An Overview of Cellular Respiration
For more detail, see Figures 6.13, 6.15, and 6.16.

6.3 Photosynthesis: Energy from Sunlight

The next time you walk outside, look at the plants around you and consider the critical role they play in supporting the web of life. Through photosynthesis, plants support humans and a great variety of other organisms, which depend on them for both food and oxygen. Before we examine photosynthesis in greater detail, let's consider the nature of light, the driving force for these reactions.

The color of an object is determined by the wavelengths of visible light it reflects

Light can be thought of as a stream of massless particles, called **photons**, that have wavelike characteristics. Every photon contains a fixed amount of energy. The energy of a photon is related to its wavelength, the distance between one wave crest and the next. The energy, and therefore the wavelength, of photons covers

The Electromagnetic Spectrum and the Colors of Objects

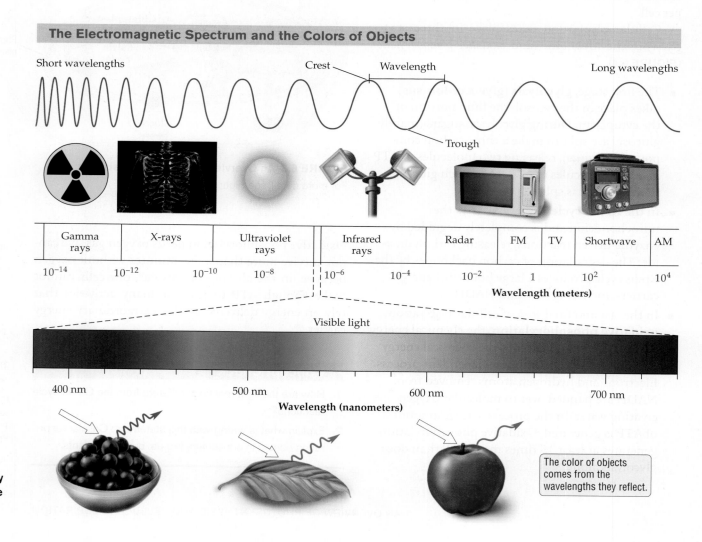

Short wavelengths
Crest
Wavelength
Long wavelengths
Trough

Gamma rays	X-rays	Ultraviolet rays	Infrared rays	Radar	FM	TV	Shortwave	AM

10^{-14} 10^{-12} 10^{-10} 10^{-8} 10^{-6} 10^{-4} 10^{-2} 1 10^{2} 10^{4}

Wavelength (meters)

Visible light

400 nm 500 nm 600 nm 700 nm

Wavelength (nanometers)

The color of objects comes from the wavelengths they reflect.

FIGURE 6.5 Why Objects Have the Colors They Do

a broad span, known as the electromagnetic spectrum (**FIGURE 6.5**). A photon with a *short wavelength* has *more energy* than a photon with a longer wavelength (compare gamma rays and radio waves in Figure 6.5).

Visible light is the portion of the electromagnetic spectrum that our eyes can perceive. It includes all photons with wavelengths between 300 and 780 nanometers (nm). When blended together, this range of the spectrum looks white to us. Isaac Newton famously split the wavelengths within white light, using a glass prism, to reveal the "seven rainbow colors," with violet at a wavelength of 300 nm and red at 780 nm.

The color of an object is determined by the wavelengths of light that bounce off it and reach our eyes. A white object *reflects* all wavelengths in the visible part of the electromagnetic spectrum. A black object *absorbs* all of these wavelengths, so no visible light bounces off it to enter our eyes.

A red apple reflects red wavelengths, and a blueberry reflects blue wavelengths. Chlorophyll, the light-absorbing pigment in leaves, absorbs almost all of the blue and red wavelengths but reflects much of the green. The absorbed light drives the light reactions.

Chloroplasts are photosynthetic organelles

In plants and protists, photosynthesis takes place inside chloroplasts. Although chloroplasts are located throughout the green parts of a plant, including young stems, leaves are especially well structured to assist in photosynthesis. The cells inside a leaf are packed with many chloroplasts, which look like flattened green footballs. The outer layers of a leaf are pockmarked with many microscopic pores, called **stomata** (STOH-**muh-tuh**; singular "stoma") that enable gas exchange (see Figure 6.3). Carbon dioxide from the air outside enters a leaf through open stomata to be used in the Calvin cycle reactions within chloroplasts. A plant absorbs water and dissolved minerals through its roots, but it gains *no energy* whatsoever from the soil.

Like mitochondria, chloroplasts are bounded by two membranes. The inner of the two membranes encloses a gel-like fluid that makes up the **stroma** (**FIGURE 6.6**). Embedded in the stroma is a network

Helpful to Know

Plants evolved to use sunlight as an energy source. However, artificial light sources can work also. Fluorescent lights are rich in the blue part of the spectrum, which is suitable for growing foliage. Red wavelengths are necessary for flowering in many plants. "Plant lights," or "grow lights," mimic the sun in the balance of blue and red wavelengths.

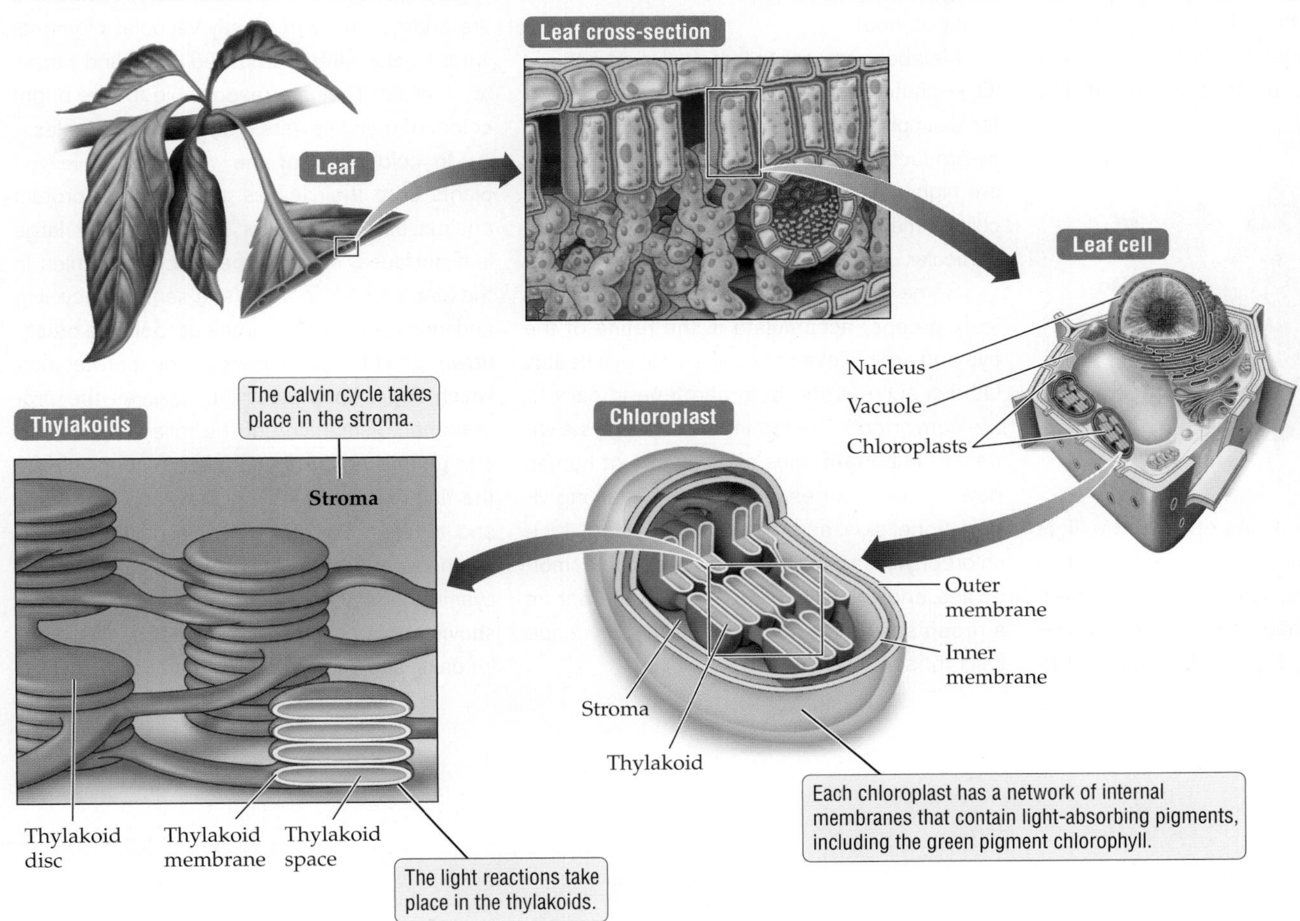

FIGURE 6.6 Chloroplasts Contain Membranes Studded with Chlorophyll

Chlorophyll absorbs light energy, which is used to drive the synthesis of energy carriers. The energy carriers fuel the synthesis of sugar in the stroma of the chloroplast.

BIOLOGYMATTERS

The Rainbow Colors of Plant Pigments

Plant pigments color our world. We like to be surrounded by greenery and take delight in the bright displays in a flower garden. What do these brightly colored pigments do for the plant, and what can they do for us?

A biological pigment is a carbon-containing molecule with a distinctive color. Chlorophylls and carotenoids are the two main families of sun-catcher pigments in the plant chloroplast. The two types of chlorophyll found in plants—chlorophyll *a* and chlorophyll *b*—are green to yellowish green; green light bounces off these molecules, while blue light and red light are strongly absorbed. Because the two chlorophylls absorb slightly different wavelengths of blue and red, collectively they harvest light over a broader range than just one pigment could.

The brightly colored pigments that plants produce have varied uses in the plant, and many of them are healthful for people

Carotenoids (**kuh-*RAH*-tuh-noyd**), which are yellow-orange pigments, also help harvest light, but they have the additional function of protecting photosynthetic cells from too much sun and from highly reactive chemicals that

are often spun off as a side effect of the light reactions. If a leaf absorbs more sunshine than it can utilize in photosynthesis, the excess energy can actually damage cellular structures—a phenomenon called photodamage. Carotenoids act as a safety valve for absorbed light energy: they pick up the excess and release it gently as heat.

Metabolic processes that involve oxygen gas (O_2)—photosynthesis and cellular respiration, for example—have the potential for generating by-products called free radicals. *Free radicals* are highly reactive chemicals that can damage cellular macromolecules—lipids and proteins, in particular.

Some carotenoids (such as lutein from leafy greens) accumulate in the retina of the eye and are believed to be good for eye health. Others, such as the beta-carotene in carrots, are converted into vitamin A, which plays a variety of important roles at all stages of human development, but especially in maintaining vision, bone, skin, and the immune system. Like chlorophyll, carotenoids are fat-soluble molecules, and you will absorb more of them from a green salad or a pumpkin pie if your recipe calls for some fat.

Most water-soluble pigments in a plant reside in the vacuole (see Section 3.4). Flavonoids are a large family of mostly vacuolar pigments whose color ranges from red, blue, and purple to … white! They are responsible for the bright colors of many flowers, fruits, and vegetables.

In cold parts of the world, broad-leaved plants drop their leaves in autumn to protect against excessive water loss from their large leaf surfaces. Most of the macromolecules in the leaf are broken down and sent for recycling and storage in the tree trunk or root. The breakdown of chlorophyll reveals the carotenoids, which linger to protect the cells as the programmed dismantling of the foliage continues. The unmasked carotenoids are responsible for the flaming yellows of nut trees, cottonwoods, and birches. Some species also step up the synthesis of protective flavonoids called anthocyanins, and these pigments help create the showstopping fall displays of red maples, scarlet oaks, and sumac, among many others.

of interconnected membrane-enclosed sacs. The sacs, known as **thylakoids** (*THYE*-**luh-koyd**), lie one on top of another in stacks. Each thylakoid consists of a *thylakoid membrane* that encloses the *thylakoid space*. The distinctive arrangement of the thylakoid membranes is crucial for light capture and for generating ATP and NADPH, the main products of the light reactions.

The Calvin cycle reactions take place in the stroma, which contains the many enzymes, ions, and molecules needed for turning carbon dioxide into sugar with the help of energy carriers.

The light reactions generate energy carriers

The thylakoid membrane is densely packed with many disclike clusters of pigments complexed with proteins. Each disclike grouping is known as an **antenna complex** (**FIGURE 6.7**) and contains different types of pigments, including chlorophylls *a* and *b* and carotenoids (see the "Biology Matters" box on page 130). The antenna complex captures light energy, especially blue and red wavelengths, and funnels it to an enzyme-chlorophyll complex known as the **reaction center**, where light reactions are initiated.

Electrons associated with certain chemical bonds in a chlorophyll molecule become more energized when they absorb light. These high-energy electrons are picked up by an **electron transport chain (ETC)**, a series of electron-accepting molecules embedded next to each other in the thylakoid membrane. As electrons pass from one component of the ETC to the next, small amounts of energy are released and are used to generate ATP.

What is the ultimate fate of the electrons that ride down the thylakoid ETCs? Eventually they are picked up by $NADP^+$, which, with the addition of a proton (H^+) from the stroma, is transformed into NADPH, the second of the two energy carriers furnished by the light reactions.

Two photosystems energize electrons

The combination of an antenna complex and its associated reaction center is called a **photosystem**. Plant chloroplasts have two interlinked photosystems—photosystems I and II—that work in tandem (see Figure

Light Harvesting and Electron Flow

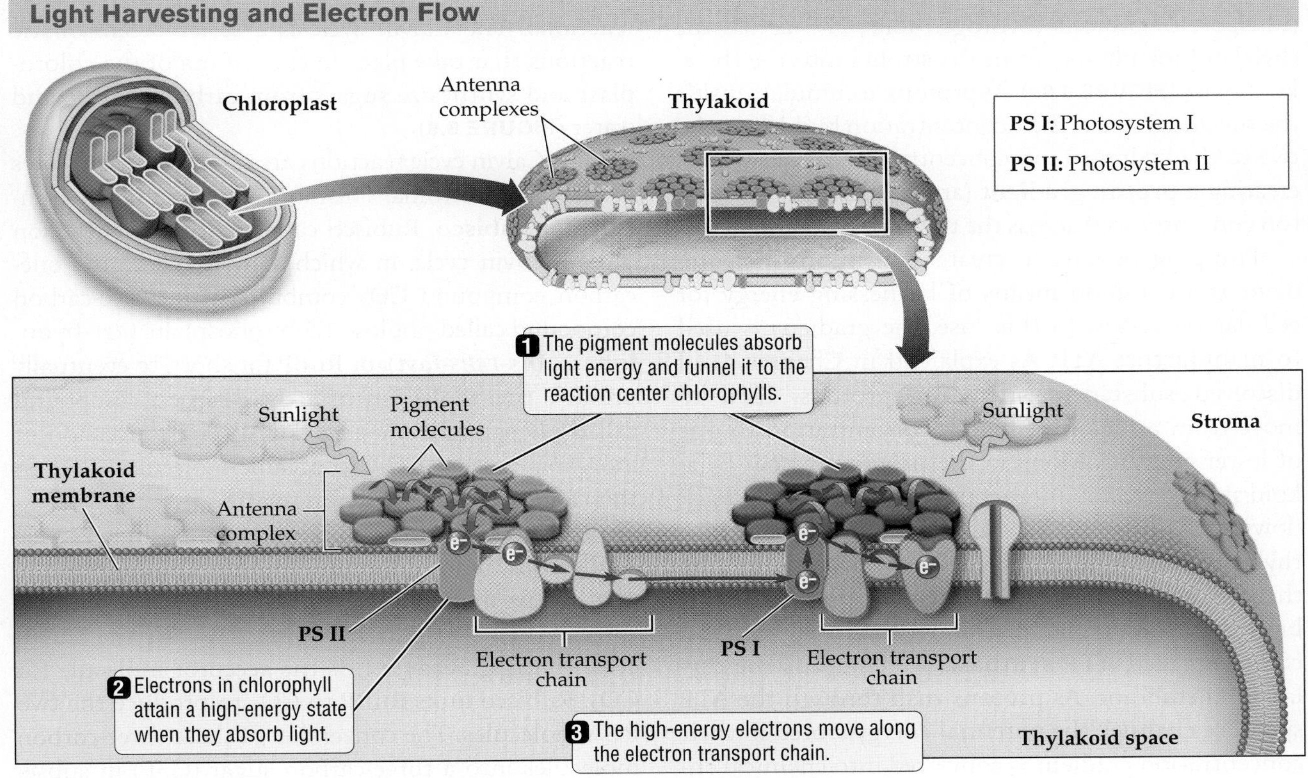

FIGURE 6.7 The Light Reactions Are Conducted by Two Linked Photosystems

EXTREME SOLAR PANELS

The leaf of the raffia palm (*Raphia regalis*), native to the moist tropics, is the largest in the world. It can grow to 24 meters (80 feet). Its fiber is used to make rope, bags, and mats. The more leaf surface a plant has, the more food it can make.

6.7). (The numbering of the two photosystems reflects the order in which each system was discovered by researchers, not the order of steps during photosynthesis.) **Photosystem II** is associated with the splitting of water (*photolysis*) and the generation of electrons, oxygen gas (O_2), and hydrogen ions (**FIGURE 6.8*a***). **Photosystem I** receives electrons from photosystem II and, after traveling down a relatively short electron transport chain, these electrons are donated to $NADP^+$ to generate NADPH (**FIGURE 6.8*b***).

The transfer of high-energy electrons is at the heart of the light reactions. The journey begins at the reaction center of photosystem II. Absorbed light energy ejects high-energy electrons from a photosystem II reaction center chlorophyll (see Figure 6.8*a*). The high-energy electrons are picked up by the first component of the ETC, which transfers them to the next component in the chain, and so on.

As the electrons travel down the ETC they lose energy, and some of that energy is used to drive the transport of protons (hydrogen ions, H^+) across the thylakoid membrane, from the stroma into the thylakoid space (**FIGURE 6.8c**). As protons accumulate inside the thylakoid space, their concentration builds up relative to the hydrogen ion concentration in the stroma, creating a **proton gradient** (an imbalance in the proton concentration) across the thylakoid membrane.

Pumping of ions to create a concentration gradient is a common means of harnessing energy for cellular processes. In this case, the gradient is used to manufacture ATP. As explained in Chapter 4, all dissolved substances, including protons, tend to move from a region of higher concentration to one of lower concentration. So the protons in the thylakoid space have a spontaneous tendency to move back down the proton gradient to the stroma. Because the thylakoid membrane will not allow protons to pass through it, the only way for them to cross the membrane is through a large, channel-containing protein complex called **ATP synthase**, which spans the thylakoid membrane. As protons rush through the ATP synthase channel, the potential energy stored in their concentration gradient is converted into chemical energy: enzymes associated with the lollipop-like head of ATP synthase catalyze the addition of a phosphate group on ADP, converting it to ATP (see Figure 6.8*c*).

Electrons lose energy as they travel down the electron transport chain from photosystem II until they reach the reaction center of photosystem I. There, light gathered by the photosystem I antenna complexes boosts each arriving electron to a very high energy level. Next, these energized elections flow through the short ETC and are then donated to $NADP^+$.

In addition to the two electrons received from the ETC, each $NADP^+$ takes up one proton (H^+) from the stroma, which converts it to NADPH. Note that these electrons and protons picked up by $NADP^+$ come ultimately from water molecules: the electrons traveling down the ETC were initially ejected from photosystem II, which replaced them by extracting electrons from water to form hydrogen ions and O_2 (as pictured in Figure 6.8*a*). In summary, the two photosystems and the two ETCs are arranged so that they synchronize perfectly to generate ATP and NADPH, making oxygen as a by-product.

The Calvin cycle reactions manufacture sugars

The energy carriers produced by the light reactions—ATP and NADPH—are used in the Calvin cycle reactions. The Calvin cycle is a series of enzymatic reactions that take place in the stroma of the chloroplast and synthesize sugars from carbon dioxide and water (**FIGURE 6.9**).

The Calvin cycle reactions are catalyzed by enzymes present in the stroma. The most abundant of these enzymes is **rubisco**. Rubisco catalyzes the first reaction of the Calvin cycle, in which a molecule of the one-carbon compound CO_2 combines with a five-carbon compound called ribulose 1,5-bisphosphate (***RYE*-byoo-lohs ... bis-*FAHS*-fayt**), or RuBP for short, to eventually produce two molecules of a three-carbon compound called phosphoglyceric acid (PGA). The conversion of inorganic carbon (CO_2) to organic molecules (PGA in this case) is known as **carbon fixation**.

This first reaction is followed by a multistep cycle catalyzed by many different enzymes. The reactions in these steps manufacture sugars and also regenerate RuBP. RuBP is absolutely necessary to keep the Calvin cycle running because it is the acceptor molecule for CO_2. Rubisco links RuBP to CO_2 to produce the two PGA molecules. The conversion of these three-carbon molecules into a three-carbon sugar (G3P) in subsequent steps requires the input of energy from ATP,

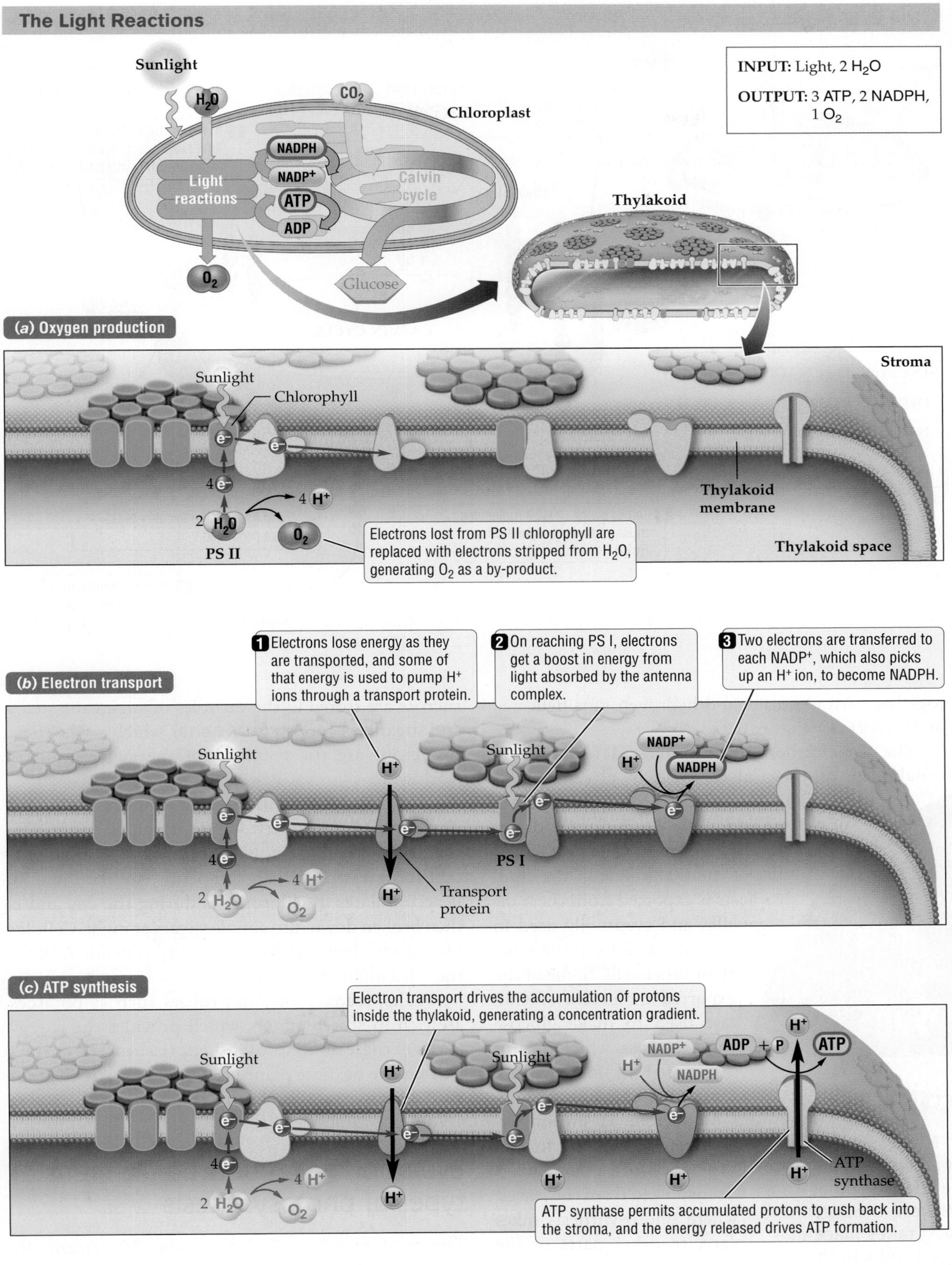

FIGURE 6.8 How the Light Reactions Generate Energy Carriers

The Light Reactions

Sunlight

H_2O

CO_2

Chloroplast

NADPH

$NADP^+$

ATP

ADP

Light reactions

Calvin cycle

O_2

Glucose

Thylakoid

INPUT: Light, 2 H_2O

OUTPUT: 3 ATP, 2 NADPH, 1 O_2

(a) Oxygen production

Stroma

Sunlight

Chlorophyll

e^-

4 e^-

4 H^+

2 H_2O

O_2

PS II

Thylakoid membrane

Thylakoid space

Electrons lost from PS II chlorophyll are replaced with electrons stripped from H_2O, generating O_2 as a by-product.

(b) Electron transport

1 Electrons lose energy as they are transported, and some of that energy is used to pump H^+ ions through a transport protein.

2 On reaching PS I, electrons get a boost in energy from light absorbed by the antenna complex.

3 Two electrons are transferred to each $NADP^+$, which also picks up an H^+ ion, to become NADPH.

Sunlight

H^+

Sunlight

$NADP^+$

H^+

NADPH

e^-

4 e^-

PS I

2 H_2O

4 H^+

O_2

H^+

Transport protein

(c) ATP synthesis

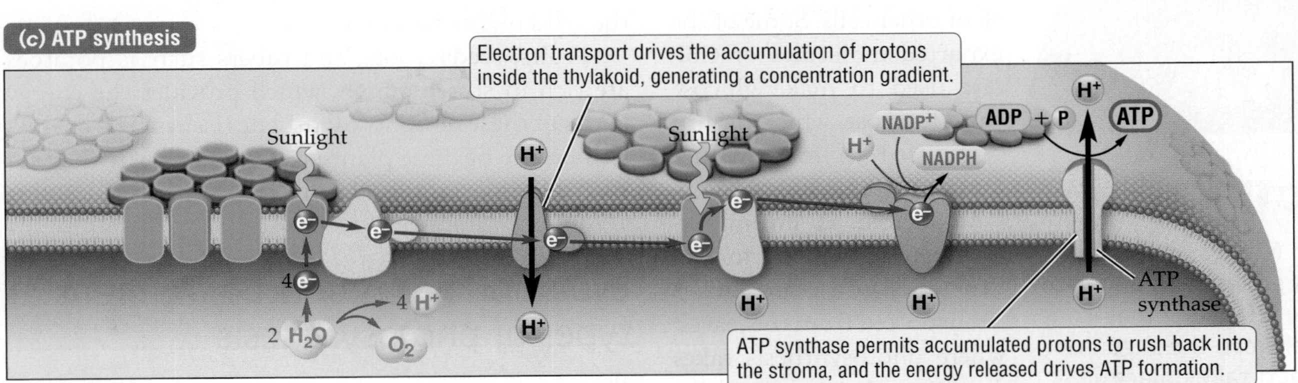

FIGURE 6.9 The Calvin Cycle Converts Carbon Dioxide into Sugar

The Calvin cycle reactions fix carbon by turning CO_2 into sugar molecules, through the use of energy delivered by ATP, and of electrons and protons delivered by NADPH.

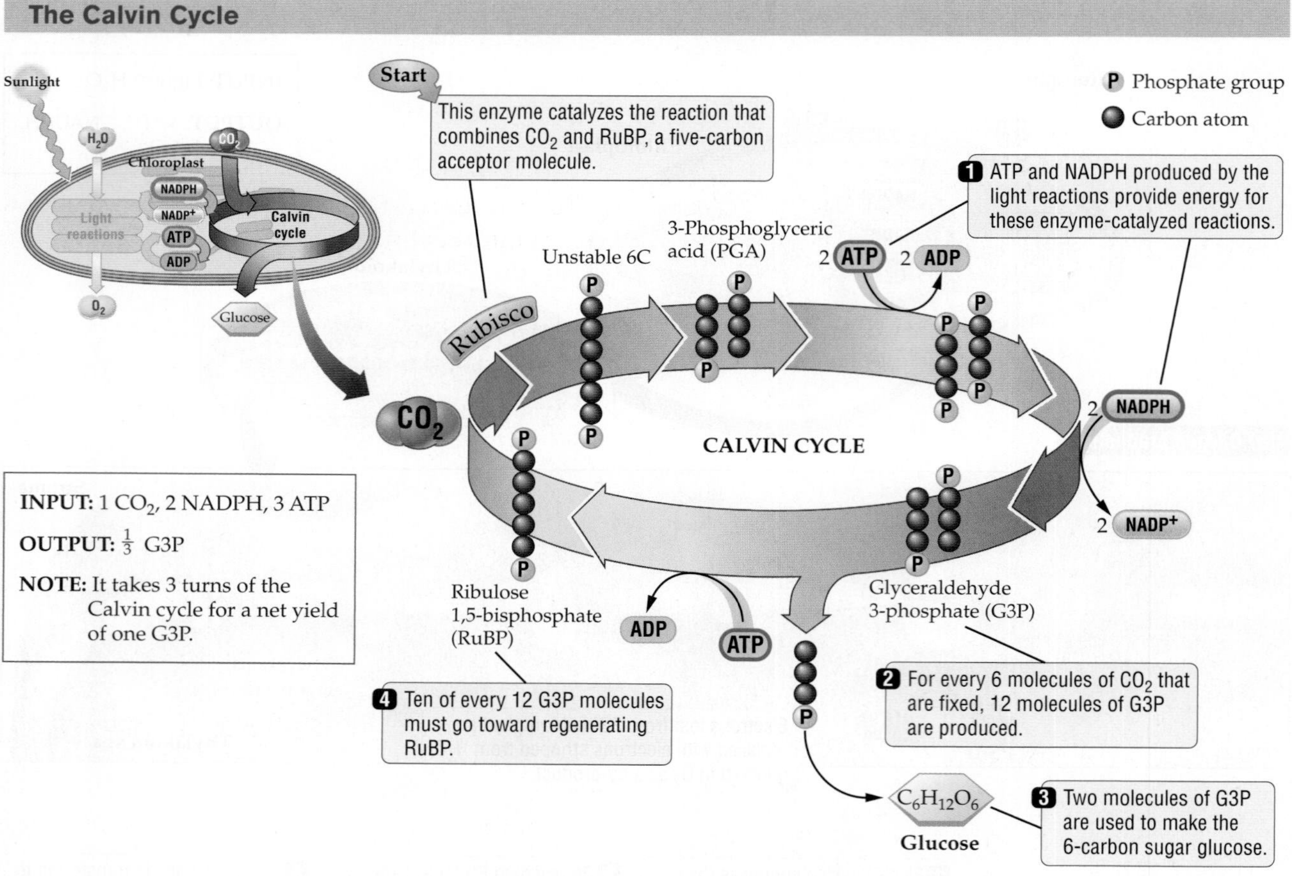

The Calvin Cycle
Sunlight
H2O
CO2
Chloroplast
NADPH
NADP+
ATP
ADP
Light reactions
Calvin cycle
O2
Glucose
Start
This enzyme catalyzes the reaction that combines CO2 and RuBP, a five-carbon acceptor molecule.
P Phosphate group
Carbon atom
1 ATP and NADPH produced by the light reactions provide energy for these enzyme-catalyzed reactions.
3-Phosphoglyceric acid (PGA)
Unstable 6C
2 ATP 2 ADP
Rubisco
2 NADPH
CO2
CALVIN CYCLE
2 NADP+
INPUT: 1 CO2, 2 NADPH, 3 ATP
OUTPUT: 1/3 G3P
NOTE: It takes 3 turns of the Calvin cycle for a net yield of one G3P.
Ribulose 1,5-bisphosphate (RuBP)
ADP
ATP
Glyceraldehyde 3-phosphate (G3P)
4 Ten of every 12 G3P molecules must go toward regenerating RuBP.
2 For every 6 molecules of CO2 that are fixed, 12 molecules of G3P are produced.
C6H12O6
Glucose
3 Two molecules of G3P are used to make the 6-carbon sugar glucose.

together with electrons and hydrogen ions delivered by NADPH (see Figure 6.9).

The three-carbon sugar generated from the final step of the Calvin cycle is glyceraldehyde 3-phosphate (G3P). G3P is the building block of glucose and all the other carbohydrates that a cell might need to manufacture. Most of the G3P made in the chloroplasts is exported from these organelles and eventually used in chemical reactions in the same cell or other cells. Some of the exported molecules of G3P are used to make glucose and fructose, which in turn are used to manufacture sucrose (table sugar).

Sucrose is an important food source for all the cells in a plant and is transported from the leaves, where photosynthesis takes place, to other parts of the plant. Significant amounts of sucrose are stored in the vacuoles of sugarcane stems and sugar beet roots (**FIGURE 6.10**), which is why these two crops are the mainstay of the sugar industry worldwide.

Not all the G3P made in the chloroplasts is shipped out. Some of it is converted into starch by enzymes in the stroma. Starch, a polymer of glucose, is an important form of stored energy in plants (see Section 2.7). It accumulates in chloroplasts during the day and is then broken down into simple sugars at night. Cellular respiration then uses these sugars to generate ATP for the cell's nighttime energy needs.

Fruits, seeds, roots, and tubers such as potatoes are rich in stored starch, which provides the energy needs of these nonphotosynthetic tissues. The energy-rich nature of these plant parts explains why they are such an important food source for animals.

EXTREME CACTUS

The cardon, found in Mexico, is the tallest cactus in the world. It grows to 19.2 meters (63 feet) and is a close relative of that icon of the American Southwest, the saguaro. Photosynthesis in cacti takes place in cells of their stems (or trunks in this case). The stems and trunks also store water.

Evolution has produced different types of photosynthesis

The great majority of plants—some 95 percent—fix CO_2 through the Calvin cycle alone. Such plants are

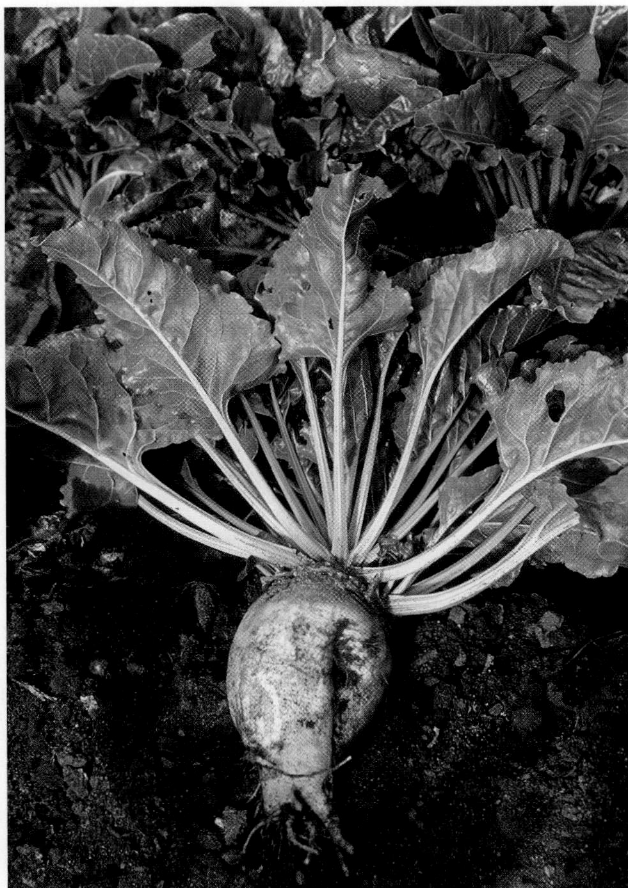

FIGURE 6.10 Sugar Fix: Plants Convert G3P into Sugars
Surplus sugars are turned into starch and are stored in the chloroplast stroma. Some plants stockpile energy in the form of sucrose stored inside vacuoles. Roughly half the table sugar used in the United States comes from the roots of sugar beets (pictured here); the rest, from sugarcane, a tropical grass that stores sucrose in the vacuoles of cells in the stem.

called **C_3 plants** because the first stable product of CO_2 fixation is a three-carbon (C_3) molecule, PGA. Photosynthesis in C_3 plants, however, can be inefficient when CO_2 levels are low and O_2 levels are high.

The reason for this low efficiency is that rubisco, the key enzyme in carbon fixation, can operate in both directions: when CO_2 is abundant, it adds carbon dioxide to RuBP; but when CO_2 concentrations are low, it will instead add oxygen to RuBP (**FIGURE 6.11**), with a subsequent release of CO_2. The series of chemical reactions that begins with rubisco adding oxygen to RuBP and ends in the release of carbon dioxide is called **photorespiration**. Photorespiration makes photosynthesis less efficient because it causes some of the fixed carbon to be lost as carbon dioxide.

Photorespiration is worst in bright sunshine, when oxygen is being produced vigorously, and in warm and dry conditions, when stomata narrow to reduce water loss, reducing the inflow of carbon dioxide. A wheat plant on a warm, sunny day can lose as much as 25 percent of its fixed carbon to photorespiration. C_3 plants—such as wheat, beans, and maple trees—simply tolerate the drain.

To counteract the effects of photorespiration, some plants that live in hot, dry climates have evolved different types of photosynthesis:

- The **C_4 pathway** (common in plants that live in hot, sunny, and relatively dry conditions)
- The **CAM pathway** (common in plants that live in desert environments)

These pathways reduce photorespiration by surrounding rubisco with a high-CO_2 environment (**FIGURE 6.12*b* and *c***). When CO_2 levels are high and O_2 levels consequently low, rubisco is more likely to combine RuBP with CO_2 than with O_2.

The C_4 pathway works by carrying out the Calvin cycle reactions in specialized tissues that maintain a high-CO_2 environment. In **C_4 plants**, carbon dioxide is converted into four-carbon (C_4) molecules in the *mesophyll cells* of the leaf (**FIGURE 6.12*b***). The C_4 molecules then move into adjacent *bundle-sheath cells*, whose

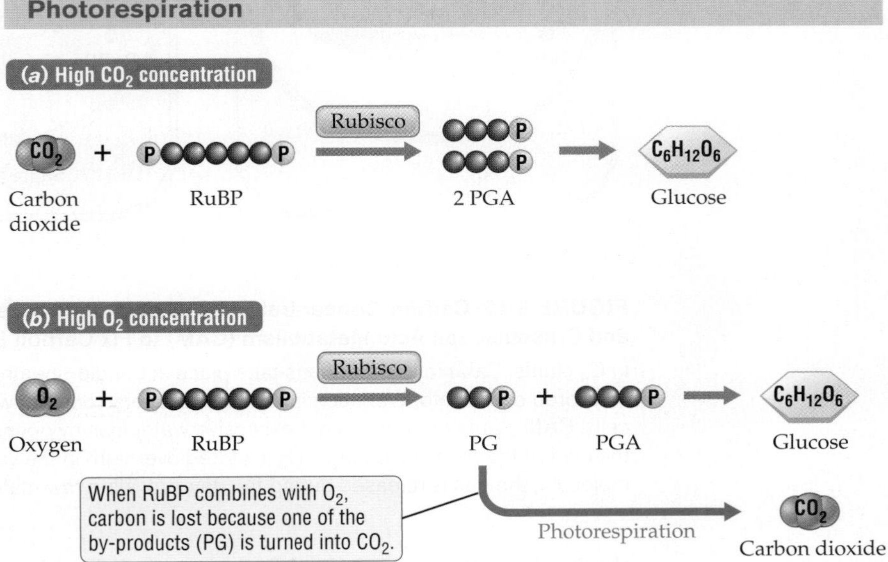

FIGURE 6.11 In Photorespiration, Fixed Carbon Is Lost as Carbon Dioxide
Rubisco has a dual nature. (*a*) With abundant CO_2, it converts RuBP into two molecules of PGA that then go on to form glucose. (*b*) When CO_2 is low and O_2 is high, rubisco converts RuBP to one PGA and one two-carbon phosphoglycolate (PG), which is lost as CO_2 via a pathway called photorespiration.

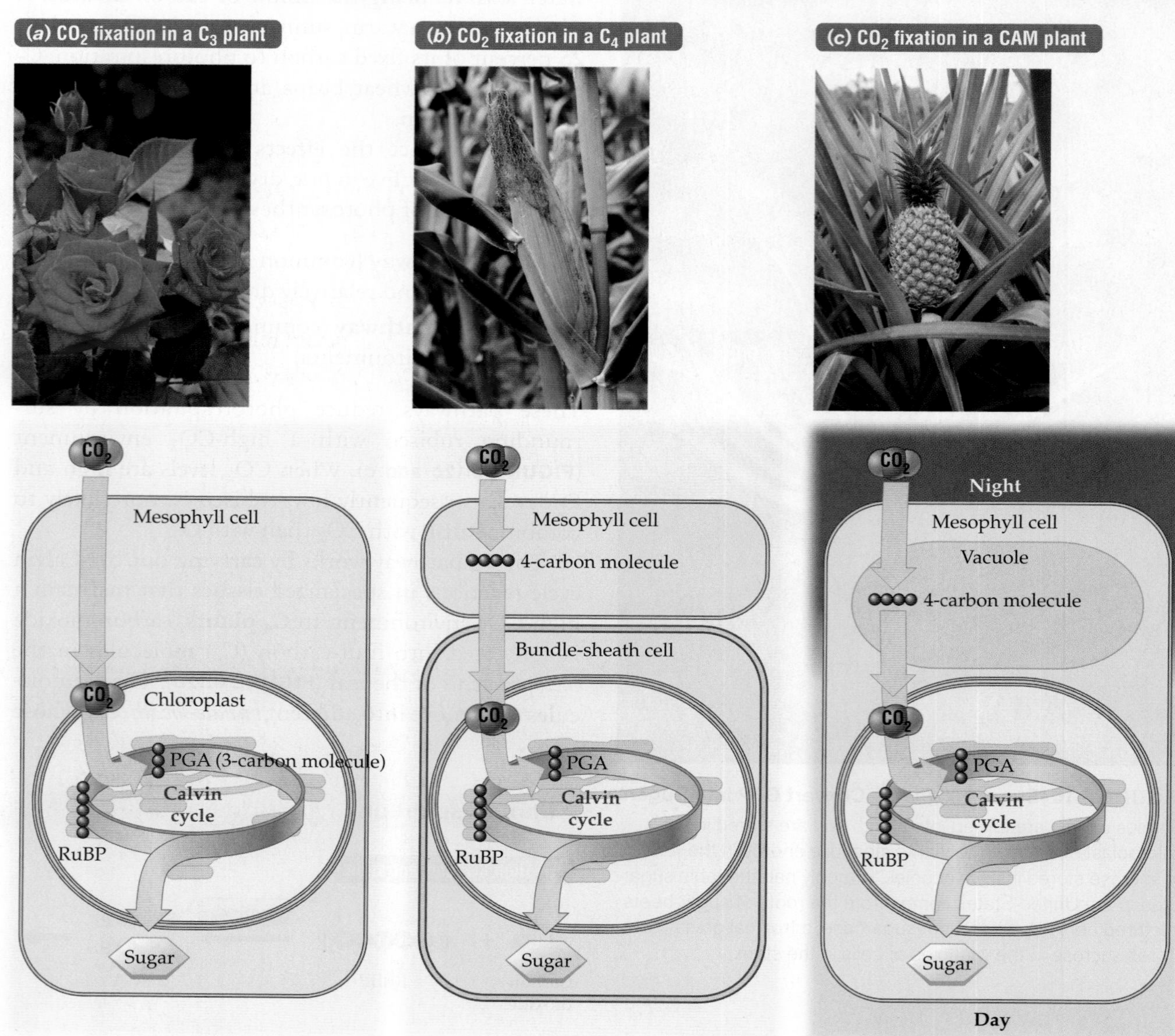

FIGURE 6.12 Carbon-Concentrating Mechanisms Enable C_4 Photosynthesis and Crassulacean Acid Metabolism (CAM) to Fix Carbon Efficiently

In C_4 plants, Calvin cycle reactions take place in bundle-sheath cells. Carbon fixation is favored over photorespiration in the high-CO_2 environment within these thick-walled cells. CAM plants protect against excessive water loss by opening their stomata only in the cool of the night. Incoming CO_2 is stored overnight in the vacuole as a four-carbon molecule; the gas is released during the day, providing raw material for the Calvin cycle.

thick cell walls are specialized for trapping gases. Here the C_4 molecules break up, releasing CO_2. In this high-CO_2 environment, rubisco has little opportunity to interact with oxygen, and therefore nearly all the CO_2 imported from mesophyll cells is converted to sugar.

The C_4 pathway requires ATP molecules at a number of steps, however. Except in hot and sunny regions, the energy cost of the C_4 pathway does not outweigh the energy cost of photorespiration. In cooler, damper climates, C_4 plants have no advantage over C_3 plants.

Corn and sugarcane are examples of C_4 plants. They manufacture glucose at an efficiency that is 4–7 times greater than the efficiency of C_3 photosynthesis. Sugarcane is often cultivated in subtropical and tropical regions (in the United States, in Florida and Louisiana), where it does best.

In plants that live in very dry regions, **crassulacean acid metabolism (CAM)** has evolved to cope with the dangers of excessive water loss through stomata. **CAM plants** open their stomata only at night, when the temperature is lower and the air is not as dry. The incoming CO_2 is converted into four-carbon molecules and stored overnight in the large vacuoles of the photosynthetic cells (**FIGURE 6.12c**).

At daybreak, CAM plants close their stomata to conserve water. The CO_2 for the Calvin cycle reactions comes not from the outside, but from the onboard "gas tank," the vacuole. The stored four-carbon molecules break down to release CO_2, which rubisco attaches to RuBP. From this point on, sugars are made just as in C_3 plants.

CAM plants grow slowly and are not as productive as C_4 plants, in part because they are not as effective in combating photorespiration. Pineapple, cacti, and jade plants are examples of CAM plants. Cacti especially are known for their adaptations to hot deserts.

Concept Check

1. Compare the functions of photosystems I and II.
2. What is rubisco, and what role does it play in photosynthesis?

6.4 Cellular Respiration: Energy from Food

Cellular respiration extracts energy from food molecules to generate ATP—a vital process in producers and consumers alike. In this section we take a closer look at each of the three major stages of cellular respiration: glycolysis, the Krebs cycle, and oxidative phosphorylation.

Glycolysis is the first stage in the cellular breakdown of sugars

Glycolysis means literally "sugar splitting." From an evolutionary standpoint, it was probably the earliest means of producing ATP from food molecules, and it is still the primary means of energy production in many prokaryotes. However, the energy yield from glycolysis is small because sugar is only partially degraded through this process. In most eukaryotes, glycolysis is just the first step in energy extraction from sugars, and additional reactions in the mitochondrion help achieve the complete breakdown of carbon backbones, typically yielding at least 15 times as much ATP as glycolysis does.

Glycolysis takes place in the cytosol (**FIGURE 6.13**). Through a series of enzyme-catalyzed reactions, glucose

Concept Check Answers

1. Photosystem II catalyzes the splitting of water molecules to release O_2, and an associated ETC creates a proton gradient to drive ATP synthesis. Electrons from the ETC enter photosystem I to generate NADPH from $NADP^+$.

2. Rubisco is an enzyme that is abundant in the stroma; it catalyzes the addition of CO_2 to RuBP, the first step in carbon fixation through the Calvin cycle.

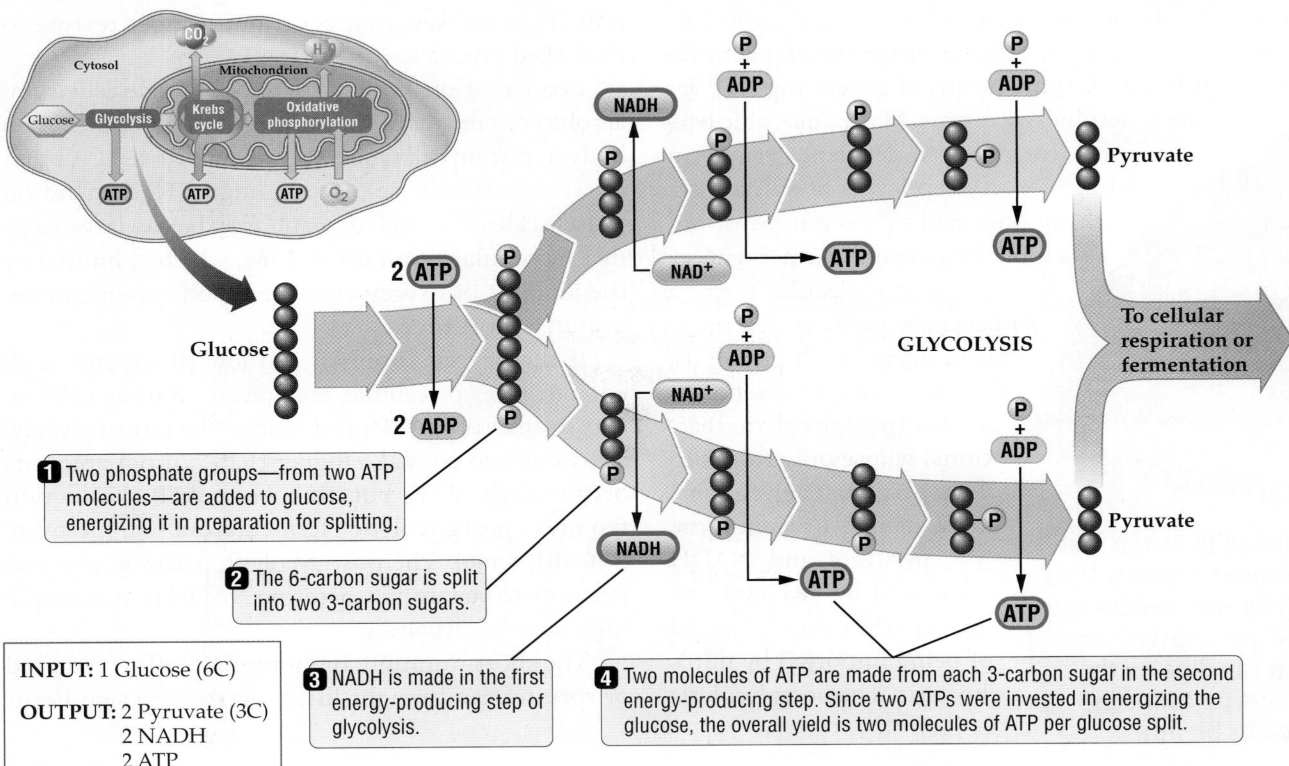

Glycolysis
Cytosol
Mitochondrion
Glucose
Glycolysis
Krebs cycle
Oxidative phosphorylation
ATP
ATP
ATP
O2
CO2
H2O
Glucose
2 ATP
2 ADP
NADH
NAD+
P + ADP
ATP
Pyruvate
GLYCOLYSIS
To cellular respiration or fermentation
1 Two phosphate groups—from two ATP molecules—are added to glucose, energizing it in preparation for splitting.
2 The 6-carbon sugar is split into two 3-carbon sugars.
3 NADH is made in the first energy-producing step of glycolysis.
4 Two molecules of ATP are made from each 3-carbon sugar in the second energy-producing step. Since two ATPs were invested in energizing the glucose, the overall yield is two molecules of ATP per glucose split.
INPUT: 1 Glucose (6C)
OUTPUT: 2 Pyruvate (3C)
2 NADH
2 ATP

FIGURE 6.13 Glycolysis Converts Glucose into Pyruvate

In glycolysis, each six-carbon glucose is converted into two molecules of pyruvate, a three-carbon organic acid.

is converted to a six-carbon intermediate, which is then split into two molecules of a three-carbon sugar, glyceraldehyde 3-phosphate (G3P). This is the same sugar that we described in photosynthesis, but in this case it is a breakdown product, not a building block. In successive steps, each G3P molecule is converted into a three-carbon organic molecule called **pyruvate** (**pye-*ROO*-vayt**).

The net energy yield for glycolysis is calculated as follows: For each molecule of glucose consumed during glycolysis, four molecules of ADP are turned into four ATP molecules, and electrons and hydrogen atoms are donated to two molecules of NAD^+, generating two molecules of NADH. Because the early steps of glycolysis consume two molecules of ATP per glucose molecule, a single glucose molecule produces a net yield of two ATP molecules and two NADH molecules (see Figure 6.13). Note that glycolysis does not require oxygen.

Helpful to Know

You will notice several biology terms include *lysis* or its related form *lytic*. The root word means "splitting" or "disintegration." This root word shows up in many biological terms. Examples include "photolysis" (splitting with light), "glycolysis" (splitting of sugars), and "hydrolysis" (breakdown with water).

Fermentation produces ATP through glycolysis when oxygen is absent

Glycolysis does not require O_2, which means it is an **anaerobic** (***AN*-ayr-*OH*-bik**) process. Glycolysis was probably the main source of energy for early lifeforms in the oxygen-poor atmosphere of primitive Earth. It is still the only means of generating ATP for some anaerobic organisms. Many anaerobic bacteria that live in oxygen-deficient swamps, in sewage, or in deep layers of soil are actually poisoned by oxygen. Most anaerobic organisms extract energy from organic molecules using a *fermentation* pathway. **Fermentation** begins with glycolysis, followed by a special set of reactions (postglycolytic reactions) whose only role is to help perpetuate glycolysis.

During fermentation, the pyruvate and NADH produced by glycolysis remain in the cytosol, instead of being imported by mitochondria. The postglycolytic reactions convert pyruvate into other molecules, such as alcohol or lactic acid, depending on the types of fermentation pathways used by the cell type. In the process, NADH is converted back to NAD^+, an adequate supply of which is necessary to keep glycolysis running. (Figure 6.13 shows the point at which NAD^+ is used in glycolysis.)

The cell has a finite pool of NAD^+, and if all of it were converted into NADH, glycolysis would cease. The postglycolytic reactions are a clever way of averting this problem: these reactions remove electrons and hydrogen ions from NADH, resulting in NAD^+ formation, which restores the cell's limited pool of this vital metabolic precursor. In regenerating NAD^+ in the postglycolytic reactions, the electrons and hydrogen ions from NADH create two-carbon or three-carbon compounds, with alcohol and lactic acid being two of the most common by-products of fermentation. The alcohol is usually ethanol, a two-carbon molecule that is the basis for alcoholic beverages.

Yeasts are single-celled fungi that resort to alcoholic fermentation when oxygen is absent or in short supply. Special yeast strains are used in the production of beer, wine, and other alcohol products. Fermentation by anaerobic yeasts converts pyruvate into ethanol, releasing CO_2 gas, which gives beer its fizz and foam (**FIGURE 6.14*a***). The gas plays an important role in bread making with baker's yeast: the CO_2 released by fermentation expands the dough by creating tiny bubbles or pockets that contribute a light texture to the baked product.

Fermentation is not limited to single-celled anaerobic organisms. It can also occur in the human body in certain cell types, such as muscle cells. A burst of strenuous exercise places a huge ATP demand on muscle cells. The ATP demand may be too large to be met by cellular respiration alone, which is limited by the amount of oxygen that the blood vessels can deliver in a short time.

If the oxygen supply is too low to support high enough rates of cellular respiration, muscle cells resort to anaerobic ATP production. The rate of glycolysis is boosted to make higher-than-normal amounts of glycolytic ATP, but this extra ATP production requires postglycolytic fermentation of pyruvate (**FIGURE 6.14*b***). The postglycolytic pathway converts pyruvate to lactate, regenerating NADH to sustain the high rates of glycolysis.

The short-duration, high-intensity effort required of sprinters and weight lifters gets a very significant

EXTREME EXHAUSTION

The cheetah can accelerate to its top speed of 120 kilometers per hour (75 miles per hour) in 4 seconds. The massive sprint requires both cellular respiration and fermentation. Before it can dine on the kill, the cheetah must rest for as much as half an hour to catch its breath.

assist from anaerobic ATP production. The burning pain felt in overtaxed muscles stems from the acidity of lactic acid, which irritates nerve endings. Lactic acid is swept to the liver by the bloodstream, where it is reconverted to pyruvate and fed into mitochondria for oxygen-dependent energy extraction. We huff and puff after strenuous exercise because we incur an "oxygen debt" when our muscles resort to lactic acid fermentation. Very soon, we must bring in extra oxygen to make up for the oxygen deficit as the chemical energy of lactic acid is salvaged through cellular respiration.

Lactic acid fermentation is of little use in prolonged exercise. "Aerobic exercise," such as long-distance running or biking, depends on the moderate but sustained ATP supply that cellular respiration provides.

Cellular respiration in the mitochondria produces much of the ATP eukaryotes need

As long as oxygen is available, most eukaryotes use cellular respiration to satisfy their relatively large ATP needs. Mitochondria break up pyruvate through a series of reactions and package the released energy into many molecules of ATP. ATP production in the mitochondrion is crucially dependent on oxygen; that is, the mitochondrial portion of cellular respiration is a strictly **aerobic** (oxygen-dependent) process.

Highly aerobic tissues tend to have high concentrations of mitochondria and a rich blood supply to deliver the large amounts of oxygen needed to support their activity. Muscle cells in the human heart, for example, have an exceptionally large number of mitochondria to produce the enormous amounts of ATP needed to keep the heart beating every second of every day of our lives.

The Krebs cycle releases carbon dioxide and generates energy carriers

The end product of glycolysis—pyruvate—is transported into the mitochondria and enters the second major stage of cellular respiration, the Krebs cycle. This cycle is a series of enzyme-driven reactions that

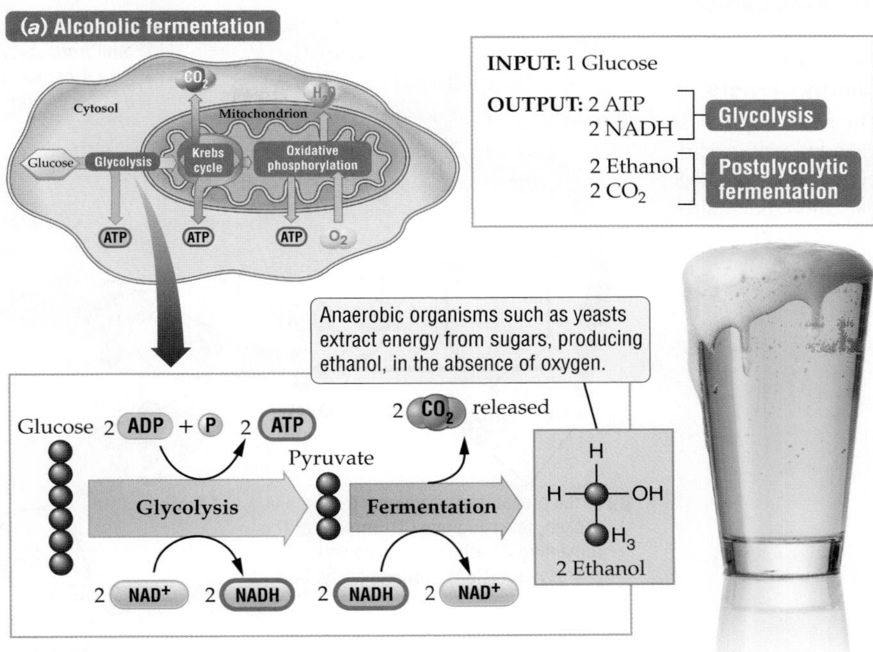

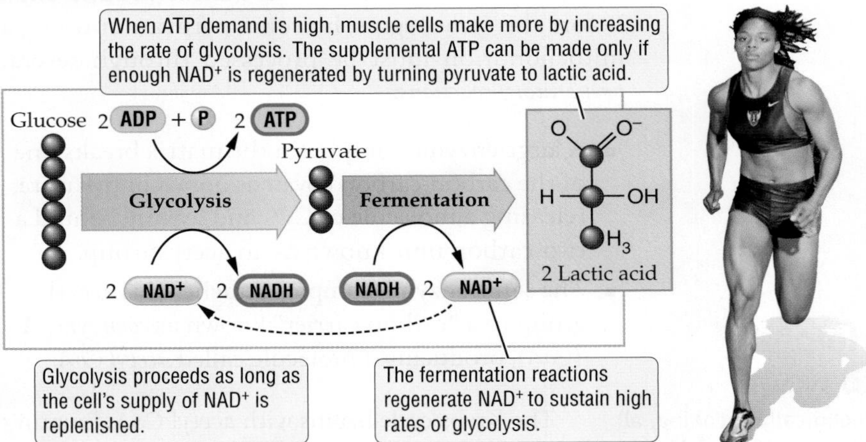

FIGURE 6.14 Ethanol and Lactic Acid Are By-products of Fermentation

When oxygen supply is too low to support ATP production through cellular respiration, fermentation provides ATP through glycolysis alone. (*a*) Strains of single-celled yeasts are used in the brewing of alcoholic beverages such as beer. When oxygen is excluded from the fermentation tanks, the yeasts resort to fermentation of sugars, producing ethanol and CO_2 as by-products. (*b*) A similar process occurs in our muscles during short bursts of strenuous exercise, except that the postglycolytic reactions turn pyruvate into lactic acid, and no CO_2 is made.

FIGURE 6.15 The Krebs Cycle Releases Carbon Dioxide and Generates Energy Carriers

The Krebs cycle, also called the citric acid cycle, occurs in the mitochondrial matrix.

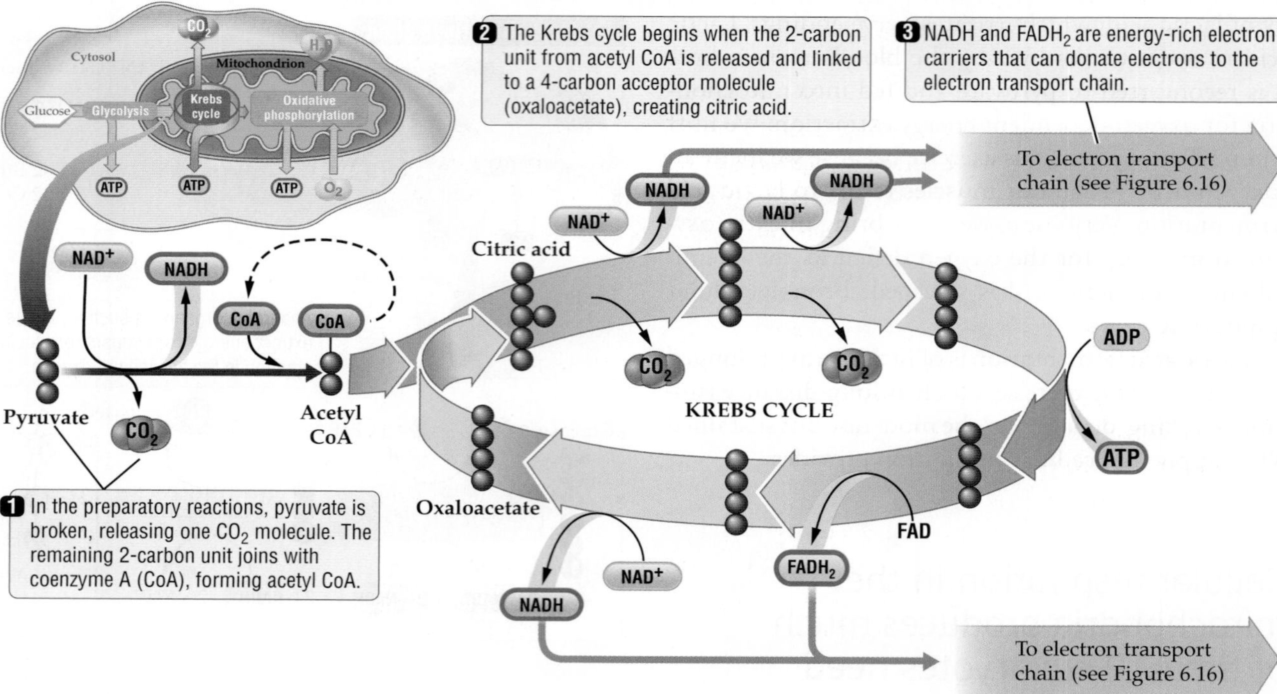

The Preparatory Reactions and Krebs Cycle
Cytosol
Mitochondrion
Glucose
Glycolysis
Krebs cycle
Oxidative phosphorylation
ATP
O2
CO2
1 In the preparatory reactions, pyruvate is broken, releasing one CO2 molecule. The remaining 2-carbon unit joins with coenzyme A (CoA), forming acetyl CoA.
2 The Krebs cycle begins when the 2-carbon unit from acetyl CoA is released and linked to a 4-carbon acceptor molecule (oxaloacetate), creating citric acid.
3 NADH and FADH2 are energy-rich electron carriers that can donate electrons to the electron transport chain.
NAD+
NADH
CoA
Pyruvate
Acetyl CoA
Citric acid
KREBS CYCLE
Oxaloacetate
ADP
ATP
FAD
FADH2
To electron transport chain (see Figure 6.16)
To electron transport chain (see Figure 6.16)

take place in the mitochondrial matrix (**FIGURE 6.15**). Before the cycle begins, however, pyruvate entering a mitochondrion must be processed through several *preparatory reactions*:

- A large enzyme complex in the matrix breaks one of the carbon-carbon covalent bonds in pyruvate, releasing a molecule of CO_2 and leaving behind a two-carbon unit known as an acetyl group.
- The same enzyme complex attaches this acetyl group to a "carbon carrier" known as *coenzyme A* (*CoA*), producing a molecule called *acetyl CoA*.

The Krebs cycle begins with acetyl CoA donating the two-carbon acetyl group to a four-carbon acceptor molecule.

The Krebs cycle is also called the **citric acid cycle** because citric acid, a six-carbon compound, is the first product formed in this looped, enzyme-driven pathway. Coenzyme A is liberated in this process to recruit yet more acetyl groups for the cycle. Citric acid is converted into a six-carbon intermediate that releases a carbon and two oxygen atoms (that is, CO_2) to become a five-carbon compound. This five-carbon molecule is then degraded to a four-carbon molecule, releasing one CO_2.

As the covalent bonds are broken, the energy stored in them is used to drive the formation of energy carriers: ATP, NADH, and a chemical cousin of NADH called $FADH_2$. As you will see shortly, the energy locked in the NADH and $FADH_2$ is used to make many molecules of ATP during the third and final stage of cellular respiration.

The Krebs cycle is like the Grand Central Station of metabolism because it is the meeting point of several different types of degradative (catabolic) and biosynthetic (anabolic) pathways (see Section 5.2). For example, carbon skeletons derived from other types of food molecules, such as lipids, can be degraded through the Krebs cycle. Triglycerides (see Section 2.9) are broken into fatty acids and glycerol in the cytosol; upon entering the mitochondrion, these molecules are converted to acetyl CoA. The lipid-derived acetyl CoA is indistinguishable from that made from pyruvate during the preparatory reactions, and its carbon atoms are broken up in the same manner by the Krebs cycle.

Oxidative phosphorylation uses oxygen to produce ATP in quantity

The largest output of ATP is generated in mitochondria during the third and last stage of cellular

Helpful to Know

Technically speaking, all physical effort by humans is aerobic, because it is absolutely dependent on oxygen. Endurance exercise is popularly called "aerobic exercise," which it is. Strength training, such as lifting weights, often involves a short burst of physical effort that depends on cellular respiration as well as glycolysis-based fermentation for ATP. Therefore, strictly speaking, strength training is aerobic *plus* anaerobic.

respiration: oxidative phosphorylation. The enzymatic reactions associated with the Krebs cycle take place in the mitochondrial matrix. Oxidative phosphorylation, however, takes place in the many folds (*cristae*) of the inner mitochondrial membrane. Folds in this membrane create a large surface area in which are embedded many electron transport chains (ETCs) and many units of ATP synthase. This ATP-manufacturing channel protein is almost identical to the ATP synthase found in the thylakoid membranes of chloroplasts (see Figure 6.8*c*).

NADH and $FADH_2$ made by the Krebs cycle, and also the NADH generated during glycolysis (see Figure 6.13), diffuse to the inner membrane and donate their high-energy electrons to the ETC. Energy is released as electrons travel down the ETC, and some of that energy drives the addition of a phosphate group on ADP (a *phosphorylation* reaction), thereby generating ATP. The phosphorylation of ADP is an oxygen-dependent process—hence the name *oxidative* phosphorylation (**FIGURE 6.16**). Let's take a closer look at the link between electron transfer through the ETC and phosphorylation of ATP.

The electron transport chains found in the inner mitochondrial membrane are similar in function to those found in the thylakoid membranes of chloroplasts. In mitochondria, electrons donated by NADH and $FADH_2$ are passed along a series of ETC components (**FIGURE 6.16*a***), and the energy released is used to pump protons (H^+) through channel proteins from the matrix into the intermembrane space, creating a proton gradient across the inner membrane (**FIGURE 6.16*b***). As in chloroplasts, the proton gradient is depleted periodically as protons gush through the ATP synthases (**FIGURE 6.16c**). The movement of protons through the ATP synthase channel activates enzymes that catalyze the phosphorylation of ADP to from ATP.

The striking similarities in the way ATP is produced in chloroplasts and mitochondria illustrate that diverse metabolic pathways can evolve through modifications of the same basic machinery. Both of these organelles use ETCs built from similar components, both use energy released during electron transfer to move protons across an inner membrane, and the two use very similar ATP synthases to harness the energy of the proton gradient to make ATP.

EXTREME FLIGHT

The bar-headed goose ascends to 30,000 feet to migrate over the Himalaya mountains. How can it fly 1,000 miles in a day, in air with only 30 percent as much O_2 as air at sea level? The birds have large, efficient lungs, an extensive blood supply to the flight muscles, and hemoglobin that binds O_2 exceptionally well.

The electrons off-loaded by NADH and $FADH_2$ travel down the ETCs of the inner mitochondrial membrane. They are accepted by oxygen (O_2), which combines them with H^+ picked up from the mitochondrial matrix to form water (H_2O). In other words, electron transfer along the ETC terminates with O_2, which serves as the final electron acceptor (see Figure 6.16*b*). Without oxygen to whisk the electrons away, the electron traffic on the ETC "highway" would come to a standstill. When electron transfer stops along the ETC, protons cannot be pumped into the intermembrane space; and without the energy of the proton gradient, ADP cannot be phosphorylated to make ATP. Like all other aerobic organisms, we need oxygen because we cannot make enough ATP without it, so anything that interferes with that process is a danger to life (see the "Biology in the News" feature on page 144).

Cellular respiration (glycolysis, the Krebs cycle, and oxidative phosphorylation) can theoretically yield 38 molecules per molecule of glucose. But because there are energy costs along the way—for example, for transporting NADH into the mitochondria—the net yield in eukaryotes is often as low as 30 ATP molecules for each glucose respired. Even so, mitochondrial respiration is much more productive than glycolysis alone, which yields only 2 ATP molecules per molecule of glucose consumed.

Concept Check

1. What is the product of glycolysis? Where else is this molecule produced?
2. Under what circumstances do cells use fermentation to produce ATP?
3. How does the ATP production of cellular respiration compare with that of glycolysis alone?
4. Why do we need oxygen (O_2) to live?

Concept Check Answers

1. Glycolysis breaks glucose into two 3-carbon sugars, G3P is also produced in the Calvin cycle of photosynthesis.
2. Under anaerobic conditions—when oxygen is absent or in short supply. Fermentation regenerates the energy carrier NAD^+ from NADH, allowing glycolysis to continue.
3. Overall, the processes of cellular respiration yield about 30–38 ATP molecules per molecule of glucose. Glycolysis alone yields only 2 ATP molecules per molecule of glucose.
4. Oxygen gas (O_2) is the final electron acceptor in the mitochondrial ETC; without it, our cells cannot make enough ATP to survive.

FIGURE 6.16 The Mitochondrial Electron Transport Chain and ATP Synthase Generate ATP through Oxidative Phosphorylation Oxidative phosphorylation is the last stage in cellular respiration, and it produces the most ATP of any metabolic pathway. (*a*) Electrons donated by NADH and $FADH_2$ enter the ETC. (*b*) A proton gradient is generated. (*c*) Proton flow through ATP synthase catalyzes the production of ATP.

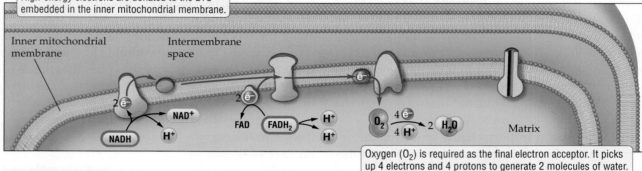

Oxidative Phosphorylation

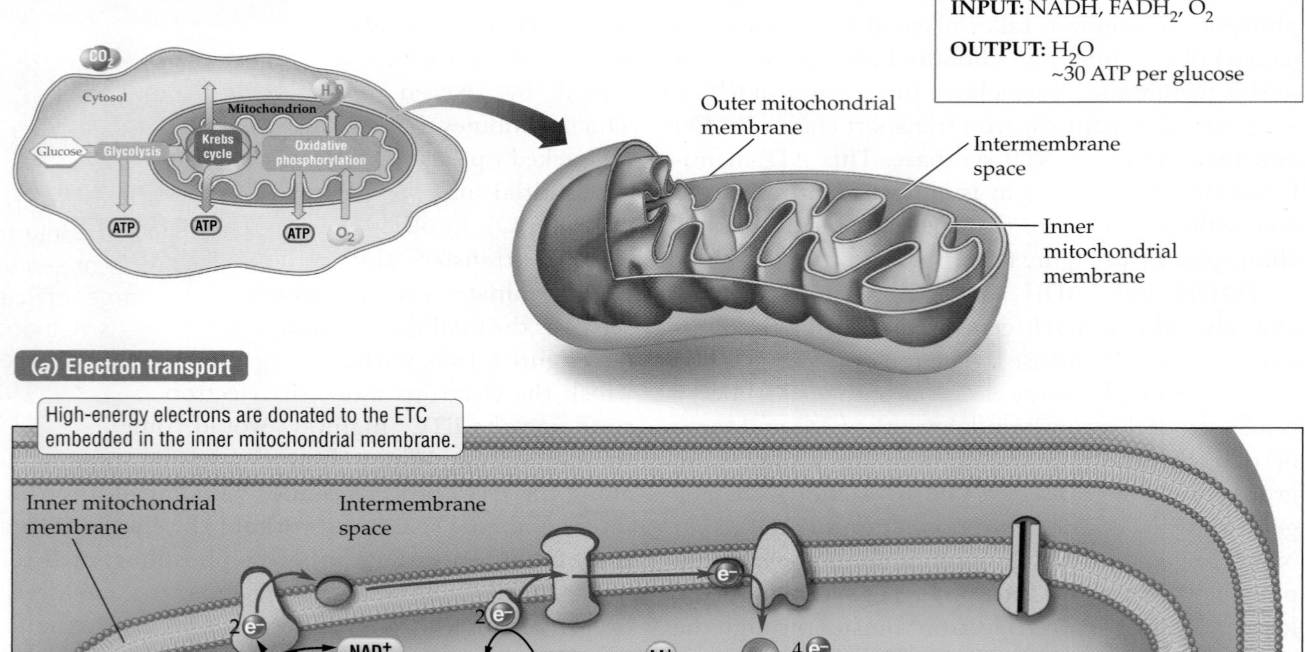

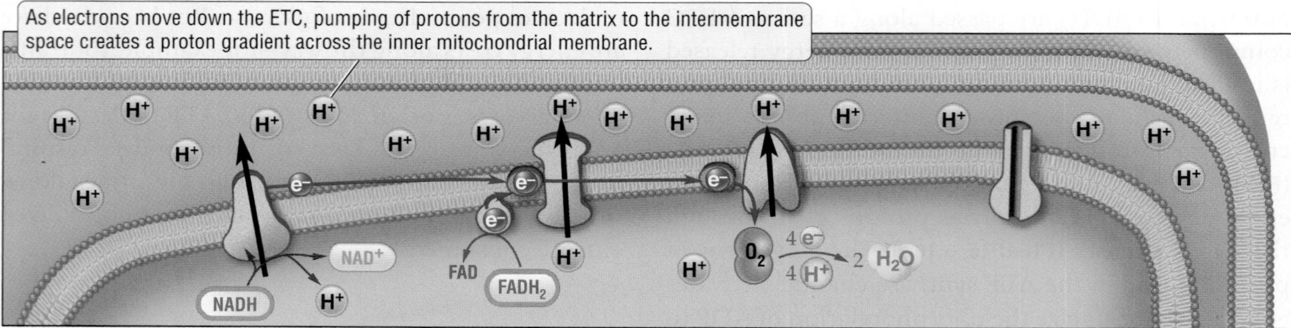

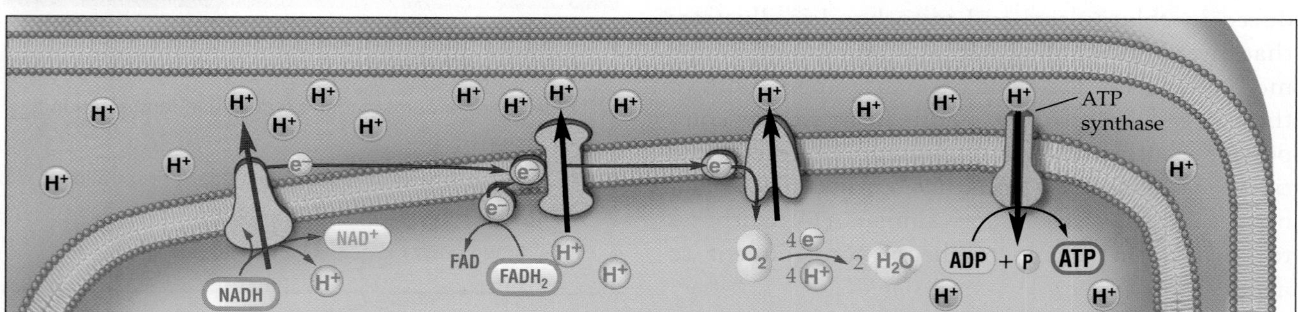

APPLYING WHAT WE LEARNED

Waiting to Exhale

Breathing is such a vital activity that if we stop breathing for even a few minutes, disaster will strike. Waste carbon dioxide from cellular respiration begins to accumulate in the body if it is not exhaled from the lungs. The buildup of carbon dioxide in turn makes the blood acidic. The brain detects the blood's increase in pH and sends signals to the body to take action. Blood pressure soars, the heart beats faster, and we gasp for air. People prone to anxiety attacks are especially sensitive and tend to experience panic at lower levels of CO_2 exposure than the average person does.

Competitive free divers must train their bodies not only to become accustomed to less oxygen but also to ignore the panic and tightness in their chests as they fight the urge to breathe. Because stressful thoughts or emotions can increase heart rate and therefore oxygen consumption, the divers often meditate to clear their minds.

Diving into cool water activates the powerful diving reflex, which diverts blood from the body to the heart and brain, preserving the function of these vital organs while reducing overall ATP use. Before a competition, breath holders reduce resting metabolic rate—and therefore oxygen consumption—by losing weight and then fasting for several hours. During the competition, they minimize ATP use by remaining as calm and still as possible.

Competitive free divers also increase their ability to withstand low oxygen levels by training at high altitudes or by sleeping in low-oxygen tents. Air at 8,000 feet has only 74 percent as much oxygen as air at sea level. The human body responds to lower oxygen levels in several ways: It increases the number and size of our red blood cells, whose job it is to ferry oxygen throughout the body. It also increases the amount of hemoglobin (the oxygen-binding protein inside the red blood cells), as well as the number of tiny blood vessels that supply the tissues. This acclimation takes a week or two to complete and disappears within 2 weeks of returning to low altitudes.

TABLE 6.1 Effects of Aerobic Exercise

- Increase in number of mitochondria in skeletal muscle cells
- Increase in blood supply to skeletal muscle cells
- Increased efficiency of heart function (more blood sent out per contraction; lower pulse rate)
- Increased lung capacity

Enormous lung capacity is another asset in both free diving and endurance sports such as long-distance running, swimming, or biking. Lung capacity varies with size, sex, age, and aerobic fitness. Taller people have larger lungs than shorter people. Women's lungs are, on average, 20–25 percent smaller than men's, and we all tend to lose lung capacity as we age. The average human lung holds about 5 liters of air. Elite Australian swimmer Grant Hacket has a lung capacity of 13 liters.

Engaging in aerobic exercise, living at high altitude, doing controlled breathing exercises, and playing a wind instrument such as a trumpet or tuba can all increase the efficiency of the respiratory system (mainly the lungs) and the circulatory system (the heart and blood vessels) and reduce age-related declines in lung capacity. Aerobic exercise, such as running or biking, and high altitudes increase the number of mitochondria in skeletal muscle cells and also the blood supply to these muscles (**TABLE 6.1**). The enhanced circulation delivers more oxygen per second.

EXTREME BREATH HOLDING

Cuvier's beaked whales are the static-apnea champions of the animal world. They can stay underwater for up to 85 minutes, diving as deep as 1,900 meters (6,230 feet). They store four times as much oxygen in their blood and muscles as humans do. Their body temperature drops as they dive, cooling the brain by 3°C to reduce its energy demands.

BIOLOGY IN THE NEWS

Cassava and Mental Deficits

BY DONALD G. MCNEIL JR. • *New York Times*, April 22, 2013

Konzo, a disease that comes from eating bitter cassava that has not been prepared properly—that is, soaked for days to break down its natural cyanide—has long been known to cripple children.

The name, from the Yaka language of Central Africa, means "tied legs," and victims stumble as if their knees were bound together.

Now researchers have found that children who live where konzo is common but have no obvious physical symptoms may still have mental deficits from the illness.

Cassava, also called manioc or tapioca, is eaten by 800 million people around the world and is a staple in Africa, where bitter varieties grow well even in arid regions. When properly soaked and dried, and especially when people have protein in their diet, bitter cassava is "pretty safe," said Michael J. Boivin, a Michigan State psychiatry professor and lead author of a study published online by Pediatrics. "But in times of war, famine, displacement and hardship, people take shortcuts."

In the Democratic Republic of Congo, Dr. Boivin and colleagues gave tests of mental acuity and dexterity to three groups of children. Two groups were from a village near the Angolan border with regular konzo outbreaks: Half had leg problems; half did not but had cyanide in their urine. The third was from a village 125 miles away with a similar diet but little konzo because residents routinely detoxified cassava before cooking it.

The children from the latter village did "significantly better" on tests of remembering numbers, identifying objects, following mazes and fitting blocks together, while healthy-looking children from the first village did almost as badly as children with obvious konzo.

The mental damage was like that done by lead exposure but more subtle, Dr. Boivin said.

The Bill and Melinda Gates Foundation is supporting efforts to create cassavas with less cyanide ... One drawback [of the low-cyanide plants], Dr. Boivin said, was that the pest-resistant and rot-resistant qualities of bitter cassava appeared to be partly due to its higher cyanide content.

Cyanide is a respiratory poison, a chemical that shuts down ATP production by mitochondria. Rotenone and carbon monoxide are other examples of respiratory poisons, also known as metabolic poisons.

Rotenone has a long history of use as a fish poison and insecticide. Rotenone is extracted by crushing the roots of various species in the bean family, including the Florida fish poison tree (*Piscidia piscipula*). The toxin binds to the first component of the electron transport chain (see Figure 6.16) and prevents electrons from being handed off to the next component in the electron transport chain.

Carbon monoxide is the most common cause of death by poisoning around the world. Faulty furnaces and fireplaces, and the use of cookstoves and power tools in an enclosed space, can cause the buildup of dangerous levels of carbon monoxide, which is odorless and tasteless. Carbon monoxide blocks electron transport by binding to the last ETC component, which is why it is toxic to all aerobic organisms. The gas also harms several other aspects of human physiology; for example, it binds to hemoglobin more strongly than oxygen does.

Cyanide is the poison of choice in murder mysteries. The smell of bitter almonds drifting from the deceased might clue the detective that cyanide was used to do in the victim. Cyanide blocks the transfer of electrons to O_2 by binding to the last component in the chain. Many plants store cyanide precursors in their seeds and roots, to deter herbivores. The precursors are converted into hydrogen cyanide by enzymes that are released when the cells are broken open. About 20 bitter almonds can produce enough cyanide to kill an average person. But don't fear the sweet almonds that you can buy at the grocery store! Cyanide was bred out of these healthful nuts long ago, when the crop was first domesticated in central Asia.

Cassava is a native of South America that was introduced to West Africa by the Portuguese some 300 years ago. It was a minor crop in Africa at first, but land deterioration over the last hundred years or so has increased the popularity of cassava, which grows well on poor soils. Cyanide can be leached out of the ground root, but sufficient access to water is necessary. As the article points out, dietary protein reduces the absorption of cyanide. Getting enough of both is a challenge for the poor.

Evaluating the News

1. Cyanide binds to the last component of the ETC. Explain how high levels of cyanide can kill a person and why lower levels are damaging to brain development.

2. Low-cyanide varieties of cassava are available in Africa. Why, then, do many farmers prefer to plant varieties with higher cyanide levels?

CHAPTER REVIEW

Summary

6.1 Molecular Energy Carriers

- Energy carriers store energy and deliver it for cellular activities.
- ATP is the most commonly used energy carrier. Photosynthesis and cellular respiration transform ADP plus phosphate into ATP.
- The energy carriers NADPH and NADH donate electrons and hydrogen ions to metabolic pathways.
- NADPH is used in biosynthetic (anabolic) pathways such as photosynthesis. NADH participates in degradative (catabolic) pathways such as cellular respiration.

6.2 An Overview of Photosynthesis and Cellular Respiration

- In chemical terms, photosynthesis is the reverse of cellular respiration.
- Photosynthesis occurs in producers only. The light reactions make ATP and NADPH, splitting water molecules and releasing oxygen gas. The energy carriers are used to convert carbon dioxide into sugar molecules during the Calvin cycle reactions.
- Cellular respiration occurs in producers and consumers. It begins in the cytoplasm and is completed in the mitochondrion.
- Small amounts of ATP and NADH are made during the first stage of cellular respiration: glycolysis. During glycolysis, sugar molecules are broken to make pyruvate, a three-carbon compound.
- The next two stages of cellular respiration take place inside the mitochondrion. Carbon dioxide is released during the degradation of pyruvate via the preparatory reactions and the Krebs cycle, which yields NADH, $FADH_2$, and ATP.
- The final stage of cellular respiration is oxidative phosphorylation, during which many molecules of ATP are made in a membrane-dependent, oxygen-utilizing process.

6.3 Photosynthesis: Energy from Sunlight

- Photosynthesis takes place in chloroplasts—light reactions in the thylakoid membrane, and Calvin cycle reactions in the stroma.
- In the light reactions, energy is absorbed using pigment molecules that include chlorophyll. Electrons are stripped from chlorophyll and replaced with electrons from water molecules, releasing oxygen (O_2). ATP is generated as electrons flow along the electron transport chain (ETC) that links photosystem II to photosystem I. In the last step, electrons are accepted by $NADP^+$, which picks up protons (H^+ ions) to become NADPH.
- Calvin cycle reactions use the ATP and NADPH produced by the light reactions to turn CO_2 into glyceraldehyde 3-phosphate (G3P), which is converted to six-carbon sugars in the cytosol. Rubisco catalyzes the fixation of CO_2 in the stroma.
- Photorespiration is a sequence of chemical reactions that "waste" fixed carbon and are triggered when rubisco adds O_2, instead of CO_2, to ribulose 1,5-bisphosphate (RuBP). Photorespiration is worse when temperatures are high.
- In C_4 plants, photorespiration is reduced because the Calvin cycle reactions are conducted in bundle-sheath cells, which maintain a high CO_2 environment. CO_2 entering through stomata is converted into a four-carbon molecule that is sent to bundle-sheath cells, where it re-releases CO_2.
- CAM photosynthesis enables plants in dry habitats to conduct photosynthesis while minimizing water loss. CAM plants open stomata only at night and store incoming CO_2 as a four-carbon molecule overnight. The molecule breaks down to release CO_2 during the day, so photosynthesis proceeds despite closed stomata.

6.4 Cellular Respiration: Energy from Food

- Cellular respiration requires oxygen and has three stages: glycolysis, the Krebs cycle, and oxidative phosphorylation.
- Glycolysis occurs in the cytosol and splits each glucose molecule into two molecules of pyruvate. It yields 2 ATP and 2 NADH.
- In fermentation, pyruvate is converted into carbon compounds such as CO_2 and alcohol (as in fermentation by yeasts) or lactic acid (as in skeletal muscles). The postglycolytic fermentation reactions regenerate NAD^+, essential for continued glycolysis when the oxygen supply is inadequate.
- In the presence of oxygen, the pyruvate from glycolysis enters the mitochondria, where it is degraded while energy carriers are generated and CO_2 is released.
- The Krebs cycle is a series of enzyme-catalyzed reactions that produces 2 CO_2, 3 NADH, 1 $FADH_2$, and 1 ATP.
- Oxidative phosphorylation generates about 30–38 molecules of ATP from each glucose. Electrons unloaded by NADH and $FADH_2$ travel an ETC, creating a proton gradient across the inner mitochondrial membrane. ATP synthase phosphorylates ADP to make ATP as the proton gradient collapses.

Key Terms

aerobic (p. 139)
anaerobic (p. 138)
antenna complex (p. 131)
ATP synthase (p. 132)
C_3 plant (p. 135)
C_4 pathway (p. 135)
C_4 plant (p. 135)
Calvin cycle (p. 126)
CAM pathway (p. 135)
CAM plant (p. 137)
carbon fixation (p. 132)
cellular respiration (p. 127)
chlorophyll (p. 126)
chloroplast (p. 126)
citric acid cycle (p. 140)
crassulacean acid metabolism (CAM) (p. 137)
electron transport chain (ETC) (p. 131)
energy carrier (p. 124)
fermentation (p. 138)
glycolysis (p. 127)
Krebs cycle (p. 127)
light reactions (p. 126)
mitochondrion (p. 127)
NADH (p. 125)
NADPH (p. 125)
oxidative phosphorylation (p. 127)
photon (p. 128)
photorespiration (p. 135)
photosynthesis (p. 126)
photosystem (p. 131)
photosystem I (p. 132)
photosystem II (p. 132)
proton gradient (p. 132)
pyruvate (p. 138)
reaction center (p. 131)
rubisco (p. 132)
stoma (p. 129)
stroma (p. 129)
thylakoid (p. 131)

Self-Quiz

1. The chemical that functions as an energy-carrying molecule in all organisms is
   a. carbon dioxide.
   b. water.
   c. RuBP.
   d. ATP.

2. The main function of the Calvin cycle reactions is to produce
   a. carbon dioxide.
   b. sugars.
   c. NADPH.
   d. ATP.

3. The oxygen produced in photosynthesis comes from
   a. carbon dioxide.
   b. sugars.
   c. pyruvate.
   d. water.

4. The light reactions in photosynthesis require
   a. oxygen.
   b. chlorophyll.
   c. rubisco.
   d. carbon fixation.

5. Glycolysis occurs in
   a. mitochondria.
   b. the cytosol.
   c. chloroplasts.
   d. thylakoids.

6. The electrons needed to replace those lost from chlorophyll in the light reactions of photosynthesis come ultimately from
   a. sugars.
   b. channel proteins.
   c. water.
   d. electron transport chains.

7. Cellular respiration
   a. converts inhaled oxygen gas into water.
   b. occurs in skeletal muscle cells in our body, but not in ordinary cells such as skin cells.
   c. converts carbon dioxide into organic molecules.
   d. breaks down sugar molecules, releasing oxygen gas as a by-product.

8. Which of the following statements is *not* true?
   a. Glycolysis produces most of the ATP required by aerobic organisms like us.
   b. Glycolysis produces pyruvate, which is consumed by the Krebs cycle.
   c. Glycolysis occurs in the cytosol of the cell.
   d. Glycolysis is the first stage of cellular respiration.

9. The Krebs cycle reactions
   a. take place in the cytoplasm.
   b. convert glucose to pyruvate.
   c. generate ATP with the help of an enzyme complex called ATP synthase.
   d. yield ATP, NADH, and $FADH_2$.

10. Which of the following is essential for oxidative phosphorylation?
    a. rubisco
    b. NADH
    c. phosphoglycolate
    d. chlorophyll

11. Oxidative phosphorylation
    a. produces less ATP than glycolysis does.
    b. produces simple sugars.
    c. depends on the activity of ATP synthase.
    d. is part of the photosystem I electron transport chain.

12. In CAM plants
    a. stomata open only at night.
    b. the first product of carbon fixation is a 3-carbon molecule.
    c. Calvin cycle reactions take place only in bundle sheath cells.
    d. photorespiration increases the efficiency of photosynthesis than a typical C_3 plant.
    e. more water is lost from the plant during photosynthesis than in a typical C_3 plant.

Analysis and Application

1. The Calvin cycle reactions are sometimes called the "dark reactions" to contrast them with the light reactions. Can the Calvin cycle be sustained in a plant that is kept in total darkness for several days? Why or why not?
2. In both chloroplasts and mitochondria, the transfer of electrons down an ETC involves hydrogen ions and leads to a similar outcome. Describe that outcome, and explain how it contributes to the production of ATP in each of these organelles.
3. Explain what is wrong with this statement: The postglycolytic fermentation reactions are a significant source of energy because they generate ATP and NADH through the degradation of pyruvate.
4. Dinitrophenol (DNP) belongs to a class of metabolic poisons known as uncoupling agents. DNP shuttles protons (H^+ ions) freely across biological membranes, thereby destroying any existing proton gradient. How would exposure to DNP affect ATP synthesis by mitochondria? ***Extra challenge:*** DNP raises body temperature and leads to rapid weight loss. Doctors prescribed DNP as a weight loss drug in the 1930s, until the death of several patients led to a ban. Explain why DNP causes weight loss in humans.
5. Compare the energy yields of glycolysis and oxidative phosphorylation. Which pathway releases more usable energy and why?
6. Explain the role of mitochondrial membranes in cellular respiration. Why do you suppose the inner mitochondrial membrane is extensively folded, but the outer membrane is not? What structure in the chloroplast performs the same function that the intermembrane space does in a mitochondrion?
7. Compare photosystem I and photosystem II by completing the table below. Write "yes" or "no," depending on whether the statement in the leftmost column applies or not. We have filled in the first line in the table as an example.

	PHOTOSYSTEM I	PHOTOSYSTEM II
Located in the stroma of the chloroplast?	No	No
Contains chlorophyll?		
Directly involved in splitting of water molecules?		
Directly involved in generating NADPH?		
Associated with an electron transport chain?		

8. How are C_4 plants different from C_3 plants and from CAM plants? Name some examples of each of these three types of plants.
9. Compare photosynthesis and cellular respiration. Name the organelles in which these processes occur. Use the diagram below to show which atmospheric gases are released and which are consumed by each of these processes. What is the role in photosynthesis of the energy carriers generated by the light reactions? Which of the three main stages of cellular respiration also generates both of these energy carriers?

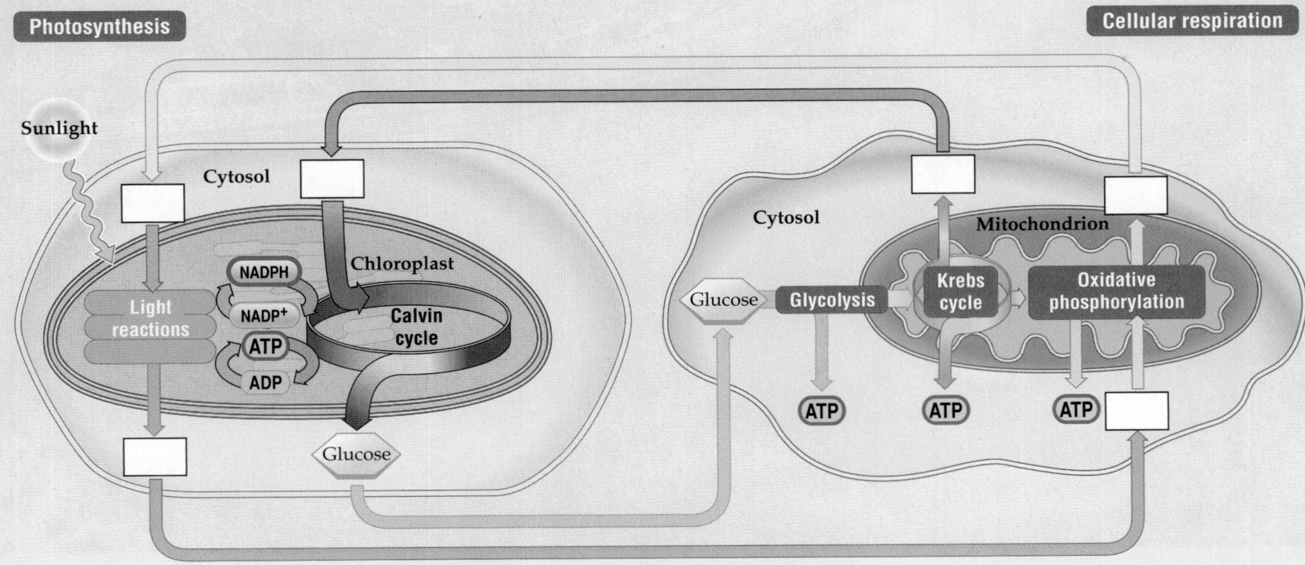

7 Cell Division

IT ISN'T GRASS A giant bloom of sticky green algae, thick and vast enough to halt sailboats and windsurfers, threatened to prevent the Chinese Olympic sailing games in 2008.

Olympic-Class Algal Bloom

In June 2008, as the People's Republic of China put the finishing touches on elaborate preparations for the Summer Olympics, a minor natural disaster struck. It was China's first chance to host the 200 countries attending the Olympics, and China had already spent nearly $2 billion to build 37 venues for 28 summer sports. Most of the events would take place in or near the inland capital city of Beijing. But one sport, sailing, would be held at Qingdao (or Tsingtao), a major port city on the Yellow Sea, 450 miles southeast of Beijing.

Olympic sailing competitions traditionally appear in the media as beautiful blue-water events, with sails bent to the wind. But starting in May, small, green mats of algae appeared here and there near the shore in Qingdao Bay. At first it was just a minor annoyance. But by the end of June, world-class sailors from all over the world had arrived to practice for the August Olympic Games, and they were beginning to complain. The problems started with dense fog and nearly windless days. But then football field–sized rafts of algae began drifting across the racecourse; mats of the sticky, green stuff wrapped around the keels of expensive racing boats, bringing them to a halt, and even entangled Olympic windsurfers. The sailing blogosphere was buzzing with complaints.

How could a handful of tiny cells multiply fast enough to cover 1,500 square miles in a few weeks? Could the masses of algae be removed in time for the Olympics?

Algae are photosynthetic organisms—eukaryotes, like us—that float near the surface of water, their cells dividing rapidly in the presence of nutrients and sunshine. In the summer of 2008, conditions in the Yellow Sea must have been ideal for these algae because, despite the best efforts of Olympic organizers, the algal cells continued to divide and multiply at a crushing pace, eventually covering a third of the planned Olympic sailing racecourse. Altogether, the algae covered 1,500 square miles of ocean. It was the world's largest drifting algal bloom.

MAIN MESSAGE Cell division is the means by which organisms grow, maintain their tissues, and pass genetic information from one generation to the next.

KEY CONCEPTS

- Cell division is necessary for reproduction in all life-forms, and for growth and repair in a multicellular body.
- The two main stages of the cell cycle are interphase (during which the cell grows, performs its life functions, and may copy its DNA) and cell division (during which the parent cell divides into daughter cells).
- A chromosome consists of one DNA molecule compacted by packaging proteins.
- Mitosis generates two genetically identical daughter cells, each of which has the total chromosome set of the parent cell.
- The main stages of mitosis are prophase, metaphase, anaphase, and telophase. In prophase, chromosomes condense and attach to the mitotic spindle, as the nuclear envelope disintegrates.
- In metaphase, the chromosomes are aligned at the cell center, and during anaphase they are segregated to the opposite sides of the cell.
- Chromosomes decondense and the nuclear envelope re-forms during telophase.
- Meiosis, necessary for sexual reproduction, produces four daughter cells, each with half of the chromosome set found in the parent cell.
- During meiosis, crossing-over and independent assortment of homologous chromosomes contribute to the genetic diversity of gametes. Random fertilization also increases genetic diversity.

OLYMPIC-CLASS ALGAL BLOOM 149

7.1 Why Cells Divide 150

7.2 The Cell Cycle 153

7.3 The Chromosomal Organization of Genetic Material 156

7.4 Mitosis and Cytokinesis: From One Cell to Two Identical Cells 159

7.5 Meiosis: Halving the Chromosome Set to Make Gametes 161

APPLYING WHAT WE LEARNED
The Great Divide 168

CELL DIVISION IS A DISTINCTIVE PROPERTY of all life-forms. Without it, there would be no eggs, no babies, no grown-ups, no circle of life. Your life began as a single fertilized egg cell, and it took many billions of cell divisions to grow you into a baby and take you from babyhood to adulthood. Even now, millions of cell divisions take place in your body every day.

Role of Cell Division

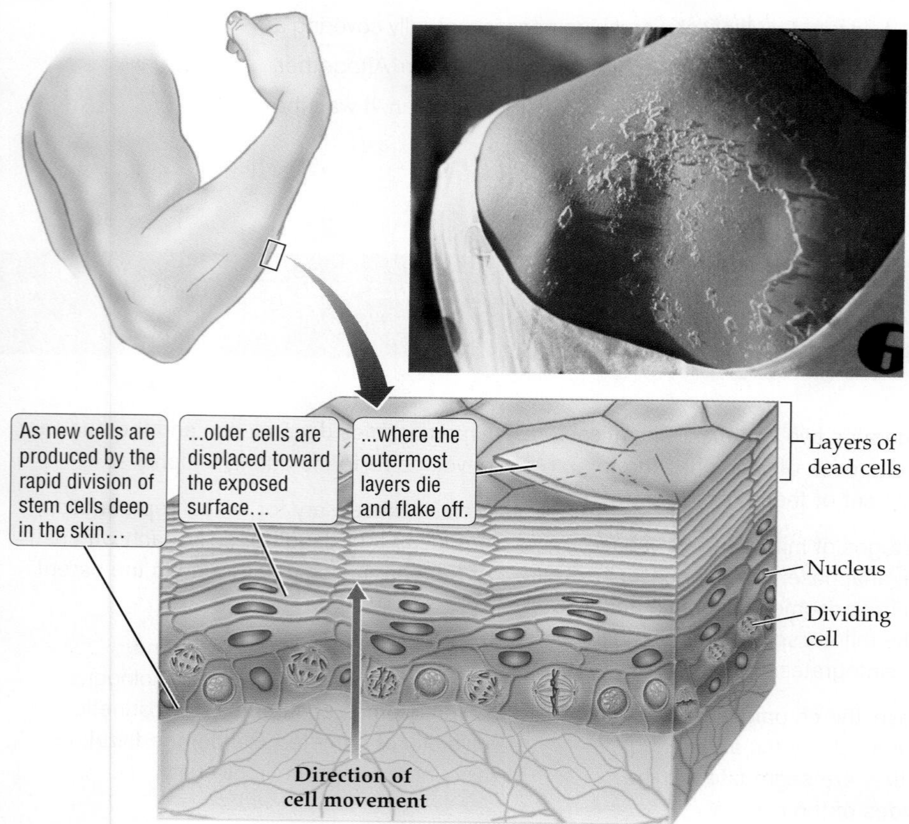

FIGURE 7.1 Cell Division Replenishes the Skin

Cell division in the deepest layer of the skin is necessary to replace skin cells (called keratinocytes) lost at the surface. The loss is due to the normal programmed death of the outer layers of mature cells, but it can also be the consequence of severe DNA damage, as in a peeling sunburn (*inset*).

The purpose of most of these cell divisions is to replace cells that have died in the line of duty—cells that have outlived their useful lives or that have been damaged for some reason. Cell divisions are also needed to increase the ranks of immune cells—those that do battle with invading organisms such as bacteria and viruses in response to an infection. A specialized type of cell division, called *meiosis*, produces the egg cells in a female and the sperm cells in a male, which join together to produce offspring.

In this chapter we discuss how cells replace themselves through cell division and the essential role of cell divisions in asexual and sexual reproduction. After examining the *cell cycle*, which describes the activities of a cell through the course of its life span, we turn the spotlight on the two main types of cell division: *mitotic division*, which replaces a cell with two identical cells, and *meiosis*, which produces eggs and sperm for sexual reproduction.

7.1 Why Cells Divide

Cells, the smallest unit of life, divide for two basic reasons:

1. In single-celled organisms, cells divide to reproduce.
2. In multicellular organisms, cells divide for growth and repair of tissues, as well as for reproduction.

Multicellular organisms like us need cell division to add new cells to the body (**FIGURE 7.1**). Cell division generates daughter cells from a parent cell, with the transfer of genetic information—in the form of DNA—from the parent to the daughter cells. All types of organisms need cell division to create the next generation of individuals. New individuals that receive DNA from the parents are called *offspring*.

Some organisms create offspring through **asexual reproduction**, which generates *clones*, offspring that are genetically identical to the parent. Bacteria and archaeans reproduce only through asexual means. Some multicellular organisms, including many plants and even some animals, reproduce asexually as well. The majority of eukaryotes, however, produce offspring through *sexual reproduction*. In **sexual reproduction**, genetic information from two individuals of opposite mating types is combined to produce offspring. Offspring resulting from sexual reproduction are similar, but not identical, to the parents.

Asexual reproduction and sexual reproduction both propagate the species (**TABLE 7.1**), but sexual reproduction adds a great deal of genetic diversity to

TABLE 7.1 Biological Relevance of Cell Division

TYPE OF CELL DIVISION	OCCURS IN	FUNCTION
Binary fission	Prokaryotes (Bacteria and Archaea)	Asexual reproduction
Mitotic division	Eukaryotes: single-celled or multicellular	Asexual reproduction
	Eukaryotes: multicellular	Growth of individual; repair and replacement of cells and tissues
Meiosis	Eukaryotes: single-celled or multicellular	Sexual reproduction

Mitosis in onion roots

the population. Genetic diversity arises because DNA from two different individuals is shuffled and combined in offspring, giving each offspring a unique mix of the genetic characteristics of its two parents.

Many bacteria use binary fission for asexual reproduction

It is likely that the first organisms to appear on Earth reproduced asexually. The strategies they used to propagate themselves probably resembled the mechanisms seen in modern-day prokaryotes (domains Bacteria and Archaea). Many prokaryotes reproduce asexually through a mechanism called **binary fission** (literally "splitting in two").

The genetic material of most bacteria and archaeans takes the form of one single loop of DNA. The first step in binary fission is the duplication of this DNA, giving rise to two DNA molecules (**FIGURE 7.2**). The cell now expands, and a partition consisting of plasma membrane and cell wall materials appears roughly at the center of the cell. The partition separates the two DNA molecules into two cytoplasmic compartments. Each compartment expands until it breaks loose of the other, so that two daughter cells now replace the parent cell. The daughter cells are identical because each inherits an exact copy of the DNA code that was present in the parent cell.

Eukaryotes use mitosis to generate identical daughter cells

Cell division in eukaryotes is more complicated than binary fission; eukaryotic cells have many molecules of DNA that have to be replicated and then distributed evenly between the two daughter cells. Eukaryotic DNA lies in the nucleus, wrapped in the double layer of membranes that make up the nuclear envelope. In most eukaryotes, the nuclear envelope is disassembled in the dividing cell and then reassembled in each of the daughter cells toward the end of cell division.

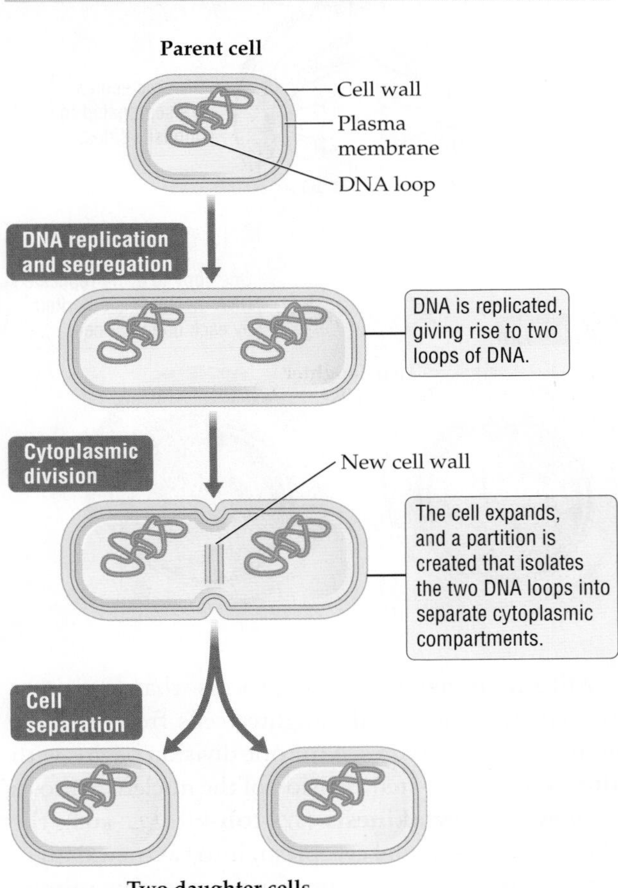
Binary Fission
Parent cell
Cell wall
Plasma membrane
DNA loop
DNA replication and segregation
DNA is replicated, giving rise to two loops of DNA.
Cytoplasmic division
New cell wall
The cell expands, and a partition is created that isolates the two DNA loops into separate cytoplasmic compartments.
Cell separation
Two daughter cells

FIGURE 7.2 Cell Division in a Prokaryote
Many prokaryotes, including many types of bacteria, propagate themselves asexually through a type of cell division known as binary fission.

FIGURE 7.3
Cell Division in a Eukaryote

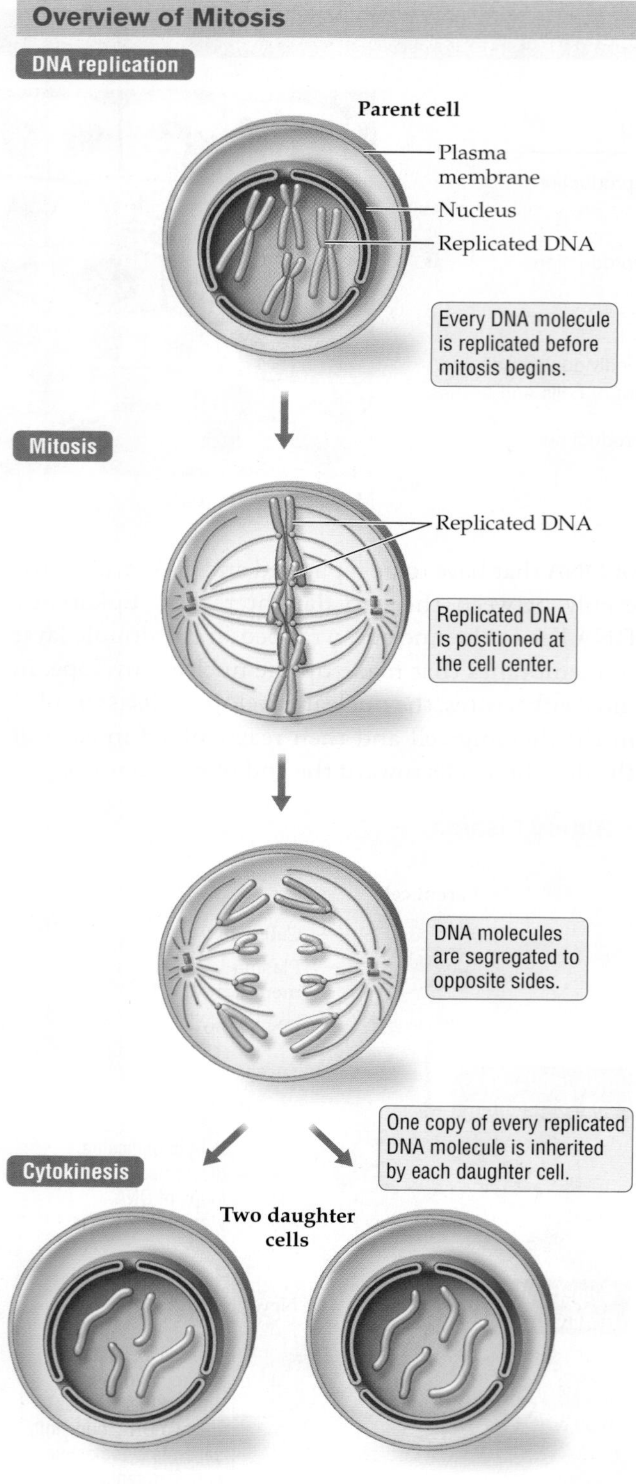
Overview of Mitosis
DNA replication
Parent cell
Plasma membrane
Nucleus
Replicated DNA
Every DNA molecule is replicated before mitosis begins.
Mitosis
Replicated DNA
Replicated DNA is positioned at the cell center.
DNA molecules are segregated to opposite sides.
One copy of every replicated DNA molecule is inherited by each daughter cell.
Cytokinesis
Two daughter cells

Mitotic division is the process that generates two genetically identical daughter cells from a single parent cell in eukaryotes. A mitotic division begins with **mitosis** (**mye-*TOH*-sus**), division of the nucleus. Mitosis is followed by **cytokinesis** (***SYE*-toh-kih-*NEE*-sus**), the splitting of the original cytoplasm into two new daughter cells. A parent cell prepares for an upcoming mitotic division by duplicating its DNA well before mitosis gets under way. During mitosis, an elaborate cytoskeletal machinery works in a precise sequence to deliver one complete copy of the replicated DNA to opposite sides of the original cell. The cytoplasm between the two copies of DNA is then divided in the process of cytokinesis, giving rise to two daughter cells (**FIGURE 7.3**).

Single-celled eukaryotes use mitotic divisions for asexual reproduction, in much the same way that prokaryotes reproduce asexually through binary fission (see Figure 7.2). Many multicellular eukaryotes use mitotic divisions to reproduce asexually as well, including most seaweed, fungi, and plants, and some animals, such as sponges and flatworms. While most multicellular organisms reproduce by sexual means, dividing their nuclei by *meiosis* rather than mitosis, all multicellular organisms rely on mitotic divisions for the growth of tissues, organs, and the body as a whole, and for repairing injured tissue and replacing worn-out cells.

Meiosis is necessary for sexual reproduction

Meiosis (**mye-*OH*-sus**) is a specialized type of cell division that makes sexual reproduction possible. In animals, meiosis in the female body generates daughter cells that mature into eggs, while meiosis in the male animal produces sperm cells. Egg and sperm are examples of sex cells, or **gametes**.

The non-sex cells in a multicellular organism are called **somatic cells** (from *soma*, Greek for "body"). The somatic cells of plants and animals have twice as much genetic information as is found in a gamete. This double set of genetic information is known as the *diploid* set, represented by $2n$. Meiosis (*meio*, "less") *reduces* the amount of genetic information transmitted by the parent cell to the daughter cells by half, so that only one set of the genetic information is inherited by each daughter cell. This single set of genetic information is called the *haploid* set, represented by the letter n.

Mating between male and female individuals can result in **fertilization**, the merging of gametes to create a single cell, the **zygote** (**FIGURE 7.4**). The zygote receives one haploid set of genetic material from the egg and the other from the sperm, and that is how fertilization restores the complete diploid set of genetic information to the offspring. From this point, the zygote undergoes mitosis to create a mass of developing cells known as an **embryo**. The embryo develops organs to become a fetus and eventually matures into a juvenile and then into an adult individual.

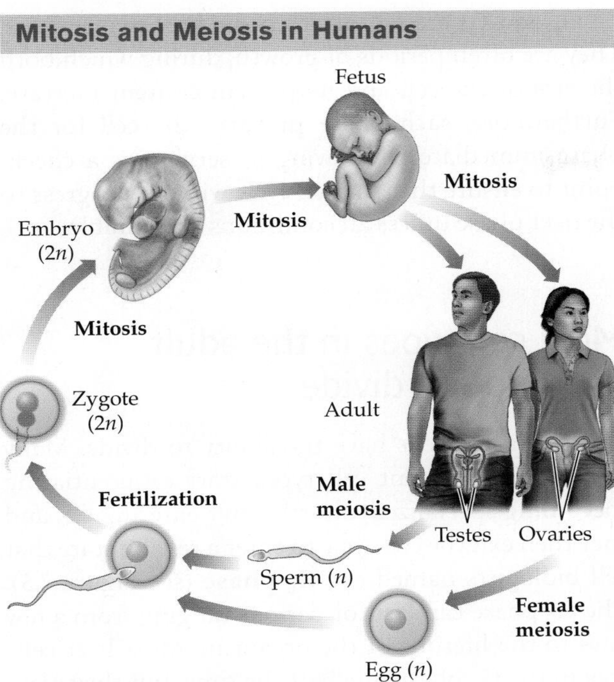

FIGURE 7.4 Mitosis and Meiosis Play Vital Roles in the Human Life Cycle

Cell divisions grow, maintain, and reproduce the human body

Your body arose from a single-celled zygote, formed by the fusion of a particular egg and a certain sperm cell. Mitotic divisions in the zygote give rise to a ball of cells, the embryo (see Figure 7.4). Cells in the very young embryo are not noticeably different from each other; but as the embryo develops, many of the cells in it acquire unique properties and highly specialized functions. The process through which a daughter cell becomes different from the parent cell is known as **cell differentiation**. Heart muscle cells and nerve cells are two examples of the 220 differentiated cell types in the adult human. A small group of cells, called **germ line cells**, is set aside very early in embryonic development. A subset of the germ line cells, called gametocytes, eventually undergo meiosis to produce gametes in adults.

Within a week after conception, germ line cells are set aside as a special lineage of cells. In about the fifteenth week after conception, the germ line cells migrate into the reproductive organs: a pair of ovaries in a developing female and a pair of testes in a developing male. In a female fetus, about a million gametocytes launch into meiosis, but then the process is put on hold until puberty! So, baby girls are born with about a million gametocytes that are arrested in an early stage of meiosis. Typically, one such arrested gametocyte *resumes* meiosis each month in one of the ovaries of a fertile female.

In males, meiosis does not begin until hormonal changes produced by puberty signal millions of gametocytes, in both testes, to begin producing sperm. Male meiosis occurs daily and continues well into old age, but in women the supply of functioning gametocytes dwindles and usually disappears by age 50. At birth a baby has most of the specialized cell types that an adult has. Although most cells in the newborn have differentiated to perform a narrow set of tasks, *stem cells* in the various organs remain unspecialized. Mitosis in stem cells contributes to the growth of the body, and to the regeneration and repair of tissues throughout our lifetime.

Concept Check

1. In what way is binary fission similar to mitotic cell division? Give one difference.
2. What is the function of mitosis?

Concept Check Answers

1. Both produce genetically identical daughter cells, but only prokaryotes undergo binary fission.

2. Mitosis is necessary for asexual reproduction in both single-celled and multicellular eukaryotes, and for growth and repair of the multicellular body.

7.2 The Cell Cycle

The **cell cycle** is a set sequence of events that make up the life of a typical eukaryotic cell that is capable of dividing. The cell cycle extends over the life span of a cell, from the moment of its origin to the time it divides to produce two daughter cells.

The time it takes to complete the cycle depends on the organism, the type of cell, and the life stage of the organism. Dividing cells in tissues that require frequent replacement, such as the skin or the lining of the intestine, take about 12 hours to complete the cell cycle. Cells in most other actively dividing tissues in the human body require about 24 hours to complete the cycle. By contrast, a single-celled eukaryote such as a yeast can complete the cell cycle in just 90 minutes.

The cell cycle has two main stages—*interphase* and *cell division*—each marked by distinctive cell activities. **Cell division** is the last stage in the life of an individual cell. Not only is it the most rapid stage of the cell cycle, but it is also the most dramatic in visual terms. In tissues with many rapidly dividing cells, such as onion root tips or fish embryos, the events of cell division can be readily seen with an ordinary light microscope.

Interphase is the longest stage of the cell cycle. Most cells spend 90 percent or more of their life span in interphase. During this stage the cell takes in nutrients, manufactures proteins and other substances, expands in size, and conducts its special functions. In cells that are destined to divide, preparations for cell division also begin during interphase. A critical event in these preparations is the copying of all the DNA molecules, which contain the organism's genetic information in the form of genes. The discussion that follows offers a closer look at the major events (*phases*) of interphase.

DNA is replicated in the S phase

In cells capable of dividing, interphase can be divided into three main phases: G_1, S, and G_2 (**FIGURE 7.5**). These phases are defined by distinctive cellular events.

- The **G_1 phase** (for "gap 1") is the first phase in the life of a "newborn" cell.
- During the **S phase**, DNA is copied (replicated), which requires synthesis of new DNA ("S" stands for "synthesis").
- The **G_2 phase** (for "gap 2") begins after the S phase and before the start of division.

Early cell biologists bestowed the term "gap" on the G_1 and G_2 phases because they believed those phases to be less important in the life of a cell, compared to the S phase and cell division. We now know that many crucial events occur during the two "gap" phases.

G_1 and G_2 are important phases for two reasons. They are often periods of growth, during which both the size of the cell and its protein content increase. Furthermore, each phase prepares the cell for the phase immediately following it, serving as a checkpoint to ensure that the cell cycle will not progress to the next phase unless all conditions are suitable.

Most cell types in the adult body do not divide

Not all of our cells have the ability to divide. Many of our 220 different cell types start differentiating (becoming specialized) shortly after entering G_1, and they then exit the cell cycle to enter a resting state that cell biologists named the **G_0 phase** (see Figure 7.5). The G_0 phase can last for periods ranging from a few days to the lifetime of the organism. Most liver cells stay in the G_0 phase much of the time, but they reenter the cell cycle about once a year, on average, to make up for cells that have died because of normal wear and tear. The liver's exceptional capacity to regenerate healthy liver tissue makes up for the cell damage the organ suffers on the front line of dealing with toxins that we ingest, from antibiotics to alcohol. The liver's regenerative capacity relies in part on a large pool of G_0 cells that can reenter the cell cycle when needed.

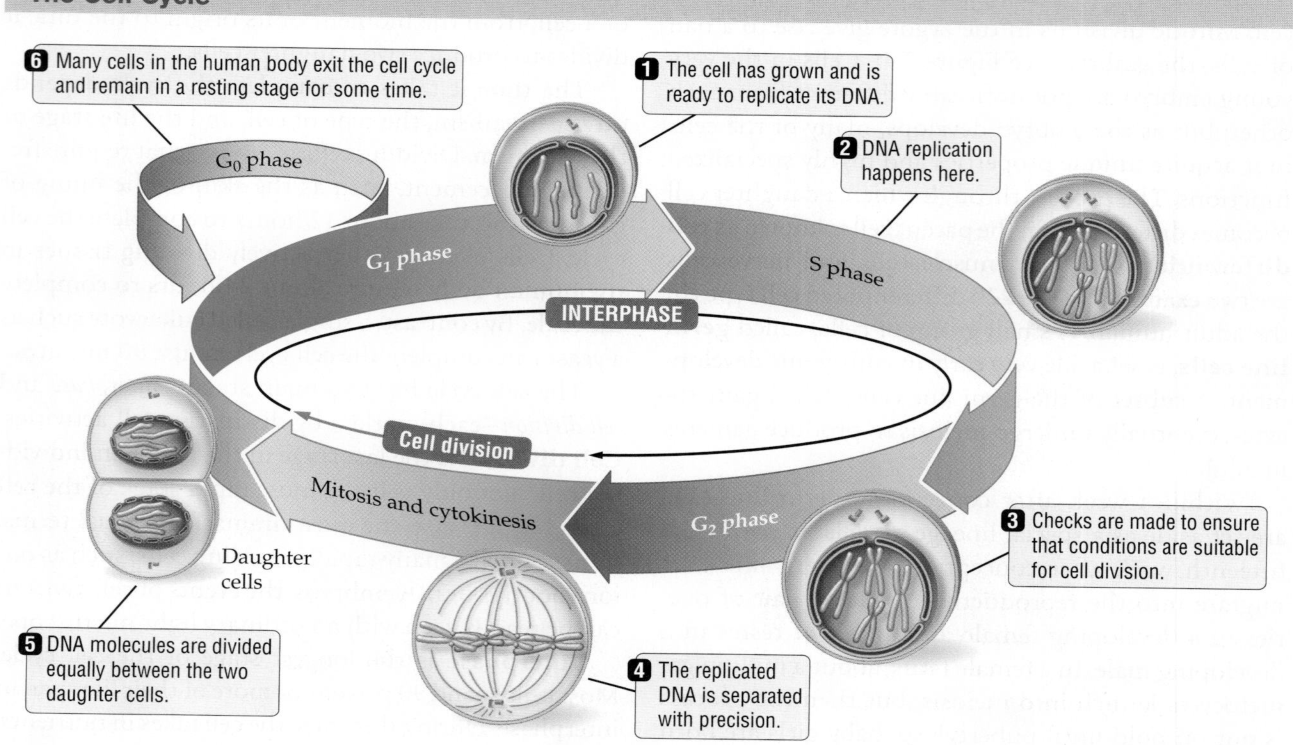

The Cell Cycle
6 Many cells in the human body exit the cell cycle and remain in a resting stage for some time.
1 The cell has grown and is ready to replicate its DNA.
2 DNA replication happens here.
G0 phase
G1 phase
S phase
INTERPHASE
Cell division
Mitosis and cytokinesis
G2 phase
3 Checks are made to ensure that conditions are suitable for cell division.
Daughter cells
5 DNA molecules are divided equally between the two daughter cells.
4 The replicated DNA is separated with precision.

FIGURE 7.5 The Cell Cycle Consists of Two Major Stages: Interphase and Cell Division

The cell prepares for division by increasing in size and producing needed proteins during the G_1 and G_2 phases, and by replicating its DNA during the S phase. Mitotic cell division consists of mitosis and cytokinesis, which result in two daughter cells that are genetically identical to the parent cell.

BIOLOGYMATTERS

Programmed Cell Death: Going Out in Style

Some cells die young. Neutrophils, the most abundant of the white blood cells, live for a day or two before they destroy themselves from the inside out. The keratinocytes in your skin (see Figure 7.1) kill themselves about 27 days after the mitotic division that produced them. Your red blood cells are demolished about 120 days after arising in the bone marrow, the destruction occurring mostly in a small organ called the spleen. And cellular suicide begins early in development; it is especially common in the fetal stage. About 50 percent of the neurons that develop in the brain of a human fetus are lost before birth.

It may seem like mayhem, but this cellular suicide is an elegantly controlled process that performs a valuable function in the body. The stepwise dismantling of a cell, controlled and conducted by the cell itself, is known as *programmed cell death* (*PCD*). In contrast to PCD, *necrosis* is the messy, disorganized death of a cell because of injury or infection that the cell is unable to resist. A paper cut kills some skin cells by rupturing them, and if bacteria enter the cut, their toxins can cause yet more cell death.

The details of PCD vary depending on the cell type. *Apoptosis* (*AP*-**up**-*TOH*-**sus** or *AP*-**uh**-*TOH*-**sus**; plural "apoptoses"), a particular form of PCD in animals, often begins with mitochondrial damage, followed by the activation of protein-destroying enzymes (called caspases) that digest the cell from the inside. The cell shrinks, and the DNA is broken into fragments as the cell begins to die. Cell remnants are typically engulfed by *phagocytes*, the immune system's cleanup crew.

Programmed cell death does away with cells that are no longer needed. In addition, PCD sculpts tissues and organs during development. For example, the fingers and toes of a human fetus are webbed initially, and were it not for the death of the webbing tissue between the digits, we would have paddlelike hands and feet.

Mitosis generates more than 200 billion nerve cells in the fetal brain, which extend fine projections (dendrites) that must link with target cells to become functional. The hit-or-miss nature of these projections means that many nerve cells fail to make an appropriate connection; superfluous nerve cells then proceed to self-destruct through PCD. However, it is a myth that we lose thousands of neurons daily; barring disease or injury, most of us keep the same population of roughly 70 billion neurons well into old age.

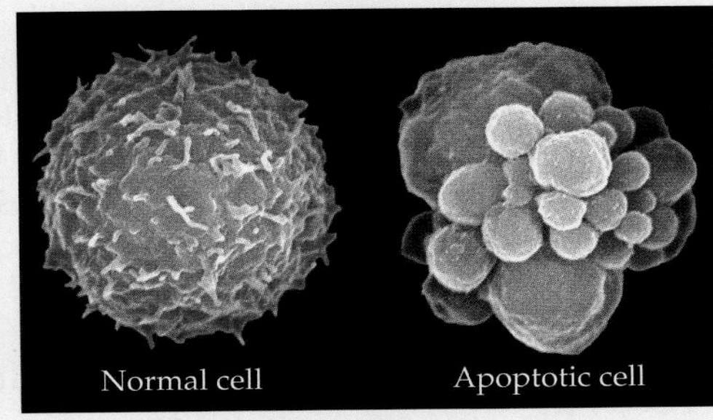

Programmed Cell Death

Some cells, such as those that form the lens of the eye, remain in G_0 for life, as part of a nondividing tissue. Many of the cell types in the brain have also exited the cell cycle, which is why neurons lost as a result of physical trauma or chemical damage are not readily replaced. Some highly specialized cell types not only exit the cell cycle, but intentionally self-destruct in a process called *programmed cell death* (described in the "Biology Matters" box on this page).

The cell cycle is tightly regulated

Cell division is metabolically expensive. It would be a reckless move for a cell to launch into cell division if food was scarce or its DNA was damaged, for example. It is little wonder, then, that the cell cycle is carefully controlled in healthy individuals.

The commitment to divide is made in the G_1 phase of the cell cycle, in response to internal and external signals. In humans, external signals that trigger cell division include hormones and proteins called growth factors. When a cell receives a signal to divide, *cell cycle regulatory proteins* are activated. These proteins "throw the switch" that enables the cell to pass the critical checkpoints and progress from one phase of the cell cycle to the next (**FIGURE 7.6**).

Cell cycle regulatory proteins also respond to negative internal or external control signals. Internal signals can pause a cell in G_1, barring entry to the S phase, under any of the following conditions: the cell is too small, the nutrient supply is inadequate, or DNA is damaged. G_2 arrests cell division in the same circumstances, as well as when the DNA duplication that begins in the S phase is incomplete for any reason. It is as if the cell cycle comes with start buttons and pause buttons, but no reverse buttons. The cell cycle can progress in only one direction—toward mitosis and the completion of cytokinesis; otherwise, it stalls indefinitely.

Despite the tight controls, cell division sometimes goes wrong. *Cancer* is a disease in which cells divide out of control and invade other tissues. A cancer begins

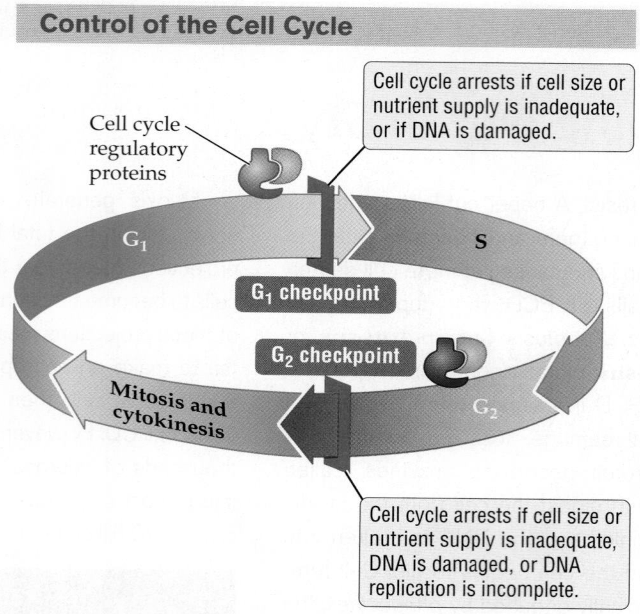

Control of the Cell Cycle

Cell cycle regulatory proteins

Cell cycle arrests if cell size or nutrient supply is inadequate, or if DNA is damaged.

G_1

S

G_1 checkpoint

G_2 checkpoint

Mitosis and cytokinesis

G_2

Cell cycle arrests if cell size or nutrient supply is inadequate, DNA is damaged, or DNA replication is incomplete.

FIGURE 7.6 Cell Cycle Regulatory Proteins Help Control the Cell Cycle

Only two of the known cell cycle checkpoints are depicted in this diagram. Checkpoints are present in the S phase and partway through mitosis as well.

Helpful to Know

"Chromosome" means literally "colored body" (*chromo*, "color"; *some*, "body"). Although the function of chromosomes was not yet known, nineteenth-century microscopists could see these bodies in dividing cells by staining them with certain dyes. Today, biologists use fluorescent dyes to produce images of specific molecules (including DNA) and particular organelles. You will see many fluorescent photographs in this and other chapters.

with a single cell that breaks loose of normal restraints on cell division and starts dividing rapidly to establish a colony of rogue cells. A clump of such cells is called a **tumor**, and tumor cells are termed **cancer cells** if they begin to invade neighboring tissues. As cancer cells spread through the body, they disrupt the normal function of tissues and organs; unchecked, cancer cells can cause death through failure of many organ systems.

Scientists believe that the G_0 state represents a "safe haven" against development into cancer. A cell in the G_0 phase and a nondividing cell in the G_1 phase may appear to behave similarly; the key difference is the complete absence of cell cycle regulatory proteins in G_0 cells. In contrast, these proteins are always present inside cells in the G_1 phase, although the proteins may be lying dormant for lack of a go-ahead signal. Because a G_0 cell *lacks* the cell cycle regulatory proteins altogether, it is further removed from the capacity to divide and is therefore less likely than a G_1 cell to turn into a rogue cancer cell.

Concept Check

1. When in the cell cycle does copying (replication) of DNA take place?
2. What is the significance of the G_0 phase?

Concept Check Answers

1. In the S (synthesis) phase of interphase.
2. A cell that leaves the cell cycle and stops dividing is in the G_0 either temporarily or permanently.

7.3 The Chromosomal Organization of Genetic Material

DNA in the nucleus is not a disorganized tangle of naked nucleotide polymers. Instead, each long, double-stranded DNA molecule is attached to proteins that help pack it into a more compact structure called a **chromosome**. The packing is necessary because each DNA molecule is enormously long, even in the simplest cells. If all of the 46 different DNA molecules in one of your skin cells were lined up end to end, they would make a double helix nearly 2 meters (about 6 feet) long. How can that much DNA be stuffed into a nucleus, which, in a human cell, has a diameter slightly less than 5 micrometers (0.005 millimeter)? Extreme packaging is the answer.

Each DNA double helix winds around DNA packaging proteins to create a DNA-protein complex known as **chromatin**. Chromatin is further looped and compressed to form an even more compact structure called a chromosome (**FIGURE 7.7**). You will

DNA Packaging

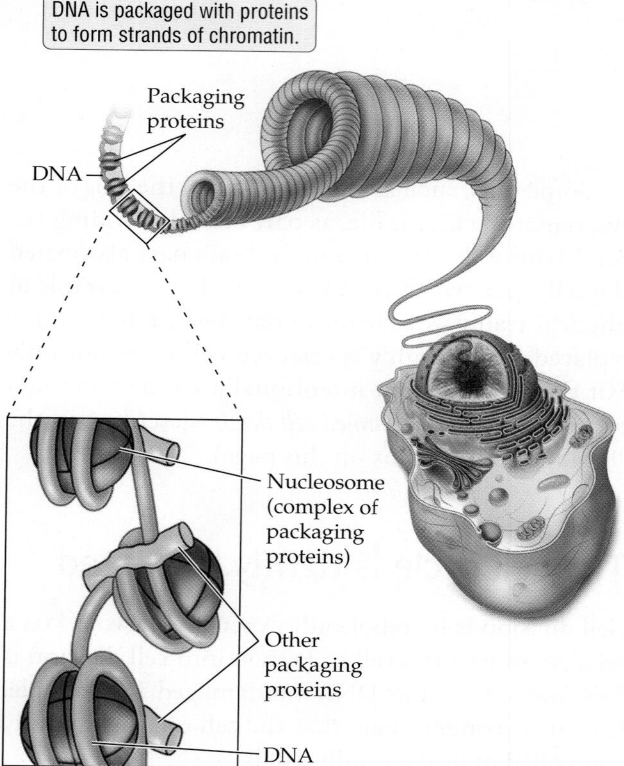

FIGURE 7.7 Each DNA Molecule in the Cell Is Packaged with Proteins to Form a Compact Structure, the Chromosome

Structure of a Replicated Chromosome

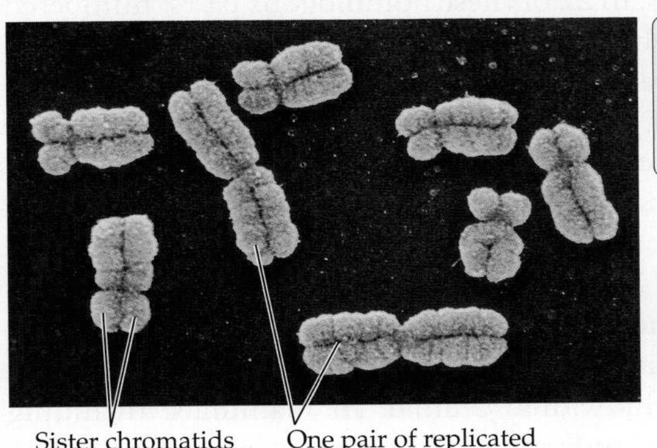

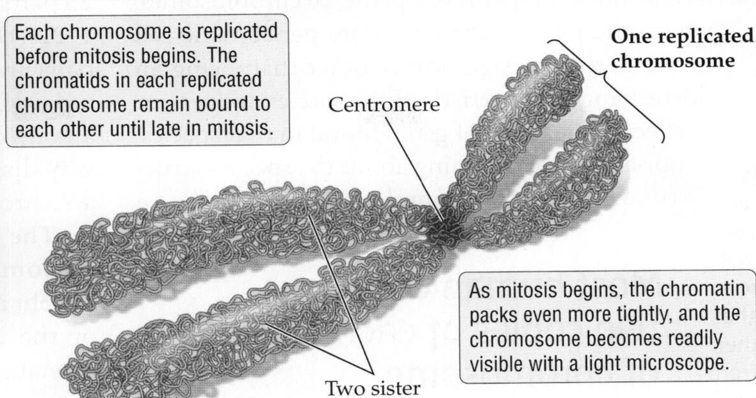

Each chromosome is replicated before mitosis begins. The chromatids in each replicated chromosome remain bound to each other until late in mitosis.
One replicated chromosome
Centromere
As mitosis begins, the chromatin packs even more tightly, and the chromosome becomes readily visible with a light microscope.
Two sister chromatids

FIGURE 7.8 Each Replicated Chromosome Consists of Two Identical Sister Chromatids

learn much more about genes, chromosomes, and DNA packaging and replication in Unit 2. For now, it is enough to know that each chromosome is a compacted DNA-protein complex, and within it is a single long molecule of DNA that bears many genes.

Before cell division can proceed, the DNA of the parent cell must be replicated so that each daughter cell can receive a complete set of chromosomes. DNA is replicated during the S phase, resulting in two identical double helices, known as **sister chromatids**. The sister chromatids remain linked to each other until the later stages of mitosis. Therefore, as mitosis begins, the nucleus of a human cell contains twice the usual amount of DNA, because each of the 46 chromosomes now consists of two identical sister chromatids, held together firmly in a region called the **centromere** (FIGURE 7.8). At the beginning of mitosis, the chromatin becomes packed and condensed even more tightly than during interphase, which is why chromosomes are most easily seen at the cell division stage.

A karyotype displays all the chromosomes in a nucleus

Every species has its own characteristic number of chromosomes in the nucleus of each of its cells. As noted earlier, somatic cells are all those cells in a multicellular organism that are not gametes (egg or sperm, in animals) or the direct precursors of gametes (germ line cells). Somatic cells of plants and animals may have anywhere from two to a few hundred chromosomes, depending on the species. During mitosis, when chromosomes are compacted to the maximum extent, the different types of chromosomes in a somatic cell can often be identified under a microscope by their size and distinctive shapes.

A display of all the chromosomes in a somatic cell is called a **karyotype** (FIGURE 7.9). Karyotypes are generally made from microscopic observations of mitotic cells, in which chromosomes are more easily seen. The karyotype of a human somatic cell shows a

The Human Karyotype

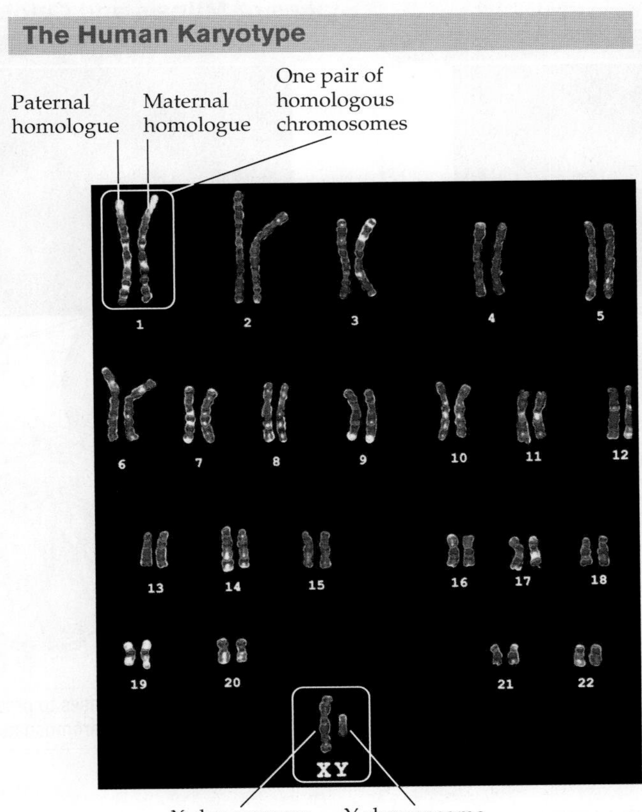

FIGURE 7.9 The Karyotype Identifies All the Chromosomes of a Species

The 46 chromosomes in this micrograph represent the karyotype of a human male. With the help of computer graphics, photos of the chromosomes have been aligned so that the two members of each homologous pair are placed next to each other. The non–sex chromosomes (known as *autosomes*) are numbered. The sex chromosomes are represented by letters (XY, in the case of this male individual).

total of 46 chromosomes. Each somatic cell in a horse has 64 chromosomes; in a corn plant, 20 chromosomes. The total number of chromosomes per somatic cell has no particular significance other than being an identifying characteristic of the species. It does not reflect the number of genes found in that species, nor does it say anything about the species' structural or behavioral complexity.

Helpful to Know

Make sure you know the difference between a pair of sister chromatids and a pair of homologous chromosomes. Sister chromatids are identical DNA molecules, replicated from a single DNA molecule, that remain bound to each other. They exist in a cell only from the S phase until anaphase of mitosis, or anaphase II of meiosis. A pair of homologous chromosomes consists of the two copies—one maternal, the other paternal—of the same type of chromosome. Homologous pairs are present at all times in diploid cells, but in a haploid cell only one chromosome of each pair is present.

Most human cells have two copies of each type of chromosome

A distinctive feature of most eukaryotes, compared to prokaryotes, is that their somatic cells contain *two* matched copies of every type of chromosome. Two matched chromosomes make up a pair of **homologous chromosomes** (**hoh-*MAH*-luh-gus**). Returning to the example of the human karyotype, our 46 chromosomes are actually a double set consisting of 23 *pairs* of homologous chromosomes. You inherited one set of 23 chromosomes from your mother and the other set, the remaining 23, from your father, to create a double set of 46, or 23 pairs. In 22 of these homologous pairs (numbered 1–22), the two *homologues* (individual members of the pair) are alike in length, shape, and the location and types of genes they carry. But the twenty-third pair is an odd couple about half the time, consisting of two very dissimilar homologues: an X chromosome and a Y chromosome (see Figure 7.9).

The X and Y chromosomes are called **sex chromosomes** because, in mammals and some other vertebrates, these chromosomes determine the sex of the individual animal. In mammals, including humans, individuals with two X chromosomes in their cells are female, and those with one X and one Y chromosome are male. The X chromosome is considerably longer than the Y chromosome and carries many more genes. Most, if not all, of the few genes found on the Y chromosome appear to be involved in controlling the development of male characteristics.

All of the many genes that are unique to the X chromosome are also present in normal males, because their cells contain one X chromosome, but males have just one copy of the X-specific genes. Normal females, with their two copies of the X chromosome (XX), have two copies of all the X-specific genes.

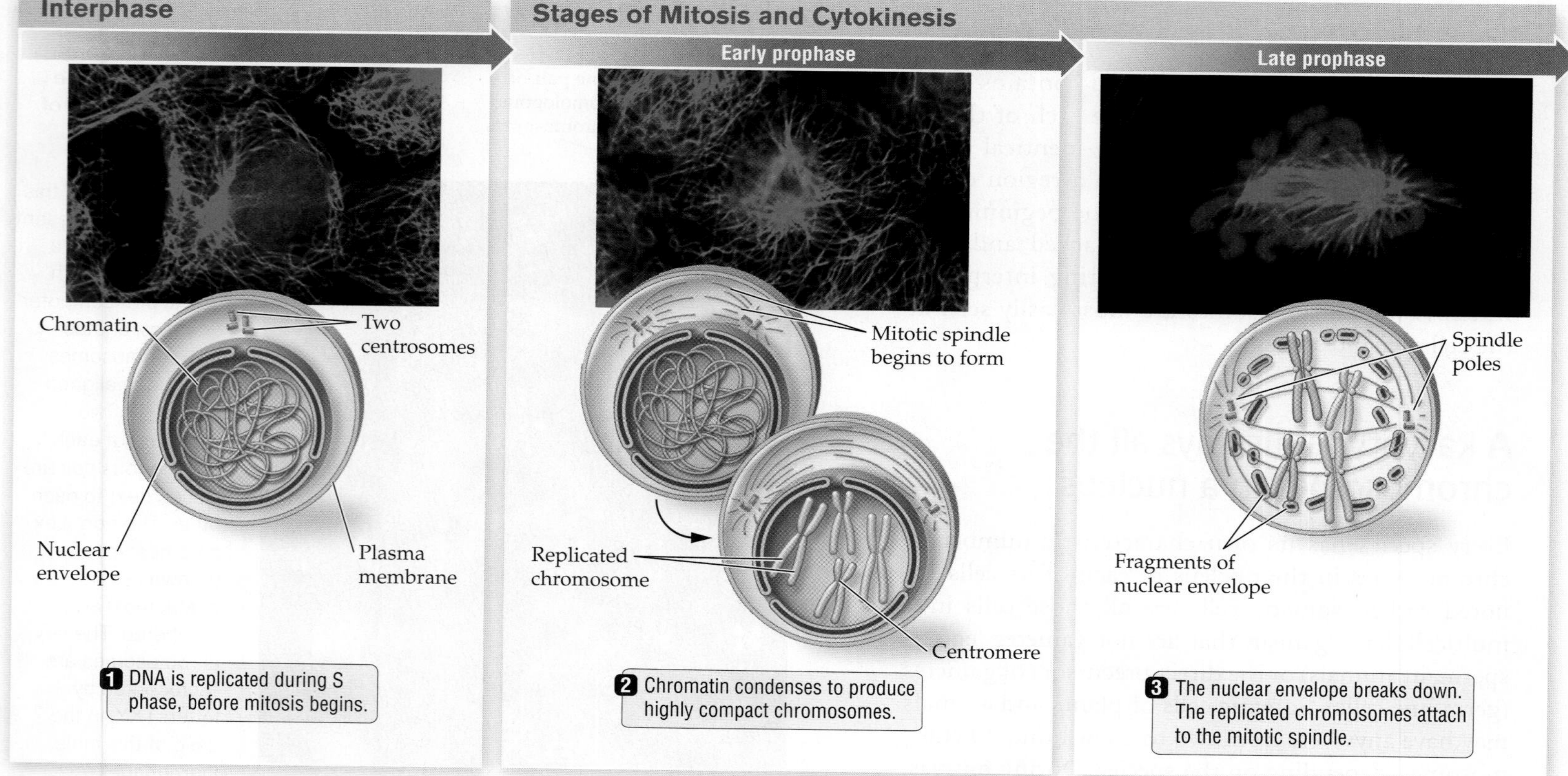

FIGURE 7.10 Mitosis and Cytokinesis Are the Two Main Stages of Mitotic Cell Division

Concept Check

1. The house cat karyotype displays a total of 38 chromosomes. How many distinct DNA molecules are present in a cat skin cell toward the end of (a) G_0 phase; (b) G_1 phase; (c) S phase; (d) G_2 phase?
2. How is the karyotype of men different from that of women?

7.4 Mitosis and Cytokinesis: From One Cell to Two Identical Cells

The climax of the cell cycle is cell division, which, in the case of mitotic divisions, consists of two steps: *mitosis* and *cytokinesis*. These steps are not separated in time; cytokinesis overlaps with the last phase of mitosis. The central event of mitosis is the equal distribution of the parent cell's replicated DNA into two daughter nuclei. This process, called DNA segregation, requires the coordinated actions of many different types of proteins, including those that make up the cytoskeleton.

Mitosis is divided into four main phases, each of them defined by easily identifiable events that are visible under the light microscope (**FIGURE 7.10**):

1. *Prophase*. Chromosomes condense and the nuclear envelope breaks down.
2. *Metaphase*. Chromosomes align at the midline of the cell.
3. *Anaphase*. Sister chromatids separate and move to opposite poles.
4. *Telophase*. New nuclear envelopes form, and chromosomes decondense.

Recall that the cell sets up for mitosis well beforehand; all the chromosomes in the nucleus have been replicated in the S phase before mitosis begins. Each replicated chromosome consists of two identical DNA molecules—the sister chromatids—held together along their length, and especially tightly at the constriction known as the centromere.

You could say that the objective of mitosis is to separate all the sister chromatids and deliver *one of each* to the opposite ends of the parent cell. The elaborate chromosomal choreography of mitosis has evolved to minimize the risk of mistakes; normally, no daughter

Concept Check Answers

1. (a) 38; (b) 38; (c) 76; (d) 76.
2. All humans have 22 pairs of autosomes and a pair of sex chromosomes. Women have a pair of identical sex chromosomes called X chromosomes. Men have two dissimilar sex chromosomes: one X and one Y.

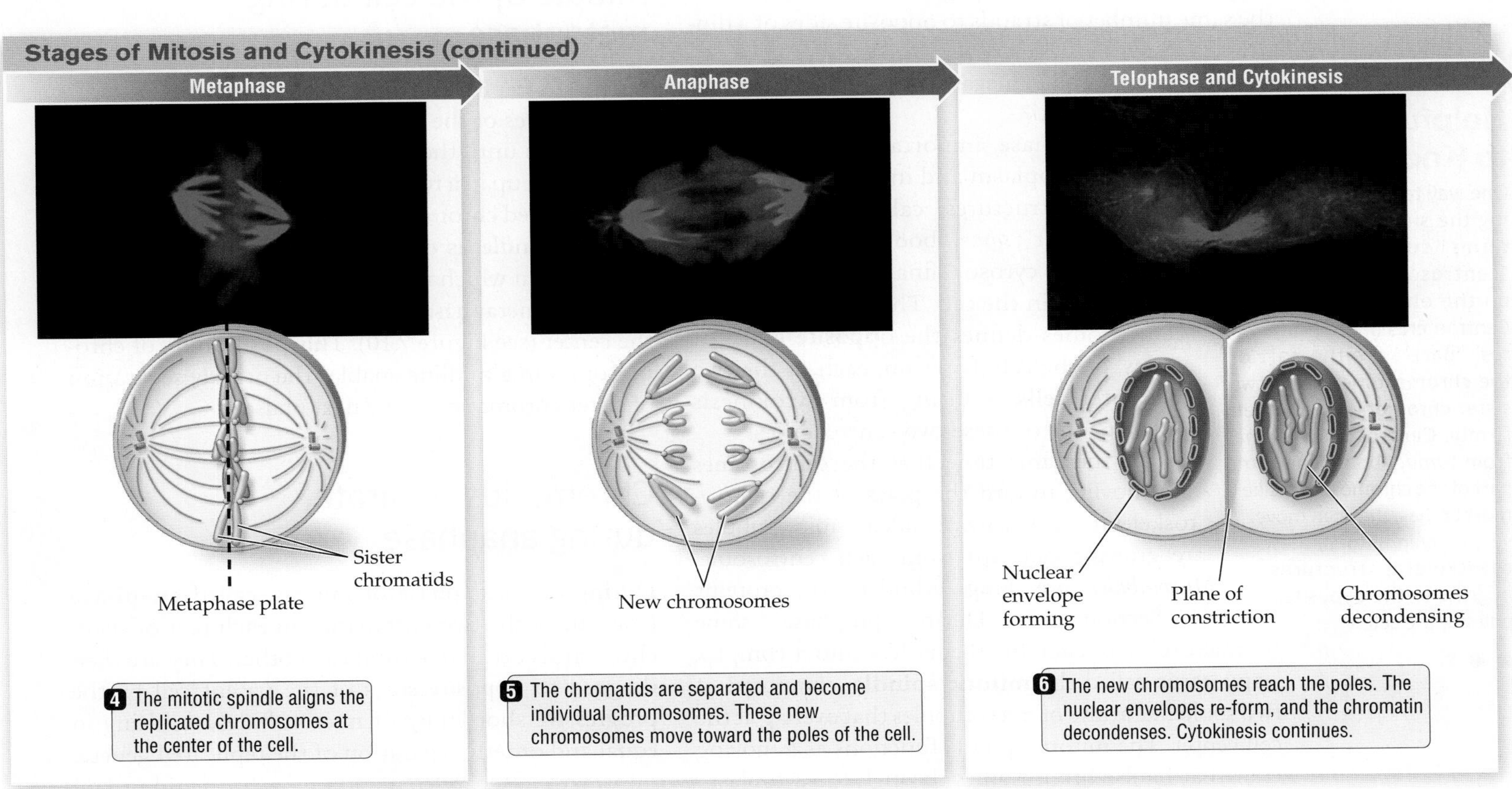

4 The mitotic spindle aligns the replicated chromosomes at the center of the cell.

5 The chromatids are separated and become individual chromosomes. These new chromosomes move toward the poles of the cell.

6 The new chromosomes reach the poles. The nuclear envelopes re-form, and the chromatin decondenses. Cytokinesis continues.

EXTREME MITOSIS

This deer, with its new horns "in velvet," sports the fastest-growing tissue in the animal world. Antlers are shed after the mating season and regrow in spring. The new antlers grow 1 inch in 2 days. Velvet is a delicate layer of skin with a rich supply of blood vessels to support rapid mitosis.

cell winds up short a chromosome, nor does it acquire duplicates. Each daughter cell inherits exactly the same information that the parent cell possessed in the G_1 phase of its life—no more, no less.

Chromosomes are compacted during early prophase

DNA molecules undergo a high level of compaction during the first stage of mitosis, **prophase** (*pro*, "before"; *phase*, "appearance"), and by the end of this phase the DNA is seven times more tightly wound than during interphase. Each chromosome gets shorter and stouter, becoming readily visible in the nucleus.

The functional value of this extra level of compaction is that the chromosomes can be lined up and sorted to the opposite poles of the cell without excessive tangling and breaking. Imagine if you had to untangle a heap of cooked spaghetti and move exactly the same number of strands to opposite sides of a dinner plate. Would the task be easier if the strands were short and stumpy, like macaroni, instead of long and floppy like spaghetti?

Helpful to Know

One way to avoid confusing the similar-sounding terms "centromere" and "centrosome" is to focus on the end of each word. Centromeres (*mere* from *meros*, "part") are the part of the chromosome where two sister chromatids are joined firmly. Centrosomes (*some* from *soma*, "body") are not part of the chromosome. Like most bodies, they can move: very early in mitosis, these cytoskeletal structures migrate to the opposite poles of the cell.

During prophase, important changes occur both in the cytoplasm and in the nucleus. Two cytoskeletal structures, called **centrosomes** (*centro*, "center"; *some*, "body"), begin to move through the cytosol, finally halting at opposite sides in the cell. This arrangement of centrosomes defines the opposite ends, or poles, of the cell. Later on, each of the two daughter cells resulting from cytokinesis inherits one of these two centrosomes.

At the same time that the centrosomes are moving toward the poles of the cell, cytoskeletal structures called microtubules are growing outward from each centrosome. *Microtubules* are long cylinders of proteins (see Section 3.5). During prophase, some microtubules assemble themselves into a complex apparatus called the **mitotic spindle**, composed of two spokelike arrays of microtubules that overlap at the cell center. The mitotic spindle functions as a moving crew that hauls chromosomes through the cytoplasm and delivers them to opposite ends of the parent cell.

Chromosomes are attached to the spindle in late prophase

The nuclear envelope breaks down late in prophase (see Figure 7.10), during a step that cell biologists call *prometaphase*. With the nuclear envelope out of the way, the microtubules of the mitotic spindle, radiating out from the centrosome at each pole, seek out and attach to the now highly condensed chromosomes. As a result, each replicated chromosome is "captured" by the mitotic spindle and becomes linked to the two centrosomes by microtubules.

Each chromosome's centromere has two patches of protein, called *kinetochores* (**kih-*NET*-uh-kohr**), that are oriented on opposite sides. Each kinetochore forms a site of attachment for at least one microtubule, so that the two sister chromatids that make up a replicated chromosome end up being linked to the centrosomes at the opposite poles of the cell. The successful "capture" of each pair of sister chromatids by the mitotic spindle sets the stage for the proper positioning of replicated chromosomes in the next phase of mitosis.

Chromosomes line up in the middle of the cell during metaphase

When each replicated chromosome has become linked to both poles of the spindle, the microtubules shrink or lengthen until the chromosomes attached to them are all lined up in a row. This stage of mitosis, when all the replicated chromosomes have been lined up by the mitotic spindle, is called **metaphase** (*meta*, "after"). The plane in which the chromosomes are arranged is called the metaphase plate, which in most cells lies at the center (see Figure 7.10). This positioning of chromosomes in a midline enables the orderly separation of sister chromatids at the next phase.

Chromatids separate during anaphase

During the next phase of mitosis, called **anaphase** (*ana*, "up"), the two chromatids in each pair of sister chromatids break free from each other. They are then dragged to opposite sides of the parent cell by the progressive shortening of microtubules, resulting in equal and orderly separation of the replicated genetic information. Once separated, each chromatid is considered a new chromosome. This segregation of the former sister chromatids to opposite poles paves the way for telophase and cytokinesis.

New nuclei form during telophase

The last phase of mitosis, **telophase** (*telo*, "end"), begins when a complete set of chromosomes arrives at a spindle pole. Major changes also occur in the cytoplasm: the spindle microtubules break down, and the nuclear envelope begins to form around the chromosomes that have arrived at each pole (see Figure 7.10). As the two new nuclei become increasingly distinct in the cell, the chromosomes within them start to unfold, becoming less distinct under the microscope. Cytokinesis begins even before telophase is quite complete.

The cytoplasm is divided during cytokinesis

Cytokinesis (*cyto*, "cell"; *kinesis*, "movement") is the process of dividing the parent cell's cytoplasm into two daughter cells. Animal cells divide by drawing the plasma membrane inward until it meets in the center of the cell, separating the cytoplasm into two compartments (**FIGURE 7.11**). The physical act of separation is performed by a ring of protein cables made of actin microfilaments. These form against the inner face of the plasma membrane like a belt at the midline of the cell. When the actin ring contracts, it acts like a drawstring, pulling the plasma membrane inward and eventually pinching off the cytoplasm and dividing it in two. Successful cytokinesis results in two daughter cells, each with its own nucleus.

A plant cell, however, is surrounded by a relatively stiff cell wall that cannot be pulled inward. Plant cells achieve cytokinesis by erecting two new plasma membranes, separated by cell wall material. Guided by cytoskeletal structures, a partition known as a **cell plate** appears where the metaphase plate had been. The cell plate, consisting mostly of membrane vesicles, starts forming in telophase (**FIGURE 7.12**). Vesicles filled with cell wall components start to accumulate in the region that was previously the metaphase plate. These vesicles join together, fusing their membranes and mingling their cell wall components (mostly polysaccharides and some protein) to create two new plasma membranes separated by a newly formed cell wall.

Cytokinesis marks the end of the cell cycle. Once the cycle is completed, the resulting daughter cells are free to enter the G_1 phase and start the process anew, or to differentiate into a specialized cell type and perhaps take a rest from cell division by entering the G_0 phase.

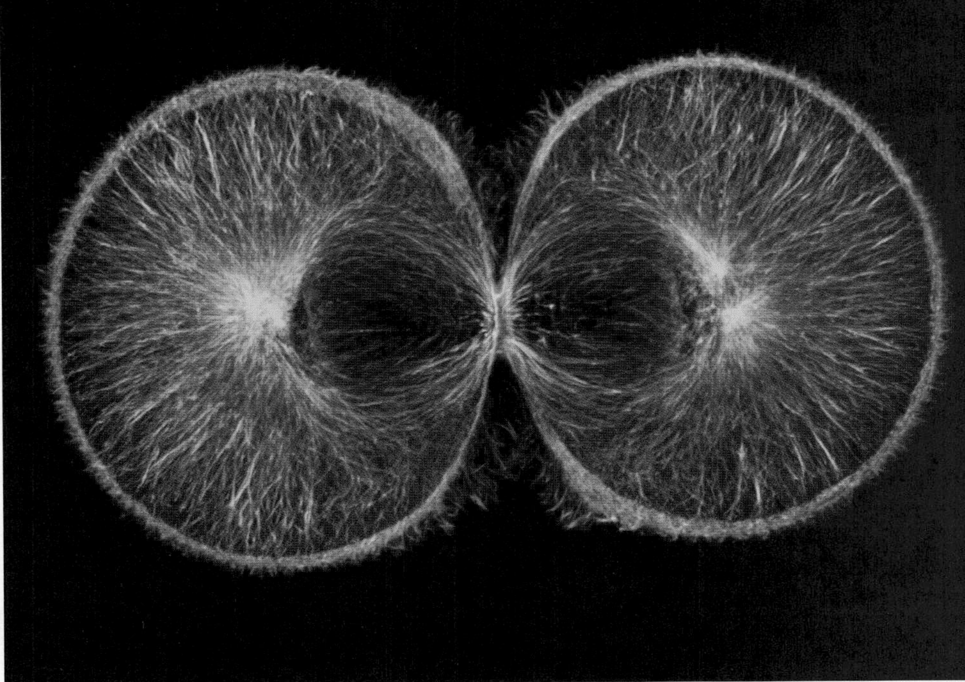

FIGURE 7.11 Cytokinesis in an Animal Cell
This fluorescence image shows cytokinesis in a sea urchin zygote that is dividing into two cells. Microtubules are orange; actin filaments, blue.

Concept Check

1. What are sister chromatids? When during mitosis do they separate to become chromosomes in their own right?
2. What is the difference between mitosis and cytokinesis?

Concept Check Answers

1. Sister chromatids are two identical DNA molecules, replicated from a single parental chromosome, that remain attached during mitosis until anaphase, when they separate to become two independent chromosomes, each headed to a different daughter cell.

2. Mitosis is nuclear division; cytokinesis is the partitioning of the cytoplasm, and ultimately the whole parent cell, into two daughter cells.

7.5 Meiosis: Halving the Chromosome Set to Make Gametes

As noted already, meiosis is a special type of cell division that produces daughter cells with half the chromosome count of the parent cell. Gametes are the only cells in the animal body that are created through meiosis.

Gametes contain half the chromosomes found in somatic cells

Sexual reproduction requires the fusion of two gametes in the process of fertilization. The successful union of an egg and a sperm creates a single-celled zygote, which develops into the multicellular embryo

FIGURE 7.12 Cell Plate Formation Is a Distinctive Feature of Cytokinesis in Plant Cells

(*a*) A microscopic view of mitosis and cytokinesis in lily pollen. The cell plate appears as a pale line in the center of the cell, in the last photograph in the series (telophase). (*b*) A diagrammatic view of the main events in mitosis and cytokinesis in a plant cell. Plant cells lack prominent centrosomes but have structures that perform the same function.

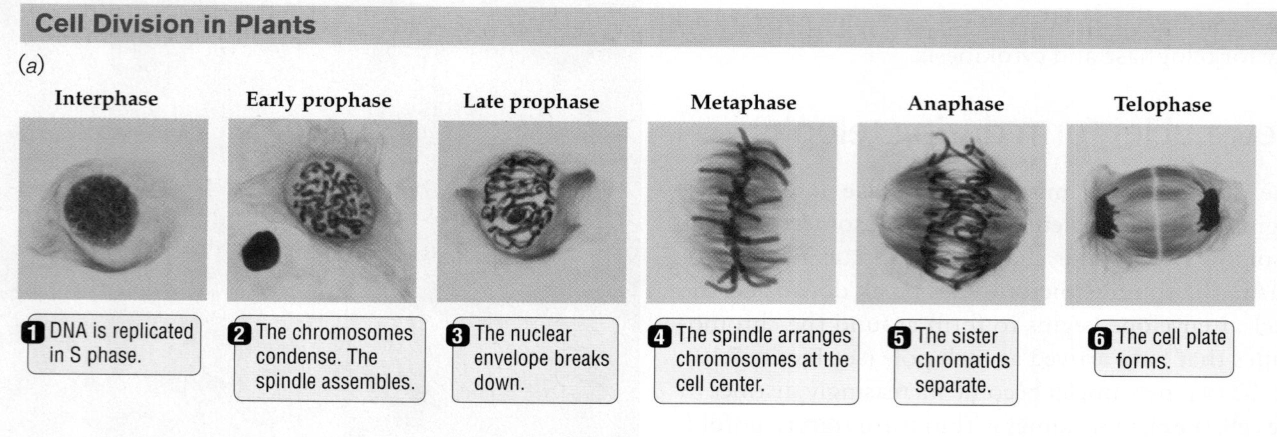

Cell Division in Plants
(a)
Interphase
Early prophase
Late prophase
Metaphase
Anaphase
Telophase
1 DNA is replicated in S phase.
2 The chromosomes condense. The spindle assembles.
3 The nuclear envelope breaks down.
4 The spindle arranges chromosomes at the cell center.
5 The sister chromatids separate.
6 The cell plate forms.

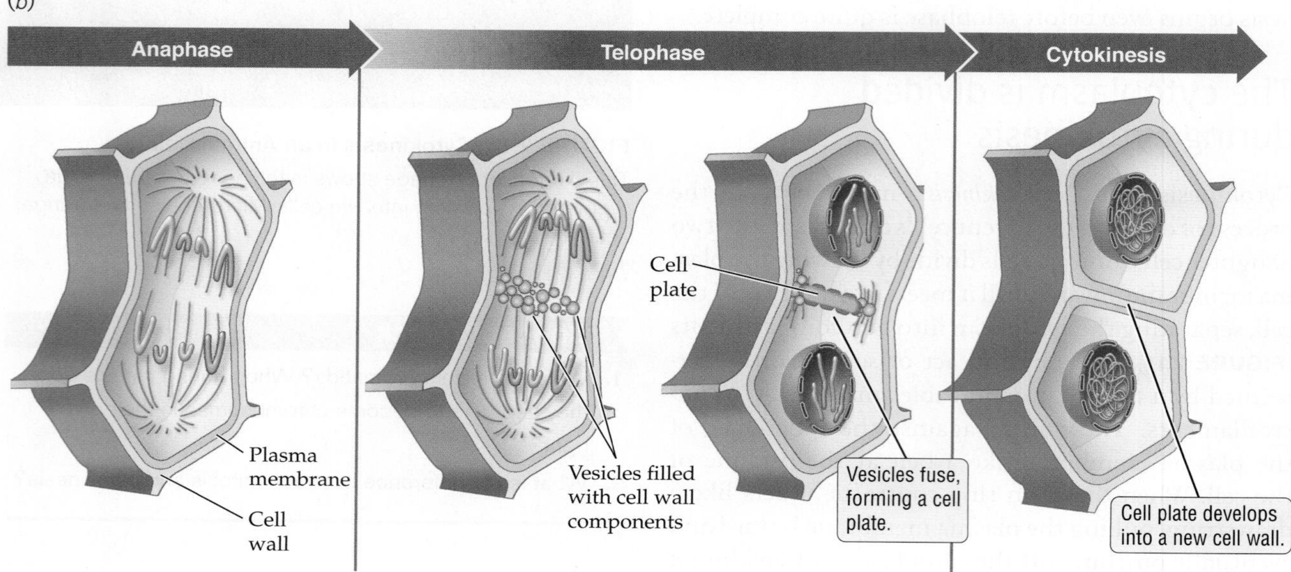

(b)
Anaphase
Telophase
Cytokinesis
Cell plate
Plasma membrane
Cell wall
Vesicles filled with cell wall components
Vesicles fuse, forming a cell plate.
Cell plate develops into a new cell wall.

(**FIGURE 7.13**). Sexual reproduction produces offspring that are genetically different from their parents and siblings.

If the sperm and the egg both contained a complete set of chromosomes (46 for humans), the resulting zygote would have twice that chromosome number (92 in the human case). The outcome of this genetic excess would be developmental chaos, generally resulting in death of the embryo. For offspring to have the same chromosome number as their parents, fertilization must yield the normal number of chromosomes in the zygote.

The simple solution to this problem is for the gametes to contain half of the full set of chromosomes found in somatic cells. Recall that somatic cells in eukaryotes possess *two copies*, or one homologous pair, for each type of chromosome found in the organism. If *only one copy* from every homologous pair is inherited by a gamete, the chromosome set is halved, and the gamete now possesses only *one copy* of all the genetic information. In humans, for example, all somatic cells contain 23 homologous pairs, for a total of 46 chromosomes. Each gamete a person produces, however, contains only one chromosome from each homologous pair, for a total of 23 chromosomes per gamete. Where the sex chromosomes are concerned, all eggs produced by a woman normally contain a single X chromosome, while 50 percent of the sperm produced by a man contain an X chromosome and the rest carry a Y chromosome.

Because gametes contain only one copy of each type of chromosome, instead of having double copies, gametes are said to be **haploid** (*haploos*, "single"). The symbol n is traditionally used to indicate the number of chromosomes in a haploid cell (see Figure 7.13). In a human gamete, $n = 23$. Somatic cells, which have twice the number of chromosomes as gametes have, are said to be **diploid** (*di*, "double")

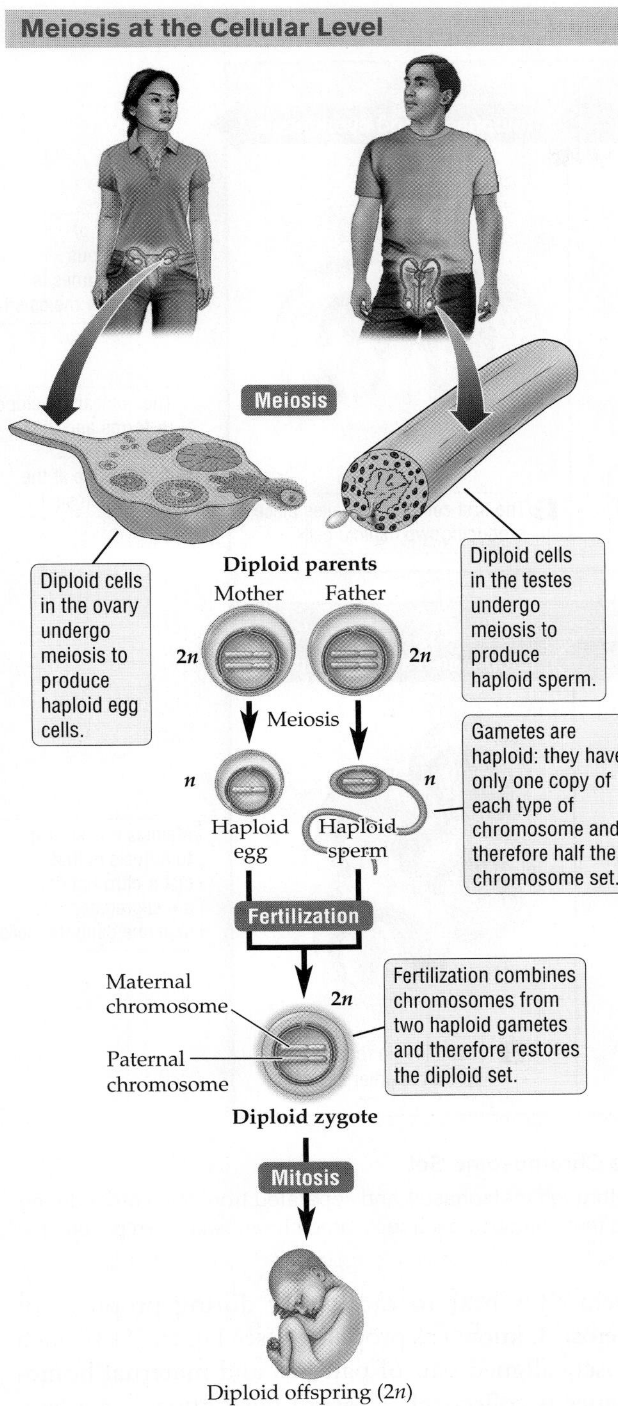

Meiosis at the Cellular Level
Meiosis
Diploid cells in the ovary undergo meiosis to produce haploid egg cells.
Diploid parents
Mother
Father
2n
2n
Diploid cells in the testes undergo meiosis to produce haploid sperm.
Meiosis
n
n
Gametes are haploid: they have only one copy of each type of chromosome and therefore half the chromosome set.
Haploid egg
Haploid sperm
Fertilization
Maternal chromosome
Paternal chromosome
2n
Fertilization combines chromosomes from two haploid gametes and therefore restores the diploid set.
Diploid zygote
Mitosis
Diploid offspring (2n)

FIGURE 7.13 Sexual Reproduction Requires a Reduction of the Chromosome Set in Gametes

The fusion of haploid sperm and egg at fertilization produces a zygote with the diploid ($2n$) chromosome set. In humans, the diploid set consists of 46 chromosomes, with two copies of each type of choromoseme. For clarity, only one homologous pair (consisting of a maternal and a paternal homologue) is shown here.

because they have $2n$ chromosomes—that is, two of every kind of chromosome.

The zygote formed by fertilization will contain $2n$ chromosomes—that is, the diploid set of chromosomes—because each gamete that contributes to the zygote contains the haploid (n) number. Furthermore, each pair of homologous chromosomes in the zygote will consist of one chromosome received from the father (**paternal homologue**) and one from the mother (**maternal homologue**), as shown in Figure 7.13. The equal contribution of chromosomes by each parent is the basis for genetic inheritance. We will investigate the details of inheritance in Unit 2.

Meiosis occurs in two stages—*meiosis I* and *meiosis II*—each involving one round of nuclear division followed by cytokinesis (**FIGURE 7.14**):

- **Meiosis I** sorts *each member of a homologous pair* into two different daughter cells, reducing the chromosome sets from $2n$ to n.
- **Meiosis II** separates *sister chromatids* in each cell produced by meiosis I into two different daughter cells.

The phases of meiosis I and meiosis II are broadly similar to those of mitosis.

Meiosis I is the reduction division

Our tour of meiosis begins with the diploid cells in reproductive tissues that are responsible for the production of gametes. Well before meiosis begins, all the chromosomes in the diploid precursor cell have been replicated during the S phase of the cell cycle. Therefore, as meiosis I begins, every replicated chromosome exists as two identical DNA molecules, with each DNA molecule constituting one chromatid. Two identical sister chromatids, bound to each other like Siamese twins, make up a *dyad*.

The first unique aspect of meiosis I, not seen at any stage of mitosis, is the coming together of each replicated homologous pair of chromosomes. In other words, early in meiosis I, each maternal homologue pairs off with its matching paternal homologue (see Figure 7.14). These homologues are sorted into two separate daughter cells at the end

EXTREME MEIOSIS

Sturgeon produce several million egg cells in each spawn. The fish can reach 2,000 pounds and don't start reproducing until they're 20 years old. The egg cells, or roe, are the product of meiosis in female fish. Beluga sturgeon roe, or caviar, often sells for $10,000 per kilogram.

The Main Events in Meiosis

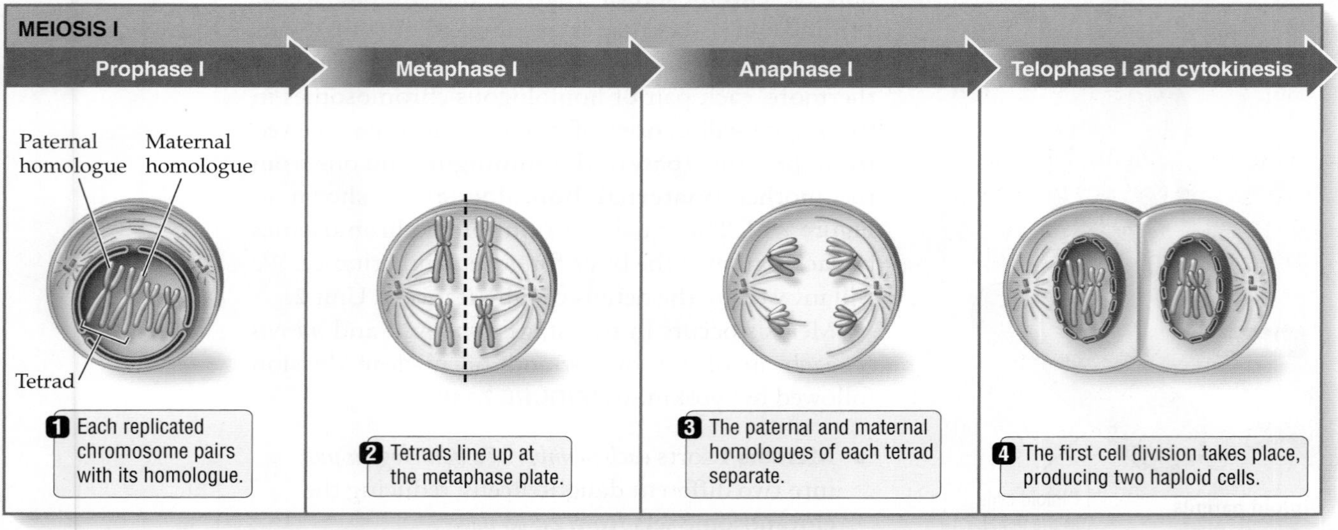

MEIOSIS I
Prophase I
Metaphase I
Anaphase I
Telophase I and cytokinesis
Paternal homologue
Maternal homologue
Tetrad
1 Each replicated chromosome pairs with its homologue.
2 Tetrads line up at the metaphase plate.
3 The paternal and maternal homologues of each tetrad separate.
4 The first cell division takes place, producing two haploid cells.
Each pair of homologous chromosomes is split up by meiosis I.
The nuclear envelope re-forms and the chromosomes decondense at the end of meiosis I.

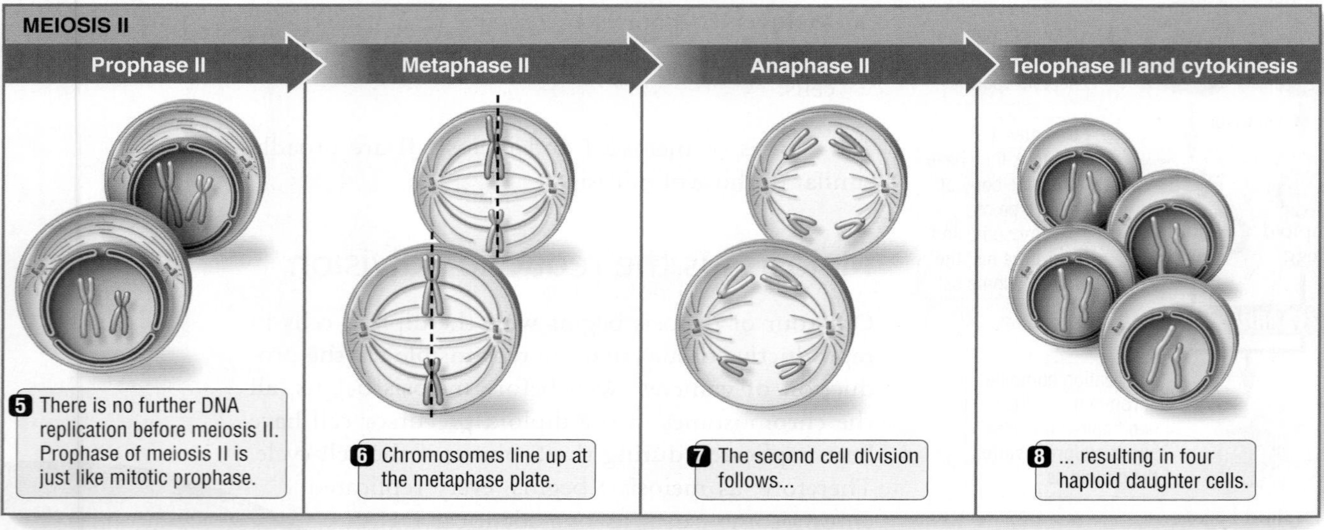

MEIOSIS II
Prophase II
Metaphase II
Anaphase II
Telophase II and cytokinesis
5 There is no further DNA replication before meiosis II. Prophase of meiosis II is just like mitotic prophase.
6 Chromosomes line up at the metaphase plate.
7 The second cell division follows...
8 ... resulting in four haploid daughter cells.
Meiosis II is similar to mitosis in that sister chromatids are segregated into separate daughter cells.

FIGURE 7.14 In Meiosis, Each Daughter Cell Receives Half the Chromosome Set
The maternal and paternal homologues are paired during prophase I through metaphase I, and separated from each other during anaphase I. Meiosis II is similar to mitosis in that the sister chromatids that compose each replicated chromosome are pulled apart.

Helpful to Know

Although anaphase I looks superficially similar to the anaphase of mitosis, remember that during anaphase I, homologous chromosome pairs—not sister chromatids—are pulled apart and deposited at opposite poles of the cell. In other words, tetrads are separated in anaphase I, but sister chromatids are separated in anaphase II and in mitotic anaphase.

of meiosis I; meanwhile, the sister chromatids of each homologue remain attached to each other. Put another way, mitosis brings about the separation and sorting of sister chromatids into different daughter cells, but meiosis I leads to the separation and sorting of each homologous chromosome pair so that daughter cells receive only one member of the pair.

The pairing off and orderly sorting of homologous chromosomes during meiosis I is what makes it possible for the resulting daughter cells to inherit exactly half the chromosome set of the parent cell. The paternal and maternal partners of each homologous chromosome pair align themselves next to each other during prophase of meiosis I, known as prophase I (see Figure 7.14). Each closely aligned pair of paternal and maternal homologues is collectively a **tetrad** (also known as a *bivalent*). This means that each tetrad, consisting of one replicated maternal chromosome and one replicated paternal chromosome, contains a total of *four* chromatids (four individual DNA molecules).

At this point, an extraordinary process unfolds: the maternal and paternal members of each pair of homologues swap pieces of themselves! The exchange of genetic material takes place between the non-sister chromatids in each pair of homologous chromosomes. This exchange is brought about by a process called *crossing-over*, a subject we will return to after completing the tour of meiosis I and II.

Late in prophase I, a meiotic spindle develops and captures each homologous pair (tetrad). The microtubules from one centrosome attach themselves to only one homologue—either the maternal homologue or the paternal homologue, as shown in Figure 7.14. Next, in metaphase I, each homologous pair is positioned at the metaphase plate (the cell's midline). Anaphase I begins after all homologous pairs have been captured and positioned in one plane at the midline. As the spindle microtubules begin to shorten during anaphase I, the paternal and maternal partners in each homologous pair are pulled to opposite poles of the cell. Which homologue of each pair goes to which pole is essentially random.

After anaphase I of meiosis, the events of telophase I follow the same patterns seen in mitosis, with the spindle disappearing and nuclear envelope re-forming. Cytokinesis of meiosis I results in two daughter cells, each with half the chromosome count of the parent cell.

As you have seen, the chromosome set becomes halved because each daughter cell receives only one member of each homologous pair—either the maternal homologue (shown in pink in Figure 7.14) or the paternal homologue (shown in blue). Meiosis I is a *reduction division* because it halves the chromosome set, as one diploid parent cell ($2n$) becomes two haploid daughter cells (n).

Meiosis II segregates sister chromatids into separate daughter cells

The two daughter cells formed at the end of meiosis I are not ready for prime time just yet: they cannot mature into gametes, because each of their chromosomes is still in a replicated state; that is, each consists of two identical DNA molecules bound together as sister chromatids. Separating these sister chromatids into two different daughter cells is the sole purpose of meiosis II.

Each of the two haploid cells formed at the end of meiosis I goes through a second round of nuclear and cytoplasmic divisions that makes up meiosis II. This time the phases of the division cycle are *almost exactly like those of mitosis*. In particular, sister chromatids separate at anaphase II, leading to an equal segregation of sister chromatids into two new daughter cells. In this manner, the two haploid cells produced by meiosis I give rise to a total of four haploid cells. The haploid cells then differentiate into gametes.

As noted earlier, the reduction in chromosome number achieved through meiosis I offsets the combining of chromosomes when gametes fuse during fertilization (**FIGURE 7.15**). It is nature's way of maintaining the constant chromosome number of a species during sexual reproduction.

Meiosis and fertilization contribute to genetic variation in a population

Meiosis and fertilization are the means of sexual reproduction in eukaryotes. Individuals in a population tend to be genetically different from each other because sexual reproduction leads to offspring that are genetically different—not only from their parents but also from their siblings. You may resemble one or the other, or both, of your parents, but you cannot be genetically identical to either of them. Similarly, you may resemble a brother or sister, but you are not a clone of anyone, unless you have an identical twin. Genetic diversity in a population is important because genetic variation is the raw material on which evolution acts.

Where does the genetic variation in a population come from in the first place? *Mutations*, which are accidental changes in the DNA code, are the ultimate source of genetic variation in all types of organisms.

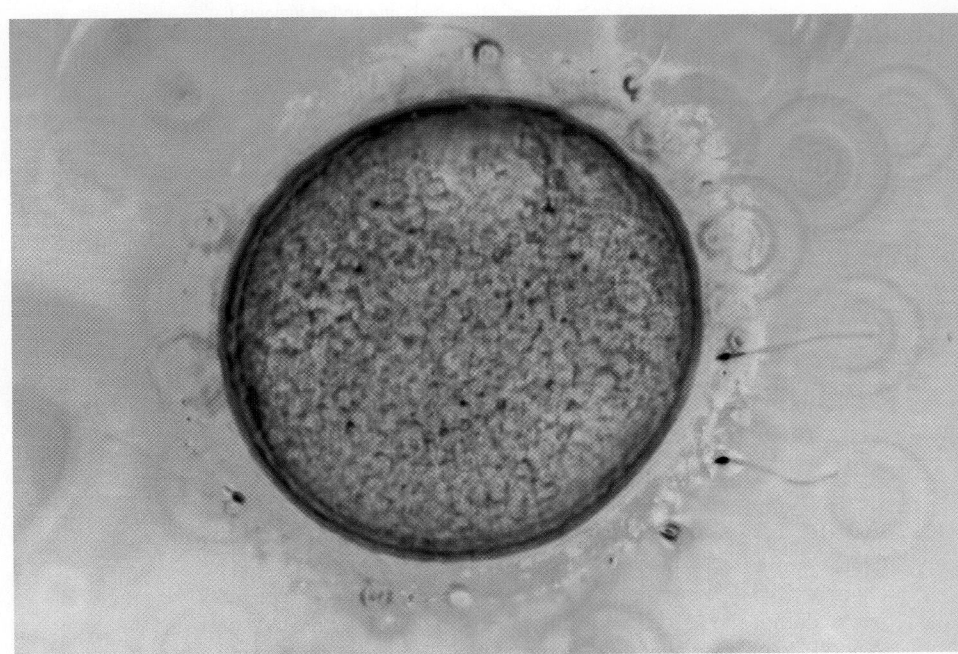

FIGURE 7.15 Fertilization of a Human Egg
Sperm attach to sugary proteins (brown) on the surface of the egg cell. Once one sperm has merged its membrane with the egg cell's, other sperm are prevented from fusing with the egg. A loose layer of cells (yellow) protects the egg and zygote and provides nutrition during the early stages of development.

The mutation of a particular gene creates a different "flavor," or genetic variant, of that gene. Different versions of a gene, created ultimately through DNA mutations, are known as *alleles*.

Meiosis is exceptionally effective at shuffling the alleles that mutations create. Meiosis in a single individual can generate a staggering diversity of gametes, by shuffling alleles between homologous pairs and then sorting these scrambled homologues *randomly* into gametes. The randomness of fertilization adds to genetic diversity in sexually reproducing populations. Entirely new combinations of genetic information are created when an egg with a unique genetic makeup fuses with one of many genetically diverse sperm cells (see Figure 7.15).

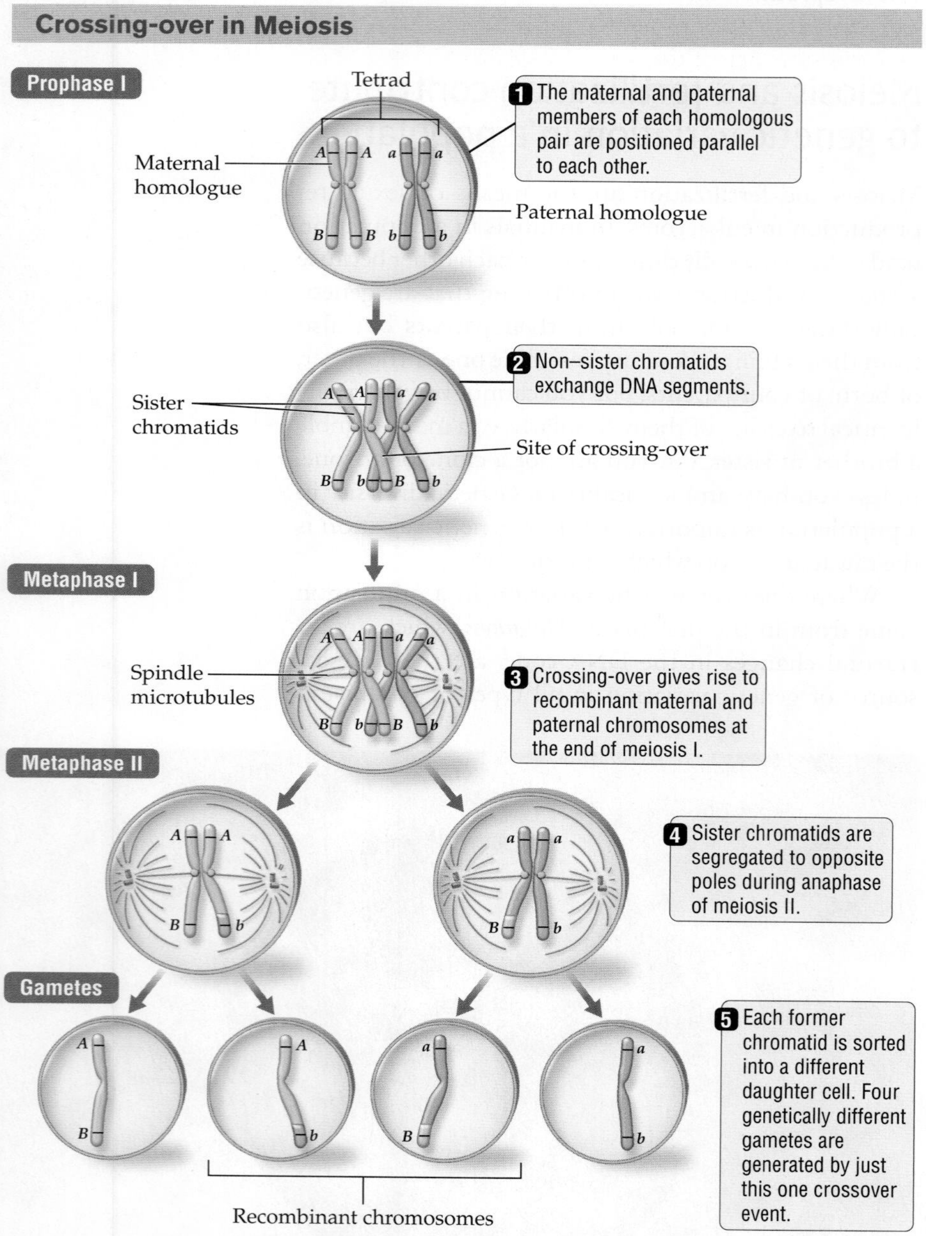

Crossing-over in Meiosis

Prophase I

Tetrad

Maternal homologue

Paternal homologue

1 The maternal and paternal members of each homologous pair are positioned parallel to each other.

Sister chromatids

Site of crossing-over

2 Non–sister chromatids exchange DNA segments.

Metaphase I

Spindle microtubules

3 Crossing-over gives rise to recombinant maternal and paternal chromosomes at the end of meiosis I.

Metaphase II

4 Sister chromatids are segregated to opposite poles during anaphase of meiosis II.

Gametes

5 Each former chromatid is sorted into a different daughter cell. Four genetically different gametes are generated by just this one crossover event.

Recombinant chromosomes

FIGURE 7.16 Crossing-over Produces Recombinant Chromosomes
Crossing-over is the physical exchange of segments between the non–sister chromatids in a pair of homologous chromosomes during prophase I. For clarity, only one maternal and one paternal chromatid are depicted here. A crossover site can be located at any point along the length of the paired homologues, not just at the tips. The letters *A*/*a* and *B*/*b* represent alternative alleles of two genes, *A* and *B*. Note that the parental combinations of these alleles have been shuffled in the recombinant chromosomes.

Crossing-over shuffles alleles

Meiosis generates genetic diversity in two ways: *crossing-over* between the paternal and maternal members of each homologous pair, and *independent assortment* of the paternal and maternal homologues during meiosis I. We will consider crossing-over first.

Crossing-over (**FIGURE 7.16**) is the name given to the physical exchange of chromosomal segments between non–sister chromatids in paired-off paternal and maternal homologues. Early in prophase I, paternal and material homologues pair and line up parallel to each other (see Figure 7.16, prophase panel). Crossing-over is initiated when a chromatid belonging to one homologue (say, the paternal one) makes contact with the chromatid across from it (belonging to the maternal homologue in this case). These *non*–sister chromatids contact each other at one or more random sites along their length. Specialized proteins at the crossover sites act in such a way that the non–sister chromatids exchange segments (see Figure 7.16).

The swapped segments contain the same genes positioned in the same order. But, as we have seen, genes can exist in different versions, called alleles. Crossing-over exchanges alleles; therefore, the chromatids produced by crossing-over are genetic mosaics, bearing new combinations of alleles compared to those originally carried by the homologous chromosomes in the diploid parent cell. The mosaic chromatid is said to be *recombined*, and the creation of new groupings of alleles through the exchange of DNA segments is known as **genetic recombination**. Without crossing-over, every chromosome inherited by a gamete would be just the way it was in the parent cell. Crossing-over between *just one* pair of homologous chromosomes enables meiosis to produce at least four genetically distinct gametes, as shown in Figure 7.16.

The independent assortment of homologous pairs generates diverse gametes

The possibilities for creating genetically diverse gametes are not restricted to crossing-over. The **independent assortment of chromosomes**—that is, the random distribution of the different homologous chromosome pairs into daughter cells during meiosis I—also contributes to the genetic variety of the gametes produced. It comes about because each homologous chromosome pair orients itself independently—without regard to the alignment of any other homologous pair—when it lines up at the metaphase plate during meiosis I.

To understand why the independent alignment of homologous pairs produces random patterns, consider a cell with just two homologous chromosome pairs ($n = 2$, $2n = 4$). During metaphase I, there are two ways of arranging each homologous pair at the metaphase plate: Option A places the maternal homologue "to the left" and the paternal homologue "to the right" for *both* chromosome pairs. Option B keeps the first homologous pair in the same orientation as in option A, but reverses the orientation of the second homologous pair. As **FIGURE 7.17** shows, option A would sort the maternal and paternal homologues in one particular pattern, creating two different types of gametes. Option B would also produce two different types of gametes, but the combinations of paternal and maternal homologues in these two gamete types would be different from the combinations created in option A.

What this means is that four types of gametes, each with a different combination of paternal and maternal homologues, can potentially be generated through

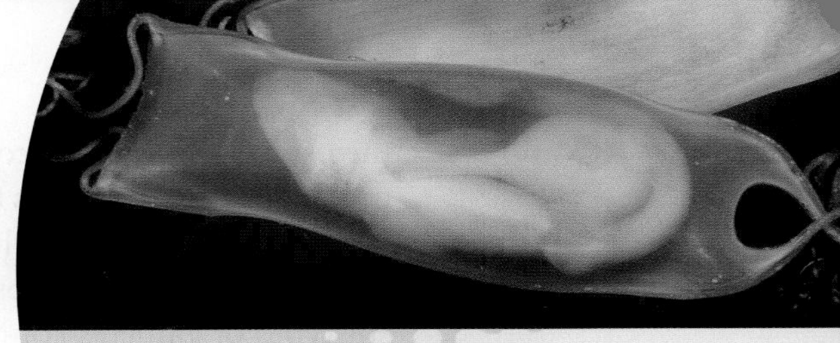

EXTREME OFFSPRING

Sharks deposit "egg cases" consisting of multiple zygotes wrapped in a tough layer of collagen protein. The biggest hatchling cannibalizes its siblings, so that only one baby shark develops within each egg case in the end. Empty cases wash up on beaches sometimes; beachcombers call them "mermaid's purses."

Independent Assortment of Chromosomes

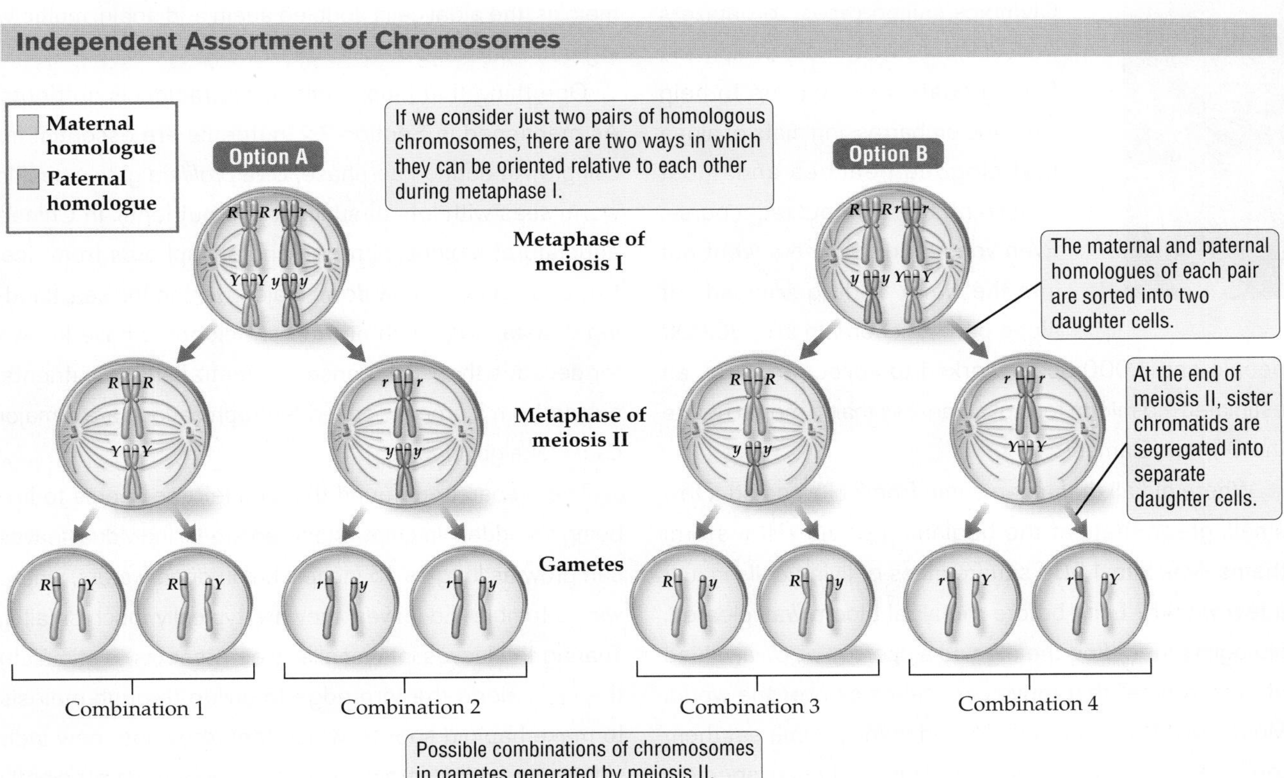

Maternal homologue
Paternal homologue
Option A
If we consider just two pairs of homologous chromosomes, there are two ways in which they can be oriented relative to each other during metaphase I.
Option B
Metaphase of meiosis I
The maternal and paternal homologues of each pair are sorted into two daughter cells.
Metaphase of meiosis II
At the end of meiosis II, sister chromatids are segregated into separate daughter cells.
Gametes
Combination 1
Combination 2
Combination 3
Combination 4
Possible combinations of chromosomes in gametes generated by meiosis II.

FIGURE 7.17 The Random Assortment of Homologous Chromosomes Generates Chromosomal Diversity among Gametes

Concept Check Answers

1. Meiosis II separates the two sister chromatids of each replicated chromosome, to form gametes with 23 chromosomes each.

2. Meiosis generates genetically diverse gametes through two different mechanisms: recombination of alleles via crossing-over, and random assortment of homologous chromosomes.

meiosis in a diploid cell containing just two pairs of homologous chromosomes ($n = 2$). Because there are two ways to arrange each of the three pairs of homologues, 2^3 (or 8) patterns are possible; therefore, eight different types of gametes could be created. In human cells, with $n = 23$, there are 2^{23}, or 8,388,608, different ways of combining homologues in the gametes.

Finally, fertilization has the potential to add tremendous genetic variation to the variation already produced by crossing-over and independent assortment of chromosomes. Because every sperm cell and every egg represents one out of more than 8 million types, there are more than *64 trillion* genetically different possibilities for offspring (8 million possible sperm $\times$ 8 million possible eggs) each time a sperm fertilizes an egg. (This number does not include the variation due to crossing-over.)

Concept Check

1. Meiosis I reduces the diploid chromosome set. What role does meiosis II play in gamete formation?
2. How does meiosis contribute to genetic variation in a population?

APPLYING WHAT WE LEARNED

The Great Divide

Just weeks before the 2008 Summer Olympics sailing races, organizers frantically mobilized heavy equipment, fishing boats, and soldiers to help clear the embarrassing mats of algae that clogged beaches and, most important, the Olympic racecourse. Even volunteer swimmers went out into the water to drag armloads of algae back to shore. In all, 130,000 people and 1,000 boats worked to collect and bury an estimated 1.5 million tons of algae less than a month before the Olympic Games.

Where did all that algae come from? How could a few small, green mats at the beginning of June transform themselves into 1,500 square miles of living cells in just a few weeks? Even before the algal bloom was cleared, biologists identified the alga as a species of *Ulva*, a kind of "sea lettuce" that thrives in oceans all over the world. Most sea lettuces are edible, and many people eat them regularly in soups and salads. This particular species, *Ulva prolifera*, had grown from a few green mats in late May to 22 tons of biomass by early July. Most of the growth, researchers said, occurred in less than 2 weeks. How did these tiny marine organisms do it? The answer is: cell division. Through a combination of mitosis and meiosis, the algal cells doubled again and again, multiplying geometrically.

One thing that helps cells divide rapidly is nutrients. As mentioned in Section 7.2, nutrients are essential for cell growth during interphase. *Ulva prolifera* grows best in warm seas with lots of sunlight and nutrients. In China's agricultural regions, nitrogen and phosphorus from rice, fish, and pig farms flow down major rivers to the sea, flooding coastal seas with nutrients. Scientists have known for decades that such dense concentrations of nutrients, arising from a process called "eutrophication," are a major cause of algal blooms.

The researchers found that sea lettuce seems to like being shredded. In quiet, sheltered areas, individual leaves can grow as long as 15 inches, but in open waters, where waves tumble the leaves, they are typically much smaller. Tearing the leaves into smaller pieces seems to stimulate the cells along the torn edge to divide through meiosis to make haploid spores, which then grow into new individuals. In fact, *U. prolifera* sheds spores most efficiently when it is torn up into bits and pieces no bigger than the asterisk above the number 8 on a computer keyboard—whether by pounding waves, the propellers of boats, or sea lettuce–munching animals.

BIOLOGY IN THE NEWS

The Science of Making Babies Becomes Commonplace

BY ERIN ELLIS • *Vancouver Sun*, May 18, 2013

A whole generation has been born since Canada's first "test tube baby" saw daylight in Vancouver almost 30 years ago ... Now almost 10,000 babies are born in Canada each year through in vitro fertilization, or IVF, as the unheard of has transformed into the unremarkable.

The latest twist is the possibility of preserving human eggs long enough for a woman to outrun her own biological clock ... [The] price is about $10,000 for each round of egg retrieval. It requires [women] injecting themselves with various hormones to force multiple eggs to mature in their ovaries. Not to mention rafts of tests and the final harvesting of the eggs, under sedation, with a needle guided by ultrasound through the vagina to the ovaries. Somewhere between eight and 15 eggs are typically harvested per ovulation cycle and some women go through several rounds to get the optimal number of eggs needed to give them a decent chance of having a child.

Will the latest technology in freezing human eggs—called vitrification—now release women from their body's limitations? It allows a younger woman to freeze 10 or 20 eggs when she is single and then use them to produce her own biological child when she finds "Mr. Right"—up to age 50 at some clinics ... "It's not a guarantee. People like to think of it as more of an insurance policy. You don't want to freeze them, forget about it and think they're automatically going to work when it comes to use them," [says Dr. Sonya Kashyap of Vancouver's Genesis Fertility Centre].

That's because by age 35, there's about a 30 per cent chance that the implantation of a fertilized egg will result in a live birth. Still, the scenario started to look a bit more believable late last year when the American Society for Reproductive Medicine—an organization that sets guidelines for the sector—removed its "experimental" label from the procedure. A committee of doctors found that the limited research available showed similar rates of live birth from freshly harvested eggs and ones preserved using vitrification.

Meiosis in male humans begins at puberty. Driven by testosterone, diploid germ line cells divide through mitosis to expand the number of meiotic precursor cells. About 100 million of these diploid cells undergo meiosis every day for the rest of a man's life.

Meiosis in a human female begins even before she is born. Germ line cells inside the ovaries begin meiosis when the female fetus is just 15 weeks old. But they don't get past prophase of meiosis I. Meiosis resumes when a girl reaches puberty. Once a month, under the influence of female sex hormones, meiosis starts up again and proceeds to metaphase of meiosis II. Then it comes to a halt again. The stalled cell—technically known as a *secondary oocyte*—is released into the tubes that connect to the uterus in a process called *ovulation*. If the cell is not fertilized within 2–3 days, the secondary oocyte will die. Fusion with a sperm cell triggers completion of meiosis II finally, and merger of the two nuclei produces the first diploid cell of the next generation, the zygote.

The "eggs" the news article refers to are secondary oocytes. Women seeking to freeze their eggs are given high doses of estrogen to trigger ovulation of multiple oocytes instead of the single oocyte that is normally released. A minimally invasive outpatient technique is used to harvest the multiple oocytes (eggs).

Just a few years ago, it was impossible to store harvested human oocytes. They had to be immediately fertilized with sperm in a lab dish, and the resulting embryo (50–100 cells all told) could then be frozen. The frozen embryos were later implanted in the mother-to-be, and this was the standard way to make "test tube babies" through in vitro fertilization (IVF).

Oocytes are the most massive cells in the human body. About 0.12 millimeter in width, they would be visible to anyone with good eyesight. But poised at metaphase II, and full of fluid, they are exceedingly fragile. The new technique of *vitrification* makes it possible to freeze and thaw the oocytes without damage from ice crystals. More than 2,000 healthy babies have been born worldwide using this technique.

Evaluating the News

1. How many DNA molecules are found in each of the "eggs" harvested for in vitro fertilization (IVF)?

2. How is the new vitrification-based technology different from conventional IVF?

3. Reproductive technology such as IVF has been controversial because some think it amounts to "playing God." Others consider it the "slippery slope" to cloning humans. What do you think? Explain your reasoning.

CHAPTER REVIEW

Summary

7.1 Why Cells Divide

- Cell division is necessary for growth and repair in multicellular organisms, and for asexual and sexual reproduction in all types of organisms.
- Many prokaryotes divide through binary fission, a form of asexual reproduction.
- Mitotic divisions produce daughter cells that are genetically identical to each other and to the parent cell.
- Meiosis is critical for sexual reproduction. In animals, meiosis produces gametes, which fuse during fertilization to make offspring.

7.2 The Cell Cycle

- The cell cycle refers to events over the life span of a eukaryotic cell.
- Interphase and cell division are the two main stages of the cell cycle. Interphase is the longest, and consists of G_1, S, and G_2. DNA is replicated in the S phase.
- The cell cycle is carefully regulated. Checkpoints ensure that the cycle does not proceed if conditions are not right.

7.3 The Chromosomal Organization of Genetic Material

- Each chromosome contains a single DNA molecule bearing many genes and compacted by packaging proteins.
- The somatic cells of eukaryotes have two of each type of chromosome, forming matched pairs called homologous chromosomes.
- One homologue in each homologous pair is inherited from the maternal parent, the other from the paternal parent. In mammals, the sex chromosomes (X and Y) determine sex. Females have two copies of the X chromosome; males have one X and one Y chromosome.

7.4 Mitosis and Cytokinesis: From One Cell to Two Identical Cells

- During mitotic cell divisions, the parent cell's replicated DNA is distributed equally between daughter cells.
- DNA replication produces two identical sister chromatids that are held together firmly at the centromere.
- Mitosis has four main phases: prophase, metaphase, anaphase, and telophase. Through these phases, the chromosomes of a parent cell are condensed, positioned appropriately, and separated to opposite ends.
- During cytokinesis, the cytoplasm of the parent cell is physically divided to create two daughter cells.

7.5 Meiosis: Halving the Chromosome Set to Make Gametes

- Meiosis—consisting of two rounds of nuclear and cytoplasmic divisions—produces haploid (*n*) gametes containing only one chromosome from each homologous pair.
- During meiosis I, the maternal and paternal members of each homologous pair are sorted into two different daughter cells.
- Meiosis II is similar to mitosis in that sister chromatids are segregated into separate daughter cells at the end of cytokinesis.
- Meiosis produces genetically diverse gametes through two means: crossing-over and the independent assortment of homologous chromosomes.
- Meiosis and fertilization introduce genetic variation in a population.

Key Terms

anaphase (p. 160)
asexual reproduction (p. 150)
binary fission (p. 151)
cancer cell (p. 156)
cell cycle (p. 153)
cell differentiation (p. 153)
cell division (p. 153)
cell plate (p. 161)
centromere (p. 157)
centrosome (p. 160)
chromatin (p. 156)
chromosome (p. 156)
crossing-over (p. 166)
cytokinesis (p. 152)
diploid (p. 162)
embryo (p. 152)
fertilization (p. 152)
G_0 phase (p. 154)
G_1 phase (p. 154)
G_2 phase (p. 154)
gamete (p. 152)
genetic recombination (p. 166)
germ line cell (p. 153)
haploid (p. 162)
homologous chromosome (p. 158)
independent assortment of chromosomes (p. 167)
interphase (p. 153)
karyotype (p. 157)
maternal homologue (p. 163)
meiosis (p. 152)
meiosis I (p. 163)
meiosis II (p. 163)
metaphase (p. 160)
mitosis (p. 152)
mitotic division (p. 152)
mitotic spindle (p. 160)
paternal homologue (p. 163)
prophase (p. 160)
S phase (p. 154)
sex chromosome (p. 158)
sexual reproduction (p. 150)
sister chromatids (p. 157)
somatic cell (p. 152)
telophase (p. 161)
tetrad (p. 164)
tumor (p. 156)
zygote (p. 152)

Self-Quiz

1. In the cell cycle, DNA is duplicated in the
   a. G_1 phase.
   b. S phase.
   c. G_2 phase.
   d. division stage.

2. Which of the following statements is true?
   a. Chromatin is more compacted in prophase than during the G_2 phase.
   b. The key event of the S phase is the segregation of sister chromatids.
   c. The mitotic spindle first appears during anaphase.
   d. The cell increases in size during metaphase.

3. Which of the following statements is *not* true?
   a. DNA is packaged into chromatin with the help of proteins.
   b. All chromosomes in the somatic cell of a particular species have the same shape and size.
   c. Each chromosome contains a single DNA molecule.
   d. Somatic cells in the animal body are diploid.

4. Which of the following correctly represents the order of the phases in the cell cycle?
   a. mitosis, S phase, G_1 phase, G_2 phase
   b. G_0 phase, G_1 phase, mitosis, S phase
   c. S phase, mitosis, G_2 phase, G_1 phase
   d. G_1 phase, S phase, G_2 phase, mitosis

5. In fertilization, gametes fuse to form a __________ zygote.
   a. tetrad
   b. haploid
   c. diploid
   d. triploid

6. Human gametes contain
   a. twice the number of chromosomes that our skin cells have.
   b. only sex chromosomes.
   c. half the number of chromosomes that our skin cells have.
   d. only X chromosomes.

7. The reduction division is
   a. prophase of mitosis.
   b. anaphase II of meiosis.
   c. metaphase II of mitosis.
   d. meiosis I.

8. Meiosis results in
   a. four haploid cells.
   b. two diploid cells.
   c. four diploid cells.
   d. two haploid cells.

Analysis and Application

1. What is the functional value of the regular checkpoints that are a feature of the cell cycle?
2. Horses have a karyotype of 64 chromosomes. How many separate DNA molecules are present in a horse cell that is in the G_2 phase just before mitosis? How many separate DNA molecules are present in each cell produced at the end of meiosis I in a horse?
3. During mitosis, can anaphase take place before metaphase? Why or why not?
4. Compare and contrast meiosis I and meiosis II. Which of the two is more similar to mitosis, and why?
5. For sexually reproducing organisms, how would the chromosome numbers of offspring be affected if gametes were produced by mitosis instead of meiosis?
6. Explain how crossing-over gives rise to recombinant chromosomes. At what stage of meiosis does crossing-over occur? Explain the biological significance of crossing-over.
7. The photo here shows cancer cells (known as HeLa cells) undergoing mitosis in a lab dish. DNA is stained pink, and the mitotic spindle and centrosome are yellow. The multiple pink dots within the nuclei are regions called *nucleoli* where ribosomal RNA is made. Identify cells in prophase, metaphase, anaphase, and telophase, and a cell about to complete cytokinesis. (*Hint:* Nucleoli are absent during mitosis, and you would expect to see two centrosomes in prophase, along with highly condensed chromosomes.)

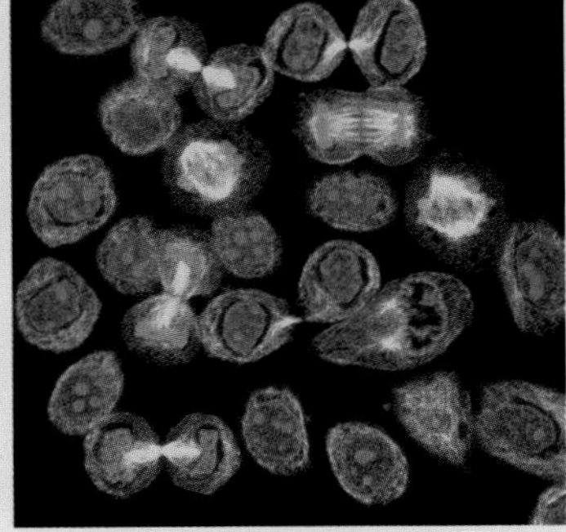

8. Compare mitotic cell division to meiosis by completing the table. Write "true" or "false" in each column as appropriate, following the example in the first row.

	MITOTIC CELL DIVISION	MEIOSIS
1. In humans, the cell undergoing this type of division is diploid.	True	True
2. A total of four daughter cells are produced when one parent cell undergoes this type of division.		
3. Our skin makes more skin cells using this type of cell division.		
4. The daughter cells are genetically identical to the parent cell in this type of division.		
5. This type of division involves two nuclear divisions.		
6. This type of division involves two rounds of cytokinesis.		
7. Maternal and paternal homologues pair up to form tetrads at some point during this type of division.		
8. Sister chromatids separate from each other at some point during this type of cell division.		

8 Cancer and Human Health

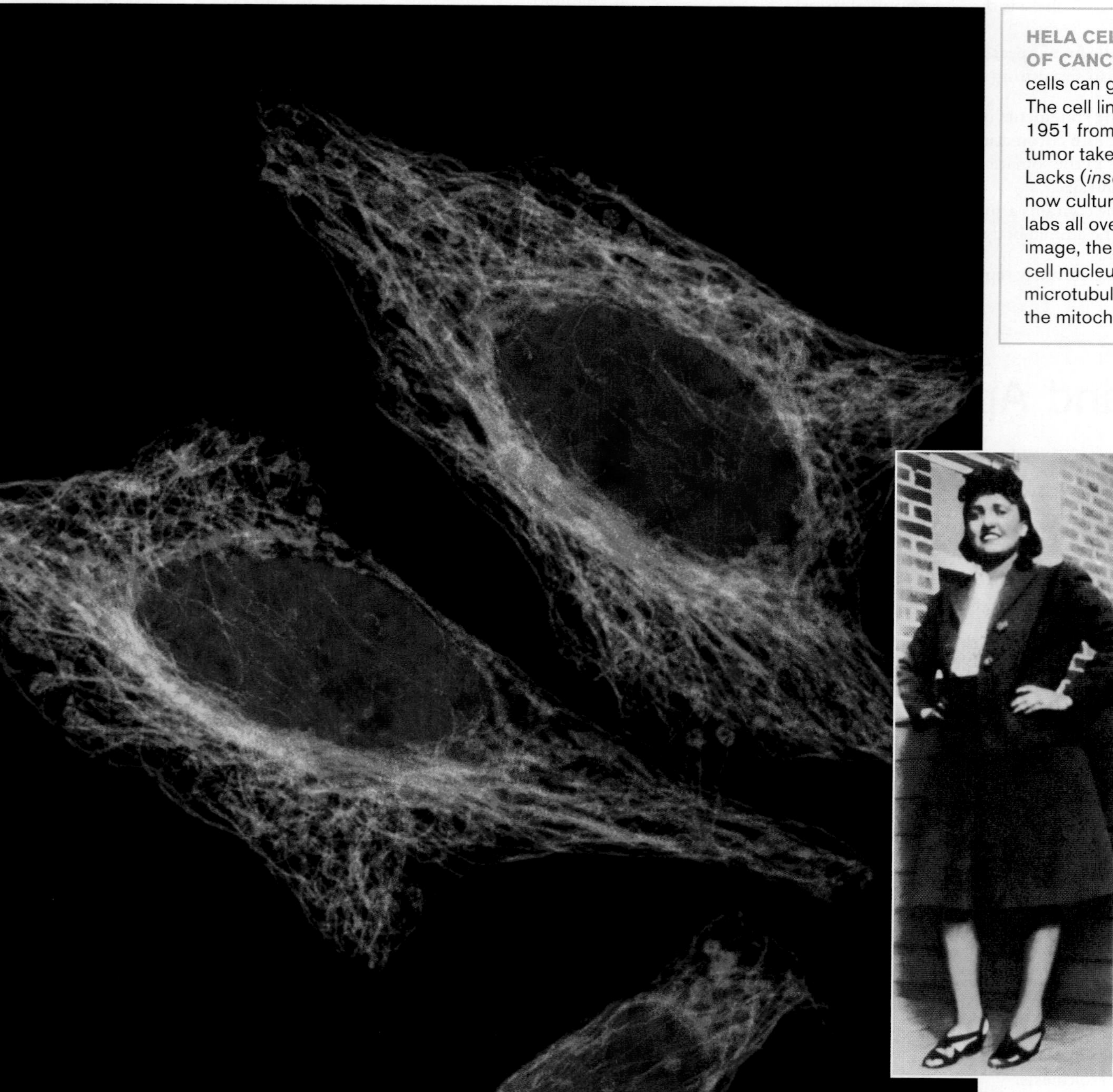

HELA CELLS ARE A LINE OF CANCER CELLS. HeLa cells can grow in a lab dish. The cell line was established in 1951 from a slice of a cervical tumor taken from Henrietta Lacks (*inset*). HeLa cells are now cultured in thousands of labs all over the world. In this image, the DNA in the HeLa cell nucleus is stained blue, the microtubules are green, and the mitochondria are red.

Henrietta Lacks's Immortal Cells

In 1951, a working-class, stay-at-home mother named Henrietta Lacks arrived at Johns Hopkins Hospital to ask about bleeding at the wrong time of the month. Her doctor took one look at the bright purple tumor on Lacks's cervix, the entrance to the uterus, and feared the worst. He soon determined that she had malignant cervical cancer and broke the news to her.

Just 31 and mother to five children, Henrietta Lacks had little idea that she would not live to see her children grow up. Most of the cells in her body were dividing at the slow, highly regulated pace that is normal for human cells—but her cancer cells were different.

When Lacks returned to the hospital for treatment, a surgeon sewed envelopes of radioactive radium into her cervix, the state-of-the-art treatment of the day. These were removed after several days, and follow-up treatments with X-rays were scheduled.

Also at Johns Hopkins were two researchers, George and Margaret Gey, who had been trying for years to "culture," or grow, human cells in test tubes so that they could study their behavior. In 1951, all of the Geys' attempts had so far ended in complete failure, and they were eagerly trying to grow any human cells they could get their hands on. To help out, Lacks's surgeon took a slice of Lacks's tumor and delivered it to the Geys.

What made Lacks's cells immortal? How have those cells contributed to our understanding of cancer biology?

The cells from the tumor, called "HeLa" cells after Henrietta Lacks, were a dramatic success, turning out to be among the fastest-dividing human cells ever known and, more important, the first human cells ever grown in the lab outside a human body. But besides being able to survive in a test tube and divide quickly, the cells had one more amazing attribute: they were immortal; they would keep dividing seemingly forever.

The treatments Henrietta Lacks received could not hold back the aggressive cancer cells. In less than a year, she died of the cancer that by then had spread throughout her body. In laboratories around the world, however, HeLa cells live on.

MAIN MESSAGE Cancer cells divide without restraint and invade other tissues. Treatment must focus on inhibiting cancer cells without harming normal cells.

KEY CONCEPTS

- When normal restraints on cell division fail, a cell starts dividing rapidly and produces a benign tumor.
- Tumor cells that gain the ability to invade surrounding tissue are cancer cells. Cancer cells can metastasize—spread to other organs—and cause further damage.
- Gene mutations are the root cause of all cancers.
- Proto-oncogenes are normal genes that stimulate cell division. When mutations make them hyperactive, they trigger cancerous change and are then called oncogenes.
- Tumor suppressor genes repair DNA and block cell division and cell migration. The risk of cancer rises when mutations make the genes nonfunctional.
- The great majority of human cancers are not hereditary; instead they are caused by the accumulation of many somatic mutations over the course of life.
- Standard cancer therapies attempt to kill rapidly dividing cells, most commonly with toxic chemicals and radiation. The treatments produce serious side effects because they also kill healthy cells.
- New cancer therapies aim to kill cancer cells selectively, most commonly by destroying them with targeted antibodies.
- Infectious agents such as viruses can trigger cancerous change; viruses are implicated in about 15 percent of human cancers.
- Environmental and lifestyle factors play a large role in human cancers and are the focus in cancer prevention.

HENRIETTA LACKS'S IMMORTAL CELLS 173

8.1 Cancer: Good Cells Gone Bad 174

8.2 Cancer-Critical Genes 178

8.3 The Progression to Cancer 180

8.4 Treatment and Prevention 181

APPLYING WHAT WE LEARNED 186
How HeLa Cells Changed Biomedicine

CANCER IS A STORY OF GOOD CELLS GONE BAD. We noted in the preceding chapter that life propagates itself through cell divisions, and cell divisions are also crucial for maintaining and growing the multicellular body. In contrast to the regulated processes of normal cell division, cancer cells divide with wild abandon, overrunning their neighbors and commandeering resources to the point of starving the hardworking normal cells. The rogue cells not only behave differently, but they are also so much larger and so abnormal-looking that lab technicians can readily pick them out from normal cells under the microscope (**FIGURE 8.1**).

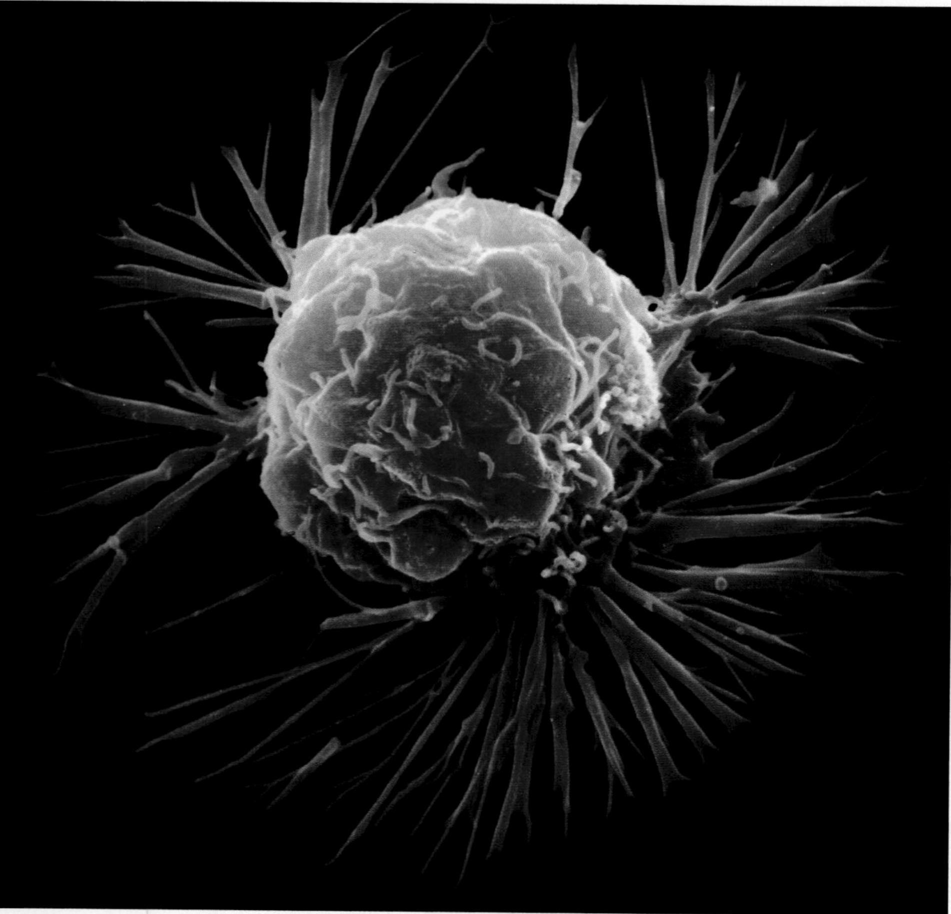

FIGURE 8.1 A Homegrown Monster
This color-enhanced photograph, captured with a scanning electron microscope, shows a breast cancer cell. A lab technician or pathologist can recognize it as a cancer cell. In what ways is a cancer cell different from a normal cell in the body? Why do normal cells become destructive monsters sometimes?

In this chapter we explore the biology of cancer by considering how a cancer is launched when normal restraints on cell division and cell movement are lost. The vast majority of cancers arise from the gradual accumulation of mutations within a cell, often as a result of DNA damage caused by environmental factors. Before we begin, here is a brief review of relevant concepts:

- **DNA** stores genetic information as a precise sequence of four nucleotides, which are the building blocks of the genetic code (see Section 2.10). We could say DNA is the software that runs all the hardware of the cell.
- A **gene** is a segment of DNA that codes for a distinct, inherited characteristic. The roughly 25,000 genes in human cells could be seen as different applications built into one software package.
- **Cell differentiation** is the developmental process by which a cell acquires a specialized function. During development, a cell is prompted by its internal genetic program and by the external cues it receives to differentiate into a particular cell type.
- **Gene expression** refers to whether a gene is read and its information turned into a product, such as a protein. All cells in the human body carry the same DNA sequence, but a different set of genes is read out, or expressed, in every unique cell type. Extending the computer analogy, the same DNA software is installed in all cells, but different cell types run a different mix of the available applications.

8.1 Cancer: Good Cells Gone Bad

Cancer accounts for more than 500,000 deaths in the United States each year. Over the course of a lifetime, an American male has a nearly one in two chance of being diagnosed with cancer. American women fare slightly better, with a one in three chance of developing cancer.

TABLE 8.1 Selected Human Cancers in the United States

TYPE OF CANCER	OBSERVATION	ESTIMATED NEW CASES IN 2013	ESTIMATED DEATHS IN 2013
Lung cancer	Accounts for 28 percent of all cancer deaths and kills more women than breast cancer does	228,190	159,480
Prostate cancer	The second leading cause of cancer deaths in men (after lung cancer)	238,590	29,720
Breast cancer	The second leading cause of cancer deaths in women (after lung cancer)	234,580	40,030
Colon and rectal cancer	The number of new cases is leveling off as a result of early detection and polyp removal	142,820	50,830
Malignant melanoma	The most serious and rapidly increasing form of skin cancer in the United States	76,690	9,480
Leukemia	Often thought of as a childhood disease, this cancer of white blood cells affects more than 10 times as many adults as children every year	48,610	23,720
Ovarian cancer	Accounts for 3% of all cancers in women	22,240	14,030

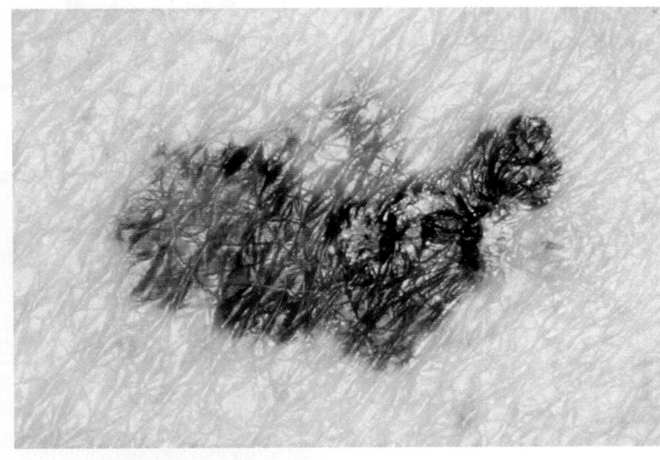

Melanoma is the most serious type of skin cancer. It is caused by the growth and spread of melanocytes, pigment-carrying skin cells that move about as a normal part of their function (delivering pigment to other cells). A melanoma on the skin can be recognized by applying the ABCDE principle:

- A is for asymmetry, the uneven shape of the tumor mass.
- B is for border, which is irregular.
- C is for color, which consists of multiple hues.
- D is for diameter, which becomes a concern if it exceeds a fourth of an inch (6 millimeters).
- E is for evolving, meaning any observed change such as a change in shape, size, color, or if there is bleeding or itchiness.

More than 200 different types of cancer have been described, but the big four—lung, prostate, breast, and colon cancers—account for more than half of all cancers combined (**TABLE 8.1**). Although the past decade has seen huge improvements in treatment and prevention, more than a 1.6 million Americans are diagnosed with some form of cancer every year, and about 1,600 die from it each day. Only heart disease kills more people in the United States. The National Cancer Institute estimates that the direct medical costs for diagnosis and treatment of the various forms of cancer amount to over $77 billion per year.

Cancer develops when cells lose normal restraints on division and migration

Every cancer begins with a single rogue cell that starts to undergo out-of-control division, giving rise to a mass of abnormal cells. The cell mass formed is known as a **tumor**. Tumors that remain confined to one site are **benign tumors** (**FIGURE 8.2**). Because they can be surgically removed in most cases, benign tumors are generally not a threat to survival. However, a benign tumor that is growing actively is like a cancer-in-training. With the passage of time, the descendants of these abnormal cells can become increasingly abnormal: they change shape, increase in size, and quit their normal job.

Tumor cells often produce proteins—known as *tumor markers*—that normal cells of that type either do not make or make in much lower quantity. The presence of tumor markers in blood, urine, and other fluids and tissues is used to screen for the *possibility* of some types of cancer.

Most cells in the adult animal body are firmly anchored in one place. Tumor cells on the path to cancer start producing enzymes that break up the adhesion proteins that attach a cell to the extracellular matrix or to other cells in the tissue. Most human cells stop dividing if they are detached from their surroundings—a phenomenon known as **anchorage dependence**. But

FIGURE 8.2
Cancers Start with a Single Cell That Begins Dividing in an Uncontrolled Manner

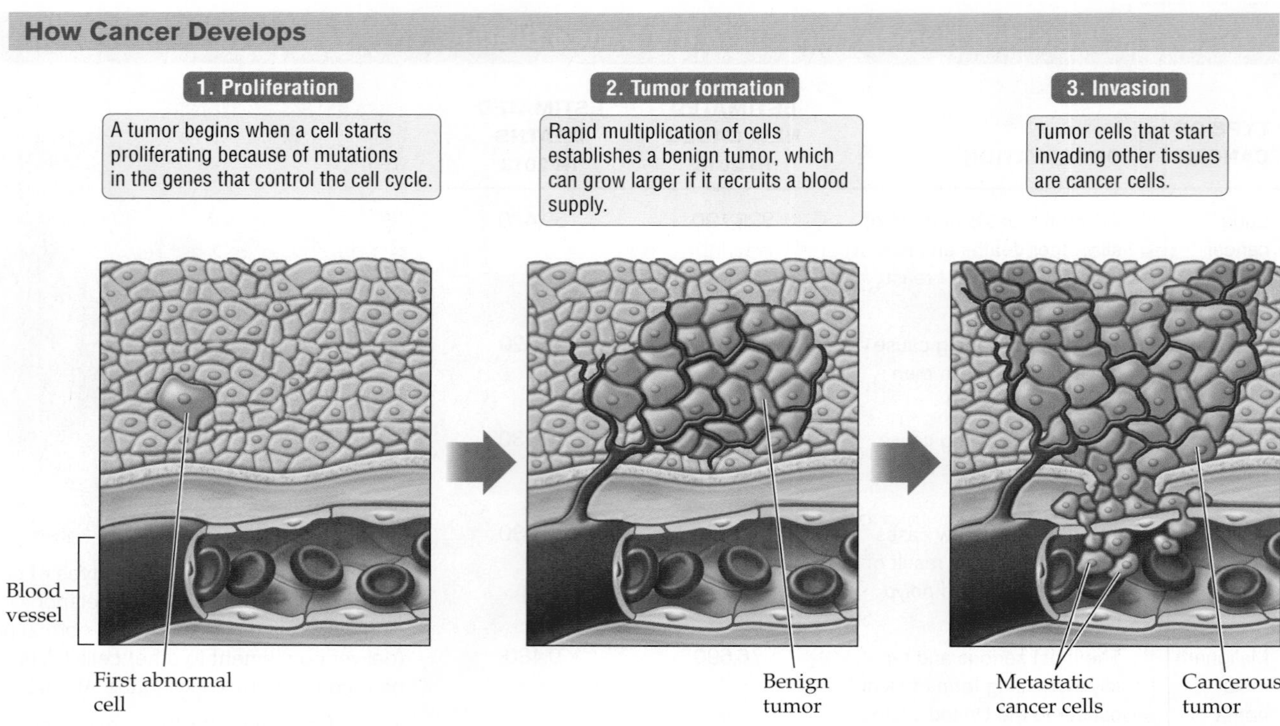
How Cancer Develops
1. Proliferation
A tumor begins when a cell starts proliferating because of mutations in the genes that control the cell cycle.
2. Tumor formation
Rapid multiplication of cells establishes a benign tumor, which can grow larger if it recruits a blood supply.
3. Invasion
Tumor cells that start invading other tissues are cancer cells.
Blood vessel
First abnormal cell
Benign tumor
Metastatic cancer cells
Cancerous tumor

Helpful to Know

The word "cancer" comes from the Old English for "spreading sore," which in turn comes from the Latin for "crab," probably because the enlarged blood vessels on a cancerous tumor were seen as resembling the legs of a crab.

some tumor cells that have broken loose from their moorings may acquire anchorage *in*dependence, the ability to divide even when released from their normal attachment sites (**TABLE 8.2**).

As tumor cells progress toward a cancerous state, they start secreting substances that cause new blood vessels to form in their vicinity (a process known as *angiogenesis*). Angiogenesis (**AN-jee-oh-JEN-uh-sus**; *angio*, "vessel"; *genesis*, "creation of") recruits a blood supply for the tumor, important for delivering nutrients and whisking away waste. Without these new blood vessels, solid tumors would not be able to grow larger than about 1 or 2 millimeters, because in a larger tumor, the cells buried inside would be starved for nutrients and poisoned by waste products. Angiogenesis, however, enables malignant tumors to grow bigger. As we explain next, the network of fine blood vessels in and around the tumor is also an easy exit route for tumor cells escaping the home port.

TABLE 8.2 Characteristics of Cancer Cells

- Cell proliferation
- Increase in cell size, change in shape
- Loss of cell adhesion
- Loss of anchorage dependence (cell continues to divide even when detached)
- Tissue invasion (local spread)
- Angiogenesis (recruiting blood supply)
- Metastasis (spread to other organs)

Cancer cells can spread to distant sites

When tumor cells gain anchorage independence and start invading other tissue layers, they are considered **cancer cells**, also known as malignant cells. At their very worst, cancer cells break loose from their neighborhood and head to other organs to set up satellite colonies there. The rogue cells enter the bloodstream or invade the lymphatic system, a network of vessels in which immune cells are made, stored, and move about. Cancer cells squeeze between the cells in the walls of blood vessels or lymphatic vessels to emerge in distant locations throughout the body, where they grow into new tumors. The tumor at the initial location is called a **primary tumor**, and the new tumors it spawns at distant sites are known as **secondary tumors** (**FIGURE 8.3**).

The spread of a disease from one organ to another is known as **metastasis (meh-TAS-tuh-sus**; *meta*, "to change"; *stasis*, "a set condition"). Metastasis typically occurs at later stages in cancer development. Because tumors are present in multiple organs, each of them capable of further metastasis, a cancer that has metastasized is difficult to fight. Some types of cancers, such as basal cell or squamous cell skin cancers, rarely metastasize and are therefore relatively easy to treat.

Some cancers typically spread to fairly predictable locations because certain routes offer less resistance to the movement of malignant cells than do others. A cancer that arises in the colon spreads most easily to the liver, for example (see Figure 8.3). Breast cancer cells commonly metastasize to the bones, lungs, liver, or brain. Because the cells that make up an organ have distinctive properties, pathologists can examine the malignant cells in a secondary tumor and tell which organ they originally came from. Knowing the origin of metastatic tumors is often valuable in choosing among treatment options.

Without restraints on their growth and migration, cancer cells take over, steadily destroying tissues, organs, and organ systems. The normal function of these organs is seriously impaired, and cancer deaths are ultimately caused by the failure of vital organs.

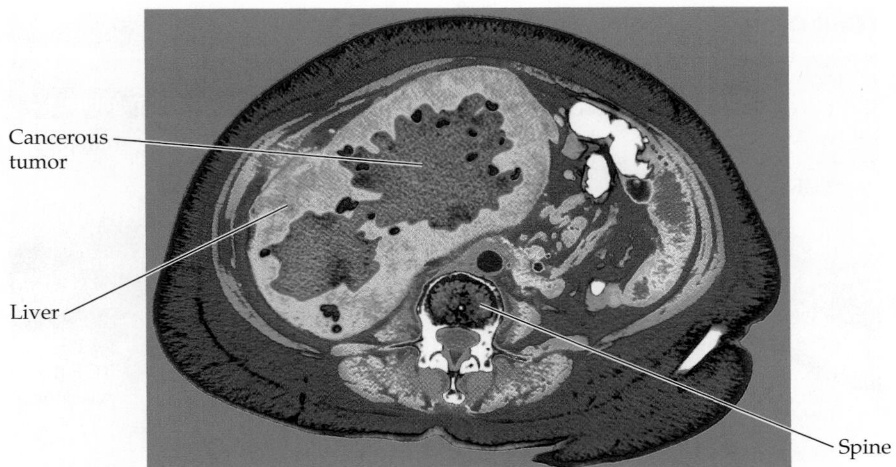

FIGURE 8.3 Cancer Metastasis

This color-enhanced CT (computed tomography) scan shows a large cancerous tumor (dark green) in the liver (light green) of a woman who was originally diagnosed with colon cancer. The cancerous cells metastasized from the large intestine to the liver to create the secondary tumors seen here. The computer-generated image shows a "slice" through the body. The patient's back and spine are at the bottom center, and her toes would be pointing toward the observer.

Cell division is controlled by positive and negative growth regulators

Rampant cell division, or proliferation, is the first abnormal behavior a cell displays on its way to becoming cancerous. As noted in Chapter 7, the cell cycle is controlled by a variety of external and internal signals. Hormones, and proteins called growth factors, are common regulators of cell division in the body. **Positive growth regulator** is a blanket term for any signal that stimulates cell division. For example, human growth hormone induces cell division at the ends of the long bones in children, and estrogen stimulates cell division in breast tissue. **Negative growth regulators** are any and all signals that put the brakes on cell division. The hormone cortisone, for example, inhibits cell division in immune cells and certain other cell types. Growth factors, too, can be positive or negative growth regulators, either pushing the cell cycle forward or restraining it (**FIGURE 8.4**).

Many of the external signals that regulate cell division do so by binding to **receptor** proteins (see Section 3.2), located either in the cell membrane or inside the cell. When a growth regulator binds to its receptor, a stepwise sequence of cellular activities, called a **signal transduction pathway**, is activated, which relays the external signal from one molecule to the next (see Figure 8.4). At the end of the pathway, cell division is either triggered (positive regulation) or repressed (negative regulation). Because they, too, control the cell cycle, components of the signaling pathways are also positive or negative growth regulators.

Under normal circumstances, positive and negative growth regulators are released only when and where they are needed. A paper cut on your finger triggers the activity of positive growth regulators that causes skin cells to divide until the wound is closed. How do cells at the wound edge "know" when they have divided enough and it is time to quit? Negative growth regulators halt the cell cycle when the newly made cells find themselves pushing up against other cells. Negative growth regulators can even prompt a potentially dangerous cell, such as one that has irreparable DNA damage, to undergo *programmed cell death* (see Chapter 7).

Cell Cycle Regulation by Growth Factors

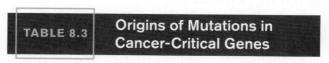

FIGURE 8.4 Growth Factors Can Be Positive or Negative Regulators of the Cell Cycle
Growth regulators work by binding to specific receptors, initiating signal transduction. Positive regulators, such as epidermal growth factor (EGF), promote cell division. Negative regulators, such as transforming growth factor beta (TGF-β), stop cell division and may lead to programmed cell death (apoptosis).

As these examples illustrate, the life of a cell is managed by a delicate interplay between a variety of positive and negative growth regulators. But what happens when this system fails? Runaway cell division is the consequence if the cell cycle is excessively stimulated through the pathways controlled by positive growth regulators. But cell division can also be triggered in a cell that ignores the message of negative growth regulators to stop dividing. In either case, excessive cell proliferation sets the stage for the development of cancer. But what would cause a cell to behave in such a reckless way, endangering the whole organism? We will see in the next section that malfunctioning genes are responsible for the wayward behavior of tumor cells.

Concept Check

1. How are cancer cells different from the cells in a benign tumor?
2. What are some characteristics of cancer cells?
3. How are positive growth regulators and negative growth regulators different?

Concept Check Answers

1. Both types of cells proliferate without restraint, but cancer cells also invade other tissues.
2. Large size, abnormal shape, loss of cell adhesion, and anchorage independence, sometimes with angiogenesis and metastasis as well.
3. Positive growth regulators trigger cell division. Negative growth regulators stop cell division, and may also trigger programmed cell death.

8.2 Cancer-Critical Genes

A **mutation** is a change in DNA sequence caused by external agents or by errors in internal processes such as DNA copying (**TABLE 8.3**). We discuss mutations in greater detail in Chapter 11. For now, keep in mind that genes typically code for proteins, and that a gene with a mutation is likely to produce an altered protein.

Most commonly, gene mutations reduce or eliminate the activity of the protein encoded by that gene, but if a mutation is severe enough, protein production might fail altogether. Such mutations are like a corrupted computer application: the code is garbled, so the application cannot run properly.

TABLE 8.3 Origins of Mutations in Cancer-Critical Genes

Inheritance

- Mutated genes inherited from one or both parents; such hereditary cancers account for less than 10% of human cancers

External mutagens (agents that cause DNA mutations)

- Chemical mutagens
- Ionizing radiation (UV, X-rays, other high-energy radiation)
- Viruses
- Chronic injury (physical, chemical)
- Lifestyle factors (e.g., diet and exercise)

Internal processes that generate mutations

- Genetic instability due to errors in DNA replication or repair; accidental chromosome breaks and rearrangements

Sometimes a gene mutation can alter the encoded protein in such a way that its activity is actually *increased*. These mutations can cause certain proteins to be made in larger-than-normal amounts, to be manufactured in cells that do not normally make them, or to be made at the wrong time. Mutations that put gene expression in "hyperdrive" are more like stealthware, unwanted applications that do things the original software was not designed to do.

Mutations in cancer-critical genes are the root cause of cancer

Genes that have been implicated in the development of cancer are dubbed *cancer-critical genes*. These genes fall into two main classes: *proto-oncogenes* and *tumor suppressor genes*.

Genes whose action results in cell proliferation are broadly classified as **proto-oncogenes**. All genes that code for positive growth regulators are proto-oncogenes. Mutations in these genes may trigger runaway cell proliferation. Proto-oncogenes are perfectly normal genes with essential roles in the body; it is only when they become *inappropriately* active, as a consequence of mutations, that they lead to excessive cell division and tumor development.

Proto-oncogenes that have become hyperactive as a result of DNA mutations are known as **oncogenes**. An oncogene is analogous to a stuck gas pedal: one hyperactive copy of the gene is all it takes to push the cell cycle forward in high gear, or to stimulate cell division in the wrong cell or at the wrong time.

Genes that protect against mutations and cancerous change are called **tumor suppressor genes**. All genes that code for negative growth regulators are tumor suppressor genes. The normal activity of these genes is to inhibit the cell cycle, stimulate repair of damaged DNA, promote cell adhesion, enforce anchorage dependence, prevent angiogenesis, or trigger cell suicide. Mutations in tumor suppressor genes can reduce or eliminate these activities, in which case the cell bearing the mutations becomes capable of uncontrolled division and possibly invasiveness as well.

As with most genes, a cell has two copies of each tumor suppressor gene. The failure of one copy can reduce the effectiveness of tumor suppression, but the loss of both copies is usually much more serious because it means total failure of control over cell proliferation. A tumor suppressor gene is analogous to the brakes in a car, and the loss of *both* copies of a tumor suppressor gene is like losing the main brake pedal and also the emergency brakes.

Most human cancers are not hereditary

Hereditary cancers are those that are linked to inherited gene mutations. A child receiving an oncogene or a mutated tumor suppressor gene from either parent has a hereditary risk of cancer. Inheriting a gene linked to cancer, however, does *not* mean that cancer is inevitable in that child's future. It simply means that he or she has an elevated risk of cancer compared to people who do not carry the faulty gene.

Only 1–5 percent of all human cancers can be traced to inherited gene defects. Some forms of breast cancer and colorectal cancer are among the handful of cancers that are based in heredity. Mutations in two tumor suppressor genes, *BRCA1* and *BRCA2*, are associated with increased risk of breast and ovarian cancer (see the "Biology in the News" feature on page 187).

Hereditary mutations are suspected in people who have a family history of a particular type of cancer. A woman who has a harmful mutation in the *BRCA1* gene, for example, is likely to have a number of close family members who have suffered from breast or ovarian cancers. Cancers caused by hereditary mutations tend to appear earlier in life than nonhereditary cancers. Genetic tests are available to screen individuals at risk for some types of hereditary cancer, including hereditary breast cancer and familial adenomatous polyposis, a hereditary condition that almost always leads to colon cancer.

EXTREME INHERITENCE

Siblings Beth and Michael inherited a cancer risk from their mother—one mutated copy of a tumor suppressor gene called *cdh1*. So did many other relatives, whose photos decorate the refrigerator. The mutation places them at very high risk of an otherwise rare type of stomach cancer. The siblings opted for surgery to completely remove the stomach.

Helpful to Know

The office of the surgeon general offers an online tool that lets you evaluate your health risks, taking into account your family history. Look up "My Family Health Portrait" on the Internet.

The risk of cancerous change increases with age

About 95 percent of the people diagnosed with cancer seem to have no inherited risk of the disease. That means their cancer was caused by a series of unfortunate changes in their DNA, resulting from environmental agents such as viruses or toxic chemicals, from unavoidable cellular accidents such as DNA copying or cell division errors, or from a combination of these two sources. Nonhereditary mutations are termed **somatic mutations**.

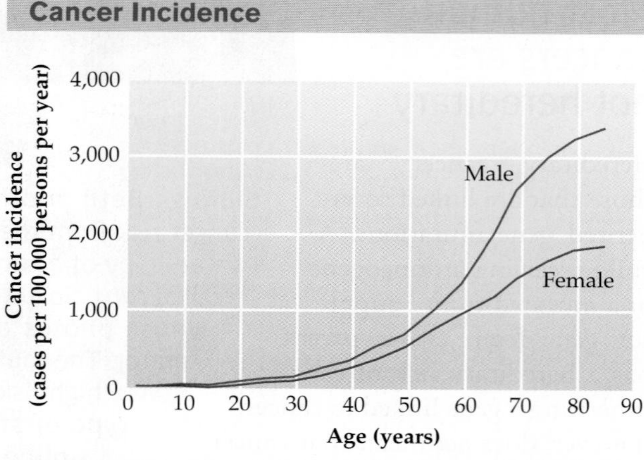

FIGURE 8.5 The Incidence of Cancer Rises Sharply as We Grow Older

The graph shows incidence rates for all cancers combined, by age group and gender. The incidence of a disease refers to the number of newly diagnosed cases per year for a defined unit of the population—in this case, per 100,000 men or women in each age group.

As we discuss in the next section, many if not all cancers are a result of multiple mutations accumulating within a single cell line. This observation explains why cancer is rare among the young, becomes more common as we get older, and steeply increases in frequency past middle age (**FIGURE 8.5**). The longer we live, the more opportunity there is for multiple well-behaved proto-oncogenes and tumor suppressor genes inside *one* cell to accumulate mutations. With every such mutation that accumulates, a good cell becomes steadily worse, until a full-fledged monster emerges in the form of a metastatic malignancy.

Concept Check

1. What is an oncogene, and how does it differ from a tumor suppressor gene?
2. Why is cancer more common in older people than in young people?

Concept Check Answers

1. Oncogenes are mutated versions of normal genes (proto-oncogenes) that promote cell proliferation. In contrast, tumor suppressor genes act to restrain cell division and migration.

2. For a cell to become cancerous, mutations must arise in multiple cancer-critical genes within it. Because it takes time to accumulate these multiple mutations, the odds of cancerous change increase with age.

8.3 The Progression to Cancer

More than one in three Americans will be diagnosed with cancer at some point in their lives. Given such a high incidence, you might think that the human body is exceptionally prone to cancer. Actually, we have robust defenses against cancer. A number of safeguards—cell adhesion and anchorage dependence, for example—reduce the likelihood of unchecked cell proliferation and tumor development, at least during the reproductive years. As we age, however, mutations start to accumulate in the genes that orchestrate our anticancer defenses, bringing us closer to the unlucky string of failures that result in a cancerous tumor.

Colon cancer develops from the accumulation of multiple mutations

In most cases of colon cancer, the tumor cells contain at least one oncogene and several completely inactive tumor suppressor genes. The mutations in these different genes that eventually lead to colon cancer usually occur over a period of years and often go hand in hand with stepwise changes in cell behavior that mark the progression toward cancer.

We can illustrate the step-by-step sequence of chance mutations, and the accompanying changes in cell activities, by following the disease progression that is characteristic of colon cancer. In most cases, the first cancer-promoting mutation results in a relatively harmless, or benign, growth described as a *polyp* (**FIGURE 8.6**). The cells that make up the polyp divide at an inappropriately rapid rate, and they are the descendants of a single cell in the lining of the colon that suffered one or more mutations, usually in a proto-oncogene.

If by chance a tumor suppressor gene fails in addition, cell division speeds up—like a runaway car with the gas pedal to the floor and brakes that have failed. The polyps grow faster than ever before, but even so, most of these polyps do not spread to other tissues and can be safely removed surgically at this stage. Colonoscopy, a microscopic examination of the lining of the colon, is advised for individuals over 50, earlier for those at high risk. A specialist can identify and remove polyps during the procedure.

The *p53* gene is the guardian angel of the cell

A common mutation in many cancers is the complete inactivation of an especially critical tumor suppressor gene, named *p53*, which has many roles in regulating cellular processes. In fact, the *p53* gene is so important that it is nicknamed the "guardian

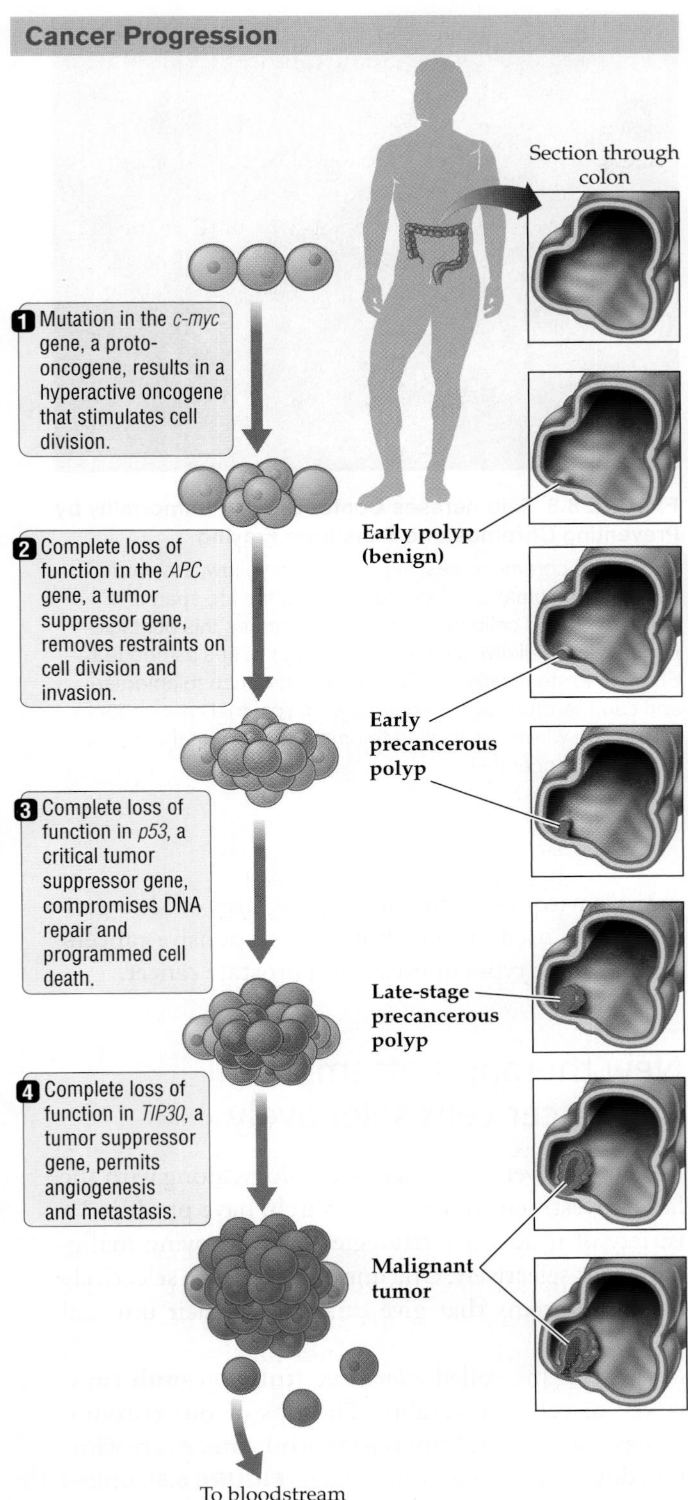

Cancer Progression
Section through colon
1 Mutation in the *c-myc* gene, a proto-oncogene, results in a hyperactive oncogene that stimulates cell division.
2 Complete loss of function in the *APC* gene, a tumor suppressor gene, removes restraints on cell division and invasion.
3 Complete loss of function in *p53*, a critical tumor suppressor gene, compromises DNA repair and programmed cell death.
4 Complete loss of function in *TIP30*, a tumor suppressor gene, permits angiogenesis and metastasis.
Early polyp (benign)
Early precancerous polyp
Late-stage precancerous polyp
Malignant tumor
To bloodstream

FIGURE 8.6 Development of Colon Cancer Is a Multistep Process

The step-by-step mutation of several genes that code for positive and negative growth regulators is responsible for the progression from a benign polyp in the colon to a malignant tumor. The order in which the proto-oncogenes and tumor suppressor genes mutate is not uniform; it can vary from one patient to another, which is why every cancer is unique.

angel of the cell." The p53 protein that this gene makes not only prevents the cell from dividing at inappropriate times, but also halts cell division when there is DNA damage in the cell. By halting the cell cycle, p53 gives the cell an opportunity to repair the damage to its DNA, keeping mutated DNA from being passed on to daughter cells. If the repair process fails—for instance, because the damage is beyond repair—p53 brings about programmed cell death. Given the important guardian functions of the p53 protein, it is not surprising that more than half of all human cancers show a complete loss of p53 activity in tumor cells. The number goes as high as 80 percent in some types of cancer, such as colon cancer.

As a polyp grows larger, the odds increase that one or more cells within this larger population of abnormal cells will acquire additional mutations. Mutations that knock out tumor suppressor genes that produce cell adhesion proteins, as well as the emergence of oncogenes that dissolve connections between cells, enable a cell to detach itself from its surroundings. Additional mutations that produce loss of anchorage dependence enable the detached cells to continue multiplying, paving the way for invasion of other tissues. The metastasis of colon cancer cells, typically to the liver, is the last and most destructive step in cancer progression.

Helpful to Know

Physicians evaluate the severity of many types of tumors on the basis of their size and spread. A stage o tumor is essentially a precancerous benign tumor. Stage I–III cancers are locally invasive, with stage II and stage III being more advanced and possibly having spread to the lymph nodes. Stage IV cancers are those that have metastasized to distant locations in the body.

Concept Check

1. Why do doctors recommend removal of benign polyps in the colon?
2. Why is the *p53* gene so crucial in the progression to cancer?

Concept Check Answers

1. A benign polyp is a group of cells carrying one or more mutations that cause excessive cell proliferation. Any additional mutations to cancer-critical genes in even one of these rapidly dividing cells could produce a cancerous cell.

2. he *p53* gene is a tumor suppressor gene responsible for halting cell division if there is DNA damage, promoting DNA repair and triggering cell death when the damage to DNA is irreparable. When both copies of the *p53* gene are nonfunctional, cells have high odds of becoming cancerous.

8.4 Treatment and Prevention

About 40 years ago, President Richard Nixon declared a war on cancer in the United States by making anticancer research a high priority. Since then, some major victories have been won, thanks to improvements in radiation and drug therapies. Whereas in the early twentieth century very few individuals survived cancer, today about 68 percent of patients are alive 5 years after treatment has begun. The 5-year survivability of some common cancers—breast and colon cancer, for example—exceeds 75 percent. Nevertheless, the war against cancer is far from over, and the need for powerful new treatments to stop or kill the malignant cells is as urgent as ever.

Conventional cancer therapies attack rapidly dividing cells

The greatest challenge in battling cancer is to selectively destroy rogue cells while sparing healthy cells. The standard plan of attack today relies on high-energy radiation (**radiation therapy**), high doses of chemical poisons (**chemotherapy**), or both in sequence to kill any and all rapidly dividing cells.

The side effects of radiation therapy and chemotherapy are terrible because this all-out assault also kills many innocent bystanders, cells necessary for the normal functions of the human body. Alopecia (hair loss) is the most visible of the many side effects of these therapies (**FIGURE 8.7**). The cells that grow and maintain hair are destroyed, but *dormant* stem cells survive the treatment, and they regrow the hair in about 3 months. Cells that divide to produce red blood cells may be killed, leading to the fatigue and weakness of anemia. The cells that line the entire digestive tract are regularly replaced by cell division, and the loss of that cell supply produces symptoms in the mouth, stomach, and intestines.

Cryosurgery makes use of extremely cold temperatures to kill abnormal cells, usually precancerous cells or cancers that are confined to a small region, such as the cervix. **Hormone therapy**, which changes the hormone environment in the body to stop or slow cancer cells, is used in some hormone-responsive cancers, such as some types of breast and prostate cancer.

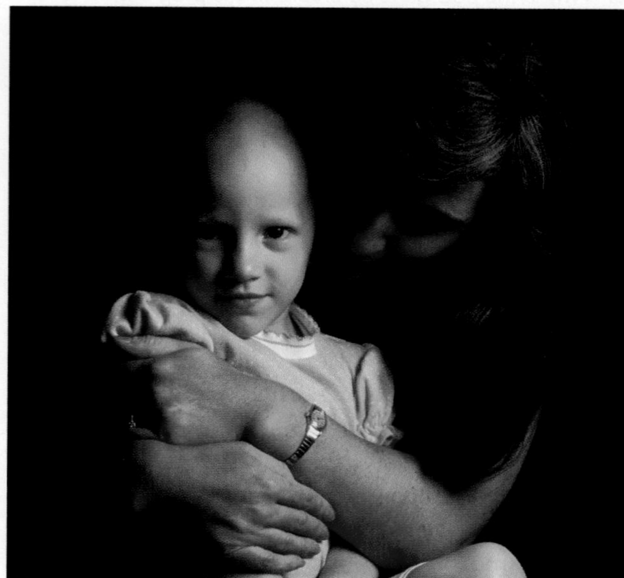

FIGURE 8.7 Hair Loss Is a Common Side Effect of Chemotherapy Chemotherapy attacks any and all rapidly dividing cells and results in collateral damage, including loss of cells that grow and maintain hair follicles. Dormant stem cells survive the treatment, and they will regrow the lost hair in about 3 months.

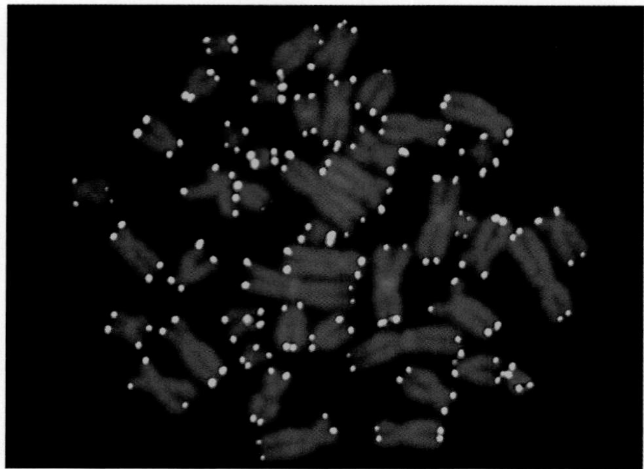

FIGURE 8.8 Telomerases Confer Cellular Immortality by Preventing Chromosome Ends from Fraying The tips of chromosomes, here shown in yellow, are eroded each time a somatic cell divides, limiting the life span of a cell. An enzyme called telomerase overcomes this obstacle to perpetual cell division by lengthening the tips as needed. Embryonic stem cells and cancer cells produce telomerase and can therefore divide endlessly. Clinical trials are under way to see whether crippling telomerase will limit the runaway growth of cancer cells.

New therapies attempt to kill cancer cells selectively

Recent discoveries in basic cell biology, along with the large investment in cancer research, have produced a variety of innovative strategies for destroying malignant cells selectively. One line of attack is to selectively disable proteins that give cancer cells their unusual immortality.

An enzyme called *telomerase* (**tuh-*LOH*-muh-rays**) is key to cell immortality. The tips of our chromosomes, known as **telomeres** (***TEH*-luh-meer**), are whittled down each time a cell divides (**FIGURE 8.8**), unless telomerase steps in to repair the ends. The few cells in our body that produce telomerase include stem cells, germ line cells (those that produce gametes), and cancer cells. Researchers have identified chemicals that knock out telomerase activity, and these are being tested in clinical trials in the hopes that wearing out the chromosome ends in cancer cells will put a stop to their proliferation.

Antibodies are powerful weapons in targeted therapy

One of the most promising new ways of fighting cancer is to use components of the body's own defense system, or immune system, in a strategy known as **cancer immunotherapy**. Genetically designed and targeted antibodies are the warhorses of cancer immunotherapy. An *antibody* is a protein that latches on tightly to highly specific targets, such as proteins that only cancer cells display on their surface.

Scientists have designed antibodies that stick to cancer cells while leaving healthy cells alone. The "designer antibodies" disable or kill cancer cells by binding tightly to cell surface proteins, such as abnormal hormone receptors in the plasma membrane (**FIGURE 8.9**). For example, trastuzumab is an antibody that homes in on proteins found on certain types of breast cancer and stomach cancer cells. The antibodies coat the cell surface, essentially smothering the signal transduction pathways of the rogue cells.

Other immunotherapies focus on priming the patient's own immune system to attack cancer cells. The binding of the therapeutic antibody to cancer cells "invites" the patient's immune cells to destroy the rogue cell (see Figure 8.9*d*).

Some therapeutic antibodies are designed to deliver chemotherapy agents or radioactive particles in a precisely targeted manner. Like a drone aircraft armed with explosives, the antibody seeks out its specific target—cancer cells—and then discharges its toxic payload.

As noted earlier, cancer cells commonly secrete chemicals that trigger angiogenesis, the growth of new blood vessels in their vicinity. One experimental approach against cancer attempts to block angiogenesis. For example, an antibody-based pharmaceutical called Avastin mops up the angiogenesis-recruiting chemicals that cancer cells release; this drug has been approved for treatment of colorectal cancer.

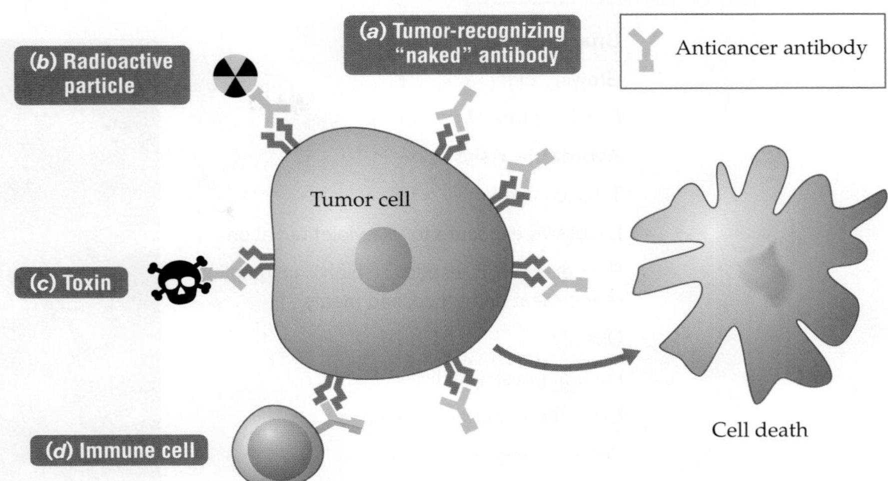

FIGURE 8.9 Cancer Immunotherapy Uses Targeted Antibodies to Destroy Cancer Cells Selectively

(*a*) Some therapeutic antibodies ("naked" antibodies) attack cancer cells by binding to cell surface proteins that cancer cells need to divide and/or stay alive; these proteins are absent or found in low amounts on healthy cells. Other targeted antibodies carry toxic cargo: either radioactive particles (*b*) or chemotherapy drugs such as taxol (*c*). Some anticancer antibodies bind to cancer cells specifically and then recruit immune cells to deliver the killing blow (*d*).

Avoiding risk factors is the key to cancer prevention

The biggest lesson learned from the past three decades of cancer research is a surprisingly simple one: instead of dealing with cancers only when they become monsters, we should try to reduce the odds that our cells will progress toward cancer in the first place. It is now abundantly clear that environmental factors, including lifestyle choices, can have a very large impact on a person's risk of developing cancer.

TABLE 8.4 lists some of the factors that are known to increase risk of cancer. Only two of the items listed—the genes we inherit and the inevitability of getting older—are factors that we cannot control. Everything else on the list, which is by no means exhaustive, is an environmental influence that we should try to avoid if we want to lower our risk.

A **carcinogen** (**kahr-*SIN*-uh-jen**) is any physical, chemical, or biological agent that elevates the risk of cancer. Asbestos, implicated in an otherwise rare cancer called mesothelioma, is an example of a physical carcinogen. The long, fine crystals of the mineral, once widely used for fireproofing and in electrical insulation, enter the cells that line the lungs and damage the cells' organelles. Hundreds of chemicals are known or suspected carcinogens in humans (see page 185).

Viruses cause some human cancers

Cancer experts believe that bacteria and viruses contribute to about 15–20 percent of all cancer cases in humans. Cancers caused by infectious agents are more common in developing nations than in developed countries. Infection with *Helicobacter pylori* (***HELL*-ih-koh-*BAK*-ter pye-*LOH*-ree**), a bacterium that thrives in the acidic environment of the stomach, is

TABLE 8.4 Common Risk Factors for Cancer

Unavoidable risks

Growing older

Family history of cancer

Avoidable risks

Tobacco use

Excessive exposure to ultraviolet radiation

Poor diet quality, especially high consumption of red meat and processed meats

Obesity

Lack of physical activity

Exposure to certain hormones

Some viruses and bacteria

Excessive alcohol consumption

Chemical carcinogens

Ionizing radiation (such as X-rays)

Skin cancers are the most common cancers in the United States. The use of tanning booths is strongly associated with elevated risk of all types of skin cancer, including the most dangerous form, malignant melanoma.

linked with increased odds of some types of stomach and intestinal cancers. Long-term infection with the hepatitis B or hepatitis C viruses increases the risk of liver cancer, which, along with stomach cancer, is common in poor parts of the world but relatively rare in the United States.

Some of the more than 100 different types of human papillomavirus (HPV) produce proteins that disable tumor suppressor genes in the cells that the virus infects. HPV infection is the main cause of cervical cancer, and HPV is still detectable in the HeLa cell line established from Henrietta Lacks's cervical cancer cells. The Pap smear is a screening test that looks for abnormal cells on the cervix, the lower end of the uterus that is exposed in the vagina. Public education and widespread screening in the United States have led to a large decline in death from cervical cancer.

A vaccine is now available that prevents infection by all the common strains of HPV. Best known by its commercial name, Gardasil, this vaccine reduces the risk of cervical, penile, and anal cancers, as well as genital warts. The U.S. Food and Drug Administration (FDA) recommends Gardasil, given as three injections over 6 months, for all persons 9–26 years old. For maximum effectiveness, the vaccination must be administered before a person becomes infected with the virus, which is why the FDA recommends vaccination even for children who are not likely to be sexually active (**FIGURE 8.10**).

EXTREME CARCINOGEN

Tobacco smoke is known to contain at least 40 different types of carcinogens. The model shows one of the most potent forms—benzopyrene or BP (yellow ring)—covalently bonded to a DNA molecule. The longer a person smokes, the more BP clumps to their DNA. Like most carcinogens, BP causes mutations by corrupting the coding information in DNA.

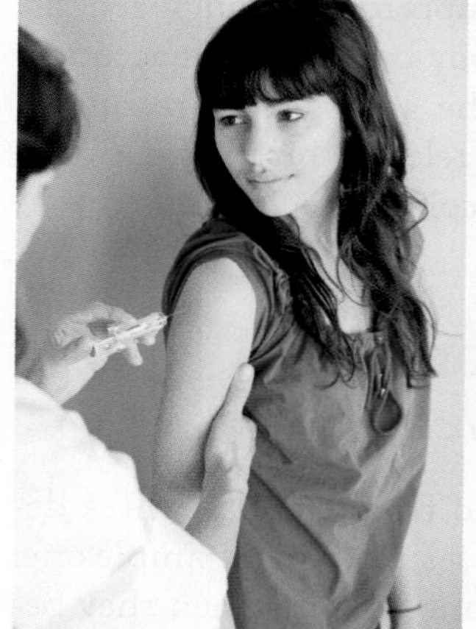

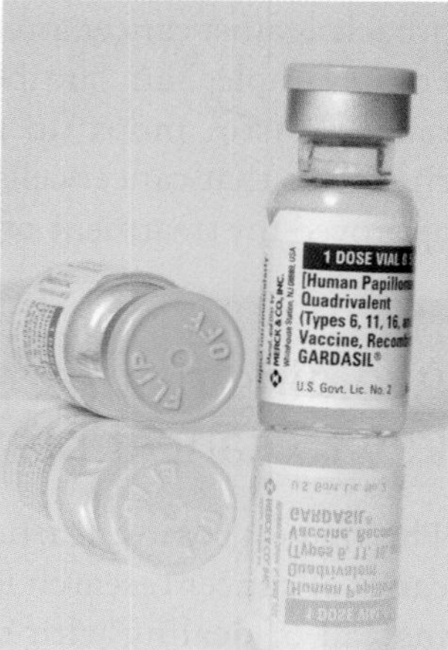

FIGURE 8.10 Preventing Cancers Linked to Human Papillomavirus (HPV)

If vaccinated before exposure, young people, both girls and boys, gain protection against cancers and other diseases caused by certain strains of HPV.

BIOLOGYMATTERS

Avoiding Cancer by Avoiding Chemical Carcinogens

Although chemical pollutants are often the most feared of carcinogens in the public mind, experts estimate that only about 2 percent of human cancers can be blamed on environmental carcinogens, whereas what we eat and drink accounts for more than 30 percent of our cancer risk. Tobacco use is responsible for about 30 percent of the cancers diagnosed in the United States.

Those who eat a lot of animal products have higher rates of some cancers than those who consume less or none. Some carcinogenic pollutants become concentrated along the food chain and are therefore found in higher concentrations in meat, fish, and dairy than in plant foods. The saturated fat in meat and dairy is an independent risk factor for cancers of the breast, prostate, and colon. Animal products also lack cancer-protective substances—such as fiber and antioxidants—that presumably help those who eat a mostly plant-based diet.

When animal flesh is cooked at high temperatures, as in grilling, broiling, or deep frying, certain amino acids that are abundant in meat, poultry, and fish are converted into a family of carcinogenic compounds known as heterocyclic amines (HCAs). High-temperature cooking of fatty foods creates yet another class of potent carcinogens, called polycyclic (*PAH*-**lee**-*SYE*-**klik**) aromatic hydrocarbons (PAHs). If you like your meat well-done, you might consider cutting back or at least resorting to safer cooking methods much of the time. Microwaving food lightly before high-temperature cooking reduces HCA and PAH formation. So does marinating in antioxidant-rich sauces—think berries, red wine, and herbs like rosemary.

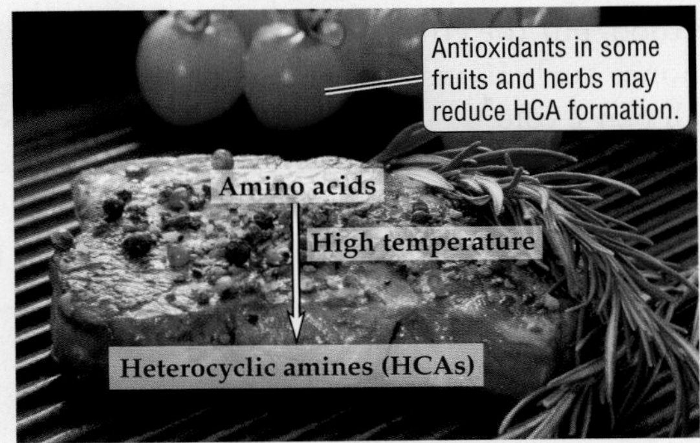

Antioxidants in some fruits, and herbs such as rosemary, may reduce HCA formation in grilled meats.

Processed meats are even worse for cancer risk than charbroiled red meat. Nitrites are commonly used to preserve the color and fresh appearance of hot dogs, sausages, bologna, cold cuts, and other processed meats. In acidic environments, such as the stomach, the nitrites react with proteins to form carcinogenic compounds called nitrosamines. Vitamin C (ascorbic acid) is often added to processed meats because this antioxidant dampens the conversion of nitrites to nitrosamines.

The link between cancer and smoking is the most dramatic illustration of how chemical exposure can transform healthy cells into dangerous ones. Lung cancer was a rare cancer just before the turn of the twentieth century, when few people smoked tobacco. Now, with nearly a third of the world's population lighting up, lung cancer is the most common and deadliest cancer worldwide, killing more than 1.2 million people annually. Both tobacco and marijuana cigarettes contain a type of PAH called benzopyrene, a powerful cancer-causing agent. Marijuana is less addictive, and those who smoke it typically consume fewer cigarettes per day and inhale less deeply. The bottom line, however, is that both types of smoke are dangerous.

Physiology and diet affect cancer risk

In addition to carcinogens, the physiological state of the body can influence our susceptibility to cancer. Exposure to hormones and growth factors affects cancer risk because these signaling molecules have such a profound influence on cell proliferation. Taller people have a slightly higher risk of some types of cancer, probably because they have higher levels of certain growth factors (in particular, insulin-like growth factor), compared to people of medium or short stature. (We hasten to add that the increased risk is small and not something that tall people should worry about.) Studies show that the more estrogen a woman is exposed to, the higher are her odds of developing breast cancer, and the risk is especially high for certain individuals and ethnic groups whose genetic profile makes them particularly vulnerable. High testosterone levels are similarly associated with prostate cancer risk in men.

Working night shifts for many decades (30 or more years) is strongly linked to an elevated risk for breast cancer. (Researchers did not find a correlation in people who did shorter stints of night work—say, on and off for 10 years.) The underlying biology is not well understood, but researchers suspect that the sleep hormone, melatonin, is involved. Melatonin levels start rising shortly before bedtime, and the hormone restores the body in many ways as we sleep. Light, especially light that is rich in blue wavelengths, inhibits melatonin production.

Obesity is strongly linked with some types of cancer, including breast, prostate, and colorectal cancers. Fat cells are known to be a potent source of hormones and growth factors, which may explain the association. Lack of physical activity is independently associated with an increase in cancer risk, meaning that even normal-weight people who are couch potatoes are at greater risk for some types of cancer, including cancers of the breast, uterus, prostate, and lung, than are people who exercise regularly. How physical activity affects tumor development and progression is not understood, but the effect of exercise on metabolism, hormones such as insulin, and the immune system may all play a role.

Diet is believed to have a substantial impact on cancer risk (see the "Biology Matters" box on page 185). Populations that eat a mostly plant-based diet—rich in whole grains, vegetables, and fruits—have lower rates of some types of cancer, including cancer of the colon, esophagus, pancreas, and kidney, than does the average American. Within a single generation, migrants from these populations who adopt a typical American diet acquire risks similar to that of the average American.

Concept Check

1. Why does chemotherapy for cancer cause alopecia (hair loss)?
2. What is the purpose of Pap smears?

Concept Check Answers

1. The cancer-fighting chemicals kill almost all rapidly growing cells, including the hair follicle cells that must divide regularly to grow hair on the scalp and elsewhere on the body.
2. Pap smears screen for precancerous or cancerous cells on the cervix.

APPLYING WHAT WE LEARNED

How HeLa Cells Changed Biomedicine

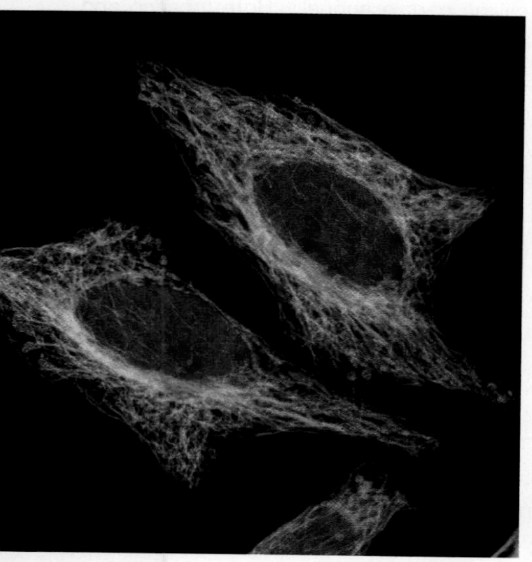

When Henrietta Lacks's tumor cells arrived in the Gey lab in early 1951, a young technician dropped them into test tubes on top of clots of chicken blood and put them in an incubator to stay warm. She didn't expect much, since all the cells she did this with seemed to die sooner or later, usually sooner.

But the Gey lab was in for a surprise. Lacks's cells had an astounding ability to divide and thrive in laboratory conditions. In just 24 hours, the cells had doubled in number. A day later, their numbers had doubled again. By day 4, the cells were outgrowing their new test-tube homes, and the lab tech had to rustle up some more test tubes for them.

It was the first time anyone had successfully cultured human cells. The Geys happily celebrated their success, and other researchers were soon demanding "HeLa cells" for their experiments. The cells have continued to grow and divide in test tubes and glass dishes for another 60 years—essentially forever, in biomedical terms.

HeLa cells are anything but normal. Henrietta Lacks had been exposed to human papillomavirus (HPV), a sexually transmitted virus that transfers its viral DNA into human cells. The result was a line of cells with DNA that was different from both the virus and the original human cells. For example, normal human somatic cells have 46 chromosomes, while HeLa cells most commonly contain about 92 chromosomes, with some chromosomes present in three or four copies. And unlike normal somatic cells, HeLa cells make the enzyme telomerase, which repairs the chromosomes' telomeres, preventing them from shortening during cell division (see Figure 8.8). Because the telomeres of HeLa cells do not shorten, HeLa cells are immortal; they can theoretically grow and divide forever.

Since 1951, HeLa cells have been used in thousands of experiments, leading to more than 60,000 scientific papers by 2013. Henrietta Lacks's cells have contributed to the rise of biomedical companies worth billions of dollars. A single anticancer drug developed using HeLa cells had revenues of more than $300 million in 2007. HeLa cells have been used to show that HPV infection can cause cervical cancer and to study other forms of cancer. HeLa cells also power the search for cures: they were used to test and develop trastuzumab (Herceptin), one of the most promising antibody-based therapies against breast cancer.

BIOLOGY IN THE NEWS

Cancer Gene Has Led Jolie and Others to Surgery

BY REX SANTUS • *Columbus Dispatch*, MAY 15, 2013

Actress Angelina Jolie is being praised by doctors and cancer advocates for disclosing yesterday that she had both breasts removed after learning she carries a gene mutation that increases the risk of cancer.

It's a choice thousands of women have had to make.

"When you have this mutation, it's basically a ticking time bomb," said Dr. Julian Kim, chief medical officer of surgical oncology at the Seidman Cancer Center at University Hospitals in Cleveland ...

But who should be tested? And if you carry the gene, is preventive surgery the only choice? The test itself can cost $3,000 to $4,000, and the surgery can run tens of thousands of dollars. Depending on your insurance, you could pay a lot out of pocket.

"My doctors estimated that I had an 87 percent risk of breast cancer and a 50 percent risk of ovarian cancer," Jolie, whose mother died at 56 after a decade-long fight with cancer, wrote. "My chances of developing breast cancer have dropped from 87 percent to under 5 percent."

She wrote that she also had reconstructive surgery.

Jolie wrote that she carries [a mutation in] the BRCA1 gene. Kim said carriers typically face an 85 percent chance of developing breast cancer.

"That's a very high risk," Kim said. "It's not uncommon for people (with the gene) to opt for prophylactic surgery."

The gene itself is not all that typical, however. About 5 to 10 percent of all breast-cancer patients have the mutation, Kim said. Experts estimate that less than 1 percent of all women inherit a BRCA mutation ...

Nationwide, more than 230,000 breast-cancer cases will be diagnosed this year, and nearly 40,000 women will die of the disease, according to the National Cancer Institute.

Only 5–10 percent of the breast cancers diagnosed in North America are hereditary in nature. As discussed in this chapter, hereditary cancers are triggered by one or more inherited gene mutations. *BRCA1* and *BRCA2* (BReast CAncer genes 1 and 2), first identified by Dr. Mary-Claire King in the 1990s, are the best known of the genes that have been implicated in hereditary breast cancer. Both are tumor suppressor genes that play an important part in repairing damaged DNA in certain tissues. Mutations that prevent these genes from working properly lead to yet more mutations, and the accumulation of these mutations in multiple genes makes the progression to cancer more likely.

In the general population, mutations in the *BRCA1/2* genes are rare. About one in 600 people in North America have a mutation in the *BRCA1* or *BRCA2* gene. But the rate is more than 10 times higher in some ethnic groups, especially Jews from eastern Europe and residents of the Bahamas. The mutated gene can be inherited from the mother's side, as in Jolie's case, or from the father's side. Women with the mutated gene have increased risk of breast cancer and ovarian cancer. Men, too, have an elevated risk for breast cancer if they carry the mutation: 65 in 1,000 men are likely to develop the disease, compared to 1 in 1,000 men who are noncarriers. The risk of prostate cancer is also higher in men who are carriers.

The genes were isolated and sequenced in 1994 by government and university researchers and a start-up company called Myriad Genetics. The company patented both genes and developed a test that detects mutations in the gene. That diagnostic test was how Angelina Jolie found out that she had a mutation in the *BRCA1* gene.

The high cost of the genetic test for *BRCA1/2* has been blamed on the monopoly that Myriad Genetics exercises over it, which led to a legal challenge. In 2013, the U.S. Supreme Court struck down the patent on *BRCA1* and *BRCA2*, ruling that "a naturally occurring DNA segment is a product of nature and not patent-eligible merely because it has been isolated." The patent on the genetic test, however, stands. The Affordable Care Act passed by Congress in 2010 requires health insurers to cover genetic tests ordered by a patient's physician.

Evaluating the News

1. Is the normal version of the *BRCA1* gene an example of a proto-oncogene, an oncogene, or a tumor suppressor gene? Explain why a faulty *BRCA1* gene was a threat to Angelina Jolie's life.

2. Who do you think should undergo testing for the *BRCA1/BRCA2* genes? Should health insurers be required to pay for the tests? Research the alternatives to surgery that might be available to women who test positive for the faulty version of the gene.

3. Angelina Jolie wrote an op-ed in the *New York Times* in which she explained her decision to get a double mastectomy (later she opted for removal of her ovaries as well). Are such highly publicized "celebrity diagnoses" helpful? Explain your viewpoint.

CHAPTER REVIEW

Summary

8.1 Cancer: Good Cells Gone Bad

- Over a lifetime, an American male has a nearly one in two chance, and an American woman a one in three chance, of developing cancer. There are more than 200 different types of cancer, but four of them—lung, prostate, breast, and colon cancers—account for more than half of all cancers combined.
- Cancer develops when cells lose normal restraints on cell division and migration.
- The cell mass formed by the inappropriate proliferation of cells is known as a tumor. Tumors grow larger when they recruit blood vessels through angiogenesis.
- Precancerous cells become cancerous (malignant) when they invade other tissues. Invasive cells detach themselves from their surroundings and lose anchorage dependence. Cancer cells can spread from a primary tumor to establish secondary tumors in other organs, in the process known as metastasis.
- Positive growth regulators stimulate cell division; negative growth regulators restrain it. Runaway cell proliferation is the consequence if the cell cycle is excessively stimulated through the pathways controlled by positive growth regulators while the antiproliferation message of negative growth regulators is ignored.

8.2 Cancer-Critical Genes

- Gene mutations are the root cause of all cancers. Most human cancers are not hereditary. Most cancers develop as multiple mutations accumulate in a single cell.
- Genes whose action results in cell proliferation are proto-oncogenes. Proto-oncogenes that have become hyperactive as a result of DNA mutations are known as oncogenes.
- Genes that hold cancerous change in check—by restraining cell division or cell migration, for example—are called tumor suppressor genes.
- The risk of cancer increases with age because as time goes by, the odds that multiple mutations will accumulate in a single cell increases.

8.3 The Progression to Cancer

- It takes mutations in multiple proto-oncogenes and tumor suppressor genes within a single cell for that cell to become cancerous.
- A cell progresses toward cancer—such as colon cancer—as a succession of proto-oncogenes become hyperactive and tumor suppressor genes fail.
- The *p53* gene is especially critical because it restrains cancerous change in several ways. When fully functional, it detects damage to DNA, triggers repair, halts the cell cycle until the repair is complete, and triggers cell death if the damage is irreparable.

8.4 Treatment and Prevention

- The traditional cancer treatments—radiation therapy and chemotherapy—attempt to kill all rapidly dividing cells. The challenge is to develop treatments that destroy malignant cells selectively, leaving healthy cells alone.
- One of the new therapies, now in clinical trials, attempts to slow runaway cell division by blocking the telomerase activity that cancer cells possess.
- Antibodies that seek out and selectively destroy cancer cells are being used against some types of cancer.
- About 5 percent of cancers are hereditary, caused at least in part by inherited mutations in one or more cancer-critical genes.
- A carcinogen is any physical, chemical, or biological agent that elevates the risk of cancer.
- Infectious agents such as viruses can trigger cancer, but in developed countries they account for only about 10 percent of diagnosed cancers.
- Avoiding tobacco, eating well, and exercising regularly are believed to reduce our risk for some types of cancer.

Key Terms

anchorage dependence (p. 175)
benign tumor (p. 175)
cancer cell (p. 176)
cancer immunotherapy (p. 183)
carcinogen (p. 183)
cell differentiation (p. 174)
chemotherapy (p. 182)
cryosurgery (p. 182)
DNA (p. 174)
gene (p. 174)
gene expression (p. 174)
hormone therapy (p. 182)
metastasis (p. 177)
mutation (p. 178)
negative growth regulator (p. 177)
oncogene (p. 179)
positive growth regulator (p. 177)
primary tumor (p. 176)
proto-oncogene (p. 179)
radiation therapy (p. 182)
receptor (p. 177)
secondary tumor (p. 176)
signal transduction pathway (p. 177)
somatic mutation (p. 179)
telomere (p. 182)
tumor (p. 175)
tumor suppressor gene (p. 179)

Self-Quiz

1. Which of the following is an example of a negative growth regulator?
   a. human growth hormone, which stimulates cell division in bone
   b. human growth hormone receptor, which is part of the signal transduction pathway for the hormone
   c. prostate-specific antigen, which is a tumor marker found on the surface of prostate cancer cells
   d. transforming growth factor beta (TGF-β), which blocks the cell cycle when there is DNA damage

2. Growth factors
   a. can only stimulate cell division, not restrain it.
   b. are signaling molecules that affect cell division.
   c. function as positive growth regulators, but not as negative growth regulators.
   d. function as negative growth regulators, but not as positive growth regulators.

3. An example of an oncogene is a gene that codes for a protein that
   a. inhibits angiogenesis.
   b. promotes anchorage dependence.
   c. pushes the cell cycle from G_1 to the S phase.
   d. triggers programmed cell death.

4. Why is cancer more common in older people than in younger people?
   a. It takes many cells to produce a cancer, and older people have more cells.
   b. The DNA of older people is more susceptible to carcinogens.
   c. Our immune system becomes stronger as we age.
   d. Multiple genes must pick up harmful mutations for cancer to develop, and that takes time.

5. Which of the following is the last step in the development of a cancer?
   a. anchorage independence
   b. loss of cell adhesion
   c. metastasis
   d. cell proliferation

6. The majority of human cancers
   a. are hereditary (caused by the inheritance of mutated genes).
   b. are the result of infection with oncogenic viruses.
   c. arise from multiple somatic mutations accumulating in a single cell.
   d. are caused by the normal activity of proto-oncogenes.

7. Tumor suppressor genes
   a. normally increase the rate of cell proliferation.
   b. normally promote cell migration.
   c. include those encoding enzymes that degrade cell adhesion proteins.
   d. include those that encode DNA repair proteins.

8. Of the following scenarios, which is *most* likely to increase the odds that a cancer will develop?
   a. One copy of a proto-oncogene is mutated such that it becomes nonfunctional.
   b. One copy of a tumor suppressor gene is mutated such that it becomes nonfunctional.
   c. One copy of an oncogene is hyperactive.
   d. Two copies of a tumor suppressor gene are hyperactive.

9. The severe side effects of chemotherapy against cancer cells come about because
   a. there is indiscriminate acceleration of the cell cycle in cells that would normally not divide because they are fully differentiated.
   b. the therapy is narrowly targeted against the biomolecule, such as a growth factor receptor, found to be responsible for the cancerous change.
   c. all rapidly dividing cells in the body are killed by the chemotherapy.
   d. all stem cells in the body are killed by the chemotherapy.

Analysis and Application

1. Describe how the interplay of positive and negative growth regulators controls cell proliferation.

2. What is angiogenesis? Explain why it is critical for tumor growth.

3. The *cdh1* gene codes for cadherin, a protein that helps anchor cells to their surroundings. Is this gene an example of a proto-oncogene, an oncogene, or a tumor suppressor gene? Explain how a mutation in this gene can increase the risk of cancer.

4. Colon cancer develops in a series of stages. Outline the stages and what happens in each stage.

5. List two unavoidable and two avoidable factors that make a person more susceptible to cancer. Evaluate your own risk using Table 8.4. How can you reduce those risks?

6. In light of the clear link between tobacco use and cancer, many have questioned the right of tobacco companies to continue selling such a deadly substance. Consider the issues of personal freedom versus public health policy and explain what restraints, if any, you think should be placed on the sale of tobacco.

7. In 2010 the federal government imposed a “tanning tax” on tanning salons in an attempt to discourage their use. Studies show there has been a decline in the use of tanning beds since then. Do you think such attempts to reduce the incidence of cancer are justified? Explain your viewpoint.

8. As environmental causes of cancer receive increasing attention, the warning labels on food have become longer and more ominous-sounding. Since many factors contribute to cancer, do you think that expanding food warning labels is an effective approach to reducing cancer risk? If so, how might one combat the public’s tendency to ignore long and complex warning labels?

9 Patterns of Inheritance

A ROYAL MYSTERY. Czar Nicholas II, seen here with the czarina and their five children, was the last czar of Russia. His youngest daughter, Princess Anastasia, is on the right in this 1910 photo.

The Lost Princess

In the early hours of July 17, 1918, Nicholas II (the last czar of Russia), his family, and his servants were awakened and directed to the ground floor of a mansion where they were being held by Bolshevik secret police. The czar had abdicated his throne the previous year, and the family had been captured and moved to a succession of hideaways. Thinking the family was about to be moved again and expecting a wait, the czar called for chairs for Czarina Alexandra, his son Alexei, and himself. Then, with his daughters—Olga, Tatiana, Maria, and Anastasia—and servants assembled, the czar turned his attention to Bolshevik officer Yakov Yurovsky. But Yurovsky abruptly announced that they were all to be executed, and seconds later soldiers opened fire, bringing a brutal end to the 300-year-old Romanov dynasty.

Or had it ended? Two years later, a young woman with scars on her head and body was admitted to a Berlin mental institution. She seemed not to know her own name and became known as Anna Anderson. A fellow patient with social connections insisted that Anna was the czar's youngest daughter, Anastasia, who had miraculously escaped execution. Hopeful friends and relatives of the czar visited Anna but sadly concluded she was not Anastasia; she could not even speak Russian. In 1927, a private detective hired by Anastasia's uncle reported that Anna was Franziska Schanzkowska, a Polish factory worker injured when a grenade exploded in a weapons factory.

What rules govern the inheritance of traits? How was genetics used to show whether Anna Anderson was Anastasia Romanov?

But the captivating legend of the Russian princess who escaped death had a life of its own, celebrated in books, magazines, and movies for more than 60 years. Minor European royalty insisted that they recognized Anna, and for the rest of her life, a succession of supporters claimed her as the lost princess Anastasia. Only a combination of detective work and genetics finally solved the mystery of Anna's background. To answer the questions raised here and to explore other aspects of inheritance, let's take a look at the principles of genetics, the focus of this chapter.

MAIN MESSAGE Inherited characteristics of organisms are governed by genes and may be influenced by environmental factors as well.

KEY CONCEPTS

- Genetics is the study of inherited characteristics (genetic traits) and the genes that affect those traits.
- A gene is a stretch of DNA that affects one or more genetic traits. It is the basic unit of inheritance.
- A phenotype is the specific version of a genetic trait that is displayed by a particular individual.
- Alternative versions of a gene (alleles) arise by mutation. The genotype is the allelic makeup of an individual.
- Diploid cells have two copies of every gene: one copy inherited from the male parent; the other, from the female parent. Homozygotes have the same two alleles for a particular gene; heterozygotes have two different alleles.
- A dominant allele controls the phenotype in a heterozygote. The recessive allele is masked.
- Mendel's laws of inheritance help us predict the phenotypes of offspring from the known genotypes of the parents.
- Mendel's laws apply broadly to most sexually reproducing organisms, but geneticists have extended those basic laws to describe more complex patterns of inheritance.
- Many aspects of an organism's phenotype are determined by multiple genes that interact with one another and with the environment, so offspring with identical genotypes can have very different phenotypes.

THE LOST PRINCESS 191

9.1 Principles of Genetics: An Overview 192

9.2 Basic Patterns of Inheritance 197

9.3 Mendel's Laws of Inheritance 199

9.4 Extensions of Mendel's Laws 203

APPLYING WHAT WE LEARNED 211

Solving The Mystery Of The Lost Princess

HUMANS HAVE USED THE PRINCIPLES of inheritance for thousands of years. Noticing that offspring tend to be similar to their parents, people raised animals and plants by mating individuals with desirable characteristics and selecting the "pick of the litter" for further breeding. Our ancestors used such methods to domesticate wild animals (**FIGURE 9.1**) and develop agricultural crops from wild plants (see Figure 1.16).

As a field of science, however, *genetics* did not begin until 1866, the year that an Augustinian monk named Gregor Mendel (**FIGURE 9.2**) published a landmark paper on inheritance in pea plants. Prior to Mendel's work, some aspects of inheritance were understood, but no one had conducted extensive and systematic experiments to explain the patterns in which inherited characteristics, or *genetic traits*, are passed from parent to offspring.

In the more than 100 years since Mendel's work, we have learned a great deal. We now know that the traits we observe in an organism are controlled by **genes**, stretches of DNA located on chromosomes, as described in Chapter 7. Genes commonly contain instructions for the manufacture of proteins, and to a very large extent, these proteins shape the individual organism. In other words, the traits we display are brought about mostly by proteins, which in turn are encoded by genes.

FIGURE 9.2 Gregor Mendel (1822–1884)

Today, Mendel's principles have been extended to reveal, in much greater detail, how genes shape the observable characteristics of organisms and how environment influences these traits. In this chapter we explain Mendel's discoveries and explore modifications and additions to his findings that came about as geneticists learned more. We begin with a broad overview of modern genetics, saving the details for the sections that follow.

FIGURE 9.1 Artificial Selection from Wolf Ancestors

You would not guess it from looks, but comparison with wolf DNA shows that the Shar-Pei is one of the oldest dog breeds still in existence. Bred as fighting dogs in China, Shar-Peis have a bristly coat and wrinkles that make them difficult to hold on to.

9.1 Principles of Genetics: An Overview

Mendel's work was largely ignored for about 30 years after it was published. Upon its "rediscovery" in the early 1900s, his principles became the foundation for the modern discipline of **genetics**, the study of genes.

Genes determine traits

A **genetic trait** is any inherited characteristic of an organism that can be observed or detected in some manner. Genetic traits are controlled at least in part by genes, although the outward manifestation of a genetic trait can also be influenced by the environment

Genetic Traits in Humans

GENETIC TRAIT	SOME PHENOTYPES	CONTROLLED BY
Blood group (ABO type)	A, B, AB, O	Single gene (*I*)
Adult lactase persistence	Lactose tolerance, lactose intolerance	Single gene (*LCT*)
Hair color	Black, brown, red, blond	Multiple genes, including *MC1R* in redheads
Risk-related personality	Risk taker, risk averse	Multiple genes (including *DRD4*) + environment
Height	Tall, medium, short	Multiple genes (including *IGF2*) + environment
Risk of type 2 diabetes	High, medium, low	Multiple genes (including *TCF7L2*) + environment

FIGURE 9.3 Many Genetic Traits Come in a Variety of Phenotypes
A phenotype is a particular version of a genetic trait in a given individual. How many genetic traits can you identify in the photo? Do you think the traits you have identified are controlled by genes exclusively, or might they be influenced by environmental factors also?

(**FIGURE 9.3**). Fruit size in pumpkins, skin looseness in dogs, and the song of the meadowlark are examples of genetic traits. How many different genetic traits can you spot among the humans pictured in Figure 9.3? Physical traits such as height or the shape of the face are among the easiest to observe. Some biochemical traits, such as hair color and skin color, are also readily observed. Other biochemical traits, such as a person's blood type or susceptibility to certain diseases, can be determined only with laboratory tests. Behaviors, too, can be genetic traits; known examples include shyness, risk taking, and extroversion.

At the molecular level, a gene consists of a stretch of DNA. Geneticists often use italicized letters, symbols, and numbers to represent a particular gene, although other conventions also exist. For example, the *Orange* gene, which leads to orange fur in cats, is designated by an italicized *O*. More than a dozen different genes affect skin color in humans, and the shorthand for one key gene is *MC1R*. A gene called *IGF2* is crucial for human development and is among the many genes that affect height and weight in humans.

Some genetic traits are invariant or nearly invariant, meaning that they are about the same in all individuals in the population. For example, all the people in Figure 9.3 have the same number of eyes (two); that is, the number of eyes in humans is essentially invariant. Other traits you see in Figure 9.3 are quite variable. The display of a *particular version* of a genetic trait in an individual is the **phenotype** of that individual. For example, fawn, sable, and apricot are all coat color phenotypes that are accepted as breed standards for the Shar-Pei dog by the American Kennel Club.

Diploid cells have two copies of every gene

As we described in Section 7.5, somatic cells in plants and animals are *diploid* (2*n*) in that they possess two copies of each type of chromosome. The two copies make up a *homologous pair* for each chromosome type. One member of the homologous pair is inherited from the male parent and is called the *paternal homologue*; the other member of the pair is inherited from the female parent and is called the *maternal homologue*. Because there are two copies of each type of chromosome (with one exception, the sex chromosomes), a diploid cell has two copies of every gene located on those chromosomes (**FIGURE 9.4**).

We also mentioned in Chapter 7 that gametes—sperm and egg cells—are *haploid* (*n*) because these cells have only one set of chromosomes (half the total number of chromosomes found in a diploid cell). Because a gamete has only one copy of each pair of homologous chromosomes, it possesses *only one copy* of every gene.

Helpful to Know

By convention, the names of genes are given in italic type, while the names of their protein products are in roman type. Often this font difference is the only thing distinguishing the protein name from the gene name. For example, in Chapter 8 you learned about the *BRCA1* gene, which is implicated in some kinds of breast cancer. Its protein product is the BRCA1 protein. This protein promotes DNA repair and triggers programmed cell death when DNA damage is too severe for reliable repair.

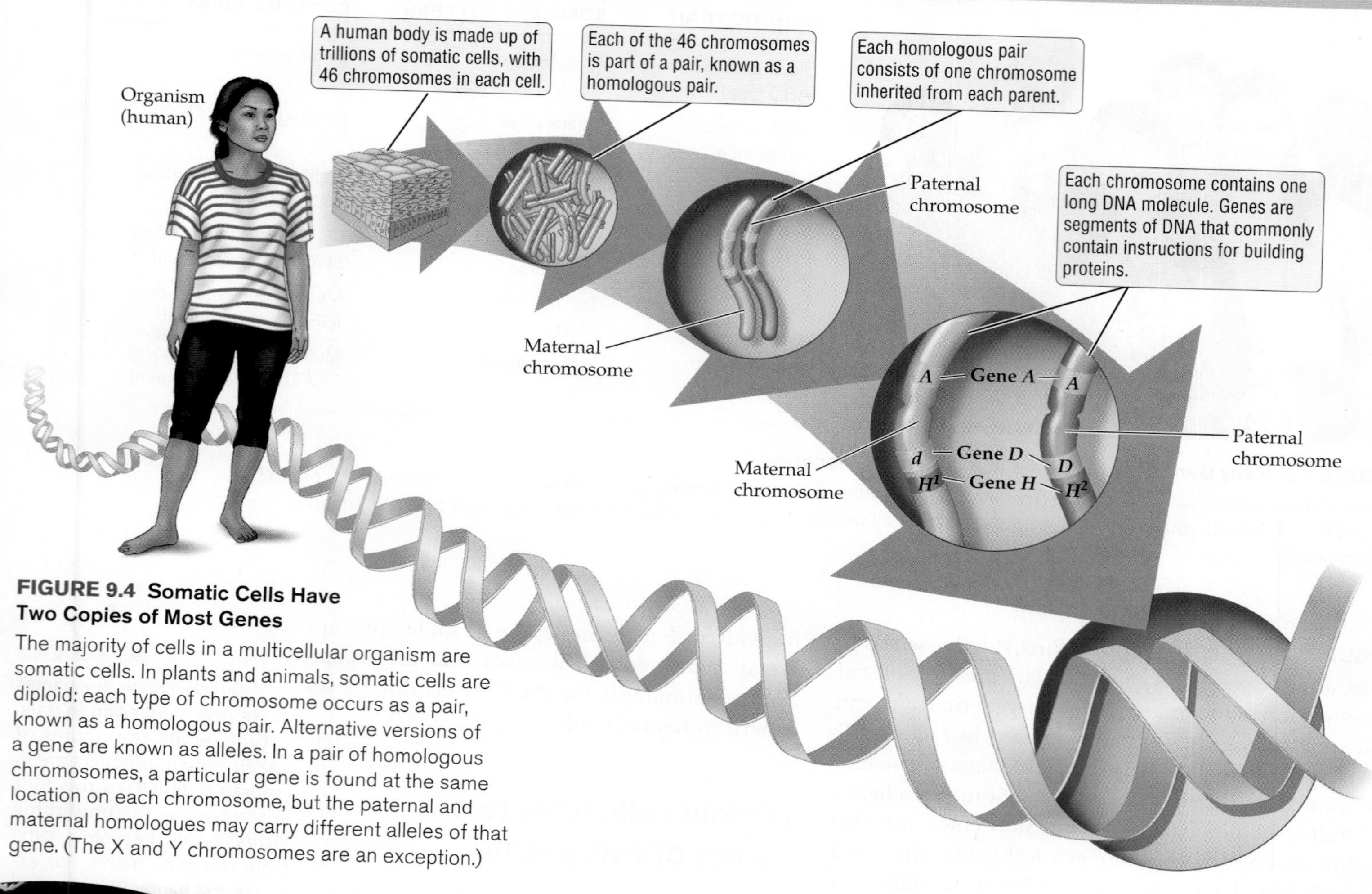

Organism (human)
A human body is made up of trillions of somatic cells, with 46 chromosomes in each cell.
Each of the 46 chromosomes is part of a pair, known as a homologous pair.
Each homologous pair consists of one chromosome inherited from each parent.
Paternal chromosome
Maternal chromosome
Each chromosome contains one long DNA molecule. Genes are segments of DNA that commonly contain instructions for building proteins.
A
Gene A
A
d
Gene D
D
H1
Gene H
H2
Maternal chromosome
Paternal chromosome

FIGURE 9.4 Somatic Cells Have Two Copies of Most Genes

The majority of cells in a multicellular organism are somatic cells. In plants and animals, somatic cells are diploid: each type of chromosome occurs as a pair, known as a homologous pair. Alternative versions of a gene are known as alleles. In a pair of homologous chromosomes, a particular gene is found at the same location on each chromosome, but the paternal and maternal homologues may carry different alleles of that gene. (The X and Y chromosomes are an exception.)

EXTREME PHENOTYPIC DIVERSITY

Male guppies come in such a rainbow of color phenotypes that in some habitats, no two males are alike. Rare color phenotypes are the sexiest, at least for the females in one Trinidad stream: researchers found unique-looking individuals were more likely to become dads.

Genotype directs phenotype

Different versions of a particular gene are known as **alleles** (**uh-**LEEL) of that gene. For example, in dogs, a single gene controls hair texture, and it has two alleles: *W* (wiry hair) and *w* (smooth hair) (**FIGURE 9.5**). As another example, ABO blood groups in humans are controlled by at least three alleles of the *I* gene: I^A, I^B, and *i*.

Note that while a diploid cell contains at most two different alleles for a given gene, and a haploid cell can have only one of all the possible alleles of a gene, a population of individuals may collectively harbor many different alleles. For example, every one of the three alleles of the *I* gene is probably represented among the students in your classroom. The genetic diversity we see in natural populations comes about because each population contains many different alleles of its many genes. The main reason just about everyone in your classroom looks recognizably different from the next person is that each of you carries different alleles for many of the roughly 25,000 genes that all humans possess.

The **genotype** of an individual is the allelic makeup of that individual for a specific genetic trait. In other words, a genotype is the pair of alleles that code for a phenotype. It is the genotype that partly, or wholly, creates the phenotype. For example, a dog with the genotype *WW* has wiry hair, as does a dog with genotype *Ww*; a dog with the genotype *ww* has smooth hair (see Figure 9.5).

Dominant and Recessive Alleles

Phenotype	Wiry hair	Wiry hair	Smooth hair
Genotype	*WW*	*Ww*	*ww*

FIGURE 9.5 Hair Texture in Dogs Is Controlled by a Single Gene
The *W* allele produces a rough-textured, or wiry, coat in dogs. Dachshunds, terriers, and schnauzers are some of the breeds in which the wiry-hair phenotype is accepted under standards set by kennel clubs. The *W* allele is dominant over the *w* allele, so dogs with *WW* or *Ww* genotype have wiry hair. The smooth-hair phenotype appears only in the *ww* homozygote.

An individual that carries two copies of the *same* allele is said to be homozygous for that gene. For example, dogs with the *WW* or *ww* genotype are **homozygotes** for the hair texture gene. An individual with a genotype consisting of two *different* alleles is said to be heterozygous for that gene. Dogs with the *Ww* genotype are **heterozygotes** for the hair texture gene.

Some phenotypes are controlled by dominant alleles

In some cases, one allele can prevent a second allele from affecting the phenotype when the two alleles are paired together in a heterozygote. The allele that exerts a controlling influence on the phenotype is said to be **dominant**. An allele that has no effect on the phenotype when paired with a dominant allele is said to be **recessive**. For example, the *W* allele is dominant over the *w* allele, which is why a heterozygous dog (*Ww*) has wiry hair (see Figure 9.5). In the heterozygote, the effect of the recessive *w* allele is masked by the presence of the dominant *W* allele. A dog must be homozygous for the recessive allele (*ww*) to display the smooth-coat phenotype, produced by the *w* allele.

Of the thousands of human genetic traits, governed by an estimated 25,000 genes, fewer than 4,000 are known or suspected to be controlled by a single gene with a dominant and a recessive allele. Cleft chin, freckles, tongue-rolling, unattached earlobes, "widow's peak," and "hitchhiker's thumb" are often given as examples of single-gene traits; the genetics of these traits, however, is actually more complex. Some of these traits (freckles, for example) are controlled by more than one gene. Others in this list are not discrete traits, meaning that their phenotypes cannot be reliably sorted into just two clear-cut categories (there are all kinds of hairline shapes and degrees of earlobe attachment, for example).

A common misconception is that a dominant allele is somehow better for an individual or is more common in a population than a recessive allele. Neither is necessarily true. For example, although the *I* allele of the gene that controls a person's ABO blood group is dominant over *i*, there is no evidence that a person with the dominant allele is better off than someone who has the recessive version of the gene. And across the globe, the *i* allele is far more common than either of the *I* alleles. In some cases, a dominant allele can be lethal, and the same is true for some recessive alleles in their homozygous state. From your own observations of neighborhood dogs, would you say the *W* allele is more common than the *w* allele?

Gene mutations are the source of new alleles

Different alleles of a gene arise by **mutation**, which we can define briefly here as any change in the DNA that makes up a gene (see Section 11.4 for a more detailed discussion). When a mutation occurs, the new allele's instructions may result in a protein whose form differs from the original version. By specifying different versions of a protein, the different alleles of a gene produce genetic differences among the individuals in a population.

Helpful to Know

The nature of mutations is often misunderstood. Two common misconceptions are, first, that mutations happen because the individual "needs" them, and second, that all mutations are inherited. In reality, mutations occur at random, with no predetermined purpose. In addition, mutations can happen at any time and in any cell of the body. In multicellular organisms, only mutations that are present in gametes can be passed on to offspring.

EXTREME HOMOZYGOSITY

The future of the Florida panther is uncertain. The present-day population of 100–160 animals is descended from about 20 individuals that survived into the 1970s. Scientists estimate that the surviving animals are homozygous at most of their genetic loci. Poor reproduction and slow growth rates have resulted.

Mutations are sometimes harmful. For example, a mutation may lead to the production of a protein that performs a vital function poorly or not at all. Harmful or nonfunctional alleles tend to be recessive; their effects do not show up unless two copies are present (that is, unless the affected individual is homozygous for the allele).

The most common mutations are those that are neither harmful nor beneficial to the individual. So-called neutral mutations arise, for instance, when a new allele specifies a protein that is nearly identical to that produced by the original allele. In some cases a cell can tolerate variation in the activity of a protein with no harm to the organism.

Occasionally, mutations produce alleles that improve on the original protein or carry out new, useful functions. These beneficial mutations are the rarest of the three mutation types.

Controlled crosses help us understand patterns of inheritance

A **genetic cross**, or just "cross" for short, is a controlled mating experiment performed to examine how a particular trait may be inherited. "Cross" can also be used as a verb, as in "individuals of genotype *AA* were crossed with individuals of genotype *aa*."

The parent generation in a genetic cross is called the **P generation**. The first generation of offspring is called the **F_1 generation** ("F" is for "filial," which describes a son or daughter). When the individuals of the F_1 generation are crossed with each other, the resulting offspring are said to belong to the **F_2 generation**.

Definitions of the important genetics terms introduced in this section are collected in **TABLE 9.1**. You may find it useful to refer to this table as you continue with the rest of the chapter.

Concept Check

1. What is an allele? How do new alleles arise? How many different alleles can a single person carry for a particular gene?
2. For a single genetic trait, what is the difference between a phenotype and a genotype?

Concept Check Answers

1. Alleles are different versions of a particular gene. New alleles arise by mutation, a change to the gene's DNA code. Normally, an individual carries only two alleles for any gene.
2. The phenotype is the particular version of a genetic trait that is displayed by an individual, such as wiry hair or smooth hair in dogs. The genotype specifies the allelic makeup that determines the phenotype. Dogs with wiry hair (a dominant trait) have either the genotype *WW* or *Ww*; those with smooth hair (recessive) have the genotype *ww*.

TABLE 9.1 Essential Terms in Genetics

TERM	DEFINITION
Allele	One of two or more alternative versions of a gene that exist in a population of organisms.
Dominant allele	The allele that controls the phenotype when paired with a recessive allele in a heterozygote.
F_1 generation	The first generation of offspring in a breeding trial involving a series of genetic crosses.
F_2 generation	The second generation of offspring in a series of genetic crosses.
Gene	The basic unit of genetic information. Each gene consists of a stretch of DNA that affects at least one genetic trait. A gene is part of a chromosome.
Genetic cross	A mating experiment, performed to analyze the inheritance of a particular trait, and also to raise crop varieties and particular breeds of animals.
Genetic trait	Any inherited characteristic of an organism that can be observed or detected; body size, color and length of fur, and aggressive behavior are known examples of genetic traits.
Genotype	The allelic makeup of an individual; more specifically, the two alleles of a gene that produce the specific phenotype displayed by an individual.
Heterozygote	An individual that carries one copy of each of two different alleles (for example, an *Aa* individual or a C^WC^R individual).
Homozygote	An individual that carries two copies of the same allele (for example, an *AA*, *aa*, or C^WC^W individual).
P generation	The parent generation in a breeding trial involving a series of genetic crosses.
Phenotype	The particular version of a genetic trait that is displayed by a given individual; for example, black, brown, red, and blond are phenotypes of the hair color trait in humans.
Recessive allele	An allele that does not affect the phenotype when paired with a dominant allele in a heterozygote.

9.2 Basic Patterns of Inheritance

Before Gregor Mendel, many people argued that the traits of both parents were blended in their offspring, much as paint colors blend when they are mixed together. According to this idea, which was known as *blending inheritance*, offspring should be intermediate in phenotype to their two parents, and it should *not* be possible for "lost" traits to reappear in later generations.

But many observations—including Mendel's—do not match these predictions. The features of offspring often are not intermediate between those of their parents, and it is common for traits to skip a generation. How can such observations be explained? During 8 years of investigation, Mendel conducted many experiments to analyze inheritance in garden pea plants, and his results led him to reject the idea of blending inheritance. Mendel proposed instead that for each trait, offspring inherit two separate units of genetic information, one from each parent.

Phenotype and Genotype

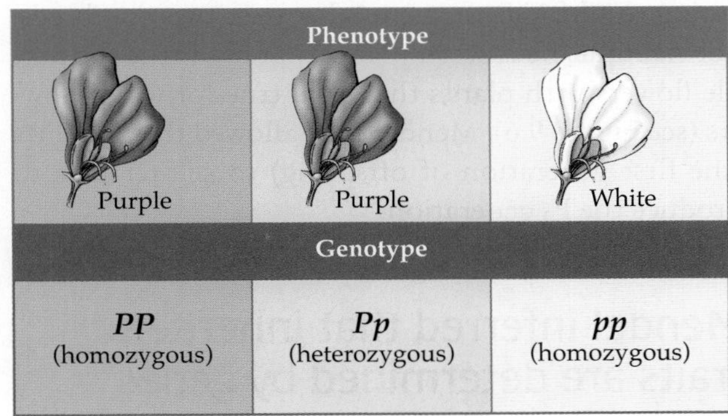

FIGURE 9.6 True-Breeding Traits Have a Homozygous Genotype

Flower color in peas is controlled by a gene with two alleles (*P* and *p*). Although there are three genotypes (*PP*, *Pp*, and *pp*), there are only two phenotypes (purple flowers and white flowers). Genotypes *PP* and *Pp* both produce purple flowers, and only *pp* produces white flowers.

Mendel's genetic experiments began with true-breeding pea plants

Mendel's extraordinary insights were made possible by his exceptional training. As a young monk, Mendel attended the University of Vienna, where he took courses ranging from mathematics to botany. Upon assuming his duties at the monastery of Saint Thomas, Mendel put his training in probability statistics and plant breeding to good use. From the patterns he observed and his careful mathematical analysis, Mendel was able to deduce the basic principles that govern how genetic information is passed from one generation to the next.

Even though Mendel's laws have been modified considerably by modern genetics, his predictions about the outcomes of certain types of mating experiments still hold true, and they apply broadly to most organisms that multiply themselves through sexual reproduction. Mendel proposed that the inherited characteristics of organisms are controlled by hereditary factors—now known as *genes*—and that one factor for each trait is inherited from each parent. Although he did not use the word "gene," Mendel was the first to propose the concept of the gene as the basic unit of inheritance.

Peas are excellent organisms for studying inheritance. Ordinarily, peas self-fertilize; that is, the flowers of an individual pea plant contain both male and female reproductive organs, so the plant is able to fertilize itself. But because peas can also be mated experimentally, Mendel was able to perform carefully controlled genetic crosses.

In addition, peas have *true-breeding varieties*, meaning that when such plants self-fertilize, *all* of their offspring have the same phenotype as the parent. For example, one variety has yellow seeds and produces only offspring with yellow seeds when it is bred with itself.

According to what we know today about how genotypes affect phenotypes, individuals that breed true for a phenotype must have a homozygous genotype. In **FIGURE 9.6**, the purple flowers with genotype *PP* and the white flowers with genotype *pp* are homozygous. Mendel based all his experiments on varieties that were homozygous—and therefore true-breeding—for traits such as plant height, flower position, flower color, and the color and shape of the seeds or of the pea pods.

Mendel crossed true-breeding lines with contrasting phenotypes (purple flowers and white flowers, for example), and he meticulously recorded the phenotypes of all the offspring over two generations. He began with a set of original, true-breeding parents

Helpful to Know

Purebred pets are simply true-breeding individuals. A mating of two purebred Labrador retrievers can be expected to produce pups that all have the distinctive qualities of this breed, known as the "breed standards." Like Mendel's true-breeding lines of peas, purebreds are homozygous for nearly all of the breed standards. The opposite of a purebred animal—a mutt—is heterozygous for many traits, and may display phenotypes that are not among the accepted standards for a particular breed.

(P generation) and tracked the phenotypes through two generations of **hybrid** offspring—that is, offspring resulting from a cross between two purebreds. For example, he crossed plants that bred true for purple flowers with plants that bred true for white flowers (see Figure 9.6). Mendel then allowed the F_1 plants (the first generation of offspring) to self-fertilize to produce the F_2 generation.

Mendel inferred that inherited traits are determined by genes

According to the theory of blending inheritance, the cross illustrated in **FIGURE 9.7** should have yielded F_1-generation plants bearing flowers of intermediate color. Instead, all the F_1 plants had purple flowers.

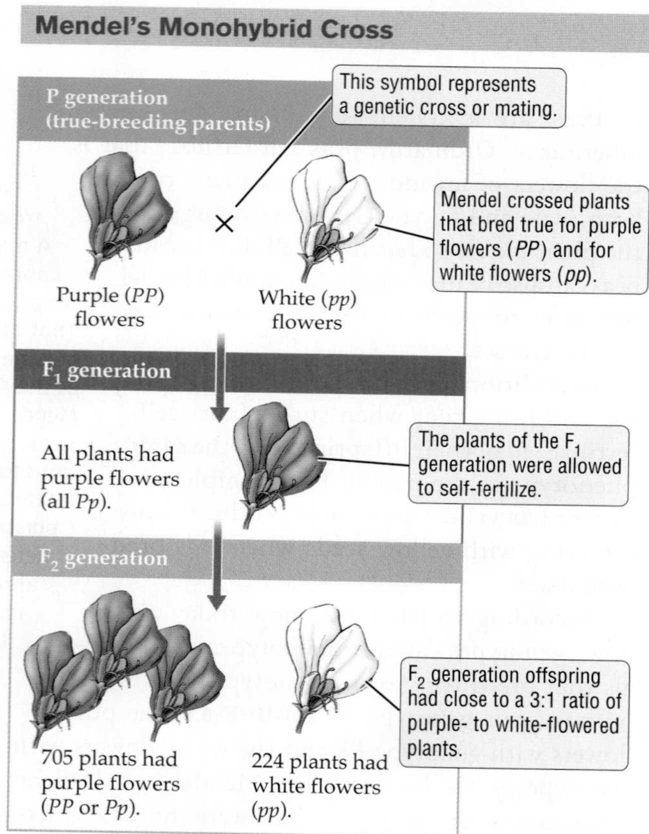

Mendel's Monohybrid Cross
P generation (true-breeding parents)
This symbol represents a genetic cross or mating.
×
Mendel crossed plants that bred true for purple flowers (PP) and for white flowers (pp).
Purple (PP) flowers
White (pp) flowers
F1 generation
All plants had purple flowers (all Pp).
The plants of the F1 generation were allowed to self-fertilize.
F2 generation
F2 generation offspring had close to a 3:1 ratio of purple- to white-flowered plants.
705 plants had purple flowers (PP or Pp).
224 plants had white flowers (pp).

FIGURE 9.7 Inheritance of a Single Trait over Three Generations

Mendel crossed parent plants that were true-breeding (homozygous) for two discrete phenotypes of a particular trait. In the example here, pea plants with purple flowers (*PP*) were crossed with pea plants with white flowers (*pp*). In the F_1 generation, all hybrids were purple (*Pp*). In the F_2 generation, the majority of the plants were purple, but some were white.

Furthermore, Mendel noted that when the F_1 plants self-fertilized, about 25 percent of the F_2 offspring had white flowers. That is, white flowers appeared in a 1:3 ratio in the F_2 generation. Mendel realized that the *reappearance of white flowers* in the second generation was not consistent with the theory of blending inheritance.

Mendel studied seven true-breeding traits in peas, repeating the experiment illustrated in Figure 9.7 for each trait. In the F_2 generation, he repeatedly observed a 3:1 ratio of the dominant to the recessive phenotype for each of the traits under study. These results led him to propose a new theory of inheritance, in which the units of inheritance (which we now call genes) exist as discrete factors that do not lose their unique characteristics when crossed, as would colors of paints blended together. Using modern terminology, we can summarize Mendel's concepts as follows:

1. *Alternative versions of genes cause variation in inherited traits.* For example, peas have one version (allele) of a certain gene that causes flowers to be purple, and another version (a different allele) of the same gene that causes flowers to be white. One individual carries at most two different alleles for a particular gene.
2. *Offspring inherit one copy of a gene from each parent.* In his analysis of crosses like the one in Figure 9.7, Mendel reasoned that white flowers could not reappear in the F_2 generation unless the white-flower allele was present in F_1 plants to pass on to F_2 plants. He deduced that the F_1 pea plant must carry two copies of the flower color gene: an allele that causes white flowers, and an allele that causes purple flowers.
3. *An allele is dominant if it has exclusive control over the phenotype of an individual when paired with a different allele.* For example, plants that breed true for purple flowers must have two copies of the *P* allele (that is, they are of genotype *PP*), because otherwise they would occasionally produce white flowers. Similarly, plants that breed true for white flowers have two copies of the *p* allele (genotype *pp*). Working back from the phenotypes of both the F_1 and F_2 generations, Mendel correctly deduced that the F_1 plants in Figure 9.7 must have genotype *Pp*.
4. *The two copies (alleles) of a gene segregate during meiosis and end up in different gametes.* Recall that each gamete receives only one copy of each gene in the process of meiosis. If an organism has two copies of a single allele for a particular trait, as in the

homozygous varieties used by Mendel, all of its gametes will contain that allele. However, if the organism has two different alleles, as an individual of genotype *Pp* has, then 50 percent of the gametes will receive one allele and 50 percent will receive the other allele.

5. *Gametes fuse without regard to the alleles they carry.* When gametes fuse to form a zygote, they do so *randomly* with respect to the alleles they carry for a particular gene.

Concept Check

1. Hair length in cats is controlled by a gene that has at least two alleles, *L* and *l*. Short-haired cats are *LL* or *Ll*. Long hair is a breed standard for Maine coon cats. Are these cats homozygous for the hair length allele? What is their genotype?
2. Which of Mendel's observations demonstrated that the theory of blending inheritance was false?

9.3 Mendel's Laws of Inheritance

Mendel summarized the results of his experiments in two laws: the *law of segregation* and the *law of independent assortment*. He was able to deduce the law of segregation from breeding experiments in which he tracked a *single trait* (such as flower color or plant height). The law of independent assortment was based on *two-trait* breeding experiments—crosses in which he tracked two different traits at the same time.

Mendel's single-trait crosses revealed the law of segregation

The **law of segregation** states that the two copies of a gene are separated during meiosis and end up in different gametes. This law can be used to predict how a single trait will be inherited, as illustrated by the experiment depicted in Figure 9.7. In that experiment, the F_1 generation was composed entirely of **monohybrids**: they were all hybrids—heterozygotes (*Pp*)—with respect to one trait, flower color. According to the law of segregation, when the F_1 plants reproduced, 50 percent of the pollen (sperm) should have contained the *P* allele; and the other 50 percent, the *p* allele. The same is true for the eggs.

We can represent the separation of the two copies of a gene during meiosis, and their random recombining through fertilization, using a checkerboard diagram called a **Punnett square** (FIGURE 9.8). The Punnett square shows all possible ways that two alleles can be brought together through fertilization.

To create a Punnett square, list all possible genotypes of the male gametes across the top of the grid, writing each unique genotype just once. List all possible genotypes of the female gametes along the left edge of the grid, again writing each unique genotype only once. Next, fill in the grid by combining in each box (or "cell") the male genotype at the top of each column with the female genotype listed at the beginning of each row (follow the blue and pink arrows in Figure 9.8).

Concept Check Answers

1. Purebred Maine coon cats are homozygous recessive for the hair length gene, with the *ll* genotype.
2. The F_1 generation did not have an intermediate phenotype, and moreover, the parental phenotypes reappeared in the F_2 generation instead of vanishing forever into a blended phenotype.

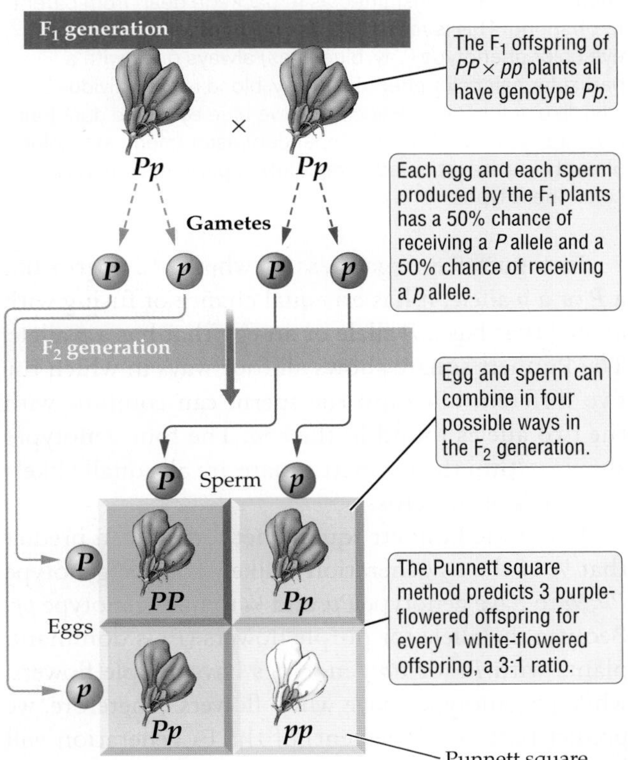

FIGURE 9.8 The Punnett Square Predicts All Possible Outcomes of a Genetic Cross

Punnett squares chart the segregation (separation) of alleles into gametes and all the possible ways in which the alleles borne by these gametes can be combined to produce offspring.

FIGURE 9.9 Inheritance of Two Distinct Traits
In his dihybrid breeding trials, Mendel was asking if phenotypes of two different traits (seed color and seed shape) would be transmitted as a "package deal" from parent to offspring. This is rather like asking whether a particular eye color phenotype (say, blue eyes) always goes with a particular hair color phenotype (say, blond hair). Individuals who, like actor Daniel Radcliffe, have blue eyes and dark hair illustrate Mendel's law of independent assortment: eye color phenotypes are inherited completely separately from hair color phenotypes.

In Figure 9.8, regardless of whether a sperm has a *P* or a *p* allele, it has an equal chance of fusing with an egg that has a *P* allele or an egg that has a *p* allele. The Punnett square shows all four ways in which the two different alleles in the sperm can combine with the two alleles found in the egg. The four genotypes shown within the Punnett square are all equally likely outcomes of this cross.

Using the Punnett square method, we can predict that ¼ of the F_2 generation is likely to have genotype *PP*, ½ to have genotype *Pp*, and ¼ to have genotype *pp*. Because the allele for purple flowers (*P*) is dominant, plants with *PP* or *Pp* genotypes have purple flowers, while *pp* genotypes have white flowers. Therefore, we predict that ¾ (75 percent) of the F_2 generation will have purple flowers and ¼ (25 percent) will have white flowers—a 3:1 ratio. This prediction is very close to the actual results that Mendel obtained. Of a total of 929 F_2 plants that Mendel raised, 705 (76 percent) had purple flowers and 224 (24 percent) had white flowers.

Mendel's two-trait experiments led to the law of independent assortment

Having demonstrated that the phenotypes of one genetic trait are inherited separately from each other, Mendel wondered whether the phenotypes of *two different traits* would also be inherited independently of each other, resulting in combinations of phenotypes in offspring that did not exist in either parent (**FIGURE 9.9**). Mendel performed experiments in which he simultaneously tracked the inheritance of *two* genetic traits by performing crosses with **dihybrids** (individuals heterozygous for two traits). For example, pea seeds can have a round or wrinkled shape, and they can be yellow or green. Two different genes control the two different traits: the *R* gene controls seed shape; the *Y* gene controls the color of the seed. Mendel wanted to know what would happen if true-breeding round, yellow-seeded individuals (genotype *RRYY*) were crossed with true-breeding wrinkled, green-seeded individuals (genotype *rryy*).

When Mendel performed the two-trait cross (**FIGURE 9.10**), all of the resulting F_1 plants had round, yellow seeds. From the phenotypes of this F_1 generation, Mendel could see that the allele for round seeds (denoted *R*) was dominant over the allele for wrinkled seeds (denoted *r*). Similarly, with respect to seed color, Mendel deduced that the allele for yellow seeds (*Y*) was dominant over the allele for green seeds (*y*).

Next, Mendel crossed large numbers of the F_1 plants (genotype *RrYy*) to raise a generation of F_2 plants. He knew that the phenotypes of the F_2 generation would answer the question that most interested him: *Is the inheritance of seed shape independent of the inheritance of seed color?* If his hunch was right, the distribution of the alleles of one gene into gametes would be independent of the distribution of the alleles of the other gene, so all possible combinations of the alleles would be found in the gametes (**FIGURE 9.11**). If this were not the case, and a particular seed color always went with a particular seed shape, we would *not* find novel combinations of phenotypes among the F_2 offspring. Instead, all of the F_2 offspring would show one of the two parental phenotypes: round, yellow seeds or wrinkled, green seeds.

Mendel tested the two possibilities by crossing the *RrYy* F_1 plants with each other (see Figure 9.10). He obtained the following results in the F_2 generation: approximately 9/16 of the seeds were round and yellow, 3/16 were *round and green*, 3/16 were *wrinkled and yellow*, and

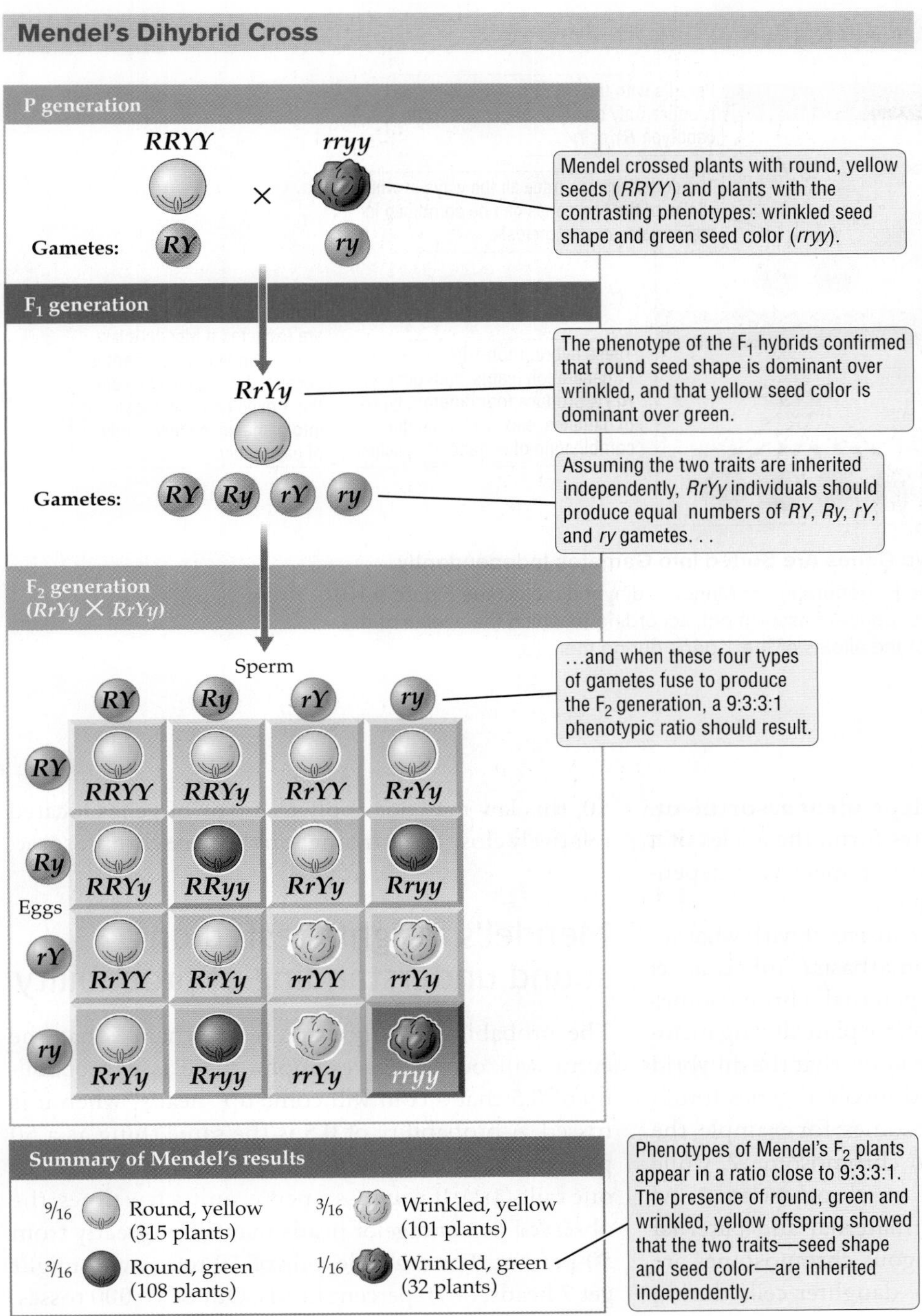

Mendel's Dihybrid Cross
P generation
RRYY
rryy
×
Gametes:
RY
ry
Mendel crossed plants with round, yellow seeds (RRYY) and plants with the contrasting phenotypes: wrinkled seed shape and green seed color (rryy).
F1 generation
RrYy
The phenotype of the F1 hybrids confirmed that round seed shape is dominant over wrinkled, and that yellow seed color is dominant over green.
Gametes:
RY
Ry
rY
ry
Assuming the two traits are inherited independently, RrYy individuals should produce equal numbers of RY, Ry, rY, and ry gametes. . .
F2 generation
(RrYy × RrYy)
Sperm
Eggs
RRYY
RRYy
RrYY
RrYy
RRYy
RRyy
RrYy
Rryy
RrYY
RrYy
rrYY
rrYy
RrYy
Rryy
rrYy
rryy
. . .and when these four types of gametes fuse to produce the F2 generation, a 9:3:3:1 phenotypic ratio should result.
Summary of Mendel's results
9/16 Round, yellow (315 plants)
3/16 Wrinkled, yellow (101 plants)
3/16 Round, green (108 plants)
1/16 Wrinkled, green (32 plants)
Phenotypes of Mendel's F2 plants appear in a ratio close to 9:3:3:1. The presence of round, green and wrinkled, yellow offspring showed that the two traits—seed shape and seed color—are inherited independently.

FIGURE 9.10 Inheritance of Two Traits over Three Generations

A two-trait breeding experiment in which the F_1 plants are double heterozygotes (heterozygous for both traits) is a dihybrid cross. Mendel used dihybrid crosses to test the hypothesis that the alleles of two different genes are inherited independently from each other. He tracked the seed shape trait controlled by the *R/r* alleles and the seed color trait controlled by the *Y/y* alleles. Two new phenotypic combinations were found among the F_2 offspring: plants that made round, green seeds (*R-yy*) and plants that made wrinkled, yellow seeds (*rrY-*). The bottom panel summarizes the ratio of the two parental phenotypes and the two novel, nonparental phenotypes.

1/16 were wrinkled and green (a 9:3:3:1 ratio). Mendel's results are what we would expect if the genes for these two traits are inherited independently of each other.

If the alleles of the *R* gene segregated independently from the alleles of the *Y* gene during gamete formation (as depicted in Figure 9.11), we would predict four phenotypes among the F_2 offspring: 9/16 displaying the dominant phenotypes for both traits, 1/16 with the recessive phenotypes of both traits, and two novel combinations of phenotypes not seen in either parent. The two nonparental combinations of phenotypes Mendel was looking for did turn up among the F_2 plants: 3/16 of the plants had round, green seeds; and another 3/16 had wrinkled, yellow seeds.

Mendel made similar crosses for various combinations of the seven traits he studied. His results led

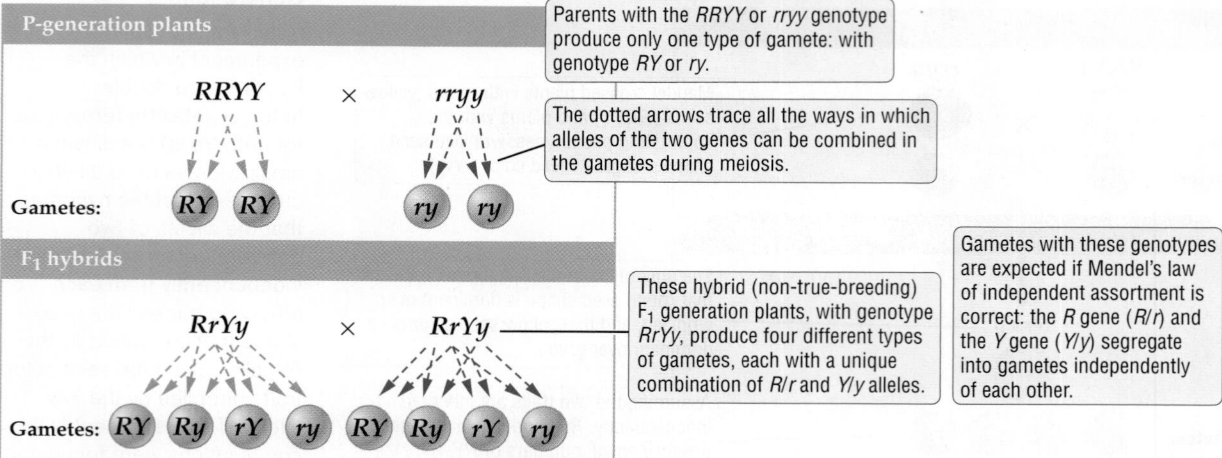

FIGURE 9.11 The Alleles of Two Genes Are Sorted into Gametes Independently
The 9:3:3:1 phenotypic ratios in the F_2 generation of Mendel's dihybrid cross (see Figure 9.10) are best explained by the law of independent assortment, according to which the alleles of the *R* gene segregate independently of the alleles of the *Y* gene during meiosis.

him to propose the **law of independent assortment**, which states that when gametes form, the alleles of a particular gene segregate during meiosis independently of the alleles of other genes.

Mendel's observations are consistent with what we now know about the chromosomal basis of inheritance: homologous maternal and paternal chromosomes line up randomly at the metaphase plate during meiosis I (see Figure 7.17). Today we know that the dihybrid crosses that Mendel published involved genes found on different pea-plant chromosomes; for example, the seed shape gene is located on chromosome 7, while the seed color gene is found on chromosome 1. As explained in Section 7.5, the maternal and paternal members of a pair of homologous chromosomes are randomly distributed into the daughter cells during gamete formation through meiosis.

Because the two members of each homologous pair are randomly sorted into gametes, the alleles on these chromosomes can also be mixed and matched in all possible allelic combinations. That is how offspring can end up with genotypes and phenotypes that were not present in either parent (such as *RRyy* and *rrYY* in Figure 9.10).

To this day, Mendel's law of independent assortment applies to the inheritance of genes that are physically separated because they lie on different chromosomes. For reasons we will explore in Chapter 10, this law may not apply to a pair of genes located relatively close to each other on the *same* chromosome.

Mendel's insights rested on a sound understanding of probability

The probability of an event is the chance that the event will occur. For example, there is a probability of 0.5 that a coin will come up "heads" when it is tossed. A probability of 0.5 is the same thing as a 50 percent chance, or ½ odds, or a ratio of one heads to one tails (1:1). If you toss a penny only a few times, the observed percentage of heads may differ greatly from 50 percent. For example, out of 10 tosses you might get 7 heads, or 70 percent heads. Out of 10,000 tosses, however, it would be very unusual to get 7,000 heads (70 percent heads). Each toss of a coin is an independent event, in the sense that the outcome of one toss does not affect the outcome of the next toss. The probability of getting two heads in a row is a product of the separate probabilities of each individual toss: 0.5×0.5, which is 0.25.

Mendel was able to deduce the patterns of inheritance because he conducted a large number of genetic crosses, which gave him data for a large number of offspring. He knew that he could not predict with certainty the phenotype of F_2 offspring from a single cross, or even a few crosses. When performing genetic crosses, the experimenter has no control over which sperm and which egg fuse to produce an offspring. Mendel knew, however, that the eggs produced by a *Pp* plant have a 0.5 probability of carrying the *P* allele and a 0.5 probability of carrying the *p* allele; the same probabilities hold for the alleles carried by the plant's sperm. When two such plants are mated, the odds that an egg with a *p* allele will fuse with a sperm carrying a *p* allele are given by the rule of multiplication outlined in the previous paragraph: 0.5×0.5, yielding a 0.25 probability of generating a *pp* offspring.

It is important to understand that the ratios predicted by Mendel (for example, the 3:1 ratio illustrated in Figure 9.8) give simply the *probability* that a particular offspring will have a specific phenotype or genotype. We cannot know with certainty what the actual phenotype or genotype of that offspring is going to be (except when true-breeding individuals are crossed). Moreover, the probability that a particular offspring will display a specific phenotype is completely unaffected by how many offspring there are. When we analyze a large number of offspring, however, the likelihood that we will see a 3:1 outcome increases.

Concept Check

1. For the offspring of a cross between an *Rr* plant and an *rr* plant, with *R* being dominant, predict the number and ratio of genotypes and phenotypes.
2. Explain Mendel's law of segregation and law of independent assortment.

9.4 Extensions of Mendel's Laws

Mendel's laws enable us to make accurate predictions of offspring phenotypes based on parental genotypes whenever a genetic trait is controlled by a single gene with two alleles—one dominant, the other recessive. But many traits are not under such simple genetic control. Geneticists refined and extended Mendel's laws through much of the twentieth century to explain more complex patterns of inheritance. They discovered, for example, that sometimes a single allele can produce a number of different phenotypes. The most complex patterns of inheritance are created when a phenotype is controlled by more than one gene, each with multiple alleles, especially if the phenotype is also affected by environmental factors rather than by the genotype alone.

Many alleles display incomplete dominance

For dominance to be complete, a single copy of the dominant allele must be enough to produce its phenotypic effect in a heterozygote; for example, one *P* allele ensures that even a *Pp* pea plant has purple flowers. But in some allele combinations, no single allele completely dominates over the other, and the heterozygote displays an "in-between" phenotype. The two alleles are said to display **incomplete dominance**. For example, two of the alleles that control flower color in snapdragons are incompletely dominant. When a homozygote with red flowers (C^RC^R) and a homozygote with white flowers (C^WC^W) are crossed, the resulting heterozygous offspring (C^RC^W) have pink flowers (**FIGURE 9.12**).

Some of the genes controlling coat color in animals also display incomplete dominance. The coat color of the palomino horse is a case in point. **FIGURE 9.13** shows how two incompletely dominant alleles, H^C and H^{Cr}, produce an intermediate phenotype in a heterozygote. (The chestnut horse [which has a reddish brown coat] is homozygous for H^C [H^CH^C]. A cream-colored horse [known as "cremello"] is produced when the horse is homozygous for H^{Cr} [$H^{Cr}H^{Cr}$].) The palomino horse, popular in parades and shows for its golden coat and flaxen mane, is heterozygous for coat

Incomplete Dominance

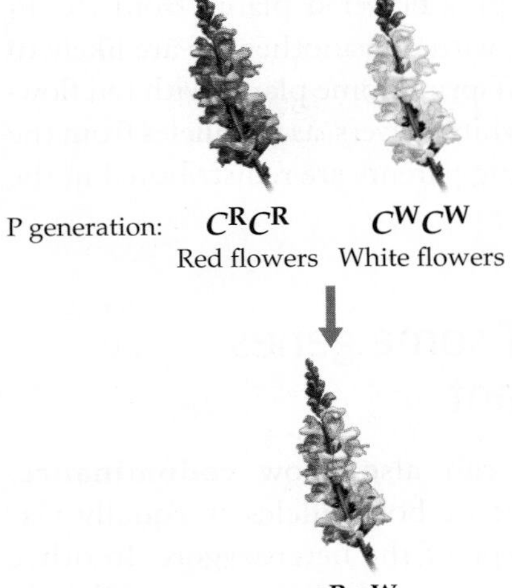

Incomplete dominance may look like blending inheritance at first glance: the pink flowers of the F_1 generation seem like a blend of red and white from the P generation. However, the fact that red and white colors reappear in the F_2 generation shows that the phenotypes have not blended together.

FIGURE 9.12 Incomplete Dominance in Snapdragons
Incomplete dominance leads to an intermediate phenotype in the heterozygote.

Concept Check Answers

1. Two genotypes (*Rr* and *rr*) and two phenotypes are predicted, each in a ratio of 1:1. Constructing a Punnett square confirms this.
2. The law of segregation states that during gamete formation, two alleles separate into different gametes so that each carries only one allele. The law of independent assortment states that alleles of two different genes are sorted into gametes independently of one another.

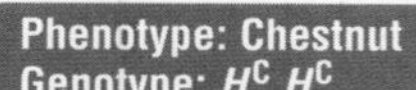

FIGURE 9.13 Incomplete Dominance in Horses
Palominos (heterozygous genotype $H^C H^{Cr}$) are intermediate in color to chestnuts ($H^C H^C$) and cremellos ($H^{Cr} H^{Cr}$) because in the heterozygote, the H^{Cr} allele "dilutes" the effect of the H^C allele to produce the intermediate phenotype.

Helpful to Know

The *I* gene that determines the ABO blood group is found on chromosome 9; however, there are many human blood groups other than the ABO group, including Rh, MNS, Kell, and Diego groups, to name just a few. The genes for these groups are found on different chromosomes, and variant alleles of these groups are rare.

color ($H^C H^{Cr}$). In the palomino, the effect of the chestnut-color allele, H^C, is "diluted" by the presence of the cream-color allele, H^{Cr}.

In cases of incomplete dominance, we can still predict the genotypes and phenotypes of F_1 and F_2 offspring using Mendelian laws of inheritance and the Punnett square method, if we assign an intermediate phenotype to the heterozygotes. For example, if two heterozygous snapdragons ($C^R C^W$) are crossed, the odds are that ¼ of the offspring will have red flowers (genotype $C^R C^R$), ½ are likely to have pink flowers (genotype $C^R C^W$), and ¼ are likely to have white flowers (genotype $C^W C^W$). Work this out for yourself using a Punnett square. You will see that Mendel's laws apply to alleles that show incomplete dominance, just as they apply to alleles that display a dominant-recessive relationship. We can also predict that when pink-flowered plants from the F_1 generation are bred with one another, we are likely to see among the F_2 offspring some plants with red flowers and some with white flowers, as the alleles from the original true-breeding parents are redistributed in the F_2 generation.

The alleles of some genes are codominant

A pair of alleles can also show **codominance**, in which the effect of both alleles is equally visible in the phenotype of the heterozygote. In other words, the influence of each codominant allele is fully displayed in the heterozygote, without being diminished or diluted by the presence of the other allele (as in incomplete dominance) or being suppressed by a dominant allele (as in the case of dominant-recessive alleles).

The ABO blood groups in humans provide an example of codominant alleles (**FIGURE 9.14**). Three alleles can determine a person's ABO blood type: the I^A, I^B, and i alleles of the I gene. As you have learned, genes commonly affect a genetic trait by coding for the manufacture of a specific protein. The I^A allele codes for an enzyme that puts an "A-type" sugar on certain proteins located on the surface of red blood cells. The I^B allele makes a different version of the enzyme, one that inserts a "B-type" sugar on the blood cells. The i allele simply fails to make a sugar-inserting enzyme that works.

The first two of these alleles, I^A and I^B, are codominant; a person carrying both alleles has the blood type AB. The AB blood type is not halfway between A and B blood types, as would be the case if these alleles were incompletely dominant. Instead, people with the AB blood type ($I^A I^B$ genotype) have *both* A and B types of sugars on the surface of their red blood cells.

In contrast, a person who is an $I^A I^A$ homozygote has only A-type sugars, and therefore an A blood type. Similarly, an individual carrying $I^B I^B$ has only B-type sugars, and therefore blood type B. The i allele is recessive to both I^A and I^B; therefore, someone with the genotype $I^A i$ has blood type A, and someone with the genotype $I^B i$ has blood type B. A person who carries two i alleles is homozygous recessive and has blood type O. The "Biology Matters" box on page 205 describes the importance of blood-type matching in transfusions.

BIOLOGY MATTERS

Know Your Type

One of the most important inherited characteristics to know about yourself and your family members is blood type, because the majority of Americans will receive donated blood at some point in their lives. Blood transfusion is the transfer of whole blood or blood products to a patient. Whole blood or its components are used for many surgical procedures, as well as for ongoing treatment of chronic diseases such as sickle-cell anemia (a hereditary disease affecting red blood cells; see Chapter 12).

Whole-blood transfusions are uncommon nowadays; a patient is more likely to receive specific cell types (such as red blood cells or platelets) or plasma, the fluid portion of blood. The immune system of the patient may recognize molecules present in transfused blood as foreign and launch an attack against the blood. The molecules that the immune system attacks are known as *antigens*. The cell surface sugars responsible for the A, B, and AB blood groups are potential antigens. If red blood cells from a person with blood type A (I^AI^A or I^Ai) is transfused into a patient with blood type B (I^BI^B or I^Bi) or blood type O(ii), the patient's immune system produces specific antigen-fighting proteins (called *antibodies*) that launch an attack against their target antigen (the A type of cell surface sugars, in this example). The transfused cells clump together when attacked, leading to life-threatening clots. To prevent transfusion incompatibility, all donated blood is extensively tested, and patients receive blood products that are matched to avoid introducing antigens. The table summarizes how a recipient's blood will respond to a blood transfusion from donors of each blood type: O, A, B, or AB.

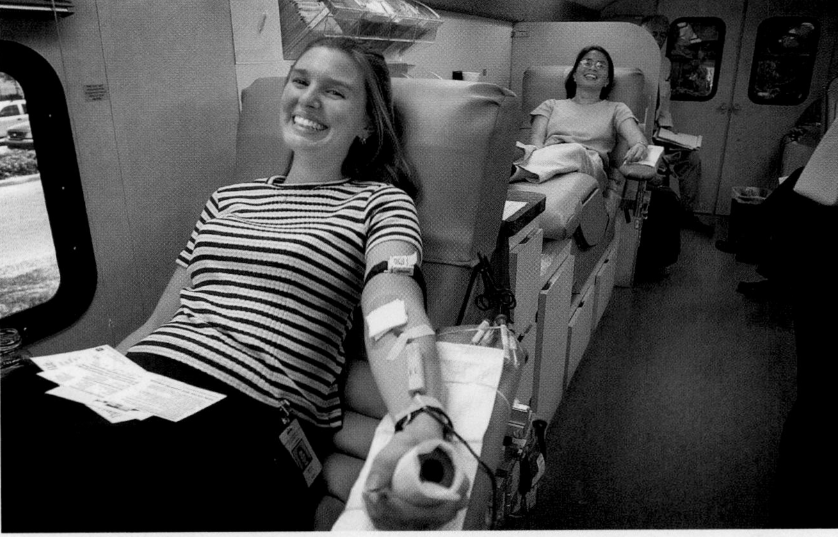

Recipient's blood			Reactions with donor's red blood cells			
ABO blood type	Antigens produced	Antibodies produced	Donor: O type	Donor: A type	Donor: B type	Donor: AB type
O	None	Anti-A Anti-B	✓	✗	✗	✗
A	A	Anti-B	✓	✓	✗	✗
B	B	Anti-A	✓	✗	✓	✗
AB	A and B	None	✓	✓	✓	✓

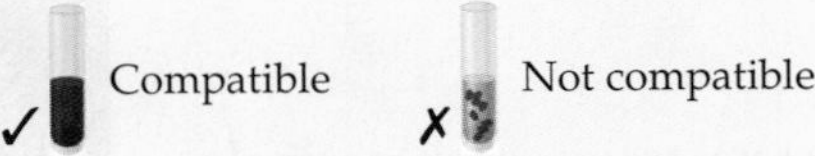

Experts estimate that 4.5 million Americans would die annually without lifesaving blood transfusions. Approximately 32,000 pints of blood are used each day in the United States, with someone receiving blood about every 3 seconds. Blood centers often run short of type O and type B blood, but shortages of all types of blood occur during the summer and winter holidays. We all expect blood to be there for us, but only a small fraction of those who can give actually do. Yet sooner or later, virtually all of us will face a situation in which we will need blood. And that time is all too often unexpected. To find out where you can donate, visit www.redcrossblood.org.

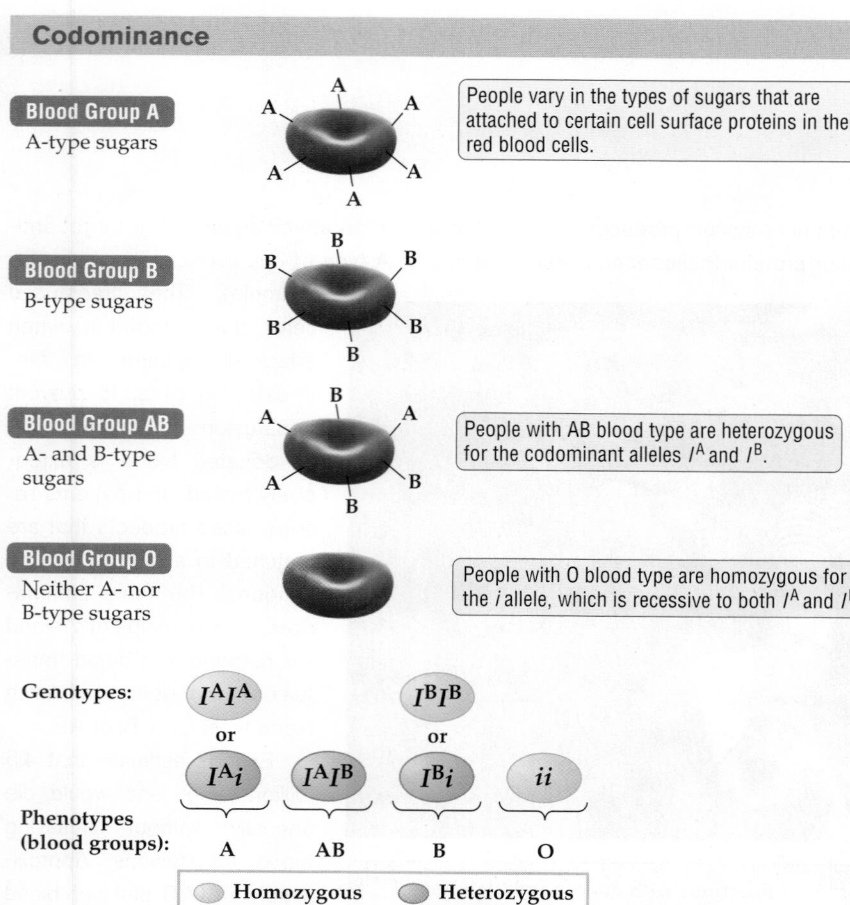
Codominance
Blood Group A
A-type sugars
People vary in the types of sugars that are attached to certain cell surface proteins in their red blood cells.
Blood Group B
B-type sugars
Blood Group AB
A- and B-type sugars
People with AB blood type are heterozygous for the codominant alleles I^A and I^B.
Blood Group O
Neither A- nor B-type sugars
People with O blood type are homozygous for the i allele, which is recessive to both I^A and I^B.
Genotypes:
I^AI^A or I^Ai
I^AI^B
I^BI^B or I^Bi
ii
Phenotypes (blood groups):
A
AB
B
O
Homozygous genotype
Heterozygous genotype

FIGURE 9.14 Genetic Basis of the ABO Blood Types in Humans

The ABO blood types are determined by chains of sugars covalently attached to certain cell surface proteins on red blood cells. The *I* gene encodes an enzyme that comes in at least two allelic forms: the form encoded by the I^A allele adds "A-type" sugars to red blood cell surfaces, and the form encoded by the I^B allele adds "B-type" sugars. The *i* allele codes for a nonfunctional version of the enzyme; individuals who have two copies of the *i* allele have neither A-type nor B-type sugars.

A pleiotropic gene affects multiple traits

Mendel studied seven discrete traits—each a single, clear-cut genetic characteristic specified by a single gene. We now know, however, that some genes control functions of such central importance that abnormal functioning of the gene in question affects many of the body's biological processes. The situation in which a single gene influences a variety of different traits is called **pleiotropy** (**plye-*AH*-truh-pee;** *pleio*, "many"; *tropy*, "change"). A pleiotropic (**plye-uh-*TROH*-pik**) gene is therefore a gene that can influence two or more distinctly different traits.

Pleiotropy is at work in the genetic condition known as albinism (***AL*-buh-*NIH*-zum** or **al-*BYE*-nih-zum; FIGURE 9.15**), in which traits as different as skin color and vision are affected by the action of a single gene. Albinism has many forms, all of them marked by the absence, or reduced production, of a brown-black pigment called melanin. About one in 17,000 Americans has albinism. Most affected individuals produce very little melanin in the skin, hair, and eyes; and most have blue eyes, although a minority have red or violet eyes. Everyone diagnosed with albinism has eye problems, ranging from being "cross-eyed" to being legally blind.

The gene involved in the most common type of albinism controls a step in the pathway for melanin production. It is easy to imagine how a malfunction in this gene would affect pigment deposition and therefore skin color, but what does that have to do with vision? The answer is that certain cells in the retina of the eye produce melanin, which is necessary for proper development of the eye, including the nerves that help the eye communicate with the brain.

Marfan syndrome is a pleiotropic disorder in which many organ systems are affected because a

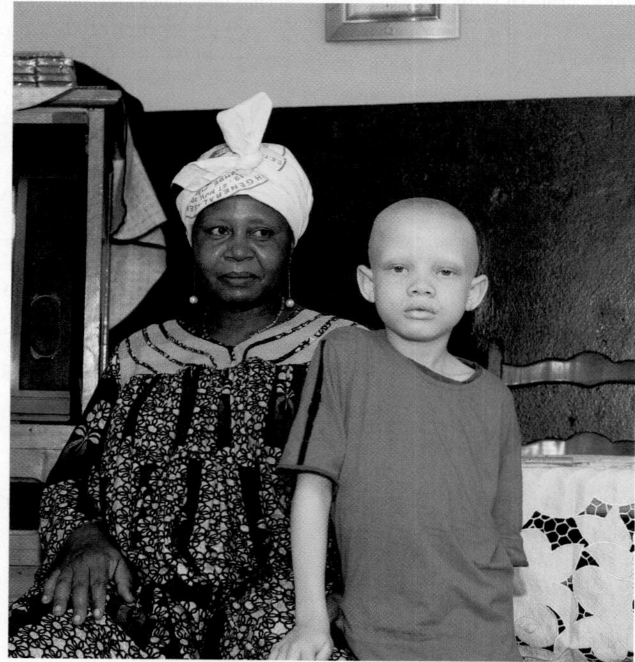

FIGURE 9.15 A Child with Albinism

A mother and her son, who has albinism, in their living room in Douala, Cameroon, Africa. Because albinism affects eyesight as well as pigmentation in skin, eyes, and hair, it is an example of a pleiotropic condition.

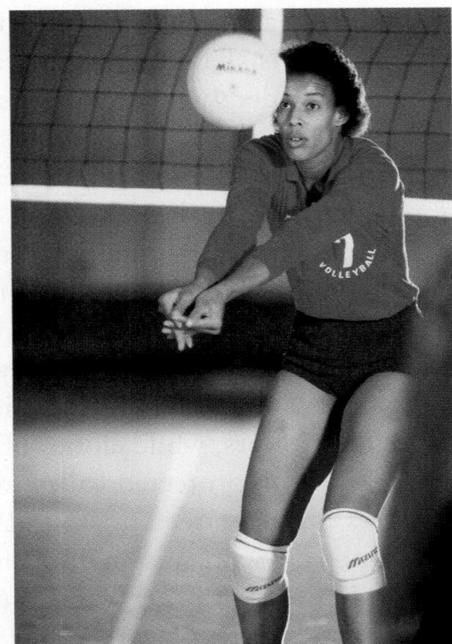

FIGURE 9.16 Flo Hyman

U.S. Olympic volleyball silver medalist Flora Jean ("Flo") Hyman, shown during practice in 1984. Complications from Marfan syndrome, a pleiotropic genetic disorder, contributed to her death in 1986.

single gene, coding for a protein called fibrillin-1, does not work properly. The protein is crucial for the normal function of connective tissues, which act as a gluing and scaffolding system for all types of organs, from bones to the walls of blood vessels. The one in 5,000 Americans diagnosed each year with Marfan syndrome show a wide range of phenotypes, depending on which allele of the fibrillin-1 gene they possess and what other genetic characteristics they have. Many are tall and gangly, with long arms, legs, fingers, and toes. Weakening of the aorta, the largest blood vessel carrying blood away from the heart, is the most serious phenotype associated with the disorder.

Flo Hyman, a volleyball player who won the Olympic silver medal with her team in 1984, was nearly 6½ feet tall (**FIGURE 9.16**). She died from aortic rupture at the age of 31, during an exhibition game in Japan. It took an autopsy to reveal that she had Marfan syndrome. Today, there is a genetic test for the condition, in addition to other tools for diagnosis and more effective care.

Alleles for one gene can alter the effects of another gene

The term **epistasis** (**epp-ee-*STAY*-sus**) applies when the phenotypic effect of the alleles of one gene depends on the presence of certain alleles for another, independently inherited gene. Such interactions among the alleles of different genes are common in all types of organisms. For example, in yeast, a single-celled fungus used in making bread and beer, each gene tested was found to interact with at least 34 other genes.

Epistasis is seen among the many genes that control coat color in mice and other animals. Many of these genes code for enzymes involved in the multistep pathway that converts the amino acid tyrosine into melanin. One such gene has a dominant allele (*B*) that leads to black fur, and a recessive allele (*b*) that produces brown fur. But the effects of these alleles (*B* and *b*) can be eliminated completely, depending on which alleles of the *C* gene are present. The *C* gene codes for an enzyme (called tyrosinase) that acts at the first step in this pathway. In mice with the *cc* genotype, the enzyme fails to do its work and the entire "assembly line" grinds to a halt. No melanin is made in the fur or eyes, resulting in an albino phenotype (**FIGURE 9.17**).

For a mouse with the *cc* genotype, it makes no difference whether the genotype with respect to the *B* gene is *BB* or *Bb* or *bb*; the albino phenotype prevails.

Helpful to Know

It is easy to confuse the terms "incomplete dominance" and "codominance." To help keep these two types of inheritance straight in your mind, remember that incomplete dominance produces an intermediate phenotype. Codominant alleles, by contrast, are like cochairs of a committee: both are active in their own right, and neither is the boss of the other.

Gene Interactions: Epistasis

3 However, mice with two c alleles of gene C produce no pigment and have white fur, regardless of their genotype for gene B (BB, Bb, or bb.)
1 Ordinarily, BB or Bb mice are black…
2 …and bb mice are brown.

FIGURE 9.17 Alleles of One Gene May Affect the Phenotype Produced by Alleles of Another Gene

In this example, the *c* allele of gene *C* masks the alleles of another gene, *B*. In mice with the *CC* or *Cc* genotype, the dominant *B* allele directs the production of melanin, resulting in black fur, while the recessive *bb* genotype "dilutes" melanin so that brown fur results. Mice with the *cc* genotype are albinos because melanin production gets blocked at an early point in the pathway, before the *B* gene enzyme comes into play.

FIGURE 9.18 The Environment Can Alter the Effects of Genes

Coat color in Siamese cats is controlled by a temperature-sensitive allele. The C^t allele of the *C* gene directs melanin production only at lower temperatures (below 37°C). Melanin therefore accumulates only in the colder extremities—the snout, tail, lower legs, and edges of the ears—and not in the main trunk, which is at the core body temperature (37°C).

The *C* gene is therefore said to be epistatic to the *B* gene. The fact that the *C* gene can obscure the effects of the *B* gene does *not* mean that the alleles of the *C* gene are dominant over those of the *B* gene. Rather, the protein encoded by the *C* gene simply acts earlier in the pathway. Without a functional *C* gene, any enzyme produced by the *B* gene has nothing to act upon.

The environment can alter the effects of a gene

The effects of many genes depend on internal and external environmental conditions, such as body temperature, carbon dioxide levels in the blood, external temperature, and amount of sunlight. For example, in Siamese cats the C^t allele of the gene that codes for the enzyme tyrosinase is sensitive to temperature (**FIGURE 9.18**). The C^t allele codes for a tyrosinase that works well at colder temperatures (35°C) but does not work at all at warmer temperatures (37°C). Because a cat's extremities tend to be colder than the rest of its body, melanin can be produced there; hence the paws, nose, ears, and tail of a Siamese cat tend to be dark. If a patch of light hair is shaved from the body of a Siamese cat and the skin is covered with a cold pack, when the hair grows back it will be dark. Similarly, if dark hair is shaved from the tail and allowed to grow back under warm conditions, it will be light-colored.

Chemicals, nutrition, sunlight, and many other environmental factors can also alter the effects of genes. In plants, genetically identical individuals (clones) grown in different environments often exhibit phenotypic differences, including variations in height and in the number of flowers they produce. Plants growing on a windswept mountainside may be short and have few flowers, while clones of the same plants grown in a warm, protected valley may be tall with many flowers. Similar effects are found in people. For example, a person who was malnourished as a child is likely to be shorter as an adult than if he or she had received good nutrition early in life.

Most traits are determined by two or more genes

Most traits are **polygenic**; that is, they are governed by the action of more than one gene—unlike the traits that Mendel studied. Examples of polygenic traits include skin color, running speed, blood pressure, and body size in humans; and flowering time and seed number in plants. Let's look in more detail at one of these examples, the inheritance of skin color in humans.

As mentioned earlier, the pigment melanin is the main source of skin color in humans and most other mammals. Geneticists estimate that more than a dozen genes are involved in controlling how much melanin is deposited in the specialized pigment-bearing cells in human skin. For simplicity, we will assume there are only three such genes (*A*, *B*, and *C*) and that each affects skin color equally. We will also assume that there are only two incompletely dominant versions of each gene: one version (A^1, B^1, C^1) that causes melanin production, and another (A^0, B^0, C^0) that prevents melanin production.

In this model, the phenotype of skin color is controlled by how many "units" of the melanin-producing allele the person has; the more such units there are, the darker the skin is. If two individuals heterozygous for each of the three genes were to marry, a child of theirs could have any one of seven potential skin color phenotypes, as shown in **FIGURE 9.19**. The Punnett square in Figure 9.19*b* shows the 64 different combinations and the skin color value that each combination produces. Notice that all combinations containing, for example, two non-melanin-producing alleles (open circles) have

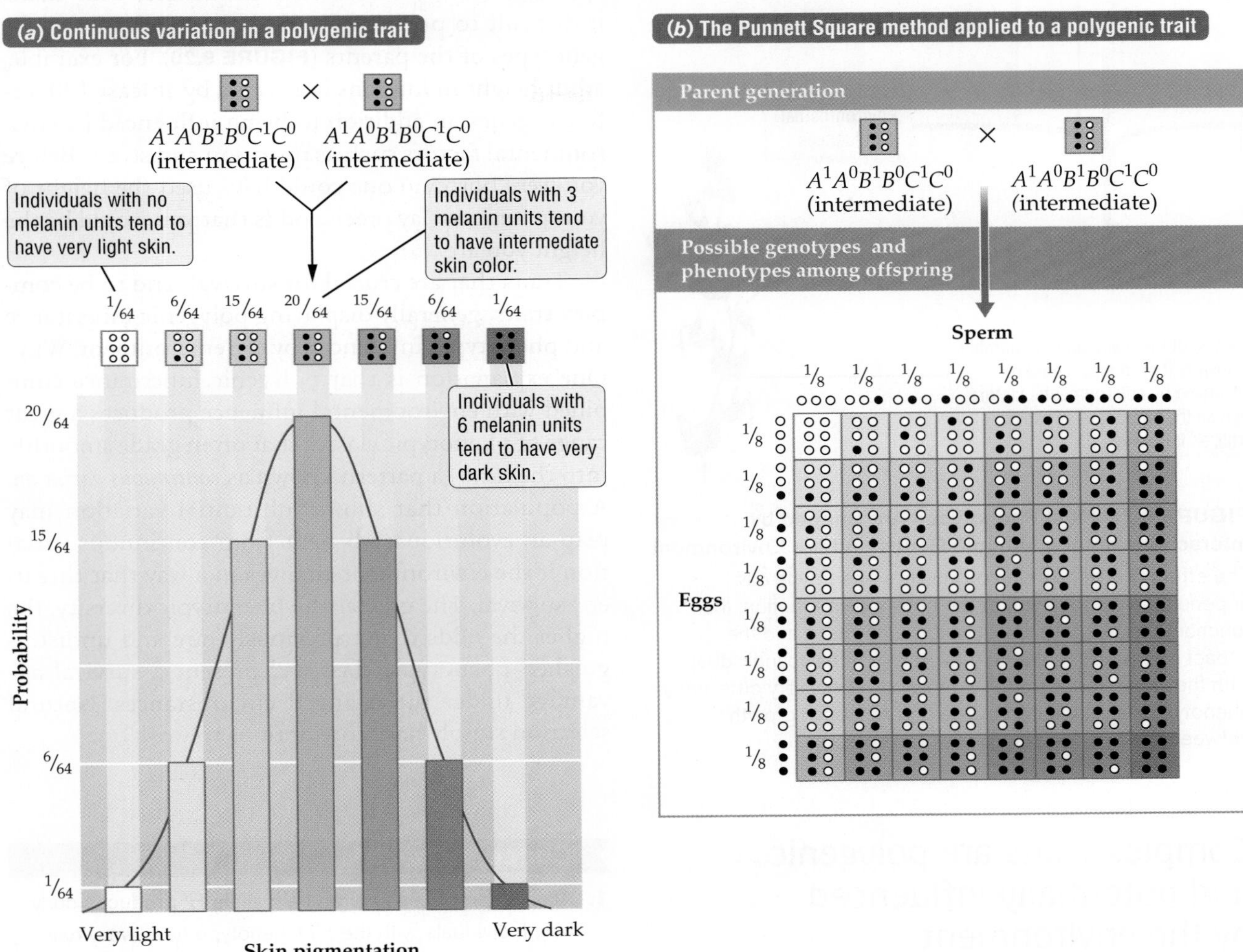

FIGURE 9.19 Three Genes Can Produce a Range of Skin Color in Humans

In this model, skin color is influenced by the total number of melanin "units" specified by the person's genotype. Alleles that do not contribute to melanin production (A^0, B^0, and C^0) are represented by open circles, while alleles that do contribute to melanin production (A^1, B^1, and C^1) are represented by solid circles. (*a*) The bar graph depicts the seven phenotypic outcomes arranged from lightest skin to darkest skin. Bar heights indicate the relative proportions of children of each phenotype. (*b*) The Punnett square shows the proportions of gametes with specific genotypes that are produced by two heterozygote ($A^1A^0B^1B^0C^1C^0$) parents and all the possible ways in which the eight different genotypes can come together during fertilization. In some cases, several different genotypes can produce the same phenotype. Additional variation in skin color would result from different levels of sun exposure.

the same color, regardless of which two genes are involved. Also notice that the darkest color value belongs to individuals with six melanin-producing alleles (solid circles); and the lightest color value, to those with six non-melanin-producing alleles. You can see from this example how complex the situation becomes with more than a dozen genes responsible for melanin production.

As demonstrated by a suntan, the skin color phenotype can be influenced by the environment. In most people, melanin production increases in response to the ultraviolet radiation in sunlight. If we added different degrees of suntanning to the seven shades of skin color predicted on the basis of genotype alone, we would get an even greater variety of skin colors—enough to fit a smooth, bell-shaped curve to the bar graph shown in Figure 9.19*a*.

In summary, human skin comes in just about every shade, from very light to very dark, because skin color in humans is a polygenic trait. In addition, skin color is affected by environmental factors, adding enough variety to the palette of skin color phenotypes that there is almost continuous variation in this trait.

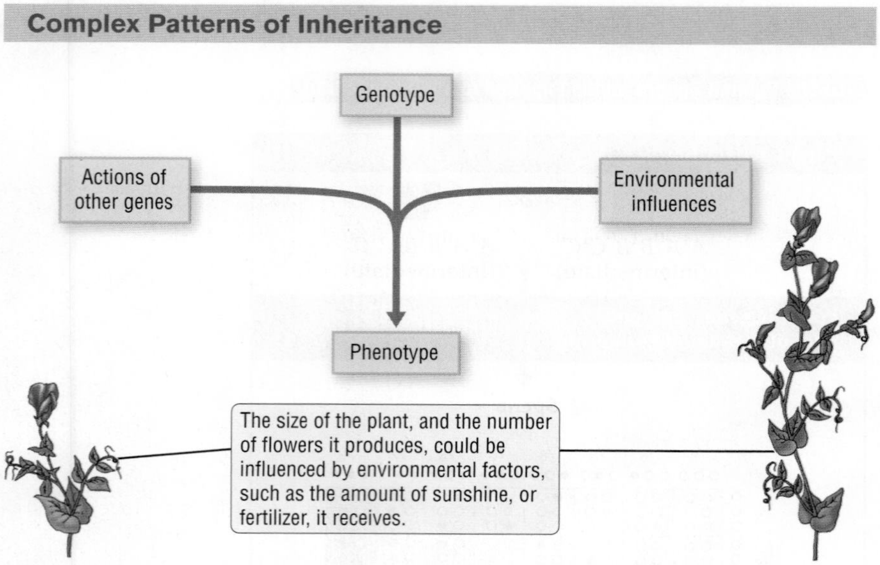

Genotype

Actions of other genes

Environmental influences

Phenotype

The size of the plant, and the number of flowers it produces, could be influenced by environmental factors, such as the amount of sunshine, or fertilizer, it receives.

FIGURE 9.20 Most Phenotypes Are Shaped by Interactions among Multiple Genes and the Environment
The effect of a gene on an organism's phenotype can depend on a combination of the gene's own function, the function of other genes with which it interacts, and the impact of environmental factors. As a result, two individuals with the same genotype for a gene may show very different phenotypes, as illustrated by the differences in growth between the two pea plants depicted here.

Complex traits are polygenic and potentially influenced by the environment

Chances are that just about all the traits you identified in Figure 9.3 are **complex traits**—that is, genetic traits whose pattern of inheritance cannot be predicted by Mendel's laws of inheritance. It is not that the inheritance of these traits violates Mendelian principles. Rather, there are additional complications, such as epistasis and environmental influences, that make it difficult to predict offspring phenotypes from the genotypes of the parents (**FIGURE 9.20**). For example, adult height in humans is affected by at least 190 different genes, in addition to being influenced by environmental factors such as nutrition and stress. Before you were born, no one could have used the height of your parents to lay precise odds that you would be the height you are.

Traits that are crucial for survival tend to be complex traits, generally displaying polygenic inheritance and phenotypes influenced by the environment. Why? One explanation is that polygenic inheritance combined with environmental influence produces a great range of phenotypic classes that often grade smoothly into the next—a pattern known as *continuous variation*. A population that shows substantial variation may reap an evolutionary benefit from continuous variation if the environment changes in a way that threatens survival. The greater the phenotypic diversity, the higher the odds that a previously rare and undistinguished phenotypic class will present a survival advantage under the changed circumstances. Natural selection simply has more options to work on.

Concept Check

1. Allele *H* produces straight hair; allele *H′* produces curly hair. Individuals with the *HH′* genotype have wavy hair, somewhere between straight and curly. Are alleles *H* and *H′* codominant?
2. What is pleiotropy?
3. The ABO blood groups are determined by several alleles of the *I* gene. Is ABO blood type a polygenic trait?

Concept Check Answers

1. No. They display incomplete dominance because they produce an intermediate phenotype in the heterozygote.
2. In pleiotropy, a single gene affects several to many different genetic traits.
3. No. "Polygenic" means "many genes;" ABO blood groups are controlled by a single gene with multiple alleles.

APPLYING WHAT WE LEARNED

Solving the Mystery of the Lost Princess

During the 1917 Russian Revolution, Czar Nicholas II abdicated his throne, leaving his country in the hands of a government that was immediately overthrown by Bolshevik revolutionaries. As long as the czar and czarina were alive, they served as a rallying point for those who opposed the Bolsheviks, but their deaths would also trigger an outcry. To prevent anyone from being certain about the fate of the family, the Bolsheviks secretly buried the czar and his family and told people that Czarina Alexandra and her daughters were in hiding. The resulting uncertainty over the fate of Alexandra and her daughters set the stage for the legend of Anastasia (**FIGURE 9.21**).

In 1991, the grave of Czar Nicholas and his family was finally discovered. To verify their identities, investigators electronically superimposed photographs of the family's skulls onto old photographs and also compared the sizes of the skeletons with their clothes, which had been saved. Such tests strongly supported the hypothesis that the skeletons in the grave were those of the Russian royal family. Still, more definitive evidence was needed.

By the late twentieth century, biologists had figured out that genes are made of DNA and that DNA can be extracted from all sorts of living and once-living tissues, including bone. Because different families have different combinations of alleles, it is possible to determine whether two people are related by comparing their alleles. When DNA from the skeletons was compared with DNA from the living descendants of the Romanovs, the results showed conclusively that the skeletal remains were those of the Russian royal family.

But missing from the Romanov grave were two sets of bones: those of Prince Alexei and those belonging to one of two princesses (actually grand duchesses)—either Maria or Anastasia. It was hard to tell which because the two sisters were about the same height. Was it possible that Anastasia had escaped? To find out whether Anna Anderson (**FIGURE 9.22**) could have been Anastasia, investigators obtained Anderson's DNA from tissue samples saved by her doctors and then compared her alleles with those of the czar and czarina.

In human DNA, five alleles of one gene—called A^1, A^2, A^3, A^4, and A^5—are common in people of European descent. The czar had genotype A^1A^2, and the czarina had genotype A^2A^3. According to Mendel's laws, for Anna to have been the daughter of the czar and czarina, she could have had any one of the following four genotypes: A^1A^2, A^1A^3, A^2A^2, or A^2A^3. If this seems confusing, draw a Punnett square with the czar's contributions (A^1 and A^2) on the top and the czarina's (A^2 and A^3) on the left side.

Anna Anderson's genotype was A^4A^5, which meant the czar and czarina could not have been her parents. Three other sets of alleles yielded the same result, showing that Anna Anderson could not be Grand Duchess Anastasia. Furthermore, in 1995, comparisons between Anna Anderson's DNA and that of the descendants of Franziska Schanzkowska's family showed that Anna was almost certainly a member of the Polish family.

The clincher came in August 2007, when an archaeologist located the last two members of the Romanov family, who had been buried some distance away from their relatives. In 2009, DNA analysis confirmed that they were the skeletons of the Romanovs' son Alexei and one of his sisters. Since there is currently no way to tell Maria and Anastasia apart by their skeletons, we may never know for certain which grave Anastasia was buried in, but we do know that she died in July 1918 with her family.

FIGURE 9.21 Anastasia and One of Her Three Sisters
Grand Duchesses Maria (left) and Anastasia visiting a hospital for soldiers.

FIGURE 9.22 Anna Anderson
Franziska Schanzkowska, also known as Anna Anderson, in about 1931. Anna Anderson was a Polish factory worker who was 5 years older than Grand Duchess Anastasia and spoke no Russian. Yet, many people believed the two women were the same person.

White Tiger Genetic Secret Unveiled: Single Mutation in Single Gene Removes Orange Color

BY ROXANNE PALMER • *International Business Times*, May 23, 2013

Chinese and South Korean scientists have discovered the secret of the white tiger's coat: a single mutation in a single gene. White tigers have been bred at zoos for decades, but have been spotted in the wild as far back as the 1500s. They're not albinos—these cats can still produce dark pigment, as seen in their stripes, but they have a genetic mutation that prevents their coats from expressing orange pigment . . .

While many traits are governed by multiple genes, tiger breeding practices have established that the white coat trait is a single recessive trait, according to Shu-Jin Luo, a researcher at the Peking-Tsinghua Center for Life Sciences.

"Thus we expect that the white tiger phenotype comes from a single mutation," Luo wrote in an email. But the actual mutation itself has long escaped discovery.

"We tried many genes that have been reported to be associated with animal coat color, but none of them is responsible for the white tiger coat," Luo wrote.

So, the researchers had to look at the whole tiger genome. First they sequenced a small piece—about 5 percent—of the genomes of 16 tigers (seven white tigers and nine orange tigers) spread across two generations to focus on a region of genes known to be associated with white coat color. Then, the researchers looked at the whole genomes of the three parent tigers (two white tigers and one orange tiger), to see where they differed from their white offspring.

That led them to a small mutation in a gene called *SLC45A2*, where just a single amino acid in an entire protein is changed. That tiny difference prevents the tiger from making the reddish and yellow pheomelanin pigments, but does not affect the production of the black pigment eumelanin. The team then confirmed that the *SLC45A2* mutation results in a white coat by scrutinizing 130 other unrelated tigers, both orange and white.

Hair color in mammals is produced by two main types of pigments: eumelanin is brown to black; pheomelanin is yellow to red. It takes a chain of several enzyme-driven chemical reactions to convert an amino acid called tyrosine into eumelanin or pheomelanin. The first few steps are the same for both pathways. The first step—turning tyrosine into a pigment precursor—is spurred by an enzyme called tyrosinase, encoded by the *C* gene. As we saw in this chapter, the *c* allele is the nonfunctional version of the *C* gene. In animals ranging from snakes to brown bears, homozygotes for the allele (*cc*) have the albino phenotype. In albinos, virtually no eumelanin or pheomelanin is made in skin, hair, or the iris of the eye.

Many more genes act farther down the production line for these pigments. The orange gene (*O*) is necessary for pheomelanin production in the wild-type Bengal tiger, as well as in house cats such as the tortoiseshell and calico breeds. Felines that are homozygous for the inactive version of the gene (*oo*) fail to make pheomelanin and have white fur. Dark markings—such as stripes—are caused by heavy deposits of eumelanin in localized regions of the body, and in a *CC* or *Cc* genotype these markings are epistatic to the *O/o* allele. In other words, the eumelanin markings look the same in *OO*, *Oo*, and *oo* genotypes because the eumelanin deposition called for by the *C* allele obscures any effect of the *O/o* allele.

The pigments are made in special cells called melanocytes, which produce *melanosomes*—little packets of pigment wrapped in membrane. The living cells at the base of a hair follicle get their color by engulfing these membrane-enclosed gifts of pigment.

The gene identified by Luo and colleagues—*SLC45A2*—codes for a tyrosine transporter in the plasma membrane of melanocytes. In tigers homozygous for the "white" allele of *SLC45A2*, tyrosine doesn't enter the cells, so little to no pheomelanin is made, which is why the fur appears white in the stripe-free zone. (The striped part of the fur has melanocytes that respond to a different set of signals, and they are able to make eumelanin through a different route.) Even before this study, the *SLC45A2* gene was known to be one of several genes involved in light coloration in horses, chickens, and people.

Evaluating the News

1. Look at the picture of the white tiger. How can you be sure that the animal is not an albino? If we represent the normal allele of *SLC45A2* as *W* and the white-coloration allele as *w*, predict the phenotypes and genotypes of a cross between a *Ww* tiger and a white tiger.
2. Is fur color in tigers a polygenic trait? Explain.

CHAPTER REVIEW

Summary

9.1 Principles of Genetics: An Overview

- Genes—the basic units of inheritance—are segments of DNA that help determine an organism's inherited characteristics, or genetic traits.
- Diploid individuals generally have two copies of each gene: one inherited from the male parent; the other, from the female parent.
- Alternative versions of a gene are called alleles. In a population of many individuals, a particular gene may have one, a few, or many alleles.
- The genotype is an individual's allelic makeup; the phenotype is the specific version of an observable trait that the individual displays.
- A dominant allele controls the phenotype of an individual even when paired with a different allele (heterozygous genotype). A recessive allele has no phenotypic effect when paired with a dominant allele.
- The different alleles of a gene arise by mutation and, by specifying different versions of the same protein, they generate hereditary differences among organisms.
- Some mutations are harmful, many have little effect, and a few are beneficial.

9.2 Basic Patterns of Inheritance

- Mendel's experiments with pea plants led him to reject the old notion of blending inheritance. Instead, they suggested that the inherited characteristics of organisms are controlled by specific units of inheritance, which we now know as genes.
- In modern terminology, Mendel's discoveries can be summarized as follows: (1) Alleles of genes account for the variation in genetic traits. (2) Offspring inherit one allele of a gene from each parent. (3) Alleles can be dominant or recessive. (4) The two copies of a gene separate into different (haploid) gametes. (5) Fertilization combines gametes in a random manner to give rise to diploid offspring.

9.3 Mendel's Laws of Inheritance

- Mendel's law of segregation states that the two copies of a gene (the two alleles) end up in different gametes during meiosis.
- The law of independent assortment says that when gametes form during meiosis, the alleles of one gene are distributed independently of the alleles of a different gene located on a different chromosome.

9.4 Extensions of Mendel's Laws

- For some traits, Mendel's laws may not predict the phenotype of the offspring because (1) many alleles show incomplete dominance or codominance; (2) one gene may affect more than one genetic trait (pleiotropy); (3) alleles for one gene can suppress the alleles of another gene (epistasis); (4) the environment can alter the phenotype of a gene; and (5) most traits are polygenic (governed by two or more genes).
- Traits that are crucial for survival tend to be complex traits. Complex traits are influenced by multiple genes that interact with one another and with the environment. The inheritance of complex traits cannot be predicted by Mendel's laws, because Mendelian traits are controlled by a single gene that is little affected by other genes or by environmental conditions.
- Complex traits often generate a great range of phenotypic classes, or continuous variation in phenotypes. A population that shows substantial phenotypic diversity may stand to reap an evolutionary benefit from continuous variation if the environment changes in a way that threatens survival.

Key Terms

allele (p. 194)
codominance (p. 204)
complex trait (p. 210)
dihybrid (p. 200)
dominant allele (p. 195)
epistasis (p. 207)
F_1 generation (p. 196)
F_2 generation (p. 196)
gene (p. 192)
genetic cross (p. 196)
genetic trait (p. 192)
genetics (p. 192)
genotype (p. 194)
heterozygote (p. 195)
homozygote (p. 195)
hybrid (p. 198)
incomplete dominance (p. 203)
law of independent assortment (p. 202)
law of segregation (p. 199)
monohybrid (p. 199)
mutation (p. 195)
P generation (p. 196)
phenotype (p. 193)
pleiotropy (p. 206)
polygenic (p. 208)
Punnett square (p. 199)
recessive allele (p. 195)

Self-Quiz

1. Alternative versions of a gene for a given trait are called
   a. alleles. c. genotypes.
   b. heterozygotes. d. copies of a gene.

2. If *A* and *a* are two alleles of the same gene, then individuals of genotype *Aa* are
   a. homozygous. c. dominant.
   b. heterozygous. d. recessive.

3. The illustration here shows the seven traits of garden peas that Gregor Mendel analyzed by conducting large numbers of crosses and recording the phenotypes of all their offspring over two generations. Which of the following statements is true?
   a. When Mendel crossed plants that were true-breeding for the seed shape trait, the F_2 plants had round seeds and wrinkled seeds in a ratio of 9:3.
   b. When Mendel crossed true-breeding tall plants with true-breeding dwarf plants, all the F_1 plants displayed the dwarf phenotype.
   c. When Mendel crossed true-breeding plants with green pods and true-breeding plants with yellow pods, the F_1 plants had pods of an intermediate color, somewhere between green and yellow.
   d. When Mendel crossed homozygous tall plants and homozygous dwarf plants, the F_2 progeny included one dwarf plant for every three tall plants.

4. Coat color in horses shows incomplete dominance (see Figure 9.13). What is the predicted phenotypic ratio of chestnut to palomino to cremello if H^CH^{Cr} individuals are crossed with other H^CH^{Cr} individuals?
   a. 3:1 c. 9:3:1
   b. 2:1:1 d. 1:2:1

5. When the phenotypes controlled by two alleles are equally displayed in the heterozygote, the two alleles are said to show
   a. codominance. c. incomplete dominance.
   b. complete dominance. d. epistasis.

6. Traits that are determined by the action of more than one gene are
   a. recessive. c. epistatic.
   b. not common. d. polygenic.

7. Which term indicates that the phenotypic effects of alleles for one gene can be suppressed by the alleles of another, independently inherited gene?
   a. phenotypic variation c. polygenic
   b. codominance d. epistasis

Trait	Dominant phenotype		Recessive phenotype	Ratio of dominant to recessive phenotypes in F_2 offspring
Flower color	Purple	×	White	3.15:1
Flower position	Axial	×	Terminal	3.14:1
Seed color	Yellow	×	Green	3.01:1
Seed shape	Round	×	Wrinkled	2.96:1
Pod shape	Inflated	×	Constricted	2.95:1
Pod color	Green	×	Yellow	2.82:1
Stem length	Tall	×	Dwarf	2.84:1

Analysis and Application

1. Describe what genes are and how they work. Include a summary of what we now know about (a) the chemical and physical structure of genes and (b) the information they encode. How many copies of each gene are found in the diploid cells in a woman's body? Explain.

2. Explain how new alleles arise, and how different alleles cause hereditary differences among organisms.

3. A purple-flowered pea plant could have genotype *PP* or *Pp*. What genetic cross could you make to determine the genotype of this plant? Explain how such a test cross enables you to deduce whether the purple-flowered plant is a homozygote or a heterozygote.

4. Can "identical" twins have different phenotypes? Why or why not?

5. Do the research to identify the four diseases responsible for more deaths than any other in North America. Is our risk for each of these diseases a genetic trait? If so, is it a complex trait? How would you explain a complex trait to your next-door neighbor? What is the practical relevance of the concept of complex traits in our everyday life?

SAMPLE GENETICS PROBLEMS: CHAPTER 9

1. If you were to repeat one of Mendel's dihybrid crosses, which of the following statements about the F_2 generation would be consistent with your results (that is, would hold true)?
   a. A total of four different genotypes is seen.
   b. A total of six different phenotypes is seen.
   c. All offspring with the same phenotype will possess the same genotype.
   d. It is not possible to tell the genotype of some offspring by looking at their phenotypes.

2. In some breeds of cats, the *L*/*l* alleles control hair length. Heterozyotes (*Ll*) have short hair; cats with the *ll* genotype have long hair. Which of these statements is true?
   a. The *L* and *l* alleles are codominant.
   b. The *L* and *l* alleles are incompletely dominant.
   c. In a mating of two heterozygotes, 25 percent of the offspring are predicted to have long hair.
   d. In a mating of *LL* and *ll* homozygotes, 50 percent of the offspring are predicted to have short hair.

3. If a child has an AB blood type, the parents
   a. must both be heterozygotes.
   b. could be A or B, but neither one of them could be AB.
   c. would both have to be AB.
   d. could not have blood type O.

4. One gene has alleles *A* and *a*, a second gene has alleles *B* and *b*, and a third gene has alleles *C* and *c*. List the possible gametes that can be formed from the following genotypes:
   a. *Aa*
   b. *BbCc*
   c. *AAcc*
   d. *AaBbCc*
   e. *aaBBCc*

5. For the same three genes described in problem 4, what are the predicted genotype and phenotype ratios of the following genetic crosses? (Following our standard notation, alleles written in uppercase letters are dominant over alleles written in lowercase letters; the phenotype produced by allele *A* is therefore dominant over the phenotype controlled by allele *a*.)
   a. *Aa* × *aa*
   b. *BB* × *bb*
   c. *AABb* × *aabb*
   d. *BbCc* × *BbCC*
   e. *AaBbCc* × *AAbbCc*

6. Sickle-cell anemia is inherited as a recessive genetic disorder in humans. With respect to this disorder, the normal hemoglobin allele (*S*) is dominant over the sickle-cell allele (*s*), so anemia is seen only in a double recessive genotype. For two parents of genotype *Ss*, construct a Punnett square to predict the possible genotypes and phenotypes of their children (do or do not have sickle-cell anemia). Also list the genotype and phenotype ratios. Each time two *Ss* individuals have a child together, what is the chance that the child will have sickle-cell anemia?

7. Alleles for a gene (*C*) that determines the color of Labrador retrievers show incomplete dominance. Black labs have genotype C^BC^B, chocolate labs have genotype C^BC^Y, and yellow labs have genotype C^YC^Y. If a black lab and a yellow lab mated, what proportions of black, chocolate, and yellow coat colors would you expect to find in a litter of their puppies?

8. For any human genetic disorder caused by a recessive allele, let *n* be the allele that causes the disease and *N* be the normal allele.
   a. What are the phenotypes of *NN*, *Nn*, and *nn* individuals?
   b. Predict the outcome of a genetic cross between two *Nn* individuals. List the genotype and phenotype ratios that would result from such a cross.
   c. Predict the outcome of a genetic cross between an *Nn* and an *NN* individual. List the genotype and phenotype ratios that would result from such a cross.

9. For any human genetic disorder caused by a dominant allele, let *D* be the allele that causes the disorder and *d* be the normal allele (where the capital "*D*" stands for "disorder").
   a. What are the phenotypes of *DD*, *Dd*, and *dd* individuals?
   b. Predict the outcome of a genetic cross between two *Dd* individuals. List the genotype and phenotype ratios that would result from such a cross.
   c. Predict the outcome of a genetic cross between a *Dd* and a *DD* individual. List the genotype and phenotype ratios that would result from such a cross.

10. If blue flower color (*B*) is dominant over white flower color (*b*), what are the genotypes of the parents in the following genetic cross: blue flower × white flower, yielding only blue-flowered offspring?

11. The fruit pods of peas can be yellow or green. In one of his experiments, Mendel crossed plants homozygous for the allele for yellow fruit pods with plants homozygous for the allele for green fruit pods. All fruit pods in the F_1 generation were green. Which allele is dominant: the one for yellow or the one for green? Explain why.

10 Chromosomes and Human Genetics

MISTY OTO WITH HER MOTHER AND DAUGHTER. Misty Oto's mother, Rosie Shaw, died of Huntington's disease, a genetic disorder passed through families as a dominant allele. Oto, who works to educate people about Huntington's, has lost several family members to the disease. When her daughter was born, Oto did not know whether she or her daughter carried the disease allele.

Family Ties

One of the first things Misty Oto remembers is her grandmother's death. Lots of kids have sad memories of losing a grandparent, but the disease that killed Misty's grandmother would profoundly shape Misty's life. Misty's grandmother died of Huntington's disease, an unusual late-onset genetic disease caused by a dominant allele. The only one of 10 children to inherit the disease, her grandmother passed the disease to all three of her daughters, including Misty's mother, Rosie. By the time Misty was 11, Rosie had begun to change from a vibrant, loving mother to an empty shell.

Symptoms of the disease usually do not show up until later in life. As Misty grew up, Huntington's decimated her family, taking her mother, both her aunts, and her brother. Yet Misty had not taken the genetic test that would tell her whether she and her children had the Huntington's gene. None of Misty's four other siblings and the nine kids in the next generation had been tested. The gene for Huntington's disease was located in 1993, and a genetic test was developed soon after. But there is still no cure. As Misty told a radio interviewer, "Why would I even put myself through that process? And to know that I have this black cloud over my head and that there's nothing I could do about it."

But in 2010, Misty's older sister began thinking about the future. If she became ill with Huntington's, she would need long-term care. Who would care for her? She asked Misty to be her caregiver, and without thinking, Misty said yes. Only then, Misty recalled, did she realize that she herself might be too ill to help. For the first time, she considered being tested.

What is the genetic basis of Huntington's disease? And why do 90 percent of at-risk people avoid the test?

To begin to understand the genetic basis of Huntington's disease, we must explore the chromosomal basis of inheritance. Along the way, you will see why some genes tend to be inherited together and why some genetic conditions are more common in males. We'll examine the patterns of inheritance that explain some common and rare human genetic disorders, including Huntington's disease.

MAIN MESSAGE The inheritance of genes is affected by their precise location on specific chromosomes.

KEY CONCEPTS

- According to the chromosome theory of inheritance, genes are located on chromosomes, which pass from parent cell to daughter cell during cell division.
- The chromosomes that determine sex in humans are the sex chromosomes, X and Y. The XY combination confers maleness, and the XX combination leads to female characteristics. The other 22 chromosome pairs are autosomes.
- The *SRY* gene, found on the Y chromosome, is a key gene for the development of male sexual characteristics.
- Genetic diversity results from crossing-over, independent assortment of chromosomes, and random fertilization of gametes.
- Genes located near one another on the same chromosome tend to be inherited together and are said to be linked. Genes located far apart on the same chromosome often behave like unlinked genes (genes on different chromosomes).
- Pedigrees can be useful tools for analyzing human genetic disorders.
- Many inherited genetic disorders in humans are caused by mutations of single genes.
- Dominant disorders with serious effects are generally rare. Harmful recessive alleles are more common because heterozygotes can transmit them from one generation to the next.
- Recessive alleles located on the X chromosome often produce gender-specific phenotypes known as X-linked traits. Recessive X-linked conditions tend to be more common among males than females.
- Some human genetic disorders result from abnormalities in chromosome number or structure. Most of these are lethal to the embryo or fetus, or cause death in infancy.

FAMILY TIES 217
10.1 The Role of Chromosomes in Inheritance 218
10.2 Genetic Linkage and Crossing-Over 221
10.3 Human Genetic Disorders 223
10.4 Autosomal Inheritance of Single-Gene Mutations 227
10.5 Sex-Linked Inheritance of Single-Gene Mutations 229
10.6 Inherited Chromosomal Abnormalities 230
APPLYING WHAT WE LEARNED 233
Testing for Huntington's Disease

OVER A HUNDRED YEARS AGO, Gregor Mendel deduced that inherited traits are governed by discrete hereditary units. As you learned in Chapter 9, these hereditary units are known today as genes. We begin this chapter with a second cornerstone of modern genetics: the *chromosome theory of inheritance*. We then explain how an individual's sex is determined in humans and other mammals, and how new combinations of alleles, different from those of either parent, can occur in offspring. This information about chromosomes, sex determination, and new allele combinations sets the stage for the discussion of human genetic disorders in the remainder of the chapter.

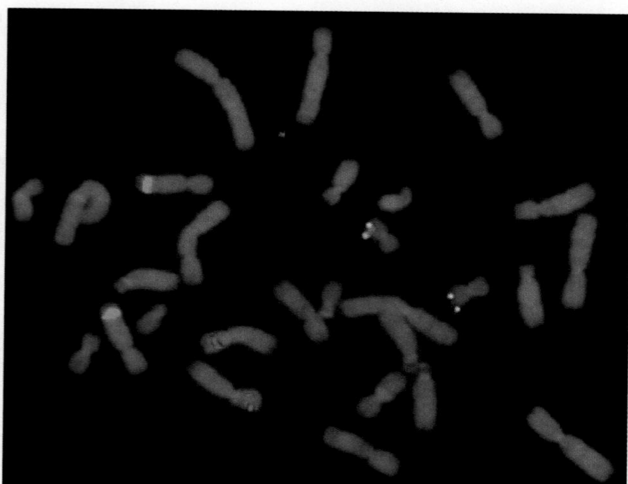

FIGURE 10.1 Genes Are Located on Chromosomes
A technique called fluorescence in situ hybridization (FISH) was used to show the location of three different genes on these chromosomes prepared from tumor cells in mitosis. Each of the three genes (*HER2*, green; *p16*, pink; *znk217*, gold) is known to be involved in cancer. These mitotic chromosomes are duplicated: each of them consists of a pair of identical sister chromatids. Can the pattern of inheritance be affected by the type of chromosome a gene is located on?

10.1 The Role of Chromosomes in Inheritance

When Mendel published his theory of inheritance in 1866, he had no idea what genes were made of or where they were located within a cell. By 1882, studies using microscopes had revealed that threadlike structures—the *chromosomes*—exist inside dividing cells. The German biologist August Weismann hypothesized that the number of chromosomes was first reduced by half during the formation of sperm and egg cells, and then restored to its full number during fertilization. The discovery of meiosis in 1887 supported Weismann's hypothesis. Weismann also suggested that the hereditary material was located on chromosomes, but at that time there was no experimental evidence for or against that idea.

Genes are located on chromosomes

Early in the twentieth century, geneticists gathered a great deal of experimental evidence in support of Weismann's hypothesis. The concept that genes are located on chromosomes came to be known as the **chromosome theory of inheritance**. Modern genetic techniques enable us to pinpoint which chromosome contains a particular gene and where on the chromosome that gene is located (**FIGURE 10.1**).

How are chromosomes, DNA, and genes related? As described in Chapters 7 and 9, chromosomes that pair during meiosis are called *homologous chromosomes* (**hoh-*MAH*-luh-gus**). One member of each pair of homologous chromosomes is inherited from the female parent (the maternal chromosome); the other, from the male parent (the paternal chromosome). Each chromosome consists of a single long DNA molecule attached to many bundles of packaging proteins (see Figure 7.7). Each gene is a small region of the DNA molecule, and many genes reside on each chromosome. For example, humans are estimated to have about 25,000 genes located on one set of our chromosomes, which consists of 23 different types of chromosomes. On average, then, we have 25,000/23 (or 1,086) genes per chromosome.

The physical location of a gene on a chromosome is called a **locus** (plural "loci"). With the exception of the sex chromosomes, every diploid cell has two copies of every gene, one on each of the chromosomes in a homologous pair, as **FIGURE 10.2** shows. Because a

Genetic Loci on Homologous Chromosomes

In a pair of homologous chromosomes, one is inherited from the male parent, the other from the female parent.

Paternal homologue

Maternal homologue

A genetic locus is the location of a particular gene on a chromosome.

At each genetic locus, an individual has two alleles, one on each homologous chromosome.

The alleles may be identical (as in *DD* or *ee* individuals)...

...or different (as in *Gg* individuals).

Three different genes, or three different genetic loci.

r *r* *H* *h* *D* *D* *e* *e* *g* *G*

FIGURE 10.2 Genes Are Located on Chromosomes
The genes shown here take up a larger portion of the chromosome than they would if they were drawn to scale. The average human chromosome has more than a thousand different genes interspersed with large stretches of noncoding DNA.

gene can come in different versions, or alleles, a diploid cell can have two *different* alleles at a given genetic locus, in which case it has a *heterozygous genotype* for the gene at that locus. But if the two alleles at a genetic locus are identical, then the diploid cell has a *homozygous genotype* for the gene at that locus. As discussed in Chapter 9, the allelic makeup, or genotype, at a particular genetic locus influences the phenotype, the detectable appearance of a genetic trait.

Autosomes differ from sex chromosomes

As noted in Chapter 9, the two chromosomes that make up a homologous pair are exactly alike in terms of length, shape, and the genetic loci they carry. But in humans and many other organisms, the chromosomes that determine the sex of the organism are dissimilar. These *sex chromosomes* are assigned different letter names. In humans, for example, males have one X chromosome and one Y chromosome (**FIGURE 10.3**), whereas females have two X chromosomes. The Y chromosome in humans is much smaller than the X chromosome, and few of its genes have counterparts on the X chromosome. Because human males have one X and one Y chromosome, they have only one copy of each gene that is unique to either the X or the Y chromosome.

Chromosomes that determine sex are called **sex chromosomes**; all other chromosomes are called **autosomes**. Human autosomes are labeled not with letters, but with the numbers 1 through 22 (for example, chromosome 4).

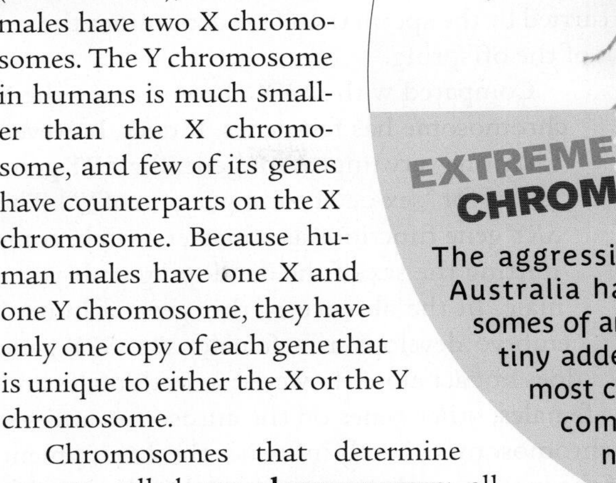

EXTREME CHROMOSOME NUMBERS

The aggressive jack jumper ant of Australia has the fewest chromosomes of any eukaryote: just 2. The tiny adder's tongue fern has the most chromosomes: 1,260. The complexity of a species is not related to how many chromosomes it has.

In humans, maleness is specified by the Y chromosome

Because human females have two copies of the X chromosome, all the gametes (eggs) they produce contain one X chromosome. Males, however, have one X chromosome and one Y chromosome, so the odds are that half of their gametes (sperm) will contain an X chromosome and half will contain a Y

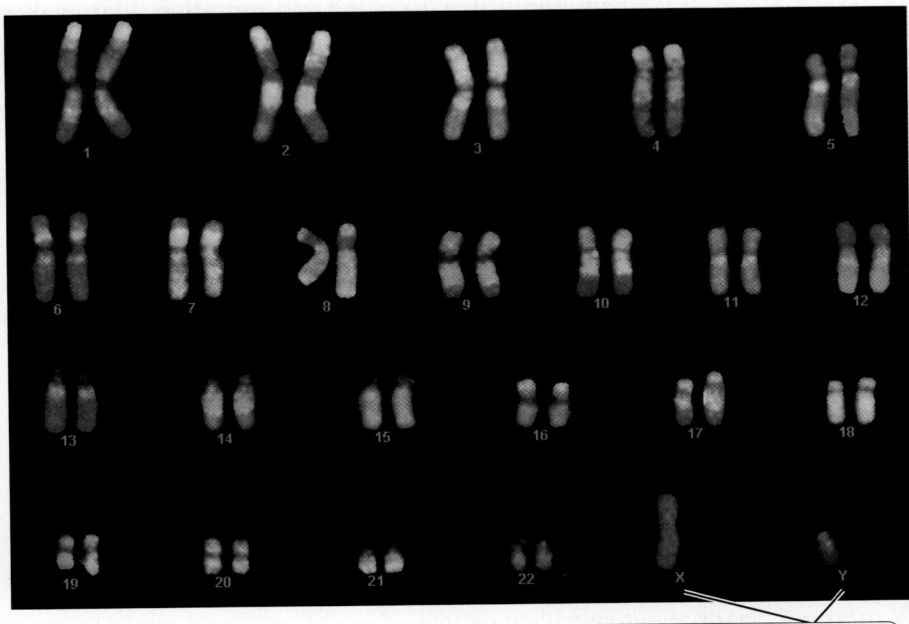

While the two autosomes in each pair are similar in shape and size, the X and Y chromosomes are not.

FIGURE 10.3 Autosomes and Sex Chromosomes
These chromosomes have been stained using the FISH technique (fluorescence in situ hybridization).

Helpful to Know

Biologists often name particles found inside cells using the suffix *some* (sohm; "body"). For example, *autosomes* ("same bodies") are chromosomes whose homologues look similar in both sexes. *Chromosomes* (literally, "colored bodies") were named for their appearance under the microscope when first discovered, before it was known what they were.

chromosome (**FIGURE 10.4**). The sex chromosome carried by the sperm therefore determines the sex of the offspring.

Compared with the X chromosome, the Y chromosome has few genes. It does, however, carry one very important gene: the ***SRY* gene** (short for "sex-determining region of Y"). The *SRY* gene functions as a master switch, committing the sex of the developing embryo to male. In the absence of this gene, a human embryo develops as a female. The *SRY* gene does not act alone, however: in both males and females, other genes on the autosomes and sex chromosomes directly influence the development of the many sexual characteristics that distinguish men and women. The *SRY* gene plays a crucial role because when it is present, it causes the other genes to produce male sexual characteristics, whereas when it is absent, the other genes produce female sexual characteristics. For example, the *SRY* gene directs the development of testes, the male reproductive organs, in place of ovaries.

The role of the non-*SRY* genes in gender determination becomes evident in disorders of sexual development, a broad range of conditions in which a person appears to be a gender different from the one predicted by his or her sex chromosomes. For example, individuals with *androgen insensitivity syndrome* (*AIS*) are female in appearance and say they "feel female," even though they are XY in their chromosomal makeup, and their ovaries fail to develop normally. AIS is most commonly due to a genetic mutation that makes these individuals unable to respond normally to male hormones such as testosterone.

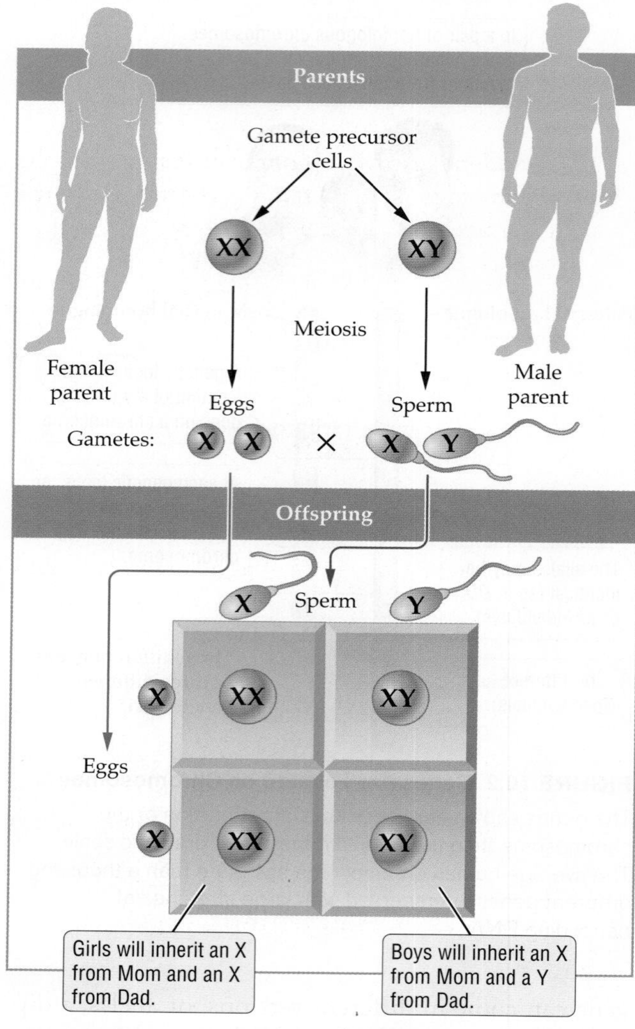

FIGURE 10.4 Dad's Chromosomes Determine Baby's Gender

Human females have two X chromosomes, while human males have one X and one Y chromosome. If a baby receives an X chromosome from the father, it's a girl! If a baby receives a Y chromosome from the father, it's a boy! The odds are 50 percent that a given sperm cell will contain an X or Y chromosome.

EXTREME CHROMOSOMAL LEGACY

The Mongol warlord Genghis Khan (about 1162–1227) conquered much of Asia. He and his sons had large harems and left so many offspring that as much as 0.5 percent of the world's population may carry his Y chromosome. That's about 35 million people. Because the Y chromosome doesn't recombine much, mutations tend to accumulate on it as linked alleles.

Crossing-over and independent assortment of chromosomes add to genetic diversity in a population

Inheritance is both a stable and a variable process. It is stable in that, most of the time, genetic info rmation is transmitted accurately from one generation to the next. Despite this stability, offspring in sexually reproducing organisms are not exact genetic copies of their parents. These genetic differences are important because they provide the genetic variation on which evolution can act.

How do genetic differences among individuals arise? As we saw in Chapter 9, new alleles arise by mutation. Once formed, those alleles are shuffled or arranged in new ways by *crossing-over*, *independent assortment of chromosomes*, and *random fertilization*.

- **Crossing-over** is a reciprocal exchange of segments of non-sister chromatids during prophase

I of meiosis. Crossing-over is part of the reason why two parents can produce offspring with a range of different phenotypes. It is one reason why siblings are not identical to one another (unless they are identical twins). **FIGURE 10.5** illustrates an example of crossing-over.

- **Independent assortment of chromosomes** results from homologous pairs lining up at random along the metaphase plate during metaphase I of meiosis (see Figure 7.14, steps 2 and 3). The paternal and maternal members of any homologous pair are sorted into gametes independently of any other pair. The larger the number of homologous pairs, the greater the variety of patterns in which homologues can be arranged at metaphase I, and the greater the variety of gametes (see Figure 7.17). For 23 pairs of chromosomes, the number of combinations would be 2^{23}, or 8,388,608 different types of gametes.
- **Random fertilization** means that any sperm can fertilize any egg, so that the genetic combinations possible for a single fertilization is an enormous number—over 64 trillion, more than the number of humans who have ever lived.

Of the three events just described, crossing-over is the one that results in novel combinations of genes on the chromosomes—combinations of genes in offspring that did not exist in the parents. Crossing-over is a random process generating patterns of inheritance that cannot be predicted using Mendel's laws.

Concept Check

1. What is the chromosome theory of inheritance?
2. How are sex chromosomes different from autosomes?

10.2 Genetic Linkage and Crossing-Over

As we saw in Chapter 9, the results of Mendel's experiments indicated that genes are inherited independently of one another (the law of independent assortment). Early in the twentieth century, however, results from several laboratories indicated that certain genes were often *inherited together*, contradicting the law of independent assortment. Much of this work was done on fruit flies, which have a short reproductive cycle. Because these insects develop from egg to reproductive adult in about 2 weeks, the inheritance of a trait can be tracked over many generations in a relatively short time span.

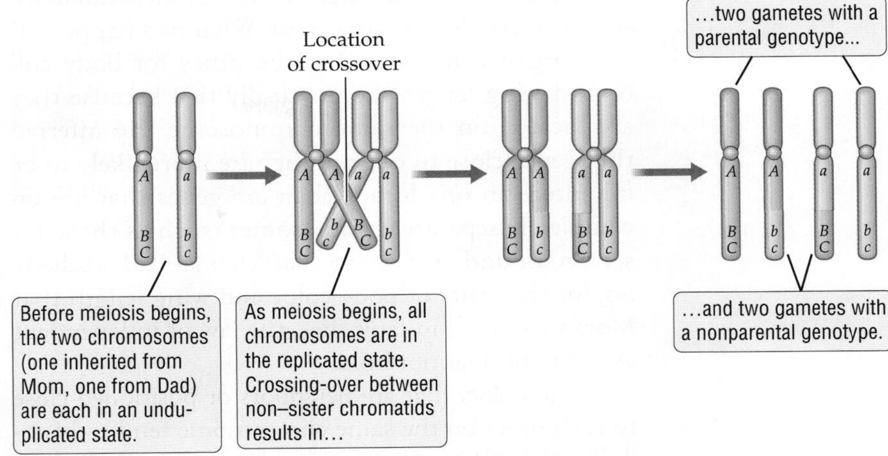

Crossing-Over
Location of crossover
...two gametes with a parental genotype...
...and two gametes with a nonparental genotype.
Before meiosis begins, the two chromosomes (one inherited from Mom, one from Dad) are each in an unduplicated state.
As meiosis begins, all chromosomes are in the replicated state. Crossing-over between non–sister chromatids results in...
A
a
B
b
C
c

FIGURE 10.5 Segments of Chromosomes Are Exchanged in Crossing-Over
Crossing-over takes place during prophase I. As a result of crossover between *A*/*a* and *B*/*b*, half of the gametes have a parental genotype (*ABC* or *abc*), while the other half have a nonparental genotype (*Abc* or *aBC*). In this example, there is no crossing-over between *B*/*b* and *C*/*c*.

Linked genes are located on the same chromosome

In his research on fruit flies, which began in 1909 at Columbia University in New York City, Thomas Hunt Morgan discovered genes that are inherited together. In one experiment, Morgan crossed a fruit fly that was homozygous for both a gray body (*G*) and wings of normal length (*W*) with another that was homozygous for both a black body (*g*) and wings that were greatly reduced in length (*w*). That is, he crossed *GGWW* flies with *ggww* flies to obtain flies of genotype *GgWw* in the F_1 generation.

Morgan then mated the *GgWw* flies with *ggww* flies, doubly recessive homozygotes. This kind of cross, in which one parent is known to be a recessive homozygote, is called a **test cross**. It enables geneticists to evaluate whether an individual with a dominant phenotype is a homozygote or a heterozygote. In this case, knowing that the F_1 flies were heterozygotes, Morgan expected to find roughly equal numbers of the four possible genotypes and phenotypes of the test-cross offspring:

GENOTYPE	PHENOTYPE
GgWw	Gray body, normal wings
ggWw	Black body, normal wings
Ggww	Gray body, short wings
ggww	Black body, short wings

Concept Check Answers

1. The theory states that genes are located on chromosomes and are therefore inherited when chromosomes are inherited by daughter cells produced when a cell divides by mitosis or meiosis.

2. Sex chromosomes differ in males and females and play a primary role in determining the sex of an individual; autosomes, the remaining chromosomes, are the same in both sexes.

As **FIGURE 10.6** shows, Morgan's results were very different from the results that the law of independent assortment had led him to expect. What had happened?

Morgan concluded that the genes for body color and wing length are physically tied because they are located on the same chromosome. He inferred that genes close to one another are more likely to be inherited "in one lump" than are genes that are on completely separate chromosomes (such as those for seed color and seed shape that Mendel had studied). So, for the traits of body color and wing length that Morgan studied in fruit flies, the law of independent assortment does not hold.

Genetic loci that are neighbors or positioned close to each other on the same chromosome tend to be inherited together and are said to be **genetically linked**. As we explain in the next section, some genes that are located far apart on a single chromosome behave as though they are not genetically linked. Genes located on different chromosomes also are not genetically linked.

Crossing-over reduces genetic linkage

If the linkage between two genes on a chromosome could never be altered, the chromosomes inherited by a gamete would never differ from those of the parent that produced that gamete. Consider, for example, the offspring of the *GGWW* × *ggww* cross shown in Figure 10.6. Since the two genes are on the same chromosome, the *GgWw* flies would have inherited a *GW* chromosome from the *GGWW* parent and a *gw* chromosome from the *ggww* parent. If linkage were absolute, however, the *GgWw* flies would have been able to make only gametes having chromosomes like those in one of their parents—namely, *GW* gametes or *gw* gametes (**FIGURE 10.7**). In that case, half of the offspring from the *GgWw* × *ggww* cross shown in Figure 10.6 would have had genotype *GgWw*, and the other half would have had genotype *ggww*. Because the majority of the offspring did have those two genotypes, Morgan

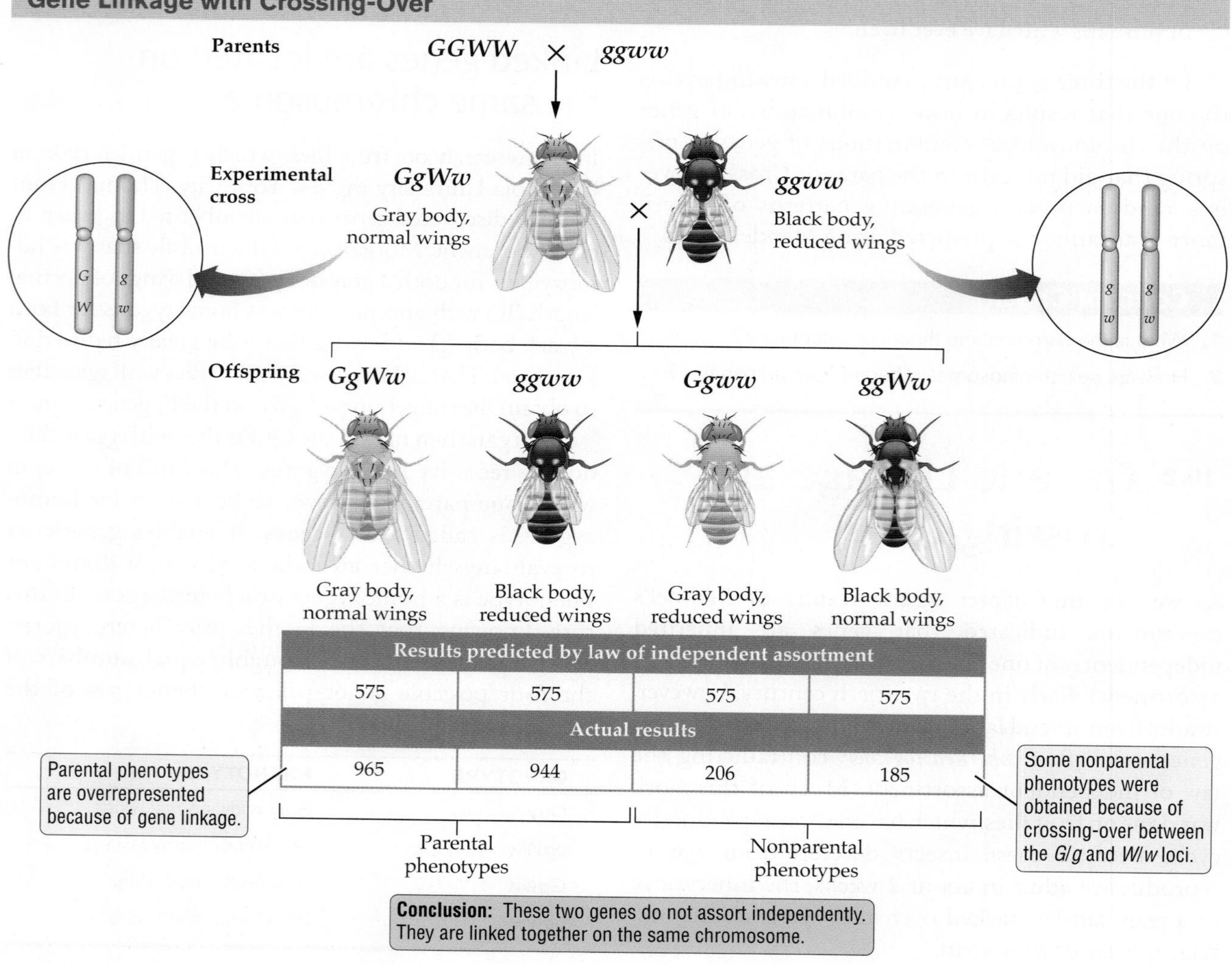

Results predicted by law of independent assortment			
575	575	575	575
Actual results			
965	944	206	185

FIGURE 10.6 Genes on the Same Chromosome May Not Assort Independently

By crossing flies of genotype *GgWw* with flies of genotype *ggww*, Morgan found that the gene for body color (*G* for gray, *g* for black) is linked to the gene for wing length (*W* for normal length, *w* for reduced length). The parental genotypes are higher in number than was predicted because the *G* and *W* genes are linked to each other; the two genes are located relatively close together on the same chromosome.

FIGURE 10.7 Without Crossing-Over, Genes on the Same Chromosome Would Be Completely Linked

If the *G/g* and *W/w* genes were completely linked in fruit flies, the offspring in Morgan's test cross (see Figure 10.6) would all have had exactly the same genotype as the parents: *GgWw* or *ggww*. But Morgan's experiment did not produce this result, indicating that some crossing-over had occurred between the two genetic loci.

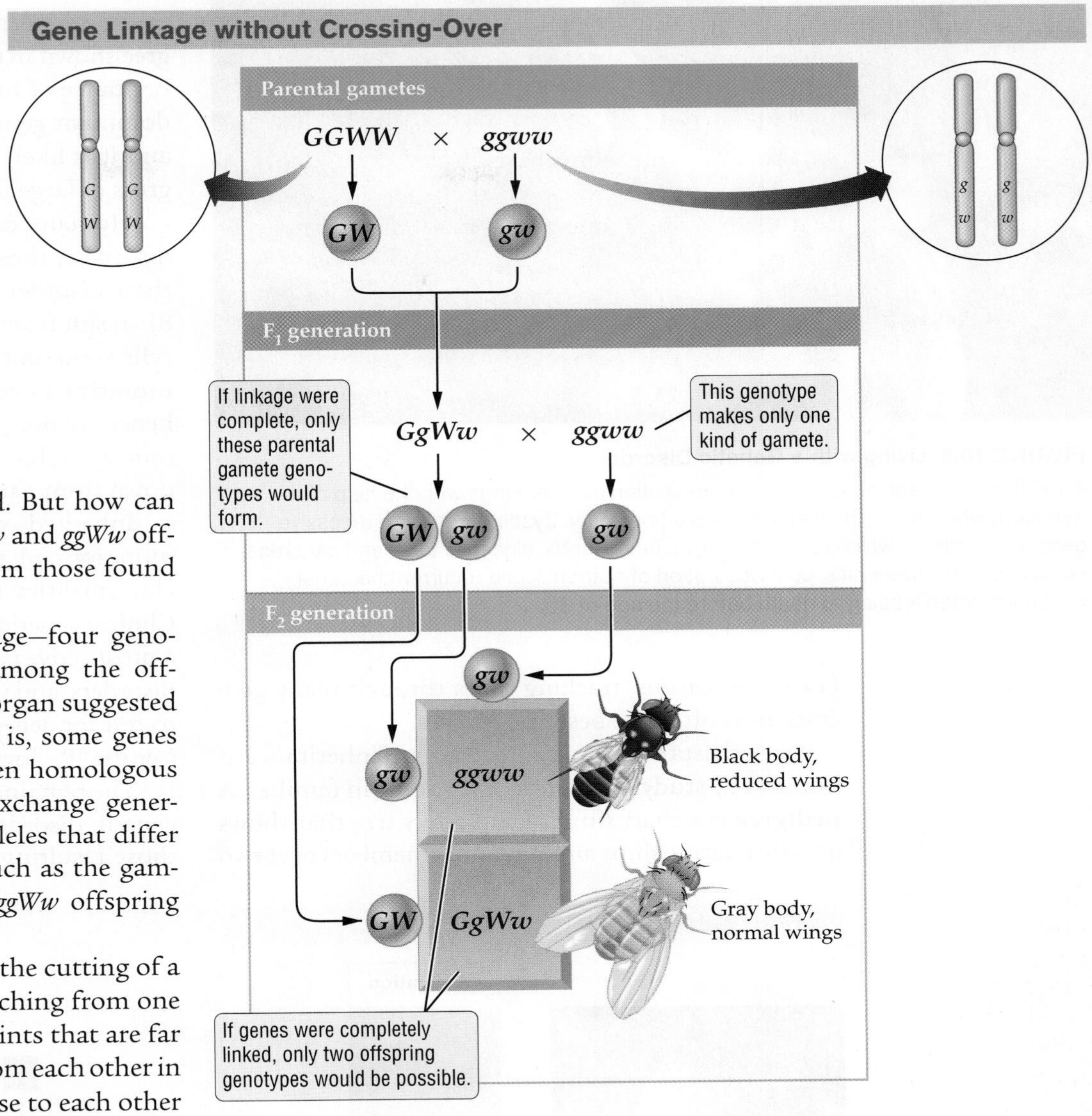

realized that the two genes are linked. But how can we explain the appearance of the *Ggww* and *ggWw* offspring, whose chromosomes differ from those found in either parent?

To explain why—in spite of linkage—four genotypes (rather than two) were seen among the offspring of the *GGWW* × *ggww* cross, Morgan suggested that crossing-over had occurred; that is, some genes had been physically exchanged between homologous chromosomes during meiosis. This exchange generates gametes with combinations of alleles that differ from those found in either parent, such as the gametes that resulted in the *Ggww* and *ggWw* offspring shown in Figure 10.6.

Crossing-over can be compared to the cutting of a string at a few random locations stretching from one end of the string to the other. Two points that are far apart on the string will be separated from each other in most cuts, whereas points that are close to each other will rarely be separated. Similarly, genetic loci that are far apart on a chromosome are more likely to be separated by crossing-over than are genes that are close together. In fact, two genes on a single chromosome that are very far apart may be separated by crossing-over so often that they behave like unlinked genes. Such genes are inherited independently, even though they are located on the same chromosome.

Concept Check

1. The offspring of Morgan's test cross between *GgWw* and *ggww* flies gave four different phenotypes, as expected. But what aspect of the result was not as expected?
2. Would two genes that are close together on a chromosome show stronger genetic linkage than a pair of genes that are far apart? Explain.

10.3 Human Genetic Disorders

Many of us know someone with a genetic disorder, such as cystic fibrosis, sickle-cell anemia, brachydactyly, or one of the many other disorders caused by inherited gene mutations (**FIGURE 10.8**). Studies of genetic disorders could lead to the prevention or cure of much human suffering, but studies in humans are beset by daunting problems: We humans have a long generation time, we select our own mates, and we decide whether to have children and how many children we want to have. Geneticists cannot perform experiments directly in humans to figure out how human genetic disorders are inherited. In addition, human families tend to be much smaller than would be ideal in a scientific study.

Concept Check Answers

1. Morgan expected the numbers of flies for each of the four phenotypes to be approximately equal. Instead, he counted many more flies with the parental phenotypes than with the other two possibilities. The simplest explanation is that the G and W genes are linked.
2. Genes located closer together show stronger genetic linkage because crossing-over that shuffles their alleles is less likely to occur between them than between more distantly separated genes.

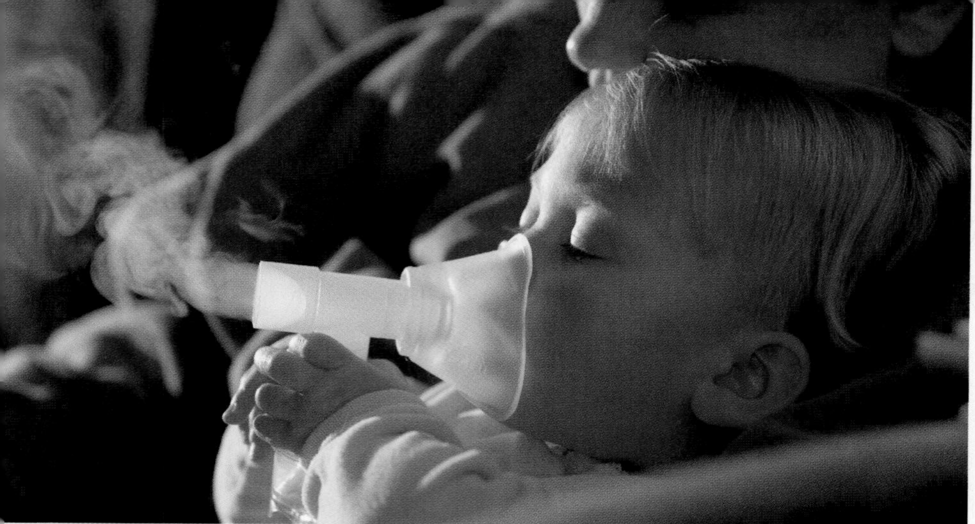

FIGURE 10.8 Living with a Genetic Disorder

A toddler with cystic fibrosis inhales medication into her lungs with the help of a nebulizer, which turns the medicine into a fine spray. Cystic fibrosis is a recessive genetic disorder in which mucus builds up in the lungs, digestive tract, and pancreas, causing chronic bronchitis, poor absorption of nutrients, and recurrent bacterial infections, often leading to death before the age of 35.

For these reasons, tracking traits through many generations is often the best way to start.

Geneticists often analyze patterns of inheritance in humans by studying traits as they occur in families. A **pedigree** is a chart similar to a family tree that shows genetic relationships among family members over two or more generations of a family's history. The pedigree shown in **FIGURE 10.9**, for example, shows the inheritance of brachydactyly. This condition involves a dominant gene, so it will be seen in all heterozygotes, and it is likely to appear in every generation in pedigrees of large families.

Humans can be afflicted by a variety of genetic disorders, those caused by gene mutations. Some of these disorders—including most cancers (see Chapter 8)—result from new mutations that occur in a person's cells sometime during that individual's life. **Somatic mutations** occur in cells other than the sex cells and hence are not passed down to offspring. However, any mutation that is present in the gametes can be passed down from parent to child.

Inherited genetic disorders can be caused by mutations in individual genes (**FIGURE 10.10**) or by abnormalities in chromosome number or structure. Clinical genetic tests can be used to determine whether a prospective parent carries an allele for certain genetic disorders, and variations on these methods can be used to test for genetic disorders long before a baby is born (see the "Biology Matters" box in Chapter 11, page 251).

The remainder of this chapter focuses on inherited genetic disorders that have relatively simple causes: those resulting from mutations of a single gene and

A Four-Generation Pedigree

Actress Megan Fox has one form of brachydactyly.

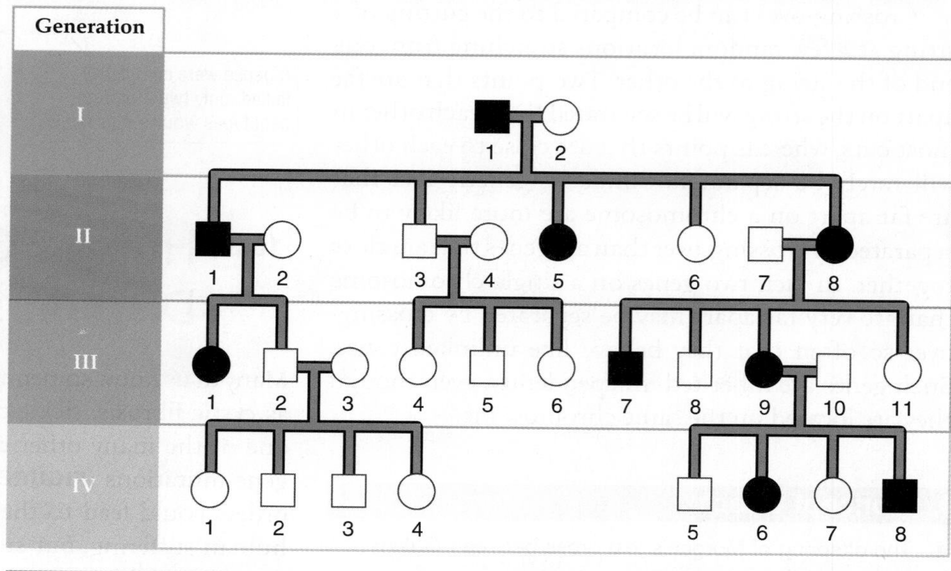

FIGURE 10.9 Patterns of Inheritance Can Be Analyzed in Family Pedigrees

The pedigree shown here illustrates symbols commonly used by geneticists. The Roman numerals at left identify different generations. Numbers listed below the symbols identify individuals of a given generation. This pedigree charts the inheritance of brachydactyly, a dominant condition marked by short or clubbed fingers and toes. Brachydactyly was the first genetic condition to be analyzed, in 1903, by tracing pedigrees.

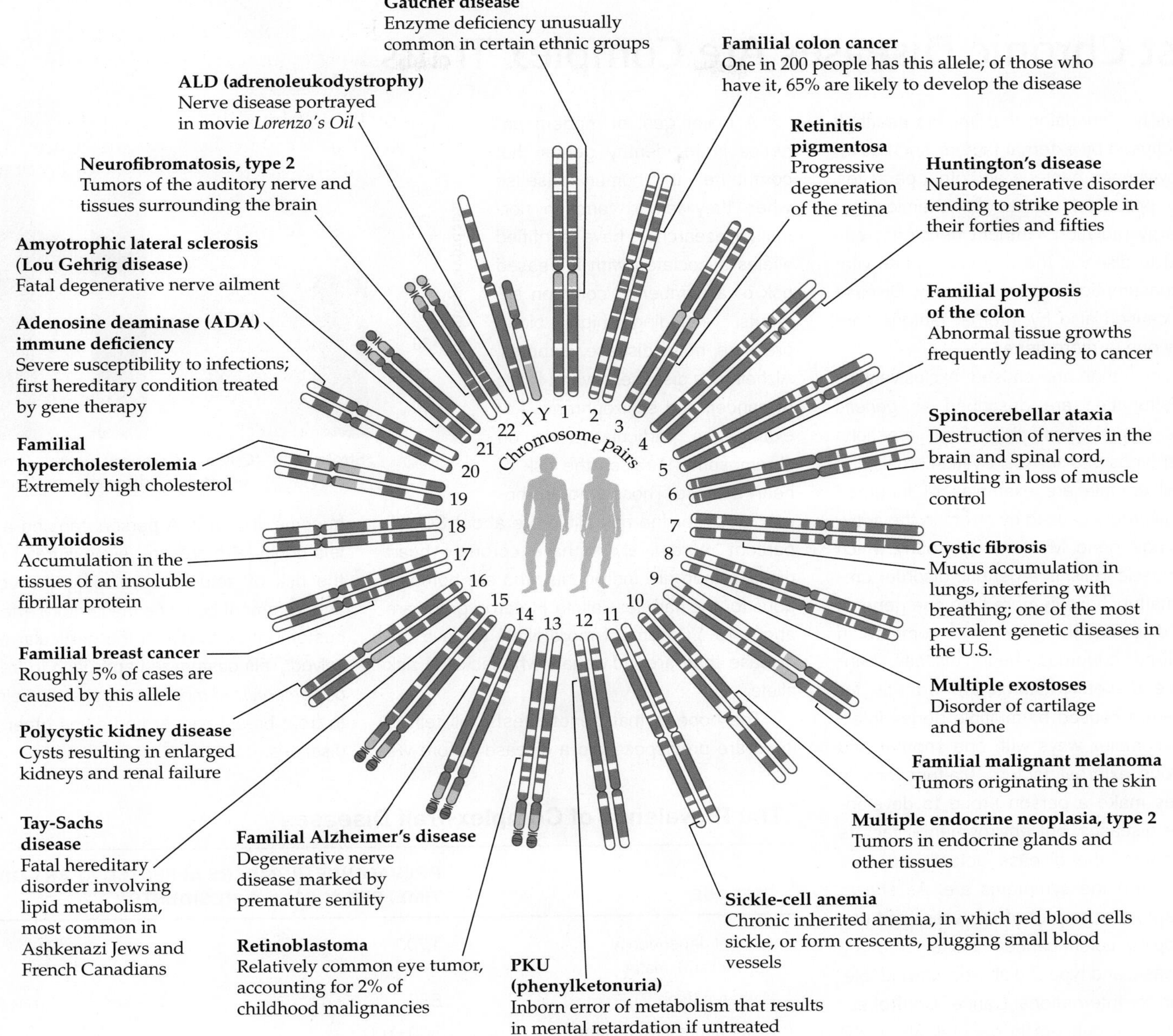

Single-Gene Disorders in Humans
Gaucher disease
Enzyme deficiency unusually common in certain ethnic groups
Familial colon cancer
One in 200 people has this allele; of those who have it, 65% are likely to develop the disease
ALD (adrenoleukodystrophy)
Nerve disease portrayed in movie *Lorenzo's Oil*
Retinitis pigmentosa
Progressive degeneration of the retina
Huntington's disease
Neurodegenerative disorder tending to strike people in their forties and fifties
Neurofibromatosis, type 2
Tumors of the auditory nerve and tissues surrounding the brain
Amyotrophic lateral sclerosis (Lou Gehrig disease)
Fatal degenerative nerve ailment
Familial polyposis of the colon
Abnormal tissue growths frequently leading to cancer
Adenosine deaminase (ADA) immune deficiency
Severe susceptibility to infections; first hereditary condition treated by gene therapy
Spinocerebellar ataxia
Destruction of nerves in the brain and spinal cord, resulting in loss of muscle control
Familial hypercholesterolemia
Extremely high cholesterol
Chromosome pairs
X Y 1 2 3 4 5 6 7 8 9 10 11 12 13 14 15 16 17 18 19 20 21 22
Amyloidosis
Accumulation in the tissues of an insoluble fibrillar protein
Cystic fibrosis
Mucus accumulation in lungs, interfering with breathing; one of the most prevalent genetic diseases in the U.S.
Familial breast cancer
Roughly 5% of cases are caused by this allele
Multiple exostoses
Disorder of cartilage and bone
Polycystic kidney disease
Cysts resulting in enlarged kidneys and renal failure
Familial malignant melanoma
Tumors originating in the skin
Multiple endocrine neoplasia, type 2
Tumors in endocrine glands and other tissues
Tay-Sachs disease
Fatal hereditary disorder involving lipid metabolism, most common in Ashkenazi Jews and French Canadians
Familial Alzheimer's disease
Degenerative nerve disease marked by premature senility
Sickle-cell anemia
Chronic inherited anemia, in which red blood cells sickle, or form crescents, plugging small blood vessels
Retinoblastoma
Relatively common eye tumor, accounting for 2% of childhood malignancies
PKU (phenylketonuria)
Inborn error of metabolism that results in mental retardation if untreated

FIGURE 10.10 Genes Known to Be Associated with Inherited Genetic Disorders in Humans

About 2,000 Mendelian traits are known, and each has been mapped to one of the 23 pairs of human chromosomes. Each disorder represented here results from mutation of a single gene. For clarity, we show only one genetic disorder per chromosome.

those resulting from chromosomal abnormalities. As you read this material, keep in mind that the tendency to develop some diseases, such as heart disease, diabetes, and some inherited forms of cancer risk, is caused by interactions among multiple genes and the environment. For most diseases caused by multiple genes, the identity of the genes involved and how they lead to disease is poorly understood.

Concept Check

1. What are some of the challenges in understanding human genetics?
2. What is a genetic pedigree?

Concept Check Answers

1. Humans have long generation times, select their own mates, and usually have a small number of offspring—all of which makes genetic analysis more challenging.

2. A pedigree is a chart that maps the inheritance of a specific genetic trait for several generations in a group of individuals related by marriage or birth.

BIOLOGYMATTERS

Most Chronic Diseases Are Complex Traits

A *disease* is a condition that impairs health. It may be caused by external factors, such as infection by viruses, bacteria, and other parasites, or injury produced by harmful chemicals or high-energy radiation. Nutrient deficiency can also lead to disease, the way inadequate vitamin C consumption produces scurvy. Disease may be caused also by internal factors, controlled by one or more genes.

Diseases that are caused exclusively by gene malfunction are described as *genetic disorders*, to distinguish them from infections and other types of diseases. Cystic fibrosis and sickle-cell anemia are examples of inherited genetic disorders caused by errors in the activity of a single gene. Myotonic dystrophy, which affects muscle cells, is a genetic disorder created by malfunctions in more than one gene.

The diseases that are most common in industrialized countries—heart disease, cancer, stroke, diabetes, asthma, and arthritis, for example—are caused by multiple genes interacting in complex ways with one another and with various external factors. Malfunctions in key genes make a person prone to developing these diseases, but environmental factors affect whether the disease actually appears or how severe the symptoms are. As shown in the graph, a large part of the estimated risk of developing colon cancer, stroke, coronary heart disease, and type 2 diabetes is avoidable. The Union for International Cancer Control estimates that for cancers in general, about 40 percent of the risk comes from environmental factors. Lifestyle factors such as good nutrition, exercising regularly, and avoiding tobacco have a significant impact on our risk of developing chronic diseases like these. The table shows the prevalence of diseases known to be controlled by multiple genes and the environment.

A major goal of modern genetics is to identify genes that contribute to human disease when they fail to function normally. Researchers have identified alleles associated with increased risk of a number of common ailments, including high blood pressure, heart disease, diabetes, Alzheimer's disease, several types of cancer, and schizophrenia. For example, a particular allele on chromosome 9 raises the risk of heart disease. Those who are homozygous for the risk allele are about 30–40 percent more likely to have coronary heart disease than are individuals who are homozygous for the harmless allele. Heterozygotes are about 15–20 percent more likely to have the disease than are individuals who lack the risk allele.

The hope is that genetic tests will tell us if we are predisposed to a disease before we become ill with it. A person carrying a risk allele could take preventive measures to reduce the risk of actually developing the condition, and treatment could be made more effective by customization to match the particular allele involved. This customized approach to treatment, or *personalized medicine*, is already being used to treat breast cancer and some other chronic diseases.

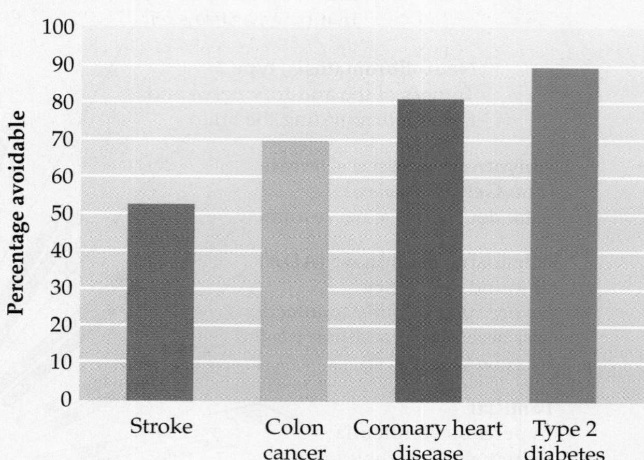

The Prevalence of Complex-Trait Diseases

DISEASE	PREVALENCE (NUMBERS AFFECTED AT ANY GIVEN TIME) PER 10,000 U.S. RESIDENTS
Alcohol dependency (alcoholism), males	1,700
Diabetes (type 2)	830
Major depressive disorder (clinical depression)	500–700
Asthma	540
Alzheimer's disease	160
Autism spectrum disorders	113
Parkinson's disease	30
Multiple sclerosis	10

10.4 Autosomal Inheritance of Single-Gene Mutations

In this discussion we characterize single-gene genetic disorders by a couple of key distinctions. First, is the gene located on an autosome or a sex chromosome? Second, if the disorder is autosomal, is the disease-causing allele recessive or dominant? As you will see, recessive genetic disorders with serious effects are much more common than dominant ones.

Autosomal recessive genetic disorders are relatively common

Several thousand human genetic disorders are inherited as recessive traits controlled by a single gene. Most of these disorders, such as cystic fibrosis, sickle-cell anemia, and Tay-Sachs disease, are caused by recessive mutations of genes located on autosomes.

Recessive genetic disorders vary in severity; some are lethal, while others have relatively mild effects. Tay-Sachs disease (see chromosome 15 in Figure 10.10) is a lethal recessive genetic disorder in which the disease-causing allele encodes a defective version of a critical enzyme. Because the enzyme does not work properly, lipids accumulate in brain cells, the brain begins to deteriorate during a child's first year of life, and death occurs within a few years. Toward the other end of the severity spectrum, adult-onset lactose intolerance is caused by a single recessive allele that leads to a shutdown in the production of lactase, the enzyme that digests milk sugar, in adolescence.

The only individuals who actually get a disorder caused by an autosomal recessive allele (say, *a*) are those that have two copies of that allele (*aa*). Usually, when a child inherits a recessive genetic disorder, both parents are heterozygous; that is, they both have genotype *Aa*. (It is also possible for one or both parents to have genotype *aa*.) Because the *A* allele is dominant and does not cause the disorder, heterozygous individuals are said to be **genetic carriers** of the disorder: they carry the disorder allele (*a*) but do not exhibit its effects.

If two carriers of a recessive genetic disorder have children, the patterns of inheritance are the same as for any recessive trait: ¼ of the children are likely to have genotype *AA*, ½ to have genotype *Aa*, and ¼ to have genotype *aa*. As **FIGURE 10.11** shows, each child has a 25 percent chance of not carrying the disorder allele (genotype *AA*), a 50 percent chance of being a carrier (genotype *Aa*), and a 25 percent chance of actually getting the disorder (genotype *aa*). Because the children of two carriers have a 75 percent chance of being *AA* or *Aa*, and therefore *not* displaying any symptoms, recessive disorders often skip a generation in a family pedigree.

These percentages reveal one way in which lethal recessive disorders such as Tay-Sachs disease can persist in the human population: although homozygous recessive individuals (with genotype *aa*) die long before they are old enough to have children, carriers (with genotype *Aa*) are not harmed

Helpful to Know

A *genetic condition* is simply a phenotype that is out of the ordinary. A *genetic disorder* is a disease caused by gene malfunction; the illness or disability it produces can range from mild to life-threatening.

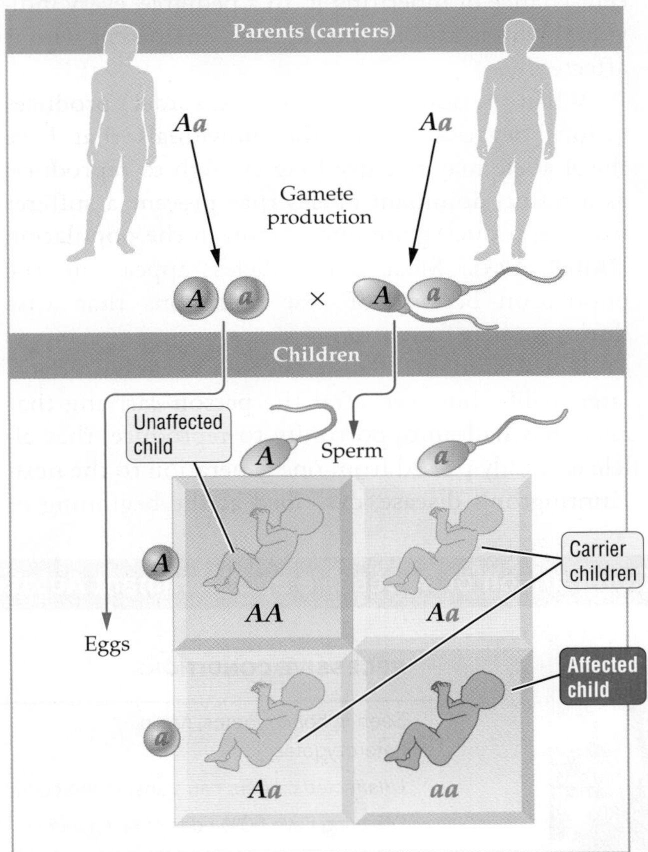

FIGURE 10.11 Genetic Carriers Transmit an Autosomal Recessive Condition without Displaying It Themselves

The patterns of inheritance for a human autosomal recessive genetic disorder are the same as for any other recessive trait (compare this figure with the pattern shown by Mendel's pea plants in Figure 9.8). Recessive, disease alleles are colored red and denoted *a*. Dominant, normal alleles are black and denoted *A*. Here, the parents are a carrier male (genotype *Aa*) and a carrier female (genotype *Aa*).

Helpful to Know

Marriage between close relatives, such as first cousins, is frowned upon in many cultures. Siblings are likely to be carriers for the same recessive disorders, and so are their offspring. When two first cousins marry, the odds of their children inheriting a homozygous recessive disorder is much higher, compared to the odds if each cousin married a nonrelative, who is much less likely to be a carrier.

by the disorder. In a sense, the *a* alleles can "hide" in heterozygous carriers, and those carriers are likely to pass the disorder allele to half of their children. Recessive genetic disorders also arise in the human population because new mutations in eggs or sperm can produce new disorder alleles.

Serious dominant genetic disorders are less common

A dominant allele (*A*) that causes a genetic disorder cannot "hide" in the same way that a recessive allele can. In this case, *AA* and *Aa* individuals get the disorder; only *aa* individuals are symptom-free (**FIGURE 10.12**). If one parent has a dominant genetic disorder, each child has a 50 percent chance of inheriting it. In a pedigree, every individual who is affected has at least one parent who is affected.

When a dominant genetic disorder produces serious negative effects, the individuals that have the *A* allele may not live long enough to reproduce; as a result, dominant alleles that prevent a sufferer from reproducing are uncommon in the population (**TABLE 10.1**). Most such alleles appear in the population because of new mutations that arise during gamete formation.

If a dominant allele expresses its lethal effects later in life, however, after the person carrying that allele has had an opportunity to reproduce, that allele is readily passed from one generation to the next. Huntington's disease, described at the beginning of

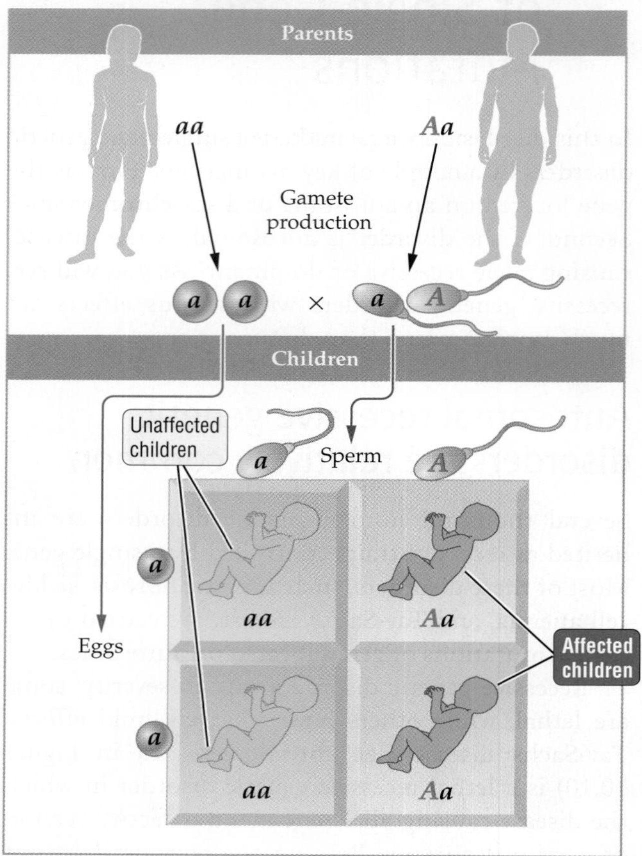

FIGURE 10.12 In an Autosomal Dominant Condition, Heterozygotes Display the Phenotype

In this figure, the dominant, disease allele is colored red and denoted *A*. Recessive, normal alleles are black and denoted *a*. Here, the parents are an unaffected female (genotype *aa*) and a heterozygous male who exhibits the disorder (genotype *Aa*).

TABLE 10.1 Comparison of Recessive and Dominant Autosomal Conditions

Actor Jason Acuña, or Wee Man, has achondroplasia. This type of dwarfism is an autosomal dominant condition.

RECESSIVE CONDITIONS	DOMINANT CONDITIONS
Seen in homozygotes, not in heterozygotes.	Seen in both heterozygotes and homozygotes.
Unaffected carriers can transmit the condition.	Only *affected* individuals transmit the condition.
Offspring have 50% odds of being carriers if one parent is a carrier and the other is an unaffected noncarrier (most common type of mating).	Offspring have 50% odds of being *affected* if one parent is an affected heterozygote and the other is unaffected (most common type of mating).
Can skip one or more generations because offspring of one affected parent and one unaffected noncarrier are not themselves affected, even though all of them are carriers.	Does not skip generations in a large pedigree sample because an affected parent has high odds (at least 50%) of having an affected child.
Condition with severe phenotype can be fairly common because it can result (with 25% odds) from a mating of two unaffected carriers.	Condition with severe phenotype and early onset is uncommon because affected persons cannot transmit the genotype, and rare mutations are the only source.

this chapter, illustrates how such a dominant lethal allele can remain in the population. The symptoms of Huntington's disease begin relatively late in life, often after victims of the disorder have already had children. Dominant disorders of this type are more commonly seen than those in which the symptoms begin before childbearing age or those that arise from new mutations alone.

10.5 Sex-Linked Inheritance of Single-Gene Mutations

Roughly 1,200 of the estimated 25,000 human genes are found only on the X chromosome or only on the Y chromosome; such genes are said to be **sex-linked**. Because sex-linked genes are found on either the X chromosome or the Y chromosome—but not both—males, who have XY chromosomes, receive only one copy of each sex-linked gene. Females, however, have two copies of every X chromosome gene, because they inherit two X chromosomes, one from each parent.

About 15 genes are shared by the X and Y chromosomes. In each of these cases, males and females receive two copies of the gene, just as they do for all autosomal genes. For this reason, these 15 genes are not termed "sex-linked," even though they are found on the sex chromosomes.

The overwhelming majority of the approximately 1,100 human sex-linked genes are located on the X chromosome, while about 50 are located on the much smaller Y chromosome. Sex-linked genes on the X chromosome are said to be **X-linked**; sex-linked genes on the Y chromosome are similarly said to be **Y-linked**. Some of the sex-linked genes on the X chromosome are known to be associated with genetic conditions that range from mild traits, such as red-green color blindness, to devastating genetic disorders like Duchenne muscular dystrophy. No well-documented cases of disease-causing Y-linked genes are known.

Recessive alleles of sex-linked genes have a different pattern of inheritance, compared to recessive alleles on autosomes. Let's consider how an X-linked recessive allele for a human genetic disorder is inherited (**FIGURE 10.13**). We label this recessive *a*, and in the Punnett square we write this allele as X^a to emphasize that it is on the X chromosome. Similarly, the dominant allele is labeled *A* and is written as X^A in the Punnett square. If a carrier female (genotype X^AX^a) has children with a normal male (genotype X^AY), each of their sons will have a 50 percent chance of having the disorder (see Figure 10.13). This result differs greatly from what would happen if the same disorder allele (*a*) were on an autosome: in that case, none of the children, male or female, would show the disorder, because both male and female children would be heterozygous.

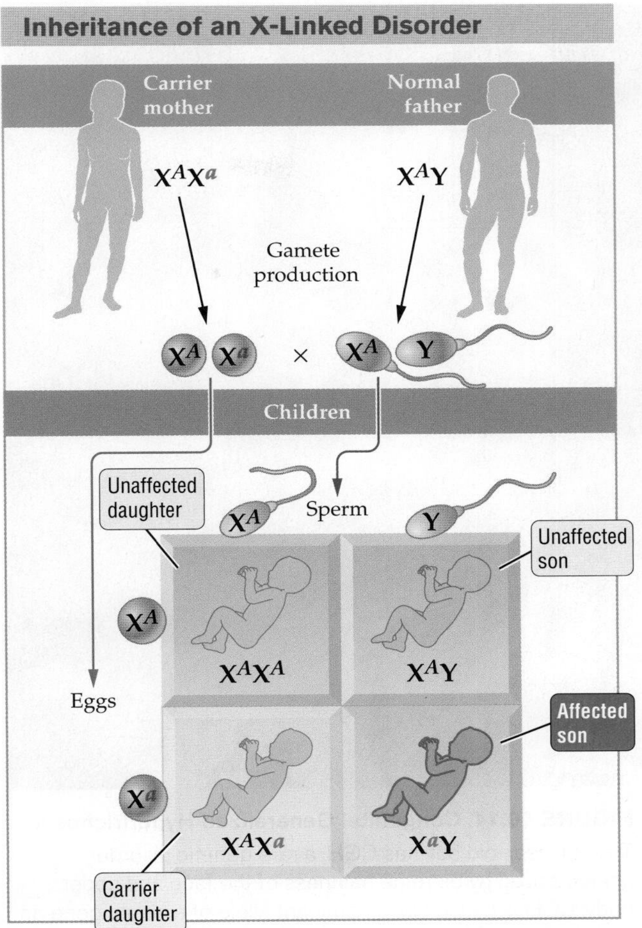

FIGURE 10.13 X-Linked Recessive Conditions Tend to Show Up More Commonly in Males Than in Females
The recessive disorder allele (*a*) is located on the X chromosome and is denoted by X^a. The dominant normal allele (*A*) on the X chromosome is denoted by X^A.

The inheritance of sex-linked disorders is unique because males who receive the recessive allele do not have another copy of the gene to offset it. Put another way, they cannot be heterozygotes—only their sisters, who receive two X chromosomes, have that possibility.

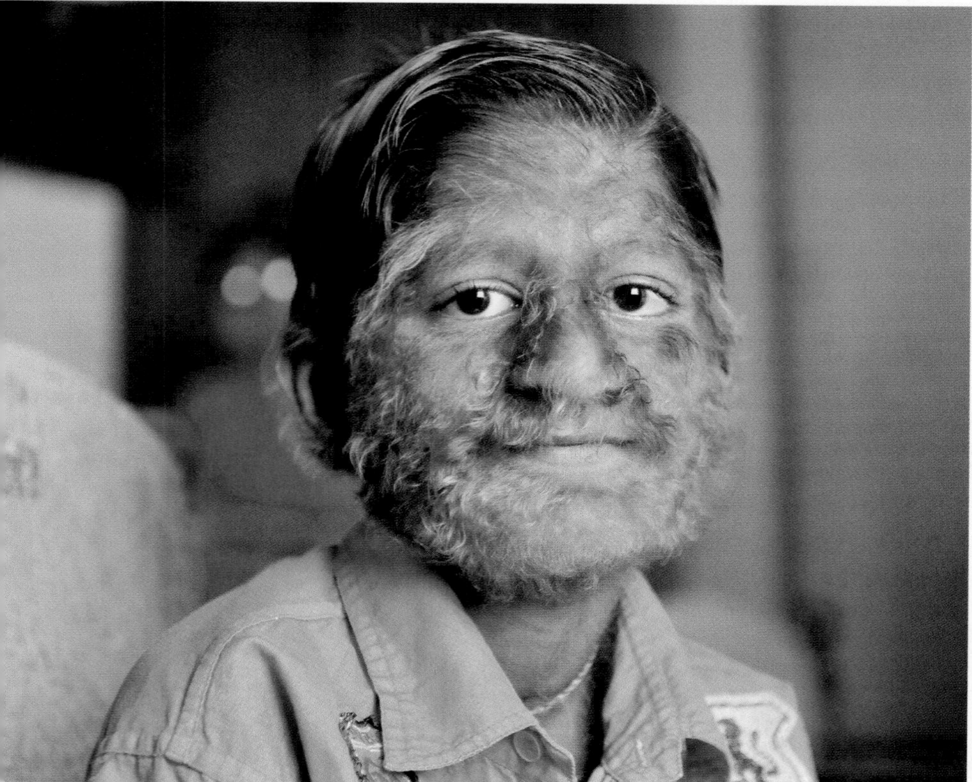

FIGURE 10.14 Congenital Generalized Hypertrichosis
This 11-year-old boy has CGH, a rare genetic disorder characterized by extreme hairiness of the face and upper body. CGH is caused by a dominant allele of a single gene on the X chromosome.

X-linked genetic disorders in humans include hemophilia, a serious bleeding disorder in which minor cuts and bruises can cause a person to bleed to death, and Duchenne muscular dystrophy, a lethal disorder that causes the muscles to waste away, often leading to death at a young age. Both of these X-linked disorders are caused by recessive alleles. An example of a dominant X-linked disorder with a daunting name—congenital generalized hypertrichosis (*HYE*-**per-trih-***KOH***-sus**), or CGH—is shown in **FIGURE 10.14**.

Note that an X-linked condition controlled by a *dominant* allele is not necessarily more common in males than in females. Male and female children both have equal odds (50 percent) of inheriting an X-linked dominant condition from an affected heterozygous mother. However, fathers transmit an X-linked condition only to their daughters, not to their sons. All the daughters of a man with an X-linked dominant trait will display that trait.

Concept Check Answers

1. Dominant autosomal disorders are less likely to be passed down multiple generations because individuals with these disorders rarely survive to produce children.

2. Males are affected if they inherit a single copy of the color blindness allele. An affected female must have two copies, which means her father must be color-blind and her mother must at least be a carrier—an uncommon occurrence.

Concept Check

1. Why are dominant autosomal disorders that produce serious negative effects less common than serious recessive autosomal disorders?
2. In North America, the X-linked form of red-green color blindness affects about 8 percent of men, but only 0.5 percent of women. Why is this condition more common in men?

10.6 Inherited Chromosomal Abnormalities

Every species has a characteristic number of chromosomes, and each chromosome has a particular structure, with specific genetic loci arranged on it in a precise sequence. Any change in the chromosome number or structure, compared to what is typical for the species, is considered a chromosomal abnormality.

Two main types of chromosomal changes are seen in humans and other organisms: changes in the overall number of chromosomes, and changes in chromosome structure, such as a change in the length of an individual chromosome. A cell is especially vulnerable to developing chromosomal abnormalities when it is dividing: chromosomes can be misaligned, misdirected, or even ripped into pieces during the delicate business of lining them up in the center of the cell and then segregating them into daughter cells.

For chromosomal abnormalities to be passed on from a parent to offspring, the chromosomal changes have to occur in the gametes or gamete-producing cells. However, relatively few human genetic disorders are caused by inherited chromosomal abnormalities, probably because most large changes in the chromosomes are lethal to the developing embryo.

The structure of a chromosome can change in several ways

When chromosomes are being aligned or separated during cell division, breaks can occur that alter one or more chromosomes.

- In **deletion**, a piece breaks off and is lost from the chromosome (**FIGURE 10.15*a***).

- In **inversion**, a fragment breaks off and reattaches at the original location but in a flipped orientation, with the genetic loci in reverse order (**FIGURE 10.15*b***).
- In **translocation**, a broken piece from one chromosome becomes attached to a different, nonhomologous chromosome (**FIGURE 10.15*c***). Translocations are frequently reciprocal, meaning that two nonhomologous chromosomes are involved in a mutual exchange of fragments. The example illustrated in Figure 10.15*c* is a reciprocal translocation.
- In **duplication**, a chromosome becomes longer because it ends up with two copies of a particular fragment (**FIGURE 10.15*d***). Errors in crossing-over are one source of duplication. If the exchange during crossing-over is not reciprocal, one chromosome can end up with two fragments and the other chromosome with none.

Changes in chromosome structure can have dramatic effects. A break in the sex chromosomes can cause a change in the expected sex of the developing fetus. A deletion in which the *SRY* gene on the Y chromosome is lost produces an XY individual who develops as a female. A translocation that results in an X chromosome to which the Y chromosome's sex-determining region is attached produces an XX individual who develops as a male. XY females and XX males are always sterile (unable to produce offspring).

Changes to the structure of autosomes can have even more dramatic effects. Cri du chat (*KREE* **doo** *SHAH*) syndrome occurs when a child inherits a chromosome 5 that is missing a particular region (**FIGURE 10.16**). *Cri du chat* is French for "cry of the cat," which describes the characteristic mewing sound made by infants with this condition. Other characteristics of this condition are slow growth and a tendency toward severe intellectual disability, a small head, and low-set ears.

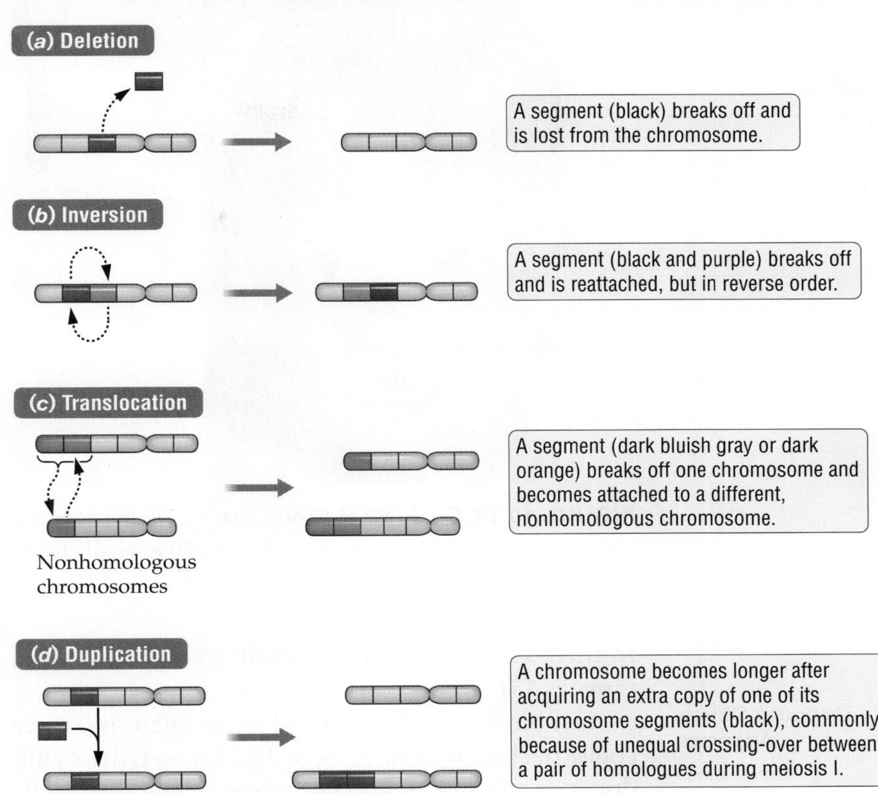

Structural Changes to Chromosomes
(a) Deletion
A segment (black) breaks off and is lost from the chromosome.
(b) Inversion
A segment (black and purple) breaks off and is reattached, but in reverse order.
(c) Translocation
A segment (dark bluish gray or dark orange) breaks off one chromosome and becomes attached to a different, nonhomologous chromosome.
Nonhomologous chromosomes
(d) Duplication
A chromosome becomes longer after acquiring an extra copy of one of its chromosome segments (black), commonly because of unequal crossing-over between a pair of homologues during meiosis I.

FIGURE 10.15 Chromosome Rearrangements Can Cause Serious Genetic Disorders

Changes in chromosome number are often fatal

Unusual numbers of chromosomes—such as one or three copies instead of the normal two—can result when chromosomes fail to separate properly during meiosis. In humans, such changes in chromosome number are often lethal to the embryo or fetus. At least 20 percent of human pregnancies abort spontaneously, largely as a result of errors in chromosome number.

In only one type of chromosome anomaly do individuals commonly reach adulthood: Down syndrome. Individuals with this condition usually have three copies of chromosome 21, one of the smallest autosomes in humans. Down syndrome is also known as trisomy 21, where **trisomy** (*TRYE*-**soh-mee**) refers to the condition of having three copies of a chromosome instead of the usual two. In a small minority (3–4 percent) of Down syndrome cases, an extra piece of chromosome 21 breaks off during cell division in a gamete or a precursor and attaches to another chromosome. Individuals with this form of Down syndrome have two copies of chromosome 21 plus part of another chromosome 21. People with Down syndrome tend to be short and intellectually disabled, and they may have defects of the heart, kidneys, and digestive tract. With appropriate medical care, most people with this condition lead healthy lives, and many live

Helpful to Know

Earlier we noted the use of the suffix *some* in the names of cellular particles, such as chromosomes. Biologists use the related suffix *somy* to name conditions involving unusual numbers of chromosomes; two examples are *trisomy* ("three bodies") and *monosomy* ("one body").

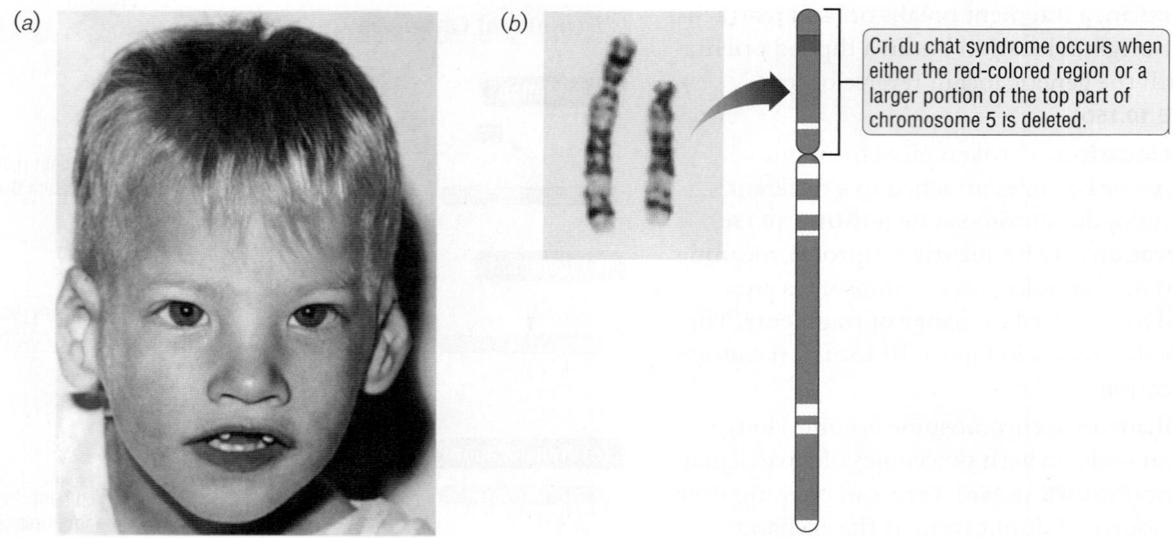

(a)
(b)
Cri du chat syndrome occurs when either the red-colored region or a larger portion of the top part of chromosome 5 is deleted.

FIGURE 10.16 Cri du Chat Syndrome
Cri du chat syndrome is caused when part of the top of chromosome 5 is deleted.

to their sixties or seventies (their average life expectancy is 55).

Live births can also occur when an infant has three copies of chromosome 9, 15, or 18. However, these children typically have severe birth defects, and they rarely live beyond their first year.

Compared with having too few or too many autosomes, changes in the number of sex chromosomes have more minor effects. Klinefelter syndrome, for example, is a condition found in males that have an extra X chromosome (XXY males). Such men have a normal life span and normal intelligence, and they tend to be tall. Many XXY males also have small testicles (about one-third normal size) and reduced fertility, and some have female characteristics, such as enlarged breasts. Females with a single X chromosome (Turner syndrome) have normal intelligence and tend to be short (with adult heights of less than 150 centimeters, about 4 feet 11 inches), to be sterile, and to have a broad, webbed neck. Other changes in the number of sex chromosomes, as in XYY males and XXX females, also produce relatively mild effects. However, when there are two or more extra sex chromosomes, as in XXXY males or XXXX females, a wide range of problems can result, including profound intellectual disability.

Concept Check

1. List two types of chromosomal abnormalities that result from the breaking and re-joining of chromosome fragments.
2. How do disorders caused by a change in chromosome number arise?

Concept Check Answers

1. Inversion, in which the re-joining fragment is flipped from its normal order; and translocation, in which a fragment joins with a different chromosome.
2. Mistakes in the segregation of chromosomes during meiosis generate some gametes with missing chromosomes and some with extra chromosomes; when these gametes participate in fertilization, the resulting zygote is deficient in chromosomes or has extra chromosomes.

APPLYING WHAT WE LEARNED

Testing for Huntington's Disease

Most genetic diseases are caused by a recessive allele, which means that any person who becomes sick must have inherited two "bad" versions of the gene (alleles)—one from the mother and one from the father. But Huntington's disease (HD) is different. Because it's caused by a dominant allele, even people who inherit only one copy of the HD allele will develop the disease unless they die from an accident or some other cause. Each child of a parent with HD has a 50 percent chance of developing the disease.

In 1983, researchers mapped the HD gene by showing that it was linked to another gene on chromosome 4. By 1993, the HD gene had been located exactly, making it the first autosomal human disease gene to be mapped through genetic linkage analysis. The gene was named *huntingtin*; the HD allele of *huntingtin* produces a defective huntingtin protein that is much longer than normal. Researchers soon developed a genetic test for the disease-causing allele. And within a handful of years, they also developed strains of laboratory mice carrying a mutated *huntingtin* allele that developed symptoms similar to those of human HD. The race was soon on to find a way to cure these mice and, after that, human patients.

Genetic counselors began to talk to the families of HD patients about being tested for the disease. Genetic testing can help reduce the number of people with genetic diseases. For example, since genetic screening for Tay-Sachs disease became available in the 1970s, the disease has nearly disappeared from families that once had it. Unlike HD, Tay-Sachs is caused by a recessive allele that affects young children, who typically die before the age of 4. In principle, both Huntington's disease and the HD allele could be eliminated in a single generation if no one carrying the HD allele had children.

But few people want to know whether they carry the allele. As Misty Oto explained to an interviewer in 2007, to find out that she definitely had HD was "not something I would want to know. I would feel guilty, not necessarily for my kids, but for [my husband]—because I would know what his future would hold, and I wouldn't want him to be trapped." Many people also worry that a positive result will affect their ability to get health insurance or keep a job. Sometimes a positive result leads to severe depression or even suicide. So, for a range of reasons, only about 10 percent of people at risk for HD have the test done.

Still, when Misty's older sister began to suspect she might have the disease, the two sisters decided to confront their fears and take the test. To Misty's relief, the test showed that she did not have an HD allele. That meant her husband and kids would not have to cope with the prospect of her developing this terrible and lingering disease. And it meant her kids would not grow up wondering if they would develop the disease. But Misty's happiness was short-lived. Her sister, who got her results later the same week, learned that she had the HD allele.

Although a complete cure is not yet in sight, our understanding of genes and gene function is advancing so rapidly that Misty's sister has much better prospects than the women's mother did. Researchers have found drugs that delay the onset of HD and reduce the severity of symptoms when they do appear, at least in mice. These drugs are currently in clinical trials, and early reports are promising: drug therapy seems to boost the survival and function of brain cells in people who have an HD allele. These advances offer new hope to the nearly 30,000 people in the United States who carry the HD allele.

BIOLOGY IN THE NEWS

Stanford University Students Study Their Own DNA

BY LISA M. KRIEGER • *San Jose Mercury News*, May 24, 2013

Most students read about genetics in a textbook. Stanford University students are reading something far more intimate: their own DNA code.

Genetics 210, a provocative class in the School of Medicine, is a pioneer in the growing movement to advance a new era in modern medicine, where genetic information allows us to not only look into our medical futures but sometimes take action ...

"Trying to use my own genetic information as a learning tool—that sounded like something I wanted," said bioscience graduate student Thomas Roos, 28, who learned he and his twin brother, Andrew, have slightly elevated risks of an Achilles tendon injury and dementia but have a reduced risk of heart disease and arthritis.

Stanford is one of a handful of universities that offers the School of Medicine course to nonphysicians, so graduate students and even undergraduates get the chance to read their very own A, C, T and G sequence of nucleotides.

"I am teaching something they need to know," said Stuart Kim, a professor of developmental biology and genetics ... "These are future scientists who need to understand the underlying concepts behind this exploding field ..."

The testing is confidential and voluntary. They use a cotton swab to get a specimen from inside their cheek, then mail it to the Mountain View company 23andMe for processing. If students don't want to study their own genes, they can use a public reference sequence.

What they learn is not a diagnosis but an estimate of risk.

It is not a complete sequence of all 3 billion nucleotides in their genome, but a snapshot of 1.1 million better-understood variants linked to thousands of conditions and traits. It is the statistics behind that analysis—and the interpretation of what the data means—that become the centerpiece of heated classroom debate ... Some of the information can be upsetting, because there aren't cures. For instance, students might find they carry a variant of the *LRRK2* gene that predicts Parkinson's disease, or variants of the apolipoprotein E gene (*ApoE*) that increase risk of developing Alzheimer's.

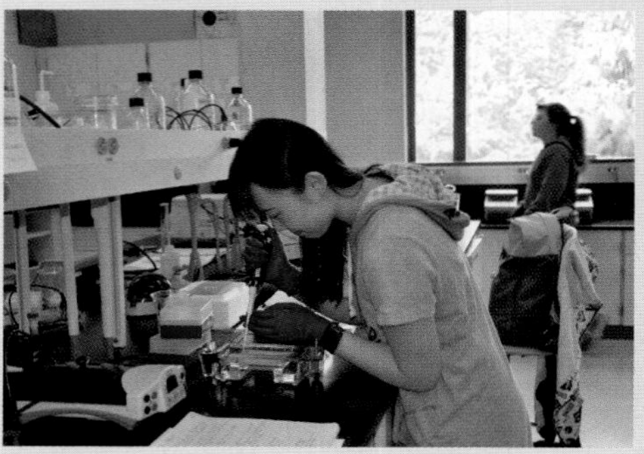

For almost a decade, some private companies have been marketing genetic information directly to individuals instead of through a doctor—a practice known as direct-to-consumer (DTC) genetic testing. The company described in the news story used to pitch the Stanford-style curriculum to other universities in the country at a discounted price. But in 2010, when the "peek into your DNA" class project was first introduced in Genetics 210, it ignited intense controversy on the Stanford campus and across the country. From lawyers to philosophers, many expressed concern that finding out they carried a high risk of a disease—especially one for which there is no cure—might cause psychological harm to the students. Despite confidentiality safeguards built into the program, some worried that a student might unwittingly reveal personal genetic information in casual conversation.

Diseases such as Huntington's, Tay-Sachs, and cystic fibrosis are strongly genetic, so knowing if you carry an allele for one of these can have a huge impact on your life. But critics say that where complex traits are concerned, information about higher-than-average risk of the disease is pretty useless. Often, the actual lifetime risk of getting a disease depends on which alleles of other genes are present and on environmental influences. Sometimes nothing at all is known about these other genes that can modify disease risk or about the environmental factors.

Another criticism of DTC genetic testing is that consumers may not be able to make sense of the data they receive. Those concerned with costs fret that consumers might be driven to unnecessary and expensive screenings and additional tests, adding to society's health care burden. Defenders of genomic testing reject what they see as a patronizing attitude, asserting instead that "knowledge is power." Late in 2013, the Food and Drug Administration (FDA) barred 23andMe from providing health-related genetic information to its customers. According to the FDA, the company failed to provide data to support its marketing claims, which promised to evaluate 254 genetic traits, including susceptibility to a host of diseases. Following the FDA order, the company must confine itself to offering DNA-based genealogies and uninterpreted "raw data" on gene sequences.

Evaluating the News

1. Explain why genomic testing cannot give us an accurate risk estimate for the three diseases—heart disease, cancer, and stroke—responsible for 60 percent of all deaths in the United States.

2. Do you support the FDA ban on genetic testing by 23andMe? Discuss the advantages and disadvantages of direct-to-consumer genetic testing.

3. If you could, would you be tested? Would you choose not to have children if you knew you had the HD allele?

CHAPTER REVIEW

Summary

10.1 The Role of Chromosomes in Inheritance

- The chromosome theory of inheritance states that genes are located on chromosomes. The physical location of a gene on a chromosome is known as its genetic locus.
- Each chromosome is composed of a single DNA molecule and many associated proteins.
- Chromosomes that pair during meiosis are homologous chromosomes. The two members of each homologous pair have the same genetic locus.
- Chromosomes that determine the sex of the organism are called sex chromosomes; all other chromosomes are called autosomes. Humans have two types of sex chromosomes: X and Y. Males have one X and one Y chromosome. Females have two X chromosomes.
- A specific gene on the Y chromosome (the *SRY* gene) is required for human embryos to develop as males.
- Genetic differences among individuals provide the genetic variation on which evolution can act. Mutations are the source of new alleles.
- Offspring differ genetically from one another and from their parents because (1) each pair of homologues exchanges random stretches of the chromosome during crossing-over, (2) the different homologous pairs are sorted into gametes independently of each other during meiosis, and (3) fertilization combines male and female genotypes randomly.

10.2 Genetic Linkage and Crossing-Over

- Genes that tend to be inherited together are said to be genetically linked.
- Two genes that are far apart on a chromosome are more likely to be shuffled by crossing-over than are genes located near one another, and they are therefore less likely to show genetic linkage.
- Mendel's law of independent assortment holds for genes located far apart on a chromosome, just as it does for a pair of genes located on two different chromosomes.

10.3 Human Genetic Disorders

- Pedigrees are useful for studying the inheritance of human genetic disorders.
- Humans suffer from a variety of genetic disorders, including some caused by mutations of a single gene and some caused by abnormalities in chromosome number or structure.

10.4 Autosomal Inheritance of Single-Gene Mutations

- Most genetic disorders are caused by a recessive allele (*a*) of a gene on an autosome. Only homozygous (*aa*) individuals get these disorders; heterozygous (*Aa*) individuals are merely carriers of the disorders.
- In dominant autosomal genetic disorders, both *AA* and *Aa* individuals are affected. If defects of the phenotype are so severe that the *AA* or *Aa* individual cannot reproduce, such a dominant genetic disorder is likely to be rare.
- Alleles that cause lethal dominant genetic disorders can remain in the population if symptoms begin late in life, as in Huntington's disease, or if new disorder alleles are produced by mutation during each generation.

10.5 Sex-Linked Inheritance of Single-Gene Mutations

- Because males inherit only one X chromosome, the patterns of inheritance for genes on sex chromosomes differ from those for genes on autosomes.
- Genes found on one of the sex chromosomes but not on both may show sex-linked patterns of inheritance. Genes found only on the X chromosome are said to be X-linked; those found only on the Y chromosome are Y-linked.
- Males are more likely than females to have recessive X-linked genetic disorders, because males need to inherit only one copy of the disorder allele to be affected, whereas females must inherit two copies to be affected. In contrast, males and females are equally likely to be affected by autosomal genetic disorders or dominant X-linked disorders.

10.6 Inherited Chromosomal Abnormalities

- The structure of an individual chromosome can change through breakage during cell division, resulting in deletion, inversion, translocation, or duplication of a chromosome fragment.
- Changes in the number of autosomes in humans are usually lethal. Down syndrome, a form of trisomy in which individuals receive three copies of chromosome 21, is an exception to this rule.
- People who have one too many or one too few sex chromosomes may experience relatively minor effects, but if there are four or more sex chromosomes (instead of the usual two), serious problems can result.

Key Terms

autosome (p. 219)
chromosome theory of inheritance (p. 218)
crossing-over (p. 220)
deletion (p. 230)
duplication (p. 231)
genetic carrier (p. 227)
genetic linkage (p. 222)
independent assortment of chromosomes (p. 221)
inversion (p. 231)
locus (p. 218)
pedigree (p. 224)
random fertilization (p. 221)
sex chromosome (p. 219)
sex-linked (p. 229)
somatic mutation (p. 224)
SRY gene (p. 220)
test cross (p. 221)
translocation (p. 231)
trisomy (p. 231)
X-linked (p. 229)
Y-linked (p. 229)

Self-Quiz

1. Genes commonly exert their effect on a phenotype by
   a. promoting DNA mutations.
   b. creating chromosomal structures such as centromeres.
   c. coding for a protein.
   d. all of the above

2. Which of the following is an autosomal dominant disorder in which the symptoms begin late in life and the nervous system is destroyed, resulting in death?
   a. Tay-Sachs disease
   b. Huntington's disease
   c. Down syndrome
   d. cri du chat syndrome

3. Crossing-over is more likely to occur between genes that are
   a. close together on a chromosome.
   b. on different chromosomes.
   c. far apart on a chromosome.
   d. on the Y chromosome.

4. Comparatively few human genetic disorders are caused by chromosomal abnormalities. One reason is that
   a. most chromosomal abnormalities have little effect.
   b. it is difficult to detect changes in the number or length of chromosomes.
   c. most chromosomal abnormalities result in spontaneous abortion of the embryo.
   d. it is not possible to change the length or number of chromosomes.

5. Nonparental genotypes can be produced by
   a. crossing-over and independent assortment of chromosomes.
   b. linkage.
   c. autosomes.
   d. sex chromosomes.

6. Sometimes a segment of DNA breaks off from a chromosome and then returns to the correct place on the original chromosome, but in reverse order. This type of chromosomal structural change is called
   a. crossing-over.
   b. translocation.
   c. inversion.
   d. deletion.

7. Which of the following can most precisely be described as a master switch that commits the sex of the developing embryo to "male"?
   a. an X chromosome
   b. a Y chromosome
   c. an XY chromosome pair
   d. the *SRY* gene

Analysis and Application

1. In terms of its physical structure, describe what a gene is and where it is located.
2. Consider the XY chromosome system by which sex is determined in humans. Do patterns of inheritance for genes located on the X chromosome differ between males and females? Why or why not?
3. In 1917, Shinoba Ishihara designed a series of colored plates for testing individuals for red-green color perception. In the images shown here, a person with red-green color blindness will not see any number in the Ishihara plate on the left, and will not be able to read the number in the plate on the right correctly. Can a woman have X-linked red-green color blindness? If so, how? Draw a three-generation pedigree of a woman with this condition. Include the woman's parents and her five children (three girls and two boys), and assume that her husband is not color-blind. Indicate all carriers in your pedigree by placing a dot in the center of the symbol representing them—a standard way of symbolizing carriers. What are the odds that this woman's sons will be color-blind? What are the odds that any of her children will be carriers for X-linked red-green color blindness?
4. Look carefully at Figure 10.6. Explain in your own words why the results shown there convinced Morgan that genes located near one another on a chromosome tend to be inherited together, or linked. What results would be expected if these genes were located on different chromosomes?
5. Explain how crossing-over occurs. Assume that genes *A*, *B*, and *C* are arranged in that order along a chromosome. From your understanding of crossing-over, do you think it will occur more often between genes *A* and *B* or between genes *A* and *C*?
6. Describe two mechanisms that give rise to nonparental genotypes in the next generation.
7. Are genetic disorders that are caused by single-gene mutations more common or less common in human populations than those caused by abnormalities in chromosome number or structure? Explain your answer.

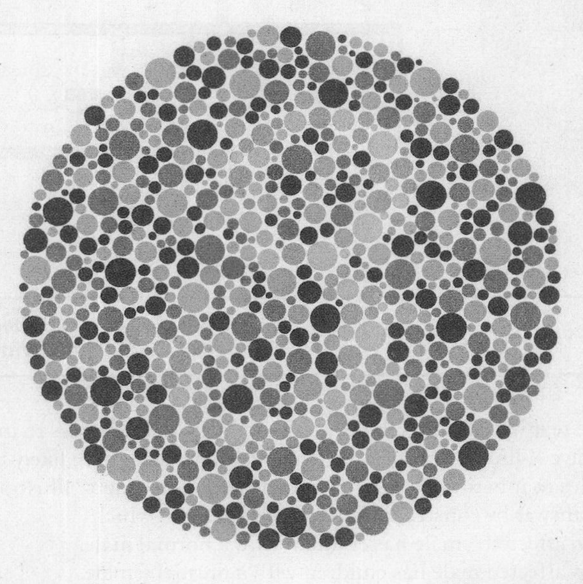

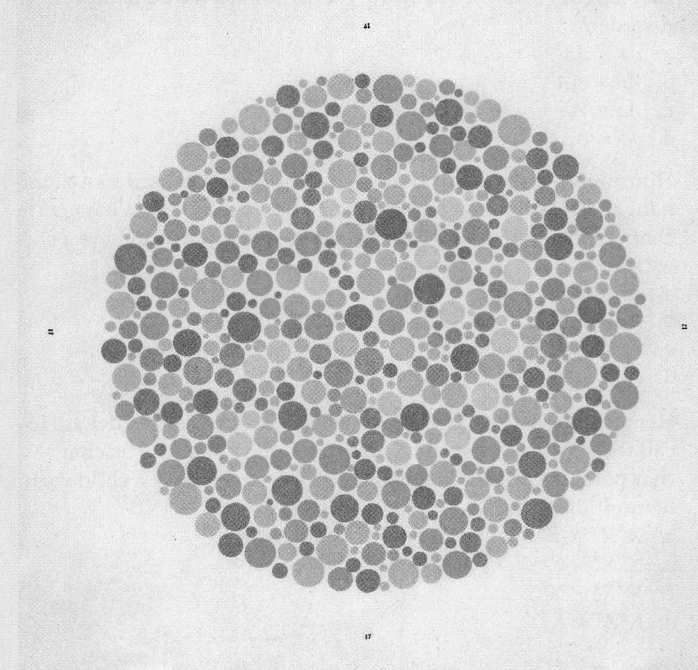

SAMPLE GENETICS PROBLEMS: CHAPTER 10

1. Recall that human females have two X chromosomes, and human males have one X chromosome and one Y chromosome.
   a. Do males inherit their X chromosome from their mother or from their father?
   b. If a female has one copy of an X-linked recessive allele for a genetic disorder, does she have the disorder?
   c. If a male has one copy of an X-linked recessive allele for a genetic disorder, does he have the disorder?
   d. Assume that a female is a carrier of an X-linked recessive disorder. With respect to the disorder allele, how many types of gametes can she produce?
   e. Assume that a male with an X-linked recessive genetic disorder has children with a female who does not carry the disorder allele. Could any of their sons have the genetic disorder? How about their daughters? Could any of their children be carriers for the disorder? If so, which sex(es) could they be?

2. Cystic fibrosis is a recessive genetic disorder. The disorder allele, which we will call *a*, is located on an autosome. What are the chances that parents with the following genotypes will have a child with the disorder?
   a. $aa \times Aa$
   b. $Aa \times AA$
   c. $Aa \times Aa$
   d. $aa \times AA$

3. Huntington's disease (HD) is a genetic disorder caused by a dominant allele—call it *A*—that is located on an autosome. What are the chances that parents with the following genotypes will have a child with HD?
   a. $aa \times Aa$
   b. $Aa \times AA$
   c. $Aa \times Aa$
   d. $aa \times AA$

4. Hemophilia is a recessive genetic disorder whose disorder allele—call it *a*—is located on the X chromosome. What are the chances that parents with the following genotypes will have a child with hemophilia?
   a. $X^AX^A \times X^aY$
   b. $X^AX^a \times X^aY$
   c. $X^AX^a \times X^AY$
   d. $X^aX^a \times X^AY$

5. Do male and female children have the same chance of getting hemophilia?

6. Explain why the terms "homozygous" and "heterozygous" do not apply to X-linked traits in males.

7. The accompanying chart shows a representative family pedigree for the inheritance of phenylketonuria. Is this disorder a dominant trait or a recessive trait? Is the disorder allele located on an autosome or on the X chromosome? What are the genotypes of individuals 1 and 2 in generation I?

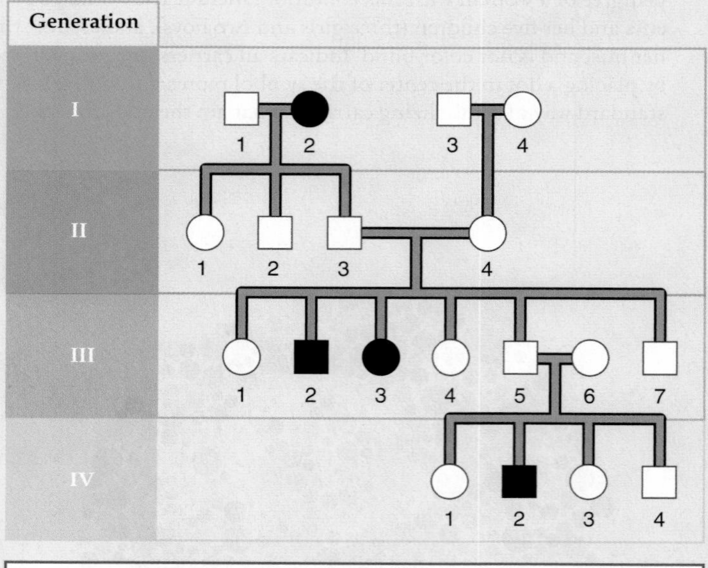

8. In the text we state that males are more likely than females to inherit recessive X-linked genetic disorders. Are males also more likely than females to inherit dominant X-linked genetic disorders? Illustrate your answer by constructing Punnett squares in which
   a. an affected female has children with a normal male.
   b. an affected male has children with a normal female.

9. Study the pedigree shown here. Is the disorder allele dominant or recessive? Is the disorder allele located on an autosome or on the X chromosome? To answer this question, assume that individual 1 in generation I, and individuals 1 and 6 in generation II, do not carry the disorder allele.

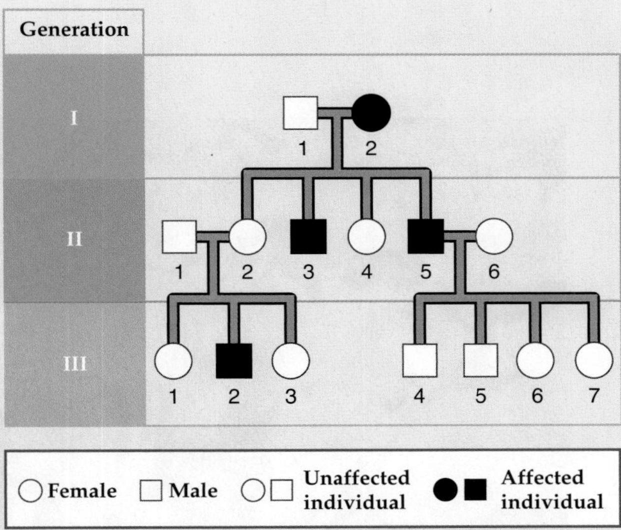

10. Imagine you are conducting an experiment on fruit flies in which you track the inheritance of two genes: one with allele *A* or *a*, the other with allele *B* or *b*. *AABB* individuals are crossed with *aabb* individuals to produce F_1 offspring, all of which have genotype *AaBb*. These *AaBb* F_1 offspring are then crossed with *aaBB* individuals. Construct Punnett squares and list the possible offspring genotypes that you would expect in the F_2 generation
    a. if the two genes were completely linked.
    b. if the two genes were on different chromosomes.

11. How many different types of gametes can be generated by a pea plant that has genotype *Aa Dd Ee gg*, where *A*/*a*, *D*/*d*, *E*/*e*, and *G*/*g* are four separate unlinked genetic loci? List all the genotypes that are possible among the gametes produced by this individual. How many different types of gametes are possible, and what would be their genotypes, if the *A* and *D* alleles were located very close together on the same chromosome and therefore completely linked?

11 DNA and Genes

ÖTZI, THE ICEMAN. In 1991, hikers discovered the body of a man who lived thousands of years ago, lying at the base of a melting glacier. The photo on the left shows a reconstruction of the Copper Age man, nicknamed Ötzi after the Ötztal Alps between Austria and Italy where he was found at an altitude of 11,000 feet.

The Man from the Copper Age

On a September day in 1991, a German couple hiking in a high mountain pass in the Italian Alps stumbled on an astonishing find: at their feet, in a pool of melting glacier water, lay a mummified corpse. Far from being even a few decades old, the unknown man turned out to be 5,200 years old—a visitor from the Copper Age promising a tantalizing peek into the past.

Extremely well preserved, the Iceman—as researchers dubbed him—was estimated to have been about 45 years old, 5 feet 5 inches tall, and 111 pounds. Near his naked, tattooed body were a beautifully made axe with a solid-copper blade, a flint dagger and woven sheath, a half-constructed longbow, a quiver full of arrows, and a full set of clothes that included a leather belt and pouch, sheepskin leggings and a jacket made of animal skin, a cape of woven grass, and calfskin shoes lined with grass. Among other items, he also carried pieces of fungus, one possibly a kind of prehistoric penicillin for treating infections, and another useful for starting a fire. He had brought berries and grains from lower elevations and was fully equipped for a mountain trek; his backpack and other gear were beautifully crafted from 18 different kinds of wood.

Frozen high in the Alps for 5,200 years, the ancient Iceman's body was perfectly suited to scientific examination and fully equipped with Copper Age tools and clothes. The discovery seemed almost too good to be true, and some people speculated that the body was an elaborate hoax; perhaps the Iceman was a transplanted Egyptian or pre-Columbian American mummy. To solve the mystery, scientists would have to study his DNA and place him in the human family tree. As unusual as the Iceman was, the way in which scientists used his DNA to learn more about him was a standard approach in biology.

How could biologists use the Iceman's DNA to determine the true identity of this long-frozen man? Were his closest relatives modern Europeans? Was he more closely related to Neandertals than modern Europeans are?

In this chapter we consider the structure of DNA and how this code-carrying molecule is copied within a cell. Despite robust mechanisms to fix "typos" in copying and to repair later damage, the DNA code can be altered, producing mutations.

MAIN MESSAGE Genes are composed of DNA sequences. DNA must be replicated before a cell divides, and mistakes during replication are one cause of gene mutations.

KEY CONCEPTS

- DNA consists of two polynucleotide strands held together by hydrogen bonds and twisted into a spiral shape.
- All cells in a multicellular body have essentially the same nucleotide sequence in their DNA. DNA sequences vary between individuals of one species and among different species.
- During DNA replication, each strand of DNA serves as a template from which a new strand is copied.
- DNA in cells is subject to damage by various physical, chemical, and biological agents. Up to a point, such damage can be repaired.
- Prokaryotes have relatively little DNA compared to eukaryotes. Eukaryotes tend to have not only more genes, but also a large amount of DNA that does not encode proteins.
- In eukaryotes, DNA is compacted with the help of proteins to organize and pack it into the tiny nucleus.
- Gene expression begins with the activation of a gene and ends with a detectable phenotype influenced by that gene. Transcription and translation are steps in the gene expression pathway.

THE MAN FROM THE COPPER AGE 241

11.1 An Overview of DNA and Genes 242

11.2 The Three-Dimensional Structure of DNA 245

11.3 How DNA Is Replicated 247

11.4 Repairing Replication Errors and Damaged DNA 248

11.5 Genome Organization 250

11.6 DNA Packing in Eukaryotes 253

11.7 Patterns of Gene Expression 254

APPLYING WHAT WE LEARNED 256
CSI: Copper Age

GENES CONTROL THE INHERITANCE OF TRAITS and are located on chromosomes, as we learned in the previous two chapters. However, this knowledge leaves several fundamental questions unanswered: How exactly does DNA store information? How is that information turned into a particular observable trait, or phenotype? What if mistakes are made in the copying of DNA? If all cells in a multicellular body have the same genes, why are heart muscle cells so very different from nerve cells in the brain?

In this chapter we describe the physical structure of DNA and how this genetic material is copied. We consider how mistakes in DNA copying that are not "fixed" by DNA repair mechanisms can result in mutations, and how mutations can give rise to genetic disorders. We tour the nucleus to inspect the orderly way in which DNA is packed inside it. We close by considering the frugal nature of cells: why not all the genes are working away in every cell all of the time, and how environmental cues and internal signals can turn genes on or off according to the needs of the organism.

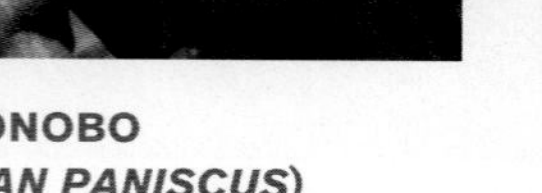

	BONOBO (*PAN PANISCUS*)	**COMMON CHIMPANZEE (*PAN TROGLODYTES*)**
HABITAT	South side of Congo River	North side of Congo River
PHYSICAL FEATURES	Slender, pink lips in dark face	More robust, face darkening with age
	Little difference between sexes	Males about 30% larger than females
VOICE	Higher pitched	Grunts, hoots, screams
BEHAVIOR	Playful throughout life	Play seen only in young animals
	Females socially dominant	Males socially dominant
	Female-female bonds important	Male-male bonds important

FIGURE 11.1 Bonobos and Chimps Have Nearly Identical Genomes
The 0.1 percent difference in the DNA sequences of bonobos and chimpanzees, though tiny, is enough to set these two species apart. How does DNA store information? Do all the cells in your body have the same DNA-based information?

11.1 An Overview of DNA and Genes

By the early 1900s, geneticists knew that genes control the inheritance of traits, that genes are located on chromosomes, and that chromosomes contain DNA and protein. The first step in the quest to understand the physical structure of genes was to determine whether the genetic material was DNA or protein. Initially, most geneticists thought that protein was the more likely candidate. Proteins are large, complex biomolecules, and it was not hard to imagine that they could store the tremendous amounts of information needed to govern the lives of cells. Over time, through a number of key experiments, that hypothesis was shown to be wrong.

- In 1928, a British medical officer named Frederick Griffith transformed a harmless strain of a bacterium (the R strain) into a deadly strain (the S strain) by exposing the R strain to heat-killed strain-S bacteria.
- In 1944, Oswald Avery and his colleagues published a landmark paper demonstrating that only the DNA from heat-killed strain-S bacteria was able to transform strain-R bacteria into strain-S bacteria.
- In 1952, Alfred Hershey and Martha Chase demonstrated that the DNA of a virus, not its proteins, was responsible for taking over a bacterial cell and producing the next generation of viruses.

With these results, nearly all biologists were persuaded that the genetic material was DNA, not protein.

DNA stores genetic information as a sequence of nucleotides

As noted in Chapter 2, deoxyribonucleic acid (**DNA**) is composed of polynucleotides. Recall that a *polynucleotide* is a long chain made up of building blocks called *nucleotides*. In DNA, two strands of polynucleotides are twisted around an imaginary axis to form a spiral structure, the double helix (see Figure 2.23). Each chromosome in your cells contains one DNA double helix. Each such helix contains many *genes*, and each gene controls at least one distinct genetic trait.

The **genome** of an organism or a particular individual is all the DNA-based information in its chromosomes. Before cell division, in S phase of the cell cycle (see Section 7.2), the entire genome of a cell is copied in a process called *DNA replication*.

The DNA sequence information in the genome of one species is different from that of another species, and this difference in sequence is what *makes* the two species different. Consider the genomes of humans and our closest living relatives, chimpanzees and bonobos (**FIGURE 11.1**). Human and bonobo genomes are 98.7 percent identical, according to researchers at the Max Planck Institute in Germany. And, as you might expect, bonobo and chimpanzee genomes are closer yet: 99.9 percent of their nucleotide sequences match. That seemingly tiny sequence variation—just 0.1 percent—is enough to produce the significant differences, especially in social behavior, between the two living apes that are most closely related to us.

Most genes code for proteins, which generate phenotypes

If the genome is the software package that runs all the hardware of the cell, then a gene is like a specific computer application. You could imagine the roughly 25,000 genes in human cells as many different applications built into one software package, the human genome.

How do we go from DNA-based information held in a gene to a particular genetic characteristic that you can see or detect in some way? In the living cell, information flows from the DNA that composes a gene to ribonucleic acid (**RNA**), another information-carrying molecule. RNA is a single-stranded nucleic acid that is similar to DNA in some ways and different from it in other ways. Most of the RNA in a cell is of a type known as **messenger RNA (mRNA)**. The molecule is called *messenger* RNA because its function is to deliver the genetic information from DNA to the ribosomes, the protein-making structures in the cytosol. **Ribosomes** manufacture proteins as directed by the code carried by mRNA molecules. When the protein carries out its particular function, either by itself or in conjunction with other molecules, the gene is expressed. *Gene expression* refers to the display of gene-based information as a phenotype (**FIGURE 11.2**).

We can summarize the process of gene expression as follows:

1. The sequence of nucleotides in DNA is copied into mRNA—a process called **transcription**.
2. Ribosomes turn the sequence of nucleotides in mRNA into a sequence of amino acids in a protein chain—a process called **translation**.
3. The protein carries out its function, producing a phenotype.

Let's look at gene expression using the example of the lactase gene, abbreviated as *LCT*. The *LCT* gene

The Flow of Genetic Information

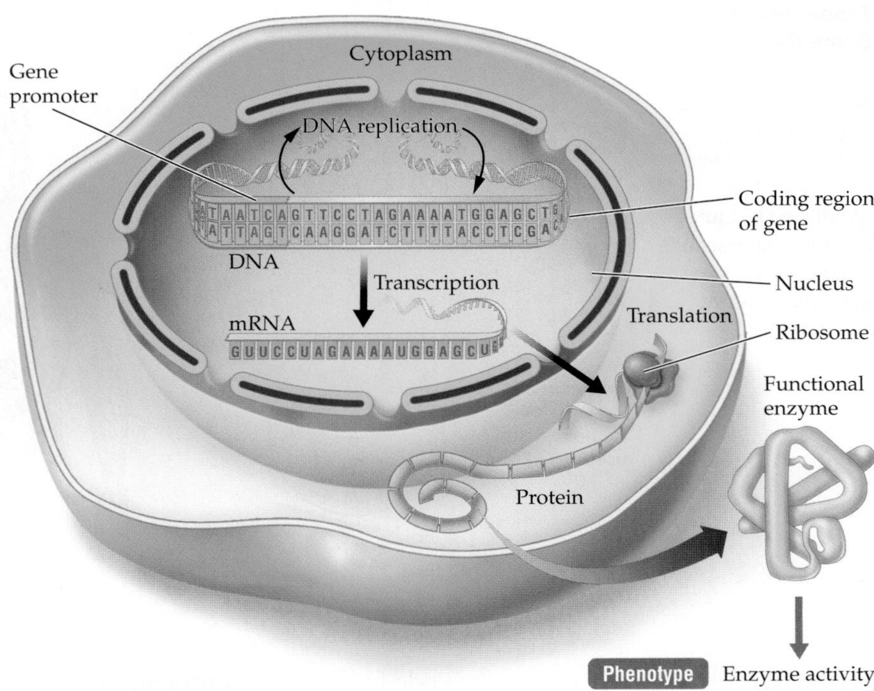

FIGURE 11.2 From Gene to Phenotype at the Molecular Level

The readout of information coded in DNA produces the phenotype, such as the activity of the lactase enzyme in human intestines. The gene promoter controls whether, and to what degree, a gene is transcribed.

Helpful to Know

No matter their race or ethnicity, two randomly selected, unrelated humans are likely to have genomes that are more than 99.9 percent identical. Identical twins have essentially the same DNA sequence information in their genomes because they arise from the same fertilized egg (zygote). But "identical" twins aren't truly identical because nongenetic factors also influence the phenotype.

FIGURE 11.3 Different Cell Types Have the Same Genes

Although all cells within a multicellular organism have the same genes, these cells can differ greatly in structure and function because different genes are active in different types of cells.

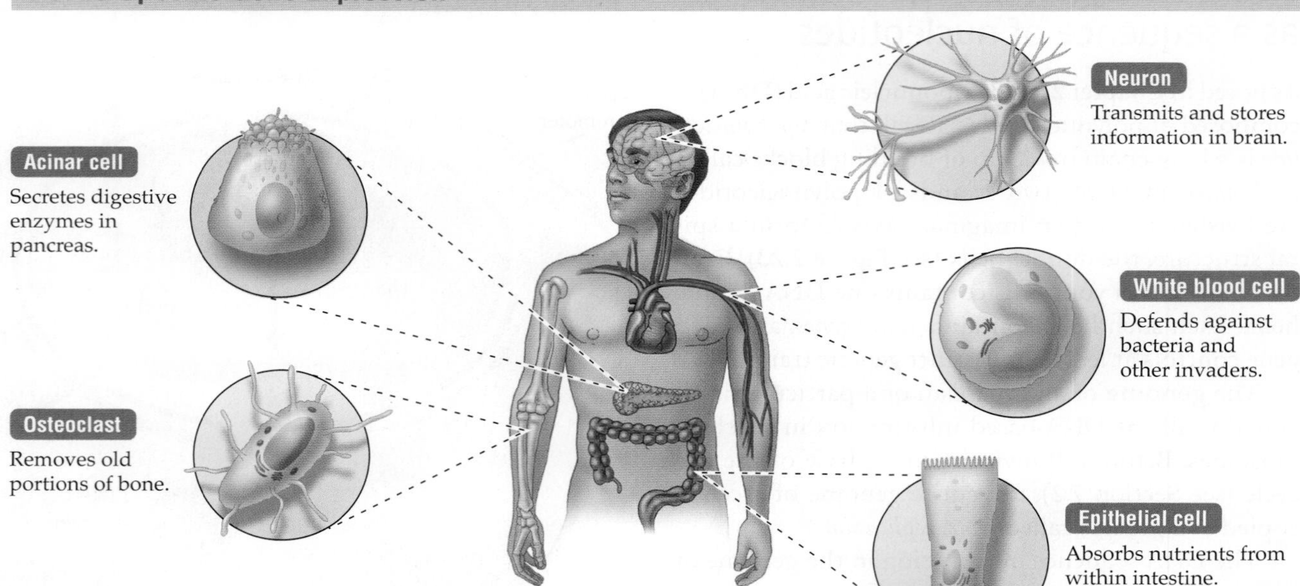

Tissue-Specific Gene Expression
Acinar cell
Secretes digestive enzymes in pancreas.
Osteoclast
Removes old portions of bone.
Neuron
Transmits and stores information in brain.
White blood cell
Defends against bacteria and other invaders.
Epithelial cell
Absorbs nutrients from within intestine.

Helpful to Know

Biologists use a variety of words and phrases to describe a gene whose product is made: the gene may be described as "activated," "turned on," or "being expressed."

carries the code for building a protein called lactase. Baby mammals need lactase to break down lactose, the most abundant sugar in milk. In the cells that line the small intestine, the *LCT* gene is transcribed into many molecules of *LCT* mRNA. The mRNA is then translated by ribosomes to produce the protein enzyme lactase, which is secreted into the small intestine. Lactase breaks up lactose to release two smaller sugars, glucose and galactose, that can be absorbed by intestinal cells.

Different sets of genes are expressed in different cell types

All cells in a multicellular individual have essentially the same DNA-based information. For example, all cells in your body have the *LCT* gene that codes for lactase. But not all cells in your body need to express the *LCT* gene. Even though every cell in a multicellular body has the same genes, specialized cell types, which differ greatly in structure and function (**FIGURE 11.3**), express very different sets of those genes. The lactase gene, for example, is expressed in only certain cells that line the intestine. Extending the computer analogy, the same DNA software is installed in all cells, but different cell types run a different mix of the available applications.

Although we have about 21,000 protein-coding genes, the average human cell expresses no more than about 10,000 of them at any one time. Most genes have a stretch of DNA, called a **gene promoter**, that functions like an on/off switch for transcription. Many genes are regulated in a tissue-specific manner, meaning that their promoter is on in some cell types but off in other cell types. The expression of a gene can also be regulated quantitatively, instead of simply being on or off: gene activity can be tweaked upward (*up-regulated*) or downward (*down-regulated*), usually in response to signals that the cell receives.

Many genes are *developmentally regulated*, meaning that their expression can change, sometimes dramatically, as an organism grows and develops. At birth, the *LCT* gene is turned on in just about all of us, but in a majority of the world's population, the expression of the gene is steadily down regulated between 5 and 10 years of age, when children were traditionally weaned from mother's milk in most cultures. The down-regulation of the *LCT* gene in adolescence and adulthood produces the condition most people call lactose intolerance, although the technical term is adult-type hypolactasia.

Concept Check

1. What is gene expression?
2. Is the genome of your neurons (nerve cells) identical to that of your liver cells?
3. We have at least 21,000 protein-coding genes. Are all of these genes expressed in each of your neurons (nerve cells)?

Concept Check Answers

1. Gene expression consists of the processes by which a gene produces a phenotype. For protein-coding genes the gene's DNA is transcribed into mRNA, mRNA is translated into protein, and protein function produces a phenotype.
2. Yes. All cells in the body have essentially the same DNA-based information, or genome.
3. No. A unique subset of the approximately 21,000 protein-coding genes is expressed in each cell type.

11.2 The Three-Dimensional Structure of DNA

Working at Cambridge University in England, the American James Watson and the Englishman Francis Crick deciphered the physical structure of DNA. In a two-page paper published in 1953, they proposed that DNA was a double helix, a structure that can be thought of as a ladder twisted into a spiral coil (**FIGURE 11.4**). Watson was 25 at the time, and Crick was 37. In 1962, Watson and Crick were awarded the Nobel Prize in Physiology or Medicine for their discovery. They shared the prize with Maurice Wilkins, who had also worked to discover the structure of DNA. Missing from the Nobel ceremony was Rosalind Franklin, a gifted young scientist whose research had provided Watson and Crick with critical data. Rosalind Franklin had died of cancer in 1958, at age 37. Because Nobel Prizes are not awarded to deceased people, or to more than three people, we will never know whether she might have received a Nobel Prize for her contributions, had she been alive in 1962.

DNA is built from two helically wound polynucleotides

As Watson and Crick described it, DNA is built from two parallel strands of repeating units called nucleotides. Each nucleotide is composed of the sugar deoxyribose, a phosphate group, and one of four **nitrogenous bases** [**nye-*TRAH*-juh-nus**]: adenine (A), cytosine (C), guanine (G), or thymine (T). The way the two strands are connected is reminiscent of the rungs that connect the sides of a ladder (see Figure 11.4). The nucleotides of each strand are connected by covalent bonds between the phosphate of one nucleotide and the sugar of the next nucleotide. The "rungs" are created by hydrogen bonds that link the bases on one strand to the bases on the other strand.

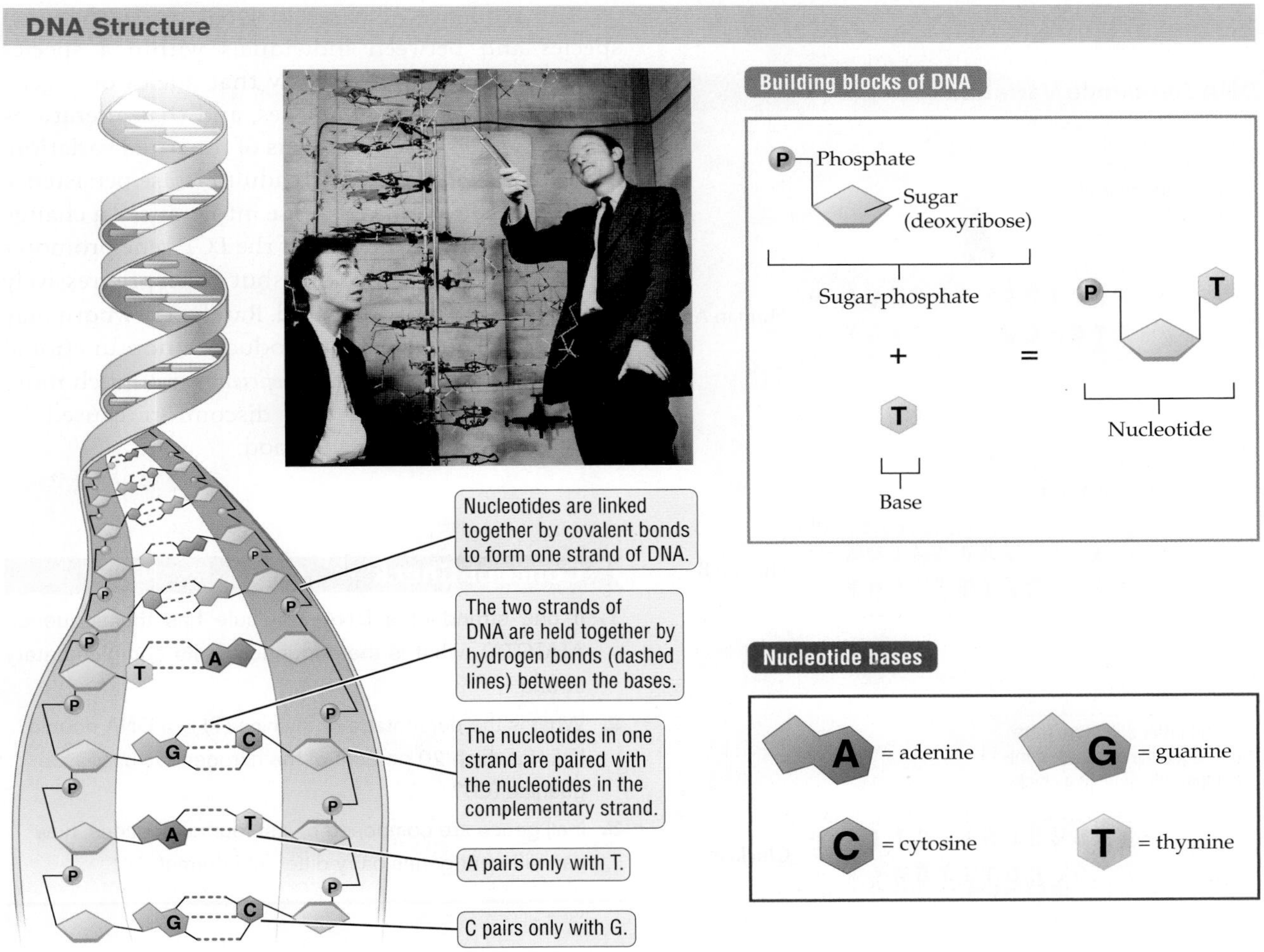

DNA Structure

Nucleotides are linked together by covalent bonds to form one strand of DNA.

The two strands of DNA are held together by hydrogen bonds (dashed lines) between the bases.

The nucleotides in one strand are paired with the nucleotides in the complementary strand.

A pairs only with T.

C pairs only with G.

Building blocks of DNA

P Phosphate

Sugar (deoxyribose)

Sugar-phosphate + T Base = Nucleotide

Nucleotide bases

A = adenine

G = guanine

C = cytosine

T = thymine

FIGURE 11.4 The DNA Double Helix and Its Building Blocks A nucleotide consists of a phosphate, a sugar, and a nitrogen-containing (nitrogenous) base. DNA contains four types of nucleotides that differ only in the type of base found in them. In DNA, two complementary strands of nucleotides are twisted into a spiral around an imaginary axis, rather like the winding of a spiral staircase. The two strands are held together by hydrogen bonds between their complementary bases. (*Inset*) James Watson (*left*) and Francis Crick, with a model of the DNA double helix.

The term **base pair** (or nucleotide pair) refers to two bases held together by hydrogen bonds in a DNA molecule.

Watson and Crick proposed a set of base-pairing rules, stating that adenine on one strand could pair only with thymine on the other strand; similarly, cytosine on one strand could pair only with guanine on the other strand (see Figure 11.4). These base-pairing rules have an important consequence: when the sequence of bases on one strand of the DNA molecule is known, the sequence of bases on the other, *complementary strand* of the molecule is automatically known. For example, if one strand consists of the sequence

ACCTAGGG

then the complementary strand has to have the sequence

TGGATCCC

Any other sequence in the complementary strand would violate the base-pairing rules.

DNA Sequence Variation

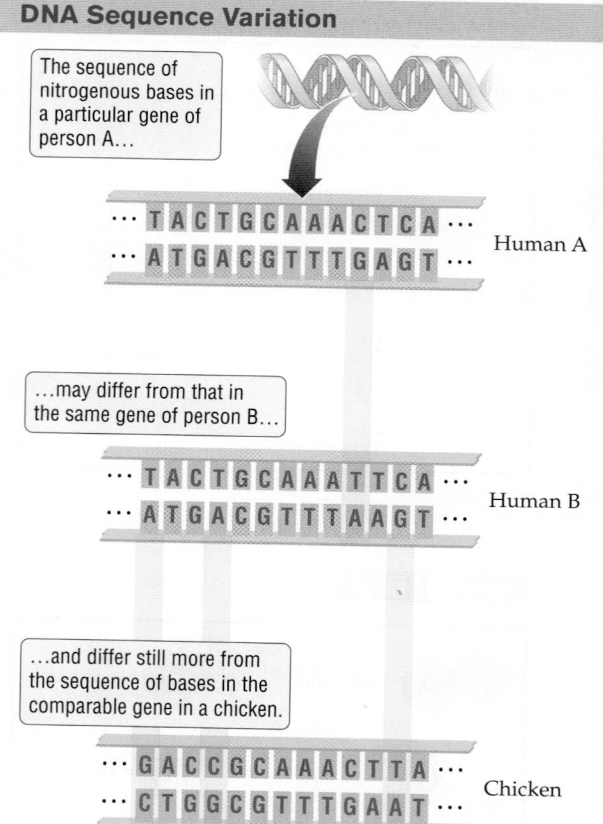

FIGURE 11.5 The Sequence of Bases in DNA Differs among Species and among Individuals within a Species

Here the sequence of bases in a hypothetical gene is compared for two humans (A and B) and a chicken. Base pairs highlighted in yellow are variant; that is, they differ between the genes of persons A and B, and between the comparable human and chicken genes.

DNA's structure explains its function

We now know that the physical structure of DNA proposed by Watson and Crick is correct in all its essential elements. This structure explains a great deal about how DNA works. For example, as you will see in Section 11.3, the fact that adenine can pair only with thymine and that cytosine can pair only with guanine suggested a simple way in which the DNA molecule could be copied: each original strand could serve as a template on which a new strand could be built.

Knowledge of the three-dimensional structure of DNA also suggested that the information stored in DNA could be represented as a long string of the four bases: A, C, G, and T. The four bases can be arranged in any order along a single strand of DNA. The fact that each DNA strand is composed of millions of these bases means that a tremendous amount of information could be stored in the order of the bases along the DNA molecule—that is, in the *DNA sequence*.

The sequence of bases in DNA differs between species and between individuals within a species (**FIGURE 11.5**). We now know that alleles of a gene have different DNA sequences, and that alterations in DNA sequence are the basis of inherited variation. For example, in people with adult lactase persistence (those who do *not* have lactose intolerance), a change in the nucleotide sequence of the *LCT* gene promoter *prevents* that gene from being shut down progressively in adolescence and adulthood. Rarely, a newborn may have an allele of *LCT* that produces a nonfunctional lactase; the resulting *lactase deficiency* is a much more serious condition than the discomfort caused by lactose intolerance in adulthood.

Concept Check

1. If one strand of a DNA molecule has the sequence ATATCTAT, what is the sequence of its complementary strand?
2. What is the percentage of thymine (T) in a DNA double helix in which 20 percent of the nitrogenous bases are guanine (G)?
3. If all genes are composed of just four nucleotides, how can different genes carry different information?

Concept Check Answers

1. A bonds with T, and C bonds with G. Therefore, the sequence of the complementary strand is TATAGATA.
2. Thirty percent. (If G = 20%, C = 20%. When G + C = 40%, then A + T = 60%; therefore T = 30%.)
3. The information encoded by a gene is determined by the precise sequence in which the four nucleotides are arranged; changing the nucleotide sequence changes the information. Because genes can differ in the total number of nucleotides, they can vary in length as well.

11.3 How DNA Is Replicated

As Watson and Crick noted in their historic 1953 paper, the structure of the DNA molecule suggested a simple way that the genetic material could be copied. They elaborated on this suggestion in a second paper, also published in 1953. Because A pairs only with T, and C pairs only with G, each strand of DNA contains the information needed to duplicate the complementary strand. For this reason, Watson and Crick suggested that **DNA replication**—the copying of DNA—might work in the following way (**FIGURE 11.6**):

1. The hydrogen bonds connecting the two strands of the DNA molecule are broken as the two strands are unwound by specialized proteins.
2. Each strand is then used as a template for the construction of a new strand of DNA. The new strand is built by an enzyme that links nucleotides together, using the sequence of the template strand as a "mold."
3. When this process is completed, there are two identical copies of the original DNA molecule, each with the same sequence of bases. Each of the two DNA copies is composed of a template strand of DNA (from the original DNA molecule) and one newly synthesized strand of DNA. This mode of replication is known as **semiconservative replication** because one "old" strand (the template strand) is retained, or conserved, in each new double helix. Put another way, half ("semi") of each new DNA double helix has been retained ("conserved") from the parent molecule.

Five years after Watson and Crick's prediction, other researchers confirmed that DNA replication does, in fact, produce DNA molecules composed of one old strand and one new strand. The key enzyme involved in the replication of DNA has now been identified as well: **DNA polymerase (puh-*LIM*-uh-rays)**.

The Watson-Crick model of DNA replication is elegant and straightforward, but the mechanics of actually copying DNA are far from simple. More than a dozen enzymes and proteins are needed to unwind the DNA, to stabilize the separated strands, to start the replication process, to attach nucleotides to the correct positions on the template strand, to "proofread" the results, and to join partly replicated fragments of DNA to one another.

Although DNA replication is a very complex task, cells can copy DNA molecules containing billions of nucleotides in a matter of hours—about 8 hours in people (over 100,000 nucleotides per second). One reason why replication of the DNA molecule is so speedy is that it begins at thousands of different places at once. Despite the speed, cells make remarkably few mistakes when they copy their DNA.

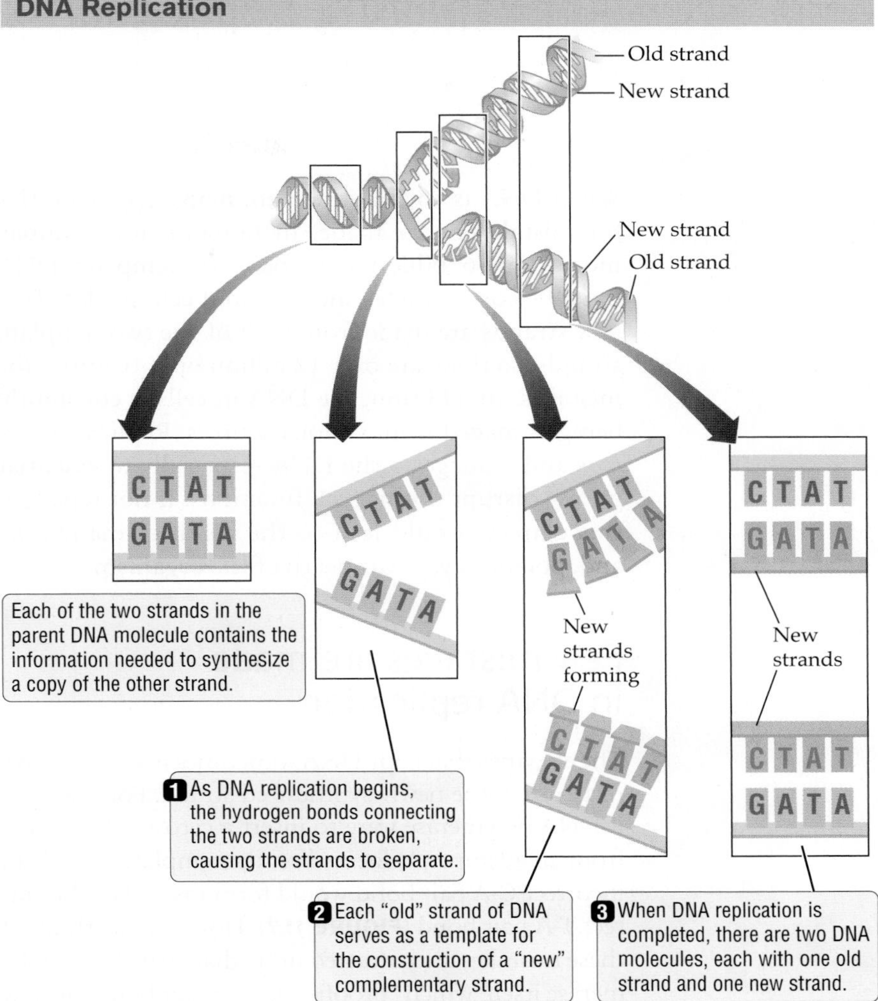

FIGURE 11.6 The Replication of DNA Is Semiconservative

In this overview of DNA replication, the template DNA strands are blue, and the newly synthesized strands are orange. One strand (blue) from the parent double helix is conserved in each newly made daughter double helix (blue and orange).

Concept Check

1. What key function is performed by DNA polymerase? Where is this enzyme found in eukaryotic cells?
2. What is meant by the "semi" in semiconservative replication?

Concept Check Answers

1. DNA polymerase catalyzes the formation of a polynucleotide strand complementary to a template DNA strand. It is localized in the nucleus in eukaryotic cells.

2. "Semi" means that one strand (or half) of each newly synthesized DNA double helix comes from the original parent DNA.

11.4 Repairing Replication Errors and Damaged DNA

When DNA is copied, there are many opportunities for mistakes to be made. In humans, for example, more than 6 billion base pairs of template DNA must be copied each time a diploid cell divides. Two new strands are made from each of the two template strands, so there are over 12 billion opportunities for mistakes. In addition, the DNA in cells is constantly being damaged from various sources. Replication errors and damage to the DNA—especially to essential genes—disrupt normal cell functions. If not repaired, this damage would lead to the death of many cells and, potentially, to the death of the organism.

Few mistakes are made in DNA replication

The enzymes that copy DNA sometimes insert an incorrect base in the newly synthesized strand. For example, if DNA polymerase were to insert a cytosine (C) across from an adenine (A) located on the template strand, an incorrect C-A pair bond would form instead of the correct T-A pair bond (**FIGURE 11.7**). However, nearly all of these mistakes are corrected immediately by DNA polymerase itself, which "proofreads" the pair bonds as they form. This form of error correction is similar to what you do when you type a paper, realize you made a mistake, and delete it right away with the backspace key.

When an incorrect base is added but escapes proofreading by DNA polymerase, a mismatch error has occurred. Mismatch errors occur about once in every 11 million bases. Cells contain repair proteins that specialize in fixing mismatch errors; these proteins play a role similar to the error checking you perform when you complete the first draft of a paper, print it, and carefully review it for mistakes. Proteins that fix mismatch errors correct 99 percent of those errors, reducing the overall chance of an error to the incredibly low rate of one mistake in every billion bases.

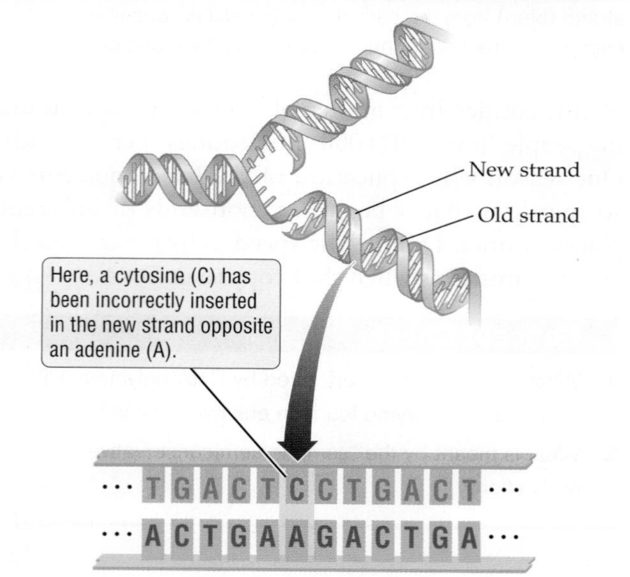

FIGURE 11.7 Mistakes in DNA Replication May Lead to Mutations
DNA repair enzymes usually detect and fix mismatch errors such as the one shown here before the cell's DNA is replicated again. If the mismatch is not repaired before the next round of DNA replication, half of the daughter helices made by this DNA will have a C-G base pair in place of the original A-T base pair. Such a change in the DNA sequence constitutes a mutation.

A mutation is a change in the DNA sequence

On the rare occasions when a mismatch error is not corrected, the DNA sequence is changed, and the altered sequence is reproduced the next time the DNA is replicated. A change to the sequence of bases in an organism's DNA is called a **mutation**. Mutations can also occur when cells are exposed to **mutagens** (substances or energy sources that alter DNA).

Mutations result in the formation of new alleles. Some of the new alleles that result from mutation are beneficial, but most are either neutral or harmful. Among the harmful alleles are those that cause cancer and other human genetic disorders, such as sickle-cell anemia and Huntington's disease. Note that our definition of mutation includes not only changes in the DNA sequence of a single gene, but also the larger-scale changes in DNA sequence that are created by chromosomal abnormalities (see Chapter 10). Changes in chromosome number or chromosome organization generally affect large numbers of genes, and such changes also amount to DNA mutations because they have the effect of adding, deleting, or rearranging nucleotide sequences.

Normal gene function depends on DNA repair

Every day, the DNA in each of our cells is damaged thousands of times by chemical, physical, and biological agents. These agents include energy from radiation or heat, collisions with other molecules in the cell, attacks by viruses, and random chemical accidents (some of which are caused by environmental pollutants, but most of which result from normal metabolic processes). Our cells contain a complex set of repair proteins that fix the vast majority of this damage. Single-celled organisms such as yeasts have more than 50 different repair proteins, and humans probably have even more.

Although our cells are very good at repairing damaged DNA, some damage far exceeds the cell's ability to repair it. One example is the damage caused by radiation. The rad is a unit for measuring the amount of absorbed radiation. Humans exposed to 1,000 rads die in a few weeks, in part because their DNA has been damaged beyond repair. Some of the people who initially survived the atomic blasts at Hiroshima and Nagasaki in 1945 were exposed to about 1,000 rads. Over the next few weeks they died from acute radiation poisoning as cells in the bone marrow and digestive system died from the severe DNA damage. (To give some idea of scale, during a dental X-ray your tissues absorb less than 0.1 rad.)

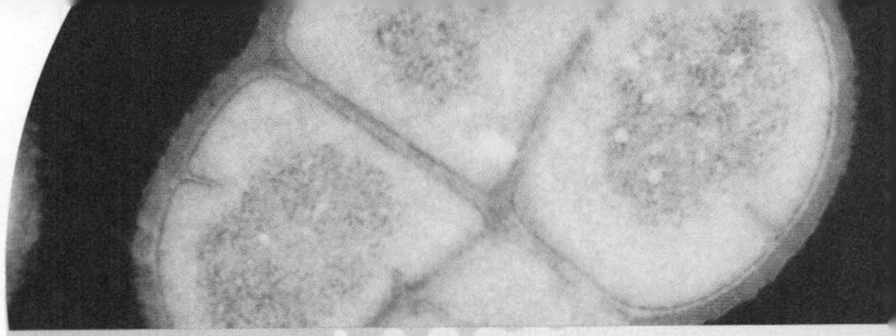

EXTREME DNA REPAIR

Although 1,000 rads of radiation kills a person, such a dose wouldn't faze the bacterium *Deinococcus radiodurans*. It's so efficient at repairing DNA damage that 1,000,000 rads merely slows its growth. Thick cell walls and pigment deposits also help shield the DNA.

There are three main steps in **DNA repair**: the damaged DNA must be (1) recognized, (2) removed, and (3) replaced. Different sets of repair proteins specialize in recognizing defects in DNA structure and in removing the damaged DNA strand (**FIGURE 11.8**). Once these first two steps have been accomplished, the final step is to add the correct sequence of nucleotides to fill the gap. This third step of the repair process uses the intact strand of DNA as template for re-creating the missing segment in the complementary strand.

Mutations are the consequence of failure in DNA repair. When an animal cell acquires many mutations in a variety of genes, it becomes dangerous because such a cell is more likely to have malfunctions in genes that normally keep cell division under tight control. Consequently, the cell is more likely to become cancerous—a condition marked by runaway cell proliferation (see Chapter 8).

Prokaryotic and eukaryotic genomes also contain "jumping genes," or **transposons**, DNA sequences that can move from one position on a chromosome to another, or even from one chromosome to another. The activity of a gene can be disrupted if a transposon inserts itself somewhere within that gene. This type of change generally cannot be detected or repaired by the DNA repair systems.

The importance of DNA repair mechanisms is also highlighted by genetic disorders in which the repair system is inactive or seriously inefficient. Xeroderma pigmentosum (XP) is a recessive genetic disorder in which even brief exposure to sunlight causes painful blisters (**FIGURE 11.9**). The allele that causes XP produces a nonfunctional version of one of the many human DNA repair proteins. The job of the normal form of this protein is to repair the kind of damage to DNA caused by UV light. The lack of this DNA repair protein in XP individuals makes them highly susceptible to skin cancer. Several inherited tendencies to develop

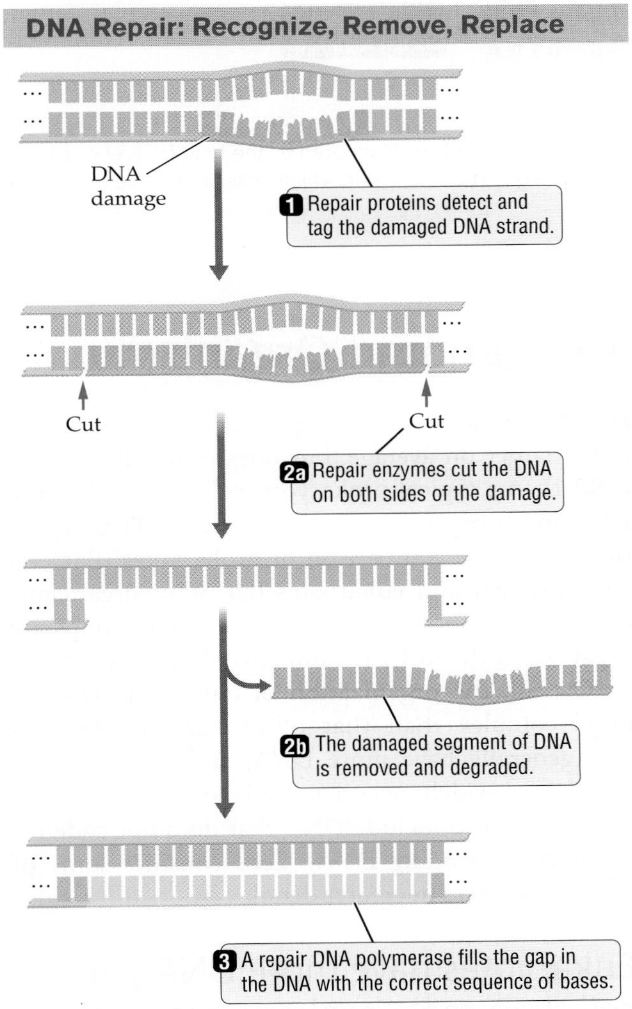

DNA Repair: Recognize, Remove, Replace

DNA damage
1 Repair proteins detect and tag the damaged DNA strand.
Cut
Cut
2a Repair enzymes cut the DNA on both sides of the damage.
2b The damaged segment of DNA is removed and degraded.
3 A repair DNA polymerase fills the gap in the DNA with the correct sequence of bases.

FIGURE 11.8 Repair Proteins Fix DNA Damage

Large complexes of DNA repair proteins work together to fix damaged DNA. They (1) recognize DNA damage, (2) remove the segment of DNA strand that is damaged, and (3) replace the missing segment through new DNA synthesis.

DNA Damage from UV Radiation

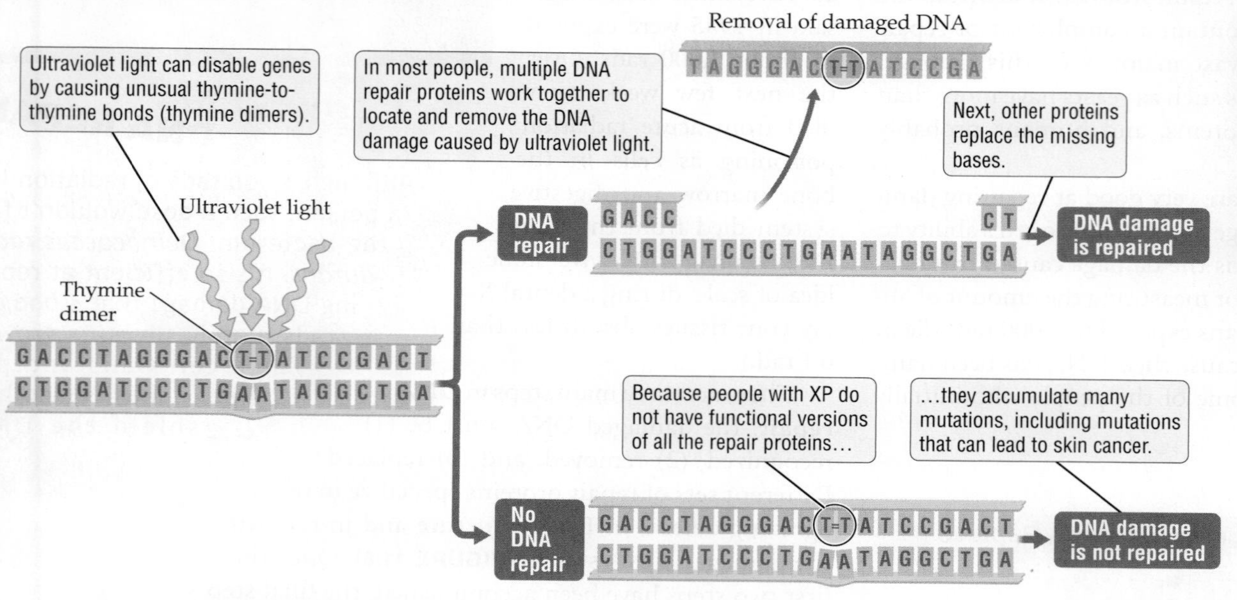

Ultraviolet light can disable genes by causing unusual thymine-to-thymine bonds (thymine dimers).
Ultraviolet light
Thymine dimer
GACCTAGGGACT-TATCCGACT
CTGGATCCCTGAATAGGCTGA
In most people, multiple DNA repair proteins work together to locate and remove the DNA damage caused by ultraviolet light.
Removal of damaged DNA
TAGGGACT-TATCCGA
DNA repair
GACC
CTGGATCCCTGAATAGGCTGA
Next, other proteins replace the missing bases.
CT
DNA damage is repaired
Because people with XP do not have functional versions of all the repair proteins…
…they accumulate many mutations, including mutations that can lead to skin cancer.
No DNA repair
GACCTAGGGACT-TATCCGACT
CTGGATCCCTGAATAGGCTGA
DNA damage is not repaired

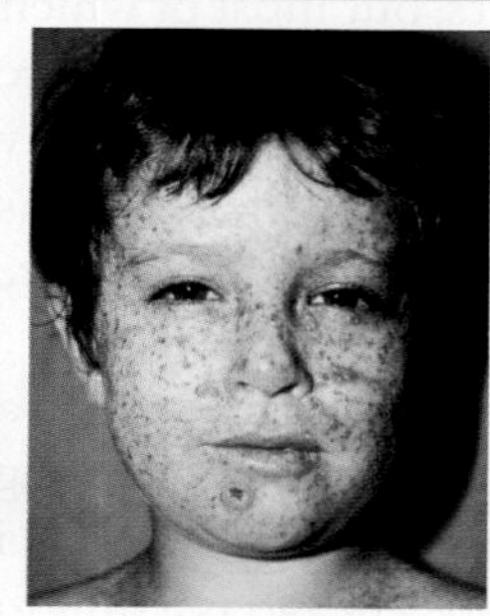

This child has XP. The growth on his chin is a skin cancer.

FIGURE 11.9 The Importance of DNA Repair Mechanisms

When DNA repair mechanisms fail to work properly, the consequences can be severe, as illustrated by the high frequency of skin cancers in people who have xeroderma pigmentosum (XP), a recessive genetic disorder in which cells are unable to make a protein used to repair DNA damage caused by ultraviolet light.

cancer, including some types of breast and colon cancer, also stem from less effective versions of genes that participate in DNA repair.

Concept Check

1. What mechanisms reduce the chance of DNA mutation? Are these mechanisms 100 percent effective?
2. What are the key steps in DNA repair?

Concept Check Answers

1. Proofreading by DNA polymerase and the DNA repair system reduce the chance of mutation. Neither mechanism is 100 percent effective.
2. There are three key steps: (1) recognition of the damaged strand, (2) removal of the damaged DNA, and (3) replacement of the removed DNA with a newly synthesized segment.

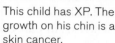

TABLE 11.1 Differences between Prokaryotic and Eukaryotic Genomes

	PROKARYOTES	EUKARYOTES
Size and organization of the genome	Several million base pairs in a single chromosome.	Hundreds of millions to billions of base pairs distributed among two to many chromosomes.
Noncoding DNA	Very little; most DNA codes for proteins.	Large amount, found both within genes and between genes.
Organization of genes	Commonly organized by function; genes for a particular pathway tend to be clustered together.	Genes with related functions may be found distant from one another, even on other chromosomes.

11.5 Genome Organization

How big is the genome of a prokaryote? A eukaryotic cell is larger on average, but does it also have more DNA than a typical prokaryotic cell? Does all of the DNA in an organism consist of genes? These questions call for a comparison of prokaryotes (Bacteria and Archaea) and eukaryotes (all other organisms) because the genomes of these two major groups are organized differently, as summarized in **TABLE 11.1**.

Overall, the organization of eukaryotic DNA is more complex than that of prokaryotes. Eukaryotes generally have more DNA, and the genes are distributed widely over a number of chromosomes. Eukaryotes also include DNA that does not code for proteins (noncoding DNA). Let's consider some of these differences in greater detail.

Eukaryotes have more DNA per cell than most prokaryotes have

Genome size is measured in base pairs. To enable comparison with prokaryotes, the genome size of eukaryotes is usually listed as the amount of DNA in a *haploid* cell. The genomes of prokaryotes vary in size from about

BIOLOGYMATTERS

Prenatal Genetic Screening

How is the baby? This is one of the first questions we ask after a child is born. Usually everything is fine, but sometimes the answer can be devastating. Today, some parents choose to make use of one of several prenatal genetic screening methods to check their baby's health long before it is born.

The practice of prenatal screening has been around a surprisingly long time. In the 1870s, doctors occasionally withdrew some of the fluid in which the fetus is suspended to obtain information about its health. Modern versions of that practice have been standard medical procedure since the early 1960s. In *amniocentesis*, a needle is inserted through the abdomen into the uterus to extract a small amount of amniotic fluid from the pregnancy sac that surrounds the fetus. This fluid contains fetal cells (often sloughed-off skin cells) that can be tested for genetic disorders. Another method is *chorionic* (**kohr-ee-*AH*-nik**) *villus sampling (CVS)*, in which a physician uses ultrasound to guide a narrow, flexible tube through a woman's vagina and into her uterus, where the tip of the tube is placed next to the villi, a cluster of cells that attaches the pregnancy sac to the wall of the uterus. Cells are removed from the villi by gentle suction and then tested for genetic disorders.

The tests are widely used by parents who know they face an increased chance of giving birth to a baby with a genetic disorder. Older parents, for example, might want to test for Down syndrome. A couple in which one parent carries a dominant allele for a known genetic disorder (such as Huntington's disease), or in which both parents are carriers for a recessive genetic disorder (such as cystic fibrosis), might also choose prenatal genetic screening.

Risks associated with amniocentesis and CVS, including vaginal cramping, miscarriage, and premature birth, have declined quite dramatically in the past 10 years because of technological advances and more extensive training. Recent studies suggest that the risk of miscarriage after CVS and amniocentesis is essentially the same, about 0.06 percent.

Until not long ago, couples who elected to have such tests performed had only two choices if their fears were confirmed: they could abort the baby, or they could give birth to a child who would have a genetic disorder. Since 1989, however, a third option has been available to couples who are willing, and can afford, to have a child by in vitro fertilization (IVF). In this procedure, fertilization occurs in a petri dish, after which one or more embryos are implanted into the mother's uterus. The advantage of this procedure is that it makes possible *preimplantation genetic diagnosis (PGD)*: one or two cells are removed from each of the developing embryos, usually 3 days after fertilization occurs while the cells are still loosely connected. Next, the cell or cells removed from the embryos are tested for genetic disorders. Finally, one or more embryos that are free of disorders are implanted into the mother's uterus, and the rest of the embryos, including those with genetic disorders, are discarded.

PGD is typically used by parents who either have a serious genetic disorder or carry alleles for one. Like all other genetic screening methods, PGD raises ethical issues. People who support the use of PGD think that amniocentesis and CVS provide parents with a bleak set of moral choices. In their view, it is morally preferable to discard an embryo at the 4- to 12-cell stage than it is to abort a well-developed fetus, or to give birth to a child that will suffer the devastating effects of a serious genetic disorder. Those opposed to the use of PGD argue that the moral choices are the same: in their view, once fertilization has occurred, a new life has formed and it is immoral to end that life, even at the 4- to 12-cell stage. What do you think?

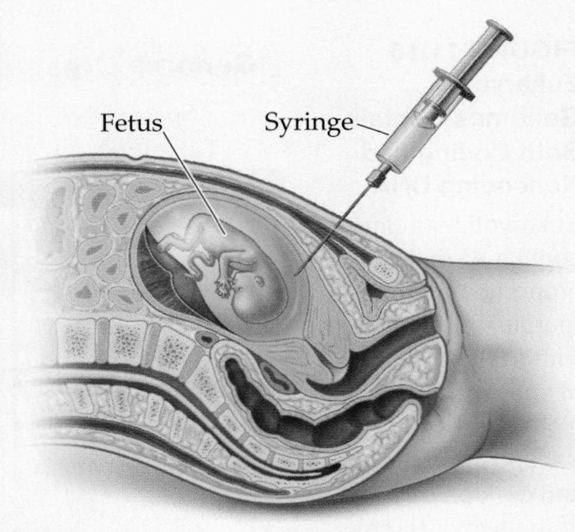

In amniocentesis, amniotic fluid, which contains fetal cells, is extracted from the uterus.

0.5 million to 30 million base pairs. The genome sizes of single-celled eukaryotes show much greater variation, ranging from 2.9 million base pairs in the haploid cells of a protist parasite called *Encephalitozoon cuniculi* to about a hundred billion base pairs in some types of amoeba. Most vertebrates have genomes that contain hundreds of millions to billions of base pairs. For example, puffer fish have only about 385 million base pairs of DNA in their haploid genome, while the marbled lungfish holds 132 billion base pairs of DNA in its haploid nucleus. **TABLE 11.2** lists additional examples.

In general, eukaryotes have much more DNA than prokaryotes have. Why? In part, the reason is that eukaryotes are structurally and behaviorally more complex than prokaryotes and hence need more genes to generate that complexity. Among eukaryotes, however, no strict relationship exists between genome size and the structural and behavioral complexity of species. The reason is that many different types of DNA are packed inside a eukaryotic nucleus, not just the bare-essential genes, and some of the "extra DNA" varies in unpredictable ways from one eukaryote to the next.

TABLE 11.2 Genome Size and Gene Number

	TYPE OF VIRUS/ORGANISM	GENOME SIZE (BP)[a]	NUMBER OF GENES[b]
Porcine circovirus type 1	Smallest eukaryotic virus; causes wasting disease in pigs	1,759	3
Mycoplasma genitalium	Smallest bacterium	580,073	517
Mimivirus	Largest known virus; infects *Amoeba*	1,181,404	1,262
Escherichia coli K-12	Most widely used lab bacterium	4,639,221	4,377
Saccharomyces cerevisiae	Yeast (fungus) species used in baking and brewing	12,110,000	5,770
Drosophila melanogaster	Fruit fly used in genetics and other types of research	130,000,000	17,000
Arabidopsis thaliana	Thale-cress, a model plant used in agricultural and other types of plant research	157,000,000	27,407
Homo sapiens	Humans	3,200,000,000	21,000
Oryza sativa	Rice plant	4,311,000,000	56,000

[a]Total DNA content in viral genome, chromosomal genome of prokaryote, or haploid nucleus of eukaryote. [b]Protein-coding genes only.

Paris heterophylla In 2011, botanists at Kew Gardens in England discovered that this rare Japanese native has the largest genome of all: 150 billion base pairs, about 50 times bigger than ours. Because its massive genome has not been sequenced, we don't know how many genes it has and what the plant is doing with all that DNA.

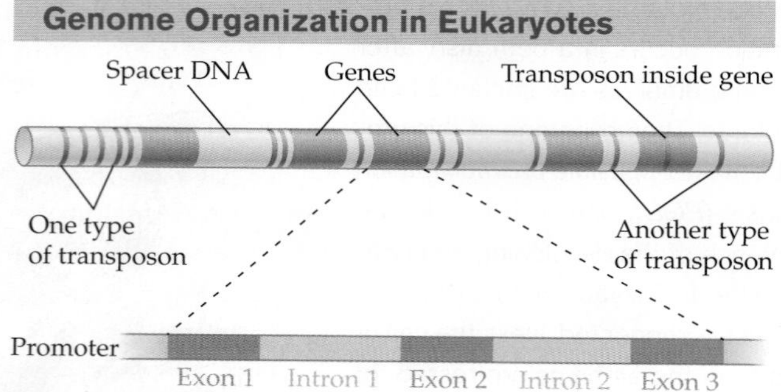

FIGURE 11.10 Eukaryotic Genomes Contain Both Coding and Noncoding DNA Eukaryotic genomes contain a small proportion of genes (purple) interspersed with a large amount of spacer DNA (light blue) and transposons (red and dark blue). Each of the two different types of transposons shown here is found in many copies throughout the human genome. Note that one transposon (dark blue) has inserted itself into one of the five genes. Most eukaryotic genes consist of coding regions (*exons*) interspersed with noncoding regions called *introns*.

Genes constitute only a small percentage of the DNA in most eukaryotes

Although eukaryotes have roughly 3–15 times as many genes as a typical prokaryote has, this difference in gene number does not fully explain why eukaryotes often have hundreds to thousands of times more DNA than prokaryotes have. Only a small percentage of the DNA in eukaryotic genomes consists of **genes**, which we can define more precisely as DNA sequences that code for RNA.

Scientists estimate that genes that encode proteins make up about 1.5 percent of the human genome. Some genes in our cells encode different types of *nonprotein RNA*, such as tRNA and rRNA. The rest of our genome consists of various types of **noncoding DNA**, defined as DNA that does not code for any kind of functional RNA.

Noncoding DNA includes *introns* and *spacer DNA* (**FIGURE 11.10; TABLE 11.3**). Most eukaryotic genes are interrupted by stretches of DNA, called **introns**. Introns do not code for the amino acid sequences of proteins and must be spliced out of a newly made mRNA molecule before translation. The segments of a gene that actually code for amino acids are called **exons**. Exons are preserved in the mature, translation-ready mRNA, and ribosomes manufacture a "made-to-order" protein according to the code carried by the exons in the mRNA. **Spacer DNA** is noncoding DNA that separates one gene from another; spacer sequences found in eukaryotic genomes are exceptionally long compared to those found in the much more compact genomes of prokaryotes.

Some of the remaining DNA has regulatory functions—for example, controlling gene expression. Some of it has architectural functions, such as giving structure to chromosomes or positioning them at precise locations within the

EXTREME PACKING These college students' 46 chromosomes are crammed into a nucleus that's only 0.005 mm (5 μm) in diameter. Placed end to end, they would be nearly 2 m (6 ft) long.

TABLE 11.3	Types of Eukaryotic DNA
TYPE	**DESCRIPTION**
Exons (in a gene)	Transcribed portions of a gene that code for the amino acid sequence of a protein
Noncoding DNA:	
Introns (in a gene)	Transcribed portions of a gene that do not code for the amino acid sequences of proteins, and whose corresponding sequence in RNA is removed before the RNA leaves the nucleus
Spacer DNA	DNA sequences that separate genes
Regulatory DNA	DNA sequences that control the expression of genes
Structural DNA	DNA sequences that have architectural function, creating structures such as the centromere of a chromosome
DNA of unknown function	DNA sequences that have no known function in the cell
Transposons ("jumping genes")	DNA sequences that can move from one position on a chromosome to another, or from one chromosome to another

The color patterns in these corncobs were caused by the activity of transposons that inserted themselves into pigment genes. A transposon that "hops" into a gene necessary for carotenoid synthesis changes the kernel color from yellow to white. The reddish color of some varieties is lost or reduced when a transposon jumps into one of the genes required for anthocyanin production.

interphase nucleus. Much of the genome, however—in fact, a majority of the genome in most eukaryotes—has no apparent function (see Figure 11.10).

At least some of the noncoding DNA appears to be nonessential and is popularly called "junk DNA" because researchers found no impact on phenotype if this DNA was lost. Although a number of hypotheses have been proposed, we do not yet understand why so many eukaryotes have large amounts of DNA with no clear-cut function.

Concept Check

1. Each human cell contains over a thousand times as much DNA as an *E. coli* bacterium has. Do we have over a thousand times as many genes as *E. coli* has? Explain.
2. Is all noncoding DNA useless "junk DNA"?

11.6 DNA Packing in Eukaryotes

To be expressed, the information in a gene must first be transcribed into an RNA molecule. A gene cannot be transcribed unless the enzymes that guide transcription are able to reach it. The task may sound simple, but it is complicated by what may be the ultimate storage problem: how to stow an enormous amount of genetic information in a small space, the nucleus, and still be able to retrieve each piece of that information precisely when it is needed.

In humans and other eukaryotes, every chromosome in each cell contains one DNA molecule and holds a vast amount of genetic information. As you learned in Chapter 7, the haploid number of chromosomes in humans is 23; these chromosomes together contain about 3.2 billion base pairs of DNA. Stretched to its full length, the DNA from all 46 chromosomes in a single human cell would be more than 2 meters long (taller than most of us). The combined length of DNA in our bodies is staggering: the human body has about 10^{13} cells, which means we have 2×10^{13} meters of DNA in the body—a length more than 130 times the distance between Earth and the sun.

How can our cells stuff such an enormous amount of DNA into such a small space? Cells use a variety of packaging proteins to wind, fold, and compress the DNA double helix, going through several levels of packing to create the DNA-protein complex we call a chromosome.

Let's examine the packing of DNA in a metaphase chromosome, beginning with the DNA double helix, which is about 2 nanometers wide (depicted at the bottom of **FIGURE 11.11**). Short lengths of this double helix are wound around "spools" of proteins, known as *histone proteins*, to create a "beads-on-a-string" structure that is

Concept Check Answers

1. No. Humans have only about 5.6 times as many genes as *E. coli* has. We have more DNA because we have a large amount of noncoding DNA.
2. No. Noncoding DNA can regulate gene expression, have structural functions, and serve as a connector (intron) between segments in a single gene. However, the function of most noncoding DNA is unknown.

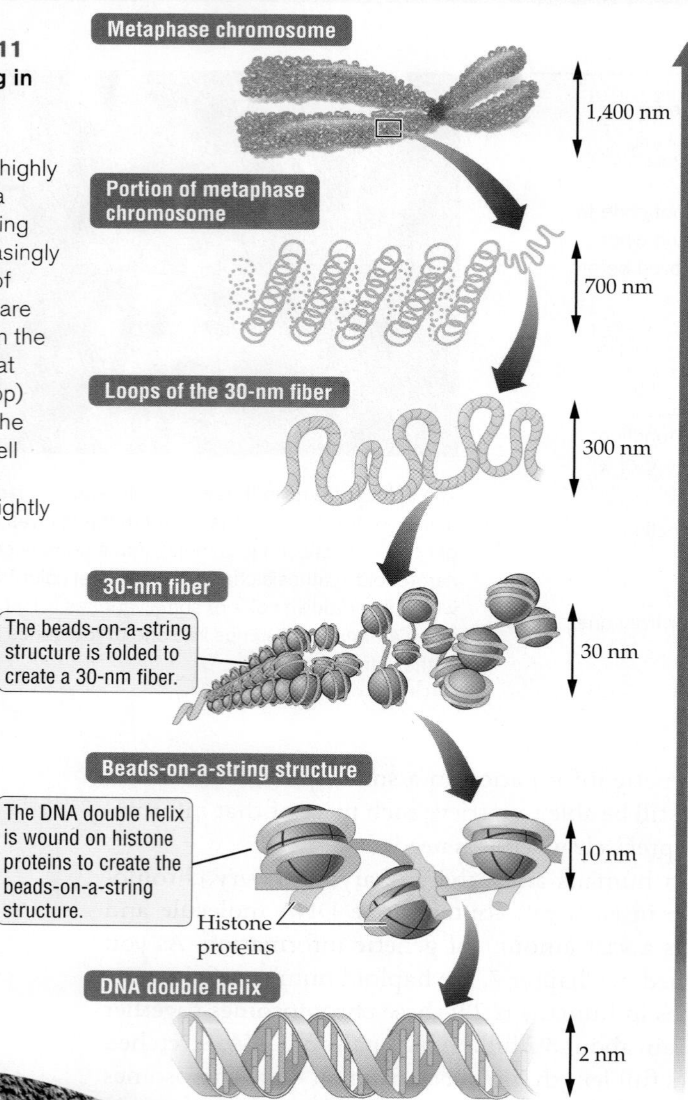

FIGURE 11.11 DNA Packing in Eukaryotes

The DNA of eukaryotes is highly organized by a complex packing system. Increasingly higher levels of DNA packing are illustrated, with the chromosome at metaphase (top) representing the stage of the cell cycle at which DNA is most tightly packed.

about 10 nanometers wide and consists of many histone "beads" linked by a DNA molecule that wraps around each bead. The beads-on-a-string structure is compressed into a more compact form, known as the 30-nanometer fiber, by yet other types of packaging proteins. This fiber is then looped back and forth to further condense the chromosome in the interior of the interphase nucleus.

During interphase, noncoding DNA is tightly condensed into a looped 30-nanometer fiber. Many of the genes that are not being transcribed in a particular cell may be likewise furled into a 30-nanometer fiber. However, any gene that is being transcribed into RNA is likely to be loosened down to the beads-on-a-string state.

The highest level of DNA packing is seen in a cell that is undergoing mitosis or meiosis. During prophase, each looped 30-nanometer fiber is further compressed into a short stub that is twice as compact as the 30-nanometer fiber. These short and stout chromosomes are easy to see even with an inexpensive microscope. The additional level of compaction protects prophase chromosomes from becoming entangled and being ripped apart during the elaborate chromosome movements that take place in dividing cells.

EXTREME BUZZ

Don't eat bananas near a beehive; they smell like an alarm pheromone that worker bees release when they sense danger. The pheromone turns on hundreds of genes in the bee brain, including those linked to aggression. European honeybee larvae reared with Africanized bees acquire a gene expression pattern like that of their more aggressive host.

Concept Check

1. Which is a more compact form of DNA: the beads-on-a-string configuration or the 30-nanometer fiber?
2. What is the functional value of the roughly twofold greater condensation that chromosomes undergo during prophase of mitosis or meiosis?

11.7 Patterns of Gene Expression

Gene expression is the means by which genes influence the structure and function of a cell or organism. In other words, gene expression enables a gene to influence a particular phenotype, the actual display of a genetic trait. Different cells express different sets of genes, and within a single type of cell the pattern of gene expression can change over time. In this section we consider some of the controls on gene expression—what factors determine which genes an organism expresses at a particular time and in a particular cell.

Organisms can turn genes on or off in response to environmental cues

Single-celled organisms such as bacteria face a big challenge because they are directly exposed to their environment. One way they meet this challenge is to express different genes as environmental conditions change. Bacteria respond to changes in nutrient availability, for example, by turning genes on or off.

If *E. coli* bacteria living in a petri dish are given lactose as the only source of energy, within a matter of minutes they turn on genes that code for lactose-digesting enzymes (**FIGURE 11.12**). When the lactose is used up, the bacteria stop producing those enzymes. In effect, these bacteria reorganize gene expression to match the food resources available to them. When a resource runs out, they turn on genes that enable them to use the next resource that becomes available (in Figure 11.12, the sugar arabinose). In this way, bacteria avoid wasting energy and cellular resources making enzymes that are not needed.

Like single-celled organisms, multicellular organisms change which genes they express in response to internal signals (those arising inside the body) or external cues (those present in the environment). For example, we humans change the genes we express when our blood sugar or blood pH levels change, enabling us to keep those levels from becoming too high or too low. Similarly, humans, plants, and many other organisms, when exposed to high temperatures, turn on genes encoding proteins that protect cells against heat damage.

In eukaryotes, different cell types express different sets of genes

Throughout the life of a multicellular organism, different types of cells express different sets of genes. Whether a cell expresses a particular gene depends on what type of cell it is, and whether that gene's function is needed in the cell's context. As a fertilized zygote develops into an embryo, a fetus, a baby, and eventually an adult, the cells, all of which started out the same, begin to follow different paths of development. Some become heart muscle cells, while others become pancreas cells, for example. To perform their developing functions, these many different types of cells turn on certain sets of genes and turn off others, sometimes permanently. All cells contain the gene for hemoglobin production, for instance, but only maturing red blood cells produce hemoglobin in an adult human (**FIGURE 11.13**). Similarly, all cells carry the gene for crystallin, but this protein is expressed only in certain cells of the developing eye, where it is the main component of the lens.

Some genes, known as **housekeeping genes**, play an essential role in the maintenance of cellular activities in all kinds of cells. These genes are therefore expressed most of the time by most cells in the body. For example, genes for rRNA, a key component of ribosomes, are expressed by almost all cells (see Figure 11.13). This is not surprising, because virtually all cells need to make proteins. Housekeeping genes tend to be highly conserved in evolutionary history, meaning that their base sequence and general function are similar in diverse groups of organisms.

Gene Expression in Response to the Environment

Environmental condition: type of food source

Supply of lactose

Supply of arabinose

Time

Gene expression in response to environment

Genes for lactose enzymes ON | Genes for lactose enzymes OFF

Genes for arabinose enzymes OFF | Genes for arabinose enzymes ON

Gene products

Enzymes for lactose utilization | Enzymes for arabinose utilization

FIGURE 11.12 Bacteria Express Different Genes as Food Availability Changes

The sugars lactose and arabinose are not always available, but both can serve as food for the bacterium *E. coli*. The presence of the sugar triggers expression of the gene that codes for enzymes needed to metabolize that sugar.

Differential Gene Expression

	Developing red blood cell	Eye lens cell (in embryo)	Pancreatic cell
Hemoglobin gene	ON	OFF	OFF
Crystallin gene	OFF	ON	OFF
Insulin gene	OFF	OFF	ON
rRNA gene	ON	ON	ON

FIGURE 11.13 Different Types of Cells Express Different Genes

Some genes, such as those encoding hemoglobin, crystallin, and insulin, are active ("on") only in the types of cells that use the protein encoded by the gene. Housekeeping genes, such as the rRNA gene, are active in most types of cells.

Concept Check Answers

1. The 30-nanometer fiber.

2. The increased condensation reduces the chance that the chromosomes will become entangled and break while being separated into different daughter cells.

Concept Check Answers

1. Gene regulation ensures that genes are expressed only when and where they are needed, preventing waste of resources. It also enables organisms to turn genes on or off to better cope with a changing environment.

2. Unlike regulated genes, housekeeping genes provide essential functions and are therefore kept on all of the time in almost all cell types.

Gene expression is the basis for the thousands of processes that take place in an organism every day. In the next chapter we will take a closer look at how genes are expressed and how their expression is controlled.

Concept Check

1. What is the adaptive value of gene regulation?
2. How are housekeeping genes different from regulated genes?

APPLYING WHAT WE LEARNED

CSI: Copper Age

Soon after the Iceman was discovered, skeptics suggested that the frozen man was a hoax, a mummy transported from Egypt or Central America. A wealth of forensic information from his body and clothes offered compelling clues about his life and the manner of his death. But in the end, it was his DNA that confirmed his identity—a genuine Copper Age European. Ötzi—as he was called, because he was found in the Ötzal Alps—shared distinctive sequences of DNA with modern Europeans (**FIGURE 11.14**).

Ötzi's pure-copper axe, high levels of copper and arsenic in his hair, and blackened lungs were all clues that he may have been a copper smelter, an early adopter of Copper Age technology. Isotopes in his bones and teeth showed that he grew up 60 miles from the 11,000-foot-high pass where he died. Frozen in his gut was a last meal containing an ancient kind of wheat. On his left hand was a deep, days-old gash. He had a recent head injury, deep bruises on his side, a slash on his right forearm, and a flint arrowhead embedded in his left shoulder. It was the arrow that likely killed him, slicing an artery, which led to massive blood loss and almost instant death.

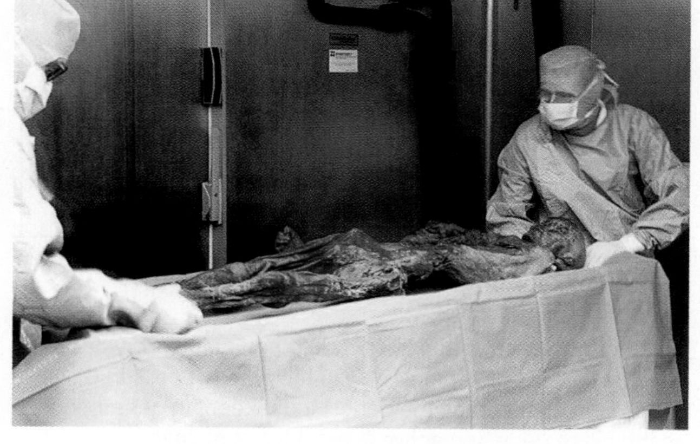

FIGURE 11.14 The Remains of Ötzi

Scientists examine the remains of Ötzi under a controlled environment designed to prevent decomposition of the well-preserved mummified body.

Within hours, a snowstorm buried him, and he remained frozen in a glacier for 5,200 years.

To find out more, researchers turned to DNA from blood that had stained Ötzi's cloak, his knife, and an arrow in his quiver. They extracted DNA from long-frozen cells in the blood and sequenced regions of the DNA that are known to vary from person to person. By comparing these "highly variable" sequences of bases, biologists can distinguish the blood of one person from another—a technique known as DNA fingerprinting (see Chapter 13). In 2008, geneticists working on Ötzi's DNA discovered that he had a variable sequence, called K, that is about 12,000 years old. K is found in peoples of North Africa, South Asia, and Europe.

In 2012, Stanford University researchers succeeded in sequencing Ötzi's full genome using DNA extracted from the hip bone. With the whole genome on hand, the researchers were able to pinpoint the modern-day Europeans that Ötzi most resembles: inhabitants of Sardinia and Corsica, rather than present-day central Europe. Ötzi's people were most likely among the Middle Eastern farmers who brought agriculture with them as they migrated into Europe. Their genomic identity faded gradually as the migrants mixed extensively with the hunter-gatherers who inhabited Europe as far north as Scandinavia. The genetic signature of the Neolithic migrant farmers is better preserved in some of the more isolated populations of southern Europe, such as Sardinians.

Recent DNA comparisons show that Ötzi's ancestors likely intermixed with Neandertals (*Homo neanderthalensis*) to a greater extent than did the ancestors of any of the present-day European groups. While the average European today has 3.5 percent of the DNA sequences that are characteristic of Neandertal genomes, Ötzi's genome harbors 5 percent of those DNA variants.

BIOLOGY IN THE NEWS

The Human Genome Project: How It Changed Biology Forever

BY JOSEPH HALL • *Toronto Star*, April 28, 2013

...The Human Genome Project—which was presented in its final form 11 years ago this month [the project was completed in 2003]—provided a map of mankind's DNA. But it also opened up a Pandora's box of boggling complexity in the biological sciences and medicine that will take decades more to unravel...

If the genome project has injected daunting complexity into biology, however, it's also proven an immensely useful research tool, many scientists say.

"Every single day, every scientist who does any biology uses that data," says Stephen Scherer, director of the Centre for Applied Genomics at The Hospital for Sick Children [Toronto, Ontario].

"It's had a huge impact."

The main benefits coming out of the project were twofold, says Scherer, also director of the [University of Toronto's] McLaughlin Centre for Molecular Medicine.

"One was the concept that the scientific community could come together in biology and generate this incredible resource that now everyone uses," says Scherer, who helped map chromosome seven during the project.

Second, he says, the original project prompted the creation of more muscular DNA sequencers that can now rapidly and cheaply generate personal genomes...

Scherer's own, post-project work has used these sequencers to show that humans actually possess a varied number of copies—Copy Number Variables or CNVs—of many of their genes.

These CNVs describe any genetic anomaly that shifts the number of copies of a gene a person has above or below the typical two they inherit—one from their mother, the other from their father.

"If it is copy number variable, you vary away from the typical two copies," Scherer explains.

"So, you either have one copy or, in some cases, zero copies, and in other cases three or four. And CNVs could help doctors to personalize the dose of a particular drug."

CNVs can have a huge impact on how an individual might handle cancer chemotherapies, for example, and will be a critical diagnostic factor in the emerging field of personalized medicine.

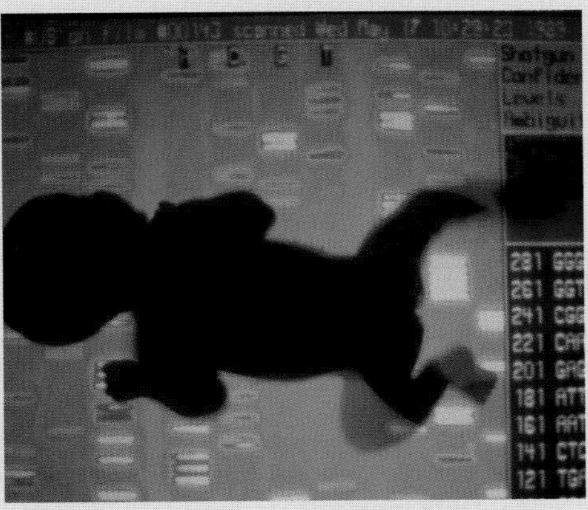

The Human Genome Project (HGP), which concluded in 2003, was sponsored mainly by the U.S. Department of Energy and the National Institutes of Health, although 18 other countries—including Canada, the United Kingdom, France, Germany, Japan, Brazil, and China—contributed. The U.S. government spent up to $3 billion on the project. What did U.S. taxpayers get for that hefty sum? Estimates vary, with some economists suggesting a threefold return on the investment and the Battelle Memorial Institute calculating a $65 economic benefit for each dollar spent. Beyond the economic benefits, there is little doubt that the project changed biology forever.

Much of what this chapter says about the organization of the human genome—the fact that only about 1.5 percent of it codes for proteins, for example—is a legacy of the HGP. The HGP's gene-predicting software estimated that we have some 21,000 protein-coding genes. The biological functions of about half of those genes are known; the race is on to decipher the functions of the rest.

The catalog of the human genome amassed by the HGP is available online to anyone with an Internet connection. Genome sequences of more than a thousand different eukaryotes are also available now, and genome comparisons have taught us a great deal about the story of life on Earth. Francis Collins, director of the NIH institute that played a leading role, described the human genome database like this: "It's a history book—a narrative of the journey of our species through time. It's a shop manual, with an incredibly detailed blueprint for building every human cell. And it's a transformative textbook of medicine, with insights that will give health care providers immense new powers to treat, prevent and cure disease."

The project led to the identification of more than 30 different types of genes linked to disorders such as breast cancer, muscle disease, deafness, and blindness. A major goal of current genomics research is to build on that success to identify genes that play an important role in health and disease. Success in these efforts will lead to better diagnosis, and possibly prevention and treatment strategies as well.

Evaluating the News

1. What have biologists learned from the Human Genome Project?

2. From what you've learned so far in this book, speculate on how copy number variables (CNVs) arise in the human genome. Where might these extra copies come from?

CHAPTER REVIEW

Summary

11.1 An Overview of DNA and Genes

- DNA stores genetic information as a sequence of four nucleotides.
- The genome is all the DNA-based information in the nucleoid of a prokaryote or the nucleus of a eukaryote.
- A gene is a segment of DNA that encodes at least one distinct genetic trait. Most genes code for proteins, which generate phenotypes.
- Genetic information flows from DNA to RNA in the process of gene transcription. In translation, ribosomes convert the genetic information in an mRNA molecule into a specific protein.
- Gene expression is the manifestation of a phenotype based on information encoded in the gene.
- Different sets of genes are expressed in different cell types. The gene promoter functions like an on/off switch for transcription.

11.2 The Three-Dimensional Structure of DNA

- Watson and Crick determined that DNA is a double helix formed by two polynucleotide strands containing the nitrogenous bases adenine (A), cytosine (C), guanine (G), and thymine (T).
- The two polynucleotide strands are held together by hydrogen bonds between each A and T, and each G and C.
- The sequence of bases in DNA, which differs among species and among individuals within a species, is the basis of inherited variation.

11.3 How DNA Is Replicated

- A complex of proteins guides the replication of DNA; the primary enzyme involved is DNA polymerase.
- The DNA helix is unwound, and DNA polymerase uses each strand as a template from which to build a new strand of DNA.
- DNA replication is semiconservative: it produces two copies of the original DNA molecule, each composed of one old strand and one newly synthesized daughter strand.

11.4 Repairing Replication Errors and Damaged DNA

- On rare occasions, mistakes occur during DNA replication. Most mistakes in the copying process are corrected, either immediately by "proofreading" or later by the mismatch repair system.
- Uncorrected mistakes in DNA replication are one source of mutations.
- The DNA in our cells is altered thousands of times every day by accidental chemical changes, and possibly by radiation and mutagens as well.
- Replication errors and damage to DNA are fixed by a complex set of DNA repair proteins.

11.5 Genome Organization

- Compared with eukaryotes, prokaryotes have a smaller genome, and it usually takes the form of a single chromosome. Most prokaryotic DNA encodes proteins, and functionally related genes in prokaryotes are grouped together in the genome.
- With regard to genome organization, eukaryotes differ from prokaryotes in several ways: (1) In eukaryotic cells, the DNA is distributed among several chromosomes.
(2) Eukaryotes have more DNA per cell than prokaryotes have, in part because they have more genes than prokaryotes have. In addition, genes constitute only a small portion of the genome in many eukaryotes; the rest consists of noncoding DNA (including introns and spacer DNA) and transposons. (3) Eukaryotic genes with related functions often are not located near one another.

11.6 DNA Packing in Eukaryotes

- Cells can pack an enormous amount of DNA into a very small space because their DNA is highly organized by a complex packing system. In eukaryotes, segments of DNA are wound around histone spools, packed together tightly into a narrow fiber, and further folded into loops.
- Chromosomes are most tightly packed during mitosis and meiosis. The packing is looser during interphase, when most gene expression occurs.
- In DNA regions that are tightly packed, genes cannot be expressed, because the proteins necessary for transcription cannot reach the genes.

11.7 Patterns of Gene Expression

- In both prokaryotes and eukaryotes, genes are turned on and off selectively in response to short-term changes in environmental conditions.
- The different cell types in a multicellular organism express different sets of genes.
- Housekeeping genes, which play an essential role in the maintenance of cellular activities, are expressed in most cells of the body.

Key Terms

base pair (p. 246)
DNA (p. 243)
DNA polymerase (p. 247)
DNA repair (p. 249)
DNA replication (p. 247)
exon (p. 252)
gene (p. 252)
gene expression (p. 254)
gene promoter (p. 244)
genome (p. 243)
housekeeping gene (p. 255)
intron (p. 252)
messenger RNA (mRNA) (p. 243)
mutagen (p. 248)
mutation (p. 248)
nitrogenous base (p. 245)
noncoding DNA (p. 252)
ribosome (p. 243)
RNA (p. 243)
semiconservative replication (p. 247)
spacer DNA (p. 252)
transcription (p. 243)
translation (p. 243)
transposon (p. 249)

Self-Quiz

1. The base-pairing rules for DNA state that
   a. any combination of bases is allowed.
   b. T pairs with C, A pairs with G.
   c. A pairs with T, C pairs with G.
   d. C pairs with A, T pairs with G.
2. DNA replication results in
   a. two DNA molecules, one with two old strands and one with two new strands.
   b. two DNA molecules, each of which has two new strands.
   c. two DNA molecules, each of which has one old strand and one new strand.
   d. none of the above
3. The DNA of cells is damaged
   a. thousands of times per day.
   b. by collisions with other molecules, chemical accidents, and radiation.
   c. not very often and only by radiation.
   d. both a and b
4. The DNA of different species differs in the
   a. sequence of bases.
   b. base-pairing rules.
   c. number of nucleotide strands.
   d. location of the sugar-phosphate portion of the DNA molecule.
5. If a strand of DNA has the sequence CGGTATATC, then the complementary strand has the sequence
   a. ATTCGCGCA.
   b. GCCCGCGCTT.
   c. GCCATATAG.
   d. TAACGCGCT.
6. Mutation
   a. can produce new alleles.
   b. can be harmful, beneficial, or neutral.
   c. is a change in an organism's DNA sequence.
   d. all of the above
7. In prokaryotes and eukaryotes, gene expression is most often controlled by regulation of
   a. the destruction of a gene's protein product.
   b. the length of time mRNA remains intact.
   c. transcription.
   d. translation.
8. The DNA of eukaryotes is packed most loosely during
   a. mitosis.
   b. meiosis.
   c. interphase.
   d. both b and c

Analysis and Application

1. A gene has two codominant alleles: A^1 and A^2. Each allele produces a different but related version of a protein found on the surface of a type of white blood cell. In physical terms, each allele is a segment of DNA. Explain how the DNA of one allele might differ from the DNA of the other allele. Describe the possible effects of these differences in the two DNA segments.
2. Explain why the structure of DNA proposed by Watson and Crick suggested a way DNA could be replicated.
3. Describe how the bases in DNA, mutations, and alleles that cause human genetic disorders are related.
4. Summarize how DNA repair works and why the repair mechanisms are essential for cells and whole organisms to function normally.
5. The total length of the DNA in a eukaryotic cell is hundreds of thousands of times greater than the diameter of the nucleus. Explain how so much DNA can be packed inside the nucleus.
6. Summarize the major differences between the genomes of prokaryotes and eukaryotes, emphasizing differences in the amount and function of DNA and the organization of genes.
7. Imagine you transferred a bacterium from an environment in which glucose was available as food to an environment in which the only source of food was the sugar arabinose. A specific enzyme is required to digest arabinose but not glucose. How do you think gene expression in the bacterium would change as a result of your action?
8. Explain how cells with the same genes can be so different (a) in structure and (b) in the metabolic tasks they perform.
9. Summarize how gene expression can be controlled at various steps in the process of generating a protein that is fully active.

12 From Gene to Protein

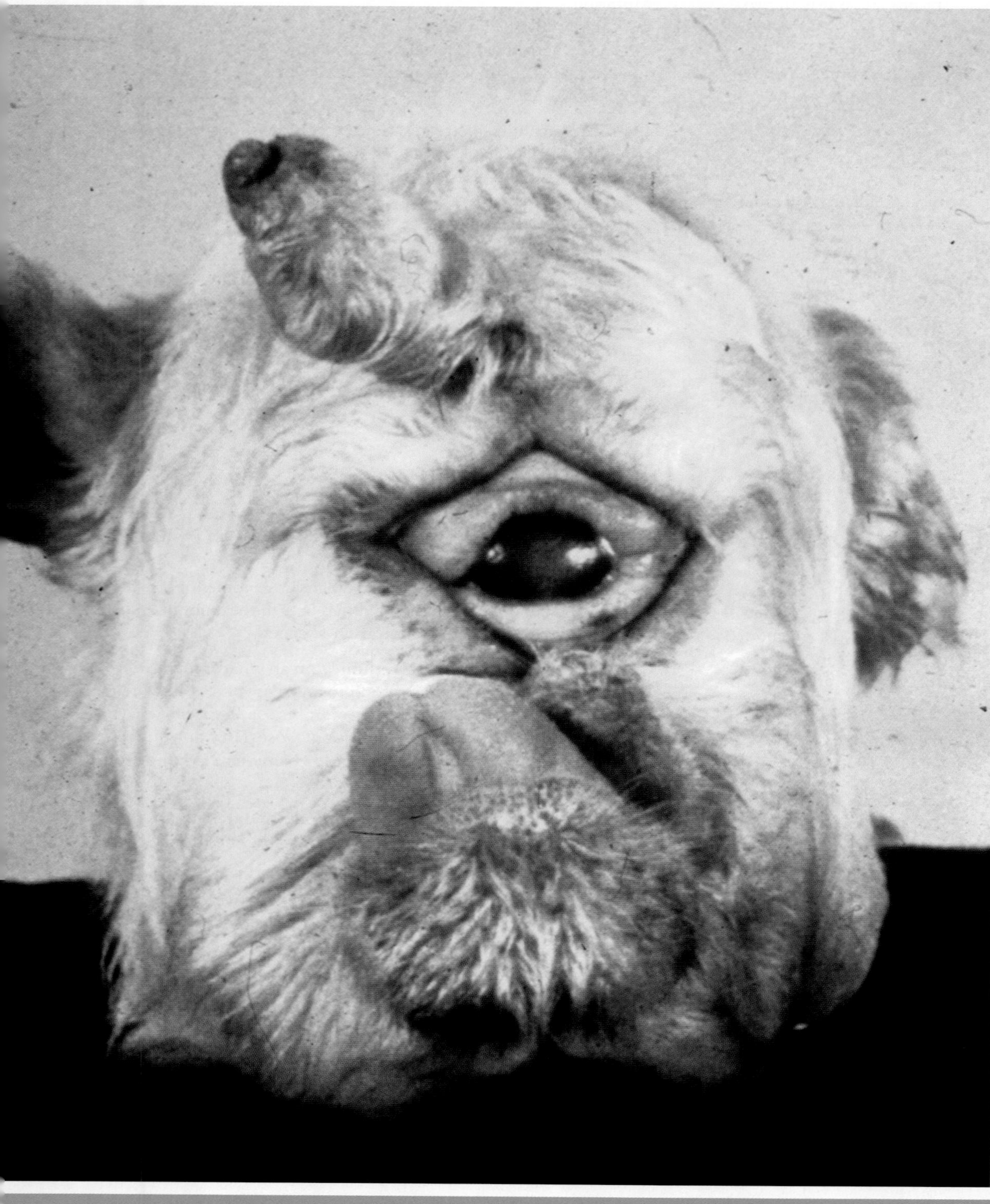

HEAD OF A LAMB WITH CYCLOPIA. This lamb suffered severe birth defects, including a single large eye and a fused, malformed brain, because its mother ate a wild plant called corn lily early in pregnancy. Corn lilies make a compound that shuts down the expression of certain genes.

Greek Myths and One-Eyed Sheep

In the epic poem *The Odyssey*, the ancient Greek author Homer tells the story of Odysseus, the king of Ithaca. Also known as Ulysses, the legendary hero was famed for his brilliance and craftiness. On his way back from the Trojan War, Odysseus and his men encountered and outwitted the Cyclops Polyphemus, a giant with a single eye in the middle of his forehead. In the myth, Polyphemus locks Odysseus and his men in a cave and commences to eat them, two at a time. The men make a daring escape by getting the Cyclops drunk on wine, putting out his eye, and hiding among his sheep.

The Cyclops is a mythical creature, but its single eye resembles a birth defect called cyclopia. Cyclopia occurs in many groups of vertebrates, including sharks, fishes, amphibians, and birds, as well as mammals such as cows, sheep, goats, cats, mice, monkeys, and rarely, even humans. This condition includes many defects in addition to the single eye, and it is invariably fatal.

Why would an animal develop with only one eye? Untangling the causes of this bizarre defect brings us to one of the most exciting and fruitful areas of modern biological research: understanding gene expression. To develop normally from a single-celled zygote, an organism must express the right genes at the right time and place, producing just the right amounts of thousands of different proteins in exactly the right cells. It is a task of monumental and bewildering complexity. By and large, organisms accomplish this feat flawlessly throughout embryonic development, and each of our many trillions of cells continues to perform the same feat every day of our lives.

However, even tiny missteps in gene expression can result in disaster: an organism may not develop properly (resulting in cyclopia or other birth defects), or a group of cells may begin multiplying out of control (becoming cancerous). At the end of this chapter we examine the consequences of gene expression gone horribly wrong.

What changes in gene expression would cause an embryo to develop into a cyclops? How do mutations and chemicals in the environment alter the expression of genes?

MAIN MESSAGE Genes carry information that affects the production of proteins, and proteins play a key role in the expression of genetic characteristics.

KEY CONCEPTS

- Some genes encode RNA molecules as their final product, but most genes contain instructions for building proteins.
- Protein-coding genes code for messenger RNA (mRNA). A noncoding RNA, such as rRNA or tRNA, is the final product of RNA-only genes.
- The flow of information from gene to protein requires two steps: transcription and translation.
- In eukaryotic cells, transcription occurs in the nucleus and produces a messenger RNA version of the sequence information stored in the gene.
- The mRNA moves from the nucleus to the cytoplasm, where translation occurs. Ribosomes convert the sequence of bases in an mRNA molecule to the sequence of amino acids in a protein, with assistance from transfer RNA.
- The ribosome-tRNA complex "reads" the mRNA in sets of three bases, and each unique sequence of three bases is a codon. The genetic code specifies which amino acid, or start/stop signal, is specified by each of the 64 possible codons.
- Gene mutations can alter the sequence of amino acids in the protein encoded by that gene. Such changes, in turn, can alter the protein's function.
- In multicellular organisms, different sets of genes are activated in different cell types, and patterns of gene expression change dramatically during developmental stages. Gene expression can be altered by internal signals or by external (environmental) conditions.

GREEK MYTHS AND ONE-EYED SHEEP 261
12.1 How Genes Work 262
12.2 Transcription: Information Flow from DNA to RNA 264
12.3 The Genetic Code 267
12.4 Translation: Information Flow from mRNA to Protein 269
12.5 The Effect of Mutations on Protein Synthesis 270
12.6 How Cells Control Gene Expression 272
APPLYING WHAT WE LEARNED 276
From Gene Expression to Cyclops

HOW DO GENES STORE THE INFORMATION needed to build RNA and proteins? How do RNA molecules direct the manufacture of proteins? Knowing how genes work can help us understand how mutations lead to new phenotypes, including disease phenotypes (**FIGURE 12.1**). We begin this chapter by describing how genetic information is encoded in genes and how the cell uses that information to build proteins. We then describe how a change in a gene can change an organism's phenotype. Next we consider how gene expression—turning genes on or off at different times or in different cells—contributes to development and enables an organism to sense and respond to its environment.

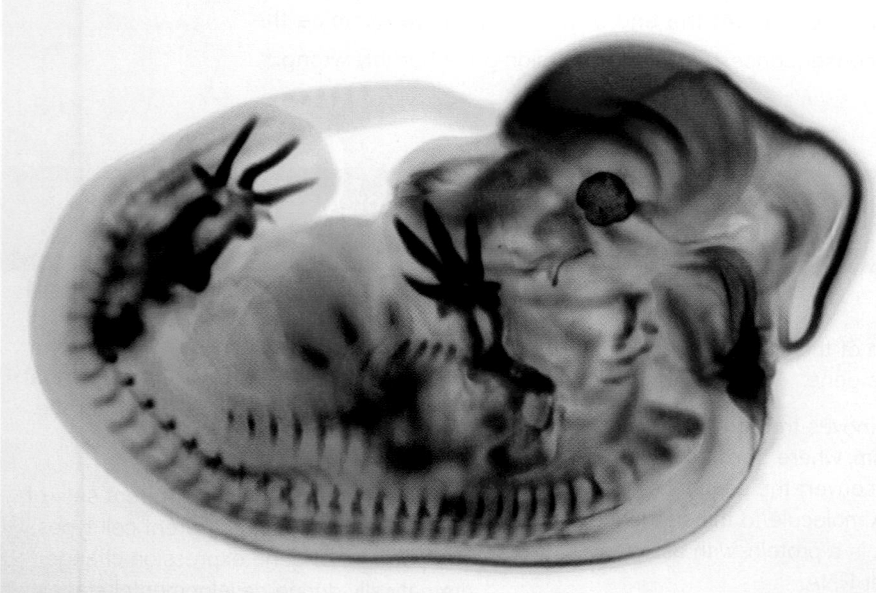

FIGURE 12.1 RNA Molecules Carry Genetic Information

A technique called in situ hybridization was used to show where mRNA (green) transcribed from the *Noggin* gene is located in a fetal mouse. The *Noggin* gene is expressed in the developing brain and in the cartilage and bones of all mammals, including humans. Serious developmental disorders result if the information in the *Noggin* gene is corrupted. How is the information in a gene actually "read out" in a cell? How does a change in the DNA sequence produce an altered phenotype?

12.1 How Genes Work

How do genes affect the phenotype of an organism? Early clues came at the beginning of the twentieth century from the work of British physician Archibald Garrod, who studied several inherited disorders in which metabolism was disrupted. In 1902, shortly after the rediscovery of Gregor Mendel's work, Garrod argued that these metabolic disorders were caused by an inability of the body to produce specific enzymes. Garrod was particularly interested in alkaptonuria, a condition in which the urine of otherwise healthy infants turns black when exposed to air. He proposed that infants with alkaptonuria have a defective version of an enzyme that, in its normal form, breaks down the substance that causes the discoloration. But Garrod did not stop there; he and his collaborator, William Bateson, went on to suggest that in general, genes are responsible for the production of enzymes.

Genes contain information for building RNA molecules

Garrod and Bateson were on the right track, but they were not entirely correct: genes control the production of all types of proteins, not just enzymes. Furthermore, not all genes contain code that specifies the order in which amino acids will be linked to make a protein. The *RNA-only genes*, also known as noncoding genes, store instructions for building any one of a number of ribonucleic acid (RNA) molecules with varied functions that do not include carrying code for the manufacture of a specific protein. In contrast, all *protein-coding genes* direct the production of messenger RNA (mRNA), which in turn directs the production of a particular protein in the cytoplasm. At the molecular level, we can say that a **gene** is any DNA sequence that is transcribed into RNA.

RNA and DNA share a number of structural similarities, as well as several important differences. Both are nucleic acids consisting of nucleotides covalently bonded to one another. But whereas DNA molecules are double-stranded, the various types of RNA molecules are all single-stranded. As in DNA, each

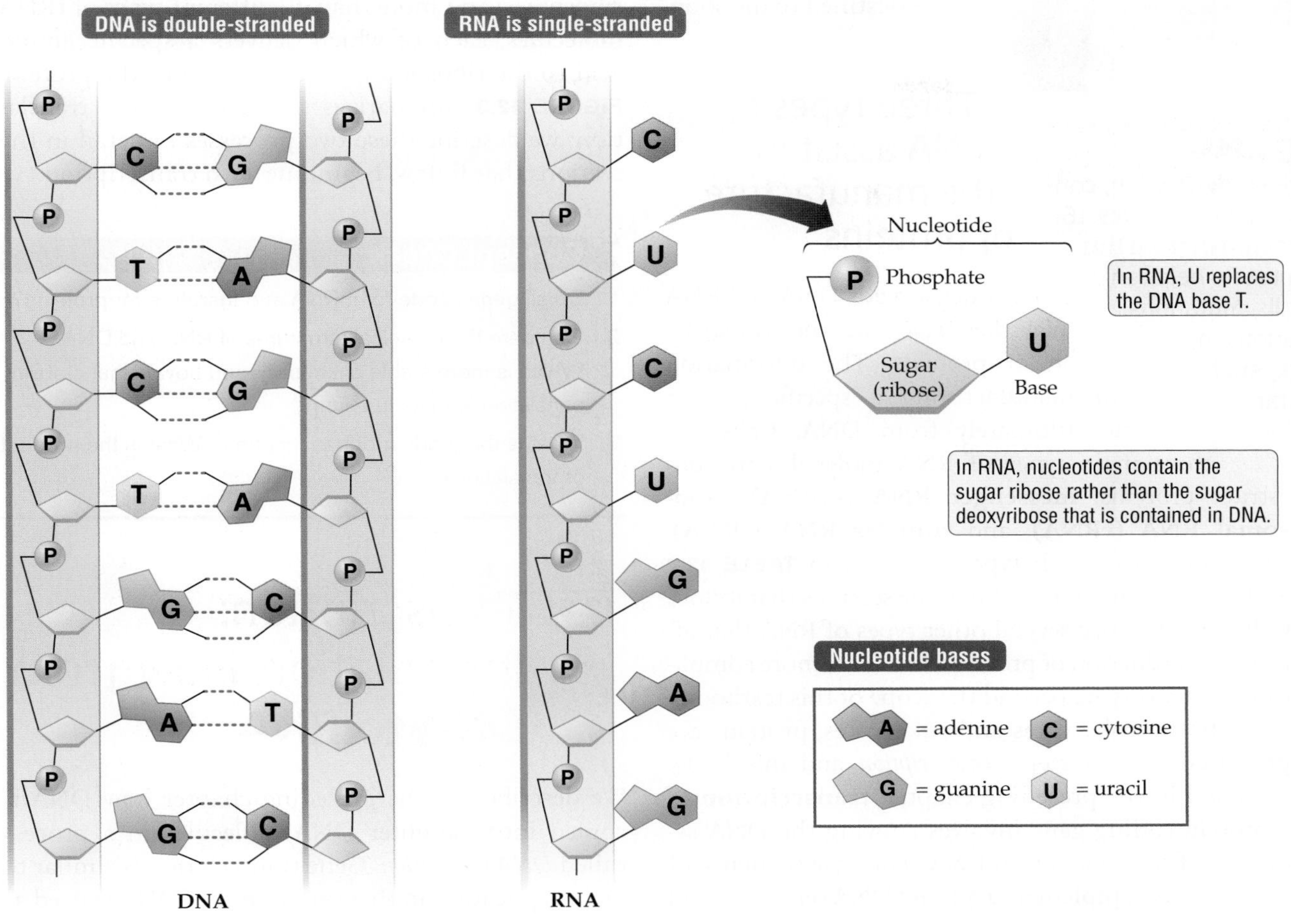
RNA Structure
DNA is double-stranded
RNA is single-stranded
P
C
G
T
A
U
DNA
RNA
Nucleotide
Phosphate
Sugar
(ribose)
Base
In RNA, U replaces the DNA base T.
In RNA, nucleotides contain the sugar ribose rather than the sugar deoxyribose that is contained in DNA.
Nucleotide bases
= adenine
= cytosine
= guanine
= uracil

FIGURE 12.2 RNA Is a Single-Stranded Chain of Nucleotides

nucleotide in RNA is composed of a sugar, a phosphate group, and one of four nitrogen-containing bases (**FIGURE 12.2**). However, the nucleotides in RNA and DNA differ in two respects: First, RNA uses the sugar ribose, whereas DNA contains the sugar deoxyribose. Second, in RNA the base uracil (U) replaces the base thymine (T). The differences between the two molecules are summarized in **TABLE 12.1**.

In general, RNA is chemically less stable than DNA, and most RNA molecules in the cell have a limited life. As the permanent store of genetic information, the DNA in the nucleus of eukaryotic cells is

TABLE 12.1 A Comparison of DNA and RNA

	DNA	RNA
Structure	Double-stranded; two polynucleotide strands wound into a helix	Single-stranded polynucleotide; may fold back on itself
Sugar	Deoxyribose	Ribose
Nucleotides	A, G, C, and T	A, G, C, and U
Function	Stores genetic information	Expresses genetic information—for example, by directing the manufacture of a specific protein
Stability	Highly stable in most cells	Generally much less stable
Location	Nucleus, chloroplasts, and mitochondria in eukaryotes; cytosol in prokaryotes	Nucleus, chloroplasts, mitochondria, and cytosol in eukaryotes; cytosol in prokaryotes

EXTREME GENE

The longest human gene, *dystrophin*, consists of 2.4 million base pairs. It takes 16 hours to transcribe one mRNA from this gene. The dystrophin protein helps anchor skeletal muscle cells. Mutations in the gene produce various muscle-wasting disorders, such as Duchenne muscular dystrophy.

Concept Check Answers

1. No. RNA-only genes are transcribed into RNA other than mRNA, and these RNAs have specialized functions other than coding for proteins.

2. RNA is single-stranded; it contains ribose and the bases A, G, C, and U. DNA is double-stranded; it contains deoxyribose and A, G, C, and T. DNA is more stable—a property it must have to serve as the storehouse of genetic information.

3. The product of transcription is an mRNA complementary to the DNA sequence of a gene. The product of translation is a polypeptide (protein chain) determined from the sequence of the mRNA.

much more stable, being destroyed only if the cell itself is destined to die soon.

Three types of RNA assist in the manufacture of proteins

The nucleic acids DNA and RNA play key roles in the construction of proteins. The information for manufacturing a specific protein comes ultimately from DNA. Cells use three main types of RNA molecules to construct proteins: **messenger RNA (mRNA)**, **ribosomal RNA (rRNA)**, and **transfer RNA (tRNA)**. The function of each type is defined in **TABLE 12.2** and discussed in more detail in the sections that follow. Cells also produce several other types of RNA that affect the production of proteins, but these more complicated functions are beyond the scope of this textbook.

In both prokaryotes and eukaryotes, proteins are produced in two steps: *transcription* and *translation*. As noted in the preceding chapter, **transcription** of a protein-coding gene involves copying the DNA sequence of the gene into mRNA. The base sequence of an mRNA is complementary to its DNA template, and it carries the code for the amino acid sequence of a specific protein. **Translation** is the process by which the mRNA code is turned into the precise sequence of amino acids that become covalently linked to form the protein product. Translation requires not only mRNA, but also rRNA, an important component of ribosomes; and more than 20 different types of tRNA molecules, each of which delivers a specific amino acid to the ribosomes for assembly into the protein. **FIGURE 12.3** summarizes transcription and translation; we describe these two processes in detail in the sections that follow, beginning with transcription.

Concept Check

1. Do all genes code for mRNA and therefore for proteins?
2. Compare the chemical structures of RNA and DNA. Which is more stable chemically, and how is that stability consistent with its function?
3. What is the product of transcription? What is the product of translation?

12.2 Transcription: Information Flow from DNA to RNA

We described in the preceding chapter how DNA is copied into daughter DNA molecules in a process called *DNA replication*. Gene transcription is similar to DNA replication in that one strand of DNA is used as a template from which a new strand—in this case, a strand of RNA—is formed. However, transcription differs from DNA replication in three important ways, as described in **TABLE 12.3**.

TABLE 12.2 RNA Molecules and Their Functions

TYPE OF RNA	FUNCTION	SHAPE
Messenger RNA (mRNA)	Specifies the order of amino acids in a protein using a series of three-base codons, where different amino acids are specified by particular codons.	
Ribosomal RNA (rRNA)	As a major component of ribosomes, assists in making the covalent bonds that link amino acids together to make a protein.	
Transfer RNA (tRNA)	Transports the correct amino acid to the ribosome, using the information encoded in the mRNA; contains a three-base anticodon that pairs with a complementary codon revealed in the mRNA.	

Transcription can be divided into three stages: *initiation*, *elongation*, and *termination*. Transcription of a gene begins when the enzyme **RNA polymerase** binds to a segment of DNA near the beginning of the gene, called the **gene promoter**. This stable binding of RNA polymerase to the promoter marks the **initiation** of gene transcription. Although the promoters of different genes vary in size and sequence, all promoters contain several specific sequences of 6–10 bases that enable the RNA polymerase to recognize and bind to them.

Once bound to the promoter, the RNA polymerase unwinds the DNA double helix at the beginning of the gene, separating a short portion of the two strands. Then the enzyme begins to construct an mRNA molecule (**FIGURE 12.4**). **Elongation** is the stage of transcription when RNA polymerase synthesizes RNA from free nucleotides.

Only one of the two DNA strands is used as a template, and this strand is called the **template strand**. If the opposite strand were used as template, it would result in a completely different sequence of amino acids, and hence a different protein. How does the RNA polymerase "choose" which strand to use as template? The answer is that RNA polymerase binds to the promoter in a specific orientation, thereby determining which strand it will be able to "read." Both the location and the orientation of the promoter sequence guide the binding of the RNA polymerase, so ultimately the positioning of the promoter is what specifies which DNA strand serves as the template.

The four kinds of bases in RNA pair with the four kinds of bases in DNA according to the rules described in Chapter 11:

THIS BASE IN DNA...	...PAIRS WITH THIS BASE IN RNA
A	U
T	A
G	C
C	G

Messenger RNA Directs Protein Synthesis

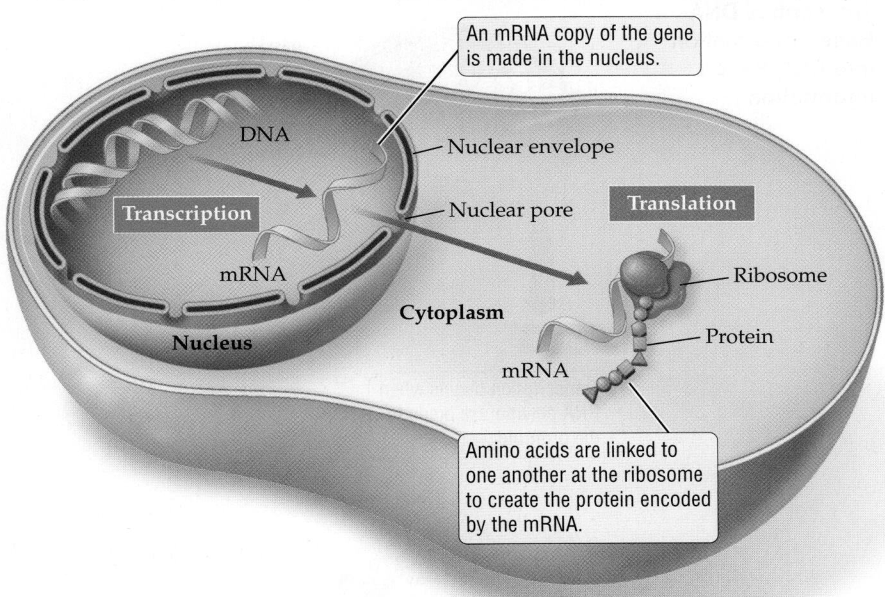

FIGURE 12.3 Genetic Information Flows from DNA to RNA to Protein during Transcription and Translation

The transcription of a protein-coding gene produces an mRNA molecule, which is then transported to the cytoplasm, where translation occurs and the protein is made with the help of ribosomes. Different amino acids in the protein being constructed at the ribosome are represented here by different colors and shapes.

As an RNA polymerase moves away from the gene promoter and travels down the template strand, another RNA polymerase can bind at the promoter and start synthesizing an mRNA fast on the heels of the previous RNA polymerase. At any given time, therefore, many RNA polymerases can be traveling down a DNA template, each synthesizing its own mRNA strand (at the rate of 60 bases per second in human cells).

In prokaryotes, **termination** of transcription occurs when the RNA polymerase

TABLE 12.3 A Comparison of Gene Transcription and DNA Replication

	GENE TRANSCRIPTION	DNA REPLICATION
Key enzyme involved	RNA polymerase	DNA polymerase
Portion of chromosome copied	Small segment	Entire DNA double helix
Product	Single-stranded RNA molecule, complementary to one DNA strand (the template strand)	Double-stranded DNA molecule

EXTREME INHIBITION

Eating a single death cap mushroom *(Amanita phalloides)* can be lethal. Alpha-amanitin, produced by the death cap mushroom, inhibits transcription in eukaryotes by binding to RNA polymerase. Some of the most potent poisons are inhibitors of transcription or translation.

FIGURE 12.4 RNA Polymerase Transcribes DNA-Based Information into RNA-Based Information

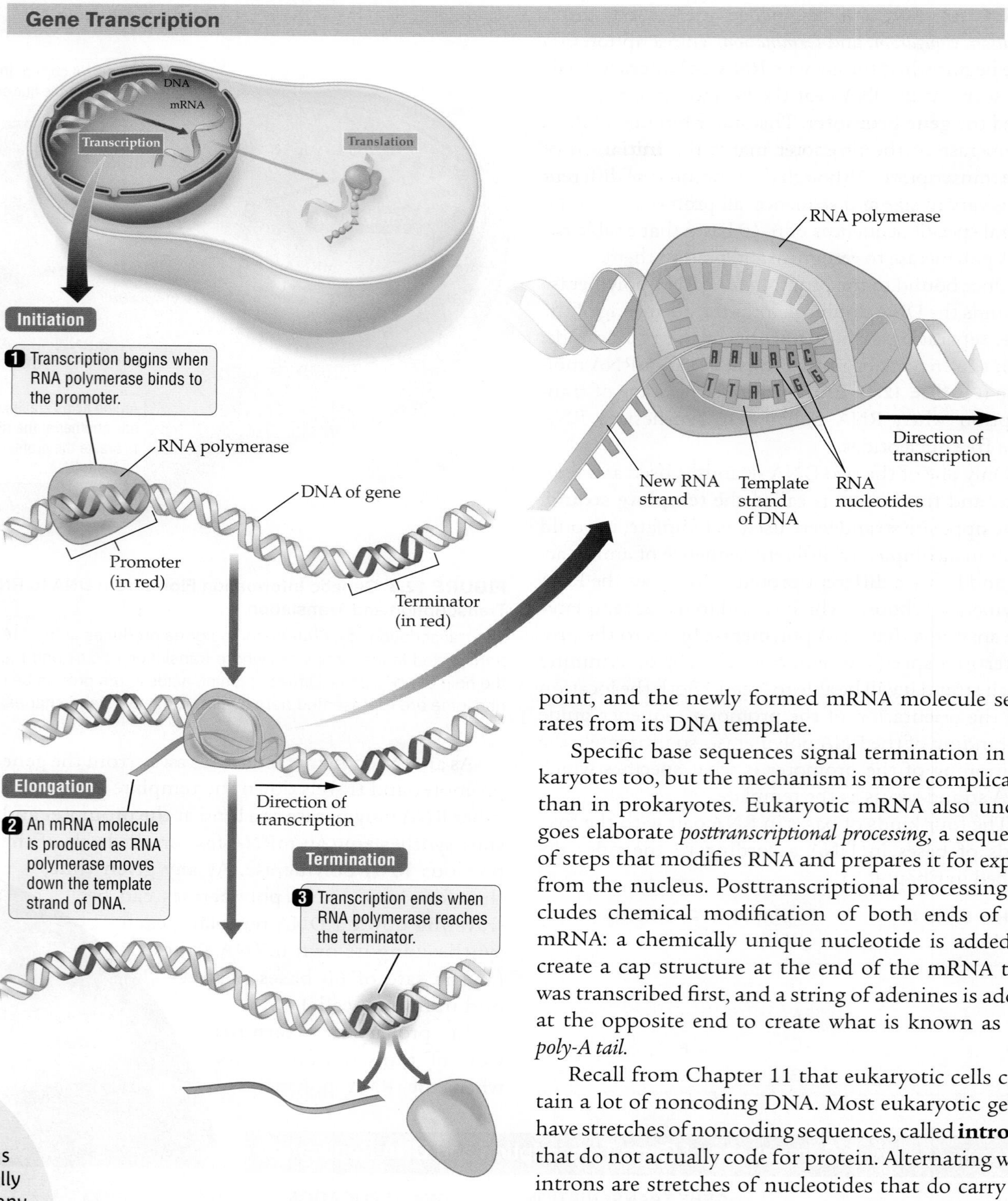

Helpful to Know

The names of proteins that are enzymes usually end with *ase*, while many names of proteins that are not enzymes end with *in*. For example, the protein RNA polymer*ase* is an enzyme, but the protein hemoglob*in* is not.

reaches a special sequence of bases called a **terminator** (see Figure 12.4). When the terminator sequence is copied into mRNA, it generates a three-dimensional shape, known as a hairpin, in the mRNA sequence. Hairpins in the mRNA destabilize the RNA polymerase and cause it to drop off the template. Transcription ends at this point, and the newly formed mRNA molecule separates from its DNA template.

Specific base sequences signal termination in eukaryotes too, but the mechanism is more complicated than in prokaryotes. Eukaryotic mRNA also undergoes elaborate *posttranscriptional processing*, a sequence of steps that modifies RNA and prepares it for export from the nucleus. Posttranscriptional processing includes chemical modification of both ends of the mRNA: a chemically unique nucleotide is added to create a cap structure at the end of the mRNA that was transcribed first, and a string of adenines is added at the opposite end to create what is known as the *poly-A tail.*

Recall from Chapter 11 that eukaryotic cells contain a lot of noncoding DNA. Most eukaryotic genes have stretches of noncoding sequences, called **introns**, that do not actually code for protein. Alternating with introns are stretches of nucleotides that do carry instructions for building the protein, and these coding segments of the gene are called **exons** (**FIGURE 12.5**). Because of this patchwork construction of eukaryotic genes, most newly transcribed mRNA (pre-mRNA) is also a patchwork of coding sequences intermixed within noncoding sequences. These pre-mRNA sequences are therefore like an uncut video recording with "extra footage."

In **RNA splicing**, the introns are snipped out of a pre-mRNA, and the remaining pieces of mRNA—the exons—are joined to generate the "processed," or mature, mRNA. Normally, only fully processed mature mRNA can exit through the nuclear pores. This mature mRNA carries the "final cut" of information, and on arriving in the cytoplasm, the code it carries is translated into a specific protein. RNA molecules that lack the distinctive features of a mature mRNA are "ignored" by ribosomes and are not translated.

Concept Check

1. The template strand of a gene has the base sequence TGAGAAGACCAGGGTTGT. What is the sequence of RNA transcribed from this DNA, assuming RNA polymerase travels from left to right on this strand?
2. The *dystrophin* gene has 78 introns. Are these introns transcribed? Do they code for amino acids?

12.3 The Genetic Code

The information in an mRNA molecule is "read" by the ribosomes in sets of three bases, and each unique sequence of three bases is called a **codon**. Because there are 64 different ways of arranging four bases to make a three-base sequence, there are 64 possible codons. By analogy, a language with only four letters (A, U, C, and G) would have 64 different words, if only three-letter words were allowed.

The **genetic code** specifies the "meaning" of every word in the language: which amino acid is encoded by a particular codon, and which codons act as signposts that communicate to the ribosomes where they should start or stop reading the mRNA. The entire genetic code is shown in **FIGURE 12.6**. Most of the 64 codons specify particular amino acids. A few amino acids are specified by only one codon; for example, only one codon (UGG) specifies the amino acid tryptophan. Other amino acids are specified by anywhere from two to six different codons.

When reading the code, a ribosome begins at a fixed starting point on an mRNA molecule, called a **start codon** (the codon AUG), and ends at one of three **stop codons** (UAA, UAG, or UGA). Having a fixed point for the start ensures that the message from a gene is read precisely the same way every time. The start codon specifies the amino acid methionine (**meh-*THYE*-oh-neen**), which is why most proteins have this amino acid at their "starting point," the portion of the protein that was translated first.

RNA Processing in the Nucleus

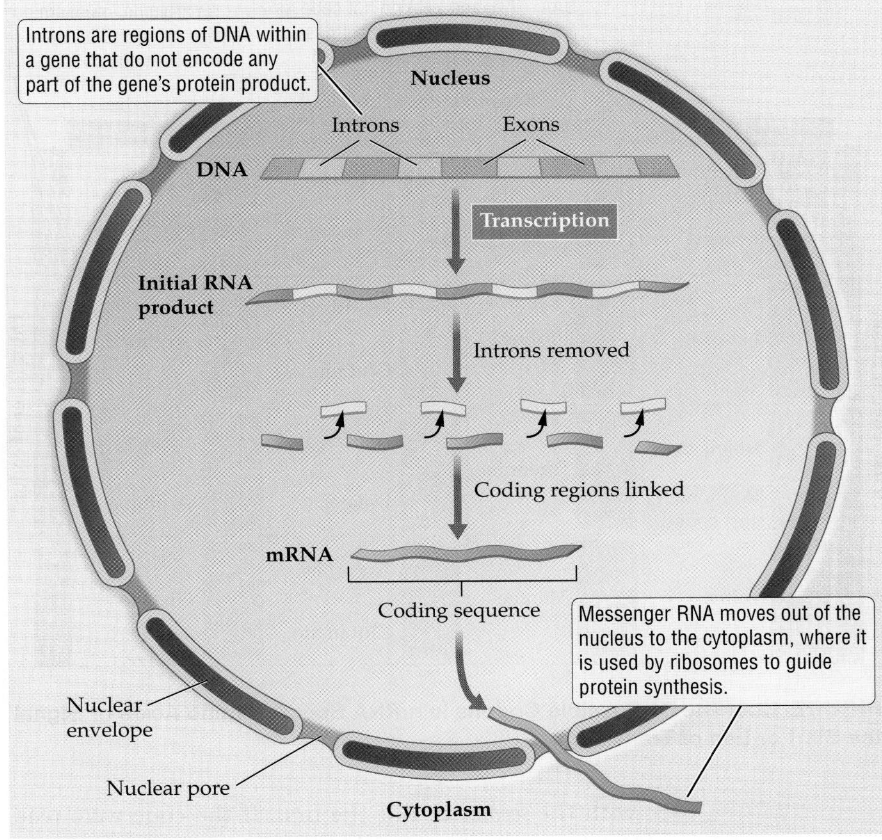

FIGURE 12.5 In Eukaryotes, Introns Must Be Removed before an mRNA Leaves the Nucleus
Most eukaryotic genes contain both coding sequences (exons) and noncoding sequences (introns). Before the mRNA transcribed from such genes can be exported to the cytoplasm, enzymes in the nucleus must remove the introns and link the remaining exons.

To examine how the rest of the amino acid–specifying codons are read, consider the example in **FIGURE 12.7**, which shows a portion of an mRNA molecule with the base sequence UUCACUCAG. Because the mRNA code is read in sets of three bases, the first codon (UUC) specifies one amino acid (phenylalanine, abbreviated as Phe), the next codon (ACU) specifies a second amino acid (threonine, abbreviated as Thr), and the third codon (CAG) specifies a third amino acid (glutamine, abbreviated as Gln). The start codon plays a crucial role in establishing which trio of bases is interpreted as a codon by the ribosomal-tRNA machinery.

Use Figure 12.6 to determine the amino acid sequence that would result if the sequence UUCACUCAG in Figure 12.7 were read in codons that began

Concept Check Answers

1. ACUCUUCUGGUCCCAACA.
2. All 78 introns are transcribed into a pre-mRNA, but they are subsequently spliced out. Because they are absent from the fully processed mature mRNA transcript, introns do not code for amino acids.

The Genetic Code

UAA, UAG, and UGA do not code for an amino acid. Translation stops when these codons are reached.

Like arginine, most amino acids are specified by more than one codon.

First letter of codon	Second letter: U	Second letter: C	Second letter: A	Second letter: G	Third letter of codon
U	UUU Phenylalanine	UCU Serine	UAU Tyrosine	UGU Cysteine	U
U	UUC Phenylalanine	UCC Serine	UAC Tyrosine	UGC Cysteine	C
U	UUA Leucine	UCA Serine	UAA Stop codon	UGA Stop codon	A
U	UUG Leucine	UCG Serine	UAG Stop codon	UGG Tryptophan	G
C	CUU Leucine	CCU Proline	CAU Histidine	CGU Arginine	U
C	CUC Leucine	CCC Proline	CAC Histidine	CGC Arginine	C
C	CUA Leucine	CCA Proline	CAA Glutamine	CGA Arginine	A
C	CUG Leucine	CCG Proline	CAG Glutamine	CGG Arginine	G
A	AUU Isoleucine	ACU Threonine	AAU Asparagine	AGU Serine	U
A	AUC Isoleucine	ACC Threonine	AAC Asparagine	AGC Serine	C
A	AUA Isoleucine	ACA Threonine	AAA Lysine	AGA Arginine	A
A	AUG Methionine; start codon	ACG Threonine	AAG Lysine	AGG Arginine	G
G	GUU Valine	GCU Alanine	GAU Aspartate	GGU Glycine	U
G	GUC Valine	GCC Alanine	GAC Aspartate	GGC Glycine	C
G	GUA Valine	GCA Alanine	GAA Glutamate	GGA Glycine	A
G	GUG Valine	GCG Alanine	GAG Glutamate	GGG Glycine	G

FIGURE 12.6 The 64 Possible Codons in mRNA Specify Amino Acids or Signal the Start or End of Translation

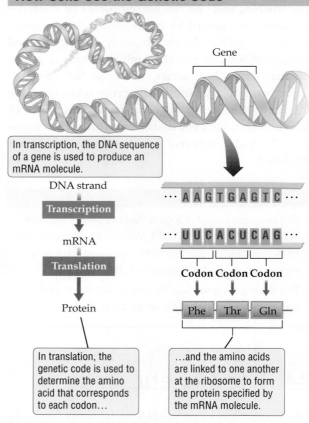

How Cells Use the Genetic Code

FIGURE 12.7 The Genetic Code Is Read in Sets of Three Bases, Each Set Constituting a Codon

with the *second* U, not the first. If the code were read starting with the second U (UUCACUCAG), we would get a very different protein chain: one containing serine followed by leucine (Ser, Leu, . . .), instead of one containing phenylalanine followed by threonine and glutamine (Phe, Thr, Gln, . . .). For a cell, this change in the protein sequence would likely be disastrous; fortunately, the start codon (AUG) prevents the scrambling of the protein sequence by establishing the start point. The bases that follow the start codon are read consecutively, with each three-base sequence being read as one codon. The start codon therefore establishes the correct order, or *reading frame*, by which the three-letter language of mRNA is translated into protein.

The genetic code has several significant characteristics:

- It is *unambiguous*. Each codon specifies only one amino acid.
- It is *redundant*. Several different codons may have the same "meaning"; that is, they may code for the same amino acid. There are four possible bases at each of the three positions of a codon, so there are a total of 64 codons ($4 \times 4 \times 4 = 64$). However, there are only 20 amino acids, so some of these codons specify the same one. For example, six different codons specify the amino acid serine (see Figure 12.6).
- It is virtually *universal*. Nearly all organisms on Earth use the same genetic code, which underscores the common descent of all organisms.

The discovery of the genetic code and its near universality revolutionized our understanding of how genes work and helped pave the way for what is now a thriving biotech industry. A few minor exceptions to the universal genetic code do exist: in certain species, some of the 64 codons are read differently than they are in most other species. In general, however, knowing the genetic code is the key to unlocking gene function in living things.

Concept Check

1. Why is the start codon, AUG, so important?
2. What does it mean to say that the genetic code is redundant?

Concept Check Answers

1. The start codon sets the reading frame; that is, it determines the grouping of the bases in the mRNA into triplets to be read as codons.
2. There are 64 possible codons, but only 20 amino acids. In most cases, a single amino acid is specified by more than one codon, and this is what is meant by redundancy. For example, tyrosine is specified by either UAU or UAC.

12.4 Translation: Information Flow from mRNA to Protein

The conversion of a sequence of bases in mRNA to a sequence of amino acids in a protein is called *translation*. It is the second major step in the process by which genes specify the manufacture of proteins (see Figure 12.3). Translation occurs at ribosomes, which are composed of more than 50 different proteins and several different strands of rRNA. Ribosomes are responsible for linking the amino acids together, using mRNA as a template.

Transfer RNA (tRNA) also plays a crucial role in the manufacture of proteins (protein synthesis). All tRNAs have a similar three-dimensional structure, shown in **FIGURE 12.8**; however, each type of tRNA specializes in binding to one specific amino acid, and it recognizes and pairs with specific codons in the mRNA. The mRNA-binding site is termed the **anticodon**, and its three-base sequence can bind only to a complementary sequence in the mRNA. For example, the tRNA that carries the amino acid serine recognizes and pairs with any AGC codon in an mRNA (see Figure 12.8).

Some tRNAs can recognize more than one codon because the base at the third position of their anticodon can "wobble," meaning it can pair with any one of two or three different bases in the codon. For example, one serine-bearing tRNA can pair with either UCU or UCC in the mRNA, while another serine-bearing tRNA pairs with either UCA or UCG. A third tRNA recognizes the two additional serine-specifying codons: AGU and AGC (consult the genetic code in Figure 12.6). This flexibility in the pairing between some anticodons and codons means that a cell does not need 61 different tRNAs, one for each of the 61 amino acid–specifying codons. In fact, most organisms have only about 40 different tRNAs, because many tRNA anticodons can recognize and pair with more than one codon in the mRNA.

For translation to occur (**FIGURE 12.9**), an mRNA molecule must first bind to a ribosome; as with transcription, this first step of translation is called **initiation**. The ribosomal machinery "scans" the mRNA until it finds a start codon, which is the first AUG codon in the mRNA sequence.

Then the ribosomes recruit the appropriate tRNAs, as determined by the codons they encounter as they proceed to read the message in the mRNA. This is the **elongation** stage of translation. With all the necessary components held together in the required three-dimensional orientation, a specialized site on the ribosome facilitates the linking of one amino acid to another. In the example in Figure 12.9, after the tRNA carrying methionine binds to the start codon, the next codon to be recognized is GGG, which corresponds to the amino acid glycine. Elongation proceeds as follows:

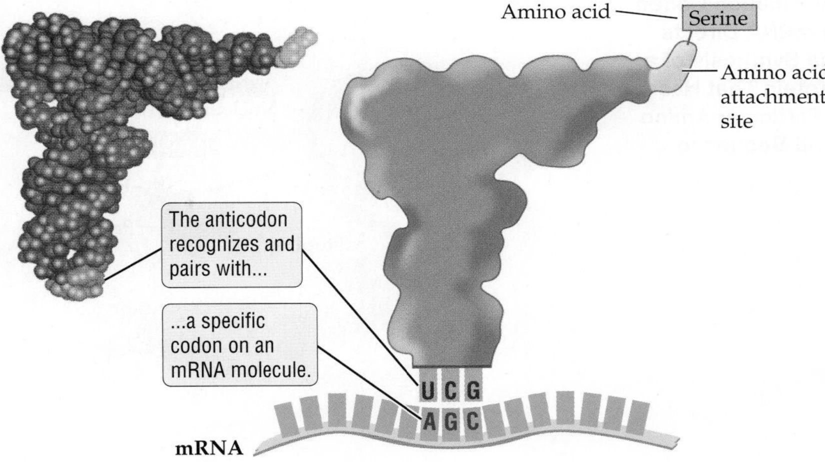

FIGURE 12.8 Transfer RNA Delivers Amino Acids Specified by mRNA Codons

A space-filling model (left) and a diagrammatic version (right) illustrate the general structure of all tRNA molecules. Similar regions are shown in matching colors in both representations. Each tRNA carries a specific amino acid (serine in this example) and has a specific anticodon sequence (UCG in this example) that binds to a complementary three-base sequence (the codon) in the mRNA.

1. A tRNA molecule that specializes in carrying glycine recognizes the GGG codon and pairs with it through its anticodon.
2. The ribosome now forms a covalent bond between the first amino acid (methionine) and the second amino acid (glycine). When the bond between the first two amino acids is formed, the first tRNA releases its amino acid (methionine).
3. This tRNA, now freed from the amino acid it had been carrying, is ejected from the mRNA-ribosome complex, and its place is taken by the next tRNA (the one carrying glycine).
4. The ribosome is now ready to read the third codon in the mRNA, which is UCC in our example. The tRNA that carries serine bears the complementary anticodon (AGG) and specifically pairs with the UCC codon.
5. This serine-specific tRNA now occupies the site in the ribosome previously occupied by the glycine codon.

FIGURE 12.9 In Translation, Information Coded in mRNA Directs the Synthesis of a Protein That Has a Particular Amino Acid Sequence

Protein Synthesis through Translation

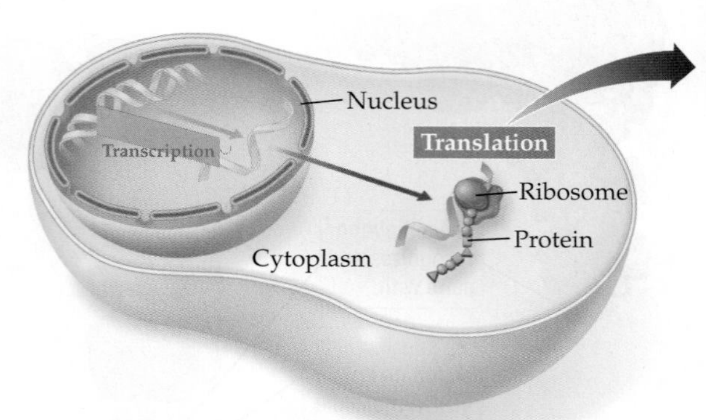

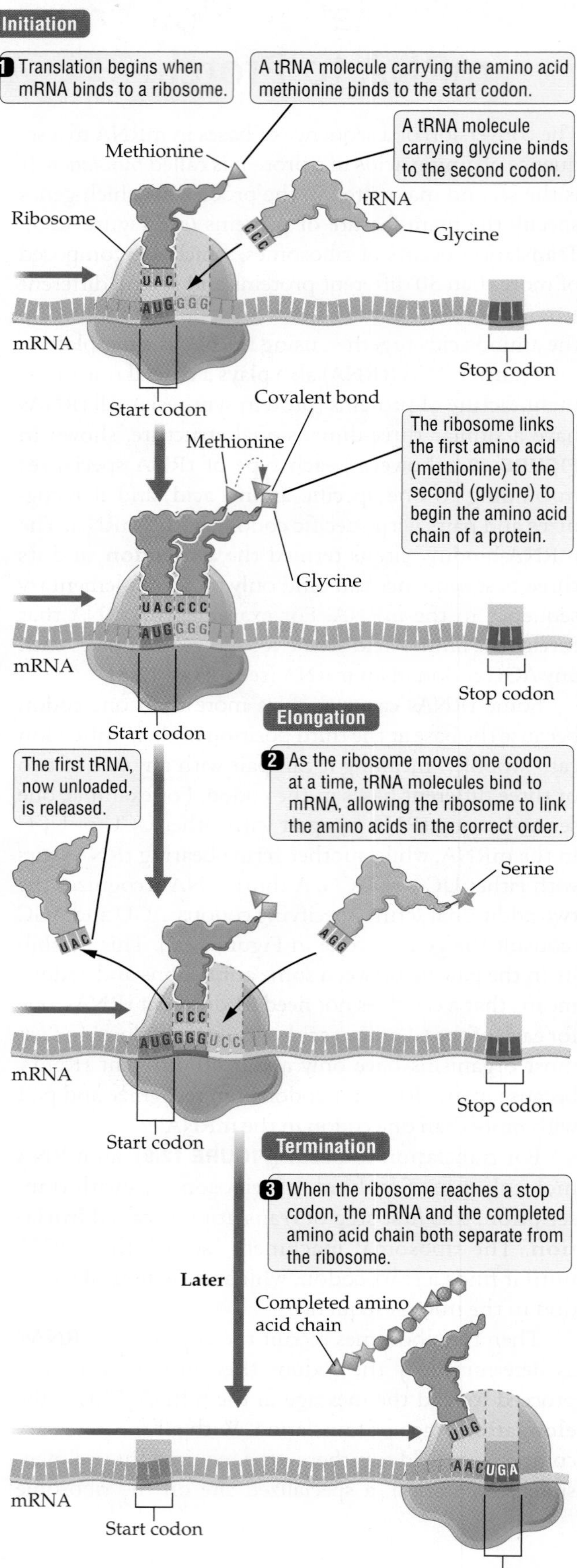

This cycle continues: each codon encountered by the ribosome is recognized by a specific tRNA, and the ribosome adds the amino acid delivered by this tRNA to the growing amino acid chain.

Finally, a stop codon is reached; this is the **termination** stage. The amino acid chain cannot be extended further, because none of the tRNAs can recognize and pair with any of the three stop codons. At this point the mRNA molecule and the completed amino acid chain both separate from the ribosome. The new protein folds into its compact, specific three-dimensional shape (discussed in Chapter 2).

Concept Check Answers

1. Translation converts a sequence of bases in mRNA to a sequence of amino acids in a protein.
2. No. The three stop codons do not specify any amino acids, and a single amino acid may be specified by as many as six different codons.

Concept Check

1. What is meant by “translation” of the genetic code?
2. Does each of the 64 codons specify a different amino acid?

12.5 The Effect of Mutations on Protein Synthesis

A *mutation* is a change in the base sequence of an organism’s DNA. As noted in previous chapters of this unit, mutations range in extent from a change in the identity of a single base pair to the addition or deletion of one or more whole chromosomes. How do mutations affect protein synthesis? In answering this question, we focus on mutations that occur within exons—the protein-encoding portions of a gene—rather than on mutations that occur in introns or on the large-scale mutations that disrupt entire chromosomes.

Mutations can alter one or many bases in a gene's DNA sequence

Three major types of mutations can alter a gene's DNA sequence: *substitutions*, *insertions*, and *deletions*. For simplicity, we first describe **point mutations**, those in which a single base is altered. We then discuss mutations that involve changes in multiple bases.

In a **substitution** mutation, one base is substituted for another in the DNA sequence of the gene. In the example shown in **FIGURE 12.10**, the sequence of the gene is changed when a thymine (T) is replaced by a cytosine (C). As the figure shows, this particular change causes the substitution of one amino acid for another: When the TAA sequence in the DNA is changed to CAA, the mRNA codon is changed from AUU to GUU; GUU is recognized by the valine-carrying tRNA, so a valine is inserted at this position in the protein, instead of an isoleucine (Ile).

Insertion or **deletion** mutations occur when a base is inserted into or deleted from a DNA sequence. Single-base insertions and deletions cause a genetic **frameshift**. Frameshift mutations, whether caused by point mutations or by large insertions or deletions, alter the resulting protein so severely that it fails to function in most cases. Consider what happens in a multiple-choice test if you accidentally record the answer to a question twice on a machine-gradable answer sheet: all your answers from that point forward are likely to be wrong, because each is an answer to the previous question. This situation is equivalent to a frameshift caused by the *insertion* of a single base. If you forget to answer a question but record the next question's answer in the overlooked question's space, all your entries from that point on will be scrambled. This situation is equivalent to a frameshift caused by the *deletion* of a base.

The insertion or deletion of one or two bases shifts all "downstream" codons by one or two bases. The entire downstream message is scrambled because the ribosomes assemble a very different sequence of amino acids from that point onward, compared to the protein encoded by the original DNA sequence and its corresponding mRNA (see Figure 12.10).

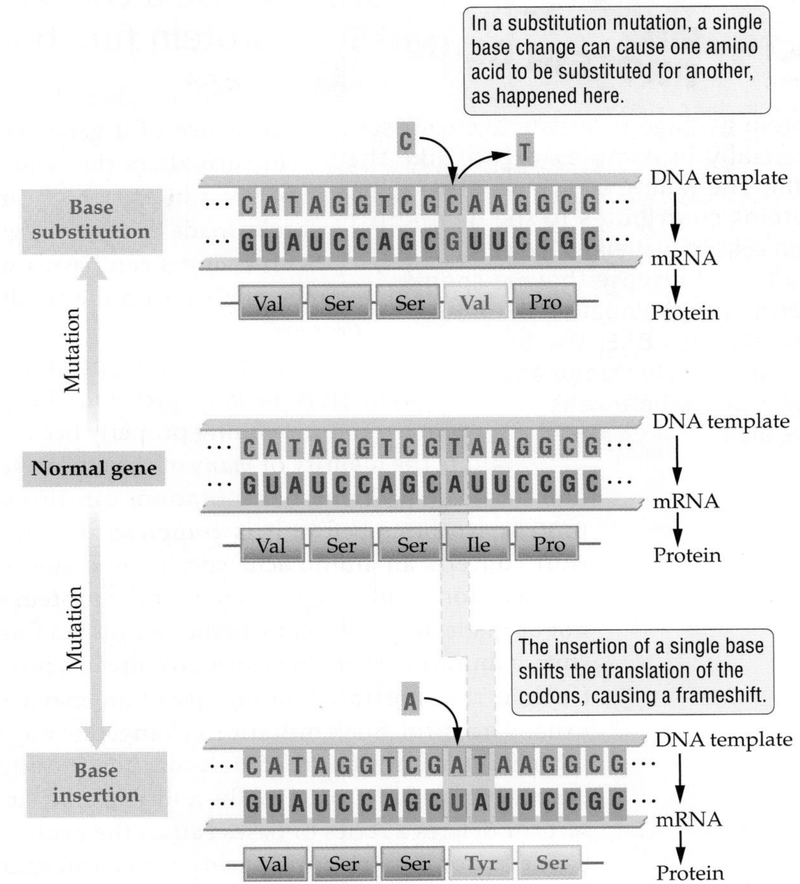

FIGURE 12.10 A Change in DNA Sequence Translates into a Change in a Protein's Amino Acid Sequence
Two kinds of mutations are shown here: a substitution and an insertion. In each case, the mutation and its effects on transcription and translation are shown in red.

What happens when *three* bases are inserted or deleted? The answer is that the reading frame does not shift, because the sequence is off by exactly one codon, and therefore the resulting protein is not changed as much as when one or two bases are added or lost. At most the protein will have one superfluous, or one missing, amino acid—a change that may or may not affect protein function. As you might guess, multiples of three codons also do not shift the reading frame.

Insertions and deletions can involve more than a few bases: sometimes thousands of bases may be added or deleted as a result of mutations. Large insertions or deletions almost always result in a protein that cannot function properly.

Helpful to Know

In rare instances, a mutation may be beneficial. Changes to the binding region of a protein, for example, might improve its efficiency—or even enable it to take on a new and useful function, such as reacting with a new substrate.

EXTREME PROTEIN CLUMPS

Protein garbage is destructive to tissues, especially in complex organs like the brain. The failure to remove damaged proteins contributes to the death of brain cells in Alzheimer's disease and "mad cow" disease (bovine spongiform encephalopathy, BSE). In patients with BSE, the brain develops protein clumps and large holes where cells have died.

Mutations can cause a change in protein function

Mutations alter the DNA sequence of a gene, which in turn alters the sequence of bases in any mRNA molecule made from that gene. Such changes can have a wide range of effects on the resulting protein.

A mutation that produces a frameshift usually prevents the protein from functioning properly because it alters the identity of *many* of the amino acids in the protein. Frameshift mutations can also stop protein synthesis before it is complete: if a frameshift converts an amino acid–specifying codon into a stop codon, a full-length version of the protein will not be made. Regardless of whether it causes a frameshift, a mutation that alters an active site in a protein (such as the substrate-binding site of an enzyme) is usually harmful. Such mutations change the way the protein interacts with other molecules, decreasing or destroying its function. Finally, a mutation that inserts or deletes a series of bases causes the protein to have extra or missing amino acids, which can change the protein's shape and hence its function.

Sometimes changing a few bases in a gene's DNA sequence has little or no effect. For example, if a single-base substitution does not alter the amino acids specified by the gene, then the structure and function of the protein are not changed. Although a change in the DNA sequence from GGG to GGA would alter the mRNA sequence from CCC to CCU, both CCC and CCU code for the same amino acid, proline (see Figure 12.6). In such cases, the substitution mutation is said to be "silent" because it produces no change in the structure (and hence the function) of the protein, and therefore no change in the phenotype of the organism.

Concept Check

1. What is a mutation? Are all mutations harmful?
2. A single-base addition or deletion in a gene is likely to alter the protein product more than a single-base substitution, such as C for T, would. Why?

Concept Check Answers

1. A mutation is a change in the base sequence of an organism's DNA. A mutation may have no detectable effects or may be harmful; in rare instances, it may even be advantageous.

2. Single-base addition or deletion shifts the reading frame, so all the amino acids downstream of such a mutation are altered. Single-base substitution alters, at most, a single amino acid.

12.6 How Cells Control Gene Expression

In **gene expression**, a gene is transcribed and its mRNA is translated into a functional protein that has an effect on phenotype. As described in Chapter 11, the study of gene expression includes not only which genes are turned on or turned off in particular cells, but also when and under what conditions, as well as the specifics of how this expression is controlled.

Cells receive signals that determine gene expression. Some of these signals are sent from one cell to another, as when one cell releases a signaling molecule that alters gene expression in another cell. Cells also receive signals from the organism's internal environment (for example, blood sugar level in humans) and external environment (for example, sunlight in plants). Overall, cells process information from a variety of signals, and that information affects which genes are expressed.

Most genes are controlled at the transcriptional level

The most common way for cells to control gene expression is to regulate the transcription of the gene. In the absence of transcription, the mRNA encoded by the gene is not made; consequently, the protein encoded by the mRNA is not made, and any phenotype directly produced by the protein is therefore absent.

In general, transcription is controlled through two essential elements: **regulatory DNA** that can activate or inactivate gene transcription, and gene **regulatory proteins** that interact with signals from the environment and also with the regulatory DNA to promote or repress gene transcription. Gene regulatory proteins are also known as *transcription factors*.

The regulation of lactose synthesis in the bacterium *E. coli* illustrates how gene regulatory DNA and gene regulatory proteins alter gene transcription in response to internal or external cues. *E. coli* can feed on a variety of different types of sugars, including the disaccharide lactose. To avoid wasting metabolic resources, the proteins needed to absorb and break down these carbohydrates are made to order: the genes encoding the proteins are expressed only when the bacterium detects the specific carbohydrate in its environment. At all other times, these genes are shut down.

The three genes needed for the uptake and degradation of lactose are sequentially arranged as an **operon**, meaning that a single promoter sequence

Gene Regulation in Bacteria: The *lac* Operon

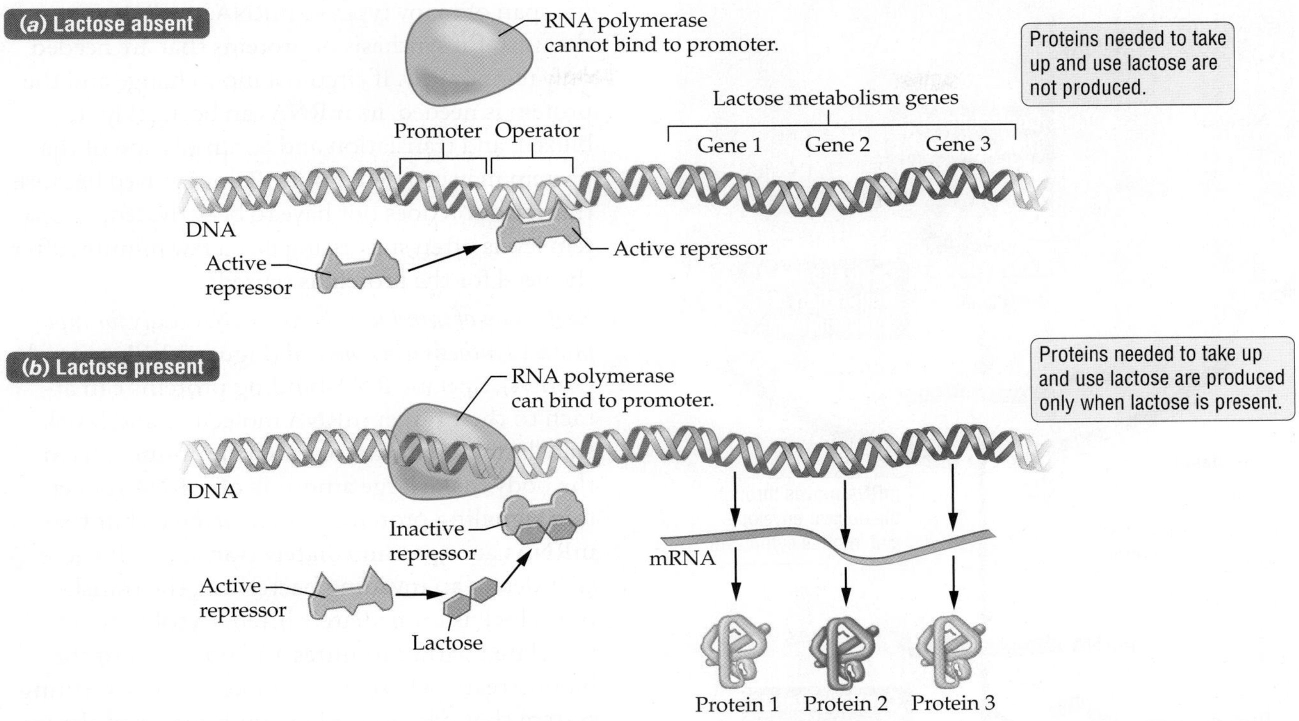

(*a*) Lactose absent

RNA polymerase cannot bind to promoter.

Proteins needed to take up and use lactose are not produced.

Lactose metabolism genes

Promoter Operator Gene 1 Gene 2 Gene 3

DNA

Active repressor

Active repressor

(*b*) Lactose present

RNA polymerase can bind to promoter.

Proteins needed to take up and use lactose are produced only when lactose is present.

DNA

Inactive repressor

Active repressor

Lactose

mRNA

Protein 1 Protein 2 Protein 3

FIGURE 12.11 Repressor Proteins Turn Genes Off
In the bacterium *E. coli*, a repressor protein interacts with an operator to control the transcription of a group of genes that encode the enzymes needed to take up and metabolize lactose. (*a*) When lactose is absent, the *lac* repressor protein binds to the operator, which blocks transcription by preventing RNA polymerase from attaching to the promoter. (*b*) When lactose is present, the sugar binds to the repressor protein, preventing it from docking with the operator. With the repressor inactivated, RNA polymerase can bind to the promoter and transcribe the entire operon.

controls the transcription of all the genes in the cluster. *E. coli* controls the three genes in the *lac* operon in the following way:

When lactose is absent, the *lac* promoter is prevented from functioning by the *lac* repressor (a gene **repressor** is a gene regulatory protein that stops the expression of one or more genes). The *lac* repressor binds to a DNA sequence called the *lac* operator. An **operator** is a gene regulatory DNA sequence that controls the transcription of an operon—in this case, the three genes needed to absorb and metabolize lactose. When bound to the operator, the *lac* repressor blocks access to the *lac* promoter, and with the repressor protein in the way, RNA polymerase cannot bind the promoter to initiate transcription(**FIGURE 12.11*a***).

When lactose is present, however, the sugar combines with the repressor protein, preventing the repressor from binding to the operator. With the repressor out of the way, RNA polymerase is free to bind to the promoter, and all the genes in the operon are transcribed together as one large *polycistronic mRNA* (**FIGURE 12.11*b***). Each gene has its own translation initiation and termination sites, and as a result, translation of the polycistronic mRNA produces three different proteins that assist in the metabolism of lactose.

Gene expression can be regulated at several levels

In eukaryotes, gene expression can be controlled at several points along the pathway from gene to protein to phenotype (**FIGURE 12.12**). Let's consider some of the ways gene expression is regulated along this pathway.

1. *Tight packing of DNA prevents it from being expressed* (Figure 12.12, control point 1). During interphase, parts of the chromosome are "unpacked" down to the beads-on-a-string state. This unpacking gives gene regulatory proteins and RNA polymerase access to gene promoters and other regulatory DNA sequences, making transcription possible. In contrast, the tightly packed regions of the chromosome are transcriptionally inactive because their gene regulatory DNA is not accessible.

FIGURE 12.12 Steps at Which Gene Expression Can Be Regulated in Eukaryotes
Each point in the pathway from gene to protein provides an opportunity for cells to regulate the production or activity of proteins.

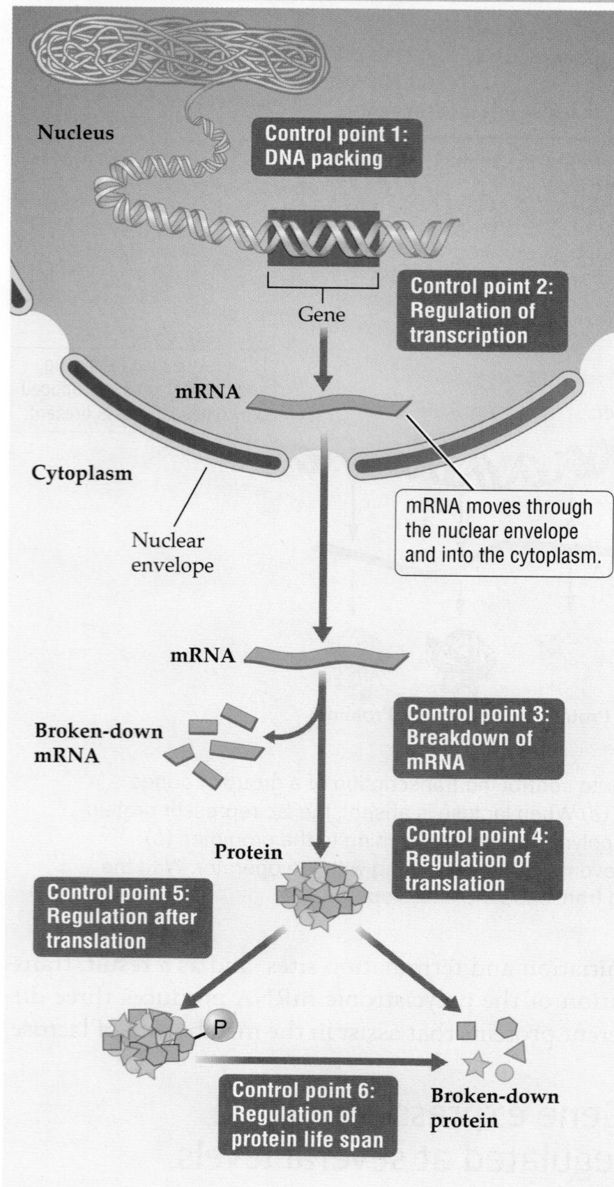

2. *Regulation of transcription conserves resources* (Figure 12.12, control point 2). Regulation of transcription is an efficient control mechanism because it enables the cell to conserve resources when it does not need the gene product. Transcriptional activation of gene expression is, however, relatively slow in eukaryotes, taking a minimum of 15–30 minutes at best.
3. *Regulation of mRNA breakdown prevents wasteful synthesis of proteins* (Figure 12.12, control point 3). Most mRNA molecules are broken down within a few minutes to hours after they are made; a few persist for days or weeks in the cytoplasm. By limiting the life span of many types of mRNA, a cell prevents the wasteful synthesis of proteins that are needed only temporarily. If circumstances change and the protein is needed, its mRNA can be quickly stabilized, and translation and accumulation of the protein may proceed rapidly. Time is saved because transcription does not have to be activated, so protein levels often start rising just a few minutes after the need for the protein is sensed.
4. *Regulation of translation keeps mRNA ready for rapid protein synthesis when needed* (Figure 12.12, control point 4). Specific RNA-binding proteins can attach to their target mRNA molecules and block translation. For example, some immune cells in the body make large amounts of mRNA for certain signaling proteins, called *cytokines*, but these mRNAs are not immediately translated. If these cells detect an invading bacterium, the translation block is immediately lifted. Cytokines are translated within minutes and poured into the bloodstream, where they act like an early-warning system that prepares other components of the immune system to defend the body.
5. *Proteins can be directly regulated by modification following translation* (Figure 12.12, control point 5). Many proteins must be chemically modified before they can exert their effect on a phenotype. Some of the blood-clotting proteins, for example, are synthesized as inactive precursors, and a segment of the protein must be cleaved off before clotting can occur. This type of control prevents premature activity of a protein, which could be disruptive, even dangerous.
6. *Regulation of protein breakdown conserves resources and prevents damage* (Figure 12.12, control point 6). The "shelf-life" of a protein provides one final opportunity to control the pathway from gene to protein. Most proteins in the cell have a limited life span, but a few, such as collagen and crystallin, last us through our lifetime. Proteins that are no longer needed, or are damaged, are taken apart and their amino acids recycled into new proteins.

Concept Check

1. Why are most genes controlled at the level of transcription?
2. If transcriptional control is the most favored method of gene regulation, why are not all genes controlled at the level of transcription?

Concept Check Answers

1. Transcriptional regulation prevents gene expression when a gene's product is not needed by a cell, enabling the cell to invest its resources elsewhere.
2. Transcriptional activation is relatively slow; controlling gene expression at a posttranscriptional step enables a cell to respond faster to environmental changes.

BIOLOGYMATTERS

One Allele Makes You Strong, Another Helps You Endure

If you have several thousand dollars to spare, you could have your entire genome sequenced. For a few hundred dollars, many genome science companies will tell you all about your genetic ancestry and also your genotype with respect to more than 50 different genes. Some of these genes may be associated with disease risk; others may be linked to less worrisome genetic traits, such as your attitude toward broccoli and other strong flavors.

Welcome to the science of *personal genomics*, the goal of which is to inspect and catalog an individual's total genetic makeup, or genome. The hope is that personal genomics will lead to personalized medicine, the practice of tailoring health care and disease prevention to a person's genotype. The field of personal genomics is not without ethical issues, however. Some opponents have expressed concern about the marketing of genomics services directly to the consumer, asserting that personal genomic information that has not been filtered through or delivered by an expert may be misunderstood, with potentially harmful consequences.

As a specific example of these concerns, an Australian company developed a commercial test for athletic potential, based on the *R* and *X* alleles of the *ACTN3* gene. Skeletal muscles contain bundles of muscle cells, also called muscle fibers. There are two main types of fibers: fast-twitch and slow-twitch. Fast-twitch fibers specialize in producing large bursts of power, but they tire quickly. Slow-twitch fibers are more efficient in extracting energy from sugars, and their power output can be sustained much longer. Most of us have roughly equal numbers of the two types of fibers in our skeletal muscles. In contrast, the muscles of elite athletes in strength-based sports, such as sprinting or weight lifting, may be 80 percent fast-twitch fibers. Those excelling in endurance sports, such as long-distance running or cycling, can have muscles that are 80 percent slow-twitch fibers.

Florence Griffith Joyner is a sprinter, a strength sport.

Paula Radcliffe runs marathons, an endurance sport.

Alpha-actinin-3, the protein encoded by the *ACTN3* gene, is made only in skeletal muscles. It anchors the contractile proteins so that muscle fibers can generate power. Australian scientists found two alleles of the *ACTN3* gene seemingly linked to athletic ability. The *R* allele codes for a functional alpha-actinin-3 protein; the *X* allele leads to the production of a shortened, nonfunctional version of the protein. The *X* allele of *ACTN3* is the result of a base substitution that changes an amino acid codon into a stop codon—a mutation that halts protein synthesis prematurely.

A study of Australian athletes showed that the *XX* genotype is rare in athletes in strength sports but is found in about 24 percent of endurance athletes (see the table). The advantage of the *X* allele in endurance sports is supported by experiments in which genetic engineering technology (see Chapter 13) was used to "knock out" the activity of both copies of the *ACTN3* gene in lab mice, creating "marathon mice."

Success in sports depends on many things, including psychological attributes such as personal drive, and top-level performance always requires extensive training and conditioning. The physical component of athletic success is likely influenced by many genes, not just one or two genes such as *ACTN3*. In one study, as many as 92 different genes were potentially associated with athletic ability and health-related fitness. Therefore, the predictive power of the *ACTN3* alleles is limited. Studies found that as many as 31 percent of the elite distance runners lacked the *X* allele, and 45 percent had only one copy.

Given the complexities in interpreting the genetic tests, would you want to know your allelic makeup? Would you want your children tested for the *ACTN3* gene, and if so, how would you use that information? Researchers justify studies of genes influencing physical performance by pointing out their relevance in conditions such as muscular dystrophy and other muscle diseases. Opponents say genetic tests of this sort will lead to abuse by overambitious athletes and pushy parents of would-be sports stars. Does the potential for medical benefit outweigh the risk for abuse?

Association between Athletic Ability and Prevalence of *ACTN3* Alleles in Strength and Endurance Athletes

	PERCENTAGE OF THE GROUP THAT HAS THE GENOTYPE		
GENOTYPE	CONTROL (NONATHLETES)	STRENGTH-SPORT ELITE ATHLETES (SPRINTERS)	ENDURANCE-SPORT ELITE ATHLETES (DISTANCE RUNNERS)
RR	30	50	31
RX	52	45	45
XX	18	6	24

APPLYING WHAT WE LEARNED

From Gene Expression to Cyclops

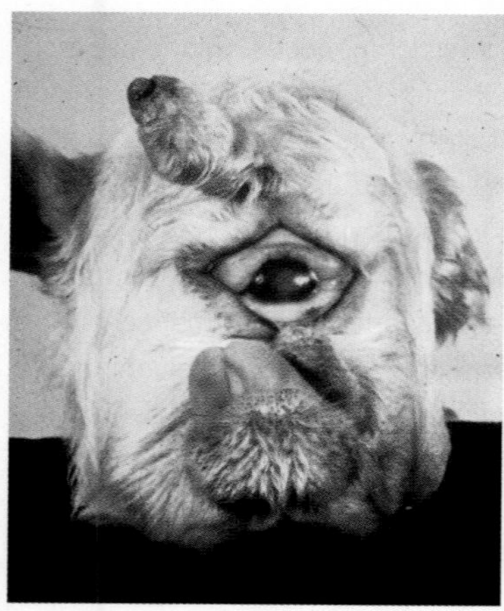

At the beginning of this chapter we described a rare birth defect called cyclopia, in which an animal is born with a single, large eye. Cyclopia may occur in as many as one in 250 embryos. Because most die before birth, the condition is rare in live births—about one in every 5,000–10,000 births. Mammals born with cyclopia usually die within a few hours of birth, but biologists have recorded some amazing exceptions—including a cyclopean skate (skates are related to sharks) that lived for several years and a trout that lived for 22 months.

Individuals with cyclopia have more problems than a missing eye; their whole face is malformed. Often they have either no nose or one that is only vaguely noselike, located above the eye or elsewhere on the head. In extreme cases, both nose and mouth are missing. Inside the skull, the two halves of the forebrain have not divided normally into separate hemispheres. This malformed brain makes cyclopia a severe and fatal birth defect.

FIGURE 12.13 Corn Lilies (*Veratrum californicum*)

A toxin produced by the plant disrupts normal gene expression during development, but for the same reason it can also disable cancer cells.

Cyclopia is sometimes common in animals that graze on wild plants. During the 1950s, 5–7 percent of all Utah lambs were born with cyclopia. Biologists eventually discovered that the lambs' mothers had eaten corn lilies on about the fourteenth day of pregnancy (**FIGURE 12.13**). Corn lilies—tall wildflowers that grow in the high mountain meadows of western North America—contain a molecule called cyclopamine.

Cyclopamine causes cyclopia by binding to a protein receptor that is part of a critical gene pathway in embryonic development. When cyclopamine binds to the receptor, it shuts down the expression of a series of genes in this pathway. The first gene is one that geneticists named *sonic hedgehog* (or *shh*), after a popular video game character. The *shh* gene expresses a protein with the same name, Sonic hedgehog (Shh), that signals the embryonic brain to divide into left and right halves—leading to left and right eyes. The Shh protein accomplishes its task by binding to a protein receptor called Smoothened (Smo), which stimulates cell division. If a mother sheep eats corn lilies, however, the plant's cyclopamine blocks the Smoothened receptor, preventing Shh from binding to the receptor. Without Shh, the brain and face don't develop normally into two symmetrical halves.

Humans almost never eat corn lilies. In rare cases, though, women have been exposed to cyclopamine from drinking the milk of goats or cows that foraged on corn lilies. Alcohol and other drugs also seem to inhibit the *sonic hedgehog* gene pathway, leading to rare cases of cyclopia in human babies. Other causes of cyclopia include diabetes during pregnancy and genetic mutations in the *hedgehog* gene pathway (including other genes besides *shh*).

Amazingly, cyclopamine is turning out to be a valuable treatment for cancer. During development, the Shh protein promotes the rapid multiplication of certain cells, helping to shape the embryo. In adults, Shh also promotes rapid cell division. But because cells should multiply slowly or not at all in most adult tissues, the *sonic hedgehog* pathway is usually quiet in adults.

Sometimes, however, a gene in the *hedgehog* pathway can mutate as a result of damage from chemicals or ultraviolet radiation, causing abnormally rapid cell division, or cancer. Cancers influenced by the *hedgehog* pathway include certain cancers of the brain and the most common form of skin cancer, basal cell carcinoma. To slow cell division in such cancers, pharmaceutical companies have developed commercial drugs from cyclopamine that suppress the *hedgehog* pathway and slow cell division.

BIOLOGY IN THE NEWS

BPA Could Affect Brain Development by Impacting Gene Regulation, Study Finds

HUFFINGTON POST • February 28, 2013

In yet another study drawing a connection between bisphenol A and potential negative health effects, researchers at Duke University have linked environmental exposure to the plastics chemical with disruption of a gene necessary for proper functioning of nerve cells.

"Our study found that BPA may impair the development of the central nervous system, and raises the question as to whether exposure could predispose animals and humans to neurodevelopmental disorders," study researcher Dr. Wolfgang Liedtke, M.D., Ph.D., an associate professor of medicine/neurology and neurobiology at Duke University, said in a statement. Liedtke's research, which was conducted in animals, is published in the journal *Proceedings of the National Academy of Sciences.*

Much of humans' BPA exposure comes through what they eat and drink, via containers used to keep the food. BPA's effects on the human body are of concern because some research suggests it is an endocrine disruptor—meaning it affects the way hormones work in the body, leading to possible reproductive and developmental effects, not to mention a possible link with a number of diseases and conditions, such as diabetes and obesity. Researchers noted that BPA is known to mimic estrogen in the body; the chemical is currently banned from baby bottles and cups in the U.S.

In this newest study in rodents, researchers found that BPA could potentially have a negative effect on development of the central nervous system by shutting down a gene necessary to the process, called the *Kcc2* gene.

If this gene is shut down, the researchers noted that it doesn't produce a protein that plays an important role removing

chloride from neurons—a fundamental step in proper functioning of brain cells.

"Our findings improve our understanding of how environmental exposure to BPA can affect the regulation of the *Kcc2* gene," Liedtke said in the statement. "However, we expect future studies to focus on what targets aside from *Kcc2* are affected by BPA. This is a chapter in an ongoing story."

As we saw in Chapters 10 and 11, genes aren't everything. The genotype directs the phenotype, but for many genetic traits the environment can also sculpt the phenotype. The concepts we discussed in this chapter show *how* this can happen. Environmental cues can tweak gene expression by altering transcription or translation.

Bisphenol A (BPA) is a human-made chemical that has been charged with a variety of harmful effects, from obesity to cancer. The chemical is everywhere: in hard plastics, CDs and DVDs, shower curtains, sports equipment, dental sealants, eyeglass lenses, lipstick and other cosmetic products, and so on. The compound is so widespread that 96 percent of Americans have measurable levels of the chemical in their blood.

The chemical structure of BPA resembles that of estrogen, the sex hormone in female mammals. BPA binds to a particular category of sex hormone receptors that are more abundant in certain tissues, such as the placenta. Not surprisingly, fetuses and newborns have been shown to be at greatest risk from BPA exposure. Evidence shows that mothers with elevated BPA levels are more likely to bear children with developmental problems, including memory deficits and sociobehavioral issues such as anxiety.

The chemical industry has responded to the alarm over bisphenol A by developing a closely related chemical called bisphenol S (BPS). Products made with BPS are being promoted as safe because they are "BPA-free." BPS is now so widespread that a 2012 study found significant amounts of the chemical substitute in nearly everyone they tested, including people from North America, China, and Japan. On a sobering note, a University of Texas study published in 2013 found that BPS disrupts gene expression in lab animals in many of the *same ways* as BPA.

Evaluating the News

1. According to the model proposed by the researchers, how does BPA affect development of the central nervous system?

2. In 2011, Maine governor Paul LePage defended the use of BPA, saying, "There hasn't been any science that identifies that there is a problem," and "the worst case is some women may have little beards." How would you respond to the governor's statement?

CHAPTER REVIEW

Summary

12.1 How Genes Work

- Genes code for RNA molecules. Protein-coding genes code for mRNA, which contains instructions for building a protein.
- An RNA molecule consists of a single strand of nucleotides. Each nucleotide is composed of the sugar ribose, a phosphate group, and one of four nitrogen-containing bases. The bases found in RNA are the same as those in DNA, except that uracil (U) replaces thymine (T).
- At least three types of RNA (mRNA, rRNA, and tRNA) and many enzymes and other proteins participate in the manufacture of proteins.

12.2 Transcription: Information Flow from DNA to RNA

- During transcription, one strand of the gene's DNA serves as a template for synthesizing many copies of mRNA.
- The key enzyme in transcription is RNA polymerase.
- Each gene has a sequence (a promoter) at which RNA polymerase begins transcription. Transcription stops after RNA polymerase encounters special transcription termination sequences.
- The mRNA molecule is constructed according to specific base-pairing rules: A, U, C, and G in mRNA pair with T, A, G, and C, respectively, in the template strand of DNA.
- Newly made, or "preliminary," eukaryotic mRNA must be processed while it is still in the nucleus to (among other things) splice out the noncoding sequences of DNA (introns) found in many genes. The remaining, protein-encoding segments of mRNA (exons) are then joined in a "mature" mRNA molecule that is exported to the ribosomes for translation into protein.

12.3 The Genetic Code

- The information encoded by an mRNA is read in sets of three bases; each three-base sequence is a codon. Of the 64 possible codons, most specify a particular amino acid, but certain codons signal the start or stop of translation. The information specified by each codon is collectively called the genetic code.
- When reading the genetic code, ribosomes begin at a fixed starting point on the mRNA (the start codon) and stop reading the code when they encounter any one of the three stop codons.
- The genetic code is unambiguous (each codon specifies no more than one amino acid), redundant (several codons specify the same amino acid), and nearly universal (used in almost all organisms on Earth).

12.4 Translation: Information Flow from mRNA to Protein

- The codon sequence in each mRNA molecule determines the amino acid sequence of the protein it encodes.
- Translation occurs at ribosomes, which are composed of rRNA and more than 50 different proteins.
- Translation calls into play transfer RNA molecules that each specialize in carrying a particular amino acid. Each tRNA has a three-base sequence, called the anticodon, that recognizes and pairs with a specific codon in the mRNA by complementary base pairing. In this way, a specific tRNA molecule delivers a specific amino acid based on the codon message present in the mRNA.
- The ribosome holds the mRNA and tRNA in a manner that allows the amino acid carried by the tRNA to be covalently bonded to the growing amino acid chain. When translation is complete, the amino acid chain folds into the three-dimensional shape of the protein.

12.5 The Effect of Mutations on Protein Synthesis

- Many mutations are caused by the substitution, insertion, or deletion of a single base in a gene's DNA sequence.
- Insertion or deletion of a single base causes a genetic frameshift, resulting in a different sequence of amino acids in the gene's protein product. Mutations causing frameshifts usually destroy the protein's function.
- Some mutations have neutral effects, and a few even have beneficial effects.

12.6 How Cells Control Gene Expression

- Most genes are regulated at the level of transcription. Transcription is controlled by regulatory DNA sequences that interact with regulatory proteins to switch genes on and off.
- Gene regulatory proteins link gene expression to internal and external signals. Some regulatory proteins (repressor proteins) inhibit transcription; others (activator proteins) promote it.

Key Terms

anticodon (p. 269)
codon (p. 267)
deletion (p. 271)
elongation (transcription, p. 265; translation, p. 269)
exon (p. 266)
frameshift (p. 271)
gene (p. 262)
gene expression (p 272)
gene promoter (p. 265)
genetic code (p. 267)
initiation (transcription, p. 265; translation, p. 269)
insertion (p. 271)
intron (p. 266)
messenger RNA (mRNA) (p. 264)
operator (p. 273)
operon (p. 272)
point mutation (p. 271)
regulatory DNA (p. 272)
regulatory protein (p. 272)
repressor (p. 273)
ribosomal RNA (rRNA) (p. 264)
RNA polymerase (p. 265)
RNA splicing (p. 267)
start codon (p. 267)
stop codon (p. 267)
substitution (p. 271)
template strand (p. 265)
termination (transcription, p. 265; translation, p. 270)
terminator (p. 266)
transcription (p. 264)
transfer RNA (tRNA) (p. 264)
translation (p. 264)

Self-Quiz

1. For a protein-coding gene, what molecule carries information from the gene to the ribosome?
   a. DNA b. mRNA c. tRNA d. rRNA

2. During translation, each amino acid in the growing protein chain is specified by how many nitrogenous bases in mRNA?
   a. one b. two c. three d. four

3. Which molecule(s) deliver the amino acid specified by a codon to the ribosome?
   a. rRNA b. tRNA c. anticodons d. DNA

4. During transcription, which molecule(s) are produced?
   a. mRNA b. rRNA c. tRNA d. all of the above

5. A portion of the template strand of a gene has the base sequence CGGATAGGGTAT. What is the sequence of amino acids specified by this DNA sequence? (Use the information in Figure 12.6 and assume that the corresponding mRNA sequence will be read from left to right.)
   a. alanine, tyrosine, proline, isoleucine
   b. arginine, tyrosine, tryptophan, isoleucine
   c. arginine, isoleucine, glycine, tyrosine
   d. none of the above

6. Which of the following is responsible for creating the covalent bonds that link the amino acids of a protein in the order specified by the gene that encodes that protein?
   a. tRNA
   b. mRNA
   c. rRNA
   d. ribosome

7. In a single mutation, the fourth, fifth, and sixth bases are deleted from a gene that encodes a protein with 57 amino acids. Which of the following would be expected to happen?
   a. The resulting frameshift would prevent protein synthesis.
   b. A protein with 56 amino acids would be constructed.
   c. A protein differing from the original one—but still with 57 amino acids—would be constructed.
   d. A protein with 54 amino acids would be constructed.

8. Most eukaryotic genes contain one or more segments that are transcribed but not translated. Each such segment is known as
   a. a start codon.
   b. a promoter.
   c. an intron.
   d. an exon.

Analysis and Application

1. What is a gene? In general terms, how does a gene store the information it contains?

2. Discuss the different products specified by genes. What are the function(s) of each of these products?

3. Describe the flow of genetic information from gene to phenotype.

4. How is the information contained in a gene transferred to another molecule? Why must the molecule that "carries" the information stored in the gene be transported out of the nucleus?

5. What is RNA splicing? Does it occur in both eukaryotes and prokaryotes? Explain your answer.

6. Describe the roles played by rRNA, tRNA, and mRNA in translation.

7. In a gene encoding a tRNA molecule, scientists have discovered a mutation that appears to be responsible for a series of human metabolic disorders. The mutation occurred at a base located immediately next to the anticodon of the tRNA—a change that destabilized the ability of the tRNA anticodon to bind to the correct mRNA codon. Why might a single-base mutation of this nature result in a series of metabolic disorders?

8. Write a paragraph explaining to someone with little background in biology what a mutation is and why mutations can affect protein function.

13 DNA Technology

THE EDUNIA. This transgenic petunia, part plant and part human, carries a gene for a human antibody extracted from the cells of artist Eduardo Kac. Kac created the "Edunia" to express some of his ideas about biotechnology.

Eduardo Kac's "Plantimal"

Brazilian artist Eduardo Kac is fascinated by biotechnology. Kac (pronounced ***KATZ***) is well known for his role publicizing a white rabbit named Alba whose cells expressed a jellyfish gene that made the rabbit glow a fluorescent green. In 2008, Kac introduced a new work of biotechnological art that dramatically expresses his thoughts about combining genes from different organisms. Kac planned and created a garden petunia that carries a gene from his own body.

The "Edunia" is a brilliant-pink petunia whose petals express a gene from Kac's own genome. The petunia's ability to express a human gene illustrates the shared heritage of plants and animals. Because plants and animals evolved from a common ancestor that lived about 1.6 billion years ago, all of them use the same cellular machinery to transcribe and translate genes into proteins. Not only can the Edunia express Kac's gene, but the plant also passes that human gene to its offspring in the form of seeds. In other words, the human gene is a permanent addition to the petunia's genome.

Kac chose a gene that would specifically illustrate another idea. The gene he selected was for an antibody from his immune system. Antibodies protect us from infections by recognizing and disarming bacteria, viruses, and other foreign entities. As Kac writes on his website, "The gene I selected is responsible for the identification of foreign bodies. In this work, it is precisely that which identifies and rejects the other that I integrate into the other, thus creating a new kind of self that is partially flower and partially human."

In this chapter we study how DNA can be manipulated using some of the tools and techniques of DNA technology. At the end of the chapter we'll also learn how Eduardo Kac created his "plantimal" Edunia.

How can biologists move DNA from one organism to another? What are the implications of this kind of transgenic technology for medicine?

MAIN MESSAGE DNA technology makes it possible to isolate a gene and produce many copies of it, and then to introduce it into whole organisms.

KEY CONCEPTS

- "DNA technology" refers to the process of manipulating DNA for research, health, forensics, agriculture and other purposes.
- We clone genes by isolating them, joining them with other pieces of DNA, and introducing such recombinant DNA into a host cell, such as a bacterial cell, that can generate many copies of the genes.
- DNA fingerprinting produces an individual's genetic profile and is widely used to establish the identities of victims, criminals, and others.
- Genetic engineering is the introduction of foreign DNA into an organism's genetic material to produce a genetically modified organism (GMO).
- The goal of reproductive cloning is to generate offspring that are genetic copies of a selected individual.
- Stem cells are a self-renewing pool of unspecialized cells. Embryonic stem cells can differentiate into any of the cell types in the adult body; in contrast, adult stem cells can differentiate into only a limited number of cell types.
- Stem cell technology finds many applications in regenerative medicine, which aims to repair or replace tissue damaged by injury or disease.
- The polymerase chain reaction (PCR) is a revolutionary technique that can make billions of copies from a few molecules of DNA.
- DNA technology provides many benefits, but its use also raises ethical concerns and poses potential risks to human society and natural ecosystems.

EDUARDO KAC'S "PLANTIMAL" 281
13.1 The Brave New World of DNA Technology 282
13.2 DNA Fingerprinting 286
13.3 Genetic Engineering 287
13.4 Reproductive Cloning of Animals 291
13.5 Stem Cells: Dedicated to Division 292
13.6 Human Gene Therapy 297
13.7 Ethical and Social Dimensions of DNA Technology 299
13.8 A Closer Look at Some Tools of DNA Technology 300
APPLYING WHAT WE LEARNED 304
How to Make a Plantimal

(a)

(b)

FIGURE 13.1 Traditional Breeding versus Direct DNA Manipulation

When it comes to strange appearances, conventional breeding methods and DNA technology can both generate some extraordinary phenotypes. (*a*) This British Belgian Blue bull was developed through conventional selective breeding practices for heavily muscled, low-fat meat. (*b*) These "naked" chickens were genetically altered using the tools of DNA technology. The goal of the research project was to develop low-fat poultry that would be cheap and environmentally friendly, in part because they have no feathers to remove and dispose of. Would you eat chicken that was genetically modified? Should all genetically modified foods be labeled as such?

PEOPLE HAVE BEEN MANIPULATING the genetic material of other organisms, indirectly, for thousands of years. The many differences between domesticated species and their wild ancestors illustrates the effects of selective breeding (**FIGURE 13.1*a***). The past 40 years have witnessed a huge increase in the power, precision, and speed with which we can alter the genes of organisms ranging from bacteria to mammals. We can now select a particular gene, produce many copies of it, and then transfer it into a living organism of our choice. In doing so, we can alter DNA directly and rapidly in ways that do not happen naturally (**FIGURE 13.1*b***).

We begin this chapter with a broad overview of DNA technology. We then turn to some of the many practical applications of this technology, including DNA fingerprinting and genetic engineering. We also consider some of the ethical issues and social and environmental risks associated with DNA technology. We close with a peek into the DNA technologist's toolkit to help you better understand the techniques and procedures used to manipulate DNA and to create genetically modified organisms.

13.1 The Brave New World of DNA Technology

DNA technology began in the 1970s, as chemists and biologists began to decipher the way DNA is modified and replicated within cells, especially by bacteria and viruses infecting their host cells. Scientists could then use nature's own toolkit—a host of DNA-modifying enzymes and other proteins—to read the DNA code, to alter it at will, and even to create it from scratch in a test tube. DNA technology has produced dramatic advances in biology, and the wide-ranging applications of this technology touch many aspects of our daily life today, from the foods we eat to the diagnosis we might hear in a doctor's office.

Revolutionary techniques are the foundation of DNA technology

The strategies and techniques that scientists use to analyze and manipulate DNA are broadly known as

DNA technology. Although the nucleotide (base) sequence of DNA can vary greatly between species, as well as among individuals of a species, the general chemical structure of the DNA molecule (see Figure 11.4) is the same in all species. This consistency means that similar laboratory techniques can be used to isolate and analyze DNA from organisms as different as bacteria and people. Here we summarize some of the technological advances made over the past 40 years. We look at many of these techniques in more detail in the last section of this chapter.

- *Use of restriction enzymes and fragment separation to identify DNA sequences*. DNA can be readily extracted from most cells and tissues. Extracted DNA can be cut into fragments with specialized enzymes (called *restriction enzymes*). These fragments can then be separated in a gel-like material, stained with certain dyes that mark known components, and finally viewed and analyzed under special lights.
- *Amplification of DNA with PCR*. A revolutionary technique known as the *polymerase chain reaction* (*PCR*) makes it possible to copy (amplify) a target DNA billions of times from a single molecule. The coding information in a piece of DNA—its nucleotide sequence—can then be determined by robotlike machines called DNA sequencers. These machines have made it possible to decipher *all* the information held in the genomes of many species of bacteria, fungi, plants, and animals.
- *Recombinant DNA and plasmids*. Fragments of DNA can be joined together with the help of enzymes, creating an artificial assembly of genetic material known as **recombinant DNA** (**FIGURE 13.2**). Many bacteria, for example, have closed loops of DNA, called **plasmids**, in addition to their single

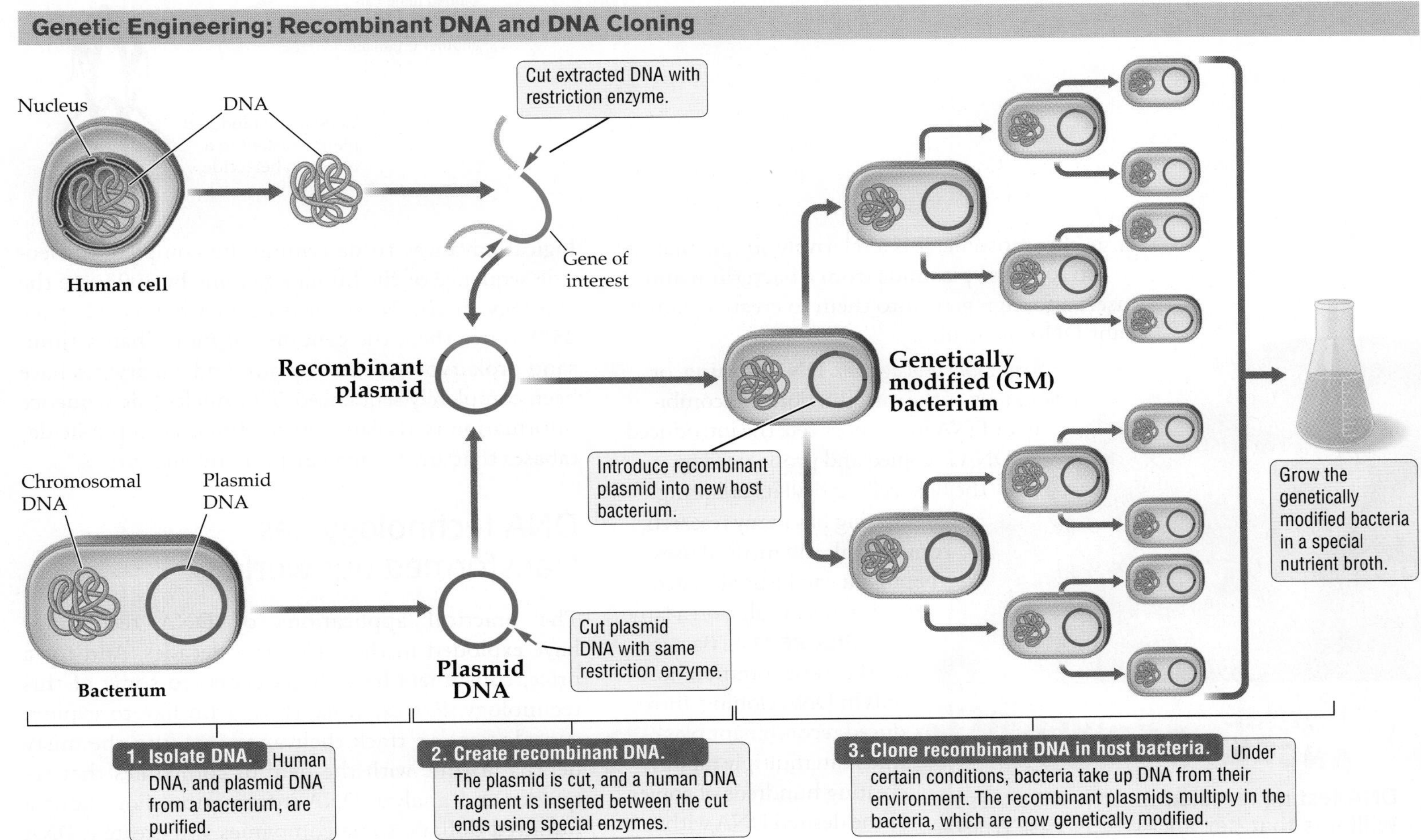
Genetic Engineering: Recombinant DNA and DNA Cloning
Nucleus
DNA
Human cell
Cut extracted DNA with restriction enzyme.
Gene of interest
Recombinant plasmid
Genetically modified (GM) bacterium
Introduce recombinant plasmid into new host bacterium.
Chromosomal DNA
Plasmid DNA
Bacterium
Plasmid DNA
Cut plasmid DNA with same restriction enzyme.
Grow the genetically modified bacteria in a special nutrient broth.
1. Isolate DNA. Human DNA, and plasmid DNA from a bacterium, are purified.
2. Create recombinant DNA. The plasmid is cut, and a human DNA fragment is inserted between the cut ends using special enzymes.
3. Clone recombinant DNA in host bacteria. Under certain conditions, bacteria take up DNA from their environment. The recombinant plasmids multiply in the bacteria, which are now genetically modified.

FIGURE 13.2 Recombinant DNA Can Be Propagated in Bacteria
This general scheme for genetic engineering uses the introduction of a human gene into a bacterium as an example. Once a recombinant plasmid has been created, it can be introduced into a desired bacterial strain in a number of different ways. The bacterium makes many copies of the recombinant plasmid, which are inherited by the daughter cells every time the bacterium divides.

FIGURE 13.3 Cloned DNA Serves Many Purposes

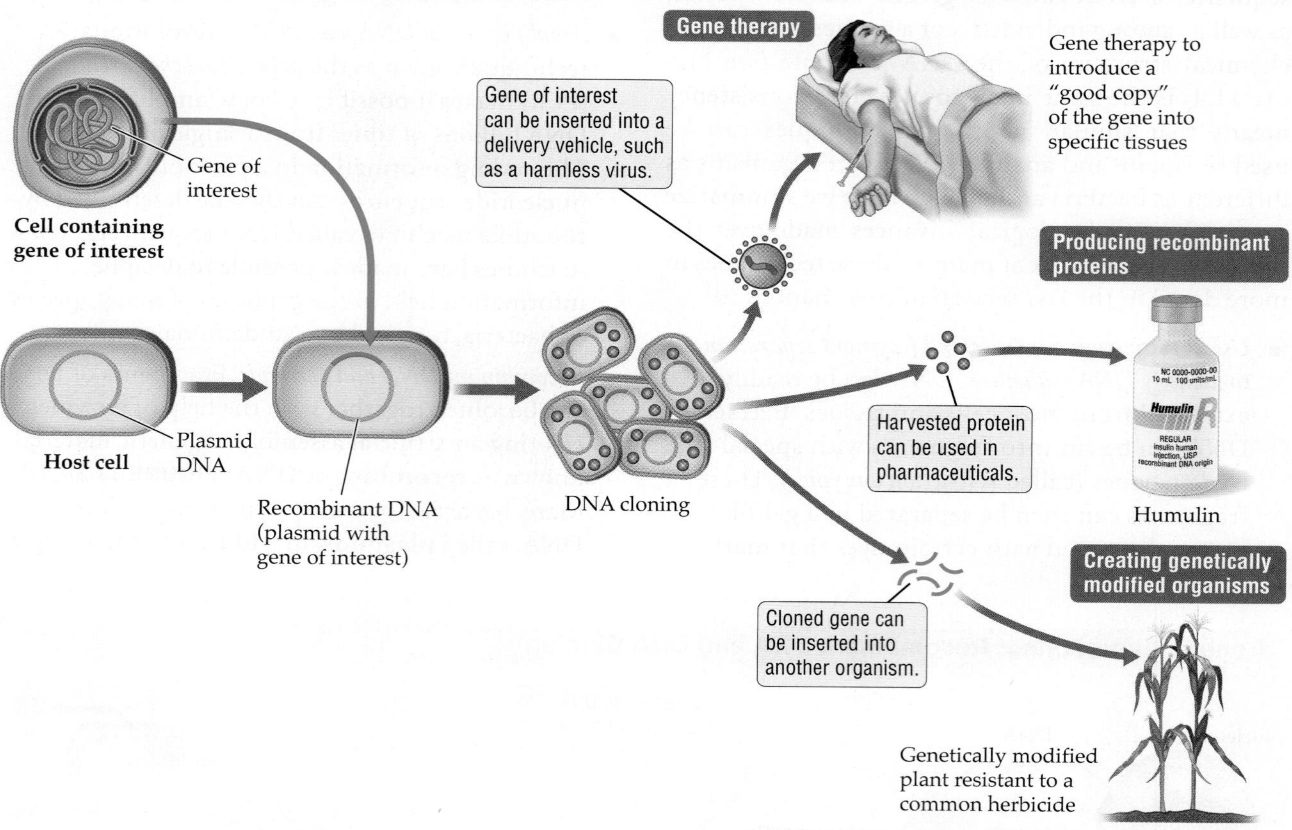

Some Applications of DNA Cloning
Gene of interest
Cell containing gene of interest
Host cell
Plasmid DNA
Recombinant DNA (plasmid with gene of interest)
DNA cloning
Gene of interest can be inserted into a delivery vehicle, such as a harmless virus.
Gene therapy
Gene therapy to introduce a "good copy" of the gene into specific tissues
Producing recombinant proteins
Harvested protein can be used in pharmaceuticals.
Humulin
Creating genetically modified organisms
Cloned gene can be inserted into another organism.
Genetically modified plant resistant to a common herbicide

large chromosome; it is a relatively simple matter to extract these plasmids from a bacterium and insert a foreign gene into them to create recombinant DNA molecules.

- *Gene cloning and propagation*. **DNA cloning**, or *gene cloning*, is the introduction of recombinant DNA into a host cell; the introduced DNA is copied and propagated by the host cells and all its offspring. Gene cloning has many research, commercial, and medical uses because it can be used to generate proteins of value on a large scale (**FIGURE 13.3**). Bacteria are the most common host cells in DNA cloning. Introduced recombinant plasmids can multiply rapidly, creating hundreds of copies of the desired DNA within the cytoplasm of a host bacterium.

The Human Genome Project used many of these technological advances to determine the complete nucleotide sequence of the human genome by 2003 (see the "Biology in the News" feature in Chapter 11, page 257). Since then, the genomes of more than a thousand prokaryotes and a few hundred eukaryotes have been completely sequenced. This nucleotide sequence information is available on the Internet in public databases that anyone may inspect and analyze.

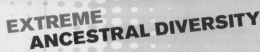

EXTREME ANCESTRAL DIVERSITY

DNA testing revealed to actress Vanessa Williams that her ancestry is 23% from Ghana, 17% from the British Isles, 15% from Cameroon, 12% from Finland, 11% from southern Europe, 7% from Togo, 6% from Benin, 5% from Senegal, and 4% from Portugal.

DNA technology has transformed our world

The practical applications of DNA technology have exploded in the past three decades. And for a price, consumers have direct access to some of this technology. For example, those who like to explore genealogies can track their ancestors into the misty depths of time with the help of companies that sequence and analyze DNA information. For about a hundred dollars, some companies will create a *DNA profile* of you; they can list your genotype for hundreds of genetic loci, including alleles associated with such traits as having a good verbal memory, or perfect pitch, or a tendency to take risks.

These advances have been made possible through a number of applications of technology; here we list some of the major developments.

- *DNA microarrays.* Informally known as a gene chip, a **DNA microarray** consists of a solid support, about the size of a microscope slide, on which known DNA sequences from a genome have been arranged in an orderly fashion. A sample can be tested against the chip to determine gene activity. DNA microarrays have made it possible to understand which sets of genes are expressed in specific cell types under certain conditions, and how the normal pattern might be altered when a person is sick. Use of this technology makes it possible to identify genes behaving in destructive ways and to test drugs that might correct the problems. **Personalized medicine**—the practice of tailoring treatments and therapies to a patient's DNA profile—depends heavily on microarray analysis.
- *Genetic engineering.* **Genetic engineering** refers to the permanent introduction of one or more genes into a cell, a certain tissue, or a whole organism, leading to a change in at least one genetic characteristic in the recipient. The organism receiving the DNA is said to be a **genetically modified organism (GMO)**, sometimes known as a genetically engineered organism (GEO). Many lifesaving medicines, such as human insulin, human growth hormone, human blood-clotting proteins, and anticancer drugs, are manufactured by genetically modified (GM) bacteria or GM mammalian cells grown in laboratory dishes. Most of the corn, soybean, canola, and cottonseed products consumed in North America are likely to be products of genetic engineering.
- *Gene therapy.* At the very forefront of DNA technology, and still in its infancy, is the strategy of curing human genetic disorders through *gene therapy*. In gene therapy, genetic engineering techniques are employed to alter the characteristics of specific tissues and organs in the human body, with the goal of treating serious genetic disorders or diseases.

DNA technology goes beyond biomedical and agricultural applications. The ability to analyze DNA from different species, and from individuals in populations of the same species, has helped us understand biological processes at all levels in the biological hierarchy. DNA analysis has played a large role in helping us reconstruct the history of life on Earth (see Chapter 14), and it has revealed, perhaps more clearly than any other approach, the common ancestry and genetic diversity of life.

FIGURE 13.4 Genome Analysis Shows That Some Neandertals Were Redheads
The genome of Neandertals (*Homo neanderthalensis*) has been sequenced and found to be 99.9 percent identical to ours. DNA analysis shows that some Neandertals were redheads, because they have a mutation in the same gene (*MCR1*)—although in a different region of that gene—that causes red hair in modern humans when its activity is reduced by mutations.

Researchers have used DNA from fossil material and from diverse groups of modern-day humans to conclude that anatomically modern humans evolved in southeastern Africa and spread out from this one area of Africa some 50,000 years ago, displacing other early humans (such as Neandertals) as they proceeded to colonize the world. Anthropologists suggest that some Neandertals may have been redheads, because DNA from the fossil bones of these human relatives contains an allele that is responsible for red hair and pale skin in modern humans (**FIGURE 13.4**).

In the rest of this chapter, we explore some of these modern technologies in detail.

Concept Check

1. What is recombinant DNA?
2. What is genetic engineering?

Concept Check Answers

1. Recombinant DNA is a single DNA molecule created by linking DNA from two different sources.

2. Genetic engineering is the permanent introduction of one or more genes into a cell or whole organism, producing a change in at least one phenotype in the recipient.

13.2 DNA Fingerprinting

Helpful to Know

"Biotechnology" is a broad term that describes the use of organisms in a wide variety of commercial processes, from beer brewing to drug manufacture. Recently, the word has also been applied to the production of GMOs or the manufacture of products from GMOs. When used in this way, the terms "biotechnology" and "DNA technology" overlap considerably, which is why many people use them interchangeably.

The process of identifying DNA that is unique to a species, or even to a specific individual within a species, is called **DNA fingerprinting**. DNA fingerprinting can be used to detect the contamination of food or water by certain harmful microorganisms, and to identify remains of endangered species (**FIGURE 13.5**). It is used to match organ donors with patients seeking organ transplants, in paternity testing, and to identify victims of mass attacks such as the one on the World Trade Center on September 11, 2001.

DNA fingerprinting is such a powerful tool in forensics today that it has traveled beyond the real world of law enforcement to be dramatized in crime novels, TV dramas, and movie thrillers. A laboratory technician can take a biological sample such as blood, tissue, or semen from a crime scene, clothing, or a victim and develop a DNA fingerprint, or profile, of the person from whom the sample came. That profile can then be compared with another profile—for example, that of a victim or suspect—to see whether they match (**FIGURE 13.6**).

DNA fingerprinting takes advantage of the fact that all individuals (except identical twins and other multiples) are genetically unique. To distinguish between different individuals, scientists examine regions of the human genome that are known to vary greatly from one person to the next. Certain highly variable regions known as **short tandem repeats (STRs)** are the ones most commonly used for DNA fingerprinting in North America.

It is theoretically possible for two people to have the same DNA profile. Therefore, a match such as the one in Figure 13.6, between a victim's DNA profile and blood stains found on a suspect's clothes, does not pro-

FIGURE 13.5 DNA Fingerprinting Can Identify Illegal Ivory

Customs officials in Thailand examine a shipment of illegal ivory. DNA fingerprinting can distinguish between elephant tusks from endangered herds and those from thriving game reserves that allow selective culling. Wildlife biologists can also use the information to identify smuggling routes and poaching hotspots.

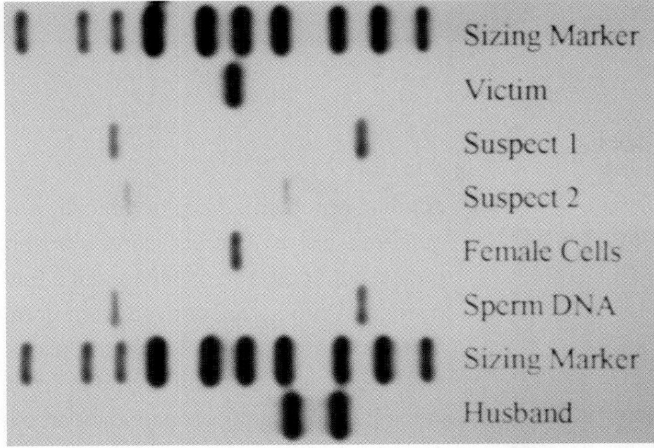

FIGURE 13.6 DNA Fingerprinting Can Be Used to Identify Criminals

With the exception of identical twins, every person's DNA is unique. By reading differences in DNA, forensic scientists can identify unique genetic profiles from crime scene evidence and match them to particular suspects. In the case above, evidence from a rape case indicates that the sperm DNA matches that of Suspect 1. "Female cells" and "Sperm DNA" refer to samples collected at the crime scene.

FIGURE 13.7 Exonerated by DNA Evidence

Damon Thibodeaux is greeted by his lawyer upon his release from prison in 2012. Thibodeaux spent 15 years on Louisiana's death row for a murder he did not commit. He became the three hundredth person to be exonerated on the strength of evidence from DNA fingerprinting.

vide absolute proof that the two samples are from the same person. However, if the investigator uses many different DNA regions to create a DNA profile, the odds that two DNA profiles would match simply by chance becomes remote. Most forensic labs in North America use at least 13 different STR loci. In most cases that involve a DNA match introduced as evidence in a courtroom, the probability of an accidental match is between one in 100,000 and one in several billion.

DNA fingerprinting can be done by various methods, although PCR-based techniques are most common today. We can use PCR to amplify STRs from a small sample of human DNA, for example. PCR-based DNA fingerprinting has been used to convict many thousands, but it has also led to the exoneration of more than 300 individuals who had been wrongfully convicted (**FIGURE 13.7**).

The Federal Bureau of Investigation (FBI) and most state law enforcement agencies use PCR-based amplification from 13 different STRs to produce a near-unique DNA profile of an individual. DNA fingerprints of missing persons, victims of unsolved crimes, and people convicted of serious crimes are maintained in a DNA database named CODIS (Combined DNA Index System). The CODIS database contains more than 5 million DNA profiles, including PCR-based fingerprints of biological material collected at the sites of unsolved crimes.

Concept Check

1. Can DNA fingerprinting identify a person with absolute certainty? Can it distinguish between identical twins?
2. What are some applications of DNA fingerprinting?

Concept Check Answers

1. No. But if at least 13 of the well-known variable DNA sequences are analyzed, the reliability of the identification is accepted as reasonable by the legal system. Identical twins cannot be distinguished, because their DNA sequence is essentially identical.

2. DNA fingerprinting can be used to identify victims of accidents or crimes, to establish a suspect's guilt or innocence, and to identify particular individuals or genetically distinct populations for research or conservation.

13.3 Genetic Engineering

As noted in Section 13.1, genetic engineering is the permanent introduction of one or more genes into a cell, a certain tissue, or a whole organism, leading to a change in at least one genetic characteristic in the recipient. The organism receiving the DNA is termed a *genetically modified organism* (*GMO*), and the gene introduced into a GMO is called a *transgene* (*trans*, "across from"), so GMO individuals are also known as **transgenic organisms**.

What are the objectives of genetic engineering? For one, if a gene can be transferred between species, it can often make a functional protein product in the new species. For example, jellyfish have a gene enabling them to produce flashes of fluorescent light that may ward off attackers. This gene, which codes for a small

BIOLOGYMATTERS

Have You Had Your GMO Today?

The use of genetically modified organisms (GMOs) in agriculture and food production has expanded dramatically since biotech crops were first commercially grown in 1996. The United States, Brazil, Argentina, and Canada are the leading producers of GMO crops. But in recent years, resource-poor farmers in Asia and Africa have taken to GMO crops, and developing countries now account for 52 percent of the worldwide acreage under GMO crops. The objectives of agricultural genetic engineering are to raise crops that are resistant to pests, herbicides, and environmental stresses such as heat, cold, or drought; and to improve the nutritional quality of food or prolong its shelf life.

Soybeans, cotton, corn, alfalfa, canola, and sugar beets dominated the 172 million acres of GMO crops planted in the United States in 2012, according to the U.S. Department of Agriculture (USDA). Other GMO crops—such as squash, potatoes, and papaya—are also grown, most of them for their enhanced resistance to plant diseases.

Because the United States has no labeling requirements for GM foods, many consumers are unaware that about 75 percent of all processed foods available in U.S. grocery stores may contain ingredients from genetically engineered plants. Breads, cereal, frozen pizzas, hot dogs, and soda are just a few of them. Corn syrup, derived from corn, is a common ingredient in many juices and sodas, and GM corn syrup is used in most brand-name sodas. Soybean oil, cottonseed oil, and corn syrup are ingredients used extensively in processed foods.

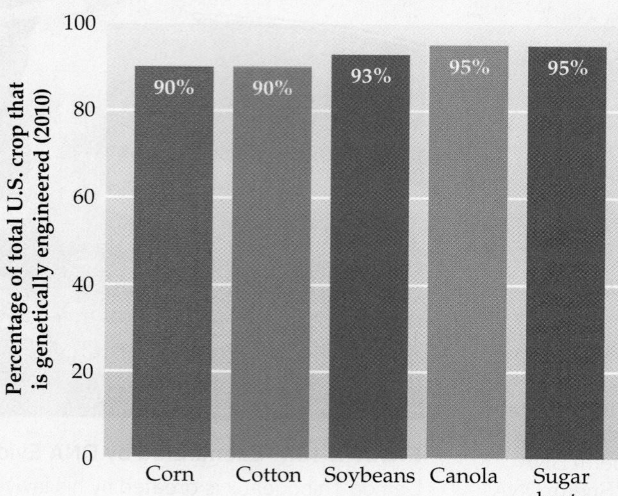

In the 1990s, much of the papaya crop in Hawaii was wiped out by the papaya ring spot virus. GM papayas resistant to the virus make up about 80 percent of the crop today. GM papayas are sold in the United States and Canada but cannot be exported to Europe.

Anti-GMO activists have campaigned for labeling of GMO foods in the United States. A ballot initiative for GMO labeling was narrowly defeated in California in 2012 (see the "Biology Matters" box in Chapter 1, page 8), but there are ongoing efforts to push for legislated labeling of GMO foods in other states. Proponents of labeling laws say people have a right to know "what's in their food." Opponents say many among the public are poorly informed about the nature of GMOs and will be unnecessarily alarmed, and the expense of complying with such regulations will drive up food prices for everyone. Besides, they say, people who want to avoid GMO foods already have an option: buying organic. Organic food is certified to be non-GMO, in addition to being raised without synthetic fertilizer and pesticides.

light-producing protein known as green fluorescent protein (GFP), has been transferred to and expressed in organisms as different as bacteria, plants, and rabbits (**FIGURE 13.8**). By attaching the GFP-producing gene to a gene of interest, researchers can track the product of the gene by the glow it creates. In other words, GFP has become invaluable as a fluorescent marker for gene expression in both basic and applied research.

The deliberate transfer of a gene from one species to another is one example of genetic engineering. As illustrated in Figure 13.2, genetic engineering involves isolating a DNA sequence (usually a gene) from one source of DNA and inserting it into a DNA molecule from another source. The inserted fragment of DNA can be introduced into the DNA of the same species or that of a different species.

Recombinant DNA or foreign genes can be introduced into a cell or whole organism in many different ways. We have already seen how plasmids can be used to transfer a gene from humans or other organisms to bacteria. Plasmids can also be used to transfer genes to plant cells or animal cells. In some species, including many plants and some mammals, genetically modified (GM) adults can be generated, or cloned, from these altered cells (see the "Biology Matters" box on page 288). Other means of gene transfer include viruses, which can be used to "infect" cells with genes from other species, and gene guns that fire microscopic pellets coated with the gene of interest into target cells.

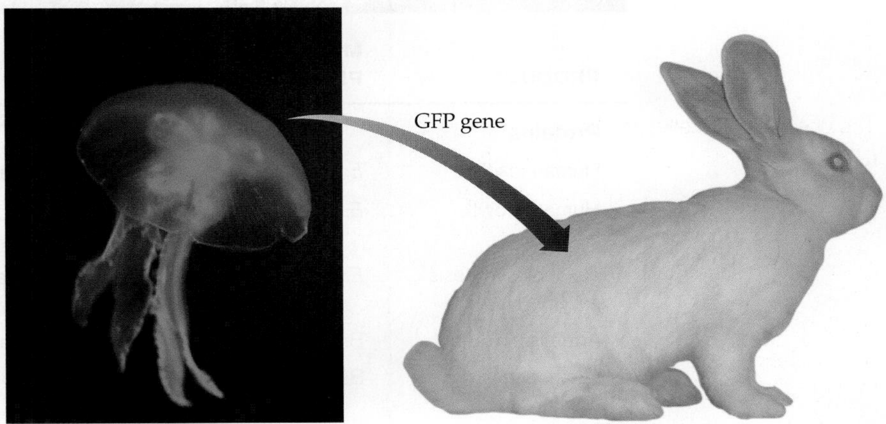

FIGURE 13.8 Alba, a White Rabbit Genetically Modified to Glow
A gene coding for a jellyfish protein called green fluorescent protein (GFP) was introduced into the fertilized egg from which this rabbit developed. Attaching this gene to genes that encode small components of the cell that would otherwise be very difficult to visualize has led to tremendous advances in our understanding of basic cell biology.

The quality of economically important species can be enhanced through genetic engineering

Genetic engineering is commonly used to alter the genetic characteristics of the recipient organism, often focusing on a particular aspect of performance or productivity. Some Atlantic salmon, for example, have been given genes that cause them to grow up to six times faster than normal (**FIGURE 13.9**).

Crop plants have been genetically engineered for a wide variety of traits, including increased yield, insect resistance, disease resistance, frost tolerance, drought tolerance, herbicide resistance (which enables crops to survive the application of weed-killing chemicals), increased shelf life, and improved nutritional value.

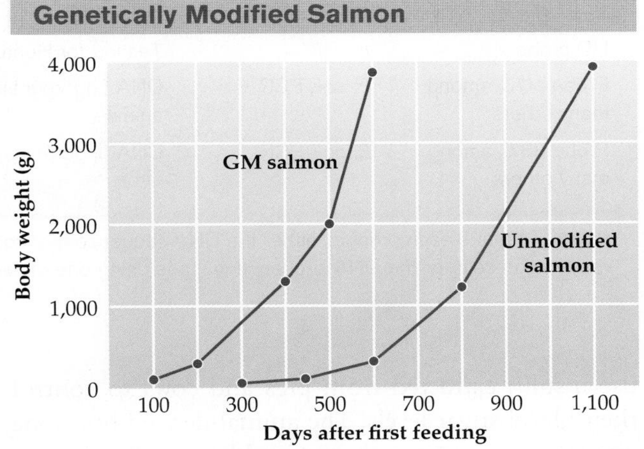

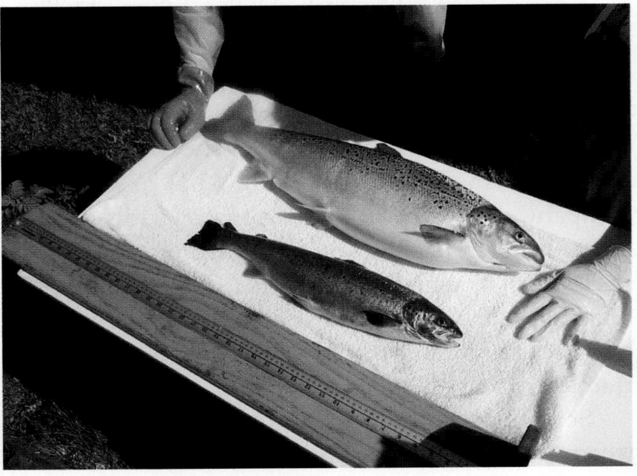

FIGURE 13.9 Genetically Modified Organisms—Good or Bad?
Genetically modified salmon eat more food and grow more rapidly on fish farms compared to unmodified fish.

Pharmaceutical products are produced in genetically engineered microbes

Another common use of genetic engineering is to churn out large amounts of a gene product, usually a protein with therapeutic or commercial value (**TABLE 13.1**). In 1978 the gene for human insulin was transferred into *E. coli* bacteria, and this protein hormone became the first genetically engineered product to be mass-produced. Before GM insulin, diabetics had to

TABLE 13.1 Methods of Production and Uses for Some Products of Genetic Engineering

PRODUCT	METHOD OF PRODUCTION	USE
Proteins		
Human insulin	*E. coli*	Treatment of diabetes
Human growth hormone	*E. coli*	Treatment of growth disorders
Taxol biosynthesis enzyme	*E. coli*	Treatment of ovarian cancer
Artemisinin	Brewer's yeast	Treatment of malaria
Luciferase (from firefly)	Bacterial cells	Testing for antibiotic resistance
Human clotting factor VIII	Mammalian cells in culture	Treatment of hemophilia
Adenosine deaminase (ADA)	Human cells in culture	Treatment of ADA deficiency
DNA sequences		
Sickle-cell probe	DNA synthesis machine	Testing for sickle-cell anemia
BRCA1 probe	DNA synthesis machine	Testing for breast cancer mutations
HD probe	*E. coli*	Testing for Huntington's disease
Probe *M13*, among many others	*E. coli*, PCR	DNA fingerprinting in plants and animals
Probe 33.6, among many others	*E. coli*, PCR	DNA fingerprinting in humans

NOTE: To make each product, either the DNA sequence that codes for the product is inserted into host cells, such as *E. coli* or mammalian cells, or the DNA sequence is made using one of several automated procedures, such as DNA synthesis or PCR.

The gene for artemisinin was transferred from the wormwood plant into yeast cells. The GM yeast produce large quantities of artemisinin, an anti-malarial drug.

use insulin extracted from pigs and cows to control their blood sugar levels. The animal-derived hormone was often in short supply and could cause allergic reactions in some people. GM insulin is safer and less expensive, and every year more than 300,000 Americans who suffer from insulin-dependent diabetes use the GM product to control their disease.

Bacteria are not the only hosts for genetically engineered medicines (see Table 13.1). A California company is producing an antimalarial drug called artemisinin in baker's yeast. Artemisinin is made by sweet wormwood (*Artemisia annua*), but extracting the drug from this plant means committing large tracts of agricultural land to drug production, and the extraction process is expensive.

Scientists isolated the artemisinin genes from wormwood and spliced them into the yeast genome. The yeast grows rapidly in fermentation tanks similar to those used in beer making, and the genetically engineered fungal cells are a cheap and reliable source of GM artemisinin. Malaria kills 655,000 people each year in the poorer parts of the world, and artemisinin is currently the most effective drug against the deadliest form of the parasite.

Vaccines can be produced in genetically engineered organisms

A vaccine is any substance that stimulates the immune system in such a way that it shields the body from future attack by a specific invading organism. Bacteria have been genetically engineered to produce large amounts of the distinctive proteins (*antigens*) that are

Reproductive Cloning of Animals

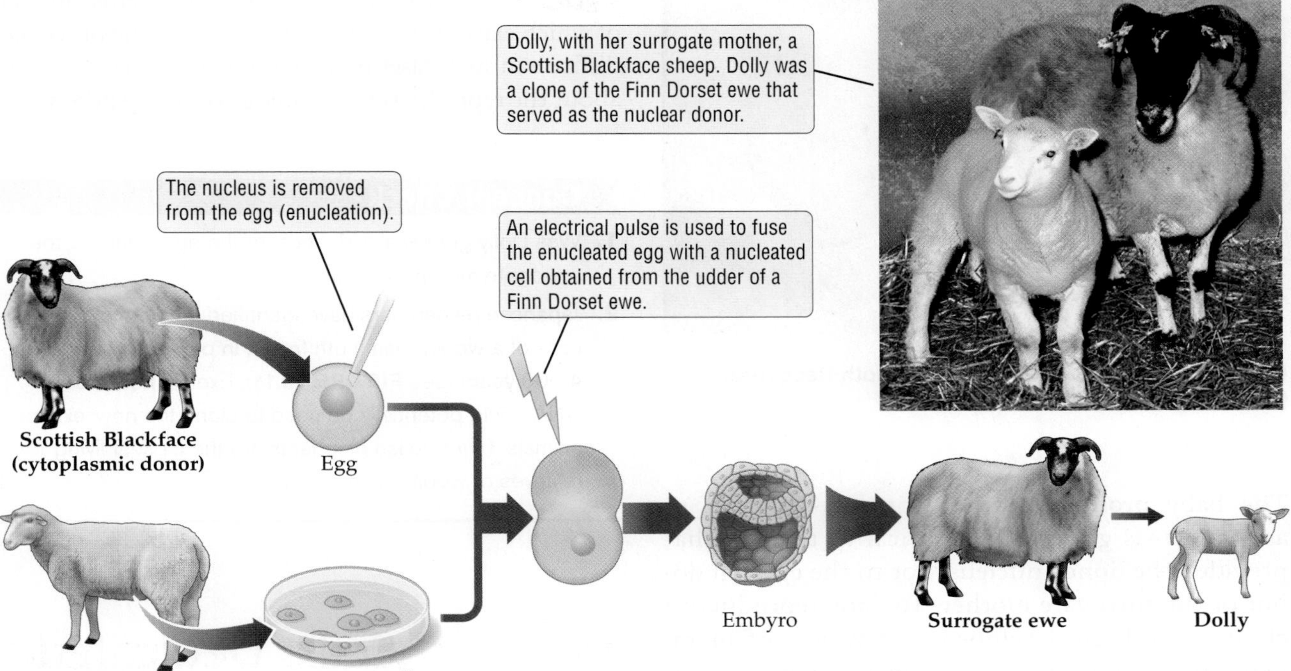

Dolly, with her surrogate mother, a Scottish Blackface sheep. Dolly was a clone of the Finn Dorset ewe that served as the nuclear donor.
The nucleus is removed from the egg (enucleation).
An electrical pulse is used to fuse the enucleated egg with a nucleated cell obtained from the udder of a Finn Dorset ewe.
Scottish Blackface (cytoplasmic donor)
Egg
Finn Dorset (nuclear donor)
Mammary cells
Embyro
Surrogate ewe
Dolly

FIGURE 13.10 Cloning Sheep: How Dolly Came to Be

present on the surface of many disease-causing organisms, including a number of viruses, bacteria, and infectious protists (such as the malaria parasite). The GM proteins produced by the host organism are usually purified and delivered through injection.

Scientists have also produced heat-stable vaccines that are edible. Bananas, tomatoes, and lettuce are among the crop plants that have been genetically modified to produce GM proteins that provide protection against infectious diseases such as cholera, measles, and hepatitis. People gain immunity against the disease simply by eating a few servings of the "vaccine fruit" each week. Edible vaccines are especially useful in poor countries because they are cheap, easy to use, and do not require refrigeration.

Concept Check

1. Name some ways in which genetic engineering can enhance food production.
2. Describe a pharmaceutical application of genetic engineering.

13.4 Reproductive Cloning of Animals

The goal of **reproductive cloning** is to produce an offspring that is a genetic copy of a selected individual. In 1996, Dolly the sheep, born on a Scottish farm, became the first mammal to be created through reproductive cloning (**FIGURE 13.10**).

There are three key steps in reproductive cloning.

1. An egg cell is obtained from a cytoplasmic donor (a Scottish Blackface ewe, in Dolly's case), and its nucleus is removed (enucleation).
2. An electrical current is used to fuse the enucleated egg with a somatic cell from a nuclear donor (for example, a Finn Dorset ewe, Dolly's genetic parent). Chemicals are used to activate this product of cell fusion—that is, to trick it into dividing so that it begins to form an embryo.
3. The embryo is transferred to the uterus of a surrogate mother, where, if all goes well, it continues to develop, ultimately resulting in the birth of a healthy baby animal.

Concept Check Answers

1. Some crop plants have been engineered to discourage attack by plant-eating pests, reducing the need for insecticides, and some have also been altered for herbicide resistance. Still others have been modified to resist frost and drought better, and to provide higher amounts of certain nutrients.
2. Human insulin, human growth hormone, and antimalarial compounds produced by the insertion of genes into bacteria or other microorganisms are all examples of pharmaceutical products made with the help of genetic engineering.

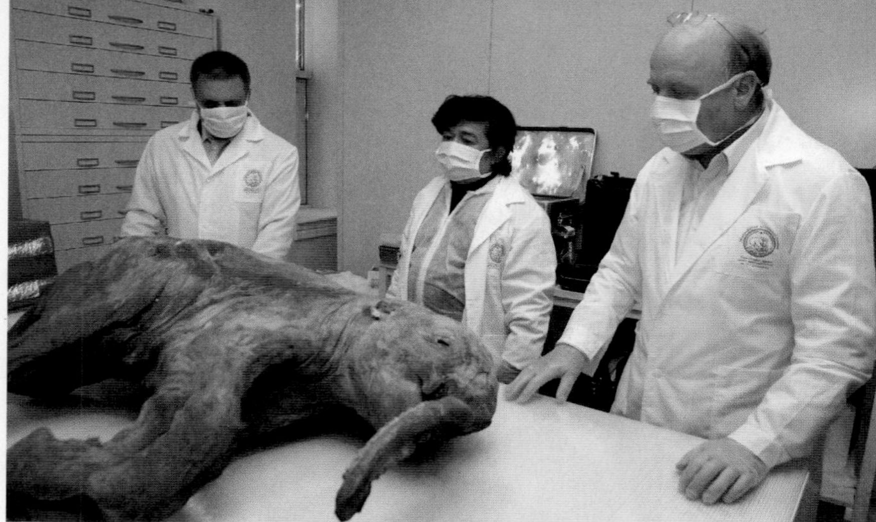

FIGURE 13.11 A Baby Woolly Mammoth Recovered from Permafrost

Concept Check Answers

1. Neither. Dolly was genetically identical to the Finn Dorset nuclear donor, whose nucleus was implanted in the enucleated egg from a Scottish Blackface ewe.

2. Assuming the frozen cells can be revived as fully functional cells, they can be fused with an enucleated egg from an Asian elephant, using an electrical pulse. The resulting embryo can be implanted in a surrogate Asian elephant.

The baby produced in this process—referred to as a clone—is genetically identical to the ewe that provided the donor nucleus, not to the egg cell donor or the surrogate mother. To date, reproductive clones have been developed in a variety of mammals, including sheep, pigs, mice, cows, horses, dogs, and cats.

Why would anyone want to clone a sheep, a pig, or a cow? Reproductive cloning can be used to produce multiple copies of an organism that has useful characteristics. For example, a company in South Dakota has cloned calves that are genetically engineered to produce human disease-fighting proteins called immunoglobulins. The ultimate aim is to create herds of genetically identical cows, each of which would serve as a "biological factory," producing large quantities of commercially valuable immunoglobulin proteins.

Reproductive cloning is also being used to produce pigs that could save the lives of people in need of organ transplants. Each year, thousands of people die while waiting for an organ transplant. Pig organs are roughly the same size as human organs and could work well in people, except for one major problem: the human immune system rejects them as foreign. In recent studies, scientists have used reproductive cloning to produce pigs whose organs lack a key protein that stimulates the human immune system to attack.

Conservationists are attempting to produce clones of highly endangered species whose numbers have dwindled to just a few individuals. Several wild cat species from Africa and Asia have been cloned using domestic cats as surrogate mothers. Wild bovines (such as an Asian wild cattle species known as the gaur) and canines (such as the gray wolf) have also been cloned using domesticated species. However, reproductive cloning of some of the most critically endangered species—such as the northern white rhinoceros, only 11 of which exist—is extremely challenging. The process is difficult in such cases partly because we know very little about the reproductive physiology of these species.

Concept Check

1. Was Dolly genetically identical to the egg donor or the surrogate mother or neither?
2. Japanese researchers have identified intact nuclei in cells of a woolly mammoth frozen in permafrost for about 4,000 years (see **FIGURE 13.11**). Explain how that DNA could potentially be used to clone the now-extinct animals. (*Hint:* Asian elephants are the closest living relatives of woolly mammoths).

13.5 Stem Cells: Dedicated to Division

The job of stem cells is to divide. Stem cells are crucial for human development and also for maintaining the body in adulthood. Stem cells helped grow you from a tiny ball of cells to an adult with about 10 trillion cells. And stem cells continue to sustain you on a daily basis by providing new cells to replace damaged or dead cells.

The manipulation of stem cells holds enormous potential for advancing human health, but stem cell research has also stirred impassioned debate and political controversy. In this section we take a closer look at the special characteristics of stem cells, their promise and potential, and also the controversy that swirls around them.

Stem cells are a source of new cells

Stem cells are undifferentiated cells with the extraordinary attribute of *self-renewal*, which means that a stem cell can undergo mitotic divisions to make more of itself. Some of the daughter stem cells generated by mitotic division become stem cells themselves to maintain or grow the stem cell population. However, other daughter cells can graduate to the status of a highly specialized cell type, such as a skin cell or a fat cell (**FIGURE 13.12**).

There are two main classes of stem cells: *embryonic stem cells* and *adult stem cells*. **Embryonic stem cells**, found only in embryos, can potentially give rise to *all* the 220 cell types known to exist in the human body. **Adult stem cells**, also called somatic stem cells, are undifferentiated cells in the adult body that produce only a limited number of different cell types. Brain stem cells, for example, give rise to neurons and two other types of brain cells (oligodendrocytes and astrocytes), but they cannot produce other cells, such as red blood cells or kidney cells.

Embryonic stem cells, and some types of adult stem cells, can be grown in a laboratory procedure known as **cell culture**. When scientists spread the stem cells in a special nutrient broth in a plastic petri dish, the cells undergo mitotic divisions to make more stem cells. Exposing these cells to particular physical and chemical signals induces them to differentiate into specific cell types, such as skin cells or neurons.

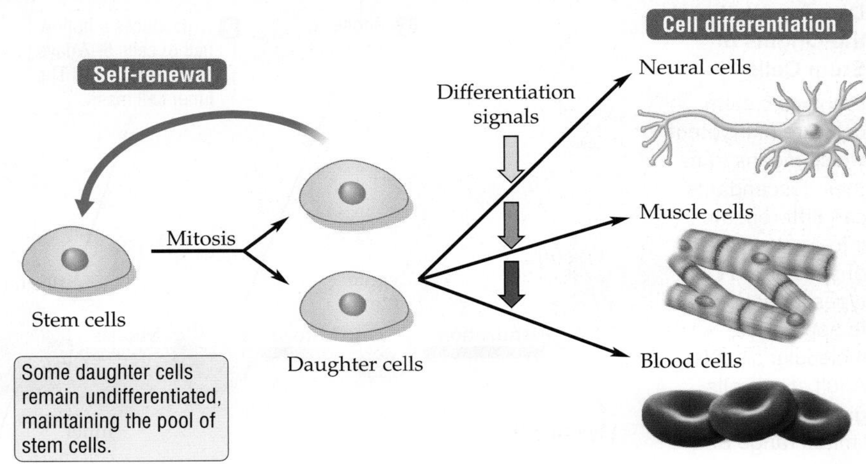

FIGURE 13.12 Stem Cells Can Renew Themselves
Stem cells are undifferentiated cells that have the unique capacity of self-renewal. Given the right mix of physical and chemical signals (differentiation signals), their descendants can differentiate into highly specialized cell types.

Embryonic stem cells have great developmental flexibility

Each of us is a product of fertilization, the union of an egg and sperm that creates a single cell, the **zygote** (**FIGURE 13.13**). It takes several mitotic divisions to convert the zygote into a ball of cells technically known as a *morula*. In the first 3–4 days after fertilization, that ball of cells is composed of **totipotent** (**toh-*TIP*-uh-tunt**) cells, so called because they can give rise to *any* cell type in the organism, including the protective birth sac (in mammals) that surrounds the developing embryo.

As cell divisions continue, the morula develops into a hollow sphere, called the *blastocyst*, with about 150 cells. Inside the blastocyst is a group of about 30 cells known as the **inner cell mass**, which gives rise to the embryo proper (see Figure 13.13). The inner cell mass is **pluripotent** (**pluh-*RIP*-uh-tunt**), which means these cells are capable of producing all of the cell types in the adult body but not, in the case of mammals, the tissues that make up the birth sac. These cells can potentially generate any cell type in the adult body.

Adult stem cells produce a limited number of cell types

Adult (somatic) stem cells are found in newborns, children, and adults. Adult stem cells are known to exist in many human organ systems, including skin, gut, liver, brain, and heart (**FIGURE 13.14**).

Adult stem cells lack both the complete developmental flexibility of the totipotent cells in the morula and the nearly complete flexibility of the pluripotent inner cell mass of the blastula. Adult stem cells are either **multipotent** (**mul-*TIP*-uh-tunt**) or **unipotent** (**yoo-*NIP*-uh-tunt**). Multipotent stem cells can generate a small repertoire of differentiated cell types. One type of bone marrow stem cell (the hematopoietic stem cell) produces at least eight different types of blood cells, including red blood cells and most of the immune cells. Unipotent stem cells produce only one cell type. Skin epidermal stem cells are unipotent because they make cells of just one type: keratinocytes, which make up most of the outer layers of skin.

Compared to embryonic stem cells, adult stem cells are smaller, fewer in number, difficult to identify, and usually difficult to grow in culture.

Helpful to Know

Despite the name, adult stem cells are found in fetuses, children, and adults. Adult stem cells are also known as somatic stem cells. Recall from Chapter 7 that somatic cells are the non-sex cells in the body of a multicellular organism.

Induced pluripotent stem cells are made from differentiated cells

The development of induced pluripotent stem cells is one of the most exciting discoveries in cell biology and biomedicine in recent years. An **induced pluripotent stem cell (iPSC)** is any cell, even a highly

FIGURE 13.13 Human Development and the Origins of Stem Cells

Embryonic stem cells are pluripotent, which means that their descendants can differentiate into a great variety of specialized cell types in response to appropriate molecular signals. Adult stem cells give rise to a more limited range of cell types.

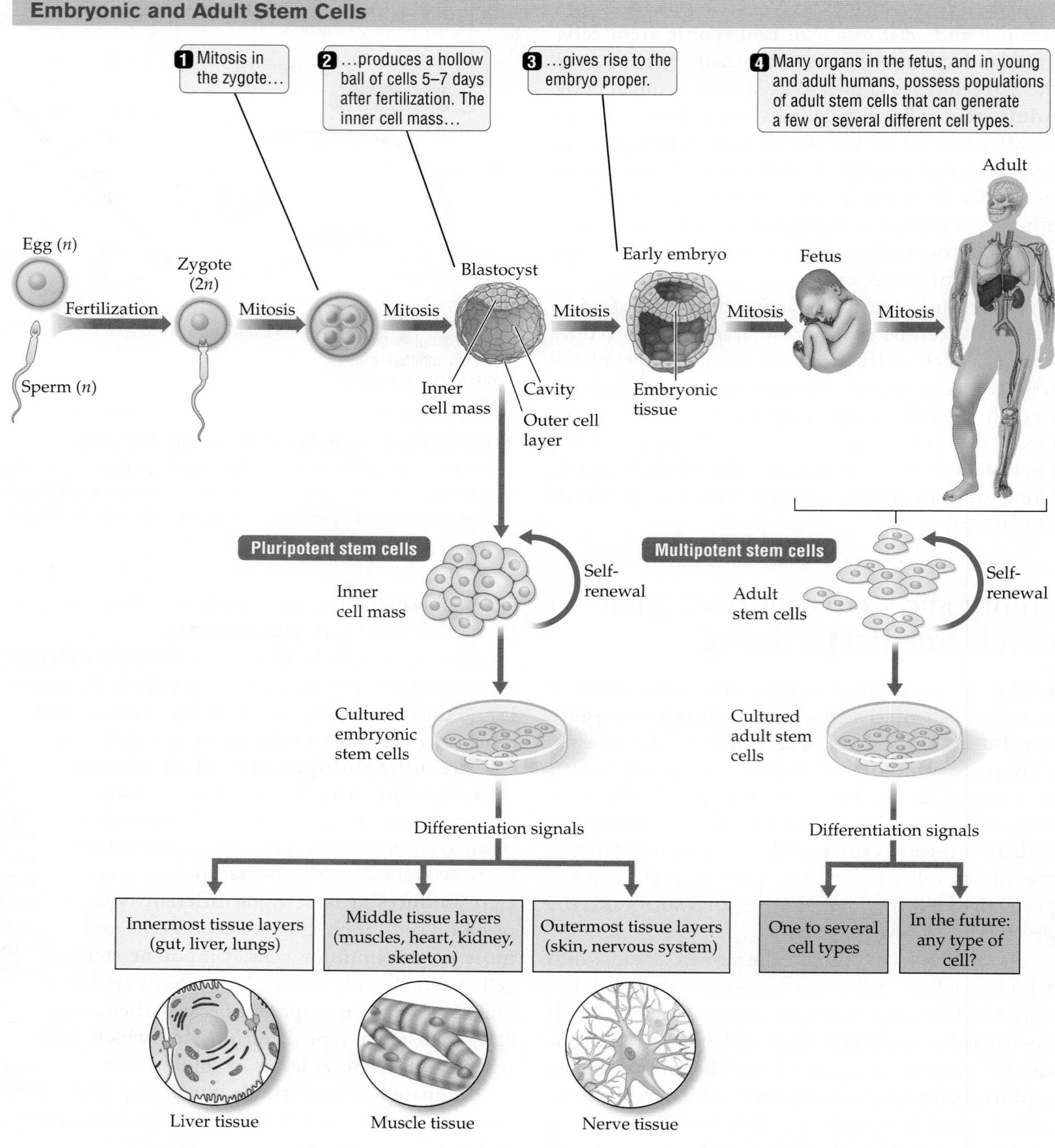

differentiated cell in the adult body, that has been genetically reprogrammed to mimic the pluripotent behavior of embryonic stem cells.

Scientists developed iPSCs by carefully studying the key genes that give an embryonic stem cell its unique properties. In 2006, researchers introduced just four critical genes into mouse skin cells, using harmless viruses as delivery vehicles. The introduced genes forced these specialized cells to start behaving almost exactly like embryonic cells. In 2007, similar techniques were used to create an iPSC line from human skin cells. More recently, iPSC lines have been raised by the introduction of key proteins, not genes, into mature skin cells.

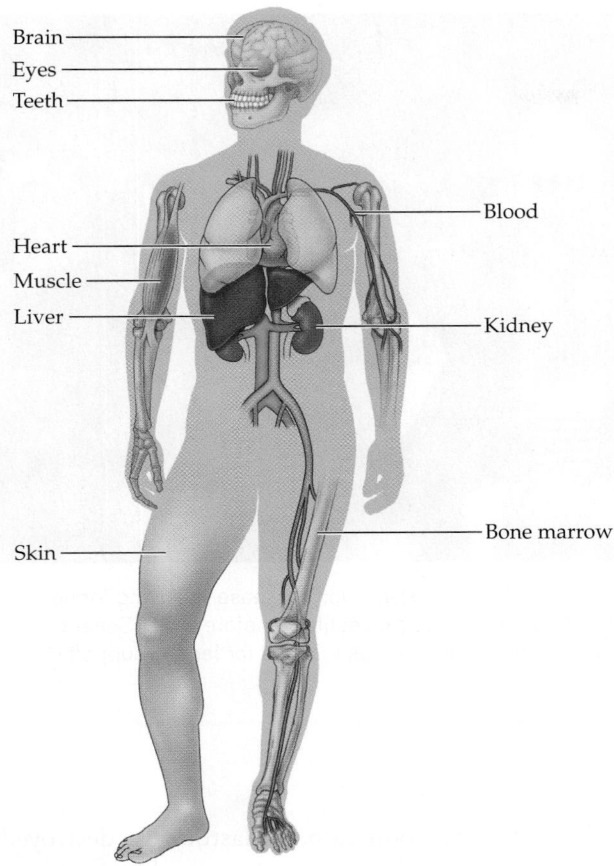

FIGURE 13.14 Small Numbers of Adult Stem Cells Are Found in Certain Tissues and Organs in the Human Body

These cells are multipotent or unipotent, rather than pluripotent like embryonic stem cells. Adult stem cells from most organs (except skin and bone marrow) are harder to identify, isolate, and culture in a lab dish, compared to embryonic stem cells derived from the inner cell mass.

Stem cell technology offers much hope and some successes

Stem cells have taught us much about cell biology and human development. For example, scientists have learned a lot about cell division and cell differentiation by culturing stem cells in a lab dish and exposing them to a variety of differentiation signals, and by tweaking the expression of specific genes using genetic engineering techniques.

Beyond basic research, human tissues made from stem cells can speed up drug discovery and lower the cost of developing new medications (**TABLE 13.2**). Human tissues raised from stem cells can be used to screen candidate drugs for effectiveness or toxicity. For example, a drug company can screen potential heart medications by trying them out on human heart tissue grown from stem cells in a lab dish. Researchers can test for potential side effects by looking for toxicity against other, nontarget, cell types generated from stem cells. Conventional drug testing in lab animals is time-consuming and expensive; the cost of testing a new drug in mice often runs to $3 million or so. Because mice are not people, drugs must also be tested in humans before a pharmacy can dispense them; the price tag for large-scale studies on human individuals often amounts to $1 billion or more. Testing on human stem cells could potentially take less time and lower costs significantly.

Stem cells offer a strategy for repairing or replacing tissues and organs damaged by injury or disease—a field known as **regenerative medicine**. Stem cell technology gives a ray of hope to patients whose tissues are damaged by accidents or a condition such as Parkinson's disease.

Adult stem cells have been used for more than 50 years to treat people with some types of blood disorders, including blood cancers such as leukemia. Adult stem cells are also used in the treatment of severe burns. Skin stem cells can be isolated from any patch of healthy skin that survives on a burn victim, and under the right conditions in the lab the cells quickly generate sheets of skin tissue that can be grafted onto the patient's body. Because it comes from the patient's own stem cells, the lab-grown skin tissue is not attacked by the immune system.

Hundreds of clinical experiments are currently under way in the United States and other countries to test various stem cell therapies, mostly using adult stem cells, although some trials using human embryonic stem cells are also in progress. Stem cell therapy for patients newly diagnosed with type 1 diabetes has been particularly encouraging. The disease damages insulin-producing cells in the pancreas. In a recent trial, children with the disease were treated with adult stem cells that had been multiplied through cell culture. Most of the children were then able to produce enough insulin on their own and no longer needed injections of this vital hormone. Hopes are high that stem cells can be used to treat Parkinson's disease, Alzheimer's disease, stroke, amyotrophic lateral sclerosis (ALS), muscular dystrophy, arthritis, Crohn disease, and some vision and hearing defects. However, much more stem cell research is needed before the hoped-for treatments can be turned into reality.

TABLE 13.2 Applications of Stem Cell Technology

Basic research

Understanding cell division and differentiation

Understanding the biology of stem cells

Understanding the biology of cancer

Understanding human development

Biomedical applications

Drug development: testing new drugs in cultured human tissues

Regenerative medicine: repairing or replacing damaged or diseased tissues

About half a million Americans suffer from Parkinson's disease, including former boxer Muhammad Ali and actor Michael J. Fox. Testifying before a U.S. Senate subcommittee in 2002, the two men joined forces to plead for federal support of stem cell research.

The use of embryonic stem cells is controversial

As we have noted, the only source of embryonic stem cells is the blastocyst derived from a zygote. All human embryonic stem cells are by-products of a form of fertility treatment known as *in vitro fertilization* (*IVF*).

In IVF, egg and sperm are combined in a lab dish and development is allowed to proceed to the blastocyst stage. A cultured blastocyst must be implanted in the woman seeking fertility treatment, or else frozen for later use, because blastocysts cannot be kept alive in a dish indefinitely. Because the process is arduous for the mother-to-be and the failure rate can be high, many eggs are removed from the woman and fertilized in vitro (literally, "in glass"), although only one or a few blastocysts can be implanted in her womb. The "extra" embryos are frozen in case they are needed later. More than 400,000 human blastocysts are held in deep freezers at fertility clinics across the United States (**FIGURE 13.15**). All human embryonic stem cells currently in use in research and therapy are derived from such "extra" embryos donated by women who used assisted reproduction. A blastocyst is destroyed when embryonic stem cells are harvested from it.

Opponents of human embryonic stem cell technology believe that the life of a human being begins at conception. They contend that a blastocyst has moral status and that it is unethical to use its cells for someone else's benefit. Advocates, in contrast, say the blastocyst is just a ball of undifferentiated cells, that the donated blastocysts would be "wasted" anyway, and that parents have the right to donate their frozen tissue to save other people's lives.

The development of induced pluripotent stem cells (iPSCs) offered the hope that these controversies would be laid to rest because iPSCs have broad developmental potential and are obtained from differentiated cells, not blastocysts. However, the technology is still in its infancy, and many advances are needed before iPSC-based therapies can be tested in human subjects. Stem cell researchers emphasize the vital importance of studying all types of stem cells, because each avenue of study has added, and continues to add, to our understanding of how, when, and where cells divide or differentiate.

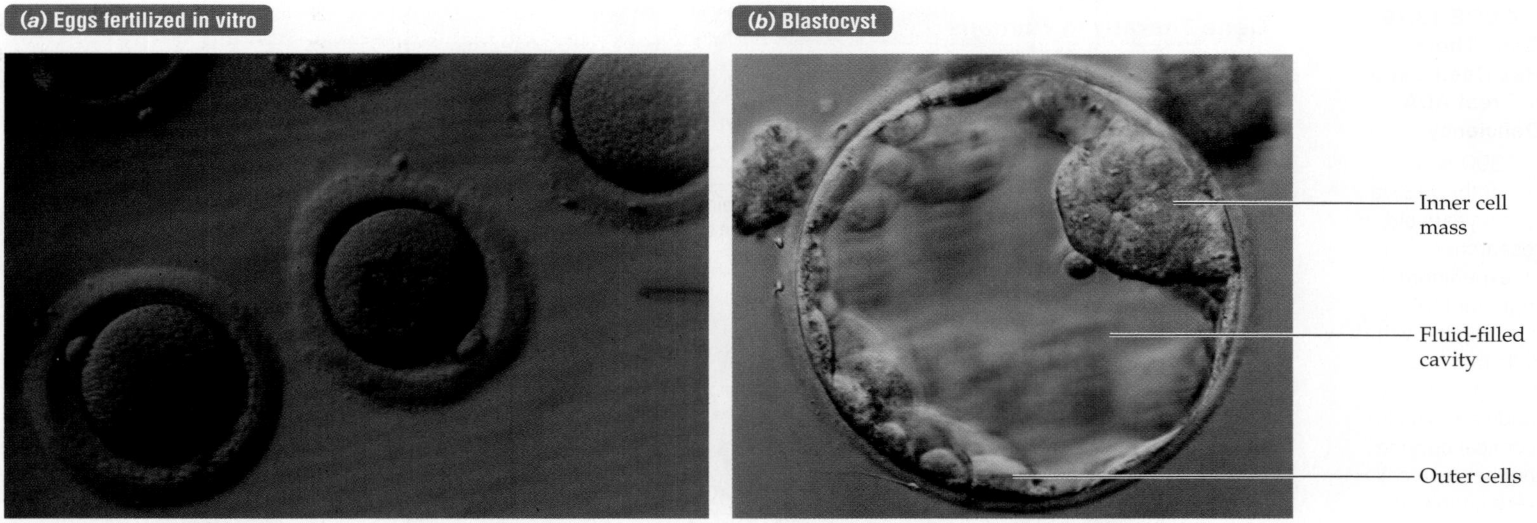

(*a*) Eggs fertilized in vitro

(*b*) Blastocyst

Inner cell mass

Fluid-filled cavity

Outer cells

FIGURE 13.15 Embryonic Stem Cells Come from Blastocysts
During in vitro fertilization (IVF), eggs are fertilized by sperm in a lab dish (*a*). After 5–6 days, mitotic divisions have transformed the zygote into a blastocyst (*b*). Blastocysts cannot survive in a lab dish beyond this point, and they must either be implanted into the uterus of the mother-to-be or else preserved by freezing. The inner cell mass is the group of about 30 cells inside the blastocyst that the fetus develops from if the blastocyst is implanted. These cells, scooped out from previously frozen blastocysts, are also the source of embryonic stem cells.

Concept Check

1. What are the unique properties of a stem cell?
2. Compare the developmental flexibility of adult stem cells to that of embryonic stem cells.

Concept Check Answers

1. A stem cell is undifferentiated, it can renew itself, and under the right conditions, its descendants can differentiate into specialized cell types.
2. An adult stem cell generates a relatively small number of specialized cell types; an embryonic stem cell can give rise to any of the cell types in the adult body.

13.6 Human Gene Therapy

On September 14, 1990, 4-year-old Ashanthi DeSilva made medical history when she received intravenous fluid that contained genetically modified versions of her own white blood cells. She suffered from adenosine deaminase (ADA) deficiency, a genetic disorder caused by a mutation in a single gene that affects the ability of white blood cells to fight off infections. Earlier, doctors had removed some of Ashanthi's white blood cells and added the normal ADA gene to them, in an attempt to fix the lethal genetic defect through genetic engineering. Ashanthi responded very well to the treatment and now leads an essentially normal life (**FIGURE 13.16**).

The treatment that Ashanthi received was the first gene therapy experiment on a person. Human **gene therapy** seeks to correct genetic disorders through genetic engineering and other methods that can alter gene function. The possibility of curing even the worst genetic disorders by reaching into our cells and restoring gene function, or turning off a troublesome gene, is a bold and captivating prospect.

Gene therapy has faced setbacks

Overall, more than 600 gene therapy experiments have been conducted worldwide, but the early trials were fraught with risk. As an example of unexpected setbacks, between 1999 and 2002, researchers attempted gene therapy as a cure for X-SCID—a disease similar to ADA deficiency that cripples the immune system. They inserted a healthy version of an X-linked gene into bone marrow cells. Of nine children with X-SCID who were treated by gene therapy alone, eight were cured. For the first time, scientists had achieved the holy grail of gene therapy: they had cured a human genetic disorder solely by fixing the gene that caused it.

FIGURE 13.16 Gene Therapy Has Been Used to Treat ADA Deficiency

In 1990, when Ashanthi DeSilva was 4 years old, researchers tried an experimental treatment (*a*), injecting her with cells that had been genetically modified to make a critical enzyme that her own cells were unable to make. (*b*) Today, DeSilva is a healthy young woman with a career in international development.

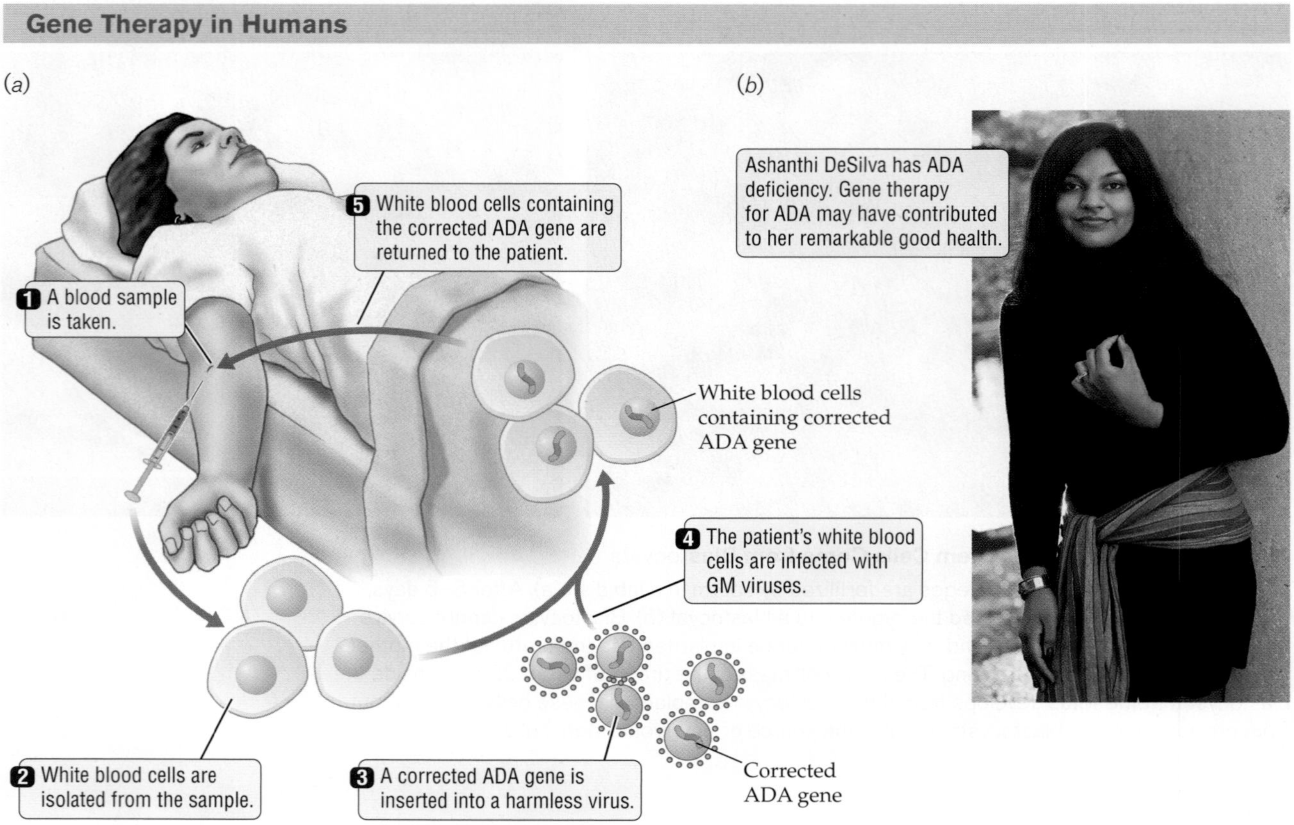

Gene Therapy in Humans

(*a*)

1 A blood sample is taken.

2 White blood cells are isolated from the sample.

3 A corrected ADA gene is inserted into a harmless virus.

4 The patient's white blood cells are infected with GM viruses.

5 White blood cells containing the corrected ADA gene are returned to the patient.

White blood cells containing corrected ADA gene

Corrected ADA gene

(*b*)

Ashanthi DeSilva has ADA deficiency. Gene therapy for ADA may have contributed to her remarkable good health.

Unfortunately, four of the children who were cured of X-SCID went on to develop leukemia, a form of cancer that strikes white blood cells. One of these children died. What went wrong? By accident, the "good copy" of the gene was inserted near and promoted the expression of a gene that causes cells to divide rapidly, increasing the risk of cancers such as leukemia.

The four cases of leukemia followed close on the heels of another terrible outcome: in 1999, a young man participating in a gene therapy experiment died from an allergic reaction to the virus used to deliver the engineered gene. The combined effect of these tragedies sent shock waves through the gene therapy field. Worldwide, many gene therapy trials were placed on hold, and there were calls to abandon gene therapy efforts altogether.

Scientists have pressed on, however, focusing their attention on the biggest hurdle: finding a way to deliver the engineered gene safely and effectively to just the right cells. The "Biology in the News" feature on page 305 describes some of these new strategies, and how these techniques have revived interest in delivering gene therapy to patients on a commercial scale.

RNA interference is a tool for shutting down problem genes

In just the past few years, *RNA interference* has emerged as a tremendously promising tool for DNA technology, including for conducting gene therapy. **RNA interference (RNAi)** selectively blocks the expression of a given gene. In RNAi, small chunks of RNA are able to silence genes that share nucleotide sequence similarity with them. The mechanism is used by many plants and animals to regulate their own genes, and it can be manipulated by DNA technologists to turn genes off in a target cell or organism.

RNAi offers a potential cure for genetic disorders caused by the inappropriate activity, or overactivity, of a gene. Clinical trials are currently under way to test the effectiveness of RNAi for shutting off viral genes in people who are infected with hepatitis B or have RSV pneumonia, a lethal viral infection. RNAi-based gene therapy is also being tested against the "wet" form of macular degeneration, which is caused by overgrowth of blood vessels in the eye.

Although viral vectors were employed in most of these trials, therapeutic RNA could also be delivered by direct injection into the cytoplasm or by packaging of the RNA into special lipids or polymers. These alternative methods for delivering gene-silencing RNA are under intense investigation because if the delivery system works, there will be no need to insert any foreign DNA into the recipient's DNA.

Concept Check

1. What is gene therapy?
2. What are the main challenges in successful gene therapy in humans?

13.7 Ethical and Social Dimensions of DNA Technology

DNA technology provides many benefits to human society. At the same time, the immense power and scope of genetic engineering raises ethical concerns in the minds of some and poses potential risks, especially to the genetic integrity of wild populations. At the most basic level, some people ask how we can assume we have the right to alter the DNA of other species. Others see no ethical conflict in altering the DNA of a bacterium or a virus but object to changing the genome of a food plant or of an animal such as a dog or a chimpanzee.

Some worry about GMO effects on human health

Few people find fault with the use of GM bacteria to produce lifesaving pharmaceuticals such as insulin for diabetics, blood-clotting proteins for hemophiliacs, and clot-dissolving enzymes for stroke victims. GM food crops, on the other hand, are bitterly opposed by some groups, especially in Europe.

In some European countries, foods containing GM products must be labeled as such, kept separate from non-GM products, and monitored through the entire food production chain. Critics of GM foods worry that the presence of a GM protein in a common food might cause a severe allergic reaction in an unsuspecting consumer. Proponents counter that no adverse reactions have been authenticated in the United States, where millions of people have been eating GM foods for more than a decade. Also, GM foods must be extensively tested for safety, including their allergic potential; these standards are not applied to new crops developed through conventional breeding.

EXTREME BURGER

Dutch scientists removed a few muscle stem cells from beef cattle and multiplied them in a lab dish. It took 3 weeks and $330,000 to grow enough of the cells to make a 5-ounce patty. Food critics who tried the grilled lab-grown fat-free burger pronounced it "close to meat, but not as juicy."

Harm to the integrity of natural ecosystems is another concern

Some environmentalists worry that engineering crops to be resistant to herbicides might promote increased use of herbicides, some of which could be harmful to the environment. Supporters of GMO technology say that herbicide-resistant plants would be good for soil health because farmers would not have to use soil-damaging tilling methods to control weeds.

Many crops, including corn and cotton, have been engineered to produce Bt toxin, named after the bacterium (*Bacillus thuringiensis*) in which it was discovered. Laboratory studies by Cornell University researchers show that corn pollen containing this protein insecticide is harmful to several insects, including monarch butterflies. However, field studies by other researchers found no evidence of ill effects on monarch butterflies or other nontarget insects.

Critics of genetic engineering have long argued that genes from genetically modified plants or animals could spread to wild species, potentially wreaking environmental havoc. Some have expressed concern that GMOs will escape from the bounds of farm fields, barnyards, and fish pens to contaminate natural ecosystems with genomes that have been altered by humans. Escaped GM salmon could threaten wild fish stocks not only by interbreeding with them and thereby reducing their natural diversity, but also by outcompeting them for resources and thereby driving them toward extinction.

Of the world's 13 most important crop plants, 12 (all except corn) can mate and produce offspring with a wild plant species in some region where they are grown.

Concept Check Answers

1. In gene therapy, a genetic disorder caused by a faulty gene is cured by alteration of the patient's DNA using the tools of genetic engineering; for example, by the insertion of a working copy of the gene into the patient's genome.

2. Challenges of gene therapy include delivering the transgene in a safe and effective way, especially preventing the gene from being inserted in the recipient's DNA in a disruptive way, and preventing allergic reactions to the vector used for gene delivery.

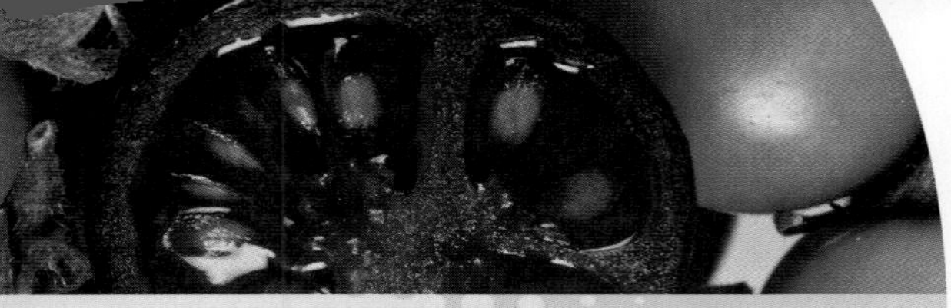

EXTREME TOMATO

Scientists in England used genetic engineering techniques to cause a purple pigment (anthocyanin) to accumulate in the flesh of tomato fruits. The GM tomatoes have twice the shelf life of regular red tomatoes. Anthocyanins are potent antioxidants, and a diet of dried purple GM tomatoes prolonged the lives of mice that were genetically susceptible to cancer.

If a crop plant is genetically engineered to be resistant to an herbicide, the potential exists for the resistance gene to be transferred (by mating) from the crop to the wild species. There is a risk that by engineering our crops to resist herbicides, we will unintentionally create "superweeds" resistant to the same herbicides. One survey of roadside canola weeds in North Dakota found that 45 percent of them were GMO escapees from farm fields.

Some oppose GMOs on sociopolitical grounds

In some circles, the heated debate over GMOs tends to center on political and socioeconomic issues. The use of "terminator genes" in a GM plant can theoretically prevent that plant from making viable seed. Proponents say that "terminator technology" would be an effective barrier against the runaway spread of GMOs. Opponents of GMOs see this technology as a veiled attempt by seed companies to control the supply of seed, because farmers would have to buy seed from the company each year instead of saving their own.

Critics of genetic engineering also point to the social costs of using such technology. They cite bovine growth hormone (BGH), which is mass-produced by GM bacteria, as a case in point. Among its other effects, BGH increases milk production in cattle. Before the introduction of genetically engineered BGH in the 1980s, milk surpluses were already common. The use of BGH by large milk producers has created even larger milk surpluses, driving down the price of milk and forcing many small producers of milk—the traditional family farms—out of business.

As some see it, the lower milk prices for consumers are not worth the social cost of driving small dairy farms into bankruptcy. Others believe that all types of commercial enterprises, including family farms, should sink or swim on their own, and that unsuccessful businesses should not be rescued because of social sentiment. As these examples illustrate, the debate over the social dimensions of GMOs often gets caught up in a much larger discussion about the politics of food and the economics of modern agriculture.

Concept Check Answers

1. The products of the transgene could produce an allergic reaction in a consumer. Defenders say GMO foods are extensively tested, more so than conventional foods, and no cases of GMO-caused food allergies have been scientifically verified.
2. Escaped GMOs could alter natural biodiversity by contaminating natural ecosystems with human-made genomes and potentially driving wild species to extinction; interbreeding of GM crops with wild relatives could lead to "superweeds" that could be difficult to eradicate.

Concept Check

1. What potential health risks of GMOs have been suggested? What arguments do defenders of GMOs put forth?
2. Explain why the release of GMOs could pose potential risks to natural ecosystems.

13.8 A Closer Look at Some Tools of DNA Technology

As researchers have come to understand the natural processes by which viruses and cells alter and maintain DNA and produce enzyme products, they have been able to "borrow" these same enzymes to manipulate DNA in a test tube. Some of these enzymes can function under extreme conditions, such as at near-boiling temperatures, because the organisms from which they are derived are adapted to living in some of the most hostile habitats on the planet. As you read the details of some of the procedures that are widely used in DNA technology, keep in mind that most of these "tricks" are inspired by nature. Just about everything in the DNA technologist's tool chest comes from living organisms or viruses and has therefore been honed by evolutionary processes.

Enzymes can be used to cut and join DNA

Each of us has 3.3 billion base pairs of DNA on 23 unique chromosomes. The DNA molecule in each chromosome is so large (140 million base pairs, on average) that after DNA has been extracted from a cell, it must be broken into smaller pieces before it can be analyzed further. DNA can be split into more manageable pieces by **restriction enzymes**, which cut DNA at highly specific sites. Discovered in the late 1960s, restriction enzymes appear to have evolved in bacteria, where they do battle against foreign DNA, such as viral DNA. Infecting viruses begin by injecting their DNA into a bacterium. The bacterium then deploys its restriction enzymes in an attempt to chop up the viral DNA, thereby "restricting" viral growth.

Researchers have isolated hundreds of restriction enzymes, each of which recognizes and cuts DNA

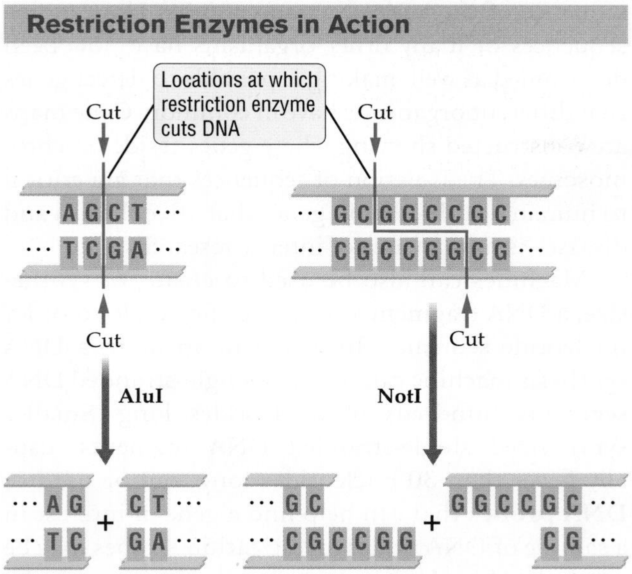

Restriction Enzymes in Action

Locations at which restriction enzyme cuts DNA

Cut

AluI

NotI

FIGURE 13.17 Restriction Enzymes Cut DNA at Specific Places

The restriction enzyme AluI specifically binds to and then cuts the DNA molecule wherever the sequence AGCT occurs. Another restriction enzyme, NotI, specifically binds to and cuts the DNA sequence GCGGCCGC. Each enzyme binds to and cuts its own special target sequence, and no other.

wherever a unique target sequence is present, and it cuts the DNA only at those sites. A restriction enzyme called AluI, for example, cuts DNA everywhere the sequence AGCT occurs, but nowhere else (**FIGURE 13.17**).

A **DNA ligase** is an enzyme that joins two DNA fragments. DNA ligases are commonly used to insert one piece of DNA, such as a human gene, into another DNA molecule, such as a plasmid extracted from bacterial cells (as illustrated in Figure 13.2), creating a recombinant DNA molecule.

The gene coding for green fluorescent protein (GFP) can be cut out of jellyfish DNA with restriction enzymes. It is then "pasted" into a bacterial plasmid with DNA ligase, creating a recombinant plasmid. The recombinant plasmid can then be inserted into bacterial cells by means of genetic engineering, and if the recombinant plasmid is constructed in the right manner, the resulting GM bacteria will glow when exposed to blue light.

Gel electrophoresis sorts DNA fragments by size

Once a DNA sample has been cut into fragments by one or more restriction enzymes, researchers often use a process called *gel electrophoresis* to help them see and analyze the fragments. In **gel electrophoresis**, fragments of DNA are placed into "wells" in a gelatin-like slab called a "gel." When an electrical current passes through the gel, it causes the DNA fragments (which have a negative electrical charge) to move toward the positive end of the gel (**FIGURE 13.18**). Long fragments of DNA pass through the gel with more difficulty, and therefore they move more slowly than short pieces.

The distance a fragment travels through the gel is related to its speed of movement. After a fixed time period, the shorter, more rapidly moving fragments are

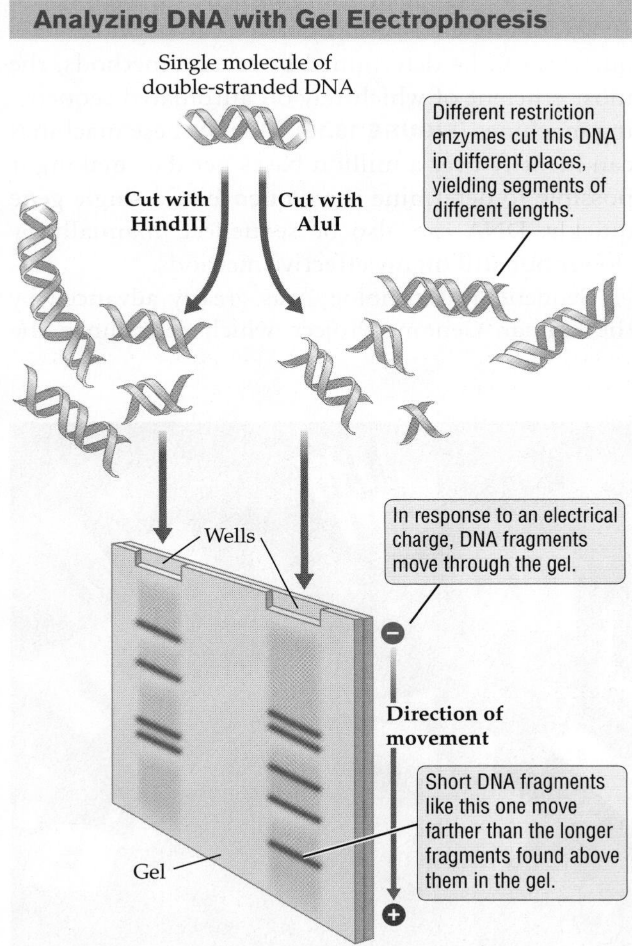

FIGURE 13.18 DNA Fragments Can Be Separated by Gel Electrophoresis

Gel electrophoresis is used to separate and visualize biomolecules. When subjected to an electrical current, DNA fragments move through a gel at different rates, depending on their size. Larger molecules move slowly, while smaller molecules move quickly. To enable visualization of the DNA, the gel is commonly stained with a DNA-binding dye, ethidium bromide, which glows pink when exposed to ultraviolet light.

found toward the positive end of the gel, and the longer, more slowly moving fragments are located closer to the negative end. Because DNA is invisible to the human eye, the fragments must be stained or labeled before they can be seen. DNA fragments having a particular length can be isolated in this way and retrieved from the gel for further analysis.

DNA sequencing and DNA synthesis are key tools in biotechnology

DNA sequencing enables researchers to identify the sequence of nucleotides in a DNA fragment, a gene, or even the entire genome of an organism. Sequences can be determined by several methods, the most efficient of which rely on automated sequencing machines (**FIGURE 13.19**). One of these machines can identify over a million bases per day, making it possible to determine the sequence of a single gene quickly. DNA can also be sequenced manually by slower but still highly effective methods.

Sequencing technology was greatly advanced by the Human Genome Project, which determined the sequence of the entire human genome. The genome sequences of many other organisms have now been determined as well, making it possible to detect genes that different organisms have in common. Gene maps are constructed showing where genes reside on chromosomes. The isolation of sequences that are critical to human health—that is, genes that affect health and disease—is now an area of intense research.

Machines can also be used to create, or synthesize, a DNA fragment with a specific, made-to-order nucleotide sequence. In less than an hour, a DNA synthesis machine can produce single-stranded DNA segments hundreds of nucleotides long. Smaller synthesized single-stranded DNA segments, usually fewer than 30 nucleotides long, can be used as **DNA probes** that can help find a gene of interest in a sample of DNA under investigation. Probes may be synthesized in the laboratory, with a sequence complementary to a target DNA sequence.

In **DNA hybridization**, the target DNA is cut into fragments by restriction enzymes and converted to a single-stranded form by treatments that break the hydrogen bonds holding the two strands of DNA together. If the single-stranded probe encounters a complementary sequence of single-stranded DNA in the target, it will bind to it (a process called hybridization). The probe can be labeled with a radioactive or fluorescent tag to make it easier to identify the DNA fragments to which it binds. PCR is used to amplify small quantities of target DNA.

The **polymerase chain reaction** (**PCR**) uses a unique, heat-stable type of DNA polymerase to make billions of copies of a targeted sequence of DNA in just a few hours. To amplify a piece of DNA by PCR, researchers must use two short segments of synthetic DNA, called **DNA primers**. Each primer is designed to bind to (hybridize with) one of the two ends of each strand of the target DNA by complementary base pairing. By the series of steps shown in **FIGURE 13.20**, DNA polymerase then produces many copies of the DNA that is flanked by the two primers. To amplify a target DNA, scientists must have at least some information about its nucleotide sequence. Without this knowledge, they cannot synthesize the specific primers required in every PCR reaction.

The power of PCR technology lies in the fact that it can amplify extremely small amounts of DNA—amounts extracted from just a few cells or a single blood stain, for example. As a result, PCR has come to be widely used in basic research and in fields as diverse as medical diagnostics, forensics, paternity test-

FIGURE 13.19 Machines Can Sequence DNA

Automated DNA sequencing machines can rapidly determine the nucleotide (base) sequence of a DNA fragment. Pictured here, a scientist examines a computer display showing part of an automated DNA sequencing gel. Each of the four chemical bases in DNA is represented by a different color (red, green, blue, or yellow).

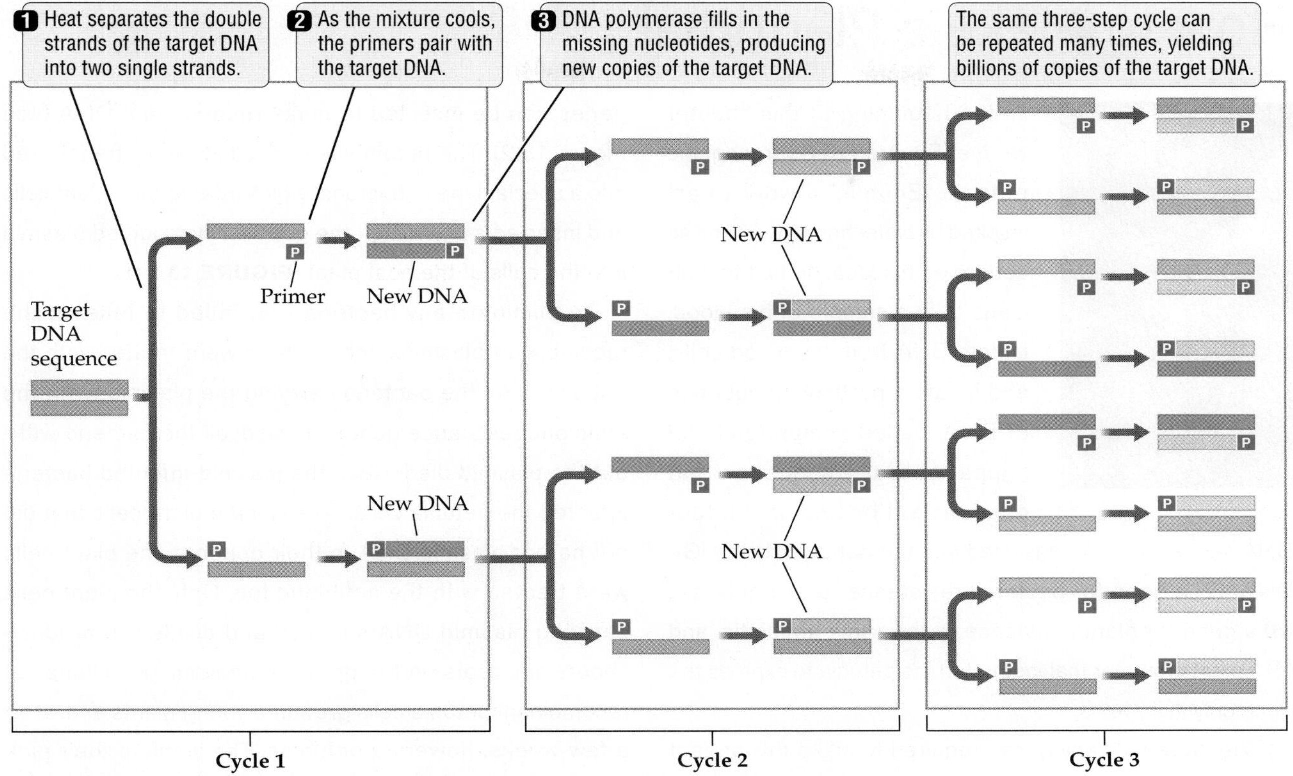

The Polymerase Chain Reaction
1 Heat separates the double strands of the target DNA into two single strands.
2 As the mixture cools, the primers pair with the target DNA.
3 DNA polymerase fills in the missing nucleotides, producing new copies of the target DNA.
The same three-step cycle can be repeated many times, yielding billions of copies of the target DNA.
Target DNA sequence
Primer
New DNA
P
Cycle 1
Cycle 2
Cycle 3

FIGURE 13.20 Small Amounts of DNA Can Be Amplified More Than a Millionfold through PCR

Short primers (red) that can pair with the two ends of a gene of interest (bluish gray) are mixed in a test tube with a sample of the target DNA, the enzyme DNA polymerase, and all four nucleotides (A, C, G, and T). A machine then processes the mixture through the three steps shown, in which the temperature is first raised and then lowered, to double the number of double-stranded versions of the template sequence. The doubling process can be repeated many times (only three cycles are shown here). For clarity, color coding is used to identify the template DNA sequence in the first cycle (bluish gray), the DNA that is newly made in each cycle (shades of orange), and the primers (red).

ing, paleoanthropology, and the authentication of delicacies such as caviar and expensive vintage wine. The technique became so successful so quickly that in 1991, only 6 years after the first paper on PCR was published, the PCR patent was sold for $300 million. Researcher Kary Mullis was awarded a Nobel Prize in 1993 for the discovery of PCR.

Concept Check

1. What roles do restriction enzymes and DNA ligases play in DNA technology?
2. Can PCR be used to amplify DNA whose nucleotide sequence is unknown?

Concept Check Answers

1. A restriction enzyme recognizes a highly specific nucleotide sequence in a target DNA and cuts the DNA at a site within that sequence. DNA ligases can join fragments of DNA to create recombinant molecules.

2. No. Some sequence information is necessary to design primers specific to the DNA target; all humans, for example, have at least some DNA sequences targeted by STR (short tandem repeat) primers.

APPLYING WHAT WE LEARNED

How to Make a Plantimal

At the beginning of this chapter we met Eduardo Kac's transgenic petunia "Edunia," a work of art inspired by biotechnology. After Kac conceived his idea, he had technicians take a sample of his blood, isolate DNA from his blood cells, and locate a particular sequence of DNA, called immunoglobulin kappa, or IGκ, a sequence that encodes antibodies. In all, four DNA sequences were inserted into the petunia: (1) the IGκ DNA; (2) a gene for bacterial resistance to an antibiotic; (3) a gene for plant resistance to the same antibiotic; and (4) a plant promoter that instructed the petunia to express the gene only in its veins.

Why were so many genes required to make this project work? Let's look at the details. The carrier for the genes was a plasmid, a small circle of DNA into which foreign genes can be inserted to make recombinant DNA (see Figure 13.2). The recombinant plasmids were transferred into a special type of bacteria, which infected the plant cells and inserted a portion of the genetically modified plasmid into the cells of the host plant (**FIGURE 13.21**).

To eliminate any bacteria that failed to take up the recombinant plasmids, the bacteria were treated with the antibiotic. All the bacteria carrying the plasmid (with the antibiotic resistance gene) survived; all the bacteria without the plasmid died. Next, the plasmid-infected bacteria infected the petunia cells. To eliminate plant cells that did not harbor plasmid DNA in their genome, the plant cells were treated with the antibiotic too. Only the plant cells carrying plasmid DNA survived and divided to produce shoots and roots on the growing medium. Tiny clumps of recombinant petunia cells grew into young plants and, after a few weeks, flowering petunias. The veins in their pink petals express Kac's immunoglobulin gene, uniting plant and human genes in a single organism.

FIGURE 13.21 Still Life with GM Petunias
Agrobacterium tumefaciens is a natural genetic engineer. When this bacterium infects a plant, it transfers a segment of plasmid DNA, called T-DNA, into the host plant's cells. Scientists can splice genes of interest into this T-DNA—even genes that code for an artist's immunoglobulins. The T-DNA is engineered to also carry a gene for resistance to an antibiotic such as kanamycin. If T-DNA integrates successfully into the nucleus of a plant cell, that cell will survive in the presence of the antibiotic, but cells lacking T-DNA will die. The photo shows the main steps in producing GM petunias using *Agrobacterium* as a vector for genetic modification: (*a*) Strips of petunia leaves were dipped in a broth of *Agrobacterium* cells that carry a genetically engineered plasmid. The leaf strips are placed on a nutritive medium with hormones that cause shoots to form. The medium also contains kanamycin, so plant cells that have received the T-DNA are likely to survive and proliferate but not so the cells that lack T-DNA. (*b*) Three weeks later, many shoots (and a few roots) have started forming on the cut edges of the leaf strips, where plant cells became infected with *Agrobacterium*. (*c*) The shoots were transferred to a medium with hormones that foster root development. After two to three weeks, the resulting plantlets can be transferred to soil.

BIOLOGY IN THE NEWS

Gene Therapy Shows New Signs of Promise

BY SCOTT KIRSNER • *Boston Globe*, June 2, 2013

Alan Smith can still recall his excitement, in the early 1990s, over early experiments in lab rats that demonstrated gene therapy's potential power to attack diseases such as cystic fibrosis.

Smith, a former chief scientific officer at Genzyme, the Cambridge biotech company, remembers feeling like the experiments were "steppingstones" to developing a whole new wave of medicines for untreatable diseases . . .

But more than two decades later, there is still no gene therapy that has won approval from the US Food and Drug Administration, and Genzyme gave up on it as a potential approach to treating cystic fibrosis . . . In 2013, though, there are signs that patients might soon start benefiting from gene therapies. A Cambridge company, Bluebird Bio, last month filed to sell stock to the public; it hopes to raise $86 million to bring to market a gene therapy that would treat a rare, fatal neurodegenerative disease known as CCALD . . .

How exactly do these new therapies work? Instead of a pill or an injection to treat a chronic ailment, many of the gene therapy approaches essentially try to install a microscopic factory inside your body. Its job is to crank out a missing enzyme or therapeutic protein continually, over the course of years. Studies in primates have seen these factories operate for more than a decade.

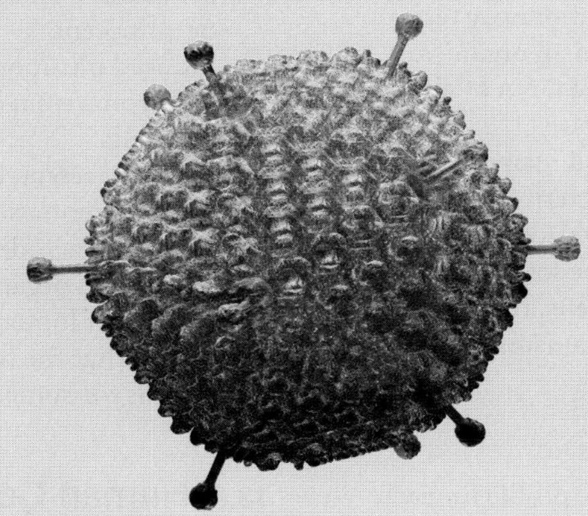

Model of adenovirus, a common gene therapy vector.

The factories themselves are created by using disarmed viruses—the same ones that might ordinarily give you the flu—that have been packed with custom-crafted DNA or RNA. They infiltrate cells in your body and tell them exactly what to make . . . These viruses can be delivered to the body by injection or inhalation, or by removing cells from the patient, exposing them to the virus, and reintroducing them to the patient.

Bluebird takes that last approach with a product it is developing for childhood cerebral adrenoleukodystrophy [CCALD], a disease that affects boys between the ages of 4 and 10, usually leading to a vegetative state and death. (It was featured in the movie "Lorenzo's Oil.") . . .

Other companies are focusing on diseases that affect larger patient populations. Genzyme . . . is working on a treatment for age-related macular degeneration, a common cause of blindness in people over 50, and for Parkinson's, a neurodegenerative disease. Between 7 million and 10 million people worldwide suffer from Parkinson's according to the Parkinson's Disease Foundation, a patient advocacy group.

In Parkinson's . . . there is already evidence that the gene treatment may have an effect for as long as five years, helping supply an enzyme lacking in those who have Parkinson's and allowing them to stay on lower doses of a drug they use called Levodopa.

As discussed in this chapter, the great challenge in gene therapy is to find a way to deliver the engineered gene safely and effectively to just the right cells. Harmless viruses are often deployed as delivery vehicles (technically, vectors), but the use of viruses for this purpose presents challenges. The human body defends itself so well against viruses that the recombinant viruses are often destroyed before they can deliver a "good copy" of the gene to enough of the target cells. Researchers at Cedars-Sinai Medical Center have developed a particularly stealthy version of a harmless virus that is almost invisible to the body's defense system. The new vector was put to the test in clinical trials aimed at treating Parkinson's disease. Worrying about unforeseen consequences, the researchers applied the gene therapy to only one side of each patient's brain. One year later, brain scans revealed improved brain activity in the treated side of the brain, while the untreated side showed a decline, compared to the pretrial status of brain function.

Scientists are also working on strategies to prevent the introduced gene from being inserted into the wrong tissues or in a wrong location within the DNA of the target cells (one that could increase the risk of cancer, for example, as it did with X-SCID, described earlier). Other scientists have focused on trying to understand how introduced genes are integrated into the DNA in human cells. This new understanding, and the improved techniques, are fueling the resurgence in gene therapy that the article describes.

Evaluating the News

1. Explain how viruses can be used "for good, not evil" in gene therapy. Should we worry that the viruses used for gene therapy will make the patient sicker? Explain.

2. There are no cures, and very little by way of treatment, for diseases as destructive and fatal as cerebral adrenoleukodystrophy. Do you think it is right to accept a higher risk level in gene therapy for such patients? Explain your views.

CHAPTER REVIEW

Summary

13.1 The Brave New World of DNA Technology

- Scientists can manipulate DNA using a variety of laboratory techniques. Because the structure of DNA is the same in all organisms, these techniques work in much the same way on the DNA of all species.
- A gene is said to be cloned if it has been isolated and many copies of it have been made in a suitable host cell, such as a bacterium.
- A recombinant DNA molecule is produced when DNA from one source is spliced into a DNA molecule from another source.
- Genetic engineering introduces cloned DNA into cells, tissues, or whole organisms to create genetically modified organisms (GMOs).

13.2 DNA Fingerprinting

- DNA fingerprinting, which creates a unique genetic profile for each individual, is widely used in criminal cases to establish the identities of victims and criminals.
- DNA fingerprinting of humans makes use of highly variable regions of the genome. PCR can be used to generate a pattern of amplified DNA fragments that is likely to vary from person to person.

13.3 Genetic Engineering

- In genetic engineering, a DNA sequence (often a gene) is isolated, modified, and inserted back into the same species or into a different species.
- Genetic engineering is used to alter the phenotype (especially the performance or productivity) of the GMO or to produce many copies of a DNA sequence, a gene, or a gene product.
- Genetic engineering is used commonly to enhance the quality of economically important species and to produce pharmaceutical products and vaccines in large quantity.

13.4 Reproductive Cloning of Animals

- The goal of reproductive cloning is to produce an offspring (baby) that is a genetic copy of a selected individual. Dolly the sheep was the first mammal to be created through reproductive cloning.
- In one method of reproductive cloning, the nucleus is removed from an egg cell, and an electrical current is used to fuse the enucleated egg with a somatic cell from a nuclear donor. The product of cell fusion forms an embryo that is then transferred to the uterus of a surrogate mother.
- Reproductive cloning can be used to produce multiple copies of an organism that has useful characteristics, or to conserve rare and endangered animals.

13.5 Stem Cells: Dedicated to Division

- Stem cells are undifferentiated cells that renew themselves and can generate descendants that differentiate into specialized cell types.
- Embryonic stem cells, derived from the inner cell mass in the blastocyst, are pluripotent, giving rise to all cell types in the body.
- Adult stem cells, which arise in the fetus and persist in small numbers in various tissues and organs in children and adults, are multipotent or unipotent.
- Induced pluripotent stem cells have the developmental flexibility of embryonic stem cells but are made by reprogramming fully differentiated cells.
- Stem cell research advances our basic understanding of cell division and differentiation, as well as drug discovery and screening. Stem cell therapies include tissue engineering to repair and replace damaged tissues and organs.
- The use of human embryonic stem cells is controversial. Opponents believe that the life of a human being begins at conception and using its cells for someone else's benefit is unethical.

13.6 Human Gene Therapy

- The goal of human gene therapy is to correct genetic disorders through genetic engineering and other methods that can alter gene function.
- Gene therapy in humans continues to present challenges, but many new advances have been made recently, especially in developing safe procedures for introducing genes into the patient; hundreds of gene therapy trials are currently under way.
- Few people object to the use of gene therapy to cure severe genetic disorders. The technology, however, faces many challenges, especially in finding safe and effective means for introducing a cloned gene into a patient.

13.7 Ethical and Social Dimensions of DNA Technology

- DNA technology offers potential benefits but also raises ethical questions and poses potential environmental risks. Opponents of genetic engineering are concerned that genes from GM plants or animals could spread to wild species, potentially wreaking environmental havoc.
- Much controversy surrounds the development of genetically modified plants, particularly in Europe. Unlike Europe, the United States does not require GM foods to be labeled as such, and they are common in the American diet.

13.8 A Closer Look at Some Tools of DNA Technology

- Restriction enzymes are used to break DNA into small pieces. Gel electrophoresis separates the resulting DNA fragments by size.
- Ligases are enzymes that join pieces of DNA.
- Automated sequencers greatly speed up the processes of sequencing and synthesizing DNA.
- To amplify a gene by PCR, primers (short segments of DNA that are complementary to the beginning and end of the target gene) are synthesized and used to produce billions of copies of the target gene in a few hours.

Key Terms

adult stem cell (p. 293)
cell culture (p. 293)
DNA cloning (p. 284)
DNA fingerprinting (p. 286)
DNA hybridization (p. 302)
DNA ligase (p. 301)
DNA microarray (p. 285)
DNA primer (p. 302)
DNA probe (p. 302)
DNA technology (p. 283)
embryonic stem cell (p. 293)
gel electrophoresis (p. 301)
gene therapy (p. 297)
genetic engineering (p. 285)
genetically modified organism (GMO) (p. 285)
induced pluripotent stem cell (iPSC) (p. 293)
inner cell mass (p. 293)
multipotent (p. 293)
personalized medicine (p. 285)
plasmid (p. 283)
pluripotent (p. 293)
polymerase chain reaction (PCR) (p. 302)
recombinant DNA (p. 283)
regenerative medicine (p. 295)
reproductive cloning (p. 291)
restriction enzyme (p. 300)
RNA interference (RNAi) (p. 298)
short tandem repeat (STR) (p. 286)
stem cell (p. 292)
totipotent (p. 293)
transgenic organism (p. 287)
unipotent (p. 293)
zygote (p. 293)

Self-Quiz

1. Which of the following cut(s) DNA at highly specific target sequences?
   a. DNA ligase
   b. DNA polymerase
   c. restriction enzymes
   d. RNA polymerase
2. The propagation of recombinant DNA in a culture of *E. coli* is known as
   a. PCR.
   b. DNA hybridization.
   c. DNA ligation.
   d. DNA cloning.
3. Genetic engineering
   a. can be used to make many copies of recombinant DNA introduced into a host cell.
   b. can be used to alter the inherited characteristics of an organism.
   c. raises ethical questions in the minds of some people.
   d. all of the above
4. Which of the following cell types displays the *least* developmental flexibility?
   a. zygote
   b. inner cell mass in a blastocyst
   c. bone marrow stem cell
   d. neuron
5. When DNA fragments are placed on an electrophoresis gel and subjected to an electrical current, which fragments move the farthest in a given time?
   a. the smallest
   b. the largest
   c. PCR fragments
   d. mRNA fragments
6. A short, single-stranded sequence of DNA whose bases are complementary to a portion of the DNA on another DNA strand is called
   a. a DNA hybrid.
   b. a clone.
   c. a DNA probe.
   d. an mRNA.
7. Small loops of nonchromosomal DNA that are found naturally in bacteria are called
   a. plasmids.
   b. primers.
   c. amplimers.
   d. clones.
8. If the DNA sequences at the beginning and end of a gene are known, which of the following methods can be used to produce billions of copies of the gene in a few hours?
   a. RNAi
   b. reproductive cloning
   c. therapeutic cloning
   d. PCR

Analysis and Application

1. Discuss the extent to which our current ability to manipulate DNA differs from what people have done for thousands of years to produce a wide range of domesticated species, such as dogs, corn, and cows.
2. What is DNA cloning? Describe how you might clone a jellyfish GFP gene into a bacterial cell.
3. Discuss some practical benefits of DNA cloning.
4. What is genetic engineering? How is it accomplished? Select one example of genetic engineering and describe its potential advantages and disadvantages.
5. Is it ethical to modify the DNA of a bacterium? A single-celled yeast? A worm? A plant? A cat? A human? Give reasons for your answers.
6. Supporters of genetic engineering claim that GM plants are likely to be safer, in many ways, than crops bred through conventional methods. They claim that GM crops undergo extensive scrutiny and environmental impact studies, while non-GM crops do not. Are you persuaded by that argument? Should new varieties of conventional crops be overseen more strictly, despite the economic cost of doing so?
7. Compare iPSCs (induced pluripotent stem cells) with embryonic stem cells and adult stem cells, in terms of both the developmental flexibility of these cells and their origins.
8. Are some modifications to the DNA of humans not acceptable? Assuming you think so, what criteria would you use to draw the line between acceptable and unacceptable changes?

14 How Evolution Works

VAMPIRE FINCH. This subspecies of the sharp-beaked ground finch lives on two of the Galápagos islands. In addition to eating insects, the birds drink nectar from flowers, the contents of eggs, and sometimes the blood of other birds. Scientists speculate that the strange diet may be driven by the scarcity of freshwater on the islands inhabited by this subspecies.

Finches Feasting on Blood

A cute little finch that drinks blood? Creepy indeed, but the sharp-beaked ground finch, also known as the vampire finch, does exactly that when it gets thirsty. Six hundred miles off the coast of South America lie the Galápagos Islands (**guh-*LAH*-puh-gohs**), an isolated group of volcanic islands that are home to hundreds of plants and animals found nowhere else on Earth. Unthinkably large tortoises weighing up to 900 pounds sway across a hot, dry landscape; 4-foot-long crested iguanas dive into the surf to feed, and then climb ashore to lie in the sun and digest the masses of seaweed in their bellies; and if the weather turns dry and there's little to drink, vampire finches hop onto the tails of large seabirds and sip their blood.

It's no wonder that Charles Darwin was fascinated by the animals of the Galápagos. Just out of college, Darwin landed a job as ship's naturalist for the British survey ship HMS *Beagle*. By the time he arrived at the Galápagos Islands 4 years into the journey, he was 26 and a seasoned naturalist who had already spent years in South America. "Nothing could be less inviting," he wrote of the dry, volcanic islands. But the Galápagos plants and animals fascinated him, and he collected them avidly—for study both on the ship and on his return home, by way of Tahiti, Australia, and South Africa.

Despite Darwin's interest, it was only on his return to England that he began to understand the significance of what he had collected in the Galápagos. Many of the species were found on only a single island and nowhere else in the world. And a scattering of odd finches and blackbirds he had collected turned out to hold the key to a mystery that had begun to intrigue Darwin.

In this chapter we will see that biological evolution explains a great deal about life on Earth, and we will explore the answers to Darwin's questions.

How do populations change? How do new species arise? Why are there so many unique species in the Galápagos Islands?

MAIN MESSAGE Evolution is a change in populations over time. Several forces can cause populations to evolve, including mutation, gene flow, genetic drift, and natural selection.

KEY CONCEPTS

- Evolution is change in the genetic composition of a population over successive generations. Populations evolve; individuals do not.
- Four mechanisms underlie evolutionary change: mutation, gene flow, genetic drift, and natural selection.
- Individuals with advantageous genetic characteristics survive and reproduce at a higher rate than other individuals—a process known as natural selection. The characteristics of individuals that produce more offspring become more common in succeeding generations.
- Natural selection leads to adaptation, the process that improves the match between a population of organisms and its environment over successive generations. Adaptive evolution can take many thousands of years to occur or can occur rapidly over the course of a few years or months.
- Mate choice may also drive evolution—a process known as sexual selection. Individuals with qualities that are desirable in mates will more successfully pass on these traits to future generations.
- An enormous amount of evidence supports evolution, including the fossil record, anatomical similarities between organisms, plate tectonics, observations of genetic change, and the ongoing formation of new species.

FINCHES FEASTING ON BLOOD 309
14.1 Evolution and Natural Selection 310
14.2 Mechanisms of Evolutionary Change 312
14.3 Natural Selection Leads to Adaptive Evolution 316
14.4 Adaptations 321
14.5 Sexual Selection 324
14.6 The Evidence for Biological Evolution 325
14.7 The Impact of Evolutionary Thought 331
APPLYING WHAT WE LEARNED 333
Darwin's Finches: Evolution in Action

EARTH TEEMS WITH LIVING THINGS. One of the most striking aspects of the planet's many organisms is the beautiful fit that so many exhibit for life in their particular environment (**FIGURE 14.1**). A hawk with its broad wings and powerful muscles can soar easily through the sky, a flower's colorful petals and sweet scent quickly attract pollinators, and an insect's body may look so much like the leaves around it that it is almost perfectly hidden from the view of hungry predators. How do organisms come to be so well matched to their surroundings? And why are there so many kinds of animals, plants, fungi, and other organisms? That is, why is there such a great diversity of life? And within that diversity, why do organisms share so many characteristics? The answer to all of these questions is the same: biological evolution.

FIGURE 14.1 Organisms Are Well Adapted to Their Environments
The natural world is filled with amazing examples of organisms seemingly perfectly suited to their environment. What kinds of adaptations can you see in these organisms? What forces may have shaped these adaptations?

This chapter introduces the concept of biological evolution, beginning with how the idea gained traction in nineteenth-century Europe, and how the contributions of two Englishmen, Charles Darwin and Alfred Wallace, advanced evolutionary thought. The mechanisms that cause evolution will then be explored, with a more in-depth look at natural selection and its impact on adaptation. Next, we present the evidence for evolutionary change from many different fields of study. The chapter closes with the impact of evolutionary thought on human society.

14.1 Evolution and Natural Selection

Through the ages, many cultures have viewed our planet, and all life on it, as fixed and unchanging. The Greek philosopher Aristotle (384–322 BC) saw the living world as unalterable and classified life-forms in 11 grades according to their level of perfection, with plants toward the bottom and humans at the pinnacle of perfection. The Greek philosophers had a profound influence on Western civilization, and the Aristotelian view of nature dominated for hundreds of years. The literal interpretation of scripture, especially the book of Genesis, shaped Judeo-Christian views about the origins of life, and these were embellished by a succession of biblical scholars. James Ussher, a seventeenth-century archbishop of Armagh in Northern Ireland, for example, claimed to know the exact date that all life was created: October 23, 4004 BC.

With intellectual and technical progress came doubts about the constancy of the world

The seventeenth and eighteenth centuries saw the dawn of a new intellectual movement, referred to as the Age of Enlightenment, during which scientific inquiry flourished, leading to rapid advances in knowledge and technological progress. The industrial revolution, launched in the mid-eighteenth century, brought with it a new understanding of landforms as geologists explored Earth's crust in search of coal and minerals. Fossils, which are the remains or imprints of past life-forms, turned up regularly as rocks were mined and quarries were dug. A number of scholars began to advance the idea that Earth was much older than the 6,000 years claimed by some, that some present-day life-forms did not exist in the early history of Earth, and that species have changed and new species have appeared over time. Sharing his ideas with just a small readership, Comte George-Louis Leclerc de Buffon wrote in his 1759 book, *Histoire Naturelle*, that species change over time either because of chance or because the changes are imposed by the environment.

Buffon's protégé, Jean-Baptiste Lamarck, was more vocal in his belief that life-forms change over time. Like other thinkers before him, Lamarck noted that many features of a particular organism enable it to function well in its particular habitat. He observed that the long necks of giraffes enable those animals to browse the tops of trees.

Lamarck missed the mark, however, when he tried to explain *why* life-forms change and *how* a group of organisms becomes better adapted to its surroundings. Lamarck claimed that the inheritance of *acquired characteristics* by offspring causes the next generation to change. Giraffes have long necks, Lamarck argued, because the ancestral animals stretched their necks to browse on high branches; and because the stretched-neck trait was passed on to offspring, necks got longer with each generation. Lamarck's compatriot Georges Cuvier (1769–1832) was among many who pointed out that Lamarck's explanation for adaptation was demonstrably wrong. Workmen with bulging muscles do not necessarily father children with large muscles, and training a border collie to run fast would not ensure that its descendants would also be fleet of foot.

People had been finding fossils for centuries, but not until the eighteenth century were fossils used as evidence that some species had become extinct. Cuvier firmly established this connection. As a vertebrate anatomist, Cuvier could see that some of the extinct forms were closely related to present-day animals, but he rejected the idea that one species can give rise to others.

In 1830 the Scottish geologist Charles Lyell published his acclaimed *Principles of Geology*, in which he explained how the surface of Earth had been shaped by the slow action of rivers, glaciers, earthquakes, and volcanoes. Darwin received the second volume of Lyell's treatise when the *Beagle* docked near Buenos Aires in 1832. Lyell's account of geologic transformation strongly influenced Darwin's theory of natural selection.

Darwin proposed natural selection as a driving force of evolutionary change

What Darwin saw in the Galápagos and in many other places during his 5-year voyage on HMS *Beagle* (**FIGURE 14.2**) led him to conclude that species were not the unchanging result of separate acts of creation. He proposed that species had descended from

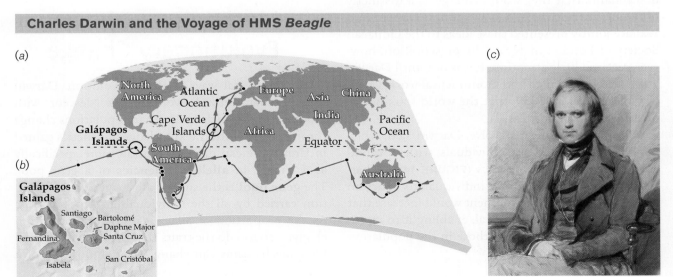

Charles Darwin and the Voyage of HMS *Beagle*

(a)

(c)

FIGURE 14.2 Charles Darwin Was Ship's Naturalist on the HMS *Beagle*

(*a*) The course sailed by the *Beagle*. (*b*) The Galápagos Islands, located 1,000 kilometers to the west of Ecuador. (*c*) Charles Darwin.

ancestral forms and had been modified to suit their environments. Darwin's compatriot Alfred Wallace had been studying the diversity of related species living on the many different islands of the Malaysian archipelago, and he independently arrived at the same conclusions as Darwin.

Darwin and Wallace went further than any scholar before them by proposing a *mechanism* for the evolution of new species. They called it **natural selection**, which we define as an evolutionary process by which a population adapts to its environment. Both men were influenced by an essay written by the Reverend Thomas Malthus, in which the clergyman stated that the growth of populations is limited by the availability of resources such as food and by adversities such as disease. According to Malthus, there is a struggle for survival when individuals produce more offspring than their environment can support, and there are winners and losers in such a struggle.

Helpful to Know

In everyday conversation, the term "fittest" describes the person who is strongest or in the best physical shape. When biologists talk about the survival of the fittest, however, they mean something quite different. The fittest are those individuals who are the bearers of advantageous inherited traits of any kind that enable them to survive and reproduce more than others. Hence, natural selection is the survival of the fittest.

Darwin and Wallace were struck by the same idea: if there was a struggle for survival, with winners and losers, then the characteristics of the winners should be better represented in the next generation because the winners would live to produce more offspring than the losers. As a consequence of such *differential reproduction*, the inherited characteristics found in the next generation would be in different proportions than they were in the original population. Specifically, because these characteristics were inherited, the ones associated with the winners would be more prevalent than the ones inherited from the losers. Over the generations, environmental pressures could so modify the descendants that they would emerge as a distinctly different form—that is, a new species. Darwin and Wallace jointly presented these ideas to the Linnaean Society of London in 1858. The presentation, however, received little attention. It was not until Darwin published his detailed and monumental work, *On the Origin of Species*, in 1859, that the world took notice of these remarkable ideas.

Groups of organisms evolve, Darwin said, when natural selection favors individuals with advantageous inherited characteristics (**FIGURE 14.3**). Characteristics that enabled the individual to function well in its particular environment would increase that individual's chances of survival, enabling it to reproduce more successfully than those in the population that lacked those advantages.

Concept Check Answers

1. Lamarck believed that evolutionary change occurred as a result of individuals passing on traits they had acquired throughout their lives. Darwin and Wallace saw that evolution occurred because those organisms with the traits suited to their environments were disproportionately able to pass on their traits.

2. No, natural selection favors individuals with characteristics that are better suited to the environment. In a population of clones, the lack of any variation would mean that no one individual would have any advantage over any other.

Evolution by Natural Selection

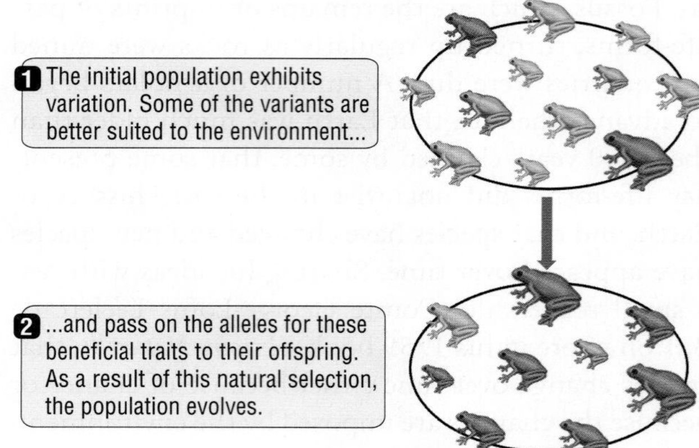

1 The initial population exhibits variation. Some of the variants are better suited to the environment...
2 ...and pass on the alleles for these beneficial traits to their offspring. As a result of this natural selection, the population evolves.

FIGURE 14.3 Darwin Proposed Natural Selection as the Mechanism of Evolutionary Change

It is hard to overstate the importance of Darwin's work. Evolution is biology's most powerful explanation for why living things exhibit the traits they do. The theory of evolution through natural selection has stood the test of time, although, as we explore next, it is not the sole means by which organisms can evolve.

Concept Check

1. Compare Lamarck's view of evolutionary change with that of Darwin and Wallace.
2. Could natural selection act on a population of clones?

14.2 Mechanisms of Evolutionary Change

There are several ways to define evolution. Darwin himself characterized evolution as "descent with modification," which means that populations change through time. With the knowledge of genetics gained in Unit 2, we can define evolution more specifically as a change in **allele frequencies** of a gene pool. The **gene pool** is the sum of all the genetic information carried by all the individuals in a population. As the proportion of alleles within the population changes, so too do the traits that these genes encode. Four mechanisms can change the composition of a population's gene pool: (1) mutation, (2) gene flow, (3) genetic drift, and (4) natural selection.

Mutations introduce genetic variation in a population

In a natural population, individuals vary in their structural, biochemical, and behavioral traits (**FIGURE 14.4**). Much of that variation is due to differences in the DNA sequences among individuals and in how those genes are *expressed*, meaning if and when they are converted into proteins. In other words, among members of a population of one species, the differences in how individuals look and behave can be traced back to differences in their DNA. We call these differences **genetic variation**. Where does genetic variation come from? **Mutations**, random changes in the DNA sequence that occur for a variety of reasons, are the original source of all genetic variation. The DNA variants produced by mutation are known as **alleles** (see Chapter 9). These types of differences serve as the raw material for evolution. In this sense, all evolutionary change depends ultimately on mutations. New mutations that are inherited cause a population to evolve because the addition of new alleles changes the gene pool.

FIGURE 14.4 Variation in Human Populations
Like all organisms, humans exhibit variation in their physical traits. Since much of this variation is controlled by a person's genes, natural selection can act to favor some traits over others.

Mutations that occur in an organism's gametes (egg or sperm cells, as opposed to mutations in muscle cells, for example) can be passed on to subsequent generations. Mutations are caused by various accidents, such as mistakes in DNA replication, collisions of the DNA molecule with other molecules, or damage from heat or chemical agents. Despite the efficiency with which repair proteins fix damage to DNA and correct errors in DNA replication (see Chapter 11), mutations occur regularly in all organisms. Humans, for example, have two copies each (one copy from each parent) of approximately 25,000 genes. On average, between two and three of these 50,000 gene copies have mutations that make them different from those of either parent. Mutations and the genetic variation they produce do not appear because an organism "needs" them; instead, mutations occur randomly and are not directed toward any goal.

Because mutations occur so infrequently in any particular gene, they generally do not have a large effect on allele frequencies on their own. Furthermore, most mutations are either harmful to their bearers or have little effect. However, by supplying new genetic variation on which natural selection can act, mutation plays a critical role in the evolution of populations. Although a beneficial mutation is rare, once it is introduced, natural selection can rapidly increase the frequency of the new allele. Take, for example, the mosquito *Culex pipiens* (KYOO-**lex** PIP-**yenz**). Genetic evidence indicates that resistance to pesticides in *C. pipiens* was caused by a single mutation that occurred in the 1960s. This mutant allele is highly advantageous to the mosquito: individuals that have the nonmutant allele die when exposed to pesticides. When the mutant allele is introduced into a population exposed to pesticides, natural selection favors the mutant, resulting in a rapid increase in the frequency of the mutant allele and leading to the evolution of resistance within the new population.

Mutations like those that enable disease agents or pests to resist our best efforts to kill them are obviously beneficial to the organisms in which they occur. In general, the effect of a mutation depends on the environment in which the organism lives. For example, certain mutations that provide houseflies with resistance to the pesticide DDT also reduce their rate of growth. In the absence of DDT, such mutations are harmful. When DDT is sprayed, however, these mutations provide an advantage great enough to

FIGURE 14.5 DDT Was Initially Described as a "Benefactor of All Humanity" Developed as the first of the modern insecticides early in World War II, DDT was initially used with great effect to combat malaria, typhus, and the other insect-borne human diseases. Often termed the "miracle" pesticide, DDT came into wide agricultural and commercial use. Here, fogging machines are shown spraying DDT at Jones Beach in New York in the 1940s. By 1972, however, the Environmental Protection Agency had banned DDT because of the devastating side effects of extensive DDT use, including insect resistance and adverse environmental impacts.

offset the disadvantage of slow growth. As a result of DDT spraying (**FIGURE 14.5**), the mutant alleles have spread throughout the housefly populations and can now be found globally.

Gene flow moves genes between populations

Gene flow is the movement of genes from one population to another. When migrants from a different population breed with a resident population, they may introduce new alleles to the resident's gene pool (**FIGURE 14.6**). Gene flow can also occur when only gametes move from one population to another, as happens when wind or pollinators like insects transport pollen from one population of plants to another. If the recipient gene pool is altered by the introduction of novel alleles, the population has evolved.

Introductions of new alleles can have dramatic effects. For example, in the case of the mosquito *Culex pipiens*, discussed in the previous section, a new allele that made the mosquito resistant to pesticides spread by gene flow across three continents. This spread of a new mutant allele enabled billions of mosquitoes to survive the application of pesticides that otherwise would have killed them.

Gene flow tends to make the genetic compositions of different populations more similar. As populations exchange alleles with one another, they become united into a single, larger gene pool. Without gene flow, the gene pools of two populations would likely

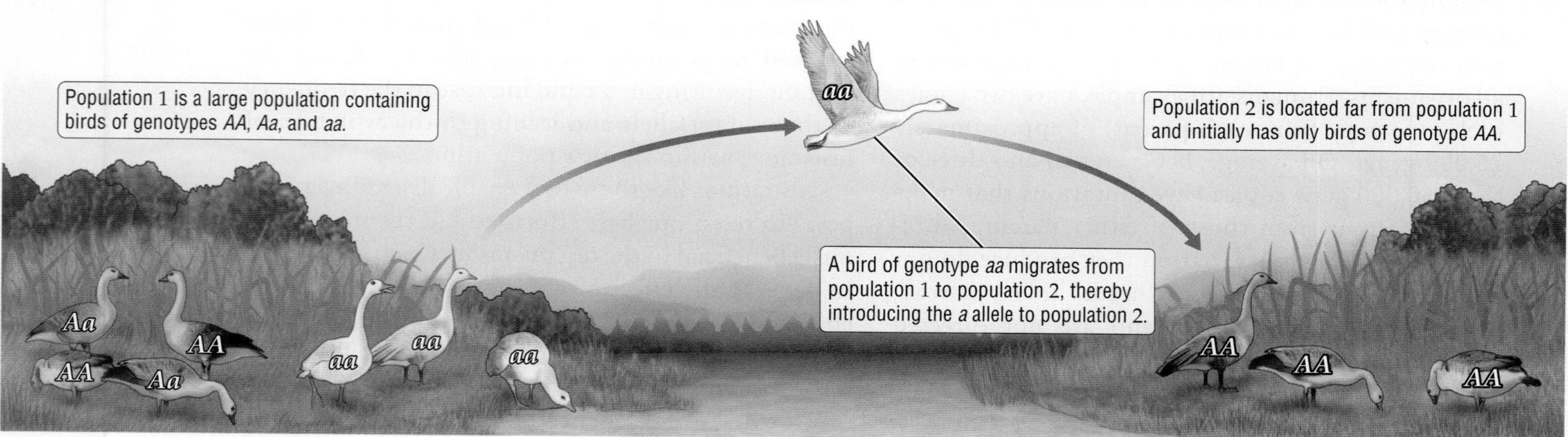

Gene Flow
Population 1 is a large population containing birds of genotypes AA, Aa, and aa.
aa
Population 2 is located far from population 1 and initially has only birds of genotype AA.
A bird of genotype aa migrates from population 1 to population 2, thereby introducing the a allele to population 2.
Aa
AA
AA
Aa
aa
aa
aa
AA
AA
AA
Population 1
Population 2

FIGURE 14.6 Migrants Can Move Alleles from One Population to Another

diverge over time. Natural selection might drive the two populations along different evolutionary paths by favoring different alleles in each habitat.

Genetic drift generates differential reproduction through accidental events

Genetic drift is a change in allele frequencies because of random differences in survival and reproduction from one generation to the next. While it is a random process, genetic drift can cause the gene pool of a population to fluctuate over time, rather than being pushed in a particular direction.

Genetic drift is more likely to alter the gene pool of a small population than that of a large population. To understand why, consider what happens when a coin is tossed. It would not be all that unusual to get four heads in five tosses; that result has about a 15 percent chance of occurring. But it would be astonishing to get 4,000 heads in 5,000 tosses. Even though the frequency of heads is the same in both cases (80 percent), the chance of getting many more, or many less, than the expected 50 percent heads is much greater if the coin is tossed a few times than if it is tossed thousands of times.

The importance of genetic drift in small populations has implications for the preservation of rare species. If the number of individuals in a population falls to very low levels, genetic drift may lead to a loss of genetic variation or to the **fixation** of harmful alleles, either of which can hasten **extinction** of the species. When a drop in the size of a population leads to such a loss of genetic variation, the population is said to experience a **genetic bottleneck**. Genetic bottlenecks often occur in nature because of the **founder effect**, which occurs when a small group of individuals establishes a new population far from existing populations (for example, on an island).

Genetic bottlenecks are thought to have occurred in populations of the Florida panther, the northern elephant seal, and the African cheetah. The endangered Florida panther population size plummeted to about 30–50 individuals in the 1980s. At that time, biologists discovered that male Florida panthers had low sperm counts and abnormally shaped sperm (**FIGURE 14.7**). The resulting low fertility in the males may have contributed to the drop in population size. The panther's numbers have increased to about 80–100 individuals in recent years, in part because of breeding programs designed to reduce the effects of genetic drift.

It is often difficult to know how much genetic variation was present before a population decreased in size. Hence, there is no way to be sure if the low levels of genetic variation observed in the Florida panther were caused by a decrease in population size. In some cases, however, it is clear that a dramatic decrease in population size was the cause of a drop in genetic variation. Recent studies on greater prairie chickens in Illinois compared the DNA of modern birds with the DNA of their pre-bottleneck ancestors, obtained from (nonliving) museum specimens. There were millions of greater prairie chickens in Illinois in the nineteenth century, but by 1993 the conversion of their prairie habitat to farmland had caused their numbers to drop to only 50 birds in two isolated populations

EXTREME GENE FLOW

Pollen grains contain the sperm of plants. Wind-pollinated plants, particularly evergreen trees, can spread their pollen over vast distances. A recent study showed that outbreaks of allergies in northern Italy were caused by pollen blown from Hungary—a distance of 200 miles.

FIGURE 14.7 Abnormally Shaped Sperm in the Rare Florida Panther

Florida panthers have more abnormal sperm than do cats from other cougar populations—a possible effect of the fixation of harmful alleles. (*Insets*) Abnormal sperm and normal sperm, for comparison.

A Genetic Bottleneck

By 1993, only 50 greater prairie chickens remained in Illinois, causing both the number of alleles and the percentage of eggs that hatched to decrease.
Pre-bottleneck Illinois (1820)
Post-bottleneck Illinois (1993)
In 1993, less than 1% of the grassland remained, and the birds could be found only in these two locations.
In 1820, the grasslands in which greater prairie chickens live covered most of Illinois.

	ILLINOIS		KANSAS	MINNESOTA	NEBRASKA
	PRE-BOTTLENECK (1933)	*POST-BOTTLENECK (1993)*	*NO BOTTLENECK*		
Population size	25,000	50	750,000	4,000	75,000 – 200,000
Number of alleles at six genetic loci	31	22	35	32	35
Percentage of eggs that hatch	93	56	99	85	96

FIGURE 14.8 A Genetic Bottleneck Can Cause a Population to Crash The Illinois population of greater prairie chickens dropped from 25,000 birds in 1933 to only 50 birds in 1993. This drop in population size caused a loss of genetic variation and a drop in the percentage of eggs that hatched. Here, the modern, post-bottleneck Illinois population is compared with the 1933 pre-bottleneck Illinois population, as well as with populations in Kansas, Minnesota, and Nebraska that never experienced a bottleneck.

(**FIGURE 14.8**). This drop in number caused a genetic bottleneck: the modern birds lacked 30 percent of the alleles found in the museum specimens, and they suffered poor reproductive success compared with other prairie chicken populations that had not experienced a genetic bottleneck (only 56 percent of their eggs hatched, versus 85–99 percent in other populations). From 1992 to 1996, 271 birds were introduced to Illinois from large populations in Minnesota, Kansas, and Nebraska, in order to increase both the size and the genetic variation of the Illinois populations. By 1997, the number of males in one of the two remaining Illinois populations had increased from a low of 7 to more than 60 birds, and the hatching success of eggs had risen to 94 percent.

Genetic drift also occurs in large populations, but in these cases its effects are more easily muted by natural selection and other evolutionary mechanisms. As in the coin toss example, a larger sample will experience less deviation from the expected value than a small sample. Likewise, in large populations, genetic drift causes relatively little change in allele frequencies over time.

Concept Check

1. What is the ultimate source of genetic diversity in a population?
2. Explain how gene flow can cause evolutionary change in a population.
3. Why does genetic drift affect smaller populations more profoundly than large populations?

Concept Check Answers

1. DNA mutations. Gene flow can also introduce genetic diversity from other populations.
2. Gene flow involves the movement of alleles between populations. If two populations have unique alleles or different allele frequencies, then the migrants will cause a change in the overall allele frequencies resulting in evolutionary change.
3. Smaller populations are subject to greater sampling error. Larger sample sizes make it more likely that actual results will approximate statistical expectations.

14.3 Natural Selection Leads to Adaptive Evolution

Evolution in populations is by no means restricted to the chance events that fuel mutation, gene flow, and genetic drift. The fourth evolutionary mechanism, natural selection, is not a random process. Unlike the other three mechanisms of evolution, natural selection is a directional process because it shifts the genetic characteristics of a population over successive generations so that advantageous traits become more prevalent in a population and the frequency of deleterious traits decreases.

BIOLOGYMATTERS

Testing Whether Evolution Is Occurring in Natural Populations

When biologists study populations in nature, it can be difficult to immediately determine whether or not evolution is occurring in a given population. In 1908, Godfrey Hardy and Wilhelm Weinberg independently developed a quick calculation based on Mendelian genetics that can reveal whether a population is evolving.

Suppose we are tracking evolution of a single gene with just two alleles: *A* and *a*. We will need to keep track of the allele frequencies of both alleles. The letter *p* traditionally is used to denote the allele frequency of the dominant allele (*A* in this case), and the letter *q* is used to denote the frequency of the recessive allele (*a*). Because this gene has exactly two alleles, we know that the sum of the two allele frequencies is $p + q = 1$.

The genetic calculation to determine whether evolution is occurring in a population relies on the *Hardy-Weinberg equation*, which has the general form

Frequency of genotype *AA*. — Frequency of genotype *aa*.

$$p^2 + 2pq + q^2 = 1$$

Frequency of genotype *Aa*.

The equation is an expression of the principle known as the *Hardy-Weinberg equilibrium*, which states that the amount of genetic variation in a population will remain constant from one generation to the next in the absence of disturbing factors. That is, the Hardy-Weinberg equation predicts the genotype frequencies of a population that is *not* evolving (in other words, a population "at equilibrium"). The assumption inherent in the Hardy-Weinberg equation is that none of the factors that cause allele frequencies to change (mutation, gene flow, genetic drift, and natural selection) affect the proportions of alleles in the gene pool (the sum of all the alleles in the population at any one time). We use the equation to test this assumption at one genetic locus having two alleles (*A* and *a*). For this locus, there are exactly three genotypes (*AA*, *Aa*, and *aa*), whose frequencies must sum to 1 (equivalent to 100 percent). As indicated in the equation above, p^2 represents the frequency of the homozygous genotype *AA*, q^2 represents the frequency of the homozygous genotype *aa*, and $2pq$ represents the frequency of the heterozygous genotype *Aa*.

The Hardy-Weinberg approach enables us to test whether actual genotype frequencies in a real population match those predicted by the equation. If the actual frequencies differ considerably from the frequencies predicted by the Hardy-Weinberg equation for a nonevolving population, we conclude that the actual population *is* evolving and that one or more of the four evolutionary mechanisms (mutation, gene flow, genetic drift, or natural selection) are at work. If the population is evolving, we may expect that allele frequencies will change over the generations until equilibrium is reached.

To find out whether genotype frequencies in a real population differ from those predicted by the Hardy-Weinberg equation, we must first determine the genotypes of individuals in the population. Suppose we have a population of 1,000 individuals—460 with genotype *AA*, 280 with genotype *Aa*, and 260 with genotype *aa*. Bearing in mind that each individual has 2 alleles for each gene, we are examining a gene pool of 2,000 alleles. The frequency of the *A* allele (denoted by the letter *p*) in this gene pool is calculated as a percentage of all the alleles in the gene pool:

$$p = [(2 \times 460) + 280]/2{,}000 = 0.6$$

Once we know the value of *p*, we can easily calculate the value of *q*, because we know that $p + q = 1$. Since $1.0 - 0.6 = 0.4$, we also know that *q*, the observed frequency of the *a* allele, is 0.4.

Next we plug allele frequencies into the Hardy-Weinberg equation to derive genotype frequencies. If the population is not evolving, the equation will hold and the observed genotype frequencies will match its predictions. With $p = 0.6$ and $q = 0.4$, the Hardy-Weinberg equation projects the following genotype frequencies:

$$p^2 = 0.6 \times 0.6 = 0.36$$
$$= \text{frequency of the } AA \text{ genotype}$$

$$2pq = 2 \times 0.4 \times 0.6 = 0.48$$
$$= \text{frequency of the } Aa \text{ genotype}$$

$$q^2 = 0.4 \times 0.4 = 0.16$$
$$= \text{frequency of the } aa \text{ genotype}$$

Given 1,000 individuals in the population, the Hardy-Weinberg equation predicts that if the population is not evolving, we should find 360 (0.36 × 1,000) *AA* individuals, 480 (0.48 × 1,000) *Aa* individuals, and 160 (0.16 × 1,000) *aa* individuals. In fact, however, the actual population in our case has 460 *AA* individuals, 280 *Aa* individuals, and 260 *aa* individuals.

The differences between the actual and expected genotype frequencies just described are large. A biologist who obtained real data like those in this example would conclude that the actual genotype frequencies differed significantly from those in the Hardy-Weinberg equation, and that the population was evolving. Next a researcher would begin to wonder what evolutionary mechanisms might be driving the population away from the predictions of the Hardy-Weinberg equation. After you finish reading this chapter, look again at the differences between the observed and the expected genotype frequencies in this example. Can you suggest one or more evolutionary mechanisms that might explain these differences?

Adaptation increases a population's chances for reproductive success

As pointed out by Malthus, and noted by Darwin and Wallace, organisms typically produce more offspring than can survive to reproduce, leading to a "struggle for existence." In this competitive environment, any individual with an advantageous inherited characteristic—one that enables the individual to function better than others in that habitat—is more likely to survive and reproduce and to pass those characteristics on. Meanwhile, an individual that lacks these advantageous characteristics, or has disadvantageous characteristics, is less likely to survive and reproduce and to pass on those characteristics. As a result, advantageous characteristics become more common among the offspring, and disadvantageous characteristics disappear or become less common over successive generations.

Natural selection causes the gene pool of descendants to become different from that of the original population, and the descendant population therefore evolves. Over time, natural selection can cause a population to evolve so that characteristics that enable individuals to function well in a competitive environment become more common among the descendants. The evolutionary process by which a population as a whole becomes better suited to its habitat is known as **adaptation**.

There are many examples of adaptation as a result of natural selection. **FIGURE 14.9*a*** depicts the evolution of pale fur in populations of beach mice. Dr. Hopi Hoekstra and her colleagues studied populations of beach mice that are found only on certain barrier islands off the Gulf Coast and off the Atlantic Coast of Florida. These mice have close relatives, called oldfield mice (*Peromyscus polionotus*), that live inland among the grassy vegetation of old farm fields. The beach mice have light coloration, while their inland cousins have darker markings (**FIGURE 14.9*b***). The hypothesis that Dr. Hoekstra started with is that the differences in coat color are the result of predation pressure by hawks, foxes, and herons that hunt the mice. Using DNA analysis, the scientists deduced that the beach mice descended from oldfield mice from the mainland over a span of about 6,000 years, when the barrier islands were formed.

To test the idea that the coat colors are the result of predation pressure, the researchers made clay replicas of mice with pale coats and dark coats. They arranged the model mice on the beach and found that the ones with dark coloration were three times more likely to have missing parts or to be missing altogether (see Figure 14.9*b*). The pattern of attack was reversed when the models were set out on dark soils, with lighter models being attacked three times more often than the paler versions. To understand how these adaptations came to be, the scientists compared the genes of beach mice and oldfield mice and discovered that beach mice have mutations in three genes, all of which depress pigment formation, causing the beach mouse fur to be light in color.

Natural selection does not craft perfection for all time and all circumstances. Light coloration in these mice is advantageous on the barrier islands, but it is a liability on the mainland. Note that the variation in the genes for coat color did not necessarily suddenly appear in the mice that colonized the barrier islands. These alleles may have existed in the oldfield mouse population, but most oldfield mice that were light in color were eaten, so they reproduced less often and had fewer babies. When mice moved from the fields to the sandy white beaches, the darker mice were suddenly at a disadvantage and the lighter ones were better camouflaged. As a result, the lighter mice survived in greater numbers than the darker mice, the lighter mice had more light-colored babies, and the barrier island populations of mice grew to be mostly light in color. This is natural selection at work.

There are three types of natural selection

Three patterns of natural selection are commonly observed: *directional selection*, *stabilizing selection*, and *disruptive selection*. Whatever the pattern, all types of natural selection operate by the same principle: individuals with certain forms of an inherited phenotypic trait tend to have better survival rates and to produce more offspring than do individuals with other forms of that trait.

In **directional selection**, individuals at one extreme of an inherited phenotypic trait have an

Helpful to Know

Adaptation is the evolutionary *process* that makes a population better suited to its habitat. However, the word "adaptation" is also used for an advantageous *characteristic*, an inherited structural, biochemical, or behavioral feature that enables an organism to function well and therefore to survive and reproduce better than do competitors lacking the characteristic.

EXTREME BOTTLENECK

Cheetah populations experienced a severe bottleneck event about 10,000 years ago, leaving only several hundred individuals. Wild cheetahs also produce only about 10 percent as much sperm as domestic cats produce. Of those sperm, 70 percent are abnormal.

Adaptation through Natural Selection

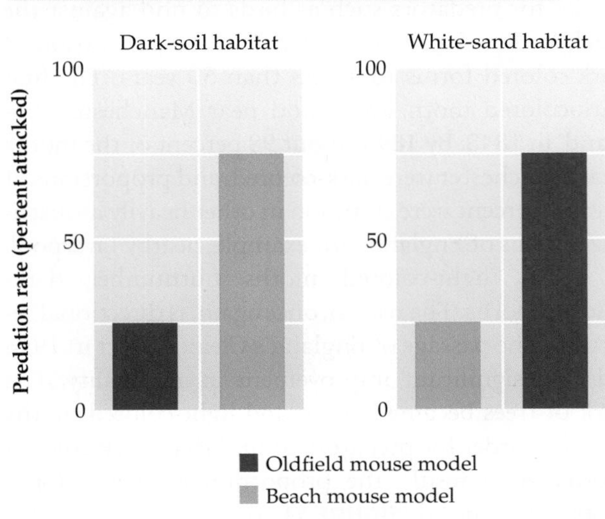

FIGURE 14.9 Natural Selection Adapts a Population to Its Environment

(*a*) Deer mice (*Peromyscus polionotus*) from the mainland colonized barrier islands off the coast of Florida. Darker pigmentation is most common among the mainland subspecies (called the oldfield mouse). The barrier island subspecies, known as the beach mouse, has a paler coat. (*b*) Researchers used clay models of oldfield mice and beach mice to test the hypothesis that predators will see and attack coat patterns that contrast with their background.

FIGURE 14.10 Directional, Stabilizing, and Disruptive Selection Affect Phenotypic Traits Differently

The graphs in the top row show the relative numbers of individuals with different body sizes in a population before selection. The phenotypes favored by selection are shown in red. The graphs in the bottom row show how each type of natural selection affects the distribution of body size in the population.

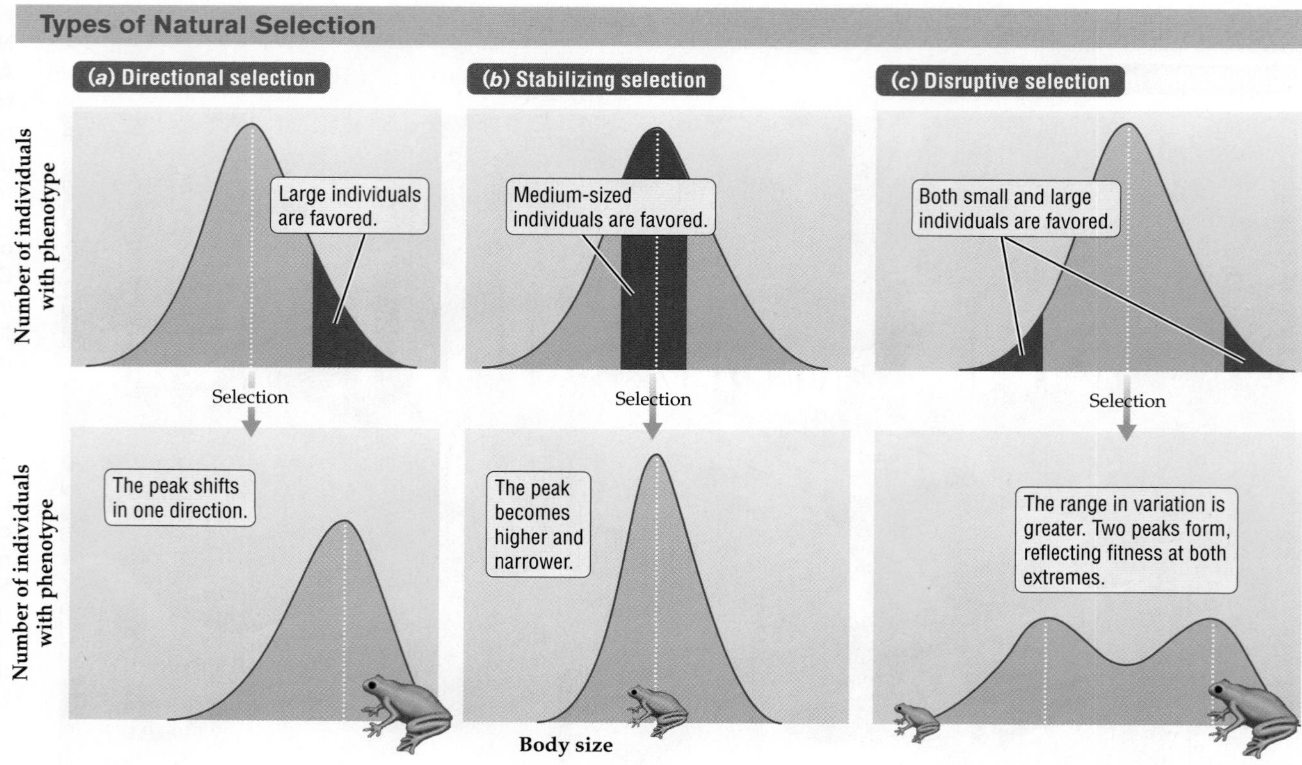

FIGURE 14.11 Directional Selection in the Peppered Moth

The frequency of dark-colored peppered moths declined dramatically from 1959 to 1995 in regions near Liverpool, England. Before 1959, dark-colored moths had risen in frequency after industrial pollution had blackened the bark of trees, causing dark-colored moths to be harder for bird predators to find than light-colored moths. A reduction in air pollution following clean-air legislation enacted in 1956 in England apparently removed this advantage, leading the dark-colored moths to decline.

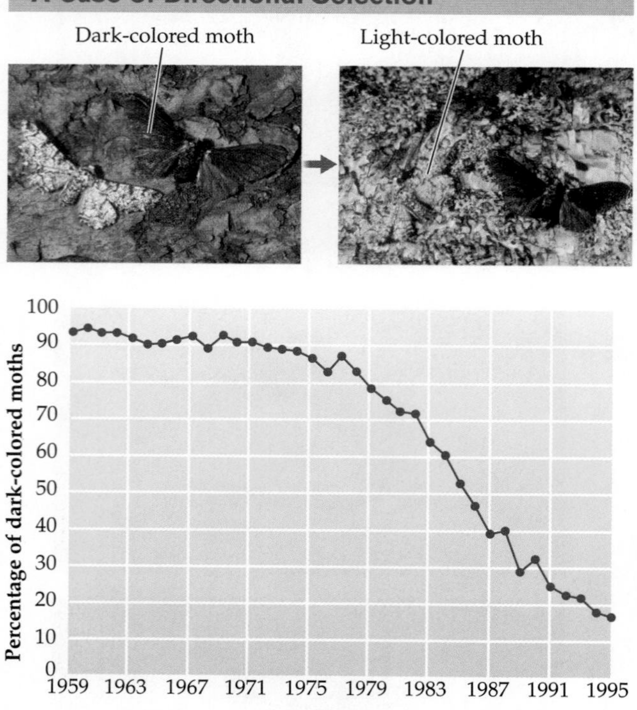

advantage over other individuals in the population. For example, if large individuals produce more offspring than do small individuals, there will be directional selection for large body size (**FIGURE 14.10*a***).

Directional selection is seen in the rise and fall of populations of dark-colored versus light-colored variants in some moth species. For example, dark-colored forms of the peppered moth were favored by natural selection when industrial pollution blackened the bark of trees in Europe and North America. The color of these moths is a genetically determined trait. The proportion of the dark-colored moths increased after industrialization, apparently because they were harder for predators such as birds to find against the blackened bark of trees. The rise in the proportion of dark-colored forms took less than 50 years. The first dark-colored moth was found near Manchester, England, in 1848. By 1895, about 98 percent of the moths near Manchester were dark-colored, and proportions of over 90 percent were common in other heavily industrialized areas of England—for example, nearby Liverpool.

Today, light-colored moths outnumber dark-colored moths. The reason, once again, is directional selection. The passage of England's Clean Air Act in 1956 led to a significant improvement in air quality. The bark of trees became lighter, and light-colored moths became harder for predators to find than dark-colored moths. As a result, the proportion of dark-colored moths plummeted (**FIGURE 14.11**).

A Case of Stabilizing Selection

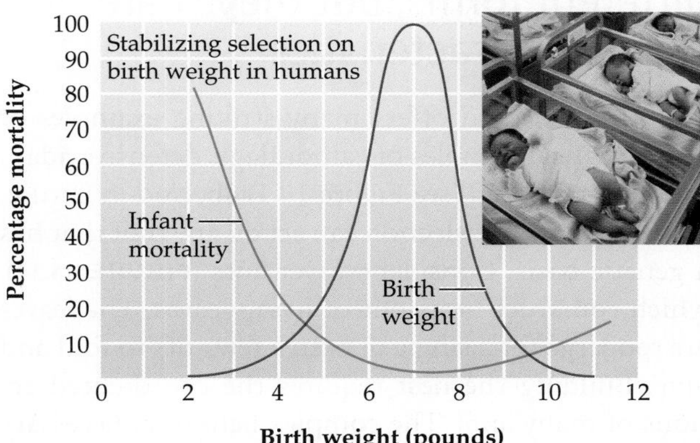

FIGURE 14.12 Stabilizing Selection for Human Birth Weight

This graph is based on data for 13,700 babies born between 1935 and 1946 in a hospital in London. In countries that can afford intensive medical care for newborns, the strength of stabilizing selection has been reduced in recent years: because of improvements in the care of premature babies and increases in the number of cesarean deliveries of large babies, a graph of such data collected today would be flatter (less rounded) at its peak than the graph shown here.

A Case of Disruptive Selection

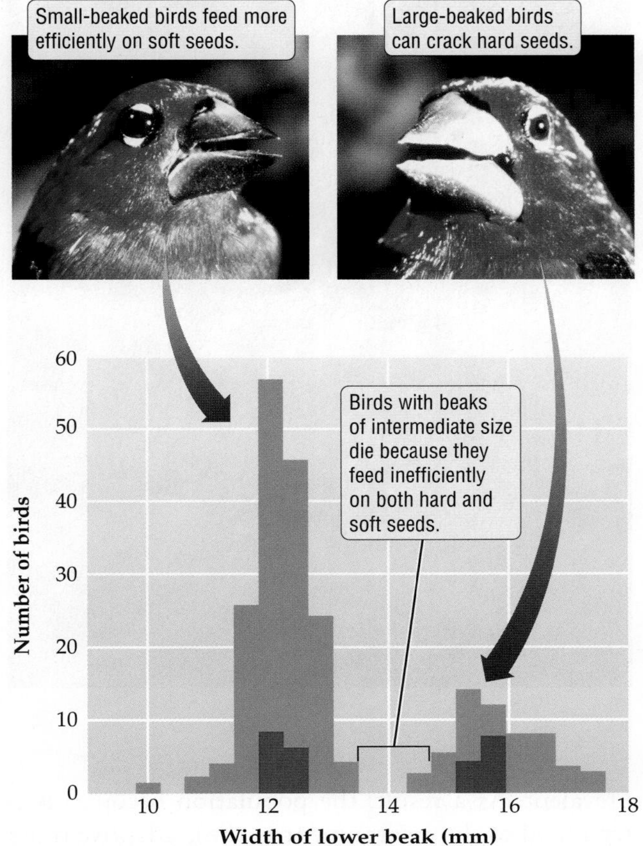

FIGURE 14.13 Disruptive Selection for Beak Size

In African seed crackers, differences in feeding efficiencies may cause differences in survival. Among a group of young birds hatched in one year, only those with a small or large beak size survived the dry season, when seeds were scarce; all the birds with intermediate beak sizes died. Therefore, natural selection favored both large-beaked and small-beaked birds over birds with intermediate beak sizes. Red bars indicate the beak sizes of young birds that survived the dry season; blue bars indicate the beak sizes of young birds that died.

In **stabilizing selection**, individuals with intermediate values of an inherited phenotypic trait have an advantage over other individuals in the population (**FIGURE 14.10*b***). Birth weight in humans provides an example. Historically, light or heavy babies did not survive as well as babies of average weight did, so there was stabilizing selection for intermediate birth weights (**FIGURE 14.12**). By the late 1980s, however, selection against small and large babies had decreased considerably in some countries with advanced medical care, such as Italy, Japan, and the United States. This reduction in the strength of stabilizing selection was caused by advances in the care of very light premature babies and by increases in the use of cesarean deliveries for babies that are large relative to their mothers (and hence pose a risk of injury to mother and child at birth).

In **disruptive selection**, individuals with either extreme of an inherited phenotypic trait have an advantage over individuals with an intermediate phenotype (**FIGURE 14.10c**). This type of selection is not common, but it appears to affect beak size within a population of African seed crackers, in which birds with large or small beaks survive better than birds with intermediate-sized beaks (**FIGURE 14.13**).

Concept Check

1. What distinguishes natural selection from the other mechanisms of evolution—mutation, gene flow, and genetic drift?
2. What are the three ways that natural selection can affect a population?

14.4 Adaptations

Natural selection is the driving force behind adaptation. As natural selection removes unfit variants from a population, favorable variants become more

Concept Check Answers

1. While mutation, gene flow, and genetic drift can all change allele frequencies within a population, these changes are random and do not lead to adaptation. Natural selection is a nonrandom mechanism that adapts a population to its environment.

2. In directional selection, one of the extremes of a distribution is favored, causing the population to shift in one direction. In stabilizing selection, extreme values are selected against, leading to a reduction in the variance of the population. Disruptive selection favors the extreme values, leading to a bimodal distribution.

FIGURE 14.14 Behavioral Adaptations in Weaver Ants Benefit All Members of the Colony
Weaver ants collaborate to pull nest leaves together.

prevalent. As a result, the population becomes better suited to the environment as these adaptive traits increase in frequency. Adaptations can take on many forms. Any trait that can be passed on to offspring—that is, any trait under genetic control—can be shaped by natural selection to be adaptive.

Adaptations can take many different forms, but they share certain characteristics

The natural world offers many striking examples of adaptations. Examples of camouflage, defensive adaptations, and speed (see Figure 14.1) abound in nature. However, natural selection can act on any trait that has a genetic basis. Consider weaver ants (**FIGURE 14.14**), which construct nests out of living leaves. The leaves are too large for a single or even a few ants to fold and glue. Building the nest requires the coordinated actions of many ants. The complex behaviors necessary for the construction of the nest are innate, or "hard-wired," rather than learned, and therefore must be under genetic control. These innate behaviors illustrate how natural selection can produce a complex *behavioral* adaptation (cooperative nest building, which benefits the ants by providing them with shelter).

In another example of adaptation, the caterpillars, or larvae (*LAHR*-**vee**; singular "larva"), of a certain moth species camouflage themselves by looking like the part of the oak tree they feed on. Larvae that hatch in the spring feed on flowers and grow to resemble oak flowers, while those that hatch in summer feed on leaves and grow to look like oak twigs (**FIGURE 14.15**). In this way the larvae develop so that they match the background they are feeding and living on, making them more difficult for predators to see.

Natural selection has also shaped some astonishing adaptations that facilitate reproduction. Instead

(*a*)

(*b*)

FIGURE 14.15 An Adaptation That Offers Protection from Predators
Although they do not look alike, the two caterpillars shown here are the same species of moth: *Nemoria arizonaria* (**nee-*MOHR*-ee-uh *AYR*-ih-zoh-*NAYR*-ee-uh**). *N. arizonaria* caterpillars differ in appearance depending on what they feed on—the flowers or leaves of oak trees. Experiments have demonstrated that chemicals in the leaves control the switch that determines whether the caterpillars will mimic flowers (*a*) or twigs (*b*).

of using nectar to lure pollinators, the flowers of some orchids release chemical attractants (called *pheromones*) that attract males. Orchids also mimic the shape, texture, and color of female wasps. Male wasps drawn to the orchid attempt to mate with the flowers (**FIGURE 14.16**). In the course of these attempts, the insects become coated with pollen, which they then transfer from one plant to another.

Look carefully at the eye of the four-eyed fish, *Anableps anableps*, in **FIGURE 14.17**. Although this fish really has only two eyes, they function as four, enabling the fish to see clearly through both air and water. The four-eyed fish is a surface feeder, so the ability to see above water helps it locate prey such as insects. Its unique eyes also enable it to scan simultaneously for predators attacking from above (such as birds) or below (such as other fish). The four-eyed fish can also walk on land, and it often escapes trouble by jumping out of the water, so although it is interesting to watch, this fish would make a poor choice for a home aquarium.

Though there are literally millions of examples of adaptations crafted by natural selection, these four examples—cooperative weaver ants, camouflaging caterpillars, wasp-mimicking orchids, and four-eyed fish—illustrate the most important characteristics of evolutionary adaptations:

- Adaptations show a close match between organism and environment.
- Adaptations are often complex.
- Adaptations help organisms accomplish important functions, such as feeding, defense against predators, and reproduction.

FIGURE 14.16 An Adaptation That Facilitates Reproduction
The flowers of this orchid have evolved to look strikingly like the females of a wasp species that can be found flying in the area. The match is so good that the orchids are able to achieve pollination by being "mated" by a fooled male wasp. The floral "female wasp" is in the center of the photo.

There are limits to adaptation

Adaptations in nature are impressive; however, natural selection rarely results in a perfect match between an organism and its environment. In many cases, genetic constraints and trade-offs between advantages and disadvantages that come with an organism's adaptation prevent further improvements in that adaptation. Scientists estimate that 99 percent of all species that have ever lived are now extinct. Every extinct species is a silent testament to an inability to adapt in the face of adversity.

For a population to become better adapted over successive generations, there must be genetic variation for traits that can enhance the match between the organism and its environment. In some cases the absence of genetic variation places a direct limit on the ability of natural selection to generate adaptations in descendant populations. For example, the mosquito *Culex pipiens* is now resistant to pesticides, but this resistance is based on a single mutation that occurred in the mosquito in the 1960s. Before this mutation, adaptation to these pesticides could not take place, and billions of mosquitoes died because their populations lacked the particular allele that could confer resistance to the pesticides.

Ecological trade-offs may also limit adaptation. To survive and reproduce, organisms must perform many functions, such as finding food and mates, avoiding predators, and surviving challenges posed by

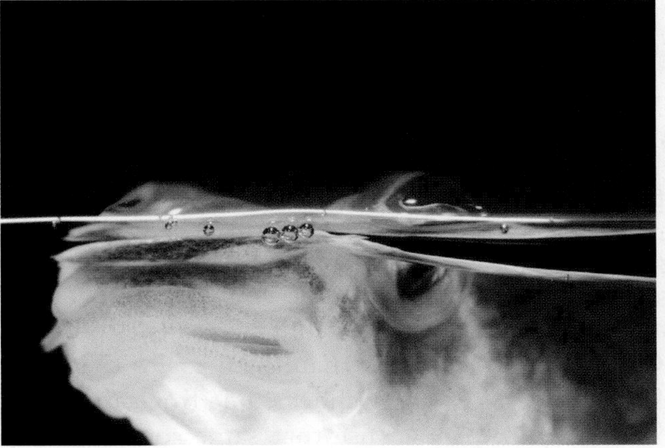

FIGURE 14.17 A Fish with Four Eyes?
The four-eyed fish (*Anableps anableps*) really has only two eyes, but each eye has a special design that lets it see clearly both in air and in water.

Concept Check Answers

1. Natural selection is a nonrandom process that removes unfit variants from the gene pool, leaving only those organisms with the best traits. The result is a population highly adapted to its environment.

2. The process of adaptation may be limited by low genetic variation, developmental constraints, or ecological trade-offs.

FIGURE 14.18 Does Love or Death Await? Male túngara frogs face an ecological trade-off: the same type of call that is most successful at attracting females also makes it easier for predatory bats to locate calling males.

the physical environment. Within the realm of what is genetically and developmentally possible, natural selection increases the overall ability of the organism to survive and reproduce. However, the conflicting demands that organisms face result in trade-offs, or compromises, in their ability to perform these essential functions.

High levels of reproduction, for example, are often associated with decreased longevity. In some cases, such a trade-off is due to relatively subtle costs of reproduction: resources directed toward reproduction are not available for other uses, such as storing energy to help the organism survive a cold winter. In red deer, for example, females that reproduced in the spring have a higher rate of death the following winter than do females that did not reproduce.

Costs associated with reproduction can sometimes be immediate and dramatic, as illustrated by the mating calls of the túngara frog. Male túngara frogs perform a complex mating call that may or may not end in one or more "chucks." Females prefer to mate with males that emit chucks, but frog-eating bats use that same sound to help them locate their prey. As a result, a frog's attempt to locate a mate can end in disaster (**FIGURE 14.18**).

EXTREME DIMORPHISM

Male elephant seals fight with one another to occupy stretches of beach where females come to mate. Sexual selection favors larger males that can dominate smaller males and secure mates. These seals exhibit significant sexual dimorphism: bulls can reach 20 feet in length and weigh 4 tons, while cows typically are about 10 feet long and weigh 1 ton.

Concept Check

1. Why do organisms show so many adaptations to life in their environment?
2. What are some reasons that adaptations are seldom perfect?

14.5 Sexual Selection

As the túngara frogs show us, an adaptation does not always confer a survival advantage to an organism. Natural selection favors individuals that can pass on their genes, even if doing so costs them their lives.

Túngara frogs exhibit a special form of natural selection, called **sexual selection**. Sexual selection favors individuals that are good at securing mates. It is often the reason behind the striking anatomical features and bizarre behaviors of courtship and mating that we see in other animals. For example, males and females of many species differ greatly in size, appearance, and behavior. In lions, for example, males are considerably larger than females; and males fight, sometimes violently, for the privilege of mating with females. Since larger males are stronger and tend to be more successful in fights for females, natural selection may have favored large size in males, leading to the substantial size difference between male and female lions. A species in which males and females look distinctly different is said to exhibit **sexual dimorphism**.

In many species, the members of one sex—often females—are choosy about whether to mate and with whom. In peafowl, for example, brightly colored males may perform elaborate displays in their attempts to woo a mate (**FIGURE 14.19**). In other species, males may attract attention by other means, such as calling vigorously; females then select as their mates the males with the loudest calls.

We would expect that if the choosy partner bases her (or, occasionally, his) choice of a mate on a trait such as color or calling vigor, then that trait should serve as a good indicator of the quality of the mate. In blackbirds, females choose males with orange beaks more often than males with yellow beaks. It turns out that orange-beaked males have had fewer infections than yellow-beaked males; so by selecting males with orange beaks, the females are select-

FIGURE 14.19 Elaborate Male Courtship Displays Influence Mate Choice

Although a peacock's elaborate plumage impairs its ability to hide or flee from predators, the trait is favored because it is attractive to females.

ing males in good health. A high-quality mate, then, might be especially healthy and more likely to produce healthy offspring or to be better at guarding a nest of young or at gathering food. In mice, females can tell from the odor of a male's urine how many parasites he is harboring, and they use this information to select their mates.

Concept Check

1. How is sexual selection a form of natural selection?
2. What mechanism of evolution is the only one that consistently leads a population toward adaptive evolution?

Concept Check Answers

1. Sexual selection favors individuals with genetic traits that are attractive to potential mates, which often include the ability to outcompete rivals.
2. Natural selection.

14.6 The Evidence for Biological Evolution

Evolution is considered by many to be a controversial topic. Kansas, notably, voted to strike evolution from its curriculum in 1999, only to reverse its decision in 2001 after severe criticism from the scientific and educational communities. Despite the misgivings of some members of the public, evolution has been a settled issue in science for nearly 150 years. Among scientists (and many nonscientists) the existence of evolution is so compelling that the concept of evolution has been elevated to the status of theory (in the scientific sense of the word).

Why do scientists find the case for evolution so convincing? As we saw in Chapter 1, a scientific hypothesis leads to predictions that can be tested, and hypotheses about evolution are no exception. Scientists have tested many predictions about evolution and have found that the evidence strongly supports the predictions. Six lines of evidence provide compelling support for biological evolution:

1. Fossils
2. Traces of evolutionary history in existing organisms
3. Similarities and differences in DNA
4. Direct observations of genetic change in populations
5. The distribution of organisms and fossils around the world
6. The present-day formation of new species

Evolution is strongly supported by the fossil record

Fossils are the preserved remains (or their impressions) of formerly living organisms (**FIGURE 14.20**). The fossil record enables biologists to reconstruct the history of life on Earth, and it provides some of the strongest evidence that species have evolved over time. For example, the fossil record shows that land-dwelling vertebrates (animals with internal skeletons) are absent in rocks that are older than 350 million years. On the other hand, fossils indicate that vertebrates lived in the oceans as far back as 525 million years ago. The rocks that entombed the oldest fossil amphibian contain fossils of fish, but none of reptiles or mammals.

The fossil record contains excellent examples of how major new groups of organisms arose from previously existing organisms. The evolutionary and ecological history of some groups is mirrored in an almost seamless fossil record that shows organisms changing from ancestral forms to forms that are still present. For example, scientists can track the evolutionary history of the horse in North America and match changes in the anatomy of descendant species with changes in the continent's climate over the past 60 million years. *Hyracotherium*, the earliest known member of the horse family (the Equidae, or equids), lived at a time when much of North America was covered in lush rainforest.

Helpful to Know

Critics of evolution often refer to evolution as "just a theory." Recall from Chapter 1, however, that a scientific theory is one that is widely accepted as fact. Other important theories widely accepted as fact include the cell theory, germ theory, and the theory of gravity. Gravity, strictly speaking, is also "just a theory."

FIGURE 14.20
Fossils Like This Triceratops Skeleton Show Evidence of Organisms Now No Longer Alive

Hyracotherium was a dog-sized animal that walked on four small toes on its front feet and three on the hind feet, and browsed on low-hanging vegetation from trees and bushes. Many of the equid species that appeared more recently had teeth adapted for grazing on tough grasses and legs adapted for fast running (**FIGURE 14.21**), consistent with a change in the climate that caused vast areas of forests to be replaced by grasslands.

About 32 million years ago, the continental interior of North America became drier, and prairies replaced forests in much of the central plains. Many of the fossil equids that first appear at this time in the central plains had longer legs than *Hyracotherium*. In addition to long legs for running, *Mesohippus* had its eye sockets farther back and higher on the head, so it had a wide field of view for spotting predators in the open grasslands. *Merychippus*, which appeared about 17 million years ago, ran on a single large toe, just as present-day horses, asses, and zebras do. Its molars had large, flat grinding surfaces for breaking up tough grasses—another similarity with modern equids. A lineage related to *Merychippus* evolved into the modern horse (genus *Equus*), which came to inhabit temperate grasslands across northern Eurasia and North America about 5 million years ago. However, North American equids became extinct toward the end of the Pliocene epoch, about 12,000 years ago. The horses we see today in North America were brought here in boats by European explorers.

Organisms contain evidence of their evolutionary history

A major prediction of evolution is that organisms should carry within themselves evidence of their evolutionary past—and they do. Related organisms often bear similar features, or the remnants of these features, because they descended from a common ancestor.

Patterns of growth in the earliest stages of life provide evidence of an organism's evolutionary past. When a sperm and egg fuse, an animal embryo begins to grow and develop (see Section 19.2). The manner in which an embryo develops, especially during the early stages, may mirror early developmental stages of its ancestors. For example, anteaters and some whales do not have teeth as adults, but as fetuses they do. Why should these organisms develop teeth during fetal development, only to reabsorb them later? Likewise, embryos of fishes, amphibians, reptiles, birds, and mammals (including humans) all develop gill pouches

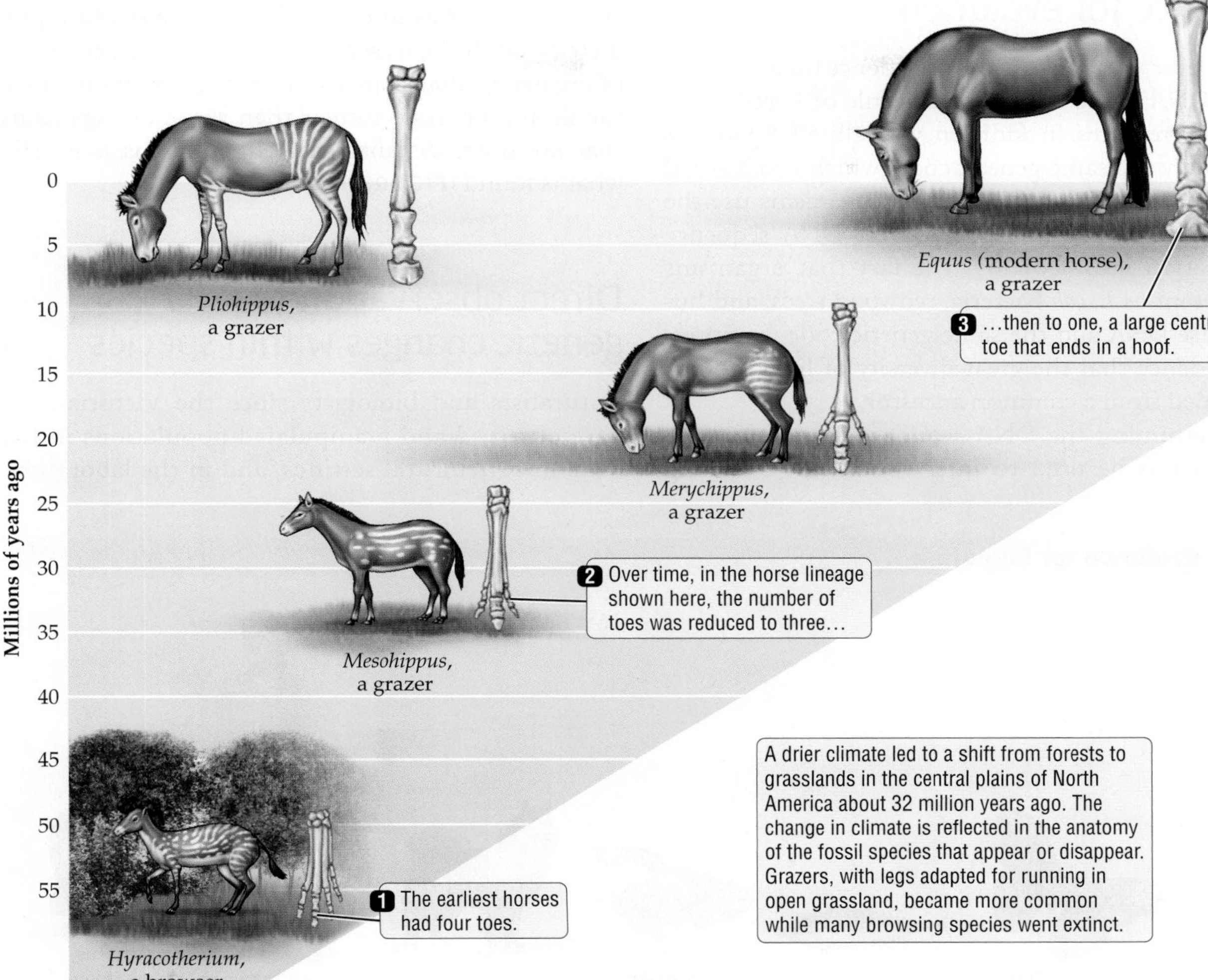

Fossil Evidence for Evolution
Millions of years ago
0
5
10
15
20
25
30
35
40
45
50
55
Pliohippus, a grazer
Equus (modern horse), a grazer
3 ...then to one, a large central toe that ends in a hoof.
Merychippus, a grazer
Mesohippus, a grazer
2 Over time, in the horse lineage shown here, the number of toes was reduced to three...
A drier climate led to a shift from forests to grasslands in the central plains of North America about 32 million years ago. The change in climate is reflected in the anatomy of the fossil species that appear or disappear. Grazers, with legs adapted for running in open grassland, became more common while many browsing species went extinct.
1 The earliest horses had four toes.
Hyracotherium, a browser

FIGURE 14.21 Fossils Reveal Descent with Modification in Horse Lineages
The evolutionary tree of the horse family (Equidae) is highly branched, as deduced from the fossil remains of more than 200 different species. Progressive changes in the anatomy, especially of the leg bones and teeth, are seen in some lineages.

(**FIGURE 14.22**). In fishes, the pouches develop into gills that the adults use to breathe underwater. But why should the embryos of organisms that breathe air develop gill pouches?

Evolution provides an answer to these puzzles: similarities in patterns of development are caused by descent from a common ancestor. Anteater and whale fetuses have teeth because anteaters and whales evolved from organisms with teeth. We see this in the fossil record. Similarly, fossil evidence indicates that the first mammals and the first birds evolved from reptiles, which are themselves descended from amphibians. The first amphibians evolved from a group of fishes. The embryos of air-breathing organisms such as humans, birds, lizards, and tree frogs all have gill pouches because all of these organisms share a common fish or fishlike ancestor.

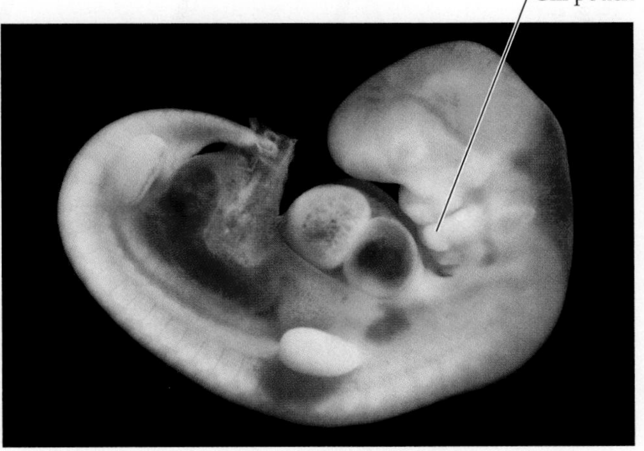

FIGURE 14.22 All Vertebrates Show Similarities in Embryological Development
A human embryo, about 30 days after fertilization. The pouches that develop into gills in fish become the voice box in humans.

DNA analysis provides some of the most compelling evidence for evolution

Within every organism is more evidence for evolution: DNA. DNA is the universal molecule of heredity for all living creatures. In addition, virtually all organisms use exactly the same genetic code (which you learned about in Chapter 12). That is, all organisms use the same cellular machinery to convert DNA sequence information into proteins. The fact that organisms as different as *E. coli* bacteria, redwood trees, and humans use DNA and the same genetic code is impressive evidence that the great diversity of living things descended from a common ancestor.

Determining the DNA sequence of different organisms can be used to infer relationships among organisms. By definition, related organisms have features in common, so related organisms are predicted to share an evolutionary history, which will be apparent as similarities in their DNA. Biologists have predicted that the DNA sequences and protein sequences of organisms that share a more recent common ancestor should be more similar than those of organisms that are more distantly related. And this is exactly what is found (**FIGURE 14.23**).

Direct observation reveals genetic changes within species

Naturalists and biologists since the Victorian era have observed and manipulated populations in the wild, in agricultural settings, and in the laboratory.

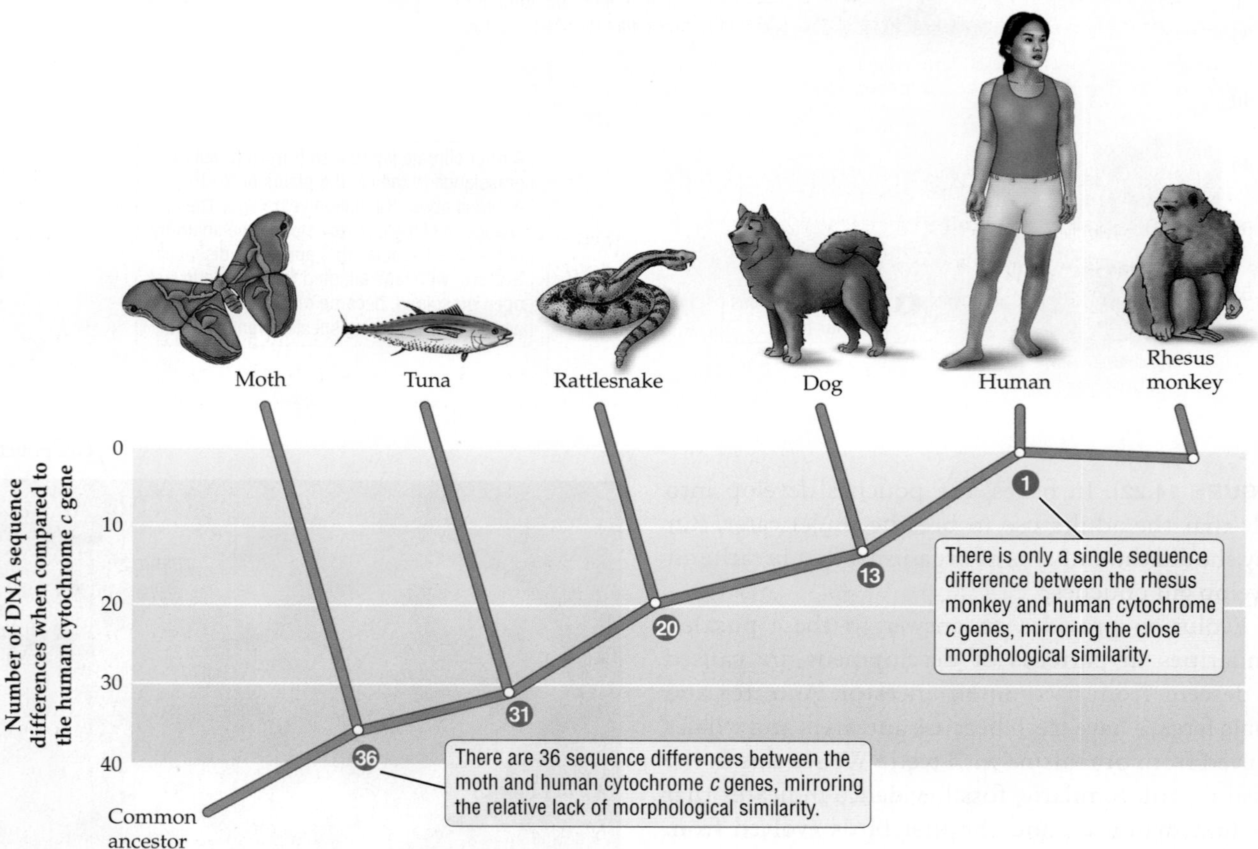

FIGURE 14.23 Independent Lines of Evidence Yield the Same Result

Cytochrome *c* is an essential metabolic enzyme that is found in all eukaryotes. From comparison of the number of DNA sequence differences between the cytochrome *c* gene found in humans and those found in other organisms, an evolutionary relationship tree can be constructed. The tree shown here, in which humans are most closely related to rhesus monkeys and least closely related to moths, matches the pattern derived independently from anatomical features.

Artificial selection is the breeding of animals and plants with specific, desirable characteristics. As Darwin recognized, breeding experiments conducted by humans provide direct, concrete evidence for evolution. Dogs, *Canis lupus familiaris,* are subspecies of the gray wolf (**FIGURE 14.24**). Humans began domesticating the gray wolf about 16,000 years ago. Individual wolves with desirable qualities, such as

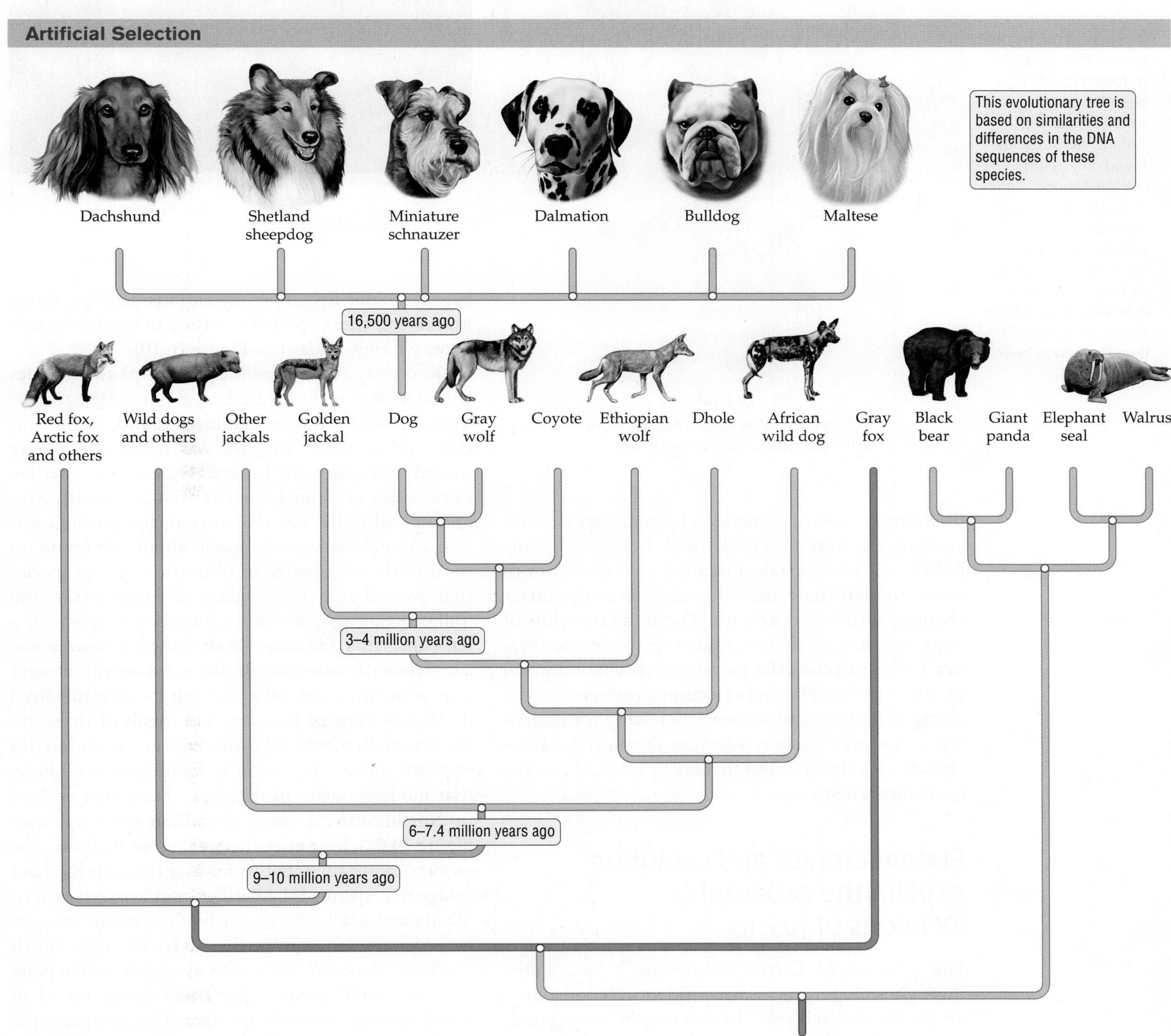

FIGURE 14.24 Artificial Selection Produces Genetic Change
Humans have directed the evolution of the gray wolf to produce the many breeds of dogs. Despite the dramatic differences exhibited by the gray wolf, the Chihuahua, and the Great Dane, all three are members of the same species: *Canis lupus.* Selective breeding for specific genetic characteristics is responsible for the difference in form and behavior among the dog breeds.

FIGURE 14.25 The Biogeography of a Lungfish Reflects Its Evolutionary Past

Ancestors of the freshwater lungfish *Neoceratodus fosteri* (*right*) lived during the time of Pangaea. *N. fosteri* fossils have been found on all continents except Antarctica. Currently found only in the orange-shaded region of northeastern Australia, this species is the only surviving member of its family. Red dots indicate places where fossils of *N. fosteri*'s ancestors have been found.

Biogeographic Evidence for Evolution

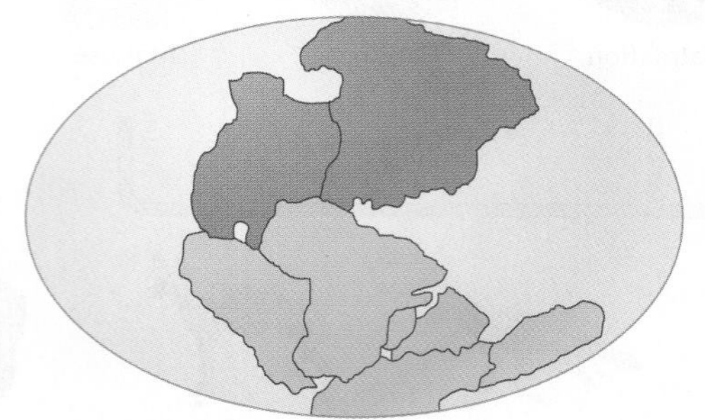

Portions of the supercontinent Pangaea began to drift apart about 200 million years ago.

friendliness toward owners and a tendency to bark at strangers, were selectively bred. By allowing only individuals with certain inherited characteristics to breed, humans have crafted enormous evolutionary changes within this species. The domestication of dogs, ornamental flowers, and crop species (see Figure 1.16) illustrates the power of artificial selection to alter species. Natural selection produces similar changes. Rather than a breeder selecting for desired traits, however, natural selection favors those individuals with the qualities that are best suited to environmental conditions.

Plate tectonics and evolution explain the geographic locations of fossils

The position of Earth's continents is not static; through geologic processes, continents are constantly on the move. Each year the distance between South America and Africa grows by about 3 centimeters, a little more than an inch. About 250 million years ago South America, Africa, and all the other landmasses of Earth formed one giant continent, called **Pangaea** (**pan-*JEE*-uh**). About 200 million years ago, Pangaea began to split up, driven by various geologic forces known together as **plate tectonics**, to form the continents we know today (see Figure 16.10).

Knowledge about evolution and plate tectonics can be used to make predictions about the geographic locations where fossils should be found. Organisms that evolved when Pangaea was intact could have moved relatively easily between regions that later became widely separated, such as what is currently Antarctica and India. For that reason, the prediction is that 250-million-year-old fossils should be found on most of the continents. In contrast, fossils of species that evolved after the breakup of Pangaea (say, 100 million years ago) should be found on only one or a few continents. For example, the lungfish *Neoceratodus fosteri* (**nee-oh-suh-*RAD*-uh-dus *FAW*-ster-ee**) is found only in northeastern Australia, but its ancestors lived during the time of Pangaea, and fossils of those ancestors are found on all continents except Antarctica (**FIGURE 14.25**). In contrast, fossil evidence shows that modern horses in the genus *Equus* first evolved in North America about 5 million years ago (see Figure 14.21), long after the breakup of Pangaea. The modern horse migrated to Eurasia through the land bridge that spanned the Bering Strait between eastern Siberia and Alaska during each of the recent ice ages. However, the land bridge that today connects North and South America was formed roughly 3 million years ago. We would predict that *Equus* fossils found in South America should be less than 3 million years old; and to date, all such discoveries have been less than 3 million years old. *Equus* fossils have not been found in Africa, Australia, or Antarctica, just as we would predict on the basis of the fossil history of *Equus* and changes in Earth's crust over that time span.

Humans Can Artificially Create New Species

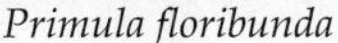

Primula floribunda

Primula kewensis

Primula verticillata

FIGURE 14.26
A New Species

In the early twentieth century, botanists at the Royal Botanic Gardens at Kew hybridized two primrose species—*Primula floribunda* (from the Himalayas, *left*) and *Primula verticillata* (from Saudi Arabia, *right*)—to form a new species, *Primula kewensis*, that is reproductively incompatible with either parent.

Formation of new species can be observed in nature and induced experimentally

Darwin believed that evolution works so slowly that humans could never observe new species arising from existing ones. Not so. Scientists have documented many new species being formed. The first experiment in which a new species was formed took place in the early 1900s, when the primrose *Primula kewensis* (*PRIM*-**yoo-luh kee-***WEN***-sis**) was produced by hybridizing two species of primrose (one from the Himalayas, the other from Saudi Arabia) to form a distinct new species that cannot interbreed with either parent species (**FIGURE 14.26**). In an example from nature, two new species of salsify (*SAL*-**suh-fee**) plants were discovered in Idaho and eastern Washington in 1950. Genetic data revealed that both of the new species had evolved from previously existing species, and field surveys indicated that this event had occurred between 1920 and 1950. The two new species continue to thrive, and one of them has become common since its discovery in 1950.

The evidence for evolutionary change is so widespread and compelling that the scientific community is united in its acceptance of evolution as a fact. Those who deny evolutionary theory must ignore or explain away the substantial body of scientific evidence that supports the fact that species change over time.

Concept Check

1. List five separate lines of evidence that life has evolved.
2. What is artificial selection?

14.7 The Impact of Evolutionary Thought

The evolution of species was a radical idea in the mid-nineteenth century, and the argument that the form and function of organisms could be explained by natural selection was even more radical. These ideas not only revolutionized biology but also had profound effects on other fields, from literature to philosophy to economics.

The idea of Darwinian evolution had a profound effect on religion as well. Evolution was viewed initially as a direct attack on Judeo-Christian religion, and this presumed attack prompted a spirited counterattack by many prominent members of the clergy. Today, however, most religious leaders and most scientists view evolution and religion as compatible but distinct fields of inquiry (**FIGURE 14.27**). The Catholic Church, for example, accepts that evolution explains the anatomical characteristics of humans but maintains that religion is required to explain our spiritual characteristics. Similarly, although the vast majority of scientists accept the scientific evidence for evolution, many of those same

Concept Check Answers

1. Fossil record, otherwise inexplicable features of existing organisms (like gill pouches in human embryos), similarities and differences in DNA, direct observation of genetic change, distribution of organisms and fossils around the world as a result of plate tectonics, and experimental formation of new species.

2. A process by which humans direct evolution in other species—like crops and domesticated animals—by allowing only individuals with certain characteristics to breed.

(a)

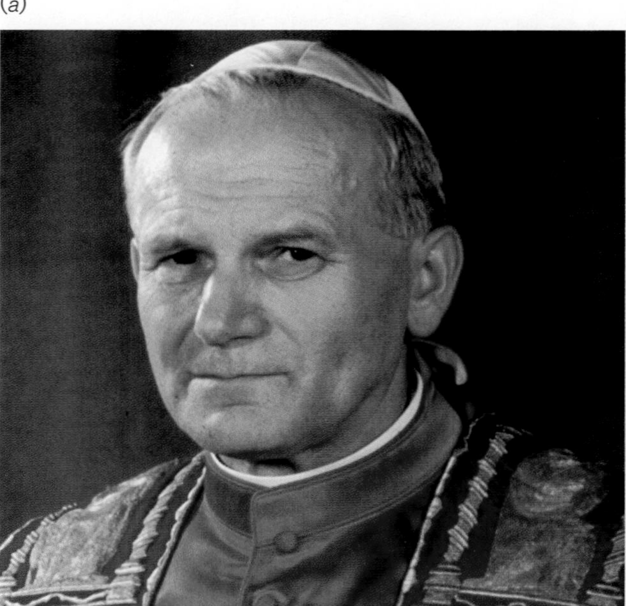

(b)

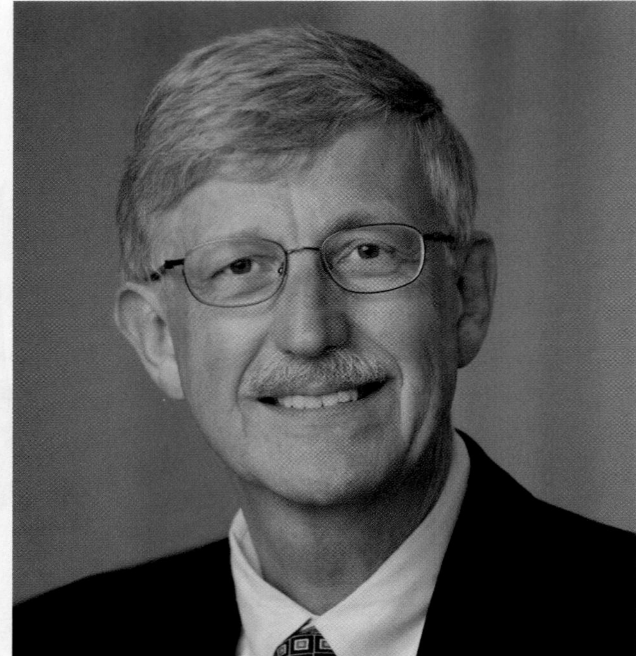

FIGURE 14.27 Evolution and Religious Beliefs Are Compatible
Although evolutionary biology is commonly perceived to be in opposition to religious orthodoxy, many scientists and religious leaders recognize that the two beliefs occupy separate spheres of knowledge. Pope John Paul II (*a*) and Francis Collins (*b*), head of the Human Genome Project and a devout Christian, have both expressed statements that religious views are fully compatible with modern science's understanding of evolutionary biology.

scientists hold religious beliefs. Overall, most scientists recognize that religious beliefs are up to the individual and that science cannot answer questions regarding the existence of God or other matters of religious importance.

The emergence of evolutionary thought has also had an effect on human technology and industry. For example, an understanding of evolution has proved essential as farmers and researchers have sought to prevent or slow the evolution of resistance to pesticides by insects. A new approach to pest control advocates the use of several different insecticides on a rotating basis. The alleles that confer resistance to one type of insecticide may increase in one year, but when the type of poison is changed, these alleles no longer confer any advantage. As a result, the population as a whole remains susceptible to the battery of insecticides available to farmers.

Evolutionary biology is sometimes called the central and unifying theme of biology. Its importance in helping to understand the biological world is so profound that in 1973, Theodosius Dobzhansky, a prominent evolutionary biologist, famously declared, "Nothing in biology makes sense except in the light of evolution."

Concept Check Answer

1. An understanding of evolutionary change has helped scientists to understand and combat phenomena such as insecticide resistance.

Concept Check

1. What are some practical ways that evolutionary thought has affected humanity?

APPLYING WHAT WE LEARNED

Darwin's Finches: Evolution in Action

During Charles Darwin's 5-year voyage around the world, he began, tentatively, to question his assumption that species could not change. There had been suggestions that species evolve—hints that simmered in the back of his mind. For example, Galápagos Islands locals had told him they could tell which island a tortoise came from by its size and shape. Every island had its own tortoise. And Darwin had noticed the same thing with the island's mockingbirds, about which he wrote, "Each variety is constant on its own island. This is a parallel fact to the one mentioned about the Tortoises." Months later, he concluded a similar discussion in his notes by saying that "such facts would undermine the stability of species." Darwin was beginning to believe that species might change over time.

When Darwin returned to England, he turned over his collection of birds to the renowned ornithologist John Gould. Darwin thought that each island had a different subspecies of mockingbird; Gould disagreed, telling Darwin that the birds were separate species. Darwin thought that the 12 birds he had brought back were a mix of blackbirds, grosbeaks, a wren, and some finches. But Gould said no, the 12 birds were all ground finches. Darwin was stunned. The birds looked so different and yet were so closely related. He later wrote, "One might really fancy that ... one species had been taken and modified for different ends." He guessed that all the finches in the Galápagos Islands were descended from a single mainland species that had split into a dozen new species.

The Galápagos Islands serve as a natural laboratory where scientists study how evolution works. The islands lie squarely on the equator in a cold ocean current. From January to May the weather is warm and rainy. The rest of the year the islands are generally dry. In 1977, Peter and Rosemary Grant, who have studied Galápagos finches every year since 1973, were on one of the islands studying medium ground finches when drought struck. Plants withered (**FIGURE 14.28**), seeds became scarce, and the seed-eating medium ground finches starved, their population plummeting from 1,200 to 180.

The plants that were able to reproduce during the dry spell were mainly drought-tolerant species that produce large seeds with hard seed coats—seeds that the finches would generally ignore in better times. Now, the large seeds of these species, especially of a plant named *Tribulus cistoides*, became the main food source for the surviving birds. Was there anything special about these survivors living mainly off *T. cistoides*? The Grants compared the average beak size of the predrought (1976) population of medium ground finches and the descendants of the drought survivors a year after the worst of the drought had passed (in 1978). Their analysis revealed that in just one generation, average beak size in the postdrought population had increased by about half a millimeter (**FIGURE 14.29**).

A thick, more powerful beak was advantageous under drought conditions because it enabled its owners to crack open the hard seed coats to get at the limited supply of food. Natural selection rapidly culled the less fit individuals—as most of the individuals with small beaks died of starvation—favoring the survival of individuals with stout beaks. Because beak size is an inherited trait, the greater reproductive success of large-beaked individuals meant that large-beaked finches became more common in the next generation; therefore, the postdrought population of finches had evolved. Exceptionally heavy rains in the early 1980s brought luxuriant plant growth, an abundance of small seeds with softer seed coats, and a decline in the large-seeded drought-adapted plants. In response to this change in the food supply, the finch population evolved yet again: the average beak size dropped to the predrought size over the rest of the 30-year study period.

FIGURE 14.28 A Drought Results in Rapid Evolutionary Change

The 1977 drought on Daphne Major in the Galápagos Islands had a dramatic effect on the plant life there, setting the stage for natural selection to cause rapid evolutionary change in birds that depended on the plants for food.

Adaptation to Drought: Change in Beak Size

(a) Medium ground finch
(b) Beak size over study period
Large
Rains
0.5
Beak size
0
−0.5
Drought
Small
1970
1980
1990
2000
Year

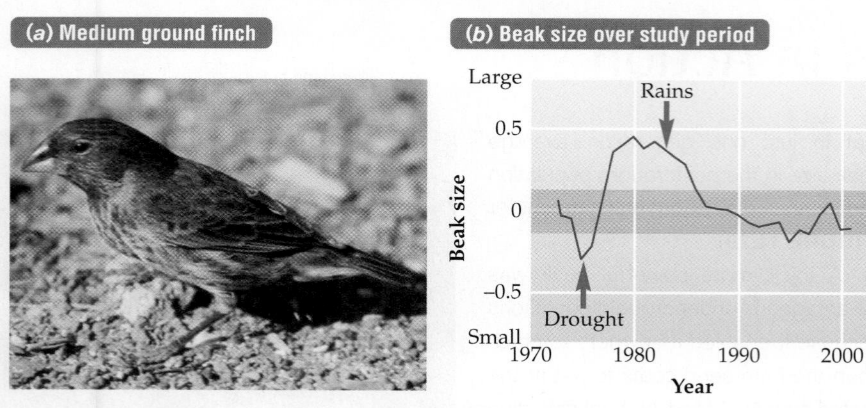

FIGURE 14.29 Evolutionary Change within a Species
Scientists have been studying the medium ground finch (*a*) on the Galápagos island Daphne Major for more than 30 years. (*b*) Average beak size increased in the population after a severe drought in 1977. When rainfall resumed, natural selection drove the evolution of smaller beaks. The orange band indicates the range of values that would be expected without evolutionary change.

Comparison of DNA from the 13 finch species in the Galápagos, and one species from the Cocos Islands to the north, have confirmed Darwin's hunch about the origins of these birds. All 14 species are descended from a single ancestral form, very likely an insect-eating warbler finch from nearby Ecuador that migrated to the islands between 2 and 3 million years ago. As the members of the ancestral population colonized diverse habitats, with varied vegetation, they adapted to local conditions, especially in their eating habits (**FIGURE 14.30**). For example, a lineage of ground finches evolved that specializes in eating seeds. Other finches took to the trees, using their blunt-ended beaks like pruning shears to snip off vegetation. The short, stubby beak of the vegetarian finch is suited to a diet of buds, flowers, and fruits. The woodpecker finch deploys a cactus spine in its tweezers-like beak to pry insects from the nooks and crannies of trees and shrubs.

A subspecies of the sharp-beaked ground finch that is found on Wolf Island and on Darwin Island has acquired a taste for blood. The behavior may have evolved from this finch's habit of picking parasites off basking marine iguanas and large seabirds such as blue-footed boobies. Sometimes, usually when food is scarce, the finches will peck at the thin skin on the back of the seabirds and lap up the blood that oozes out.

Evolutionary History of the Galápagos Finches

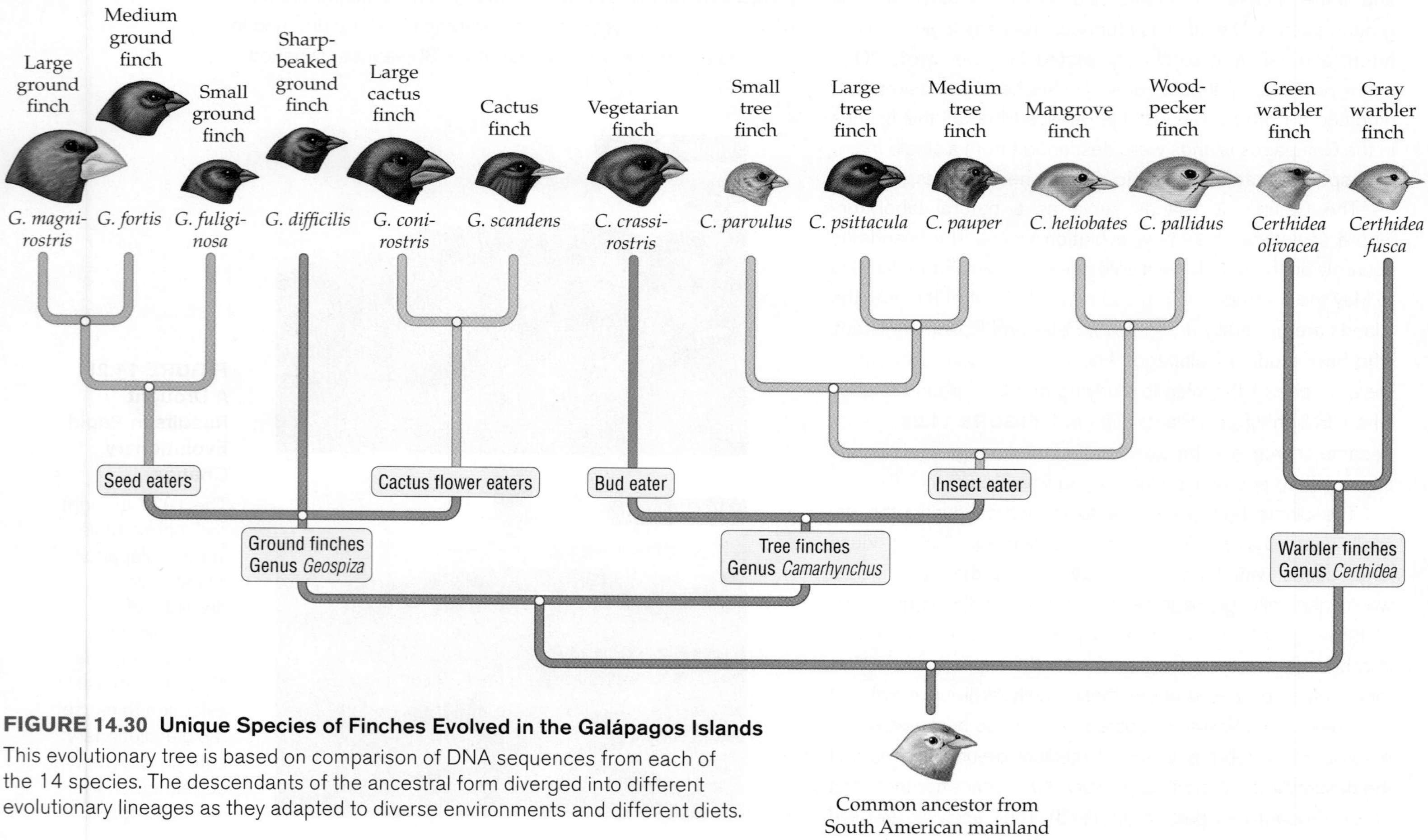

Large ground finch
Medium ground finch
Small ground finch
Sharp-beaked ground finch
Large cactus finch
Cactus finch
Vegetarian finch
Small tree finch
Large tree finch
Medium tree finch
Mangrove finch
Wood-pecker finch
Green warbler finch
Gray warbler finch
G. magnirostris
G. fortis
G. fuliginosa
G. difficilis
G. conirostris
G. scandens
C. crassirostris
C. parvulus
C. psittacula
C. pauper
C. heliobates
C. pallidus
Certhidea olivacea
Certhidea fusca
Seed eaters
Cactus flower eaters
Bud eater
Insect eater
Ground finches Genus Geospiza
Tree finches Genus Camarhynchus
Warbler finches Genus Certhidea
Common ancestor from South American mainland

FIGURE 14.30 Unique Species of Finches Evolved in the Galápagos Islands
This evolutionary tree is based on comparison of DNA sequences from each of the 14 species. The descendants of the ancestral form diverged into different evolutionary lineages as they adapted to diverse environments and different diets.

BIOLOGY IN THE NEWS

Evolution via Roadkill

BY SARAH WILLIAMS • *Science*, March 18, 2013

Cliff swallows that build nests that dangle precariously from highway overpasses have a lower chance of becoming roadkill than in years past thanks to a shorter wingspan that lets them dodge oncoming traffic. That's the conclusion of a new study based on 3 decades of data collected on one population of the birds. The results suggest that shorter wingspan has been selected for over this time period because of the evolutionary pressure put on the population by cars.

"This is a clear example of how you can observe natural selection over short time periods," says ecologist Charles Brown of the University of Tulsa in Oklahoma, who conducted the new study with wife Mary Bomberger Brown, an ornithologist at the University of Nebraska, Lincoln. "Over 30 years, you can see these birds being selected for their ability to avoid cars."

The Browns have studied cliff swallows (*Petrochelidon pyrrhonota*) in southwestern Nebraska since 1982. They return to the same roads every nesting season to perform detailed surveys of the colonies of thousands of birds that build mud nests on bridges and overpasses in the area . . .

When the researchers looked back at the numbers of swallows collected as roadkill each year, they found that the count had steadily declined from 20 birds a season in 1984 and 1985 to less than five per season for each of the past 5 years. During that same time, the number of nests and birds had more than doubled, and the amount of traffic in the area had remained steady.

The birds that were being killed, further analysis revealed, weren't representative of the rest of the population. On average, they had longer wings. In 2012, for example, the average cliff swallow in the population had a 106-millimeter wingspan, whereas the average swallow killed on the road had a 112-millimeter wingspan.

"Probably the most important effect of a shorter wing is that it allows the birds to turn more quickly," says Charles Brown. Previous studies on the dynamics of flight have illustrated the benefits of short wings for birds that perform many pivots and rolls during flying and shown that shorter wings also may allow the birds to take off faster from the ground, he adds.

When the researchers analyzed the average wing length of the living birds in the population, they discovered that it had become shorter over time, from 111 millimeters in 1982 to the 106 millimeter average in 2012. The data suggested to the Browns that roadkill deaths were a major force driving this selection. Birds with longer wings would be more likely to be killed by vehicles and less likely to reproduce, the team reports online today in *Current Biology*.

The data illustrate a "beautiful trend that never could have been predicted," says evolutionary biologist John Hoogland of the University of Maryland Center for Environmental Science in Frostburg, who was not involved in the study. "We humans, because we're changing the environment so much, are adding a new kind of natural selection to these animal populations."

Humans have long been tough on wildlife. We have hunted and fished numerous species to extinction. Farming and the development of cities and suburbs have decimated many populations of plants and animals. But as the study described in the article indicates, humans can also have more subtle impacts on the evolution and adaptation of wildlife. Cliff swallows are birds that eat insects on the wing (catching them in midair) and natural selection has selected for slender wings that allow for speed and flying over long distances.

The presence of humans and human-made structures, however, has changed the selective environment for these birds. Swallows commonly nest in artificial structures such as buildings and bridges, placing them in close contact with humans. Threats such as speeding vehicles present new challenges for the birds and their flying style. As the data indicate, shorter, broader wings with higher wing loads are selectively favored because birds with these wings would be better able to avoid cars on the roads.

As you have seen in this chapter, adaptations are never perfect. In this case, different environmental pressures—namely, the presence of speeding vehicles—have led to an evolutionary response in the swallows, favoring birds with shorter wings. Natural selection is a continuous process of refinement leading to the wondrous adaptations we see in nature.

Evaluating the News

1. Why are swallow wings normally long and slender? What evolutionary pressures made this shape adaptive?

2. What is the hypothesis for the change in wing shape? How might scientists investigate this question further?

3. In what other ways might humans be influencing the evolution of wild organisms?

CHAPTER REVIEW

Summary

14.1 Evolution and Natural Selection

- Evolution can be broadly defined as change in the genetic characteristics of populations of organisms over successive generations.
- A number of scholars had advanced the idea that present-day species are descended from more ancient forms. Charles Darwin and Alfred Wallace offered a mechanism for the evolution of new species through modification of preexisting forms: natural selection.
- According to the theory of natural selection, individuals with adaptive traits (characteristics that enable them to function well in a competitive environment) produce more offspring, and therefore their characteristics become more common in descendant generations.

14.2 Mechanisms of Evolutionary Change

- Individuals in populations differ genetically in their structural, biochemical, and behavioral characteristics. The gene pool is the sum of all the genetic information carried by all the individuals in a population. Evolution can be defined as a change in the gene pool of a population over successive generations.
- Mutations in DNA are the ultimate source of all genetic variation. Mutations are random events, and as such they are not goal directed. Once introduced into a gene pool, however, mutant alleles can be acted upon by natural selection to increase or decrease in frequency.
- Gene flow is the movement of genes from one population to another. A resident population can evolve if its gene pool is altered by the introduction or loss of novel alleles as individuals immigrate or emigrate.
- Genetic drift is a random process that can cause changes in allele frequencies over time. Genetic drift is more pronounced in small populations than in large ones.

14.3 Natural Selection Leads to Adaptive Evolution

- Natural selection is the greater reproductive success of individuals with advantageous genetic characteristics, compared to competing individuals with other characteristics. The characteristics of individuals that produce more offspring become more common in the succeeding generations.
- Natural selection is a nonrandom process, and it generates adaptation in descendant populations. Adaptation is the process by which a population becomes better matched to its environment.
- Natural selection can influence populations in three ways: by favoring one extreme of a characteristic's range (directional selection), by favoring the intermediate value (stabilizing selection), or by favoring both of the extremes (disruptive selection).

14.4 Adaptations

- An adaptive trait is an inherited characteristic—structural, biochemical, or behavioral—that enables an organism to function well in its environment and that therefore increases survival and reproductive success.
- Adaptive evolution can be limited by genetic constraints (lack of genetic variation gives natural selection little or nothing to act on), developmental constraints, and by ecological trade-offs (conflicting demands faced by organisms can compromise their ability to perform important functions).
- Adaptation does not craft perfect organisms.

14.5 Sexual Selection

- Sexual selection is a special form of natural selection in which individuals with certain anatomical characteristics differ in their ability to secure mates.
- Sexual selection underlies many differences between males and females, such as differences in size and courtship behavior.

14.6 The Evidence for Biological Evolution

- The fossil record provides clear evidence for the evolution of species over time.
- Within organisms, evidence for evolution includes remnant anatomical structures, patterns of embryonic development, the universal use of DNA and the genetic code, and molecular similarities among organisms.
- As predicted by our understanding of evolution and plate tectonics, fossils of organisms that evolved when the present-day continents were all part of the supercontinent Pangaea have a wider geographic distribution than do fossils of more recently evolved organisms.

14.7 The Impact of Evolutionary Thought

- Darwin's ideas on evolution and natural selection revolutionized biology, overturning the views that species do not change with time.
- Evolutionary biology has many practical applications in agriculture, industry, and medicine.

Key Terms

adaptation (p. 318)
allele (p. 313)
allele frequency (p. 312)
artificial selection (p. 329)
directional selection (p. 318)
disruptive selection (p. 321)
extinction (p. 315)
fixation (p. 315)
fossil (p. 326)
founder effect (p. 315)
gene flow (p. 314)
gene pool (p. 312)
genetic bottleneck (p. 315)
genetic drift (p. 315)
genetic variation (p. 313)
mutation (p. 313)
natural selection (p. 312)
Pangaea (p. 330)
plate tectonics (p. 330)
sexual dimorphism (p. 324)
sexual selection (p. 324)
stabilizing selection (p. 321)

Self-Quiz

1. In natural selection,
   a. the genetic composition of the population changes randomly over time.
   b. new mutations are generated over time.
   c. all individuals in a population are equally likely to contribute offspring to the next generation.
   d. individuals that possess particular inherited characteristics consistently survive and reproduce at a higher rate than other individuals.
2. A study of a population of the goldenrod *Solidago altissima* (**sah-luh-DAY-goh al-TIS-ih-muh**) finds that large individuals consistently survive at a higher rate than small individuals. Assuming size is an inherited trait, the most likely evolutionary mechanism at work here is
   a. disruptive selection.
   b. directional selection.
   c. stabilizing selection.
   d. natural selection, but it is not possible to tell whether it is disruptive, directional, or stabilizing.
3. Use the Hardy-Weinberg equation (see the "Biology Matters" box on page 317) to solve the following problem: If the frequency of the *A* allele is 0.7 and the frequency of the *a* allele is 0.3, what is the expected frequency of individuals of genotype *Aa* in a population that is not evolving?
   a. 0.21
   b. 0.09
   c. 0.49
   d. 0.42
4. Adaptations
   a. match organisms closely to their environment.
   b. are often complex.
   c. help the organism accomplish important functions.
   d. all of the above
5. The fossil record shows that the first mammals evolved 220 million years ago. The supercontinent Pangaea began to break apart 200 million years ago. Therefore, fossils of the first mammals should be found
   a. on most or all of the current continents.
   b. only in Antarctica.
   c. on only one or a few continents.
   d. only in Africa.
6. Differences in survival and reproduction resulting from chance events can cause the genetic makeup of a population to change randomly over time. This process is called
   a. mutation.
   b. natural selection.
   c. macroevolution.
   d. genetic drift.
7. Artificial selection is the process by which
   a. natural selection fails to act in wild populations.
   b. humans prevent natural selection.
   c. humans allow only organisms with specific characteristics to breed.
   d. humans cause genetic drift in domesticated populations.
8. Which of the following provides evidence for evolution?
   a. direct observation of genetic changes in populations
   b. sharing of characteristics between organisms
   c. the fossil record
   d. all of the above

Analysis and Application

1. Explain what evolution is and why we state that populations evolve but individuals do not.
2. A population of lizards lives on an island and eats insects found in shrubs. Because the shrubs have narrow branches, the lizards tend to be small (so they can move effectively among the branches). A group of lizards from this population migrates to a nearby island where the vegetation consists mostly of trees; the branches of those trees are thicker than the shrubs on which the lizards formerly fed. A few of the lizards that migrate to the new island are slightly larger than the others; large size is not a disadvantage for moving in the trees (because the branches are thicker) and is advantageous when males compete with other males to mate with females. Assuming that large size is an inherited characteristic, explain what is likely to happen to the average size of the lizards in their new home, and why.
3. Select one of the four evolutionary mechanisms discussed in this chapter (mutation, gene flow, genetic drift, or natural selection), and describe how it can cause allele frequencies to change from generation to generation.
4. In your own words, define and explain the following: (a) gene flow; (b) genetic drift; (c) natural selection; (d) sexual selection.
5. Genetic drift is more likely to produce evolutionary change in a small population than in a large population. Explain why.
6. Select an organism (other than humans) that is familiar to you. List two adaptations of that organism. Explain carefully why each of these features is an adaptation.
7. What is adaptive evolution? Apply your understanding of adaptive evolution to organisms that cause infectious human diseases, such as bacterial species that cause plague or tuberculosis. How do our efforts to kill such organisms affect their evolution? Are the evolutionary changes we promote usually beneficial or harmful for us? Explain your answer.
8. Why are scientists throughout the world convinced that evolution happens? Consider at least three of the lines of evidence discussed in this chapter.

15 The Origin of Species

CICHLID DIVERSITY. Cichlids are an extremely diverse group of fishes. Africa alone is believed to have at least 1,600 species of cichlids. Hundreds more have evolved in other parts of the world, from Madagascar, India, and Syria to Central America, Mexico, and the southern United States.

Cichlid Mysteries

Over time, the surface of Earth changes slowly but dramatically. Chains of islands rise from the sea; new lakes form and old ones disappear; mountains thrust upward, miles above sea level; and rivers cut massive canyons that divide continents. Such changes separate populations of organisms, alter the environments in which they live, and set the stage for evolution.

Some of the most remarkable examples of rapid evolutionary change are the cichlid (*SIK*-**lud**) fishes of Lake Victoria in East Africa. Since its formation 400,000 years ago, Victoria has dried up and filled with water again and again. It last filled with water about 15,000 years ago, and until the 1970s it was home to about 500 species of cichlids—more kinds of fishes than in all the lakes and rivers of Europe. In the 1970s, researchers reported so many fish that, in just 10 minutes, they were able to catch 1,000 fish of 100 different species. Amazingly, all of these cichlid fishes descended from just two ancestor species from nearby Lake Kivu, and all had evolved in 15,000 years.

In the 1960s, human activities began to significantly degrade Lake Victoria. Rapidly growing populations drained sewage and industrial chemicals into the lake, causing eutrophication (the overgrowth of algae and loss of oxygen in the water). Neighboring forests were logged, and with nothing holding the soil in place, it eroded, pouring sediments into the lake. People overfished the lake and introduced a predatory fish, the gigantic Nile perch, that ate cichlids faster than they could reproduce. The Nile perch ate so well that the giant fish drove 200 species of cichlids to extinction in the 1980s. Biologists wrung their hands in anguish over the destruction of so much biodiversity. Yet the lake's evolutionary history might have given them reason to hope. In this chapter we'll learn how new species arise.

How could so many species have evolved in just 15,000 years? What hope was there that the cichlids might adapt to the presence of perch and polluted lake water?

MAIN MESSAGE New species arise when populations no longer exchange alleles with one another.

KEY CONCEPTS

- The biological species concept defines a species as a group of populations that can potentially interbreed in nature to produce fertile offspring and that is reproductively isolated from other groups.
- Speciation is the process by which one species splits to form two or more new species.
- Speciation is usually a by-product of genetic divergence between populations. The genetic divergence is promoted by mutation, genetic drift, and natural selection.
- Speciation often occurs when populations of a species become geographically isolated, because isolation limits gene flow between the populations, making genetic divergence more likely.
- Speciation can also occur in the absence of geographic isolation—for example, when gene flow is blocked because of differences in mating behavior.
- Under certain conditions, a single lineage may diversify into many descendant species—a phenomenon known as an adaptive radiation.
- Two species derived from the same ancestral lineage will share traits because of their common ancestry.
- Species formation generally takes thousands of years, but it has been observed in the span of a single generation.

CICHLID MYSTERIES 339

15.1 What Are Species? 340

15.2 Speciation: Generating Biodiversity 342

15.3 Adaptive Radiations: Increases in the Diversity of Life 348

15.4 Evolution Can Explain the Unity and Diversity of Life 349

15.5 Rates of Speciation 351

APPLYING WHAT WE LEARNED 352
Lake Victoria: Center of Speciation

WHAT ARE SPECIES? It turns out the term "species" is not so easy to define. Building on the previous chapter's introduction of natural selection and the other forces that drive evolution, in this chapter we consider how species arise. We then discuss the mechanisms that cause populations to diverge to the point that they become different species. The chapter closes with how those mechanisms have led to the spectacular diversity of life on Earth.

15.1 What Are Species?

It may surprise you to learn that there are several ways to answer the question "What are species?" While the word **species** is commonly applied to members of a group that can mate with one another to produce fertile offspring, not all species can be defined by their ability to interbreed. For example, many species, including all bacteria, reproduce asexually. As a result, various "species concepts" have been advanced by biologists, which, taken together, help us understand what defines a species.

Species are often morphologically distinct

We can often clearly see the difference between two species; for example, we can readily distinguish polar bears (*Ursus maritimus*) from brown bears (*Ursus arctos*). Identifying species by their morphology, or external form, has long been a method of distinguishing species. This strategy is based on what is known as the **morphological species concept**, the notion that many species can be identified as a separate and distinct group of organisms by the unique set of morphological characteristics they possess. Indeed, morphology is sometimes the only way we can identify and distinguish fossil species.

However, the morphological species concept does not always work well. Sometimes distinct and separate species have members that look very much alike (**FIGURE 15.1**). For example, researchers from the Smithsonian Institution were startled to discover that the 3 species of *Starksia* blennies they had been studying in the Caribbean islands were really *10* different species. Each of the 10 species of these reef fishes inhabits a quite different habitat in the western Atlantic, and the species cannot interbreed; however, even experts cannot tell all 10 species apart by appearance alone.

Conversely, different populations can vary in appearance (sometimes dramatically) yet be assigned to

FIGURE 15.1 Distinct Species May Look Similar to One Another
The eastern meadowlark (*a*) and the western meadowlark (*b*) are very similar in form, but they are distinct species. How do you recognize species? What makes these two birds distinct species? Why do these birds look so similar?

the same species. All the brown bears of the world, for example—including the grizzly bear of inland Alaska and the much larger coastal brown bear—belong to one species: *Ursus arctos*. Likewise, *Heliconius* butterflies of Central and South America exhibit extraordinary morphological variation within a single species (**FIGURE 15.2**).

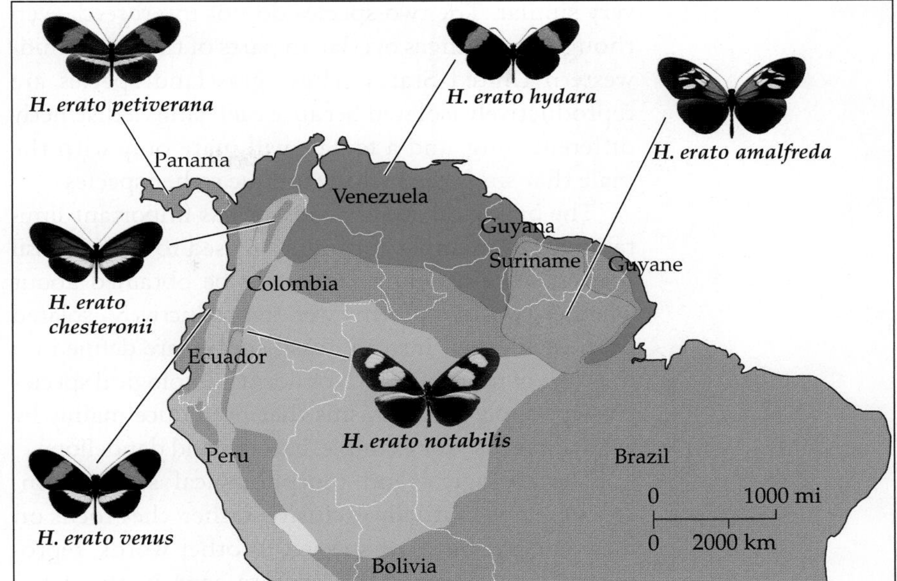

FIGURE 15.2 Members of a Single Species May Look Different from One Another
Heliconius butterflies exhibit an astonishing range of morphological variation, even within a species. Shown here are different variants of *H. erato*. Some forms bear a stronger resemblance to *Heliconius* butterflies of other species than to members of their own species.

Species are reproductively isolated from one another

The most commonly used definition of a species is the **biological species concept**. Here a species is defined as one or more populations that can interbreed to produce fertile offspring but that are reproductively isolated from other groups. In most cases, members of different species cannot reproduce with each other under natural conditions. Defining species in this manner emphasizes the effect of gene flow in maintaining the homogeneity of a gene pool (concepts that we introduced in Chapter 14). When two species are prevented from interbreeding, we say that the species are *reproductively isolated* from one another.

A wide variety of cellular, anatomical, physiological, and behavioral mechanisms generate barriers to reproduction, but they all have the same overall effect: few or no alleles are exchanged between individuals of different species. Barriers to reproduction ensure that the members of a species share a unique **genetic heritage**, a particular set of genes and alleles that is typical of the species but different from that of all other species. Because members of a species exchange alleles with one another but not with members of other species, they usually remain phenotypically similar to one another and different from members of other species.

Barriers to reproduction are often divided into two categories (**TABLE 15.1**):

- Barriers that prevent fertilization or the formation of a zygote are called **prezygotic barriers**.
- Barriers that prevent zygotes from developing into healthy and fertile offspring are **postzygotic barriers**.

Note that in the biological species concept, reproductive isolation is distinct from geographic isolation. For example, the eastern and western meadowlarks are different species, although they look

TABLE 15.1 Barriers That Can Reproductively Isolate Two Species in the Same Geographic Region

TYPE OF BARRIER	DESCRIPTION	EFFECT
Prezygotic barriers		
Temporal isolation	The two species breed at different times.	Mating is prevented.
Ecological isolation	The two species breed in different portions of their habitat.	Mating is prevented.
Behavioral isolation	The two species respond poorly to each other's courtship displays or other mating behaviors.	Mating is prevented.
Mechanical isolation	The two species are physically unable to mate.	Mating is prevented.
Gametic isolation	The gametes of the two species cannot fuse, or they survive poorly in the reproductive tract of the other species.	Fertilization is prevented.
Postzygotic barriers		
Zygote death	Zygotes fail to develop properly, and they die before birth.	No offspring are produced.
Poor hybrid performance	Hybrids survive poorly or reproduce poorly.	Hybrids are not successful.

very similar. The two species do not interbreed, even though their ranges overlap in parts of the upper midwestern United States. These grassland species are reproductively isolated because each sings a distinctly different song, and a female will mate only with the male that sings the melody unique to her species.

The biological species concept has important limitations. For example, it cannot be used to define fossil species, since no information can be obtained about whether two fossil forms were reproductively isolated from each other. Instead, fossil species are defined on the basis of morphology. Nor does the biological species concept apply to organisms that reproduce mainly by asexual means—for example, bacteria and dandelions.

The biological and morphological species concepts are not mutually exclusive; rather, they focus on different attributes of species. In other words, reproductive isolation creates separate gene pools in the isolated populations, which can then diverge, leading the two lineages to acquire unique forms that make them look different from one another. The biological species concept remains the most useful definition for most biologists, so this is the definition we will use in this book. Remember, however, that while reproductive isolation (the cornerstone of the biological species concept) is very useful in establishing a species definition, the concept of a species in nature can be considerably more complicated.

Concept Check Answers

1. The morphological species concept defines species by similarities in their physical form. The biological species concept defines species as organisms that can actually or potentially interbreed.
2. Reproductive isolation.

Concept Check

1. Distinguish between the morphological and biological species concepts.
2. What is the fundamental requirement for speciation?

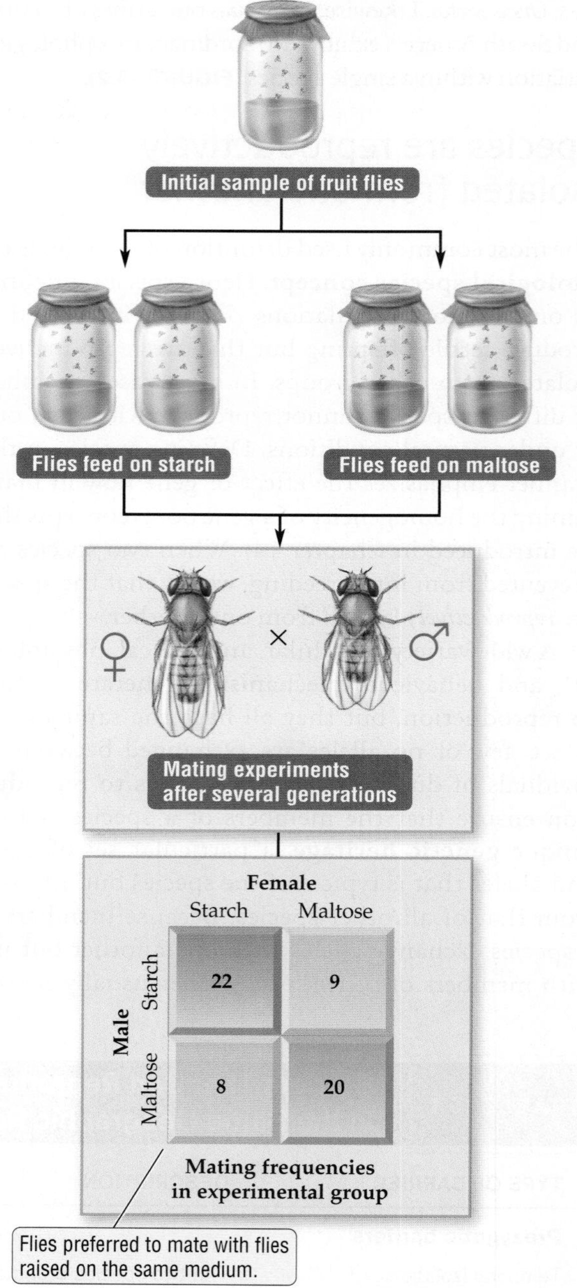

FIGURE 15.3 When What You Eat Affects Whom You Love

Fruit flies in an initial sample were separated into four populations and raised on two different kinds of food (starch or maltose) for several generations. Scientists found that flies adapted to feed on starch preferred mating with other flies adapted to starch. Flies adapted to maltose likewise preferred other flies adapted to maltose. These preferences are the early stages of reproductive isolation that can eventually lead to speciation.

15.2 Speciation: Generating Biodiversity

Speciation is the process in which one species splits to form two or more species that are reproductively isolated from one another. The study of speciation is fundamental to understanding the diversity of life on Earth.

How do new species form? The crucial event in the formation of new species is the evolution of **reproductive isolation**, which requires populations that once could interbreed to diverge enough that they are no longer able to do so. As we saw in Chapter 14, populations within a species can be linked through gene flow, which tends to keep them genetically similar to one another. In order for speciation to occur, reproductive isolation must halt this gene flow, allowing the two populations to acquire unique gene pools and develop into different species. How does reproductive isolation develop?

Speciation can be explained by the same mechanisms that cause the evolution of populations

When gene flow between two populations decreases or stops, evolutionary processes begin to act on each gene pool independently. Over time, mutation, genetic drift, and natural selection cause the two populations to evolve genetic differences from one another. These genetic differences sometimes result in reproductive isolation.

This idea can be illustrated by the results of an experiment with fruit flies. A population of flies was separated into smaller populations and placed in similar environments except that some flies were fed maltose while others were fed starch. Over time, the flies raised on these two different foods started to prefer mating with flies raised on the same food supply (**FIGURE 15.3**). This difference in mate preference arose simply because the flies adapted to living on two different kinds of food. The fly populations had changed genetically in response to natural selection for the ability to grow and develop on maltose versus starch. Those genetic changes had the side effect of causing reproductive isolation. Flies that were adapted to feed on maltose preferred to mate with flies that were also adapted to feed on maltose. The flies adapted to feed on starch likewise preferred to mate with flies adapted to feed on starch. This is an example of behavioral isolation, described in Table 15.1. Given enough time, these mating preferences can lead the two fly populations to be become separate species.

As illustrated in Figure 15.3, natural selection can cause populations to diverge genetically when populations located in different environments face different selection pressures—in the case of our fruit flies, the selection to grow on maltose versus starch. Populations can also diverge from one another as a result of mutation and genetic drift. In contrast, gene flow always operates to counteract the genetic divergence of populations. For populations to accumulate enough genetic differences to cause speciation, the factors that promote divergence must have a greater effect than does the amount of ongoing gene flow.

Speciation can result from geographic isolation

New species can form when populations of a single species become separated, or geographically isolated, from one another. This process can begin when a newly formed barrier, such as a river or a mountain chain, isolates two populations of a single species. Such **geographic isolation** can also occur when a few members of a species colonize a region that is difficult to reach, such as an island located far outside the usual geographic range of the species. For example, as we saw in Chapter 14, Darwin's finches on the separate Galápagos islands were geographically isolated from one another and from finches on the South American mainland by ocean waters.

The distance required for geographic isolation to occur varies tremendously from species to species, depending on how easily the species can travel across any given barrier. Populations of squirrels and other rodents that live on opposite sides of the Grand Canyon—a formidably deep and large barrier for a rodent—have diverged considerably (**FIGURE 15.4**).

An Example of Geographic Isolation

FIGURE 15.4 The Grand Canyon Is a Geographic Barrier for Squirrels

The Kaibab squirrel is confined to the North Rim of the Grand Canyon. Abert's squirrel lives on the South Rim and also farther south on the Colorado Plateau. The Kaibab population became isolated from the Abert's population when the Colorado River cut a canyon—as deep as 6,000 feet in some places. With gene flow between them blocked, the Kaibab population diverged from the ancestral Abert's squirrel.

Allopatric Speciation
A single plant species is distributed over a broad geographic range.
Time
The sea level rises and isolates plant populations from each other. The populations may adapt to different environments on opposite sides of the barrier, indirectly causing genetic changes that reduce their ability to interbreed.
Time
When the barrier is removed, the plants recolonize the intervening area and mingle, but do not interbreed.
Range of overlap

FIGURE 15.5 Physical Barriers Can Cause Speciation by Blocking Gene Flow

New species can form when populations are separated by a geographic barrier, such as a rising sea.

Meanwhile, populations of birds—which can easily fly across the canyon—have not diverged. In general, geographic isolation is said to occur whenever populations are separated by a distance that is great enough to limit gene flow.

Geographically isolated populations are essentially disconnected genetically; there is little or no gene flow between them. For this reason, mutation, genetic drift, and natural selection can more easily cause isolated populations to diverge genetically from one another. If the populations remain isolated long enough, they can evolve into new species. The formation of new species from geographically isolated populations is called **allopatric speciation** (*allo*, "other"; *patric*, "country"; **FIGURE 15.5**).

One fascinating example of speciation by geographic isolation is the recent discovery of the so-called hobbit people of the island of Flores in Indonesia. These tiny human relatives, also known as the "little people" of Indonesia, were discovered in 2004. Measuring 3 feet tall as adults, *Homo floresiensis* lived as recently as 17,000 years ago on their isolated island along with giant Komodo dragon lizards and pygmy elephants. The size of a typical human toddler, these people appeared to use stone tools and fire. Anatomical studies have persuaded most anthropologists that these islanders were a separate species, not members of our species, *Homo sapiens* (**FIGURE 15.6**).

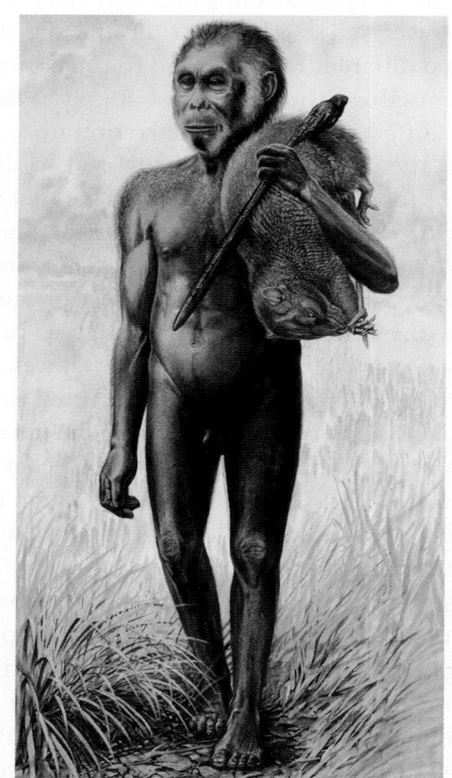

FIGURE 15.6 Island Isolation May Have Driven the Evolution of *Homo floresiensis*

Foot structure and other anatomical features suggest that *Homo floresiensis* was a distinctly different species, not a scaled-down version of *H. erectus* or *H. sapiens*.

Islands Are Centers for Speciation—and Extinction

Since the time of Darwin and Wallace, island archipelagoes, such as the Hawaiian Islands, have been recognized as places with uniquely high biodiversity. Though encompassing only 5 percent of Earth's surface, islands contain roughly 20 percent of Earth's species. The islands of Hawaii, for example, contain nearly a thousand species of the fruit fly, *Drosophila*, as well as 54 distinct species of honeycreeper.

Why are there so many species on islands? A major reason for this high level of biodiversity is that individual islands in a chain are geographically isolated from one another and provide many opportunities for allopatric speciation. Because migration between the islands is rare, there is little gene flow between populations on different islands—certainly not enough to maintain the populations as a single gene pool. When migration does occur, the new migrants are unable to interbreed with the resident populations, and (because they are now isolated from their source population) will eventually become new species.

Islands also have highly variable environments, with diverse microclimates. Coastal and interior areas, for example, will have significantly different environmental conditions. Many islands also possess mountains, where conditions at high altitude differ significantly from those at low altitude. These microclimates create opportunities for speciation through ecological isolation.

Finally, colonists to the islands sometimes experience adaptive radiations when they first arrive, particularly those colonists that arrive at relatively new, remote, or barren islands. Because there are few competitors or predators and many ecological niches available, rapid speciation can occur.

The species found on islands exhibit high endemism, meaning that often they are found nowhere else in the world. Species that evolve on islands generally do not disperse well enough to return to the mainland, so islands not only contain a wealth of species, but also are home to many unique species.

While islands are rich in biodiversity, the species found there are particularly vulnerable to extinction. Approximately 40 percent of critically endangered species reside on islands, and 80 percent of all known extinctions have occurred on islands. The small size of islands makes populations highly subject to genetic drift. On average, island species maintain about 29 percent less genetic diversity than their mainland counterparts. Furthermore, the relationships between species within an island community are easily upset by the introduction of invasive species that may act as predators or competitors. The introduction of the brown tree snake on the island of Guam in the early 1950s, for example, caused the rapid extinction of most native birds and mammals, which had never experienced predation by snakes and therefore had no defense mechanisms against them.

Hawaiian Honeycreepers

Conservation biologists have begun to recognize the unique characteristics of island biodiversity and are developing strategies to preserve it. For example, scientists can use DNA analysis to estimate the minimum population size necessary to maintain healthy levels of genetic diversity. Attempts to control invasive species also are aimed at protecting vulnerable island communities. Armed with a greater understanding of the processes that generate as well as threaten the diversity of species on islands, biologists hope to preserve these incredible centers of biodiversity.

1 Species A migrates from the mainland to an island.

2 On the island, species A diverges to become a separate species, B.

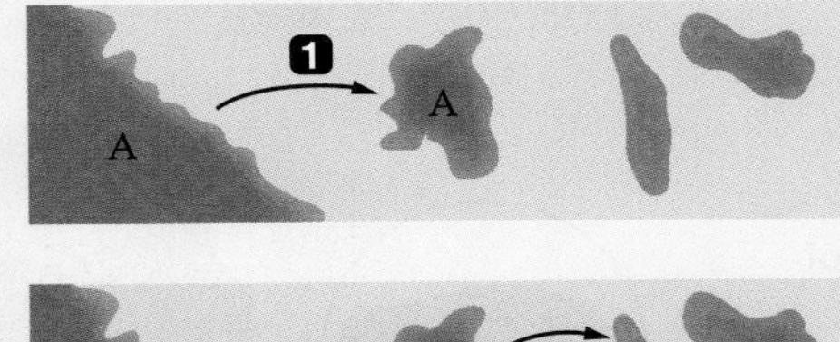

3 A rare migration event moves several members to another island.

4 Isolated from the original population, the migrants evolve into another species, C.

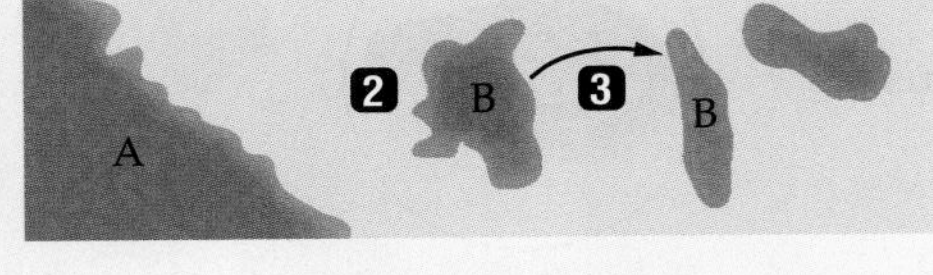

5 Migration events carry species C to the original island...

6 ...and to an unoccupied island.

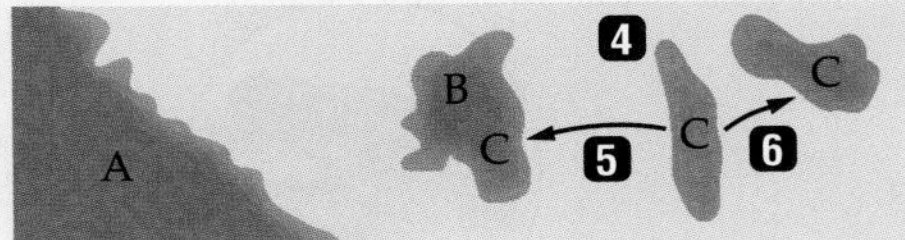

7 On that isolated island, species C evolves into species D.

8 While on the original island, C evolves into another species, E.

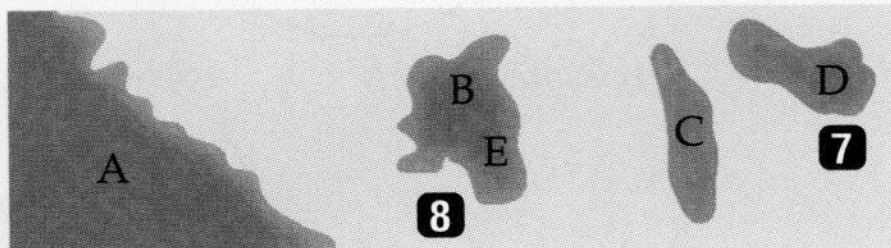

9 Another migratory event brings species D to the original two islands. Note that these migrants cannot interbreed with the resident populations and will, over time, become separate species in their own right.

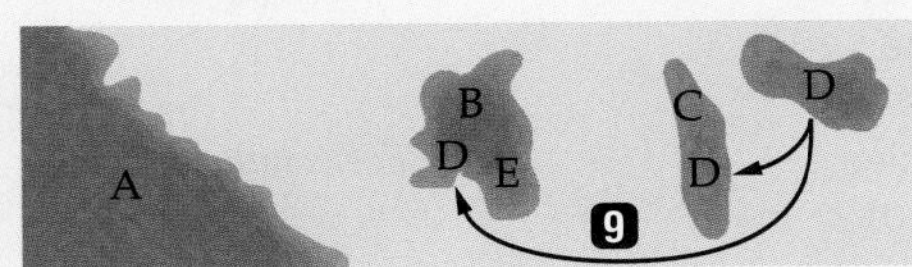

Islands provide enough geographic isolation for allopatric speciation to occur. Rare migratory events increase species diversity.

EXTREME POLYPLOIDY

Domestic and wild strawberries are reproductively isolated because of incompatibilities in chromosome number. Wild strawberries are typically diploid with small (marble-sized) fruits. Commerically grown strawberries are octoploid, with eight sets of chromosomes, and yield larger fruits. Many commerically grown crops, including peanuts and wheat, are the products of polyploidy

A second line of evidence for the importance of geographic isolation in speciation comes from cases in which individuals of a population found at one end of a species' geographic range reproduce poorly with individuals of a population at the other end of its range, even though individuals from both ends reproduce well with individuals from intermediate portions of the species' range. A special case of this phenomenon occurs when the populations loop around a geographic barrier; in this case, **ring species** form, in which populations at the two ends of the loop are in contact with one another, yet individuals from these populations cannot interbreed (**FIGURE 15.7**). Ring species have been found in salamander populations that loop around the San Joaquin Valley in California.

Speciation can occur without geographic isolation

Most speciation events are thought to occur by allopatric speciation, because speciation requires gene flow between populations to stop, and the potential for gene flow is much greater between populations whose geographic ranges overlap or are adjacent to one another than between populations that are geographically isolated from one another. However, it is well established that plants can form new species in the absence of geographic isolation, and recent work has provided convincing evidence that animals can as well. The formation of new species in the absence of geographic isolation is called **sympatric speciation** (*sym*, "together").

In plants, rapid changes in chromosome numbers can cause sympatric speciation. New plant species can form in a single generation as a result of **polyploidy**, a condition in which an individual has more than two sets of chromosomes, usually because the chromosomes did not separate during meiosis (discussed in Chapter 7). Polyploidy can also occur when a hybrid spontaneously doubles its chromosome number. Chromosome doubling can lead to reproductive isolation because the

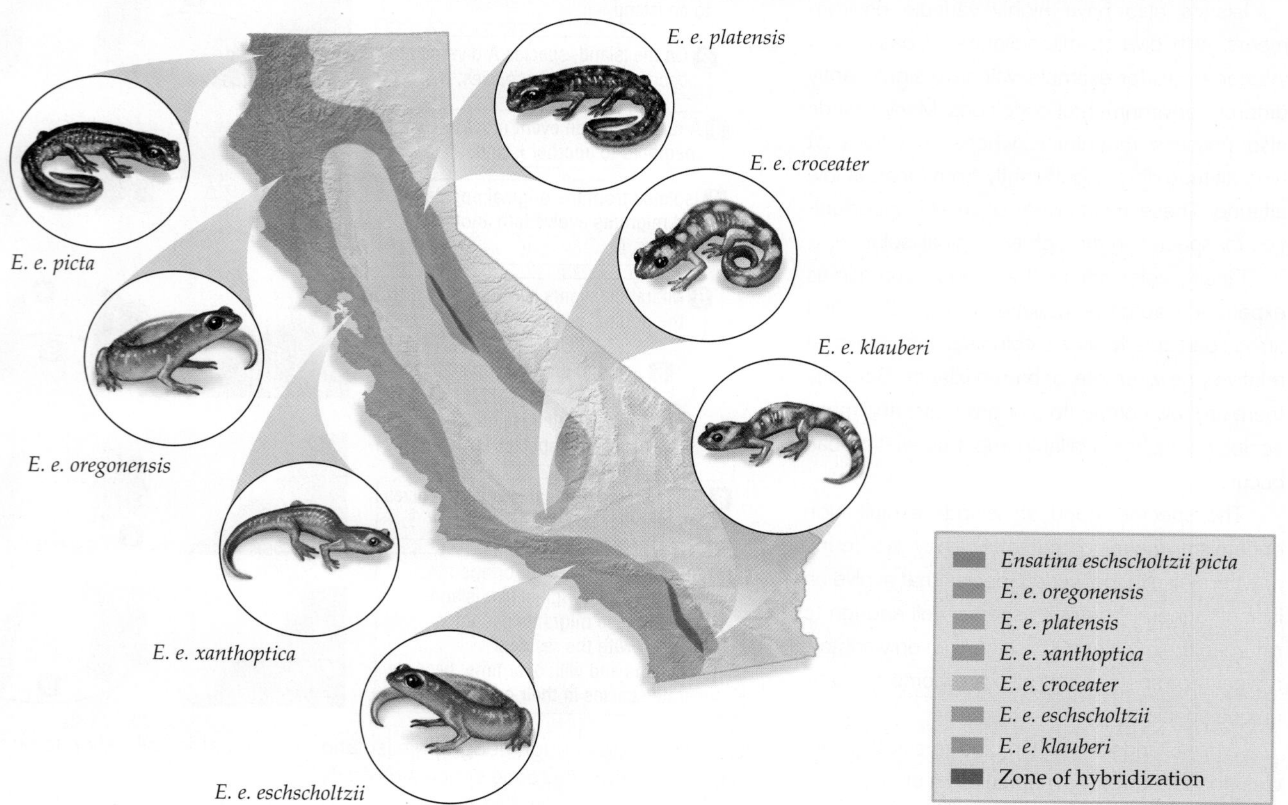

FIGURE 15.7
Ring Species of Salamanders from the Genus *Ensatina*
Although each of the adjoining populations of salamanders can interbreed, *E. croceater* and *E. eschscholtzii* cannot. Theoretically, alleles from *E. croceater* can reach *E. eschscholtzii* by going around the ring through the various subspecies, but the amount of gene flow is so low that reproductive isolation has occurred.

chromosome number in the gametes of the polyploid no longer matches the number in the gametes of either of its parents. For example, a cross between a diploid organism ($2n$) and a tetraploid one ($4n$) would result in a triploid ($3n$) offspring. After undergoing meiosis, it would not be able to produce gametes with the proper number of chromosomes to mate with either parent. Polyploidy has had a large effect on life on Earth: more than half of all plant species alive today are descended from species that originated by polyploidy.

A few animal species appear to have originated by polyploidy, including several species of lizards and fishes, and one mammal (an Argentine rat). Sympatric speciation can occur in animals by means other than polyploidy as well. More commonly, animals that experience sympatry do so without changes in their chromosome number. For example, new cichlid species have formed in the crater lakes of Cameroon without being geographically isolated. These crater lakes are formed when the tops of mountains are blown away by volcanic eruptions and then fill with water. Because these lakes are not fed by tributaries as a traditional lake is, species introductions are rare. Furthermore, each lake is relatively small. Yet a great diversity of cichlid species inhabits each lake. Each species is more related to other species in the lake than to any species in other lakes. The formation of new species within these small, environmentally uniform lakes provides strong evidence for sympatric speciation. The same pattern has been observed in crater lakes in Central America, as well as the Great Lakes of Africa (such as Lake Victoria, as discussed at the start of the chapter). Scientists believe that ecological specialization has led to the diversification of these cichlid fishes (**FIGURE 15.8**).

North American populations of the apple maggot fly, *Rhagoletis pomonella* (**rag-oh-*LEE*-tis poh-muh-*NELL*-uh**), are in the process of diverging into new species, even though their geographic ranges overlap. Historically, *Rhagoletis* ate native hawthorn fruits, but starting in the mid-nineteenth century these flies became pests to apple farmers (note that apples were an introduced nonnative species). Over time the two populations of *Rhagoletis* diverged. *Rhagoletis* populations that feed on apples are now genetically distinct from populations that feed on hawthorns. Members of the apple and hawthorn populations mate at different times of year and usually lay their eggs only on the fruit of their particular food plant. As a result, there is little gene flow between the apple-eating and hawthorn-eating fly populations. In addition, researchers have identified alleles that benefit flies that feed on one host plant but are detrimental to flies that feed on the other host plant.

Sympatric Speciation

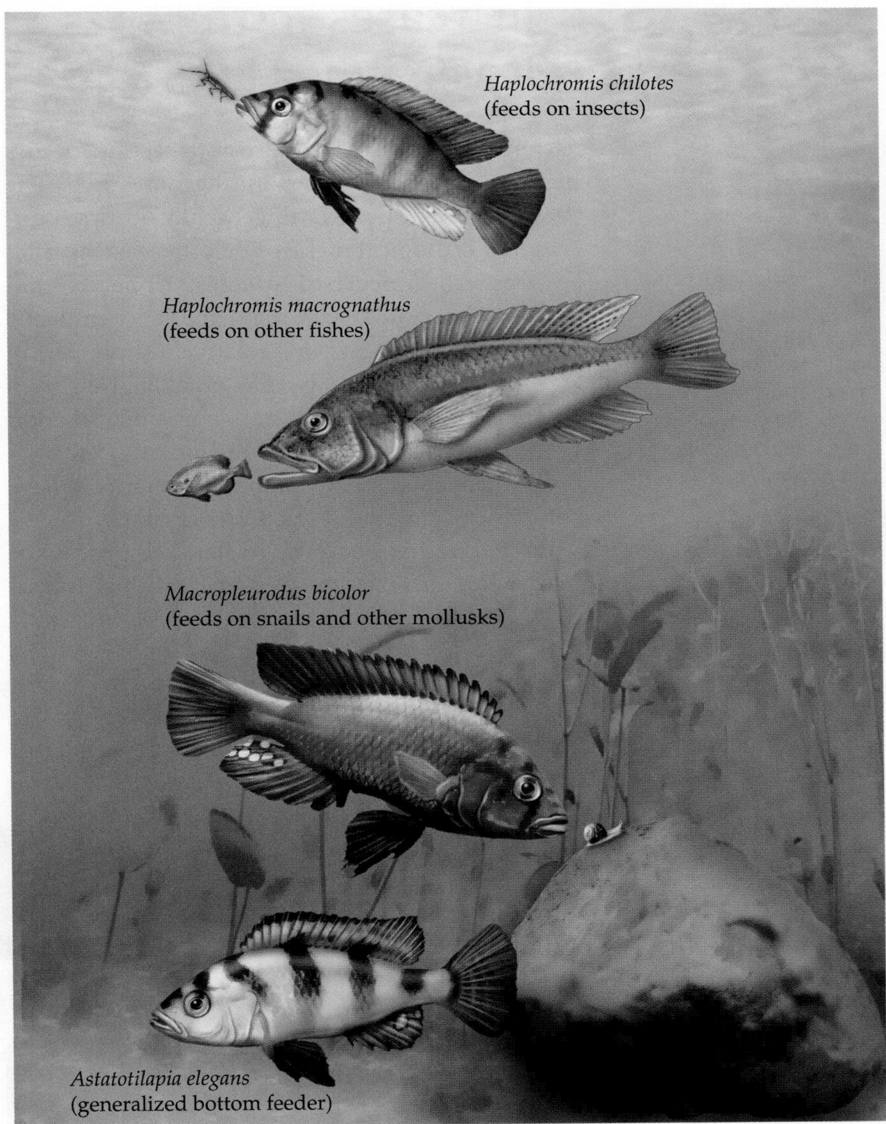

FIGURE 15.8 Food Preferences and Female Mate Preferences May Have Driven Speciation among Lake Victoria Cichlids

The four species shown here illustrate some of the differences in feeding behavior and morphology among Lake Victoria cichlids.

Natural selection operating on these alleles acts to limit whatever gene flow does occur. Over time, the ongoing research on *Rhagoletis* may well provide a dramatic case history of sympatric speciation.

Concept Check

1. What is allopatric speciation?
2. Why is sympatric speciation generally thought to be more difficult than allopatric speciation?

Concept Check Answers

1. The formation of new species that occurs when two populations become reproductively isolated because of a geographic barrier.
2. When organisms inhabit the same area, the likelihood of gene flow between them is higher. Gene flow reduces the probability of speciation.

15.3 Adaptive Radiations: Increases in the Diversity of Life

Speciation, as discussed in the previous section, occurs when one lineage diverges and becomes two. The newly derived species will generally occupy a different ecological niche or be otherwise distinct from the **ancestral species**. Under certain conditions, a lineage may experience multiple speciation events in a relatively short period of time. One lineage may quickly give rise to *many* descendant species. This phenomenon is known as an **adaptive radiation**. Throughout Earth's history, adaptive radiations have contributed greatly to the diversity of life.

Helpful to Know

Here, "radiation" is not related to radioactivity or radioisotopes. In this case, the prefix radi conveys the idea of expanding outward (like rays of light)—not just geographically but ecologically, into new roles.

Adaptive radiations occur when the process of speciation is made easier by a relaxation of natural selection. When ecological niches are more freely available, a greater variety of lifestyles can be sustained. In this way, speciation can proceed rapidly, leading to the formation of many new species. The conditions that lead to adaptive radiations include (1) the colonization of a new location, (2) mass extinctions that remove existing species, and (3) the evolution of a novel trait that confers a significant competitive advantage.

Darwin himself observed the effects of adaptive radiations in his finches (see Figure 14.30). A species of finch from the South American mainland colonized the Galápagos Islands a few million years ago. Once on the islands, however, many niches not available on the mainland were open to these colonizers, since there were no other bird species to compete with. While originally adapted for seed eating, the finches on the Galápagos were now able to feed on flowers and insects. The lack of competitors in the new environment made it possible for ecological isolation (see Table 15.1) to rapidly lead to many descendant species.

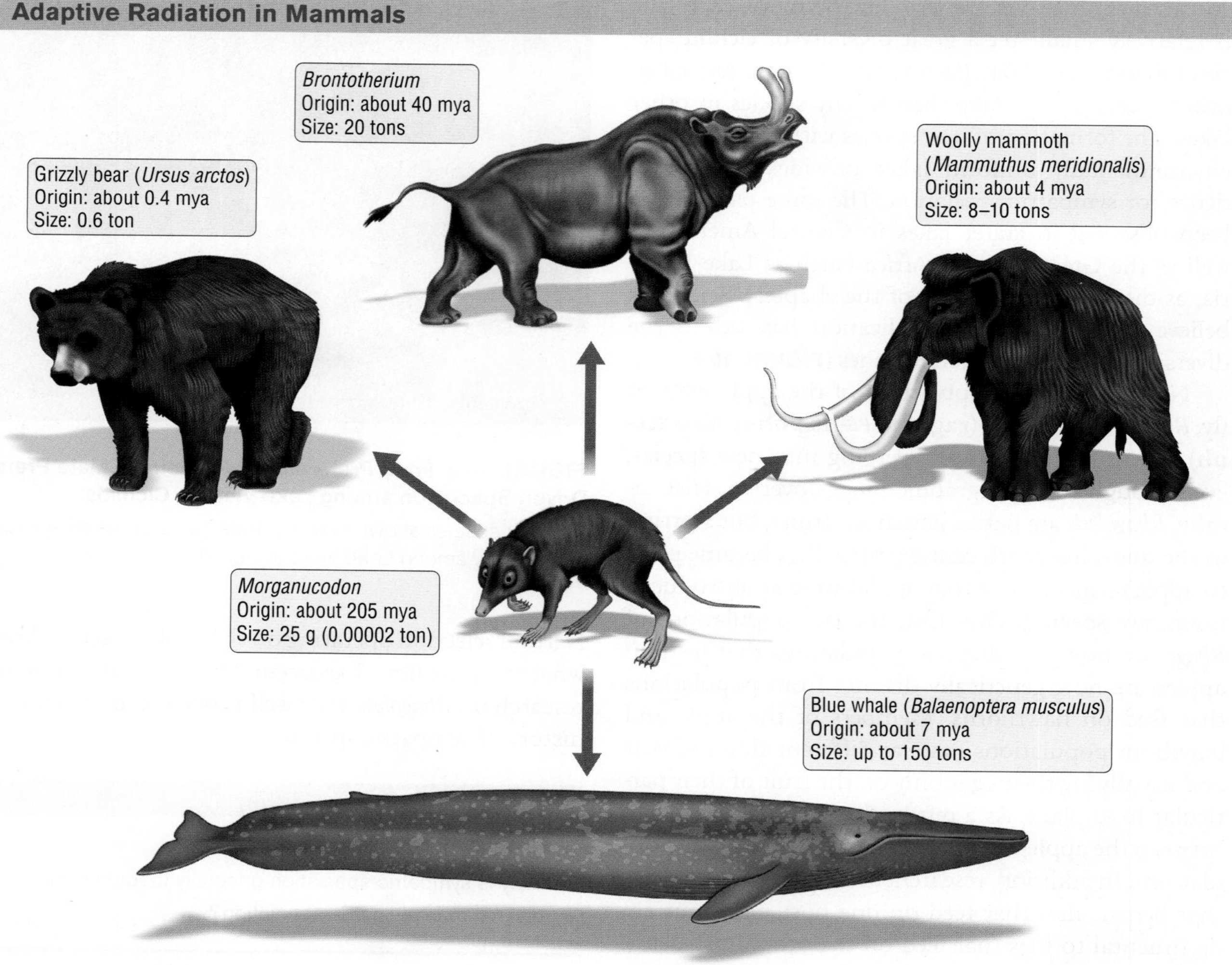

FIGURE 15.9 Extinction of the Dinosaurs Enabled Mammals to Radiate

Early mammals (such as *Morganucodon*; MOHR-**guh**-NOO-**kuh-dahn**) were small and are thought to have been nocturnal. Following the extinction of the dinosaurs, mammals radiated to occupy the ecological niches vacated by the dinosaurs. Because of the huge range in size between the smallest (*Morganucodon*, about the size of a shrew) and the largest (the blue whale, which can reach 25–30 meters long), none of these animals are drawn to scale. Note: mya = million years ago.

Extinction events may also facilitate adaptive radiations. The removal of the ecologically dominant species allows some of the surviving groups of organisms to diversify to replace those that have become extinct. Some of the great adaptive radiations in the history of life occurred after mass extinctions, as when the mammals diversified to replace the dinosaurs 65 million years ago (**FIGURE 15.9**).

Adaptive radiations may also occur after a group of organisms has acquired a new adaptation that enables it to use its environment in new ways. The first terrestrial plants, for example, possessed adaptations that helped them thrive on land, a new and highly challenging environment. The descendants of those early colonists radiated greatly, forming many new species and higher taxonomic groups that were able to live in a broad range of new environments (from desert to Arctic to tropical regions).

It is important to remember that the root cause of speciation is reproductive isolation, whether it occurs singly or as an adaptive radiation. No radiation could occur without some other isolating mechanism to prevent gene flow and drive speciation.

Homologous Traits

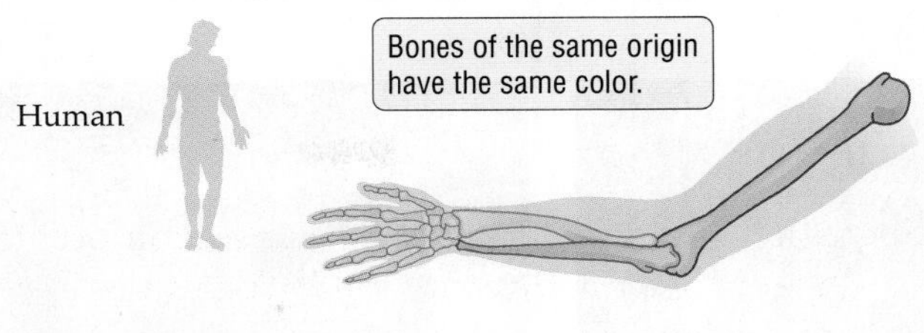

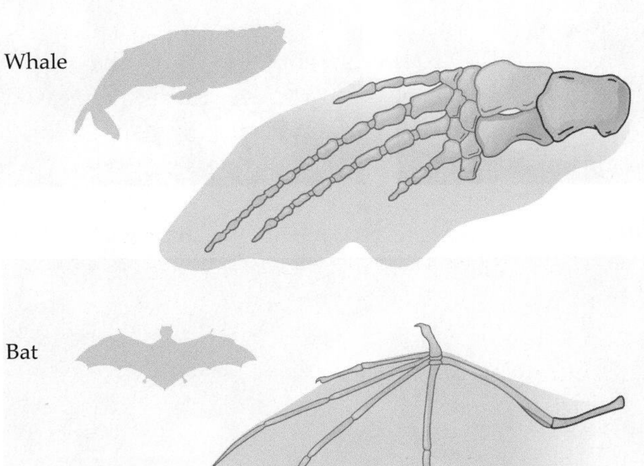

Bat

FIGURE 15.10 Shared Characteristics The human arm, the whale flipper, and the bat wing are homologous structures, all of which have what are essentially a matching set of five digits and a matching set of arm bones that have been altered by evolution for different functions.

Concept Check

1. Under what conditions might an adaptive radiation occur?
2. How might a small amount of gene flow affect speciation rates?

15.4 Evolution Can Explain the Unity and Diversity of Life

The processes that create new species can explain both the unity and the incredible diversity of life. The natural world is filled with many puzzling examples of organisms that look and behave very differently, yet share characteristics that we would not imagine they would share. Consider the wing of a bat, the arm of a human, and the flipper of a whale. Each is used for a very different purpose, yet each has five digits attached to the same set of bones (**FIGURE 15.10**). These similarities among organisms arise because the organisms evolved from a common ancestor. When one species splits into two, the two species that result still share many features. Features of organisms related to one another through common descent are said to be **homologous** (**hoh-*MAH*-luh-gus**).

Many living organisms also show the puzzling characteristic of having features that appear to be of no use to them, such as **vestigial organs** (**ves-*TIH*-jee-ul**), which are reduced or degenerate parts whose function is hard to discern. For example, why do humans have a reduced tailbone and the remnants of muscles for moving a tail? Why do whales have remnants of hip bones when they have no leg bones requiring support? And why do some snakes have rudimentary leg bones but no legs (**FIGURE 15.11**)? Again the answer is that species arise

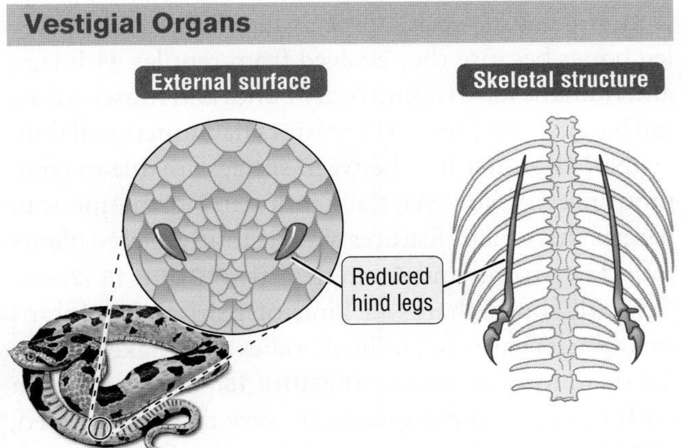

FIGURE 15.11 A Python Has Rudimentary Hind Legs Visible Externally and Internally

Concept Check Answers

1. Colonization of a new habitat, mass extinction of competitor species, or the evolution of a novel trait that confers significant evolutionary advantage.

2. Gene flow homogenizes gene pools, making speciation more difficult.

FIGURE 15.12 The Power of Natural Selection

(*a–c*) These three plants evolved from very different groups of leafy plants. They resemble one another because of convergent evolution: each was similarly driven by natural selection to adapt to life in a desert. Their shared structures (fleshy stems, spines, reduced leaves) are therefore analogous, not homologous. (*a*) *Euphorbia* (**yoo-*FOHR*-bee-uh**) belongs to the spurge family and can be found in Africa. (*b*) *Echinocereus* (**ee-*KYE*-noh-*SEER*-ee-us**) is a cactus found in North America. (*c*) *Hoodia*, a fleshy milkweed, can be found in Africa. (*d,e*) Convergent evolution can be a powerful force shaping animals as well. Here we see how natural selection has caused two distantly related animals—sharks (*d*) and dolphins (*e*)—to look very similar. Sharks are a kind of fish; dolphins are mammals.

from previous species. Some snakes have rudimentary leg bones because they evolved from reptiles with legs, and humans have rudimentary bones and muscles for a tail because our (distant) ancestors had functional tails.

Not all similarities between species are due to common ancestry, however. Cacti found in North American deserts share many features with distantly related plants found in African and Asian deserts (**FIGURE 15.12*a–c***). Similarly, both sharks (a kind of fish) and dolphins (mammals) have streamlined bodies that make it easier for them to swim rapidly (**FIGURE 15.12*d* and *e***). Yet in both examples these species are very distantly related, and their overall similarities do not represent homology. Instead, these organisms share features as a result of **convergent evolution**, which occurs when natural selection causes distantly related organisms to evolve similar structures in response to similar environmental challenges. The characteristics that species share because of convergent evolution, not common descent, are said to be **analogous** (**uh-*NAL*-uh-gus**).

Concept Check

1. What is the relationship between speciation and homologous traits?
2. What is the main force that drives two separate species to exhibit convergent evolution?

Concept Check Answers

1. Speciation occurs when a single lineage diverges to become two or more. The descendant lineages share similarities due to their common ancestry. These traits are called homologous traits.

2. Natural selection.

15.5 Rates of Speciation

How quickly can a new species evolve? When speciation is caused by polyploidy or other types of rapid chromosomal change, new species can form in a single generation. New species also appear to have formed relatively rapidly in the case of some African cichlid fishes. As we saw at the beginning of this chapter, scientists have described some 500 species of cichlids from Lake Victoria, and genetic analyses indicate that all of them descended from just two ancestor species over the past 100,000 years.

Many cichlids in Lake Victoria use color as a basis for mate choice: females prefer to mate with males of a particular color (an example of sexual selection; see Chapter 14). Furthermore, cichlids have unusual jaws that can be modified relatively easily over the course of evolution to specialize on new food items. This feature of their biology causes the cichlids in Lake Victoria to vary greatly in form and feeding behavior (see Figure 15.8). If female mate choice caused two populations to begin to be reproductively isolated from each other, the resulting lack of gene flow could enable the populations to specialize on different sources of food, which would make it increasingly likely that they would continue to diverge and form new species.

The cichlid rate of speciation is unusual. DNA evidence suggests that in most plants and animals, speciation occurs more slowly. Freshwater fishes can speciate in a time span ranging from 3,000 years (in pupfishes) to over 9 million years in characins (*KAYR*-**uh-sun**), a group that includes carp, piranhas, and many aquarium fishes. As we saw in the cichlids, rates of speciation and rates of extinction are affected by various factors, including how fast reproductive isolation occurs and whether changes in the physical environment are sudden or gradual.

Once gene flow between populations of a species has been interrupted, speciation rates can be affected by factors other than the reproductive barrier itself, such as the frequency and severity of genetic events (for example, polyploidy affects speciation much more quickly than random mutation does), doubling time (the time it takes for a population to double in size), and behavior (such as the processes by which mates are chosen).

Some populations can be geographically isolated for a long time without evolving reproductive isolation. North American and European sycamore trees, for example, have been separated for more than 20 million years, yet the two populations remain morphologically similar and can interbreed. Fossil evidence and DNA analysis show that polar bears (*Ursus maritimus*) and the brown bear (*Ursus arctos*) diverged into two distinctly different species about 200,000 years ago. Reports of hybrid bears have been increasing in recent years, and DNA tests on odd-looking bears shot by hunters confirm that they are hybrids between polar bears and grizzly bears. The ongoing loss of summer sea ice in the Arctic, one of many effects of climate change, may be driving the polar bear farther south in search of food, while grizzlies appear to be expanding their range northward. A gene flow that all but ceased 200,000 years has resumed, and the mighty hunter of the frozen north may come to be replaced by pizzly bears, prizzly bears, and grolar bears.

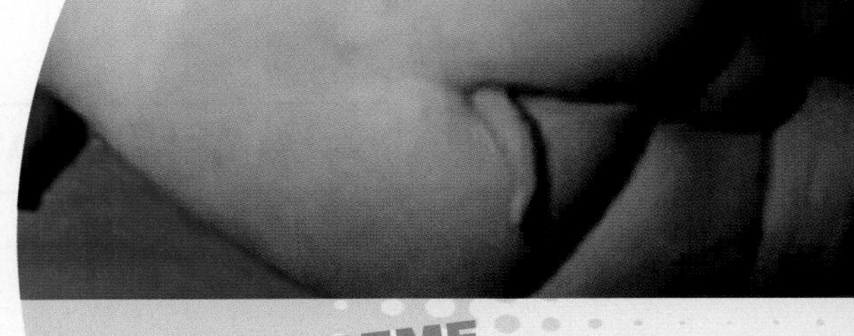

EXTREME VESTIGE

Although humans do not have tails, our DNA contains the information for this trait as a genetic vestige of our descent from organisms with tails. In normal human development, these genes are repressed. In rare cases, however, these genes become activated, leading to tail development.

Concept Check

1. What factors govern the rate of speciation?
2. How might a small amount of gene flow affect speciation rates?

Concept Check Answers

1. Speciation is thought to happen most often when populations are geographically isolated. However, some species exhibit more rapid speciation because of ecological specialization, while others may be separated for long periods of time without significant genetic differentiation.
2. Since speciation requires reproductive isolation, a small amount of gene flow will slow speciation rates.

APPLYING WHAT WE LEARNED

Lake Victoria: Center of Speciation

In the 1980s a population explosion of introduced Nile perch led to the extinction or near extinction of two-thirds of cichlid fish species in Lake Victoria, in East Africa. This catastrophic mass extinction drastically reorganized the lake's ecology. Cichlids, which had made up 80 percent of all the fish in the lake, now made up just 1 percent.

This was not the first time the lake had experienced such a disaster. Despite its vast size (nearly 27,000 square miles), Lake Victoria has dried up three times since it first formed 400,000 years ago. Each time the lake dried up, its fish species either went extinct or survived in only a few small ponds. Yet every time the lake refilled, the cichlids recolonized the lake and diversified into hundreds of new species in a classic example of adaptive radiation.

Like other fishes, cichlids are masters of speciation. Half of all species of vertebrates are fish. Of the roughly 30,000 fish species, nearly 10 percent are cichlids. How could 500 species of cichlids evolve in one lake in just 15,000 years? If the fish had evolved in a set of separate lakes, as in allopatric speciation, it would be easier to understand, but cichlids seem to speciate while living side by side, in sympatry.

One key aspect of cichlid biology partly explains the fishes' rapid diversification. All animals tend to evolve the ability to see best in the color of light that bathes their environment. At the surface of a lake, red light dominates, while deeper down, blue light dominates. The eyes of fish species that feed deeper in the water are most sensitive to blue light. These blue-seeing fish usually stay deep, where they can see best. In addition, deepwater, blue-seeing females tend to choose bluish males as mates because bluer males are easier to see than reddish ones (which look black). In contrast, fish that feed near the surface see reds better, and likewise the females choose redder mates. The combination of cichlids' specialized color vision and the range of light color in the water helps to reproductively isolate each cichlid species, setting the stage for speciation.

But human development around Lake Victoria has changed all this. The water has become so filled with pollutants, sediment, and algae that the fish can hardly see. Instead of separating into distinct mating groups according to depth and light color, the fish swim wherever they like. Unable to see one another's colors, the fish crossbreed with other species—a process called hybridization.

Evolutionary biologists think that this species blending may be a first step in the process of adaptive radiation. Because about two-thirds of Lake Victoria's cichlid species are extinct, the lake is awash in empty niches that an evolving fish can exploit. One species, *Yssichromis pyrrhocephalus*, seems to have jumped at this opportunity. Before the lake-wide extinctions, this cichlid fed on plankton floating on the surface of the lake. It had nearly gone extinct when, in 2008, biologists discovered a surviving population that fed by rooting around in the mud at the bottom of the lake. The new *Yssichromis pyrrhocephalus* had gills for absorbing oxygen that were two-thirds larger than before—possibly because eutrophication had reduced oxygen levels in the water. *Y. pyrrhocephalus* 2.0 also had a strikingly different head shape, which apparently helped it prey on tough invertebrates in the muddy lake bottom. The fish had also changed its behavior: instead of avoiding the voracious Nile perch, it somehow managed to live in close proximity to this dangerous predator; how, biologists still didn't know.

BIOLOGY IN THE NEWS

First Love Child of Human, Neanderthal Found

BY JENNIFER VIEGAS • *Discover News*, March 27, 2013

The skeletal remains of an individual living in northern Italy 40,000–30,000 years ago are believed to be that of a human/Neanderthal hybrid ...

If further analysis proves the theory correct, the remains belonged to the first known such hybrid, providing direct evidence that humans and Neanderthals interbred. Prior genetic research determined the DNA of people with European and Asian ancestry is 1 to 4 percent Neanderthal.

The present study focuses on the individual's jaw, which was unearthed at a rock-shelter called Riparo di Mezzena in the Monti Lessini region of Italy. Both Neanderthals and modern humans inhabited Europe at the time ...

[Silvana] Condemi is the CNRS [France's National Center for Scientific

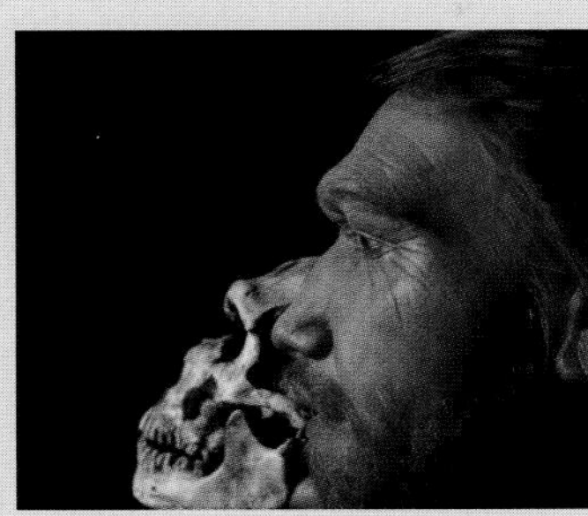

Research] research director at the University of Ai-Marseille. She and her colleagues studied the remains via DNA analysis and 3D imaging. They then compared those results with the same features from *Homo sapiens* ...

By the time modern humans arrived in the area, the Neanderthals had already established their own culture, Mousterian, which lasted some 200,000 years. Numerous flint tools, such as axes and spear points, have been associated with the Mousterian. The artifacts are typically found in rock shelters, such as the Riparo di Mezzena, and caves throughout Europe.

The researchers found that, although the hybridization between the two hominid species likely took place, the Neanderthals continued to uphold their own cultural traditions.

That's an intriguing clue, because it suggests that the two populations did not simply meet, mate and merge into a single group.

As Condemi and her colleagues wrote, the mandible supports the theory of "a slow process of replacement of Neanderthals by the invading modern human populations, as well as additional evidence of the upholding of the Neanderthals' cultural identity."

Humans (*Homo sapiens*) and Neandertals (*Homo neanderthalensis*) are the tips of two lineages that evolved in Africa and separated half a million years ago. The ancestors of Neandertals migrated north to Europe and Asia, where they lived and evolved into Neandertals over hundreds of thousands of years. Our own ancestors stayed in Africa until about 100,000 years ago, when they, too, began migrating north out of Africa.

Neandertals didn't die out until about 30,000 years ago, which means both they and humans lived in Europe and Asia at the same time for perhaps thousands of years. But exactly when and where remains controversial. Some researchers argue that the two species lived side by side in the Middle East about 60,000 years ago, and perhaps in Europe as well, as recently as 24,000 years ago. Other researchers doubt that there was much overlap at all.

In the study reported here, scientists examined the skeletal structure of a newly discovered specimen found in Italy. By comparing the sample to Neandertals and modern humans, the researchers found evidence of hybrid features. Combining biological evidence with archaeological clues, the researchers surmise that as modern humans moved out of Africa into Europe, they came into contact and mated with European Neandertals.

Previous research has shown that humans today still have some traces of Neandertal genes. Any human whose ancestral group evolved outside of Africa carries between 1 and 4 percent of genes that are recognizably Neandertal. What this means is that non-African humans were mating with Neandertals at some time in our past. Why only non-Africans? Although Neandertals were descended from a lineage that came out of Africa, Neandertals themselves never lived in Africa, so it would have been unlikely for the two groups to interbreed. And in fact, Africans carry no Neandertal genes.

We know that two separate species of humans—*Homo sapiens* and *Homo neanderthalensis*—hybridized. Does this mean that modern humans and Neandertals are the same species? The answer is more complicated than a simple yes or no. Recall that gene flow counteracts divergence by homogenizing gene pools. If there were a significant mixing of alleles from modern humans and Neandertals, then we could clearly conclude they were the same species. However, it is evident that Neandertals had their own culture, which remained intact even after the contact with modern humans. The amount of gene flow between the two groups was probably not sufficient for a complete fusion of the two gene pools. However, the existence of these hybrids quite clearly suggests that we are not so different from our Neandertal cousins.

Evaluating the News

1. Do you think Neandertals and anatomically modern humans should be classified as separate species? Offer evidence to support your case.

2. Thinking about what you have learned in this and the previous chapter, what kind of population-wide situations might lead to two species hybridizing?

CHAPTER REVIEW

Summary

15.1 What Are Species?

- The morphological species concept recognizes that species can often be distinguished by external appearance alone.
- The biological species concept defines species as a group of interbreeding natural populations that is reproductively isolated from other such groups.
- The biological species concept has important limitations: it does not apply to fossil species (which must be identified by morphology) or to organisms that reproduce mainly by asexual means.

15.2 Speciation: Generating Biodiversity

- The crucial event in the formation of a new species is the evolution of reproductive isolation.
- Speciation usually occurs as a by-product of the genetic divergence of populations from one another caused by natural selection, genetic drift, or mutation.
- Most new species are thought to arise through allopatric speciation, which occurs when populations are geographically isolated from one another long enough for reproductive isolation to evolve. Extended periods of reproductive isolation do not guarantee allopatric speciation, however.
- Speciation that occurs in the absence of geographic isolation is called sympatric speciation, which can occur when part of a population diverges genetically from the rest of the population.
- Polyploidy is one way that many plants and some animals evolve new species during a single generation.

15.3 Adaptive Radiations: Increases in the Diversity of Life

- In an adaptive radiation, a group of organisms diversifies greatly and takes on new ecological roles.
- Adaptive radiations can be caused by colonization of a new area, mass extinctions, or the evolution of novel traits that confer significant evolutionary advantage.

15.4 Evolution Can Explain the Unity and Diversity of Life

- Shared characteristics suggest either descent from a common ancestor or convergent evolution. Shared characteristics that result from common descent are said to be homologous; those that result from convergent evolution are said to be analogous.
- Vestigial organs, such as the human tailbone, are features that are reduced but represent homology between related organisms.

15.5 Rates of Speciation

- Rates of speciation vary. Speciation occurs rapidly in some cases but requires hundreds of thousands to millions of years in other cases.
- Rates of speciation depend both on how fast reproductive isolation is established (how long it takes to prevent gene flow between populations) and on factors that influence how quickly the species branches into a new lineage.

Key Terms

adaptive radiation (p. 348)
allopatric speciation (p. 344)
analogous (p. 350)
ancestral species (p. 348)
biological species concept (p. 341)
convergent evolution (p. 350)
genetic heritage (p. 341)
geographic isolation (p. 343)
homologous (p. 349)
morphological species concept (p. 340)
polyploid (p. 346)
postzygotic barrier (p. 341)
prezygotic barrier (p. 341)
reproductive isolation (p. 342)
ring species (p. 346)
speciation (p. 342)
species (p. 340)
sympatric speciation (p. 346)
vestigial organ (p. 349)

Self-Quiz

1. Species that have overlapping geographic ranges but do not interbreed in nature are said to be
   a. geographically isolated.
   b. reproductively isolated.
   c. influenced by genetic drift.
   d. hybrids.

2. Which of the following evolutionary mechanisms acts to slow down or prevent the evolution of reproductive isolation?
   a. natural selection
   b. gene flow
   c. mutation
   d. genetic drift

3. The splitting of one species to form two or more species most commonly occurs
   a. by sympatric speciation.
   b. by genetic drift.
   c. by allopatric speciation.
   d. suddenly.

4. The time required for populations to diverge to form new species
   a. varies from a single generation to millions of years.
   b. is always greater in plants than in animals.
   c. is never less than 100,000 years.
   d. is never more than 1,000 years.

5. Traits that are similar because of common ancestry are called
   a. homologous.
   b. evolved.
   c. analogous.
   d. adaptive.

6. Prezygotic and postzygotic barriers to reproduction have the effect of
   a. reducing genetic differences between populations.
   b. increasing the chance of hybridization.
   c. preventing speciation.
   d. reducing or preventing gene flow between species.

7. Evidence suggests that sympatric speciation may have occurred or may be in progress in all of the following cases *except*
   a. the apple maggot fly.
   b. squirrels on opposite sides of the Grand Canyon.
   c. cichlid fishes.
   d. polyploid plants (or their ancestors).

8. The diploid number of chromosomes in plant species A is 8; the diploid number in plant species B is 16. If plant species C originated when a hybrid between A and B spontaneously doubled its chromosome number, what is the most likely number of diploid chromosomes in C?
   a. 8
   b. 12
   c. 24
   d. 48

9. The biological species concept
   a. can be applied to organisms that reproduce asexually.
   b. can be applied to fossil life-forms.
   c. would classify two natural populations, A and B, as separate species if A and B were separated by a geographic barrier.
   d. would classify two natural populations, A and B, as separate species if A and B were unable to exchange genes even if they co-occurred.

10. Which of the following is a good example of an analogous trait?
    a. A horse limb and a human limb have the same bones.
    b. Brown bears are distributed around the world.
    c. European and American vultures are not related but look very much alike.
    d. All organisms have ribosomes.

Analysis and Application

1. Distinguish between the morphological species concept and the biological species concept. For each species concept, list three organisms to which that concept would not apply.

2. Why is reproductive isolation such a critical component of speciation? Explain the relationship between gene flow and speciation. What other evolutionary factors play a role in the process of speciation?

3. Imagine that a species legally classified as rare and endangered is discovered to hybridize with a more common species. Since the two species interbreed in nature, should they be considered a single species? Since one of the two species is common, should the rare species no longer be legally classified as rare and endangered?

4. Many plant species, such as grasses, hybridize in nature. Should species that hybridize naturally be considered one species or two? What factor should define species in these cases?

5. High winds during a tropical storm blow a small group of birds to an island previously uninhabited by that species. Assume that the island is located far from other populations of this species, and that environmental conditions on the island differ from those experienced by the birds' parent population. Is natural selection or genetic drift (or both) likely to influence whether the birds on the island form a new species? Explain your answer.

6. Hundreds of new species of cichlids evolved within the confines of a large lake in Africa known as Lake Victoria. Some of these species live in different habitats within the lake and rarely encounter one another. Would you consider such species to have evolved with or without geographic isolation?

7. How can new species form by sympatric speciation? Why is it harder for speciation to occur in sympatry than in allopatry?

16 The Evolutionary History of Life

THE ROSS ICE SHELF, ANTARCTICA. The Antarctic continent is so cold and dry that few organisms can survive here on land year-round. Yet millions of years ago the continent supported an ecosystem with giant ferns, amphibians, dinosaurs, and flightless birds.

Puzzling Fossils in a Frozen Wasteland

Antarctica is a crystal desert, an ice-covered land in which warmth and liquid water are nearly absent. Few organisms can survive in this cold and arid landscape, and those that do are small or live along the coast or are able to survive months of being frozen solid. Antarctica is almost entirely covered by an ice sheet that is up to a mile thick. In the few places that are not permanently covered with ice, just two species of flowering plants eke out a living—a modest grass and a mosslike pearlwort. Mosses, lichens, and tiny invertebrates also survive in the ice and cutting winds. The frigid coasts of the continent support a thriving community of plankton, fish, whales, seals, and penguins. But most are visitors that eat fish or plankton and fly or swim away when winter comes. On land, the largest year-round terrestrial animal is a flightless 5-millimeter fly.

In the interior of the continent, organisms are even tinier: in most places the only living things are microscopic bacteria and protists. A few interior valleys are nearly ice-free and seem less forbidding. But dry, freezing winds of up to 200 miles per hour blast away any remaining traces of water. In the McMurdo Dry Valleys, even bacteria survive by living inside of rocks where a bit of water remains or within ice beneath the ground surface.

Despite the near lifelessness of modern Antarctica, paleontologists have uncovered fossils of trees as tall as 22 meters, as well as fossils of dinosaurs, other reptiles, mammals, and terror birds (fast-running flightless birds that stood 3.5 meters tall), not to mention ferns, freshwater fishes, large amphibians, and aquatic beetles. In this chapter we'll take a look back at the spectacular history of life on Earth. We'll find out how whole continents have moved, how the global climate has changed from warm and dry to frozen to tropical and back again, and how those dramatic changes have stimulated the evolution of life on Earth.

How could animals and plants adapted to warm, wet conditions have survived on a continent that is so barren today? What happened in Antarctica?

MAIN MESSAGE The climate and geology of our planet have changed again and again over hundreds of millions of years, causing dramatic changes in the life-forms on Earth.

KEY CONCEPTS

- Macroevolution involves large-scale changes in organisms, generally occurring over millions of years.
- The fossil record documents the history of life on Earth.
- Early photosynthetic organisms released oxygen to the atmosphere, setting the stage for the evolution of eukaryotes and multicellular organisms.
- The Cambrian explosion was an astonishing increase in animal diversity in which most of the major living animal phyla appeared in the fossil record in a relatively short time span.
- The colonization of land by plants, fungi, and animals marked the beginning of another major increase in the diversity of life.
- The history of life can be summarized by the rise and fall of major groups of organisms. This history has been greatly influenced by plate tectonics, mass extinctions, and adaptive radiations.
- Advances in developmental biology and gene expression have shown that macroevolutionary change can happen very rapidly.
- The diversity of life can be classified and organized according to evolutionary principles.

PUZZLING FOSSILS IN A FROZEN WASTELAND 357
16.1 Macroevolution: Large-Scale Body Changes 358
16.2 The Fossil Record: A Guide to the Past 359
16.3 The History of Life on Earth 361
16.4 The Effects of Plate Tectonics 365
16.5 Mass Extinctions: Worldwide Losses of Species 366
16.6 Rapid Macroevolution through Differential Gene Expression 368
16.7 Phylogenetics: Reconstructing Evolutionary Relationships 371
APPLYING WHAT WE LEARNED 374
When Antarctica Was Green

EARTH ABOUNDS WITH LIFE. About 1.7 million species have been described, and millions more await discovery. Though these numbers are large, the species alive today are thought to represent far less than 1 percent of all the species that have ever lived. How did they come to be—these "endless forms, most beautiful and most wonderful," as Darwin described them? We track that story of life in this chapter.

We begin with an introduction to large-scale evolutionary change, called macroevolution, and then continue with a look at how the history of life on Earth is documented in the fossil record. Next we summarize the major events in the history of life. We then consider some of the forces—both geologic and genetic—that alter the amount of biodiversity that has lived and is living on Earth. We conclude with a discussion of how scientists use the tools and principles of evolution to classify these "endless forms."

16.1 Macroevolution: Large-Scale Body Changes

Speciation, as we saw in Chapter 15, is the process of generating new species through the reproductive isolation of lineages. This process contributes greatly to the diversity of life observed in nature. Life's diversity, however, is more complex than the sheer number of species. Organisms differ radically in form, in embryological development, in behavior, and in ecology. Consider the whale, a mammal that has been so altered by its adaptations to aquatic life that it scarcely retains any similarity to its terrestrial ungulate (hoofed) ancestors (**FIGURE 16.1**). Clearly, the process

Evolution of the Whales

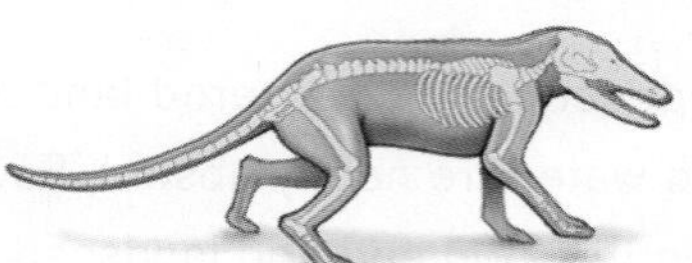

The oldest whale ancestor, *Pakicetus*, lived on land 53 million years ago.
Pakicetus
1.8 meters long, 53 mya

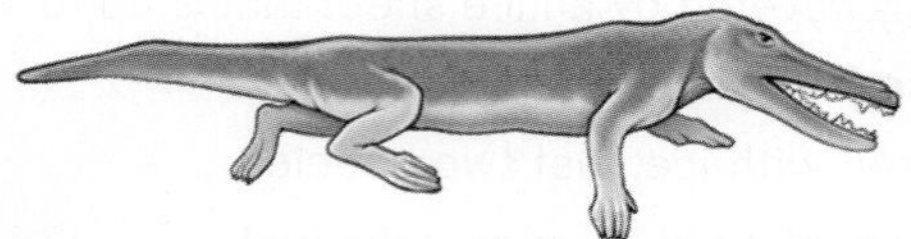

Ambulocetus
4.2 meters long, 49 mya
Ambulocetus had strong, well-developed legs and probably was semiaquatic, living at the water's edge and hunting in much the same way that a crocodile does today.

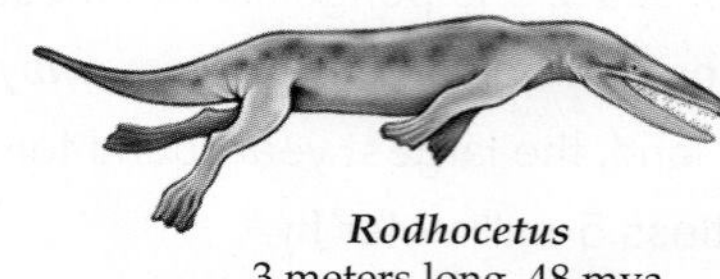

In *Rodhocetus*, the body was more streamlined and the front legs were shaped more like flippers.
Rodhocetus
3 meters long, 48 mya

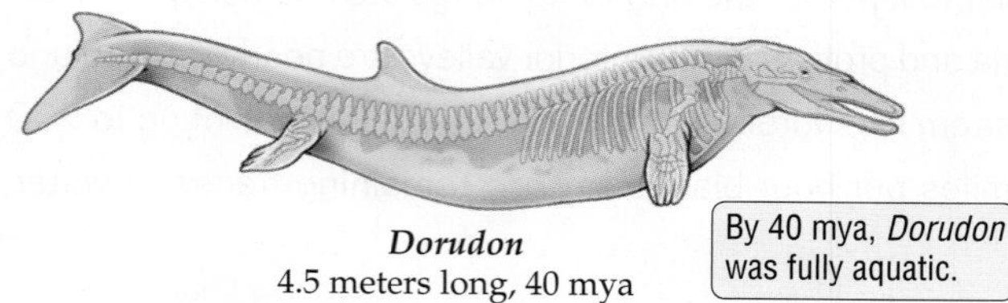

Dorudon
4.5 meters long, 40 mya
By 40 mya, *Dorudon* was fully aquatic.

FIGURE 16.1 Shape-Shifters
It took roughly 15 million years for whale ancestors to make the transition from life on land to life in water. These drawings are based on reconstructed fossil skeletons, which are superimposed on two of the whale ancestors. Compare the whale ancestors in this sequence with *Orcinus*, a modern toothed whale. What processes could have led to such a dramatic shift in body form?

of evolution can result in significant changes to body plans. How do such large-scale changes occur? Such changes through time are the focus of **macroevolution**, which explores the processes involved in the dramatic alteration of forms.

Darwin recognized that while the force of natural selection is potent enough to generate adaptations in organisms, significant timescales would be necessary to explain the great diversity of body plans observed in nature. In the late eighteenth century, James Hutton proposed that the forces that change Earth's geologic landscape are extremely slow-acting, so Earth is far older than was previously thought. Seventy years later, Charles Lyell (whose book Darwin read on his voyage on the *Beagle*) expanded on these theories. Estimates of Earth's age during Darwin's time ranged greatly, but it was generally thought not to exceed 100 million years, an estimate proposed by Lord Kelvin in 1862. This estimate deeply troubled Darwin, as he had serious doubts that natural selection could act sufficiently in that relatively short amount of time to produce the massive diversity of life.

Supporters of Darwin's theory of natural selection were vindicated in the early twentieth century when studies of radioactive isotopes placed the age of Earth at several billion years—sufficient time for the process of natural selection to create large-scale changes in organisms. **Radioisotopes** (*RAY*-**dee-oh**-*EYE*-**soh-tohp**) are unstable forms of elements that decay to more stable forms at a constant rate over time (see Section 2.1). The application of isotope studies to geology brought the current estimate of Earth's age to approximately 4.6 billion years. This scale of time, known as **geologic time**, may be difficult for most people to grasp. As humans, we tend to think of timescales in years, decades, or centuries. Geologic time spans millions of years or even eons. To put this scale of time into perspective, imagine the entire history of Earth as a 24-hour day. Dinosaurs evolved just before 11:00 PM, and humans, just a minute before midnight (**FIGURE 16.2**).

While studies of the geologic properties of Earth had a major impact on evolutionary thought by determining Earth's age, advocates of macroevolutionary change had another piece of major evidence buried within the very rocks they were studying: the fossil record.

Concept Check

1. What is macroevolution?
2. Why is the concept of geologic time so important to our understanding of macroevolution?

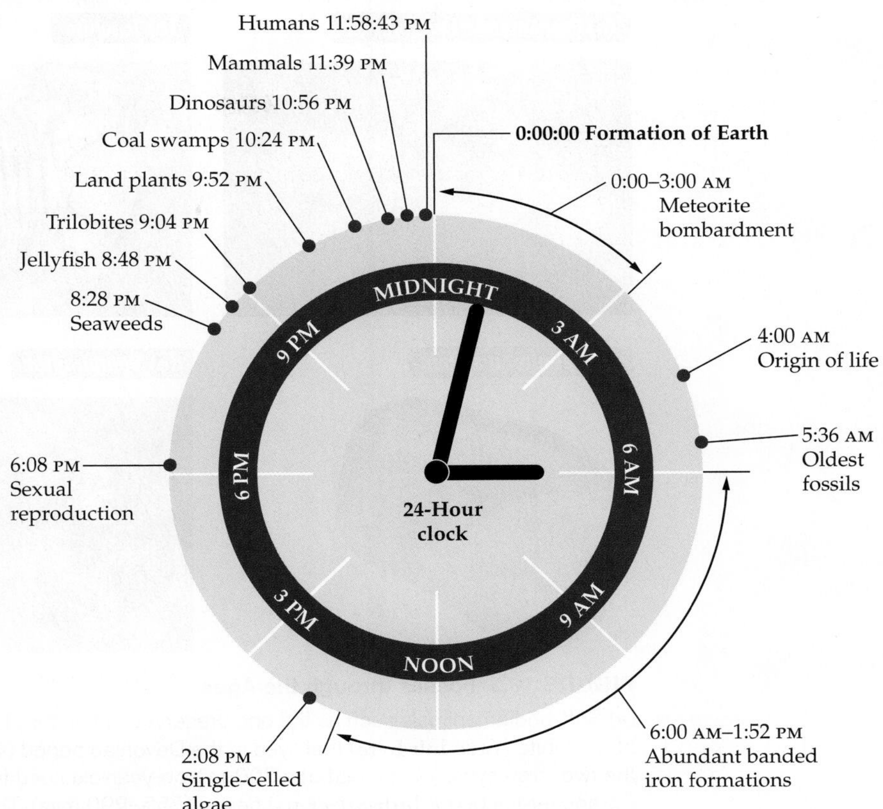

FIGURE 16.2 Earth's 4.6-Billion-Year History Represented as a 24-Hour Day
The history of Earth and major biological events laid out in a single 24-hour day. Notice the evolution of humans at 1 minute 17 seconds before midnight.

16.2 The Fossil Record: A Guide to the Past

Fossils are the preserved remains or impressions of individual organisms that lived in the past (**FIGURE 16.3**). Most commonly, fossils form when hard parts of animals, such as bones, teeth, and shells are preserved in sediments. Over time, the sediments harden into rock, and minerals replace organic materials, resulting in the formation of fossils. Fossils can also be formed by other means, such as when insects are trapped in amber (hardened tree sap; see Figure 16.3*d*). Although fossil preservation is a rare event, particularly for organisms that have few hard parts or that do not live near areas where sediments typically form, fossils are an invaluable resource for glimpsing the history of life on Earth.

The fossil record is central to the study of evolution. Fossils provided the first compelling evidence

Concept Check Answers

1. Macroevolution is large-scale evolution in which the effect of natural selection is seen as radical changes in body plan, not simply a change in allele frequencies.
2. A short timescale would make it difficult to argue convincingly that natural selection could shape the entire biodiversity seen on Earth.

FIGURE 16.3 Fossils through the Ages

(*a*) Soft-bodied animals such as the one preserved in this fossil dominated life on Earth 600 million years ago (mya). (*b*) A fossil of a trilobite (*TRYE*-**loh-byte**) that lived in the Devonian period (410–355 mya). Note the furrowlike rows of lenses on each of the two large eyes. (*c*) The leaf of a 300-million-year-old seed fern found near Washington, DC. The fossil formed during the Carboniferous (*KAHR*-**buh**-*NIF*-**er-us**) period (355–290 mya). The great forests of this period led to formation of the fossil fuels (oil, coal, and natural gas) that we use today as sources of energy. (*d*) This 20-million-year-old termite is preserved in amber, the fossilized resin of a tree. (*e*) A fossil of a *Velociraptor* entangled with a *Protoceratops*, which bit down on the predator's claw, locking both in a death grip. (*f*) Petrified wood. Here we see how what was once solid wood has fossilized into solid rock.

that past organisms were unlike organisms alive today, that many forms have disappeared from Earth completely, and that life has evolved through time.

As sedimentary rocks form, distinct layers are created, each containing the remains of organisms that lived at the time of formation. Older fossils are found in deeper, older rock layers. The order in which organisms appear in the fossil record agrees with our understanding of evolution based on other evidence, providing strong support for evolution. For example, analyses of the morphology (external form and internal structure), DNA sequences, and other characteristics of living organisms indicate that bony fishes gave rise to amphibians, which later gave rise to reptiles, which still later gave rise to mammals. This is also the order in which fossils from these groups appear in the fossil record. The fossil record also documents the evolution of major new groups of organisms, such as the evolution of mammals from reptiles, which occurred 220 million years ago. Mammal fossils are unknown in rocks from before this time.

Helpful to Know

The prefix *radi* (or *radio*), when used in biological terms, can have several meanings, none having to do with radios. Often it signals a word related in some manner to radioactivity, as here, with "radioisotopes."

Knowing the order of various species in the fossil record is helpful, but it can provide only *relative* ages of fossils. That is, it can reveal which fossils are older than others. In some cases we can approximate a fossil's age better by using radioisotopes. For example, for a given amount of the radioisotope carbon-14 (^{14}C), half of the total decays to the stable element carbon-12 every 5,730 years. By measuring the amount of ^{14}C that remains in a fossil, scientists can estimate the age of the fossil. Carbon-14 can be used to date only relatively recent fossils: too little ^{14}C remains to date fossils formed more than 70,000 years ago. But elements such as uranium-235, which has a half-life of 700 million years, can be used to date much older materials. If, as commonly occurs, a fossil does not contain any radioisotopes, methods like carbon or uranium dating enable us to determine an approximate date for the fossil by dating rocks found above and below the fossil.

The fossil record is incomplete

The fossil record shows clearly that there have been great changes in the groups of organisms that have dominated life on Earth over time. Although great numbers of fossils have been found, the fossil record contains many gaps. Because most organisms decompose rapidly after death, very few of them form fossils. Even if an organism is preserved initially as a fossil, a variety of common geologic processes, such as erosion and extreme heat or pressure, can destroy the rock in which it is embedded. Furthermore, fossils can be difficult to find. Given the unusual circumstances that must occur for a fossil to form, remain intact, and be discovered by scientists, a species could evolve, thrive for millions of years, and become extinct without our ever finding evidence of its existence in the fossil record. Nevertheless, each year new discoveries fill in some of those gaps. One evolutionary event that has long been of interest is the evolution of sea-dwelling whales from land-dwelling mammals. We look at this example next.

Fossils reveal that whales are closely related to a group of hoofed mammals

Most mammals live on land, and it is hard to imagine how a land mammal could be transformed into something as seemingly different as a whale. But recently discovered fossils provide a glimpse of how that adaptive transformation occurred.

Whales have a relatively complete fossil record, which enables scientists to observe many transitional forms that document their evolutionary past (see Figure 16.1). *Pakicetus* is an early whale ancestor that spent most of its time on land but does not particularly resemble a whale. Its overall body form, especially bones in its ankle, show that it was related to the hoofed mammals of today, such as the hippopotamus, camel, and giraffe (a group called the artiodactyls [***AHR*-tee-oh-*DAK*-tul**]). However, incomplete fossils of *Pakicetus* recovered in the 1980s did show some similarities to whales, such as bones in the inner ear that were similar to those found in modern whales. In 2001, the anklebones of several whale ancestors, including *Pakicetus* and *Rodhocetus*, were discovered. In all artiodactyls (as well as whales, which do have anklebones, despite their lack of hind legs), the anklebone has an unusual shape, with both the top and bottom surfaces of the bone resembling those of a pulley (**FIGURE 16.4**). This shape is an adaptation for running on land, so it is highly unlikely that whale ancestors developed such bones as a result of convergent evolution (see Section 15.4). Instead, these fossils strongly suggest that whale ancestors had such bones because they shared a common ancestor with the artiodactyls.

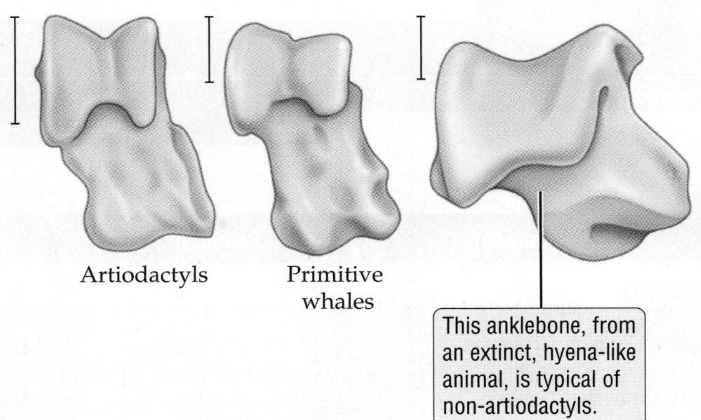
Changes in Anklebone Structure
Artiodactyls
Primitive whales
This anklebone, from an extinct, hyena-like animal, is typical of non-artiodactyls.

FIGURE 16.4 Whale Anklebones Illustrate Relatedness to Artiodactyls
The anklebones of two whale ancestors—*Pakicetus* and *Rodhocetus*—are similar in shape to those of artiodactyls (hoofed mammals), but very different from those found in most other mammals. Note: scale bar = 1 cm.

These fossil discoveries confirmed the results of genetic analyses that had been carried out on the relationship between whales and artiodactyls. Whales show the greatest DNA similarity to their artiodactyl cousins.

Concept Check

1. What is a fossil?
2. Why is the fossil record so incomplete?

Concept Check Answers

1. A fossil is the preserved traces of an organism. Fossils may include petrified bone, impressions cast in stone, or any other remnant of a once-living organism.
2. The conditions for fossil preservation are relatively rare. Some organisms are more likely to be fossilized than others (for example, bony versus soft-bodied animals).

16.3 The History of Life on Earth

With our increasing understanding of the geologic and biological history of Earth, scientists are able to piece together the major events in the history of life on Earth: the origin of cellular organisms, the beginning of multicellular life, and the colonization of land. **FIGURE 16.5** provides a sweeping overview of this history.

The first single-celled organisms arose at least 3.5 billion years ago

Our solar system and Earth formed 4.6 billion years ago. The oldest known rocks on Earth (3.8 billion years old) contain carbon deposits that hint at life. Cell-like structures have been found in layered mounds called stromatolites (**stroh-*MAT*-uh-lyte**) that formed

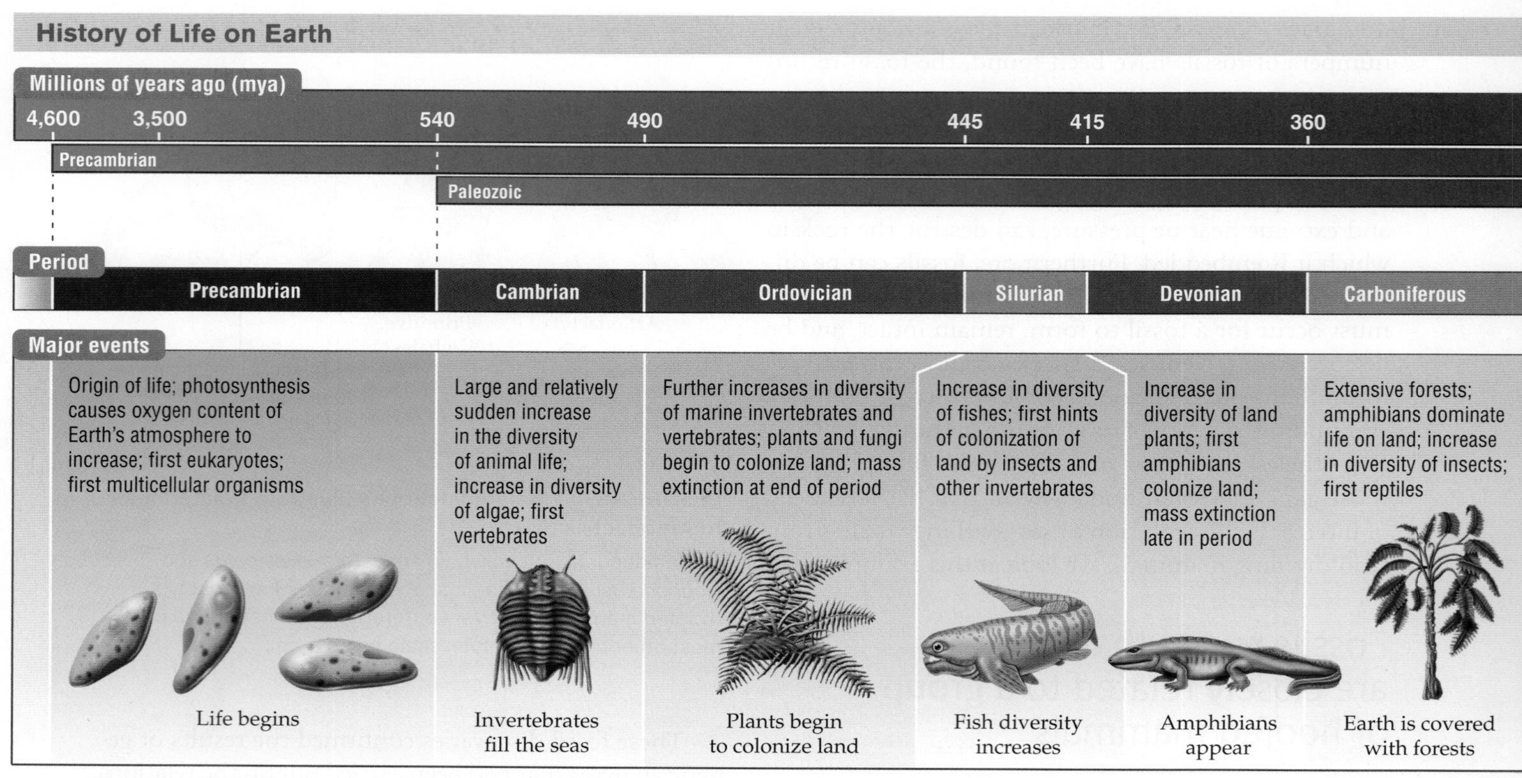

History of Life on Earth
Millions of years ago (mya)
4,600
3,500
540
490
445
415
360
Precambrian
Paleozoic
Period
Precambrian
Cambrian
Ordovician
Silurian
Devonian
Carboniferous
Major events
Origin of life; photosynthesis causes oxygen content of Earth's atmosphere to increase; first eukaryotes; first multicellular organisms
Large and relatively sudden increase in the diversity of animal life; increase in diversity of algae; first vertebrates
Further increases in diversity of marine invertebrates and vertebrates; plants and fungi begin to colonize land; mass extinction at end of period
Increase in diversity of fishes; first hints of colonization of land by insects and other invertebrates
Increase in diversity of land plants; first amphibians colonize land; mass extinction late in period
Extensive forests; amphibians dominate life on land; increase in diversity of insects; first reptiles
Life begins
Invertebrates fill the seas
Plants begin to colonize land
Fish diversity increases
Amphibians appear
Earth is covered with forests

FIGURE 16.5 The Geologic Timescale and the Major Events in the History of Life
The history of life can be divided into 12 major geologic time periods, beginning with the Precambrian (4,600–540 mya) and extending to the Quaternary (1.8 mya to the present). This timescale is not drawn to scale; to do so while including the Precambrian would require extending the diagram off the book page to the left by more than 5 feet.

3.5 billion years ago (**FIGURE 16.6**). Projections based on DNA analysis also support the idea that life had appeared on Earth by 3.5 billion years ago.

Those first life-forms were prokaryotes. After they appeared, it took well over a billion years for the first eukaryotes to appear. Eukaryotes are first seen in the fossil record at about 2.1 billion years ago. During this long period between the origin of prokaryotes and eukaryotes, the evolution of eukaryotes may have been constrained in part by low levels of oxygen in the atmosphere. Chemical analyses of very old rocks indicate that Earth's atmosphere initially contained almost no oxygen. Roughly 2.8 billion years ago, however, a group of bacteria evolved a type of photosynthesis that releases oxygen as a by-product. As a result, oxygen began to accumulate in the atmosphere, and its concentration increased over time (**FIGURE 16.7**).

FIGURE 16.6 Stromatolites Contain Evidence of Ancient Cellular Life
These mounds were formed from compacted sediments that trap the remains of bacterial cells. Stromatolites similar to these have been dated to 3.5 billion years, indicating that life on Earth is very old.

Eukaryotic cells are larger than most prokaryotic cells. Because of their relatively large size, eukaryotic cells would not have been able to get enough oxygen to meet their needs until the atmospheric concentration of oxygen reached at least 2–3 percent of present-day levels (see Section 17.3). Once those levels were reached, about 2.1 billion years ago, the first single-celled eukaryotes—organisms that resembled some modern algae—appeared. When oxygen levels reached their current levels, the evolution of larger and more complex multicellular organisms became possible.

Oxygen can be toxic. And it was to many early forms of life, so as the oxygen concentration in the

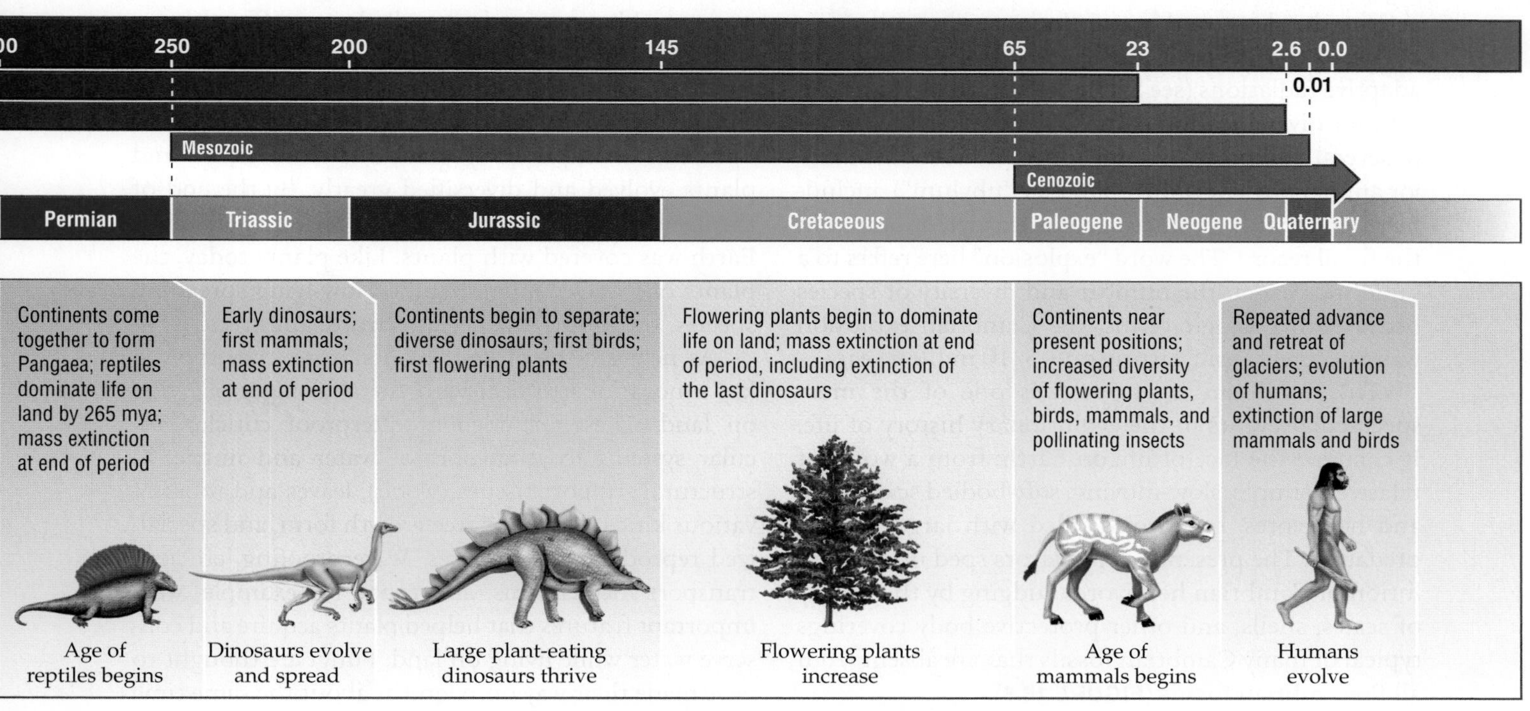

300
250
200
145
65
23
2.6
0.0
0.01
Mesozoic
Cenozoic
Permian
Triassic
Jurassic
Cretaceous
Paleogene
Neogene
Quaternary
Continents come together to form Pangaea; reptiles dominate life on land by 265 mya; mass extinction at end of period
Early dinosaurs; first mammals; mass extinction at end of period
Continents begin to separate; diverse dinosaurs; first birds; first flowering plants
Flowering plants begin to dominate life on land; mass extinction at end of period, including extinction of the last dinosaurs
Continents near present positions; increased diversity of flowering plants, birds, mammals, and pollinating insects
Repeated advance and retreat of glaciers; evolution of humans; extinction of large mammals and birds
Age of reptiles begins
Dinosaurs evolve and spread
Large plant-eating dinosaurs thrive
Flowering plants increase
Age of mammals begins
Humans evolve

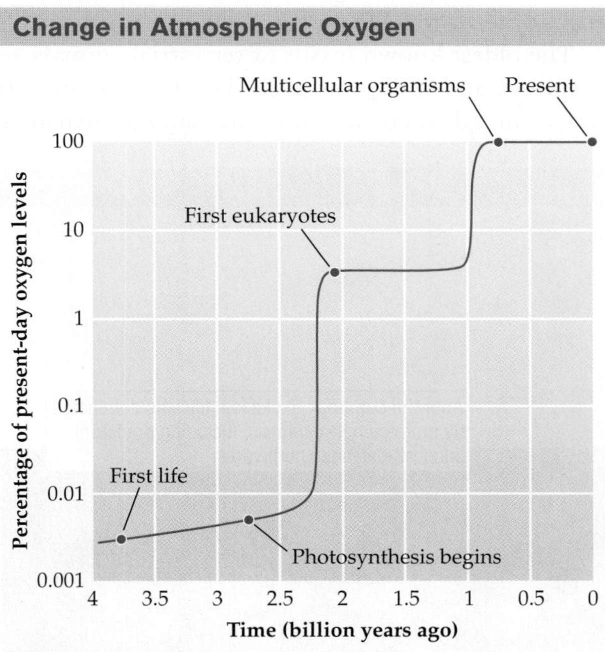

Change in Atmospheric Oxygen
Multicellular organisms
Present
100
10
1
0.1
0.01
0.001
First eukaryotes
First life
Photosynthesis begins
Percentage of present-day oxygen levels
4
3.5
3
2.5
2
1.5
1
0.5
0
Time (billion years ago)

FIGURE 16.7 Oxygen on the Rise

The release of oxygen as a waste product by photosynthetic organisms has caused its concentration in Earth's atmosphere to increase greatly over the last 3–4 billion years, facilitating the evolution of eukaryotes and multicellular organisms. Shown here are the hypothesized levels of oxygen at key points in the history of life, but much is still unknown about the dynamics of the rise of oxygen over time.

atmosphere increased, numerous early prokaryotes went extinct or became restricted to environments that contained little or no oxygen. Because the biologically driven increase in the oxygen concentration of the atmosphere drove many early organisms extinct while simultaneously setting the stage for the origin of multicellular eukaryotes, this increase in oxygen was one of the most important events in the history of life on Earth.

Multicellular life evolved about 650 million years ago

During the Precambrian period, about 650 million years ago (mya), the number of organisms appearing in the fossil record increased. At that time, much of Earth was covered by shallow seas filled with plankton. Protists, small multicellular animals, and single-celled and multicellular algae floated freely in the water. Later in the Precambrian, by about 600 mya, larger, soft-bodied multicellular animals had begun to appear (see Figure 16.3*a*). These flat animals crawled or stood upright on the seafloor, probably feeding on living plankton or their remains. Many of these early multicellular animals belonged to groups of organisms that are no longer found on Earth.

Later, during the Cambrian period (about 540 mya), there was an astonishing burst of evolutionary activity that led to a dramatic increase in the diversity of life, known as the **Cambrian explosion**. Multiple adaptive radiations (see Section 15.3) resulted in a significant diversification of animal life. Over the course of several million years, large forms of most of the major animal phyla (*FYE*-**luh**; singular "phylum"), including many that have since become extinct, appeared in the fossil record. The word "explosion" here refers to a rapid increase in the number and diversity of species because, in geologic terms, the Cambrian explosion was extremely rapid, lasting only 5–10 million years.

The Cambrian explosion was one of the most spectacular events in the evolutionary history of life. It changed the face of life on Earth: from a world of relatively simple, slow-moving, soft-bodied scavengers and herbivores, to a world filled with large, mobile predators. The presence of predators sped up the evolution of Cambrian herbivores, judging by the variety of scales, shells, and other protective body coverings typical of many Cambrian fossils that are absent from all Precambrian fossils (**FIGURE 16.8**).

Colonization of land followed the Cambrian explosion

Up to this point in time, all living things had lived in water. The land was barren, since colonizing it posed enormous challenges for living organisms. Many of the functions basic to life, like structural support, movement, reproduction, and the regulation of ions, water, and heat, must be handled very differently on land than in water. About 500 mya, descendants of green algae were the first organisms to meet these challenges. These early terrestrial colonists had few cells and a simple body plan, but from them land plants evolved and diversified greatly. By the end of the Devonian (**duh-*VOH*-nee-un**) period (360 mya), Earth was covered with plants. Like plants today, the plants of the Devonian included low-lying spreading species, short upright species, shrubs, and trees.

As new groups of land plants arose, they evolved key innovations to deal with the challenges of living on land. These included a waterproof cuticle, vascular systems for transport of water and nutrients, structural support tissues (wood), leaves and roots of various kinds, seeds, the tree growth form, and specialized reproductive structures. Waterproofing, efficient transport mechanisms, and roots, for example, were important features that helped plants acquire and conserve water while living on land. Fungi are thought to have made their way onto land at about the same time as plants. In fact, mutualistic associations between plants and fungi called *mycorrhizae* (see Section 18.5) may have helped both groups contend with the challenges of terrestrial life.

The oldest known fossils of terrestrial animals are of spiders and millipedes that date from about 410 mya, although there are hints that land animals may

Diversification of Animals

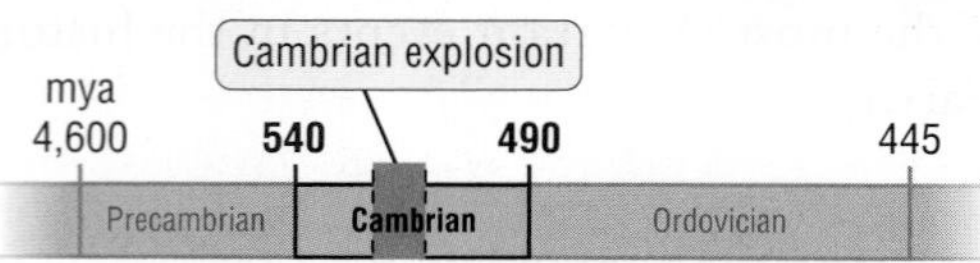

FIGURE 16.8 The Cambrian Explosion Transformed the History of Life on Earth

have lived as early as 490 mya. Many of the early animal colonists on land were predators; others, such as millipedes, fed on living plants or decaying plant material. Insects, which are currently the most diverse group of terrestrial animals, first appeared roughly 400 mya, and by 350 mya they were playing a major role in terrestrial ecosystems.

The first vertebrates to colonize land were amphibians, the earliest fossils of which date to about 365 mya. Early amphibians resembled, and probably descended from, lobe-finned fishes (**FIGURE 16.9**). Amphibians were the most abundant large organisms on land for about 100 million years. In the late Permian period, the reptiles, which had evolved from a group of reptile-like amphibians, rose to become the most common vertebrate group. Reptiles were the first group of vertebrates that could reproduce without returning to open water (for example, to lay eggs). The evolution of the amniotic egg was a major event in the history of life because it established a new evolutionary branch, the amniotes, which includes all reptiles, birds, and mammals. Reptiles dominated vertebrate life on land for 200 million years (265–65 mya). Dinosaurs, a group of reptiles, arose about 230 mya and dominated the planet from about 200 mya to about 65 mya. Mammals, the vertebrate group that currently dominates life on land, evolved from reptiles roughly 220 mya (see Figure 16.5).

Concept Check

1. Why was it significant that early bacteria evolved the ability to carry out photosynthesis?
2. How did the Cambrian explosion change life on Earth?

16.4 The Effects of Plate Tectonics

The enormous size of the continents may lead us to think of them as immovable. However, the continents do move slowly relative to one another, and over hundreds of millions of years they travel considerable distances (**FIGURE 16.10**). This movement of the continents over time is called **plate tectonics**.

How can something as big as a continent move from place to place? Continents are not anchored to the center of the planet. They "float" on Earth's **mantle**, a hot layer of semisolid rock. Two forces cause the continental plates to move. First, hot plumes of liquid rock rise from Earth's mantle to the surface and push the continents away from one another. This process can cause the seafloor to spread, as it is doing right now between North America and Europe, which are separating at a rate of 2.5 centimeters per year. This process can also cause bodies of land to break apart, as is currently happening in Iceland and East Africa. Second, where two plates collide, one plate sometimes slips underneath the other and begins to sink into the mantle below. As the now hidden end of the continental plate sinks down and slowly melts, it gradually pulls the rest of the plate down along with it, causing the rest of the plate to continue to move.

Patterns of plate tectonics have had dramatic effects on the history of life. Continents have in the past collided and joined into large landmasses called **supercontinents**. The latest supercontinent, Pangaea, began to break apart early in the Jurassic period (about 200 mya), ultimately separating into the continents we know today (see Figure 16.10). As the continents drifted apart, populations of organisms that once were connected by land became isolated from one another.

As we noted in Chapter 15, geographic isolation reduces or eliminates gene flow, thereby promoting speciation. The separation of the continents was geographic isolation on a grand

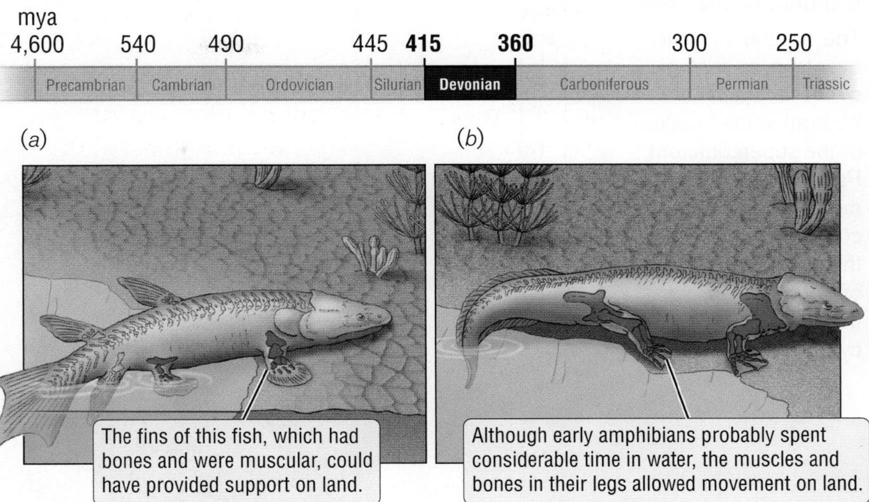

Evolution of Tetrapods
mya
4,600
540
490
445
415
360
300
250
Precambrian
Cambrian
Ordovician
Silurian
Devonian
Carboniferous
Permian
Triassic
(a)
(b)
The fins of this fish, which had bones and were muscular, could have provided support on land.
Although early amphibians probably spent considerable time in water, the muscles and bones in their legs allowed movement on land.

FIGURE 16.9 The First Amphibians
(*a*) Amphibians probably evolved over long periods of evolutionary time from a lobe-finned fish ancestor like the one shown here. (*b*) This early amphibian was reconstructed from a 365-million-year-old (late Devonian) fossil.

Concept Check Answers

1. Oxygen is a by-product of photosynthesis, and increasing oxygen in the atmosphere set the stage for the evolution of eukaryotes, a group that includes all multicellular life.
2. The Cambrian explosion marks the appearance of most of the major living animal phyla, many of which were the world's first predators.

EXTREME TECTONICS
Each year, tectonic forces pull the plates apart, and Iceland grows wider by about 5 centimeters. That's because tectonic forces are pulling the North American and European plates apart, with Iceland right in the middle.

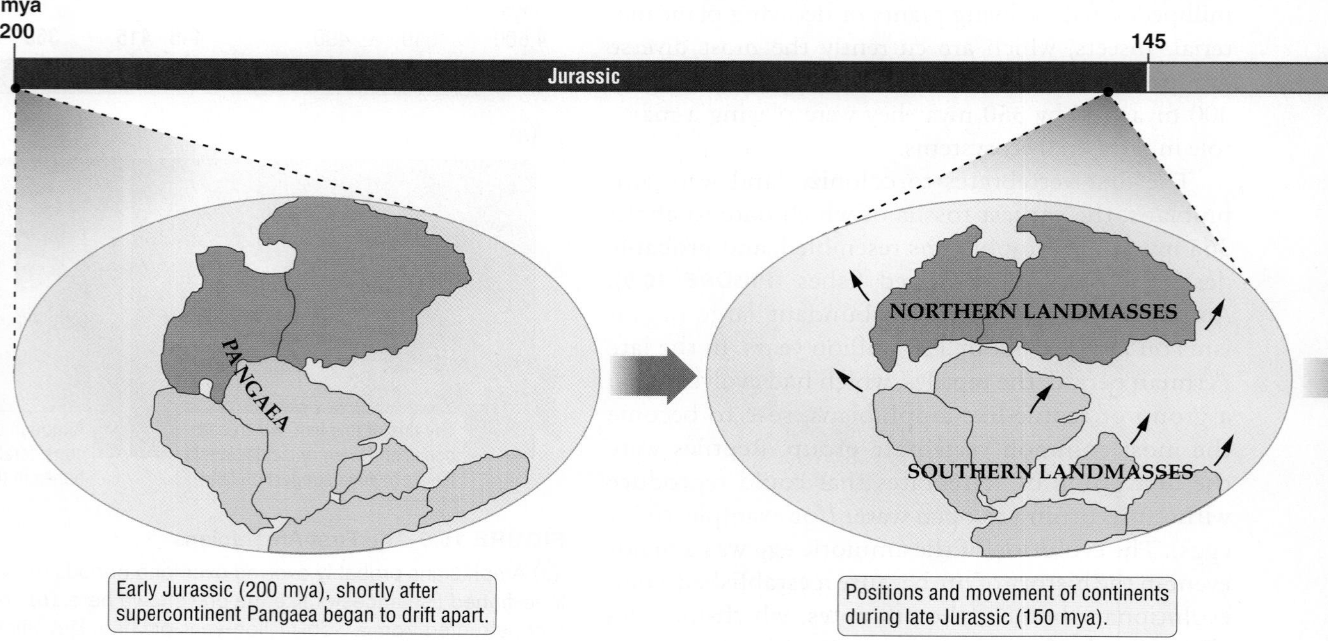

Plate Tectonics
mya
200
145
Jurassic
PANGAEA
NORTHERN LANDMASSES
SOUTHERN LANDMASSES
Early Jurassic (200 mya), shortly after supercontinent Pangaea began to drift apart.
Positions and movement of continents during late Jurassic (150 mya).

FIGURE 16.10 Movement of the Continents over Time

The continents move over time, as shown by these "snapshots" illustrating the breakup of the supercontinent Pangaea. Earlier movements of the continents had led to the gradual formation of Pangaea, a process that was complete by 250 mya.

scale, and it led to the formation of many new species. Among mammals, for example, kangaroos, koalas, and other marsupials unique to Australia evolved in geographic isolation on that continent, which broke apart from Antarctica and South America about 40 mya.

Plate tectonics also affects climate. The position of the continents influences the flow of Earth's oceans, which distributes heat and moisture to different parts of the Earth. These climate trends have a profound effect on the evolution of life by altering what kinds of organisms natural selection will favor. Organisms adapted to life in warm tropical climates will face significant challenges if their landmass moves to a much colder climate. The continent of Antarctica, for example, harbors little plant life currently, but fossil discoveries have shown that the continent was once heavily forested with beech and ginkgo trees—indicators that Antarctica once exhibited a subtropical climate similar to that of New Zealand today. The continent also hosted several dinosaur species before the combination of plate tectonics and climate change made the continent the harsh, ice-covered environment we see today.

Concept Check

1. Why is the age of Earth a key factor in discussions of plate tectonics?
2. What are two ways that plate tectonics can affect biodiversity?

Concept Check Answers

1. Because the continents move so slowly, the effects of plate tectonics are apparent only over long timescales.

2. The movement of continents can geographically isolate populations, leading to allopatric speciation (see Section 16.2). It can also alter climate by changing ocean currents.

16.5 Mass Extinctions: Worldwide Losses of Species

As the fossil record shows, species have come and gone throughout the history of life. Usually the extinction rate, which is a measure of how fast species disappear, is fairly steady. At several points in Earth's history, however, the extinction rate has soared. The fossil record shows that there have been five **mass extinctions**, periods of time during which great numbers of species went extinct all across the planet. Each of these upheavals left a permanent mark on the history of life, driving more than 50 percent of Earth's species to extinction. **FIGURE 16.11** shows the effects of these mass extinctions on animal life alone. Though difficult to determine, the causes of these extinctions are thought to include such factors as climate change, massive volcanic eruptions, asteroid impacts, changes in the composition of marine and atmospheric gases, and changes in sea levels.

The largest mass extinction occurred at the end of the Permian period, 250 mya. The **Permian extinction** radically altered life in the oceans. Among marine invertebrates (animals without backbones), an estimated 50–63 percent of existing families, 82 percent of genera, and 95 percent of species went extinct. The Permian mass extinction also devastated life on land. It removed

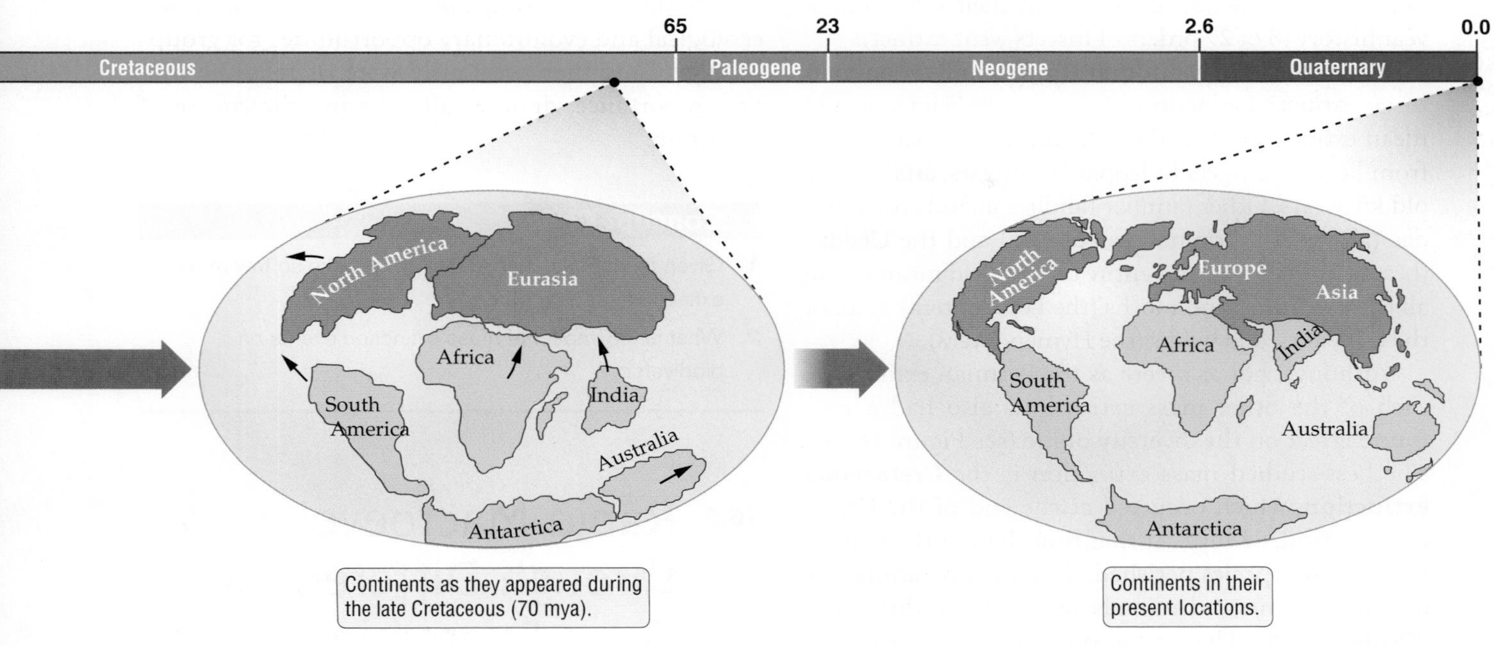

Mass Extinctions

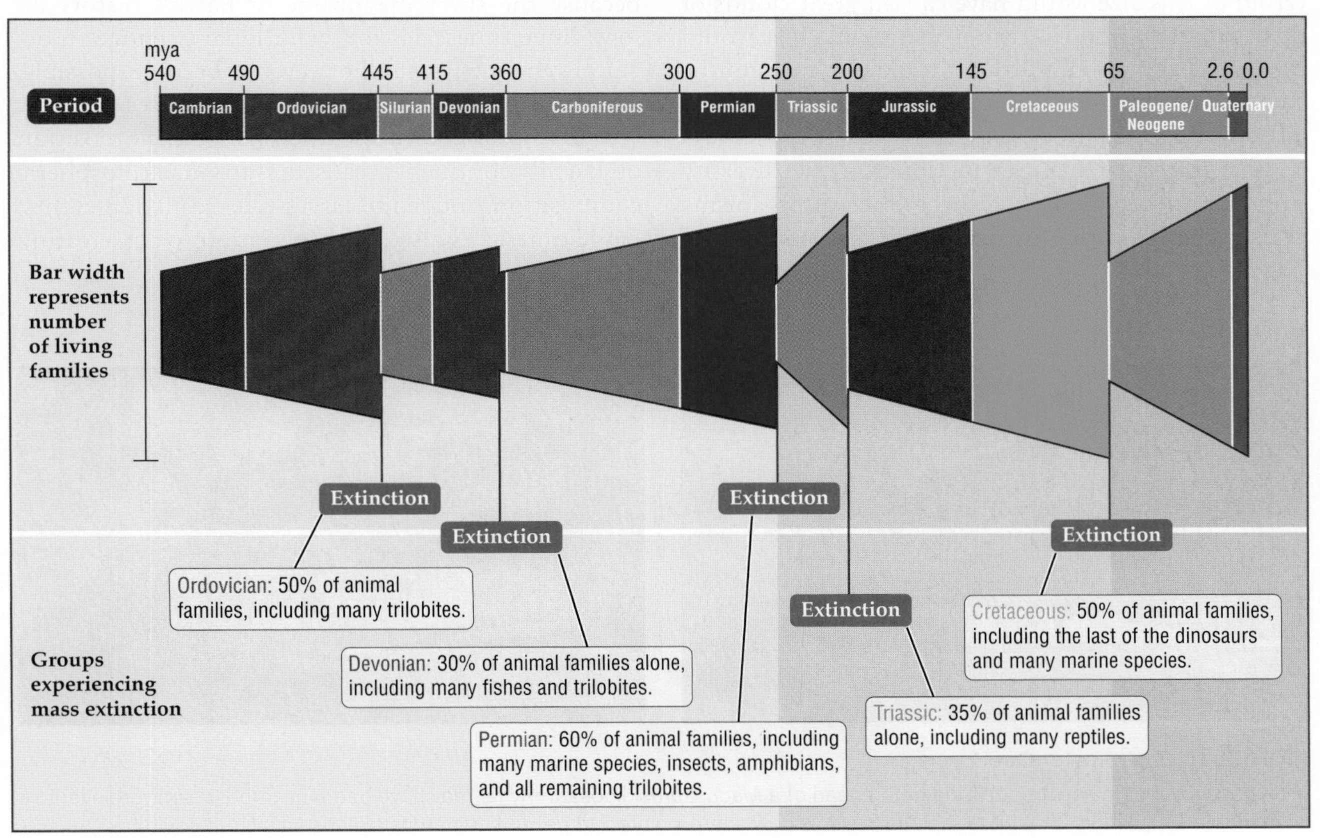

FIGURE 16.11 The Five Mass Extinctions Drastically Reduced the Diversity of Animals

In addition to both marine and terrestrial animals, plant groups (not shown) were severely affected by the five mass extinctions that have occurred in Earth's history. After each extinction, life again diversified.

Concept Check Answers

1. The causes of mass extinctions are global in scale and catastrophic in magnitude. Such events are inherently rare.
2. Many species are lost in mass extinction events, but their loss opens up ecological niches permitting the adaptive radiation of successor species.

62 percent of the existing terrestrial families, brought the reign of the amphibians to a close, and caused the only major extinction of insects in their 400-million-year history (8 of 27 orders of insects went extinct).

What does it mean for an entire family of animals to go extinct? Extinction of the family Felidae would mean extinction of all the wild and domesticated cats, from lions and tigers to leopards, cougars, and regular old kitty cats. Other familiar families include the Canidae (all wild and domesticated dogs) and the Ursidae (bears). Today, losing an entire order could mean losing all the butterflies and moths (the Lepidoptera) or all of the bees, ants, and wasps (the Hymenoptera).

Although not as severe as the Permian extinction, each of the other mass extinctions also had a profound effect on the diversity of life (see Figure 16.11). The best-studied mass extinction is the **Cretaceous extinction**, which occurred at the end of the Cretaceous period, 65 mya. At that time, half of the marine invertebrate species perished, as did many families of terrestrial plants and animals, including the dinosaurs (**FIGURE 16.12**). The Cretaceous mass extinction was caused at least in part by the collision of an asteroid with Earth. A 65-million-year-old, 180-kilometer-wide crater lies buried in sediments off the Yucatán coast of Mexico; this crater is thought to have formed when an asteroid 10 kilometers across struck Earth. An asteroid of this size would have caused great clouds of dust to hurtle into the atmosphere; this dust would have blocked sunlight around the globe for months to years, causing temperatures to drop drastically and thereby driving many species extinct.

The effects of mass extinctions on the diversity of life are twofold. First, entire groups of organisms perish, changing the history of life forever. Second, mass extinctions enable surviving species to experience *adaptive radiations* (see Section 15.3), as the extinction of dominant groups of organisms provides new ecological and evolutionary opportunities for groups of organisms that previously were of relatively minor importance, dramatically altering the course of evolution.

Concept Check

1. Given the long history of Earth, why have so few mass extinctions been recorded?
2. What is the impact of mass extinction events on biodiversity?

16.6 Rapid Macroevolution through Differential Gene Expression

For many years, evolutionary biologists relied exclusively on the concept of geologic time to explain macroevolutionary change. As you have seen in previous sections, dramatic changes in body form can occur because the sheer magnitude of Earth's history allows huge timescales for evolutionary change to occur. Even one-in-a-million events become possible, and even likely, under such vast expanses of time. Geologic processes also have changed the position of Earth's continents, caused shifts in atmospheric composition, and led to large-scale extinction events. Large timescales also allow speciation to act, creating

(*a*)

(*b*)

FIGURE 16.12 Gone for Good
(*a*) A dog sits next to a reconstruction of the head of a *Mapusaurus* dinosaur—what may have been the largest carnivore ever to walk the Earth. (*b*) This *Allosaurus* skeleton reveals how sharp, pointed, large, and numerous a dinosaur's teeth could be.

BIOLOGY MATTERS

Is a Mass Extinction Under Way?

The International Union for Conservation of Nature (IUCN), known also as the World Conservation Union, maintains what it calls its Red List, which identifies the world's threatened species. To be defined as such, a species must face a high to extremely high risk of extinction in the wild. The 2013 Red List contains 20,934 species threatened with extinction, of a total of 70,293 species assessed. That means about 30 percent of the species evaluated by the IUCN are threatened. Because this assessment accounts for only about 4 percent of the world's 1.7 million described species, the total number of species threatened with extinction worldwide would actually be much larger. In 2003, the IUCN list showed 12,259 species threatened with extinction. The number of threatened species has therefore increased nearly 71 percent in 10 years.

The Iberian lynx (*Lynx pardinus*) is the most critically endangered feline in the world. About 100 individuals survive in small populations in Spain.

The Red List is based on an easily understood system for categorizing extinction risk; it is also objective, yielding consistent results when used by different people. These two attributes have earned the Red List international recognition as an effective method to assess extinction risk.

For many taxonomic groups, only a few of the described species have been evaluated. For example, survival risk has been assessed for only 771 out of 950,000 described insect species. If the data already gathered are indicative of similar patterns in unassessed species, then current species loss will approach the rates seen in some of the previous mass extinctions.

For more information about the Red List, visit www.iucnredlist.org.

Number of Species at Risk of Extinction in Some Major Groups of Organisms

	MAMMALS	BIRDS	REPTILES	AMPHIBIANS	FISHES	MOLLUSKS	OTHER	PLANTS	TOTAL
Canada	11	16	5	1	36	5	10	2	86
United States	36	78	37	56	236	301	265	270	1,279

NOTE: Based on the IUCN Red List, 2013.

the diversity of life observed today. Combined with the effects of natural selection, organisms have been constantly modified over billions of years to their current forms.

While the geologic timescale is certainly an important factor in explaining macroevolutionary change, scientific advances, particularly in the fields of evolutionary developmental biology (or **evo-devo**), have shed new light on the process. In the late twentieth century, the field of developmental biology demonstrated that gene expression has as much impact on body plans as do the genes themselves. An organism may bear a different form because of how and when a particular gene is expressed in the developing embryo.

Changes in gene expression can radically influence the body plan of an organism. In particular, developmental biologists identified a class of genes, called **homeotic genes**, that cause the development of structures in embryological development. While it is not correct to say that there is a "gene" for limbs (since limb development requires many proteins to complete the process of, say, forming a leg), there is a homeotic gene that regulates limb development by controlling the expression of all the other genes necessary for limb growth in an embryo. Although every cell has the same DNA (and therefore the same homeotic genes), not every cell expresses the same genes; that is, different cells will "turn on" different sets of genes. Homeotic genes are expressed in *specific* parts of the developing embryo, which is why your limbs are located where they are and not placed randomly over your body.

Experiments have demonstrated that altering the expression of homeotic genes can have significant impacts on body plans in only one or a few generations.

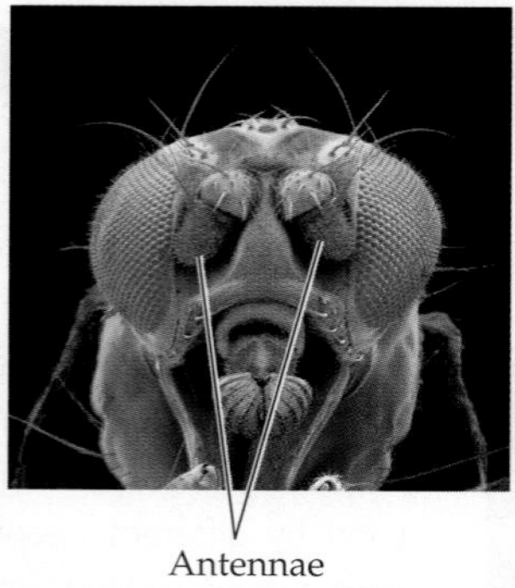

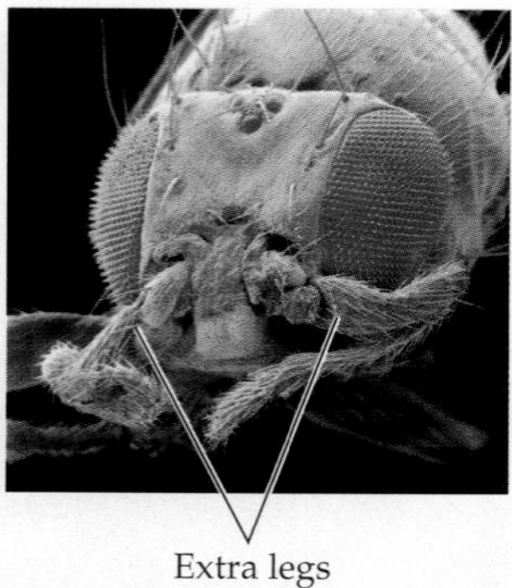

FIGURE 16.13 Homeotic Gene Expression Can Rapidly Lead to Different Forms
The homeotic gene for limb development was activated in the head region of a fly, leading to the growth of legs from the facial region.

For example, expressing the homeotic genes that control limb development in the head region of a fly embryo leads to the development of a fly with legs growing from its head (**FIGURE 16.13**). If such a change were adaptive, natural selection would increase its frequency in the population, and it could rapidly become the most common allele. While this particular example was a laboratory experiment, it is not hard to imagine how such changes in homeotic genes would act in wild populations. For example, a legless lizard looks to most people like a snake (**FIGURE 16.14**). It is the product of a change in homeotic gene expression, where the genes for limb development were suppressed. Such a change did not require millions of years of progressive and incremental evolution. A small alteration of homeotic gene expression rapidly led to a significant phenotypic change and introduced a completely novel vertebrate body plan. Macroevolution, as it turns out, does not absolutely require vast timescales in order to function.

Atavistic traits and vestigial traits provide evidence that many existing body plans are the product of developmental changes (see Section 15.4). An atavistic trait is one that represents a reversion to an ancestral state. Being descended from organisms with tails, humans are sometimes born with tails because we possess the genes for tail development within our genomes. Our current form lacks tails simply because the homeotic genes that control tail development are suppressed. Similarly, horses are occasionally born with three toes instead of one because ancestral horses had multiple toes (see Figure 14.21).

Differential gene expression can occur in any gene, not just homeotic genes. The human genome and chimpanzee genome, for example, are approximately 98 percent similar, yet the two species differ significantly in form. Certainly, the 2 percent of genetic difference contributes to the differences between the two species, but the bulk of the differences arise from the way in which the remaining 98 percent of the

FIGURE 16.14 A Legless Lizard
Although similar in appearance and lifestyle to a snake, a legless lizard cannot unhinge its jaw like a snake can. It also possesses the distinct lizard ability to detach its tail when threatened.

genes are expressed. Humans, for example, have the same genes for hair growth that a chimpanzee has, but these genes are expressed only in certain areas of the body. Likewise, the same genes responsible for facial elongation during growth are found in both species, but they are expressed for a longer period of time in chimps than in humans, leading to their relatively longer faces (**FIGURE 16.15**).

Concept Check

1. What is a homeotic gene?
2. Explain why the discovery of homeotic genes had a significant impact on evolutionary thought.

16.7 Phylogenetics: Reconstructing Evolutionary Relationships

Earth's biodiversity is staggering. Speciation and macroevolution have created millions of species that are diverse in form and function, yet all of life is related. Organizing and classifying this array of biodiversity is the focus of a discipline called **taxonomy**. Scientists classify organisms into groups based on morphological or genetic similarities. These similarities reflect the evolutionary histories of the organisms, and can be used to construct evolutionary trees that illustrate the patterns of species evolution. This is the goal of a field called **phylogenetics**.

Similarities between organisms can be used to infer their evolutionary histories

Since life on Earth began, living things have diversified into many lines of descent, or **lineages**, that have evolved into millions of different types of organisms, many of them now extinct.

An evolutionary tree maps the relationships between ancestral groups and their descendants, the way your family tree describes your relationships to your mother, grandmother, and great grandmother. An evolutionary tree clusters the groups that are most closely related on neighboring branches, the way you and your siblings would be depicted on one branch of your family tree and your cousins on another branch.

In an evolutionary tree, the organisms under consideration are depicted as if they were leaves at the tips of the tree branches (**FIGURE 16.16**). A **node** marks the moment in time when an ancestral group split, or diverged, into two separate lineages (such as brown bears and polar bears). The node represents the **most recent common ancestor** of the two lineages in question—that is, the most *immediate* ancestor that *both* lineages share.

How do we know which groups of organisms belong where on an evolutionary tree—which "leaves" belong

EXTREME ATAVISM

The Gomez brothers, Larry and Danny, exhibit congenital hypertrichosis, a rare condition in which hair grows over the entire body. This trait is thought to be an atavistic trait, indicative of our descent from animals with more fur than we currently possess.

Concept Check Answers

1. A homeotic gene controls the embryological development of a body structure.

2. Prior to the discovery of homeotic genes, macroevolutionary change was thought to occur solely over long periods of time. Changes in homeotic gene expression, however, have shown that significant alterations in body plans can occur in a single generation.

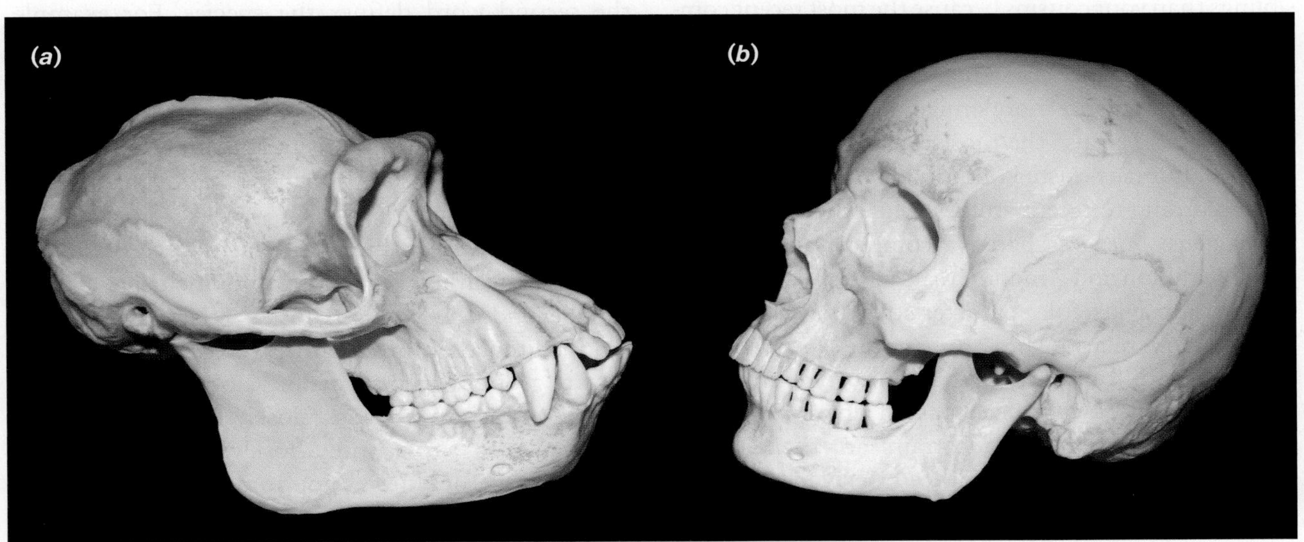

FIGURE 16.15 Same Gene, Different Timing
A chimpanzee's face (*a*) is longer than a human's (*b*) because the genes that are responsible for facial elongation are expressed for a longer period of time in chimps.

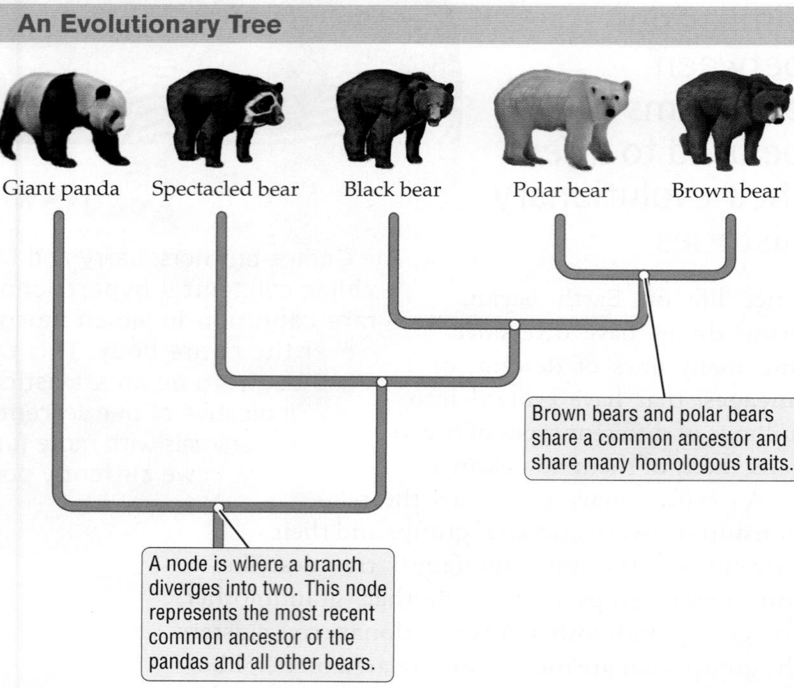
An Evolutionary Tree
Giant panda
Spectacled bear
Black bear
Polar bear
Brown bear
Brown bears and polar bears share a common ancestor and share many homologous traits.
A node is where a branch diverges into two. This node represents the most recent common ancestor of the pandas and all other bears.

FIGURE 16.16 Evolutionary Trees Depict Relationships between Organisms

This tree shows the evolutionary relationships among several members of the bear family. At the root of the tree is the common ancestor of these organisms. One can see that the polar bear and brown bear are most closely related because they share a more recent common ancestor.

next to each other in the same cluster? Recall from Section 15.4 that when two lineages diverge to become separate species, they retain certain similarities because of their shared ancestry. These similar characteristics are called *homologous traits*. Two organisms that share a very recent common ancestor are likely to exhibit more homologous traits, compared to more distantly related organisms. For example, you more closely resemble your siblings than your cousins because the most recent common ancestors of siblings are parents, while the most recent common ancestors of cousins are grandparents. Likewise, humans are more similar to each other than to chimpanzees because the most recent common ancestor of all humans existed only about 2,000–3,000 years ago, according to recent studies, while the human and chimpanzee lineages shared a common ancestor between 5 million and 7 million years ago.

The most useful characteristics for discerning evolutionary relationships are unique (*derived*) features that are found in a group's most recent common ancestor. Having been passed down from that ancestor, these homologous traits are then shared by the descendant groups. **Shared derived traits** are evolutionary novelties shared by an ancestor and its descendants but not seen in groups that are not direct descendants of that ancestor. For example, fur and mammary glands are shared derived traits that are unique to mammals and the most recent common ancestor of mammals; no nonmammals on the tree of life display these traits.

Evolutionary trees imply that time has passed when going from ancestors to the descendants at the tips of the branches. The time sequence may have a scale—in millions of years, for example—but trees depicting large-scale evolution may lack a precise time scale because the information simply is not available. In evolutionary trees of this type, however, the base of the tree represents the point furthest in the past, and the "leaves" of the tree are in the present.

The Linnaean system of biological classification reflects evolutionary history

Modern biologists have devised different classification systems for organizing biodiversity. The scientific standard still in use is based on a hierarchical classification system that was first introduced in the 1700s. The **Linnaean hierarchy** (**lih-*NEE*-un**) is a system of biological classification devised by a Swedish naturalist named Carolus Linnaeus (**lih-*NEE*-us**). The **species** is the smallest unit (lowest level) of classification in the Linnaean hierarchy (**FIGURE 16.17**). Closely related species are grouped together to form a **genus** (plural "genera"). Using these two categories in the hierarchy, every species is given a unique, two-word Latin name, called its **scientific name**. The first word of the name identifies the genus to which the organism belongs; the second word defines the species. For example, humans are called *Homo sapiens*: *Homo* ("man") is the genus to which we belong, and *sapiens* ("wise") is our species name. We are the only living species in our genus. Other species in the genus include *Homo erectus* ("upright man") and *Homo habilis* ("handy man"), both of which are extinct.

In the Linnaean hierarchy, each species is placed in successively larger and more inclusive categories beyond the genus. Closely related genera are grouped together into a **family**. Closely related families are grouped into an **order**. Closely related orders are grouped into a **class**. Closely related classes are grouped into a **phylum**. Closely related phyla are grouped together into a **kingdom**. Finally, each kingdom falls into one of three **domains**.

Linnaean Taxonomy

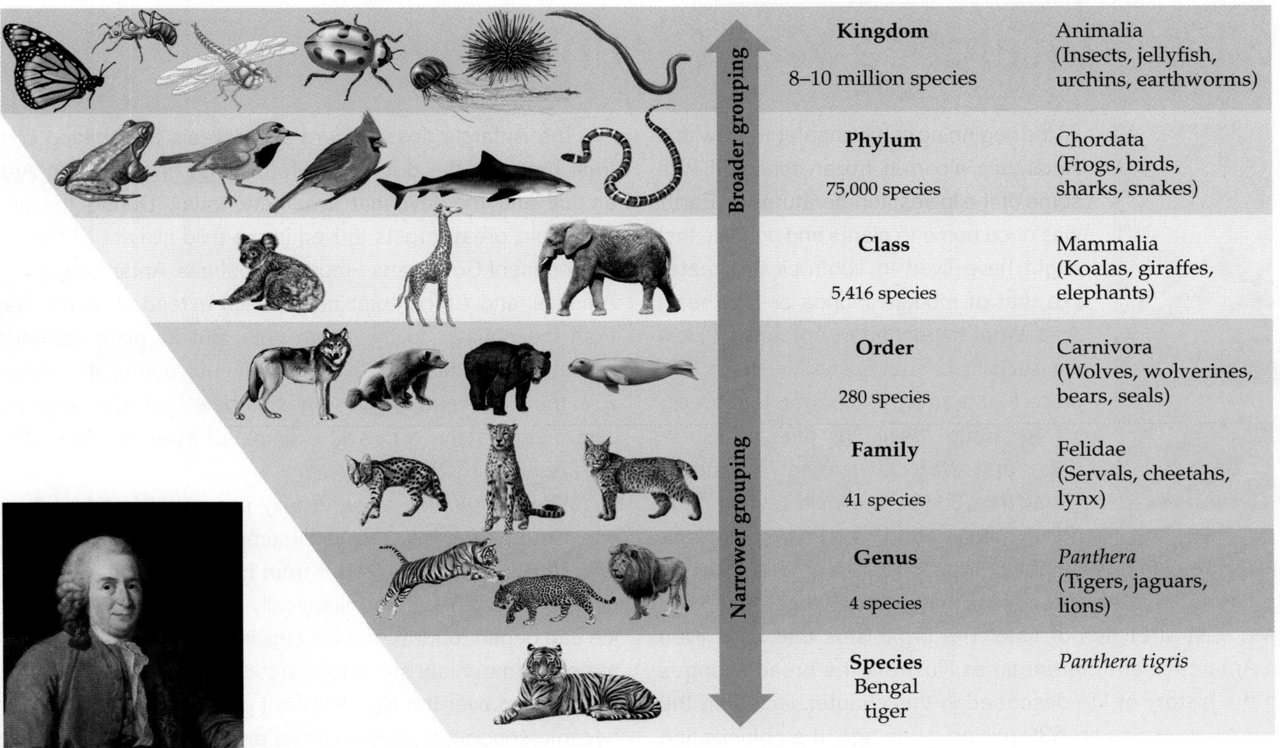

CAROLI LINNÆI

SYSTEMA NATURÆ

PER REGNA TRIA NATURÆ,

SECUNDUM CLASSES, ORDINES, GENERA, SPECIES,

CUM CHARACTERIBUS, DIFFERENTIIS, SYNONYMIS, LOCIS.

TOMUS I.

HOLMIÆ, IMPENSIS DIRECT. LAURENTII SALVII, 1758.

FIGURE 16.17 The Linnaean Hierarchy Places Organisms in Successively Larger Categories

The smallest unit of classification is the species (here, the Bengal tiger, whose scientific name is *Panthera tigris*). This species belongs to the genus *Panthera*, which includes other large cats. The genus *Panthera* lies within the family Felidae, which lies within the order Carnivora, within the class Mammalia, within the phylum Chordata and the kingdom Animalia. We can use the same categories—from species to kingdom—to classify all organisms. This classification system was first devised by the Swedish naturalist Carolus Linnaeus (*inset*).

Taxonomy is the branch of biology that deals with the naming of organisms and with their classification in the Linnaean hierarchy. Biologists refer to a group of organisms at any of these various levels of classification as a taxonomic group or, more simply, a **taxon** (plural "taxa"). Using Figure 16.17 as an example, you can see that the species *Panthera tigris* is a taxon, but so are the higher levels of classification to which it belongs: Felidae, Carnivora, Mammalia, and so on, up to Animalia. Each of these is a taxon, or taxonomic group, to which the Bengal tiger belongs, along with all the other organisms grouped with it in that taxon.

In the next several chapters we will explore life's biodiversity. An understanding of evolutionary classification and the underlying evolutionary processes that shaped this diversity will help you to make sense of the wide variety of organisms you will meet.

Concept Check

1. In evolutionary terms, what happens at a node?
2. List the eight taxonomic tiers of Linnaean classification, beginning with the most inclusive.

Concept Check Answers

1. A node represents a speciation event in which one lineage splits into two.
2. Domain, kingdom, phylum, class, order, family, genus, species.

APPLYING WHAT WE LEARNED

When Antarctica Was Green

At the beginning of the chapter we saw that Antarctica, a barren, frozen continent, with some of the lowest temperatures on Earth, was once home to plants and animals that might have lived in subtropical climates like that of modern Florida or Southeast Asia. What happened in Antarctica? How did such lush, subtropical life thrive in a place that is so icy and barren today?

Antarctic fossils of dinosaurs, forests, and tropical marine organisms (**FIGURE 16.18**) are a vivid testimony to our dynamic world. These fossils, ranging from Cambrian marine organisms to early land plants to birds and mammals, reveal that great changes occurred over vast stretches of time. The organisms that have lived in Antarctica at different times illustrate the broad changes in the history of life described in this chapter, including the Cambrian explosion 530 million years ago, the colonization of land 400 million years ago, and the different periods of domination by amphibians, reptiles, and mammals.

FIGURE 16.18 Antarctica Was Once Home to Subtropical Flora and Fauna

The Antarctic fossil record also reveals the striking contrast between the diverse life-forms that once lived in Antarctica and the few that live there today. During the late Paleozoic, great forests thrived in the mild climate of the supercontinent Gondwana—modern Australia, Antarctica, South America, and Africa combined—which extended all the way from the equator to the south pole. But as plate tectonics broke Gondwana into several continents during the Mesozoic, the plants and animals on each new continent began to evolve separately—a classic example of allopatric speciation (see Section 15.2).

When Antarctica became fully separated from Australia and South America, a cold circumpolar current developed around Antarctica, isolating it from the rest of the world. The new continent became dramatically colder, and an immense ice cap began to form. The ice cap, in turn, caused Earth's climate to grow colder still. As Antarctica moved toward its present position over the South Pole, it grew ever colder. Once it became separated from Australia and South America, about 40 million years ago, the warm-adapted ferns, dinosaurs, and mammals of Antarctica were trapped on a continent drifting south. As the continent continued to cool, reaching its present position over the South Pole about 25 million years ago, most of these forms of life died out.

Recent research has revealed that as recently as 14 million years ago, alpine meadows and mosses filled the Dry Valleys of Antarctica. Then, in a short time, the temperature dropped precipitously—by 8°C—freezing virtually every living thing. In 14 million years, none have ever thawed again, and paleontologists marvel at the detail preserved in these frozen and fossilized Dry Valley organisms. Today, the Dry Valleys can barely support a few bacteria and are considered a good model for what life on Mars might be like.

The same continental movements that brought destruction to terrestrial life in Antarctica sowed the seeds of evolutionary diversity elsewhere. The rerouting of ocean currents, which contributed to the formation of the Antarctic ice cap, produced the largest difference in temperature between the poles and the tropics that Earth has ever known. The wide range of new habitats that resulted from this temperature difference set the stage for adaptive radiations in many organisms, including humans.

BIOLOGY IN THE NEWS

Fish's DNA May Explain How Fins Turned to Feet

BY NICHOLAS WADE • *New York Times*, April 17, 2013

In the hope of reconstructing a pivotal step in evolution—the colonization of land by fish that learned to walk and breathe air—researchers have decoded the genome of the coelacanth, a prehistoric-looking fish whose form closely resembles those seen in the fossils of 400 million years ago.

Often called a living fossil, the coelacanth (pronounced SEE-**luh-canth**) was long believed to have fallen extinct 70 million years ago, until a specimen was recognized in a fish market in South Africa in 1938. The coelacanth has fleshy, lobed fins that look somewhat like limbs, as does the lungfish, an air-breathing freshwater fish. The coelacanth and the lungfish have long been battling for the honor of which is closer to the ancestral fish that first used fins to walk on land and give rise to the tetrapods, meaning all the original vertebrates and their descendants, from reptiles and birds to mammals.

The decoding of the coelacanth genome, reported online Wednesday in the journal *Nature*, is a victory for the lungfish as the closer relative to the first tetrapod. But the coelacanth may have the last laugh because its genome—which, at 2.8 billion units of DNA, is about the same size as a human genome—is decodable, whereas the lungfish genome, a remarkable 100 billion DNA units in length, cannot be cracked with present methods. The coelacanth genome is therefore more likely to shed light on the central evolutionary question of what genetic alterations were needed to change a lobe-finned fish into the first land-dwelling tetrapod ...

Dr. Amemiya's team [Chris Amemiya is from the University of Washington in Seattle] has sifted through the coelacanth's genome for genes that might have helped its cousin species, the ancestor to the first tetrapod, invade dry land some 400 million years ago. They have found one gene ... that enhances the activity of the genes that drive the formation of limbs in the embryo. The Amemiya team focused on the enhancer DNA sequence because it occurred in the coelacanth and animals but not in ordinary fish. They then inserted the coelacanth enhancer DNA into mice.

"It lit up right away and made an almost normal limb," said Neil Shubin, meaning that the coelacanth gene enhancer successfully encouraged the mouse genes to make a limb. Dr. Shubin, a member of the team, is a paleontologist at the University of Chicago.

The transition from aquatic fish to terrestrial tetrapod is one of the most significant events in the biological history of vertebrates. Several transitional forms have been found in the fossil record that document the evolutionary changes that occurred between fishes and terrestrial tetrapods. The fact that this transition occurred has been well supported for many years. *How* it occurred, on the other hand, has been an open question for nearly as long. Amemiya's findings offer an intriguing suggestion as to the mechanism of this transition. The study demonstrates that the evolution of terrestrial tetrapods was likely facilitated by the evolution of certain homeotic genes.

Limb development in organisms is controlled by homeotic genes, and the function of these genes is highly conserved across animal species, from insects to mice to sea stars. Even though the limbs of these animals are vastly different in form, the underlying genetic toolbox that constructs the limbs is the same. In that respect, it should not be surprising that a coelacanth's fin genes can trigger limb development in mice. A similar experiment had been done previously in which *PAX6*, the homeotic gene that controls eye development in mice and humans, was inserted into an eyeless fly embryo and led to the development of eyes!

While Amemiya's study is important in elucidating the genetic basis of the transition from water to land in vertebrates, the truly amazing implication is that the developmental tools to construct a body are more or less shared by very different species. Our basic developmental program is versatile enough to create the entire spectrum of forms observed in nature. Understanding our genetic commonality makes it seem less absurd to speak of the relatedness of a mouse to a mushroom because, in fact, both fungi and mammals possess homeotic genes.

Evaluating the News

1. What is the significance of sequencing the coelacanth genome?
2. One of the more exciting aspects of this study is that the genes for fin development in the coelacanth stimulated limb development in mice. Why might this have occurred?
3. What are the evolutionary implications of the results of this study? What might be the significance for the future evolutionary patterns of living things?

CHAPTER REVIEW

Summary

16.1 Macroevolution: Large-Scale Body Changes

- Macroevolution examines how dramatic changes in body form occur.
- One important aspect of macroevolutionary change is the immense age of Earth. It is approximately 4.6 billion years old. This scale of time is called geologic time.

16.2 The Fossil Record: A Guide to the Past

- The fossil record documents the history of life on Earth. Fossils reveal that past organisms were unlike living organisms, that many species have gone extinct, and that the dominant groups of organisms have changed significantly over time.
- The order in which organisms appear in the fossil record is consistent with our understanding of evolution gained from other kinds of evidence, including morphology and DNA sequences. Sometimes the approximate age of a fossil can be determined through analysis of radioisotopes.
- Although the fossil record is not complete, it provides excellent examples of the evolution of major new groups of organisms.

16.3 The History of Life on Earth

- The first single-celled organisms resembled bacteria and probably evolved about 3.5 billion years ago.
- The release of oxygen by photosynthetic bacteria caused oxygen concentrations in the atmosphere to increase. Rising oxygen concentrations made possible the evolution of single-celled eukaryotes about 2.1 billion years ago. Multicellular eukaryotes followed about 650 million years ago (mya).
- Life on Earth changed dramatically during the Cambrian explosion (530 mya), when large predators and well-defended herbivores suddenly appear in the fossil record.
- The land was first colonized by plants (about 500 mya), fungi (about 460 mya), and invertebrates (about 410 mya), which were followed later by vertebrates (about 365 mya).

16.4 The Effects of Plate Tectonics

- The shifting of Earth's crust, or plate tectonics, has had profound effects on the history of life on Earth.
- The separation of the continents over the past 200 million years has led to geographic isolation on a grand scale, promoting the evolution of many new species.
- At different times, climate changes caused by the movements of the continents have led to the extinctions of many species.

16.5 Mass Extinctions: Worldwide Losses of Species

- There have been five mass extinctions during the history of life on Earth.
- The extinction of a dominant group of organisms can provide new opportunities for other groups.

16.6 Rapid Macroevolution through Differential Gene Expression

- Evolutionary developmental biology (evo-devo) is a relatively new field of science that studies how differences in gene expression can cause different body plans during embryological development.
- Homeotic genes are genes that determine body structures. They are expressed only in certain body segments to determine the organism's overall form. Altering the expression of these homeotic genes can yield dramatically altered body plans.
- Differential expression in regular genes can also create differences in body form. Two organisms with the same genes may express them in different amounts, leading to different morphology.

16.7 Phylogenetics: Reconstructing Evolutionary Relationships

- Biologists use evolutionary trees to model ancestor-descendant relationships among different organisms. The tips of branches represent existing groups of organisms, and each node represents the moment when an ancestor split into two descendant groups.
- Closely related groups of organisms share distinctive features that originated in their most recent common ancestor. These shared derived traits are used to identify lineages of closely related organisms.
- The Linnaean hierarchy is a classification system for organizing life-forms. In this scheme, every species of organism has a two-part scientific name indicating its genus and species.
- The lowest level of the Linnaean hierarchy is the species. Each species falls into ever more inclusive groups: genera, families, orders, classes, phyla, kingdoms, and domains.

Key Terms

atavistic trait (p. 370)
Cambrian explosion (p. 364)
class (p. 372)
Cretaceous extinction (p. 368)
domain (p. 372)
evo-devo (p. 369)
family (p. 372)
genus (p. 372)
geologic time (p. 359)
homeotic gene (p. 369)
kingdom (p. 372)
lineage (p. 371)
Linnaean hierarchy (p. 372)
macroevolution (p. 359)
mantle (p. 365)
mass extinction (p. 366)
most recent common ancestor (p. 371)
node (p. 371)
order (p. 372)
Permian extinction (p. 366)
phylogenetics (p. 371)
phylum (p. 372)
plate tectonics (p. 365)
radioisotope (p. 359)
scientific name (p. 372)
shared derived trait (p. 372)
species (p. 372)
supercontinent (p. 365)
taxon (p. 373)
taxonomy (p. 371)

Self-Quiz

1. The fossil record
   a. documents the history of life.
   b. provides examples of the evolution of major new groups of organisms.
   c. is not complete.
   d. all of the above
2. Mass extinctions
   a. are always caused by asteroid impacts.
   b. are periods of time when many species go extinct worldwide.
   c. have little lasting effect on the history of life.
   d. affect only terrestrial organisms.
3. The Cambrian explosion
   a. caused a spectacular increase in the size and complexity of animal life.
   b. caused a mass extinction.
   c. was the time during which all living animal phyla suddenly appeared.
   d. had few consequences for the later evolution of life.
4. The history of life shows that
   a. biodiversity has remained constant for about 400 million years.
   b. extinctions have little effect on biodiversity.
   c. macroevolution is greatly influenced by mass extinctions and adaptive radiations.
   d. macroevolution can be understand solely in terms of the evolution of populations.
5. Large-scale evolution characterized by the rise and fall of major groups of organisms and significant alterations of body forms is called
   a. macroevolution.
   b. microevolution.
   c. mass extinction.
   d. adaptive radiation.
6. Homeotic genes
   a. are present in only a few cells.
   b. determine the body plan of an organism.
   c. are found only in humans.
   d. must all be suppressed in order for an organism to develop properly.
7. A node (represented in Figure 16.16 by a circle) represents
   a. species that are extinct.
   b. the descendant lineage.
   c. the most recent common ancestor of two or more descendant lineages.
   d. the shared derived trait.
8. According to the Linnaean system of classification,
   a. all life-forms can be divided into two domains: prokaryotes and eukaryotes.
   b. members of the same class are also members of the same order.
   c. phylum is a broader, more inclusive category than order.
   d. the scientific name of each unique organism consists of two parts: the genus and the family.

Analysis and Application

1. The fossil record provides clear examples of the evolution of new groups of organisms from previously existing organisms. Describe the major steps of one such example.
2. How did the evolution of photosynthesis affect the history of life on Earth?
3. What is the Cambrian explosion, and why was it important?
4. Life arose in water. Explain why the colonization of land represented a major evolutionary step in the history of life. What challenges—and opportunities—awaited early colonists of land?
5. Mass extinctions can remove entire groups of organisms, seemingly at random—even groups that possess highly advantageous adaptations. How can this be?
6. Homeotic genes are highly conserved between species. Scientists have been able to use homeotic genes from mice to induce developmental changes in fruit flies. Do you think it would be possible to use the homeotic genes from insects to, say, give pigs wings? Why or why not?
7. Humans and chimps share 98 percent of their DNA and yet look drastically different from one another. Likewise, identical twins (with 100 percent genetic similarity) will also look slightly different from one another. What nongenetic factors might be behind these differences in form?
8. Name one of your favorite wild animals and one of your favorite wild plants. Find the scientific names of each, and research their taxonomic classification (that is, determine the phylum, class, order, and family they belong to). Name two other species that are classified in the same order, but *not* the same family, as your favorites. Are there any species that belong to the same family as your favorites but *not* to the same order? Explain.

17 Bacteria, Archaea, and Viruses

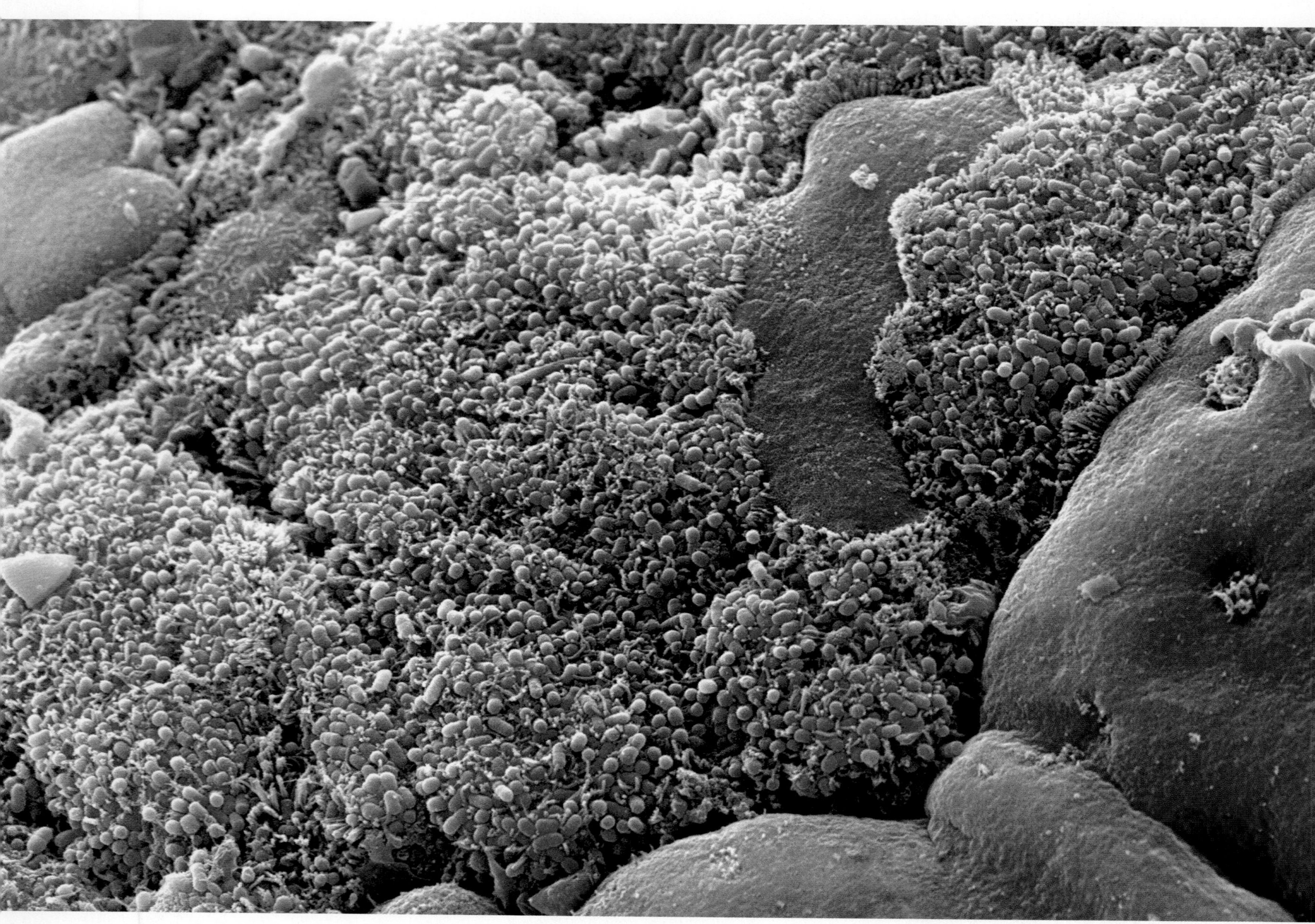

GUT BACTERIA. Microscopic bacteria called *Escherichia coli* (*E. coli*), shown here in blue, grow in the human intestine (green) as part of a community of hundreds of kinds of microorganisms.

A Hitchhiker's Guide to the Human Body

There's more to you than meets the eye. The next time you're feeling lonely, think about the fact that of the trillions of cells in your body, about 90 percent belong to other individuals, living everywhere on your skin and deep inside your intestines. In the farthest reaches of your gut live some 10 trillion individual bacteria, archaeans, and fungi.

Even more live higher in the gut, as well as in your mouth and in every crevice of your skin. Some of these organisms inhabit moist tropical regions such as your underarms, while others specialize in oily habitats like the sides of your nose. The greatest biodiversity of skin bacteria and fungi lies not in any of the places you might think, but on your forearms. The least diverse region is the ecological desert that is the back of your ears.

Taken together, about a thousand different species of microbes live in you and on you—several hundred species in your mouth alone. Most of these tiny companions are bacteria or archaeans. But some are yeasts (fungi), not to mention a smattering of protozoans and even arthropods like the tiny mites that live, mate, and die in the roots of your eyelashes.

Virtually all organisms share their bodies with other organisms, in various kinds of *symbiosis*—the shared existence of two or more species. Most of the bacteria and other organisms that live with us are harmless, and many are essential to good health. For example, many of those in the gut help digest food and make vitamins. And many bacteria act as bouncers, shouldering aside harmful organisms.

All these organisms constitute the human *microbiome*, all the tiny organisms that live on and in us. To better understand the difference between microorganisms that we need and microorganisms that may harm us, let's take a look at the huge diversity of prokaryotic organisms, as well as the viruses that plague them and every other life-form on Earth.

Where does this huge diversity of organisms hide in our bodies? How do we know which are good and which are bad?

MAIN MESSAGE Of the three domains of life, Bacteria and Archaea are the most ancient, most diverse, and most abundant.

KEY CONCEPTS

- The three-domain system organizes all life on Earth into three large categories, or domains: Bacteria, Archaea, and Eukarya.
- Biodiversity embraces all the world's living things, as well as all the variety in their interactions with the living and nonliving components of the ecosystems they inhabit.
- Bacteria and Archaea are single-celled organisms informally known as prokaryotes.
- Prokaryotes are the most numerous in terms of number of individuals and diversity of types.
- Prokaryotes are the most widespread of all living groups, and some flourish in extremes of heat, cold, acidity, or saltiness.
- Prokaryotes have the most diverse modes of nutrition and are essential to the web of life as both producers (autotrophs) and consumers (heterotrophs). Bacteria are especially important as decomposers. Cyanobacteria release oxygen as a by-product of photosynthesis, but photosynthetic archaeans are unknown.
- Some bacteria cause disease in animals and other organisms. However, no archaean is known to be a pathogen.
- Antibiotics are molecules produced by one organism that kill or slow the growth of another organism. Many antibiotics produced by fungi are effective against bacterial pathogens, but viruses are unaffected by antibiotics.
- Viruses are infectious agents containing genetic material, either DNA or RNA, wrapped in protein layers. Viruses lack cellular organization and independent metabolism.
- Viruses evolve rapidly because their DNA mutates quickly.

A HITCHHIKER'S GUIDE TO THE HUMAN BODY 379

17.1 The Diversity of Life 380

17.2 Bacteria and Archaea: Tiny, Successful, and Abundant 383

17.3 How Prokaryotes Affect Our World 391

17.4 Viruses: Nonliving Infectious Agents 395

APPLYING WHAT WE LEARNED 398
All of Us Together

YOU MAY HAVE ENCOUNTERED SOME LIVING THING somewhere that made you wonder, what in the world is that? If you cannot recall such a moment, we invite you to look at **FIGURE 17.1** and play a game of "Animal, Vegetable, Mineral, or Thing." Can you identify the living organisms in the portrait gallery? Which ones would you classify as plants? Which are animals? What criteria do you apply to decide that something is a plant, not an animal? Do you think some of these organisms should be placed in yet other categories, apart from plants or animals? What might those categories be? How would you set the boundaries of those categories; in other words, what characteristics must an organism display to be placed in the groupings you have in mind? And in lumping organisms into a particular category, are you going by their appearance, by the way they obtain energy from their environment, or by their evolutionary descent from a recent common ancestor?

Biologists are faced with much the same puzzle when they observe the bewildering variety of life. In this chapter we begin with some thoughts about the dazzling diversity of life on Earth. Then we take a closer look at two of the earliest and most awe-inspiring branches on the tree of life: Bacteria and Archaea. We conclude with a description of viruses, infectious agents that parasitize every known life-form.

17.1 The Diversity of Life

The enormous variety of life on Earth is still far from being completely known, counted, or named. Of the known forms of life, we can describe only a tiny fraction—focusing on the main groups—in the space available in this and the next two chapters. Let's begin our survey of life by considering what biodiversity is and why it matters.

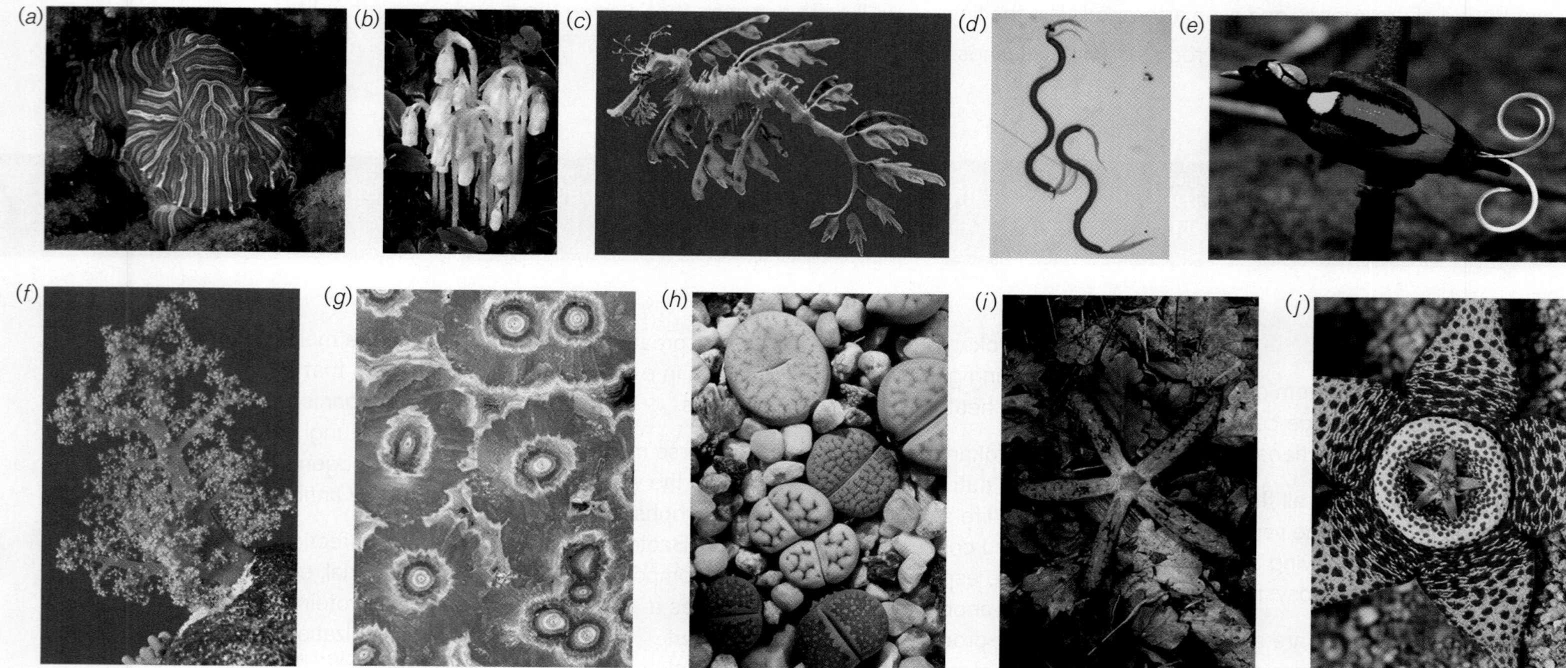

FIGURE 17.1 How Many Organisms Do You See Here?
Can you distinguish the nonliving from the living in these photos? Can you classify any of these organisms as plants or animals? Some of the organisms pictured here are single-celled; others are multicellular. Some are microscopic (not readily seen without a microscope), but most are visible to the naked eye. For the answers, turn to page 384.

The extent of Earth's biodiversity is unknown

The term **biodiversity** embraces all the world's living things, as well as all the variety in their interactions with the living and nonliving components of the ecosystems they inhabit. Biodiversity can be described at the level of genes or species or whole ecosystems. In other words, biodiversity on our planet has three main components:

- Genetic diversity
- Species diversity
- Ecosystem diversity

For example, you could inventory biodiversity at the level of DNA variation by estimating the full range of genetic information that is held collectively in the DNA of all the grasshoppers in a patch of prairie. You could also measure the biodiversity of a place by totaling up the species of organisms found there—all the animal species in a prairie ecosystem, for example. Biodiversity also embraces the variety in the interactions of organisms with each other and with the nonliving part of their environment—that is, all components of an ecosystem. To measure biodiversity on the prairie at this broader level, you could determine the variety in food webs and the interactions among the food webs by making a comprehensive record of who eats whom in that ecosystem.

In spite of intense worldwide interest, scientists do not know the exact number of species alive today. Estimates range from 3 million to 100 million species. Most estimates, however, fall in the range of 3–30 million. So far, a total of about 1.7 million species have been collected, identified, named, and placed in the most basic system for biological classification, the Linnaean hierarchy (see Chapter 16).

Earth's biodiversity is disappearing even as scientists race to catalog it. Why should we care? The "Biology Matters" box on page 382 explains the value of biodiversity for functioning ecosystems and for our own well-being. In Chapter 18 we examine the many threats to Earth's biodiversity.

Life on Earth can be sorted into three distinct domains

To organize the mind-numbing diversity of life in our biosphere, biologists start by sorting all life-forms into the broadest of all groupings: domains (**FIGURE 17.2**).

There are three domains of life:

- **Bacteria**, which includes familiar disease-causing bacteria.
- **Archaea** (**ahr-*KEE*-uh**), which consists of single-celled organisms (archaeans) best known for living in extremely harsh environments.
- **Eukarya** (**yoo-*KAYR*-ee-uh**), which includes all the rest of the living organisms (eukaryotes), including us. Domain Eukarya is divided into four kingdoms: Protista (including amoebas and algae); Plantae (all plants, from mosses to maple trees); Fungi (including molds and mushrooms); and Animalia (all animals).

Even though bacteria and archaeans belong to two different domains, and archaeans are in some ways more like eukaryotes than like bacteria, the two non-Eukarya domains have traditionally been lumped under a common label: *prokaryotes*. **Prokaryotes** is an informal label for Bacteria and Archaea; that is, the term is not part of the modern system of biological classification. As used in this textbook, "prokaryotes" means simply "organisms that are not in the domain Eukarya."

EXTREME DIVERSITY SHIFT

About 100 years ago, there were 300 species of plants, 60 different mammals, 300 types of birds, and many thousands of insect species in the short-grass prairie that is replaced by corn in much of Iowa today. A recent survey of an Iowa cornfield turned up 7 other species visible to the naked eye in a sea of corn: 1 mushroom, 4 insects, 1 spider, and 1 tiny red mite.

Concept Check

1. What are some ways for estimating the biodiversity of a habitat?
2. Do all prokaryotes belong to the same domain?

Concept Check Answers

1. Cataloging genetic diversity, number of species, and interactions among organisms and between organisms and their environment.
2. No. The informal term encompasses two domains: Bacteria and Archaea.

Domains and Kingdoms of Life

(*a*) Three-domain system

Bacteria	Archaea	Eukarya

(*b*) Six-kingdom system

Bacteria	Archaea	Protista	Plantae	Fungi	Animalia

FIGURE 17.2 The Three Domains of Life Are Divided into Six Kingdoms
This book employs both the three-domain system (*a*) and the widely used six-kingdom system (*b*) for classifying life. The domain Bacteria is equivalent to the kingdom Bacteria, and the domain Archaea is equivalent to the kingdom Archaea. The domain Eukarya encompasses four kingdoms in the six-kingdom scheme: Protista (protists, which include organisms such as amoebas and algae), Plantae (plants), Fungi (including yeasts and mushroom-producing species), and Animalia (animals).

BIOLOGY MATTERS

The Importance of Biodiversity

Many people wonder whether the loss of one beetle here or a buttercup there really makes a difference to humanity. To answer these questions, it helps to look at the situation from the perspective of biologists who have long wrestled with how to assess the value of species diversity within particular habitats. One question that biologists are particularly interested in answering is how, if at all, biodiversity affects the forests, wetlands, oceans, rivers, and other wild ecosystems of the world. Since the 1990s, researchers have been studying how biodiversity can contribute to the health and stability of ecosystems.

The Madagascar rosy periwinkle (*Catharanthus roseus*) is a source of anticancer drugs.

Biodiversity Can Improve the Function of Ecosystems

From tiny experimental ecosystems in an English laboratory to experimental prairies in the midwestern United States, researchers have found that the more species an ecosystem has, the healthier it appears to be. For one thing, species diversity maximizes the use of available resources because different species are good at using varied resources; for example, some thrive in the sun-drenched portions of a habitat, whereas others can make the best use of the areas that lie mostly in the shade.

Evidence also suggests that the more species an ecosystem has, the more resilient that ecosystem is. For example, the greater the number of species present in a patch of prairie, the more easily that area can return to a healthy state following a drought. Scientists have also found that an increased diversity of species in an area leads to a lower incidence of disease and lower rates of invasion by introduced species. Furthermore, diversity can even lead to more diversity: researchers have found that greater plant diversity in a plot of ground leads to greater insect diversity as well.

Biodiversity Provides People with Goods and Services

Why should we care whether various ecosystems are healthy? The reason is that the biosphere provides us with many goods and services. Even the most basic requirements of human life are provided by other organisms. Plants produce the oxygen we breathe; they also provide us with food and many other necessities. One-fourth of all prescription drugs dispensed by pharmacies are extracted from plants. Quinine, used as an antimalarial drug, comes from a plant called yellow cinchona (**sing-*KOH*-nuh**). Taxol, an important drug for treating cancer, was originally found in the Pacific yew tree (it is now synthesized rather than harvested). Bromelain (***BROH*-muh-lun**), a substance that controls tissue inflammation, comes from pineapples.

Whole ecosystems full of species can provide what are known as ecosystem services. For example, the coastal redwood trees in northern California intercept fog, mist, and rain, and channel the water onto and into the ground. Where these forests have been cut down, moisture-laden air masses are much less likely to linger, and the moisture is more likely to evaporate in the sun, so that much less water ends up entering the ground and reservoirs. Without trees on the ground, hillsides erode more easily and rivers become filled with sediments.

The value of seemingly useless marshes full of reed and grass species became all too apparent when Hurricane Katrina barreled through the Gulf Coast in 2005. A healthy marsh acts much like a barrier island, blocking storm surges while also creating areas for fish to spawn and birds to nest. But the wetlands outside New Orleans had been so reduced by humans—for flood control, navigation, agriculture, oil drilling, and other human activities—that storm surges easily destroyed the rest. Hurricane Katrina caused some marshy areas east of the Mississippi River to lose 25 percent of their land areas, thereby not only interfering with natural fisheries and birdlife, but also reducing the power of remaining marshes to act as natural filters for the water moving through them, providing water-cleaning services to growing human populations.

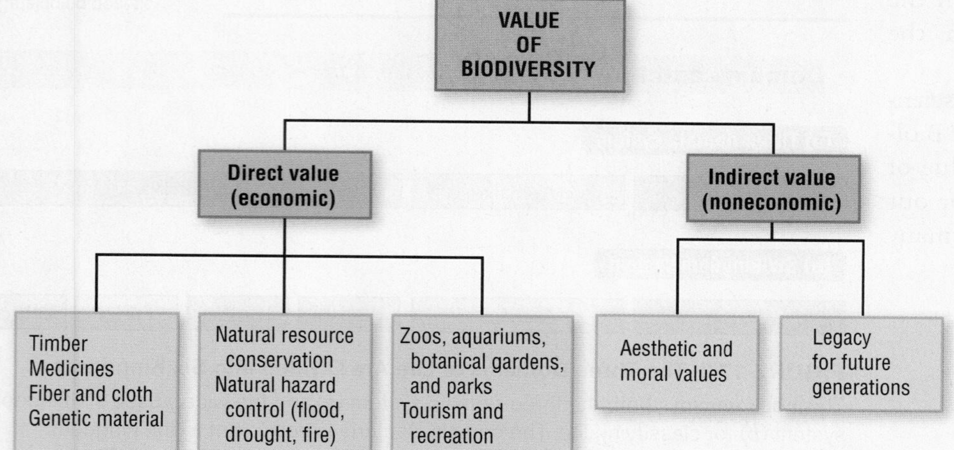

17.2 Bacteria and Archaea: Tiny, Successful, and Abundant

The first cell on Earth, the presumed universal ancestor of us all, probably resembled a modern-day bacterium. Now, about 3.8 billion years later, the domain Bacteria still rules Earth, if we go by sheer abundance and diversity of lifestyles.

Chemical clues in fossil remains suggest that the domain Archaea had evolved by about 2.7 billion years ago, but the most ancient prokaryotic fossils are difficult to interpret and assign conclusively to one or the other domain. Therefore, biologists are still uncertain about the placement of Archaea in the tree of life. The evolutionary tree in **FIGURE 17.3** presents one hypothesis; it shows descendants of the universal ancestor splitting into Bacteria and another lineage, and the other lineage itself diverging into Archaea and Eukarya later in the history of life.

Discovered in the 1970s, archaeans are superficially like bacteria in that both are microscopic and single-celled, and both can reproduce only asexually. However, archaeans have several unique features, such as the distinctive lipids in their plasma membranes. But DNA comparisons, more than anything else, have convinced biologists that archaeans belong in a domain of their own.

Because of their status as "non-Eukarya," the prokaryotes are a good place to begin our introduction to the major groups of life. We will point out significant differences between the two domains (Archaea and Bacteria) as they come up.

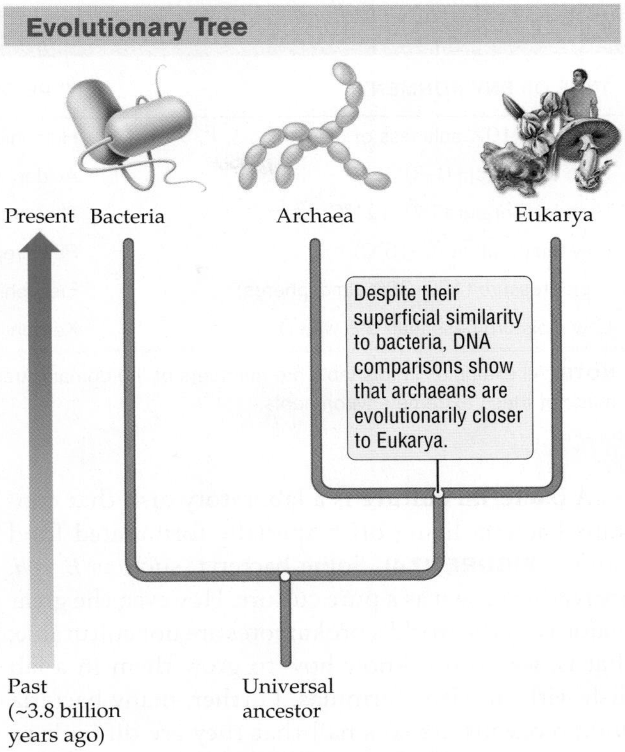

Evolutionary Tree

Present
Bacteria
Archaea
Eukarya
Despite their superficial similarity to bacteria, DNA comparisons show that archaeans are evolutionarily closer to Eukarya.
Past (~3.8 billion years ago)
Universal ancestor

FIGURE 17.3 Evolutionary Tree of Domains
This tree is one model of how the three domains might be related. At the root of the tree is the universal ancestor, from which all living things descended. Of the three surviving lineages, the first split came between Bacteria and the lineage that would give rise to Archaea and Eukarya. The next split was between Archaea and Eukarya, making Archaea and Eukarya more closely related to each other than either group is to Bacteria.

Prokaryotes represent biological success

Most people tend to think of the living world in terms of butterflies, tigers, and orchids, giving little thought to microscopic organisms, even though the vast majority of life on Earth is in fact single-celled and prokaryotic. Scientists estimate that the number of prokaryotes on Earth is about 5,000,000,000,000,000,000,000,000,000,000 (5 nonillion = 5×10^{30}). The success of prokaryotes is due, in part, to how quickly they reproduce. Prokaryotes typically reproduce by splitting in two—a process called binary fission (see Chapter 7). Overnight, a single bacterium of the common species *Escherichia coli* (usually referred to by the abbreviated form *E. coli*) can divide to produce a population of 16 million bacteria.

Prokaryotes are also the most widespread of organisms, able to live in many places where few other forms of life can exist. Prokaryotes that are **extremophiles** (literally, "lovers of the extreme") are known to flourish in boiling-hot geysers in Yellowstone National Park, the freezing-cold seas off Antarctica, and 2 miles deep in the lightless depths of a South African gold mine (see **TABLE 17.1** for examples of archaeal extremism).

Prokaryotes are also abundant in milder habitats, from doorknobs to kitchen sinks. In one teaspoon of garden soil, there are many millions of several thousand different types of bacteria. Scientists estimate that 1 square centimeter of healthy human skin is home to between 1,000 and 10,000 different types of bacteria.

Helpful to Know

Microscopic organisms, or microbes, are invisible to the naked eye because they are less than 0.17 millimeter (170 micrometers, or μm) in diameter. Most prokaryotes are microbes, but so are many members of the kingdom Protista, some fungi, and even some animals. The science of microbes is microbiology.

ANSWERS TO FIGURE 17.1
(*a*) Psychedelic frogfish; (*b*) Indian pipe (*Monotropa uniflora*), a parasitic plant; (*c*) leafy sea dragon, a type of fish; (*d*) *Spirillum volutans*, a bacterium; (*e*) Wilson's bird of paradise; (*f*) soft coral; (*g*) mineral crystal geode (nonliving mineral deposit); (*h*) stone plant (*Lithops hookeri*); (*i*) devil's finger fungus (*Clathrus archeri*); (*j*) carrion flower (*Stapelia variegata*), which smells like rotting meat.

TABLE 17.1 Lifestyles of the Extreme

TYPE OF ENVIRONMENT	TYPE OF EXTREME ORGANISM	EXAMPLE
High salt (10× saltiness of ocean)	Halophiles	*Halobacterium salinarum*
High acidity (pH 0–3)	Acidophiles	*Acidianus infernus*
High temperature (80–121°C)	Thermophiles	*Pyrodictium abyssi*
Low temperature (0–15°C)	Psychrophiles	*Methanogenium frigidum*
High pressure (100–380 atmospheres)	Piezophiles	*Pyrolobus fumarii*
Low moisture (less than 5% water)	Xerophiles	*Pyrococcus furiosus*

NOTE: All examples in this table are members of the domain Archaea; however, many bacterial species are also known to inhabit some of these extreme environments.

A **bacterial culture** is a laboratory dish that contains bacteria living off a specially formulated food supply (**FIGURE 17.4**). Some bacteria, such as *E. coli*, are readily grown as a pure culture. However, the great majority of the world's prokaryotes are not culturable; that is, we do not know how to grow them in a lab dish with nutrient formulas. Further, many bacteria and archaeans are so small that they are difficult to see except with high-powered microscopes. Scientists therefore resort to DNA analysis not only to detect these organisms' presence, but also to study and classify the many millions of unseen and unculturable species of bacteria and archaeans that we now know populate every imaginable habitat on land and in the ocean.

Prokaryotes are single-celled

Bacteria and Archaea are quite variable in shape, and biologists sometimes categorize them by these shapes.

- *Cocci* (*KAH*-**kye**; singular "coccus") are spherical.
- *Bacilli* (**buh-*SILL*-eye**; singular "bacillus") are rod shaped.
- *Spirilla* (**spye-*RILL*-uh**; singular "spirillum") are corkscrew shaped.

Regardless of shape, all bacteria and archaeans share a basic structural plan (**FIGURE 17.5**). Most have a protective cell wall that surrounds the plasma membrane. Some have an additional wrapping, called a *capsule*.

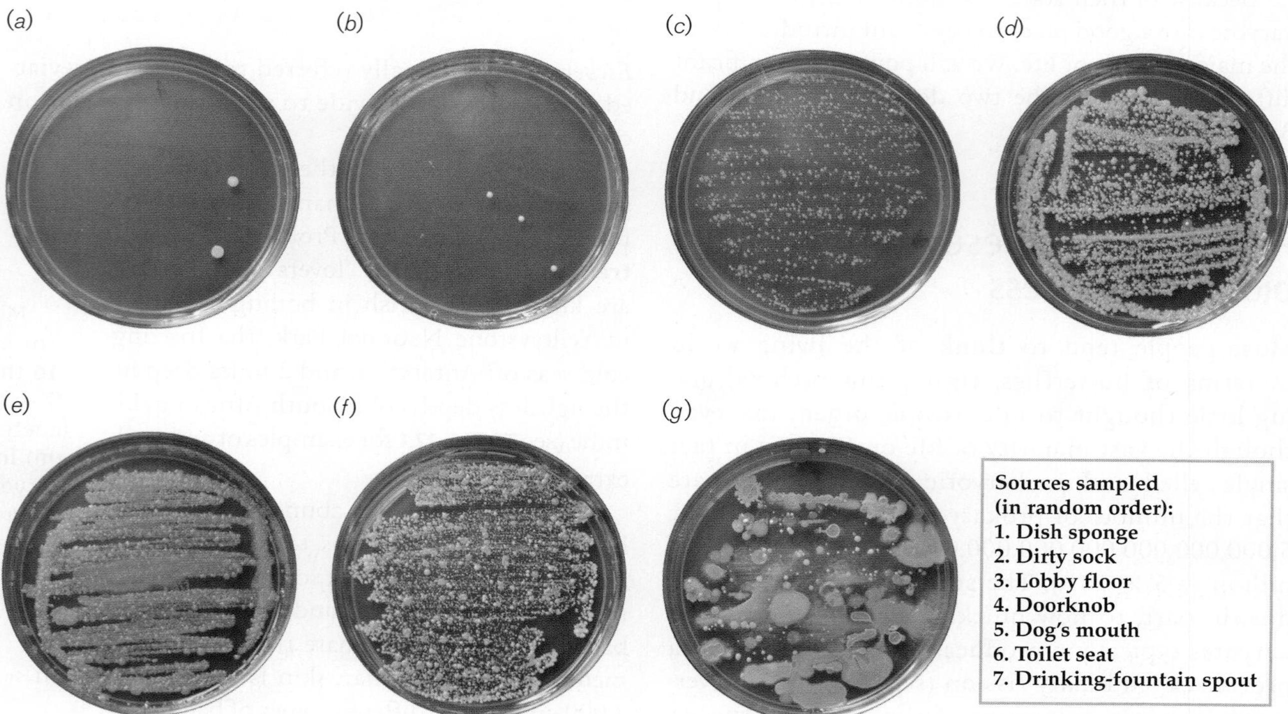

FIGURE 17.4 Some, but Not All, Microbes Can Be Cultured in an Artificial Medium
In this experiment, students collected microbial samples from various sources by swabbing them with a moistened sterile cotton swab and inoculating petri dishes containing a nutrient medium called LB agar. The photos show the colonies of microbes that appeared after 2 days of incubating the plates at 37°C. Most of the smooth, shiny colonies are bacteria, and the fuzzy ones are fungi. Try matching the microbial cultures (*a–g*) with their sources (1–7). For answers, turn to page 392.

Structure of a Prokaryote

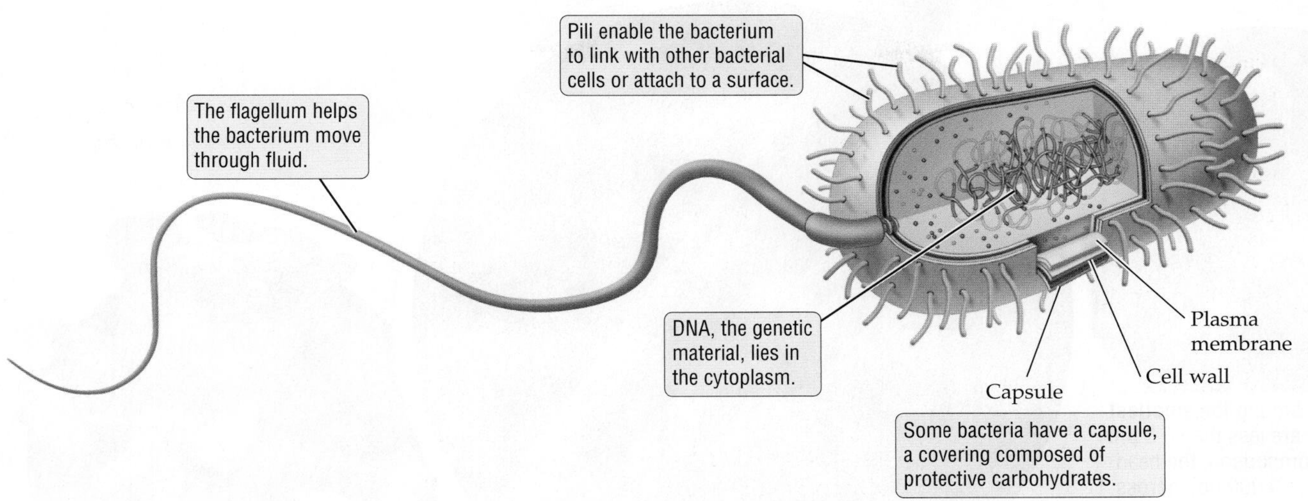

Pili enable the bacterium to link with other bacterial cells or attach to a surface.
The flagellum helps the bacterium move through fluid.
DNA, the genetic material, lies in the cytoplasm.
Plasma membrane
Cell wall
Capsule
Some bacteria have a capsule, a covering composed of protective carbohydrates.

FIGURE 17.5 All Prokaryotic Cells Lack a Nucleus

Prokaryotic cells tend to be about 10 times smaller than eukaryotic cells and generally have much less DNA.

The capsule, made of slippery polysaccharides, helps disease-causing bacteria evade the immune system, which protects the animal body from foreign invaders.

The surface of some bacteria is covered in many short, hairlike projections called **pili** (*PYE*-**lye**; singular "pilus"). Bacteria use their pili to link together to form bacterial mats, or to attach to surfaces in their environment, including the cells inside a host animal they have infected. Some bacteria have one or more long, whiplike structures called **flagella** (**fluh-*JELL*-uh**; singular "flagellum") that spin like a boat's propeller to push the bacterium through liquid.

As noted in Chapter 3, prokaryotes do not have a true nucleus. They typically have much less DNA than do cells of eukaryotic organisms. Eukaryotic genetic material often includes a large amount of extra, noncoding DNA that serves no clear function. In contrast, prokaryotic genetic material is composed mainly of coding DNA that is actively used for the survival and reproduction of the bacterial cell.

Although prokaryotes are regarded as single-celled organisms, some form colonies, produced by the splitting of one original cell. Many bacteria and archaeans form long chains of cells, called filaments, produced by the splitting of an original cell. For example, the cyanobacterium *Nostoc* forms long filaments of photosynthetic cells. Some of the cells in the filament are even specialized for unique functions. In *Nostoc*, a specialized cell, called the *heterocyst*, converts atmospheric nitrogen gas into organic molecules (**FIGURE 17.6**). Some scientists see colonial forms as a "semi-multicellular" lifestyle.

Some prokaryotes display social behavior

Many prokaryotes have a busy social life. Individual cells release chemical signals that can be picked up by other cells of the same or even different species. Like a flash mob coordinated by text messages, these cells can swim toward each other to form a swarm.

The communication system that enables microbes to coordinate their behavior in response to cell density is called **quorum sensing**. Free-swimming bacteria latch on to a living or nonliving surface with the help of their pili or their sticky capsules. The attached bacteria cover themselves in a protective slime. They secrete signaling molecules that recruit more bacteria. The chemical signals also enable the swarming bacteria to sense their numbers: a threshold population size (quorum) is reached when the concentration of the signal rises to a critical level. The bacteria now divide explosively, forming a tough, slime-covered aggregate known as a **biofilm** (**FIGURE 17.7**).

Many disease-causing bacteria use quorum sensing as a means of timing an all-out assault on the host's immune defenses. The cells lie low until they detect a strength in numbers. Upon reaching a quorum, they multiply so quickly and suddenly that they can overwhelm an immune system that could have demolished them singly or at low density.

Helpful to Know

Dental plaque is a biofilm that forms on the teeth. Plaque can contribute to tooth decay and periodontal disease, which can lead to tooth loss. Regular brushing and flossing, along with professional cleaning, controls plaque by breaking up biofilms. Sugary foods and soft drinks feed the bacteria, and use of a mouthwash alone is largely ineffective once a biofilm has formed.

Structure

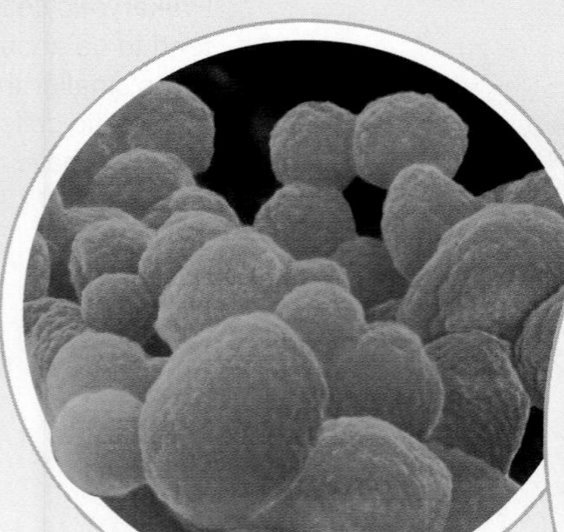

Mycoplasmas are among **the smallest organisms**; many are less than 0.5 μm in diameter. For comparison, the head of a pin is 2 mm, or 2,000 μm, across. Most mycoplasmas are parasites of plants and animals.

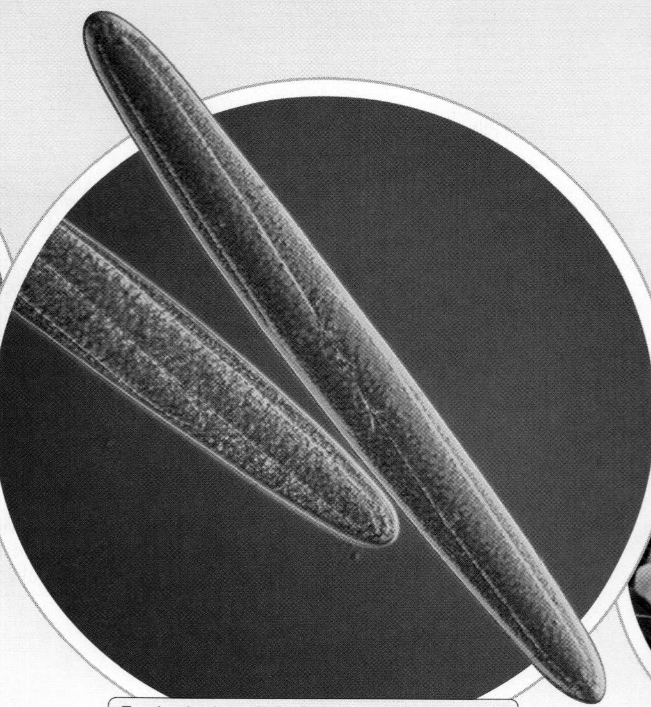

Epulopiscium fishelsoni is **a giant among prokaryotes**, measuring about 600 μm in length. The bacterium lives inside the gut of the surgeonfish.

Bacteria and Archaea range from balloonlike forms to flattened triangles. Shown here are **rod and coccal bacteria** found on a dollar bill. A wide variety of bacteria are found on paper money, and they may spread infections.

Habitat

Psychrophiles live at temperatures **close to freezing**. *Psychrobacter* is just one of many bacteria found in this subglacial stream in the Swiss Alps.

Hyperthermophiles inhabit some of the **hottest places** on Earth. *Pyrodictium abyssi* is an archaean that lives at about 110°C at extremely high pressure around deep-sea hydrothermal vents like this one.

Halophiles can live in water 10 times **saltier than the sea**. Seawater is evaporated to make salt in these lagoons by San Francisco Bay. The pink and purple tints come from an abundance of *Halobacterium*, an archaean that uses colored pigments to turn sunshine into chemical energy.

FIGURE 17.6 Diversity in the Domains Archaea and Bacteria

Energy Acquisition

Producers

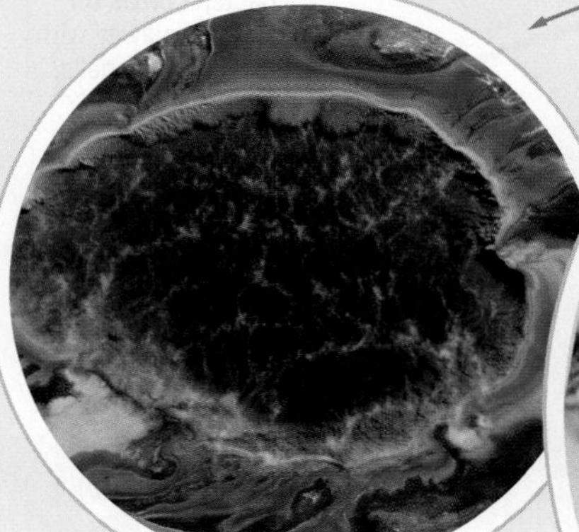

Chemolithotrophs tap the energy in minerals to make food. The "**rock eaters**" in these highly acidic hot springs in Yellowstone National Park obtain energy from hydrogen sulfide and arsenite.

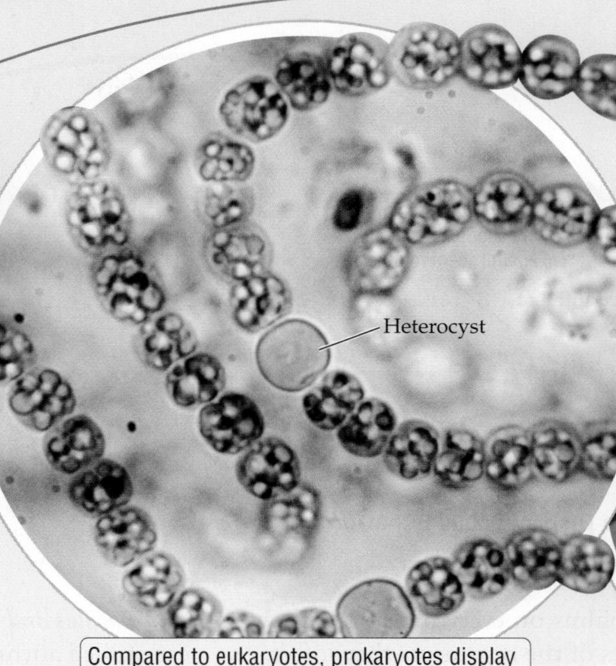

Compared to eukaryotes, prokaryotes display more metabolic diversity. *Nostoc*, the cyanobacterium in this photo, **uses light energy** to make food through photosynthesis. It can also turn nitrogen gas from the atmosphere into organic molecules—something no eukaryote can do.

Some prokaryotes **can live completely independently from sunshine**. *Desulfotomaculum*, discovered about 2 miles underground in a South African gold mine, is a bacterium that taps sulfur minerals, as well as hydrogen gas released from water by radioactivity, as sources of energy for converting carbon dioxide into food.

Decomposers are consumers that break down the remains of **dead organisms** and absorb the chemicals that are released. Millions of species of Bacteria and Archaea are hard at work in these holding tanks at a sewage treatment plant in Houston, Texas.

Some prokaryotes are **collaborators**. The root nodules on these bean plants are colonized by *Rhizobium* bacteria that obtain sugar from the host plant. The bacteria return the favor by converting atmospheric nitrogen gas into a form that the plant can use to build proteins and other vital molecules.

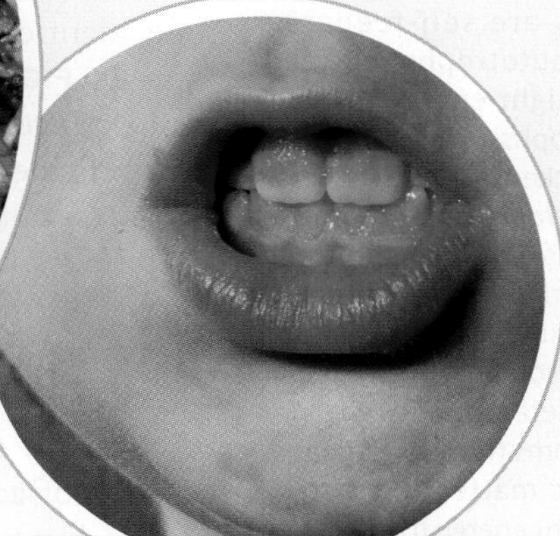

More than 500 species of bacteria live **in the human mouth**. Dentists use a red dye to reveal the presence of dental-plaque, caused by tooth decay bacteria such as *Streptococcus mutans*.

Consumers

Quorum Sensing and Bacterial Biofilms

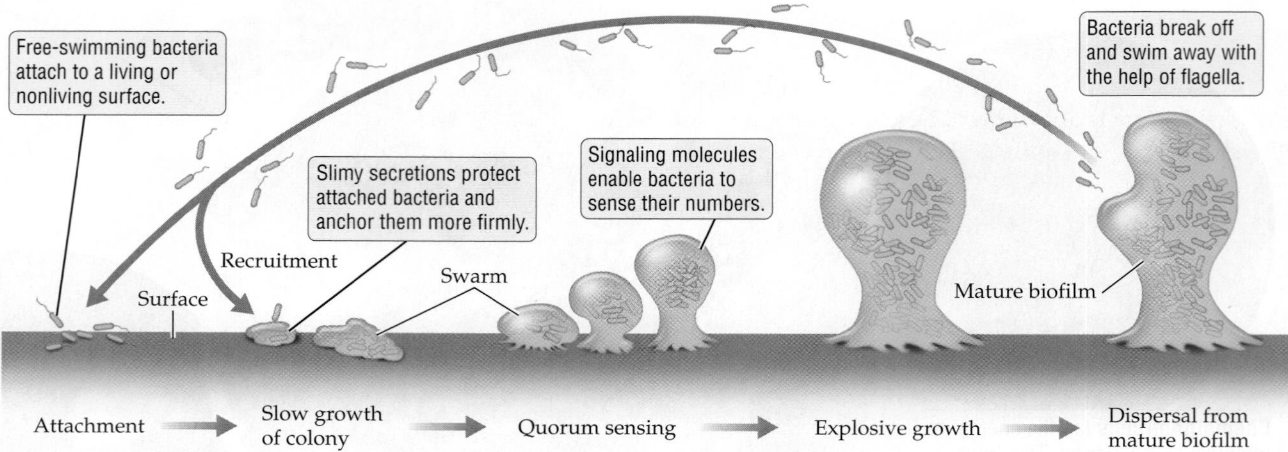

Free-swimming bacteria attach to a living or nonliving surface.
Bacteria break off and swim away with the help of flagella.
Slimy secretions protect attached bacteria and anchor them more firmly.
Signaling molecules enable bacteria to sense their numbers.
Recruitment
Swarm
Surface
Mature biofilm
Attachment
Slow growth of colony
Quorum sensing
Explosive growth
Dispersal from mature biofilm

FIGURE 17.7 Bacteria Display Social Behavior

Bacteria form biofilms on everything from shower curtains to the tissues of the organisms they infect. As they build up, biofilms become more and more resistant to dislodging by physical forces, and could even repel antibacterial drugs. Biofilms formed by antibiotic-resistant strains of human pathogens, such as the dreaded MRSA (methicillin-resistant *Staphylococcus aureus*), kill tens of thousands of Americans every year.

Prokaryotes reproduce asexually

Prokaryotes typically reproduce by splitting in two in the process called *binary fission*. The DNA in the parent cell is copied before fission, and one copy is transferred to each of the resulting offspring, known as daughter cells (see Chapter 7). The genetic information in the daughter cells is virtually identical to that of the parent cell.

As **FIGURE 17.8** illustrates, bacteria can reproduce quickly. A single *E. coli* cell, which can divide in two in just 20 minutes, can become a billion (1,000,000,000) cells in about 10 hours, as long as the cells continue to receive all the nutrients they need. The main reason the world is not completely buried in bacteria is that under natural conditions the growth of bacteria is rapidly curtailed by the lack of adequate resources. Some types of bacteria, but not any archaeans, can form *spores*, thick-walled dormant structures that can survive boiling and freezing. Both *Clostridium tetani* (which causes a life-threatening infection known as tetanus) and *Bacillus anthracis* (which causes an often fatal disease called anthrax) form spores.

Helpful to Know

Troph comes from the Greek for "feeding." *Autotrophs* are self-feeders. Photoautotrophs self-feed using light energy; chemoautotrophs use chemical energy locked in certain minerals. *Heterotrophs* feed off other organisms. Chemoheterotrophs gain carbon atoms *and* energy by consuming other organisms or their organic by-products. Photoheterotrophs gain carbon atoms from consuming organic matter, but also trap light energy to make ATP the way plants do.

Prokaryotes can take up genetic material from their environment

Although sexual reproduction has not been seen in prokaryotes, these enormously diverse organisms are capable of capturing bits of DNA from their environment

Cell Division in Prokaryotes

(*a*) Binary fission

Many prokaryotic cells divide by binary fission.

Daughter cells

(*b*) Exponential growth

Number of bacteria: 0, 1,000, 2,000, 3,000, 4,000

Time (hours): 1, 2, 3, 4

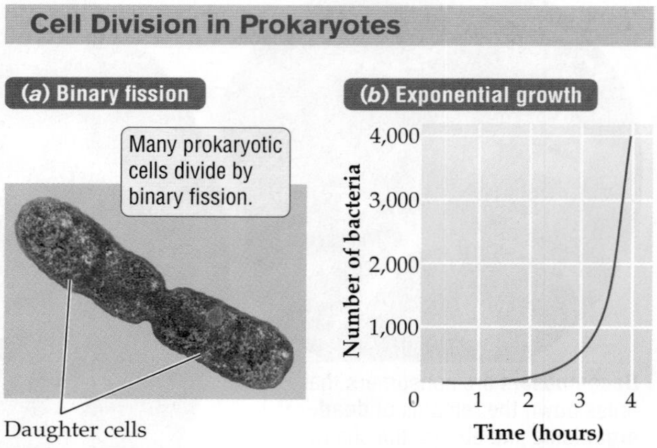

FIGURE 17.8 Through Binary Fission, *E. coli* Multiplies at an Astounding Rate

(*a*) *E. coli* divides into two every 20 minutes, given the right conditions, especially the ready availability of food and the normal temperature of the human body (37°C). (*b*) This graph shows how the population of *E. coli* increases, starting from a single cell that turns into 1,000 cells after 3 hours and 20 minutes. This type of growth, in which population doubles every 20 minutes, is an example of exponential growth.

and incorporating them into their own genetic material. The transfer of genetic material between different species under natural conditions is known as **lateral gene transfer** (or horizontal gene transfer).

One mechanism through which Bacteria and Archaea trade genetic material involves **plasmids** (*PLAZ*-**mid**), which are loops of extra DNA in the cytoplasm of prokaryotes (and some eukaryotes as well). A bacterium may actively trade DNA with another bacterium, usually of the same species, through a process known as **bacterial conjugation** (**FIGURE 17.9**). Both plasmids and chromosomal DNA may be transferred through conjugation.

When a bacterium dies, the cell may burst open and the released DNA—plasmid or chromosomal DNA—may be taken up by another bacterium of the same, or even a different, species. The process is termed *transformation*, and the bacteria that take up the DNA are said to be transformed. Genes for bacterial resistance may move from one species to another in this way.

Lateral gene transfer through these and other mechanisms appears to have sped up the rate of evolution in prokaryotes. As with sexual reproduction in eukaryotes, lateral gene transfer in prokaryotes increases genetic diversity.

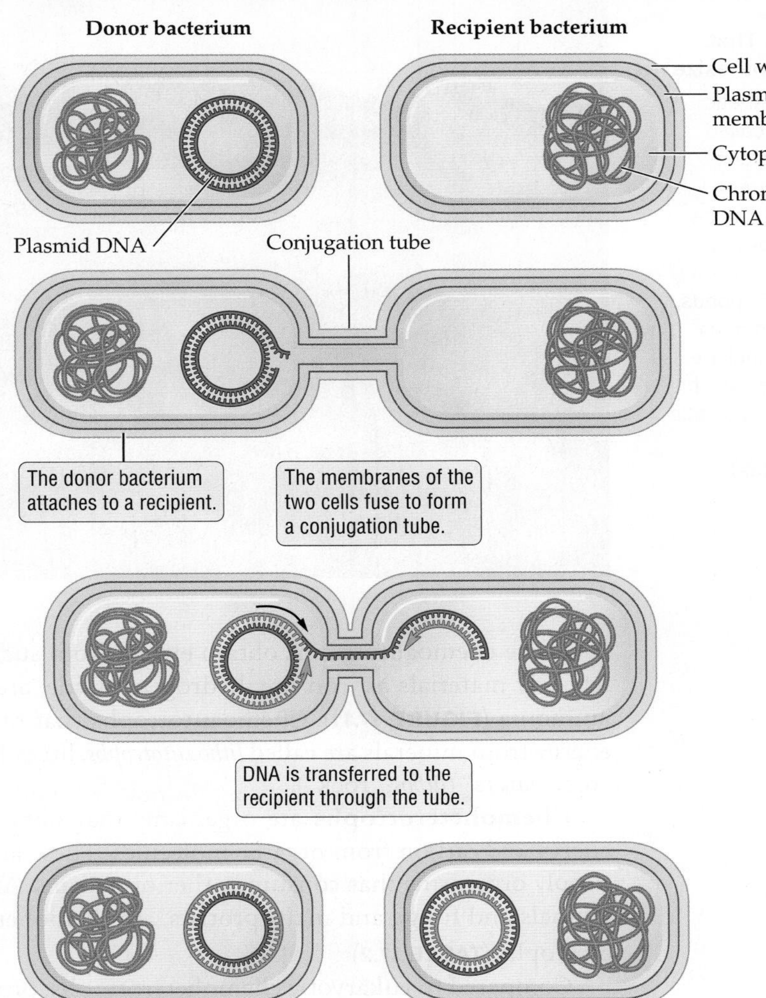
Bacterial Conjugation
Donor bacterium
Recipient bacterium
Cell wall
Plasma membrane
Cytoplasm
Chromosomal DNA
Plasmid DNA
Conjugation tube
The donor bacterium attaches to a recipient.
The membranes of the two cells fuse to form a conjugation tube.
DNA is transferred to the recipient through the tube.

FIGURE 17.9 Lateral Gene Transfer Accelerates the Rate of Evolution in Prokaryotes

Bacterial conjugation is one mechanism by which DNA is transferred from one bacterium to another. This diagram depicts the transfer of plasmid DNA, but chromosomal DNA can also be transferred through conjugation. For clarity, the structures are not drawn to scale. A plasmid is much smaller than the chromosomal DNA in a bacterial cell.

Prokaryotes are unrivaled in metabolic diversity

As noted in Chapter 1, some prokaryotes are producers, or **autotrophs**, meaning that they obtain energy from the nonliving part of their environment, such as sunlight or inorganic chemical compounds. Others are consumers, or **heterotrophs**: they obtain energy from the living or once-living part of their environment, such as by consuming other organisms or organic matter. Prokaryotes exhibit the greatest variety in both autotrophic and heterotrophic modes of life.

Like plants, some bacteria are photosynthetic autotrophs (**photoautotrophs**; *photo* means "light," *auto* means "self," *troph* means "to eat"). They absorb the energy of sunlight and take in carbon dioxide to conduct photosynthesis. Cyanobacteria (*SYE*-**an-oh-bak**-*TEER*-**ee-uh**) are commonly found in the green slime that people call pond scum, but they are known to inhabit just about every environment on Earth. Cyanobacteria are the only prokaryotes that produce oxygen gas, as a by-product of photosynthesis (**FIGURE 17.10**).

Curiously, some autotrophic prokaryotes get their energy not from light but from inorganic chemicals in their environment. They are known as **chemoautotrophs** (*chemo*, "chemical"), organisms that make food from carbon dioxide and energy extracted from chemicals in their environment. Examples include some of the bacteria and archaeans that inhabit hydrothermal vents in the lightless depths of the ocean (see Figure 17.6). These chemoautotrophs are food for an extraordinary ocean floor ecosystem that includes animals such as vent clams, tube worms, and even octopi.

FIGURE 17.10 Pond Scum Contains Bacteria That Photosynthesize Photosynthetic bacteria, called cyanobacteria or "blue-green bacteria," can be found growing as slimy mats on freshwater ponds. The green mats may also include true algae, which are photosynthetic protists (eukaryotes).

Some chemoautotrophs obtain energy from such unlikely materials as iron ore, hydrogen sulfide, and ammonia (**FIGURE 17.11**). Chemoautotrophs that tap energy from minerals are called *lithoautotrophs*, literally "rock eaters" (*lithos*, "rock").

Chemoheterotrophs are organisms that obtain energy *and* carbon from organic molecules. These are simply organisms that consume other organisms. All animals and fungi, and many protists, are chemoheterotrophs (**TABLE 17.2**).

Compared to eukaryotic chemoheterotrophs, prokaryotes can use a greater variety of carbon sources. Some prokaryotes can live off the carbon-rich molecules in petroleum. Petroleum, as you may know, represents the fossilized remains of ancient organisms; these organisms have been transformed over hundreds of thousands of years into liquid by intense heat and pressure generated by geologic processes. Some heterotrophs need oxygen to extract energy from organic material, and are therefore *aerobic*. *Anaerobic* heterotrophs can tap the chemical energy in organic molecules even in the absence of oxygen, and sometimes *only* when oxygen is absent.

Prokaryotes vary not only in type of autotrophy (photoautotrophy or chemoautotrophy), but also in type of heterotrophy. While some prokaryotes are *chemo*heterotrophs, just like us, others are *photo*heterotrophs. **Photoheterotrophs** (*photo*, "light"; *hetero*, "other"; *troph*, "to eat") use light as an *energy* source (as do plants) but get their carbon from organic material (not from carbon dioxide, the way plants do). For example, the salt-tolerant halobacteria (which are archaeans, despite their name) absorb sunlight using a pigment called bacteriorhodopsin, the way plants

TABLE 17.2 Modes of Nutrition among Prokaryotes

TYPE OF NUTRITION	SOURCE OF ENERGY	SOURCE OF CARBON	EXAMPLE[a]
Photoautotroph	Light	Carbon dioxide	Cyanobacteria
Chemoautotroph	Inorganic chemicals (such as iron ore)	Carbon dioxide	*Thiobacillus ferrooxidans*
Chemoheterotroph	Organic molecules	Organic molecules	*Escherichia coli*
Photoheterotroph	Light	Organic molecules	*Heliobacterium chlorum*

NOTE: The real key to the success of prokaryotes is the great diversity of ways in which they obtain and use nutrients.

[a]All the examples in this table are members of the domain Bacteria. The domain Archaea has all the listed modes of nutrition *except* photoautotrophy.

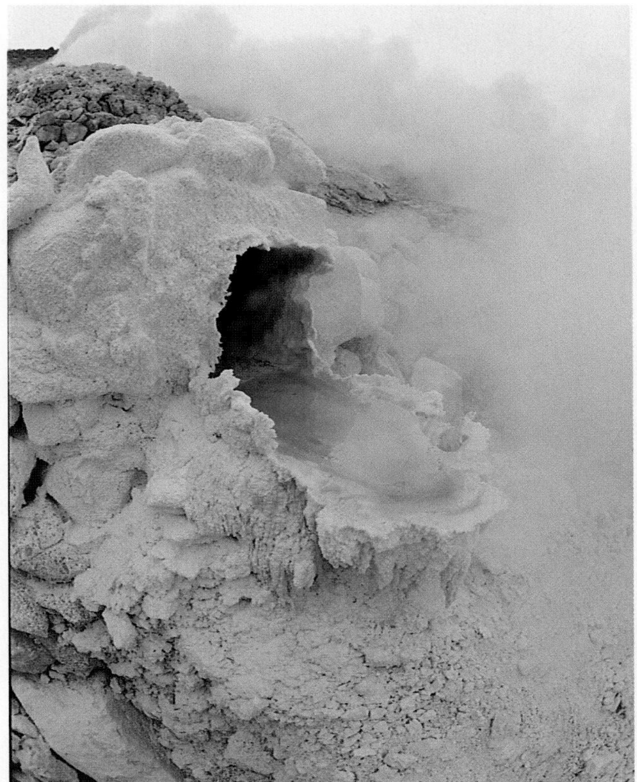

FIGURE 17.11 This Chemoautotroph Has an Appetite for Metal

The crusty orange and yellow puddle is a colony of the organism known as *Sulfolobus*, an archaean that gets its carbon from carbon dioxide, as plants do. This archaean, however, acquires its energy in an unusual way—not by harnessing sunlight (as plants do), or by eating other organisms (as animals do), but by chemically processing inorganic chemicals such as sulfur. This chemoautotroph is living in a volcanic vent in Japan.

absorb light energy with their green-colored chlorophyll pigment. The absorbed light energy is used to make ATP that then drives the manufacture of food molecules using carbon atoms released from the remains of once-living organisms. Bacteriorhodopsin is red to purple in color, and large *Halobacterium* populations often lend a colorful tint to salt lakes and to the evaporation ponds used in salt manufacturing (see Figure 17.6).

Concept Check

1. What characteristics of prokaryotes make them so numerous and so successful?
2. What are some differences between archaeans and bacteria?

Concept Check Answers

1. Most prokaryotes are small and reproduce quickly. They also have diverse modes of nutrition, tolerate environmental extremes, and therefore occupy a greater variety of habitats in larger numbers than does any eukaryote.
2. Archaeans have unique membrane chemistry, are not photoautotrophic, do not form spores, and are not known to be pathogenic. DNA comparisons show that they are actually closer to Eukarya than to Bacteria.

17.3 How Prokaryotes Affect Our World

There was very little oxygen gas on the early Earth, which was a hothouse thick with gases like carbon dioxide, methane, and ammonia. The evolution of oxygen-generating photosynthesis, probably about 2.5 billion years ago, changed the chemistry of Earth and, in so doing, changed the world forever. The increase in atmospheric oxygen is believed to have spurred the evolution of eukaryotes, which use a more efficient and oxygen-dependent form of metabolism—cellular respiration—to power the energy needs of their larger cells (see Chapter 6).

Prokaryotes play important roles in the biosphere

Because of the wide range of evolutionary innovations they possess—particularly those that enable them to obtain nutrients in a variety of ways—prokaryotes play many important roles in ecosystems and in human society. Bacteria that are producers, like cyanobacteria, are at the base of the food chain in many aquatic ecosystems, such as the open ocean.

Many heterotrophic bacteria and archaeans are **decomposers**, consumers that extract nutrients from the remains of dead organisms and from waste products such as urine and feces (see Figure 17.6). Decomposers play a crucial role in **nutrient recycling**: by breaking down dead organisms or waste products, decomposers release the chemical elements locked in the biological material and return them to the environment, where they can be used by autotrophs and eventually heterotrophs as well. Decomposer prokaryotes include oil-eating bacteria used to clean up ocean oil spills and bacteria that live on sewage, breaking down waste so that it can be released into the environment in a safer form.

Bacteria can directly aid plants as well. Plants need nitrogen in the form of ammonia or nitrate, which they cannot make themselves. Plants benefit

EXTREME BURP

The white "pancakes" are deposits of methane hydrate frozen in an Arctic lake. They were made by chemoheterotrophic archaeans that ferment dead plants and animals under anaerobic conditions. As the lake thaws, the deposits pop and fizz, releasing methane gas into the air. Gas-producing archaeans lurk in cow guts and human intestines too.

FIGURE 17.12
We Use Bacterial Fermentation to Preserve Foods
Cookbook author and natural-foods "guru" Sandor Katz teaches workshop participants how to pickle and preserve nature's bounty with the help of bacterial fermentation.

from bacteria that can take nitrogen, a gas in the air, and convert it to ammonia, in a process known as **nitrogen fixation**. Most nitrogen-fixing bacteria live free in the soil or water. But some, such as species of *Rhizobium*, form elaborate and intimate associations with certain plants, which house them inside special outgrowths of the roots (root nodules; see Figure 17.6).

Prokaryotes have economic value

Because prokaryotic metabolism is so enormously diverse, prokaryotes produce an astonishing variety of metabolic by-products, ranging from antibiotics used as medicine to the acetone in nail polish remover. Some oxygen-utilizing heterotrophic bacteria resort to **fermentation** to produce ATP, usually in low-oxygen environments (see Section 6.4). Whether conducted by prokaryotes or by eukaryotes such as yeasts, fermentation results in the accumulation of a variety of end products, depending on the species and the food molecules being broken down. The breakdown products range from acetic acid, the main acid in vinegar, to butyric acid, which is responsible for the distinctive flavor of Swiss cheese. Yogurt, buttermilk, soy sauce, some cheeses, and pickled vegetables (such as kimchi) are some of the foods that are made with the help of bacterial fermentation (**FIGURE 17.12**).

The extraordinary metabolic diversity of prokaryotes explains their ability to live in an astonishing diversity of habitats, and that metabolic versatility often comes to our aid in resolving mistakes made by humans. **Bioremediation** is the use of organisms to clean up environmental pollution. Prokaryotes have been especially useful in cleaning up oil spills.

Some chemoheterotrophic bacteria and archaeans can use the organic molecules in petroleum as a source of both energy and organic molecules, turning the contaminant molecules into harmless carbon dioxide and water in the process. Oil spill sites are often sprayed with fertilizer (usually containing nitrogen and sulfur) to encourage the explosive growth of bioremediator prokaryotes (**FIGURE 17.13**). The fertilizer components are nutrients that, in nature, are usually not available in sufficient quantity to support rapid proliferation of prokaryotic cells.

Some bacteria cause disease

Although the great majority of bacteria are harmless, and many are actually beneficial to humans, some

ANSWERS TO FIGURE 17.4
a: 7 (drinking-fountain spout); *b*: 4 (doorknob); *c*: 3 (lobby floor); *d*: 2 (dirty sock); *e*: 6 (toilet seat); *f*: 5 (dog's mouth); *g*: 1 (dish sponge). Note that sources that tend to be moist have the greatest abundance of microbes.

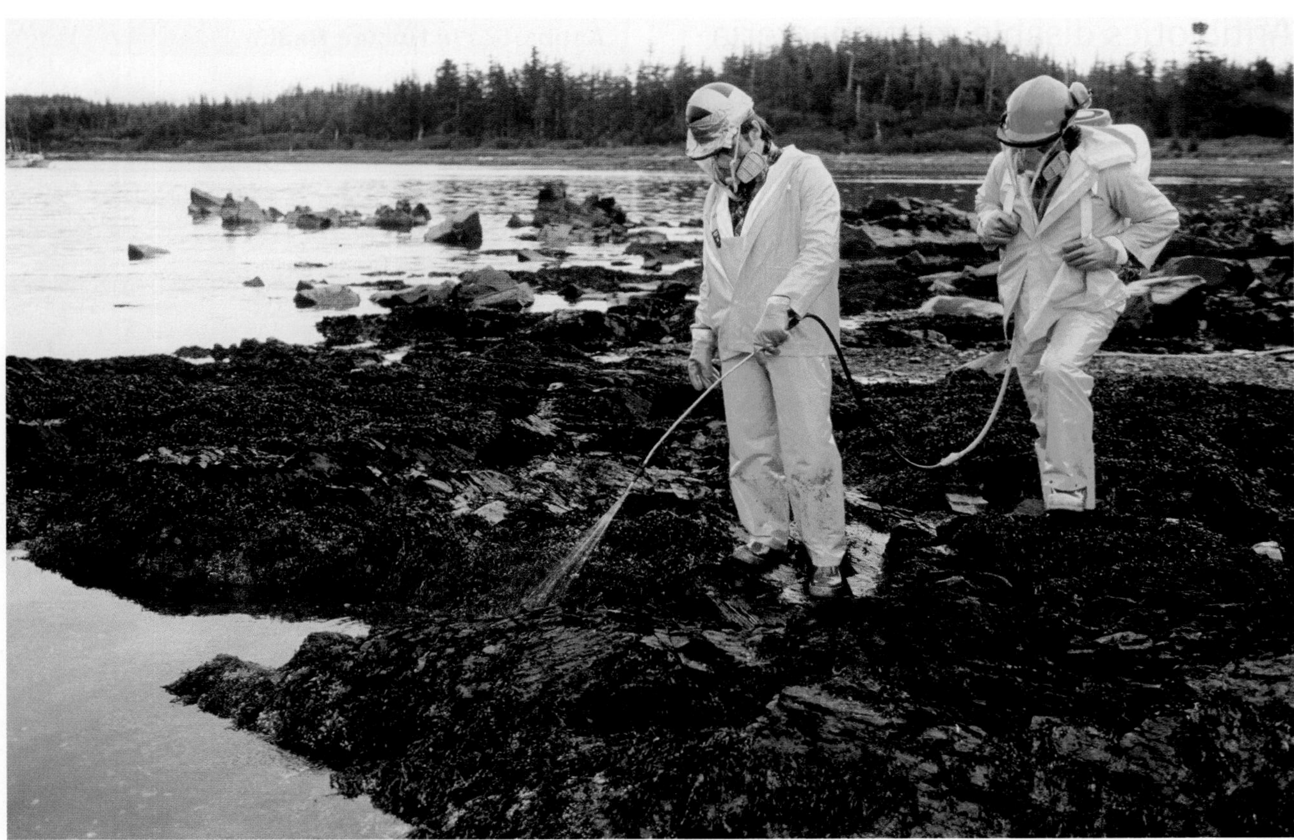

FIGURE 17.13 Bioremediation by Prokaryotes

Workers spray fertilizer on an oil-contaminated shore to stimulate the growth of oil-degrading bacteria.

cause diseases ranging from mild to deadly. Organisms that cause disease in other organisms are called **pathogens**. Interestingly, archaeans do not appear to be pathogens of any organism.

With their ability to use almost anything as food, bacteria infiltrate crops, stored foods, and domesticated livestock, in addition to sickening or killing humans. Like most other pathogens, pathogenic bacteria tend to be quite host-specific, meaning they infect a specific type of organism. Most bacteria that infect plants, for example, do not affect humans. Some bacterial infections, such as anthrax, can be communicated to humans from another animal species (cattle, in the case of anthrax). However, most bacterial species infect just one or a few closely related species.

Across the world, bacterial infections kill more than 2 million people each year, mainly from tuberculosis, typhoid, and cholera. In the developed world, better sanitation makes bacterial infections less common. Bacterial food poisoning, bacterial pneumonia, and "strep throat" (caused by *Streptococcus* species) are the most common bacterial infections in the United States.

Federal agencies estimate that 5 million people per year are sickened by food-borne bacteria, mainly species of *Campylobacter* and *Salmonella*. Certain bacteria, including some that cause food poisoning, such as *Salmonella* and *Vibrio cholerae*, produce endotoxins. An **endotoxin** is a component of the bacterial cell wall that triggers illness. Endotoxins can produce fever, blood clots, and toxic shock (a sudden, sometimes fatal, drop in blood pressure).

Some bacteria, such as *Clostridium botulinum*, produce an **exotoxin**, which is a poison that an organism releases into its surroundings. Exotoxins disrupt cellular processes or damage cell structures such as the plasma membrane. Botulinum, the exotoxin produced by *C. botulinum*, blocks nerve transmission to muscles. It is more commonly known as Botox. Unlike endotoxins, most exotoxins are destroyed by heat. Improperly canned meat contaminated with *C. botulinum* causes the deadly food poisoning known as botulism. Thorough cooking of meat breaks down endotoxins such as botulinum.

EXTREME PATHOGEN

As a grad student, Aimee Copeland contracted necrotizing fasciitis, commonly called "flesh-eating bacteria disease." She described her heroic struggle on Katie Couric's talk show. *Streptococcus pyogenes* is among the bacteria known to cause the rare, fast-spreading infection that can kill in a couple of days if left untreated. Scientists don't know why some people are susceptible to this common bacterium.

Antibiotics disable many bacteria but have no effect on viruses

Antibiotics are commonly used to combat bacterial infections. Although scientists can also make antibiotics from scratch nowadays, the term **antibiotic** refers to a molecule secreted by one microorganism to kill or slow the growth of another microorganism. Antibiotics are naturally produced by a variety of fungi. As decomposers and pathogens, bacteria and fungi often occupy the same ecological *niche*—that is, have the same habitat and food requirements. The two are therefore in direct competition for resources, and antibiotics are the fungal weapons of choice against the bacterial world. However, some species of bacteria also deploy antibiotics in warfare against other bacteria, and a number of commercial antibiotics (such as streptomycin) come from bacteria rather than fungi.

In 1928, Alexander Fleming described antibiotics from a mold called *Penicillium notatum* (**FIGURE 17.14**).

Discovery of Penicillin

Penicillium (mold)
Penicillin
Dead bacteria
Dying colonies
Normal staphylococcal colony

FIGURE 17.14 The Discovery of Penicillin, an Antibiotic Produced by *Penicillium notatum*, Transformed Medicine
Alexander Fleming noticed that the disease-causing bacterium *Staphylococcus aureus* died when it was near a mold colony (green) that had contaminated his bacterial plates.

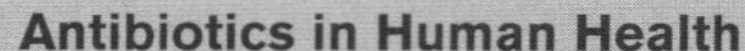

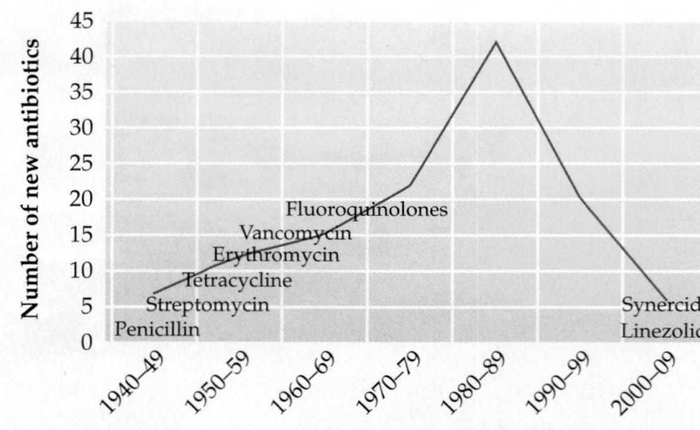

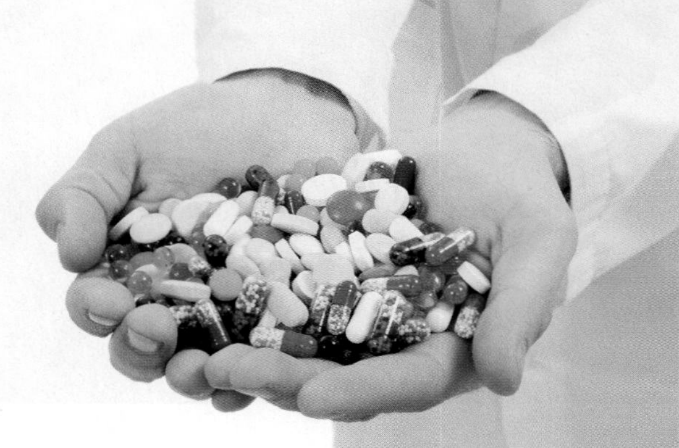

FIGURE 17.15 Antibiotic Misuse Leads to Resistant Strains That Are Hard to Treat
Fewer new antibiotics are being developed for pharmaceutical use. Only two new antibiotics have been introduced in the past 5 years. That means there will be very few, possibly no, options that we can fall back on if pathogens become resistant to multiple antibiotics that are currently available.

In 1939, Ernst Chain and Howard Florey developed methods for mass-producing the antibiotic, penicillin, and launched the antibiotic era. Since then, many natural and synthetic (human-made) antibiotics have been put to use in fighting bacterial infections. To appreciate the social transformation brought about by antibiotics, visit an old cemetery and observe the large numbers of young children that families lost to infectious diseases before the 1940s.

Using antibiotics inappropriately—for example, not completing a prescribed course, or overusing them in farm animals—can lead to natural selection for antibiotic resistance in bacteria, reducing the options available for fighting bacterial infections (**FIGURE 17.15**). It is important to realize that antibiotics are ineffective against viruses (discussed next), which can cause diseases similar to bacterial infections (some types of pneumonia and some types of food poisoning, for example). Therefore, prescribing and using antibiotics in cases of simple viral disease, such as a cold, is not recommended.

Concept Check

1. What are some roles of prokaryotes in ecosystems?
2. What are antibiotics? Should they be used to fight a viral infection?

17.4 Viruses: Nonliving Infectious Agents

A **virus** is a microscopic, noncellular infectious particle. Most viruses are little more than genetic material wrapped in proteins, yet they can attack and devastate organisms in every kingdom of life—bacteria, archaeans, protists, fungi, plants, and animals.

Viruses lack cellular organization

Like living organisms, viruses can have DNA, they can reproduce, and they evolve. Yet viruses lack some of the key characteristics of life, which is why, after many years of head-scratching, most scientists today regard viruses as nonliving infectious particles. A virus is much smaller and simpler than a cell and usually consists of genetic material—DNA or RNA—wrapped in a coat of proteins (**FIGURE 17.16**). Some viruses also have an envelope, a lipid layer somewhat like a cell's plasma membrane, enclosing the central core of genetic material and protein.

Another difference, compared to living organisms, is that viruses lack the many structures within cells that are necessary for critical cellular functions such as homeostasis, reproduction, and metabolism. To gain these functions, viruses make the cells of other organisms do the work for them. They accomplish this feat by invading those cells, releasing their genetic material into the cytoplasm, and essentially "hijacking" the host cell's metabolism.

Viruses are classified by structure and type of infection

A special classification system, similar to the Linnaean hierarchy, is used to categorize viruses. Viruses are generally classified by the type of genetic material they possess (type of DNA or RNA molecule), their shape and structure, the type of organism (host) they infect, and the disease they produce. **FIGURE 17.17** shows some of the variations in viral shape and structure.

The variant forms of a particular type of virus are called **viral strains**, or serotypes. The common cold, for example, is caused most often by one of the many strains of rhinovirus, a member of the picornavirus family. Viral strains are sometimes named after the place where they were identified. For example, the Ebola virus is named after the Ebola Valley in the Democratic Republic of the Congo, in Africa. More often, newly discovered strains of otherwise well-known viruses are named simply with letters and numbers (such as the H1N1 strain of the influenza virus).

Concept Check Answers

1. As photosynthesizers, cyanobacteria add oxygen to the atmosphere; as decomposers, prokaryotes cycle nutrients; prokaryotes are the only organisms that can fix atmospheric nitrogen, converting it into a form that plants can use.
2. Antibiotics are molecules secreted by one microorganism to kill or slow the growth of a competing microorganism; today, they can also be made artificially. Antibiotics are not effective against viruses.

Structure of the Influenza Virus

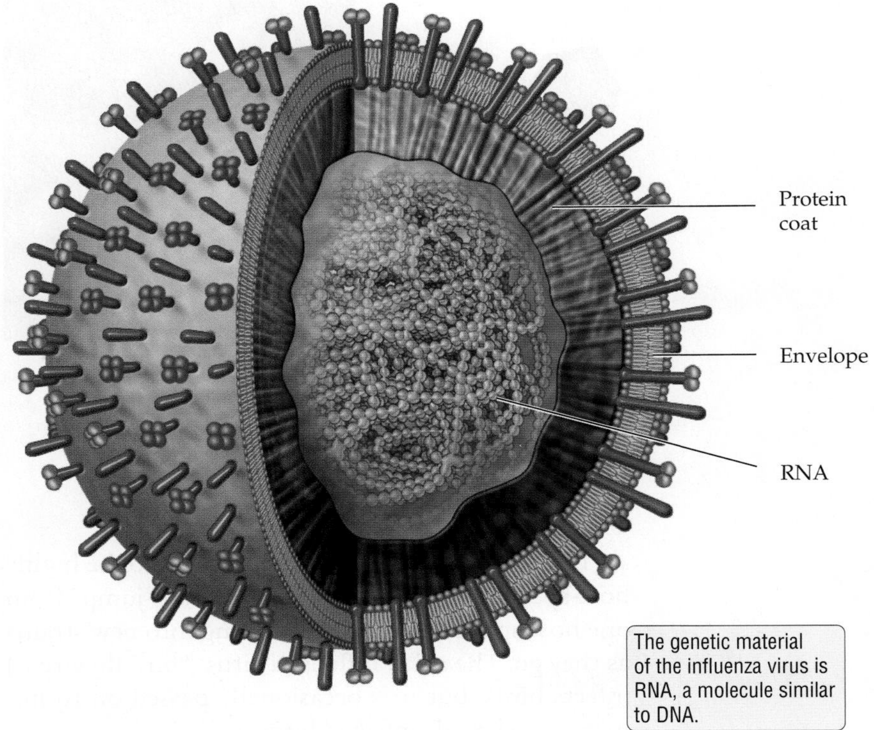

Protein coat
Envelope
RNA
The genetic material of the influenza virus is RNA, a molecule similar to DNA.

FIGURE 17.16 Viral Structure

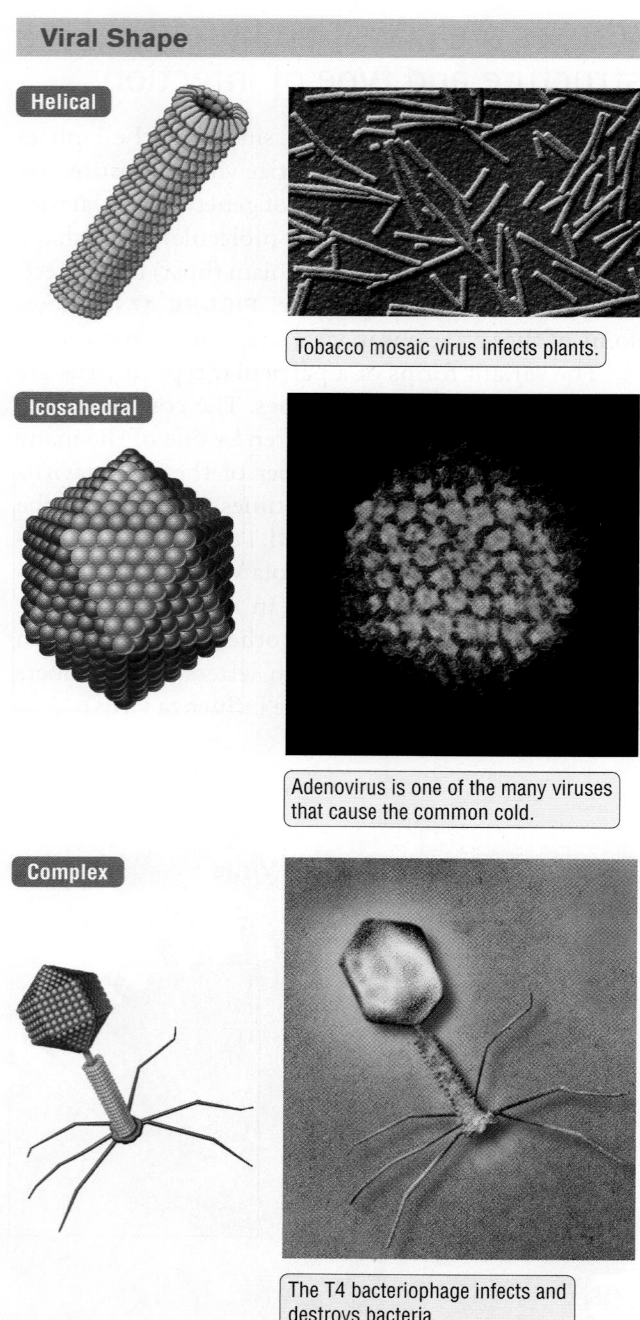

Viral Shape
Helical
Tobacco mosaic virus infects plants.
Icosahedral
Adenovirus is one of the many viruses that cause the common cold.
Complex
The T4 bacteriophage infects and destroys bacteria.

FIGURE 17.17 Viruses Can Be Classified by Their Shape

Like bacterial pathogens, viruses tend to be highly host-specific. However, some viruses can "jump" from one host species to another, evolving into new strains as they go. The avian influenza virus ("bird flu virus") infects birds, but it is occasionally passed on to humans who handle infected birds.

Some viruses, known as *retroviruses*, can insert their genetic material into a host cell's DNA and then lie dormant for long periods of time, even for the life span of the infected organism. Pathogenic retroviruses usually become active after a while, multiply to huge numbers, and then escape the host cell. What triggers this sequence of events is largely unknown. The viral offspring escape from a host cell either by causing it to burst open or by budding off from the cell wrapped in a layer of the host cell's plasma membrane.

The HIV-1 virus, which causes HIV-AIDS in humans, is a retrovirus that invades immune cells by fusing with the host cell's membrane and releasing the genetic material (which, in the case of HIV, is RNA) into the cytoplasm. The viral genetic material is converted into DNA and then integrated into the host cell's DNA.

In most infected adults, the immune system beats back the virus enough that no disease symptoms are seen for the next 2–20 years. However, the continued proliferation of the virus during this time means that in most people the virus gains the upper hand eventually, as the critical immune cells begin dying faster than they can be replaced. Once the number of these immune cells has fallen below a certain threshold, the body's defense capacity declines sharply, making the infected individual extremely vulnerable to opportunistic infections by other pathogens, including other viruses, bacteria, and fungi.

Flu viruses evolve rapidly

Many people are laid up each winter by the influenza virus. Influenza viruses are RNA viruses, but they do not insert their genetic material into the host cell's DNA. However, the virus takes over the host cell's protein and RNA-making machinery, directing the cell to make many copies of virus particles. The particles then exit the cell enclosed in a "bubble" made by the plasma membrane, without destroying the host cell (**FIGURE 17.18**). Most people start "shedding" the virus into the environment 2–3 days after becoming infected, which is usually a day before symptoms appear, and they remain infective for about 7 days after first catching the virus. The virus can survive on a doorknob for a few days, and on a moist surface for about 2 weeks.

Influenza viruses attach to cells in the nose, throat, and lungs of humans and in the intestines of birds. Flu sufferers cough and sneeze, develop a fever, and ache all over. Runny noses suit the virus very well because, through sneezing, they help disseminate the virus from one victim to the next. Some flu symptoms are the result of the actions of immune cells on a mission to attack and destroy the virus. Fever, for example, is a defensive strategy: turning up body temperature slows viral reproduction. But very high fever is often an overreaction that can damage the body; doctors therefore have to exercise judgment in deciding whether to fight a fever with medication or let nature take its course.

Why are people susceptible to the influenza virus year after year? The reason is that as this microscopic virus proliferates throughout the body's cells—whether for days or for weeks—it is evolving rapidly. Viruses evolve into new strains so quickly that sometimes an antiviral drug developed to fight an older strain becomes useless against a new strain.

Remember, antibiotics are useless against viral infections. Life-threatening viral infections, such as HIV-AIDS, are treated instead with drug cocktails that disable the reproduction of viruses by interfering with the copying of their genetic material. The antiviral drug cocktails can have harsh side effects, however, which is why they are reserved for serious illness.

Concept Check

1. Are viruses living organisms?
2. Why must health agencies release new flu vaccines nearly every year?

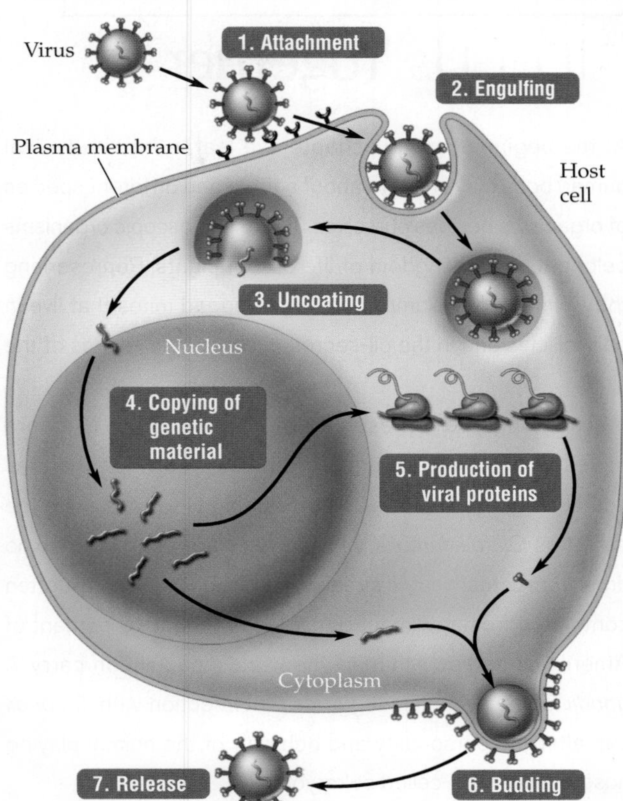

FIGURE 17.18 Viruses Reproduce inside Their Host Cells

Concept Check Answers

1. Viruses are infectious agents. Most biologists regard them as nonliving organisms because they lack cellular organization and independent metabolism.

2. The flu virus evolves rapidly, producing new strains that an older vaccine may not protect against. Each year, new vaccine must be formulated using the most recent available strains.

APPLYING WHAT WE LEARNED

All of Us Together

At the beginning of this chapter we learned that a typical human body plays host to about a thousand different species of organisms besides ourselves. These microscopic organisms belong to every kingdom of life except plants. Representing the Animalia, for example, are eight-legged mites that live in hair follicles and in the oil-secreting sebaceous glands of the skin.

The Fungi are represented by a dozen or more genera of yeasts and molds that inhabit the skin between the toes and other nutritious zones. Representing the Protista are the parasites *Giardia lamblia*, which gives diarrhea to hikers who drink from contaminated streams, and *Toxoplasma gondii*, often contracted from raw meat or cat feces. About 10 percent of Americans and one-third of the world's population carry *T. gondii*. Some evidence suggests that infection with *T. gondii* can alter the personality and behavior of the animal playing host to this single-celled eukaryote.

The overwhelming majority of organisms that live in and on humans, however, are prokaryotes—mostly bacteria, but a few archaeans as well. Archaeal methanogens live in our guts, generating methane gas, and also in our gums. And then there are the bacterial species, hundreds of them—including *E. coli*, which lives in the colon; numerous kinds of *Bacteroides*, a genus that makes up about a third of all gut bacteria; and others, such as *Streptococcus mutans*, that live on our teeth and make slimy plaque.

Each of us has a unique microbiome "aura." However, your microbiome is more similar to that of other people in your region than to the microbial community that lives on and within people on the other side of the country. In fact, University of Oregon researchers could distinguish the members of East Coast and West Coast roller derby teams just by running DNA tests on the microbial samples they collected from the upper arms of the players.

Couples who live together have similar microbiomes. Children have microbial communities most similar to those of their parents, and this is more true of older children than younger ones, simply because older children have shared one roof with the parents longer. Dogs share part of their microbiomes with their owners, which is why a dog owner's microbiome has more in common with the microbiome of other dog owners than with people who don't own a dog (**FIGURE 17.19**).

FIGURE 17.19 Dog Owners Acquire Part of Their Microbiome from Their Pets

Bacteria and other microbes can definitely be good. Mice raised in germ-free environments have to eat 30 percent more calories than regular mice to make up for the absence of "germs" that help digest food. Babies born by C-section (surgically) have different microorganisms than those born vaginally, and the C-section babies are more susceptible to dangerous infections. Children treated with antibiotics early in life seem more likely to develop asthma. The microorganisms found in overweight adults differ from those in slimmer people, and when people lose weight, the biodiversity of their gut changes.

Yet, even though microbes are clearly essential to good health, some of those that live with us every day sometimes turn on us. Some 40 kinds of staphylococci (STAF-**ih-loh-**KAH-**kye**) harmlessly inhabit our skin and mucous membranes, but one strain of *Staphylococcus **aureus*** stands ever ready to invade a break in the skin or elsewhere and initiate an infection. Yet this exact strain of staph is a normal resident of our bodies, occurring naturally in 25 percent of humans. People who take powerful antibiotics, causing them to lose much of their community of friendly bacteria, become more prone to infection by *Clostridium difficile*, a decidedly unfriendly bacterium that is difficult to fend off, even with modern medicines.

BIOLOGY IN THE NEWS

Gut Microbiome Largely Stable over Time, Feces Study Finds

BY ERYN BROWN • *Los Angeles Times*, July 5, 2013

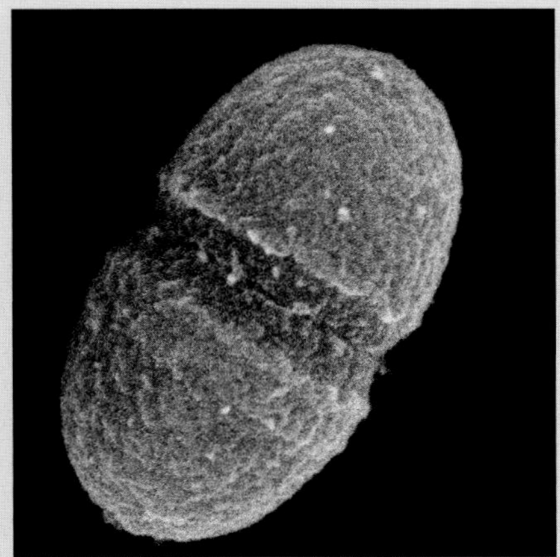

Each of us has a core community of bacterial strains that lives in our lower digestive tract—a personal set of microbes that remains relatively stable in the gut over the course of decades, scientists said Thursday.

In an examination of the bacteria contained in dozens of fecal samples collected from 37 healthy American adults, researchers at the Center for Genome Sciences and Systems Biology at the Washington University School of Medicine in St. Louis found that every person has about 100 species and 200 strains of bacteria in the gut.

On average, 60% of the bacterial strains detected were retained over a period of five years, suggesting that the mix of bacteria in the digestive tract, also known as the gut microbiome, is largely stable ...

Scientists want to know whether each person's gut microbiome changes or remains stable over time so that they can better interpret what shifts in microbial makeup mean, said Washington University biologist Dr. Jeffrey Gordon, senior author of a study describing the research, published Thursday in the journal *Science* ...

Rather than isolate all of the bacteria in a sample and culture them to see what's there, scientists look at the entire community of bacteria at once, focusing on the differences in a single gene in each microbe, 16s rRNA, to separate out the different species that are present ...

... The team performed a number of analyses, calculating the numbers of species and strains in each sample, comparing samples from a single individual over time, comparing samples between family members (who had been shown to share microbiomes) and seeing how weight loss or gain affected bacterial makeup, in subjects who were following a special diet as well as subjects who ate as they pleased ...

[The group] estimated that, were physicians to begin collecting fecal samples to monitor patient health, doing so just once a year would be sufficient to reveal significant changes in the core microbiome. That could provide a useful tool for doctors, Gordon said.

Stanford's [Dr. David] Relman, who has also studied how human gut microbiome changes over time, said that understanding stability in microbial communities was "fundamentally important," and praised the new study as "an important contribution to the story." He agreed that there could be clinical value in monitoring "yearly poop samples" in patients.

Evidence is growing that the community of microbes we harbor—our microbiome—has a big impact on our health. Conditions such as diabetes, obesity, allergies, Crohn disease, and even mood disorders are believed to be influenced by the gut microbiome, the prokaryotes and fungi that live in the digestive system.

Take weight gain. Researchers observed a difference between the gut bacteria of obese and normal-weight animals, including mice and humans. Since that discovery, scientists have been conducting experiments to tease apart the complex ways in which genetics, diet, gut bacteria, and disease susceptibility are related. The studies show how these might be linked: in a particular strain (genotype) of mice, a high-sugar and high-fat diet fostered the growth of certain *Firmicutes* bacteria over the *Bacteroidetes* type; the favored bacteria produced chemicals that increased the absorption and storage of fats and carbohydrates in the body, leading to weight gain. Fecal transplant experiments have established cause and effect quite clearly. When scientists moved gut microbes from heavy mice to lean ones, the latter gained weight.

Studies on human subjects corroborate the mouse studies, with participants who have a sugary, fatty, "Western" diet showing lower microbial diversity and an increase in the actinobacteria population. Some microbial by-products were found to aggravate the immune system suggesting that the microbes we harbor might trigger immune system dysfunction.

It seems each of us is a walking habitat for microbes, and which microbes flourish on and in us, as well as how they interact with each other and the body, affects many aspects of our health. The goal of the study described in this article was to do a census of the gut microbiome of 37 volunteers and to track the microbial population over 5 years. Although most of the approximately thousand species of gut bacteria remained stable, some did change over time. The challenge now is to understand when and why that change occurred, and what it means for the health and well-being of the individual.

Evaluating the News

1. Why is it useful to know whether our microbiome is largely stable over the years?

2. How did scientists distinguish cause and effect in the association between certain types of microbial communities and weight gain in mice?

3. How would you explain the clinical value of yearly fecal monitoring to a skeptical and "grossed out" friend?

CHAPTER REVIEW

Summary

17.1 The Diversity of Life

- The most basic and ancient branches of the tree of life define three domains: Bacteria, Archaea, and Eukarya. All life-forms fall into one of these three domains. The domains are further divided into six kingdoms: the prokaryotic Bacteria and Archaea; and the eukaryotic Protista, Fungi, Plantae, and Animalia.
- The universal ancestor arose about 4.5 billion years ago and most likely resembled bacteria. Archaeans are superficially similar to bacteria, but DNA analysis shows that Archaea is a domain in its own right and evolutionarily closer to Eukarya than to Bacteria.
- Biodiversity includes all the world's living things, as well as all the variety in their interactions with the living and nonliving components of the ecosystems they inhabit. It can be described at the level of genes or species or whole ecosystems.
- Earth's biodiversity is crucial for healthy ecosystems and also benefits humanity in direct and indirect ways. The world's biodiversity is disappearing even as scientists race to catalog and understand it.

17.2 Bacteria and Archaea: Tiny, Successful, and Abundant

- The non-Eukarya, commonly called prokaryotes, are microscopic, single-celled organisms, but Bacteria and Archaea differ in significant ways, such as in their plasma membrane chemistry and metabolism.
- Prokaryotes can reproduce extremely rapidly and are the most numerous life-forms on Earth. They also have the most widespread distribution. Some prokaryotes, including many archaeans, thrive in extreme environments. Thermophiles, for example, live in extremely hot places, and halophiles inhabit very salty places.
- Prokaryotes do not reproduce through sexual means but can acquire genetic information from other organisms by picking up DNA from their environment—a process known as lateral gene transfer.
- Prokaryotes exhibit unmatched diversity in methods of getting and using energy and nutrients. Prokaryotes can be photoautotrophs, chemoautotrophs, chemoheterotrophs, or photoheterotrophs.

17.3 How Prokaryotes Affect Our World

- Prokaryotes perform key tasks in ecosystems, including photosynthesizing, fixing atmospheric nitrogen into a form that plants can use, and decomposing dead organisms.
- Prokaryotes are useful to humanity in many ways (for example, helping in the preparation of fermented foods, cleaning up oil spills, and assisting our digestion).
- Some bacteria (but no known archaeans) cause diseases in animals and other eukaryotes.
- Endotoxins are cellular components of the bacterium, such as cell wall polysaccharides, that trigger a severe overreaction by the immune system of animals like us. Exotoxins are molecules released by invading bacteria that can damage cellular processes and destroy cell components such as the plasma membrane.
- An antibiotic is a molecule that kills or slows the growth of a microorganism. Inappropriate use of antibiotics can lead to microorganisms becoming resistant to the molecules.

17.4 Viruses: Nonliving Infectious Agents

- A virus is a microscopic, noncellular infectious particle.
- Viruses lack some of the characteristics of living organisms: they are not made of cells, and they lack the structures necessary to perform certain activities essential to life. Because viruses exhibit only some of the characteristics of living organisms, most biologists consider them to be nonliving.
- Viruses do have genetic material, and they do evolve. Some viruses use DNA as genetic material; others use a related molecule called RNA.
- Many viruses, such as the flu virus, accumulate mutations in their genetic material, enabling them to evolve rapidly.

Key Terms

antibiotic (p. 394)
Archaea (p. 381)
autotroph (p. 389)
Bacteria (p. 381)
bacterial conjugation (p. 389)
bacterial culture (p. 384)
biodiversity (p. 381)
biofilm (p. 385)
bioremediation (p. 392)
chemoautotroph (p. 389)
chemoheterotroph (p. 390)
decomposer (p. 391)
endotoxin (p. 393)
Eukarya (p. 381)
exotoxin (p. 393)
extremophile (p. 383)
fermentation (p. 392)
flagellum (p. 385)
heterotroph (p. 389)
lateral gene transfer (p. 389)
nitrogen fixation (p. 392)
nutrient recycling (p. 391)
pathogen (p. 393)
photoautotroph (p. 389)
photoheterotroph (p. 390)
pilus (p. 385)
plasmid (p. 389)
prokaryote (p. 381)
quorum sensing (p. 385)
viral strain (p. 395)
virus (p. 395)

Self-Quiz

1. Which of the following groupings lists only domains?
   a. Eukarya, Bacteria, and Animalia
   b. Plantae, Protista, and Archaea
   c. Archaea, Bacteria, and Eukarya
   d. Bacteria, Archaea, and Fungi

2. Most bacteria
   a. have a nucleus.
   b. cannot be cultured on an artificial medium in a lab dish.
   c. reproduce using RNA as the hereditary material.
   d. are multicellular.

3. Prokaryotes
   a. display a greater diversity of metabolism, compared to eukaryotes.
   b. have larger cells, on average, than most eukaryotes.
   c. lack a self-generated plasma membrane.
   d. use sexual reproduction to generate offspring.

4. Bacteria
   a. are different from the organisms in other kingdoms of life in that they are known to have arsenic instead of phosphorus as a component of their DNA.
   b. live in more diverse environments than Archaea.
   c. are not known to conduct photosynthesis, whereas most archaeans are photosynthetic.
   d. are more distantly related to the eukaryotes than are the Archaea.

5. Archaeans
   a. can be autotrophs but not heterotrophs.
   b. are responsible for a number of human diseases.
   c. lack cellular organization.
   d. can be found in the human body.

6. Some prokaryotes
   a. can obtain energy from minerals, such as iron, that are found in some rocks.
   b. are consumers that use sunlight to synthesize (manufacture) food.
   c. obtain energy from water molecules, which are abundant in most habitats on Earth.
   d. are producers that get their energy by eating other organisms.

7. Quorum sensing
   a. is the transfer of plasmid DNA from one bacterium to another.
   b. enables bacteria to form biofilms.
   c. is the formation of thick-walled dormant structures, called spores, under conditions unfavorable for growth.
   d. enables bacteria to switch from cellular respiration to fermentation when they sense that oxygen levels are low.

8. Antibiotics
   a. are toxins produced by viruses.
   b. are an example of endotoxins produced by archaeans.
   c. produced by fungi work by disrupting cellular processes or cell structures in specific bacterial species.
   d. are effective against a wide range of viruses and are commonly used to treat viral influenza.

9. Viruses
   a. divide by a form of cell division known as binary fission.
   b. are considered nonliving because they contain no hereditary material, such as DNA.
   c. infect humans but not bacteria.
   d. lack the ability to acquire energy independently.

10. Figure 17.18 shows the life cycle of the flu virus. According to the figure, the flu virus
    a. inserts its genetic material into the host cell's DNA.
    b. becomes wrapped in an envelope derived from the host cell's plasma membrane during the budding stage (step 6).
    c. breaks the host cell open, destroying it in the process, during the release stage (step 7).
    d. replicates itself through binary fission in the host cell's cytoplasm.

Analysis and Application

1. Describe the key features of a bacterial cell and compare it with a typical eukaryotic cell.
2. Explain the ecological importance of bacteria and archaeans.
3. What are extremophiles? Describe some of the habitats known to be home to prokaryotic extremophiles.
4. Compare exotoxins and endotoxins. Can one or both be destroyed by high heat?
5. Explain the life cycle of the flu virus. The human body usually responds to a flu infection with a fever. Why is the fever response useful, within limits?
6. Examine the flowchart here depicting the four main types of nutrition found among bacteria. Complete the diagram by labeling the two empty boxes.

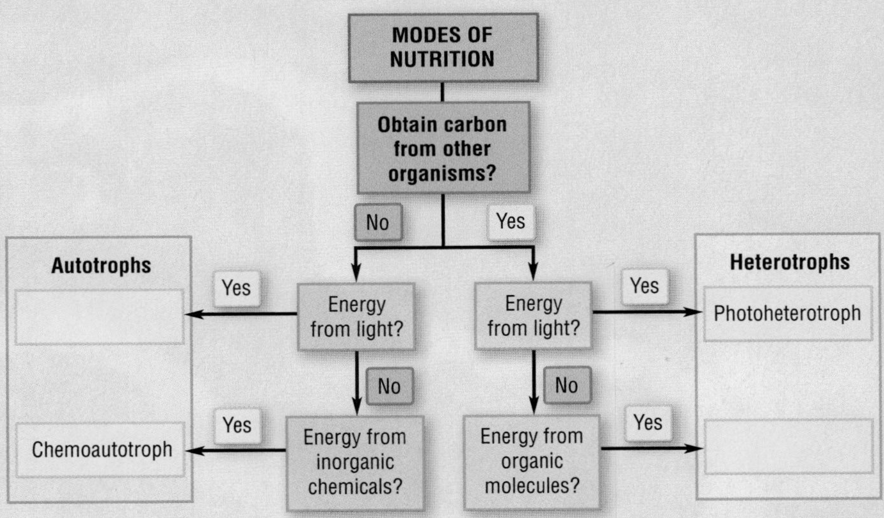

18 Protista, Plantae, and Fungi

MEANDERING RIVER. A river displays distinctive meanders as it winds through the upper Amazon basin in Peru.

Did Plants Teach Rivers to Wander?

Look at a river flowing across a gentle landscape and one of the first things you'll notice about it is that it meanders from side to side, winding in ever-wider loops. At every loop, one side is steeply cut and the other side has a low beach of sand or rocks.

Over geologic time, the meanders themselves sweep from side to side, touching first one side of a valley, then the other. Gradually the meanders move downstream, like the coils of a snake moving from head to tail. River meanders wander across floodplains, creating new deposits of sand and gravel, tree-lined sloughs, and lakes that teem with wildlife. Rivers gradually create rich, flat bottomland where marshes, meadows, and farms appear. As rivers sculpt the landscape, they promote the diversity of plants, fungi, animals, and other organisms. Meandering rivers support more bird species than straight rivers do. Rivers are a valuable part of the ecology of Earth.

Rivers that have been damaged by development, deforestation, or floods sometimes lose their curves. Instead of helping to create fertile soils, they wash them away. As a result, re-creating a river's meandering shape has been a major goal of river restoration projects that bring back natural vegetation and habitat. Yet amazingly, only recently have scientists discovered what makes rivers meander.

Until now, scientists didn't know whether rivers' meanders were caused by things like the volume of water or by qualities of the rocky sediments that make up the soil. Some scientists tried to explain meandering with mathematical equations. More hands-on researchers built artificial indoor streams using different kinds of gravel, sand, and mud—all of which refused to meander. Out in the field, restoration ecologists tried digging winding channels that looked like meandering streams through fields, only to have them form wide, shallow streams in the next big storm.

What were scientists missing? Why do natural rivers meander, and in what way could plants be involved?

The answer has to do with how plants with roots came to dominate on land 425 million years ago—a topic we explore in this chapter. The greening of Earth's continents through the evolution of plants is one of the most significant events in the history of Earth.

MAIN MESSAGE Eukaryotic cells have a nucleus and other membrane-bound components; sexual reproduction is among the evolutionary innovations of the Eukarya.

KEY CONCEPTS

- Eukaryotic cells are generally larger than prokaryotic cells. Effective predation, escape from predation, and greater metabolic capacity are some of the adaptive benefits of larger cell size.
- Eukaryotes are distinguished by having a nucleus and complex subcellular organization.
- The many membrane-enclosed compartments enable eukaryotic cells to function efficiently through division of labor among specialized organelles.
- Sexual reproduction, which generates genetic diversity within a population, is a key evolutionary innovation of eukaryotes.
- The domain Eukarya encompasses animals, plants, fungi, and a catchall category, the protists.
- Some protists, most fungi, and all plants and animals are multicellular; multicellularity makes large individuals possible.
- The kingdom Protista is an artificial grouping of mainly single-celled eukaryotes that includes algae and protozoans.
- Plants are descended from green algae; the evolutionary innovations of different plant groups include vascular tissue, seeds, and flowers.
- Fungi include yeasts, molds, and mushrooms; they acquire nutrients by absorption and are important decomposers.
- Some fungi form symbiotic relationships with other organisms, such as with a green alga in a lichen, and with plant roots in mycorrhizae.

DID PLANTS TEACH RIVERS TO WANDER? 403

18.1 The Dawn of Eukarya 405

18.2 Protista: The First Eukaryotes 407

18.3 Plantae: The Green Mantle of Our World 413

18.4 Fungi: A World of Decomposers 419

18.5 Lichens and Mycorrhizae: Collaborations between Kingdoms 424

APPLYING WHAT WE LEARNED 426
The Root of the Problem: Why Rivers Meander

THE MIND-BOGGLING DIVERSITY OF LIFE began with the original prokaryotic ancestor, which arose about 3.8 billion years ago. We noted in Chapter 17 the amazing array of bacteria and archaeans that have descended from that universal ancestor and that now colonize nearly every imaginable habitat, from ocean depths to mountaintops. The evolution of the Eukarya, the third domain of life, was the next momentous event in the history of life.

In this chapter we explore three of the four kingdoms of Eukarya: Protista, Plantae, and Fungi (**FIGURE 18.1**). We will review the fourth eukaryotic kingdom—Animalia—in the next chapter. The catchall kingdom of Protista is the first stop in our journey through the world of Eukarya. Then we survey the kingdoms Plantae and Fungi, which appear at nearly the same time in the fossil record, working sometimes as friends and sometimes as foes to transform the land. We begin by describing how the eukaryotes arose and what sets them apart from prokaryotes.

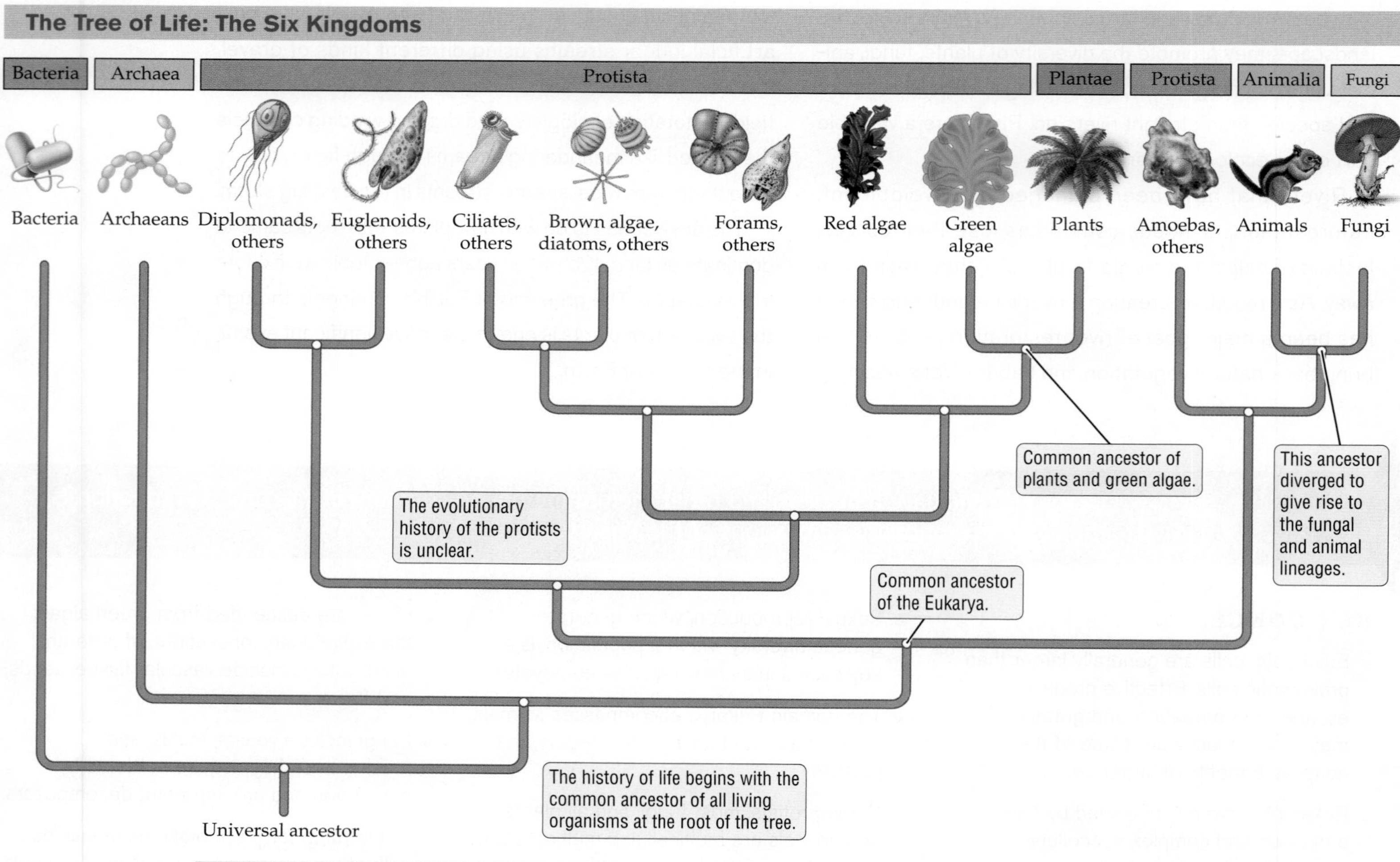

FIGURE 18.1 The Six Kingdoms of Life
This evolutionary tree shows the hypothesized relationships among the six kingdoms of life (Bacteria and Archaea are domains as well as kingdoms). As fellow eukaryotes, we share certain characteristics with plants. What are those shared traits? Are we more closely related to plants or to fungi?

18.1 The Dawn of Eukarya

If you study the chart "Milestones in the History of Life on Earth," which is printed inside the back cover of this book, you will notice that prokaryotes dominated the scene for about a billion years after life first appeared. On the basis of chemical clues from ancient rocks, some scientists suggest that eukaryotes—organisms with a nucleus—had evolved by 2.7 billion years ago. But the oldest fossil that is indisputably a eukaryote comes from iron-rich mud shale that formed 2.1 billion years ago. The fossil *Grypania spiralis* (**FIGURE 18.2**) is similar to modern-day red algae that form long ribbons of relatively large cells.

The evolution of Eukarya was a major milestone in the history of life, not only because this domain spawned organisms like us, but also because eukaryotes represent new ways of organizing cell structure and novel strategies for propagating life. These are the key evolutionary innovations of eukaryotes:

- Larger cell size compared to prokaryotes
- The presence of many membrane-enclosed internal compartments, including the nucleus
- Sexual reproduction
- Multicellularity (in some groups)

FIGURE 18.2 Fossil Eukaryote
Grypania spiralis, the oldest fossil eukaryote currently known, dates to 2.1 billion years ago. It resembles certain modern-day multicellular red algae.

Eukaryotic cells are larger than prokaryotic cells

On average, eukaryotic cells are 10 times wider than prokaryotic cells, and their cell volume is a thousandfold greater. The great majority of prokaryotes measure about 1 micrometer or less in width, and generally speaking, prokaryotes cannot afford to get much larger than that. Why?

The answer is that as cell diameter increases, cell volume increases more dramatically than surface area does. Surface area increases as a squared function, but volume increases as a cubic function (see Figure 3.5). A bigger cell has a larger volume of cytoplasm, so its metabolic capacity is greater. But a larger cell also needs more nutrients and produces more waste, both of which must enter and leave across the cell surface. For prokaryotes, which live in a highly competitive environment, a relatively large surface area is more valuable than a relatively large cell volume.

In the right ecological context, however, being big can be advantageous. Within limits, a larger cell volume translates into higher metabolic capacity for a cell, meaning that the cell can acquire and store more food. Being larger is often an advantage for predatory organisms, heterotrophs that live off other organisms. The amoeba in **FIGURE 18.3** can readily engulf whole bacterial cells and even other eukaryotes, such as single-celled algae.

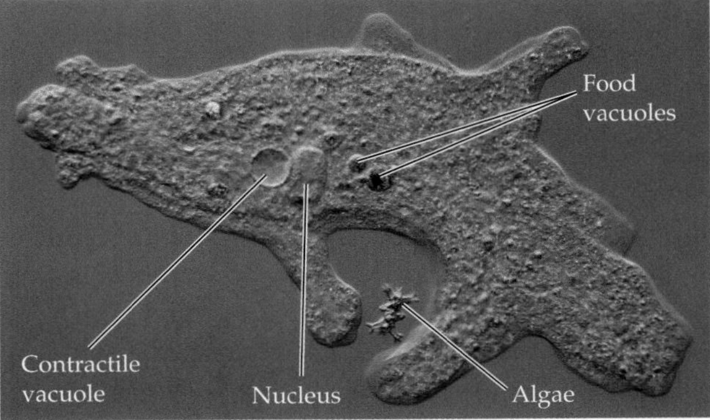

FIGURE 18.3 Internal Compartments in an Amoeba
This live image shows *Amoeba proteus* ingesting desmids, a type of single-celled algae. Intracellular compartments, such as food vacuoles, enable the amoeba to digest the ingested food. The amoeba expels excess water with the help of its contractile vacuoles, another example of intracellular compartments at work.

FIGURE 18.4 Internal Compartments in an Alga

Euglena gracilis, seen in a colorized TEM image here, is a relatively large single-celled alga. The protist uses a long, whiplike structure (the flagellum) to swim about. The flagellum is not visible in this color-enhanced electron microscope photograph. The reservoir is a membrane-lined pocket in which the flagellum is anchored.

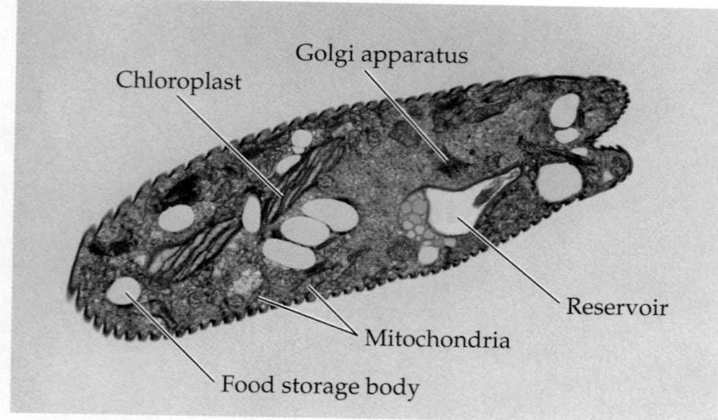

Compartmentalization of the cell interior also enabled the evolution of novel cell functions not seen in prokaryotes. Consider, for example, that unlike prokaryotes, many heterotrophic eukaryotes can engulf their prey and digest them internally, as the amoeba in Figure 18.3 is doing. These eukaryotes possess an elaborate system of membrane-enclosed organelles for digesting engulfed prey, ridding the cell of waste, and storing surplus food. Prokaryotes, in contrast, digest their prey externally—a more wasteful way of obtaining nutrients because some are lost to the surrounding environment.

Being large can also benefit potential prey species. The *Euglena* in **FIGURE 18.4** is too big a meal for the amoeba, for example. Incidentally, the researchers who discovered *Epulopiscium*, the giant fish-gut bacterium (see Figure 17.6), hypothesize that large size protects this unusual prokaryote from being eaten by other prokaryotic and eukaryotic residents that share its home in the intestines of the surgeonfish.

Eukaryotic cells have many membrane-enclosed internal compartments

Compared to prokaryotes, eukaryotes have a great variety of membrane-enclosed cytoplasmic compartments, or subcellular compartments. Some prokaryotes do have cytoplasmic compartments, including spherical structures, called vesicles, that store nutrients and other substances. However, as illustrated by the amoeba in Figure 18.3 and the alga *Euglena* in Figure 18.4, eukaryotic cells display much greater complexity in subcellular compartmentalization. Each type of internal compartment in *Euglena* specializes in conducting a unique set of functions. Through specialization and division of labor, these subcellular compartments can function with greater efficiency.

The largest subcellular compartment of most eukaryotic cells is also the defining feature of the Eukarya: the *nucleus*. Instead of lying free in the cytoplasm, eukaryotic chromosomes are enclosed in two concentric layers of cell membranes that together make up the *nuclear envelope*. The membranes isolate nuclear processes from the quite different cytoplasmic activities, and that separation makes both types of cellular functions more efficient.

Helpful to Know

In general, larger organisms have an advantage. But sometimes being small and fast is a better adaptation. As an analogy, think of the difference between a short running back and a tall receiver on a football team. Both can successfully help move the ball downfield. It's the same with evolution and biodiversity—there's more than one way to be successful.

Sexual reproduction increases genetic diversity

Perhaps the most outstanding evolutionary innovation of the Eukarya is sexual reproduction, the fusion of gametes to produce offspring that are genetically different from each other and from both parents. It is one means by which natural populations become genetically diverse. As we noted in Chapter 1, genetic variation is the raw material for evolution by natural selection. If the environment changes, a genetically diverse population is more likely to evolve adaptively than is a genetically uniform population.

In many algae, most plants, and even some animals, a single individual—known as a *hermaphrodite*—can produce both male and female gametes. For example, in most plants a single flower produces both eggs and sperm. Most hermaphroditic species have mechanisms that prevent self-breeding, ensuring that an individual's male gametes fertilize another individual's female gametes, not its own. **FIGURE 18.5** illustrates the roles of asexual and sexual reproduction in the life cycle of a brown alga known as rockweed. The rockweed life cycle resembles the way sexual reproduction works in animals, but protists actually display many variations on the basic theme of sexual reproduction that is depicted in Figure 18.5.

Asexual reproduction, which generates genetically identical offspring, is also common among the Eukarya. As with sexual reproduction, protists exhibit a variety of asexual reproduction methods. *Euglena* and its relatives can produce *clones*, genetically identical offspring, by partitioning a single cell into many minicells, and then releasing the minicells as free-swimming offspring called *zoospores*. Protists like *Amoeba* split into two in a process similar to binary fission in prokaryotes (see Chapter 17). Many large multicellular algae ("seaweeds") can fragment into pieces, each piece developing into a new individual the way cuttings from some plants can grow into whole new plants (see Figure 18.5).

Multicellularity evolved independently in several eukaryotic lineages

Most protists are single-celled, but multicellular forms evolved several times among different lineages of the eukaryotes, including some groups that are currently lumped under Protista. The fungi include some single-celled species and many multicellular forms. Plants and animals are both exclusively multicellular.

Multicellular organisms enjoy the benefits of cell specialization—namely, more efficient functioning through division of labor. Multicellularity also enables the individual organism to grow large, which can be advantageous for evading potential predators. Although there are many exceptions, a bigger individual can often gather resources from its environment more effectively than can a smaller individual. Having more resources, such as light or food, usually translates into producing more surviving offspring, the ultimate measure of biological success.

The giant kelp, which can grow to 60 meters, illustrates the adaptive benefits of multicellular organization (**FIGURE 18.6**). Cells at the base form a specialized tissue, the holdfast, that anchors the giant seaweed and keeps it from being washed out to sea. Seaweed blades are broad and flat, the better to capture light for photosynthesis.

Loosely arranged cells in the seaweed body form swollen bladders (technically called *pneumatocysts*) that trap air and give buoyancy to the kelp. Without these flotation devices, the kelp's large body would collapse to the seafloor, where it would die from lack of sunlight. Being large presents a challenge: how to quickly deliver food from one part of the body to another. Food-conducting tubes in the interior of this brown alga enable transport of photosynthetic sugars from the blades to all other parts of the kelp.

Concept Check

1. Name at least three evolutionary innovations of the Eukarya.
2. What is the adaptive value of membrane-enclosed subcellular compartments?
3. What are some advantages of multicellularity?

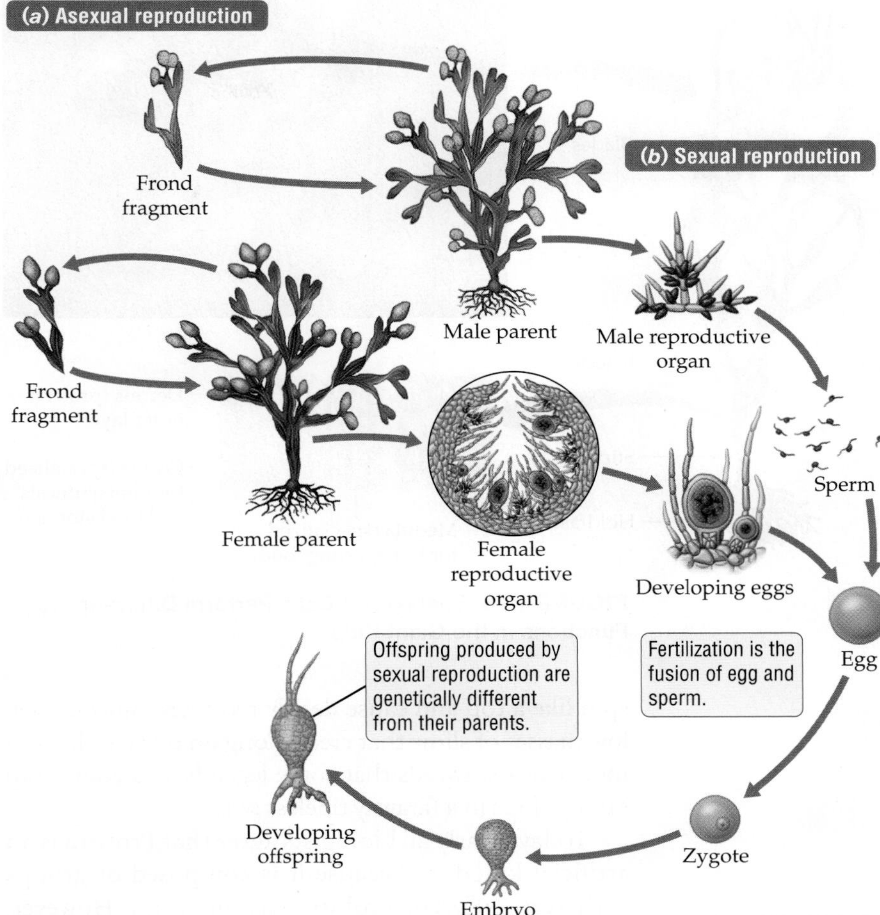

FIGURE 18.5 Asexual and Sexual Reproduction in a Eukaryote

Some eukaryotes, such as the rockweed illustrated here, can clone themselves through a type of asexual reproduction known as fragmentation. However, sexual reproduction is one of the outstanding evolutionary innovations of the Eukarya. The rockweed, a brown alga found on many rocky shores, also reproduces sexually. It produces eggs and sperm that unite to create offspring, in a life cycle that resembles that of animals. Other seaweeds have different, and more complex, life cycles.

18.2 Protista: The First Eukaryotes

The kingdom **Protista** is an artificial grouping, defined only by what members of this group are *not*: protists are not plants, animals, or fungi, nor are they bacteria or archaeans. The kingdom was first proposed by nineteenth-century biologists attempting to deal with puzzling and poorly understood forms of life—species as disparate as amoebas that make people sick, dinoflagellates that

Concept Check Answers

1. Among the novel features of eukaryotes are larger cell size and complex subcellular compartments, as well as sexual reproduction and multicellularity in some forms.
2. Greater efficiency through specialization and division of labor among subcellular compartments, and novel cellular functions such as intracellular digestion of food.
3. Just as subcellular compartments enable individual cells to divide labor, multicellularity enables individual organisms to divide labor among different cell types. Multicellularity also enables increased organism size, which can be an advantage in avoiding predators and collecting resources.

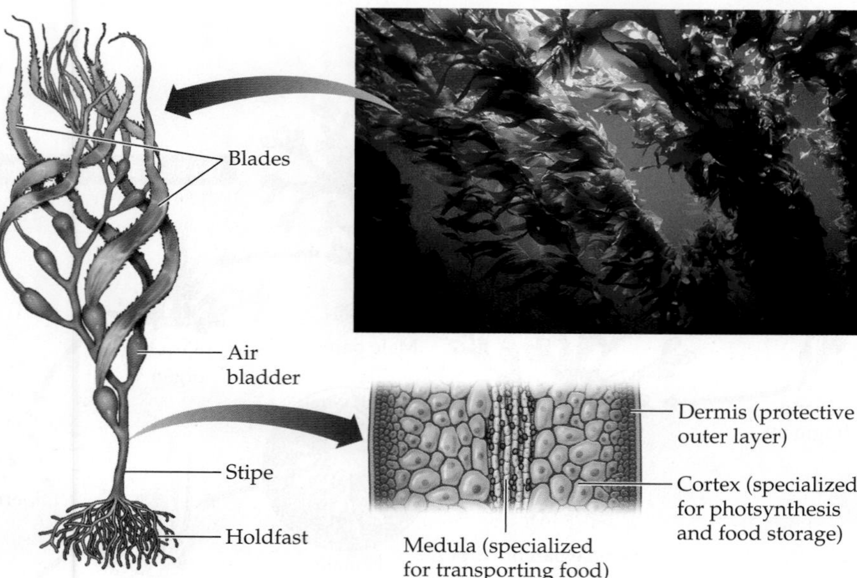

FIGURE 18.6 Specialized Cells Perform Different Functions in the Giant Kelp

spin like a top and cause deadly red tides, mustard-yellow masses of slime that creep along on tree trunks, and monstrous seaweeds that some legends said could trap sailing ships in a faraway tideless sea.

Today, nearly all biologists agree that Protista is an artificial kingdom because it is composed of groups with very different evolutionary histories. However, no consensus has been reached yet on how many new kingdoms should be carved out of Protista, which groups should be placed in them, and what the new kingdoms should be called. We therefore retain "Protista" as a label of convenience. For our purposes, Protista is a catchall category of eukaryotic organisms that have not been formally assigned to other kingdoms or to separate kingdoms of their own.

The evolutionary history of protists is unclear

Much remains unclear about the evolutionary relationships of the protists to one another and to other living organisms. As a result, biologists have proposed a number of competing hypotheses and different evolutionary trees for the protists. The Protista section of the evolutionary tree in Figure 18.1 shows a recent interpretation of the main protist lineages. This scheme is based on many different lines of evidence, including comparisons of cell structure, metabolic chemistry, and the DNA code; and it makes clear that protists do not make up just one distinct branch (clade) on the tree of life.

Some groups traditionally placed in Protista are actually more closely related to plants or animals than they are to other protists. For example, red and green algae share a most recent common ancestor with land plants, and these three groups therefore form one distinct branch on the evolutionary tree. The lineages that gave rise to animals and fungi diverged from each other more recently than the lineages that gave rise to animals, fungi, *and* amoebozoans (which include amoebas and slime molds). Because they shared a common ancestor more recently, animals and fungi are closer to each other evolutionarily than either of them is to amoebas and slime molds.

The major groups of protists have traditionally been imagined as falling into two broad categories: the **protozoans**, which are nonphotosynthetic and motile (capable of moving); and the **algae** (singular "alga"), which are photosynthetic and may or may not be motile. The evolutionary tree in Figure 18.1 reveals that the grouping into protozoans and algae is also artificial and is evolutionarily meaningless. For example, species of *Euglena* are often green and photosynthetic, but this group is about as distantly related to green algae and plants as it is to amoebas and slime molds. Instead, *Euglena* is more closely related to diplomonads such as *Giardia*, a colorless, single-celled parasite that lives in animal guts and causes painful diarrhea, as campers who drink untreated water from contaminated streams know all too well. In other words, *Euglena* and *Giardia* are in a distinct clade that is neither plantlike nor animal-like; they belong in a group of their own and are likely to gain their own kingdom when the dust finally settles on the reorganization of the protists.

Most protists are single-celled and microscopic

In part because they are not a single evolutionary lineage, protists are diverse in size, shape, cellular organization, modes of nutrition, and life cycle. Most protists are single-celled and microscopic, and these forms are often pigeonholed with prokaryotes under the tag of "microbes." Most microbial protists are motile. They can swim with the help of one or more *flagella* (singular "flagellum"), or by waving a carpet of tiny hairs called *cilia* (singular "cilium"). Some microbial protists, like the amoeba in Figure 18.3, crawl on a solid surface with the help of cellular projections called *pseudopodia* (literally, "false feet"; singular "pseudopodium").

EXTREME ALGA

Some algae are comfortable in hot springs as corrosive as battery acid (pH 0.8). The photo shows mats of *Cyanidinium caldarum* (green) on Whirligig Geyser (135°F) in Yellowstone National Park.

Some protists, especially the ones that live as parasites within the bodies of multicellular organisms, are single cells bounded by nothing more than a flexible plasma membrane. Others are covered in protective sheets, heavy coats, chalky plates, or other types of armor. Diatoms, for example, are renowned for the exquisite beauty of their glassy cell walls containing silicon dioxide (**FIGURE 18.7**). When these organisms die, their cell walls can create thick sediments on the ocean floor.

The white color of the famed white cliffs of Dover, on the southern shores of England, comes from the casings of fossil coccolithophores, which are related to diatoms. The intricately patterned armor plates of coccolithophores are made of calcium carbonate, or chalk (**FIGURE 18.8**).

Red, green, and brown seaweeds are protists with multicellular bodies. Some groups have evolved from free-living single cells into multicellular associations that seem to function remarkably like a single individual. Among the more interesting of these multicellular-like associations are the slime molds, protists that were originally mistaken for fungi. Commonly found on rotting vegetation, slime molds are protists that eat bacteria and live their lives in two phases: as independent, single-celled creatures, and as members of a multicellular association (**FIGURE 18.9**).

Protists are autotrophs, heterotrophs, or mixotrophs

The protists known as algae play a vital role as producers, especially in oceans, lakes, rivers, and streams (see Figure 18.9). Algae are autotrophic protists that carry out oxygen-generating photosynthesis. Roughly half of the photosynthesis that occurs on Earth takes place in the oceans, and algae, together with photosynthetic bacteria, are responsible for nearly all of that activity. In other words, algae and photosynthetic bacteria carry out about as much photosynthesis in the world's oceans as all the crops, forests, and other types of vegetation do on land.

Helpful to Know

"Algae" is an everyday term for any photosynthetic eukaryote other than a plant. "Algae" is not evolutionarily meaningful, because it lumps together protists as different as *Euglena* (which can live as a heterotroph) and green algae (closely related to land plants). Large, multicellular algae—which are broadly classified as brown, red, or green algae—are commonly called "seaweeds."

Seaweeds create food-rich habitats in coastal areas, but much of the photosynthesis in open water–both freshwater and salt water–is carried out by free-floating, single-celled algae called **phytoplankton**. **Plankton** (from the Greek *planktos*, "drifting") is a general term for microbes that drift at or near the surface of water bodies. Diatoms and coccolithophores, described earlier, are some of the most abundant marine phytoplankton.

Heterotrophic prokaryotes and protists, together with microscopic animals, form **zooplankton**.

FIGURE 18.7 Variety in Diatom Shapes
Diatoms are common phytoplankton in both fresh and salt water. Their silica-containing cell walls are usually transparent and highly ornamented. Different species vary in size, shape, and structure.

FIGURE 18.8 Coccolithophores Are Phytoplankton with Chalky Shells
The protective shells of coccolithophores dissolve in acidic environments, which is why the acidification of the oceans, caused by carbon dioxide spewed by human activities, is a serious threat to these photosynthetic protists.

Life Strategies

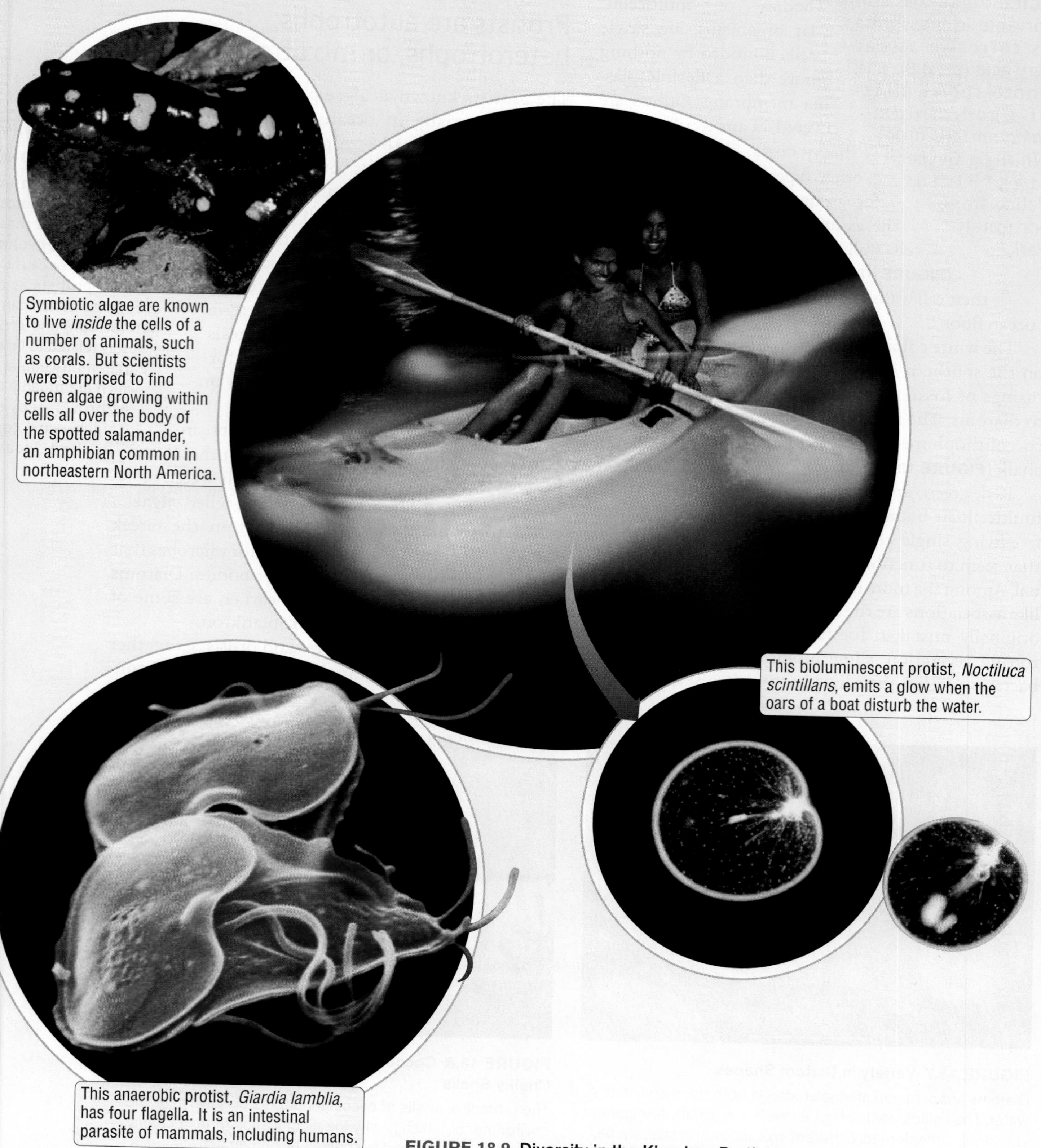

Symbiotic algae are known to live *inside* the cells of a number of animals, such as corals. But scientists were surprised to find green algae growing within cells all over the body of the spotted salamander, an amphibian common in northeastern North America.
This bioluminescent protist, *Noctiluca scintillans*, emits a glow when the oars of a boat disturb the water.
This anaerobic protist, *Giardia lamblia*, has four flagella. It is an intestinal parasite of mammals, including humans.

FIGURE 18.9 Diversity in the Kingdom Protista

Structure

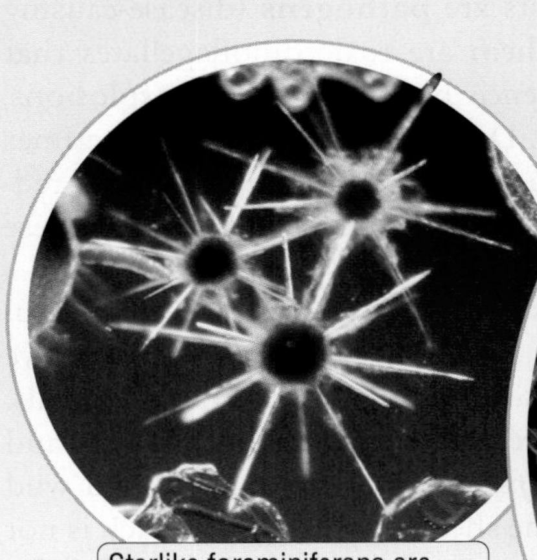

Starlike foraminiferans are seen here with other plankton.

This multicellular red alga, *Antithamnion plumula*, has special cells that attach it to coastal rocks. The small, clublike side branches are reproductive organs.

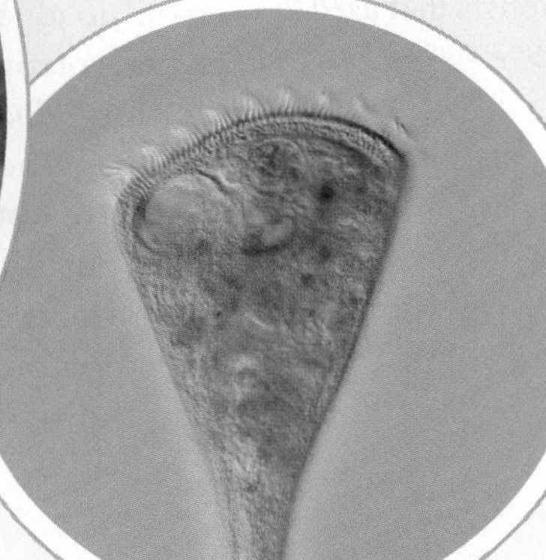

Stentor coeruleus is a ciliated protozoan common in freshwater worldwide. Up to 2 mm in length, it is one of the largest single-celled organisms.

Energy Acquisition

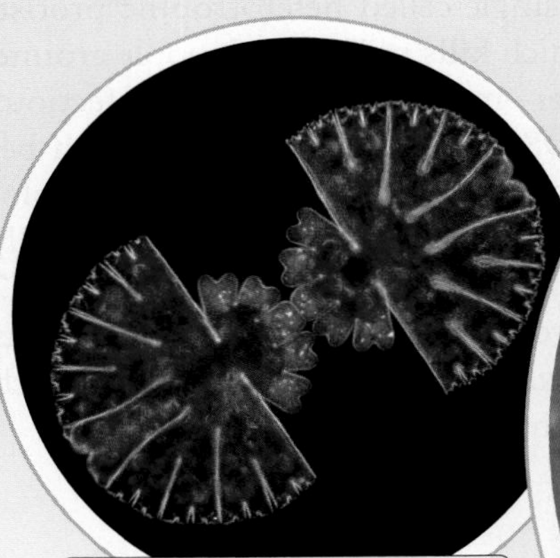

Photosynthesis in *Micrasterias thomasiana*, a single-celled desmid (a type of alga), is enabled by its many chloroplasts (green). One cell has divided to produce the two cells seen here.

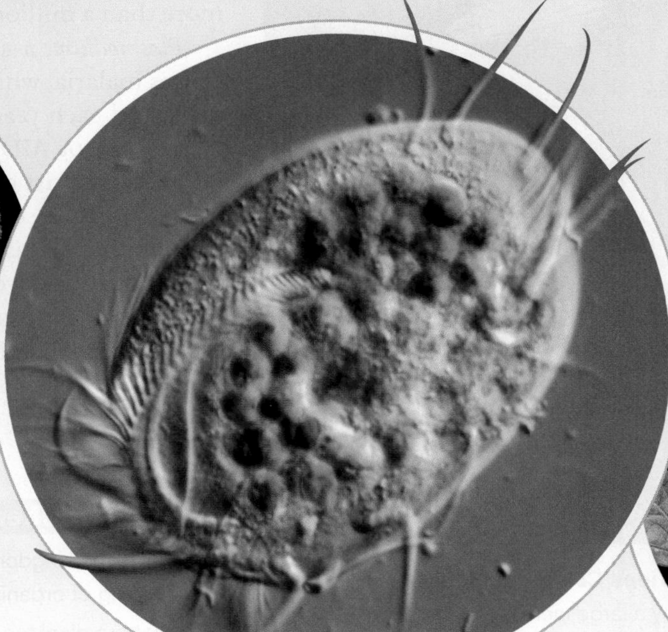

Ingestive heterotrophy in *Euplotes*. This ciliated protist has engulfed more than a dozen algal cells (green) that are at various stages of degradation.

Absorptive heterotrophy is how water molds (oomycetes) obtain energy. A water mold called *Phytophthora* destroyed the potato crop and contributed to the Great Famine in Ireland in the mid-1800s.

Helpful to Know

Diatomaceous earth, used in swimming pool filters, is mined from deposits of ancient diatoms. If you examine scrapings from diatomaceous earth under a microscope, you will see a kaleidoscope of glassy shapes, the fossil remains of protists that lived millions of years ago.

Foraminiferans (or, more commonly, "forams") and radiolarians are among the important planktonic protists in the ocean, occurring in such vast numbers that their remains are a major component of seafloor sediments. Many heterotrophic protists function as decomposers in terrestrial or aquatic ecosystems.

Some protists are nutritional opportunists, or **mixotrophs**, organisms that can use energy and carbon from a variety of sources to fuel their growth and reproduction. Mixotrophic algae function as photoautotrophs *or* as heterotrophs, depending on environmental conditions. *Euglena* and some of its relatives are mixotrophs. When light and mineral nutrients such as nitrogen are abundant, they may live photoautotrophically; but if the resources become scarce, these versatile protists switch to engulfing prey organisms such as bacteria or absorbing organic molecules from their surroundings.

FIGURE 18.10 Red Tides Can Close Beaches

Red tides are caused by a large increase in the population of pigmented protists such as dinoflagellates. Some dinoflagellates, such as the *Gymnodinium* (*inset*), produce nerve toxins that can poison mammals.

Some protists are pathogens

Although most protists are harmless, many of the best-known protists are **pathogens** (disease-causing agents). Among them are toxic dinoflagellates that sometimes experience huge population explosions, known as blooms. Dinoflagellate blooms sometimes cause the water to turn red ("red tides") from a reddish pigment concentrated inside the cells of these protists (**FIGURE 18.10**).

Dinoflagellates can produce a variety of harmful chemicals (toxins), including some that cause nerve and muscle paralysis in humans and other mammals. Some of these toxins accumulate in shellfish and cause *paralytic shellfish poisoning* in humans and wild animals who eat the shellfish. The toxin, which is not destroyed by cooking, can remain in the shellfish for weeks and even for a year or more in some species of clams. Scientists do not have a full understanding of what causes the sudden blooms, but the frequency of such blooms has been rising around the world, and pollution from fertilizer runoff and sewage are thought to be among the culprits.

Protists left their mark on human history forever when a water mold (an oomycete, mistakenly referred to as a fungus sometimes) attacked potato crops in Ireland in the 1800s, causing the disease known as potato blight. The resulting widespread loss of potato crops caused a devastating famine and a major emigration of Irish people to the United States (among other countries) in the 1840s. It is estimated that more than a million people died of starvation.

Plasmodium, a single-celled heterotrophic protist, causes malaria, which kills millions of people around the world each year—more than any other infectious disease except AIDS. *Trichomonas vaginalis*, which belongs to a protist group most closely related to diplomonads, causes one of the most common sexually transmitted diseases in the United States. About 7 million men and women have the infection, known as trichomoniasis (*TRIH*-**kuh-muh**-*NYE*-**uh-sus**), which is readily cured with medications.

Concept Check

1. Why is the kingdom Protista described as an artificial grouping of organisms?
2. What are plankton, and what is their ecological significance?

Concept Check Answers

1. Rather than forming one distinct branch on the tree of life, Protista includes groups that are descended from many different lineages and that are more closely related to organisms in other kingdoms of life, such as plants, animals, or fungi, than to other protists.
2. Plankton are microscopic organisms that drift in the water in aquatic ecosystems. They include algae (phytoplankton) and eggs and larval stages of animals (zooplankton). They form the base of most aquatic food webs.

18.3 Plantae: The Green Mantle of Our World

Life on Earth began in the water, where it stayed for nearly 3 billion years. It was only when the kingdom **Plantae**—the plants—evolved that life took to land in a big way. The first plants appeared about 470 million years ago, having most likely evolved from a lineage of multicellular green algae that made the move to land and became successfully established in freshwater habitats. In colonizing the land, plants turned barren ground into a green paradise in which a whole new world of land-dwelling organisms, including humans, could then evolve.

Plants are multicellular autotrophs that are mostly terrestrial (land-dwelling). Like photosynthetic protists (algae), plants use chloroplasts in order to photosynthesize. Most photosynthesis in plants takes place in plants' leaves, which typically have a broad, flat surface—a design that maximizes light interception. Because plants are producers, they form the basis of essentially all food webs on land.

Plants reproduce both asexually and sexually. Although their life cycle is distinctly different from that of animals, plants are like animals in that they produce embryos: the fusion of egg and sperm produces a single cell, called a *zygote*, which then divides to produce a multicellular structure called an *embryo*.

Today the diversity of members of the kingdom Plantae ranges from the most ancient lineages—liverworts and mosses—to ferns, which evolved next; to gymnosperms; and finally, to the most recently evolved plant lineage, the angiosperms, or flowering plants (**FIGURE 18.11**).

Mosses, liverworts, and hornworts—informally known as **bryophytes**—were among the earliest land plants. These "amphibians of the plant world" still thrive in moist habitats throughout the world, and some can even withstand drying out or freezing during the non-growing season.

Gymnosperms are familiar to us as the conifers, or cone-bearing trees, such as pines and firs, that dominate in the colder regions of the world. Tropical gymnosperms, such as the palmlike cycads in Florida, are probably less familiar to people from nontropical regions. The *angiosperms*, or flowering plants, are familiar to everyone because we depend on them for food, clothing, building material, paper, medicines, and many other products.

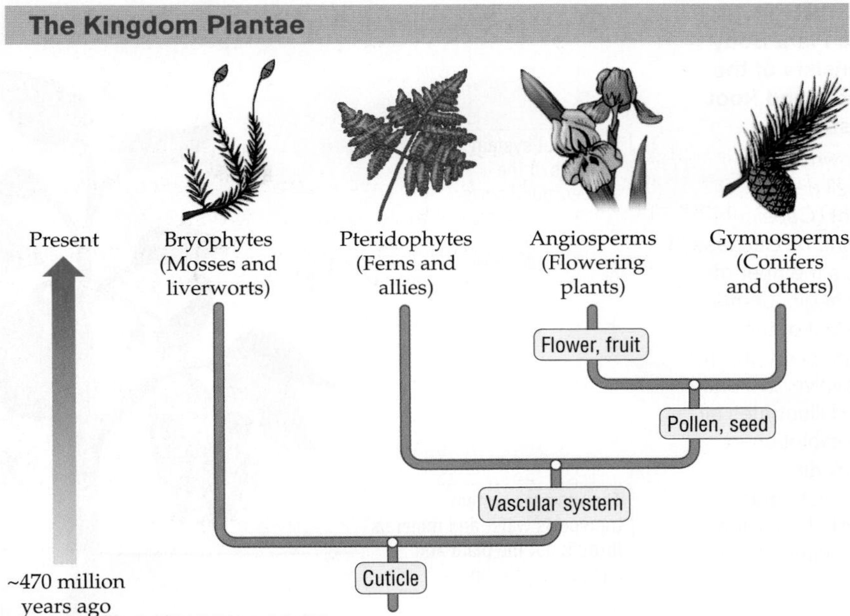

FIGURE 18.11
Evolutionary Tree of the Plantae

Plants had to adapt to life on land

Organisms on land had to adapt to challenges not faced by organisms living in water. The biggest challenge was how to obtain and conserve water. Plants have a waxy covering, known as the **cuticle**, that covers their aboveground parts (**FIGURE 18.12**). A waxy cuticle holds in moisture and thereby keeps plant tissues from drying out, even when exposed to sun and air all through the day. Plants that live in relatively dry climates tend to have exceptionally thick cuticles.

As described in Chapter 6, carbon dioxide in air enters leaf cells through many minute openings in the cuticle called **stomata** (singular "stoma"). Each stoma is bordered by a pair of **guard cells**, which can inflate or deflate like water balloons to open or close the stoma. Although they are essential for photosynthesis, stomata also put a plant at risk of drying out because they allow moisture-laden air to escape from leaves.

Lignin enabled plants to grow tall

The first plants, the ancestors of present-day liverworts and mosses, grew as ground-hugging carpets of greenery. To this day, reproductive structures—such as capsules raised on stiff stalks—are the only vertical structures that these groups produce. Like green algae, plant cells have strong but flexible cell walls composed of *cellulose* (see Section 2.7). Cellulose cell walls give structural

FIGURE 18.12 The Plant Body Consists of the Shoot and Root Systems

Shown here is a bell pepper plant (*Capsicum annuum*). Because it is a member of the angiosperms, the last of the major plant groups to evolve, this one plant illustrates all the evolutionary innovations that distinguish plants from other kingdoms of life.

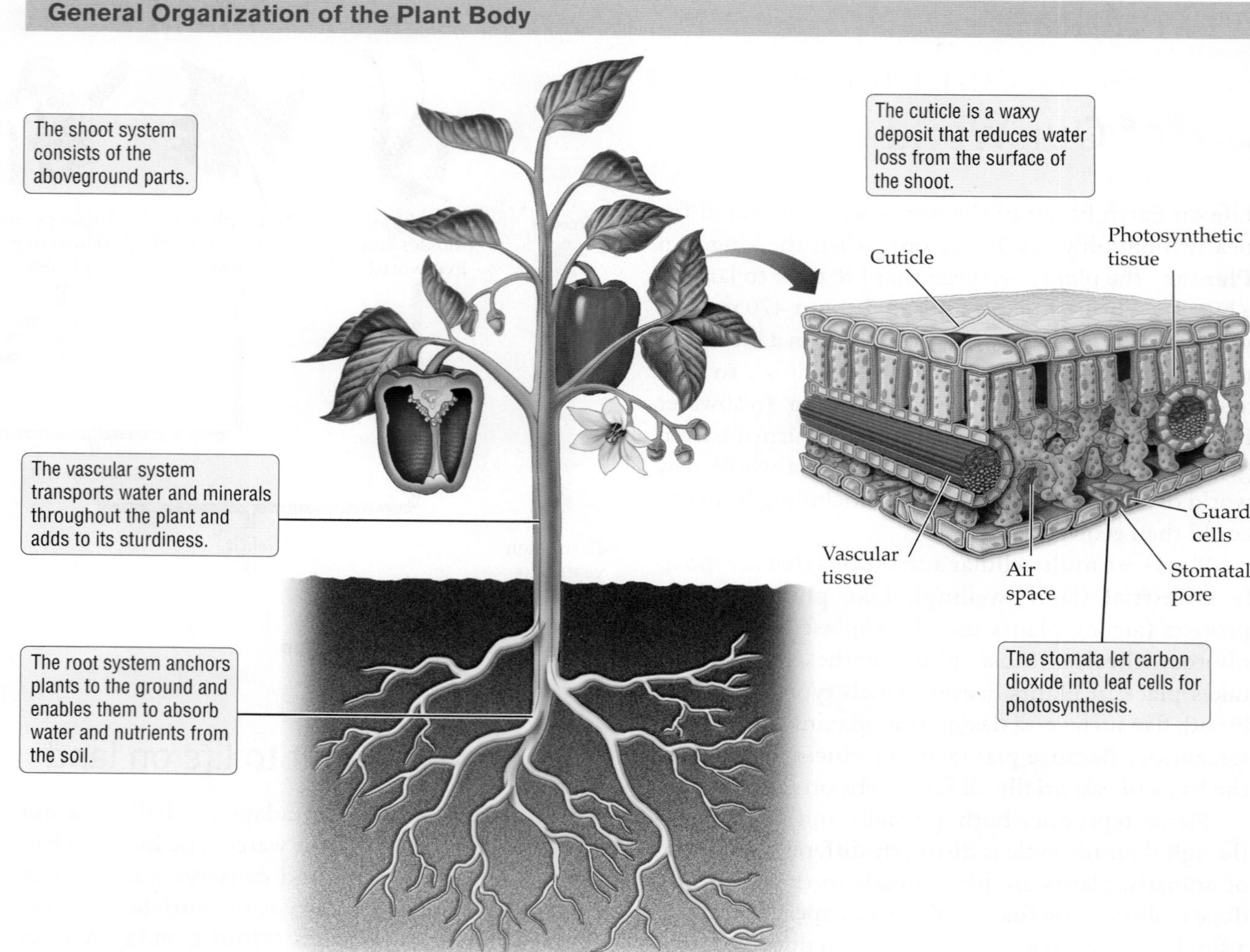

General Organization of the Plant Body
The shoot system consists of the aboveground parts.
The vascular system transports water and minerals throughout the plant and adds to its sturdiness.
The root system anchors plants to the ground and enables them to absorb water and nutrients from the soil.
The cuticle is a waxy deposit that reduces water loss from the surface of the shoot.
Cuticle
Photosynthetic tissue
Guard cells
Stomatal pore
Air space
Vascular tissue
The stomata let carbon dioxide into leaf cells for photosynthesis.

strength to all types of plant cells, including those of the low-growing mosses and liverworts.

The evolution of strengthening material enabled plants to hold their bodies upright, which gave them better access to sunlight. Plants other than mosses and liverworts stiffen themselves with *lignin*, one of the strongest materials in nature. **Lignin** is a polymer that links cellulose fibers in the cell wall of plants to create a rigid three-dimensional network. Wood is strong because it is made of cells whose cell walls are reinforced with lignin. Lignified tissues enabled plants to reach for the sky, like the 300-foot redwoods that soar into the ocean mists along the coast of northern California.

Growing tall poses its own set of challenges, however; one of these is how to raise fluids, such as water, from ground level to the crown of a bush or tree. As we discuss next, the evolution of lignin—in fernlike plants that lived about 425 million years ago—went hand in hand with the evolution of good plumbing.

The vascular system enables plants to move fluids efficiently

Bryophytes have relatively thin bodies, often just a few cells thick, and sprawling on wet surfaces enables these plants to absorb water through a wicking action. Many bryophytes have tufts of threadlike cells (*rhizoids*) on their lower surface that grow into the soil and soak up water a few centimeters deeper in the ground. These simple strategies are effective in delivering water, and the mineral nutrients dissolved in it, to a relatively thin plant body that is close to the ground. However, absorption by direct contact cannot transport fluids effectively in a plant that rises a foot (0.3 meter) or more aboveground.

About 425 million years ago, plants evolved a network of tissues, called the **vascular system**, that includes tubelike structures specialized for transporting fluids. Vascular tissues that specialize in transporting food molecules, such as sugars, are called **phloem** (*FLOH*-**em**). Vascular tissues that specialize in transporting water and dissolved nutrients are known as **xylem** (*ZYE*-**lem**). Xylem and phloem are usually bundled together in branching strands that snake throughout the plant body in a way similar to our own network of blood vessels.

The inside of a tree trunk is taken up mostly by bundles of xylem—which you recognize as wood—with phloem in a ring under the corky layers that make up the bark. The water-conducting tubes of xylem are reinforced with lignin, which is why wood is strong enough for building things.

Roots—the water-absorbing organ system found in all plants except bryophytes—also have an extensive vascular system. Root xylem brings water from the soil to the aboveground parts of the plant, and root phloem delivers sugars made in the leaves to the nonphotosynthetic tissues below ground.

The evolution of seeds contributed to the success of gymnosperms

Gymnosperms (*JIM*-**noh-sperm**) evolved about 365 million years ago. Conifers (cone-bearing plants) are the most diverse and abundant gymnosperms today: spruce, fir, pine, and larch are the dominant vegetation in large swaths of the northern lands, including much of Canada, northern Europe, and Siberia.

Gymnosperms were the first plants to produce **pollen**, a microscopic structure that contains and protects sperm cells (**FIGURE 18.13**). All plant lineages that appeared before the gymnosperms produce flagellated sperm, which need at least a film of water in order to swim to a nearby egg cell. Pollen, in contrast, is dry and powdery and can be lofted into the air in massive quantities. The evolution of pollen freed the gymnosperms and their close cousins, the angiosperms, from a dependence on water for fertilization.

Gymnosperms were also the first plants to develop the **seed**, which consists of the plant embryo and a supply of stored food, encased in a protective covering called

EXTREME FERN

The resurrection fern (*Pleopeltis polypodioides*) loses as much as 97 percent of its water under drought conditions. But sprinkling some water on the dead-looking growth on this oak tree in Florida yields a lush bed of ferns within a few hours. In 1997, a dried-up specimen shipped out on the space shuttle *Discovery*, and astronauts watched its resurrection in zero gravity.

The Life Cycle of a Gymnosperm

Scales
Female cone
Mature tree
Fertilization
Seed coat
Male cone
A cone is a cluster of modified leaves (scales) bearing male or female reproductive organs.
Pollen contains sperm cells.
The seed contains an embryo, along with a food supply.
Food supply
Embryo
Seedling

FIGURE 18.13
Gymnosperms Were the First Plants to Produce Pollen and Seeds

FIGURE 18.14
The Flower
A flower is really a meeting place for plant gametes. Pollen contains sperm cells. The egg cells are held inside roughly oval structures, called ovules, that turn into seeds after fertilization.

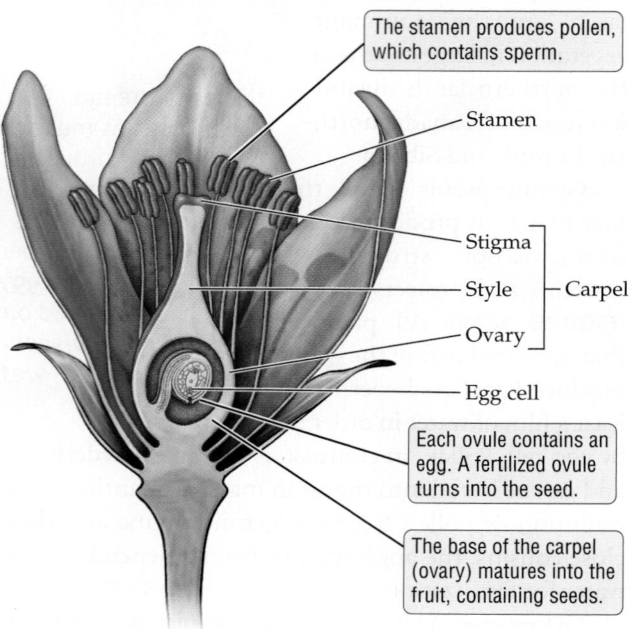

The Reproductive Structures of an Angiosperm
The stamen produces pollen, which contains sperm.
Stamen
Stigma
Style
Ovary
Carpel
Egg cell
Each ovule contains an egg. A fertilized ovule turns into the seed.
The base of the carpel (ovary) matures into the fruit, containing seeds.

the seed coat (see Figure 18.13). Seeds could be disseminated farther away, so they would not be in competition with the mother plant for sunlight or for water and mineral nutrients in the soil. Pine, fir, spruce, and hemlock produce winged seeds that ride wind currents when they break out of the cones and drift away from the mother tree. Seeds contain stored food that plant embryos use to grow before they are able to make their own via photosynthesis. Seeds also provide embryos with protection from drying or rotting, and from attack by predators.

Helpful to Know

What is the difference between a fruit and a veggie? A fruit is the mature *ovary* of an angiosperm and normally contains seeds. "Vegetable" is not a scientific term. In everyday life, people tend to use the word "vegetables" to refer to plant structures that do not taste sweet. This popular usage includes storage roots such as carrots and radishes, plant leaves such as lettuce and spinach leaves, but also some true fruits—such as tomatoes and cucumbers—that happen to be nonsweet.

Angiosperms produce flowers

The flowering plants, or **angiosperms** (*AN*-**jee-oh-sperm**), are a relatively recent development in the history of life. The first flowering plants evolved about 145 million years ago, shortly after another upstart group, the mammals, appeared on the Mesozoic scene. Today, with about 250,000 species, angiosperms are the most dominant and diverse group of plants on our planet. Angiosperms include orchids, grasses, corn plants, and trees such as apple and maple. Highly diverse in size and shape, angiosperms live in a wide range of habitats—from mountaintops to deserts to salt marshes and freshwater habitats.

The crucial evolutionary innovation of angiosperms, and the key to their success, is the **flower**, a structure that evolved through modification of the conelike reproductive organs of gymnosperms. Flowers are structures that enhance sexual reproduction in angiosperms by bringing sperm to the egg in highly efficient ways. In the most dramatic example, animals are enlisted to carry pollen from flower to flower.

The evolution of flowers has been shaped by the availability of specific animal pollinators, especially insects, and the evolution of the pollinators has likewise been driven by the structure and function of the flowers. The lockstep evolution of two groups—with the evolution of one influencing and being influenced by the other—is called *coevolution* (see Chapter 23). The coevolution of flowering plants and their pollinators is one of the great stories in the evolutionary history of life. Most flowers are hermaphroditic in that the male and female structures, known as the stamen and carpel, respectively, are present in the same flower. Stamens consist of sacs, often borne on long stalks, that produce pollen (**FIGURE 18.14**). Pollen grains are most familiar as the dustlike particles that some plants release into the air and that can cause allergies in some people. The receipt of pollen by the carpel is known as *pollination*.

Pollen grains germinate on the receptive surface of the carpel (the *stigma*), producing a long extension called the *pollen tube*. Sperm cells travel in the growing pollen tube to the egg cells inside the base of the carpel (the *ovary*), where fertilization takes place. As in gymnosperms, the embryo of angiosperms is enclosed in protective layers, which collectively form the seed.

Some angiosperms are like the overwhelming majority of gymnosperms in that they rely on wind currents for pollination. Angiosperms that depend on wind pollination—all grasses, and trees such as birch and oak—tend to have small, dull-colored flowers that spew prodigious quantities of tiny pollen grains into the air. In contrast, angiosperms that produce brightly colored or strongly scented flowers, usually with a nectar reward, depend on animals such as insects to transport their pollen. Although animal-pollinated plants do not have to invest energy in producing massive amounts of tiny pollen, they must devote a good deal of energy to making special products—bright petals, odors, and the sugary liquid known as nectar—to lure their pollinators.

Pollen delivery by animals is highly effective because many pollinators specialize in the plant species they visit and therefore tend to carry pollen to the carpel of the same species, instead of wasting it on nontarget species. Wind pollination works best in species that grow in thick, near-pure stands, but animal-pollinated species can afford to be scattered across the landscape.

Angiosperms produce fruit

In angiosperms, the ovary of the flower encloses the egg-bearing structures, or *ovules*, and protects them as they develop. After fertilization, the ovules develop into seeds, and the ovary wall that enclosed them becomes the fruit wall. A **fruit**, therefore, is a mature ovary with seeds inside it.

What is the adaptive value of the fruit, which encloses the seeds of angiosperms? In addition to protecting immature seeds from would-be seed predators, the angiosperm fruit wall is a seed dispersal device. In some species, the fruit wall dries out and flings the seeds from the spent carpel with considerable force. For example, the fruit of the squirting cucumber, *Ecballium elaterium*, explodes when it becomes ripe and discharges its seeds a distance of 3–6 meters (10–20 feet) in a stream of mucilage. The fruit wall that surrounds the coconut has a leathery jacket that resists salt water, as well as fibrous, air-filled layers that make the heavy seed buoyant and act as a flotation device (**FIGURE 18.15**), permitting seeds to travel in the ocean to colonize new habitats. In maple trees, the fruit has flattened wings that slow its descent because the wings generate lift as they spin like helicopter blades; a slower descent increases the likelihood that the twirling fruit will drift away on a wind current, instead of falling straight to the ground.

The most effective seed dispersal strategies are those that employ animals as a delivery service. Some fruit walls, such as those of burweeds, have sharp hooks or Velcro-like barbs that enable the fruits to hitch a ride on animal fur. Fleshy fruits become palatable when they ripen, after the seeds inside have matured and developed a tough, protective seed coat. Animals that eat fruit often excrete the seeds whole in their feces, well away from the mother plant. The nutrient-rich wastes provide a good place for excreted seeds to begin their new life.

FIGURE 18.15 Plants Get Around in a Variety of Ways
Plants have evolved many ways of spreading to new areas. The seed inside this coconut fruit can float for hundreds of miles until it reaches a new beach where it can take root and grow.

Plants are the basis of land ecosystems and provide many valuable products

It is difficult to overstate the significance of plants, which are major producers. Nearly all organisms on land ultimately depend on plants for food, either directly by eating plants or indirectly by eating other organisms (such as animals) that eat plants. Many organisms live on or in plants, or in soils largely made up of decomposed plants. Aquatic plants, such as the water lilies and duckweed in **FIGURE 18.16**, provide food and shelter for organisms that range from bacteria to adult animals and their larvae.

Humans have used flowering plants in a variety of ways, ranging from sources of materials such as cotton for clothing to pharmaceuticals such as morphine. Nearly all agricultural crops are flowering plants, and the entire floral industry rests on the reproductive structures of angiosperms. Gymnosperms such as pine, spruce, and fir are the basis of forestry industries that provide wood and paper.

As much value as plants have when harvested, they are also valuable when left in nature. By soaking up rainwater in their roots and other tissues, for example, plants prevent runoff and erosion that can contaminate streams. Plants also recycle carbon dioxide and produce oxygen. The importance of plants not only to humans but to all of Earth's ecosystems makes conserving plant biodiversity an important goal. However, humans have drastically and rapidly

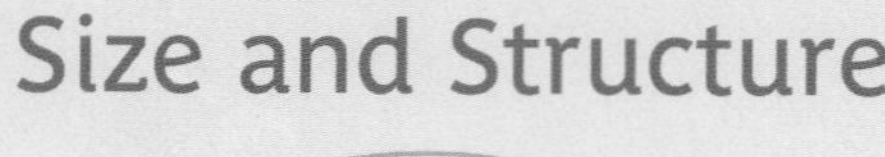

Size and Structure

The leaf of the Amazon lily, *Victoria regia*, is the largest in the world.

The duckweed, *Wolffia globosa*, is the smallest plant. Its entire body is just under 1 mm in length.

The baobab tree (*Adansonia rubrostipa*) from Madagascar grows up to 150 feet in circumference. It can hold more than 10,000 gallons of water in its hollow trunk, which serves as a water storage tank for local people.

Sequoiadendron giganteum, the giant sequoia, can reach 300 feet tall, with a circumference of 56 feet.

Reproduction

Measuring about 3 feet across, the flower of *Rafflesia arnoldii* is the largest. *Rafflesia* is a parasite on other plants, and the massive flower is the only structure it exposes aboveground.

Orchids, like this brown bee orchid (*Ophrys fusca*), produce the tiniest seeds. You could fit a million of the dustlike seeds of some tropical parasitic species in a quarter teaspoon.

The unbranched flower stalk (inflorescence) of *Amorphophallus titanum* is the largest in the world. The flowers smell like a rotting corpse.

FIGURE 18.16 Diversity in the Kingdom Plantae

modified Earth's plant landscapes with cities, agriculture, and industries (see the "Biology Matters" box on page 420).

> **Concept Check**
>
> 1. Name some key evolutionary innovations of plants.
> 2. Which plant groups produce pollen, and what is the adaptive value of pollen?

18.4 Fungi: A World of Decomposers

Most people are familiar with fungi as the mushrooms on their pizza or in their lawns. Fungi also include single-celled yeasts that ferment beer and make bread rise, and the threadlike mold that grows on cheese and bread. Because much of the fungal body remains hidden from view, however, most people fail to realize the extent to which fungi permeate our world, and fungi remain poorly understood organisms.

The kingdom **Fungi** consists of *absorptive* heterotrophs: fungi digest organic material outside the body and then absorb the molecules that are released as breakdown products. In contrast, all animals and most protist consumers are *ingestive* heterotrophs: they bring prey organisms or organic material into the body, or into the cell, and break it down internally.

Unlike animal cells, fungal cells have a protective cell wall that wraps around the plasma membrane and encases the cell. But like *some* animals, such as insects and lobsters, fungi produce a tough material called **chitin** that strengthens and protects the body. Fungi are similar to animals in that they store surplus food energy in the form of *glycogen*, the same molecule that serves as an energy reserve in our liver and skeletal muscle cells. Before the tools of DNA analysis became widely available, many biologists believed that fungi were more closely related to plants than to animals. However, DNA comparisons show that fungi diverged from animals more recently than both groups diverged from plants.

Because fungi do not fossilize well, their early evolutionary history is shrouded in mystery. Reconstructing the evolutionary history of eukaryotes from DNA data, scientists estimate that the common ancestor of fungi and animals diverged from all other eukaryotes about 1.5 billion years ago, and fungi diverged from their closest cousins, the animals, about 10 million years after that. Fungi were mostly likely aquatic for much of their evolutionary history and made the journey from water to land about 500 million years ago, about the time that plants appeared on land.

The majority of fungal species fall into three main groups (**FIGURE 18.17**):

- **Zygomycetes**, the first fungal group on land, produce diploid spores (zygospores).
- **Ascomycetes** are informally known as sac fungi because they produce spores in sacs.
- **Basidiomycetes**, or club fungi, produce the fruiting bodies we call mushrooms. They evolved about 400 million years ago.

Each of these groups exhibits—and is named for—unique reproductive structures.

Fungi play several roles in terrestrial ecosystems. Many are decomposers that live off nonliving organic material. Playing the role of garbage processor and recycler, these fungi speed the return of the nutrients in dead and dying organisms to the ecosystem. Some fungi live in or on other organisms and harm them (that is, they are *parasites*); others benefit from, and provide benefits to, organisms they associate with,

Concept Check Answers

1. Cuticle to prevent water loss, stomata to allow gas exchange, and vascular system to carry water and nutrients throughout the plant body.
2. Gymnosperms and angiosperms produce dry pollen grains, which carry sperm cells via wind or animal pollinators to the female organs of other individuals. This adaptation frees plants from depending on water for fertilization.

The Kingdom Fungi

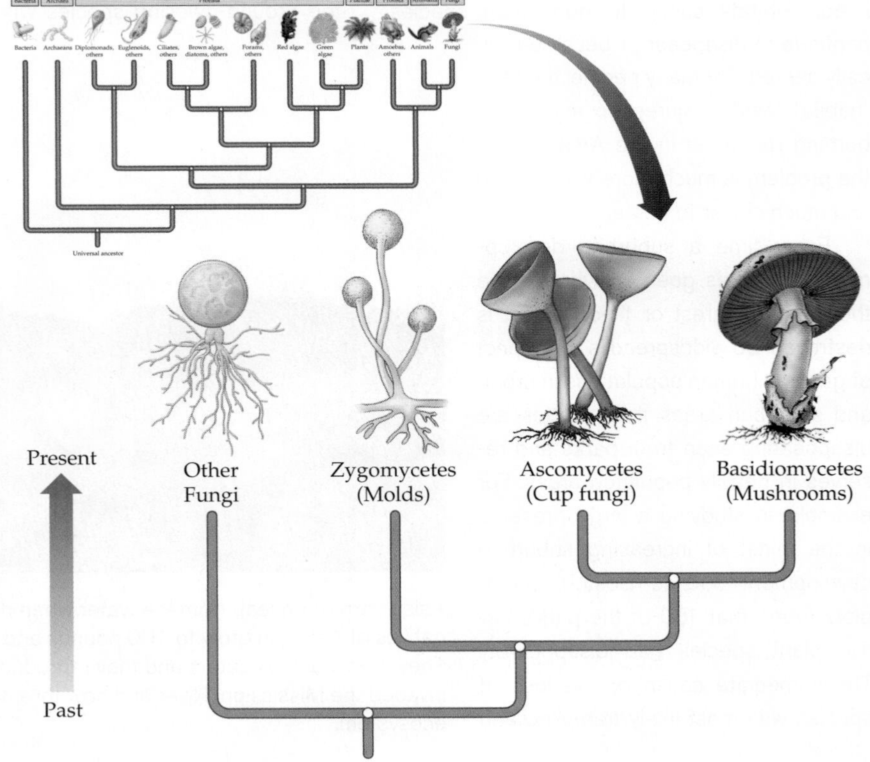

FIGURE 18.17 Evolutionary Tree of the Fungi

BIOLOGYMATTERS

The Many Threats to Biodiversity

The history of life on Earth includes a handful of drastic events, known as *mass extinctions*, during which huge numbers of species went extinct. Today, even as researchers struggle to get a total species count, many biologists assert that we are on our way toward a new mass extinction. In fact, many biologists say that the ongoing extinction—if it continues unabated—will lead to the most rapid mass extinction in the history of Earth. As with the total number of species on the planet, extinction rates are estimates. But even using conservative calculations, species are being lost at a staggering pace. The cause of the contemporary extinction is clear: the activities of the ever-increasing number of people living on Earth.

Habitat Loss and Deterioration Are the Biggest Threats to Biodiversity

Foremost among the direct threats to biodiversity is the destruction or deterioration of habitats. As human homes, farms, and industries spring up where natural areas once existed, habitats suited to nonhuman species continue to disappear or become radically altered. For many people, the term "habitat loss" conjures up images of burning rainforest in the Amazon, but the problem is much more widespread and much closer to home.

Every time a suburban development of houses goes up where once there was a forest or field, habitat is destroyed. So widespread is the impact of growing human populations in urban and suburban areas that species are disappearing even from parks and reserves in heavily populated areas. For example, in studying a large preserve in the midst of increasing suburban development outside Boston, ecologists found that 150 of the park's native plant species had disappeared. The immediate cause of the loss of species was most likely trampling and other disturbances, as more and more people—probably including many nature lovers—used the park. But the increasing number of homes in the area and the decreasing number of nearby natural areas—from which seeds could have come to repopulate the park—also played an important role in the loss of species.

Introduced Foreign Species Can Wipe Out Native Species

Another threat to existing species is the introduction of nonnative species. Researchers estimate that 50,000 such *introduced species* have entered the United States since Europeans arrived. Some of these are *invasive species*, so called because they sweep through a landscape, competing with and wiping out native species.

In Hawaii, introduced pigs that have escaped into and are living in the wild are devouring native plant species. Domesticated cats and mongooses, also introduced species, have killed many of Hawaii's native birds, especially the ground-dwelling species whose nests are easy targets. Purple loosestrife, eucalyptus trees, and Scotch broom are invasive plant species that are choking out native plants in various parts of the United States.

Asian carp often leap from the water when disturbed. These natives of Asia can grow to 100 pounds and more than 4 feet long. They are voracious eaters and they reproduce rapidly. They have invaded the Mississippi River and now threaten the Great Lakes ecosystem.

Climate Changes Also Threaten Species

Recent changes in climate, which the vast majority of scientists now agree are caused largely by human activities, constitute another threat affecting many species. In Austria, for example, biologists have found whole communities of plants moving slowly higher in the Alps. Apparently this change in their range is a response to global warming; because these plants are adapted to cooler temperatures, they are able to survive only at ever-higher elevations as lower elevations warm up. These plant communities are migrating at an average rate of about a meter per decade. If the climate continues to warm, these alpine plants—which exist nowhere else in the world—will eventually run out of mountaintop and go extinct. And in areas where organisms do not have cool mountains to climb, many species have begun moving to and living in increasingly northern latitudes to escape the heat.

Human Population Growth Underlies Many, If Not All, of the Major Threats to Biodiversity

The biggest threat overall to nonhuman species is the growth of human populations. Our growth is what spurs continuing habitat deterioration as natural areas are converted to the farms, roads, housing developments, and factories needed to support human life. The effects of our growing population are further magnified by the fact that more resources are being used *per person* now than in the past. In the "Biology Matters" box in Chapter 17, page 382, we described what we stand to lose when we lose biodiversity.

and for that reason are termed *mutualists*. Chapter 23 describes the many relationships in which organisms interact in nature.

Fungi are adapted for absorbing nutrients

Most fungi are multicellular, but there are some single-celled species, which are collectively known as "yeasts." A key evolutionary innovation of the multicellular fungi is their body form, which is exceptionally suited for absorbing nutrients from their surroundings as a source of energy. The main body of a multicellular fungus is called a **mycelium** (**mye-*SEE*-lee-um**; plural "mycelia"), which consists of an extensive mat of highly branched strands (**FIGURE 18.18**). Each mycelial strand, known as a **hypha** (***HYE*-fuh**; plural "hyphae" [*HYE*-fee]) is a slender thread-like filament consisting of multiple cells arranged in a row. In some species, the hyphal cells are incompletely separated. Instead of a complete partition, adjacent cells in the hyphae of basidiomycete fungi are separated by a structure called a *septum* (plural "septa"). Openings in the septum allow organelles—even the nucleus sometimes—to move from one cell compartment to another.

Mycelia extend deep into the medium the fungus is growing in, weaving into the soil, a rotting tree stump, or the compost that commercial mushroom growers commonly use. The mycelia of parasitic fungi send fine branches through the living tissues of their hosts. The hyphae, which are typically 10 micrometers (0.01 millimeter) wide, are in intimate and extensive contact with the medium they permeate. Mycelia therefore have an enormous surface area for taking up food and nutrients from their environment.

Like animals and all chemoheterotrophs, fungi rely entirely on other organisms for both energy and carbon. Fungal hyphae release special digestive proteins to break down organic material, and even living tissues, through which they grow. The hyphae then absorb the nutrients for the fungus to use. Some fungi have evolved to become predators, actively trapping small animals, such as the tiny worms known as nematodes, with sticky secretions or a noose made from three cells arranged in a ring.

Fungi have unique ways of reproducing

Fungi can reproduce both asexually and sexually. Some species, such as the yeasts, appear to multiply only asex-ually. Baker's yeast, *Saccharomyces cerevisiae* (***SAK*-uh-roh-*MYE*-seez sayr-uh-*VEE*-see-eye**), is a single-celled fungus that divides in a lopsided manner, known as budding, to produce new daughter cells that are genetically identical to the parent cell. Most multicellular fungi can reproduce asexually through fragmentation—that is, by simply breaking off from the mother colony.

Many fungi, like the greenish blue *Penicillium* that is so common on blocks of cheese, produce asexual *spores* (**FIGURE 18.19**). A **spore** is a reproductive structure, usually thick walled, that can survive for long periods of time in a dormant state and will sprout under favorable conditions to produce the body of the organism (mycelium, in the case of fungi). Fungal spores may be single-celled or multicellular and are usually encased in a thick wall that is resistant to decay and keeps the cytoplasm

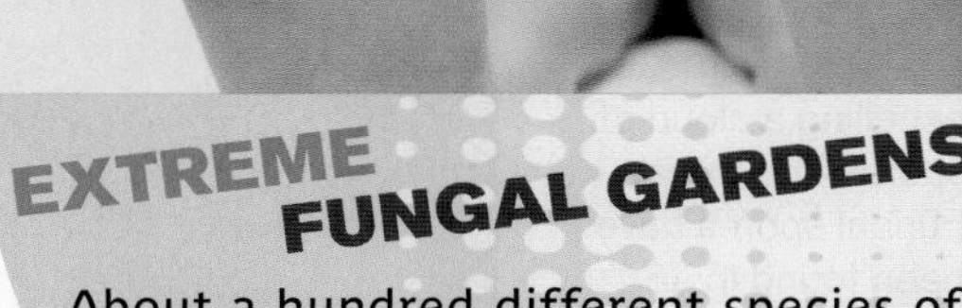

EXTREME FUNGAL GARDENS

About a hundred different species of fungi live on and in the average person. The richest fungal gardens? Our feet. Researchers found 60 different species in toe clippings and 40 in swabs between the toes. The great majority are harmless, but some cause skin infections such as athlete's foot.

The Fruiting Body of a Basidiomycete Fungus

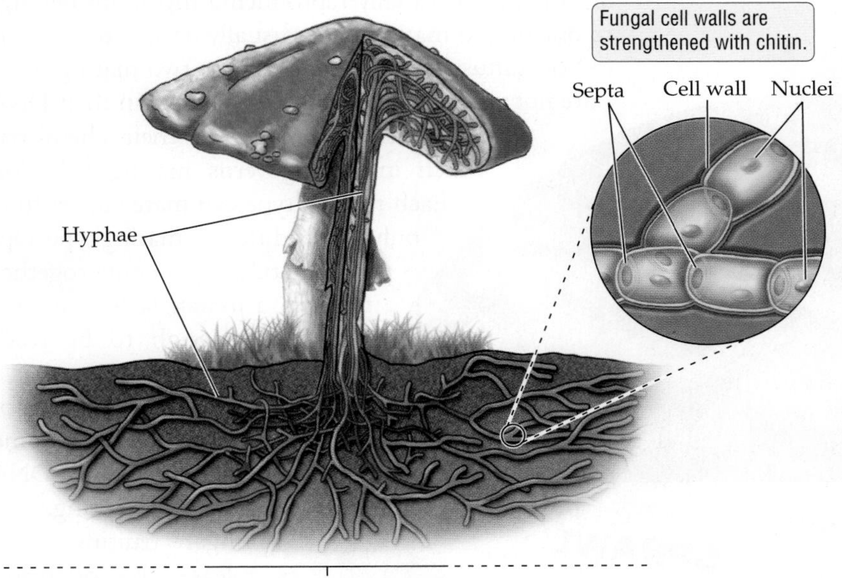

FIGURE 18.18 The Basic Structure of a Multicellular Fungus
Mats of hyphae, known collectively as a mycelium, form the main body of a fungus. Each hypha is a row of cells separated by septa. Openings in the septa allow organelles to move from one compartment to another. The fungal cell walls encasing the hyphae contain chitin, the same material that makes up the hard outer skeleton of insects.

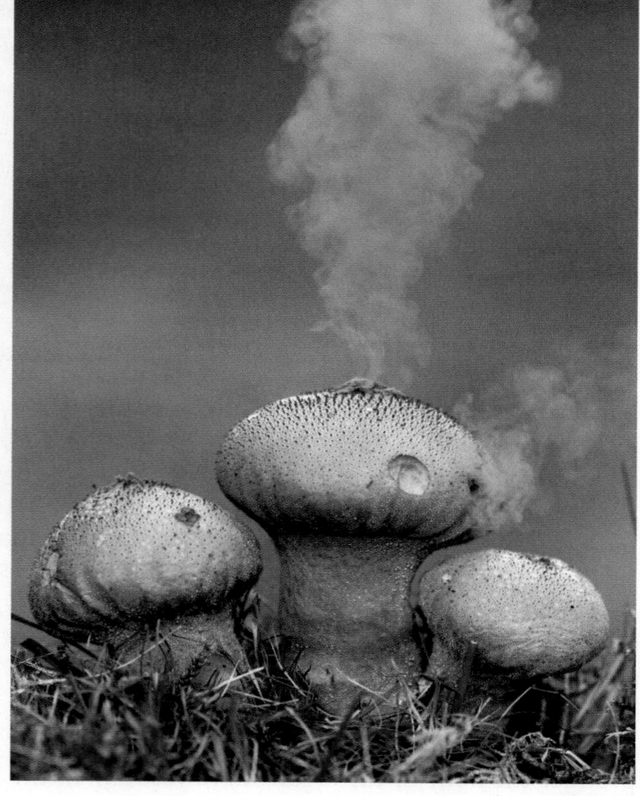

FIGURE 18.19
Fungi Spread via Spores
This puffball fungus, a basidiomycete, is expelling a cloud of spores into the air. Fungal spores have been found in air samples collected at 30,000 feet.

inside from drying out. The powdery material that speckles moldy bread or the walls of a damp basement consists of fungal spores.

Sexual reproduction in fungi is varied and complex. Fungi do not have distinct male and female individuals. Instead, a sexually reproducing mycelium belongs to one of two mating types, usually designated as plus (+) or minus (−) mating types. The two mating types are not visibly different, but differences in their DNA translate into differences in their chemistry, which in turn governs mating behavior. Each mating type can mate successfully only with a different mating type. Opposite mating types come together to form a *fruiting body* that may be large enough to be readily observed. A mushroom is a fruiting body in which opposite mating types have come together to fuse their DNA and produce offspring.

Fungal fruiting bodies release the offspring as sexual spores that are scattered into the world by wind, water, and animals, the way asexual spores are. Spores released from a fruiting body raised above the growing medium are better able to catch a ride on wind currents or to attract animals (as with flies drawn to stinkhorn mushrooms) that can carry them far and wide (**FIGURE 18.20**). Once carried to new locales, these sexual spores can begin growing as new, separate individuals that are genetically diverse because they contain DNA shuffled and blended from two different parent mycelia.

EXTREME SPRAWL

Armillaria ostoyae is the most humongous fungus among us. DNA tests show that the mycelium of a single individual of this tree-killing species snakes through 3.4 square miles of forest soil in the Blue Mountains of eastern Oregon. The mycelium—still growing—is estimated to be at least 2,400 years old.

Fungi play a key role as decomposers

Fungi are among the most important decomposers in terrestrial ecosystems, recycling a large proportion of the dead and dying organisms on land. They play a crucial role in nutrient recycling because they break down organic matter, such as leaf litter or the dead bodies of other organisms, releasing the nutrients as inorganic chemicals. Fungi are particularly effective in breaking down plant remains, including lignin, which no other organism except certain bacteria can decompose. The inorganic chemicals released become available to producers such as autotrophic bacteria, algae, and plants, which use them to manufacture food "from scratch."

What would the world look like if all the fungi disappeared? Great piles of leaf litter and brush and whole tree trunks would accumulate on the forest floor, and even animal carcasses would decay more slowly, with mainly bacteria and some protists available to decompose them.

What would the world look like if fungi were around but there was a sudden, colossal die-off among *other* organisms, especially plants? Paleobiologists, who study ancient life-forms, believe that this scenario already occurred in the past. For example, the mass extinctions that mark the end of the Permian period, which ended about 250 million years ago, were often accompanied by a *fungal spike*—a sudden and massive increase in the abundance of fossil spores belonging to decomposing fungi. The explanation is that the death of large numbers of organisms set up a feast for decomposing fungi, which multiplied explosively.

Fungi can be dangerous parasites

Parasitic fungi grow their hyphae through the tissues of living organisms, causing diseases in the organisms they infect. Mammals have a complex immune system that appears to defend them well against fungi. In humans with a healthy immune system, fungi cause no more than mild disease, such as athlete's foot or ringworm.

Life Strategies

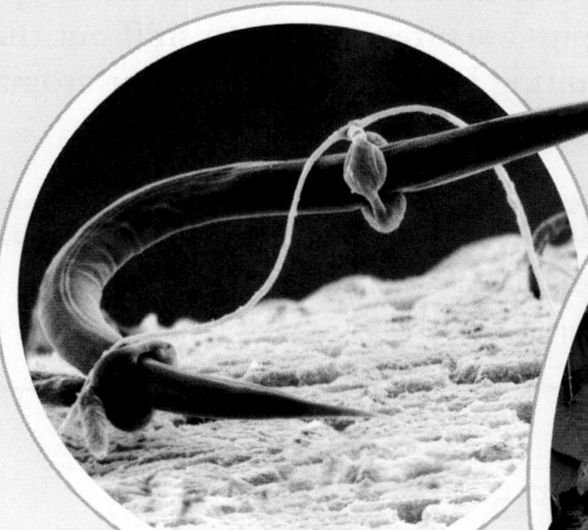

This predatory fungus, *Arthrobotrys anchonia*, is a killer with a lasso. It has trapped a nematode worm in a lasso made of three inflatable cells.

Like the more famous *Psilocybe*, this showy flame cap (*Gymnopilus spectabilis*) produces hallucinogenic chemicals to deter animals from nibbling it.

This bioluminescent fungus, *Mycena lampadis*, grows in Australia.

The death cap fungus, *Amanita phalloides*, contains a deadly toxin. Ingesting just half of one cap can kill a person.

Reproduction

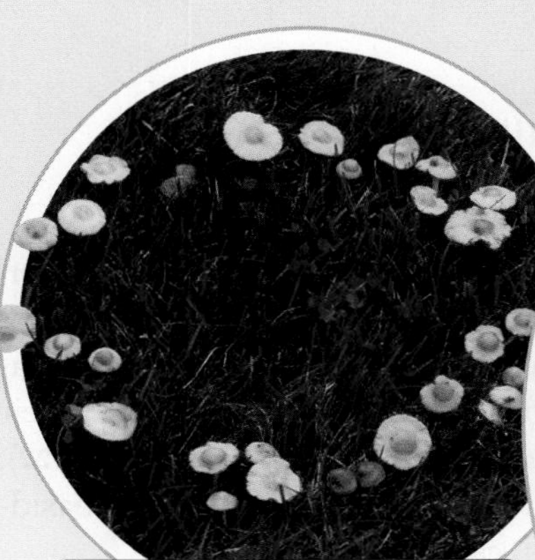

This fairy ring was produced by *Marasmius oreades*, a basidiomycete. Fairy rings start from a single spore at the center that sprouts mycelia that continue to grow outward until they reach a barrier. The largest fairy ring, located in France, measures nearly half a mile in diameter.

The best shot in the fungal world is *Pilobolus*, a zygomycete that lives in dung. Certain worm larvae that complete their development in the guts of herbivores are known to climb up the 1-centimeter-long stalk of the fruiting body. They hitch a ride on the spore mass as it rockets out of the dung heap and lands on adjacent vegetation, where it can be ingested by a cow or a horse.

The aptly named stinkhorn mushroom, *Phallus impudicus*, is a basidiomycete that attracts flies. The insects get covered with the sticky spores and scatter them as they fly to other locations.

FIGURE 18.20 Diversity in the Kingdom Fungi

FIGURE 18.21
Fungal Parasites
Some fungi are parasites, making their living by attacking the tissues of other living organisms. This beetle, a weevil in Ecuador, has been killed by a *Cordyceps* (**KOHR-duh-seps**) fungus, the stalks of which are growing out of the beetle's back.

But in humans with compromised immune systems, fungal diseases can be deadly, such as the pneumonia caused by the fungus *Pneumocystis carinii*, the leading killer of people suffering from AIDS.

Fungi are the most significant parasites of plants: they are responsible for two-thirds of all plant diseases, causing more crop damage than do bacteria, viruses, and insect pests combined. *Ceratocystis ulmi*, a fungal import from Asia that causes Dutch elm disease, has destroyed a majority of the American elm trees in the United States, which once formed arching canopies over streets all across the country. Likewise, *Cryphonectria parasitica*, another Asian import, has all but wiped out the American chestnut, a handsome native hardwood that can grow to 200 feet.

Rusts and smuts are serious fungal pests of crop plants, especially grain crops such as wheat and rice. Insects are another group of organisms that appear to be especially susceptible to fungal pathogens. Agricultural scientists are trying to use fungi that specialize in killing insects to protect crop plants from insect pests (**FIGURE 18.21**).

Fungi can benefit human society

Although some fungi can be costly to human society, others are beneficial, providing us with pharmaceuticals, including antibiotics such as penicillin. Yeasts such as *Saccharomyces cerevisiae* can feed on sugars and produce two important products: alcohol and the gas carbon dioxide—crucial to the rising of bread, the brewing of beer, and the fermenting of wine. Species of *Aspergillus* are used to ferment soybean extracts in the manufacture of soy sauce and tamari.

Fungi also are the source of highly sought-after delicacies such as the edible mushrooms and the fungi known as truffles. Truffles grow underground, and when the closed, saclike fruiting bodies are fully mature, they release a distinctive aroma to attract animals that dig up and ingest the reproductive structures whole and disseminate the spores in their droppings. Dogs or pigs are often trained to sniff out the subterranean fruiting bodies, whose pungent aroma is also appreciated by many food connoisseurs.

Concept Check

1. What are some distinctive characteristics of fungi?
2. What is a fungal spore?

Concept Check Answers

1. Fungi are eukaryotes that digest food externally and absorb it; they strengthen their cell walls with chitin and use glycogen as an energy storage molecule.
2. A fungal spore is a thick-walled dormant structure that, under the right conditions, can sprout to produce a mycelium.

18.5 Lichens and Mycorrhizae: Collaborations between Kingdoms

The long-term and intimate association of two different types of organisms is known as **symbiosis** (*sym*, "with"; *biosis*, "life"). Symbiosis may be beneficial to both organisms, to neither, or to one but not the other.

A **mutualism** is a close association between two species that benefits both symbiotic partners. Despite their fearsome image as destroyers of the world, fungi have evolved mutualisms with nearly every kingdom of life: with photosynthetic bacteria, with photosynthetic protists (algae), with plants, and also with some animals.

Lichens contain a fungus and a photosynthetic microbe

A **lichen** (**LYE-kun**) is a mutualistic association between a photosynthetic microbe and a fungus. The lichen-forming microbe can be a single-celled green alga, a cyanobacterium, or both together. Ascomycetes, with their distinctive cup-shaped fruiting bodies (**FIGURE 18.22**), are the most common fungi in lichens, although basidiomycetes (club fungi) can also form lichens.

Much of the body of the lichen is created by packed mycelial strands, with algal or cyanobacterial cells embedded in the mycelial mat. The fungus receives sugars and other carbon compounds from its photosynthetic partner. In return, the fungus produces lichen acids, a mixture of chemicals that scientists believe may function to protect both the fungus and the alga from being eaten by predators.

FIGURE 18.22
A Lichen Is a Mutually Beneficial Relationship between a Fungus and a Photosynthetic Prokaryote or Protist
The three different growth forms of lichens—crustose, foliose, and fruticose—are seen here. The photosynthetic partner in lichens is a cyanobacterium or a green alga, and occasionally both.

FIGURE 18.23
Mycorrhizae Are Mutualisms between Fungi and Plant Roots

Lichens grow very slowly, typically increasing in size by less than a centimeter per year. They can multiply by fragmentation, or by disseminating dry powdery packages, called *soredia* (singular "soredium"), that consist of a few photosynthetic cells wrapped in fungal mycelia.

The lichen body is thin and has no protective sheath like the cuticle of a plant or the "armor plating" of some protists. Nor does it have any mechanism for ridding the body of wastes or toxic substances. Therefore, lichens readily absorb and accumulate air- or water-borne pollutants. Lichens are destroyed by acid rain, heavy-metal pollutants, and organic toxins, which is why they tend to disappear in heavily industrialized areas with poor pollution controls.

Lichens are especially abundant in the Arctic tundra, where they form an important food source for herbivores such as reindeer. Lichens are often pioneers in barren environments. Acids produced by the fungal symbionts in lichens wear down a rocky surface, facilitating soil formation. Soil particles build up from the slow weathering of rock, and over time other life-forms, including pioneering plant species, gain a toehold in the newly made soil.

Mycorrhizae are beneficial associations between a fungus and plant roots

Plants probably would not have been as successful on land if they had not entered into a mutualistic relationship with fungi almost immediately on arrival. Some of the earliest fossil plants—liverworts that lived about 460 million years ago—appear to have had fungal endosymbionts in their roots. Today, the vast majority of plants in the wild have mutualistic fungi, known as mycorrhizal fungi (*mykos*, "fungus"; *rhiza*, "root"), associated with their root systems. Truffles, morels, and chanterelles—highly prized as food by some—are the reproductive structures of mycorrhizal fungi.

Mycorrhizae (*MYE*-**koh**-*RYE*-**zee;** singular "mycorrhiza") are mutualistic associations between fungal mycelia and the root system of a plant. Mycorrhizal fungi form thick, spongy mats of mycelium on and in the roots of their plant hosts (**FIGURE 18.23**) and also extend into the surroundings. A mycelial mat draws up far more water and mineral nutrients, such as phosphorus and nitrogen, than the plant's root system could absorb on its own. In return for sharing absorbed water and mineral nutrients, the fungus obtains sugars that the plant manufactures through photosynthesis. **FIGURE 18.24** illustrates the impact of mycorrhizal associations on the growth of tomato plants.

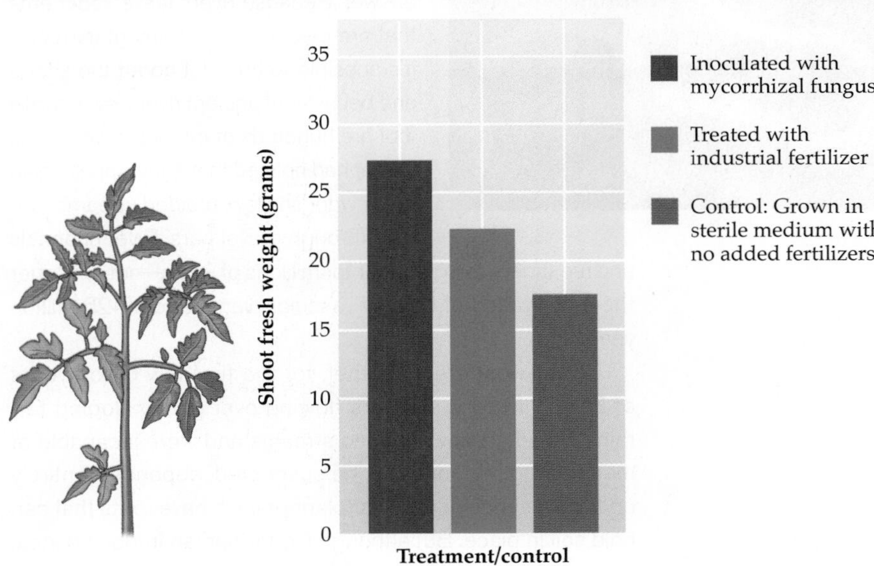

FIGURE 18.24 Mycorrhizal Associations Benefit the Growth of Tomato Plants

Concept Check Answers

1. Lichens produce acids that slowly break down rocks, creating soil; they are an important food source for many organisms, including reindeer.

2. Mycorrhizal fungi live in and on plant roots, helping the plants to absorb water and mineral nutrients and receiving sugars from the plants in exchange. Most wild plants have mycorrhizal associations, which boost growth in many plant species.

Mycorrhizal fungi also improve soil quality by changing its chemical and physical properties. The mycorrhizal mantle protects roots from potentially harmful soil pathogens, including other types of fungi, bacteria, and root-nibbling animals. A predatory mycorrhizal fungus called *Laccaria* attracts and kills small arthropods called springtails, which are related to insects, and it shares with its plant host the nitrogen that the mycelia absorb from the digested animal.

Certain practices in intensive agriculture, such as soil fumigation, destroy native mycorrhizal fungi. Plants grow very poorly in fumigated fields, unless large amounts of industrial fertilizer are added. Mycorrhizal fungi, dependent as they are on plant roots, cannot be grown and multiplied in a lab dish, but biologists and organic farmers have cultured fungal isolates in soil that are sold commercially for use by home gardeners and organic farmers.

Concept Check

1. Explain the ecological significance of lichens.
2. What are mycorrhizal fungi, and what is their significance?

APPLYING WHAT WE LEARNED

The Root of the Problem: Why Rivers Meander

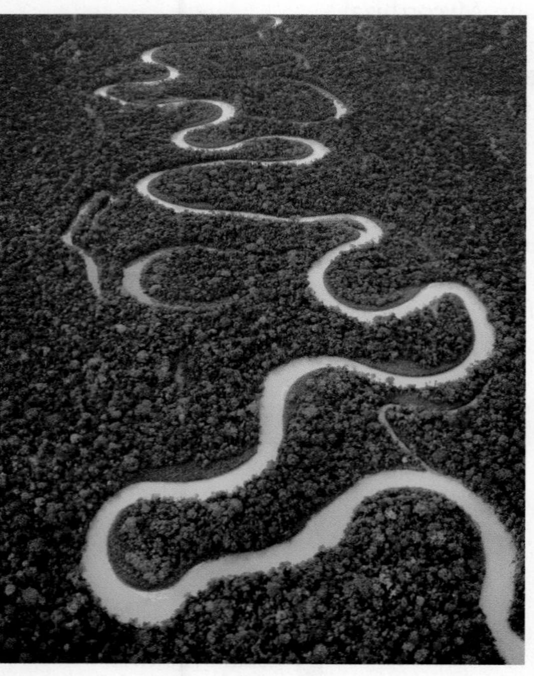

At the beginning of this chapter we learned that the meanders of rivers are important to the formation of flat valleys, rich sediments, and other features of Earth's surface that promote the diversity of plants and other organisms. But until recently, it wasn't clear what caused rivers to meander.

Back in the 1980s, a Bucknell University geologist named Edward Cotter, who was studying ancient sediments (250–450 million years old) in the Appalachian Mountains, guessed the answer. Because rivers leave sediments that are preserved in sedimentary rocks, it's possible to tell a lot about the shape and behavior of ancient rivers, even those that are hundreds of millions of years old. Cotter had noticed that early Appalachian rivers were shallow, braided streams running through gravel bars. River channels and meanders didn't appear for millions of years—not until after the colonization of land by vascular plants about 425 million years ago.

Cotter wondered whether, for the first few billion years of Earth's history, all rivers flowing over gently sloping terrain formed shallow, braided streams and were incapable of meandering. Meandering, he suspected, depended entirely on the presence of vascular plants, which have roots that can hold soil in place. But although Cotter had an intriguing idea, he didn't have any strong evidence.

In recent decades, researchers have begun looking a lot harder at river meandering. In 2009, Christian Braudrick, a graduate student at UC Berkeley, built a miniature river in a 20-by-50-foot box. To his delight, it meandered not just for a few hours or days, but for an entire year. By using just the right slope, just the right sized sand, and one exciting new ingredient—alfalfa sprouts—he succeeded where countless others had failed. The sprouts' roots stabilized the sand enough to keep the water from cutting through it. Over the course of a year, the tiny river formed five new meanders, each of which migrated downstream as it grew and disappeared. It was a dramatic breakthrough.

Meanwhile, at Dalhousie University, geologists Martin Gibling and Neil Davies decided to evaluate the evidence for and against the theory that vascular plants sculpt modern rivers. Davies and Gibling read and analyzed 144 scientific papers that discussed the shapes of ancient rivers. The two researchers concluded that until vascular plants evolved and came to dominate on land, ancient rivers left thin, wide blankets of gravel that were consistent with shallow, braided streams, not meandering rivers. Some of these ancient rivers were as much as a thousand times wider than they were deep. (For comparison, the Mississippi River is only about fifty times as wide as it is deep.) To further test their theory, the two researchers are now looking to see whether rivers changed back to braided streams after a mass extinction that wiped out many terrestrial plants 250 million years ago.

Mounting evidence suggests that the living and nonliving features of our planet dynamically shape one another. When plants colonized the land, their roots stabilized soils perhaps for the first time, creating tough, erosion-resistant banks capable of channeling wide, shallow streams into beautiful, meandering rivers that help create rich soil and diverse vegetation. Besides promoting the diversity of terrestrial life, these deep rivers later gave rise to major human civilizations by allowing humans to travel along waterways, carrying goods and information.

BIOLOGY IN THE NEWS

Nectar That Gives Bees a Buzz Lures Them Back for More

BY JAMES GORMAN • *NEW YORK TIMES*, March 7, 2013

Nothing kicks the brain into gear like a jolt of caffeine. For bees, that is.

And they don't need to stand in line for a triple soy latte. A new study shows that the naturally caffeine-laced nectar of some plants enhances the learning process for bees, so that they are more likely to return to those flowers.

"The plant is using this as a drug to change a pollinator's behavior for its own benefit," said Geraldine Wright, a honeybee brain specialist at Newcastle University in England, who, with her colleagues, reported those findings in *Science* ...

Several varieties of coffee and citrus plants have toxic concentrations of caffeine in leaves and other tissues, but low concentrations, similar to that in weak coffee, in the nectar itself. The toxic concentrations help plants fend off predators ...

[Dr. Wright and her colleagues] conducted learning experiments with bees to see if they associated a reward with an odor, the reward being either sugar water or a combination of sugar water and caffeine in the same concentrations found in the nectar of coffee and citrus plants ...

"If you put a low dose of caffeine in the reward when you teach them this task, and the amount is similar to what we drink when we have weak coffee, they just don't forget that the odor is associated with the reward," she said.

After 24 hours, three times as many bees remembered the connection between odor and reward if the reward contained caffeine. After 72 hours, twice as many remembered. They then tested the effect of caffeine on neurons in the bee brain and found that its action could lead to more sensitivity in neurons called Kenyon cells, which are involved in learning and memory ...

Insect and human brains are vastly different, and although caffeine has many effects in people, like increasing alertness, whether it improves memory is unclear. But the excitation of the Kenyon cells was similar to the action of caffeine on cells in the hippocampus in a recent experiment on rats, Dr. Wright said.

Showy flowers—roses, snapdragons, and magnolias—have evolved for one reason: to lure pollinators. Shiny petals, vivid colors, sweet perfumes, or stinky odors—they're all advertising gimmicks.

Some plants sweeten the deal by offering nectar, a liquid rich in sugars, amino acids, and other nutritive molecules. Pollen—which is rich in protein and antioxidants such as carotenoids—is an easy sell to pollinators such as beetles. Although the insects gobble up a good deal of a plant's pollen production, their bumbling around in the flowers transfers just enough pollen that it's a fair trade all in all.

Some flowers extend an open invitation to a whole host of pollinators—bees, syrphid flies, butterflies, and more. Most, however, are quite selective in their clients. The delightful variety in shape, size, color, and odor of flowers has evolved to enable plants to cater to specific pollinators. For example, moth-pollinated flowers are strongly scented and have a corolla tube about as long as the proboscis (feeding tube) of the moth species that has coevolved to pollinate them; because the preferred pollinators fly at night, color is of little use, and moth-pollinated flowers are almost always white. Unlike most insects, birds see red, but unlike most insects and mammals, birds aren't good at smelling. That explains why most bird-pollinated flowers are red and unscented.

Why such choosiness in the pollinator clientele? A pollinator that visits one or only a few species of flowers is more likely to deliver the real goods, instead of a whole lot of "junk mail," to the female parts of the flower. Researchers have known for some time that coevolutionary processes enhance fidelity in the relationship between plant and pollinator. Some flowers, for example, seal the deal by producing nectar with just the right kinds of amino acids to suit the needs of their specific bumblebee pollinator. But using drugs like caffeine to keep the pollinator coming back for more? That's a new one even for botanists who think they've heard it all in this business of the birds and the bees.

Evaluating the News

1. Give two explanations for why a coffee plant produces caffeine. (Keeping people alert doesn't count.)

2. How did the researchers demonstrate that caffeine-laced nectar actually affects learning in bees?

CHAPTER REVIEW

Summary

18.1 The Dawn of Eukarya

- The Eukarya have traditionally been divided into four major kingdoms: Protista, Plantae, Fungi, and Animalia.
- Eukaryotes possess a true nucleus, and they have complex subcellular compartments, which enable larger cell size.
- Sexual reproduction is another evolutionary innovation of the Eukarya; however, many eukaryotes reproduce asexually as well.
- The Eukarya evolved multicellularity separately in several lineages, which endowed these groups with the benefits of functional specialization among different cell types.

18.2 Protista: The First Eukaryotes

- The Protista (protists), the most ancient category of eukaryotes, is a highly diverse group.
- Protista is a catchall category that lumps together many evolutionarily distinct lineages, some of which are only distantly related to others in the category.
- Most protists are single-celled and microscopic.
- Protists can be photosynthetic autotrophs, heterotrophs, or mixotrophs.
- Algae are photosynthetic protists. Nonphotosynthetic motile protists are commonly called protozoans.
- Sexual reproduction, a key evolutionary innovation of the protists, combines genetic information from two parents to produce offspring that are genetically different from each other and from both parents. Many protists can reproduce asexually as well.
- Some protists are poisonous, and some are pathogens of other organisms, including humans.

18.3 Plantae: The Green Mantle of Our World

- The Plantae (plants) are multicellular photosynthetic eukaryotes that are adapted for life on land.
- Plants evolved a waxy covering, the cuticle, that reduces water loss.
- Plant cells have cell walls stiffened with a strong but flexible biomolecule called cellulose. The evolution of another cell wall–strengthening biomolecule, lignin, enabled plants to resist the pull of gravity.
- The vascular system, containing bundles of water-conducting xylem tubes and food-conducting phloem tubes, was also critically important in enabling plants to grow tall. All plants except bryophytes have a vascular system.
- Pollen and seeds first evolved among the gymnosperms. Pollen delivers sperm cells to the female reproductive structures. Each seed contains an immature plant, or embryo, along with a food supply enclosed in a protective seed coat.
- Angiosperms evolved flowers, reproductive structures with numerous innovations to aid in pollen dispersal. The fruit wall protects the seeds inside it and often develops special elaborations that aid in seed dispersal. Many angiosperms recruit animals to deliver pollen and also to disperse their seeds.
- As producers, plants are the ultimate food source for nearly all terrestrial organisms.

18.4 Fungi: A World of Decomposers

- The Fungi (fungi) are eukaryotic absorptive heterotrophs with chitin-containing cell walls. DNA comparisons show that fungi are more closely related to animals than to any other kingdom of life.
- Although some fungi are single-celled, most are multicellular.
- Multicellular fungi have a unique body plan: threadlike branched hyphae that penetrate organic material. The organic material is digested externally, and the breakdown products are absorbed as food. Fungi are important as decomposers in terrestrial ecosystems.
- Fungi reproduce both asexually and sexually. The products of fertilization become encased in thick walls to form spores, which are remarkably resistant to harsh conditions and able to disperse long distances via animals, wind, and water.
- There are at least three main groups of fungi—zygomycetes, ascomycetes, and basidiomycetes—each characterized by distinctive reproductive structures, or fruiting bodies.
- Some fungi are parasites and cause serious disease, especially in plants and insects. Others are beneficial.

18.5 Lichens and Mycorrhizae: Collaborations between Kingdoms

- Mutualisms are close associations between two species that benefit both organisms.
- In lichens, algae or cyanobacteria live in a mutually beneficial association with a fungal partner.
- Mycorrhizal fungi live in and on the roots of most wild plants, and they assist their plant hosts in absorbing water and mineral nutrients from the soil. In return, the plants supply sugars to the fungi.

Key Terms

alga (p. 408)
angiosperms (p. 416)
ascomycetes (p. 419)
basidiomycetes (p. 419)
bryophytes (p. 413)
chitin (p. 419)
cuticle (p. 413)
flower (p. 416)
fruit (p. 417)
Fungi (p. 419)
guard cell (p. 413)
gymnosperms (p. 415)
hypha (p. 421)
lichen (p. 424)
lignin (p. 414)
mixotroph (p. 412)

mutualism (p. 424)
mycelium (p. 421)
mycorrhiza (p. 425)
pathogen (p. 412)
phloem (p. 414)
phytoplankton (p. 409)
plankton (p. 409)
Plantae (p. 413)
pollen (p. 415)
Protista (p. 407)
protozoan (p. 408)
seed (p. 415)
spore (p. 421)
stoma (p. 413)
symbiosis (p. 424)
vascular system (p. 414)
xylem (p. 415)
zooplankton (p. 409)
zygomycetes (p. 419)

Self-Quiz

1. Compared to prokaryotes, eukaryotes
   a. do not have membrane-enclosed compartments in their cells.
   b. exhibit a much greater diversity in modes of nutrition.
   c. have a nucleus.
   d. are more widespread.
2. Which of the following evolutionary innovations enabled larger cell size?
   a. autotrophic mode of nutrition
   b. multicellularity
   c. sexual reproduction
   d. subcellular compartmentalization
3. Which of the following groups contains *only* multicellular species?
   a. algae
   b. protists
   c. bryophytes
   d. yeasts
4. Which of the following groups consists *entirely* of autotrophic species?
   a. fungi
   b. protists
   c. gymnosperms
   d. animals
5. Protists
   a. are all descended from a single evolutionary lineage and are more closely related to each other than to any other kingdom of life.
   b. are mostly multicellular, with a few single-celled species.
   c. are more diverse in modes of nutrition and life cycle characteristics than fungi are.
   d. were the first pioneers on land and dominate the terrestrial environment to this day.
6. Fungi grow by extending their
   a. hyphae.
   b. septa.
   c. basidiomycetes.
   d. coelomic cavities.
7. Which of the following constitutes an evolutionary innovation that enabled plants to become taller?
   a. a cuticle
   b. a mycelial network
   c. the presence of cell walls reinforced with lignin
   d. the presence of mycorrhizal associations
8. Mycorrhizal fungi are
   a. beneficial to plants because they help plants stay dry.
   b. harmful to plants because they secrete acids.
   c. beneficial to plants because they help in absorbing minerals.
   d. harmful to plants because they degrade lignin.

Analysis and Application

1. Is sexual reproduction seen among the prokaryotes? Is it known among each of the three kingdoms of Eukarya described in this chapter? What might be the adaptive benefits of sexual reproduction compared to asexual reproduction?
2. Name a multicellular protist. How does this protist benefit from being multicellular instead of single-celled?
3. What are red tides? Could humans be contributing to red tides? Explain.
4. Fungi are like us in that they are heterotrophs. How is their heterotrophic mode of nutrition different from ours? In terms of evolutionary relatedness, are fungi closer to plants or to animals? What evidence can you cite in support of your answer?
5. Imagine that an especially potent strain of a broad-spectrum fungal virus (mycovirus) destroys almost all the major fungi in your part of the world. What consequences would you expect? Describe the scene as if you were a TV news reporter touring your county a year after the massive fungal die-off.
6. Two of the major challenges facing plants when they colonized land were (a) obtaining and retaining water and (b) fighting gravity to grow taller. What evolutionary innovations enabled plants to deal with these challenges?
7. Lichens are so sensitive to pollution that they can be used as indicators of air quality. What is a lichen, and why are lichens like canaries in a coal mine when it comes to environmental pollution?

19 Animalia

SPONGES ARE THE SIMPLEST ANIMALS. Like other animals, sponges develop from an embryo and feed off other organisms. But sponges lack both symmetry and complex tissues and organs.

Who We Are

Take a swim at any beach and you will likely swallow a gulp or two of water that contains a clue to the origin of animals. It may sound far-fetched, but most water contains tiny microorganisms that are distant cousins to all animals.

A single gallon of seawater can contain millions of single-celled choanoflagellates, whose whiplike flagella not only push them through the water but also beat bacteria and food particles into a collar that traps food. Lakes and oceans are filled with tiny microorganisms—plankton—that drift with currents and serve as food for larger organisms. But choanoflagellates have some traits that set them apart from other plankton. Many of them have an ability to form colonies—cooperative groups that live and work together. Cooperation is a central theme in the evolution of life.

For the first few billion years of life on Earth, all living organisms were single cells. But sometime between about a billion years ago and 700 million years ago, cells began forming cooperative groups, or colonies. From these early colonies evolved multicellular organisms such as plants and animals. Scientists believe one such colonial organism was the common ancestor of present-day colonial choanoflagellates, as well as all the animals that have ever lived.

The earliest animals were soft, multicellular eukaryotes similar to sponges and jellyfishes. Later came worms, clams, squids, sharks, and bony fishes, followed on land by insects, amphibians, reptiles, and mammals. Like all these other animals, we are descendants of tiny colonial swimmers resembling the "choanos" that still thrash around in duck ponds all over the world. But unlike the choanos, all animals have multicellular bodies.

How do biologists know animals are related to choanoflagellates? What can humans and other animals possibly have in common with plankton?

Your own body consists of trillions of individual cells that somehow manage to communicate and cooperate second by second for an entire lifetime. Your nerve and muscle cells coordinate with your brain to help you swallow water or allow you to breathe at the right time while swimming laps. Amazingly, the communication systems that human cells use to cooperate are similar to those that choanos use. In this chapter we tour the world of animals and consider the distinctive characteristics of the major groups. At the end of the chapter we'll find out how choanos help us understand how we, and all other members of the kingdom Animalia, came to be.

MAIN MESSAGE

Animals are multicellular ingestive heterotrophs that display a remarkable diversity in form and behavior.

KEY CONCEPTS

- Animals are multicellular, ingestive heterotrophs that evolved by about 700 million years ago.
- Animal cells lack cell walls, but most are bound to an extracellular matrix and, through cell junctions, to each other.
- Key evolutionary innovations of animals include specialized tissues, organs, and organ systems; complete body cavities; body segmentation; and an astounding range of behaviors.
- Insects are the most species-rich group of all organisms. Mollusks are the most diverse animals in the sea.
- Fish were the first vertebrates, and amphibians were the first tetrapods.
- Reptiles and birds produce an amniotic egg, a major adaptation for terrestrial living.
- The success of mammals is attributed to an exceptional investment in nourishment and care of the young.

WHO WE ARE		431
19.1	The Evolutionary Origins of Animalia	432
19.2	Characteristics of Animals	433
19.3	The First Invertebrates: Sponges, Jellyfish, and Relatives	438
19.4	The Protostomes	440
19.5	The Deuterostomes	448
19.6	Chordates like Us: The Vertebrates	450
APPLYING WHAT WE LEARNED Clues to the Evolution of Multicellularity		460

WHAT ARE THE QUINTESSENTIAL QUALITIES of an animal? As you consider this question, examine the photos in **FIGURE 19.1**. Would you say that all animals are capable of moving? Can all animals see? Do they all have a brain? Which of the animals in Figure 19.1 is our closest relative? Why?

(*a*) Yeti crab, *Kiwa hirsuta*

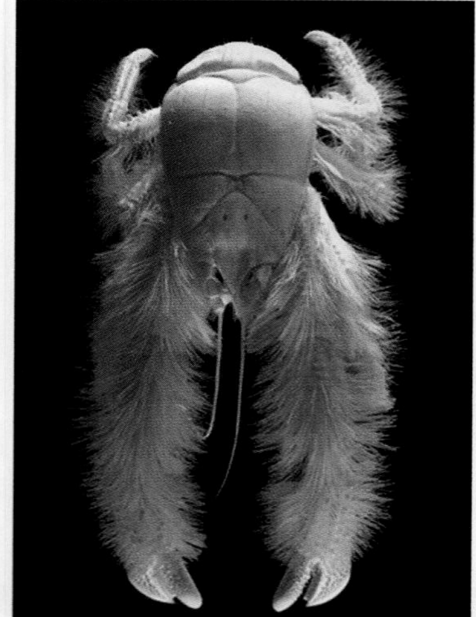

(*b*) Sea squirt, a tunicate

(*c*) Sea slugs or nudibranchs

(*d*) Christmas tree worm

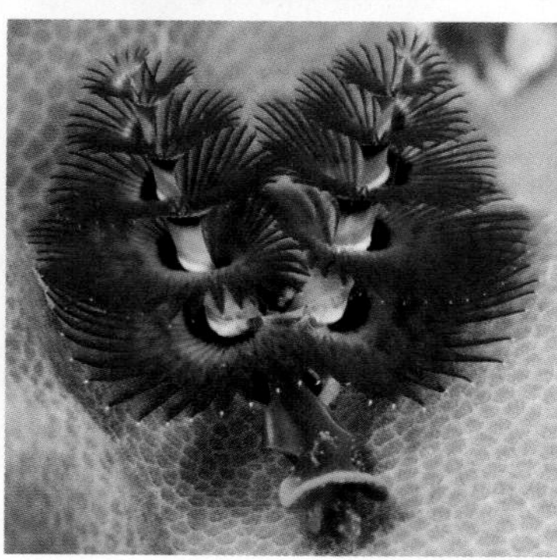

FIGURE 19.1 A Gallery of Animalia
Can you guess which of these animals is most closely related to us?

In this chapter we explore the captivating diversity of the world of animals. We will see that animal bodies are organized in many different ways and will contemplate the pros and cons of these varied body plans. We will note that the evolution of organ systems has given rise to complex behaviors, including our own ability to reflect on the nature of all life.

19.1 The Evolutionary Origins of Animalia

The kingdom **Animalia** comprises a diverse variety of organisms, including sponges, corals, worms, snails, insects, sea stars, and flashy beasts like Komodo dragons, Bengal tigers, and us. Animals are multicellular *ingestive* heterotrophs, obtaining energy and carbon by *ingesting* food—that is, by bringing food inside their bodies and digesting and absorbing it internally. As you learned in the preceding chapter, fungi are different in that they are *absorptive* heterotrophs: they digest food externally and then absorb the nutrients that are released across the body surface.

The oldest animal fossils are vaselike sponges, dating to 760 million years ago, that were unearthed recently in Etosha National Park, in Namibia. Cnidarians (**nye-*DAYR*-ee-un**), a group that includes jellyfish, sea anemones, and corals, evolved next (**FIGURE 19.2**). The remaining animal phyla fall into two broad groups: the *protostomes* and *deuterostomes*, which are distinguished by the way their embryos develop.

Protostomes are divided into at least 20 phyla, including mollusks (such as snails and clams), annelids (segmented worms), and arthropods (including crustaceans, spiders, and insects). Deuterostomes (***DOO*-ter-oh-*STOHM***) include echinoderms (**ee-*KYE*-noh-derm**), which are sea stars and their relatives; and the chordates, which include us.

The chordates form a large phylum that encompasses all animals with backbones, such as fishes, birds, and humans. The phylum also includes a few subgroups of less familiar animals, such as sea squirts and lancelets, that have a nerve cord along the back of the body but no backbone.

Chordates that possess a backbone are known as vertebrates, and all other phyla of animals are informally lumped together under invertebrates. However, as Figure 19.2 illustrates, "invertebrates" is not an evolutionarily meaningful label, because it merges lineages with divergent evolutionary histories and different sets of evolutionary innovations.

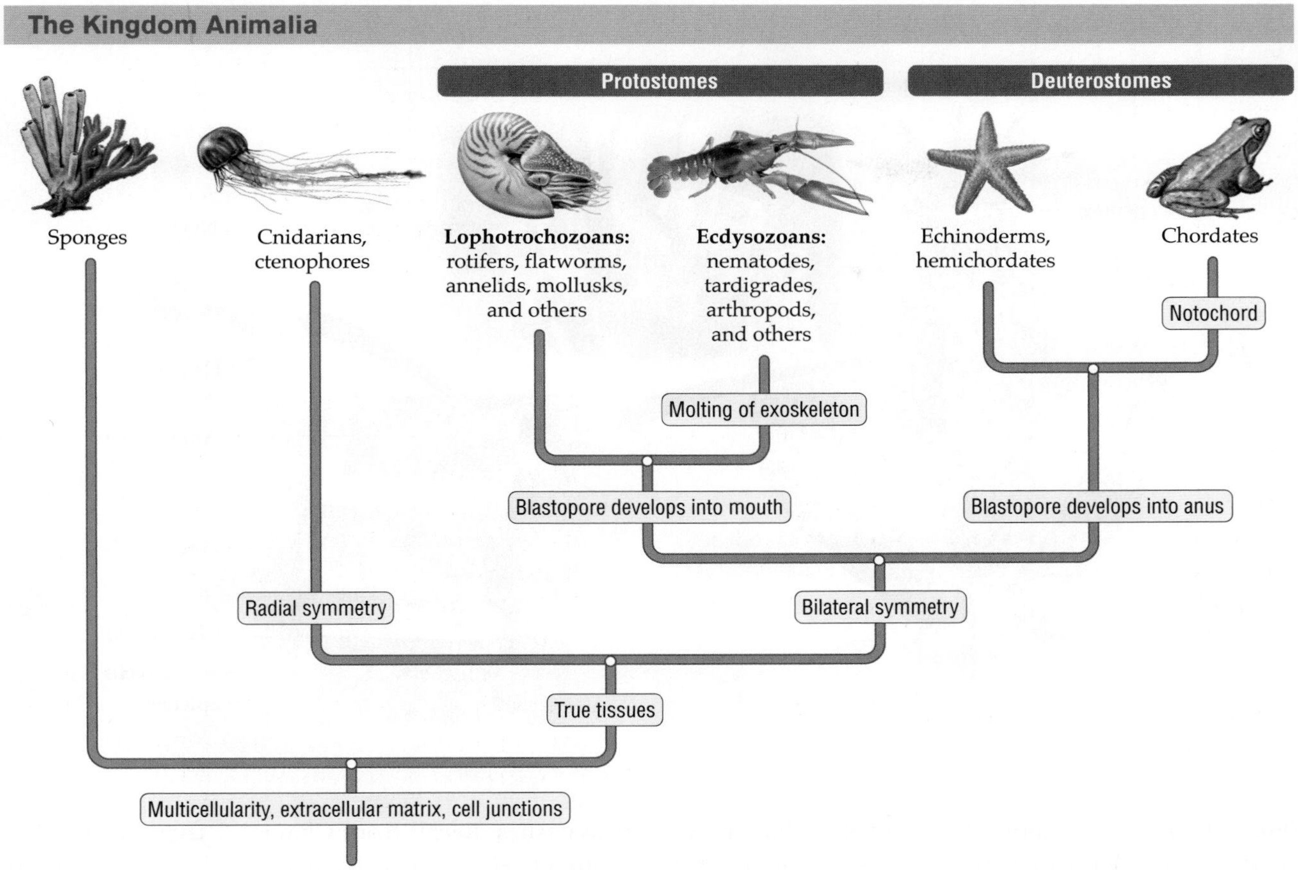

The Kingdom Animalia

Protostomes

Deuterostomes

Sponges

Cidarians, ctenophores

Lophotrochozoans: rotifers, flatworms, annelids, mollusks, and others

Ecdysozoans: nematodes, tardigrades, arthropods, and others

Echinoderms, hemichordates

Chordates

Notochord

Molting of exoskeleton

Blastopore develops into mouth

Blastopore develops into anus

Radial symmetry

Bilateral symmetry

True tissues

Multicellularity, extracellular matrix, cell junctions

FIGURE 19.2 Evolutionary Tree of the Animalia

The unique evolutionary innovations of each lineage are shown with boxed labels.

Concept Check

1. In what way are animals similar to fungi, and in what key way are they different?
2. Which animals are the most ancient?

19.2 Characteristics of Animals

All multicellular organisms must have a way of gluing the cells in the body together. Cell specialization is a major adaptive benefit of multicellularity (see Chapter 18). Most cells in the animal body are enveloped in, or attached to, a feltlike layer known as the *extracellular matrix*. Most cells in the animal body are also firmly attached to one another by Velcro-like patches of proteins located at special sites called *cell junctions* (see Chapter 4).

All animals except sponges have true tissues, and embryo development in animals includes a stage, called *gastrulation*, in which cells are rearranged to produce a multilayered embryo. The major groups of animals differ in their *body plan*, or the overall organization of the body. These features provide a road map for our discussion of the characteristics of various groups of animals:

- Presence of true tissues
- Pattern of gastrulation during embryo development
- Body symmetry
- Organs and organ systems
- Complex body cavities
- Segmentation of the body

With this road map in hand, let's look at the loose community of specialized cells that makes up the simplest of all animal bodies: the sponge body.

Sponges have the most basic body plan

Among animals, sponges have the simplest organization of the body—the least complex body plan

Helpful to Know

Zoion is Greek for "animal" and the root word in "zoology," the science of animals (*logia*, "study of"). Animals are also referred to as "fauna," just as plants are generically known as "flora." "Metazoans" is another general label for animals.

Concept Check Answers

1. Both are multicellular heterotrophs, but whereas fungi digest food externally, animals ingest food and then digest and absorb it internally.

2. Sponges. They had evolved by 760 million years ago.

FIGURE 19.3 Sponges Have Specialized Cells but Lack Tissues

Sponges are loose associations of cells that each perform a special function. Choanocytes create water currents and snare food particles. Amoebocytes defend against pathogens and can also turn into other cell types.

Specialized Cells in Sponges

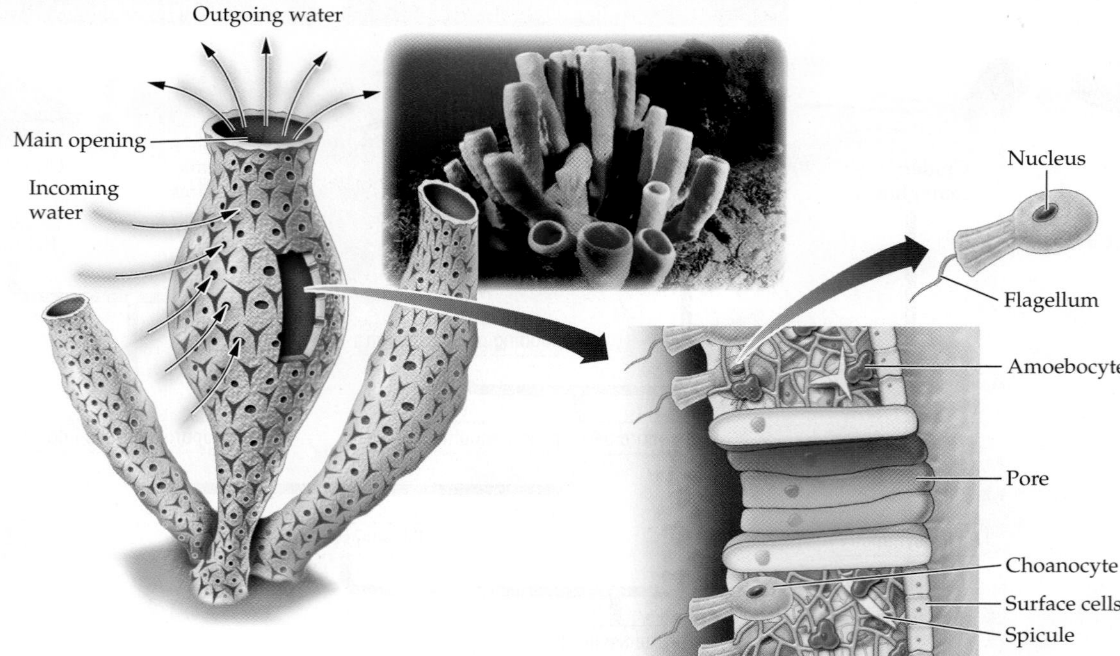

(**FIGURE 19.3**). Many sponges consist of nothing more than an outer layer of flattened cells surrounding a pouchlike cavity lined with the ciliated cells. Sponges with more complex organization have a multilayered body wall that includes a network of canals.

The cells that line the central cavity and the canals are called *choanocytes*, or "collar cells" (in Greek, *choane* is "collar" and *cyte* is "cell"), and they strongly resemble single-celled choanoflagellates. Choanocytes use sticky secretions to snare prey, such as bacteria, drifting by in the water current.

Most sponges have spiny cells, called *spicules*, coated in a glassy or chalky extracellular matrix. The sharp structures, which can resemble shards of glass, are a deterrent for would-be grazers and also shape and strengthen the animal's body. A sponge's amazing powers of regeneration are made possible by crawling cells (*amoebocytes*) that can regrow the entire body of the animal.

Although they have specialized cell types, sponges lack true tissues. Recall from Chapter 1 that a tissue is a group of cells that works in a coordinated fashion to perform a unique set of functions. A sponge is a loose collection of cells, with each cell functioning largely independently from other cells (see Figure 19.3). Most sponges have irregular shapes, with no distinct pattern of symmetry. There are no muscles, no nerves, and no coordination of activities among the different cell types in the body. The upside of this minimalist body plan is developmental flexibility: a small broken-off fragment can regenerate a whole new sponge body.

Most animals have true tissues

The cnidarians are among the first animal groups to have evolved true tissues. Jellyfish (**FIGURE 19.4**) exhibit a two-layered pattern of tissue organization: an outer layer of tissues makes up the body wall, and an inner layer lines the gut cavity. These cnidarians have a network of interconnected nerve cells that enable rapid communication from one part of the body to another. Although they do not have eyes like ours that can form images (camera eyes), cnidarians do have light-sensitive regions, known as *eyespots*, that can sense light and shade.

Muscle tissue resides between the inner and outer layers in most species of cnidarians. Muscle tissue is

EXTREME LIFE CYCLE REVERSAL

The immortal jellyfish (*Turritopsis nutricula*) can reverse its life cycle. If the bell-shaped adult form (seen here) is damaged, the remaining cells will turn into a blob that then sprouts many polyps, the juvenile form of the species. Because the adult-juvenile switch seems to happen indefinitely, some say this jellyfish is immortal.

True Tissues in Cnidarians

Gut cavity
The body wall tissues, derived from ectoderm, are protective.
A network of nerve cells extends through the jellylike mesoglea and into the adjacent tissue layers.
The cells lining the gut cavity are derived from the endoderm and facilitate digestion by releasing proteins that break down food.
Tentacles

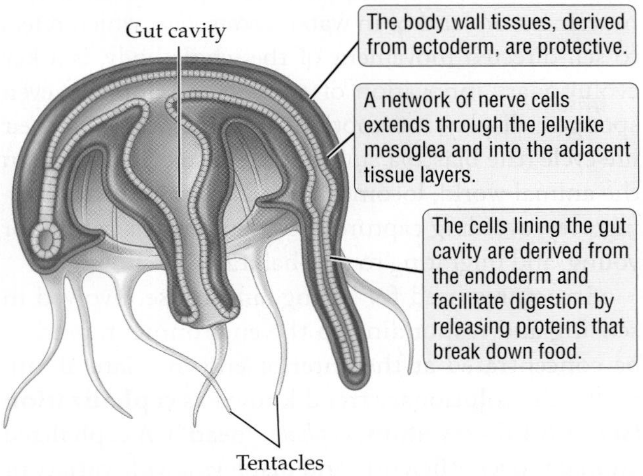

FIGURE 19.4 Jellyfish Have True Tissue Layers
Cnidarians, which include jellyfish, were one of the earliest groups to evolve true tissues. These tissues include the ectoderm (*ecto*, "outer"; *derm*, "skin") and the endoderm (*endo*, "inner"). For clarity, these two layers are color-coded blue and yellow here, respectively. Sandwiched between them is an inner (red) layer of secreted material known as the mesoglea (*meso*, "middle"; *glea*, "jelly").

unique to animals and found in all phyla of living animals except the sponges and placozoans ("flat animals," which constitute a puzzling phylum with a single species).

The coordinated action of the specialized tissues—the nerve net and muscle tissue, for example—makes complex behaviors possible. For example, some jellyfish impale small animals with special stinging cells located on their tentacles, and then use the tentacles to bring the captured prey to the mouth.

Animals exhibit unique patterns of embryo development

Although some animals can propagate themselves asexually, most animals produce offspring through sexual reproduction. In animals, the single-celled *zygote* that results from fertilization divides repeatedly to form a **blastula**, a hollow, fluid-filled sphere with a single layer of cells on the surface.

The movement of cells converts the single-layered blastula into a **gastrula**, an embryo with at least two cell layers (as in sponges). This embryonic reorganization—known as *gastrulation*—varies from simple (in sponges and jellyfish) to extremely complex (as in humans). In most animals, gastrulation begins with cells in the surface layer "diving" into the interior. The movement of cells into the interior of the hollow sphere creates a dent, called the *blastopore* (**FIGURE 19.5**).

In most protostomes (*proto*, "first"; *stoma*, "mouth"), tissues at or near the blastopore give rise to the mouth of the adult animal. In protostomes with a complete gut, a second opening that forms on the opposite side from the blastopore develops into the anus. In deuterostomes, by contrast, the blastopore generally gives rise to the anus, not the mouth (see Figure 19.5). A secondary opening, often but not always at the opposite end from the blastopore, gives rise to the mouth in deuterostomes (*deutero*, "second").

In animals with true tissues, the cells on the outer surface of the gastrula become the *ectoderm*, the embryonic tissue layer that generates the outer tissues of the animal body, such as the body wall and the nerve network of the jellyfish in Figure 19.4. The innermost tissue layer, created by the inward migration of cells at the blastopore, forms the *endoderm* and gives rise to the digestive system (shown in yellow in Figures 19.4 and 19.5).

Most protostomes and all deuterostomes have a third embryonic tissue layer, called the *mesoderm*, that arises from clumps of cells near the blastopore in most protostomes or, in many deuterostomes, from pouches that form at the far end of the pocket created by ingrowth of the blastopore. The mesoderm gives rise to tissues such as muscle and reproductive structures; in chordates, the mesoderm gives rise to the skeleton as well.

Embryo Development in Animals

Egg
Zygote
Blastula
Sperm
Gastrula
In most protostomes, the blastopore gives rise to the mouth; in deuterostomes, the anus usually develops at this end.

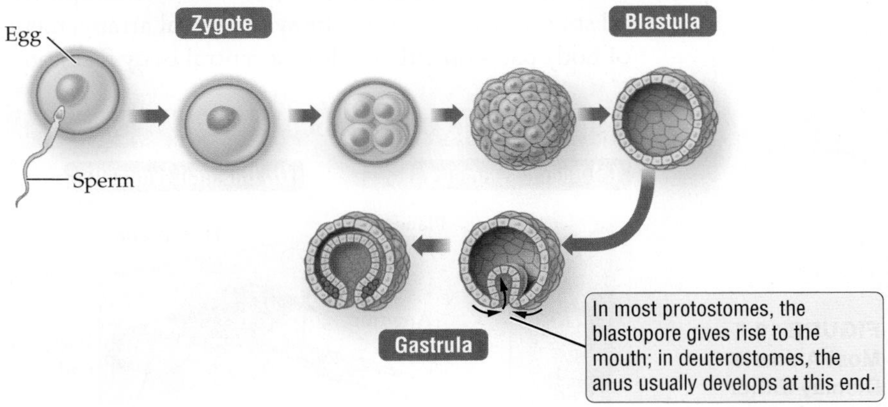

FIGURE 19.5 Early Embryo Development Distinguishes Protostomes and Deuterostomes

Most animals have symmetrical bodies

All animals except the sponges have a distinct body symmetry. Animals other than sponges can be divided into two main groups: those with *radial symmetry*, and those with *bilateral symmetry*.

The body of an animal with **radial symmetry** can be sliced symmetrically along any number of vertical planes that pass through the center of the animal (**FIGURE 19.6*a***). A sea anemone has radial symmetry because it can be cut, like a pie, to produce body parts that are nearly identical. Cnidarians (sea anemones, jellyfish, and corals, among others) and ctenophores (comb jellies) are among the animals that display radial symmetry.

Radial symmetry gives an animal sweeping, 360-degree access to its environment. The animal can snare food drifting in from any direction of the compass, and it can also sense and respond to danger from any of these directions. Radial symmetry is adaptive for sessile animals and those that drift in currents without being able to propel themselves in a preferred direction; because these animals cannot track down food or flee from a predator, they must be prepared to deal with whatever comes their way from any direction.

In animals with **bilateral symmetry**, only one plane passes vertically from the top to the bottom of the animal, dividing the body into two halves that mirror each other (**FIGURE 19.6*b***). Bilateral animals have distinct right and left sides, with near-identical body parts on each side. The top of a bilateral animal is the *dorsal* side, and its bottom surface is the *ventral* side. Bilateral animals almost always have a clear-cut front end (the *anterior* end) and a distinct back end (the *posterior* end).

Bilateral symmetry is seen in virtually all protostomes and deuterostomes, at least at some developmental stage in the life cycle. The symmetrical arrangement of body parts on either side of a central body facilitates movement in bilateral animals. The paired arrangement of limbs or fins, for example, enables quick and efficient movement on land or in water. *Locomotion*, which refers to self-directed movement of the whole body, is a key evolutionary innovation of animals. All animals, even sponges, display locomotion at some stage of their life cycle (the blastula stage, in the case of sponges). In the animal world, locomotion is important for capturing prey, avoiding capture, attracting mates, caring for young, and migrating to new habitats.

Structures used for eating, and those involved in sensing and responding to the environment, tend to be concentrated at the anterior end in bilateral animals—an evolutionary trend known as **cephalization** (*SEFF*-**uh-luh**-*ZAY*-**shun**; *kephale*, "head"). A cephalized animal moves efficiently and displays a wide variety of adaptive behaviors because it can receive and process information rapidly from the most important direction: the direction in which it is traveling. In other words, cephalization enables animals to look where they are going!

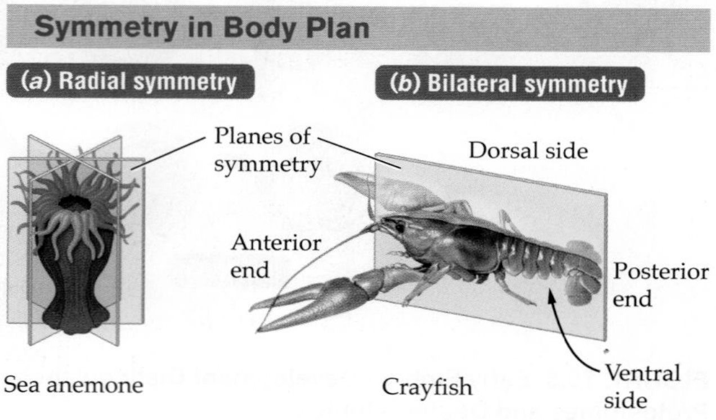

Symmetry in Body Plan

FIGURE 19.6 Most Animals Display Either Radial or Bilateral Symmetry

Most animals have organs and organ systems

Most protostomes and all deuterostomes evolved organs and organ systems, which enabled animals to function yet more efficiently. Recall that organs are body parts composed of two or more tissues that work together to carry out certain body functions. Usually organs have a defined boundary and a characteristic size, shape, and location in the body (see Chapter 1). **FIGURE 19.7** shows some of the organs found in the body of a flatworm, a protostome with one of the simplest body plans. The flatworm is a hermaphrodite, meaning that it contains both sperm-producing organs, known as *testes*, and egg-producing organs, known as *ovaries*. Clusters of nerve cells in the anterior are organized into a simple brain, which is a centralized information-processing organ found only in cephalized animals. The flatworm brain is networked with nerves that extend in a ladderlike pattern through the length of the body.

As you learned in Chapter 1, an organ system is composed of two or more organs that perform a set of related functions in the body. The flatworm brain is networked with nerves that extend in a ladderlike pattern through the length of the body. The simple brain and the nerves together constitute the *nervous system*. As minimal as it seems, the nervous system of flatworms is more complex than the nerve net of

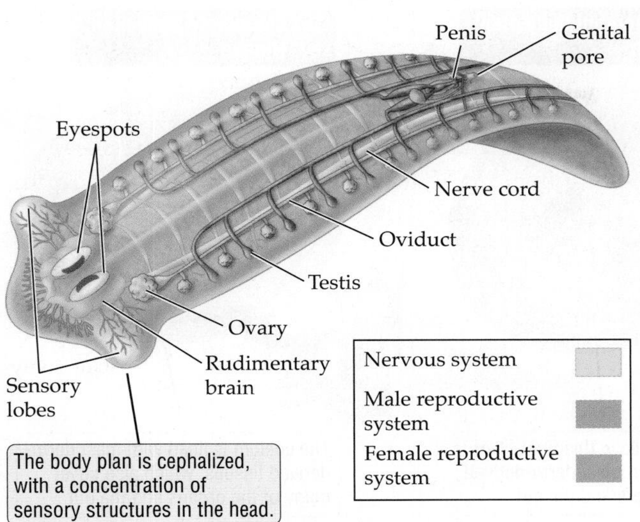

FIGURE 19.7 Organ Systems in a Protostome
The flatworm is a hermaphrodite, meaning that it contains both male and female reproductive structures.

cnidarians. Certain nerves deliver sensory input to the brain, which processes the information. Nerves that specialize in relaying output signals rapidly carry the response generated in the brain to the rest of the body for action.

Like the nervous system, the *reproductive system* of the flatworm is composed of organs linked to accessory structures. Eggs produced in the ovaries move through tubes called oviducts and exit the animal through the genital pore, for example. Flatworms also have a simple *excretory system* (not shown in Figure 19.7), which removes excess water and waste. The waste fluid is collected by specialized cells, called flame cells, routed through a system of collecting tubes, and then discharged to the outside.

Flatworms have a saclike gut (not shown in Figure 19.7) that is rather like the simple pouch of cnidarians and ctenophores; food and waste must enter and leave through the same opening in such dead-end guts. In contrast, most other protostomes and all deuterostomes have a complete *digestive system*, the organ system specialized for breaking down food, absorbing usable nutrients, and expelling undigested material. For example, the digestive system of vertebrates, such as fishes, consists of a one-way tube that includes a stomach and two functionally different intestinal regions, as well as accessory organs such as the pancreas and liver.

The *respiratory system*, which facilitates uptake of oxygen and removal of carbon dioxide gas, is structured in varied ways among the different groups of protostomes and deuterostomes. Many protostomes and all deuterostomes have a *circulatory system*, which delivers essential gases and nutrients to, and whisks waste away from, all the cells in the body. The *muscular system* works in conjunction with an external skeleton in some protostomes, such as the arthropods, or with the bony internal *skeletal system* of vertebrates, to facilitate movement. The defensive system, or *immune system*, varies from simple to complex across the animal phyla, as does the *endocrine system*, which consists of organs that produce signaling molecules called hormones.

Some animals have evolved complex body cavities

The evolution of complex body cavities was another major step in the evolution of the animal body. In animals like us, many of the internal organs are protected within a well-organized cavity called the *coelom* (*SEE*-**lum**). In contrast, flatworms are *acoelomate* (lacking a coelom); their internal organs lie crowded between the body wall and the gut, embedded in tightly packed layers of mesoderm-derived tissues, rather than in a true cavity.

Pseudocoelomate animals, which include roundworms, do have a fluid-filled internal cavity, the *pseudocoelom*, that surrounds many organs, including the gut. However, the pseudocoelom develops in a different way from a true coelom: it arises as a space between the mesoderm and the endoderm (**FIGURE 19.8**).

Coelomate animals have a true **coelom**, a body cavity that lies *within* the mesoderm. Tissues derived from the mesoderm therefore line the internal surface of the coelom and also wrap around all the organs suspended in the cavity. Many protostomes and all deuterostomes have a true coelom.

The evolution of body cavities made it possible for an animal's internal organs to grow more freely and function independently, as the organs became liberated from the body wall on the outside and the gut on the inside. Body cavities also provide padding and protection for the organs, as well as turgidity and mechanical support for the entire body.

Helpful to Know

Why does it feel like your stomach drops when you descend a roller coaster? Some organs in the abdominal cavity are held in place by ligaments, but the stomach and intestines can shift slightly inside the coelom. With your body strapped down, your nervous system senses the changes as your gut floats free in your coelom for a split second before returning to its more familiar position.

FIGURE 19.8 Some Protostomes and All Deuterostomes Have a True Coelom

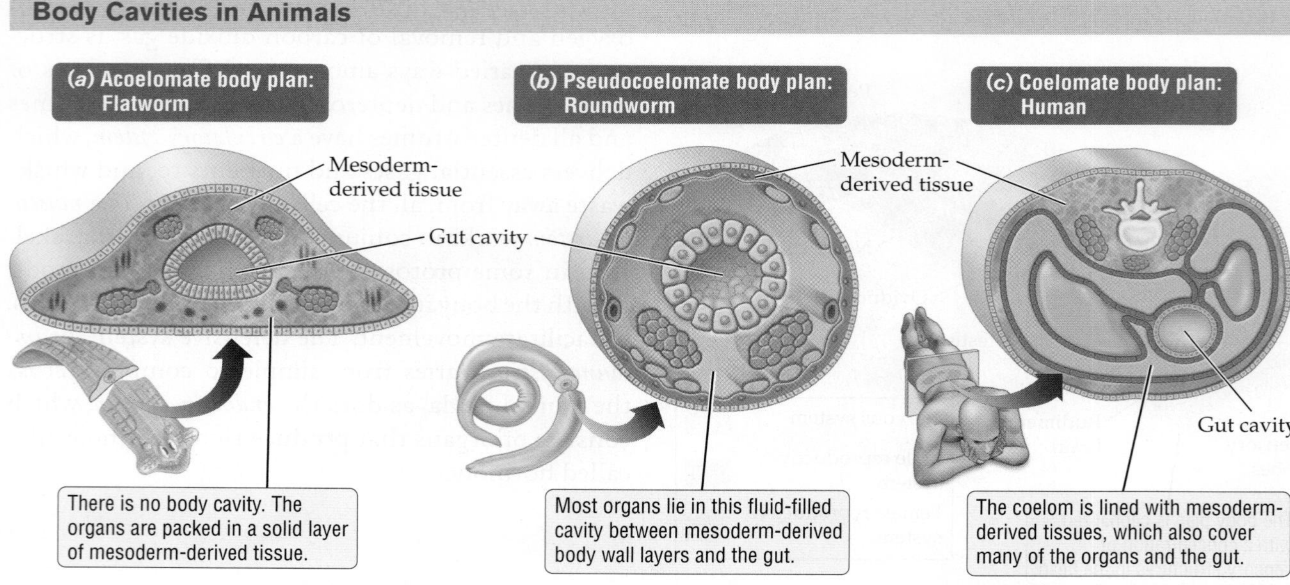
Body Cavities in Animals
(a) Acoelomate body plan: Flatworm
Mesoderm-derived tissue
Gut cavity
There is no body cavity. The organs are packed in a solid layer of mesoderm-derived tissue.
(b) Pseudocoelomate body plan: Roundworm
Mesoderm-derived tissue
Most organs lie in this fluid-filled cavity between mesoderm-derived body wall layers and the gut.
(c) Coelomate body plan: Human
Gut cavity
The coelom is lined with mesoderm-derived tissues, which also cover many of the organs and the gut.

Segmentation enables division of labor among body parts

Many animals have segmented bodies; that is, their body plan consists of repeated units known as *segments*. Specialized body parts, known as *appendages*, originate, often in pairs, from specific segments of the body. Over evolutionary time, the segments and the appendages that spring from them have acquired diverse form and function, enabling the animal body to adapt to new habitats or acquire new modes of life. The many uses of segments and their appendages are beautifully exemplified by the lobster in **FIGURE 19.9**.

The evolution of just the posterior segments of arthropods illustrates how evolution can take a basic structure, such as the segmented body plan, and modify it to produce many variations over time. The last segment in the body of arthropods has evolved into the delicate abdomen of the butterfly, the piercing abdomen of the wasp (which has a needlelike structure for inserting and laying eggs deep in another animal's body), and the delicious tail of the lobster. The front appendage of vertebrates is another good example of variation on a basic structure. This appendage has evolved as an arm in humans, a wing in birds, a flipper in whales, an almost nonexistent nub in snakes, and a front leg in salamanders and lizards.

Concept Check

1. Name four evolutionary innovations of animals.
2. Are humans and jellyfish bilateral? What are some advantages of bilaterality?

Concept Check Answers

1. All of the following are innovations found in animals: true tissues, unique patterns of embryo development, body symmetry, organs and organ systems, complex body cavities, and segmentation.
2. Humans are bilaterally symmetrical; jellyfish are not. Having paired parts on either side of the body's central axis enhances locomotion.

19.3 The First Invertebrates: Sponges, Jellyfish, and Relatives

The Kingdom Animalia

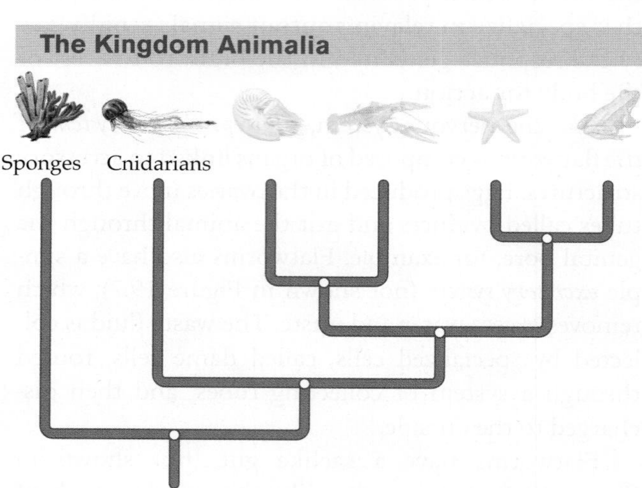
Sponges
Cnidarians

Examine the evolutionary tree of animals in Figure 19.2 and you will see that we share our world with some ancient lineages of animals that diverged very early in animal evolution, before the most recent common ancestor of protostomes and deuterostomes put in an appearance. These ancient animals include the sponges (phylum Porifera); jellyfish, corals, hydras, and their relatives (phylum Cnidaria); and the comb jellies (phylum Ctenophora). In this section we will take a closer look at these ancient animal lineages.

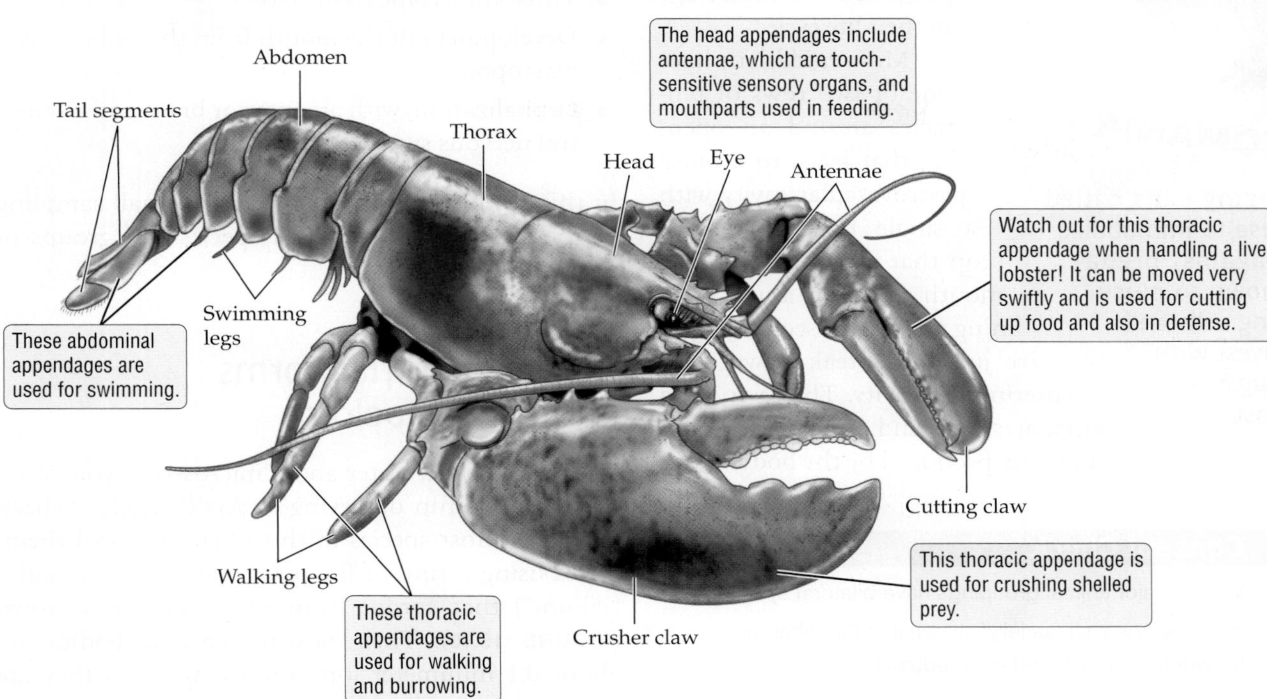
Adaptive Value of Segmentation
Abdomen
Tail segments
Thorax
The head appendages include antennae, which are touch-sensitive sensory organs, and mouthparts used in feeding.
Head
Eye
Antennae
Watch out for this thoracic appendage when handling a live lobster! It can be moved very swiftly and is used for cutting up food and also in defense.
Swimming legs
These abdominal appendages are used for swimming.
Cutting claw
Walking legs
These thoracic appendages are used for walking and burrowing.
Crusher claw
This thoracic appendage is used for crushing shelled prey.

FIGURE 19.9 Body Segments Evolved Diverse Functions

Segmentation, a body plan in which segments repeat, enabled the evolution of diverse uses of appendages, as shown here in the body of a lobster.

Sponges are the simplest animals

Sponges are found in shallow tidal zones and at great depths in the open ocean, and they live from the poles to the tropics. Some sponges are absorbent and yellow and porous, but sponges are much more varied than that. Nearly every color of the rainbow is seen in living sponges.

There are three main groups of sponges (desmosponges, glass sponges, and calcareous sponges), which differ in the type of extracellular strengthening material they produce. Desmosponge species, which lack the sharp spicules but have an abundance of the tough but flexible *spongin* (an extracellular matrix protein) were once widely used as bath sponges. Today, most commercial sponges for use in the home are made from synthetic (human-made) material to protect natural sponges from being overharvested.

Most members of the phylum Porifera (literally, "pore-bearers") are filter feeders: the choanocytes lining the canals and interior chambers beat their flagella, creating a current that draws water through the pores on the surface (see Figure 19.3). Sponges feed on bacteria, amoebas, and other tiny organisms in their aquatic environment, filtering a ton of water just to get enough food to grow one ounce of tissue.

Cnidarians and ctenophores have radial symmetry

Cnidarians include corals and sea anemones (anthozoans), hydrozoans, and jellyfish (scyphozoans). *Ctenophores* look superficially like jellyfish, but they differ from cnidarians in having a complete gut, meaning a tube with a mouth at one end and anal pores at the other end. Cnidarians and ctenophores have the following characteristics in common:

- Radial symmetry
- Two distinct tissue layers separated by a jellylike extracellular matrix (mesoglea)
- A nerve network consisting of interlinked nerve cells
- Lack of organs and organ systems

The adult form of some cnidarians is a sessile cylindrical *polyp*; the adult anemone is a polyp, as is the adult coral animal. Other adult cnidarian species have a motile, bell-shaped *medusa* (plural "medusae"), which is essentially an upside-down, free-floating polyp (see page 434). The medusae drift in currents but may move actively by contracting the muscle cells in the "bell."

EXTREME KLEPTOMANIA

Cnidarians have stinging cells called *cnidocytes* that they use to immobilize prey and to defend against predators. Sea slugs are known to pluck the harpoonlike stinging cells and stick them on themselves! With the stolen arms, this slug has gained protection against its own enemies.

Some species have both a polyp and a medusa stage in their life cycle.

Most cnidarians have one or two rows of tentacles around an opening that leads to a blind gastrovascular cavity with a single opening at the top that serves as both the mouth and the anus. Cells lining the cavity secrete digestive juices to break down food entering the cavity. The cavity also circulates food and oxygen and flushes out waste produced by the body.

Concept Check

1. Which major animal groupings have bilateral symmetry?
2. In cnidarians such as jellyfish, what is the difference between the polyp and the medusa?

Concept Check Answers

1. Protostomes and deuterostomes.
2. The poly is the cylindrical sessile stage of the animal; a medusa is bell shaped and motile.

19.4 The Protostomes

The Kingdom Animalia

Protostomes

Protostomes form the largest branch on the evolutionary tree of animals, encompassing a staggering diversity that ranges from the feathery Christmas tree worms to the colorful sea slugs (nudibranchs) seen in Figure 19.1*c*. Protostomes range in size from microscopic rotifers, which are smaller than many single-celled protists, to mollusks like the colossal squid, which can grow to 14 meters (46 feet). **Protostomes** share the following characteristics:

- Bilaterality at some stage in the life cycle
- Three embryonic tissue layers
- Development of the mouth from the embryonic blastopore
- Cephalization, with an anterior brain and a ventral nervous system

In this section we will explore just a small sampling of the more widespread and species-rich groups of protostomes.

Rotifers and flatworms lack a true coelom

With some pond water and a microscope, you could have a lot of fun observing *rotifers* (literally, "wheel-bearers"). Most species in this phylum propel themselves using a ring of fine projections (*cilia*; singular "cilium") that sprout from the head like a crown (**FIGURE 19.10**). Their near-transparent bodies are about 0.1 millimeter long on average, and they are sometimes mistaken for ciliated protists. Rotifers are important decomposers, using their crown of cilia to whisk organic particles and fish waste into the mouth. They also graze on algae and are very effective at keeping fish tanks clean. Rotifers are preyed on by many aquatic animals, including jellyfish and sea stars.

Flatworms (phylum Platyhelminthes) are acoelomate animals with a dead-end gut (one that lacks an anus). They look nothing like rotifers, but DNA comparisons show that the two phyla are closely related. Flatworms are either free-living (such as the several thousand species of freshwater planarians) or parasitic (such as flukes and tapeworms). Planarians are distinctly cephalized, with a triangular head bearing sensory cells and a pair of eyespots (pigments in the eyespots give the animal its cross-eyed appearance, which you can see in Figure 19.7). They lack respiratory and circulatory systems, and gases are exchanged across the skin. The thin, flattened shape of the animal facilitates gas exchange because the cells are close to the skin surface, where they can take up oxygen and release waste carbon dioxide.

A parasitic blood fluke that migrates from snails to humans is responsible for schistosomiasis, a devastating illness found in parts of Asia, Africa, and South America. Tapeworms, another serious parasite of humans, attach to the intestinal walls of their hosts with hooks and suckers on their anterior end. Fertilized eggs are shed in feces, and when eaten by animals such as pigs and cattle, the eggs become juvenile forms that

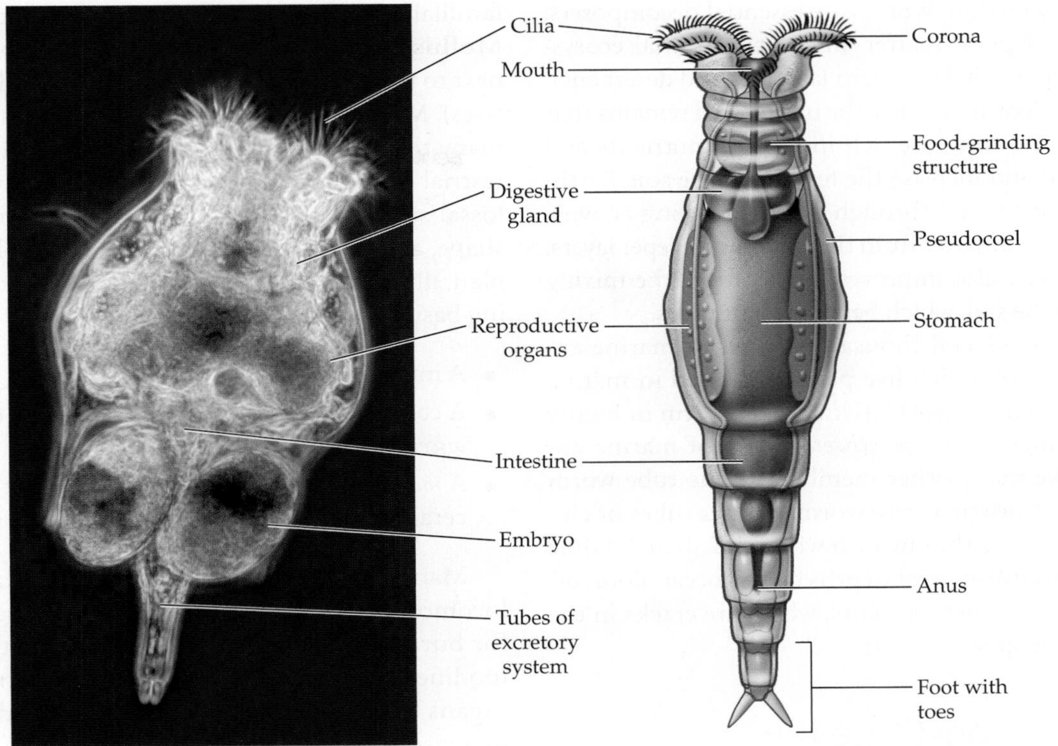

FIGURE 19.10 Rotifers Are Smaller Than Some Protists

migrate through the bloodstream to lodge in muscle tissue. Humans eating raw or underdone meat can acquire the parasite, which proceeds to reproduce in the intestines, continuing the infection cycle.

Annelids are coelomate worms with segmented bodies

All worms with true coeloms and segmented bodies are placed in the phylum **Annelida** (literally, "ringed ones"). The earthworms that emerge from their flooded underground burrows after a hard rain are the most familiar members of this phylum. Like most annelids, earthworms have a simple brain connected to two *nerve cords* (bundles of nerves) that run along the ventral side of the body (**FIGURE 19.11**). The skin of earthworms is thin and moist and is used for gas exchange. Earthworms have a closed circulatory system, which means the blood moves in closed tubes (blood vessels) at all times, instead of draining into a body cavity at any point.

Earthworms have a well-developed digestive system, with a *crop* for storing food and a *gizzard* in which ingested leaf litter is ground with swallowed grit for better

The Annelid Body Plan

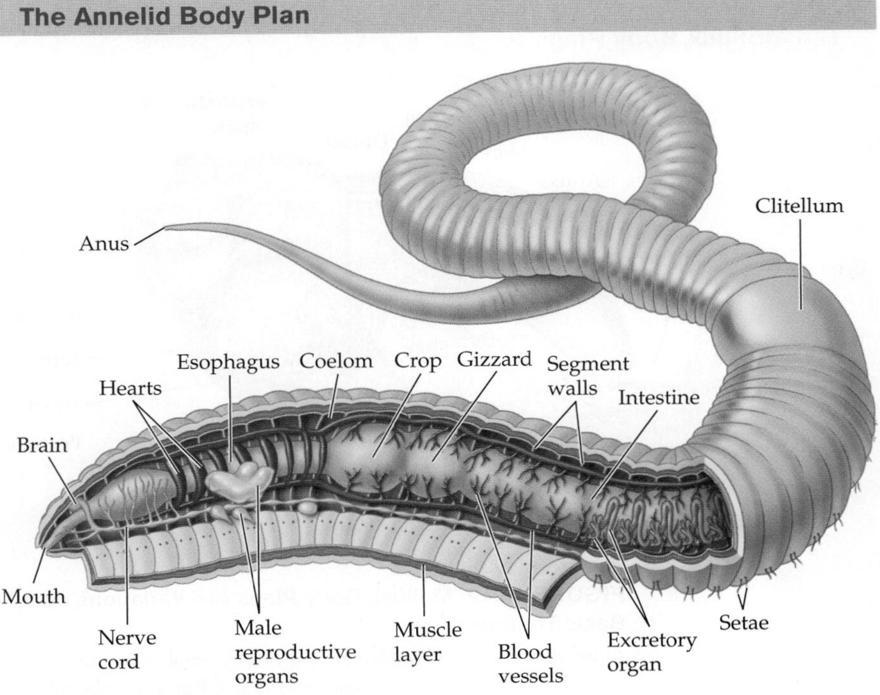

FIGURE 19.11 Segmentation Is a Key Feature of the Annelids

Annelids have a true coelom and well-developed organ systems.

nutrient extraction. Worms are essential decomposers, especially of plant matter, in most terrestrial ecosystems, except in cold northern latitudes and desert environments. Worm castings, the undigested remains that exit from the anus, are rich in mineral nutrients and humic acid and increase the fertility of the soil. Earthworms turn the soil through their burrowing activity, bringing plant matter from the surface to deeper layers. Worm activity also improves soil aeration, the mixing of air into the soil, which benefits plant roots.

There are several thousand species of marine annelids, most of which live partially buried in marine sediments. The showy Christmas tree worm in Figure 19.1*d* belongs to the *polychaete* group of marine annelids. Like many other members of the tube worm group, the Christmas tree worm secretes tubes of chitin that it can withdraw into when threatened. Some tube worms live at great depths on the ocean floor, often at deep-sea thermal vents, which are cracks in the ground that spew hot water.

Mollusks constitute the largest marine phylum

One of the largest and most diverse group of protostomes consists of the mollusks, which include familiar shellfish, snails, slugs, squid, and octopi. **Mollusca** is the most diverse phylum of animals, next to the arthropods (spiders, insects, and their relatives). Most mollusks are saltwater animals, although many thrive in freshwater and some live in moist terrestrial environments. From chitons to conchs to colossal squid, mollusks are enormously varied in size, shape, and details of the life cycle. The mollusk body plan, illustrated in **FIGURE 19.12**, includes the following basic characteristics:

- A muscular *foot* at the base of body
- A compact grouping of internal organs called the *visceral mass*
- A *mantle* enclosing the body cavity and the visceral mass

Many mollusks use their thick, muscular foot for locomotion. Mollusks like clams also use the foot for burrowing, and in squid and octopi, the foot is modified into several tentacles. Most of the internal organs of a typical mollusk are bunched together in a visceral mass. The outer layers of the body wall form a flap of tissue called the mantle that encloses a body cavity (*mantle cavity*) and also forms a protective cover over the visceral mass. In some species the mantle secretes a protective shell.

The *gills*, which are specialized for gas exchange, protrude into the mantle cavity; the gills absorb oxygen from and release carbon dioxide into the water that is pumped into and out of the cavity. Mollusks have a circulatory system made up of a simple heart and blood vessels, as well as a fluid-filled cavity.

Mollusks have a complete digestive system and an excretory system with structures called *nephridia* (singular "nephridium") that filter waste substances from blood, combine them with excess water, and expel both from the body in a fluid called urine. The nervous system varies from relatively simple in bivalves (clams, oysters) to highly complex in octopi.

Some marine experts estimate that over 23 percent of marine species are mollusks. We introduce here just three large—and probably familiar—groups: the *bivalves* (shellfish), the *gastropods* (which include snails), and the *cephalopods* (which include squid and octopi).

The Mollusk Body Plan

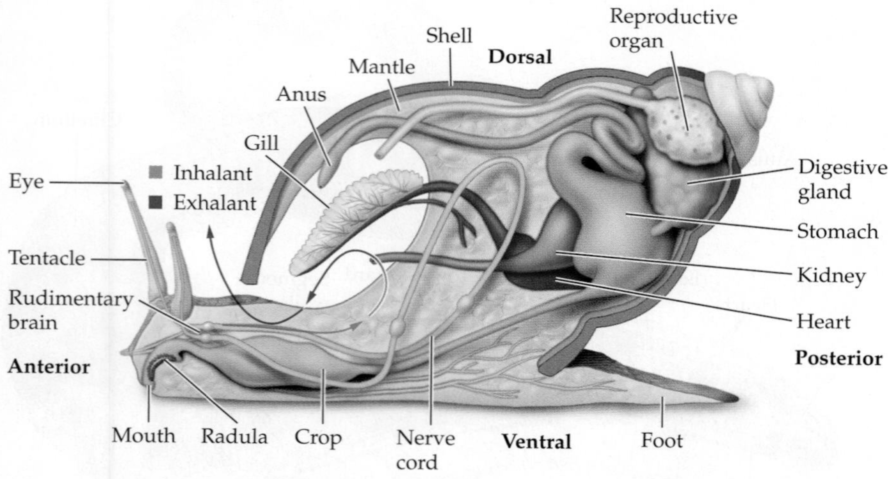

FIGURE 19.12 Mollusk Body Plans Are Variations on a Basic Theme

Although they show astounding diversity, mollusks have certain commonalities in their body plans. For example, all mollusks are bound by a taut sheet of tissue called the mantle, which wraps around the visceral mass, the compact lump formed by the internal organs. Most mollusks have a muscular foot used in locomotion.

BIVALVES *Bivalves* are mollusks that shelter their soft bodies inside a hard, hinged shell (**FIGURE 19.13*a***). Bivalves include familiar shellfish such as clams, oysters, scallops, cockles, and mussels. A bivalve animal opens its shell just enough to draw in water, often through

Mollusk Diversity

(*a*) Bivalve

(*b*) Gastropod

(*c*) Cephalopod

FIGURE 19.13 Mollusks Dominate the Sea

(*a*) Notice the rows of simple eyes near the edges of the shell. This bivalve, an Atlantic bay scallop, can detect motion by sensing changes in light levels. (*b*) At 40 pounds and 30 inches long, the Australian trumpet snail is the largest gastropod in the world. (*c*) Cephalopods like Paul the octopus are the brainiest invertebrates. Paul is credited with correctly predicting the outcomes of all international games played by the German national soccer team. Paul died in 2010 of natural causes.

a specialized intake tube, the *incurrent siphon*. The incoming water passes over the gills, enabling the animal to absorb oxygen and rid itself of waste carbon dioxide.

In filter-feeding species, which are the majority among bivalves, fine cilia on the surface of gills secrete mucus that traps incoming plankton and send the food-laden mucus into the mouth. The water is routed through the mantle cavity and then exits the shell, sometimes through a specialized *excurrent siphon*. Some bivalves, such as scallops, can move by jet propulsion: they shoot water out the rear by clapping their valves forcefully, thereby moving the animal forward like a jet.

GASTROPODS Most *gastropods* have a spiral calcareous shell on the dorsal side of the animal (**FIGURE 19.13*b***). Gastropods (literally, "belly-foot" animals) include slugs, snails, periwinkles, limpets, whelks, abalones, and the sea slugs (nudibranchs) pictured in Figure 19.1*c*. Some gastropods, such as garden slugs and sea slugs, lack a shell. In land snails and slugs, the gill is modified into a lung with an extensive network of fine blood vessels (capillaries). In some cases the bright coloration of nudibranchs (literally, "naked gills") is camouflage; in other cases it is a warning to would-be predators that the nudibranchs are poisonous.

Some gastropods use their thick foot for gliding on a slime trail. In others, the foot does double duty as a digging tool, and in some species it assists in swimming as well. Many gastropods scrape food, such as algae, with the rasping action of a grooved, tonguelike structure called the *radula* (plural "radulae"; see Figure 19.12). The radula has evolved additional functions in some species. In cone shells, for example, the radula is modified into a venom-injecting tube. Stings from these tropical gastropods can be fatal.

CEPHALOPODS *Cephalopods* (literally, "head-foot" animals) are exclusively marine and include shelled species such as the nautilus; species with reduced shells, such as squid and cuttlefish; and species that have no shell at all, such as the octopus. In most species the foot is modified into several clasping arms, or tentacles.

The octopus (**FIGURE 19.13*c***) has a pair of large, image-forming eyes (camera eyes) that rival human eyes in complexity, although vertebrate eyes evolved independently from those of any mollusk. The paired eyes have lenses to focus light and an inner surface (*retina*) capable of forming detailed images. In contrast to vertebrate eyes, which focus by changing the shape of the lens, the typical cephalopod eye focuses by moving the lens away from or toward the retina.

The sophisticated nervous system of cephalopods is an adaptation to their predatory lifestyle, particularly

Helpful to Know

Have you noticed that clams and mussels gape open when they are cooked? Strong elastic tissues hold the two halves of the shell closed much of the time for protection. Quite conveniently, the shell springs open when cooking destroys the ligaments that help close the two valves.

Helpful to Know

Cephalopods are thought to be the brainiest invertebrates. They have a well-developed brain and an extensive system of branched nerves. Octopi can learn to navigate a maze and have been known to use tools (such as carrying away a coconut shell and later using it as a hideout). An octopus at a German zoo learned how to open screw-top jars with its tentacles after watching zoo staff unscrewing the lids on jars containing shrimp, a favorite food.

the need to swim fast and track down prey, which range from crabs to schools of fish. Many cephalopods can change color, either for camouflage or to communicate with other individuals, such as potential mates.

Most cephalopods rely on jet propulsion for speedy locomotion, using strong contractions of the mantle cavity to shoot out water. Many cephalopods expel large clouds of ink, which they store in a sac below the gills, to help them escape from predators; in addition, some species blanch as they shoot out the ink—a disappearing act that must further confuse the predator.

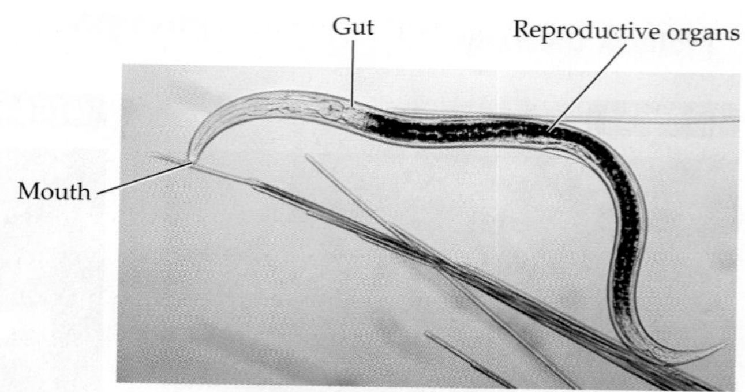

FIGURE 19.14 The Nematode *Caenorhabditis elegans*
Nematodes are often hermaphrodites, commonly reproduce about 4 days after hatching, and live for 3–4 weeks. Much of the body is filled with reproductive structures, which include two ovaries and two sperm-producing structures.

Some protostomes shed their outer covering to enable growth

The members of one group of protostomes, the ecdysozoans (*ecdysis*, "getting out of"), have the habit of shedding their outer covering on a regular basis. Ecdysozoans have a protective noncellular layer, the cuticle, composed of organic material secreted by the outermost layer of skin cells. The cuticle is relatively thin in some phyla but forms a thick, platelike exoskeleton in other phyla. Juveniles encased in the cuticle cannot grow unless they shed the covering in the process known as **molting** (ecdysis), grow rapidly, and then secrete a new cuticle to protect themselves.

Arthropoda is the largest phylum grouped under the ecdysozoan label, and it is also the largest phylum among all eukaryotes. Arthropods (*arthro*, "jointed"; *pod*, "foot") are animals with jointed body parts, including crustaceans (shrimp and lobsters), chelicerates (horseshoe crabs and spiders), millipedes and centipedes (myriapods), and insects and their relatives (hexapods). We consider some interesting nonarthropod ecdysozoans—the tardigrades and nematodes—before turning our attention to the dramatically diverse arthropods.

Tardigrades, commonly known as water bears, are soft-bodied, usually microscopic, segmented animals with three to four pairs of fleshy, unjointed legs. They are found in deep oceans, beach sand, freshwater sediments, garden soil, and the film of water that covers lichens and mosses. Two million or more of these animals may occur in one square meter of a mossy bank.

Organisms in the phylum Nematoda (roundworms) are widely distributed in freshwater, marine, and terrestrial environments. Most of the species are free-living soil dwellers, but many are parasitic. Soil nematodes often live at extremely high density: billions of them may be found in a single acre of fertile farmland. Nematodes range in size from microscopic to gigantic: the largest nematode known measures 9 meters and lives inside the placentas of sperm whales.

A soil nematode called *Caenorhabditis elegans* (**FIGURE 19.14**) has become a star in the modern study of animal development. The entire DNA of the species has been decoded, and the tools of DNA technology make it very easy to manipulate the animal's genes. The transparent body of the adult worm, about 1 millimeter long, contains exactly 959 cells, including 302 nerve cells, and researchers have been able to track the development of every one of these cells from the zygote stage.

Arthropoda is the most species-rich phylum

More than a million species of arthropods, mainly crustaceans and insects, have been described. Members of the group are also extremely abundant: about 10^{18} (a billion billion) arthropods are estimated to be alive at any given time. Arthropods have a hard outer cuticle, the **exoskeleton** (*exo*, "outer"), that is made of *chitin* (*KYE*-**tin**), the same biomolecule that is found in the cell walls of fungi.

Arthropods shed their exoskeleton in order to grow. Enzymes secreted by the underlying skin cells break down the base of the cuticle, which splits, allowing the animal to crawl out. The new cuticle that forms is soft and pliable initially, enabling the animal to grow. Until the growth stops and the cuticle hardens, the animal is vulnerable to predation and desiccation. Between molts, most arthropods stockpile food stores to fuel the growth that occurs immediately after the exoskeleton is shed.

Why are arthropods so successful? One characteristic that has contributed to the evolutionary diversity and success of this phylum is the segmented body plan. Over time, individual body segments have evolved different combinations of legs, antennae, and other specialized appendages, resulting in a huge number of different types of animals, many of them exceptionally well adapted to their habitat (see Figure 19.9).

The tough exoskeleton provides waterproofing on land and protection against many potential predators in all types of habitats. The rigid exoskeleton also anchors muscles, which in turn enable rapid and precise movements of the different types of appendages, including the jointed legs that facilitate efficient locomotion. The arthropods share the following general features:

- Jointed appendages that facilitate quick and precise movements of body parts
- A cuticle that forms a hard exoskeleton
- A segmented body plan at some stage of the life cycle
- A three-part body plan, consisting of an anterior head and thorax and a posterior abdomen

Four major arthropod groups are illustrated in **FIGURE 19.15**. We will discuss each of them in turn.

CRUSTACEANS **Crustaceans** are aquatic arthropods that are especially diversified in the marine environment (**FIGURE 19.15*a***), although freshwater species

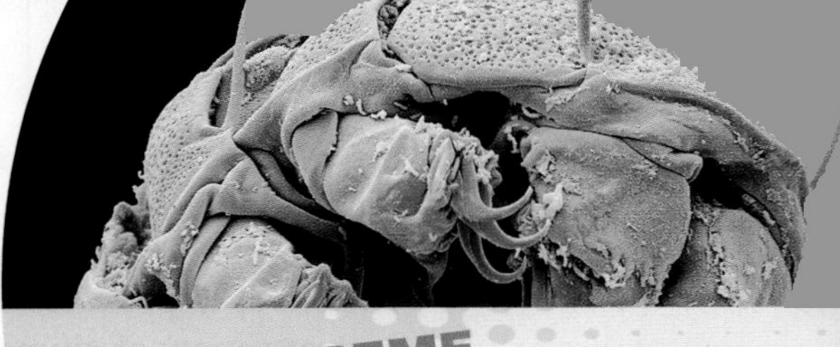

EXTREME SURVIVORS

The near vacuum of outer space is just a walk in the park, if you're a tardigrade. Popularly known as "water bears," these microscopic arthropods have endured temperatures between 151°C (304°F) and −200°C (−328°F), and they've survived 100 years as dehydrated, dormant balls that revive when sprinkled with water.

(*a*) Crustacean

(*b*) Insect

(*c*) Arachnid

(*d*) Myriapod

FIGURE 19.15 Arthropods

(*a*) Crustaceans (lobsters, shrimp, and crabs) are primarily water dwellers with 10 or more legs. (*b*) Insects have 6 legs. (*c*) Arachnids (mites, ticks, spiders, and scorpions) have 8 legs. (*d*) Myriapods (millipedes and centipedes) live on land and have many more than 10 legs.

are also common. Shrimp, lobsters, and crabs are among the most familiar crustaceans. Copepods, and the shrimplike krill, are microscopic crustaceans of enormous importance in aquatic food chains because a great variety of animals, including whales, feed on these abundant and prolific animals.

As noted earlier, the many different appendages of crustaceans such as the lobster perform a great variety of functions (see Figure 19.9), from sensing their world to capturing food and eating it. Many crustaceans have a flap of exoskeleton—the carapace—that covers and protects the head and thorax. Most crustaceans produce free-swimming planktonic larvae that are important food for larger animals.

INSECTS No animal group is more species-rich than the insects (**FIGURE 19.15*b***). More than half of the nearly 1.7 million known species of eukaryotes are insects. All the remaining animals make up only about 300,000 species (**FIGURE 19.16**). Insects are mainly terrestrial and are distinguished by the ability of many species to fly. Including grasshoppers, beetles, butterflies, and ants, among others, the insects are probably the best-known arthropod group. They have a three-part body plan with six legs attached to the thorax, but none to the abdominal segments, in contrast with many crustaceans (**FIGURE 19.17*a***).

Insects were among the earliest animals on land, and they exhibit many adaptations for a terrestrial existence. The exoskeleton, which prevents the soft body tissues from drying out, is one such adaptation. In contrast to the external gills of crustaceans, the gas exchange surfaces for insects are internal, so the moist surfaces are protected from desiccation (**FIGURE 19.17*b***).

As an adaptation to life on dry land, insects evolved a series of breathing tubes that branch inward from openings (*spiracles*) on the surface of the body. The largest tubes, called *tracheae* (singular "trachea"), branch repeatedly; the smallest of these, called *tracheoles*, allow gases to move directly to and from the bloodlike fluid (hemolymph) that fills the coelom. The valvelike outside openings of the tracheae can be opened and closed to regulate oxygen uptake (see Figure 19.17*b*). Larger insects can also pull air into the tracheal system, or push it out, by using muscles to alternately expand and contract the abdomen. Insects have well-developed visual systems. Some insects can detect motion far better than we can. Dragonflies can see and snatch mosquitoes (their prey) when both are flying in midair—no mean feat, if you recall from experience how difficult it is to swat a fly. One basic visual system found in many arthropods consists of **simple eyes** that can distinguish light and dark, and sometimes distance and color.

In addition to simple eyes, insects have a pair of **compound eyes** that can form images. Honeybees, for example, use their compound eyes to see patterns of color on flowers. A compound eye consists of many individual light-receiving units, each with its own lens and small cluster of photoreceptors. In dragonflies, the light-receiving units are densely clustered on certain portions of the eye, which, like the human retina, can form especially sharp images.

Flight has very likely been a big factor in the extraordinary success of insects. Wings enable insects to escape predators and unfavorable conditions, to locate food and mates far and wide, and to disperse

Insects: The Most Species-rich Group

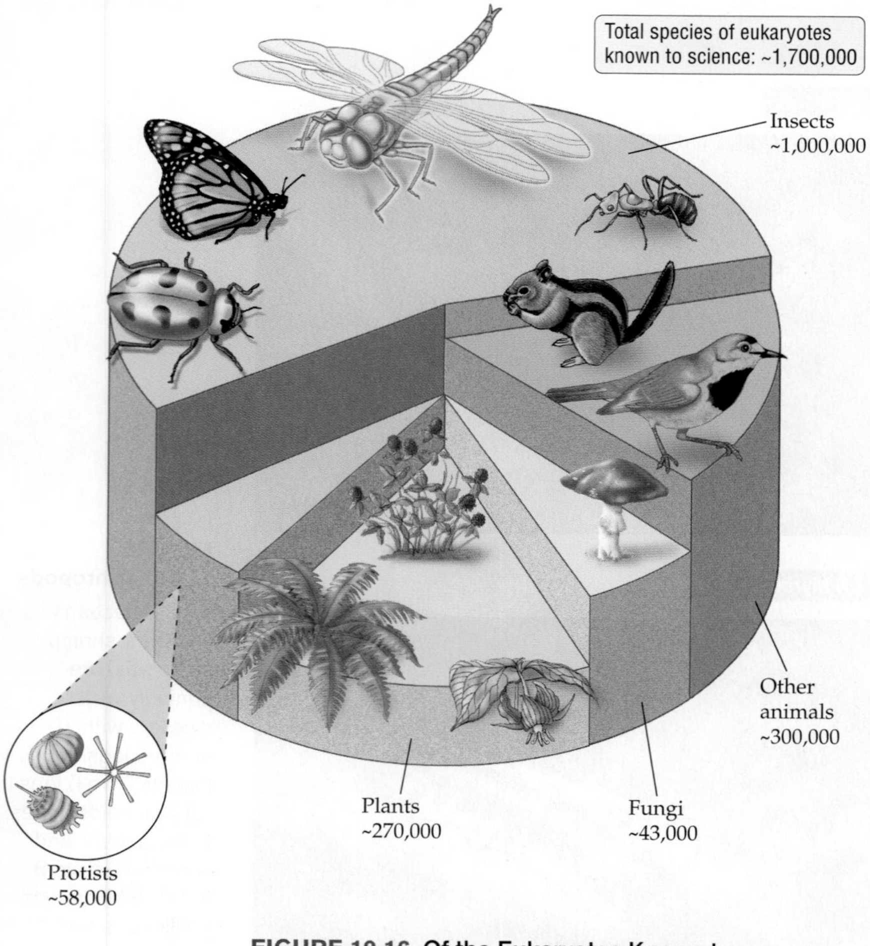

FIGURE 19.16 Of the Eukaryotes Known to Science, Insects Are the Most Diversified

The Insects

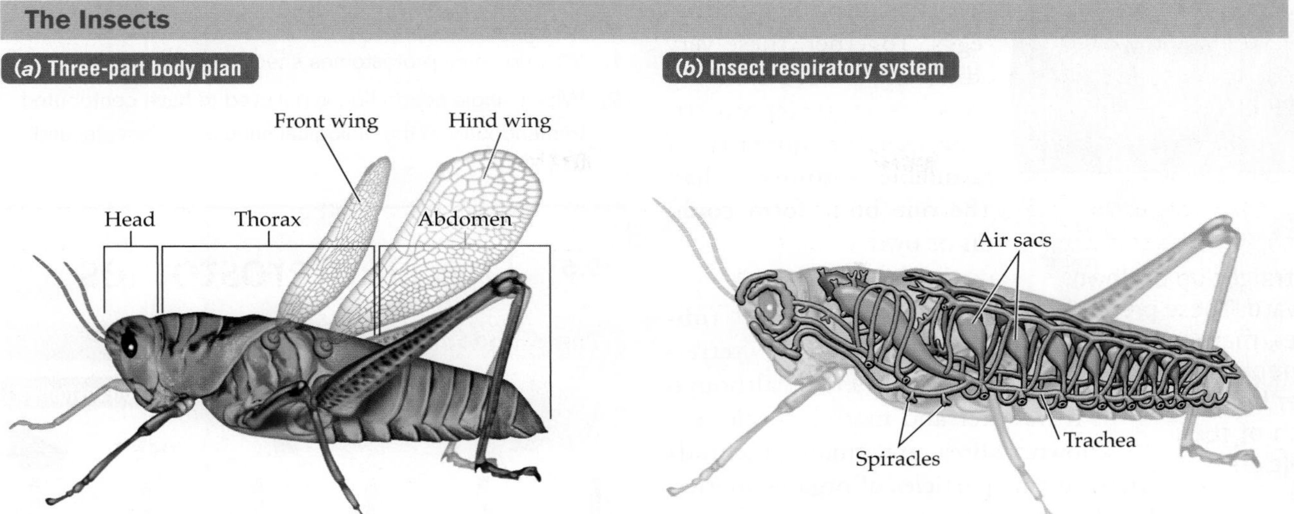

FIGURE 19.17 Insects Have a Three-Part Body Plan and an Enclosed Respiratory System

offspring over a broad landscape. Most insects have two pairs of stiff, gauzy wings. Flies have only one pair of wings, with the second pair modified into stubby balancing organs (called haltares). In beetles, the thin and gauzy second pair of wings is tucked under thick and sturdy forewings. Some insects, such as lice, had wings at an earlier point but then lost them over evolutionary time.

Many insects have complex life cycles in which the immature insect looks nothing like the adult. Caterpillars look very different from their adult forms, for example. The multistep developmental changes through which immature forms of animals are transformed into adults are collectively called **metamorphosis** (**FIGURE 19.18**). In arthropods, the transition from one developmental stage to the next is generally accompanied by molting. In some species, including grasshoppers and cockroaches, the developmental changes are gradual; such species are said to have *incomplete metamorphosis*. In other species, such as butterflies, the transition from one developmental stage to the next is dramatic, with the immature forms bearing little resemblance to the adults; these insects are said to have *complete metamorphosis*, as illustrated in Figure 19.18.

Why do some animals produce markedly different forms at various stages of their life cycle? Why do butterflies live much of their lives as the stub-legged leaf-munching machine that is the caterpillar, and then transform into a nectar-sipping, gossamer-winged adult? The answer is that this change has an evolutionary advantage. By having two very different body forms in their life cycle, metamorphic insects can pack two very different but highly successful and highly specialized modes of living into the life cycle of one animal. The body plan of the larval stage (the caterpillar) is well suited for voracious eating in a limited area, whereas the winged adult (the butterfly) is best suited for seeking mates and the best

Complete Metamorphosis

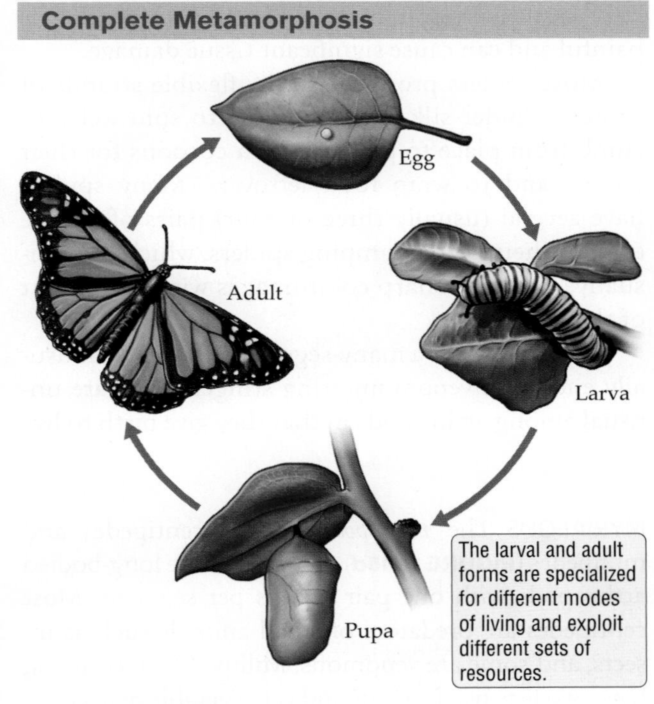

FIGURE 19.18 Butterflies Are among the Insects That Undergo Complete Metamorphosis

EXTREME FLYER

Dragonflies can fly straight up or down, and forward or backward. These predatory insects can hover, mate in midair, and in some species, clock speeds of 36 miles per hour. The 30,000 lenses in each of the compound eyes enable an almost 360-degree field of view.

locations for depositing eggs. Together, these very different modes of living acquire a greater variety, and therefore quantity, of available resources than the one body form could on its own.

ARACHNIDS *Arachnids* (uh-*RAK*-**nid**) are mostly terrestrial (**FIGURE 19.15*c***), although freshwater and marine species are known. Mites are small arachnids that eat tiny particles of organic matter, and although some are noxious parasites, especially of plants, most are beneficial as decomposers. The dust mites that inhabit most homes live mainly on the skin cells we shed regularly (some people are allergic to these mites' droppings). Ticks are tiny, spiderlike arachnids that are carriers for a number of infectious organisms, such as the bacterium that causes Lyme disease.

The sensory hairs that cover the body of most spiders (see Figure 19.15*c*) help the animals feel their way through the world. Spiders partially digest their prey by injecting digestive juices into them and then sucking up the liquefied remains with tubelike mouthparts. Many spiders inject venom to subdue prey or deter predators. No spider produces enough venom to kill a healthy person, but the bite of some spiders is painful and can cause significant tissue damage.

Most spiders produce strong, flexible strands of protein—spider silk—that they use to spin webs, to climb from place to place, to spin cocoons for their young, and to wrap food leftovers. Many spiders have several (usually three or four) pairs of simple eyes on their heads. Jumping spiders, which hunt visually, can form sharp color images with the largest of their eight eyes.

Scorpions have a many-segmented trunk that usually ends in a venom-injecting stinger. They are unusual among arthropods in that they give birth to live young.

MYRIAPODS The *myriapods* include centipedes and millipedes (**FIGURE 19.15*d***). Centipedes are long-bodied arthropods with one pair of legs per segment. Most centipedes are predators of small animals such as insects, and some are venomous. Millipedes, in contrast, have two legs per segment and eat vegetable matter.

Concept Check Answers

1. Protostomes that are ecdysozoans have a tough exoskeleton that strengthens and protects them, but also constrains growth. Young animals shed the exoskeleton periodically so that they can grow.
2. Flight, because it enabled insects to escape at least some potential predators, track down food and mates, and disperse offspring.

Concept Check

1. Why do some protostomes shed their skin?
2. Which single adaptation is believed to have contributed tremendously to the biological success of insects, and why?

19.5 The Deuterostomes

The Kingdom Animalia

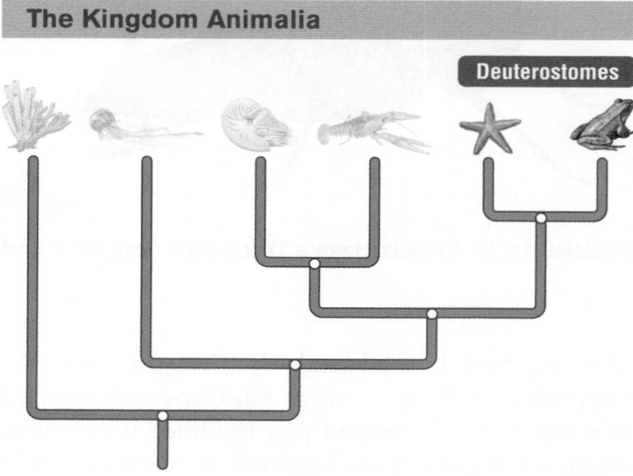

What do sea stars, sea squirts, and people have in common? Although we look very different, live in very different places, and behave in diverse ways, we are all deuterostomes. The deuterostome heritage of these groups is not readily apparent at the level of the whole animal, but **deuterostomes** share the following characteristics:

- The embryonic blastopore gives rise to the anus, and the mouth forms secondarily.
- A hollow nerve cord is located on the dorsal side (rather than ventrally).

DNA comparisons offer the strongest support for the idea that all deuterostomes descended from a single branch on the tree of life. All deuterostomes have a true coelom, and the adult individual is generated by three distinct embryonic tissue layers (germ layers) that arise during gastrulation. Deuterostomes with skeletal support systems have these structures on the inside (endoskeleton), in contrast to the external support structures (exoskeleton) of some protostomes.

The deuterostomes include a number of species-poor groups—for example, the phylum Hemichordata (literally, "half chordates"), which includes only the filter-feeding acorn worms. In the remainder of this section we will focus on the more familiar deuterostomes: echinoderms and chordates.

Echinoderms use a water vascular system for locomotion and gas exchange

Echinoderms such as sea stars, sea urchins, and sand dollars are radially symmetrical as adults, but their larvae are bilaterally symmetrical. The loss of bilateral symmetry may have been an adaptation to a slow-moving lifestyle in adulthood, when access to the environment from all sides would be more adaptive than a head-to-tail body plan.

Echinoderms such as sea stars and brittle stars have a mouth on the ventral side, an anus on the dorsal side, and five or more spokelike arms. Like some other echinoderms, sea stars have thick plates of calcified tissue that function like an internal skeleton just under the skin.

Below the branched sacs of the digestive system is a system of water-filled canals; it consists of a central ring, with branches radiating out into the arms and into the stumpy, flexible protrusions known as *tube feet*. Seawater entering the water vascular system is squeezed by the contraction of the surrounding muscles, and that compression, in turn, flexes the body, enabling the animal to move or to attach to its surroundings. The tube feet can latch on to a choice prey item through suction, and the body muscles can generate enough force to pry open a bivalve. The circulation of water through the canals also facilitates gas exchange, ushering in oxygen-rich water and spewing out the carbon dioxide–laden wastewater.

Chordates possess a dorsal notochord at some stage of their life cycle

You might find it hard to believe that the brightly colored, sessile sea squirt in Figure 19.1*b* is our closest relative in that photo gallery (see also **FIGURE 19.19*a***), but you might be more convinced if you could peer inside the juvenile sea squirt (**FIGURE 19.19*b***). Like all other **chordates**, the tadpole-like larvae of sea squirts (more formally known as tunicates) have the following characteristics:

- A dorsal rod of strengthening tissue, the notochord
- Pharyngeal pouches, which develop on either side of the throat in the embryo
- A post-anal tail

The **notochord** is composed of large cells that collectively form a strong but flexible bar running dorsally along the length of the animal. The notochord provides support for the rest of the body, the way a ridge beam supports the structure of a frame house. So, where is the human notochord? In some chordates, and all vertebrates, the notochord is lost during early development, and its function is transferred to stronger skeletal structures, such as the backbone (spine). In humans, traces of the notochord survive in the flattened pads of tissue—the intervertebral discs—that act as cushions between the vertebrae, the bones that make up our spine.

Like the notochord, *pharyngeal pouches* are found in the early embryos of all chordates. The **pharyngeal pouches** first appear as pockets of tissue on either side of the embryonic pharynx. The pharynx, commonly known as the throat, is the passageway posterior to (behind) the mouth. In fish and larval amphibians, the pharyngeal pouches deepen into the mesoderm and ectoderm layers until they merge with

EXTREME ARMS

When the going gets hot, the ochre sea star drops its arms. If exposed to high temperatures (35°C plus), this echinoderm draws heat away from its main body into the arms and sheds them! Thanks to the animal's amazing regenerative abilities, the lost arms easily regrow.

The Chordate Body Plan

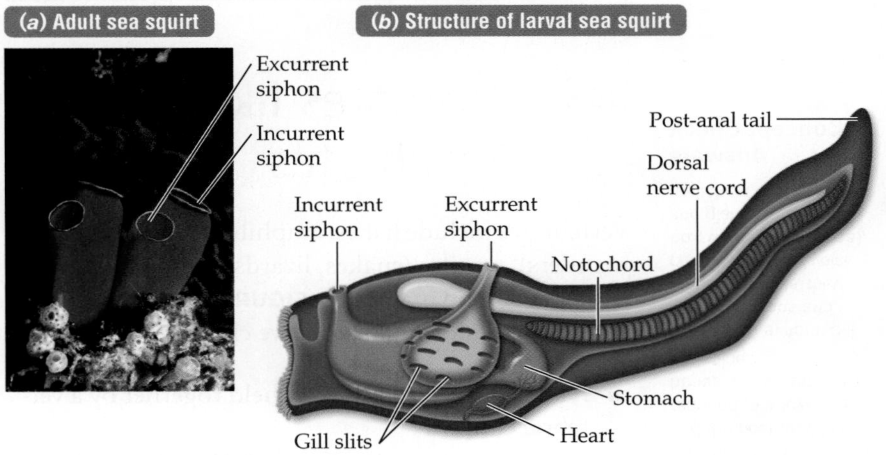

(*a*) Adult sea squirt

(*b*) Structure of larval sea squirt

Excurrent siphon
Incurrent siphon
Incurrent siphon
Excurrent siphon
Post-anal tail
Dorsal nerve cord
Notochord
Stomach
Heart
Gill slits

FIGURE 19.19 The Notochord Is Unique to the Chordates

Although the adult form of the sea squirt *Ciona* (*a*) looks very different from that of other chordates, such as people, the chordate hallmarks are evident in the larval form of the animal (*b*).

EXTREME JAWS

Some say the megalodon was the most formidable carnivore ever. The jaws of the prehistoric shark, reassembled from fossil fragments, are 9 feet high and 11 feet across. Megalodons, which died out about 150 million years ago, grew to 52 feet long and weighed up to 100 tons.

the body surface to create openings, called gill slits, in which the gills develop. In other vertebrates, the pharyngeal pouches fail to extend all the way to the body surface, and instead cells derived from these pouches give rise to a diversity of structures in the developing embryo. In mammalian embryos, for example, the pharyngeal pouches give rise to parts of the larynx (voice box) and trachea (windpipe), as well as the thyroid and thymus glands in the neck and chest.

The phylum Chordata includes several small subgroups of marine organisms, but we will focus on the vertebrates, animals with an internal *vertebral column* (backbone or spine) composed of a series of strong, hollow, cylindrical sections, known as *vertebrae* (singular "vertebra"). The vertebral column encloses the nerve cord that all deuterostomes have on the dorsal side of the body.

Concept Check

1. What do sea squirts have in common with us?
2. Pharyngeal pouches are found in the early embryos of all chordates. What becomes of them in fish and in humans?

Concept Check Answers

1. Both sea squirts and humans are deuterostomes and chordates. Sea squirts exhibit bilateral symmetry, a post-anal tail, and a dorsal notochord only in their juvenile form. In humans, the notochord all but vanishes in adulthood, and the post-anal tail is reduced to a few fused bones at the bottom of the spine.
2. In fish they form gill slits. In humans and other mammals, they contribute to the voice box, windpipe (trachea), and glands in the neck and chest.

19.6 Chordates like Us: The Vertebrates

Vertebrates include fishes, amphibians (frogs and salamanders), reptiles (snakes, lizards, turtles, and crocodiles), birds, and mammals (**FIGURE 19.20**). Vertebrates exhibit the following distinctive characteristics:

- A strong internal skeleton held together by a vertebral column
- An anterior braincase, or skull
- A closed circulatory system, with a pumping organ, the heart

The skeleton of the first vertebrates—the jawless fishes—was made from a strong but flexible tissue, called *cartilage*, that also extended over the brain in the form of a protective braincase, or skull. A muscular heart pumped blood through a network of blood vessels. Fine extensions of the blood vessels (capillaries) permeated the gills, which in these early fish were supported on an archlike framework of cartilage. The lamprey is one representative of jawless fishes found in today's oceans (**FIGURE 19.21*a***).

Jaws and a bony skeleton represent major steps in vertebrate evolution

The next great leap in vertebrate evolution came when two or three of the anterior gill arches became modified into hinged jaws. Jaws were a major evolutionary advance because jawed predators can grab, overpower, and swallow prey more efficiently than can jawless animals. The evolution of teeth made jaws yet more effective because they enabled animals to seize, tear, and cut up food. With their fearsome rows of serrated teeth (**FIGURE 19.21*b***), sharks had appeared on the world stage by the mid Paleozoic. The evolution of jawed fishes ushered the demise of the once-mighty jawless fishes, which are represented today by just a few members, such as hagfishes and lampreys.

Another major step in the evolution of fish, and vertebrates in general, was the replacement of the cartilage-based skeleton with a denser tissue strengthened by calcium salts: bone. Although the descendants of cartilaginous fishes—sharks, skates, and rays—are still with us today, *bony fishes* are far more diversified and widespread in both marine and freshwater environments. With more than 40,000 species, a majority of them marine, fish are the most diversified vertebrates today.

Most fish have two sets of paired fins, and typically several unpaired fins as well. The fins, which are modified appendages, are used for stability and swimming. *Ray-finned fishes* evolved slender, spikelike extensions of their bony skeleton to serve as a support framework for their fins (**FIGURE 19.21c**). *Lobe-finned fishes* are a third group of jawed fishes distinguished by paired fins having a thick muscular lobe supported by jointlike bones (**FIGURE 19.21*d***).

The body surface of bony fishes is typically covered by thin, flattened *scales*, which, along with the slime secreted by skin glands, reduces resistance to water flow over their streamlined bodies. The gills of ray-finned fishes are covered by a movable flap, the *operculum* (plural "opercula"). Fanning the operculum improves the flow of water over the gills, which enhances gas

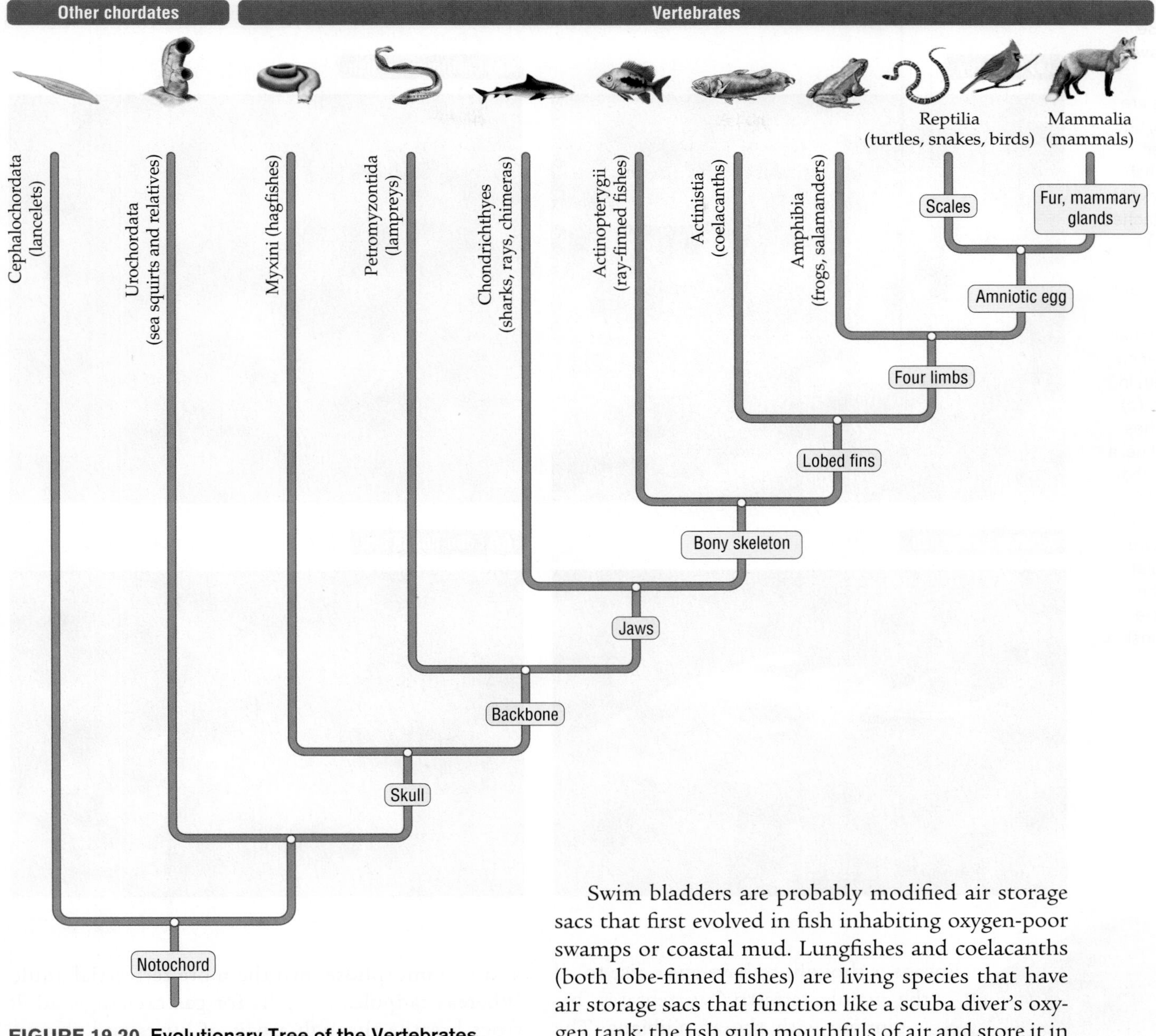

FIGURE 19.20 Evolutionary Tree of the Vertebrates

exchange. Sharks and other cartilaginous fishes do not have an operculum, and many of them must gulp regularly or swim constantly to pump enough oxygen-rich water over their gills.

Because bone is heavier than cartilage, gas-filled swim bladders have evolved in the bony fishes to help them stay buoyant with a minimum expenditure of energy (**FIGURE 19.22**). By controlling the amount of gas in the swim bladders, bony fishes can stay at a preferred depth without having to swim actively to maintain that position.

Swim bladders are probably modified air storage sacs that first evolved in fish inhabiting oxygen-poor swamps or coastal mud. Lungfishes and coelacanths (both lobe-finned fishes) are living species that have air storage sacs that function like a scuba diver's oxygen tank: the fish gulp mouthfuls of air and store it in their lunglike sacs to supplement the oxygen supply from the gills. Scientists believe that a pair of air storage sacs in a bony vertebrate ancestor evolved into a pair of dorsal swim bladders in the bony-fish lineage, and the same two sacs were modified into a pair of lungs in the lineage that gave rise to *tetrapods*, the four-legged vertebrates that colonized land.

Amphibians breathe through both the lungs and the skin surface

Terrestrial vertebrates with four limbs are known as **tetrapods** (*tetra*, "four"; *pod*, "foot"). This Devonian

FIGURE 19.21 Fishes May Be Jawless or Jawed
(*a*) Lampreys and hagfishes are jawless fishes, but the three other categories of fish shown here have a hinged jaw attached to the skull. (*b*) Sharks have thousands of teeth. The large main teeth are continually shed and replaced by smaller teeth in the inner rows. (*c*) Ray-finned fishes have a lateral line, a sensory organ that enables them to detect movement by sensing pressure. (*d*) The muscular lobed fins of the coelacanth have leglike bones inside.

Fish Diversity

(*a*) Lamprey

(*b*) Cartilaginous fish

(*c*) Ray-finned fish

(*d*) Lobe-finned fish

EXTREME BURROWER

The spadefoot toad of the Sonoran Desert develops a thick skin and burrows underground, where it remains without food or drink for 10 months. When the rains come, it emerges to reproduce—quickly! From egg to a little toadlet takes just 2 weeks.

lineage is thought to have given rise to all the mainly terrestrial vertebrates, beginning with the amphibians. The coelacanth, a "living fossil" that was thought to have become extinct in the Devonian period, has just such jointed, lobelike fins, attached to the skeleton by one large bone.

Many amphibians, such as frogs, have a life cycle characterized by complete metamorphosis: fertilized eggs (zygotes) hatch into tailed tadpoles that are entirely aquatic until they develop limbs, lose the tail, and metamorphose into the more terrestrial adult. Whereas tadpoles use gills for gas exchange, adult frogs breathe through a pair of lungs and also exchange gases through the moist surface of the skin and mouth. The advent of lungs was, of course, a crucial milestone in vertebrates' transition onto land.

Reptiles exhibit adaptations to a drier environment

The amphibious life limits vertebrates to habitats in which bodies of water are found at least seasonally. Reptiles were the first vertebrates to head into drier environments, and a number of adaptive traits evolved in this group to cope with dehydration. The

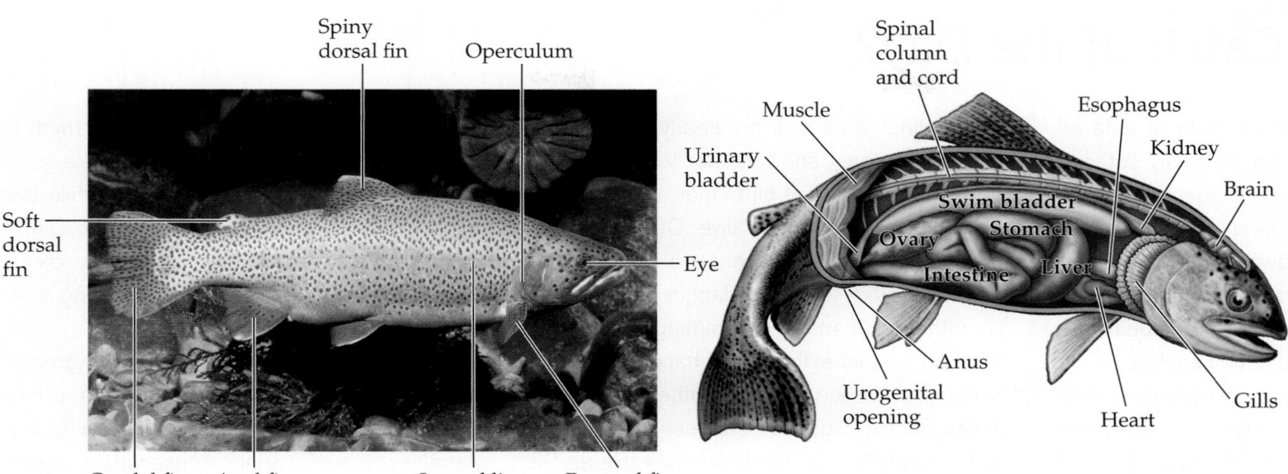

FIGURE 19.22 Bony Fishes Have a Strong Internal Skeleton
The swim bladder provides buoyancy, reducing the energy required to stay afloat.

major evolutionary innovations that first appeared among the reptiles include the following:

- Skin covered in waterproof scales
- A water-conserving excretory system
- The amniotic egg, with stored food and a waterproof shell
- Internal fertilization

A lineage of tetrapods with these characteristics gave rise to the reptiles in the Carboniferous period, about 354 million years ago. Reptiles dominated on land, in salt water and freshwater, and even in the skies, for much of the Mesozoic era (**FIGURE 19.23**), which is therefore known as the Age of Reptiles.

One line of reptiles, the dinosaurs, became spectacularly successful in the middle of the Mesozoic, although their fortunes were to decline toward the end of the era, 65 million years ago. Dinosaurs ranged greatly in size: the horned-faced *Microceratops* stood about 0.5 meter tall (about a foot and a half), while another plant eater, *Argentinosaurus*, was a 100-ton behemoth that measured 37 meters (about 120 feet) from nose to tail. Although *Tyrannosaurus rex*, at 40 feet in length and 40 tons in weight, tends to steal the limelight, some of the smaller predators that hunted in packs, like the 20-foot-long *Utahraptor* and the chicken-sized *Velociraptor*, were every bit as ferocious.

Turtles (which are aquatic) and tortoises and box turtles (which are terrestrial) are among the oldest reptiles. The dorsal side of the rib cage is modified into a tough, leathery protective shell in these reptiles. Snakes are a lineage of legless reptiles.

The **amniotic egg** is the most magnificent evolutionary innovation to appear among the reptiles (**FIGURE 19.24**). The developing embryo is surrounded and protected by layers of extraembryonic membranes that also promote gas exchange and store waste. The calcium-rich protective shell retards moisture loss but is porous enough to allow the entry of life-giving oxygen and the release of waste carbon dioxide. The amniotic egg hoards food in the form of a large *yolk* mass, which enables the young to achieve a relatively advanced level of development before they emerge (hatch) from the shell.

FIGURE 19.23 Jurassic Park
Dinosaurs reached the height of their dominance during the Jurassic, the middle period of the Mesozoic era.

BIOLOGYMATTERS

Good-bye, Catch of the Day?

Biologists have amassed a wealth of data on species already gone or on their way out, and even conservative analyses suggest that huge numbers of species have been, and continue to be, lost. According to Edward O. Wilson, a biologist at Harvard University, at the current rate of rainforest destruction 27,000 additional species will be doomed to extinction each year—an average of 719 per day, or 3 every hour. And although rainforests are particularly rich in biodiversity, they are just one of many different habitats that are threatened.

Despite the difficulty of estimating extinction rates for the world as a whole, scientists have definitively documented the extinction, caused by humans, of many hundreds of particular species in the last few thousand years. Twenty percent of the freshwater fish species known to be alive in recent history either have gone extinct already or are nearly extinct. One large-scale study showed that 20 percent of the world's species of birds that existed 2,000 years ago are no longer alive. Of the remaining bird species, 10 percent are estimated to be endangered—that is, in danger of extinction.

Although it may be tempting to assume that little of this extinction is happening close to home, evidence suggests otherwise. In the United States, frogs are disappearing in Yosemite National Park. In North America overall, 29 percent of freshwater fishes and 20 percent of freshwater mussels are endangered or extinct.

Humans eat many different species, but not all meals have the same impact on the planet's biodiversity. In particular, a number of fish and other marine species that humans consume are being overharvested and are in rapid decline. Eating these species only pushes them into steeper declines and could drive them to extinction. In addition, some species, like salmon, are raised on aquatic farms. But while that might seem like the perfect solution to overfishing wild species, these farms sometimes create high levels of ocean pollution, threatening other species.

So, when we are shopping at the grocery store or scanning a menu at the restaurant, how can we decide what might be both delicious and harmless to eat? Various conservation organizations have put together lists of seafood that are best to eat and best to avoid, often printed on handy wallet-sized cards. Shown here are some of the recommendations from the wallet card of the marine conservation organization known as Blue Ocean Institute. Just by choosing wisely the next time you crave seafood, you can help preserve the planet's biodiversity.

ENJOY	BE CAREFUL	AVOID
Arctic Char	Crabs (Blue, Snow, and Tanner)	Chilean Sea Bass
Clams, Mussels, and Oysters (farmed)	Monkfish	Cod (Atlantic)
Mackerel	Rainbow Trout (farmed)	Halibut (Atlantic)
Mahi-Mahi (pole- and troll-caught)	Sea Scallops	Salmon (farmed)
Salmon (wild Alaskan)	Swordfish	Sharks
Striped Bass	Tuna: Albacore, Bigeye, Yellowfin, and Skipjack (canned or longline-caught)	Shrimp (imported)
Tilapia (U.S. farmed)		Tuna: Bluefin (Atlantic)

SOURCE: Blue Ocean Institute, "Guide to Ocean Friendly Seafood," September 2007.

The body of reptiles is covered in overlapping layers of scales (**FIGURE 19.25**), made chiefly of a protein called *keratin*, that reduce water loss from the skin surface. Gases are exchanged exclusively through lungs, which have a larger surface area than the lungs of amphibians.

Living reptiles do not generate extra metabolic heat to warm their bodies. Instead, they are **ectotherms**, meaning that their body temperature matches the temperature of their environment. In cool temperatures, reptiles like lizards bask in the sun or sprawl on sun-warmed rocks; when the temperature rises, they seek shade.

Birds are adapted for flight

In the mid Mesozoic, about 175 million years ago, a lineage of feathered theropod dinosaurs, which

Structure of the Amniotic Egg

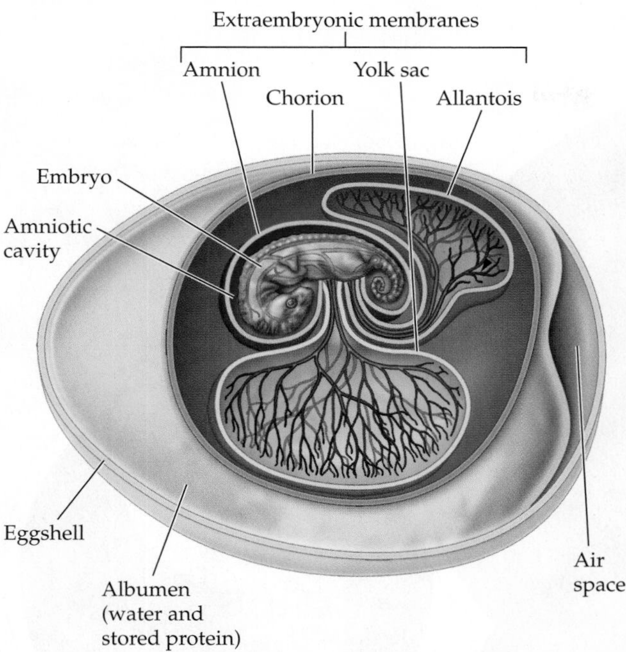

FIGURE 19.24 The Amniotic Egg Was a Crucial Evolutionary Innovation for Life on Land

include raptors and ornithomimids (*OHR*-**nih-thoh-***MYE*-**mid**; "bird mimics"), evolved into birds. Like other theropods, this lineage had an upright (*bipedal*) stance with large, powerful legs and smaller forelimbs. Their scales were modified into feathers made of keratin. The feathers evolved initially not for flight, but as a layer of insulation that kept the animals warm. In time, the feathers attached to the forelimbs became modified for stronger, sustained flight, with the tail feathers acting as stabilizers.

Transitional fossils, such as the 150-million-year-old *Archaeopteryx* discovered in a Bavarian quarry, offer a rare and wonderful glimpse of the evolutionary transformation of feathered dinosaurs into birds (**FIGURE 19.26**). Like its theropod relatives, *Archaeopteryx* had teeth in its beak, although modern birds have lost them as a weight-minimizing adaptation. The three claws on the forelimbs have also been lost in most modern birds, although the hoatzin, a puzzling species from the Amazon, has two claws that it uses for clambering in trees as a chick. *Archaeopteryx* perched on tree branches with clawed hind limbs covered in scales that resemble those of modern birds, and its feathers seem virtually identical to those of modern birds in their system of branching and interlocking barbs. Beyond hollow bones and light, toothless jaws (bills or beaks), modern birds exhibit other adaptations for flight, such as a reduction in internal organs; birds, for example, have only one ovary, instead of the two that are characteristic of all other tetrapods.

Unlike the living amphibians and reptiles, birds have a four-chambered heart that resembles ours: oxygenated blood and nonoxygenated blood are kept in separate chambers, and the blood that is pumped to the rest of the body therefore contains high levels of oxygen, supporting a high metabolic rate.

The respiratory system of birds is considered even more efficient than ours. Birds have a one-way flow of air through their respiratory passages, so that incoming air never mixes with outgoing air. In contrast, the air we inhale and exhale takes the same route (although in opposite directions), so there is some mixing of the incoming and outgoing airstreams. Like the mammals we discuss next, birds generate extra metabolic heat to warm their bodies (that is, they are **endotherms**) and they maintain a near-constant internal temperature (they are **homeotherms**).

Mammals came into prominence when dinosaur populations declined

Through much of the Mesozoic, when dinosaurs roamed all the continents, there were small, hairy creatures that kept a low profile as they scurried about in the vegetation living mostly on insects. The mammalian lineage split off from the reptilian line about 225 million years ago, in the early Mesozoic. With the decline of the dinosaurs toward the end of Mesozoic, mammals came into their own and diversified into a spectacular diversity of forms—from koalas to giant sloths—in the last 70 million years of Earth's history. The following characteristics of mammals have contributed to their success:

- Hair on the body and endothermy
- Sweat glands that cool the body through evaporation
- Young nourished with milk from mammary glands
- Internal fertilization and parental care

Size

Pygmy marmosets are the smallest primates.
This tiny carp from Indonesia is the world's smallest vertebrate. It is about 8 millimeters (0.3 inch) long.
The blue whale is the largest animal that has ever lived. It can grow to 110 feet and weigh 190 tons.
The colossal squid is the largest invertebrate. This rare cephalopod is believed to grow to 14 meters (46 feet) and can weigh 500 kilograms (1,100 pounds).

Locomotion

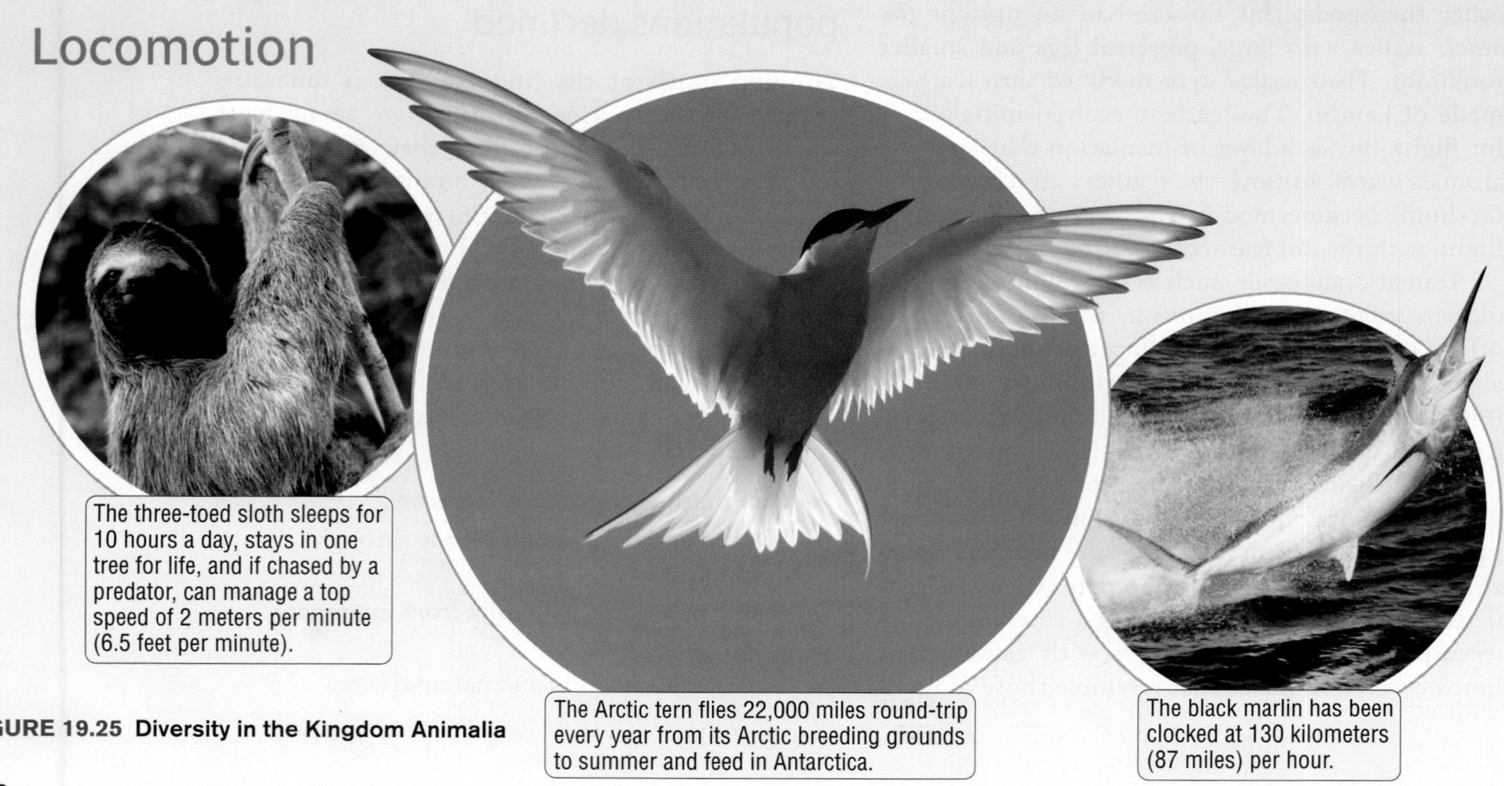

The three-toed sloth sleeps for 10 hours a day, stays in one tree for life, and if chased by a predator, can manage a top speed of 2 meters per minute (6.5 feet per minute).
The Arctic tern flies 22,000 miles round-trip every year from its Arctic breeding grounds to summer and feed in Antarctica.
The black marlin has been clocked at 130 kilometers (87 miles) per hour.

FIGURE 19.25 Diversity in the Kingdom Animalia

Reproduction

Clownfishes are born male; dominant individuals change gender to become female if a resident female dies.

Asian elephant mothers have the longest gestation (pregnancy): about 22 months. The long duration enables the newborn to become tall enough to reach with its trunk and nurse. Mothers nurse their calves for 2–3 years.

Komodo dragons are among the reptiles that can produce young without mating and fertilization. This phenomenon, known as parthenogenesis, has been reported in sharks, as well as in other fishes and in many invertebrates.

Life Span

The mayfly has no gut and lives for about 1 day.

Galápagos turtles are known to live to 150 years. They can go for 18 months without food or water.

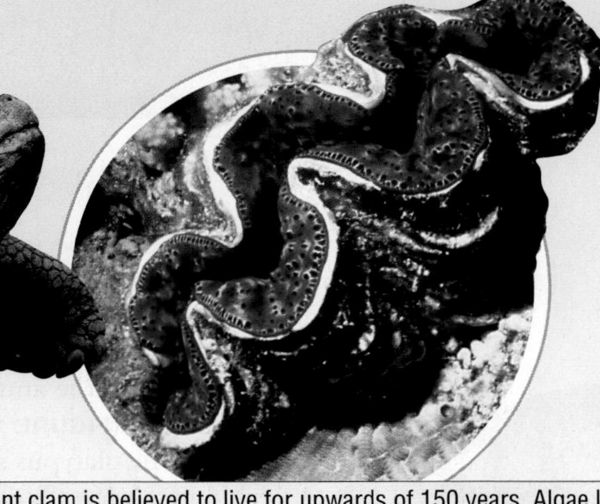

The giant clam is believed to live for upwards of 150 years. Algae live inside its mantle, and the clam obliges its mutualistic partners by keeping the valves open during the day for maximum photosynthesis.

Eating

Koalas live on the leaves of just a handful of *Eucalyptus* species. There are more than 600 species of *Eucalyptus* in Australia.

This new species of catfish from Peru eats wood from trees that fall into the rainforest rivers where it makes its home.

This deep-sea anglerfish dangles a bioluminescent lure that grows out from its head. The light, produced by bacteria that colonize the fleshy lure, attracts other predatory fish, which become a meal for the angler.

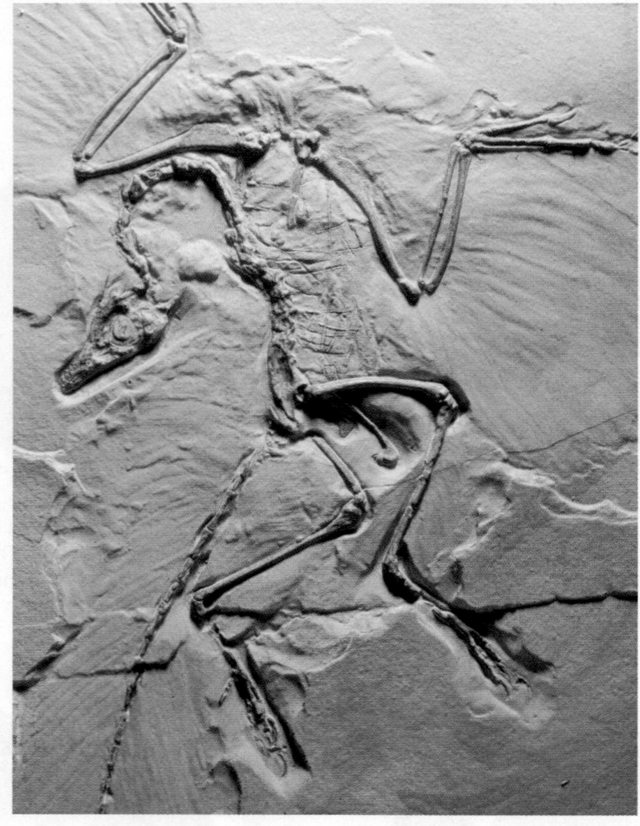

FIGURE 19.26
Archaeopteryx
This fossil from a German quarry shows the clear imprint of feathered wings and a feathered tail.

There are about 5,000 species of mammals, divided into three broad categories (**FIGURE 19.27**). The nonplacental, egg-laying mammals are classified as **monotremes** (*mono*, "one"; *treme*, "hole," referring to a common opening for the anus and urinary system). Monotremes (**FIGURE 19.27*a***) today consist of just one platypus species and several echidna species, all confined to Australia and New Guinea. The **marsupials** are animals that protect and feed their newborns with milk in an external pocket or pouch (*marsupium*, "pouch"). Marsupials (**FIGURE 19.27*b***) are found mainly in Australia and New Zealand, with a few species in the Americas. The North American, or Virginia, opossum is the only marsupial in North America.

More than 95 percent of mammals alive today are **eutherians** (*eu*, "true"; *therion*, "beast") a category that includes us. A unifying characteristic of eutherians (**FIGURE 19.27*c***) is that the developing young are nourished inside the mother's body through a special organ called the placenta and are therefore born in a relatively well-developed state.

EXTREME NEWBIE!

The olinguito is the newest ball of fur. Discovered in 2013 in the cloud forests of the Andes, this shy, tree-dwelling member of the raccoon family had earlier been confused with a related species, the olingo. The furriness of the olinguito, the "little olingo," is a legacy of its mammalian lineage.

Mammals can live in diverse habitats because they regulate body temperature

All birds and mammals are endotherms and homeotherms: they use metabolic energy to warm up the body, and they maintain a near-constant body temperature. The reptilian scales of the mammalian ancestor evolved into long, keratin-containing strands that we recognize as hair (fur is especially thick and luxuriant hair, such as that of nonhuman animals like foxes and mink). Muscles in the skin can raise the hair to trap a thicker layer of air next to the skin, thereby increasing the insulating properties of body hair.

Endothermy combined with hair has enabled mammals to colonize cold regions of the world and to remain active at temperatures too low for most other animal groups, except some birds. Hair is reduced on some aquatic mammals, such as whales, but the fetuses of even these species sport a good head of hair. Many adult whales have hair on the chin or snout (in right whales, for example), and the bumps (tubercles) on the snout of a humpback whale are large hair follicles, the living hair roots embedded in skin.

Only mammals have sweat glands. Evaporation of sweat cools terrestrial mammals and enables them to maintain a moderate body temperature even in extremely hot and dry environments, such as the desert. Aquatic animals and many fur-bearing mammals lack sweat glands.

Parental care contributed to the success of mammals

Like all reptiles and birds, and in contrast to most amphibians, mammals have internal fertilization, meaning that the male deposits the sperm inside the female's body and fertilization occurs within. In monotremes, the embryo develops within a shelled egg that is deposited outside the body. In marsupials, gestation (pregnancy) lasts for a short period—about 4 weeks in kangaroos—and the offspring are born in a relatively immature state. The joeys, as the infant marsupials are called, have strong limbs that they use to clamber into a pouch on the ventral side of the mother's body.

The Mammals

(*a*) Monotreme

(*b*) Marsupial

(*c*) Eutherian

FIGURE 19.27 Three Categories of Mammals

(*a*) Monotremes like the platypus are characterized by milk-producing mammary glands in females. Unlike all other mammals, however, monotremes lay eggs instead of giving birth to live young. (*b*) Kangaroos are marsupial mammals that give birth to relatively immature young that finish developing in a pouch (marsupium). (*c*) Eutherians give birth to well-developed young that suckle from the mammary glands.

All eutherians, and some marsupials, are placental animals. Embryonic tissues and maternal tissues combine to form a very special structure, the **placenta** (**pluh-*SEN*-tuh**), with an extensive blood supply. As the offspring develops in the special chamber known as the *uterus* (womb), it receives nutrients and oxygen across the placenta, which also removes waste chemicals and carbon dioxide. Nourished and protected inside the mother's body, eutherian offspring are born at a much more advanced stage of development than are marsupial young.

Mammary glands, which are modified sweat glands, are the most distinctive feature of mammals. They produce a liquid rich in fat, proteins, salts, and other nutritive substances that nourish the newborn. Monotremes secrete the fluid from the glands directly on the fur, where it is lapped up by the newborns after they hatch from their shells. Marsupial females have a nipple in each pouch that the newborn attaches to and feeds from. Eutherians nurse their newborns from two or more nipples on the ventral side of the body. The fat and sugar composition of eutherian milk varies: whale milk and seal milk have 10 times as much fat as cow's milk and almost no milk sugar (lactose).

Eutherians are highly successful in terms of the variety of habitats they occupy and the range of sizes they display, from shrews that weigh less than a gram to elephants that tip the scales at more than 5 tons. They have replaced dinosaurs as the top predators in most terrestrial habitats, and they thrive in both marine and freshwater habitats. Bats are the only mammals that can fly, although a number of eutherians and marsupials can glide.

Learning is developed to the highest degree among mammals, but many birds also transmit learned behaviors (such as song). Some eutherians are social animals, living in groups or herds in which guard duty and even rearing of the young are shared. Social eutherians tend to have comparatively large brains that enable complex behaviors, which in turn pave the way for exploiting a greater range of resources and habitats.

Concept Check

1. What is the adaptive value of the amniotic egg?
2. What are the key characteristics of mammals?

Concept Check Answers

1. The amniotic egg enables offspring to achieve a more advanced state of development before hatching by supplying the offspring with food and oxygen, removing waste, and protecting the offspring from drying out with extraembryonic layers.
2. Hair, endothermy and homeothermy, mammary glands, extensive parental care, and social living in some groups.

APPLYING WHAT WE LEARNED

Clues to the Evolution of Multicellularity

At the beginning of this chapter we met our tiny cousins the choanoflagellates. Just as we share a grandparent with a cousin (and a great grandparent with a second cousin), humans and all other animals share a distant ancestor with the choanoflagellates. In this ancestor, which lived between 700 million and 1 billion years ago, the ability to form colonies of communicating cells evolved. But while animals are multicellular all of the time, with specialized cells (and usually tissues and organs), choanoflagellates stayed single-celled and kept their colony-forming abilities as just an option—like a biology study group.

One way we know that choanoflagellates are our kin is that their individual cells look and work just like the special feeding cells of the simplest of all animals, the sponges (see Figure 19.3). Sponges are marine animals that attach themselves to a rock or other solid base and filter tiny organisms out of the water. Sponges lack a nervous system, blood circulation, or even a one-way digestive tract like ours, but they do have an interior space lined with *choanocytes* ("collar cells"). The choanocytes lining the interior cavity of a sponge look amazingly like choanoflagellates and function in a similar way, waving their whiplike flagella to pump water and food where they're needed.

Another clue to our kinship with choanoflagellates is that the choanoflagellates that form colonies do so using the same process that animals use to form embryos. A fertilized egg—whether it's from a human, a hamster, a fish, or a fly—divides into two cells, which stay in contact with each other. Those 2 cells divide again into 4 cells (and then 8, 16, and 32). But even though the cells divide, they stick together in a clump. They don't normally break away and go their separate ways. In the same way, colonial choanoflagellates repeatedly divide and stick together to create a cooperative group. Primitive colonies form in other ways as well. For example, in slime molds, separate cells come together to form a sluglike mass. The divide-and-stick-together development of animals and colonial choanoflagellates is a shared trait supporting the theory that they're related.

Still another clue that choanoflagellates are related to animals is that animals as different as flies and mice share dozens of nearly identical proteins with choanoflagellates. Intriguingly, nearly all of those proteins are involved in helping animal cells stick to one another or talk to one another. It's an ability that is very useful in a multicellular organism.

Surprisingly, some choanoflagellates that have these sticky proteins are not colonial. What do they use these proteins for? One possibility is that they use the sticky proteins to capture food particles. In some species, a long time ago, the food-snaring proteins may have been repurposed to become the glue that binds a colony of individual cells. At least one such colonial lineage evolved into a close-knit community of interdependent cells with division of labor: a multicellular animal. All animals, from sponges to people, are likely descended from such an ancestor.

BIOLOGY IN THE NEWS

Prime Number Cicadas

BY BARRY EVANS • *North Coast Journal* (Humboldt County, CA), August 1, 2013

... The male of the species [*Magicicada septendecim*] has but four weeks or so to attract females, mate, mate and mate ... and die ...

The eggs hatch about eight weeks later, and the newborn nymphs fall to the ground where they burrow down about a foot to start [17 years of underground life, sustained by] ... fluids found in the roots of deciduous woodland trees.

Finally, on a single spring evening when the soil temperature is just right, the nymphs emerge in their multitudes—up to two million per acre! ... A few days later, after a series of moltings, the adult males congregate in vast choruses of lust. Unlike crickets, which produce their signature chirping by rubbing their wings together, cicadas vibrate their entire corrugated exoskeletons, or timbals. The noise of millions of these creatures can be thunderous; they produce some of the loudest decibel ratings (up to 107 dB) of any insects.

The question, of course, is why 17 years? Or 13 years, in related species of periodic cicadas? The most common answer given is predator saturation. When all the nymphs in one area emerge simultaneously, birds, squirrels and other predators soon gorge themselves, leaving the survivors to breed in peace. The prime number pattern (13 and 17 are both prime) optimizes this saturation strategy by keeping the cicadas out of step with predators. If instead cicadas emerged every, say, 12 years, a predator might well evolve a strategic periodicity of its own, synchronizing its breeding cycle to two, three, four or six years (factors of 12); the 13- and 17-year cycles avoid a matching predator periodicity. Predators on the same cycle would decimate the cicadas. Also, the two cycles of 13 and 17 years helps keep "rival" cicada species separate, again optimizing survival rates by minimizing hybridization, which could throw off their timing, again putting them at greater risk from predators. The 13-year cicadas are found mostly in southern states, and the 17-year variety stays in northern states, with very little overlap.

Outside of the tropical rainforests—where not much changes over the course of the year—most organisms are seasonal breeders, timing their reproduction for when food will be most plentiful and conditions best suited to the survival of their young. Some plants and animals also coordinate their breeding so that an entire population reproduces at the same time.

Synchronized reproduction in trees is called seed masting. Even different species of plants—spruce, maple, and beech, for example—will time their flowering and fruiting for the same year, and just grow and stockpile food in their nonmasting "off" years.

In animals, synchronized reproduction is most common in prey species, from lemmings in the high Arctic to wildebeest in Tanzania. The leading hypothesis explains synchronized reproduction in the same way for plants and animals: it keeps predator populations from building up, because predators face either feast or famine from one season to the next. Predators gorge themselves on the bumper crop of prey in an "on" year, but because of their sheer numbers, enough of the prey survive to reproduce.

There are about 170 species of cicadas in North America, but only 7 of them—all belonging to the *Magicicada* genus—are periodic. Three of them emerge on a 17-year cycle, the other four on a 13-year cycle. Because there's genetic variation in populations, there are also scramblers and stragglers: Among 17-year cicadas, scramblers tend to appear 4 years earlier than their predicted emergence. Among 13-year cicadas, stragglers tend to pop up 4 years behind schedule.

As noted in Chapter 15, adaptive strategies often have trade-offs, and no strategy is perfect. Although periodic cicadas prevent their many predators from wiping them out completely, the flip side of that strategy is that they must put off reproduction for many years. And in the never-ending coevolutionary struggle between predator and prey, a fungus called *Massospora cicadina* appears to have evolved a reproductive cycle that synchronizes with its favorite prey, the 17-year cicada.

Evaluating the News

1. To which group of animals labeled in Figure 19.2 do cicadas belong?

2. What is the adaptive value of synchronized reproduction, and of reproducing on a prime number cycle, for a prey species such as the cicada?

3. If you live in eastern North America, explore the local lore of periodic cicadas in your region—for example, by looking up old newspapers or by interviewing older people. If you live elsewhere, research the culinary uses of some of the many hundreds of cicada species around the world.

CHAPTER REVIEW

Summary

19.1 The Evolutionary Origins of Animalia

- The Animalia are multicellular, ingestive heterotrophs. Most animals are capable of locomotion at some stage in the life cycle.
- Sponges are the most ancient animal lineage. Cnidarians evolved next. Sponges lack distinct symmetry; cnidarians have radial symmetry.
- Animals with bilateral symmetry are divided into two main groups—protostomes and deuterostomes—on the basis of their pattern of embryo development. Protostomes include mollusks, annelids, and arthropods. Among the deuterostomes are echinoderms and the chordates, which include us.
- The chordates include all animals with a dorsal nerve cord. Chordates with a backbone are vertebrates, and all other phyla of animals are informally designated as invertebrates.

19.2 Characteristics of Animals

- All animals except sponges have true tissues. All animals other than sponges and cnidarians have organs and organ systems.
- Animal development is characterized by cell migration and the formation of embryonic layers that generate all the different tissue types in the adult body.
- After fertilization, the single-celled zygote turns into a hollow ball, which forms a blastopore as cell migration begins. In most protostomes, the blastopore end gives rise to the mouth. In most deuterostomes, the anus develops from the blastopore end.
- Although some animals lack body cavities (are acoelomate), most have either a pseudocoelom or a true coelom.
- Many animals have segmented bodies; segments and their various appendages have become adapted to perform diverse functions in different species.

19.3 The First Invertebrates: Sponges, Jellyfish, and Relatives

- Sponges have specialized cells, but no tissues or distinct body symmetry.
- Cnidarians and ctenophores have true tissues and are radially symmetrical.

19.4 The Protostomes

- The evolutionary innovation of complete body cavities first evolved in the protostomes.
- Rotifers and flatworms lack a true coelom. Annelids are coelomate worms with segmented bodies.
- Most mollusks have a hard shell. Bivalves have hinged shells; gastropods have a dorsally located spiral shell. Cephalopods have a well-developed nervous system and are capable of complex learning tasks.
- Arthropods have jointed body parts and must shed their protective layer, the cuticle, in order to grow.
- Arthropods include crustaceans, insects, arachnids, and myriapods. Crustaceans are the most diversified aquatic animals, whereas insects are the most species-rich class of animals on Earth. Crustaceans and insects have a tough exoskeleton that protects them from desiccation and possibly from predation.

19.5 The Deuterostomes

- Deuterostomes have a dorsal nerve cord, and in most, the blastopore gives rise to the anus.
- Echinoderms are deuterostomes that use a water vascular system for locomotion and gas exchange.
- Chordates have a dorsal notochord at some stage of their life cycle.

19.6 Chordates like Us: The Vertebrates

- Vertebrates (which include fishes, amphibians, reptiles, birds, and mammals) are distinguished by having an internal backbone and an anterior braincase.
- The evolution of jaws and a bony skeleton were major steps in vertebrate evolution. These innovations first occurred in fishes.
- Fishes have paired fins that enable rapid swimming maneuvers.
- Amphibians breathe through both lungs and the skin surface. Their life cycles are often characterized by complete metamorphosis.
- Reptiles have evolved adaptive traits that reduce the risk of desiccation: waterproof scales, a water-conserving excretory system, and an amniotic egg.
- Birds evolved from theropod dinosaurs. They are adapted for flight, with a bipedal stance, feathers modified from scales, and hollow bones.
- Mammals are subdivided into three categories: monotremes, marsupials, and eutherians. Mammals have hair on the body, endothermy and homeothermy, mammary glands, and sweat glands.
- The success of mammals can be attributed to long gestation periods and enhanced investment in parental care.

Key Terms

amniotic egg (p. 453)
Animalia (p. 432)
Annelida (p. 441)
Arthropoda (p. 444)
bilateral symmetry (p. 436)
blastula (p. 435)
cephalization (p. 436)
chordate (p. 449)
coelom (p. 437)
compound eye (p. 446)
crustacean (p. 445)
deuterostome (p. 448)
ectotherm (p. 454)
endotherm (p. 455)
eutherian (p. 458)
exoskeleton (p. 444)

gastrula (p. 435)
homeotherm (p. 455)
mammary gland (p. 459)
marsupial (p. 458)
metamorphosis (p. 447)
Mollusca (p. 442)
molting (p. 444)
monotreme (p. 458)
notochord (p. 449)
pharyngeal pouch (p. 449)
placenta (p. 459)
protostome (p. 440)
radial symmetry (p. 436)
simple eye (p. 446)
tetrapod (p. 451)

Self-Quiz

1. Which animal group is the most abundant in number of individuals and number of species?
   a. insects
   b. birds
   c. protists
   d. mammals
2. Which of the following statements about animals is *not* true?
   a. Animals are ingestive heterotrophs.
   b. Animal cells are enclosed in a cell wall made of polysaccharides.
   c. All animals have at least some specialized cell types.
   d. Most animal cells are attached to an extracellular matrix.
3. Which of the following groups was the first to take to the air?
   a. bats
   b. birds
   c. protostomes
   d. certain reptiles
4. In deuterostomes,
   a. bilateral symmetry is absent.
   b. the mouth does not develop from the blastopore.
   c. the notochord is absent.
   d. the body is acoelomate.
5. Segmentation is beneficial
   a. to insects because it helps them stay dry.
   b. to arthropods because it facilitates specialization among body parts.
   c. to annelids because it is necessary for coelomate organization.
   d. to sponges because it led to cephalization.
6. True tissues
   a. are found in all animals.
   b. are thought to be absent in sponges.
   c. consist of two or more organs that work together in an integrated manner to carry out specific functions.
   d. consist of loose collections of cells that function independently from each other.
7. Chordates are distinguished from all other animals in that all of them
   a. have a dorsal nerve cord and a post-anal tail.
   b. possess mammary glands.
   c. have an anterior skull and well-developed jaws.
   d. have a backbone.
8. An amniotic egg
   a. is a characteristic of all tetrapods.
   b. is believed to have first evolved in jawless fishes.
   c. is found in birds, but not reptiles such as snakes and crocodiles.
   d. contains membranes that facilitate gas exchange.

Analysis and Application

1. What were the major challenges facing insects when they colonized land? What evolutionary innovations did insects use to deal with these challenges?
2. Why are sponges considered the simplest of animals? Compare the body plan of a sponge with that of a cnidarian such as a jellyfish.
3. What is the evolutionary significance of segmentation in the animal body? Name an animal that illustrates the adaptive value of segmentation.
4. Animals are typically mobile. What tissue types and organ systems enabled locomotion in animals? Compare locomotion in fishes and tetrapods. What is the adaptive value of locomotion?
5. What evolutionary innovations adapted birds for flight? List the characteristics shared by birds and reptiles such as crocodiles and dinosaurs. Which of these characteristics are also found in mammals?
6. List the main structures in a hen's egg. What is the function of each of these structures?
7. Did feathers first evolve as an adaptation for flight? Cite evidence that supports your answer.
8. How is complete metamorphosis different from incomplete metamorphosis? What is the adaptive rationale for complete metamorphosis (in other words, why do some animals go through such elaborate and dramatic changes in their life cycle)?
9. What *unique* adaptations do mammals possess for coping with cold and hot environments?
10. Compare how female monotremes, marsupials, and eutherians nurture their young.

20 The Biosphere

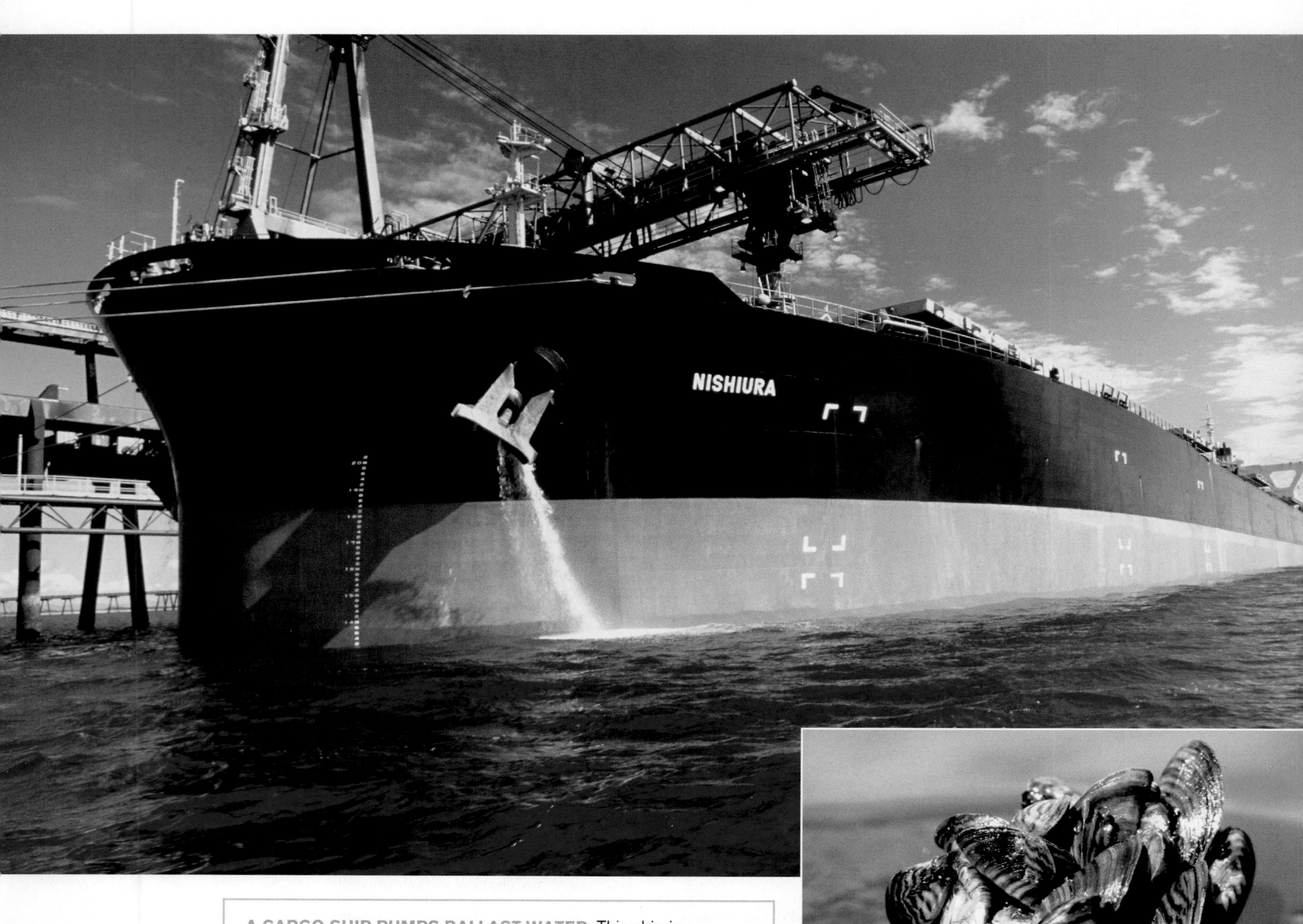

A CARGO SHIP PUMPS BALLAST WATER. This ship in Queensland, Australia, is pumping ballast water while being loaded with coal bound for Japan. Dumping ballast water into local waters introduces aquatic species from other parts of the world. Zebra mussels introduced in this way are one of the most economically destructive invasive species in North America. Today, most ships replace ballast water while still at sea, where the saltiness of the ocean seawater kills most freshwater stowaways.

Invasion of the Zebra Mussels

Walk outside on a spring day and the wild riot of life is plain to see in even the biggest city: weeds slowly lift chunks of sidewalk, dandelion seeds float by, and insects buzz through the air, while in every yard worms squirm through the soil beneath our feet. The rich abundance of life that surrounds us is a tiny fraction of the biosphere, a global system of life including all of Earth's organisms and their physical environment. From the depths of the oceans to the tops of the highest peaks, all of the trillions of individual organisms on Earth are part of the biosphere.

Human actions can have enormous impacts on the biosphere. Consider the seemingly harmless practice in which ships take up ballast water in one port and discharge it in another. If you live near a shipping port, you may have seen ships come into port and discharge ballast water by the ton, water gushing from the sides of the ship in giant streams. Whether the ships carry new laptops or furniture, they need ballast to keep from tipping over and to maintain the right depth in the water. Water is great ballast because it's easy for ships to load or discard as needed.

But each time a ship dumps ballast water, it may also dump hundreds of aquatic species—everything from bacteria and viruses to small schools of fish. The result has been the transfer of aquatic species all over the world. One especially destructive species is the zebra mussel, a freshwater shellfish from Eurasia that arrived in the Great Lakes in the 1980s. Zebra mussels have flourished so well in their North American home away from home that they have displaced native species and caused billions of dollars in damage by clogging pipes and covering the hulls of boats. In this chapter we will discover how organisms interact with one another and with their environment. These interactions occur at many different levels: populations, communities, ecosystems, and the broadest level of them all, our biosphere.

Why does the zebra mussel cause more trouble in North America than in its original home? And how can we prevent this and other similar problems in the future?

MAIN MESSAGE Organisms interact with one another and their physical environment, forming a web of interconnected relationships that is heavily influenced by climate.

KEY CONCEPTS

- Ecology is the study of interactions between organisms and their environment.
- The biosphere consists of all living organisms on Earth, together with the environments they inhabit.
- Climate has a major effect on the biosphere. Climate is determined by incoming solar radiation, global movements of air and water, and major features of Earth's surface.
- The biosphere can be divided into large terrestrial and aquatic areas, called biomes, based on their climate and ecological characteristics.
- Terrestrial biomes are usually named for the dominant plants that live there. The locations of terrestrial biomes are determined mainly by climate.
- The major terrestrial biomes include tundra, boreal forest, temperate deciduous forest, grassland, chaparral, desert, and tropical forest.
- Aquatic biomes are usually characterized by the environmental conditions that prevail there, especially salt content. They are heavily influenced by the surrounding terrestrial biomes and by climate.
- Each of the two major aquatic biomes—freshwater and marine—encompasses a variety of different ecosystems.
- Because components of the biosphere interact in complex ways, human actions that affect the biosphere can have unexpected side effects.

INVASION OF THE ZEBRA MUSSELS 465

20.1 Ecology: Understanding the Interconnected Web 466

20.2 Climate's Large Effect on the Biosphere 468

20.3 Terrestrial Biomes 471

20.4 Aquatic Biomes 477

APPLYING WHAT WE LEARNED 484
How Invasive Mussels Can Harm Whole Ecosystems

Helpful to Know

Ecology and *economy* are intertwined in word derivation as well as in practice. *Eco* is Latin for "household," so "ecology" literally means "study of the household," and "economy" literally means "management of the household."

A VIEW OF EARTH FROM SPACE highlights the beauty and fragility of the **biosphere**, which consists of all organisms on Earth, together with the physical environments in which they live (**FIGURE 20.1**). All life on Earth, including humanity, depends on the biosphere. Earth provides food and the raw materials we use to manufacture things. In this unit we discuss *ecology*, the branch of science devoted to understanding how the biosphere works.

FIGURE 20.1 The Biosphere

Despite its diversity, Earth is a single entity with interconnected physical and biological components. What kinds of factors influence the distribution of organisms on Earth? What impact do the organisms have on each other? How does Earth sustain life? The field of ecology examines these kinds of questions to gain an understanding of the relationship between the living world and its environment.

Ecology can be defined as the scientific study of interactions between organisms and their environment. The environment of an organism includes both biotic factors (other organisms) and abiotic (nonliving) factors. Ecologists are interested in how the two parts of the biosphere—organisms and the environments in which they live—interact with and affect each other. Ecology is an expansive subject that looks at the world from many different perspectives. In the chapters of this unit, our study of ecology covers several levels of the biological hierarchy: individual organisms, populations, communities, ecosystems, biomes, and the biosphere (**FIGURE 20.2**).

We begin this chapter by discussing why ecology is important and how it helps us understand the relationships between organisms and their environments. We go on to discuss climate and other factors that shape the biosphere, since all ecological interactions, at whatever level they occur, take place in the biosphere. We then briefly characterize the terrestrial and aquatic regions that are found on Earth.

20.1 Ecology: Understanding the Interconnected Web

The science of ecology helps us understand the natural world in which we live. Such understanding is vitally important because we are changing our world in ways that are not always for the better. Some of these changes can be expensive, and in some cases difficult or even impossible, to fix. Consider species such as zebra mussels that are accidentally or deliberately moved by people from place to place around the globe. In the United States, people have introduced thousands of nonnative species, some of which have become pests (invasive species) that collectively cause an estimated \$120 billion in economic losses each year—a cost similar to the annual economic impact of smoking (\$150 billion per year) in the United States. By studying the ecology of invasive species, we can understand how people help them spread to new regions, why they increase dramatically in abundance once they are relocated, how they affect natural communities, and how they cause economic disruption—all of which can be helpful in limiting the damage that these species cause.

Every organism affects its environment (as when a beaver builds a dam that blocks the flow of a stream and creates a pond), just as the environment affects

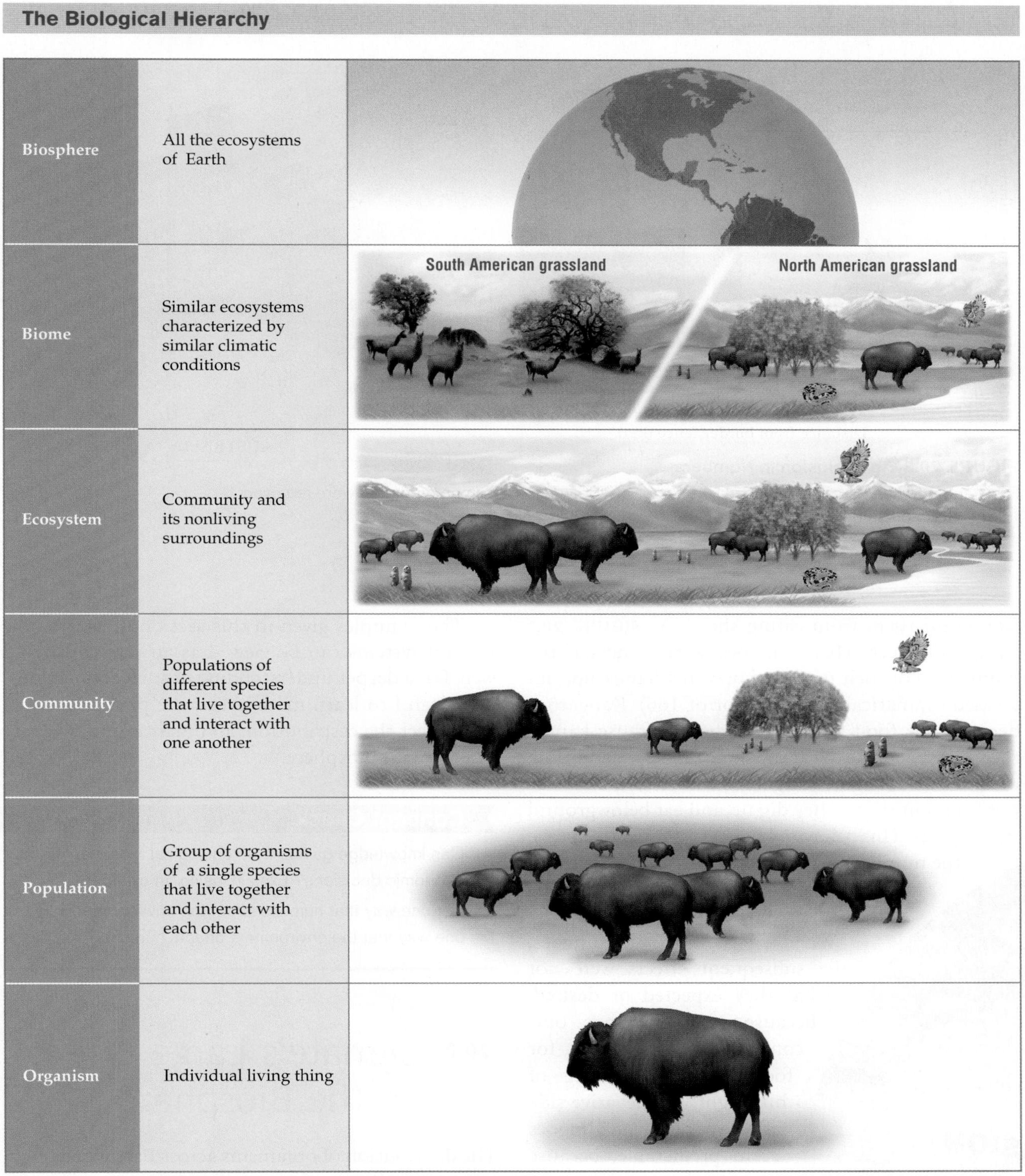

The Biological Hierarchy

Level	Description	
Biosphere	All the ecosystems of Earth	
Biome	Similar ecosystems characterized by similar climatic conditions	South American grassland / North American grassland
Ecosystem	Community and its nonliving surroundings	
Community	Populations of different species that live together and interact with one another	
Population	Group of organisms of a single species that live together and interact with each other	
Organism	Individual living thing	

FIGURE 20.2
Ecology Is Studied at Different Levels
Ecologists study organisms and their relationship to the environment.

the organism (as when an extended drought limits the growth of plant species that the beaver depends on for food). With interactions going back and forth between organisms and their physical environments, the biosphere can be thought of as forming a web of interconnected relationships.

Thinking of organisms and their environment as an interconnected web helps us understand the consequences of human actions. Consider what happened when people used fencing, poison, and hunting to remove dingoes from a large region of rangeland in Australia. Dingoes, a type of wild dog, were removed

Predator-Prey Relationships

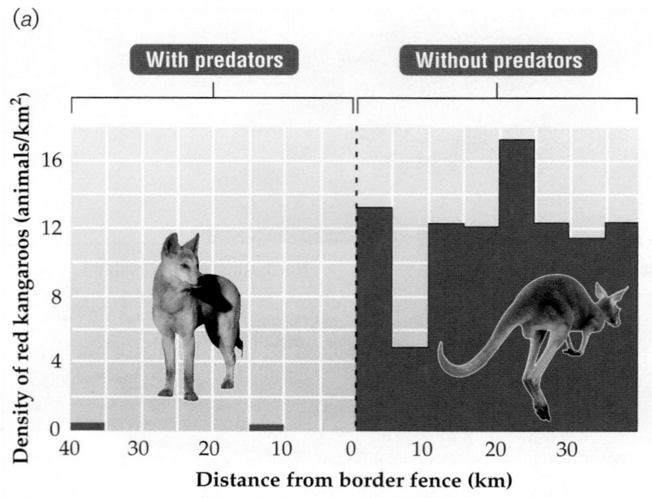

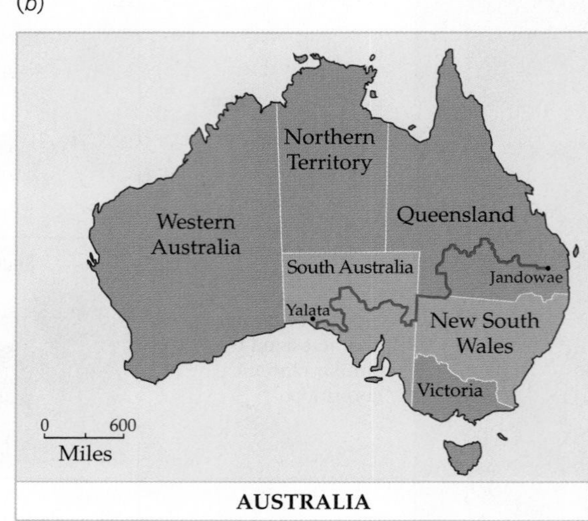

FIGURE 20.3 An Explosion in Numbers

Red kangaroo numbers (*a*) increased by a factor of 166 when their dingo predators were removed from the Australian rangelands located south of the world's largest fence (red line in *b*).

Concept Check Answers

1. Yes. Studying the ecology of invasive species enables us to understand how and why these organisms spread, how they affect ecological communities, and how they cause economic disruption.

2. Humans can affect the environment in many different ways. Climate change, the relocation of species, and the destruction of habitats for consumption or resource extraction are a few examples.

to prevent them from eating sheep. As **FIGURE 20.3** shows, in areas where dingoes were removed, the population of their preferred prey, red kangaroos, increased dramatically (by a factor of 166). Kangaroos decrease the food available to sheep because kangaroos and sheep like to eat the same plants. In addition, in times of drought, kangaroos resort to a behavior not found in sheep: they dig up and eat belowground plant parts. This behavior has the potential to change the number and types of plant species found in rangeland, thereby further increasing the effect that kangaroos have on sheep.

When people removed dingoes, the subsequent effects were not what they expected or desired, because red kangaroos outcompeted their sheep for food. With the advantage of hindsight, the negative side effects of removing dingoes seem predictable, because changes that affect one part of the biosphere (such as removal of dingoes) can have a ripple effect to produce changes elsewhere (such as increases in red kangaroos and decreases in the food available to sheep).

EXTREME INVASION

Rabbits were introduced into Australia in the mid-nineteenth century. With no predators, a high fertility rate, and a highly adaptable diet, the population rapidly expanded. From an initial population of 24 wild rabbits in 1859, the population exploded to 10 billion by the 1920s, threatening crops, native species, and the natural environment.

The examples given in this section illustrate how natural systems can be viewed as an interconnected web. For a deeper understanding of such connections, it is useful to learn more about the physical factors that affect the distribution, abundance, and diversity of life in our biosphere.

Concept Check

1. Can knowledge gained from ecological research affect economic decisions? Explain, using an example.
2. List one way that humans affect the environment, and one way that the environment affects humans.

20.2 Climate's Large Effect on the Biosphere

The distribution of organisms across Earth is strongly affected by climate, the prevailing environmental conditions of an area. The terms "weather" and "climate" should not be used interchangeably. **Weather** refers to the temperature, precipitation (rainfall and snowfall), wind speed and direction, humidity, cloud cover, and other physical conditions of Earth's lower atmosphere at a specific place over a short period of time. **Climate** refers to the average weather conditions experienced in a region over relatively long periods of time. Weather changes quickly and is hard to predict. Climate, on the other hand, is more predictable, because it describes the state of the atmosphere over larger parts of Earth and typically over decades or centuries.

Climate has major effects on ecological communities because organisms are more strongly influenced by climate than by any other feature of their environment. While modeling climate is difficult, several basic factors play a major role in shaping climatic conditions in a given geographic region. In the discussion that follows we look at some of these factors.

Incoming solar radiation shapes climate

Energy from the sun is a major factor in shaping climate on Earth. This energy, however, is not uniformly distributed across the planet, simply because Earth is more or less spherical. Sunlight strikes Earth directly at the equator, but at a slanted angle near the North and South poles. This difference causes more solar energy to reach the equator, as well as the regions on either side of it (the tropics), than reaches the poles (**FIGURE 20.4**). Tropical regions show small seasonal fluctuations in temperature, so organisms that live there experience a relatively warm, stable climate throughout the year. Generally speaking, sunlight and warmth promote photosynthesis and thereby increase the productivity of plants and other light-dependent producers.

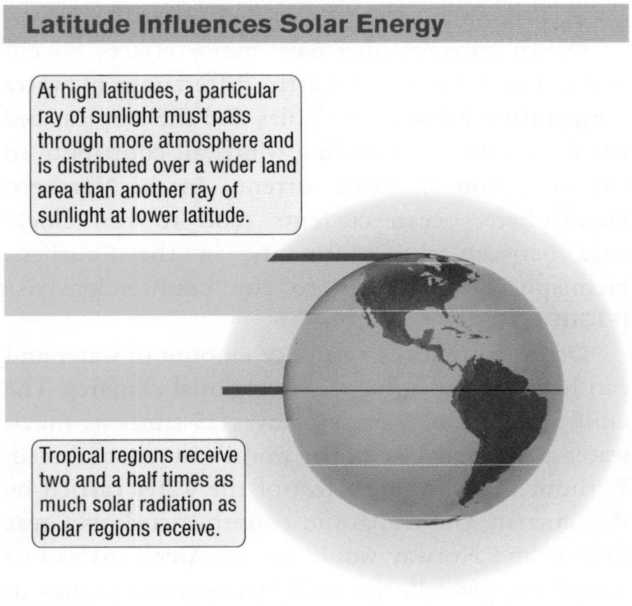

Latitude Influences Solar Energy

At high latitudes, a particular ray of sunlight must pass through more atmosphere and is distributed over a wider land area than another ray of sunlight at lower latitude.
Tropical regions receive two and a half times as much solar radiation as polar regions receive.

FIGURE 20.4 Earth's Spherical Shape Causes an Uneven Distribution of Solar Radiation

Helpful to Know

Productivity is a measure of the energy that producers are able to store in the form of biological material, or biomass. Consumers, including animals and decomposers, depend on the productivity of producers.

Wind and water currents affect climate

Near the equator, intense sunlight causes water to evaporate from Earth's surface. The warm, moist air rises because heat causes it to expand and therefore to be less dense, or lighter, than air that has not been heated. The warm, moist air cools as it rises. Because cool air cannot hold as much water as warm air can, much of the moisture is released from a cooling air mass and falls as rain.

Usually, cool air sinks. The cool air above the equator cannot sink, however, because of the warm air rising beneath it. Instead, the cool air is drawn to the north and south, tending to sink back to Earth at about 30° latitude. Part of the air mass flows back toward the equator, and as it does so it absorbs moisture from Earth's surface. By the time it reaches the equator, the air is once more warm and moist, so it rises, repeating the cycle. This grand cycle of air movement is called a **convection cell**. Earth has six giant convection cells in which warm, moist air rises and cool, dry air sinks (**FIGURE 20.5**). Two convection cells are located in tropical regions and two in polar regions, where they generate relatively consistent wind patterns. In temperate regions (roughly 30–60° latitude), winds are more variable but generally move in the opposite direction.

Precipitation occurs when moisture-laden warm air cools and releases its water as rain or snow. This happens near the equator as the warm equatorial air ascends and cools. It also occurs near 60° latitude, when cool, dry air from polar regions collides with warm, moist air moving toward the poles.

The winds produced by the giant convection cells do not move straight north or straight south relative to the land spinning below. Instead, because of Earth's rotation, these winds appear to curve as they travel near Earth's surface (**FIGURE 20.6**). This phenomenon is known as the **Coriolis effect**. Winds traveling toward the equator appear to curve to the left. To an Earth-bound observer, these winds seem to blow from the east; hence, such winds are called easterlies. Similarly, winds that travel toward the poles curve to the right; and because they seem to blow from the west, these winds are

Helpful to Know

One's position on the Earth can be stated in latitude and longitude. Latitude is an indication of your position along a north–south axis, while longitude is a measure of your east–west position.

Global Air Circulation Patterns

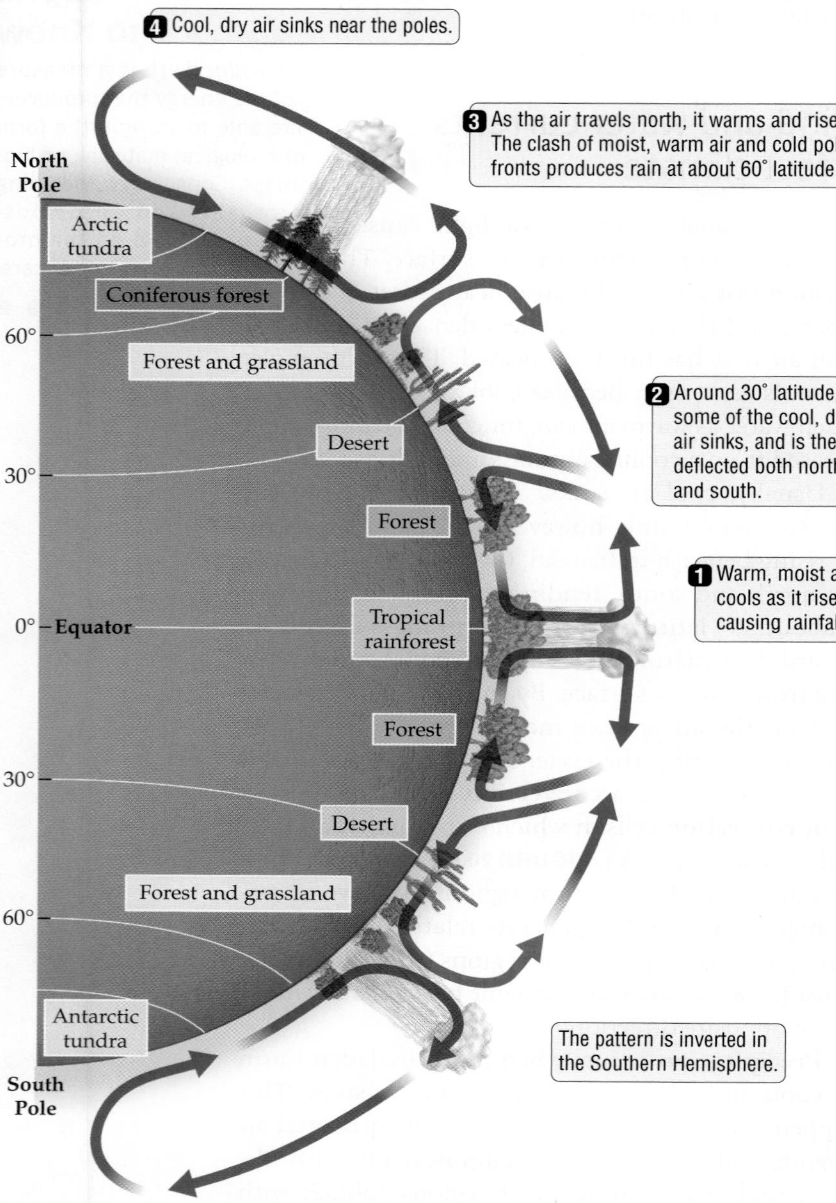

4 Cool, dry air sinks near the poles.
3 As the air travels north, it warms and rises. The clash of moist, warm air and cold polar fronts produces rain at about 60° latitude.
2 Around 30° latitude, some of the cool, dry air sinks, and is then deflected both north and south.
1 Warm, moist air cools as it rises, causing rainfall.
The pattern is inverted in the Southern Hemisphere.
North Pole
Arctic tundra
Coniferous forest
60°
Forest and grassland
Desert
30°
Forest
0° Equator
Tropical rainforest
Forest
30°
Desert
Forest and grassland
60°
Antarctic tundra
South Pole

FIGURE 20.5 Earth Has Six Convection Cells

Air flows in six large, circular cells (three in each hemisphere) that are driven by the relative temperature of the air. Warm, moist air rises and cools, releasing its moisture as rain or snow. Cool air is denser and descends.

Global Wind Patterns

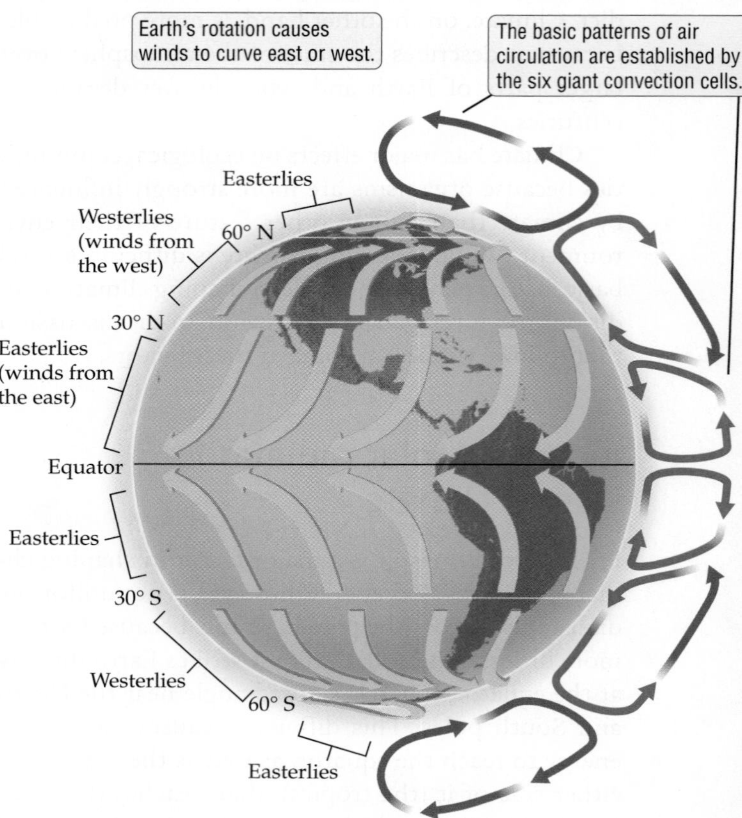

Earth's rotation causes winds to curve east or west.
The basic patterns of air circulation are established by the six giant convection cells.
Easterlies
Westerlies (winds from the west)
60° N
30° N
Easterlies (winds from the east)
Equator
Easterlies
30° S
Westerlies
60° S
Easterlies

FIGURE 20.6 Prevailing Winds Are Determined by Global Patterns of Air Circulation

Earth's rotation causes winds to curve to the east or west. The direction they curve depends on their latitude, but for most regions on Earth, the seasonal winds blow from a consistent direction.

Helpful to Know

A common misconception based on the Coriolis effect is that toilets flush clockwise in the Northern Hemisphere and counterclockwise in the Southern Hemisphere. While in ideal circumstances this would be the case, the Coriolis effect is too weak to counteract the effects of the flowing water of a toilet.

called westerlies. At any given location the winds usually blow from a consistent direction, and these predictable patterns of air movement are known as prevailing winds. In southern Canada and much of the United States, for example, winds blow mostly from the west, so storms in these regions usually move from west to east.

Ocean currents also have major effects on climate. The rotation of Earth, differences in water temperature between the poles and the tropics, and the directions of prevailing winds all contribute to the formation of ocean currents. In the Northern Hemisphere, ocean currents tend to run clockwise between the continents; in the Southern Hemisphere, they tend to run counterclockwise (**FIGURE 20.7**).

Ocean currents carry a huge amount of water and can have a great influence on regional climates. The Gulf Stream, for example, moves 25 times as much water as is carried by all the world's rivers combined. Without the warming effect of the water carried by this current, the climate in countries such as Great Britain and Norway would be sub-Arctic instead of temperate. Overall, the Gulf Stream causes cities in western Europe to be warmer than cities at similar latitudes in North America.

Major Ocean Currents of the World

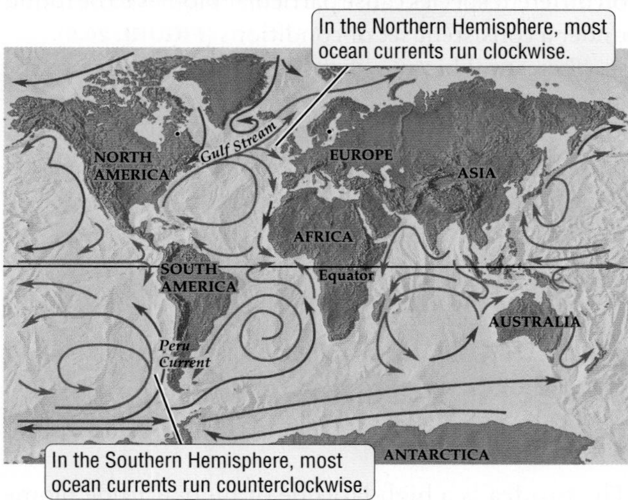
In the Northern Hemisphere, most ocean currents run clockwise.
Gulf Stream
NORTH AMERICA
EUROPE
ASIA
AFRICA
SOUTH AMERICA
Equator
AUSTRALIA
Peru Current
ANTARCTICA
In the Southern Hemisphere, most ocean currents run counterclockwise.

FIGURE 20.7 Why Do Ocean Temperatures Vary? Ocean currents can be cold (blue) or warm (red), depending on a combination of factors, including water depth and latitude.

The Rain Shadow Effect

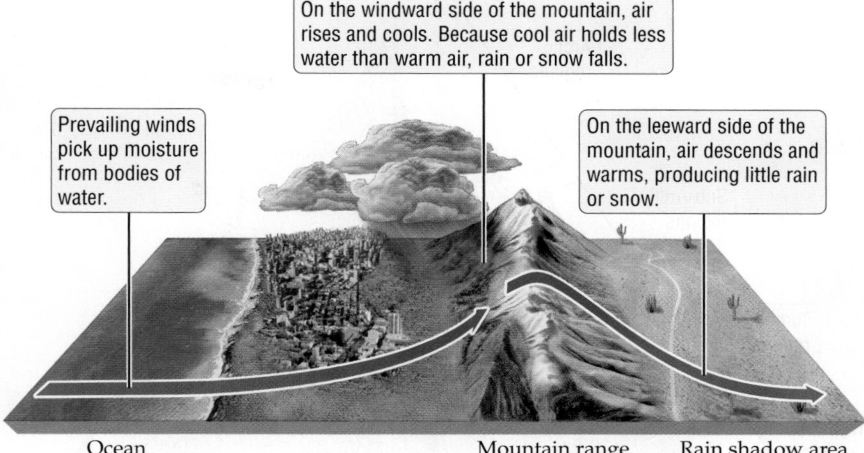
On the windward side of the mountain, air rises and cools. Because cool air holds less water than warm air, rain or snow falls.
Prevailing winds pick up moisture from bodies of water.
On the leeward side of the mountain, air descends and warms, producing little rain or snow.
Ocean
Mountain range
Rain shadow area

FIGURE 20.8 The Leeward Side of a Mountain Is Usually Dry The side of a high mountain that faces the prevailing winds (the windward side) receives more precipitation than the side of the mountain that faces away from the prevailing winds (the leeward side). The leeward side is therefore said to be in a rain shadow.

The major features of Earth's surface also shape climate

The climate of a place may also be affected by the presence of large lakes, the ocean, and mountain ranges. Heat is absorbed and released more slowly by water than by land. Because they retain heat comparatively well, large lakes and the ocean moderate the climate of the surrounding lands. Mountains can also have a large effect on a region's climate. For example, mountains often produce a **rain shadow**, in which little precipitation falls on the side of the mountain that faces away from the prevailing winds (**FIGURE 20.8**). As moisture-laden air rises over the windward slope of a mountain range, it cools, releasing that moisture. Once the air reaches the leeward side of the range, it has lost its moisture and causes arid conditions. In the Sierra Nevada of North America, five times as much precipitation falls on the western side of the mountains (which faces toward winds that blow in from the ocean) as on the eastern side, where the lack of precipitation contributes to the formation of deserts. Mountain ranges in northern Mexico, South America, Asia, and Europe also create rain shadows.

EXTREME CORIOLIS EFFECT

During World War I, German long-range artillery such as the super-gun "Big Bertha" had to correct for Earth's rotation in order to accurately hit their targets.

Concept Check

1. What causes convection cells to form in Earth's atmosphere, and what effects do they have on climate?
2. What impact do ocean currents have on Earth's climate?

Concept Check Answers

1. Convection cells form where moist, warm air rises and cool, dry air descends. Rain falls as ascending moist air releases moisture. Convection cells also determine the direction of prevailing winds.
2. Because water retains so much heat, ocean currents can distribute thermal energy over long distances.

20.3 Terrestrial Biomes

Ecologists categorize large areas of the biosphere into distinct regions, called **biomes**, based on the unique climatic and ecological features of each such region. Because climate strongly influences the life-forms that can live in a particular place, each biome is associated with a characteristic type of plant and animal species. While the species themselves may be different, similar prevailing climatic conditions lead to evolutionary convergence, resulting in similar assemblages of organisms. A biome may encompass more than one ecosystem and typically stretches across large swaths of the globe. Biomes on land—terrestrial biomes—are usually named after the dominant vegetation in the area.

Climate is the single most important factor controlling the location of natural terrestrial biomes. The climate of the area—primarily the temperature and the amount and timing of precipitation—allows some species to thrive and prevents other species from living

Typical Distribution of Biomes

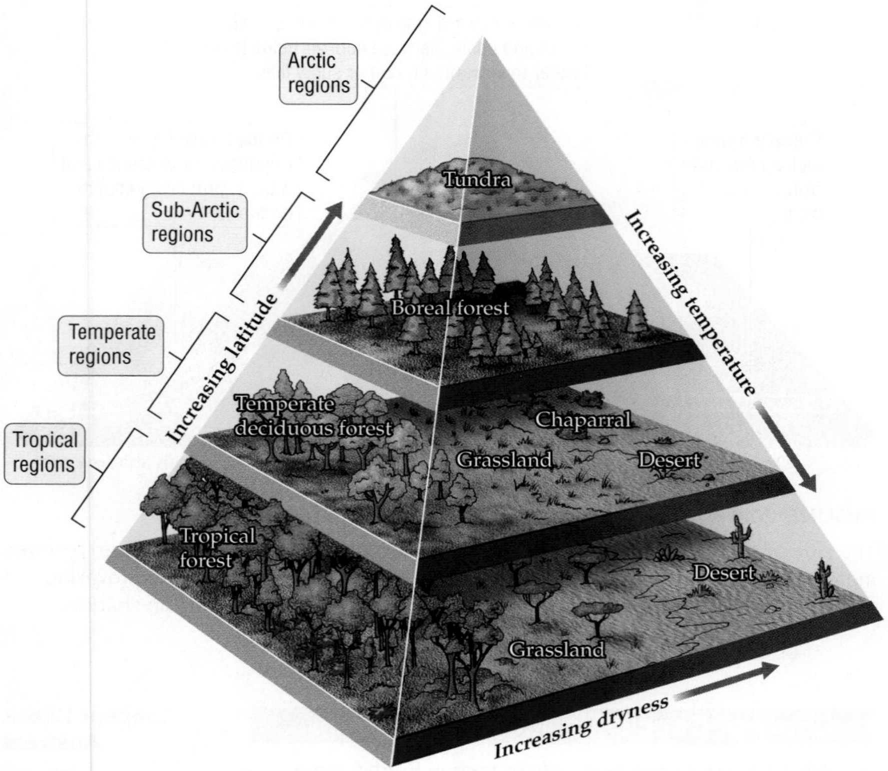

FIGURE 20.9 The Location of Terrestrial Biomes Depends on Temperature, Precipitation, and Altitude

there. Overall, the effects of temperature and moisture on different species cause particular biomes to be found under a consistent set of conditions (**FIGURE 20.9**).

Terrestrial biomes change from the equator to the poles, and from the bottoms to the tops of mountains. In this section we explore some of Earth's major terrestrial biomes: tundra, boreal forest, temperate deciduous forest, grassland, chaparral, desert, and tropical forest (**FIGURE 20.10**).

The tundra is marked by cold winters and a short growing season

The **tundra** is a high-latitude or high-altitude biome characterized by cold, windswept plains. The Arctic tundra covers nearly one-fourth of Earth's land surface, encircling the North Pole in a vast sweep that includes half of Canada and Alaska and sizable portions of northern Europe and Russia. A similar habitat, known as the alpine tundra, is encountered above the tree line in high mountains.

Winter temperatures in the Arctic tundra can dip to −50°C (−58°F), although they average about −34°C (−29°F). Even in summer, temperatures

Major Terrestrial Biomes

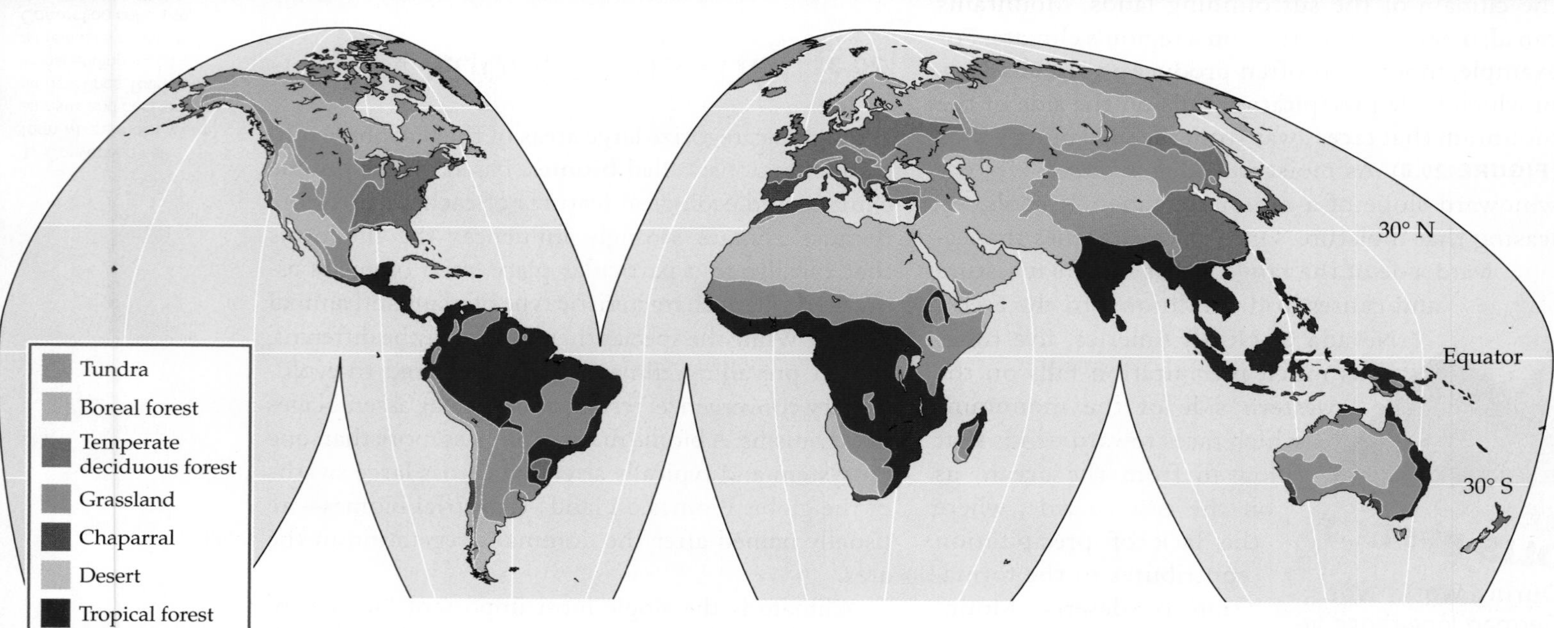

FIGURE 20.10 The Distribution of Biomes Is Affected by Climate, Latitude, and Disturbance
Biomes do not begin and end abruptly. Instead, they generally transition into one another. Disturbances such as storms, fires, and human activities can alter biomes.

seldom climb above 12°C (54°F), and freezing weather is not uncommon. The ground is frozen for 10 months of the year, thawing to a depth of no more than a meter (about 3 feet) during the short summer. Below these surface layers of soil is **permafrost**, permanently frozen soil that may be a quarter of a mile deep. Precipitation in the Arctic tundra ranges from 15 to 25 centimeters (up to 10 inches) per year—lower than in many of the world's deserts. However, evaporation is low because of the cold. In the summer, ice melt and thawing of the upper layers of soil create an abundance of bogs, ponds, and streams, which are prevented from draining because of the underlying permafrost.

Because of the short growing season and because the permafrost is a barrier to the deep taproots of woody plants, trees are absent or scarce in the tundra (**FIGURE 20.11**). The vegetation is dominated by low-growing flowering plants, such as grasses, sedges, and members of the heath family. The rocky landscape is covered in mosses and lichens, which are an important food source for herbivores such as reindeer (known as caribou in North America). Rodents such as lemmings, voles, and Arctic hare provide food for carnivores like Arctic foxes and wolves. Bears and musk oxen are among the few large mammals. Insects abound in the summer, along with the migratory birds that feed on them. There are few amphibians and even fewer reptiles.

FIGURE 20.11 Tundra: Denali National Park, Alaska
Tundra is found at high latitudes and high elevations, and it is dominated by low-growing shrubs and nonwoody plants that can cope with a short growing season.

A few coniferous species dominate the boreal forest

The **boreal forest** (from *borealis*, "northern") is the largest terrestrial biome. It is also known as taiga (*TYE*-**guh**), based on the Mongolian word for coniferous forests. It includes the sub-Arctic landmass immediately south of the tundra, covering a broad belt of Alaska, Canada, northern Europe, and Russia, approximately between 60° and 50° north latitude. Winters in the boreal forest are nearly as cold as in the tundra and last about half the year. Summers are longer and warmer than in the tundra, with temperatures reaching as high as 30°C (86°F) in some boreal forests of the world. Typically, the soil is thin and nutrient-poor. Precipitation in the form of rain or snow is low in most boreal forests, but some areas, most notably the Pacific Coast rainforests of North America, receive large amounts of precipitation. Because evaporation rates are low in these cold, northern latitudes, plants in the boreal forest generally receive adequate moisture during the growing season.

FIGURE 20.12 Boreal Forest: Banff National Park, Alberta
Boreal forests are dominated by coniferous (cone-bearing) trees that grow in northern or high-altitude regions with cold, dry winters and mild summers.

Conifers, which are cone-bearing trees with needle-like leaves, dominate the boreal forest vegetation (**FIGURE 20.12**). Spruce and fir are the most common species in the North American boreal forest; pine and larch are abundant in Scandinavia and Russia. Broad-leaved species such as birch, alder, willow, and aspen are also found, especially in the southern ranges of this

Helpful to Know

Fire is a significant environmental factor in the chaparral. Several chaparral plant species produce seeds that lie dormant in the soil and germinate only after a fire destroys larger, more established plants.

biome. Plant diversity is relatively low, except in the rainforests of the Pacific Coast, where the seaside climate is milder and soils are richer. The large herbivores of the boreal forest include elk and moose. Small carnivores, such as weasels, wolverines, and martens, are common. Larger carnivores include lynx and wolves. Bears, which are omnivores, are found throughout the world's boreal forests.

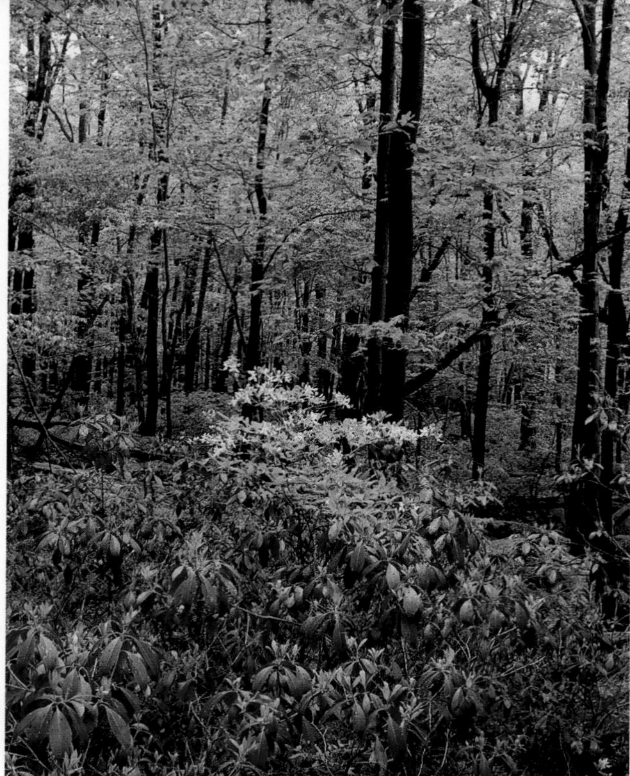

FIGURE 20.13 Temperate Deciduous Forest: Pocono Mountains, Pennsylvania

Temperate deciduous forests are dominated by trees and shrubs adapted to relatively rich soil, snowy winters, and moist, warm summers.

Temperate deciduous forests have fertile soils and relatively mild winters

Temperate deciduous forests are a familiar forest type for most people who live in North America. They also constitute the dominant biome in large areas of Europe and Russia and parts of China and Japan. Temperate deciduous forests occur in regions with a distinct winter, with subfreezing conditions that may last 4 or 5 months. However, winter temperatures are not as harsh as in the Arctic and sub-Arctic biomes, and summers are typically much warmer. Temperatures commonly range from lows of −30°C (−22°F) in winter to summertime highs above 30°C (86°F). Annual rainfall varies from 60 centimeters (nearly 24 inches) to more than 150 centimeters (59 inches), and precipitation is distributed evenly through much of the year.

Deciduous trees—that is, trees that drop their leaves in the cold season—are the dominant vegetation in this biome (**FIGURE 20.13**). Temperate deciduous forests display greater species diversity than do the tundra and boreal biomes. Oak, maple, hickory, beech, and elm are common in these woodlands, which also harbor an understory of shade-tolerant shrubs and herbaceous plants forming a ground cover. Coniferous species, such as pine and hemlock, also occur, but they are not the dominant trees. The fauna includes squirrels, rabbits, deer, raccoons, beavers, bobcats, mountain lions, and bears. There are many different fishes, and amphibians and reptiles are common.

EXTREME PRESERVATION

The extremely cold temperatures of the tundra provide excellent preservation conditions. In 2012, Russian scientists claimed to have regenerated a viable plant from the tissues of a 30,000-year-old specimen, *Silene stenophylla*, recovered from the Siberian tundra.

Grasslands appear in regions with good soil but relatively little moisture

The **grassland** biome is characteristic of regions that receive about 25–100 centimeters (about 10–40 inches) of precipitation annually. The moisture levels are insufficient for vigorous tree growth, but they are not as low as in a desert. Grasslands are found in both temperate and tropical latitudes. The prairies of North America, the steppes in Russia and central Asia, and the pampas of South America are examples of temperate grasslands. Tropical and subtropical grasslands are known as savanna. Grasses dominate in this biome, although the landscape may be dotted with shrubs and small trees, as in the African savanna. Soils in some grasslands, especially the prairies of North America and the pampas of Argentina, are exceptionally deep and fertile. As a result, most of these areas have been converted to agriculture, and today they are some of the most productive grain-growing regions of the world.

Before they succumbed to the plow, the grasslands of central North America formed the largest stretch of grassland biome in the world. The northern and central parts of the Great Plains receive moderate rainfall—about 100 centimeters annually—enough to support the growth of "hat-high" grasses like big bluestem, which averages about 6 feet in height. The tallgrass prairies (**FIGURE 20.14**), as these grasslands are called, once stretched from Manitoba in Canada, through the Dakotas and Nebraska, south to Kansas and Oklahoma. They also extended into neighboring states to the east, as far as the western edge of Indiana.

Only about 1 percent of the original prairie remains, most of it in protected areas. Grasses, and herbaceous plants such as coneflower and shooting star, predominate in these grasslands. That predominance is maintained by prairie fires, which destroy shrubs and trees but not the underground roots and stems of prairie plants. Burrowing rodents like voles and prairie dogs (a type of ground squirrel) aerate the soil, thereby improving growing conditions for the root systems of prairie plants. There are many butterflies and other insects, and the greatest diversity of mammals in North America is found here.

Rainfall declines farther west, and the prairie is replaced by mixed grassland and then short grassland along the eastern foothills of the Rocky Mountains from Montana, Wyoming, and Colorado south to New Mexico. Drought-resistant grasses, such as buffalo grass, are the dominant vegetation, and bison and pronghorn are the larger herbivores of the short grasslands.

FIGURE 20.14 Grassland: Eastern Kansas
Grasslands are common throughout the world and are dominated by grasses, although scattered trees are found in some, such as the tropical grasslands known as the savanna. Pictured here is bluestem grass (*Andropogon gerardii*) in a tallgrass prairie.

Chaparral is characteristic of regions with wet winters and hot, dry summers

The **chaparral** (*SHAP*-**uh**-*RAL*) is a shrubland biome dominated by dense growths of scrub oak and other drought-resistant plants (**FIGURE 20.15**). It is found in regions with a "Mediterranean climate," characterized by cool, rainy winters and hot, dry summers. Annual rainfall ranges from 25 to 100 centimeters (about 10–40 inches). Chaparral occurs in regions of southern Europe and North Africa bordering the Mediterranean Sea, and also in coastal California, southwestern Australia, and along the west coast of Chile and South Africa.

The soil is relatively poor in these habitats, and most species are adapted to hot, dry conditions. Many chaparral plants have thick, leathery leaves that reduce water loss. Low moisture and high temperatures in summer make the chaparral exceptionally susceptible to wildfires. Common vegetation in the California chaparral includes scrub oak, pines, mountain mahogany, manzanita, and the chemise bush.

FIGURE 20.15 Chaparral: North of San Francisco, California
Chaparral is characterized by shrubs and small, nonwoody plants that grow in regions with cool, rainy winters and hot, dry summers.

FIGURE 20.16
Desert: Near Phoenix, Arizona
Deserts form in regions with low precipitation, usually 25 centimeters (10 inches) per year or less. The photo shows saguaro cacti (tall green columns) and other plants in the Sonoran Desert.

The California quail is an iconic bird of the chaparral. Rodents such as jackrabbits and gophers are common, and there are many species of lizards and snakes. Mammals include deer, the peccary (a piglike animal), lynx, and mountain lions.

EXTREME TEMPERATURES

Characterized by a lack of atmospheric moisture, desert temperatures are the most variable among all terrestrial biomes. Both the coldest (Vostok Station, Antarctica, −89.2°C [−128.6°F]) and warmest (Death Valley, CA, 56.7°C [134°F]) temperatures recorded on Earth occurred in deserts.

The scarcity of moisture shapes life in the desert

The **desert** biome makes up one-third of Earth's land surface. The defining feature of a desert is a scarcity of moisture, not high temperatures. A desert experiences less than 25 centimeters (10 inches) of precipitation annually. Antarctica, which receives less than 2 centimeters of precipitation per year, is the largest cold desert in the world. The Sahara desert in northern Africa is the largest hot desert. Because desert air lacks moisture, it cannot retain heat and therefore cannot moderate daily temperature fluctuations. As a result, temperatures can be above 45°C (113°F) in the daytime and then plunge to near freezing at night.

There are numerous evolutionary adaptations to living in the desert. Desert plants have small leaves; the reduced surface area minimizes water loss. Succulents, such as cacti, store water in their fleshy stems or leaves (**FIGURE 20.16**). Some desert plants produce enormously long taproots that are able to reach subsurface water. Most animals in the desert are nocturnal, hiding in burrows during the heat of the day and emerging at night to feed. Jackrabbits have large ears that act like a radiator to help dissipate heat. Desert mammals, such as the desert fox, have light-colored fur, which deflects some of the radiant energy from the sun. The kangaroo rat loses very little water in its urine because its kidneys recover water with exceptional efficiency. The rodent's respiratory passages wring moisture from exhaled air by cooling it. A camel uses a similar mechanism in its nose to recover moisture from the air it exhales.

Tropical forests have high species diversity

The **tropical forest** biome is characterized by warm temperatures and about 12 hours of daylight all through the year. Rainfall may be heavy year-round or occur only during a pronounced wet season. Tropical rainforests, which are tropical forests that remain wet all year, may receive in excess of 200 centimeters (80 inches) of rain annually. Because organic matter decomposes rapidly at warm temperatures, there is little leaf litter in tropical forests. Soils in this biome tend to be nutrient-poor for two main reasons: First, a large percentage of nutrients are locked up in the living tissues (biomass) of organisms, especially large trees. Second, heavy rains tend to leach out soil nutrients, depleting certain minerals in particular.

The abundance of sunshine and moisture makes tropical rainforests the most productive terrestrial habitat (**FIGURE 20.17**). They are also hot spots for diversity of life-forms (biodiversity). Tropical rainforests currently occupy about 6 percent of Earth's land area but harbor nearly 50 percent of its plant and animal species. South America has the largest tracts of tropical rainforest, especially in Brazil, Peru, and Bolivia. Large areas of Southeast Asia and equatorial Africa are also covered in tropical rainforests.

Tropical rainforests are under severe threat from logging and clearing for livestock grazing and other types of agricultural activity. More than half of the original tropical rainforest has been lost. About 2 acres of tropical rainforest is lost every second. Loss of rainforest biomes does not just mean loss of habitat for much of the world's biodiversity, but also loss of a significant carbon dioxide (CO_2) sink. A CO_2 sink is a means of sequestering carbon dioxide. The main natural sinks are the oceans, and plants and other organisms that use photosynthesis to remove carbon dioxide from the atmosphere by incorporating it into biomass. The loss of the rainforests is likely to worsen global warming, because global warming is associated with increasing CO_2 in Earth's atmosphere.

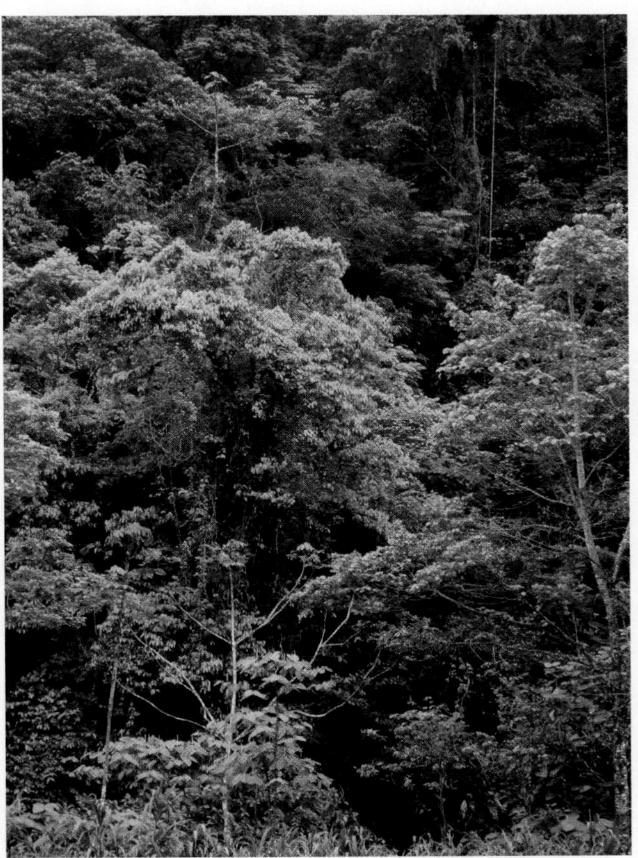

FIGURE 20.17 Tropical Forest: El Yunque National Forest, Puerto Rico

Tropical forests form in warm regions with either seasonally heavy rains or year-round rain. Tropical rainforests, which receive abundant moisture throughout the year, are some of the most productive ecosystems on Earth. They have a rich diversity of trees, vines, and shrubs.

EXTREME MEDICINES

Rainforest species are important sources of new medical drugs. Approximately 7,000 medicines originate from rainforest species, and this number is expected to grow as scientists identify new species that produce unique biological compounds.

Concept Check

1. Why do deserts experience such extreme daily temperature differences?
2. What two factors explain why tropical rainforests have such nutrient-poor soil?

Concept Check Answers

1. Because water has such high thermal retention, it is an important factor in mitigating temperature changes. Deserts, however, are characterized by a lack of moisture.

2. A large proportion of nutrients are stored away in the living tissues of the forest vegetation, and heavy rains wash away minerals from the soil.

20.4 Aquatic Biomes

Aquatic ecosystems cover about 75 percent of Earth's surface. Just as the terrestrial environment can be divided into discrete biomes, Earth's waters exhibit distinct regions. Aquatic biomes are shaped primarily by the physical characteristics of their environment, such as salt content, water temperature, water depth, and the speed of water flow. Within an aquatic biome, we can recognize various habitats, or ecological zones—defined by their nearness to the shore, the depth of the water, and the extent of light penetration.

Two main aquatic biomes can be distinguished on the basis of salt content: freshwater and marine. Lakes, rivers, and wetlands are examples of ecosystems within the freshwater biome. Estuaries, coral reefs, the coastal region, and the open ocean are examples of ecosystems within the marine biome.

Aquatic biomes are influenced by terrestrial biomes and climate

Ecosystems in aquatic biomes, especially lakes, rivers, wetlands in the freshwater biome, and the coastal portions of the marine biome, are heavily influenced by the terrestrial biomes that they border or through which their water flows. High and low points of the land, for example, determine the locations of lakes and the speed and direction of water flow. In addition, when water drains from the land into an aquatic biome, it brings with it dissolved nutrients (such as nitrogen, phosphorus, and salts) that were part of the terrestrial ecosystem. Rivers and streams carry nutrients from terrestrial environments to the ocean, where they may stimulate large increases in the abundance of phytoplankton.

Aquatic ecosystems also are strongly influenced by climate. Climate helps determine the temperature, movement, and salt content of the world's oceans, for example. Such physical conditions of the ocean have dramatic effects on the organisms that live there, so climate has a powerful effect on marine life. Consider the El Niño (**ell** *NEEN*-**yoh**) events that are often reported in the news. These events begin when warm waters from the west deflect the cold Peru Current along the Pacific coast of South America (**FIGURE 20.18**). The results of this change are spectacular, including dramatic decreases in numbers of fish, die-offs of seabirds, storms along the Pacific coast of North America that destroy underwater "forests" of a brown alga called kelp, crop failures in Africa and Australia, and drops in sea level in the western Pacific that kill huge numbers of coral reef animals.

Aquatic biomes are also influenced by human activity

Like terrestrial biomes, aquatic biomes are strongly influenced by the actions of humans. Portions of some aquatic biomes, such as wetlands and estuaries in the freshwater biome, are often destroyed to allow for development projects. Rivers, wetlands, lakes, and coastal marine ecosystems are negatively affected by pollution. Aquatic biomes also suffer when humans destroy or

El Niño Events

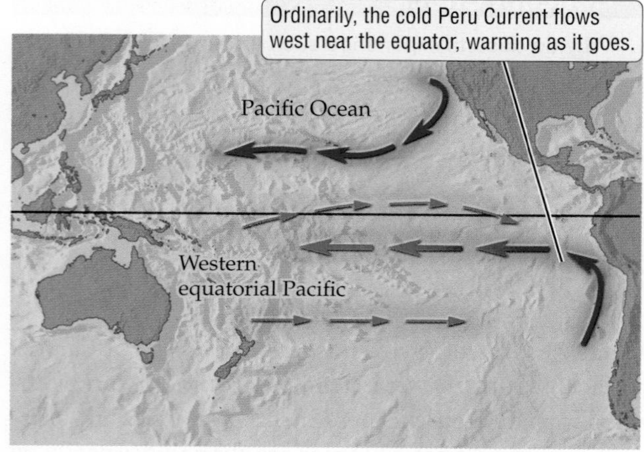

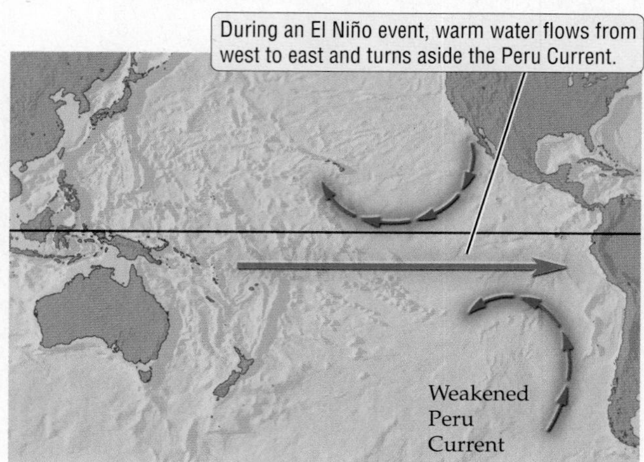

FIGURE 20.18 Ocean Currents Change during El Niño Events

During an El Niño event, winds from the west push warm surface water from the western Pacific to the eastern Pacific. The resulting changes in sea surface temperatures cause changes in ocean currents (shown here in blue for cold, red for warm). El Niño events cause many additional changes (not shown here), altering wind patterns, sea levels, and patterns of precipitation throughout the world.

BIOLOGY MATTERS

The Great Pacific Garbage Patch

In the twentieth century, the use of plastics became so commonplace that it is now difficult to find any product that does not utilize the material. Because of the low cost of production and the fact that plastics can be molded into a variety of shapes, they are used to make everything from disposable eating utensils to consumer electronics to automobile parts. Worldwide production of plastic products is approximately 300 million tons per year.

In 1997, sailor and oceanographer Charles Moore was returning to California from a yacht race in Hawaii when he stumbled upon a patch of plastic debris in the middle of the Pacific Ocean. Now labeled the Great Pacific Garbage Patch, what Moore had discovered was a collection of refuse that had been gathered by the circular flow of Earth's oceans. Since then, Moore and other scientists have examined the extent of this patch, sought out other patches, and explored the impact of this debris on marine ecosystems.

Because plastic products neither float nor sink, they can remain in the water column for long periods of time. Animals may mistake items such as plastic bags for prey species like jellyfish and ingest them, or they may get entwined in debris and either starve or drown. Marine birds have been found dead with stomachs full of plastic material. Plastics are not biodegradable and persist for extremely long periods of time. Over time, ultraviolet light will break plastic into progressively smaller bits, but even these so-called microplastics have a devastating impact on marine ecosystems. One study of water quality in the Great Pacific Garbage Patch revealed that microplastics outnumbered plankton (the base of the food chain) by a ratio of 6 to 1. Another study estimates that fish in the North Pacific ingest 24,000 tons of plastic per year.

Shoreline communities very often see the impact of this accumulation of trash. Storms will wash up tons of debris onto places like Hawaii's North Shore. Microplastics are also affecting these beach areas. On Kamilo Beach in Hawaii, microplastic pieces outnumber sand granules. As the beach grows, the plastic content of the sand will continue to increase.

Experts contend that efforts to remove the trash from the ocean are all but impossible and have focused on decreasing the amount of plastic debris that is deposited into the ocean. While recycling efforts have reduced the amount of plastic entering landfills and oceans, disposal of plastic materials still represents a significant ecological challenge.

(*a*) (*b*)

(*c*)

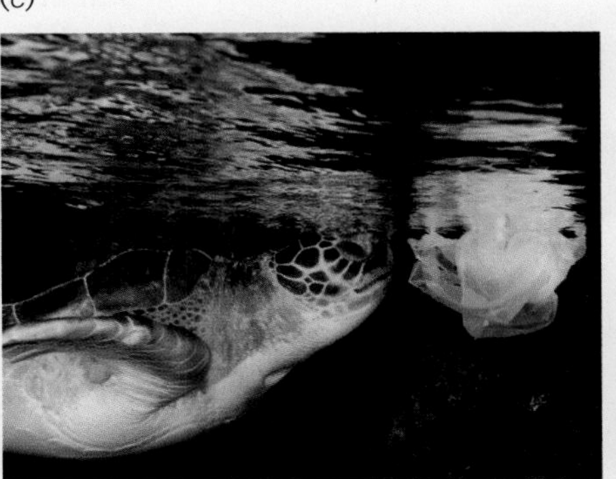

(*d*)

The Effects of Plastic-Based Trash in the Ocean

(*a*) Albatross chick with assorted debris in its stomach. Note the bottle caps and cigarette lighter. (*b*) Microplastics mix with sand granules to create plastic sand beaches. (*c*) Sea turtles often mistake suspended plastic bags for jellyfish. (*d*) Storms wash large amounts of marine litter onto beaches.

modify neighboring terrestrial biomes. For example, when forests are cleared for timber or to make room for agriculture, the rate of soil erosion increases dramatically because trees are no longer there to hold the soil in place. Increased erosion can cause streams and rivers to become clogged with silt, which harms or kills invertebrates, fishes, and many other species.

Next we examine six ecosystems that are common in the freshwater or marine biome: lake, river, wetland, estuary, coastal region, and oceanic region.

(a)

(b)

(c)

FIGURE 20.19 Ecosystems in the Freshwater Biome

Lakes (*a*) are landlocked bodies of standing water. They vary in size from one-fiftieth of a square kilometer (5 acres) to thousands of square kilometers. In contrast, rivers (*b*) are bodies of freshwater that move continuously in a single direction, and wetlands (*c*) are characterized by shallow waters that flow slowly over lands that border rivers, lakes, or ocean waters.

Lakes, rivers, and wetlands are part of the freshwater biome

Lakes are standing bodies of water that are surrounded by land and, according to some authorities, are at least 2 hectares (5 acres) in size (**FIGURE 20.19*a***). The productivity of a lake, as well as the abundance and distribution of its life-forms, is strongly influenced by nutrient concentrations, water depth, and the extent to which the lake water is mixed. Northern lakes tend to be clear because they usually have low nutrient concentrations and therefore do not support vigorous growth of photosynthetic plankton (floating microscopic organisms). Lakes with higher nutrient concentrations appear more turbid because plankton thrive there. In temperate regions, seasonal changes in temperature cause the oxygen-rich water near the top of a lake to sink in the fall and the spring, bringing oxygen to the bottom of the lake. This seasonal turnover also delivers nutrients from the bottom sediments to the surface layers, where they enhance the growth of photosynthetic organisms. In tropical regions, seasonal differences in temperature are not great enough to cause a similar mixing of water. This lack of mixing causes the deep waters of tropical lakes to have low oxygen levels and relatively few forms of life.

Rivers are bodies of freshwater that move continuously in a single direction (**FIGURE 20.19*b***). The physical characteristics of a river tend to change along its length. At its source—whether glacier, lake, or underground spring—the current is stronger and the water colder. The waters in these upper reaches are highly oxygenated because O_2, like most other gases, dissolves more readily in cold water and the turbulence created by rapids and riffles causes more of the gas to dissolve in water. As they approach their emptying point, rivers become wider, slower, warmer, and less oxygenated.

Wetlands are characterized by standing water shallow enough that rooted plants emerge above the water surface (**FIGURE 20.19*c***). A bog is a freshwater wetland with stagnant, acidic, oxygen-poor water. Because bogs are nutrient-poor, their productivity and species diversity are low. In contrast, marshes and swamps are highly productive wetlands. A marsh is a grassy wetland, but swamps are dominated by trees and shrubs.

Estuaries and coastal regions are highly productive parts of the marine biome

The marine biome is the largest biome on our planet, since it includes the open oceans, which cover about three-fourths of Earth's surface. This biome is critically important for us, for all other biomes, and for our planet in general. Photosynthetic plankton in the oceans release more oxygen (and absorb more CO_2) through photosynthesis than do all the terrestrial producers combined. Much of the rain that falls on terrestrial biomes comes from water evaporated off the surface of the world's oceans.

Estuaries are the shallowest of the marine ecosystems. An estuary is a region where a river empties into the sea. It is marked by the constant ebb and flow of freshwater and salt water, and all organisms that thrive on its bounty have to be able to tolerate daily and seasonal fluctuations in salinity (salt levels). The water depth fluctuates with ocean tides and river floods, but most of the time it is shallow enough that light can reach the bottom. The plentiful light, the abundant supply of nutrients delivered by the river system, and the regular stirring of nutrient-rich sediments by water flow create a rich and diverse community of photosynthesizers. Grasses and sedges are the dominant vegetation in most estuaries (**FIGURE 20.20**). The producers provide food and shelter for a varied and prolific community of invertebrates and vertebrates, including crustaceans, shellfish, and fishes. The abundance and diversity of life make estuaries one of the most productive ecosystems on our planet. Many species use estuaries as breeding grounds because of this immense productivity.

As on land, the ocean is divided into distinct regions with different conditions, supporting specific types of ecological communities (**FIGURE 20.21**). The **coastal region** stretches from the shoreline to the edge of the continental shelf, which is the undersea extension of a continent (**FIGURE 20.22**). The coastal region is among the most productive marine ecosystems because of the ready availability of nutrients and

FIGURE 20.20 Estuaries

Estuaries are tidal ecosystems where rivers flow into the ocean. They are usually classified as part of the marine biome. This photo shows a salt marsh at sunrise in Rachel Carson National Wildlife Refuge, Maine.

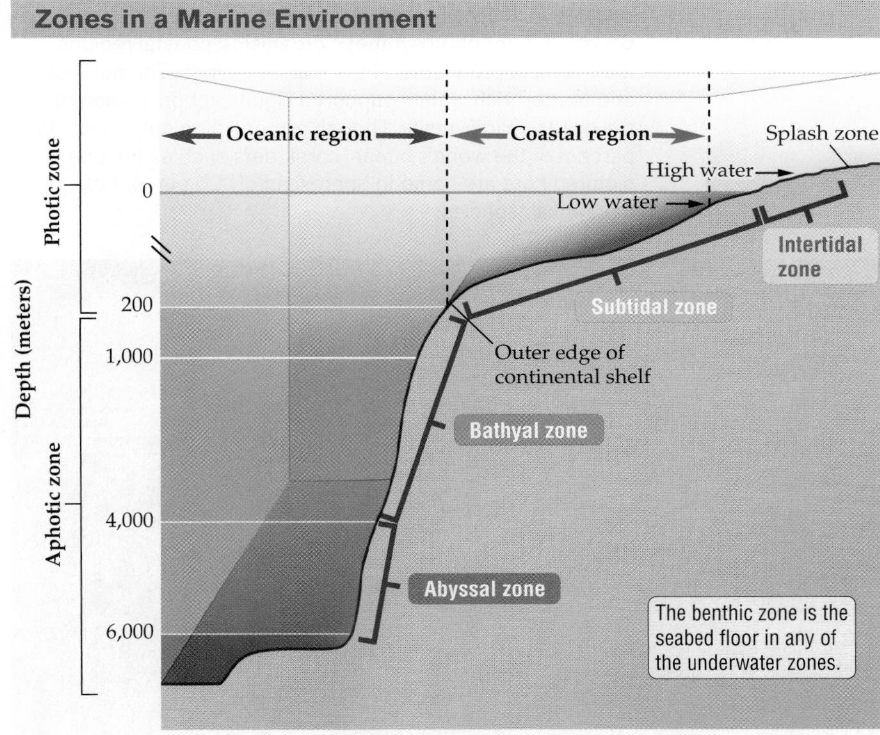

FIGURE 20.21 Ecological Zones in the Marine Biome

This cross-sectional view of the land and water progresses from the shoreline toward the open ocean (oceanic region). The coastal region stretches out to sea as far as the continental shelf, which is the underwater extension of a continent's rim. Productivity, and the abundance of life-forms, declines with increasing water depth because sunlight, which producers need for photosynthesis, diminishes with depth. The sunlit zone, or photic zone, extends to a depth of 200 meters (656 feet). The waters are dimly lit as deep as 1,000 meters, but they lie in complete darkness at greater depths. Productivity also decreases with distance away from the shore because nutrient levels typically decline farther out to sea. The well-lit waters of the open ocean are less productive than the well-lit regions of coastal areas, and the deepest layers of the ocean (the abyssal zone) are typically the least productive of all aquatic habitats.

FIGURE 20.22 Coastal Regions Support a High Level of Biodiversity

Located in shallow waters that allow sufficient light penetration for photosynthetic organisms, coastal regions are the most productive of the marine biomes. The nutrient- and oxygen-rich waters support a significant proportion of the ocean's biodiversity. Though representing only about 0.2 percent of the world's ocean, coral reefs such as the one pictured here are home to approximately 25 percent of all marine fish species.

FIGURE 20.23 The Intertidal Zone

Intertidal zones are found in coastal regions where tides rise and fall on a daily basis, periodically submerging a portion of the shore.

oxygen. Nutrients delivered by rivers and washed off the surrounding land accumulate in the coastal region. Nutrients that settle to the bottom are stirred up by wave action, tidal movement, and the turbulence produced by storms. The nutrients support the growth of photosynthetic producers, which inhabit the well-lit upper layers of coastal waters to a depth of about 80 meters (260 feet). The vigorous mixing by wind and waves also adds atmospheric oxygen to the water. A majority of Earth's marine species live in the coastal region. Not surprisingly, most of the world's highly productive fisheries are also located along coasts.

The **intertidal zone** is the part of the coast that is closest to the shore, where the ocean meets the land (**FIGURE 20.23**). This ecological zone spans the uppermost reaches of ocean waves during the highest tide and the shoreline that remains submerged at the lowest of low tides. In other words, the intertidal zone extends from the highest tide mark to the lowest tide mark. The intertidal zone is a challenging environment for plants and animals because they must cope with being submerged and exposed to dry air on a twice-daily basis, in addition to being pounded by surf and sand. The organisms that inhabit the upper regions of the intertidal zone are subject to predation by shorebirds and other animals when the tide is out. Despite these challenges, a diverse community of seaweeds, worms, crabs, sea stars, sea anemones, mussels, and other species is adapted to living in this zone.

The **benthic zone**, or ocean floor, of coastal regions may lie as deep as 200 meters (656 feet) below the water surface (**FIGURE 20.24**). The coastal benthic zone is a relatively stable habitat, rich in sediments containing the dead and decaying remains of organisms (detritus). The detritus forms the basis of a food web that supports a wide variety of consumers, including sponges, worms, sea stars, sea fans, sea cucumbers, and many fishes.

Productivity in the oceanic region is limited by nutrient availability

The open ocean, or **oceanic region**, begins about 40 miles offshore, where the continental shelf, and therefore the coastal region, ends. The open oceans form a vast, complex, interconnected ecosystem that we know relatively little about. Although they are well lit and have sufficient oxygen, the surface layers of the

FIGURE 20.24 The Benthic Zone
Located on the floor of the ocean, benthic communities can exist in shallow waters near the shore, or in deep waters, where they feed on dead organic matter drifting down from upper, better-lit zones.

open ocean are much less productive than estuarine and coastal waters, because they are relatively nutrient-poor. Nutrients are lacking because detritus tends to settle on the seafloor, and the nutrients locked in it are not readily stirred up from the great depths of the open ocean.

Where the continental shelf ends, the seafloor plunges steeply to a depth of approximately 6,000 meters (almost 20,000 feet). The cold, dark waters at these great depths constitute the **abyssal zone**. Few organisms can survive at the great pressures and low temperatures (about 3°C, or 37°F) of the abyssal zone. However, complex communities of archaeans and invertebrates are known to be associated with hydrothermal vents in geologically active regions of the deep oceans. A submarine hydrothermal vent is a crack in the seafloor that releases hot water containing dissolved minerals. Certain archaeans can extract energy from these minerals, and these single-celled prokaryotes form the basis of a food web that supports invertebrates such as giant tube worms, clams, and shrimp.

Concept Check

1. What are some of the physical conditions that influence life in aquatic biomes?
2. Why are open oceans generally less productive than coastal waters?

Concept Check Answers

1. Water temperature, water depth, speed of water flow, and salt levels.

2. Open oceans are relatively nutrient-poor because nutrients settle to the ocean floor and are not easily brought to the surface.

APPLYING WHAT WE LEARNED

How Invasive Mussels Can Harm Whole Ecosystems

International cargo ships that discharge ballast water in American ports introduced the Eurasian zebra mussels to the waterways of North America. Millions of the little inch-long mussels colonized lakes and rivers beginning in the Great Lakes and rode all the way to California on recreational fishing boats. Zebra mussels compete with native species and block or damage the cooling pipes of factories and power plants, causing both ecological and economic damage.

Mussels are filter feeders that live by filtering algae and other organic matter out of the water and consuming it. Zebra mussels have made Great Lakes waters so clear that there's little for other aquatic species to eat. They also attach themselves by the thousands to the backs of native mussels, preventing the natives from filter-feeding. In Lake St. Clair and Lake Erie, native mussels have all but vanished.

Zebra mussels are an example of an invasive species. Ecologists have cataloged hundreds of examples of destructive invasive species all over the world. But not all introduced species become invasive, causing destruction to the native species in an area.

The biosphere is an interconnected web of relations among different species and ecosystems. When we alter one component of the biosphere, we inevitably alter other parts. When the United States and China burn coal, for example, we raise carbon dioxide levels in the atmosphere throughout the world, contributing to global warming and the acidification of oceans all over the planet. Currents of air and water can carry pollutants around the world, causing damage to distant ecosystems. Agricultural runoff enters rivers, which carry pollutants into rich coastal ecosystems.

Humans, with our large numbers and heavy exploitation of natural resources, inevitably damage the natural world. Still, we are beginning to understand the biosphere. How can we use what we know about how the biosphere works to solve current environmental problems and prevent future ones? We have no easy answers to such questions, but in the next several chapters we'll examine how a sound understanding of ecological principles, combined with public awareness, citizen pressure, and political will, can help correct environmental mistakes of the past and prevent such blunders in the future.

BIOLOGY IN THE NEWS

Kenyan Company Turns Old Sandals into Colorful Array of Toys and Safari Animals

BY JOE MWIHIA • Associated Press, May 8, 2013

The colorful handmade giraffes, elephants and warthogs made in a Nairobi workshop were once only dirty pieces of rubber cruising the Indian Ocean's currents.

Kenya's Ocean Sole sandal recycling company is cleaning the East African country's beaches of used, washed-up flip-flops and other sandals.

About 45 workers in Nairobi make 100 different products from the discarded flip-flops. In 2008, the company shipped an 18-foot giraffe to Rome for display during a fashion week.

Company founder Julie Church says the goal of her company is to create products that people want to buy, then make them interested in the back-story.

Workers wash the flip-flops, many of which show signs of multiple repairs. Artisans then glue together the various colours, carve the products, sand and rewash them.

Church first noticed Kenyan children turning flip-flops into toy boats around 1999, when she worked as a marine scientist for WWF [the World Wildlife Fund] and the Kenya Wildlife Service on Kenya's coast near the border with Somalia.

Turtles hatching on the beach had to fight their way through the debris on beaches to get to the ocean, Church said, and a plan to clean up the debris and create artistic and useful items gained momentum. WWF ordered 15,000 key rings, and her eco-friendly project took off.

The world's oceans have historically been dumping grounds for human waste and debris, with various transmission routes delivering these products to the ocean. Tons of human-manufactured waste and trash are deliberately dumped from ships and boats and washed into oceans from freshwater rivers. Some items are lost during transit. At any given moment, the world's shipping industry is moving 5–6 million shipping containers, and on average, one of them falls off a ship every hour.

Plastic debris is a particular concern, because of its longevity. Things we drop in the ocean don't necessarily just go away, especially if they're made of plastic. The plastic pieces can be broken down into smaller pieces, but since they're nondegradable, these pieces will persist for centuries. As we have seen in this chapter, ocean currents carry marine litter over long distances. Because of the circular movement of water, this debris accumulates in particular areas of the world's oceans.

Within the past 20 years, both scientists and the general public have become increasingly aware of the issue of marine pollution. Community cleanup efforts have removed large amounts of rubbish that washed onto beaches after storms, scientists have studied the impact of garbage on ecosystems and food webs, and recycling efforts have attempted to reduce the production of waste plastics.

This article highlights an effort by a Kenyan company to address the problem of marine pollution. By producing products from waste gathered from the ocean, the founder hopes to raise public consciousness of a worldwide issue that has a significant impact on our biosphere.

Evaluating the News

1. Bacteria and other organisms break down organic matter into individual molecules that go to building new plants and animals. In contrast, most plastics undergo photodegradation, a process in which ultraviolet light from the sun breaks the plastic into tiny pieces, which may be consumed by aquatic animals. What kind of aquatic animals would consume plastic particles? Which animals farther up the food chain could be affected?

2. The Great Pacific Garbage Patch was not actually observed until after scientists predicted its existence in a 1988 paper. Despite its enormous size and density, it is invisible in satellite photos because it consists of millions of tiny particles of plastic suspended in miles of seawater. Discuss how researchers might determine the size and depth of the Great Pacific Garbage Patch.

3. Discuss the value of efforts like the one profiled in this article to raise public awareness of the marine pollution issue. Would an argument based on economics be more effective than an argument based on biology? For example, should companies like this Kenyan start-up focus more on producing more profitable consumer goods with their recycled materials?

CHAPTER REVIEW

Summary

20.1 Ecology: Understanding the Interconnected Web

- Ecology, the scientific study of interactions between organisms and their environment, helps us understand the natural world around us and our relationship with it.
- Organisms affect—and are affected by—their environment, which includes not only their physical surroundings but also other organisms. As a result, a change that affects one organism can affect other organisms and the physical environment.
- It is helpful to think of organisms and their environment as forming a web of interconnected relationships.
- Because components of the biosphere depend on one another in complex ways, human actions that affect the biosphere can have surprising side effects.

20.2 Climate's Large Effect on the Biosphere

- Weather describes the physical conditions of Earth's lower atmosphere in a specific place over a short period of time. Climate describes a region's long-term weather conditions.
- Climate depends on incoming solar radiation. Tropical regions are much warmer than polar regions because sunlight strikes Earth directly at the equator, but at a slanted angle near the poles.
- Climate is strongly influenced by six giant convection cells that generate relatively consistent wind patterns over much of the Earth.
- Human actions can affect climate, which influences almost all aspects of the biosphere.
- Ocean currents carry an enormous amount of water and can have a large effect on regional climates. Regional climates are also affected by major features of Earth's surface, as when mountains create rain shadows.

20.3 Terrestrial Biomes

- Climate is the most important factor controlling the potential (natural) location of terrestrial biomes. Climate can exclude a species from a region directly (if it finds the climate intolerable) or indirectly (if other species in the region outcompete it for resources).
- Human activities heavily influence the actual location and extent of terrestrial biomes.
- Some of the major terrestrial biomes are tundra, boreal forest, temperate deciduous forest, grassland, chaparral, desert, and tropical forest.

20.4 Aquatic Biomes

- Aquatic biomes are usually characterized by physical conditions of the environment, such as temperature, salt content, and water movement.
- Aquatic biomes are strongly influenced by the surrounding terrestrial biomes, by climate, and by human actions.
- Two major aquatic biomes are recognized on the basis of salt content: the freshwater and marine biomes. The freshwater biome includes lakes, rivers, and wetlands; estuaries, the coastal region, and the open ocean are ecosystems that lie within the marine biome.
- Human activities, such as dams on rivers and the application of fertilizer and pesticides on lawns and farm fields, can affect aquatic biomes.

Key Terms

abyssal zone (p. 483)
benthic zone (p. 482)
biome (p. 471)
biosphere (p. 466)
boreal forest (p. 473)
chaparral (p. 475)
climate (p. 468)
coastal region (p. 481)
convection cell (p. 469)
Coriolis effect (p. 469)
desert (p. 476)
ecology (p. 466)
estuary (p. 481)
grassland (p. 474)
intertidal zone (p. 482)
lake (p. 480)
oceanic region (p. 482)
permafrost (p. 473)
precipitation (p. 469)
rain shadow (p. 471)
river (p. 480)
temperate deciduous forest (p. 474)
tropical forest (p. 477)
tundra (p. 472)
weather (p. 468)
wetland (p. 480)

Self-Quiz

1. Which of the following properties of Earth explains why solar radiation is most intense at the equator?
   a. Earth is spinning on its axis.
   b. Earth is revolving around the sun.
   c. Earth is round.
   d. Earth's surface is mostly water.

2. Earth has six stable regions ("cells") in which warm, moist air rises, and cool, dry air sinks back to the surface. Such cells are known as __________ cells.
   a. temperate
   c. rain shadow
   b. latitudinal
   d. giant convection

3. The biosphere consists of
   a. all organisms on Earth only.
   b. only the environments in which organisms live.
   c. all organisms on Earth and the environments in which they live.
   d. none of the above

4. What aspect(s) of climate most strongly influence the locations of terrestrial biomes?
   a. rain shadows
   b. temperature and precipitation
   c. only temperature
   d. only precipitation

5. Winds that blow from the west across warm waters in the Pacific Ocean become warm and moist. By analogy to what happens in a rain shadow, what do you think would happen if such warm, moist winds blew across the cold Peru Current (see Figure 20.18)?
   a. An El Niño event would occur.
   b. The winds would continue to pick up moisture from the ocean currents.
   c. The warm, moist winds would cool, causing rain to fall.
   d. The warm, moist winds would cool, but rain would not fall.

6. Which of the following represents a large area of the globe with unique climatic and ecological features?
   a. population
   b. community
   c. biosphere
   d. biome

7. Wetlands, ponds, and streams are common in summertime in the Arctic tundra because
   a. the permafrost thaws completely, releasing large amounts of water.
   b. water from melted snow and ice is prevented from draining by the underlying permafrost.
   c. precipitation in the Arctic is high, exceeding 80 inches per year.
   d. tundra trees grow best in wet, boggy ground.

8. Temperate deciduous forests
   a. experience cool, rainy winters and hot, dry summers.
   b. are maintained by frequent fires.
   c. are not dominated by coniferous trees.
   d. were the predominant biome in Great Plains states such as Nebraska before they were settled by Europeans.

9. Estuaries are highly productive because
   a. they are dominated by trees and shrubs, which have a large biomass.
   b. they receive nutrients delivered by rivers and stirred up by tide action.
   c. they lie between the high-tide mark and low-tide mark in the coastal region.
   d. archaeans that extract energy from minerals are especially abundant in this type of ecosystem.

Analysis and Application

1. This chapter suggests that the organisms and environments of the biosphere can be thought of as forming an "interconnected web." Why do ecologists think of the biosphere like this? Give an example illustrating the interconnections between organisms and their environment.

2. Explain in your own words how global patterns of air and water movement can cause local events to have far-reaching ecological consequences. Give an example that shows how local ecological interactions can be altered by distant events.

3. Name seven of the major terrestrial biomes. How many of those biomes are located within 100 kilometers (about 60 miles) of your home? Describe the chief climatic and ecological characteristics of one terrestrial biome, and explain how those characteristics help determine which life-forms are found in that biome.

4. Using examples, explain the following statement: The defining characteristic of a desert is low moisture, not high temperature.

5. Explain why desert plants generally have smaller leaves than plants that are native to tropical rainforests. Describe some adaptations of desert animals.

6. How does spring and fall turnover contribute to the productivity of temperate lakes?

7. What accounts for the high productivity of estuaries and coastal regions? Explain why the open ocean has lower productivity than these two regions of the marine biome.

21 Growth of Populations

EASTER ISLAND WAS ONCE COVERED BY A LUSH FOREST. Long ago, the Rapanui people cut down all the trees on Easter Island—to build houses, to fuel cooking fires, and to move giant stone statues, visible in this photograph. The unsustainable use of resources led to an ecosystem collapse, and the island could no longer support the thousands of Rapanui who once lived there.

The Tragedy of Easter Island

Imagine living on a beautiful subtropical island in the South Pacific, 2,000 miles from the nearest continent and 1,300 miles from the next island. The temperature and weather are idyllic, with soft ocean breezes. Thirty species of seabirds nest on the island, and a forest of palms and flowering trees is home to parrots, herons, and other birds. Frigate birds wheel overhead, stealing fish from other seabirds. The ocean is rich with sea life, the soil is fertile, and human families raise vegetables and chickens.

It may sound romantic, but now imagine that the island is only 15 miles long and the population of humans is growing out of control. There's not enough food for everyone; the trees have nearly all been cut down; and without wood, islanders cannot even build boats to escape. This place was Easter Island.

Today, Easter Island is a small, barren grassland, just 166 square kilometers in area, about the size of St. Louis, Missouri. The island has lost its forest, lost its fertile soil, and even lost its generous rainfall. Scattered around the island are nearly a thousand ancient stone statues representing ancestors, some of them 30 feet tall and weighing 75 tons. Some of the statues, called Moai, stand upright but unfinished, as if the sculptors dropped their tools midstroke. Other Moai are complete but lie toppled in the grass, often broken. Hundreds more lie scattered along the coast.

Who carved these statues? What happened to the people who made them? What happened to the forest and the birds? And what lessons can we apply to the human population of Earth?

In this chapter we examine the characteristics of populations, including how and why populations increase or decrease in size—topics that are the main focus of population ecology. Toward the end of the chapter we'll return to the story of Easter Island and think about what happens when humans use more resources than the biosphere can support. Will the biosphere continue to support all of us?

MAIN MESSAGE

No population can continue to increase in size for an indefinite period of time.

KEY CONCEPTS

- A population is a group of interacting individuals of a single species located within a particular area.
- Populations increase in size when birth and immigration rates exceed death and emigration rates, and they decrease in size when the reverse is true.
- A population that increases by a constant proportion from one generation to the next exhibits exponential growth.
- Eventually, the growth of all populations is limited by environmental factors such as lack of space, food shortages, predators, disease, and habitat deterioration. These limits will determine the carrying capacity of the environment.
- Population growth may start exponentially, but typically levels off at carrying capacity, in a pattern called logistic growth.
- The world's human population is increasing exponentially. Rapid human population growth cannot continue indefinitely; either we will limit our own growth, or the environment will do it for us.
- Population ecology has broad practical application, including managing pests, regulating harvesting, and protecting endangered species.

THE TRAGEDY OF EASTER ISLAND		489
21.1	What Is a Population?	490
21.2	Changes in Population Size	491
21.3	Exponential Growth	492
21.4	Logistic Growth and the Limits on Population Size	494
21.5	Applications of Population Ecology	498
APPLYING WHAT WE LEARNED What Does the Future Hold?		501

POPULATION ECOLOGISTS CONCERN THEMSELVES with the number of organisms in a particular place, exploring questions like "How many organisms live here, and why?" Investigations in **population ecology** not only provide insight into the natural world, but are essential for solving real-world problems, such as controlling pests or protecting rare species (**FIGURE 21.1**). To set the stage for our study of the factors that influence how many individuals are in a population, we begin by defining what populations are. We then describe how populations grow over time, and we consider the limits to growth that all populations face, including the human population.

21.1 What Is a Population?

A **population** is a group of interacting individuals of a single species located within a particular area. The human population of Easter Island, for example, consists of all the people who live on the island.

Ecologists usually specify the number of individuals in a population by **population size** (the total number of individuals in the population) or by **population density** (the number of individuals per unit of area). To calculate population density, we divide the population size by the total area. As an illustration,

FIGURE 21.1 How Many Pandas Are Left in the Wild?
The giant panda is an endangered species native to China. The fact that it lives in remote bamboo forests makes it difficult to accurately estimate the size of the population (estimates range from 1,500 to 3,000). Which aspects of the panda population might interest an ecologist? How might understanding panda population structure contribute to the conservation of pandas?

let's return to the Easter Island example: In the year 1500, the population size was 7,000 people. If we divide that number by 166 square kilometers (the size of the island), we get a population density of 42 people per square kilometer (7,000/166 = 42).

Easter Island is an easy example for determining what constitutes a population because islands have well-defined boundaries, and human individuals are relatively easy to count. But often it is more difficult to determine the size or density of a population. Suppose a farmer wants to know whether the aphid population damaging a crop is increasing or decreasing (**FIGURE 21.2**). Aphids are small and hard to count. More important, it is not obvious how the aphid population should be defined. What do we mean by "a particular area" in this case? Aphids can produce winged forms that can fly considerable distances, so how do we know which aphids to count? Should we count only the aphids in the farmer's field? What about the aphids in the next field over?

In general, what constitutes a population is often not as clear-cut as in the Easter Island example. Overall, the area appropriate for defining a particular population depends on the questions being asked and on aspects of the biology of the organism of interest, such as how far and how rapidly it moves.

Aphids insert their mouthparts into a plant and withdraw nutrients, thus damaging the plant.
Aphids have infested this rose in large numbers.

FIGURE 21.2 Populations of Aphids Can Cause Extensive Crop Damage

Aphids are small insects with sucking mouthparts. They are pests on many plant species, which they infest in such large numbers that they can be difficult to count.

Concept Check

1. What is a population?
2. What is population density? Explain why it can be difficult to measure.

Concept Check Answers

1. A population consists of organisms of the same species interacting with each other in a given area.

2. Population density is the total number of interacting individuals of a single species, divided by the particular area they inhabit. It may be difficult to measure because individuals may be hard to detect, may move outside of or into the area being counted, and may inhabit a complex, hard-to-define area.

21.2 Changes in Population Size

Populations tend to change in size over time—sometimes increasing, sometimes decreasing. In 2002, monarch butterfly populations in North America fluctuated wildly. Each spring, monarchs make a spectacular migration that begins in the mountains west of Mexico City, where they overwinter, and ends in eastern North America (**FIGURE 21.3**). On January 13, 2002, the monarchs' overwintering sites were hit with an unusual storm that first drenched the butterflies with rain and then subjected them to freezing cold. The combination of wet and cold proved lethal: an estimated 70–80 percent of the butterflies—roughly 500 million of them—died overnight, the worst die-off in 25 years.

Helpful to Know

Remembering the meanings of two common prefixes can help you distinguish three similar-sounding terms. "Migration" comes from the Latin meaning "to move." The prefix *in* or *im* can mean "into," while *ex* or *e* means "out of, away from." So "immigration" is movement into an area, and "emigration" is movement out of an area.

Fortunately for the butterflies, this huge increase in winter death rates was followed by an equally spectacular rise in birth rates during the summer of 2002. Monarch birth rates shot up that year because it turned out to be a great summer for the monarch's primary food plant, milkweed. Because of this chance good fortune, butterflies that had survived the winter produced so many young in the summer of 2002 that monarch numbers quickly rebounded to almost the historic average.

Whether a population increases or decreases in size depends on the number of births and deaths in the population, as well as on the number of individuals that immigrate to (enter) or emigrate from (leave) the population. We can express this relationship in equation form:

$$\text{Growth rate} = (\text{birth} + \text{immigration}) - (\text{death} + \text{emigration})$$

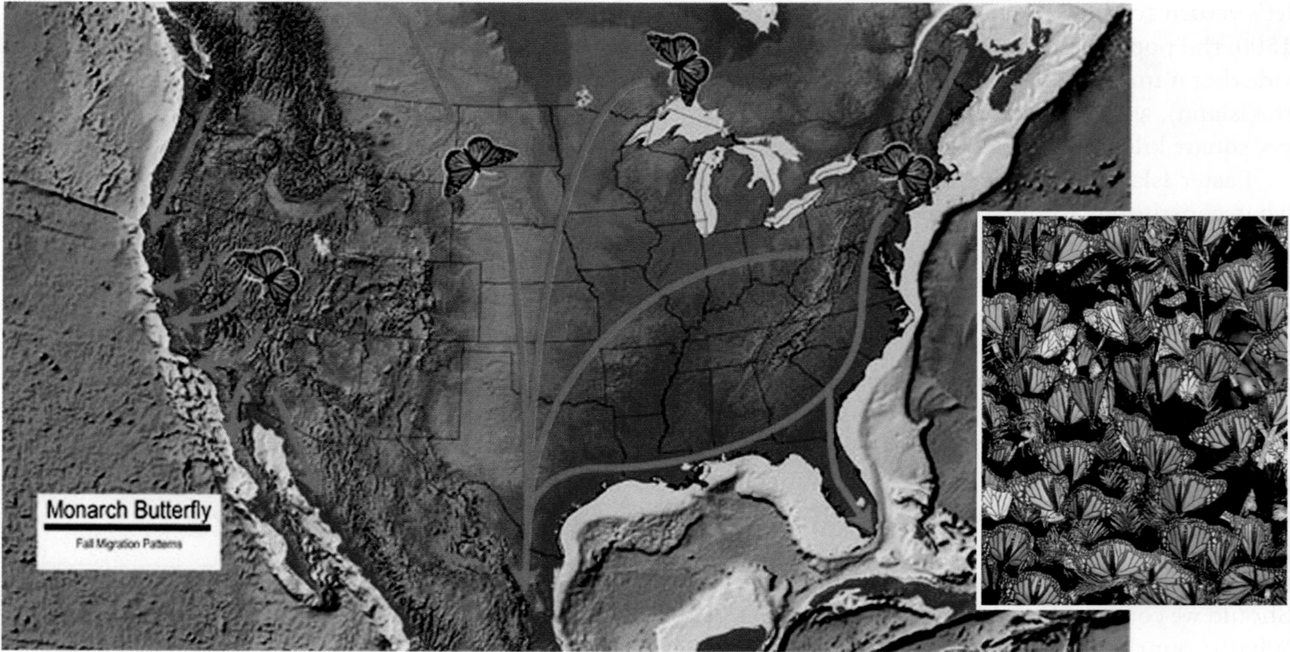

FIGURE 21.3
Before the Crash
Huge numbers of monarch butterflies overwinter in mountains west of Mexico City and then migrate each spring to eastern North America. In 2002, 70–80 percent of the overwintering butterflies died in an unusual winter storm.

For example, a population increases in size whenever the number of individuals entering (by birth and immigration) is greater than the number of individuals leaving (by death and emigration).

The environment has a major impact on birth, death, immigration, and emigration rates of a population. In one year, abundant rainfall and plant growth may cause mouse populations to increase; in the next, drought and food shortages may cause mouse populations to decrease dramatically. Such changes in the population sizes of organisms can have important consequences for people. For example, an increase in the number of deer mice, carriers of hantavirus, is thought to have been responsible for the 1993 outbreak of this deadly disease in the southwestern United States.

Concept Check

1. What factors influence a population's growth rate?
2. How does the environment influence a population's growth rate?

Concept Check Answers

1. Birth rate, death rate, immigration rate, and emigration rate.
2. Environmental fluctuations can lead to increased death rates or birth rates. Regional variability may also contribute to migration rates.

21.3 Exponential Growth

As with monarch butterflies, many organisms produce vast numbers of young. If even a small fraction of those young survive to reproduce, a population can grow extremely rapidly.

An important type of rapid population growth is **exponential growth**, which occurs when a population increases by a constant rate (r) over a constant time interval, such as 1 year (**FIGURE 21.4**). We can represent exponential growth from one year to the next by the equation

$$\frac{dN}{dt} = rN$$

where N is the number of individuals in the population, t is the time interval, dN/dt is the change in the population size during that time interval, and r is the rate of population growth. For example, if $r = 0.5$ and the current population size is 40, then the change in population size for that year (dN/dt) is $40 \times 0.5 = 20$, making the population in the next generation equal to the original population plus the change in the population, which is $40 + 20 = 60$ individuals.

In the next year the new value of N is 60. But r, the rate of population growth is unchanged; it is still 0.5. So for the year when the population begins with 60 individuals, the number of new individuals equals $60 \times 0.5 = 30$. At the end of this year, then, the population size increases to 90. In the third year, the starting population is 90, so the number of new individuals that year equals 45, and the total population at the end of the year is 135.

In exponential growth, the growth rate (r) is constant, but the numerical increase—the number of individuals added to the population—becomes larger with each generation. For example, the population in Figure 21.4 increases by 50 percent every generation (that is,

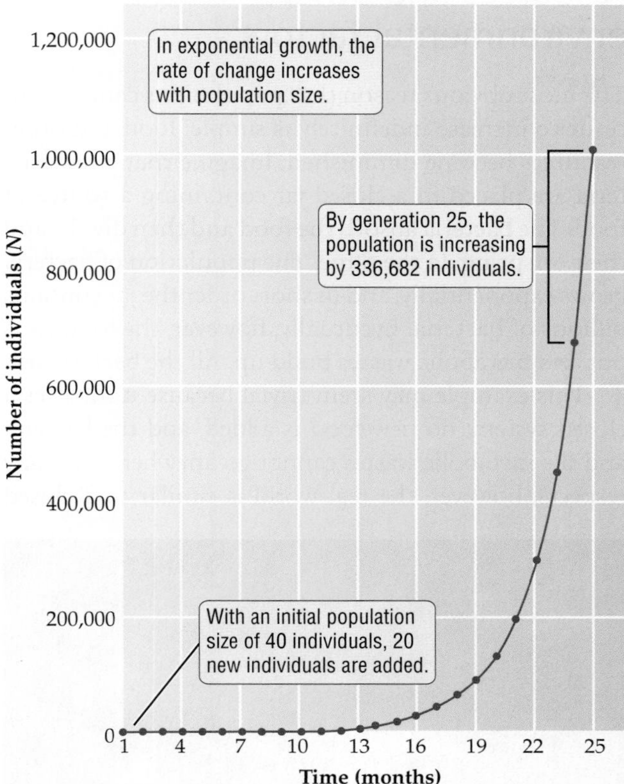
Exponential Growth
1,200,000
1,000,000
800,000
600,000
400,000
200,000
0
Number of individuals (N)
In exponential growth, the rate of change increases with population size.
By generation 25, the population is increasing by 336,682 individuals.
With an initial population size of 40 individuals, 20 new individuals are added.
1
4
7
10
13
16
19
22
25
Time (months)

FIGURE 21.4 The Size of a Population That Is Growing Exponentially Increases at a Constant Rate
In this hypothetical population, the growth rate of the population is 0.5. The number of individuals added to the population increases each generation, resulting in the J-shaped curve that is characteristic of exponential growth. Exponential-growth curves are always J shaped, but the curve's steepness varies depending on the rate of increase.

$r = 0.5$). With respect to its *numerical* increase, however, the population increases vary: the population increases by only 20 individuals between generations 1 and 2, but by 101 individuals between generations 5 and 6. As the population size increases, the rate of growth also increases. When plotted on a graph, exponential growth forms a **J-shaped curve**, as seen in Figure 21.4. Any population exhibiting a constant r value greater than 0 will exhibit exponential growth.

Endless exponential growth of a population is not seen under natural conditions, because sooner or later populations run out of vital resources, such as food or living space, that are needed to sustain them. (In terms of our equation, this means that the value of r begins to get smaller.) However, a population can increase exponentially over a short window of time. In 1839, a rancher in Australia imported a species of *Opuntia* (prickly pear cactus) and used it as a "living fence" (a thick wall of this cactus is nearly impossible for human or beast to cross). Unlike a metal or wooden fence, however, the *Opuntia* cactus did not stay in one place; it spread rapidly throughout the landscape. As the cactus spread, whole fields were turned into "fence," crowding out cattle and destroying good rangeland.

In 90 years, *Opuntia* cacti spread exponentially across eastern Australia, covering more than 243,000 square kilometers (over 60 million acres) and causing great economic damage. All attempts at control failed until 1925, when scientists introduced a moth species, appropriately named *Cactoblastis cactorum*, whose caterpillars feed on the growing tips of the cactus. This moth killed billions of cacti, successfully bringing the cactus population under control (**FIGURE 21.5**).

(*a*)

(*b*)

FIGURE 21.5 Blasting a Troublesome Cactus
In Australia, the moth *Cactoblastis cactorum* was introduced in 1925 to halt the exponential growth of populations of an introduced cactus species. (*a*) Two months before release of the moth, *Opuntia* cacti were growing in a dense stand. (*b*) Three years after introduction of the moth, the same stand had been almost completely eliminated.

EXTREME DENSITY

Bracken Cave near San Antonio, Texas, houses the largest known concentration of mammals. Twenty million Mexican free-tailed bats reside in this single cave. As many as 500 bat pups have been reported in a single square foot of space.

Introducing nonnative species to control a problematic invader (biological control) is risky because the control agent (the cactus moth in this case) could multiply exponentially and become a problem itself. Fortunately, the cactus moth is a specialist feeder, preferring cacti over any other plant the Australian outback had to offer. The moth's success in eradicating the cactus led to its own demise because decimating the prickly pear cactus population meant depleting the moth's own food supply. Moth numbers plummeted as a result. Today, both the cactus and the moth exist in eastern Australia in low numbers.

Concept Check

1. In exponential growth, what two factors contribute to the change in population size in a given generation?
2. Why does a population experiencing exponential growth exhibit a J-shaped curve?

Concept Check Answers

1. The rate of growth and the population size.
2. As the population increases in size, its *rate* of increase (dN/dt) also gets larger. Increases therefore become larger with every generation.

21.4 Logistic Growth and the Limits on Population Size

A giant puffball mushroom can produce up to 7 trillion offspring (**FIGURE 21.6**). If all of these offspring survived and reproduced at this same (maximal) rate, the descendants of a giant puffball would weigh more than Earth in just two generations. Humans and *Opuntia* cacti have much lower growth rates than giant puffballs, but given enough time, they, too, can produce an astonishing number of descendants. Obviously, Earth is not covered with giant puffballs, *Opuntia* cacti, or even humans, which illustrates an important general point: No population can increase in size indefinitely. Limits exist.

Growth is limited by essential resources and other environmental factors

The most obvious reason that populations cannot continue to increase indefinitely is simple: food and other resources become diminished. Imagine that a few bacteria are placed in a closed jar containing a source of food. The bacteria absorb the food and then divide, and their offspring do the same. The population of bacteria grows exponentially, and in short order the jar contains billions of bacteria. Eventually, however, the food runs out and metabolic wastes build up. All the bacteria die.

This example may seem trivial because it involves a closed system: no new food is added, and the bacteria and the metabolic wastes cannot go anywhere. In many respects, however, the real world is similar to a closed

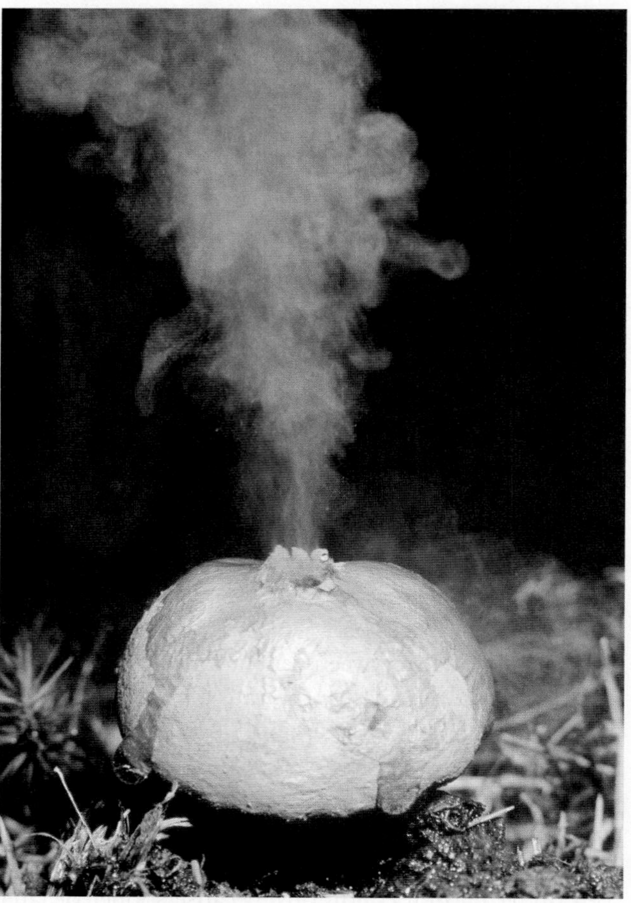

FIGURE 21.6 Will They Overrun Earth?
Given the number of spores it produces, a giant puffball mushroom has the potential to produce 7 trillion offspring in a single generation. However, relatively few of those spores land in a habitat suited for their growth. Large giant puffballs weigh 40–50 kilograms each; a medium-sized example is shown here.

system. Space and nutrients, for example, exist in limited amounts. In the *Opuntia* example of the previous section, even if humans had not introduced the *Cactoblastis* moth, the cactus population could not have sustained exponential growth indefinitely. Eventually, growth of the cactus population would have been limited by an environmental factor, such as a lack of suitable **habitat** (the type of environment in which an organism lives).

Logistic population growth is the norm in the real world

The exponential-growth model assumes that resources are unlimited, which is impossible over the long run. The logistic-growth model takes into consideration changes in growth rates that occur as resources become limited. **Logistic growth** is represented by an **S-shaped curve**, in which a population grows nearly exponentially at first but then stabilizes at the maximum population size that can be supported indefinitely by the environment. This maximum population size that can be sustained in a given environment is known as the **carrying capacity** (**FIGURE 21.7**), and is denoted by the variable K. The growth rate of the population decreases as the population size nears the carrying capacity because resources such as food and water begin to be in short supply. At the carrying capacity, the change in population size over time is zero.

Logistic growth can be modeled with this equation:

$$\frac{dN}{dt} = r_{max} N\left(\frac{K - N}{K}\right)$$

where r_{max} refers to the population's maximum rate of growth. The factor $(K - N)/K$ very simply is a measure of how far away from carrying capacity a particular population is. Consider what happens when a population is very small relative to the carrying capacity. In this case, $(K - N)/K$ effectively reduces to 1, and the rate of population change is similar to exponential growth. Now consider the case when a population is very near carrying capacity. As N approaches K, the factor $(K - N)/K$ becomes zero, making the rate of population change zero as well.

In the 1930s the Russian ecologist G. F. Gause carried out experiments on *Paramecium caudatum*, a common protist. He found that laboratory populations of paramecia increased to a certain size and then remained there (see Figure 21.7). In these experiments, Gause added new nutrients to the protists' liquid medium at a steady rate and removed the old solution at a steady rate. At first, the population increased rapidly in size. But as the population continued to increase, the paramecia used nutrients so rapidly that food began to be in short supply, slowing the growth of the population. Eventually the birth and death rates of the protists equaled each other and the population size stabilized. In contrast to natural systems, there was no immigration or emigration in Gause's experiments. In natural systems, populations reach and remain at a constant population size when birth plus immigration equals death plus emigration for extended periods of time.

Like laboratory populations of bacteria and paramecia, natural populations also experience limits (**FIGURE 21.8**). Their growth can be held in check by a number

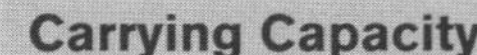

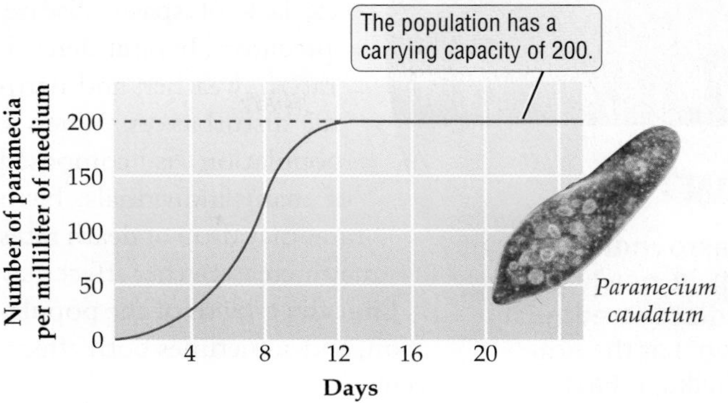

FIGURE 21.7 Carrying Capacity Is the Maximum Population Size a Particular Environment Can Sustain

A laboratory population of the single-celled protist *Paramecium caudatum* increases rapidly at first and then stabilizes at the maximum population size that can be supported indefinitely by its environment—that is, at its carrying capacity. This growth pattern can be graphed as an S-shaped curve.

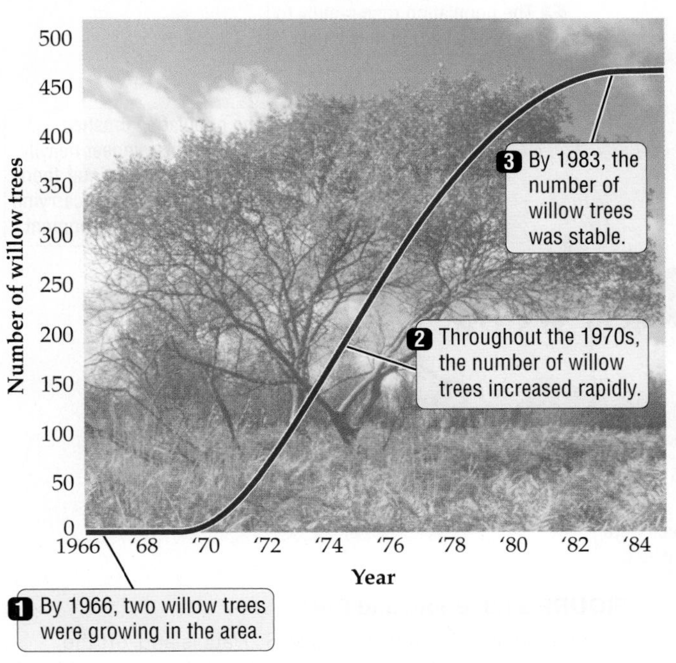

FIGURE 21.8 A Logistic Curve in a Natural Population

At a site in Australia, rabbits heavily grazed young willow trees, preventing willows from growing in the area. The rabbits were removed in 1954. By 1966, two willows had taken root in the area, presumably from seed blown in or carried in by animals. They increased rapidly in number, and the population then leveled off at about 475 trees.

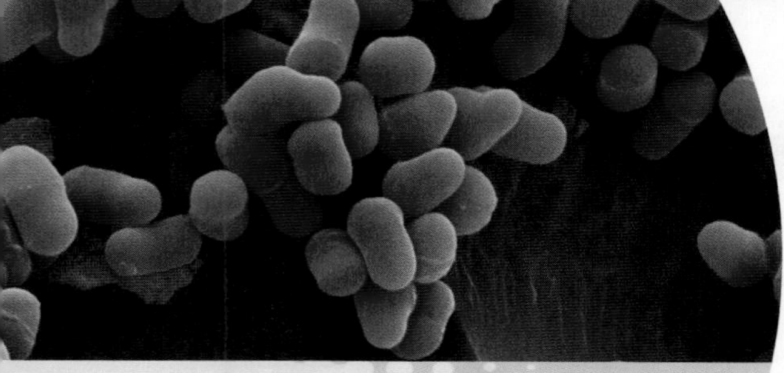

EXTREME GROWTH

Bacteria have an astounding rate of population growth. A newly found bacterial species, discovered near the hydrothermal vents at the bottom of Lake Tanganyika in East Africa and as yet unnamed, can produce 500 million cells in 5 hours.

of environmental factors, including food shortages, lack of space, disease, predators, habitat deterioration, weather, and natural disturbances. When a population is composed of many individuals, birth rates may drop or death rates may increase; either effect may limit the growth of the population, and sometimes both effects occur.

Large populations can also damage or deplete their resources. If a population exceeds the carrying capacity of its environment, it may damage that environment so badly that the carrying capacity is lowered for a long time. A drop in the carrying capacity means that the habitat cannot support as many individuals as it once could. Such habitat deterioration may cause the population to decrease rapidly (**FIGURE 21.9**). In 1911, 25 reindeer were introduced to the island of St. Paul in the Bering Sea. With no predators on the island, the population increased to over 2,000 individuals by 1938, well exceeding the carrying capacity of the island. The large number of reindeer resulted in severe degradation of the vegetation on the island, leading to a massive die-off so that by 1950, only 8 reindeer remained on the island. A similar pattern was seen in reindeer imported to St. Matthew and St. George islands.

Density-Dependent Growth

FIGURE 21.10 The Effects of Overcrowding in Plantain Weed, *Plantago major*

The number of seeds produced per plant drops dramatically under increasingly crowded conditions in plantain, a small herbaceous plant.

Habitat Deterioration Limits Growth

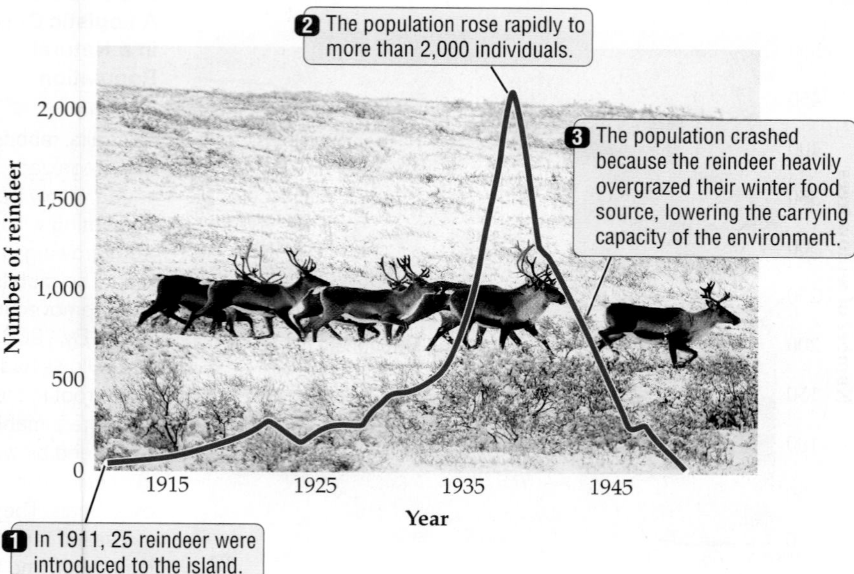

FIGURE 21.9 Boom and Bust

When reindeer were introduced to St. Paul Island, off the coast of Alaska, in 1911, their population increased rapidly at first and then crashed. By 1950, only eight reindeer remained.

Some growth-limiting factors depend on population density; others do not

Food shortages, lack of space, disease, predators, habitat deterioration—all of these factors influence a population more strongly as it grows and therefore increases in density. The birth rate, for example, may decrease or the death rate may increase when the population has many individuals. When birth and death rates change as the density of the population changes, such rates are said to be **density-dependent** (**FIGURE 21.10**).

In other cases, populations are held in check by factors that are not related to the density of the population; such factors cause the population to change in a **density-independent** manner. Density-independent factors can prevent populations from reaching high densities in the first place. Year-to-year variation in weather, for example, may cause conditions to be suitable for rapid population growth only occasionally. Poor weather conditions may reduce the growth

BIOLOGYMATTERS

How Big Is Your Ecological Footprint?

Humankind's use of resources and our overall impact on the planet have increased even faster than our population size. For example, from 1860 to 1991 the human population increased four-fold, but our energy consumption increased 93-fold. Many people think that modern societies of the developed world, like that of Easter Island, are not based on the *sustainable* use of resources. The term "sustainable" describes an action or process that can continue indefinitely without using up resources or causing serious damage to the environment.

One measure of sustainability is the *Ecological Footprint*, which is the area of biologically productive land and water that an individual or a population requires in order to produce the resources it consumes and to absorb the waste it generates. Scientists compute the Ecological Footprint using standardized mathematical procedures and express it in global hectares (gha; a hectare is a measure of land area equal to 10,000 square meters or approximately 2.5 acres). According to recent estimates, the Ecological Footprint of the average person in the world is 2.7 gha, which is about 30 percent higher than the 2.1 hectares that would be needed for each of the world's 7 billion people if they were supported in a sustainable manner. Overall, such estimates suggest that, since the late 1970s, people have been using resources faster than they can be replenished—a pattern of resource use that, by definition, is not sustainable. As the world population grows, the amount of biologically productive land available per person continues to decline, increasing the speed at which Earth's resources are consumed.

The Ecological Footprint of individuals in some countries, such as the United States (8.0 gha per person) and the United Kingdom (5.45 gha per person), is three to five times what is sustainable. As the chart indicates, the per capita consumption of Earth's resources by different countries is most directly related to energy demand, affluence, and a technology-driven lifestyle. However, population size also has a large overall impact on sustainability. For example, in 2007 the *total* Ecological Footprint of the Chinese population (1.3 billion) was 2,959 gha, compared to 2,468 for the U.S. population of 308 million people. Like overconsumption, overpopulation has a severe negative impact on ecosystems, and pollution and habitat degradation are often most severe in densely populated countries that otherwise have a low per capita Ecological Footprint. As people in populous countries such as China and India become wealthier, their overall and per capita footprints are growing rapidly.

What is *your* Ecological Footprint? If you are a typical college student, your footprint is probably close to the U.S. average of 8.0 gha (about 20 acres). It would take about 4.5 planet Earths to support the human population if everyone on Earth enjoyed the same lifestyle that you do. Your Ecological Footprint depends on four main types of resource use:

1. *Carbon footprint*, or energy use
2. *Food footprint*, or the land, energy, and water it takes to grow what you eat and drink
3. *Built-up land footprint*, which includes the building infrastructure (from schools to malls) that support your lifestyle
4. *Goods-and-services footprint*, which includes your use of everything from home appliances to paper products

If you drive a gas guzzler, live in a large suburban house, routinely eat higher up on the food chain (more beef than vegetables), and do not recycle much, your footprint is likely to be higher than that of a person who uses public transportation, shares an apartment, eats mostly plant-based foods, and sends relatively little to the local landfill. You can estimate your impact on the planet by using one of the many "footprint calculators" on the Internet, such as one offered by the Global Footprint Network. Most of us can significantly reduce our Ecological Footprint with little reduction in our quality of life, while bestowing an outsized benefit on our planet.

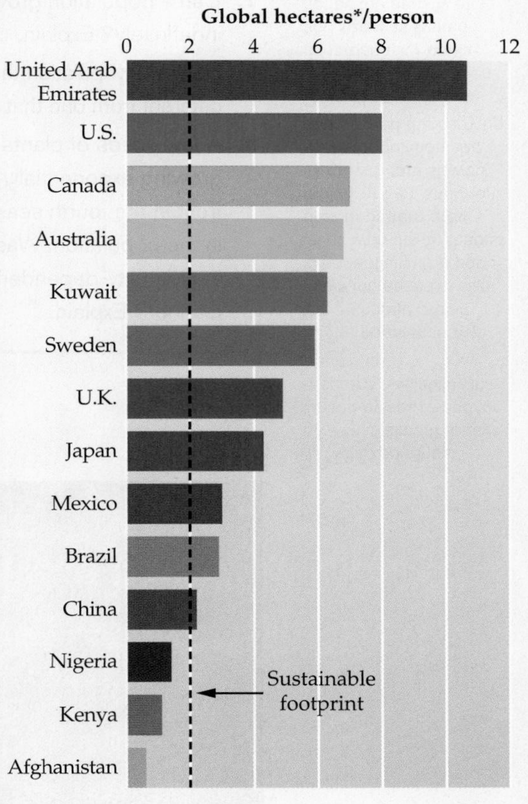

Ecological Footprints across the World

of a population directly (by freezing the eggs of an insect, for example) or indirectly (by decreasing the number of plants available as food to that insect). Natural disturbances such as fires and floods also limit the growth of populations in a density-independent way. Finally, the effects of human activities such as

Concept Check Answers

1. No. Exponential growth is almost always limited by space and/or resource availability in the habitat.

2. The population with the J-shaped growth curve increased in size exponentially. The population with the S-shaped growth pattern grew rapidly at first, and then its growth rate slowed as the population size approached the carrying capacity.

3. In a density-independent manner. Population size was limited not by the density of the population but by a natural disturbance that prevented the population from reaching the carrying capacity.

deforestation are density-independent; destroyed habitats are unsuitable for all organisms, no matter what their density is.

Concept Check

1. Can a population grow in an exponential manner indefinitely? Explain.
2. How is a population that shows an S-shaped growth curve different from one that exhibits a J-shaped growth curve?
3. A few seeds of plantain arrive in a vacant lot and start growing exponentially over the next three seasons. A late frost in the fourth season destroys most of the individuals in the population. Was the size of this population limited in a density-dependent manner or a density-independent manner? Explain.

21.5 Applications of Population Ecology

Not all populations will exhibit the idealized J- or S-shaped growth patterns discussed in the previous two sections. **Irregular fluctuations** are far more common in nature than is the smooth rise to a stable population size shown in Figure 21.8. Populations may experience booms and busts as a result of varying environmental conditions. Consider, for example, swarms of locusts (**FIGURE 21.11**). When densely packed together, grasshoppers experience behavioral and physiological changes that cause them to associate in huge, flying swarms of voracious locusts.

FIGURE 21.11 Locusts Exhibit Irregular Fluctuations in Population Size
Locust swarms occur when grasshoppers are densely packed together. Environmental factors, such as drought, may decrease the vegetation in an area and concentrate the grasshoppers into a smaller area, triggering the swarming behavior.

In March 2013, a swarm of locusts numbering an estimated 30 million grasshoppers ravaged Egypt and Israel, destroying valuable crops in these desert nations.

Understanding the reasons behind a population's growth pattern can provide critical information for monitoring and managing species: for conservation, for pest control, or for protecting economically important species. In the mid-twentieth century, populations of raptors (predatory birds) declined dramatically, and the die-off associated with the widespread use of the insecticide DDT. Many bird species, including the bald eagle, were declared endangered. Use of DDT was banned in 1972, and population ecologists have documented a steady rise in eagle populations since then (**FIGURE 21.12**).

Population studies can also be used to document the decline of species, particularly those that are commercially harvested. In the 1990s, Chilean sea bass experienced a significant population decline because of a dramatic increase in fishing pressure. Now considered an unsustainable fishery, Chilean sea bass is on many "do not eat" lists published by conservation groups (**FIGURE 21.13**).

Besides merely documenting population growth or decline, population ecologists can also help make informed decisions on the management of species and human interests. For managed fisheries, scientists regularly set limits on how many fish can be harvested in order to keep populations from declining. There are usually strict rules about the number and size of fish that can be caught, as well as the season in which they can be caught. In another example, during the 1980s forest managers needed to decide where and how much (if any) mature or old-growth forest could be cut without harming the rare spotted owl. For each owl population, researchers first gathered data on the birth rate and the amount of habitat used by each individual. The researchers then used these data to predict how the growth of spotted owl populations would be affected by the number, size, and location of patches of the bird's preferred habitat: old-growth forest. (Patches are portions of a particular habitat that are surrounded by a different habitat or habitats.) The researchers discovered that the total area of these forest patches, and how they are arranged in the landscape, has a big impact on the growth rates

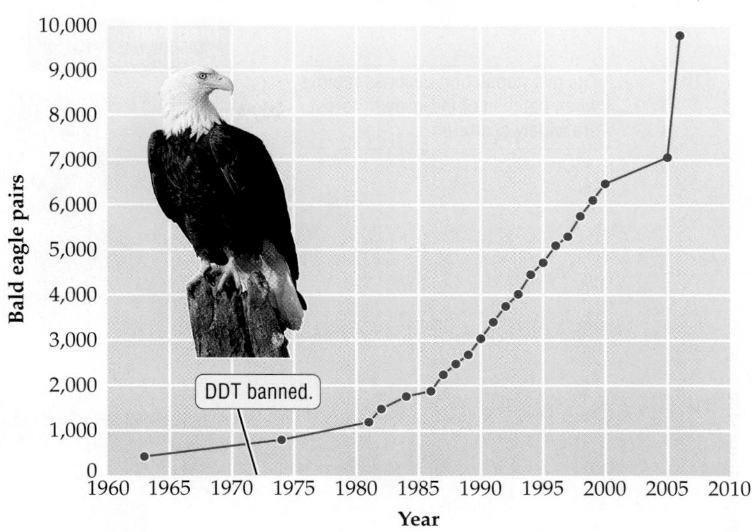

FIGURE 21.12 Bald Eagle Populations Increased after DDT Was Banned

Bald eagles are relatively easy to count; they have large, conspicuous nests to which they return year after year. By the early 1960s, population counts indicated that only 417 breeding pairs of eagles remained in the lower 48 states—a huge drop from the estimated 100,000 breeding pairs present in 1800. It was determined that DDT poisoning was directly responsible for declining eagle populations, which prompted a ban on DDT in 1972.

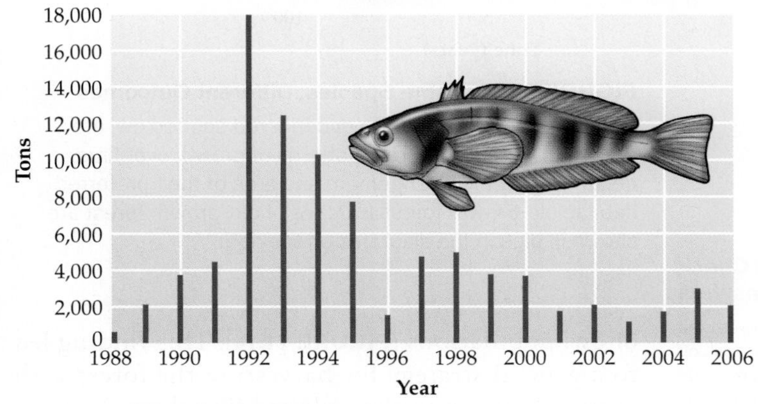

FIGURE 21.13 Popularity of the Chilean Sea Bass Caused Its Decline

An increase in the culinary popularity of Chilean sea bass in the 1990s led to overfishing, causing the population size to decline. Ecologists can utilize data such as these to assess the health of fisheries.

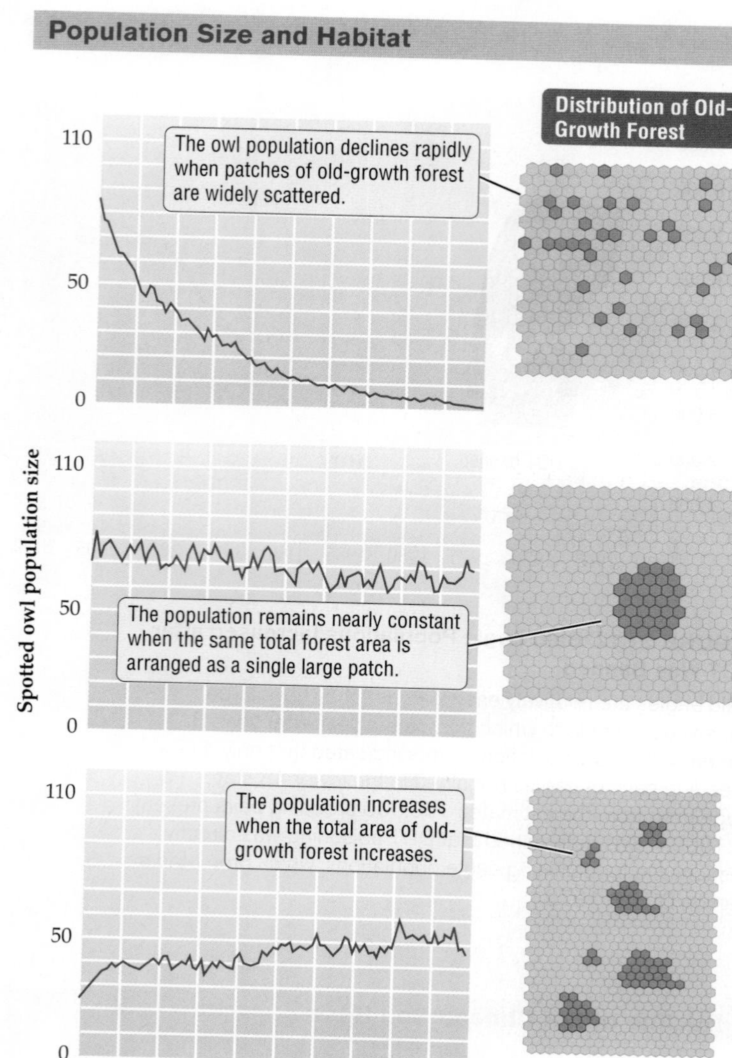

Population Size and Habitat
Distribution of Old-Growth Forest
The owl population declines rapidly when patches of old-growth forest are widely scattered.
The population remains nearly constant when the same total forest area is arranged as a single large patch.
The population increases when the total area of old-growth forest increases.
Spotted owl population size
110
50
0
Time (years)
50
100

FIGURE 21.14 Same Species, Different Outcomes
Different populations of the endangered spotted owl are predicted to show different patterns of growth over time, depending on the arrangement and area of their preferred habitat: old-growth forest. Patches of old-growth forest are shown in blue in the diagrams on the right.

of owl populations (**FIGURE 21.14**). This finding led to a general strategy for harvesting the forest with minimal impact to the owls residing there.

The principles of population ecology are also shaping modern agricultural practices. Traditionally, farmers controlled crop-damaging insects by spraying fields with insecticides to reduce their numbers as much as possible. Insect biology, however, made this strategy only marginally effective at controlling crop losses: the reproductive capacity (r) of insects is so high that even with the depressed numbers (N), populations were able to rebound quickly, making constant spraying a necessity. Extensive use of insecticides led to increased rates of resistance among the insects (see Section 14.2), diminishing the effectiveness of these treatments.

Ecologists quickly recognized that pest numbers can be lowered effectively by reductions in the carrying capacity (K) of the environment and the reproductive rate (r) of the insects. Modern pest control methods now take a comprehensive approach to insect control. Farmers clear out breeding areas, encourage the presence of predators or competitors, and introduce sterile males so that birth rates decline.

Nowhere are the applications of population ecology more relevant than to human population growth. For millennia, the human population was stable and small. However, the advent of modern medicine in the nineteenth century led to a meteoric rise in human populations. In a scant two centuries, humankind has exhibited exponential growth, growing from fewer than 1 billion individuals to well over 7 billion.

As we learned in Section 21.4, populations cannot maintain exponential growth indefinitely. Natural environments have a finite number of resources. The human population, like all other natural populations, must taper off at carrying capacity or else suffer a fate similar to that of the reindeer on St. Paul Island. This inevitable effect leads to an important question: What is the carrying capacity of Earth? Unfortunately, there is no simple answer. The chapter's closing essay discusses some of the challenges for future human population growth.

Concept Check

1. Why might two species exhibit interdependent population cycles?
2. Why are irregular fluctuations so common in nature?
3. What purpose does the study of population growth patterns serve?

Concept Check Answers

1. Two species that are intimately linked ecologically will strongly influence one another's abundance.
2. Because variable environmental conditions influence populations in unpredictable ways.
3. Proper management of a species (for example, for harvest or conservation) requires knowing how its population size is influenced by environmental factors.

APPLYING WHAT WE LEARNED

What Does the Future Hold?

As on Easter Island, or Rapa Nui, many of the current problems facing humans relate to excess population growth and destruction of the environment. The human population is growing at a spectacular rate, passing the 7 billion mark in 2011 and growing by about 200,000 people a day (**FIGURE 21.15**). Are we in danger of suffering the same fate that befell the Rapanui?

Following the excessive population growth and environmental destruction that ravaged it, Easter Island today is a nearly barren volcanic island. Before human colonization, Easter Island was a rich—but fragile—forest ecosystem. What happened and what role humans played in this dramatic change is still being debated. But a rich mine of archaeological evidence suggests that a combination of rapid population growth and habitat destruction led to the extinction of major groups of plants and animals and a human population crash. According to archaeological evidence, humans most likely first colonized Easter Island between AD 1000 and AD 1200, when a group of Polynesians arrived in canoes carrying domesticated chickens and crop plants such as yams, bananas, and taro root.

Almost as soon as the islanders arrived, they began cutting down the native palm forest. The trees were useful for building boats and houses and for megaprojects such as rolling 75-ton statues around the island. By about AD 1200, nearly half the trees were gone and the population of the island had reached about 2,000 people. By about 1300, the population is thought to have peaked at between 7,000 and 20,000. Archaeologists have calculated that such numbers must have existed to build and transport the hundreds of giant statues that still populate the island. By 1450, the palm trees were extinct. And by 1650, 20 other species of trees and shrubs had vanished as well.

Without wood, the islanders could not build wood houses; to cook, they burned grass or other plants. Before they cut down the forest, the islanders had relied on the palms and other trees to make baskets, sails, mats, roofing, rope, cloth, and wood carvings. The palm forests had also protected the soil,

Human Population Growth over the Millennia

(*a*)

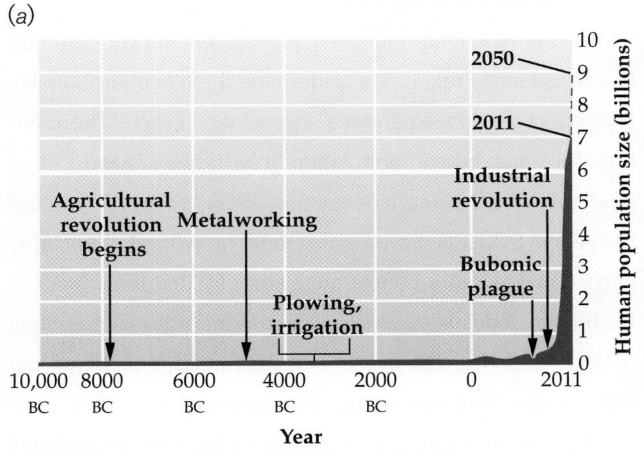

(*b*)

FIGURE 21.15 Rapid Growth of the Human Population
An indication of our growing numbers on Earth, as charted in (*a*), this image from space (*b*) shows the planet brightly lit by its most populous cities' lights. The brightness correlates well with population densities in industrialized countries but underrepresents population densities in countries such as China and India, which have sizable populations but poor access to modern amenities. The image was created by researchers at NASA using satellite image data.

FIGURE 21.16 Our Rising Ecological Impact

The global ecological impact of people has increased steadily. This graph compares human demands on the biosphere in each year between 1961 and 1999 with the capacity of the biosphere to regenerate itself. Human demand has exceeded the biosphere's entire regenerative capacity since the late 1970s.

Human Demand on the Biosphere

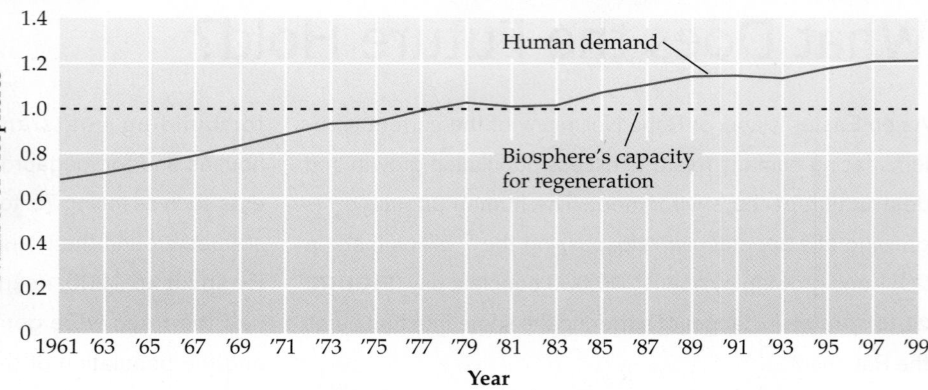

shaded crops, as well as providing nuts to eat and sap to drink. With the gradual removal of the forest, annual rainfall declined, limiting what could grow. But despite the decline in rainfall, soil erosion increased. The island's soil, exposed by the loss of forest, washed away during winter storms at a rate of 3 meters per year starting along the shoreline and gradually working up the mountain slopes. Landslides buried houses and gardens.

The thousands of Rapanui people on the tiny island were in dire straits. Without wood to build new boats, they could not leave the island; thousands were trapped on an island with dwindling resources. They struggled to deal with the drastic changes by moving inland to areas that still had soil. By AD 1400, they were building mountainside terraces for growing taro root; they struggled to protect the remaining soil by paving it with stones—laying approximately a billion stones averaging 2 kilograms each.

By some estimates, the Rapanui cut 6 million trees in just 300 years, severely reducing the island's carrying capacity. As food production dropped, families probably starved; the death rate increased and the birth rate declined. By 1600, the population had dropped back to about 2,000 people. The island no longer had the resources to support the huge populations required to carve and move the gigantic statues. Indeed, according to oral tradition, the Rapanui stopped carving statues by 1680. During fierce intertribal wars over the next 200 years, the people toppled most of the hundreds of statues and abandoned their former homes to hide in caves.

By the year 2025, the global human population is projected to surpass 8 billion people. As the global human population grows, we use more resources—including burning more fossil fuels, cutting forests faster, emptying the oceans of marine life more quickly—and generate more pollution and waste. Freshwater is becoming scarcer throughout the world, global fisheries are collapsing from too much fishing, and if current rates of logging continue, scientists estimate that, just as on Easter Island, all of Earth's tropical forests will be gone in less than 150 years.

All the evidence suggests that the current human impact on Earth is *un*sustainable. Since the 1970s, the human population has been using resources faster than they can be replenished—a pattern of resource use that, by definition, cannot be sustained (**FIGURE 21.16**). If people spend more than they earn, they have unsustainable spending habits. Humans, and especially Americans, are using global resources unsustainably.

People in wealthy countries leave a larger footprint on the planet than do those in poor countries. The United States has the third largest population in the world, and because its per capita footprint is so immense, it uses more than twice the total resources that India uses.

Will humans limit their own growth? Or will we continue to use too many resources, undermine the biosphere's ability to support us, and experience a population crash? There are hopeful signs: human population growth has slowed from a peak rate of 2.2 percent per year in 1963 to only 1.1 percent per year in 2009. But even a 1 percent growth rate translates into 70 million more people each year. To limit the impact of the human population, we must address the interrelated issues of population growth, overuse of resources, environmental deterioration, and sustainable development.

Hope for the future of our species (and the biosphere) lies in realistically evaluating the problems we face, and then addressing those problems. In the end, it is up to all of us to help ensure that humankind does not repeat on a grand scale the tragic lessons of Easter Island.

BIOLOGY IN THE NEWS

Eating Bugs: Would You Dine on Cicadas? Crickets? Buttered Beetles?

BY RENE LYNCH • *Los Angeles Times*, May 16, 2013

Mmmm. Just look at that plump little cicada. Can you imagine plucking it off its leaf and popping it in your mouth? Too much? How about after it's flash fried with a little butter, garlic and sea salt?

Face it, America. We're inch-worming our way closer to a dinner plate piled high with crickets, grasshoppers, grubs, cicadas, and more.

Don't believe us? The United Nations' Food and Agriculture Organization issued a report just this week aimed at raising "the profile of insects as sources of food and feed" as experts wonder how the world will feed a population that is expected to explode to 9 billion men, women, and children by 2050.

"To meet the food and nutrition challenges of today—there are nearly 1 billion chronically hungry people worldwide—and tomorrow, what we eat and how we produce it needs to be reevaluated," the report states.

Enter insects. Specifically, the hundreds of species of insects that are widely deemed as edible, including beetles, caterpillars, wasps, ants, grasshoppers, crickets, cicadas, termites, dragonflies ... stop us when you've had your fill.

Insects are considered a viable alternative because they are so plentiful, they pack a protein punch, they require little preparation, and they do not take a heavy toll on the environment.

"Insects as food and feed emerge as an especially relevant issue in the twenty-first century due to the rising cost of animal protein, food and feed insecurity, environmental pressures, population growth, and increasing demand for protein among the middle classes," the report says.

"Thus, alternative solutions to conventional livestock and feed sources urgently need to be found. The consumption of insects, or *entomophagy*, therefore contributes positively to the environment and to health and livelihoods."

Billions of people on the planet already include bugs as part of their diet. But Westerners, the report notes, "view entomophagy with disgust and associate eating insects with primitive behaviour." ...

Are you biting? Can you envision a day when you'll tuck into a plate of beetles?

The rapid growth of human populations in the past two centuries is perhaps the most significant ecological crisis confronting Earth. With populations expected to exceed 9 billion by the mid-twenty-first century, humans presently are exhibiting an exponential rate of growth. As noted in this chapter, exponential growth is an unsustainable condition. Either populations degrade their environment—as happened to the reindeer on St. Paul Island and the humans on Easter Island—or growth rate slows as a population reaches carrying capacity. Either scenario is troubling for the future of humanity. Life for a population at carrying capacity is highly competitive, with mortality rates equal to birth rates. That is, for every child that is born, someone will die.

As with any population, as humanity approaches carrying capacity, we will be confronted with limits, including limits on food and water. Presently, starvation and famine are the result of an unequal distribution of food that is caused by social, political, and economic shortcomings, not a deficiency in global food production. Food production has increased significantly since the industrial revolution, actually outpacing human population growth over the same period. With our excessive growth rate, however, it is estimated that production will simply not be able to keep up. If production does not increase, then new food sources must be found.

Insects are abundant and have a rapid growth rate. Furthermore, they contain enough protein to make them a viable option, should our population continue to grow beyond the present capacity of Earth.

Evaluating the News

1. Why is human growth rate such a concern? What are the ecological impacts of our ever-increasing population? What are some possible outcomes? What ecological limits might humanity confront as we approach carrying capacity?

2. How might limits to human population growth be overcome? What lifestyle changes can be exercised in order to mitigate our ecological impact on Earth? Should we advocate policies such as China's one-child rule in order to prevent reaching carrying capacity?

3. What factors make insects an attractive choice as a nutritional source?

CHAPTER REVIEW

Summary

21.1 What Is a Population?

- A population is a group of interacting individuals of a single species located within a particular area.
- Two basic concepts used in studying populations are population size (the total number of individuals in the population) and population density (the number of individuals per unit of area).
- What constitutes an appropriate area for determining a population depends on the questions of interest and the biology of the organism under study.

21.2 Changes in Population Size

- All populations change in size over time. Populations increase when birth and immigration rates are greater than death and emigration rates, and they decrease when the reverse is true.
- Because birth, death, immigration, and emigration rates are all affected by environmental factors, the environment plays a key role in changing the size of a population.

21.3 Exponential Growth

- A population grows exponentially when it increases by a constant proportion from one generation to the next. Exponential growth produces a J-shaped curve.
- Populations may grow at an exponential rate when organisms are introduced into or migrate to a new area.

21.4 Logistic Growth and the Limits on Population Size

- Because the environment contains limited amounts of space and resources, no population can continue to increase in size indefinitely.
- Some populations increase rapidly at first and then level off and stabilize at the carrying capacity, the maximum population size that their environment can support. This logistic growth pattern is represented by an S-shaped curve.
- Density-dependent environmental factors limit the growth of a population more strongly when the density of the population is high. Such factors include food shortages, diminishing space, disease, predators, and habitat deterioration.
- Density-independent factors, such as weather and natural disturbances, limit the growth of populations without regard to their density.

21.5 Applications of Population Ecology

- In natural systems, a growth pattern of irregular fluctuations is much more common than a logistic growth pattern.
- Understanding why different populations have different patterns of growth can provide critical information on how best to manage endangered species, pest species, and species that are of commercial interest to humans.

Key Terms

carrying capacity (p. 495)
density-dependent (p. 496)
density-independent (p. 496)
exponential growth (p. 492)
habitat (p. 495)
irregular fluctuation (p. 498)
J-shaped curve (p. 493)
logistic growth (p. 495)
population (p. 490)
population density (p. 490)
population ecology (p. 490)
population size (p. 490)
S-shaped curve (p. 495)

Self-Quiz

1. A group of interacting individuals of a single species located within a particular area is
   a. a biosphere.
   b. an ecosystem.
   c. a community.
   d. a population.
2. A population of plants has a density of 12 plants per square meter and covers an area of 100 square meters. What is the population size?
   a. 120
   b. 1,200
   c. 12
   d. 0.12
3. A population that is growing exponentially
   a. increases by the same number of individuals each generation.
   b. increases at a constant rate (r) each generation.
   c. increases in some years and decreases in other years.
   d. none of the above
4. In a population with an S-shaped growth curve, after an initial period of rapid increase the number of individuals
   a. continues to increase exponentially.
   b. drops rapidly.
   c. remains near the carrying capacity.
   d. cycles regularly.
5. The growth of populations can be limited by
   a. natural disturbances.
   b. weather.
   c. food shortages.
   d. all of the above
6. Factors that limit the growth of populations more strongly at high densities are said to be
   a. density-dependent.
   b. density-independent.
   c. exponential factors.
   d. sustainable.
7. The maximum number of individuals in a population that can be supported indefinitely by the population's environment is called the
   a. exponential size.
   b. J-shaped curve.
   c. sustainable size.
   d. carrying capacity.
8. A population that initially has 40 individuals grows exponentially at an annual rate (r) of 0.6. What is the size of the population after 3 years? (*Note:* Round down to the nearest individual.)
   a. 16
   b. 163
   c. 192
   d. 102,400

Analysis and Application

1. Explain why it can be difficult to determine what constitutes a population.
2. Populations increase in size when birth and immigration rates are greater than death and emigration rates. Keeping this basic principle in mind, what actions do you think a scientist or policy maker might take to protect a population threatened by extinction?
3. Assume that a population grows exponentially, increasing at a constant rate of 0.5 per year. If the population initially contains 100 individuals, it will contain 150 individuals in the next year. Graph the number of individuals in the population versus time for the next 5 years, starting with 150 individuals in the population.
4. Population growth cannot increase indefinitely.
   (a) What environmental factors prevent unlimited growth?
   (b) Why is it common for populations of species that enter a new region to grow exponentially for a period of time?
5. Describe the difference between density-dependent and density-independent factors that limit population growth. Give two examples of each.
6. Different populations of a species can have different patterns of population growth. Explain how an understanding of the causes of these different patterns can help managers protect rare species or control pest species.
7. List five specific actions you can take to limit the growth or impact of the human population.

22 Animal Behavior

DOMESTICATED SIBERIAN SILVER FOXES. Not just hand-raised descendants of wild foxes, these foxes have been bred since 1959 to respond only positively to humans. In the process of genetic domestication, the foxes also evolved multicolored coats, curly tails, broader heads, and other traits typical of domesticated animals. Domesticated foxes, unlike their wild cousins, make great pets.

The Evolution of Niceness

Dogs are famous for their affectionate and loyal nature. Playful and relaxed around people, "man's best friend" will approach total strangers, lick faces, and welcome impromptu belly rubs. In contrast, wild wolves and coyotes avoid humans and even at their least shy are wary, unfriendly, and often aggressive. It's easy to assume that dogs are friendly just because they've been raised by humans from puppies. But coyotes, wolves, and foxes that have been raised by humans are not like dogs. The submissive and friendly behavior of most domesticated dog breeds is fundamentally different from the wary alertness of their wild counterparts, and the difference is genetically based.

One of the most dramatic demonstrations of the genetic basis of domestic behavior came from an unlikely source—a Soviet research program that had been virtually stripped of genetics. Beginning in the 1930s, the powerful Soviet scientist Trofim Lysenko launched a campaign to wipe out the science of genetics—at least in the USSR. Lysenko rejected the very idea of genes and, with the support of the ruthless Soviet premier Joseph Stalin, eliminated departments of genetics throughout the Soviet Union. An entire generation of Soviet geneticists died in Soviet prisons, were executed, or else fled to the West. One canny survivor was geneticist Dmitry Belyaev. His own brother, also a geneticist, died in a concentration camp. After Dmitry lost his job as head of an animal breeding lab in Moscow in 1948, he accepted a transfer to Siberia and reported that he was switching to research in physiology.

Inside, however, Belyaev remained a geneticist, hanging on until Stalin's death in 1953, when Lysenko's hold on Soviet science began to weaken. In 1959, Belyaev began a multidecade effort to domesticate silver foxes, valued for their beautiful fur. Belyaev was convinced that behavior had a genetic basis, and he planned to select for just one behavior—tameness—and no other traits.

Did Belyaev succeed in taming the foxes? What did his experiment tell us about the domestication of dogs and other animals and about the genetics of behavior? Is behavior controlled by genes? Is behavior learned?

Before we address these questions, we will examine the main types of behaviors displayed by animals, including humans.

MAIN MESSAGE Animal behavior is the way an animal responds to information obtained from its environment; patterns of behavior can be fixed or learned.

KEY CONCEPTS

- Behavior is a predictable response to external stimuli.
- Behavior can be fixed (inborn) or learned. Most animals have some fixed behaviors, which are usually triggered by simple external stimuli.
- Learned behaviors enable individual animals to respond to their environment flexibly.
- Group living and social behavior offer many advantages, including extra protection from predators and greater access to resources.
- Social behavior involves interaction among members of a group, usually of the same species.
- Altruistic individuals increase the adaptive fitness of the recipient at an actual or potential cost to themselves.
- Altruism is usually extended toward close relatives and therefore increases the fitness of the social group as a whole.
- Communication is an important type of behavior. Communication ranges from simple chemical signals to elaborate displays and songs to the intricate languages of humans.
- Mating behaviors are influenced by differences in parental investment. Males and females often have different interests that lead to different mating strategies.
- Mating systems are the result of differences in parental contributions. Organisms can be monogamous, promiscuous, or polygamous.

THE EVOLUTION OF NICENESS 507

22.1 Sensing and Responding: The Nature of Behavioral Responses 508

22.2 Fixed and Learned Behaviors in Animals 510

22.3 Social Behavior in Animals 513

22.4 Facilitating Behavioral Interactions through Communication 515

22.5 Mating Behaviors 517

APPLYING WHAT WE LEARNED 520
The Genetics of Domestication

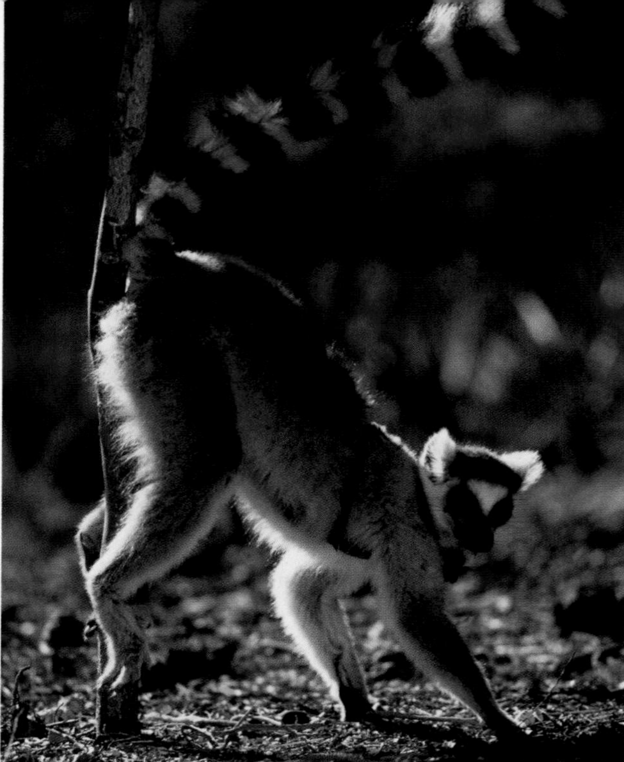

FIGURE 22.1 Scent Marking Is a Common Behavior in Mammals
Like many other mammals, this ring-tailed lemur is using its urine to scent-mark a tree. Why might mammals engage in this behavior? What genetic or physiological factors might underlie this behavior?

ANIMALS ACQUIRE INFORMATION from the environment and respond to it in predictable ways. The sum total of these responses constitute an animal's **behavior**, and the environmental cues that elicit behavioral responses are called **stimuli** (singular "stimulus"). This broad definition of animal behavior encompasses responses that are familiar—for example, courtship in songbirds, and scent marking by territorial animals (**FIGURE 22.1**)—but also more subtle and nuanced actions, such as alcohol abuse or sexual promiscuity. Behavior enables animals to respond quickly to changes in the environment, such as the presence of a predator or the changing of the seasons. Moreover, it enables animals to coordinate their own activities with those of other individuals of the same or different species. Through appropriate behavioral responses, honed by natural selection, animals can find food, defend territory and other resources, avoid predators, choose mates, and care for their young. Behavioral scientists are interested in learning how a specific behavior increases or decreases an animal's *adaptive fitness*—its ability to survive and reproduce in its particular habitat.

Understanding how and why an organism reacts to a stimulus in a particular way can be a complex undertaking. Some behaviors can be influenced by genes; others, by an organism's experience. *Fixed behaviors* are expressed from birth; *learned behaviors* are those that animals acquire as they gain experience with their surroundings.

This chapter begins with an examination of the roles of genes and environment in shaping behavioral responses. We then discuss different types of behaviors, including both fixed and learned behaviors. We explore sociality in organisms, and how communication facilitates social interactions. Finally, we discuss mating behaviors and the factors that influence them.

22.1 Sensing and Responding: The Nature of Behavioral Responses

Unlike physical traits, such as flower color or wing shape, behaviors are complex manifestations of both genetic and learned responses. Scientists have many ways to approach the study of behavior. Behavioral ecologists are interested in how behaviors may be adaptive to an organism, since the way an animal interacts with its environment has important consequences for its survival and reproduction. Behavioral scientists also aim to dissect the genetic basis of behavior and understand how nongenetic influences, such as the stress experienced by an individual, might affect behavior. Biologists also want to know the functional basis of behavior: how the stimulus is perceived, what information is processed by the nervous system, and how organ systems execute a coordinated

BIOLOGYMATTERS

Drinking and the Dark Sides of Human Behavior

Human behavior changes dramatically under the influence of alcohol, as does our ability to regulate that behavior. Among college students, the consequences of abusing alcohol go far beyond the fatalities and injuries caused by drinking and driving. Each year in the United States, approximately 600,000 physical assaults and 70,000 cases of sexual assault involving college students are alcohol related. In addition, 400,000 students have unprotected sex, and 100,000 students report that they were too intoxicated to know whether they had consented to sexual intercourse. About one in 100 students attempt suicide because of alcohol or other drugs. On the academic side, students who drink to excess miss more classes, get lower grades, and are more likely to fall behind in their work.

Furthermore, consequences of drinking spill over into the lives of those who are not even doing the drinking. Among students who live on campus or in sororities or fraternities and do not drink, or drink moderately, 60 percent have had sleep interrupted because of drinking behavior by others, almost half have had to take care of a drunken fellow student, and 15 percent have had property damaged as a result of drunken behavior. One in three has been insulted or humiliated by or has had a serious quarrel with a drunken student, and almost one in 10 has been pushed or hit in an alcohol-related incident. One study found that alcohol is one of the most significant contributing factors to sexual aggression by male college students. One in 100 college-age women have been victims of alcohol-related sexual assault or acquaintance rape. The damage to the psychological well-being and academic performance of these victims is difficult to measure.

Alcohol, when consumed in excess, very often causes drastic and dangerous changes in human behavior. An awareness of the costs and personal consequences associated with those changes can build a better appreciation for the advice we often see in the media: "If you drink, drink responsibly."

DUI Arrest

An officer from the Miami Beach Police Department arrests a motorist who failed a Breathalyzer test. Two readings with the device measured her blood alcohol concentration (BAC) at 0.19 and 0.183, about twice the legal limit in Florida.

response. Finally, reconstructing the evolutionary history of specific behaviors, such as tool use, and the role of natural selection in shaping those behaviors, is an important goal in the study of animal behavior.

Although single-celled organisms, such as amoebas, exhibit some basic behaviors, behavioral responses are most obvious in complex animals because their nervous, muscle, and skeletal systems assist them in rapidly processing sensory information and in generating behaviors that are readily observed.

Behavior is a crucial part of human life, which is why it is studied not just by biologists, but also by many others, from behavioral psychologists to political analysts. Humans have a complex, highly developed nervous system that can receive and process diverse external stimuli and launch diverse and complex responses to them. As a general principle, the most complex animal behaviors are found among species that have the most complex nervous systems.

Complex behaviors are formed from a combination of genetic and environmental factors. Careful studies of identical and fraternal twins in humans, for example, have revealed that alcoholism has a strong genetic component. However, unknown environmental factors control whether a genetically susceptible person will actually display the condition. Traits like alcoholism that run in families clearly have a genetic basis (around 50 percent in the case of alcoholism), yet no *single* gene encodes this behavior (**FIGURE 22.2**). How, then, does genetics play a role in this type of behavior?

When alcohol is consumed, it becomes metabolized by the body and affects brain chemistry. These metabolic processes are all mediated by enzymes (which are proteins). Individuals with different versions of the same enzyme may process alcohol in different ways. Some, for example, may experience greater feelings of euphoria or may be less able to break down alcohol. These biochemical responses may result in behaviors such as alcohol abuse or aversion.

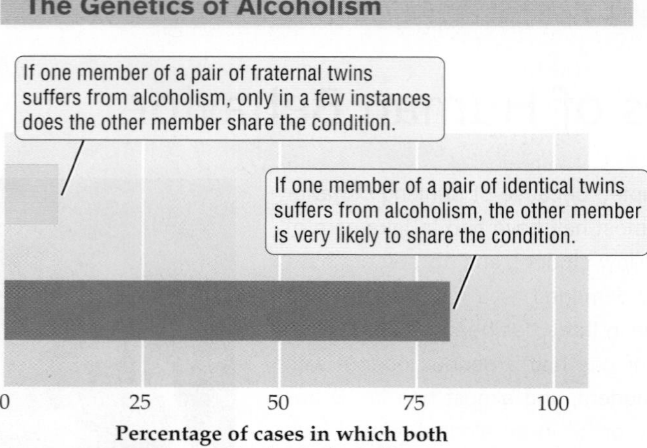

The Genetics of Alcoholism
If one member of a pair of fraternal twins suffers from alcoholism, only in a few instances does the other member share the condition.
If one member of a pair of identical twins suffers from alcoholism, the other member is very likely to share the condition.
0
25
50
75
100
Percentage of cases in which both twins suffer from alcoholism

FIGURE 22.2 Behavioral Studies of Twins Can Reveal a Genetic Basis of Behavior
Alcohol abuse and dependence shows some genetic basis. However, the environment also affects the expression of the genes causing alcohol dependence, because even among genetically identical twins, in many cases only one twin develops the condition.

Concept Check Answers

1. Behavioral scientists are interested in many different aspects of behavior, including its adaptive value, its genetic basis, its physiological and functional basis, and its evolutionary history.

2. No single gene encodes a complex behavior like alcoholism. However, alcohol's digestion, metabolism, and impact on brain chemistry are all influenced by genes. Differences in these genes might influence an organism's response to alcohol and determine whether it might be prone to alcoholic behavior.

Scientists have made significant progress in identifying many genes that are responsible for complex behaviors. Most laboratory rats, for example, can learn to negotiate mazes, but some rats learn faster than others. By mating fast learners with other fast learners, researchers can produce offspring that also learn to negotiate mazes quickly. Mating slow learners with other slow learners produces rats that have a difficult time learning to find their way through a maze.

Some behaviors that would seem to defy scientific explanation are also influenced by the genetic composition of an organism. Fidelity, for example, is caused by a chemical response in the brain to a partner. Studies on voles have shown differences in brain function (and presumably genetics) between monogamous species and promiscuous species (**FIGURE 22.3**). Even the maternal instinct has some root in the genetics of mothers. Studies of mice in which a single gene was suppressed led to mice that failed to properly care for their offspring.

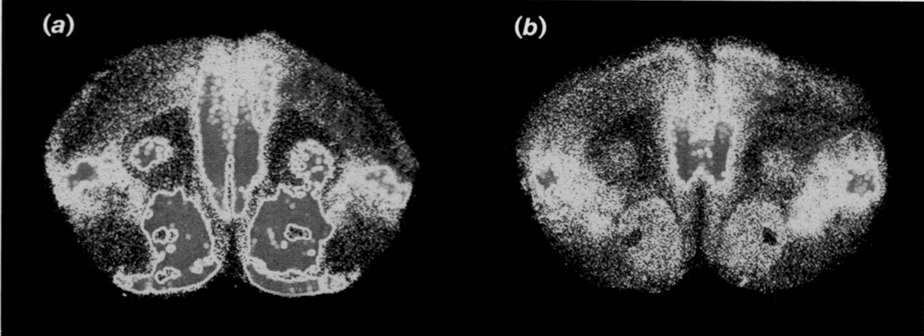

FIGURE 22.3 Complex Behaviors Can Have a Genetic Basis
Brain scan of the monogamous prairie vole (*a*) shows a high concentration of oxytocin receptors (red), which are linked to emotional attachments. The montane vole, which does not form long-term pair bonds, possesses fewer of these receptors (*b*).

Concept Check

1. What are the major goals for the scientific study of behavior?
2. Are there "genes" for specific behaviors, such as alcoholism?

22.2 Fixed and Learned Behaviors in Animals

Behavioral patterns can be either fixed (innate) features of an animal's behavior or developed through the animal's experiences. For instance, males of most bird species sing a distinctive song to attract females. Males of some species know their courtship songs from birth, but others must learn them. Male brown-headed cowbird chicks that are raised in captivity and never hear an adult male cowbird sing still give a flawless rendition of their species' courtship song when they mature. In contrast, male white-crowned sparrows raised in captivity fail miserably in their attempts to sing a song that is attractive to females. To learn their courtship song, male white-crowned sparrow chicks must hear other males of their species sing.

Fixed behaviors are elicited at the first encounter with a stimulus

A **fixed behavior** is an innate behavior in which the first encounter with the appropriate stimulus leads predictably to a well-defined response. Prior experience of the stimulus is not a prerequisite for an animal to display a fixed behavior. Such behaviors are strongly rooted in genetics and are commonly described as instinctive or "hardwired." In humans, fixed behaviors are most readily seen in very young babies and include some behaviors that are displayed only in the first few weeks or months after birth. Pediatricians commonly test newborns for some of these behaviors, known as neonatal reflexes, to determine whether a baby's nervous system is healthy. Healthy newborns display the **grasp reflex**,

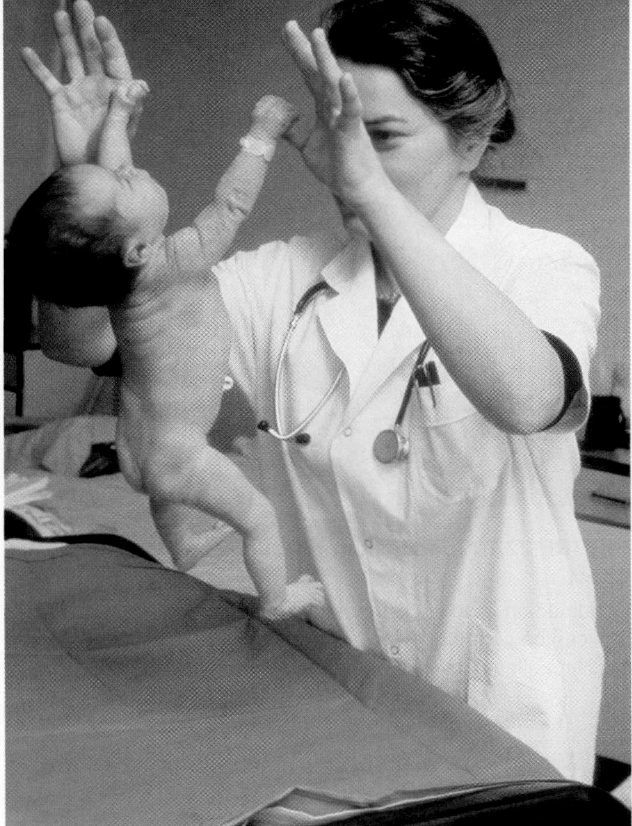

FIGURE 22.4 The Grasp Reflex in a Human Newborn
Healthy newborns will cling tightly to any object placed in the palm. This innate behavior disappears 5–6 months after birth.

for example, which enables them to hold on tightly to anything that will fit in their tiny fists (**FIGURE 22.4**).

Fixed behaviors are common in the young of other animals as well. Kittens reflexively cover their droppings when first introduced to a litter box—even kittens removed from their mothers at too early an age ever to have seen their mothers demonstrate this behavior. Pipevine swallowtail caterpillars can distinguish between food plants that are suitable for them and those that are not.

Fixed behaviors enable animals to behave appropriately when they have no chance to learn from experience and the risks associated with the wrong behavior are great. Consider female cowbirds, which lay their eggs in the nests of other bird species (**FIGURE 22.5**). Because of the nest-parasitizing behavior of their mothers, male cowbird chicks grow up hearing only the courtship songs of their foster parents, and they have no opportunity to learn their own species' song from their fathers. Male cowbirds have to be able to sing the cowbird-specific courtship song to have any chance of attracting female cowbirds as mates, so they must ignore any singing lessons they might receive while being reared in another species' nest and instead depend on their fixed, genetic programming to produce their song. The ability to sing the "cowbird song" is therefore a fixed behavior that is automatically displayed when male cowbirds become sexually mature and receive the appropriate stimulus, such as the lengthening days of spring.

The stimuli that trigger fixed behaviors can be quite simple. Herring gull chicks, for example, beg eagerly for food whenever their parents return to the nest. In a famous series of experiments, Nobel Prize–winning behavioral scientist Niko Tinbergen showed that the chicks "aim" their begging behavior at a conspicuous red dot on their parents' bills. Oddly colored or oddly shaped models of herring gull heads trigger begging behavior just as effectively as lifelike models do, as long as they feature the red dot (**FIGURE 22.6**). A simple stimulus that causes a fixed behavior, such as the red dot in herring gulls, is called a **releaser**.

Learned behaviors add flexibility to animal responses

In **learned behaviors**, the response to a stimulus depends on an animal's past experience. For instance, we can train a dog to sit in response to a whistle or to the command "Sit." A dog trained to the oral command sits in response to "Sit"; a dog trained to the whistle also sits, but in response to a different stimulus. The two dogs learn to respond appropriately to their different learning experiences.

The feeding behavior of rats illustrates some advantages of learned behavior. Rats in laboratory cages can learn to press a lever that releases food pellets.

FIGURE 22.5 Brown-Headed Cowbirds
Nest parasites, such as the cowbird, lay their eggs in other songbird nests, leaving their offspring to be raised by the surrogate parents. Without any opportunity to learn behaviors from parents, cowbird courtship songs and rituals must be genetically controlled.

A Fixed Behavior

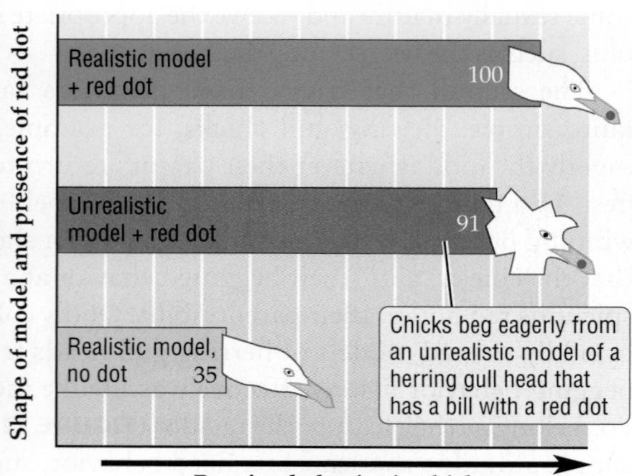

FIGURE 22.6 Simple Stimuli Trigger Fixed Behaviors

Herring gull chicks will beg almost as eagerly from an unrealistic model of an adult gull head with the critical red dot (the releaser for begging behavior) as they will from a realistic one. The red dot on the adult's bill (photo), not the appearance of the whole head, is what triggers begging behavior.

FIGURE 22.7 Conditioning Is Associating Cause and Effect

Conditioning is an important learned behavior in animals. Lab rats can be conditioned to perform specific actions, such as pressing a lever, by rewarding them with food.

When first placed in the cage, a rat will sometimes accidentally bump into the lever and discover the food pellet. The rat soon learns to press the lever deliberately to get more pellets. This association between cause and effect is called **conditioning**, and it is perhaps the most important learned behavioral response in animals (**FIGURE 22.7**).

Because of the diversity of their diet, rats cannot have a set of fixed rules about what to eat and what to avoid. Instead, they combine a specific behavior with the capacity to learn. When rats encounter a food for the first time, they sample only an amount small enough to avoid serious consequences, should the material prove inedible. If the food sample sickens them, they avoid that food in the future. Rats will sniff the mouths of littermates returning to the nest and stay away from any food that has an odor they smelled on the breath of a sick rat. If a food causes no harm, they add it to their learned list of good things to eat. By learning what to avoid from their experiences, rodents can cope with new and unexpected kinds of food–something they could not do with inflexible fixed behaviors.

Helpful to Know

Behavior researchers must be careful to avoid *anthropomorphism*—that is, ascribing humanlike attitudes or thought processes to nonhuman animals when there is no evidence to support such ascriptions. Modern behavior researchers use standardized protocols to minimize subjectivity and individual or group bias.

An interesting type of learning, called **imprinting**, occurs in animal species in which parental involvement in rearing the young is important. You may have seen a mother goose with her goslings trailing along behind her. Within a short time after hatching, the goslings develop a "vision" of who their mother is and what a goose is supposed to look like. This sense of "mother goose" is critical to the survival of young animals that must spend time with parents in order to learn how to survive on their own. Imprinting usually takes place only during a specific period of time in the early life of offspring. In ducklings, for example, imprinting occurs during (but not before or after) the period about 7–23 hours after hatching, with a peak sensitivity to imprinting at about 15 hours (**FIGURE 22.8**).

Habituation is another type of learned behavior in which animals learn to filter out uninformative stimuli. The sheer amount of sensory information that can be taken in by an organism would overwhelm the brain without the ability to filter out information that is unimportant. When a scarecrow is first introduced to a garden, birds are frightened off. Within a short period of time, however, the scarecrow loses its effectiveness as the birds learn that the presence of the scarecrow is nonthreatening and therefore unnecessary information to process.

Animals with higher cognitive functions are able to figure out solutions to challenging issues. **Problem solving** has been observed in rats, primates, even octopi. A particularly interesting example was seen in Akita, Japan. Crows and other birds drop hard objects such as shellfish from a significant height in order to break them open. This technique, however, failed to break open walnuts because of the walnuts' soft outer shell. In the early 1990s, crows were noticed dropping walnuts onto a road at a driving school and waiting for the cars to run over the nuts. Since that time, these Japanese crows have refined their technique. They drop them on roadways near crosswalks, and wait until the signal lights stop the traffic before retrieving their nuts. If passing cars do not immediately break open a nut, the crows will fly down and adjust the position of the nut until a car finally drives over it (**FIGURE 22.9**).

Concept Check

1. Compare fixed (innate) behaviors and learned behaviors.
2. Can learned behaviors be acted upon by natural selection?

22.3 Social Behavior in Animals

Many animals live in closely interacting groups. **Social behavior**—behavioral interactions among members of a group, usually of the same species—offers its members advantages not available to solitary animals. Organisms in groups may cooperate in food gathering, defending against predators, and child rearing. Although living in groups increases competition among individuals, may spread disease more easily, and makes organisms more visible to predators, the many benefits of group living can more than compensate for these costs.

Imprinting

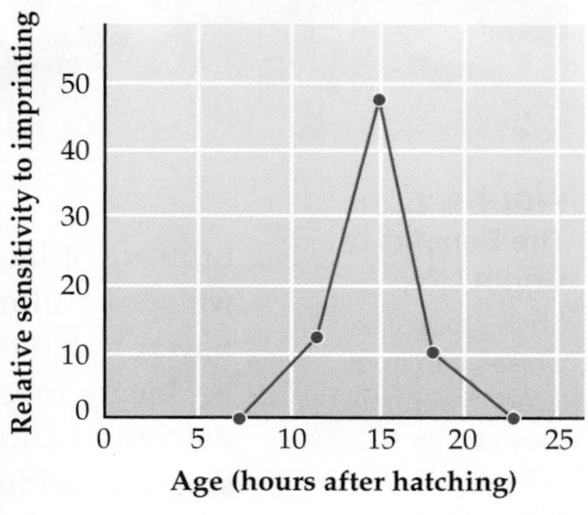

FIGURE 22.8 Some Animals Learn Who Their Parents Are by Imprinting
Imprinting usually takes place only during a brief "window" of sensitivity. A study of ducklings found that their peak sensitivity to imprinting occurred at about 15 hours after hatching. Young geese that imprinted on the famous behaviorist Konrad Lorenz (photo) considered him their "mother" and followed him everywhere.

Group living is an effective survival strategy

One important benefit of group living is that often groups can stir up and capture prey more effectively than individuals can, and unsuccessful hunters or hunters-to-be can watch and learn while others in their group demonstrate successful hunting behaviors. The

FIGURE 22.9 Problem Solving Is a Complex Learned Behavior
Crows in Akita, Japan, drop walnuts onto the roadways, letting the cars crack them open.

Concept Check Answers

1. Fixed behaviors are predictable, invariant, and immediately manifested in a young animal because prior exposure to the relevant stimulus is not required. Learned behaviors are more flexible, can be modified by experience, and are therefore adaptive in a changeable environment.

2. Many learned behaviors are influenced by an organism's genetic composition. Since natural selection acts only on heritable traits, learned behaviors can be acted on by natural selection.

FIGURE 22.10 The Benefits of Group Life

(*a*) In groups, wolves can hunt larger prey than they can as individuals. (*b*) Although a single musk ox may be vulnerable to predators such as wolves, a group that forms a circle makes a difficult target. (*c*) Family groups of scrub jays work together to feed young, defend nests, and hold on to valuable breeding territories.

hunting of large mammals by groups of cooperating wolves or humans illustrates such collaborative behavior well (**FIGURE 22.10*a***).

Living in groups also gives animals two effective defenses against predators that are not available to single individuals. Several prey individuals acting together may be able to repel attacks from predators (**FIGURE 22.10*B***). Large groups of prey may also be able to provide better warning of a predator's attack. Because more individuals can watch for predators, a large flock of wood pigeons detects the approach of a goshawk (*GAHS*-**hawk;** a predatory bird) much sooner than a single pigeon does. The success rate of goshawk attacks drops from nearly 80 percent when they are attacking single pigeons to less than 10 percent when they are attacking flocks of more than 50 birds (**FIGURE 22.11**).

In some cases, living in a group enables animals to use scarce resources in a more efficient manner. Florida scrub jays live in shrubby oak thickets that are dry for much of the year and that have only a limited supply of both food and nest sites. The jays form lifelong monogamous pairs, and DNA tests show that there is little "cheating" among mated scrub jays (**FIGURE 22.10*c***). After fledging, both the male and female offspring of a breeding pair may remain with the parents as nonbreeding members of a cooperative group. The nonbreeding members help care for the newly hatched young and join in defending the breeding territory against intruders. When they cannot find territories of their own, it is in the genetic interest of the nonbreeding birds to promote the survival of their closest relatives. Belonging to a group makes the nonbreeding birds less vulnerable to predators than a solitary bird would be. Furthermore, belonging to a cooperative group puts the helpers on a "waiting list" to inherit the territory when the breeding pair becomes too old to reproduce or if one or both birds are lost to predators. The fledglings produced by a former helper are themselves likely to display the helping behavior, boosting the reproductive success of the bird that inherits the parental territory.

Individuals in groups may act to benefit other members more than themselves

Animals that live in groups often display **altruism**; that is, they do things that help other members of their group survive or reproduce while decreasing their own chances of doing so. The scrub jays mentioned earlier provide a good example of altruism: nonbreeding members of a group help the breeding pair raise their young while possibly forgoing reproduction themselves. At first glance, altruism seems to contradict Darwin's idea that only traits that improve the *individual's* reproductive success can spread through a population. However, a deeper examination

Benefits of Group Living

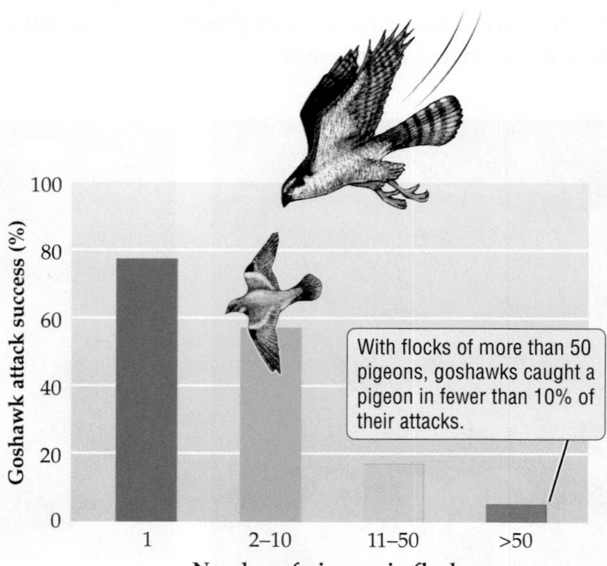

FIGURE 22.11 Safety in Numbers

The success of goshawk attacks on wood pigeons decreases greatly when there are many pigeons in a flock.

EXTREME GROUPS

European starlings can fly in flocks of more than a million individual birds. These large groups, called *murmurations*, fly in a coordinated manner to create elaborately fluid movements of the entire flock.

reveals that altruistic individuals further their genetic interests by improving the reproductive success of individuals that are genetically related to them.

In general, social groups consist of closely related individuals. A pride of lions, for example, centers around a group of closely related females. Similarly, among scrub jays the helpers are most often older offspring of the breeding pair, so the young birds they are helping to raise are their younger siblings. In these cases, the individuals benefiting from altruistic behavior carry many of the same genes that the altruistic individuals carry, so altruistic behavior often helps to spread many of the same genes that the altruist carries.

Social insects such as ants, bees, and termites include some of the most successful species on Earth. Although these insects evolved social behavior independently of one another, they share some remarkable traits. Each of these three groups lives in large colonies containing individuals belonging to distinct classes that serve distinct functions within the colony (**FIGURE 22.12**). The workers collaborate to build complex nests, forage widely for food, and defend the colony against predators in ways that individuals acting by themselves never could. The queen spends her life producing massive numbers of eggs. The workers in a hive are the female offspring of the queen and are therefore closely related to her and to one another. The workers are sterile; they do not reproduce, but they contribute to reproduction by carefully tending to the queen and her larvae. Their altruistic behaviors, including defending to the death the queen and the colony or hive, do not reduce the reproductive capacity of the colony but rather increase its chances of survival.

FIGURE 22.12 Cooperation in an Ant Colony
An ant colony is made up of many different types of individuals, most of whom are sterile. Typically, a single queen (*a*) is responsible for reproduction, while the other ants have specialized roles including soldiers (*b*), workers (*c*), and nurses (*d*).

Concept Check

1. What are the potential benefits of group living for social animals?
2. Why might an organism choose to forgo its own reproduction? How can such altruistic behaviors be favored by natural selection?

22.4 Facilitating Behavioral Interactions through Communication

Social interactions rely on effective communication between members of the social group. **Communication** is a type of behavior that enables one individual to exchange information with another, thereby making it possible for animals to coordinate their activities (**FIGURE 22.13**). Communicating facilitates group behavior, such as fending off large predators, capturing large prey, or delivering emergency assistance.

One individual produces signals that stimulate responses in others

The communication behavior of animals varies widely in its complexity and includes just about any type of signal that other animals can sense: sounds, visual signals, odors, electrical pulses, touch, and tastes.

At the simple end of the spectrum, the release of a chemical signal, called a **pheromone**, by one individual informs others of the same species about its identity, its location, its physical condition, or a situation in the environment. For example, female silkworm moths release bombykol, a sex pheromone, to attract males. The pheromone not only reveals to a male that the individual releasing the chemical is a female of his species and that she is interested in finding a mate; it also tells him where she is. The male can

Concept Check Answers

1. Hunting and defending as a group are more likely to increase survival than are hunting and defending by lone individuals. Scarce resources, such as nesting sites, can be shared through cooperative breeding. Group members can assist close relatives in raising offspring through altruistic behaviors.

2. The recipients of altruistic acts are typically close relatives of the altruists. Because close relatives share many of the same genes as the altruist, an altruist's actions ensure that copies of its genes are passed on to subsequent generations.

Concept Check Answers

1. Animals communicate in a variety of ways. Pheromones are chemical signals between organisms. Communication can also be accomplished through visual or audible signals.

2. Communication enables social organisms to relay information such as position and physical condition, as well as information such as interest in mating, coordination for group efforts, or aggression.

FIGURE 22.13 Communication Is Vital to Animals

(*a*) Howling is important for pack cohesion among wolves and may warn outsiders to stay clear. (*b*) The "jump-yip" display of black-tailed prairie dogs maintains group cohesion in these social ground squirrels. (*c*) Verbal and nonverbal communication is vital to these traders in Brazil's Bovespa exchange.

Helpful to Know

One intriguing study of the role of pheromones in humans asked men to wear T-shirts for three consecutive days without deodorant, cologne, or heavily scented soaps. Females sniffed these shirts and were asked to rate the attractiveness of the scent. Overall, women preferred the scent of males with different immunity genes from their own—which would allow their offspring to possess a wide variety of these genes and presumably have more robust immune systems

locate the female by simply moving toward an area with a higher amount of the pheromone (**FIGURE 22.14**). Evidence suggests that we humans also produce airborne signals that may influence sexual behavior.

Pheromones communicate other conditions besides sexual readiness. For example, when a colony or hive is disturbed, ants, honeybees, and other social insects release alarm pheromones that inform other individuals in the community about the situation. Those individuals interpret the disturbance as an attack and respond accordingly.

Pheromones are relatively simple signals. An example of a more complex signal is the intricate dance used by honeybees to communicate the distance and location of food to other members of their hive (**FIGURE 22.15**). And at the extreme end of the complexity spectrum of communication lies human **language**. Most human languages consist of thousands of words representing everything from objects to actions to abstract ideas.

Bombykol and Moths

FIGURE 22.14 Pheromones Can Signal Potential Mates

The hormone bombykol is a sex pheromone secreted by female silkworm moths that indicates readiness to mate. By flying upwind toward increasing concentrations of bombykol, a male is able to find the receptive female.

Identification is a key function of animal communication

Animals most often communicate to identify themselves to other animals, to avoid conflict, or to coordinate their activities. Signaling an individual's identity is probably the most common function of communication.

An animal uses a variety of signs and signals to inform other members of its species that it is a member of the same species, as well as to indicate its sex, its physical condition, and its location. When a dog sprays a fire hydrant with its urine, the scent communicates the dog's sex, breeding condition, health, and status to other dogs. The mating songs and behaviors of male birds, frogs, and crickets tell females their species, location, and potential quality as mates (**FIGURE 22.16*a***).

A second important function of communication in animals is to avoid potentially harmful conflicts. Physical conflict over food or mates can lead to serious injury for both the winner and the loser. To reduce the risk of injury in such encounters, many species communicate their fighting ability through ritual displays. When male red-winged blackbirds flash red wing patches at one another, for example, their brightness and size signal a male's quality as a fighter to other males (**FIGURE 22.16*b***). This ritual allows the less able fighter to back down without engaging in a potentially dangerous fight. The males of many mammalian species mark their territory by depositing their scent on objects throughout their territory. The scent marking serves as a deterrent to potential competitors.

Animals can even communicate across species lines, as some birds do when they mob a predator. The loud mobbing calls of a chickadee will attract other chickadees, but they may also draw other tiny birds, such as warblers, all of whom will join in harassing and dive-bombing a hawk that has come too close to their nesting sites. Mobbing behavior deters or distracts predators, but it may also be a teaching tool: mobbing by adults may help young nestlings learn to identify the species they need to be wary of.

Concept Check

1. List some common modes of animal communication.
2. Why is communication such a vital aspect of social living?

22.5 Mating Behaviors

By its very nature, sexual reproduction requires two distinct partners who contribute differently to the creation of offspring. Males contribute sperm, which is energetically simple and cheap to make; females produce eggs, which are more nutritionally demanding. In animals that provide a significant amount of parental care, this burden typically falls to the female. Mammals, for instance, require milk from their mothers, further increasing the biological costs of reproduction for the female. These biological differences contribute to distinct mating behaviors.

In sexual organisms, reproductive behaviors may differ between males and females

In the production of offspring, females are typically more invested than males. Because of this difference in parental contribution, males and females often exhibit different reproductive strategies. For males, the best strategy to maximize reproductive output is to emphasize quantity in their search for mates. In contrast, because of the significant investment a female makes in her offspring, she is generally more selective about her mate choice.

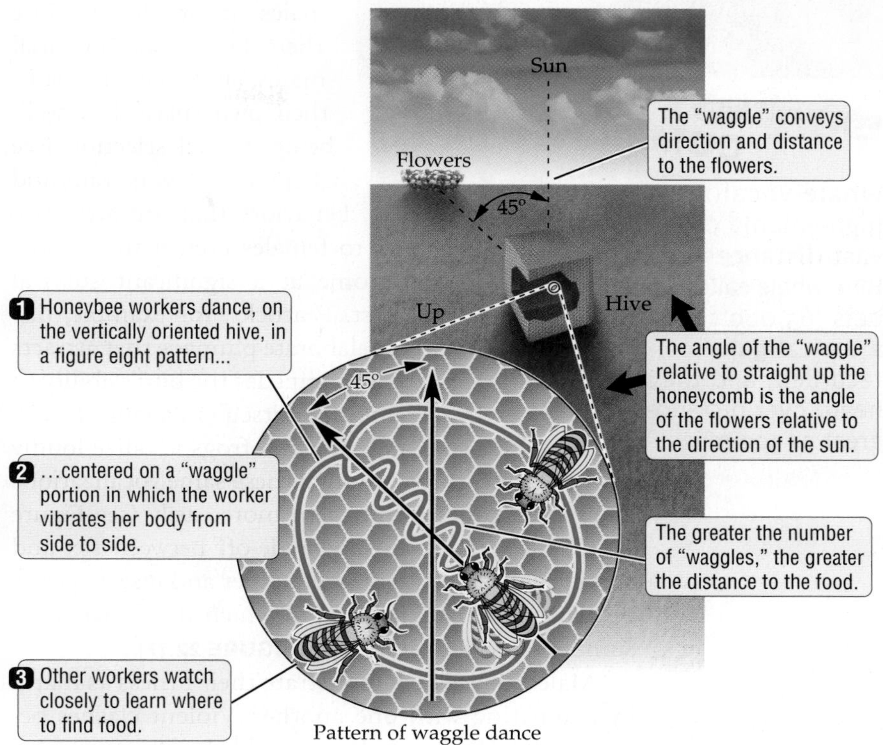

The Honeybee Waggle Dance
Sun
Flowers
45°
The "waggle" conveys direction and distance to the flowers.
Up
Hive
1 Honeybee workers dance, on the vertically oriented hive, in a figure eight pattern...
45°
The angle of the "waggle" relative to straight up the honeycomb is the angle of the flowers relative to the direction of the sun.
2 ...centered on a "waggle" portion in which the worker vibrates her body from side to side.
The greater the number of "waggles," the greater the distance to the food.
3 Other workers watch closely to learn where to find food.
Pattern of waggle dance

FIGURE 22.15 Honeybees Dance to Communicate

Bees use a dance language to communicate complex information about the location of distant food sources to other worker bees in their hive.

FIGURE 22.16 Showing Off to Attract Mates Is a Form of Communication

(*a*) A singing male frog advertises to listening females its species, its location, and its quality as a mate. (*b*) Male blackbirds display red patches on their wings, indicating territorial aggression.

EXTREME COMMUNICATION

Whale vocalizations are low-pitch, high-volume calls that can carry for vast distances in the ocean water. Blue whale calls as high as 188 decibels (65,000 times louder than a typical gunshot) have been recorded, and they can be heard over distances as great as 500 miles.

When females are choosy about their mates, males must demonstrate their fitness as potential mates, often at great cost to their own survival or well-being. Sexual selection (see Chapter 14) favors traits and behaviors that are attractive to females even if these traits come at a significant survival cost. Peacocks, for example, display elaborate plumage that attracts mates but impedes the bird's ability to fly or camouflage itself (see Figure 14.19). Likewise, male túngara frogs vocalize loudly to attract mates, though these same vocalizations allow predators to find them more easily (see Figure 14.18). In some cases, the trade-off between sex and survival is even greater. Some spider and insect species display sexual cannibalism, in which the female consumes the male after mating (**FIGURE 22.17**).

Males may also demonstrate their fitness as mates by competing with one another. Violent clashes between male bighorn sheep establish dominance relationships between the males, allowing females to easily select the strongest males to mate with (**FIGURE 22.18**). Likewise, male lions will battle one another for access to the females of a pride. Although males can be injured or killed in these contests, sexual selection favors these displays as females elect to mate with the champions, allowing their genes to be passed on.

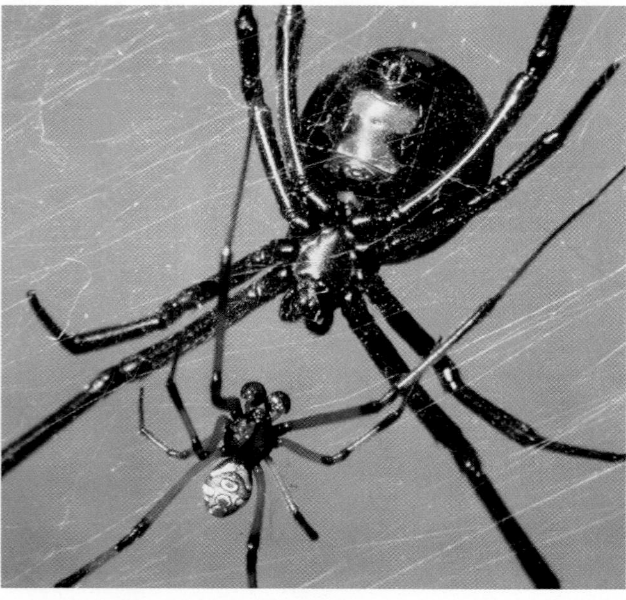

FIGURE 22.17 Love Hurts

Male black widows run a significant risk of death when mating, as the female (larger) will attempt to eat the male (smaller) after copulation.

FIGURE 22.18 Males Fight with One Another to Demonstrate Dominance

Male bighorn sheep communicate their genetic quality to potential mates by engaging in head-butting contests. They hit with an estimated 800 kilograms of force, and the sounds of the impact can be heard up to 2 kilometers away.

Males can also curry favor with females by providing them with **nuptial gifts**. These are rewards (typically food items) presented to females for sexual attention. Male snowy owls, as part of their courtship behavior, will catch lemmings and present them to females (**FIGURE 22.19**). A female owl will consent to mating only after she consumes her lemming. Some spider and insect species also utilize nuptial gifts to increase the likelihood of sexual union.

In some organisms, sexual roles are reversed. Male sea horses, for example, exhibit more parental care than females. In these cases, males are the more selective species, and it is the females that compete for mates.

Mating systems are generally the result of differences in parental contributions

Organisms exhibit a wide range of mating systems. Some organisms are **monogamous**, in which a male and female are bonded to one another solely. Others are **promiscuous**, where males and females mate with multiple partners. Finally, organisms can be **polygamous**, in which (usually) a single male mates with multiple females. Which mating system an organism exhibits depends strongly on the amount of parental investment each partner contributes (**FIGURE 22.20**).

FIGURE 22.19 A Male Snowy Owl Presents a Nuptial Gift
Nuptial gifts are given to females to entice them to mate.

Monogamy is observed when both parents contribute more or less equally to child rearing. Many songbird species are monogamous; chicks are much more likely to survive when two parents are available to provide food for them. In monogamous mating systems, males and females are equally selective in choosing their partners.

In promiscuous mating systems, typically one or both parents exhibit little parental care. Males and females come together for mating and then separate to find other mates. Some species with parental care, such as rodents or cats, do exhibit promiscuity, but it is more commonly observed in species with less parental care. Mate choice in a promiscuous system is usually less rigorous than in other mating systems.

In polygamous organisms, typically one male has sexual access to many females, called a **harem**. The parental contribution of the male is considerably less than that of the female, and females opt to mate with the single most dominant male and raise the offspring by themselves. Females in most polygamous systems are the choosier sex; by mating with the dominant male, they ensure the genetic quality of their offspring.

FIGURE 22.20 Mating Systems
(*a*) Monogamous organisms, such as these lovebirds, form pair bonds—often for life. Both parents contribute to child rearing. (*b*) Promiscuous species, like many rodents, mate multiple times with multiple partners. (*c*) Polygamous organisms typically have one male mating with a harem of females.

Concept Check

1. What is the principal factor that determines the choosiness of mates?
2. Under what conditions would a species be likely to exhibit monogamy?

A male anglerfish is small and underdeveloped compared to its better-known female counterpart. To initiate mating, a male begins nibbling a female. Its lips begin to dissolve and fuse with the female's body. Over time, its body and organs will be absorbed into the female's body, leaving nothing but its testes as outgrowths on the female.

Concept Check Answers

1. The amount of effort that each respective parent puts into childbearing and child rearing.

2. When both parents contribute significantly to child rearing.

APPLYING WHAT WE LEARNED

The Genetics of Domestication

In 1959, a Russian geneticist committed to the idea that genes influence behavior began an experiment to breed a strain of friendly foxes from 130 farm-raised silver foxes (*Vulpes vulpes*). In simple terms, the selection for friendly foxes was ruthless. Each litter of fox puppies was tested several times for friendliness. The 10 percent of fox puppies that didn't growl, bare their teeth, or show too much fear were selected for breeding; for the rest, it was off to the fur coat factory. So that the animals didn't become tame—that is, unafraid of humans because of associating with humans—all the foxes had minimal contact with people.

Over the years, Dmitry Belyaev and his colleagues selected the friendliest foxes from 45,000 foxes, breeding those that showed the least aggression and fear. In just 10 generations, 18 percent of foxes were extremely friendly, whining for attention and licking experimenters just like dogs do. By 1985, all the foxes were as tame as dogs. Although less than 100 years removed from wild foxes, today Belyaev's domesticated Siberian foxes are as friendly as dogs, and their behavior is completely unlike that of wild foxes, which—even when hand-raised by humans—are skittish, liable to bite, and not eager to please.

In less than 40 years, Soviet researchers succeeded in fully domesticating a wild animal—a process that has generally been assumed to have taken thousands of years in dogs, horses, and other domestic animals.

To check whether the change in behavior really was genetic, Belyaev's successor to the breeding program (Lyudmila N. Trut) and her colleagues transplanted embryos from genetically domesticated foxes into genetically wild fox mothers. They also performed the reverse experiment—transplanting wild fox embryos into genetically domesticated mothers. In both cases, the embryos developed into foxes like their original mothers, not their surrogate mothers, showing that genes caused the difference in behavior. The researchers calculated that about 35 percent of a typical domestic fox's friendliness is genetic, meaning that the other 65 percent is environmental.

Although the researchers selected only for friendliness, the domesticated foxes differ from wild foxes in other ways as well. Many have white stars on their heads, white feet and chests, spots, floppy ears, broad heads that give them more puppyish faces, and curly tails. What's remarkable about this suite of traits is how many other domesticated animals share similar color changes. Dogs, cats, horses, goats, and pigs all share this suite of traits, including spotted or piebald coats, white stars on their heads, floppy ears, and broader heads than their wild cousins. It is as if all domesticated animals experience a similar selection program.

Belyaev expected this result. Charles Darwin had noticed that all domestic animals show these traits to some degree and wild animals almost never do. Belyaev and Trut believed that selecting for genetically based tameness was the same as selecting for a major difference in animal development.

When they looked more closely, the researchers found that domesticated foxes show a lower activity of the adrenal glands, which produce the "stress" hormone adrenaline. As puppies, domesticated foxes develop a fear response later than regular foxes do. Their brains also produce more serotonin, a neurotransmitter connected to happiness and appetite. Many of the traits that domesticated animals display are likely related to delays in development—a phenomenon called *pedomorphosis* (keeping juvenile traits into adulthood). Some researchers hypothesize that humans are themselves "domesticated" in similar ways, with a suite of traits that enable large groups of strangers to assemble without aggression.

BIOLOGY IN THE NEWS

Study Discovers DNA That Tells Mice How to Construct Their Homes

BY JAMES GORMAN • *New York Times*, January 16, 2013

The architectural feats of animals—from beaver dams to birds' nests—not only make for great nature television, but, since the plans for such constructions seem largely inherited, they also offer an opportunity for scientists to tackle the profoundly difficult question of how genes control complicated behavior in animals and humans.

A long-term study of the construction of burrows by deer mice has the beginnings of an answer ... the report, in the current issue of *Nature*, identifies four regions of DNA that play a major role in telling a mouse how long a burrow to dig and whether to add an escape tunnel.

The research could eventually lead to a better understanding of what kind of internal reward system motivates mice to dig, or tells them to stop. And although humans do not dig burrows, that, said the leader of the three-person research team, Hopi E. Hoekstra of Harvard, could "tell us something about behavioral variation in humans."

Dr. Hoekstra started with a species called the oldfield mouse (*Peromyscus polionotus*), the smallest of the deer mice. For 80 years or more, field scientists have documented its behavior, including excavating characteristically long burrows with an escape tunnel, which the mice will dig even after generations of breeding in cages in a laboratory.

Dr. Hoekstra treated tunnel length and architecture as a physical, measurable trait, much like tail length or weight, by filling burrows with foam that would produce a mold easily measured and cataloged—behavior made solid.

She and her students did this in the field and repeated it in the laboratory by putting the mice in large, sandbox-like enclosures, letting them burrow and then making molds of the burrows. They did the same with another deer mouse species, *Peromyscus maniculatus*, that digs short burrows without escape tunnels ...

Then the scientists matched variations in tunnel architecture to variations in DNA. What they found were three areas of DNA that contributed to determining tunnel length, and one area affecting whether or not the crossbred mice dug an escape tunnel. That was a separate behavior inherited on its own, so that the mice could produce tunnels of any length, with or without escape tunnels.

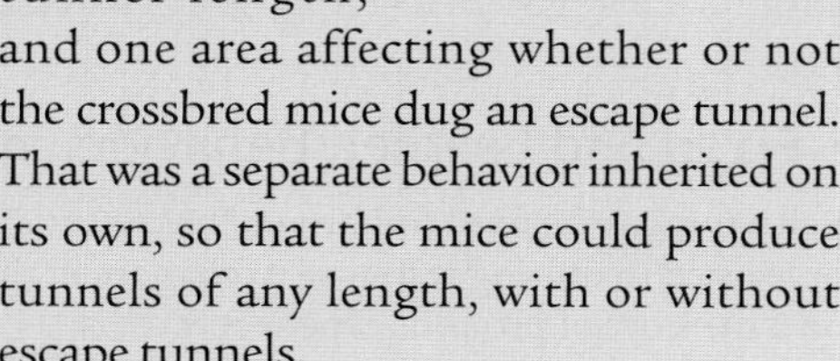

All complicated behaviors are affected by many things, Dr. Hoekstra said, so these regions of DNA do not determine tunnel architecture and length by themselves. But tunnel length is about 30 percent inherited, she said, and the three locations account for about half of that variation. The rest is determined by many tiny genetic effects. As for the one location that affected whether or not mice dug an escape tunnel, if a short-burrow mouse had the long-burrow DNA region, it was 40 percent more likely to dig a complete escape tunnel.

Complicated behaviors, such as web building in spiders and burrowing in ants and mice, have long been known to have at least some genetic basis, since these organisms are instinctively able to perform these actions even without the benefit of having observed parents or other species members performing the behavior. However, uncovering the genetic basis for these behaviors has eluded scientists in most cases. Behaviors are typically encoded not by single genes, but by a suite of genes that interact and contribute to the biochemical structures that influence decision making and therefore behavior.

In the study described in this article, scientists were able to link certain regions of the *genome* (the entire genetic information of an individual) to a specific, quantifiable behavior, illustrating the extent of its genetic basis. The next step would be to identify those genes, understand their role in brain chemistry, and link these biological traits to behavioral responses. The task is daunting, but not without precedent. Scientists have identified genes that are responsible for monogamy versus promiscuity in voles, and even genes that dictate maternal instinct in mice. The difficulty in understanding the relationship between genes and behavior comes from the fact that genes do not directly encode behaviors, but rather encode proteins that work together within a cognitive system to ultimately determine a behavioral response.

The correlation of genes and behaviors has important implications for understanding human behaviors as well. Scientists have already identified as many as 50 regions of DNA that influence alcoholism, as well as behaviors such as obsessive-compulsive disorder and schizophrenia. Understanding the biochemical basis of these behaviors may lead to better treatment for these conditions.

Evaluating the News

1. The researchers in the reported study suggest that tunnel length is approximately 30 percent inherited. What other factors might explain why this behavior is not perfectly correlated with genetics? Is this behavior learned or innate?

2. Explain how this technique could be useful in determining the genetic basis of human behaviors. What challenges would confront such an attempt?

CHAPTER REVIEW

Summary

22.1 Sensing and Responding: The Nature of Behavioral Responses

- A behavior is a coordinated response made by one animal in reaction to another organism or to the physical environment. Behavioral scientists are interested in how behaviors are adaptive, how they are shaped by genes and cognitive functions, and how they develop through evolutionary history.
- Complex behaviors are often a combination of genetic and environmental factors. Genes that affect the physiology or brain chemistry of an animal may govern its response to a particular stimulus.

22.2 Fixed and Learned Behaviors in Animals

- Behaviors can be fixed (inborn or "hardwired," requiring no learning) or learned (requiring response to changes in the environment based on a memory of previous experiences).
- Fixed behaviors have a simple pattern: a single (usually simple) stimulus (called a releaser) leads to a single response. Fixed behaviors enable animals to behave appropriately when they have no chance to learn by experience or when the risks associated with the wrong behavior are great. Fixed behaviors may be stimulated only under certain conditions.
- Learned behaviors are responses based on the past experience of the individual animal. Different types of learned behaviors include conditioning, imprinting, habituation, and problem solving

22.3 Social Behavior in Animals

- Social behavior—the interactions among members of a group of the same species—provides several different advantages. For prey animals, it provides added protection against predation. Social behavior also enables animals to gain access to resources, including food and breeding territories, more effectively—in some cases enabling animals to obtain foods not available to individuals working alone.
- Individuals living in groups often act in ways that benefit other group members more than themselves. Such altruistic behavior can evolve within groups of closely related animals. Some individuals sacrifice their lives or reproductive success to increase the overall survival and success of the group.

22.4 Facilitating Behavioral Interactions through Communication

- Communication enables individuals to coordinate their activities with those of other individuals. Information is exchanged through the production of signals by one animal that stimulate a response in another.
- Animals communicate to identify themselves (as individuals and as members of a species), to avoid conflict, to coordinate their activities in order to perform shared tasks, to indicate their sex and sexual readiness, and to convey information about their physical condition and location.
- Communication can range from simple to highly complex. Pheromones are a simple means of communication that can be used to communicate identity, location, physical condition, or a situation in the environment. Language is the most complex means of animal communication.

22.5 Mating Behaviors

- Because of the differences in parental investment in reproduction, males and females often exhibit different reproductive strategies. Males are generally less particular about the quality of their mates than females are; they focus instead on quantity of mates. Females seek the fittest males to mate with.
- Choosy females result in male traits and behaviors that are attractive to females, displays of physical fitness such as combat between other males, and the presentation of nuptial gifts to entice females to mate.
- Mating systems of animals include monogamy, in which a male and female are bonded to one another; promiscuity, in which organisms have multiple sexual partners; and polygamy, in which one sex (usually the male) has multiple partners.

Key Terms

altruism (p. 514)
behavior (p. 508)
communication (p. 515)
conditioning (p. 512)
fixed behavior (p. 510)
grasp reflex (p. 510)
habituation (p. 512)
harem (p. 519)
imprinting (p. 512)
language (p. 516)
learned behavior (p. 511)
monogamous (p. 518)
nuptial gift (p. 518)
pheromone (p. 515)
polygamous (p. 518)
problem solving (p. 512)
promiscuous (p. 518)
releaser (p. 511)
social behavior (p. 513)
stimulus (p. 508)

Self-Quiz

1. Behavior
   a. is seen in vertebrates, but not in invertebrates.
   b. always involves communication.
   c. enables animals to evolve.
   d. enables animals to respond quickly to changes in their environment.

2. Genes
   a. have no influence on behavior.
   b. influence only social behavior.
   c. affect only fixed behaviors.
   d. can affect fixed and learned behaviors.

3. Fixed behaviors
   a. derive from experience.
   b. are always imprinted.
   c. are genetically inherited.
   d. require language.

4. Learning
   a. occurs only in humans.
   b. overrides all genetic control of behavior.
   c. depends on an animal's past experience.
   d. requires pheromones.

5. Which of the following animals would rely most on learned behaviors?
   a. termite
   b. frog
   c. rat
   d. spider

6. The dance of a honeybee worker
   a. communicates the location of a food source.
   b. communicates the location of a mate.
   c. communicates the sex of the dancer.
   d. has no influence on fixed behaviors.

7. Animals that live in groups
   a. have no fixed behaviors.
   b. usually rely on language for communication.
   c. can reap benefits that compensate for increased competition among group members.
   d. have no genetically controlled behaviors.

8. Which of the following is not an example of altruistic behavior?
   a. A butterfly lays an egg.
   b. A scrub jay chases a snake away from its parents' nest.
   c. A sterile worker ant defends the nursery of its anthill from invading ants.
   d. A lioness regurgitates food for her sister's cubs.

9. An organism in which females compete for male partners would most likely have
   a. promiscuous males.
   b. males that perform most of the child care.
   c. a monogamous mating system.
   d. extended care of offspring performed by females.

10. In most species, females are the choosier sex because
    a. females are more invested in reproduction.
    b. males are more abundant than females.
    c. males are more invested in reproduction.
    d. most species are monogamous.

Analysis and Application

1. Give one example of a fixed behavior, and one example of a learned behavior, in humans.
2. Do you think humans are born with an innate fear of spiders? Explain and defend your answer.
3. Some people are born with athletic or musical talent. But even the most talented must learn and practice in order to excel. Pick a sport or a musical instrument, and describe in detail the dimensions of learning that must take place in order to fully develop a talent. Keep in mind that learning includes more than just practice.
4. Language is a major component of the human ability to communicate. But we also communicate without words. What nonverbal forms of human communication have you responded to or used?
5. What kinds of fixed behaviors might be advantageous to animals that live in a group, and why?
6. Consider the customary dating and mating rituals of humans—such as males paying for dinner on the first date or buying an engagement ring in order to propose. Do you think these rituals are societal constructs, or do they have a biological basis?

23 Ecological Communities

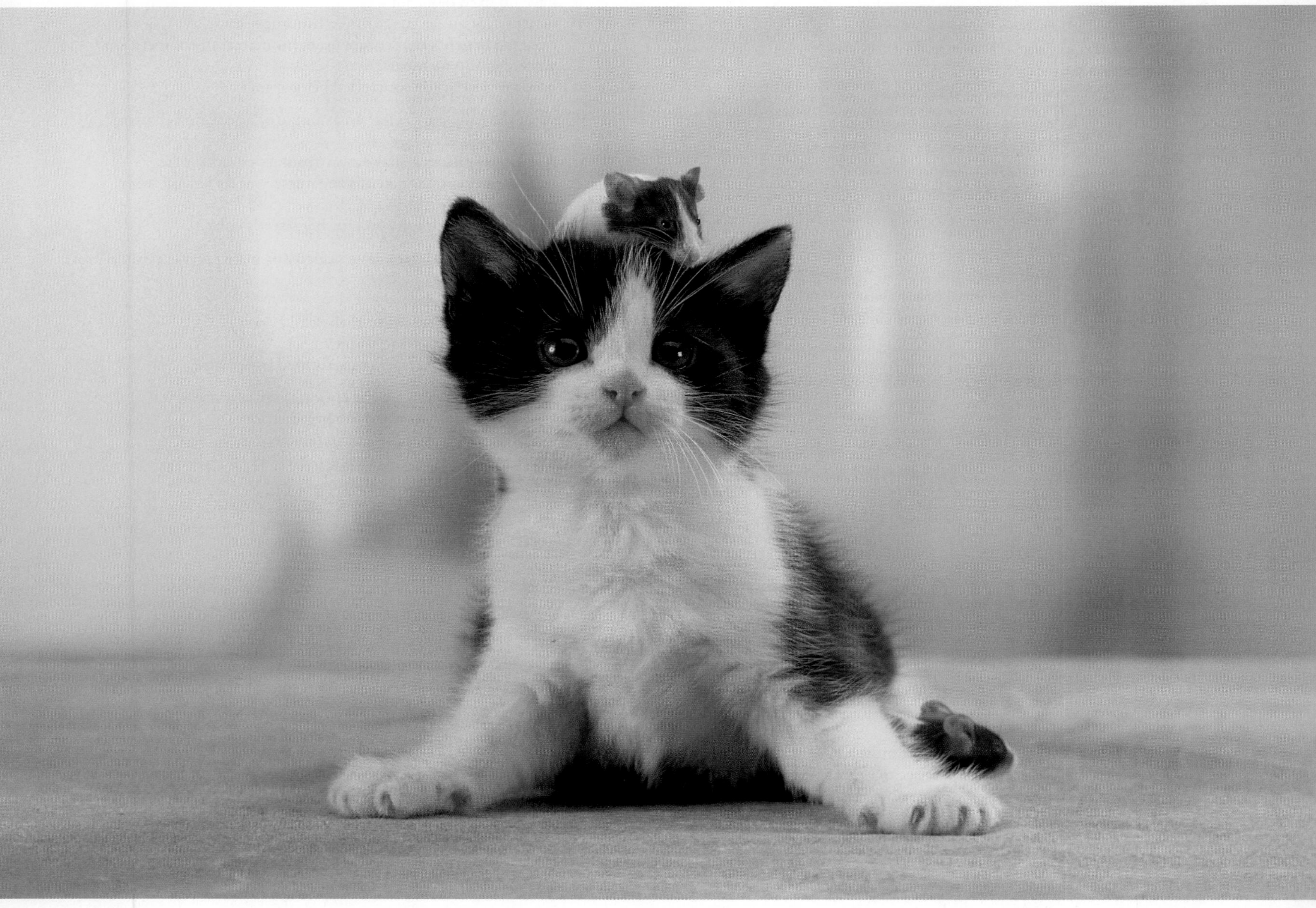

A BABY RAT SITS ATOP A KITTEN'S HEAD. When the parasite *Toxoplasma gondii* infects a rat, the rat loses its natural fear of cats and will readily approach a cat. The cat eats the rat and itself becomes infected by the parasite. Very young animals like the ones in this photo tend to be less fearful than adults; these two are probably not infected.

Fatal Feline Attraction

Imagine an alien from outer space that enters people's brains and changes their behavior, sometimes even driving them crazy. Millions of people secretly harbor the infection. It kills babies in the womb and infects so easily that a person can get it just by making a hamburger. Does this sound like a bad plot for a science fiction story? Amazingly enough, it's all true, except that the "alien" is a parasite that lives right here on Earth.

Toxoplasma gondii is a single-celled parasite, a relative of the parasite that causes malaria. "Toxo," as it's informally called, can infect nearly any warm-blooded animal, including birds and mammals. Most often contracted from raw meat, Toxo causes more deaths and illness than any other food-borne pathogen except *Salmonella*. And 60 million Americans are infected with the parasite.

Toxo's primary hosts, cats, get the parasite by eating rodents or, like us, raw meat. When rats and mice accidentally contact cat feces, they become infected. But then *Toxoplasma gondii* does something unexpected; it enters the rats' brains and makes them unafraid of cats. In addition, male rats (but not female rats) actually become sexually excited by the smell of cat urine and go out of their way to find it. It's a bad move for the rats. But for the parasite—which ends up in a new cat, where it can reproduce—it all ends happily.

How can *Toxoplasma gondii* hijack the chemistry of the brain and force humans and other animals to do things they wouldn't normally do? How does the parasite benefit from altering the behavior of its hosts? Is this the way different species normally interact?

So, does the parasite affect our behavior too? You won't like the answer. Most people infected with the parasite either don't notice any symptoms or have mild flulike symptoms for a few weeks, from which they recover completely. But just as in rats, the parasite can take up residence in the brain and alter our behavior.

As we'll see in the course of this chapter, the species that make up a community interact in all kinds of ways. But the species interactions involving *T. gondii* are a bit bizarre.

MAIN MESSAGE Communities change naturally as a result of both interactions among species and interactions between species and their physical environment.

KEY CONCEPTS

- A community is an association of populations of different species that live in the same area.
- The diversity of a community has two components: the number of different species that live in the community, and the relative abundance of each species.
- Community structure is influenced by a wide range of factors, such as climate, disturbance, and species interactions.
- Species interactions are classified by whether they help, harm, or have no effect on each of the species involved.
- The four primary types of species interactions are mutualism, commensalism, exploitation (predation and parasitism), and competition.
- Species interactions drive natural selection and evolution. Evolutionary change caused by these interactions is called coevolution.
- Species interactions help determine where organisms live and how abundant they are, and therefore strongly influence community structure.
- All communities change over time. As species colonize new or disturbed habitat, they tend to replace one another in a directional and fairly predictable process called succession.
- Humans transform natural habitats in many ways. Communities can reassemble after some forms of human-caused disturbance, but the process may take from a few years to many centuries.

FATAL FELINE ATTRACTION 525
23.1 Species Interactions 527
23.2 How Species Interactions Shape Communities 534
23.3 How Communities Change over Time 536
23.4 Human Impacts on Community Structure 539
APPLYING WHAT WE LEARNED 542
How a Parasite Can Hijack Your Brain

AN ASSOCIATION OF DIFFERENT SPECIES that live in the same area is known as an ecological **community**. Communities vary greatly in size and complexity, from the community of microorganisms that inhabits a small, temporary pool of water, to the community of plants that lives on the floor of a forest, to a forest community that stretches for hundreds of kilometers (**FIGURE 23.1**). Whatever its size or type, an ecological community can be characterized by its species composition, or *diversity*. The **diversity** of a community has two components: **species richness**, the total number of different species that live in the community; and **relative species abundance**, the number of individuals of a species in a given community compared to individuals of other species in that community. **FIGURE 23.2** compares the diversity of two communities that have the same number of species (that is, the same species richness). Because community A is dominated by a single, highly abundant species, it is considered less diverse than community B, in which all species are equally common.

The study of an ecological community is a study of the relationships among organisms in a defined area. Most communities contain many species, and the interactions among those species can be complex. Ecologists seek to understand how interactions among organisms influence natural communities, because these interactions have huge effects. For example, as we saw in Chapter 21, the moth *Cactoblastis cactorum*, by feeding on the cactus *Opuntia*, caused *Opuntia* populations to crash throughout a large region of Australia. Overall, interactions among organisms have an influence at every level of the biological hierarchy at which ecology is studied.

Ecologists also study how human actions affect communities. People are having profound effects on

Ecological Communities

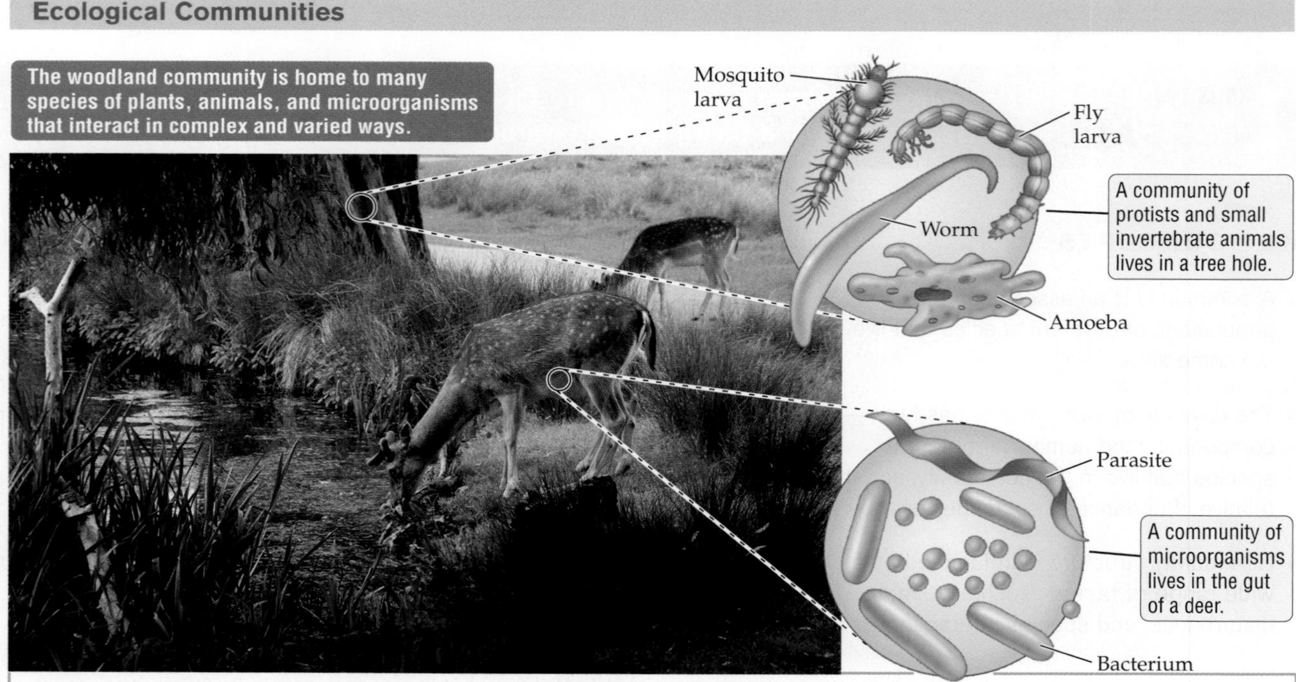

The woodland community is home to many species of plants, animals, and microorganisms that interact in complex and varied ways.
Mosquito larva
Fly larva
Worm
A community of protists and small invertebrate animals lives in a tree hole.
Amoeba
Parasite
A community of microorganisms lives in the gut of a deer.
Bacterium

FIGURE 23.1 Community Ecology Studies Relationships between Species
A community is an association of species that interact with each other in a given area. How many different communities are represented in this scene? Choose any two organisms and describe the relationship between them. Is the interaction positive or negative for each species? How many species do you interact with in your day-to-day life?

Species Diversity
Community A
In this community, there are four species, but this one is more abundant than any others.
Community B
In this community, there are four species, and all are equally abundant.

FIGURE 23.2 Which Community Has Greater Diversity?
Community A is dominated by a single species; in community B, all four species are equally represented. Therefore, ecologists consider community A less diverse than community B.

many different kinds of ecological communities. When we cut down tropical forests we destroy entire communities of organisms, and when we give antibiotics to a cow we alter the community of microorganisms that live in its digestive tract. To prevent such actions from having effects that we do not anticipate or want, we must understand how ecological communities work. In this chapter we describe how species interact with each other, and we discuss the factors that influence which species are found in a community. We pay particular attention to how communities change over time and how they respond to disturbance, including disturbance caused by people.

23.1 Species Interactions

The millions of species on Earth interact in many different ways. Communities generally are too complex to describe every relationship that occurs between the various species within it. Community ecologists therefore focus on interactions between pairs of species and classify these relationships into four broad categories, based on whether the interaction is beneficial (+), harmful (−), or without benefit or harm (0) to each of the interacting species:

1. *Mutualism* (+/+). Both species benefit.
2. *Commensalism* (+/0). One species benefits at no cost to the other.
3. *Exploitation* (+/−). One species benefits and the other is harmed.
4. *Competition* (−/−). Both species may be harmed.

Each type of interaction plays a key role in determining where organisms live and how abundant they are. Two species that interact may trigger evolutionary change in each other as a consequence of their interactions—a concept known as **coevolution**. Species interactions can also alter the composition of communities over both short and long spans of time. We will see how community stability is a function of species interactions and how changes in interactions among organisms can alter ecological communities.

EXTREME DIVERSITY

Though comprising only 7 percent of Earth's surface, rainforests harbor about 50 percent of all species. A single hectare of rainforest may be home to 1,500 species of flowering plants, 750 species of trees, 400 species of birds, and 125 different mammal species.

Mutualistic relationships benefit both species involved

Mutualism is an association between two species in which both species benefit more than it costs them to interact with each other. The benefits of mutualism increase the survival and reproduction of both of the interacting species. Often, mutualistic species live together—an association known as **symbiosis** (plural "symbioses"). Insects such as aphids, which feed on the nutrient-poor sap of plants, often have a mutualistic, symbiotic association with bacteria that live within their cells. The bacteria receive food and a home from the insects, and the aphids receive nutrients that the bacteria (but not the aphids) can synthesize from sugars in the plant sap.

NATURE ABOUNDS WITH VARIETIES OF MUTUALISM. Organisms can benefit from symbiotic relationships in various ways. Here we describe only some of the most common types of mutualism. In *gut inhabitant mutualism*, organisms that live in an animal's digestive tract receive food from their host and benefit the host

Helpful to Know

"Symbiosis" (from *sym*, "with"; *biosis*, "life") describes some (but not all) mutualistic and parasitic interactions. Individuals of one species, for at least part of their life cycle, must live close to, in, or on individuals of another species for the interaction to be considered symbiotic.

by digesting foods, such as wood or cellulose, that the host otherwise could not use. Termites, for example, are well known for eating wood. However, the digestion of the cellulose is performed by mutualistic protists that reside in the termite's gut. In fact, the mutualism includes another species: within the protists live bacteria that are the actual digesters of the cellulose. We humans, too, benefit from gut inhabitant mutualism because some of the hundreds of species of bacteria that colonize our large intestines manufacture beneficial nutrients, such as vitamin K.

Mutualism in which each partner has evolved to alter its behavior to benefit the other species is called *behavioral mutualism*. The relationship between certain shrimp and fishes is a good example of behavioral mutualism (**FIGURE 23.3**). Shrimp of the genus *Alpheus* live in an environment with plenty of food but little shelter. They dig burrows to hide in, but they see poorly, so they are vulnerable to predators when they leave their burrows to feed. These shrimp have formed a fascinating relationship with some goby fishes in the genera *Cryptocentrus* and *Vanderhorstia*. When a shrimp ventures out of its burrow to eat, it keeps an antenna in contact with an individual goby with which it has formed a special relationship. If a predator or other disturbance causes the fish to make a sudden movement, the shrimp darts back into the burrow. The goby acts as a "seeing eye" fish for the shrimp, warning it of danger. In return, the shrimp shares its burrow with the goby, thereby providing the fish with a safe haven.

Behavioral Mutualism

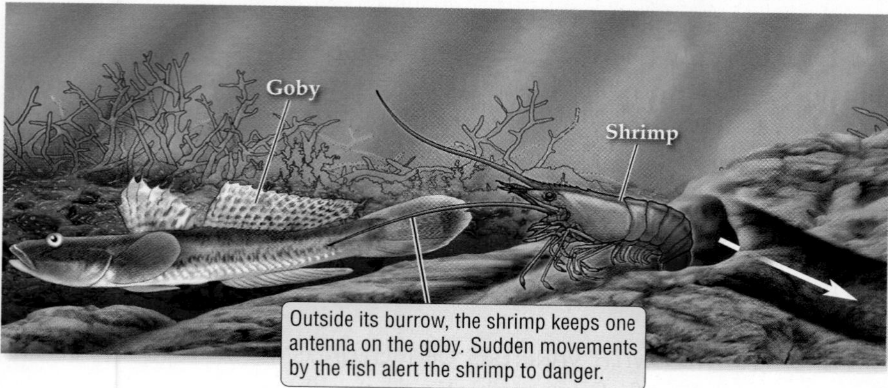

Goby
Shrimp
Outside its burrow, the shrimp keeps one antenna on the goby. Sudden movements by the fish alert the shrimp to danger.

FIGURE 23.3 Friends in Need
Each *Alpheus* shrimp builds a burrow for shelter, which it shares with a goby fish. The fish provides an early-warning system to the nearly blind shrimp when the shrimp leaves the burrow to feed.

Aggressive ants that defend their hosts from predators provide some of the best-known examples of *protection mutualism*. Certain species of *Acacia* trees play host to fierce ants that fend off herbivores. The trees secrete a nectarlike food from the leaf tips and produce swollen structures to shelter the insects.

In *seed dispersal mutualism*, an animal such as a bird or mammal eats a fruit that contains plant seeds and then later defecates the seeds far from the parent plant. Such dispersal by animals is the primary way that many plant species reach new areas of favorable habitat. For example, most of the plant species that live on isolated oceanic islands (those that are farther than 1,000 kilometers from land) are thought to have arrived there by bird dispersal of their seeds.

In *pollinator mutualism*, an animal such as a honeybee transfers pollen (which contains sperm) from one flower to the female reproductive organs (carpels) of another flower of the same species. Without such **pollinators**, many plants could not reproduce. To ensure that pollinators come to their flowers, plants offer a food reward, such as pollen or nectar, and both species benefit from the interaction. Pollinator mutualism is critical in both natural and agricultural ecosystems. For example, the apples we buy at the supermarket are available only because honeybees pollinate the flowers of apple trees, enabling the trees to produce their fruit.

MUTUALISTS ARE IN IT FOR THEMSELVES. Although both species in a mutualism benefit from the relationship, what is good for one species may come at a cost to the other. For example, a species may use energy or increase its exposure to predators when it acts to benefit its mutualistic partner. From an evolutionary perspective, mutualism evolves when the benefits of the interaction outweigh the costs for both species.

Consider the pollinator mutualism between the yucca plant and the yucca moth. A female yucca moth collects pollen from yucca flowers, flies to another group of flowers, and lays her eggs at the base of the carpel in a newly opened flower. After she has laid her eggs, the female moth climbs up the carpel and deliberately places the pollen she collected earlier onto the stigma of the flower. By this act she fertilizes the eggs of that second yucca plant (**FIGURE 23.4**). (For a review of the reproductive parts of a flower, see Figure 18.14.) When the moth larvae hatch, they feed on the seeds of the yucca plant.

In this mutualism, the plant gets pollinated (a reproductive benefit provided to the plant by the moth) and the moth eats some of its seeds (a food benefit provided to the moth by the plant). In fact, plant and pollinator each depend absolutely on the other: the yucca is the moth's only source of food, and this moth is the only species that pollinates the yucca. But there are costs for both species. In a cost-free situation for the plant, the moth would transport pollen but would not destroy any of the plant's seeds. In a cost-free situation for the moth, the moth would produce as many larvae as possible, and they would consume many of the plant's seeds. In actuality, an evolutionary compromise has been reached: the moth usually lays only a few eggs per flower, and the plant tolerates the loss of a few of its seeds. Yucca plants have a defense mechanism that helps keep this compromise working: if a moth lays too many eggs in one of the plant's flowers, the plant can selectively abort that flower, thereby killing the moth's eggs or larvae.

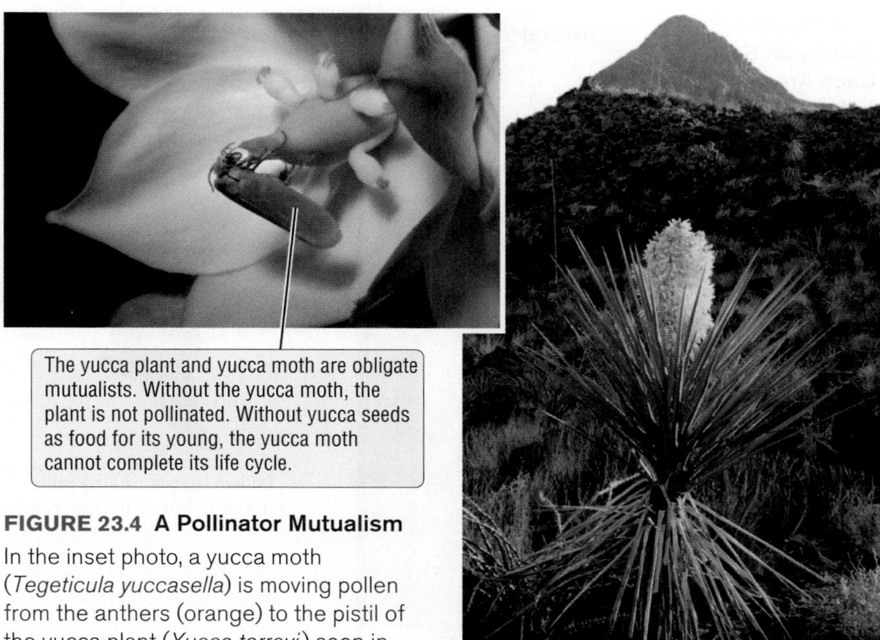

The yucca plant and yucca moth are obligate mutualists. Without the yucca moth, the plant is not pollinated. Without yucca seeds as food for its young, the yucca moth cannot complete its life cycle.

FIGURE 23.4 A Pollinator Mutualism
In the inset photo, a yucca moth (*Tegeticula yuccasella*) is moving pollen from the anthers (orange) to the pistil of the yucca plant (*Yucca torreyi*) seen in the larger photo.

Only one partner benefits in commensalism

In **commensalism**, one partner benefits while the other is neither helped nor harmed. Man-of-war fishes (Nomeidae)—which have acquired immunity from the deadly tentacles of jellyfish—evade predators by congregating among jellyfish. The man-of-war fishes clearly depend on this interaction with jellyfish, but the jellyfish get nothing out of it. Barnacles, as another example, attach themselves to whales (**FIGURE 23.5**). The whales are not harmed, but the filter-feeding hitchhikers get ferried around the ocean and may find more food than if they were stuck in one place.

FIGURE 23.5 Commensalism
This gray whale's rostrum (snout) is covered in barnacles.

FIGURE 23.6 Spines on Some Cacti Are an Induced Defense

On three islands off the coast of Australia, the percentage of cacti with spines is higher on the island that has cattle than on the two islands that do not. Field and laboratory experiments show that grazing by cattle directly stimulates the production of spines in this species of cactus.

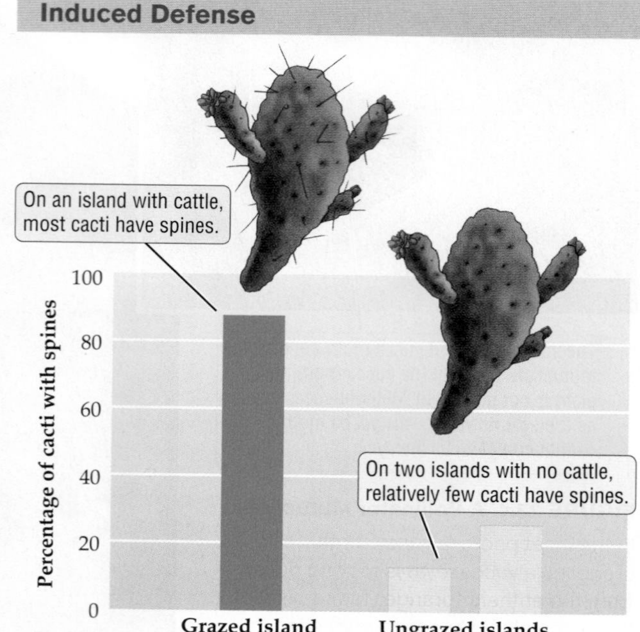

In exploitation, one member benefits while another is harmed

Exploitation encompasses a variety of interactions in which one species (the exploiter) benefits and the other (the species that is exploited, usually for food) is harmed. Exploiters are generally consumers falling into one of three main groups:

1. **Herbivores** are consumers that eat plants or plant parts.
2. **Predators** are animals (or, in rare cases, plants) that kill other animals for food; the animals that are eaten are called **prey**.
3. **Parasites** are consumers that live in or on the organisms they eat (which are called **hosts**). An important group of parasites are **pathogens**, which cause disease in their hosts.

The three major types of exploitation are very different from one another. Whereas predators (such as wolves) kill their food organisms immediately, parasites (such as fleas) usually do not. Although the different types of exploitation have obvious and important differences, we focus here on some general principles applying to all three.

EXTREME PARASITE

Cymothoa exigua is a marine crustacean that enters through the gills of fish. This parasite attaches to the base of the fish's tongue and draws blood from it. Starved of oxygen, the tongue will eventually atrophy and die. The fish, however, will not. It will continue to live with *C. exigua*, still attached, acting as a surrogate tongue.

CONSUMERS AND THEIR FOOD ORGANISMS CAN EXERT STRONG SELECTION PRESSURE ON EACH OTHER. The presence of consumers in the environment has caused many species to evolve elaborate strategies to avoid being consumed—yet another example of species interactions affecting evolutionary outcomes. Many plants, for example, produce spines and toxic chemicals as defenses against herbivores. Some plants rely on **induced defenses**, responses that are directly stimulated by an attack from herbivores. Spine production is an induced defense in some cactus species; an individual cactus that has been partially eaten, or grazed, is much more likely to produce spines than is an individual that has not been grazed (**FIGURE 23.6**).

Many prey organisms have evolved bright colors or striking patterns, known as **warning coloration**, to warn potential predators that they are heavily defended (**FIGURE 23.7*a***). Such warning coloration can be highly effective. Blue jays, for example, quickly learn not to eat brightly colored monarch butterflies, which contain chemicals that, in birds (and people), cause nausea and, at high doses, cause sudden death from heart failure. Other prey have evolved to avoid predators by being hard to find or hard to catch (**FIGURE 23.7*b***). **Mimicry** is a type of adaptation arising from predator-prey interactions in which a species evolves to imitate the appearance of something unappealing to its would-be predator (**FIGURE 23.7*c***).

Exploitative relationships tend to be under intense natural selection; after all, whether an organisms eats or is eaten will strongly influence its reproductive success. Often the selection pressure results in coevolution between the exploiter and the exploited. We have just looked at several examples of defensive adaptations that reduce the chances of being eaten. If a plant, host, or prey species evolves a particularly powerful defense against attack, its consumers, in turn, experience strong selection pressure to overcome that defense—an evolving situation sometimes called an *evolutionary arms race*. Take for example the cheetah and the Thomson's gazelle. Cheetahs are famous for their speed, which is a product of natural selection favoring individuals that could chase down gazelles. For their part, gazelles have been pressured by natural selection to become faster in order to avoid predation by cheetahs. As a consequence of this evolutionary arms race, these two animals are vastly faster runners than any other organism on the African savanna.

Defensive Adaptations

(a) Warning coloration

The poison dart frog is among the most toxic animals on Earth.

(b) Camouflage

Can you find the insect in this photo?

(c) Mimicry

FIGURE 23.7 Adaptive Responses to Predation
(*a*) The bright colors of this poison dart frog warn potential predators of the deadly chemicals contained in its tissues. (*b*) With its long legs outstretched, this lichen-mimic katydid lies motionless on lichen-covered branches to escape detection in daytime by lizards, birds, and other predators. A relative of the cricket, the katydid forages at night. (*c*) The viceroy butterfly (*left*) mimics the color and pattern of the monarch butterfly (*right*), which contains toxic compounds. Confusing it with the monarch, predators tend to leave the viceroy alone.

Population Cycles

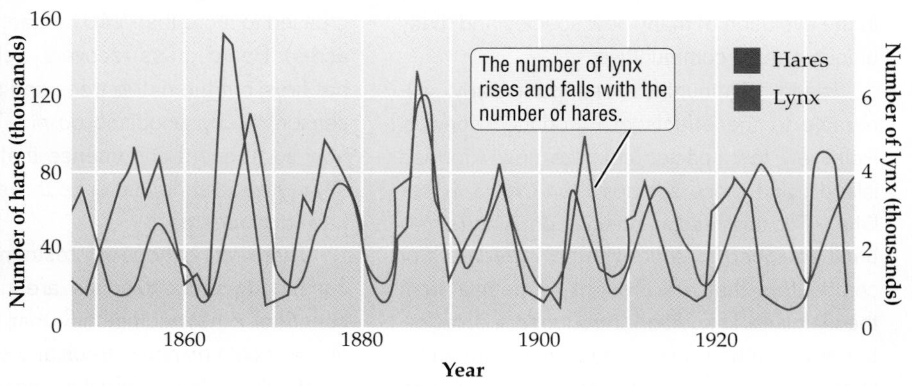

FIGURE 23.8 Populations of Two Species Occasionally Increase and Decrease Together
The Canadian lynx depends on the snowshoe hare for food, so the number of lynx is strongly influenced by the number of hares. Experiments conducted in the early twentieth century indicate that hare populations are limited by their food supply and by their lynx predators.

CONSUMERS CAN STRONGLY INFLUENCE THE ABUNDANCE AND DISTRIBUTION OF THEIR FOOD ORGANISMS. Sometimes a consumer exerts such a significant impact on a prey species that the population sizes of two species change together in a tightly linked cycle, which is known as a **population cycle**. The Canadian lynx, for example, depends on the snowshoe hare for food. Lynx populations increase when hare populations increase, but the increased number of lynx drive the hare populations down. As a consequence, their own numbers decrease until there are few enough lynx that the hare populations can increase again (**FIGURE 23.8**).

The impact of consumers of prey species can be intensive enough to drive species to extinction in certain local areas. The American chestnut used to be a dominant tree species

Helpful to Know

An evolutionary arms race ultimately will end with one species winning. North America once had a cheetah-like large cat that preyed on pronghorn antelope in the American Southwest. Over time, the pronghorn evolved to be such fast runners that the American cheetah was driven to extinction. Today, the pronghorn have no natural predators that can match their speed.

BIOLOGYMATTERS

Introduced Species: Taking Island Communities by Stealth

The Hawaiian Islands are the most isolated chain of islands on Earth. Because the islands are so remote, entire groups of organisms that live in most communities never reached them. For example, there are no native ants or snakes in Hawaii, and there is only one native mammal (a bat, which was able to fly to the islands).

The few species that did reach the Hawaiian Islands found themselves in an environment that lacked most of the species from their previous communities. The sparsely occupied habitat and the lack of competitor species resulted in the evolution of many new species and many unique natural communities.

Island communities are particularly vulnerable to the effects of introduced species. Relatively few species colonize newly formed islands, and those species then evolve in isolation. For this reason, species on islands may be ill equipped to cope with new predators or competitors that are brought by people from the mainland. In addition, introduced species often arrive without the predator and competitor species that held their populations in check on the mainland. On islands the potential exists for populations of introduced species to increase dramatically and become invasive.

In some cases, invasive species can destroy entire communities. The introduction of beard grass to Hawaii (as forage for cattle) is a case in point. By the late 1960s, beard grass had invaded the seasonally dry woodlands of Hawaii Volcanoes National Park. Before that time, fires had occurred there every 5.3 years on average, and each fire had burned an average of 0.25 hectare (about five-eighths of an acre). Since the introduction of beard grass, fires have occurred at a rate of more than one per year, and the average burn area of each fire has increased to more than 240 hectares (about 600 acres). Beard grass recovers well from large, hot fires, but the native trees and shrubs of the seasonally dry woodland do not. The fires are now so frequent and intense that the seasonally dry woodlands that once thrived in the park have disappeared.

There is no hope of restoring the native community, but ecologists are now trying to construct a new community that is tolerant of fire yet contains native trees and shrubs. This is a difficult challenge, and it is uncertain whether the effort will succeed. If not, what was once woodland is likely to remain indefinitely as open meadows filled with introduced grasses.

Great Diversity from a Single Ancestor Hawaiian silverswords are found only in the Hawaiian Islands. Genetic evidence indicates that this diverse group of plant species evolved from a single ancestor (a tarweed from California). Although the three silversword species shown here are closely related, they live in very different habitats and differ greatly in form.

across much of eastern North America. Within its range, anywhere from one-quarter to one-half of all trees were chestnuts. They were capable of growing to large size: trunks up to 10 feet in diameter were noted by settlers in colonial times. In 1900, however, a fungus that causes a disease called chestnut blight was introduced into the New York City area. This fungus spread rapidly, killing most of the chestnut trees in eastern North America. Today the American chestnut survives throughout its former range only in isolated patches, primarily as sprouts that arise from the base of otherwise dead trunks. With few exceptions, the new sprouts die back from reinfection with the fungus before they can grow large enough to generate new seeds.

In competition, both species are negatively affected

An ecological **niche** is the sum total of the conditions and resources a species or population needs in order to survive and reproduce successfully in its particular habitat. When the niches of two species overlap, competition may ensue. **Competition** is most likely when two species share an important limited resource, such as food or space. When two species compete, each has a negative effect on the other because each uses resources that could have been used by its competitor. This is true even when one species is so superior as a competitor that it ultimately drives the other species out of a given area.

In such **competitive exclusion**, the inferior competitor eventually loses so much ground to the superior rival that it becomes locally extinct.

There are two main types of competition:

1. In **interference competition**, one organism directly excludes another from the use of a resource. For example, individuals from two species of birds may fight over the tree holes that they both use as nest sites.
2. In **exploitative competition**, species compete indirectly for a shared resource, each reducing the amount of the resource available to the other. For example, two plant species may compete for a limited amount of nitrogen in the soil.

COMPETITION CAN LIMIT THE ABUNDANCE AND DISTRIBUTION OF SPECIES. Competition between species often has important effects on natural populations. These effects, as shown by a great deal of field evidence, include limiting the distribution and abundance of species. Let's explore two examples.

Along the coast of Scotland, the larvae of two species of barnacles—*Semibalanus balanoides* (**SEM-ee-buh-LAY-nus BAH-luh-NOY-deez**) and *Chthamalus stellatus* (**thuh-MAY-lus stell-AY-tus**)—both settle on rocks on high and low portions of the shoreline. However, as adults, *Semibalanus* individuals appear only on the lower portion of the shoreline, which is more frequently covered by water; and *Chthamalus* individuals are found only on the higher portion of the shoreline, which is more frequently exposed to air (**FIGURE 23.9**). In principle, the distributions of these two barnacles could have been caused either by competition or by environmental factors. In an experimental study, however, ecologists discovered that *Chthamalus* could thrive on low portions of the shoreline, but only when *Semibalanus* was removed. Hence, competition with *Semibalanus* ordinarily prevents *Chthamalus* from living low on the shoreline. This interaction is an example of interference competition because *Semibalanus* individuals often crush the smaller and more delicate *Chthamalus* individuals. The distribution of *Semibalanus*, on the other hand, depends mainly on environmental factors: the increased heat and dryness found at higher levels of the shoreline prevent *Semibalanus* from surviving there.

A second case of competition affecting distribution and abundance concerns wasps of the genus *Aphytis* (**ay-FYE-tus**). These wasps attack scale insects, which can cause serious damage to citrus trees. Female wasps lay eggs on a scale insect, and when the wasp larvae hatch, they pierce the scale insect's outer skeleton and then consume its body parts.

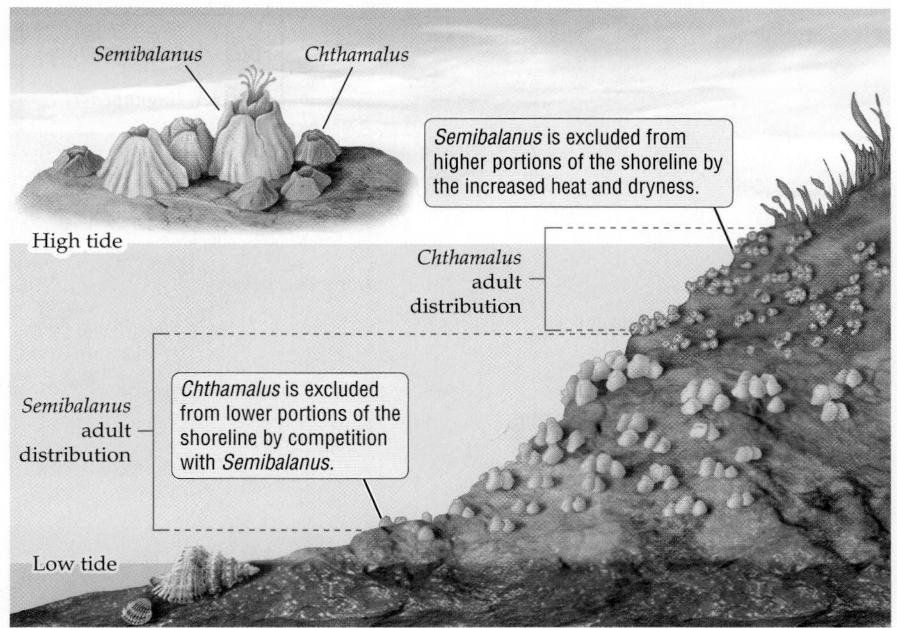

FIGURE 23.9 What Keeps Them Apart?
On the rocky coast of Scotland, the larvae of *Semibalanus* and *Chthamalus* barnacles settle on rocks on both high and low portions of the shoreline. However, adult *Semibalanus* barnacles are not found on high portions of the shoreline, and adult *Chthamalus* individuals are not found on low portions.

In 1948, the wasp *Aphytis lingnanensis* was released in southern California to curb the destruction of citrus trees caused by scale insects. A closely related wasp, *A. chrysomphali*, was already living in that region at the time. *A. lingnanensis* was released in the hope that it would provide better control of scale insects than *A. chrysomphali* did. *A. lingnanensis* proved to be a superior competitor (**FIGURE 23.10**), in most locations driving *A. chrysomphali* to extinction by exploitative competition. As hoped for, *A. lingnanensis* also provided better control of scale insects.

Although competition between species is very common, it does not necessarily follow when two species share resources or space. This is especially true when the resources are abundant. Competition among leaf-feeding insects, for example, is relatively uncommon for this reason. A huge amount of leaf material is available for the insects to eat, and usually there are too few insects to cause their food to be in short supply. As long as their food remains abundant, little competition occurs.

Natural selection can also drive competing species to use their common niche in different ways—a phenomenon known as **niche partitioning**. The differential use of space or resources minimizes competition,

Exploitative Competition

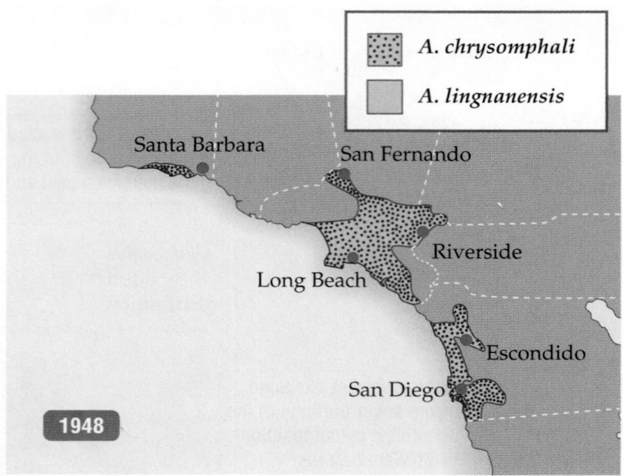

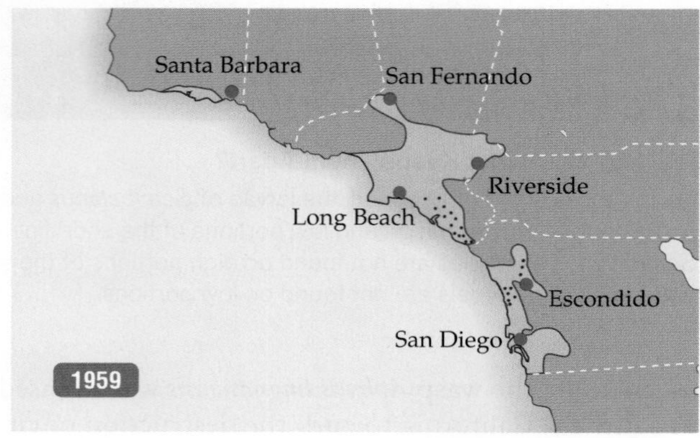

FIGURE 23.10 A Superior Competitor Moves In

After being introduced to southern California in 1948, the wasp *Aphytis lingnanensis* rapidly drove its competitor, *Aphytis chrysomphali*, extinct in most locations. Both species of wasps prey on scale insects that damage citrus crops (such as lemons and oranges).

Concept Check Answers

1. It is a mutualism. There are costs for both: the moth's reproduction is limited because the plant restricts the number of eggs the moth can lay successfully, and the plant in turn loses some of its seeds to the moth larvae.
2. This is commensalism. The birds benefit because they get better access to their insect prey. The grazers are neither harmed nor benefited.
3. *Chthamalus* would thrive low on the shoreline as long as *Semibalanus* was absent.

Helpful to Know

In genetics, specific traits of organisms, such as height, beak size, or the chemical structure of proteins, are sometimes referred to as "characters." In character displacement, a specific trait (such as beak size) becomes more different over time in populations of two species that compete for resources.

enabling species to coexist despite their potential for competition (**FIGURE 23.11**). For example, many prairie plants that grow together in dense stands can coexist through niche partitioning. A shallow-rooted prairie grass mines the soil close to the surface to obtain water and mineral nutrients, while a neighboring coneflower plant taps deeper layers of soil with its deep root.

COMPETITION CAN INCREASE THE DIFFERENCES BETWEEN SPECIES. As Charles Darwin realized when he formulated the theory of evolution by natural selection, competition between species can be intense when the two species are very similar in form. For example, birds whose beaks are similar in size eat seeds of similar sizes and therefore compete intensely, whereas birds whose beaks differ in size eat seeds of different sizes and compete less intensely. Intense competition between similar species may result in **character displacement**, in which the forms of the competing species evolve to become more different over time. By reducing the similarity in form between species, character displacement reduces the intensity of competition.

Concept Check

1. The yucca moth pollinates the yucca plant and depends on it for food. How would you classify this type of interaction? Is it cost-free for both moth and plant? Explain.
2. Cattle egrets trail livestock—sometimes perching on their backs—to pick up insects stirred up by the grazing animals. How would you classify this type of interaction? Who benefits in this type of interaction?
3. *Chthamalus* (a barnacle that lives high on the shoreline) and *Semibalanus* (a barnacle that lives in the low intertidal zone) exhibit interference competition. How would the survival and reproduction of a *Chthamalus* colony be affected if it were relocated to the low intertidal zone either in the absence of *Semibalanus* or intermixed with *Semibalanus*?

23.2 How Species Interactions Shape Communities

We have seen how interactions among organisms help determine their distribution and abundance. Interactions among organisms also have large effects on the communities and ecosystems in which those organisms live.

When dry grasslands are overgrazed by cattle, for example, grasses may become less abundant and desert shrubs may become more abundant. These changes in the abundances of grasses and shrubs can change the physical environment. The rate of soil erosion may increase because shrubs do not stabilize soil as well as grasses do. Ultimately, if overgrazing is severe, the ecosystem can change from a dry grassland to a desert.

A variety of factors other than species interactions can affect species diversity in a community. A fire can wipe out trees and the species they harbor and allow

(a)

(b)

(c)

FIGURE 23.11 Niche Partitioning Minimizes Competition
Along coastal shores, different species of seabirds wade through the water and use their bills to dig for crustaceans. To minimize competition, each species has its own specialized niche. (*a*) The dowitcher wades through shallow waters. (*b*) The whimbrel can reach deeper waters with its longer legs and bill. (*c*) The curlew can forage in even deeper waters

grasses and other sun-loving plants to flourish for the first time. Urbanization reduces habitat, diverts water, and introduces pollution that affects species composition and density in a community. Any event—whether caused naturally or by humans—that changes the habitat in which species live will affect their chances of survival. Any change in species diversity in a community will have a ripple effect throughout the community.

Keystone species have profound effects on communities

Certain species have a disproportionately large effect, relative to their own abundance, on the types and abundances of the other species in a community; these influential species are called **keystone species**. Keystone species are usually noticed when they are removed or they disappear from an ecosystem, resulting in dramatic changes to the rest of the community.

In an experiment conducted along the rocky Pacific coast of Washington State, ecologist Robert Paine demonstrated that the sea star *Pisaster ochraceus* (**pih-*ZAS*-ter oh-*KRAY*-see-us**) is a keystone species in its intertidal-zone community. He removed sea stars from one site and left an adjacent, undisturbed site as a control. In the absence of sea stars, all of the original 18 species in the community, except mussels, disappeared (**FIGURE 23.12**). When the sea stars were present, they ate the mussels, thereby keeping the number of mussels low enough that the mussels did not crowd out the other species.

In general, the term "keystone species" can include any producer or consumer of relatively low abundance that has a large influence on its community. The most abundant or dominant species in a community (such as the corals in a coral reef or the mussels in Paine's intertidal zone) also have large effects on their communities, but because their abundance is not low, they are not considered keystone species.

Helpful to Know

In architecture, a keystone is the central, topmost stone that keeps an arch from collapsing. A keystone species serves a similar role with respect to the "architecture" of a community: if a keystone species is removed, the composition of the entire community will change dramatically.

Invasive species can overtake communities

Nonnative species introduced by people to new regions sometimes disrupt the ecological communities there. Members of mature communities typically have long-established evolutionary relationships between them that create stable distributions, abundances, and ecological roles. Predators keep prey species in

A Keystone Species

How a predator maintains diversity

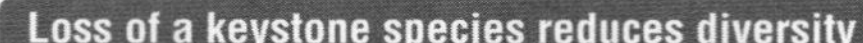

Loss of a keystone species reduces diversity

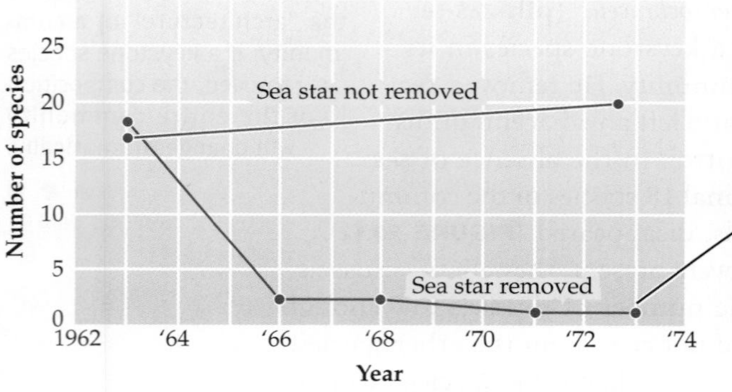

Sea stars completely eliminated mussels in submerged areas of this marine community, enabling other intertidal species to thrive there.
When the sea star *Pisaster* was removed from a community experimentally, the number of species dropped from 18 to 1, a mussel.
Sea star not removed
Sea star removed
Number of species
25
20
15
10
5
0
1962
'64
'66
'68
'70
'72
'74
Year

FIGURE 23.12 The Star of the Community

The sea star *Pisaster ochraceus* is a predator that feeds on mussels and thereby prevents the mussels from crowding out other species in their community.

Populations Grow When Predation Is Low

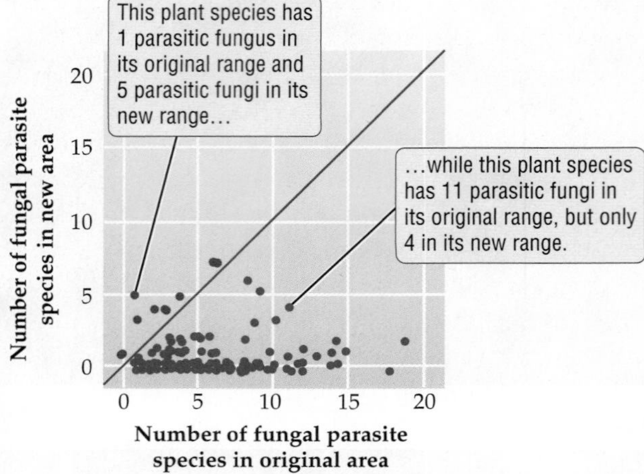

This plant species has 1 parasitic fungus in its original range and 5 parasitic fungi in its new range...
...while this plant species has 11 parasitic fungi in its original range, but only 4 in its new range.
Number of fungal parasite species in new area
20
15
10
5
0
0
5
10
15
20
Number of fungal parasite species in original area

FIGURE 23.13 Introduced Species Leave Their Predators at Home

Generally, introduced species have fewer parasites and other predators in their new homes than in their original homes. Each point on the graph represents a different plant species that has been introduced to a new area. The blue points below the diagonal line represent plants with fewer fungal parasites in their new home than in their old home; points above the red line represent the opposite.

Concept Check Answers

1. Any organism of relatively low abundance that has a disproportionately large influence on the diversity of a community is a keystone species.
2. Because they lack biological controls on their population growth, such as predators or parasites, invasive species can potentially grow rapidly and displace, outcompete, or prey on native species.

check, mutualists benefit each other, and competitors evolve means to coexist. When nonnative species are introduced to a community, the invasive species lack any evolutionary relationship with existing members, such as predators, competitors, or parasites, and can multiply to such an extent that they overtake the entire community (**FIGURE 23.13**). We have already seen examples, such as the zebra mussels and *Cactoblastis cactorum* in previous chapters, of invasive species devastating communities.

After World War II, brown tree snakes were accidentally introduced on Guam, a 209-square-mile island located about 3,800 miles west of Hawaii. Guam had only one native snake, which was blind and insectivorous. The fauna of Guam therefore had no experience with predatory snakes and had evolved no defense against them. Within 50 years, nearly all of the native birds and many of the native small mammals and reptiles went extinct. The abundance of defenseless prey, combined with the absence of predators, allowed brown tree snakes to multiply exponentially on the island. Current estimates of the population abundance range from 12,000 to 15,000 snakes per square mile.

Concept Check

1. What is a keystone species?
2. Why are invasive species such a significant problem?

23.3 How Communities Change over Time

All communities change over time. The number of individuals of different species in a community often changes with the seasons. For example, although butterflies might be abundant in summer, we would not find any of them flying in a North Dakota field in the middle of winter. Similarly, every community shows year-to-year changes in the abundances of organisms, as we saw in Chapter 21. In addition to such seasonal and yearly changes, communities show broad, directional changes in species composition over longer periods of time.

A community may begin when new habitat is created, as when a volcanic island like Hawaii rises out of the sea. New communities may also form in regions that have been disturbed, as by a fire or hurricane. Some species arrive early in such new or disturbed habitat. These early colonists tend to be replaced later by other species, which in turn may be replaced by still other species. These later arrivals replace other species because they are better able to grow and reproduce under the changing conditions of the habitat.

Succession establishes new communities and replaces disturbed communities

The process by which species in a community are replaced over time is called **succession**. In a given location, the order in which species will replace one another is fairly predictable. Such a sequence of species replacements sometimes ends in a *mature community*, which, for a particular climate and soil type, is a community whose species composition remains stable over long periods of time. But in many—perhaps most—ecological communities, disturbances such as fires or windstorms occur so frequently that the community is constantly changing in response to a disturbance event, and a mature community never forms.

Primary succession occurs in newly created habitat, as when an island rises from the sea or when rock and soil are deposited by a retreating glacier (**FIGURE 23.14**). In such a situation, the process begins with a habitat that contains no species and little to no soil. The first species to colonize the new habitat usually have one of two advantages over other species: either they can disperse more rapidly (and hence reach the new habitat first), or they are better able to grow and reproduce under the challenging conditions of the newly formed habitat. In some cases of primary succession, the first species to colonize the area alter the habitat in ways that enable later-arriving species to thrive. In other cases, the early colonists hinder the establishment of other species.

Secondary succession is the process by which communities regain the successional state that existed before a disturbance, as when natural vegetation recolonizes a field that has been taken out of agriculture, or when a forest grows back after a fire (**FIGURE 23.15**). In contrast to primary succession, habitats undergoing secondary succession often have well-developed soil containing seeds from species that usually predominate late in the successional process. The presence of such seeds in the soil can considerably shorten the time required for the later stages of succession to be reached.

Primary Succession

(*a*) Stages of succession

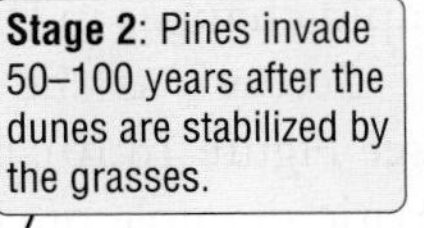
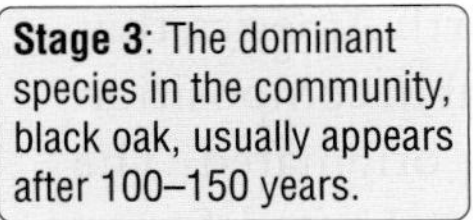
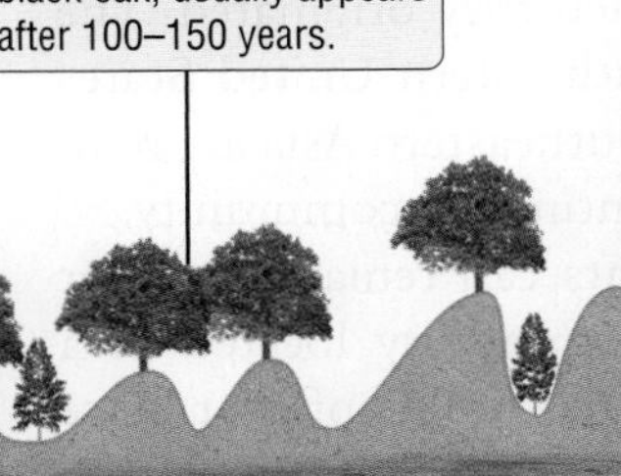

Stage 1: Bare sand is first colonized by dune-building grasses, such as marram grass, which spread rapidly and stabilize the moving sand of the dunes.
Stage 2: Pines invade 50–100 years after the dunes are stabilized by the grasses.
Stage 3: The dominant species in the community, black oak, usually appears after 100–150 years.
Lake Michigan
Older sand dunes

(*b*) A mature community

FIGURE 23.14 Sand Becomes Woodland

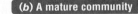

When strong winds cause sand dunes to form at the southern end of Lake Michigan, succession often leads to a community dominated by black oak. Succession on such dunes occurs in three stages and forms black-oak communities that have lasted up to 12,000 years. Under different local environmental conditions, succession on Michigan sand dunes (*a*) can lead to the establishment of stable communities as different as grasslands, swamps, and sugar maple forests (*b*).

Secondary Succession

FIGURE 23.15 The Slow but Steady Regrowth of Lodgepole-Pine Forest in Yellowstone National Park

These photographs, taken in different locations, show the regrowth of lodgepole-pine forest following the large fire that struck Yellowstone National Park in 1988 (*a, b*), compared to unburned, mature forest (*c*).

Communities change as climate changes

Some groups of species stay together for long periods of time. For example, an extensive plant community once stretched across the northern parts of Asia, Europe, and North America. As the climate grew colder during the past 60 million years, plants in these communities migrated south, forming communities in Southeast Asia and southeastern North America that are similar to one another—and similar in composition to the community from which they originated. The iconic magnolias of the southeastern United States and very similar species in southeastern Asia are remnants of that ancient intercontinental community.

Although groups of plants can remain together for millions of years, the community located in a particular place changes as the climate of that place changes. The climate at a given location can change over time for two reasons: global climate change and continental drift.

First, consider the climate of Earth as a whole, which changes over time. What we experience today as a "normal" climate is warmer than what was typical during the previous 400,000 years. Over even longer periods of time, the climate of North America has changed greatly (**FIGURE 23.16**), causing dramatic changes in the plant and animal species that live there. For example, fossil evidence indicates that 35 million years ago, the areas of southwestern North America that are now deserts were covered with tropical forests. Historically, changes in the global climate have been due to relatively slow natural processes, such as the advance and retreat of glaciers. However, evidence is mounting that human activities are now causing rapid changes in the global climate (see Chapter 25).

Second, as the continents move slowly over time (see Figure 16.10), their climates change. To give a dramatic example of continental drift, 1 billion years ago Queensland, Australia, which is now located at 12° south latitude, was located near the North Pole. Roughly 400 million years ago, Queensland was at the equator. Since the species that thrive at the equator and in the Arctic are very different, continental drift has resulted in large changes in the communities of Queensland over time.

Concept Check

1. Describe the distinctive characteristics of species that tend to be the first to colonize a new habitat.
2. Compare primary succession with secondary succession.

Concept Check Answers

1. They disperse more rapidly than other species, and/or they are better able to survive and reproduce in challenging environments.
2. Primary succession is the order in which species replace others in a new habitat (one that contains no species). Secondary succession is the sequence of species that appear in a preexisting community disturbed by natural processes (such as fire) or human actions (such as deforestation).

Climate Change over Time

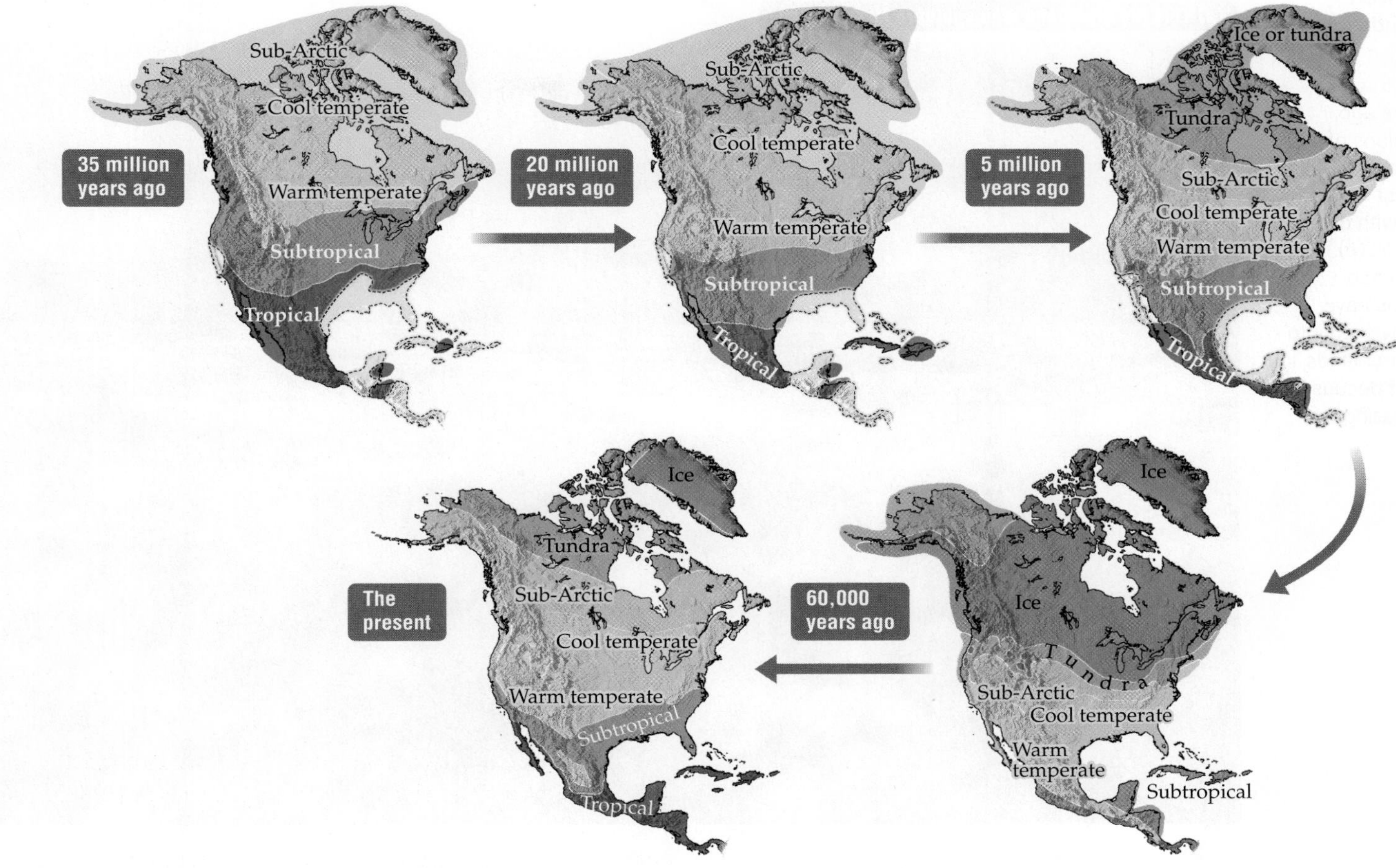

FIGURE 23.16 Climates Change, Communities Change

The climate of North America has changed greatly during the past 35 million years. As the climate has changed, the communities found in particular places have changed as well. The white regions surrounded by dotted lines were below sea level at the time indicated.

23.4 Human Impacts on Community Structure

Ecological communities are subject to many natural forms of disturbance, such as fires, floods, and windstorms. As we saw with the example of Yellowstone Park (see Figure 23.15), following a disturbance, secondary succession can reestablish the previously existing community. In this way communities can and do bounce back from some forms of disturbance. Depending on the community, the time required to regain a prior state varies from years to decades to centuries.

Communities have been exposed over long periods of time to natural forms of disturbance, such as windstorms. In contrast, people may introduce entirely new forms of disturbance, such as the dumping of hot wastewater into a river by a nuclear power plant. Human actions may also alter the frequency of an otherwise natural form of disturbance—for example, causing a dramatic increase or decrease in the frequency of fires or floods.

Communities can reassemble after some human-caused disturbances

Can communities regain their previous state after disturbances caused by people? For some forms of human-caused disturbance, the answer is: yes. Throughout

FIGURE 23.17 Overgrazing Can Convert Grasslands into Deserts

(*a*) More than 200 years ago, large regions of the American Southwest were covered with dry grasslands. (*b*) Most of these grasslands have been converted into desert shrublands, in large part because of overgrazing by cattle.

Overgrazing

(*a*) This dry grassland is in southern New Mexico.

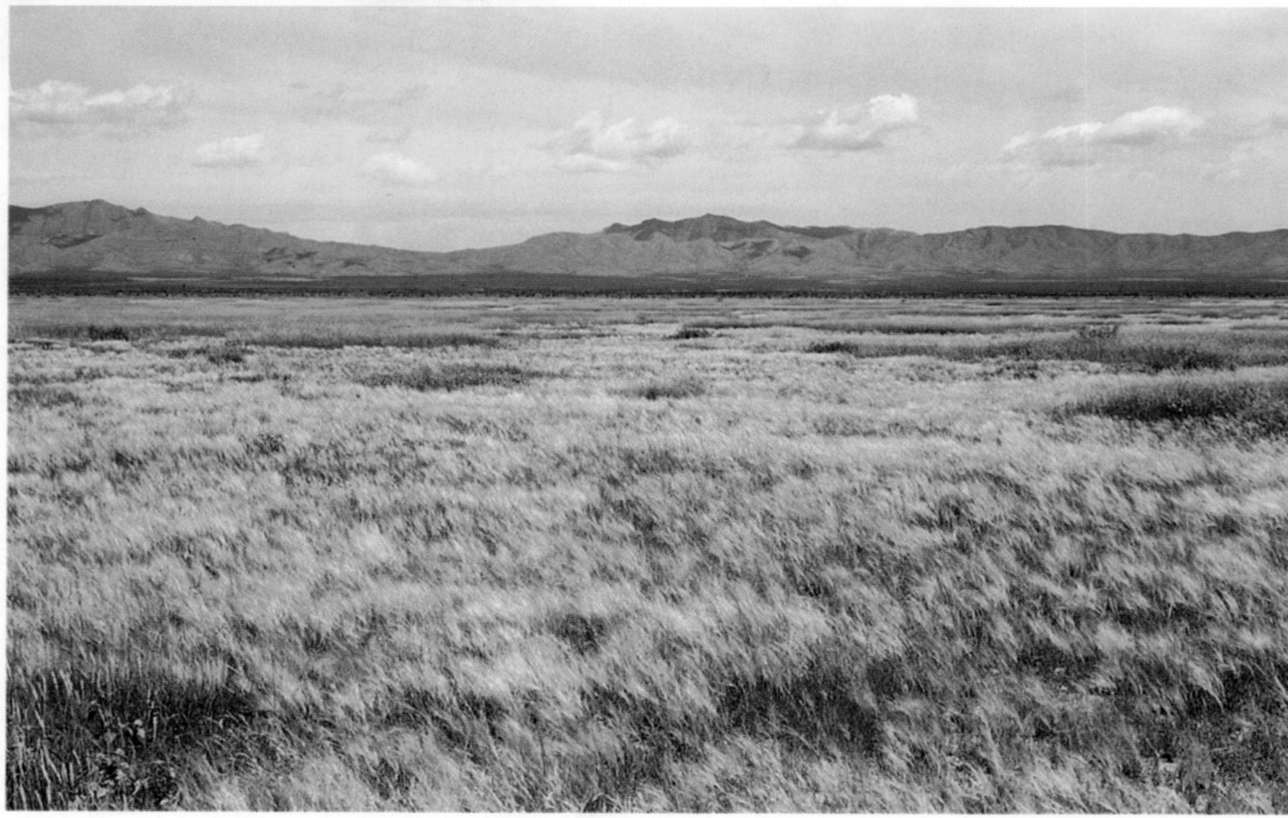

the eastern United States, for example, there are many places where forests were cut down and the land used for farmland; years later, the farmland was abandoned. Second-growth forests have grown on these abandoned farms, often within 40–60 years after farming stopped. Second-growth forests are not identical to the forests that were originally present. The sizes and abundances of the tree species are different, and fewer plant species grow beneath the trees of a second-growth forest than beneath a virgin forest (one that has never been cut down). However, the second-growth forests of the eastern United States already have recovered partially from cutting. If current trends continue, over the next several hundred years there will be fewer and fewer differences between such forests and the original forests.

People can cause long-term damage to communities

It is encouraging that complex ecological communities like northeastern forests can recover rapidly from disturbances caused by people. However, communities do not always recover so quickly from human-caused disturbances, as a few examples will show.

- Northern Michigan once was covered with a vast stretch of white-pine and red-pine forest. Between 1875 and 1900, nearly all of these trees were cut down, leaving only a few scattered patches of virgin forest. The loggers left behind large quantities of branches and sticks, which provided fuel for fires of great intensity. In some locations, the pine forests of northern Michigan have never recovered from the combination of logging and fire.
- Throughout South America and Southeast Asia, large tracts of tropical forest have been converted into grasslands by a combination of logging and fire. Scientists estimate that it will take tropical forest communities hundreds to thousands of years to recover from such changes.

(*b*) This nearby, former dry grassland is now a desert.

- In some areas of the American Southwest, grazing by cattle has transformed dry grasslands into desert shrublands (**FIGURE 23.17**). How do cattle cause such large changes? Overgrazing and trampling by cattle decrease the abundance of grasses in the community. With less grass to cover the soil and hold it in place, the soil dries out and erodes more rapidly—a process known as **desertification**. Desert shrubs thrive under these new soil conditions, but grasses do not. These changes in soil characteristics can make it very difficult to reestablish grasses, even when the cattle are removed.

Grazing can have a dramatic effect on dry grasslands, but how do you think the effects of grazing would compare to the effects of an atomic bomb? The first aboveground explosion of an atomic bomb occurred on July 16, 1945, at the Trinity site in New Mexico. Fifty years later, dry grasslands destroyed by the bomb blast (but never grazed) had recovered. In contrast, nearby dry grasslands that had been heavily grazed (but not destroyed by the bomb blast) had not recovered, even though they had not been grazed since the time of the blast. The plant community recovered more rapidly from the effects of a nuclear explosion than from the effects of grazing—a dramatic example of how strongly ecological interactions can affect natural communities.

Concept Check

1. What kinds of human activities influence community structure?
2. Can human activity ever induce primary succession?

Concept Check Answers

1. Logging, fire, agriculture, and grazing are among the many human activities that affect communities.

2. Not likely. Human activity may damage or degrade communities, but it does not remove all species and soil.

APPLYING WHAT WE LEARNED

How a Parasite Can Hijack Your Brain

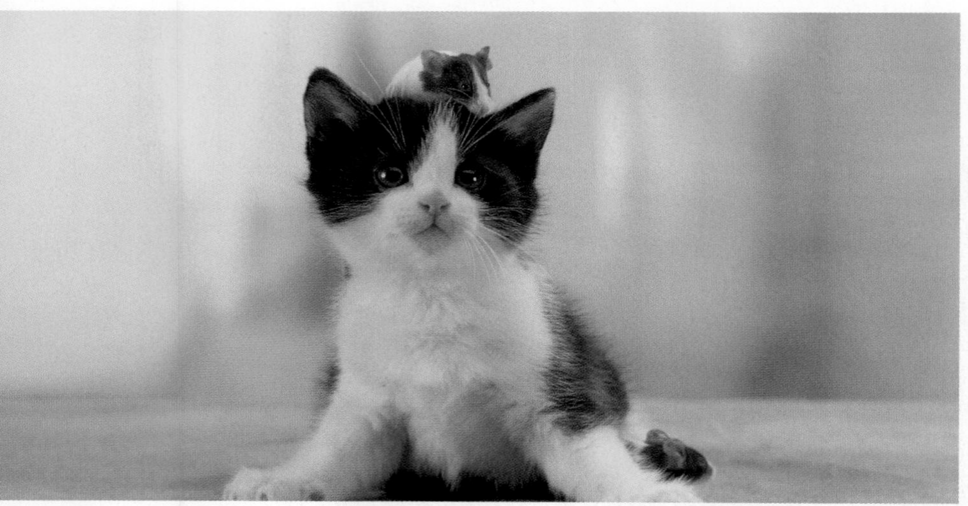

At the beginning of this chapter we learned about a parasite that changes the behavior of its host. Specifically, male rats infected with *Toxoplasma gondii* find the smell of cat urine irresistible. When the unfortunate rat locates a cat, the cat eats the rat and—voilà—the cat is infected by the parasite. Cats (including lions and leopards) are the only animals in which *T. gondii* can sexually reproduce, making cats the parasite's primary host. When an infected cat passes feces containing the parasite, the parasite remains infectious for up to 2 years.

T. gondii also infects humans—causing two forms of toxoplasmosis. In the first form, an initial ("acute") infection can cause (1) no symptoms; (2) mild, flulike symptoms such as fever and enlarged lymph nodes; or (3) serious illness such as fever, seizures, and eye inflammation in those with compromised immune systems. Pregnant women are at special risk from acute Toxo infections, since the parasite can cause birth defects.

Most people do not even notice the initial infection. Once an untreated infection is over, the parasite enters the brain and muscle and forms cysts, creating a permanent "latent" infection. Despite the lack of medical symptoms, people with latent toxoplasmosis are measurably different. Early studies of people carrying antibodies to the parasite show that they have, on average, slower reaction times, higher rates of automobile accidents, increased rates of anxiety, and resistance to new situations. One study has even shown that women with a latent infection are more likely to give birth to sons.

Just as in male and female rats, men and women react differently to *T. gondii* infection. Some studies suggest that infected men are more likely than uninfected men to ignore rules and to be suspicious, jealous, or strongly opinionated. Infected men have three to four times the rate of automobile accidents that normal men have. In contrast, infected women are said to be more warmhearted, outgoing, and conscientious than are uninfected women. *T. gondii* infection has also been tied to higher rates of the mental disorder schizophrenia.

How does Toxo alter the brain? The verdict is still out, but very recently, researchers may have discovered some answers. The brains of many animals, including mammals, secrete a chemical called dopamine that is involved in the sense of pleasure. In response to a "rewarding" experience such as food, sex, or cocaine, the brain releases dopamine, which makes an animal feel good. (Dopamine is also involved in other important activities, such as thought, movement, and sleep.)

Single-celled organisms do not need dopamine and do not normally carry genes for making dopamine. Yet the single-celled *T. gondii* carries a gene for making an enzyme that makes dopamine. Once *T. gondii* has infected an animal's brain, the parasite floods the brain with dopamine, apparently causing poor choices in both male humans and male rats. But rats at least don't lose their fear of all dangerous situations—just of cats. In fact, *T. gondii* goes a step further and makes infected rats actually crave the smell of cat urine.

It's obvious how changing the behavior of rats improves *T. gondii*'s chances of spreading to more cats. It's less obvious if the parasite's effect on humans is just a side effect that doesn't help the parasite or if, in our evolutionary past, when our ancestors were prey for leopards, *T. gondii* in the brain made humans take risks, such as approaching leopards.

Today, nearly a quarter of the U.S. population is infected by *T. gondii*. Although most media accounts say that the parasite is contracted from cleaning cat litter boxes, the Department of Agriculture reports that half of all *T. gondii* cases are contracted from handling raw meat or from consuming undercooked meat, unpasteurized milk, and unwashed fruits and vegetables.

BIOLOGY IN THE NEWS

Beyond the Kiss, Mistletoe Helps Feed Forests, Study Suggests

BY ALANNA MITCHELL • *New York Times*, December 17, 2012

For years, mistletoe has suffered from a split reputation: either the decorative prelude to a sweet Christmas kiss or the tree-killing parasite that must be mercilessly excised for the good of the forests.

David Watson . . ., known in academic circles as "the mistletoe guy," had long suspected that his favorite plant was a keystone species ... but even he was unprepared for the results. He had supposed that creatures that fed or nested on mistletoe would be affected by its removal. Instead, he found that the whole woodland community in the mistletoe-free forests declined.

Three years after the mistletoe vanished, so had more than a third of the bird species, including those that fed on insects. Bird diversity is considered an indicator of overall diversity. Where mistletoe remained, bird species increased slightly. It was a similar story for some mammals and reptiles, but, in another surprise, particularly for those that fed on insects on the forest floor . . .

Analysis showed that species of mistletoe play an important role in moving nutrients around the forest food web. That has to do with their status as parasites.

Nonparasitic plants suck nutrients out of their own leaves before they let them fall, sending dry containers to the ground. But because the vampiric mistletoe draws water and nutrients from the tree stem or branch it attaches to, it is more nonchalant about leaving that nutrition in falling leaves. That means the fallen leaves still contain nutrients that feed creatures on the forest floor.

Not only that, but mistletoes make and drop leaves three or four times as rapidly as the trees they live off of, said Dr. Watson. As evergreens, they also do it throughout the year, even when trees are dormant. It is like a round-the-calendar mistletoe banquet . . .

Dr. Watson said it was possible that introducing mistletoe into a damaged forest could help restore it to health.

But introducing mistletoe onto trees could prove controversial. While the parasites are like Robin Hood, stealing from rich trees to feed the forest poor, they can spoil individual trees for lumber . . .

Still, Dr. Watson's findings add a touch of science to the folkloric view of mistletoe as a tantalizer, inducing people to wait under it for a kiss at Christmas. The custom stems from the ancient Druids, who believed mistletoe could work magic because it grew high in bare oak trees in midwinter where nothing else did, seemingly out of thin air. They cut it down with golden sickles, never letting it touch the ground, and hung it in homes to foster fertility.

Mistletoe is a well-known tree parasite, but as this study shows, its impact on the community goes beyond its relationship with its host. As a parasite, mistletoe attaches to trees and extracts water and nutrients. Heavy infestations of mistletoe can impair tree growth and even cause death. Nonetheless, mistletoe is also an important food species for birds, which help to disperse it to other trees. This study indicates that mistletoe's highly inefficient and abundant leaf shedding distributed nutrients from the trees to the forest floor, enabling the growth of insect communities that, in turn, affected the birds, reptiles, and mammals that fed on the insects. In places where mistletoe was removed, community diversity decreased, indicating that mistletoe is a keystone species in forest communities.

This study illustrates the complex relationships that can form within a community. While the mistletoe is very clearly a parasite on trees, it is also prey for herbivorous organisms such as birds. Furthermore, by shedding its nutrient-rich leaves, it acts as a food source for soil insect communities, whose abundance influences many other species that prey on them. Such complex relationships between species is the rule rather than the exception in communities. Even well-known relationships, such as the one between the various clownfish species and certain anemones, are seldom as straightforward as they appear. The clownfish usually aids its anemone partner by cleaning up and helping to aerate the anemone, but sometimes it also steals the anemone's food.

Understanding these complex community relationships can help people manage ecosystems better. Policy makers and conservationists must make policy and management decisions regarding the natural world. An appreciation for the complexities of relationships between species is essential to good decision making.

Evaluating the News

1. Given that the presence of mistletoe increases mammal numbers by increasing the number of soil insects, can you say that mistletoe and mammals have a community-level interaction?

2. What life history traits do mistletoe possess that make them keystone species in forest ecosystems?

3. The study reported here suggests that adding mistletoe to a forest may actually increase its diversity. What are the potential perils of such a move? Would you support such an effort?

CHAPTER REVIEW

Summary

23.1 Species Interactions

- An ecological community is an association of populations of different species that live in the same area. Communities vary greatly in size and complexity and are characterized by both species richness and species abundance.
- Species interactions are classified by whether they hurt, harm, or have no effect on the species involved. These interactions fall into four categories: mutualism, commensalism, exploitation, and competition.
- Mutualism evolves when the benefits of the interaction to both partners are greater than its costs for both partners.
- In commensalism, one species benefits while the other is neither harmed nor benefited.
- In exploitation, one species benefits while the other is harmed. Exploiters are consumers. Consumers include herbivores, predators, and parasites.
- Consumers can be a strong selective force, leading their food organisms to evolve various ways to avoid being eaten. Many plants have evolved induced defenses, such as the growth of spines, that are directly stimulated by attacking herbivores. Food organisms, in turn, exert selection pressure on their consumers, which evolve ways to overcome the defenses of the species they eat.
- Consumers can restrict the distribution and abundance of the species they eat, thereby strongly affecting community structure.
- Competition between species negatively affects both. Competition can strongly influence the distribution and abundance of species.
- In interference competition, one species directly excludes another species from the use of a resource. In exploitative competition, species compete indirectly, each reducing the amount of a resource available to the other species.

23.2 How Species Interactions Shape Communities

- Community structure can be shaped by the organisms within it. Some species are necessary for the stability of the community. Others disrupt communities and may lead to their collapse.
- A keystone species has a large effect on the composition of a community relative to its abundance within that community. Keystone species alter the interactions among organisms in a community, and they change the types or abundance of species in the community.
- Invasive species have no evolutionary relationship with the organisms within a community. Invasive species may have no predators or competitors, allowing them to take over entire communities, leading to the extinction of species.

23.3 How Communities Change over Time

- All communities change over time. Directional changes that occur over relatively long periods of time have two main causes: succession and climate change.
- Primary succession occurs in newly created habitat. Secondary succession occurs as a community regains its previous state after a disturbance.

23.4 Human Impacts on Community Structure

- Humans transform natural habitats in many ways, from turning forests into farmland, to diverting water, polluting air and water, introducing new species, and altering the global climate.
- Communities can bounce back from some forms of natural and human-caused disturbance. The time required for restoration varies from years to centuries.
- Understanding the consequences of community change can help people to make decisions that take community stability into account.

Key Terms

character displacement (p. 534)
coevolution (p. 527)
commensalism (p. 529)
community (p. 526)
competition (p. 532)
competitive exclusion (p. 533)
desertification (p. 540)
diversity (p. 526)
exploitation (p. 530)
exploitative competition (p. 533)
herbivore (p. 530)
host (p. 530)
induced defense (p. 530)
interference competition (p. 533)
keystone species (p. 535)
mimicry (p. 530)
mutualism (p. 527)
niche (p. 532)
niche partitioning (p. 533)
parasite (p. 530)
pathogen (p. 530)
pollinator (p. 528)
population cycle (p. 531)
predator (p. 530)
prey (p. 530)
primary succession (p. 537)
relative species abundance (p. 526)
secondary succession (p. 537)
species richness (p. 526)
succession (p. 537)
symbiosis (p. 537)
warning coloration (p. 530)

Self-Quiz

1. The advantages received by a partner in a mutualism can include
   a. food.
   b. protection.
   c. increased reproduction.
   d. all of the above

2. Defensive adaptations exhibited by prey species include all of the following *except*
   a. warning coloration.
   b. camouflage.
   c. induced defense.
   d. character displacement.

3. The shape of a fish's jaw influences what the fish can eat. Researchers found that the jaws of two fish species were more similar when they lived in separate lakes than when they lived together in the same lake. The increased difference in jaw structure when the fishes live in the same lake may be an example of
   a. warning coloration.
   b. character displacement.
   c. mutualism.
   d. exploitation.

4. Experiments with the barnacle *Semibalanus balanoides* showed that
   a. where this species was found on the shoreline was not influenced by physical factors.
   b. competition with *Chthamalus* restricted *Semibalanus* to high portions of the shoreline.
   c. competition with *Chthamalus* restricted *Semibalanus* to low portions of the shoreline.
   d. this species restricted *Chthamalus* to high portions of the shoreline.

5. Niche partitioning is a way for two species to minimize which of the following community-level interactions?
   a. commensalism
   b. exploitation
   c. mutualism
   d. competition

6. A low-abundance species that has a large effect on the composition of an ecological community is called a
   a. predator.
   b. herbivore.
   c. keystone species.
   d. dominant species.

7. A directional process of species replacement over time in a community is called
   a. global climate change.
   b. succession.
   c. competition.
   d. community change.

8. The conversion of grasslands into desert shrublands in the American Southwest was caused by
   a. excessive water usage.
   b. overgrazing.
   c. the removal of trees.
   d. invasive species.

Analysis and Application

1. Mutualism typically has costs for both of the species involved. Why, then, is mutualism so common?

2. Consumers affect the evolution of the organisms they eat, and vice versa. Explain how this interaction occurs, and illustrate your reasoning with an example described in the text.

3. Rabbits can eat many plants, but they prefer some plants over others. Assume that the rabbits in a grassland containing many plant species prefer to eat a species of grass that happens to be a superior competitor. If the rabbits were removed from the region, which of the following do you think would be most likely to happen?
   a. The plant community would have fewer species.
   b. The plant community would have more species.
   c. The plant community would remain largely unchanged.

   Explain and justify your answer.

4. Describe how each of the following factors influences ecological communities: (a) species interactions; (b) disturbance; (c) climate change; (d) continental drift.

5. Provide an example of how the presence or absence of a species in a community can alter a feature of the environment, such as the frequency of fire.

6. Consider two forms of human disturbance to a forest:
   a. All trees are removed, but the soils and low-lying vegetation are left intact.
   b. The trees are not removed, but a pollutant in rainfall alters the soil chemistry to such an extent that the existing trees can no longer thrive.

   Which form of disturbance do you think would require the longest recovery time before a healthy forest community could once again be found at the site? Explain your assumptions in answering this question, and justify the conclusion you reach.

7. Do you think it is ethically acceptable for people to change natural communities so greatly that it takes thousands of years for the communities to recover? Why or why not?

24 Ecosystems

A MASSIVE OIL SLICK. In 2010, an explosion at the BP *Deepwater Horizon* deep-sea oil-drilling rig in the Gulf of Mexico resulted in a gigantic oil spill that coated thousands of square miles of ocean and destroyed marine life throughout the northern Gulf of Mexico and along 125 miles of coastline.

Deepwater Horizon: Death of an Ecosystem?

On April 20, 2010, an oil-drilling rig in the Gulf of Mexico exploded, killing 11 people and initiating the largest oil spill in U.S. history. A fire on the *Deepwater Horizon* raged for 36 hours before the state-of-the-art oil rig sank, extinguishing the flames. From April to July, when the well leased by BP was finally capped, the gushing well a mile below the surface and 40 miles off the Louisiana coast released nearly 5 million barrels of crude oil. Within days, the oil that floated to the surface spread over thousands of square miles. The well was a rich source of natural gas (which had caused the explosion), and a 22-mile-long underwater plume of oil, gas, and organic solvents such as benzene and xylene spread a half mile beneath the surface of the ocean. Fishery experts predicted major damage to the Gulf shrimp and fishing industries, and ecologists predicted unprecedented damage to the Gulf ecosystem food chains.

The 1989 crash of the *Exxon Valdez* oil tanker in Alaska had spilled only 5 percent as much oil, yet it killed 250,000 seabirds, thousands of marine mammals, and billions of salmon and herring eggs—leading to long-term declines in fish populations that supported humans, as well as sea otters, killer whales, and other marine mammals. Twenty years later, oil still coats Alaskan rocks and sand, and it continues to leach slowly into the ocean, contaminating filter-feeding animals such as mussels, as well as animals farther up the food chain.

Will the long-term damage from the *Deepwater Horizon* oil spill be as bad as that from the *Exxon Valdez* oil spill in Alaska? How are ecosystems affected by such catastrophes?

Knowing all this, Gulf Coast residents in 2010 feared the worst. Coast Guard personnel wept at the destruction, and fishermen said they'd never be able to fish the Gulf again. Before we address the long-term impact of catastrophes such as the *Deepwater Horizon* spill, we will look at how healthy ecosystems function. In particular, we'll see how energy flows through ecosystems, driving the cycling of carbon, nitrogen, water, and other materials. Understanding how healthy ecosystems function will enable us to better understand how an insult like a major oil spill disrupts an ecosystem. And we'll see how much we depend on the services provided by ecosystems.

MAIN MESSAGE Ecosystem ecology studies the interaction between organisms and the physical environment, including the flow of energy through ecosystems, and the cycling of nutrients within them.

KEY CONCEPTS

- An ecosystem consists of a community of organisms together with the physical environment in which those organisms live. Energy, materials, and organisms can move from one ecosystem to another.
- Energy enters an ecosystem when producers capture it from an external source, such as the sun. A portion of the energy captured is lost as heat at each step in a food chain. As a result, energy flows in only one direction through ecosystems.
- Earth has a fixed amount of nutrients. If nutrients were not recycled between organisms and the physical environment, life on Earth would cease. Human activities affect the cycling of some nutrients.
- Ecosystems provide humans with essential services, such as nutrient recycling, at no cost. Our civilization depends on these and many other ecosystem services.
- Human activities frequently damage ecosystem processes, reducing the value of the services that the ecosystem provides. Loss of these services incurs both environmental and economic costs.

DEEPWATER HORIZON: DEATH OF AN ECOSYSTEM? 547

24.1 How Ecosystems Function: An Overview 548

24.2 Energy Capture in Ecosystems 550

24.3 Energy Flow through Ecosystems 552

24.4 Biogeochemical Cycles 554

24.5 Human Actions Can Alter Ecosystem Processes 559

APPLYING WHAT WE LEARNED 564
What Happens When The Worst Happens?

TO SURVIVE, ALL ORGANISMS NEED ENERGY, as well as materials to construct and maintain their bodies. For their energy needs, most organisms depend directly or indirectly on solar energy, an abundant supply of which reaches Earth each day. By contrast, materials such as carbon, hydrogen, oxygen, and other elements are added to our planet in relatively small amounts (in the form of meteoric matter from outer space). Earth, therefore, has essentially fixed amounts of materials for organisms to use. This simple fact means that for life to persist, natural systems must recycle materials. In this chapter we focus on these two essential aspects of life—energy and materials—as we discuss *ecosystem ecology*, the study of how energy and materials are used in natural systems.

24.1 How Ecosystems Function: An Overview

An **ecosystem** consists of communities of organisms together with the physical environment they share. The assemblage of interacting organisms—prokaryotes,

FIGURE 24.1 Ecosystems in the Snake River Valley, Near Jackson Hole, Wyoming
An ecoststem consists of communities of organisms interacting with the physical environment. How many ecosystems can you discern in this photo? What is the energy base in each one? Name some abiotic factors that are likely to affect the organisms living in the ecosystems you identified.

protists, animals, fungi, and plants—constitutes the **biotic** world within an ecosystem ("biotic" means "having to do with life"). The physical environment that surrounds the biotic community—the atmosphere, water, and Earth's crust—is the **abiotic** part of an ecosystem. An ecosystem is therefore the sum of its biotic and abiotic components. Ecosystems do not always have sharply defined physical boundaries. Instead, ecologists recognize an ecosystem by the distinctive ways in which it functions, especially the means by which energy is acquired by the biotic community (**FIGURE 24.1**).

Ecosystems may be small or very large; a puddle teeming with protists is an ecosystem, as is the Atlantic Ocean. Smaller ecosystems can be nested inside larger, more complex ecosystems. In fact, global patterns of air and water circulation (discussed in Chapter 20) may be viewed as linking all the world's organisms into one giant ecosystem, the biosphere.

For a broad overview of how energy and nutrients move through ecosystems, trace the paths of the arrows in **FIGURE 24.2**. **Producers** are organisms that capture light energy and convert it into chemical forms through the process of photosynthesis. **Consumers** obtain their energy and nutrients by eating other organisms. **Decomposers** break down the tissue from once-living organisms to obtain their energy and nutrients. As organisms consume or decompose one another, only a portion of the energy within the food source is available to the consumer. The majority is lost as metabolic heat (red arrow in Figure 24.2). Metabolic heat is a by-product of chemical reactions within a cell—most important, cellular respiration (discussed in Chapter 6). Organisms lose a lot of energy as metabolic heat. Think about how quickly a small room crowded with people warms up. Because of this steady loss of heat, *energy flows in only one direction through ecosystems.* Most of it enters Earth's ecosystems from the sun and leaves as metabolic heat.

Helpful to Know

In discussing the biology of humans and other animals, "nutrients" refers to vitamins, minerals, essential amino acids, and essential fatty acids (see Chapter 27). In the ecosystem context of this chapter, however, "nutrients" has a more restricted meaning, referring only to the essential elements required by producers.

In contrast to energy, **nutrients**—the chemical elements required by living organisms—are largely recycled between living organisms and the physical environment. Although it receives a constant stream of light energy from the sun, Earth does not acquire more nutrients on a daily basis. Finite amounts of chemical elements cycle through the land, water, and air, within and between ecosystems. These elements may pass from rocks and mineral deposits to the soil and water and then to producers, various consumers, and decomposers—and back again to the abiotic world (blue arrows in Figure 24.2). Ecologists and earth scientists describe the passage of chemical elements through the abiotic and biotic worlds as *nutrient cycles* or *biogeochemical cycles.*

The physical, chemical, and biological processes that link the biotic and abiotic worlds in an

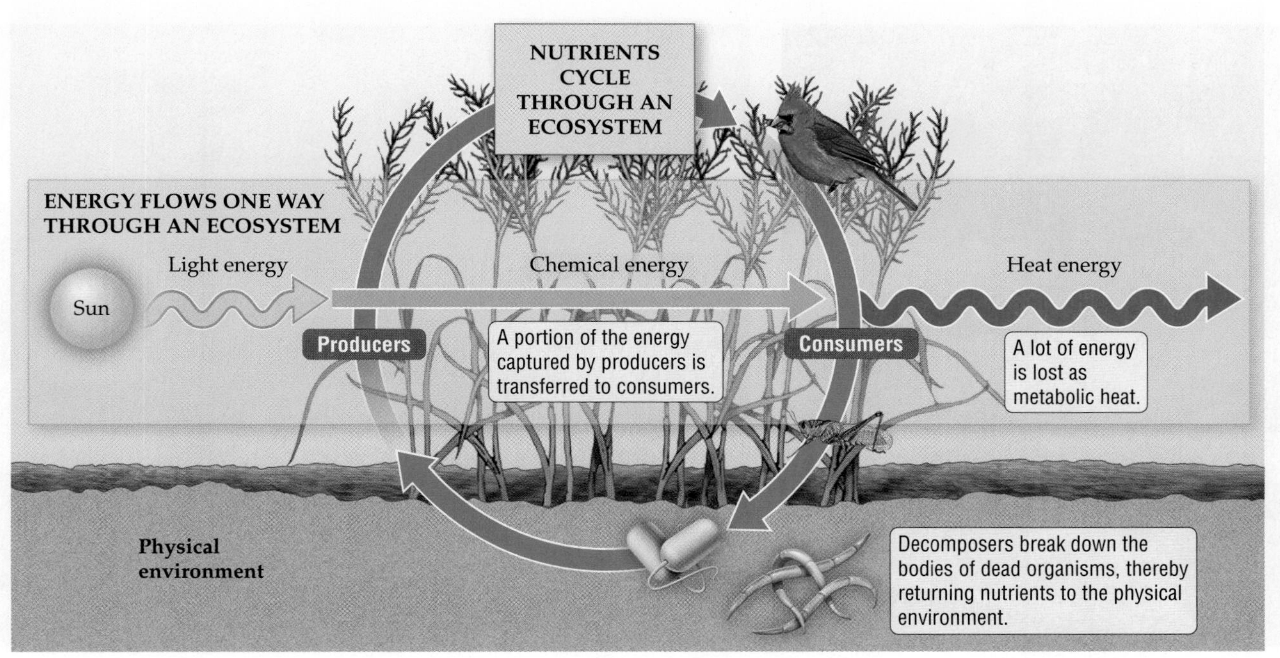
Energy and Nutrient Flow in an Ecosystem
NUTRIENTS CYCLE THROUGH AN ECOSYSTEM
ENERGY FLOWS ONE WAY THROUGH AN ECOSYSTEM
Light energy
Sun
Chemical energy
Heat energy
Producers
A portion of the energy captured by producers is transferred to consumers.
Consumers
A lot of energy is lost as metabolic heat.
Physical environment
Decomposers break down the bodies of dead organisms, thereby returning nutrients to the physical environment.

FIGURE 24.2 How Ecosystems Work
At each step in a food chain, a portion of the energy captured by producers is lost as metabolic heat (indicated by the red arrow). Therefore, energy flows through the ecosystem in a single direction and is not recycled (yellow, orange, and red arrows). In contrast, nutrients such as carbon and nitrogen cycle between organisms and the physical environment (blue arrows).

Concept Check Answers

1. Energy flows through an ecosystem in only one direction. Nutrients generally cycle between living organisms and their physical environment.

2. Producers directly convert the sun's energy into organic forms of chemical energy (carbohydrates). Consumers obtain their energy from the foods they eat.

ecosystem are known as *ecosystem processes.* Energy capture through photosynthesis, release of metabolic heat, decay of biomass through the action of decomposers, and movement of nutrients from living organisms to the abiotic world are examples of ecosystem processes. The activity of producers, in particular, profoundly influences ecosystem processes and therefore the characteristics of an ecosystem (see Figure 24.3). Ecologists often demarcate ecosystems according to the types of producers they contain and the communities of consumers that these producers support. A duckweed-covered pond, a saltwater marsh, a tallgrass prairie, and a beech-maple woodland are all examples of ecosystems that can be defined by the specific types of producers that capture energy and supply it to a characteristic assemblage of consumers.

Concept Check

1. Compare the direction of energy transfer with the movement of nutrients in an ecosystem.
2. What is the distinction between producers and consumers in terms of their acquisition of energy?

24.2 Energy Capture in Ecosystems

Almost all life on Earth depends on the capture of solar energy by producers. There are some exceptions, such as deep-sea vents and hot springs, where the main producers are prokaryotes that can extract chemical energy from inorganic substances that bubble out of Earth's mantle. Here we focus on ecosystems that depend on solar energy, because these are by far the most common in all regions of the world.

Ecosystems depend on energy captured by producers

Plants and other photosynthetic organisms capture energy and store it in the form of chemical compounds, such as carbohydrates. Herbivores (which eat plants and other producers), carnivores (which eat herbivores and other animals), omnivores (which eat everything), and decomposers (which consume the remains of dead organisms) all depend on the solar energy originally captured by plants, photosynthetic bacteria, and algae.

Imagine that all the plants in a terrestrial ecosystem suddenly vanished. Although bathed in sunlight, the animals living there would eventually starve because they could not use the sun's energy to produce food. Without plants the herbivores would perish, leaving the carnivores and omnivores with a dwindling supply of food. This thought experiment should also help you understand why the tropics can support more animals than temperate forests or tundra can. In a tropical forest there is an abundance of sunlight (along with enough rain) to support the growth of many plants (**FIGURE 24.3*a***). As a result, a large amount of

FIGURE 24.3 Producers Are the Energy Base in an Ecosystem
(*a*) In tropical rainforests, the abundant producers (plants) store a lot of chemical energy, which in turn is available to consumers. (*b*) In deserts, because of the sparse plant life, relatively little chemical energy is available for consumers.

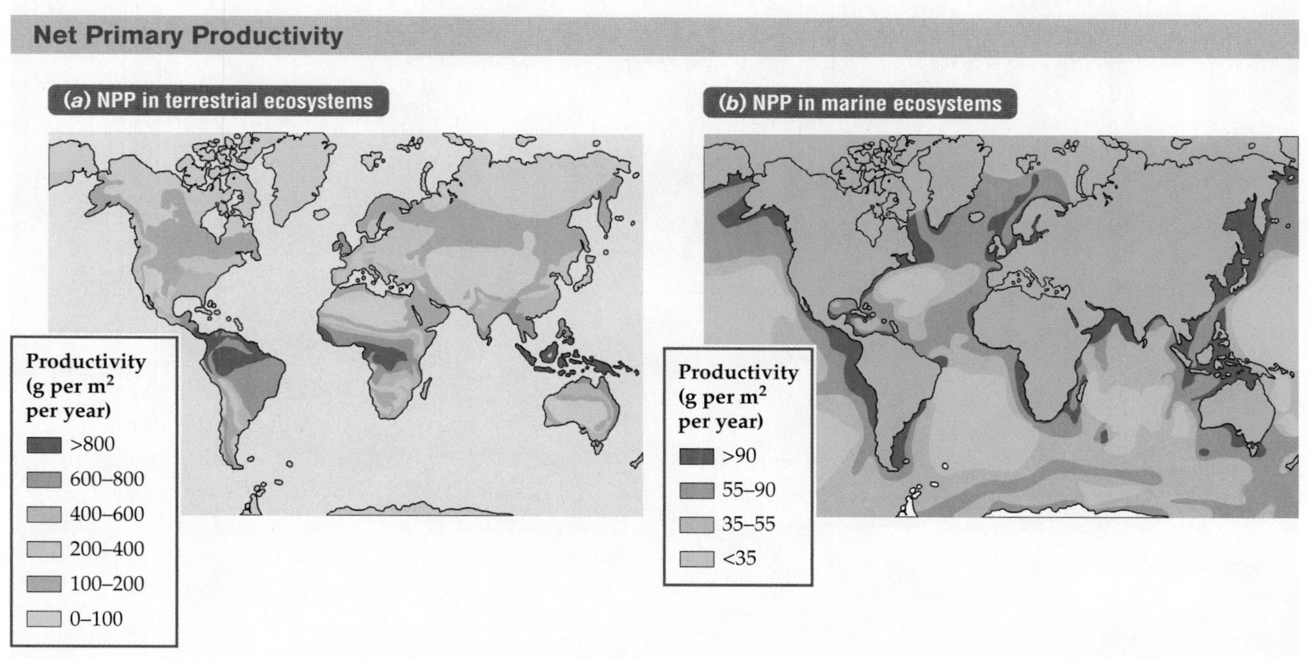

FIGURE 24.4 Net Primary Productivity Varies Greatly in Both Terrestrial and Marine Ecosystems NPP is commonly measured as grams of new biomass made by plants or other producers each year in a square meter of area.

energy from the sun is stored in chemical forms that can be used as food by animals. In contrast, deserts cannot support as many plants, because of the limited amount of available water (**FIGURE 24.3*b***), so relatively little energy is captured from the sun. Hence, less food is available in such environments, and fewer animals can live there.

Assessing the overall amount of energy captured by plants is an important first step in determining how a terrestrial ecosystem works. The amount of energy influences the amount of plant growth and hence the amount of food available to other organisms. Each of these factors, in turn, influences the type of terrestrial ecosystem found in a region and how that ecosystem functions.

The rate of energy capture varies across the globe

The amount of energy captured by photosynthetic organisms minus the amount they expend on cellular respiration and other maintenance processes is called **net primary productivity (NPP)**. You can think of NPP as the energy acquired through photosynthesis that is available for growth and reproduction in a producer. Although NPP is defined in terms of energy, it is usually easier to estimate it as the amount of new **biomass** (the mass of organisms) produced by the photosynthetic organisms in a given area during a specified period of time. In a grassland ecosystem, for example, ecologists would estimate NPP by measuring the average amount of new grass and other plant matter produced in a square meter each year. Such NPP estimates based on biomass can be converted to units based on energy.

NPP is not distributed evenly across the globe. On land, NPP tends to decrease from the equator toward the poles (**FIGURE 24.4*a***). This decrease occurs because the amount of solar radiation available to plants also decreases from the equator toward the poles (as we saw in Chapter 20). But there are many exceptions to this general pattern. For example, there are large regions of very low NPP in northern Africa, central Asia, central Australia, and the southwestern portion of North America. Each of these regions is a site of one of the world's major deserts.

The low NPP in deserts underscores the fact that sunlight alone is not sufficient to produce high NPP; water is also required. In addition to water and sunlight, productivity on land can be limited by temperature and the availability of nutrients in the soil. The most productive terrestrial ecosystems are tropical rainforests; the least productive are deserts and tundra (including some mountaintop communities).

The global pattern of NPP in marine ecosystems (**FIGURE 24.4*b***) is very different from that on land. Marine NPP decreases very little from the equator

FIGURE 24.5 The Most Productive Ecosystems
Wetlands like this estuary in Virginia can harness energy from sunlight and oxygen and translate it into high net primary productivity on a par with terrestrial systems like tropical forests and agricultural fields.

toward the poles. Instead, the productivity of marine ecosystems is typically high in ocean regions close to land and relatively low in the open ocean, which is, in essence, a marine "desert." In all parts of the ocean, the producers live within 150–200 meters of the ocean surface, because that is as far as sunlight penetrates into the water. In coastal areas, wind and ocean currents drive cold, nutrient-rich layers of water to the surface to replace warm, nutrient-depleted water—a process known as **upwelling**. The combination of nutrients and abundant sunlight in coastal surface waters provides ideal conditions for plankton to flourish, contributing to a high NPP in these areas. For example, coral reefs, which are found in warm, sunny, relatively shallow seas, are among Earth's most productive ecosystems, rivaling tropical forests in their output of NPP (see Figure 20.23). Upwelling is largely absent from the deeper open ocean. As a result, the open ocean has a limited supply of the nutrients needed by aquatic photosynthetic organisms. Nutrients released by the death and decay of organisms tend to settle on the deep-ocean floor, so they are not immediately available to photosynthetic producers, which live near the surface.

Streams and rivers also deliver nutrients to coastal areas. Nutrients drained off the land stimulate the growth and reproduction of phytoplankton, the small photosynthetic producers that form the foundation of aquatic food webs. Estuaries—regions where rivers empty into the sea—are some of the most productive habitats on the planet precisely because their rich supply of nutrients supports large populations of producers, which in turn nourish large populations of consumers (**FIGURE 24.5**). Wetlands such as swamps and marshes can also be highly productive. They trap soil sediments rich in nutrients and organic matter, thereby promoting the growth of flood-tolerant plants and phytoplankton, which in turn feed a complex community of consumers.

Concept Check

1. What is net primary productivity? What factors limit NPP in terrestrial ecosystems?
2. Explain why NPP is lower in the open ocean than in coastal areas.

Concept Check Answers

1. Net primary productivity is the amount of new biomass produced by photosynthetic organisms in a given area in a given period of time. NPP in terrestrial ecosystems is limited by temperature, light, moisture, and nutrients.
2. The open ocean is nutrient-poor because the remains of dead organisms settle on the ocean floor, where the nutrients they contain are inaccessible to the producers living thousands of meters above the ocean floor. Coastal areas receive nutrients from upwelling and from rivers and streams that carry nutrients drained off the land. The rich nutrient supply supports photosynthetic producers, giving the ecosystem a high NPP.

24.3 Energy Flow through Ecosystems

As we discussed in the previous section, the sun is the principal source of energy for most living organisms.

Its energy is captured by producers, which in turn pass that energy and nutrients along to the consumers that eat them. In this section we focus on the transfer of energy between organisms in an ecosystem, and in Section 24.4 we will look at how nutrients cycle through ecosystems.

Food chains transfer energy through an ecosystem

A **food chain** describes a linear sequence of who eats whom. Most dietary relationships, however, are not simply linear. Ecologists often refer instead to **food webs**, which show the various food chains of a community, where they overlap, and how they are connected (**FIGURE 24.6**). Food chains help us to see the dietary relationships between organisms, as well as the passage of energy within the biotic components of ecosystems. Energy flows into ecosystems as producers convert solar energy into organic forms, such as glucose. As consumers eat organisms lower on the food chain, the energy contained in those prey species is passed upward. Eventually, however, every unit of energy captured by producers is lost from the biotic part of the ecosystem as heat. Therefore, energy cannot be recycled within an ecosystem.

Consumers obtain energy by eating all or parts of other organisms or their remains. But not all consumers are the same. In the food chain highlighted in Figure 24.6, krill are consumers that eat producers directly. These **primary consumers** are one step removed from producers. **Secondary consumers** are organisms, such as the seal in Figure 24.6, that feed on primary consumers. This sequence of organisms eating organisms that eat other organisms can continue: a killer whale that eats seal that ate krill that ate plankton is an example of a **tertiary consumer**. The different steps of a food chain are called **trophic levels** and reflect the number of energy conversions that occur from the ultimate source of energy (usually the sun).

An energy pyramid shows the amount of energy transferred up a food chain

When primary consumers eat producers, and then secondary consumers eat the primary consumers, the energy contained in the prey is transferred up the food chain. However, the transfer of energy is never 100

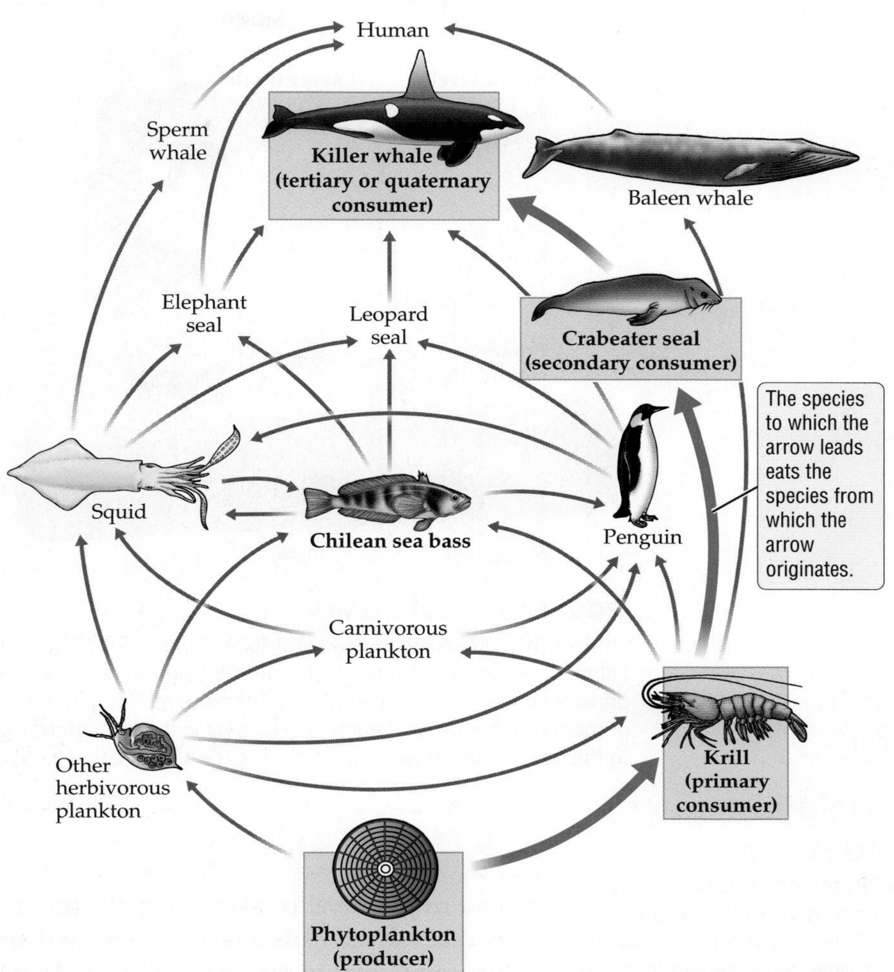

FIGURE 24.6 Food Webs Summarize the Movement of Energy through an Ecosystem

Food webs are composed of many specific sequences, known as food chains, that show one species eating another. To make it easier to follow a single sequence of feeding relationships, one of the food chains in this food web is highlighted with red arrows and orange boxes.

percent efficient. In fact, on average, only about 10 percent of the energy at one trophic level is transferred to the next trophic level. In other words, only about 10 percent of the energy captured by the grasses is transferred to the herbivores that eat the grasses, and then 10 percent of the energy in the

EXTREME INEFFICIENCY

A panda may spend 12–16 hours a day eating. Pandas are descended from carnivores and have the digestive systems of meat eaters, which makes it difficult for them to efficiently extract energy from their bamboo diet.

An Energy Pyramid

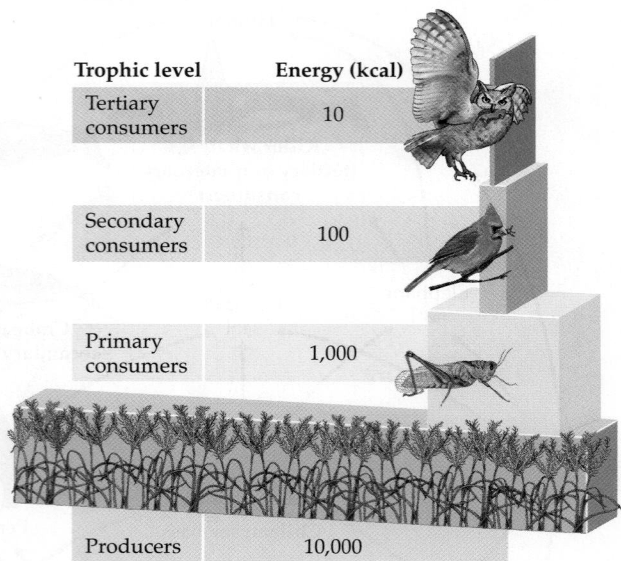

Trophic level
Energy (kcal)
Tertiary consumers
10
Secondary consumers
100
Primary consumers
1,000
Producers
10,000

FIGURE 24.7 Trophic Levels

Of each 10,000 kilocalories (kcal; 1 kilocalorie = 1,000 calories) of energy from the sun captured by producers, primary consumers store only about 10 percent. Roughly 10 percent of the energy at each trophic level is then transferred to the next trophic level.

Helpful to Know

"Reservoir" has two meanings. It can refer to a body of water held aside for use by people. More generally, a reservoir is any place or substance that holds something (not necessarily water) in storage—which is what we mean with regard to nutrient cycling.

second trophic level is taken up by the predators that eat the herbivores. The energy that is not transferred to the next trophic level was not consumed (for example, when we eat an apple we eat only a small part of the apple tree), was not taken up by the body (for example, we cannot digest the cellulose that is contained in the apple), or was lost as metabolic heat.

As we follow the movement of energy up a food chain, we see that the available energy at each trophic level decreases. The energy obtained by primary consumers passes through only two conversions (source to producer, and producer to primary consumer). On the other hand, for energy to reach tertiary consumers, it must pass through an additional two conversions, with losses of energy at each step. The amounts of energy available to organisms in an ecosystem are often represented in what is called an **energy pyramid**. **FIGURE 24.7** shows an energy pyramid containing four trophic levels; a grassland is on the first level, grasshoppers are on the second, insect-eating birds are on the third, and bird-eating birds are on the fourth. Note that each trophic level contains only about 10 percent of the energy of the level below. Top-level consumers have far less available energy than those feeding at lower trophic levels.

Concept Check Answers

1. Of the energy present in primary producers, 10 percent will find its way to the primary consumers, 1 percent to the secondary consumers, and only 0.1 percent to the tertiary consumers.
2. It is not consumed, is not absorbed, or is lost as metabolic heat.
3. Primary producers will have the greater biomass because more energy is available at the lower trophic levels.

Decomposers return nutrients to the abiotic world

Some organisms, called decomposers, obtain their energy from the remains of dead organisms. Organic tissues contain high-energy molecules that can serve as a source of energy for these organisms (**FIGURE 24.8**). In some ecosystems, 80 percent of the biomass produced by plants is used directly by decomposers. Eventually, since all organisms die, all biomass made by producers, herbivores, predators, and parasites is consumed by decomposers. In some instances, people bypass the decomposers, as when we use crops or agricultural refuse to produce fuels.

Concept Check

1. How much energy present in primary producers will find its way up the food chain to the tertiary consumers?
2. If only about 10 percent of the energy from one trophic level is transferred to the next, where does the rest of it go?
3. In a given ecosystem, which trophic level will have the greater biomass: primary producers or primary consumers? Why?

24.4 Biogeochemical Cycles

All living things are made up of a limited set of chemical elements, which in ecological terms we call *nutrients*. The main nutrients are carbon, hydrogen, nitrogen, oxygen, and phosphorus. Producers obtain these and other essential nutrients from the soil, water, and air in the form of ions such as nitrate (NO_3^-) and inorganic molecules such as carbon dioxide (CO_2). Consumers obtain them by eating producers or other consumers. Because these nutrients are essential for life, their availability and movement through ecosystems influence many aspects of ecosystem function.

Nutrients are transferred between organisms (the biotic community) and the physical environment (the abiotic world) in cyclical patterns called **nutrient cycles** or **biogeochemical cycles**. **FIGURE 24.9** illustrates how nutrient cycles work in general.

Nutrients can be stored for long periods of time in abiotic reservoirs such as rocks, ocean sediments, or fossilized remains of organisms. The nutrients stored in such abiotic reservoirs are not readily accessible to producers. Weathering of rocks, geologic uplift, human actions, and other forces can move nutrients back and forth between these reservoirs and **exchange pools**, abiotic sources such as soil, water, and air where nutrients *are* available to producers.

Once captured by a producer, a nutrient can be passed from the producer to an herbivore, then to one or more predators or parasites, and eventually to a decomposer. Decomposers break down once-living tissues into simple chemical components, thereby returning nutrients to the physical environment. Without decomposers, nutrients could not be repeatedly reused, and life would cease because all essential nutrients would remain locked up in the bodies of dead organisms.

Abiotic conditions, especially temperature and moisture, influence the length of time it takes for a nutrient to cycle from the biotic community to the exchange pool and back again to a producer. Warmer temperatures, for instance, increase the activity of decomposers, which in turn speeds up nutrient release from biomass. By influencing such processes as weathering and runoff, rainfall may control nutrient release from nutrient reservoirs and also determine whether significant amounts of nutrients are lost from one ecosystem (a forest community, for example) to another (such as a stream).

How long it takes for a nutrient to cycle from a producer to the physical environment and back to another producer depends on the element. Elements that have an **atmospheric cycle**, such as carbon, hydrogen, oxygen, nitrogen, and sulfur, can be transported over great distances and often very rapidly. Once in the atmosphere, air currents move these nutrients from one region of Earth to another, where they can affect nutrient cycles in distant ecosystems. Phosphorus is one of the few major nutrients that does not have an atmospheric cycle, but rather displays a **sedimentary cycle**, moving mostly through land and water, rather than the atmosphere. Nutrients with sedimentary cycles tend to move slowly and are not dispersed as widely as those with atmospheric cycles.

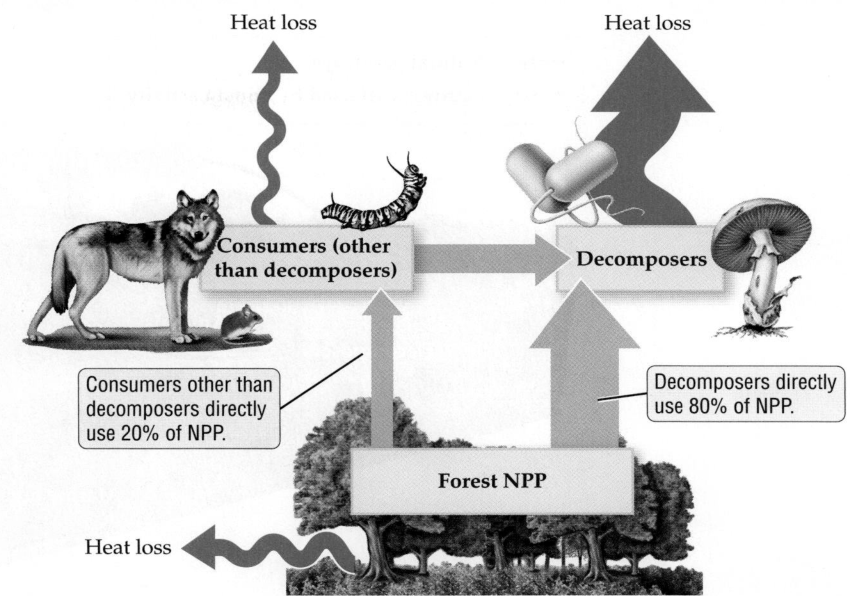

FIGURE 24.8 Decomposers Return Nutrients to the Environment
Without decomposers, not only would nature's waste pile up, but the soil would become devoid of nutrients. Decomposers such as bacteria and fungi recycle more than 50 percent of net primary productivity in ecosystems of all types. In the forest represented here, 80 percent of NPP is used directly by decomposers, and the remaining 20 percent is used by other consumers (such as herbivores and predators).

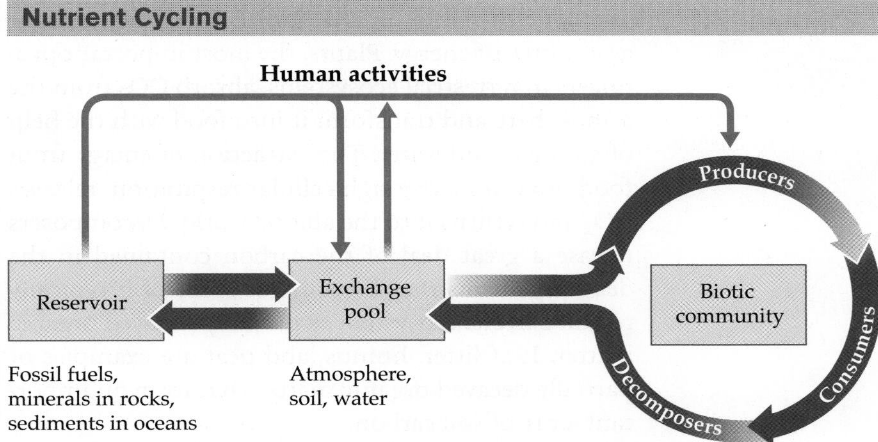

FIGURE 24.9 Nutrient Cycling
Nutrients cycle among reservoirs inaccessible to producers, exchange pools that are available to producers, and living organisms. Human activities, such as fossil fuel use and fertilizer synthesis, can move nutrients between reservoirs and exchange pools, altering their availability to producers and changing nutrient cycles.

Carbon cycling between the biotic and abiotic worlds is driven by photosynthesis and respiration

Living cells are built mostly from organic molecules, which are molecules that contain carbon atoms bonded to hydrogen atoms. The process of photosynthesis converts inorganic carbon in the form of atmospheric CO_2 into organic molecules. The transfer of carbon within biotic communities, between living organisms and their physical surroundings, and within

FIGURE 24.10
The Carbon Cycle

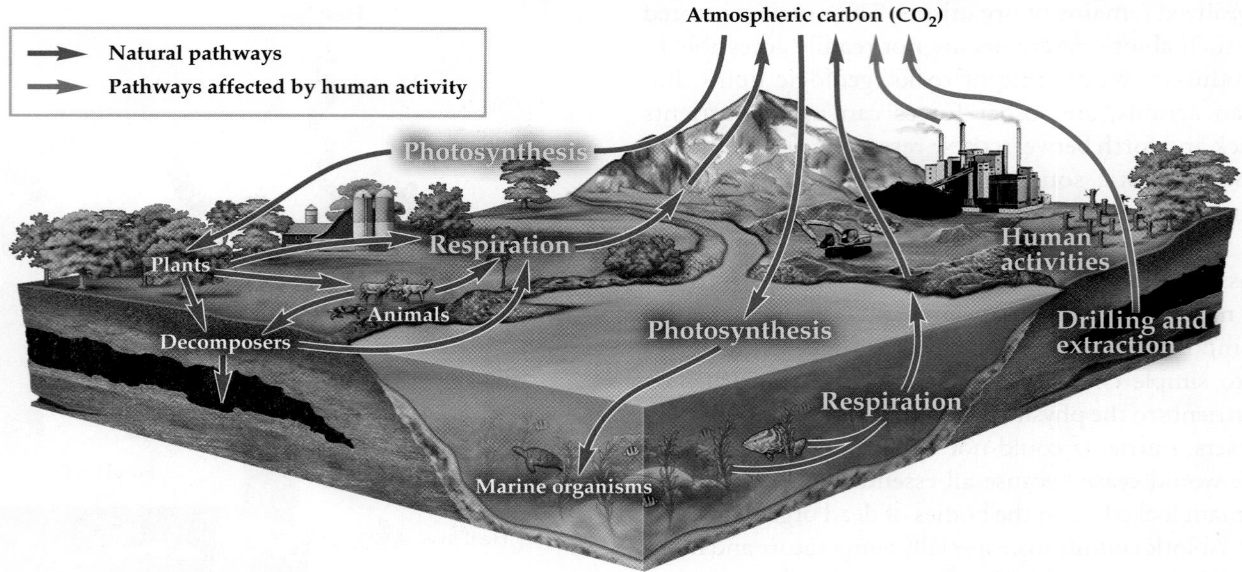

the abiotic world is known as the global **carbon cycle** (**FIGURE 24.10**).

Living organisms, in both aquatic and terrestrial ecosystems, acquire carbon mostly through photosynthesis. Aquatic producers, such as photosynthetic bacteria and algae, can absorb dissolved carbon dioxide (in the form of bicarbonate or carbonate ions) and convert it into organic molecules using sunlight as a source of energy. Plants, the most important producers in terrestrial ecosystems, absorb CO_2 from the atmosphere and transform it into food with the help of sunlight and water. The extraction of energy from food molecules, through cellular respiration, releases CO_2 and returns it to the abiotic world. Decomposers release a great deal of the carbon contained in the dead organisms they live on, but some of it typically remains in the ecosystem as partially decayed organic matter. Leaf litter, humus, and peat are examples of partially decayed organic matter that form an important store of soil carbon.

Although carbon makes up a large part of biomass, the element is not abundant in the atmosphere. Carbon, in the form of CO_2 gas, makes up only about 0.04 percent of Earth's atmosphere, although that percentage has been creeping upward every year for the last 100 years or so (the effect of increasing CO_2 on global warming is covered in Chapter 25). The oceans represent the largest store of carbon on our planet. Most of this carbon is dissolved inorganic carbon (such as bicarbonate ions, HCO_3^-), and a minor portion is held in the biomass of marine organisms. Earth's crust generally contains only about 0.038 percent carbon by weight, but it also has regions with carbon-rich sediments and rocks formed from the remains of ancient marine and terrestrial organisms (see Figure 24.10). Some of the organic matter from ancient organisms has been transformed by geologic processes into deposits of fossil fuel, such as petroleum, coal, and natural gas. When we extract these fossil fuels and burn them to meet our energy needs, the carbon locked in these deposits for several hundred million years is released into the atmosphere as carbon dioxide (shown by a red arrow in Figure 24.10).

Biological nitrogen fixation is the most important source of nitrogen for biotic communities

Nitrogen is a key component of amino acids, proteins, DNA, and RNA. It is therefore an essential element for all living organisms. The fact that this nutrient is not naturally abundant in soil and water limits the growth of producers in most ecosystems.

Earth's atmosphere is the largest store of nitrogen in the biosphere: N_2 gas (molecular nitrogen) makes up 78 percent of the air we breathe. However, this gaseous form of nitrogen is not useful to most living organisms. Most of the nitrogen in soil and water that is available for plants to use originates in a remarkable metabolic process known as biological **nitrogen fixation**, in which bacteria convert molecular nitrogen (N_2) into ammonium ions (NH_4^+) (**FIGURE 24.11**).

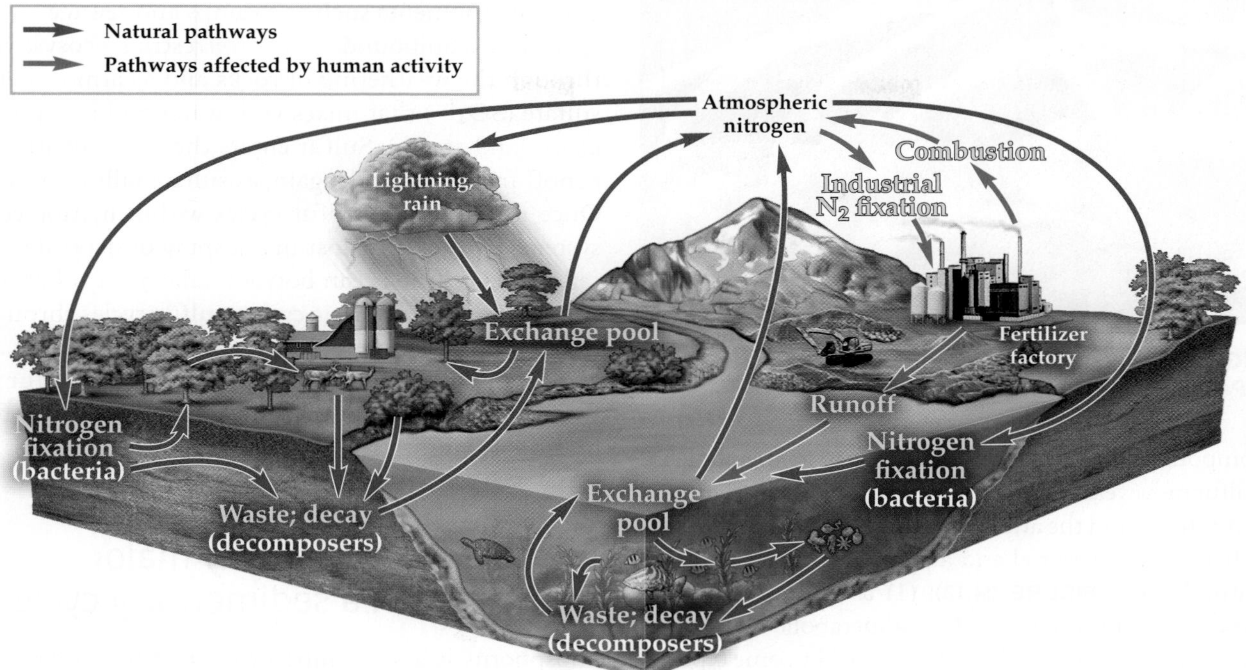

FIGURE 24.11
The Nitrogen Cycle

Some nitrogen-fixing bacteria live free in the soil, and others fix nitrogen in a mutualistic symbiosis with certain plants, including legumes (members of the bean family), as pictured in **FIGURE 24.12**; alder trees; and an aquatic fern called *Azolla*. The plant host obtains ammonium from its nitrogen-fixing partner and supplies food energy derived from photosynthesis to the bacteria. Nitrogen fixed by bacteria is added to soil or water when these organisms, and any host species they might associate with, die and decompose. Consumers acquire nitrogen when they eat plants that have acquired nitrogen from the soil or through their association with nitrogen-fixing bacteria. Decomposers transfer nitrogen from the biotic community to the abiotic component of an ecosystem.

Energy from lightning converts a small amount of atmospheric N_2 into nitrogen compounds that mix with rainwater to form nitrate ions (NO_3^-). Nitrate ions are the most common form of nitrogen in soil and water, but nitrate is highly soluble and therefore is easily lost from an ecosystem through runoff. Waste produced by animals, and the death and decay of organisms, adds ammonium ions (NH_4^+) to soil and water (see Figure 24.11). Bacteria rapidly convert ammonium ions to nitrate, which is why ammonium ions do not accumulate in most habitats. Denitrifying bacteria, common in oxygen-poor environments, convert nitrate into molecular nitrogen (N_2) or into N_2O (nitrous oxide), returning nitrogen to the vast atmospheric pool.

Humans manufacture fertilizer through industrial nitrogen fixation, a process that combines nitrogen gas (N_2) with hydrogen gas (H_2) to make ammonium compounds. The manufacturing process relies on high temperature and pressure generated by the burning of fossil fuels. Applying nitrogen fertilizer to crop plants boosts their productivity, but the runoff can stimulate NPP in aquatic ecosystems in ways that are harmful to the biotic community. Excess nitrogen can disturb terrestrial communities as well, sometimes decreasing diversity (species number and abundance). This shift in diversity has been noted in some grassland communities, in which a few species that responded with exceptional vigor to supplemental nitrogen were able to outcompete the other species.

Sulfur is one of several important nutrients with an atmospheric cycle

Sulfur is a component of certain amino acids, many proteins, some polysaccharides and lipids, and other organic

EXTREME NITROGEN

On April 17, 2013, a fertilizer plant in the town of West, Texas, exploded with a force so great that it registered 2.1 on the Richter scale. Although potentially hazardous, synthetic nitrogen production is commonplace for the production of explosives and fertilizer. Human activities now contribute more nitrogen to the atmosphere than do all natural sources combined.

FIGURE 24.12 Nitrogen-Fixing Nodules on the Roots of a Pea Plant

compounds that have important roles in metabolism. Sulfur moves easily among terrestrial ecosystems, aquatic ecosystems, and the atmosphere. Sulfur enters the atmosphere from terrestrial and aquatic ecosystems in three natural ways (**FIGURE 24.13**): (1) as sulfur-containing compounds in sea spray; (2) as a metabolic by-product (the gas hydrogen sulfide, H_2S) released by some types of bacteria; and (3) as a result of volcanic activity.

About 95 percent of the sulfur that enters the atmosphere from the world's oceans does so in the form of strong-smelling sulfur compounds (such as dimethyl sulfide) that are breakdown products of organic molecules made by phytoplankton. The odorous compounds are lofted into the air by wave action and contribute to the smell we associate with seaside air. Hydrogen sulfide, another smelly gas, is generated by metabolic reactions in bacteria that inhabit oxygen-poor environments such as swamps and sewage.

Sulfur compounds enter terrestrial ecosystems through the weathering of rocks and as atmospheric sulfate (SO_4^{2-}) that mixes with water and is deposited on land as rain. Sulfur enters the ocean in stream runoff from land and, again, as sulfate falling as rain. Once in the ocean, sulfur cycles within marine ecosystems before being lost in sea spray or deposited in sediments on the ocean bottom. Like most other nutrients with atmospheric cycles, sulfur cycles through terrestrial and aquatic ecosystems relatively quickly. As we will see shortly, human activities also add sulfur to the atmosphere, often with adverse consequences for biotic communities and human economic interests.

Phosphorus is the only major nutrient with a sedimentary cycle

Phosphorus is a vital nutrient because it is a component of DNA and RNA. Phosphorus strongly affects net primary productivity, especially in aquatic ecosystems. NPP usually increases, for example, when phosphorus is added to lakes. As we will see in Section 24.5, when such an increase in productivity occurs too rapidly it can lead to the death of aquatic plants, fish, and invertebrates.

Among the major nutrients that cycle within ecosystems, phosphorus is the only one with a sedimentary

The Sulfur Cycle

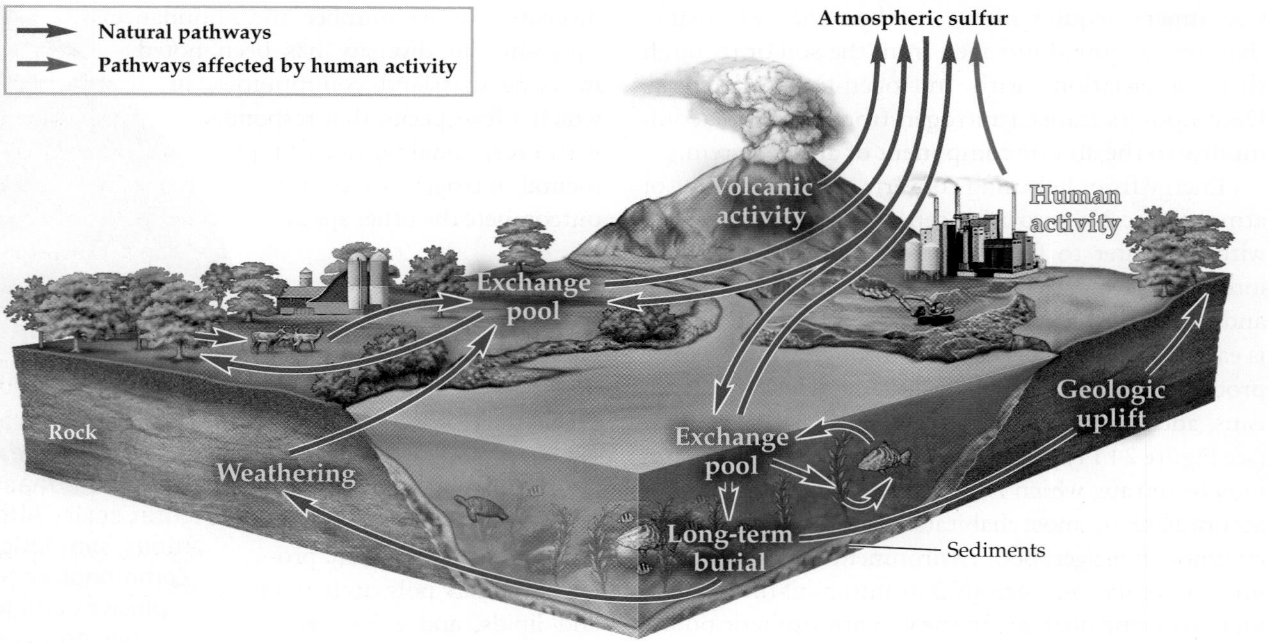

FIGURE 24.13
The Sulfur Cycle

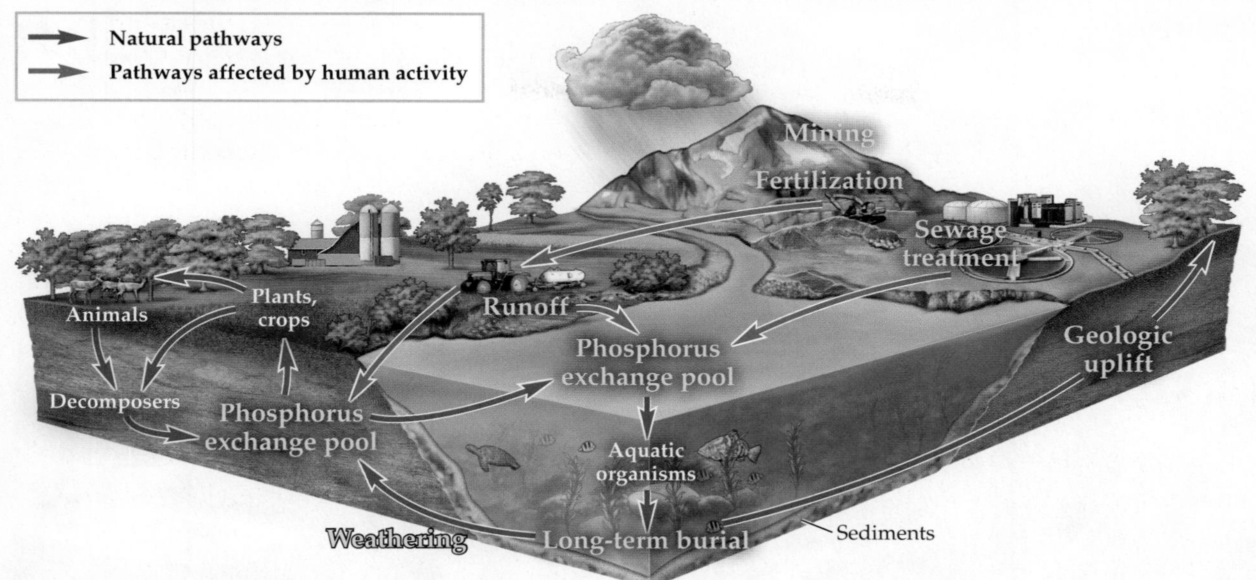

FIGURE 24.14 The Phosphorus Cycle

cycle (**FIGURE 24.14**). Nutrients like phosphorus first cycle within terrestrial and aquatic ecosystems for variable periods of time (from a few years to many thousands of years); then they are deposited on the ocean bottom as sediments, unavailable to most organisms for hundreds of millions of years. Eventually, however, the bottom of the ocean is thrust up by geologic forces to become dry land, and once again the nutrients in the sediments may be available to organisms. Sedimentary nutrients usually cycle very slowly, so they are not replaced easily once they are lost from an ecosystem.

Concept Check

1. What are fossil fuels? How can the consumption of fossil fuels influence the carbon cycle?
2. Plant fertilizers are often rich in nitrogen and phosphorus. What aspects of these nutrient cycles make them especially important for enriching soil?

24.5 Human Actions Can Alter Ecosystem Processes

Humans have been disrupting ecological communities for many hundreds, perhaps thousands, of years. The history of the Moai-building Polynesian settlers of Easter Island (see Chapter 21) demonstrates that such disruptions can have tragic consequences for humans. The disruptions of preindustrial times are dwarfed, however, by the scale of the ecological changes that humans have brought about in the past 200 years. In this section we examine some of the ecosystem processes that are readily altered by human activities, reserving a broader discussion of the human impact on the biosphere for the next chapter.

Human activities can increase or decrease NPP

Human activities can change the amount of energy captured by ecosystems on local, regional, and global scales. For example, rain can cause fertilizers to wash from a farm field into a stream, which then flows into a lake. The addition of extra nutrients to lakes or streams or offshore waters leads to **eutrophication** (literally, "overfeeding"; from *eu*, "well"; *troph*, "nourished"). In a eutrophic lake, NPP increases because the added nutrients cause photosynthetic algae to become more abundant, so they absorb more energy from the sun.

It is not necessarily a good thing when human actions increase NPP. Nutrient-rich waters provide fertile environments for algal population, which rapidly multiply. As the algae die, they drift into deeper waters, where bacteria decompose their bodies. A larger-then-normal "rain" of dead algae triggers a population boom among decomposers, which utilize all of the available dissolved oxygen in the water (**FIGURE 24.15**). Oxygen levels may drop so low that virtually all animals die or flee, creating a large "dead zone."

Concept Check Answers

1. Fossil fuels, such as petroleum, coal, and natural gas, are repositories of organic carbon that were formed from the remains of ancient organisms. Extracting and burning these fossil fuels releases carbon into the atmosphere that had been locked away by photosynthesis for millions of years.
2. The nitrogen cycle relies on bacteria for converting inorganic, atmospheric nitrogen into organic forms. Phosphorus has a sedimentary cycle, relying on geologic processes to replenish terrestrial supplies. The slow rate of natural replenishment for both of these nutrients means that farmers seeking to enrich soils need to add both nitrogen and phosphorus.

FIGURE 24.15 Eutrophication Can Be Too Much of a Good Thing

Large inputs of nutrients such as nitrogen and phosphorus can cause a population boom in photosynthetic plankton (algae). Dying algae in turn trigger explosive growth of oxygen-consuming decomposers, choking off the oxygen supply for animals.

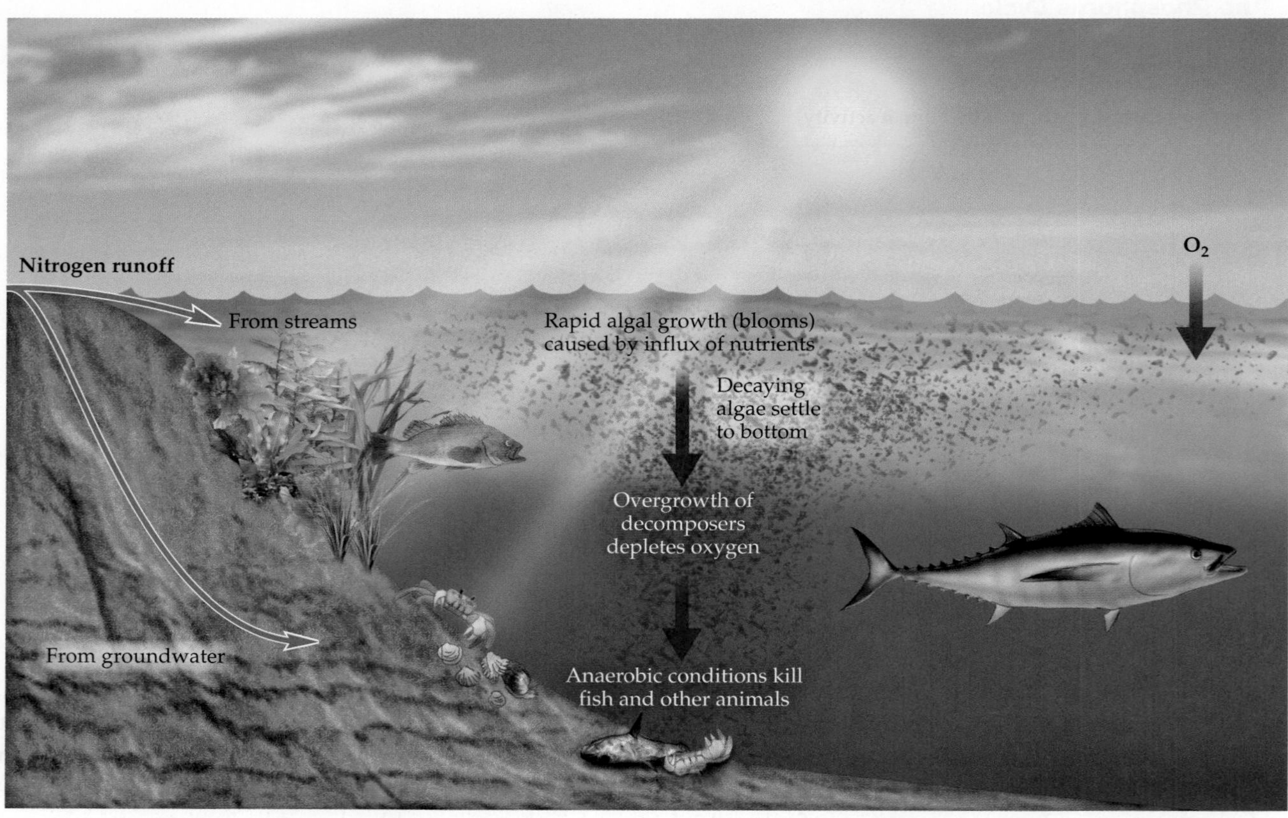

Each summer, large amounts of nitrogen and phosphorus are carried by the Mississippi and other rivers to the Gulf of Mexico (**FIGURE 24.16**). This eutrophication creates a widespread dead zone that threatens to diminish the fish and shellfish industry in the Gulf, which produces an annual catch worth about $500 million. In the summer of 2002, the dead zone reached its largest size ever—about 22,000 square kilometers (8,500 square miles)—covering an area greater than the state of Massachusetts.

Human activities can also change NPP on land. For example, NPP decreases when logging and fire convert tropical forest to grassland. Globally, scientists estimate that human activities leading to such land conversions have decreased NPP in some regions while increasing it in other regions, but the net effect is a 5 percent decrease in NPP worldwide.

Helpful to Know

Deforestation has significant implications for the global carbon cycle. The Amazon rainforest, the world's largest forest, is experiencing rapid rates of deforestation. As trees are cut, the biological material contained within them decomposes or is burned. The carbon from these trees is then released into the atmosphere as CO2. Recent reports suggest that deforestation rates may exceed the total photosynthetic activity of the forest, so that the Amazon is now a net emitter of carbon dioxide.

Human activities can alter nutrient cycles

Human activities can have major effects on nutrient cycles. Ecologists have shown, for example, that clear-cutting a forest, followed by spraying with herbicides to prevent regrowth, causes the forest to lose large amounts of nitrate, an important source of nitrogen for plants (**FIGURE 24.17**).

Many human activities release chemicals into the water or air, which then are moved over long distances. Consider sulfur dioxide (SO_2), which is released into the atmosphere when we burn fossil fuels such as oil and coal. Most human inputs of sulfur into the atmosphere come from heavily industrialized areas such as central Europe and eastern North America. Burning fossil fuels has greatly altered the sulfur cycle: annual human inputs of sulfur into the atmosphere are more than one and a half times the inputs from all natural sources combined. Once in the atmosphere, SO_2 combines with oxygen and water to become sulfuric acid (H_2SO_4), which then returns to the land in rainfall. Rainfall normally has a pH of 5.6, but sulfuric acid (as well as nitric acid, HNO_3, caused by nitrogen-containing pollutants) has caused the pH of rain to drop to values as low as 2 or 3 in the United States, Canada, Great Britain, and Scandinavia (see Chapter 2 for a review of pH). Rainfall with a low pH is called **acid rain**.

Dead Zone in the Gulf of Mexico

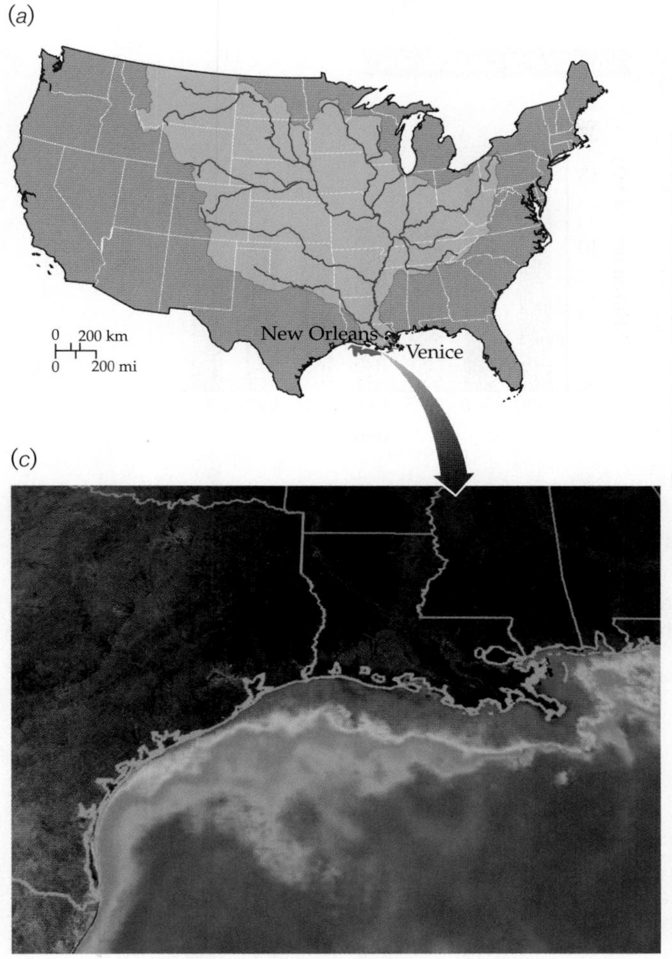

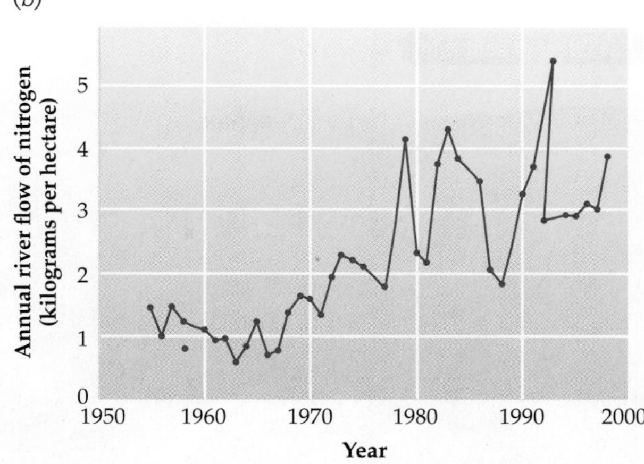

FIGURE 24.16 Nutrient Loading Can Deplete Oxygen in Aquatic Ecosystems

The Mississippi and Atchafalaya river basins drain nearly 40 percent of U.S. land area (*a*). Since the 1970s, large amounts of nitrogen (N) from fertilizer, sewage, and industrial by-products have entered waters within these watersheds. The resulting nutrient addition can be measured as the amount of nitrogen flowing past particular points on the Mississippi per year (*b*). This nitrogen then drains into the Gulf of Mexico, where each summer the extra nutrients create a large dead zone (*c*) in which virtually all animals are killed. The image, based on satellite photos from NASA, shows the concentrations of sediment and chlorophyll in the ocean. Red and orange represent high concentrations of phytoplankton and river sediment.

Vegetation Reduces Nutrient Runoff

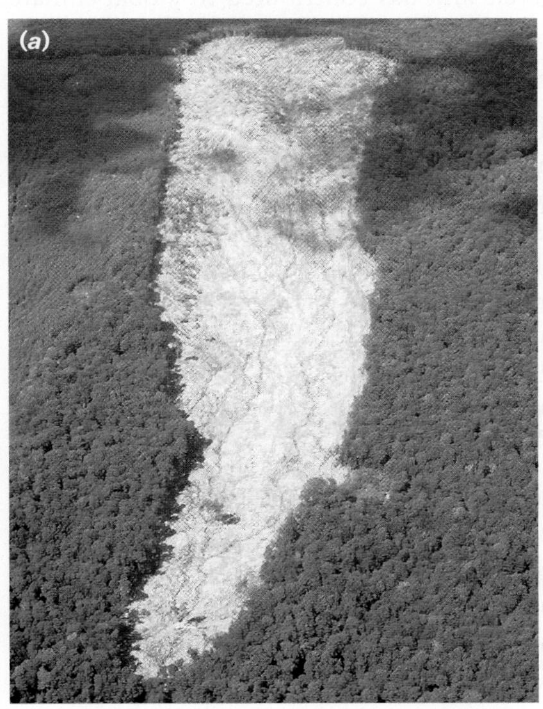

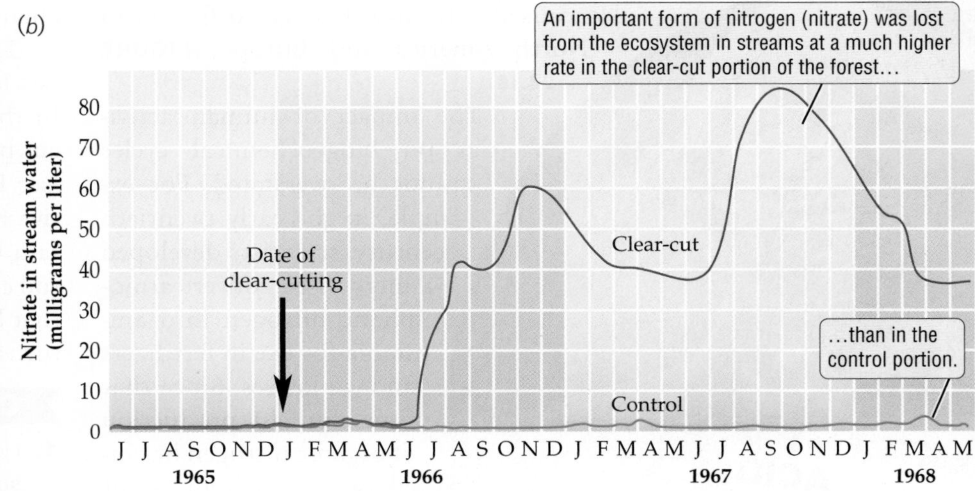

FIGURE 24.17 Altering Nutrient Cycles in a Forest Ecosystem

(*a*) This portion of the Hubbard Brook Experimental Forest in New Hampshire has been clear-cut. (*b*) A portion of the forest was first clear-cut and then sprayed with herbicides for 3 years to prevent regrowth. A second portion of the forest, which was not clear-cut or sprayed, served as a control plot. Nitrate, a form of nitrogen that is important to plants, was lost from the ecosystem in streams at a much higher rate in the clear-cut portion of the forest than in the control portion. (The loss of nitrate was measured in milligrams of nitrate in the stream per liter of water.)

Acid Rain

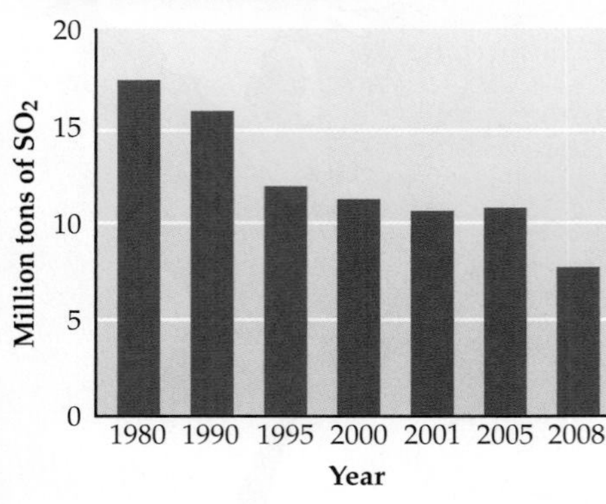

FIGURE 24.18 Acid Rain Can Damage Many Ecosystems
(*a*) This spruce forest in the Jizerské Mountains of the Czech Republic was killed by acid rain. (*b*) Thanks to regulations imposed by the 1990 Clean Air Act, the amount of sulfur dioxide, a major contributor to acid rain, emitted into the atmosphere each year in the United States has fallen by more than 50 percent since 1980.

Concept Check Answers

1. Nutrients in fertilizer and sewage are delivered by the Mississippi River to the Gulf of Mexico, where they stimulate NPP and an algal bloom. Oxygen-consuming decomposers feed on algae that die and settle in deeper waters; dissolved oxygen is depleted as the decomposer population soars, depriving invertebrates and fish of essential oxygen and causing them to die in large numbers.

2. Acid rain is caused when sulfur and nitrogen pollutants combine with water to form sulfuric and nitric acids. These pollutants are associated with areas of heavy industrialization.

Acid rain can have devastating effects on human-made structures (such as statues) and on natural ecosystems. Acid rain has drastically reduced fish populations in thousands of Scandinavian and Canadian lakes. Much of the acid rain that falls in these lakes is caused by sulfur dioxide pollution that originates in other countries (such as Great Britain, Germany, and the United States). Acid rain has also caused extensive damage to forests in North America and Europe (**FIGURE 24.18**).

The impact of human activities on biogeochemical cycles cannot be overstated. For example, in the early twentieth century, scientists developed a method to convert atmospheric nitrogen into ammonia for use in fertilizers and explosives. Since that time, human production of organic nitrogen has increased to such an extent that we now artificially fix more nitrogen than do all natural sources combined. Likewise, humans have mined phosphorus and sulfur so rapidly that the high levels of these elements in ecosystems are unparalleled in Earth's history. As we have seen, the disruption of these nutrient cycles has led to phenomena such as eutrophication and acid rain. Perhaps most important, the massive consumption of fossil fuels during the past century has significantly affected the carbon cycle and has contributed to global climate change, a topic we will visit in the next chapter.

The international nature of the acid rain problem has led nations to agree to reduce sulfur emissions. In the United States, annual sulfur emissions were cut by more than 50 percent between 1980 and 2008 (see Figure 24.18*b*). Such reductions are a very positive first step, but the problems resulting from acid rain will be with us for a long time: acid rain alters soil chemistry and therefore has effects on ecosystems that last for many decades after the pH of rainfall returns to normal levels.

EXTREME ACID

Acid rain can be very corrosive. This photo shows a statue of George Washington in Washington Square Park, New York City. The statue was erected in 1944, and its decay illustrates the effects of acid rain on buildings and monuments.

Concept Check

1. How does runoff from farm fields and outflow from sewage treatment facilities contribute to the "dead zone" in the Gulf of Mexico?
2. How does acid rain form? What human factor strongly influences acid rain production?

Is There a Free Lunch? Ecosystems at Your Service

Next time you drink a glass of water, think for a moment about where your water comes from. If you are like many of us, you may not know. Does it come from surface waters, such as rivers, lakes, or reservoirs? Or does it come from deep underground? Consider New York City. The 8 million people who live there get about 90 percent of their water from the Catskill Watershed, with the remainder coming from the Croton Watershed. Together, these watersheds store 580 billion gallons of water in 19 reservoirs and 3 controlled lakes. Over 1.3 billion gallons of this water is delivered to New York each day. The water flows by gravity to the city in a vast set of pipes, some of them large enough to drive a bus through.

For years, New Yorkers drank high-quality water, essentially for free; their water was kept pure by the root systems, soil microorganisms, and natural filtration processes of forests in the Catskill and Croton watersheds. By the late 1980s, however, pollutants such as sewage, fertilizers, pesticides, and oil had begun to overwhelm these purification processes, causing the quality of the water to decline. In the early 1990s the city embarked on an ambitious but simple plan: protect the watershed's environment so that natural systems could resume supplying the city with clean water, for about a tenth of the cost that a conventional water treatment project would have cost. The city is buying land that borders rivers and streams in the Catskills, and protecting the land from development to minimize the flow of fertilizers, pesticides, and other pollutants into the water. New York City's decision to invest in the long-term health of the watershed shows that what is good for the environment can also be sound economic policy.

Ecological communities serve human communities in varied and valuable ways. The ecosystem processes and resources that benefit humankind are called *ecosystem services*. For example, floodplains provide us with a free service: they act as safety valves for major floods, preventing even larger floods. Floodplains normally function as huge sponges: when streams and rivers overflow, floodplains absorb the excess water, thereby preventing even more severe floods from occurring farther downstream. Dikes and levees are often constructed, and rivers are often diverted in an attempt to protect homes or industrial areas located in what were once floodplains. But in preventing rivers from overflowing into floodplains, we reduce the ability of the ecosystem to handle periods of heavy rainfall.

Although they are "free," ecosystem services should never be taken for granted. Critical ecosystem services include removal of pollutants from the air by plants, pollination of plants by insects (essential for about 30 percent of U.S. food crops), maintenance of breeding grounds for shellfish and fish in marine ecosystems, prevention of soil erosion by plants, screening of dangerous ultraviolet light by the atmospheric ozone layer, and moderation of the climate by vegetation. Intact and healthy natural ecosystems hold cultural, recreational, aesthetic, and spiritual value for many.

Ecosystem services are essential for maintaining healthy ecological communities, and they also provide people with enormous economic benefits. Globally, the value of ecosystem services provided by lakes and rivers alone has been estimated by some researchers to be a staggering $1.7 trillion each year. Other examples of the value of ecosystem services include the world's fish catch (valued at between $50 billion and $100 billion per year) and the billions of dollars' worth of crops that could not be produced if plants were not pollinated by insects.

Albany
Oneonta
Catskill Watershed
Hudson River
MASSACHUSETTS
Kingston
Liberty
Poughkeepsie
CONNECTICUT
Ellenville
Catskill Aqueduct
Croton Watershed
PENNSYLVANIA
NEW YORK
NEW JERSEY
Catskill Watershed area
Croton Watershed area
Rivers and reservoirs
— Aqueducts and tunnels
- - - State borders
Croton Aqueduct
New York City

New York City's Water Supply System

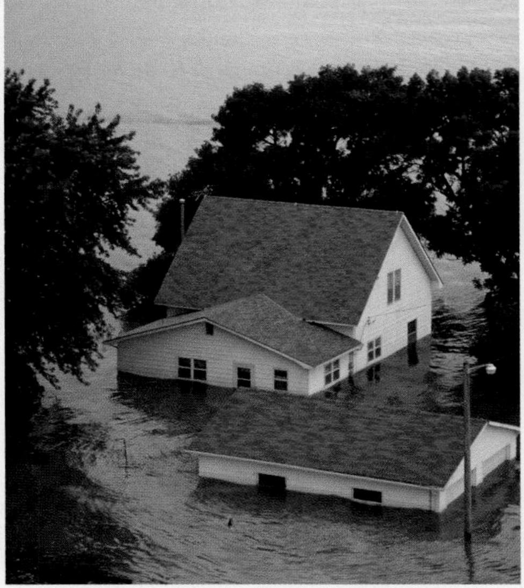

Flood Devastation in the Midwest

APPLYING WHAT WE LEARNED

What Happens When the Worst Happens?

The Gulf of Mexico is one of the most productive fisheries in the world, a maritime treasure that provides major ecosystem services to humans. The Gulf supports a multibillion-dollar fishing industry that produces shrimp, crabs, oysters, and fish. Gulf marshlands protect coastal cities from flooding, clean wastes from water, and nurture marine life. Scientists estimate that 75–90 percent of animal species in Gulf Coast waters—including fish, oysters, anchovies, and shrimp—spend at least part of their lives in estuaries and marshes. The oil released by the *Deepwater Horizon* spill could potentially kill thousands of young marine animals and marsh plants.

A year after the disastrous oil spill, scientists took stock. One government official optimistically reported that the Gulf would be fully recovered by 2012. Indeed, the methane from the well seemed to have vanished altogether, and researchers reported that microbes had consumed it. And just days after the gushing well was finally capped, the surface oil had all but vanished from satellite photos.

Where had the oil gone? Some of it apparently dispersed naturally through the churning action of waves and ocean currents. By some estimates, the bright Gulf sunshine had evaporated a quarter of the oil. Some went up in flames after BP set fire to it. And BP poured about 2 million gallons of a chemical "dispersant" into the ocean; the dispersant broke the thick globs of oil into droplets that spread through the water column instead of congealing at the surface. Large amounts of oil fell to the seafloor, where, biologists reported, thick mucuslike layers of oil covered ocean beds miles from the spill site.

Damage to beaches and marshes could have been far worse. Ocean currents had carried the bulk of the oil away from the coast. And Louisiana had released Mississippi River water into coastal marshes, possibly pushing the oil offshore. Crab and shrimp fisheries, some fishery biologists announced, would be fine within a year or two, as would most fish populations.

Yet, other biologists remained pessimistic. To be sure, a months-long fishing moratorium had taken the pressure off marine populations. But laboratory work demonstrated that the mixture of oil and dispersant could kill, stunt, or sicken juvenile and adult fish and crabs. With fewer of these animals near the bottom of the food chain, there would be less food later for red snapper and bluefin tuna to eat. The food chain appeared to have some damaged links.

Other scientists argued that the oil was still around, buried under beaches or coating the seafloor, where it would be suffocating tube worms, corals, sea urchins, crustaceans, and other marine creatures that play vital roles in the Gulf ecosystem. One biologist described offshore reefs, where fish breed, as "dead"; and another compared parts of the Gulf seafloor to a "graveyard." Half of Louisiana's oyster beds were destroyed—partly by oil and partly by the surge of fresh Mississippi River water that had flooded the oyster beds for months—and could take a decade to recover.

Ecosystem damage from the *Exxon Valdez* spill continued for years after, and oil damage may continue in the Gulf as well. After both spills, researchers recorded unusual numbers of fish with skin sores, parasites, strange pigmentation patterns, damaged organs, and oil compounds in their organs. After the *Exxon Valdez* oil spill, a herring fishery near the spill collapsed, and 20 years later it still had not recovered. Meanwhile, oil sitting on the Gulf seafloor is not degrading as quickly as expected, and ecologists are still studying the aftermath of the spill.

BIOLOGY IN THE NEWS

Too Much Deer Pee Changing Northern Forests

BY BECKY OSKIN • *LiveScience.com*, June 5, 2013

The booming deer population in the northern United States is bad for the animal's beloved hemlocks, a new study finds.

During Michigan winters, white-tailed deer converge on stands of young hemlocks for protection from winter chill and predators. The same deer return every year to their favorite clumps of the bushy evergreens, called deeryards. The high concentration of deer in a small space saturates the soils with nitrogen from pee, according to a study published online in the journal *Ecology*. While deer pee can be a valuable source of nitrogen, a rare and necessary nutrient for plants, some deeryards are now too rich for the hemlocks to grow.

"Herbivores like deer interact with the ecosystem in two ways. One is by eating plants and the other is by excreting nutrients," said Bryan Murray, an ecologist and doctoral student at Michigan Tech University. "Urine can be a really high nitrogen resource, and hemlock can be out-competed by other species in really high nitrogen environments."

Slow-growing hemlocks prefer low-nitrogen soil, and the prolific pee results in nitrogen-loving species like sugar maple outgrowing the hemlocks, the researchers found.

Hemlocks are already struggling to recover from logging and other ecosystem changes that reduced their numbers to 1 percent of pre-settlement populations in some parts of Michigan, Murray said. "At the moment, it's difficult to find hemlock stands where there are saplings in the understory that are going to replace the hemlocks in the overstory when they die," he told OurAmazingPlanet. The lack of regeneration could be due to a number of issues, but deer overpopulation is a factor, he added.

With the reduced hemlock cover available for deer, the booming white-tailed deer population means more deer crowd into the remaining forest. The researchers found more than 100 deer per square mile (2.6 square kilometers) in popular deeryards. And young hemlocks have a tough time recovering from the deer nibbling and browsing.

In the eastern United States, an invasive sap-sucking bug called the adelgid is also killing off hemlocks.

"The Upper Midwest represents one of the last strongholds of hemlocks," Murray said.

Nitrogen is an essential element in ecosystems, which normally rely on the metabolic activities of bacteria to convert atmospheric nitrogen into organic forms that can be taken up by plants. But some plants—for example, hemlocks—prefer environments with lower nitrogen content, which limits the growth of competitor species. The additional nitrogen being deposited by booming deer populations in the upper Midwest is restricting the growth of these plants. Northern hemlock forests are already facing a host of other problems, including the invasive adelgid bug and a decrease in soil nutrients caused by increased runoff due to deforestation. The latest challenge highlighted in the article illustrates the delicate nature of nutrient cycling and the impact that organisms can have on ecosystems.

A single species like the deer can affect the nutrient cycling of local ecosystems, but human activities can disrupt ecosystem cycling on a global scale. In the early twentieth century, German scientists developed a method to convert molecular nitrogen into chemically useful forms. Since then, human production of nitrogen has steadily and rapidly increased to the point that it now rivals or perhaps exceeds all natural organic nitrogen production.

This disruption of the nitrogen cycle has resulted in increased levels of atmospheric nitrates that contribute to acid rain, as well as in eutrophication in aquatic ecosystems due to nitrate runoff, which causes dead zones. Nitrogen dioxide (NO_2, a major air pollutant) and nitrous oxide (N_2O, a potent greenhouse gas) also have increased in abundance as a result of human activities. Air and water circulation can carry human-produced nitrogen by-products across vast distances, making the overall impact of human activities global in scope.

Evaluating the News

1. What is the normal pattern of nitrogen cycling in ecosystems, and how does the increase in deer populations influence this biogeochemical cycle? What impact does this disruption have on other members of the community?

2. When populations become too abundant, normal ecosystem patterns can be affected. Wildlife managers often will suggest culling (hunting or removal) programs to reduce population sizes in order to mitigate the impacts of overabundant populations, such as bison in Yellowstone National Park or elephants in South Africa. Discuss the merits and dangers of this approach. What are some alternative management strategies? What other factors could managers consider in order to restore ecosystem function?

CHAPTER REVIEW

Summary

24.1 How Ecosystems Function: An Overview

- Energy and materials can move from one ecosystem to another.
- A substantial portion of the energy captured by producers is lost as metabolic heat at each step in a food chain. Therefore, energy moves through ecosystems in just one direction.
- Nutrients are recycled in ecosystems. They pass from the environment to producers to various consumers, and then back to the environment when the ultimate consumers—decomposers—break down the bodies of dead organisms.

24.2 Energy Capture in Ecosystems

- Energy capture in an ecosystem is measured as net primary productivity; assessing the amount of NPP in an area is an important first step in determining the type of ecosystem found there and how it functions.
- On land, NPP tends to decrease from the equator toward the poles. Temperature, moisture, and nutrient availability all influence NPP.
- In marine ecosystems, NPP tends to decrease from relatively high values where the ocean borders land to low values in the open ocean (except where upwelling provides scarce nutrients to marine organisms). Aquatic ecosystems on land (such as wetlands) can also show high NPP.

24.3 Energy Flow through Ecosystems

- Food webs describe the dietary relationships between organisms and depict the flow of energy throughout an ecosystem.
- Producers obtain their energy from an external source, such as the sun. Consumers get their energy by eating other organisms. Primary consumers eat producers, secondary consumers feed on primary consumers, and so on.
- An energy pyramid depicts the amounts of energy available to organisms at different trophic levels in an ecosystem, with each successive level harvesting only about 10 percent of the energy from the level below.

24.4 Biogeochemical Cycles

- Nutrients are transferred between organisms and the physical environment in what are called nutrient cycles or biogeochemical cycles.
- Decomposers return nutrients from the bodies of dead organisms to the physical environment.
- Nutrients that enter the atmosphere easily (in gaseous form) have atmospheric cycles, which occur relatively rapidly and can transfer nutrients between distant parts of the world.
- Nutrients that do not enter the atmosphere easily (such as phosphorus) have sedimentary cycles, which usually take a long time to complete.
- The carbon cycle is mediated by the processes of photosynthesis and respiration. Photosynthesis draws CO_2 from the atmosphere and converts it into organic forms, while respiration and decomposition release CO_2 back to the environment.
- Bacteria play an important role in the nitrogen cycle. In biological nitrogen fixation, N_2 is converted to NH_4^+ by bacteria.
- Sulfur has an atmospheric cycle, entering land and water through weathering of rocks. Sulfur-containing compounds produced by phytoplankton are lofted into the atmosphere in sea spray.
- Phosphorus is a component of vital macromolecules, such as DNA. Phosphorus inputs usually boost NPP in an ecosystem. Phosphorus is the only major nutrient that has a sedimentary (not atmospheric) cycle.

24.5 Human Actions Can Alter Ecosystem Processes

- Human activities can alter nutrient cycles, as well as net primary productivity, on local, regional, and global scales.
- The addition of extra nutrients (especially nitrogen and phosphorus) to lakes, streams, and coastal waters leads to eutrophication, marked by a population boom among producers.
- Human inputs to the sulfur cycle exceed those from all natural sources combined, creating problems of international scope, such as acid rain.

Key Terms

abiotic (p. 549)
acid rain (p. 560)
atmospheric cycle (p. 555)
biogeochemical cycle (p. 554)
biomass (p. 551)
biotic (p. 549)
carbon cycle (p. 556)
consumer (p. 549)
decomposer (p. 549)
ecosystem (p. 548)
energy pyramid (p. 554)
eutrophication (p. 559)
exchange pool (p. 555)
food chain (p. 553)
food web (p. 553)
net primary productivity (NPP) (p. 551)
nitrogen fixation (p. 556)
nutrient (p. 549)
nutrient cycle (p. 554)
primary consumer (p. 553)
producer (p. 549)
secondary consumer (p. 553)
sedimentary cycle (p. 555)
tertiary consumer (p. 553)
trophic level (p. 553)
upwelling (p. 552)

Self-Quiz

1. The amount of energy captured by photosynthesis, minus the amount lost as metabolic heat, is
   a. primary consumption.
   b. consumer efficiency.
   c. net primary productivity.
   d. photosynthetic efficiency.

2. The movement of nutrients between organisms and the physical environment is called
   a. nutrient cycling.
   b. ecosystem services.
   c. net primary productivity.
   d. a nutrient pyramid.

3. Free services provided to humans by ecosystems include
   a. prevention of severe floods.
   b. prevention of soil erosion.
   c. filtering of pollutants from water and air.
   d. all of the above

4. Each step in a food chain is called
   a. a trophic level.
   b. an exchange pool.
   c. a food web.
   d. a producer.

5. What type of organism consumes 50 percent or more of the NPP in all ecosystems?
   a. herbivore
   b. decomposer
   c. producer
   d. predator

6. Sources of nutrients that are available to producers, such as soil, water, or air, are
   a. called essential nutrients.
   b. called exchange pools.
   c. considered eutrophic.
   d. called limiting nutrients.

7. Which of the following is the most representative term for an organism that gets its energy by eating all or parts of other organisms or their remains?
   a. decomposer
   b. predator
   c. consumer
   d. producer

8. Nutrients that cycle between terrestrial and aquatic ecosystems and are then deposited on the ocean bottom
   a. have a short cycling time.
   b. have an atmospheric cycle.
   c. are more common than nutrients with a gaseous phase.
   d. have a sedimentary cycle.

9. Eutrophication refers to
   a. reduced NPP because of low nutrient levels.
   b. increased NPP because of a low concentration of nitrogen.
   c. increased numbers of secondary and tertiary consumers.
   d. a population boom in producers because of increased nutrient levels.

Analysis and Application

1. Some people think the current U.S. Endangered Species Act should be replaced with a law designed to protect ecosystems, not species. The intent of such a law would be to focus conservation efforts on what its advocates think really matters in nature: whole ecosystems. Given how ecosystems are defined, do you think it would be easy or hard to determine the boundaries of what should and should not be protected if such a law were enacted? Give reasons for your answer.

2. What prevents energy from being recycled in ecosystems?

3. Why do coastal areas have higher NPP than the open ocean? The commercial fisheries in the Gulf of Alaska are among the most productive in the world. Alaskan fishing boats haul in more than a billion dollars worth of salmon, pollack, herring, halibut, cod, crabs, and shrimp each season. How can such cold, northern waters be so productive?

4. What essential role do decomposers play in ecosystems?

5. Explain why human alteration of nutrient cycles can have international effects.

6. Describe some key ecosystem services and discuss the extent to which human economic activity depends on such services.

7. The table here shows the land area needed to raise 1 kilogram (35 ounces) of edible portions of the foods listed. Explain why it takes more land to produce a kilogram of chicken meat than a kilogram of wheat.

FOOD	LAND AREA NEEDED (m^2) TO PRODUCE 1.0 kg EDIBLE PORTION	CALORIES PROVIDED BY 1.0 kg EDIBLE PORTION
Milk	9.8	610
Beef (grain-fed)	7.9	2,470
Eggs	6.7	1,430
Chicken (broiler)	6.4	1,650
Wheat	1.5	3,400

25 Global Change

TOP OF THE FOOD CHAIN, TOP OF THE WORLD. As global warming melts the Arctic ice cap, receding summer ice is increasingly leaving polar bears stranded on land or ice floes. Some bears are moving south into Canada and Alaska, but global warming may also affect the bears in a surprising way—by damaging marine food chains from the bottom up.

Is the Cupboard Bare?

Polar bears are in trouble. The big, white bears live by hunting seals on vast stretches of sea ice covering the Arctic Ocean. Yet each summer a larger area of the ice cap melts, leaving the bears stranded on land, often with hungry cubs. Until the ice returns in winter, the giant bears fast or eat garbage at Canadian and Russian garbage dumps. As the planet heats up, the ice melts earlier each spring and freezes later each fall, extending the bears' months of fasting by weeks. Unless they abandon the Arctic and begin migrating south into Canada and Siberia to compete with grizzly bears each summer, polar bears may go extinct.

Surprisingly, global warming could also wallop the bears from a completely different direction. In the summer of 2010, ecologists reported that global populations of tiny but critically important organisms called phytoplankton had declined by 40 percent since the 1950s. Like plants, phytoplankton live by photosynthesizing—that is, making sugar molecules from carbon dioxide and water using energy from the sun. On land, the big photosynthesizers are trees and other plants. But in the world's oceans, photosynthesis is carried out by phytoplankton—a mix of bacteria, algae, diatoms, and other protists.

Floating at the surface of millions of square miles of ocean, phytoplankton perform half of all the photosynthesis on the planet and supply half of all the new oxygen in Earth's atmosphere. Photosynthetic phytoplankton are the foundation for all the great marine food webs. Every marine animal—from tiny shrimplike krill to fish, seals, and polar bears—depends on the energy and building blocks stored in phytoplankton. A 40 percent decline in phytoplankton would lead to a drastic decline in krill and marine fish. Without fish, there would be no seals, to say nothing of polar bears, poised precariously at the top of the energy pyramid.

What could cause such a decline in phytoplankton? Could a shortage of phytoplankton lead to a collapse of marine food chains?

In 2010, researchers reported that phytoplankton had declined in 8 out of 10 regions of the ocean since 1950. In this chapter we'll learn how the expanding human population is dramatically affecting the biosphere.

MAIN MESSAGE Global change caused by human actions is occurring at a rapid pace. The consequences for humans and other species are real and immediate.

KEY CONCEPTS

- The main causes of the current high rate of species extinctions are likely human activities that transform Earth's land and water.
- Human activities have added natural and synthetic chemicals to the environment, and these additions in turn have altered how natural chemicals cycle through ecosystems.
- Toxins such as heavy metals that do not break down readily in the environment can become concentrated in living tissues over time in a phenomenon called bioaccumulation, or at higher trophic levels because of biomagnification.
- The concentration of carbon dioxide (CO_2) gas in the atmosphere is increasing at a dramatic rate, largely because of the burning of fossil fuels.
- Strong evidence suggests that human activity has increased the concentrations of CO_2 and certain other gases in the atmosphere. These increases have led to the observed rise in average global temperature over the past century.
- The warming of Earth's climate has led to melting of polar and glacial ice, rising sea levels, acidification of the oceans, and shifts in where many species live.
- Climate models predict an increase in the frequency of severe weather and changes in rainfall patterns.
- Many species are likely to become extinct over the next 100 years.

IS THE CUPBOARD BARE? 569
25.1 Land and Water Transformation 570
25.2 Changes in the Chemistry of Earth 573
25.3 Human Impacts on the Global Carbon Cycle 575
25.4 Climate Change 576
25.5 Timely Action Can Avert the Worst-Case Scenarios 583
APPLYING WHAT WE LEARNED 584
Bye-Bye, Food Chain?

ON LAND AND SEA AND SKY, our world is changing, and the change is occurring at a pace unprecedented in human history. Much of this change has been caused by humans, and most of it has taken place in just the last 100 years. Statements by politicians, talk show hosts, and others can give the impression that worldwide change in the environment—**global change**—is a controversial topic. Such statements cause many in the general public to think that global change may not really be occurring, or cause them to wonder whether anything really needs to be done about it (**FIGURE 25.1**).

This impression of controversy is unfortunate because we know with certainty that global change *is* occurring. We can readily demonstrate that invasions of nonnative species have increased worldwide, large losses of biodiversity have occurred, and pollution has altered ecosystems on a global scale (see examples in Chapters 21, 23, and 24). The data show that increased accumulation of greenhouse gases has led to global warming, which in turn has changed the climate of Earth.

In this chapter we describe how people influence global change. We first discuss changes in land and water use, and changes in the cycling of nutrients through ecosystems. Next we focus on one of the most serious ecological issues of our time: **climate change**, which is a large-scale and long-term alteration in Earth's climate. Although Earth's average climate has gone through many changes over its 4.5-billion-year history, the speed of the change that has taken place in the past 100 years is without precedent in the climate record. Moreover, there is compelling evidence that this current climate change is, to a large extent, caused by human actions. Humanity is therefore responsible for the most rapid climate change in Earth's history, the effects of which may negatively impact people and ecosystems around the world.

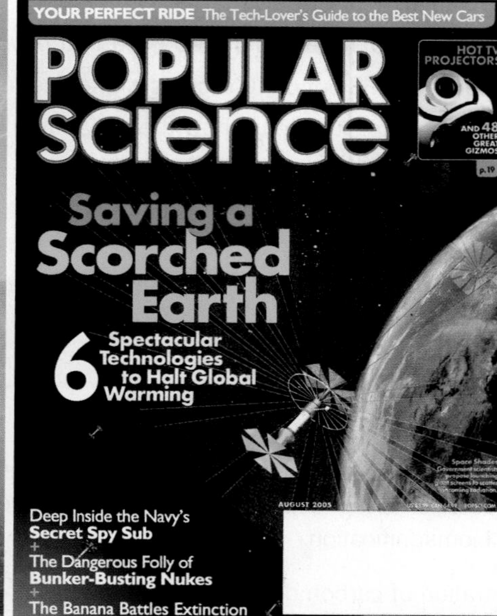

FIGURE 25.1 Climate Change in the Public Media
The subject of global climate change is a matter of considerable debate in the popular press, although the scientific community is generally in agreement that it is occurring. Why is there such a disconnect between the scientific and public spheres? How do scientists and journalists differ in their approach to a topic like global climate change?

25.1 Land and Water Transformation

People make many physical and biotic changes to the land surface of Earth, which collectively are referred to as **land transformation** or land-use change. Such changes include agriculture, urban and suburban growth (**FIGURE 25.2**), and the destruction of natural habitat to obtain resources (as when a forest is clear-cut for lumber). Land transformation also includes many human activities that alter natural habitat less dramatically, such as grazing cattle on grasslands.

Land Transformation

(*a*) 1973

(*b*) 2006

FIGURE 25.2 Human Activities Can Dramatically Alter the Terrestrial Environment

Rapid conversion of desert areas to urban and suburban regions near Las Vegas, Nevada, between 1973 (*a*) and 2006 (*b*).

Similarly, **water transformation** refers to physical and biotic changes that people make to the waters of our planet. For example, we have drastically altered the way water cycles through ecosystems (**FIGURE 25.3**). People now use more than half of the world's accessible freshwater, and we have altered the flow of nearly 70 percent of the world's rivers. Water is essential to all life, so our heavy use of the world's waters has many far-reaching effects, including changing where water is found and consequently altering the species makeup of communities and ecosystems.

Many of our effects on the lands and waters of Earth are local in scale, as when we cut down a single forest or pollute the waters of one river. However, these local effects add up to have a global impact because of the interconnectedness of the world's ecosystems.

There is ample evidence of land and water transformation

Aerial photos, satellite data, and changing urban boundaries show how humanity is changing the face of Earth. In modifying land and water for our own use, we have dramatically affected many ecosystems. The ongoing destruction of tropical rainforests (**FIGURE 25.4**) and the conversion of once vast grasslands in the American Midwest to cropland are two examples of human effects on ecosystems.

Wetlands have been hit particularly hard by human activities. Half of the world's wetlands, from mangrove swamps to northern peat bogs, have been lost in the last 100 years. From the 1780s through the 1980s, wetlands declined in every state in the United States. Thanks to protective legislation, widespread public outreach, and initiatives that encourage landowners and public groups to conserve and restore wetlands, the loss of wetlands was sharply curtailed in the last decade of the twentieth century.

Estuaries, saltwater marshes, mangrove swamps, and coastal shelf waters are among the most productive ecosystems on Earth. Yet, with about 50 percent of the world's population living within 3 miles of a coastline, coastal ecosystems are extremely vulnerable to human activity. That enormous population

Helpful to Know

The terms "global warming," "climate change," and "global change" are related but not synonymous. *Global warming* is a significant increase in the average surface temperature of Earth over decades or more. *Climate change* is a long-term alteration in Earth's climate, and it includes such phenomena as global warming, change in rainfall patterns, and increased frequency of violent storms. *Global change* is an even broader concept, encompassing all types of worldwide environmental change, including large-scale pollution and loss of biodiversity.

FIGURE 25.3 Human Activities Alter the Global Water Cycle Water circulates through Earth's ecosystems in a global cycle that involves evaporation, precipitation, and runoff. Humans influence the natural water cycle in significant ways.

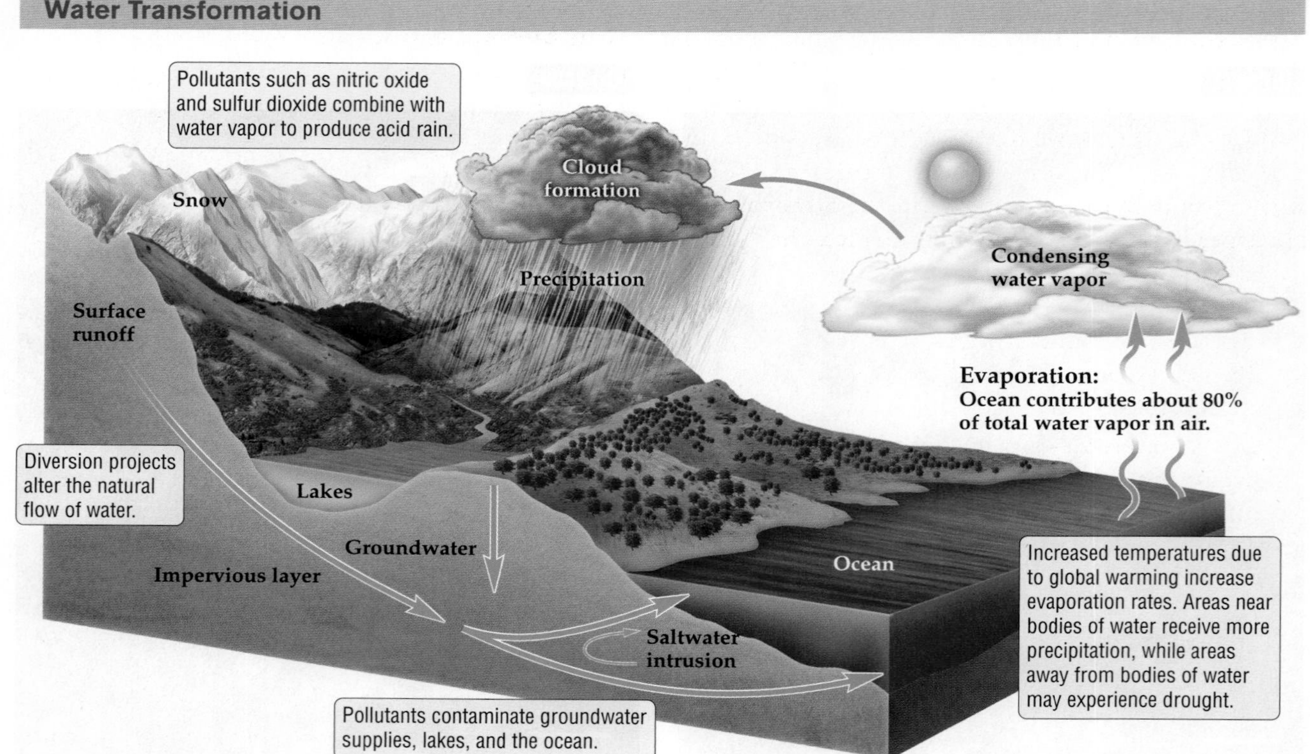

Water Transformation
Pollutants such as nitric oxide and sulfur dioxide combine with water vapor to produce acid rain.
Snow
Cloud formation
Condensing water vapor
Precipitation
Surface runoff
Evaporation:
Ocean contributes about 80% of total water vapor in air.
Diversion projects alter the natural flow of water.
Lakes
Groundwater
Impervious layer
Ocean
Saltwater intrusion
Increased temperatures due to global warming increase evaporation rates. Areas near bodies of water receive more precipitation, while areas away from bodies of water may experience drought.
Pollutants contaminate groundwater supplies, lakes, and the ocean.

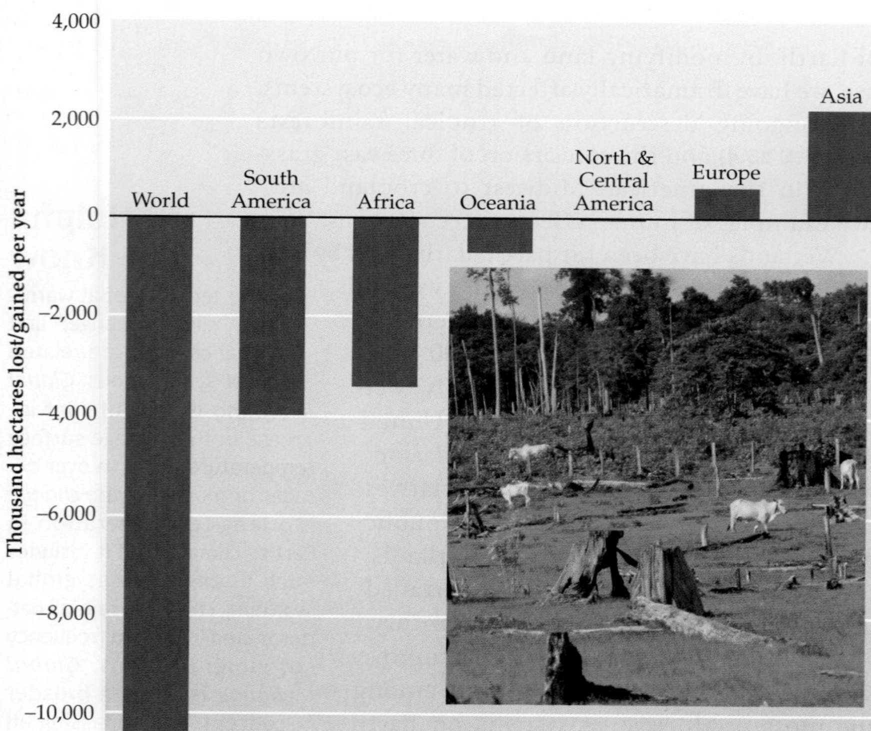

4,000
2,000
0
-2,000
-4,000
-6,000
-8,000
-10,000
Thousand hectares lost/gained per year
World
South America
Africa
Oceania
North & Central America
Europe
Asia

FIGURE 25.4 Disappearing Forests
This graph, based on data from the United Nations Food and Agriculture Organization (FAO), shows that forest cover has shrunk in most regions of the world, except Europe. Asia's relatively good standing is due largely to extensive reforestation efforts in China over the past few years. (*Inset*) Cattle grazing on land previously covered by Amazonian rainforest, in Para, Brazil.

density—more than 3.5 billion people—puts most of the world's coastline under siege today from urban development, sewage, runoff from farm fields, chemical pollution, and unsustainable harvesting of marine life.

Land and water transformation have important consequences

As we alter the land and water in the service of an ever-increasing number of people, we use a large share of the world's resources. Estimates suggest that humans now control roughly 30–35 percent of the world's total net primary productivity (NPP) on land. As described in Chapter 24, NPP is the new growth that producers generate in a unit area per year. By controlling such a large portion of the world's land area and resources, we reduce the amounts of resources and land available to other species. Water transformation has similar effects. Overfishing and pollution have dramatically affected marine ecosystems, threatening organisms such as whales and coral reef communities.

Land and water transformation can also change local climate. For example, when a forest is cut down, the local temperature may increase and the humidity may decrease, since the lack of trees exposes the ground to direct sunlight and leads to an increase in evaporation rates. Such climatic changes can make it less likely that the forest will regrow even if the logging stops. In addition, as we will see shortly, the cutting and burning of forests increases the amount of carbon dioxide in the atmosphere—an aspect of global change that can alter the climate worldwide.

Concept Check

1. Give some examples of human activities that lead to land transformation.
2. Describe some causes of the degradation of coastal ecosystems.

25.2 Changes in the Chemistry of Earth

In Chapter 24 we learned that life on Earth depends on and participates in the cycling of nutrients in ecosystems. Net primary productivity often depends on the amount of nitrogen and phosphorus available to producers, for example, and an overabundance of sulfuric acid in rainfall lowers the pH of lakes and rivers, destroying fish populations. In the next two sections we look at how human activities, like manufacturing, motorized transportation, and overpopulation, have altered the natural nutrient cycles (see also Figures 24.10, 24.11, 24.13, and 24.14).

Bioaccumulation concentrates pollutants up the food chain

Humans release many synthetic chemicals and pollutants into the air, water, and soil that then cycle through ecosystems. These human-made chemicals can be ingested, inhaled, or absorbed by organisms. If a chemical binds to cells or tissues and stays there, then we say it **bioaccumulates** in an individual. As a substance bioaccumulates, its concentration within an organism exceeds the amount found in the environment.

Many organic molecules found in pesticides, plastics, paints, and solvents tend to bioaccumulate in cells and tissues. Long-lived organic molecules of synthetic origin that bioaccumulate in organisms, and that can have harmful effects, are broadly classified as **persistent organic pollutants (POPs)**. Some of the most damaging POPs that are widespread in our biosphere include different types of PCBs (polychlorinated biphenyls, used in the production of electronics) and dioxins (a by-product of many industrial processes, such as the bleaching of paper pulp). Because many of these pollutants have an atmospheric cycle, they can be transported over vast distances across the globe to contaminate food chains in remote places where the chemicals have never been used.

Heavy metals such as mercury, cadmium, and lead can also bioaccumulate in a wide variety of organisms. Mercury enters the food chain when bacteria absorb it from soil or water and convert it to an organic form known as methylmercury. Methylmercury is much more toxic than inorganic forms of mercury, in part because the organic form bioaccumulates more readily, being stored in muscle tissues of shellfish, fish, and humans. Methylmercury bioaccumulated by bacteria is passed on to consumers, such as zooplankton (microscopic aquatic animals), that feed on mercury-accumulating bacteria. In this way, the methylmercury is progressively transferred to other consumers throughout the food web. The FDA has issued an advisory suggesting that pregnant women, in particular, abstain from eating mackerel, shark, swordfish, and tilefish, because these predatory fishes tend to accumulate higher levels of mercury.

The increase in the tissue concentrations of a bioaccumulated chemical at successively higher trophic levels in a food chain is known as **biomagnification**. Bioaccumulation and biomagnification might seem similar at first glance. Bioaccumulation is the accumulation of a substance in an individual within a trophic level, and biomagnification is the increase in tissue concentrations of a chemical as organic matter is passed from one trophic level to the next in a food chain.

Chemicals that are biomagnified persist in the body and in the environment. PCBs, for example, are hydrophobic (see Chapter 2) molecules that combine with fat and become locked within fatty tissues. Predators in the next trophic level acquire the chemical when they eat the fatty tissues of their prey.

EXTREME DEFORESTATION

The Amazon rainforest has been particularly influenced by human activities. Since 1970, 270,000 square miles of the forest has been destroyed—an area roughly the size of Texas. Deforestation in the Amazon in some years has exceeded 10,000 square miles.

Concept Check Answers

1. Destruction of habitats for resources or agriculture; alterations of habitats due to activities such as grazing.
2. Coastal ecosystems are affected by urban development, sewage dumping, excessive nutrient runoff, chemical pollution, and overfishing.

Because predators consume large quantities of prey, and lose little if any of the chemical, its concentration builds up in their tissues over time. This is why top predators—those that feed at the end of a food chain—usually have the highest tissue concentration of biomagnified chemicals. **FIGURE 25.5** illustrates the 25-million-fold biomagnification of PCBs that has been recorded in some northern lakes. An important aspect of biomagnification is that pollutants that are present in minuscule amounts in the abiotic environment, such as the water in a lake, can build up to damaging, even lethal, concentrations in the top predators of a food chain.

Helpful to Know

The FDA's advisory concerning mercury consumption extends to women of reproductive age as well as nursing mothers. While methylmercury does break down in the body, the process takes several months, so women who plan to become pregnant should abstain from fish with high mercury content well before pregnancy. Nursing mothers similarly should avoid these fish in order to avoid the possibility that their breast milk will become contaminated with methylmercury.

The pesticide DDT is an example of a POP that is bioaccumulated and biomagnified along a food chain. Until its use was banned in 1972, DDT was extensively sprayed in the United States to control mosquitoes and protect crops from insect pests. The pesticide ended up in lakes and streams, where it was taken up by phytoplankton, such as algae, which were in turn ingested by zooplankton. As the pesticide moved up the food chain, from zooplankton to shellfish to birds of prey such as ospreys and bald eagles, its tissue concentrations increased by hundreds of thousands of times. DDT disrupts reproduction in a variety of animals, but predatory birds were hit especially hard. The chemical interferes with calcium deposition in the developing egg, producing thin, fragile eggshells that break easily. The result was huge losses in the populations of peregrine falcons, California condors, and bald eagles.

DDT is an example of an **endocrine disrupter**, a chemical that interferes with hormone function, resulting in reduced fertility, developmental abnormalities, immune system dysfunction, and increased risk of cancer. Bisphenol A (found in many plastic water bottles) and phthalates (found in everything from soft toys to cosmetics) are examples of endocrine disrupters that can be readily detected in the tissues of most Americans. In laboratory animals, bisphenol increases the risk of diabetes, obesity, reproductive problems, and various cancers. Phthalate exposure is associated with lowered sperm counts and defects in development of the male reproductive system. There is much to be learned about endocrine disrupters, but for now there is no assurance that long-term exposure to multiple endocrine disrupters, even at low doses, is safe for us.

Biomagnification

FIGURE 25.5 PCB Levels Become More Concentrated in Consumers Higher in the Food Chain

Many pollutants cause changes in the biosphere

The effects of POPs extend beyond the organisms of Earth; some POPs have also been shown to affect the physical environment itself. **CFCs**, **chlorofluorocarbons** (*KLOHR*-**oh**-*FLOHR*-**oh**-*KAHR*-**bun**), are chemicals used as refrigerants or propellants. The addition of CFCs to the atmosphere is one of the most wide-ranging changes that humans have made to the chemistry of Earth. CFCs have eroded the thickness of the atmospheric ozone layer across the globe, and contributed to the ozone hole above Antarctica. Because the ozone layer shields the planet from harmful ultraviolet light (which can cause mutations in DNA), damage to the ozone layer poses a serious threat to all life.

Fortunately, the international community responded quickly to this threat by phasing out the use of CFCs, and the ozone layer has recently begun to show signs of a recovery. Clearly, in some cases we have succeeded in slowing down or undoing the harm caused by chemical pollution or the alteration of nutrient cycles (the mitigation of acid rain, discussed in Chapter 24, is another example). But in other cases, such as the global nitrogen and carbon cycles, great challenges lie ahead.

Concept Check

1. Compare and contrast bioaccumulation and biomagnification. What are some distinctive characteristics of chemicals that tend to bioaccumulate?
2. In a food chain, which organisms are most affected by biomagnification?

25.3 Human Impacts on the Global Carbon Cycle

Nearly all of us have had a hand in changing the world's nutrient cycles, at least a tiny bit. We affect nutrient cycles when we sprinkle fertilizer on our lawns and gardens, and when we send our waste to landfills, sewage plants, or septic tanks. The cheap and abundant food that people in rich countries take for granted comes for the most part from intensive farming, with its heavy input of fertilizer and energy from nonrenewable sources. We add huge amounts of carbon dioxide, nitrogen, phosphorus, and sulfur to our environment. Of particular concern in the context of climate change is our disruption of the global carbon cycle.

Atmospheric carbon dioxide levels have risen dramatically

Although CO_2 makes up less than 0.04 percent of Earth's atmosphere, it is far more important than its low concentration might suggest. As we saw in earlier chapters, CO_2 is an essential raw material for photosynthesis, on which most life depends. CO_2 is also the most important of the atmospheric gases that contribute to global warming. Therefore, scientists took notice in the early 1960s when new measurements showed that the concentration of CO_2 in the atmosphere was rising rapidly.

Scientists have been measuring the concentration of CO_2 in the atmosphere since 1958. By also measuring CO_2 concentrations in air bubbles trapped in ice, scientists have been able to estimate the concentration of CO_2 in the atmosphere over the last several hundred thousand years (**FIGURE 25.6**). Both types of measurements show that CO_2 levels have risen dramatically during the past two centuries. Overall, of the current yearly increase in atmospheric CO_2 levels, about 75 percent is due to the burning of fossil fuels. Logging and burning of forests are responsible for most of the remaining 25 percent, but industrial processes such as cement manufacturing also make a significant contribution.

The recent increase in CO_2 levels is striking for two reasons. First, the increase happened quickly: the concentration of CO_2 increased from 280 to 380 parts per million (ppm) in roughly 200 years. Measurements from ice bubbles show that this rate of increase is greater than even the most sudden increase that occurred naturally during the past 420,000 years. Second, CO_2 levels are higher than those estimated for any time during that same period. In the middle of 2013, global carbon dioxide concentrations stood at 397 ppm, with the levels increasing at the rate of about 3 ppm per year.

Helpful to Know

The FDA banned the use of bisphenol A (BPA) in children's bottles and sippy cups in 2012, and from infant formula packaging in 2013. However, BPA is still found in a wide variety of products, including children's toys, canned goods, and register receipts. Its use in everyday products is so widespread that the CDC has found detectable levels of BPA in 93 percent of the Americans it tested.

Increased carbon dioxide concentrations have many biological effects

An increase in the concentration of CO_2 in the air can have large effects on plants (**FIGURE 25.7**). Many plants increase their rate of photosynthesis and use water more efficiently, and therefore grow more rapidly, when more CO_2 is available. When CO_2 levels remain high, some plant species keep growing at higher rates, but others drop their growth rates over time. As CO_2 concentrations in the atmosphere rise, species that maintain rapid

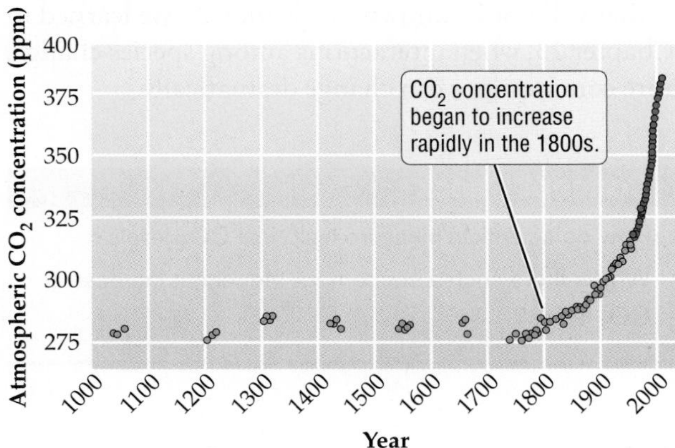

FIGURE 25.6 Atmospheric CO_2 Levels Are Rising Rapidly Atmospheric CO_2 levels (measured in parts per million, or ppm) have increased greatly in the past 200 years. The red circles are direct measurements at the Mauna Loa Observatory in Hawaii, at 11,135 feet above sea level. The green circles indicate CO_2 levels measured from bubbles of air trapped in ice that formed many hundreds of years ago.

Concept Check Answers

1. In bioaccumulation, a chemical accumulates within an organism's tissues at concentrations higher than in the surroundings; it takes place *within a trophic level*. In biomagnification, accumulated chemicals are passed on from one consumer to the next—that is, *from one trophic level to the next*. Chemicals that bioaccumulate are long-lived and not easily secreted, often because they bind to proteins or fats.

2. Top predators, because toxins accumulate between each trophic level.

FIGURE 25.7 High CO_2 Levels Can Increase Plant Size
Rice plants were grown at different CO_2 concentrations, ranging from 30 ppm (parts per million) to 800 ppm. Increased CO_2 levels resulted in higher growth rates. Note that the control plant, grown at current CO_2 levels (390 ppm), appears at the right.

growth at high CO_2 levels might outcompete other species in their current ecological communities or invade new communities.

Differences in how individual species respond to higher CO_2 levels may cause changes to entire communities. However, it is difficult to predict exactly how communities will change under higher CO_2 levels. Increased CO_2 levels in the atmosphere have contributed to the warming of Earth's climate, as we will see in Section 25.4. As both temperatures and CO_2 levels change, many different competitive and exploitative interactions may also change, but usually in ways that will not be known in advance. As we learned in Chapter 23, when interactions among species change, entire communities can change dramatically.

Helpful to Know

Changes in atmospheric CO_2 concentrations are predicted to have ecological consequences in marine ecosystems as well. Coral reefs, for example, are highly sensitive to changes in temperature and ocean acidity, which are both strongly influenced by rising CO_2 levels.

Concept Check

1. How do scientists measure historical CO_2 levels?
2. Why is there such a concern for the recent increase in CO_2 levels?

Concept Check Answers

1. By measuring CO_2 concentrations in air bubbles trapped in ice.
2. The current speed of CO_2 accumulation in the atmosphere is unparalleled in nearly half a million years. Furthermore, the levels of CO_2 are higher than scientists have seen for 420,000 years.

25.4 Climate Change

Some gases in Earth's atmosphere, such as carbon dioxide (CO_2), water vapor (H_2O), methane (CH_4), and nitrous oxide (N_2O), absorb some of the heat that radiates from Earth's surface to space. These gases are called **greenhouse gases** because they function much as the walls of a greenhouse or the windows of a car do: they let in sunlight but trap heat. **FIGURE 25.8** illustrates how these gases contribute to the **greenhouse effect** that warms the surface of Earth.

About one-third of the solar radiation received by Earth bounces back off the upper layers of the atmosphere. The rest is absorbed by the land and oceans, and to a lesser degree by the air. The warmed Earth emits some of its heat as long wavelengths of energy, known as *infrared radiation*. Some infrared radiation escapes Earth's atmosphere, but a good deal is absorbed by the greenhouse gases. Much of the heat absorbed by greenhouse gases is effectively trapped on Earth because when it is reemitted, it does not have sufficient energy to pass through the atmosphere and escape into outer space. Greenhouse gases have existed in Earth's atmosphere for more than 4 billion years and have played an important part in maintaining temperatures that are warm enough for life to thrive over most of Earth's surface. Scientists reconstructing the past climate of our planet have noted a near-perfect correlation between the levels of carbon dioxide (the most potent of the greenhouse gases) and the surface temperature on Earth. These data show that as the concentration of greenhouse gases in the atmosphere increases, more heat is trapped, raising temperatures on Earth.

Global temperatures are rising

Carbon dioxide is the most important of the greenhouse gases because so much of it enters the atmosphere as

The Greenhouse Effect

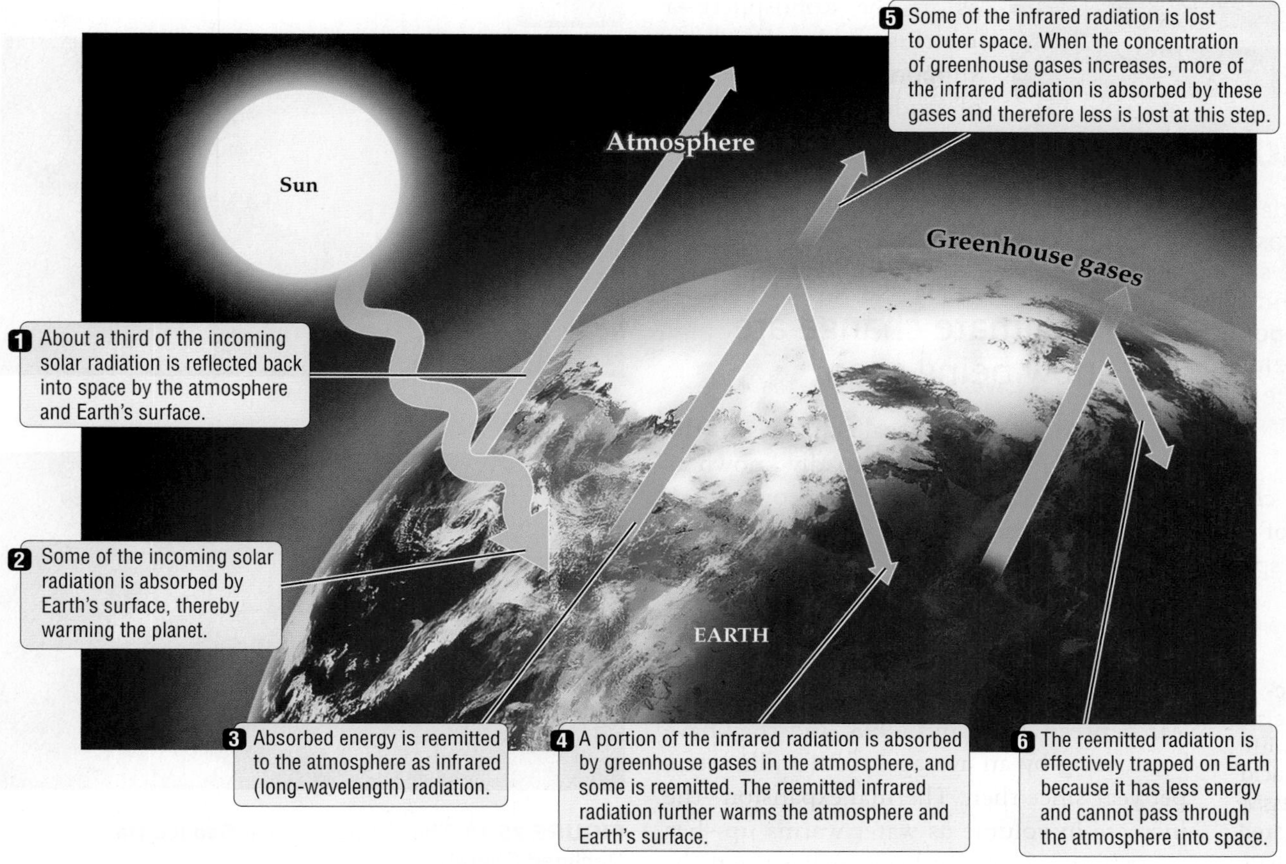

Sun
Atmosphere
Greenhouse gases
EARTH
1 About a third of the incoming solar radiation is reflected back into space by the atmosphere and Earth's surface.
2 Some of the incoming solar radiation is absorbed by Earth's surface, thereby warming the planet.
3 Absorbed energy is reemitted to the atmosphere as infrared (long-wavelength) radiation.
4 A portion of the infrared radiation is absorbed by greenhouse gases in the atmosphere, and some is reemitted. The reemitted infrared radiation further warms the atmosphere and Earth's surface.
5 Some of the infrared radiation is lost to outer space. When the concentration of greenhouse gases increases, more of the infrared radiation is absorbed by these gases and therefore less is lost at this step.
6 The reemitted radiation is effectively trapped on Earth because it has less energy and cannot pass through the atmosphere into space.

FIGURE 25.8 How Greenhouse Gases Warm Earth's Surface

a result of human activities. Since the 1960s, scientists have predicted that the ongoing increases in atmospheric CO_2 concentrations would cause temperatures on Earth to rise. This aspect of global change, known as **global warming**, has provoked controversy in both the media and the political arena.

Although year-to-year variation in the weather can make it hard to persuade everyone that the climate really is getting warmer, the overall trend in the data (**FIGURE 25.9**) has convinced the great majority of the world's climatologists and other scientists. A 2007 report from the United Nations–sponsored Intergovernmental Panel on Climate Change (IPCC) concluded that global surface temperatures rose by an average of 0.75°C between 1906 and 2005, with land warming more than the oceans, and higher rates of warming in the northern latitudes compared to the more tropical and equatorial regions of the planet. The IPCC also concluded that the increase in global temperatures since the mid-twentieth century is very likely a result of human-caused (anthropogenic) increases in

Global Temperature Change

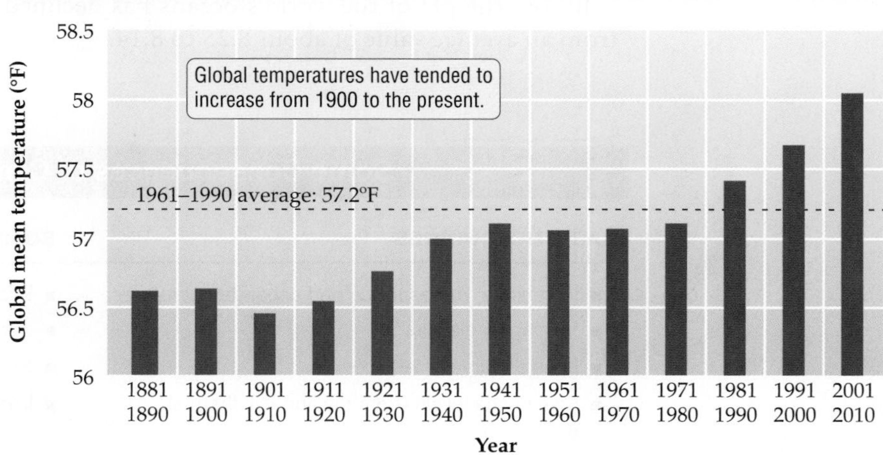

FIGURE 25.9 Global Temperatures Are on the Rise Global air temperatures are plotted relative to the average temperature between 1961 and 1990 (dashed line). The past three decades have seen significantly higher-than-average temperatures, with the last decade recording the highest temperatures since recording began in the 1850s.

EXTREME ICE CORE

Scientists are able to deduce global prehistoric climate patterns and CO_2 levels by collecting ice core samples from around the world and analyzing tiny air bubbles trapped within the ice. Cores are removed from ice sheets—some as much as 11,000 feet deep.

Helpful to Know

Ocean acidification occurs with higher CO_2 levels because CO_2 combines with water to form carbonic acid (H_2CO_3), which is why excessive consumption of carbonated beverages leads to erosion of tooth enamel. On a global scale, as CO_2 levels rise, levels of carbonic acid in the ocean increase.

the concentration of CO_2 and other greenhouse gases in the atmosphere—a conclusion that has been supported by hundreds of studies published since 1995.

Some predicted consequences of climate change are now being seen

Long-term and large-scale changes in the state of Earth's climate are broadly known as climate change. Global warming is one component of climate change, and some of its effects on the biosphere are now evident (**TABLE 25.1**). Consistent with the warming trend, satellite images show that Arctic sea ice has been declining by 2.7 percent per decade since 1978 (**FIGURE 25.10**). Sea levels rose by an average of 1.8 millimeters per year between 1961 and 1993, and they have been rising by an average of 3.1 millimeters per year since then. Thermal expansion—the increase in volume as water warms up—has contributed to sea level rise, as has the melting of glaciers (**FIGURE 25.11**) and polar ice. As atmospheric carbon dioxide levels rise, more of the gas is absorbed by the oceans, leading to ocean acidification. Since the industrial revolution, the pH of the world's oceans has declined from an average value of about 8.25 to 8.14.

Decline of the Arctic Ice Cap

(*a*) 1980

(*b*) 2012

FIGURE 25.10 The Extent of Polar Sea Ice Has Declined Sharply

Summer sea ice in the Arctic has declined by almost 25 percent compared to preindustrial levels. Climate change has affected wind and ocean currents in different ways across the globe, explaining why the Antarctic ice sheet is relatively stable. The satellite-based illustrations show the extent of the polar ice cap in the Arctic and the ice sheet on Greenland in 1980 (*a*) and 2012 (*b*).

TABLE 25.1 Some Consequences of Climate Change

ABIOTIC CHANGES	SOME BIOTIC CONSEQUENCES
Increase in near-surface and ocean temperatures	Ecosystem disruption, loss of ecosystem services; species extinction
Melting of glaciers	Spring floods, summer drought in glacier-fed regions
Loss of summer sea ice	Species extinction, loss of cultural and economic resources
Rise in sea levels (from melting ice, thermal expansion)	Loss of habitat, human habitation, and livelihood
Ocean acidification	Loss of marine organisms with calcified structures, coral bleaching; damage to fisheries
Increased frequency of severe weather	Habitat destruction; loss of human life, economic damage
Change in rainfall pattern, drought in some regions	Ecosystem degradation; severe agricultural and other economic loss

A Melting Glacier

(*a*) 1913

(*b*) 2009

FIGURE 25.11 Many Glaciers Are in Retreat
The extent of Shepard Glacier in Glacier National Park, Montana, in 1913 (*a*) versus 2009 (*b*). Most of the world's glaciers are in retreat, although some, especially in parts of South America and Central Asia, are either stable or growing slightly.

The additional heat energy that warmer temperatures generate, especially over the tropical oceans, is increasing the frequency of severe weather and lengthening the storm season. Since the middle of the twentieth century, the number of tropical storms sweeping into North America has not changed significantly, but the number of class 3 and class 4 hurricanes has nearly tripled. Rainfall patterns have changed: there is more rain in the eastern United States and northern Europe, and less in parts of the Mediterranean, northeastern and southern Africa, and parts of South Asia. Some recent climate simulations predict that global warming will worsen ozone depletion—with the highest increase in UV radiation in tropical rather than polar regions—because of alterations in wind flow patterns in the upper atmosphere.

Climate change has brought many species to the brink

Recent temperature increases have also changed the biotic (living) component of ecosystems. Many northern ecosystems are shifting their range poleward at a rate of about 0.42 kilometer (a quarter of a mile) per year, as species migrate north in an attempt to find their "comfort zone." For example, as temperatures increased in Europe during the twentieth century, dozens of bird and butterfly species shifted their geographic ranges to the north (see Figure 1.3). Similarly, the length of the growing season has increased for plants in northern latitudes as temperatures have warmed since 1980. However, some species—Arctic and alpine plants and animals, for example—have nowhere else to go (**FIGURE 25.12**). Canadian researchers have recorded a 60 percent decline in caribou and reindeer populations worldwide. There is higher calf mortality among the herds, and the animals suffer more from attacks by biting insects, whose populations have climbed.

In tropical waters, high temperatures combined with lower pH result in *coral bleaching*, caused by a loss of the algal symbiotic partner and often resulting in the death of the coral animal as well. About a third of the tropical coral reefs have been destroyed in the last few decades, succumbing to the collective onslaught of coral bleaching, pollution, and physical damage from an increase in severe storms.

Although the magnitude of warming is much larger in the northern latitudes, scientists expect a more severe impact on tropical ecosystems. Plants and animals in the moist tropics are adapted to a stable habitat and therefore live very close to the limits of their tolerance. Any change in that previously stable environment—increased temperature and reduced moisture, for example—puts them in jeopardy. In general, species with specialized habitat requirements are the most vulnerable. Experts studying species vulnerability warn that only 18–45 percent of the plants and animals native to the moist tropics are likely to survive beyond 2100. According to the International Union for Conservation of Nature

EXTREME SHORTCUT

The receding Arctic ice cap is bad news for polar bears but good news for transport companies. Climate scientists anticipate that within the next few decades, ships will be able to ferry goods between Asia and Europe by sailing through the Arctic Ocean, rather than making the long transit around Africa or through the Suez Canal.

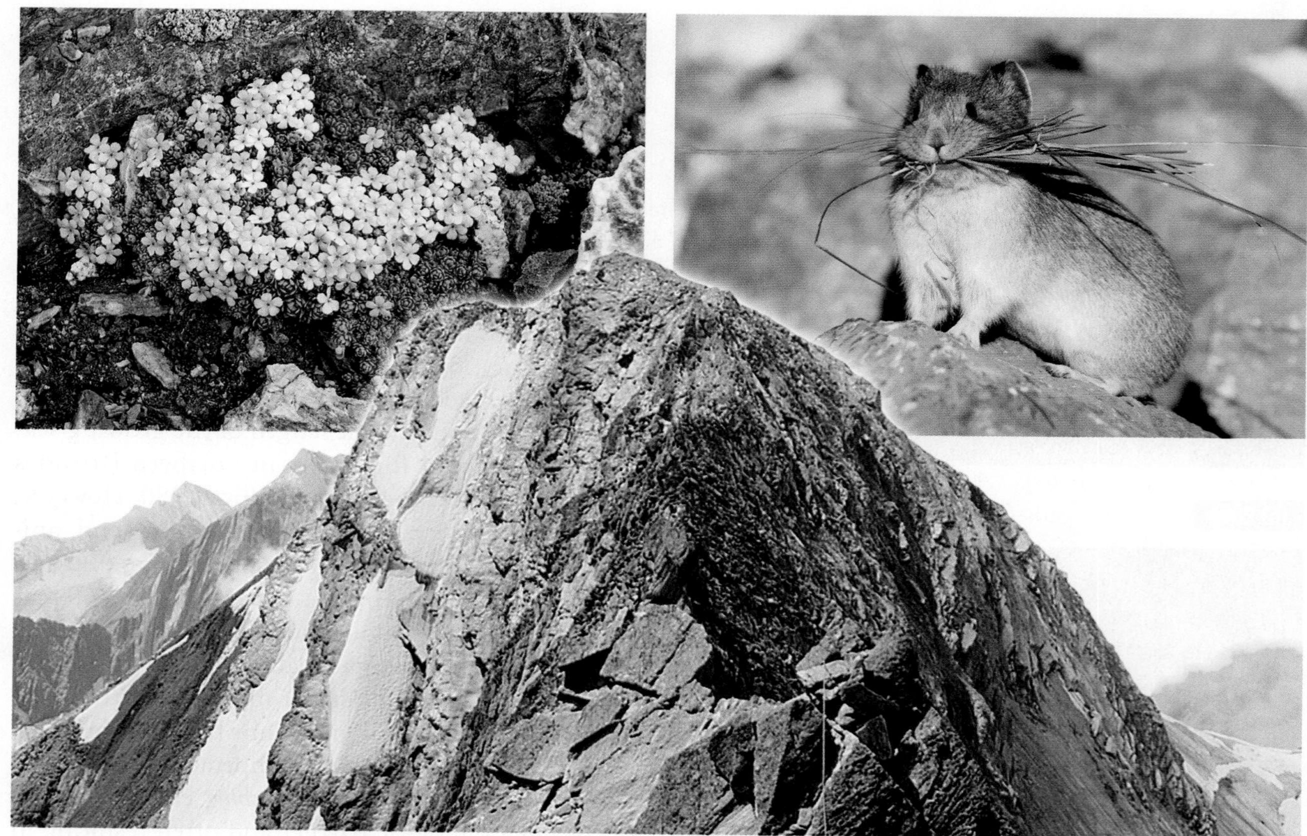

FIGURE 25.12 Running Out of Altitude

Alpine species such as the mountain *Androsace* (**an-DRAH-suh-kee**; *left*) and pika (*right*) are adapted to cold, high-elevation habitats. These species have been increasing their range upslope in the past few decades, but scientists fear that they will soon run out of options as they reach the tops of mountain peaks. Pikas live in boulder-strewn talus slopes in the highest reaches of the alpine zone. With their dense fur, these rodents suffer heat stress at temperatures that we would consider cool.

Helpful to Know

In one of the most alarming models of climate change, a rise in temperature of approximately 3°C is projected to cause "runaway" global warming. At this temperature, plants will no longer be able to properly photosynthesize, engaging instead in a process called photorespiration (discussed in Chapter 6), which will release CO_2 instead of absorbing it as they would normally do with photosynthesis.

(IUCN), 35 percent of the world's birds, 52 percent of amphibians, and 71 percent of warm-water reef-building corals have characteristics that make them especially vulnerable to the impacts of climate changes.

Will some species emerge as winners? Organisms that have broader tolerances and can live in a variety of habitats are most likely to emerge unscathed and even to expand their range. Versatility and wide tolerances are the calling cards of weedy plants and animal pests. Duke University researchers found that supplementing a forest ecosystem with extra carbon dioxide not only led to a boom in the growth of poison ivy but made the plants "itchier" because they produced a more potent form of the rash-inducing chemical urushiol. Another study predicts that the venomous brown recluse spider will continue to expand its range northward. Mosquito species that carry dengue fever virus and West Nile virus, and that were previously confined to the tropics, are also spreading northward.

Climate change will likely have severe consequences

Because there is no end in sight to the rise in CO_2 levels, the current trend of increasing global temperatures seems likely to continue. Even if humans stopped adding CO_2 to the atmosphere today, we would not see an immediate decline in global temperatures, because of the already high levels. How will increased temperatures affect life on Earth in the future? Not surprisingly, the effects will depend on how much, and how fast, global warming occurs.

Computer models predict that by the end of the twenty-first century, average temperatures on Earth will have risen by anywhere from 1.1°C to 6.4°C

BIOLOGYMATTERS

Toward a Sustainable Society

Many different lines of evidence suggest that the current human impact on the biosphere is not sustainable (see Figure 1). An action or process is *sustainable* if it can be continued indefinitely without serious damage being caused to the environment. Consider our use of fossil fuels. Although fossil fuels provide abundant energy now, our use of these fuels is not sustainable: they are not renewable, and hence supplies will run out, perhaps sooner rather than later (see Figure 2). Already, the volume of new sources of oil discovered worldwide has dropped steadily from over 200 billion barrels during the period from 1960 to 1965, to less than 30 billion barrels during 1995–2000. In 2007, the world used about 31 billion barrels of oil, but only 5 billion barrels of new oil was discovered in that year.

Actions that cause serious damage to the environment are also considered unsustainable, in part because our economies depend on clean air, clean water, and healthy soils. People currently use over 50 percent of the world's annual supply of available freshwater, and demand is expected to rise as populations increase. Many regions of the world already experience problems with either the amount of water available or its quality and safety. Declining water resources are a serious issue today, and experts are worried that matters may get much worse.

To illustrate the problem, let's look at water pumped from underground sources, or *groundwater*. How does the rate at which people use groundwater compare with the rate at which it is replenished by rainfall? The answer is that we often use water in an unsustainable way: we pump it from *aquifers* (underground bodies of water, sometimes bounded by impermeable layers of rock) much more rapidly than it is renewed.

In Texas, for example, for 100 years water has been pumped from the vast Ogallala aquifer faster than it has been replenished, causing the Texan portion of the aquifer to lose half its original volume. If that rate of use were to continue, in another 100 years the water would be gone, and many of the farms and industries that depend on it would collapse. Texas is not alone. Rapid drops in groundwater levels (about 1 meter per year) in China pose a severe threat to its recent agricultural and economic gains; and at current rates of use, large agricultural regions in India will completely run out of water in 5–10 years. In Mexico City, pumping has caused land within the city to sink by an average of 7.5 meters (more than 24 feet) since 1900, damaging buildings, destroying sewers, and causing floods.

Sustainability is one aspect of ecology where each of us has a role. We can build a more sustainable society by supporting legislation that fosters less destructive and more efficient use of natural resources; by patronizing businesses that take measures to lessen their negative impact on the planet; by supporting sustainable agriculture; and by modifying our own lifestyle to reduce our Ecological Footprint (see the "Biology Matters" box in Chapter 21, page 497). For example, we can:

- Increase our use of renewable energy and energy-efficient appliances;
- Reduce all unnecessary use of fossil fuels (for instance, by biking to work or using public transportation);
- Support organic farming; buy seafood from sustainable fisheries;
- Use "green" building materials; and reduce, reuse, and recycle waste.
- Support aid efforts that provide education, health care, and family-planning services in poor countries.

Experts estimate that more than 200 million women around the world wish to limit their family size but have no access to family planning.

FIGURE 1 The Most Inconvenient Truth: Climate Change Is Caused by Overpopulation and Overconsumption

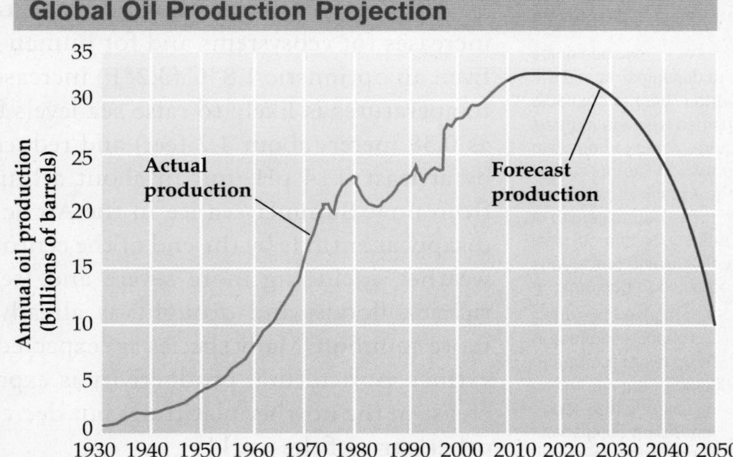

FIGURE 2 Running Out of Oil
Many experts predict that the annual global production of oil will peak, and then decline, sometime before 2020.

Rise in Sea Levels

(a)

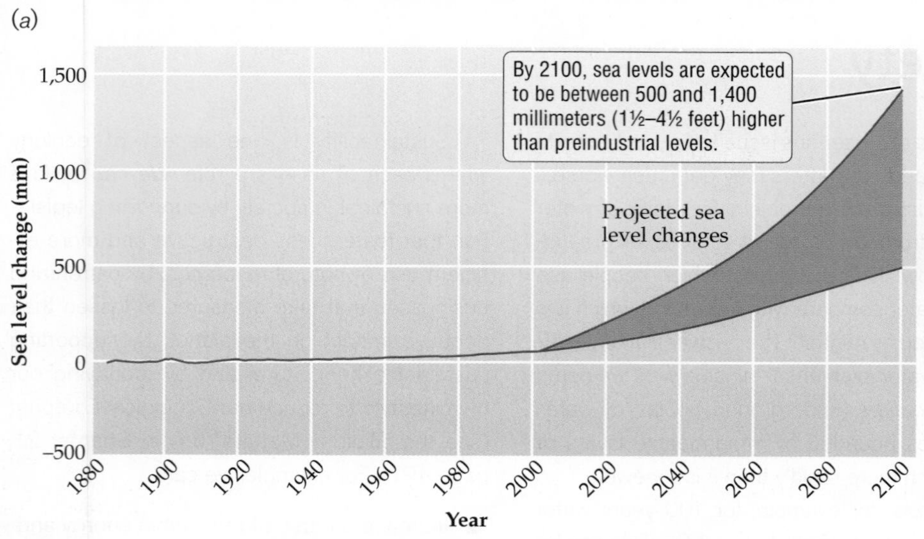

(b)

FIGURE 25.13 Thermal Expansion Combined with Ice Melt Has Raised Sea Levels

(*a*) Average sea levels have risen by about 200 millimeters since the start of the industrial revolution in 1880 and are expected to continue to rise. (*b*) The prime minister of the Maldives held an underwater cabinet meeting in 2009 to draw international attention to the threat that climate change presents to his island nation. The Maldive Islands lie so low that they are likely to be submerged by the year 2100 if sea levels continue to rise at their current rate.

(2–11.5°F) above the average global temperatures that prevailed between 1980 and 1990. The projections are based on a "business as usual" scenario, with no checks on the current trends in greenhouse gas emissions. The broad range reflects best estimates based on differing assumptions with regard to some aspects of climate change that scientists are still uncertain about.

What are the implications of such temperature increases for ecosystems and for human well-being? Even an optimistic 1.8°C (3.2°F) increase in surface temperatures is likely to raise sea levels by as much as 0.38 meter (about 1.2 feet) and reduce ocean pH by at least 0.14 pH unit in about a hundred years from now. Summer sea ice in the Arctic is likely to disappear entirely by the end of the century. Extreme weather, including more severe and frequent hurricanes, floods, and drought, is already becoming more common. Many species are expected to become extinct. Agricultural productivity is expected to increase in the northern latitudes but decrease in most other parts of the world.

A global temperature increase of 4°C or so will intensify the severity of these effects. Sea levels, for instance, are likely to rise as much as 0.59 meter (almost 2 feet). That may not sound like much, but combined with storm surges, it is enough to wipe out some island nations and destroy many coastal communities across the globe (**FIGURE 25.13**). A 4°C rise in global surface temperatures will wreak large-scale alterations in Earth's biomes. Some species will migrate, others will adapt, but a very large number will probably become extinct.

The world's agricultural systems will be severely strained, and it is unlikely that there will be enough food to nourish the extra 4–5 billion who are expected to join the human population between now and the end of this century. A recent analysis projects that by 2050, climate change will result in $125 billion in economic losses, will displace 26 million people, and will contribute to 300,000 deaths each year.

Concept Check

1. How do greenhouse gases contribute to global warming?
2. Which predicted effects of global warming are already apparent?

Concept Check Answers

1. Increases in the atmospheric levels of greenhouse gases, including CO_2, increase global temperatures by intensifying the greenhouse effect, which is caused by the trapping of infrared radiation by greenhouse gases in Earth's atmosphere.

2. Shrinking of polar ice sheets and glaciers; sea level increases; change in rainfall patterns; increased frequency of severe storms; poleward migration of many warm-weather species; earlier blooming of many northern plants; extinction of some species.

25.5 Timely Action Can Avert the Worst-Case Scenarios

Although climate change is already under way, experts say the worst-case scenarios can be averted by timely action using technology that is currently available. Battling climate change will require reduced use of fossil fuels, increased energy efficiency, and increased reliance on renewable energy such as cellulose-based ethanol and solar power. Innovative carbon capture methods have been developed, and more are under development, to reduce atmospheric CO_2 levels. In one such strategy, CO_2 from factories and power plants is turned into oil by algae, and the oil is converted to biodiesel (**FIGURE 25.14**). Improvements in waste management—reducing the release of greenhouse gases by landfills, for example—will be necessary. Agricultural practices will have to change, placing greater emphasis on sustainability, improved manure management to minimize the release of methane, and fertilizer application techniques that reduce the emission of N_2O (nitrous oxide, a greenhouse gas that is released in the breakdown of synthetic fertilizers). Halting deforestation in tropical countries and increasing forest cover worldwide will be crucial.

Some renewable-energy technologies may need government support (such as tax credits) to compete with conventional energy sources. New regulations, such as higher fuel economy standards for vehicles and stringent energy efficiency codes for appliances, are needed, and these will require political will, resting on public support. Efforts to curb global warming will have social and economic costs, but any delay will most likely lead to even greater costs in the future. Because some of the effects of climate change are inevitable, we must also develop plans to cope with the floods, storm surges, violent weather, increased fire risk, water shortage, reduced crop yields, and harm to human health that are the expected outcomes of climate change even in the best-case scenario.

Concept Check

1. What efforts can be undertaken to reduce carbon levels in the atmosphere?
2. Why might some new technologies require government support?

Concept Check Answers

1. CO_2 levels in the atmosphere could be reduced with improved agricultural methods, increased energy efficiency, new sources of energy, carbon capture technology, and improvements in waste management.

2. Conventional energy sources benefit from existing infrastructure and proven technology. Novel sources would initially be more costly and therefore would require government support to be profitable.

FIGURE 25.14 Technological Innovations Can Produce Alternative Energy Sources
This facility in France uses household waste to produce clean energy. The waste is first composted to produce methane, which is burned to produce electricity. The CO_2 from burning the methane is captured and pumped into ponds like the one above to promote the growth of algae, which is used to synthesize biofuels.

APPLYING WHAT WE LEARNED

Bye-Bye, Food Chain?

Vast ocean populations of tiny phytoplankton are responsible for about half of all the photosynthesis on Earth. These tiny organisms supply all of the energy for marine food chains and about half of all newly generated oxygen. Every organism on Earth depends on the work of phytoplankton—not just polar bears and fish, but humans too. Every time we take a breath, some of the oxygen comes from phytoplankton. Earth's atmosphere of 20 percent oxygen comes largely from the work of photosynthetic ocean bacteria that lived 2.5 billion years ago. And much of the oil that fuels our vehicles and industry comes from phytoplankton that lived millions of years ago.

So, when researchers at Dalhousie University in Nova Scotia reported in 2010 that marine phytoplankton were in sharp decline, ecologists were shocked. Without phytoplankton, life in the oceans would die and atmospheric oxygen levels would begin to decline. Were things that bad?

Oceanographers have known for a long time that rising temperatures hurt phytoplankton. As ocean waters warm up, they become more "stratified," with warmer water sitting on top of colder water. These static layers of water don't turn over, so there's no upwelling of nutrients from the seafloor. And without nutrients, phytoplankton can't grow. Data collected from satellite images suggest that warming oceans have caused a 6 percent drop in phytoplankton numbers since the early 1980s.

How does global warming hurt phytoplankton? We know that burning fossil fuels adds carbon dioxide to the atmosphere and that a lot of the extra carbon dioxide is absorbed by the oceans. Since carbon dioxide is a building block used for making sugars during photosynthesis, you might think that more carbon dioxide would be good for phytoplankton. And in some parts of the ocean, populations of phytoplankton *are* increasing. Unfortunately, carbon dioxide dissolved in water also makes water acidic, and acidic water contains less iron, an essential nutrient for phytoplankton.

Different parts of the ocean are warming up faster or slower, and some parts are becoming more acidic while other parts are not. Oceanographers aren't sure where all this is going. To find out whether the changes they had observed were short-term (and perhaps not serious) or long-term (and serious), the Nova Scotia researchers looked at half a million measurements of ocean clarity and color collected since the late 1800s. In general, cloudier water and greener water contain more phytoplankton. After analyzing the data, these researchers concluded that phytoplankton had declined by 40 percent since 1950.

Not so fast, said researcher Mark Ohman, of the Scripps Institution of Oceanography. Ohman didn't object to the data, but he claimed that the other researchers' analysis contained errors. For example, those researchers had reported that 59 percent of local ocean measurements had shown declines in phytoplankton. But Ohman pointed out that only 38 percent of those declines were statistically significant. The researchers had also used measurements from only the top 20 meters of the ocean, but in some parts of the ocean, phytoplankton live much deeper than that.

Although ocean scientists agree that phytoplankton have declined in number, and that this decline is most likely related to global warming and ocean acidification, just how extensively phytoplankton populations have declined is, for now, unsettled. Marine fisheries *are* collapsing, but mostly because of overfishing. That said, we need to halt ongoing declines in phytoplankton. Otherwise, it will be bye-bye, marine food chain and bye-bye, polar bears.

BIOLOGY IN THE NEWS

Blind, Starving Cheetahs: The New Symbol of Climate Change?

BY ADAM WELZ • *Guardian*, June 21, 2013

The world's fastest land animal is in trouble. The cheetah, formerly found across much of Africa, the Middle East and the Indian subcontinent, has been extirpated from at least 27 countries and is now on the Red List of threatened species.

Namibia holds by far the largest remaining population of the speedy cat. Between 3,500 and 5,000 cheetahs roam national parks, communal rangelands and private commercial ranches of this vast, arid country in south-western Africa, where they face threats like gun-toting livestock farmers and woody plants.

Yes, woody plants. Namibia is under invasion by multiplying armies of thorny trees and bushes, which are spreading across its landscape and smothering its grasslands.

So-called bush encroachment has transformed millions of hectares of Namibia's open rangeland into nearly impenetrable thicket and hammered its cattle industry...

Bush encroachment can also be bad news for cheetahs, which evolved to use bursts of extreme speed to run down prey in open areas. Low-slung thorns and the locked-open eyes of predators in "kill mode" are a nasty combination. Conservationists have found starving cheetahs that lost their sight after streaking through bush encroached habitats in pursuit of fleet footed food.

... An emerging body of science indicates that rapidly increasing atmospheric carbon dioxide may be boosting the onrushing waves of woody vegetation.

Savanna ecosystems, such as those that cover much of Africa, can be seen as battlegrounds between trees and grasses, each trying to take territory from the other. The outcomes of these battles are determined by many factors including periodic fire, an integral part of African savannas.

In simple terms, fire kills small trees and therefore helps fire-resilient grasses occupy territory. Trees have to have a long-enough break from fire to grow to a sufficient size—about four metres high—to be fireproof and establish themselves in the landscape. The faster trees grow, the more likely they are to reach four metres before the next fire.

Lab research shows that many savanna trees grow significantly faster as atmospheric CO_2 rises, and a new analysis of satellite images indicates that so-called "CO_2 fertilisation" has caused a large increase in plant growth in warm, arid areas worldwide.

... Increased atmospheric CO_2 seems to be upsetting many savanna ecosystems' vegetal balance of power in favour of trees and shrubs.

If increasing atmospheric carbon dioxide is causing climate change and also driving bush encroachment that results in blind cheetahs, should blind, starving cheetahs be a new symbol of climate change?

The astounding predatory behavior of cheetahs is well known among the general public. Incredible bursts that propel the cheetah to speeds of 60 miles per hour have been captured by videographers for decades. This speed, however, is becoming a liability as climate change alters the savanna landscape. Woody vegetation is encroaching on once open plains and presents dangerous obstacles to the famously fleet African cat. Trying to negotiate dense vegetation at high speed has led many cheetahs to become blinded by woody growth.

This study illustrates the practical consequences of a changing climate. The issue of climate change is largely discussed on a global scale by climatologists, but biologists are gaining more insight into the direct effects of such changes on organisms and ecosystems. For example, the melting Arctic sea ice has significantly affected populations of polar bears, which need the ice sheets for hunting grounds. Tropical frogs are experiencing increased fungal infections as temperatures warm. Coral reefs have been decimated by bleaching events, in which corals eject their photosynthetic symbionts as ocean temperatures warm.

Just as climate scientists are unable to predict the exact effects of global climate change, biologists are unable to deduce the impact of climate change on species. However, these examples of how warming temperatures are affecting biological organisms and communities illustrate that the effects of global climate change resonate throughout all levels of biological organization.

Evaluating the News

1. From your reading of this chapter, what factors do you think are influencing the distribution of woody vegetation, and therefore the cheetahs?

2. While the expansion of woody vegetation is bad for the cheetah, some species may benefit from the change in vegetative structure. How might the savanna community change in the face of global climate change?

3. Do you think studies such as the one reported in this article or the numerous studies on the impact of warming on polar bear populations may motivate people to address global climate change?

CHAPTER REVIEW

Summary

25.1 Land and Water Transformation

- Many lines of evidence show that human activities are changing land and water worldwide.
- Land and water transformation has caused extinctions of species and has the potential to alter local and global climate.

25.2 Changes in the Chemistry of Earth

- Human activities are changing the way many chemicals, both natural and synthetic, are cycled through ecosystems.
- Bioaccumulation is the tendency of some chemicals to be deposited in the tissues of an organism at concentrations higher than those found in the surrounding environment. Methylmercury and PCBs are among the many persistent organic pollutants (POPs) that are bioaccumulated.
- In biomagnification, tissue concentrations of pollutants increase as biomass is transferred from one trophic level to another. Fishes, birds, and mammals that feed at the highest trophic levels in a food chain tend to accumulate the highest tissue concentrations of biomagnified chemicals.
- The release of chlorofluorocarbons (CFCs) into the atmosphere has thinned the ozone layer over Earth, posing a serious threat to all life.

25.3 Human Impacts on the Global Carbon Cycle

- Concentrations of atmospheric CO_2 have increased greatly in the past 200 years and are higher now than in the past 420,000 years. These CO_2 increases are caused by the burning of fossil fuels and the destruction of forests.
- Increased CO_2 concentrations can alter the growth of plants in ways that will probably cause changes in many ecological communities.

25.4 Climate Change

- Carbon dioxide and other greenhouse gases in the atmosphere trap heat that radiates from Earth's surface. As the concentration of greenhouse gases increases, average temperatures on Earth are also rising.
- Human activities have contributed to the global warming that has occurred in the past 100 years.
- Some predicted effects of global warming are already being witnessed, including the melting of polar ice sheets, sea level rise, ocean acidification, and migration of some species toward the poles.
- Climate models predict changes in rainfall patterns, increased frequency of severe weather, and species extinctions.

25.5 Timely Action Can Avert the Worst-Case Scenarios

- Despite the challenges confronting humanity, the worst effects of climate change can be mitigated if action is taken in time.
- Proposed measures include reduced use of fossil fuels, increased energy efficiency, and increased reliance on renewable energy such as solar power.

Key Terms

bioaccumulation (p. 573)
biomagnification (p. 573)
chlorofluorocarbon (CFC) (p. 574)
climate change (p. 570)
endocrine disrupter (p. 574)
global change (p. 570)
global warming (p. 577)
greenhouse effect (p. 576)
greenhouse gas (p. 576)
land transformation (p. 570)
persistent organic pollutant (POP) (p. 573)
water transformation (p. 571)

Self-Quiz

1. Which of the following is most directly responsible for global warming?
   a. increased CO_2 concentration in the atmosphere
   b. sunspot activity
   c. climate change
   d. biomagnification

2. CO_2 absorbs some of the ________ that radiates from the surface of Earth to space.
   a. ozone
   b. infrared energy
   c. ultraviolet light
   d. smog

3. Most scientists think that three of the following four statements related to global warming are correct. Which one is *not* correct?
   a. The concentration of greenhouse gases in the atmosphere is not increasing.
   b. Dozens of species have shifted their geographic ranges to the north.
   c. Plant growing seasons are longer now than they were before 1980.
   d. Human actions, such as the burning of fossil fuels, contribute to global warming.
4. Substances that bioaccumulate
   a. are found in organisms at lower concentrations than in their abiotic surroundings.
   b. are readily eliminated by excretion—for example, in urine.
   c. are invariably inorganic, rather than organic, substances.
   d. tend to be chemically stable, and are not degraded in the body or in the environment.
5. In a simple food chain, which of the following would be most strongly affected by biomagnification?
   a. primary producers
   b. primary consumers
   c. secondary consumers
   d. top predators
6. Mercury that enters waterways
   a. can undergo biomagnification, but not bioaccumulation.
   b. will always be more abundant in a larger animal than in a smaller one.
   c. occurs in increasingly higher concentrations at successively higher trophic levels in the food chain.
   d. is found in secondary consumers but not in primary consumers.
7. Which of the following is *not* a method to address atmospheric carbon levels?
   a. carbon capture
   b. reduction in fossil fuel consumption
   c. improved manure management
   d. moving away from ethanol-based fuels

Analysis and Application

1. Summarize the major types of global change caused by human activities. What consequences do such types of global change have for species other than humans?
2. Compare examples of human-caused global change with examples of global change not caused by people. What is different or unusual about human-caused global change?
3. How does the current atmospheric CO_2 concentration compare with concentrations over the past 420,000 years? How do scientists know what Earth's CO_2 concentrations were hundreds of thousands of years ago?
4. The future magnitude and effects of global warming remain uncertain. Do you think we should take action now to address global warming, despite those uncertainties? Or do you think we should wait until we are more certain what the ultimate effects of global warming will be? Support your answer with facts already known about global warming.
5. Would you be willing to pay a gasoline tax to help fund aggressive actions to reduce global warming? If so, how much tax per gallon would you be willing to pay—50 cents, one dollar, two dollars? If not, why not?
6. What changes to human societies would have to be made for people to have a sustainable impact on Earth?
7. The graph shows the change in ocean pH since the industrial revolution, computed on the basis of carbon dioxide emissions over this time span. Average ocean pH is projected to decline by about 0.5 pH unit below the preindustrial levels (pH 8.2) by the end of the twenty-first century. Imagine you are taking tourists on an underwater dive to see Australia's Great Barrier Reef in the year 2100. What kinds of changes are the tourists likely to see, compared to the state of this underwater wonderland at the turn of the twentieth century? What biological mechanisms would explain the observed changes? What can we do now to ensure that these tourists-of-the-future will have a more rewarding experience of nature under the waves?

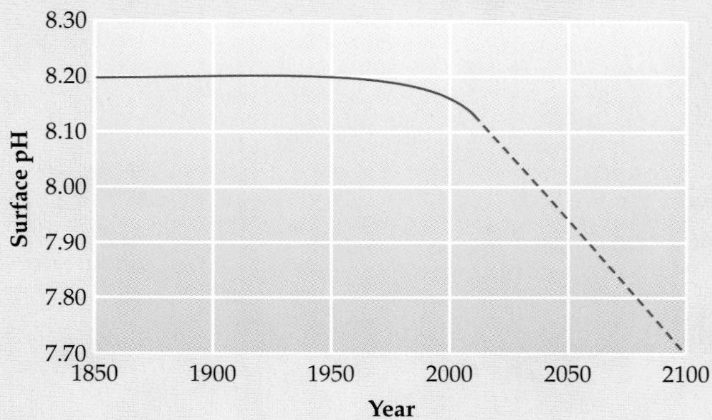

8. What are endocrine disrupters? Name two endocrine disrupters that are found in the tissues of most Americans. What industrial products or consumer items do these pollutants come from, and what harmful effects are known to be associated with exposure to these chemicals?

Appendix
The Hardy–Weinberg Equilibrium

In this appendix we describe the conditions under which populations do not evolve. Specifically, we discuss the conditions for the Hardy–Weinberg equation, a formula that allows us to predict genotype frequencies in a hypothetical nonevolving population. As described in Chapter 14 (see the box on page 317), this equation provides a baseline with which real populations can be compared in order to figure out whether evolution is occurring.

A population can evolve as a result of mutation, gene flow, genetic drift, or natural selection. Put another way, a population does *not* evolve when the following four conditions are met:

1. There is no net change in allele frequencies due to mutation.
2. There is no gene flow. This condition is met when new alleles do not enter the population via immigrating individuals, seeds, or gametes.
3. Genetic drift does not change allele frequencies. This condition is met when the population is very large.
4. Natural selection does not occur.

The Hardy–Weinberg equation is derived from the assumption that all four of these conditions are met. In reality, these four conditions are rarely met completely in natural populations. However, many populations meet these conditions well enough that the Hardy–Weinberg equation is approximately correct, at least for some of the genes within the population.

To derive the Hardy–Weinberg equation, consider a hypothetical population of 1,000 moths. The dominant allele for orange wing color (W) has a frequency of 0.4, and the recessive allele for white wing color (w) has a frequency of 0.6. What we seek to do now is predict the frequencies of the WW, Ww, and ww genotypes in the next generation for a population that is not evolving.

If mating among the individuals in the population is random (that is, if all individuals have an equal chance of mating with any member of the opposite sex), and if the four conditions just described are also met, we can use the approach described in the accompanying figure to predict the genetic makeup of the next generation. This approach is similar to mixing all the possible

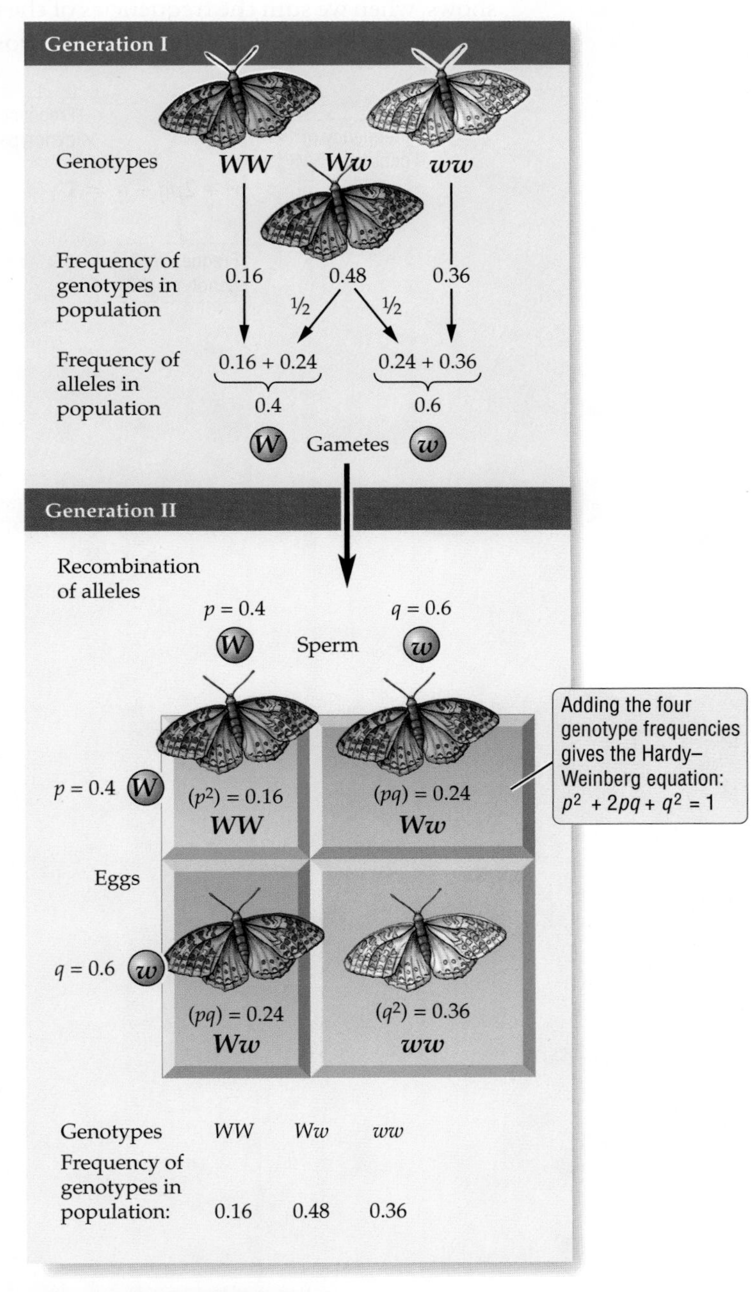

The Hardy–Weinberg Equation

When mating is random and certain other conditions are met, allele and genotype frequencies in a population do not change. p = frequency of the W allele; q = frequency of the w allele.

gametes in a bag and then randomly drawing one egg and one sperm to determine the genotype of each offspring. With such random drawing, the allele and genotype frequencies in our moth population do not change from one generation to the next, as the figure shows.

Because the *WW*, *Ww*, and *ww* genotypes are the only three types of zygotes that can be formed, the sum of their frequencies must equal 1. As the figure shows, when we sum the frequencies of the three genotypes, we get the Hardy–Weinberg equation:

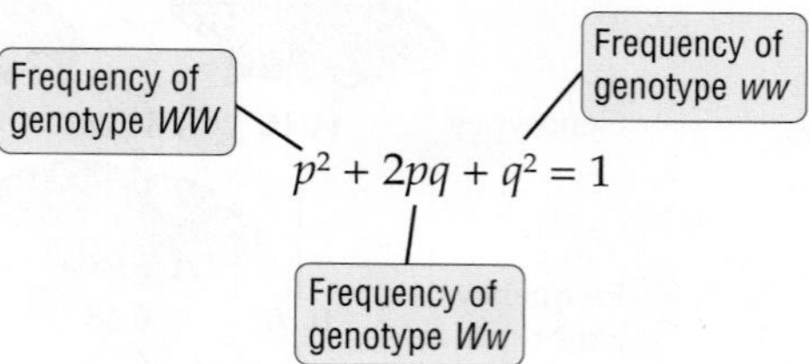

In this equation, the frequency of the *W* allele is labeled p and the frequency of the w allele is labeled q.

In general, once the genotype frequencies of a population equal the Hardy–Weinberg frequencies of p^2, $2pq$, and q^2, they remain constant over time if the four specified conditions continue to be met. A population in which the observed genotype frequencies match the Hardy–Weinberg predicted frequencies is said to be in Hardy–Weinberg equilibrium

Table of Metric–English Conversion

Common conversions

LENGTH		TO CONVERT	MULTIPLY BY	TO YIELD
nanometer (nm)	0.000000001 (10^{-9}) m	inches	2.54	centimeters
micrometer (μm)	0.000001 (10^{-6}) m	yards	0.91	meters
millimeter (mm)	0.001 (10^{-3}) m	miles	1.61	kilometers
centimeter (cm)	0.01 (10^{-2}) m			
meter (m)	—	centimeters	0.39	inches
kilometer (km)	1,000 (10^{3}) m	meters	1.09	yards
		kilometers	0.62	miles

WEIGHT (MASS)				
nanogram (ng)	0.000000001 (10^{-9}) g	ounces	28.35	grams
microgram (μg)	0.000001 (10^{-6}) g	pounds	0.45	kilograms
milligram (mg)	0.001 (10^{-3}) g			
gram (g)	—	grams	0.035	ounces
kilogram (kg)	1,000 (10^{3}) g	kilograms	2.20	pounds
metric ton (t)	1,000,000 (10^{6}) g (=10^{3} kg)			

VOLUME				
microliter (μl)	0.000001 (10^{-6}) l	fluid ounces	29.57	milliliters
milliliter (ml)	0.001 (10^{-3}) l	quarts	0.95	liters
liter (l)	—			
kiloliter (kl)	1,000 (10^{3}) l	milliliters	0.034	fluid ounces
		liters	1.06	quarts

TEMPERATURE				
degree Celsius (°C)	—	To convert Fahrenheit (°F) to Celsius (°C): $°C = \frac{5}{9}°F - 32°$ To convert Celsius (°C) to Fahrenheit (°F): $°F = \frac{9}{5}°C + 32°$		

PERIODIC TABLE OF THE ELEMENTS

Elements essential for living organisms

Key: 1 — Atomic number; H — Symbol; Hydrogen — Name; 1.00794 — Average atomic mass

Metals | Metalloids | Nonmetals

Period	1 1A	2 2A	3 3B	4 4B	5 5B	6 6B	7 7B	8 8B	9 8B	10 8B	11 1B	12 2B	13 3A	14 4A	15 5A	16 6A	17 7A	18 8A
1	1 **H** Hydrogen 1.00794																	2 **He** Helium 4.002602
2	3 **Li** Lithium 6.941	4 **Be** Beryllium 9.012182											5 **B** Boron 10.811	6 **C** Carbon 12.0107	7 **N** Nitrogen 14.0067	8 **O** Oxygen 15.9994	9 **F** Fluorine 18.9984032	10 **Ne** Neon 20.1797
3	11 **Na** Sodium 22.98976928	12 **Mg** Magnesium 24.3050											13 **Al** Aluminum 26.9815386	14 **Si** Silicon 28.0855	15 **P** Phosphorus 30.973762	16 **S** Sulfur 32.065	17 **Cl** Chlorine 35.453	18 **Ar** Argon 39.948
4	19 **K** Potassium 39.0983	20 **Ca** Calcium 40.078	21 **Sc** Scandium 44.955912	22 **Ti** Titanium 47.867	23 **V** Vanadium 50.9415	24 **Cr** Chromium 51.9961	25 **Mn** Manganese 54.938045	26 **Fe** Iron 55.845	27 **Co** Cobalt 58.933195	28 **Ni** Nickel 58.6934	29 **Cu** Copper 63.546	30 **Zn** Zinc 65.38	31 **Ga** Gallium 69.723	32 **Ge** Germanium 72.64	33 **As** Arsenic 74.92160	34 **Se** Selenium 78.96	35 **Br** Bromine 79.904	36 **Kr** Krypton 83.798
5	37 **Rb** Rubidium 85.4678	38 **Sr** Strontium 87.62	39 **Y** Yttrium 88.90585	40 **Zr** Zirconium 91.224	41 **Nb** Niobium 92.90638	42 **Mo** Molybdenum 95.96	43 **Tc** Technetium [98]	44 **Ru** Ruthenium 101.07	45 **Rh** Rhodium 102.90550	46 **Pd** Palladium 106.42	47 **Ag** Silver 107.8682	48 **Cd** Cadmium 112.411	49 **In** Indium 114.818	50 **Sn** Tin 118.710	51 **Sb** Antimony 121.760	52 **Te** Tellurium 127.60	53 **I** Iodine 126.90447	54 **Xe** Xenon 131.293
6	55 **Cs** Cesium 132.9054519	56 **Ba** Barium 137.327	57 **La** Lanthanum 138.90547	72 **Hf** Hafnium 178.49	73 **Ta** Tantalum 180.94788	74 **W** Tungsten 183.84	75 **Re** Rhenium 186.207	76 **Os** Osmium 190.23	77 **Ir** Iridium 192.217	78 **Pt** Platinum 195.084	79 **Au** Gold 196.966569	80 **Hg** Mercury 200.59	81 **Tl** Thallium 204.3833	82 **Pb** Lead 207.2	83 **Bi** Bismuth 208.98040	84 **Po** Polonium [209]	85 **At** Astatine [210]	86 **Rn** Radon [222]
7	87 **Fr** Francium [223]	88 **Ra** Radium [226]	89 **Ac** Actinium [227]	104 **Rf** Rutherfordium [261]	105 **Db** Dubnium [262]	106 **Sg** Seaborgium [266]	107 **Bh** Bohrium [264]	108 **Hs** Hassium [277]	109 **Mt** Meitnerium [268]	110 **Ds** Darmstadtium [271]	111 **Rg** Roentgenium [272]	112 **Cn** Copernicium [285]	113 **Uut** Ununtrium [284]	114 **Fl** Flerovium [289]	115 **Uup** Ununpentium [288]	116 **Lv** Livermorium [292]	117 **Uus** Ununseptium [294]	118 **Uuo** Ununoctium [294]

6 Lanthanides	58 **Ce** Cerium 140.116	59 **Pr** Praseodymium 140.90765	60 **Nd** Neodymium 144.242	61 **Pm** Promethium [145]	62 **Sm** Samarium 150.36	63 **Eu** Europium 151.964	64 **Gd** Gadolinium 157.25	65 **Tb** Terbium 158.92535	66 **Dy** Dysprosium 162.500	67 **Ho** Holmium 164.93032	68 **Er** Erbium 167.259	69 **Tm** Thulium 168.93421	70 **Yb** Ytterbium 173.05	71 **Lu** Lutetium 174.967
7 Actinides	90 **Th** Thorium 232.03806	91 **Pa** Protactinium 231.03588	92 **U** Uranium 238.02891	93 **Np** Neptunium [237]	94 **Pu** Plutonium [244]	95 **Am** Americium [243]	96 **Cm** Curium [247]	97 **Bk** Berkelium [247]	98 **Cf** Californium [251]	99 **Es** Einsteinium [252]	100 **Fm** Fermium [257]	101 **Md** Mendelevium [258]	102 **No** Nobelium [259]	103 **Lr** Lawrencium [262]

We have used the U.S. system as well as the system recommended by the International Union of Pure and Applied Chemistry (IUPAC) to label the groups in this periodic table. The system used in the United States includes a letter and a number (1A, 2A, 3B, 4B, etc.), which is close to the system developed by Mendeleev. The IUPAC system uses numbers 1–18 and has been recommended by the American Chemical Society (ACS). While we show both numbering systems here, we use the IUPAC system exclusively in the book. Elements with atomic numbers higher than 112 have been reported but not yet fully authenticated.

Self-Quiz Answers

Chapter 1

1. *a*
2. *b*
3. *d*
4. *c*
5. *d*
6. *a*
7. *c*
8. *b*
9. *c*

Chapter 2

1. *a*
2. *c*
3. *d*
4. *a*
5. *c*
6. *b*
7. *c*
8. *b*
9. *d*
10. *c*
11. *c*
12. *d*
13. *a*
14. *b*

Chapter 3

1. *b*
2. *a*
3. *c*
4. *d*
5. *b*
6. *a*
7. *b*
8. *a*
9. *b*
10. *c*

Chapter 4

1. *d*
2. *b*
3. *a*
4. *c*
5. *c*
6. *d*
7. *b*
8. *a*
9. *d*
10. *b*

Chapter 5

1. *c*
2. *a*
3. *b*
4. *c*
5. *d*
6. *a*
7. *c*
8. *d*
9. *d*
10. *c*

Chapter 6

1. *d*
2. *b*
3. *d*
4. *b*
5. *b*
6. *c*
7. *a*
8. *a*
9. *d*
10. *b*
11. *c*
12. *a*

Chapter 7

1. *b*
2. *a*
3. *b*
4. *d*
5. *c*
6. *c*
7. *d*
8. *a*

Chapter 8

1. *d*
2. *b*
3. *c*
4. *d*
5. *c*
6. *c*
7. *d*
8. *c*
9. *c*

Chapter 9

1. *a*
2. *b*
3. *d*
4. *d*
5. *a*
6. *d*
7. *d*

Chapter 10

1. *c*
2. *b*
3. *c*
4. *c*
5. *a*
6. *c*
7. *d*

Chapter 11

1. *c*
2. *c*
3. *b*
4. *a*
5. *c*
6. *d*
7. *c*
8. *c*

Chapter 12

1. *b*
2. *c*
3. *b*
4. *d*
5. *a*
6. *d*
7. *b*
8. *c*

Chapter 13

1. *c*
2. *d*
3. *d*
4. *d*
5. *a*
6. *c*
7. *a*
8. *d*

Chapter 14

1. *d*
2. *b*
3. *d*
4. *d*
5. *a*
6. *d*
7. *c*
8. *d*

Chapter 15

1. *b*
2. *b*
3. *c*
4. *a*
5. *a*
6. *d*
7. *b*
8. *c*
9. *d*
10. *c*

Chapter 16

1. *d*
2. *b*
3. *a*
4. *c*
5. *a*
6. *b*
7. *c*
8. *c*

Chapter 17

1. *c*
2. *b*
3. *a*
4. *d*
5. *d*
6. *a*
7. *b*
8. *c*
9. *d*
10. *b*

Chapter 18

1. *c*
2. *d*
3. *c*
4. *c*
5. *c*
6. *a*
7. *c*
8. *c*

Chapter 19

1. *a*
2. *b*
3. *d*
4. *b*
5. *b*
6. *b*
7. *a*
8. *d*

Chapter 20

1. *c*
2. *d*
3. *c*
4. *b*
5. *c*
6. *d*
7. *b*
8. *c*
9. *b*

Chapter 21

1. *d*
2. *b*
3. *b*
4. *c*
5. *d*
6. *a*
7. *d*
8. *b*

Chapter 22

1. *d*
2. *d*
3. *c*
4. *c*
5. *c*
6. *a*
7. *c*
8. *a*
9. *b*
10. *a*

Chapter 23

1. *d*
2. *d*
3. *b*
4. *d*
5. *d*
6. *c*
7. *b*
8. *b*

Chapter 24

1. *c*
2. *a*
3. *d*
4. *a*
5. *b*
6. *b*
7. *c*
8. *d*
9. *d*

Chapter 25

1. *a*
2. *b*
3. *a*
4. *d*
5. *d*
6. *c*
7. *d*

Analysis and Application Answers

Chapter 1

1. Science is a body of knowledge about the natural world and an evidence-based process for generating that knowledge. The characteristics of science include these: (1) Science deals with the natural world, which can be detected, observed, and measured. (2) Scientific knowledge is based on evidence that can be demonstrated through observations and/or experiments. (3) Scientific knowledge is subject to independent validation and peer review. (4) Science is open to challenge based on evidence by anyone at any time. (5) Science is a self-correcting enterprise.

   Science cannot answer all types of questions that humans might raise. Science is restricted to seeking natural causes to explain the workings of our world. Science cannot tell us what is morally right or wrong, or speak to the existence of God or any supernatural being. Science can only attempt to find the objective truth; it cannot address subjective issues such as what is beautiful or ugly.
2. "Correlation" means that two or more aspects of the natural world behave in an interrelated manner: if one variable shows a particular value, we can predict a particular value for the other. However, correlation between two variables does not necessarily imply that one is the cause of the other.

   As one example, studies show that the incidence of skin cancer has gone up sharply since the 1950s, after sunscreens were introduced and their use climbed in the decades that followed. Does that mean that the use of sunscreen causes skin cancer? Further analysis revealed that post–World War II generations spend much more time in the sun, and that people who sunburn easily and would have spent less time in the sun in the past are more exposed to the sun now because of a false sense of security. Further, dermatologists say that most people use sunscreen inappropriately, usually applying too little of it and too infrequently for maximum protection.

   As another example, a Hungarian study reported that men who carry their cell phones in their pants pockets have a 30 percent lower sperm count than do men who keep their cell phones in a jacket pocket. The media ran with the assumption that cell phones reduce sperm count. Shortly thereafter, a physician pointed out that men who use their pants pockets for cell phones are disproportionately smokers, who commonly save their jacket pockets for their cigarette packs because there they are less likely to get crushed. That smoking reduces sperm count is already well known.
3. **Observation:** Some of the lowest rates of heart disease are found among communities that eat a lot of fish and other seafood, such as native Alaskans and Greenland Inuits.

   Hypothesis: Fish oil in the diet reduces the risk of death from heart disease.

   Experiment: About 2,000 men diagnosed with heart disease were enrolled in a controlled study in which the independent variable was fish oil supplementation. About half of the study participants were randomly assigned to the control group; these men were not directed to change their usual diet in any way. The men in the treatment group took 900 milligrams (mg) of purified fish oil each day. At the end of the 2-year study period, there were 62 percent fewer deaths in the treatment group compared to the control group. The experiment therefore supported the hypothesis that fish oils reduce heart disease mortality.
4. A scientific hypothesis (1) is an educated guess that seeks to explain observed phenomena; (2) makes clear predictions that can be arranged in "if ... then" statements; (3) must be testable repeatedly and independently; (4) must be potentially refutable; (5) can never be proved, but only supported or refuted.
5. A scientific fact is a direct and repeatable observation of any aspect of the physical world. A scientific theory is a component of scientific knowledge that has been repeatedly confirmed in diverse ways and is provisionally accepted by the experts of the discipline because it has stood the test of time.
6. Energy flows from the sun to photosynthesizing organisms such as grasses. The grasses, which are producers, use the sun's energy to produce chemical energy in the form of sugars and starches. Antelope, which are consumers, feed on the grasses to produce energy for their own use. Lions, also consumers, then eat the antelope. Ticks, consumers as well, feed on both the antelope and the lions. The grasses are producers because they capture sunlight and convert it to energy. The antelope, lions, and ticks are consumers because they eat either plants or other organisms that derive energy from plants. The food chain looks like this:

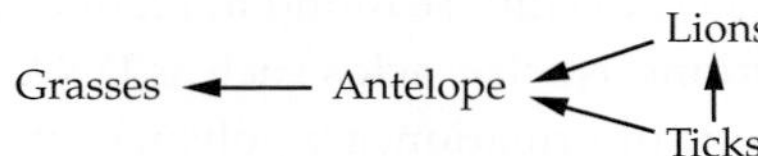

7. Levels of the biological hierarchy, from smallest to largest (with examples), are: atom (carbon); molecule (DNA); cell (bacteria); tissue (muscle tissue); organ (heart); organ system (stomach, liver, and intestines in the digestive system); individual (human); population (field mice in one field); community (different species of

insects living in a forest); ecosystem (a river and the communities of organisms living in and around it); biome (the Arctic tundra or a coral reef); biosphere (Earth).

8. The American Council on Exercise (ACE) and *Consumer Reports* have both found fault with the claims made by makers of toning shoes. Check ACE's website to see how it designed and conducted experiments to test the claims.

Chapter 2

1. Monomers are small molecules that serve as repeating units in a larger molecule (macromolecule). Macromolecules that contain monomers as building blocks are known as polymers. Lipids, such as triglycerides and sterols, are macromolecules, but since they are not built from discrete units that are repeated multiple times, they are not usually regarded as polymers.
2. The pH of pure water should be 7. Units on the pH scale represent the concentration of free hydrogen ions in water. In the presence of a base, the pH of a solution will be above 7, indicating that there are more hydroxide ions than hydrogen ions, so the solution is basic. In the presence of an acid, the pH will be below 7, indicating more free hydrogen ions than hydroxide ions, so the solution is acidic. Pure water has equal amounts of hydrogen and hydroxide ions and is therefore neutral.
3. A hydrogen bond is a noncovalent bond created by the electrical attraction between a hydrogen atom with a partial positive charge and any other atom that has a partial negative charge. Hydrogen bonds are weaker than ionic bonds, which are weaker than covalent bonds. Water molecules are polar: the region around the oxygen atom is slightly negative, and the regions around the two hydrogen atoms are slightly positive. This property provides for the formation of hydrogen bonds between water molecules, since each partially charged hydrogen atom in a water molecule is attracted to any atom with a partial negative charge, including the oxygen atom of a nearby water molecule.
4. Each carbon atom can form strong covalent bonds with up to four other atoms, including other carbons, creating large molecules that contain hundreds, even thousands, of atoms. These molecules play many different roles critical to life.
5. Cells use carbohydrates as a readily available energy source. Some carbohydrates, such as the cellulose found in plant cell walls, have structural functions. Nucleic acids such as DNA and RNA, which carry genetic information, are polymers of nucleotides. Some nucleotides act as energy carriers. Proteins make up the physical structures of organisms, as well as the enzymes that catalyze biochemical reactions. Lipids such as triglycerides are common means of long-term energy storage, and lipids such as phospholipids are important components of cell membranes.
6.

FOOD INGREDIENT	MACRO-MOLECULE	BUILDING BLOCK(S)	FUNCTION
Hamburger bun	Starch (polysaccharide)	Glucose (sugar)	Energy source; building block for many other biomolecules
Lettuce, onion, tomatoes	Cellulose (polysaccharide)	Glucose (sugar)	Non-nutritive because undigested; important for intestinal health
Meat patty	Protein	Amino acids	Building block for proteins, as well as other biomolecules, such as neurotransmitters
Cheese	Triglyceride (fat)	Fatty acids; glycerol	Energy source; building block for other biomolecules, including membrane phospholipids

7. Elements that are abundant in the human body but scarce in rocks are carbon and nitrogen. Amino acids contain both of these elements. Proteins are polymers of amino acids.
8. The solvent in black coffee is water. In a cup of coffee that is sweetened and topped with whipped cream, the hydrophilic solutes are the hundreds of organic molecules found in coffee (including caffeine) and sugar (sucrose, a disaccharide). The hydrophobic substances, perhaps partially mixed with the coffee but mainly floating on the top, are the whipped-cream constituents, which are mostly lipids (triglycerides and fatty acids).
9. Ice floats on water because it is 9 percent less dense than liquid water. If ice were denser than liquid water, it would sink in this lake in winter, and the lake would then freeze from the bottom up. Instead, the ice that forms on the lake surface acts like an insulating blanket, enabling aquatic organisms to survive the winter in the liquid water below.

   Water molecules in liquid water move about vigorously, constantly forming and breaking hydrogen bonds with neighboring water molecules. The water molecules in ice have less energy and cannot move about as vigorously; they form a more stable network of hydrogen bonds. A given mass of liquid water occupies more space when it turns into ice because the molecules are spaced farther apart in ice, locked into an orderly pattern known as a crystal lattice.

Chapter 3

1. One feature of all cells, the plasma membrane, provides a necessary boundary between a cell and its surrounding environment. The plasma membrane is selectively permeable, controlling what gets in and what flows out. Both prokaryotic and eukaryotic

cells also contain DNA, cytosol, and ribosomes. DNA contains the information for producing the proteins needed by each cell. Cytosol is the watery medium in which biochemical reactions take place. Ribosomes are the workbenches for producing proteins.

2. The major components of a plasma membrane are a phospholipid bilayer and an assortment of proteins. The phospholipid molecules are oriented so that their hydrophilic heads are exposed to the watery environments both inside and outside the cell. Their hydrophobic fatty acid tails are grouped together inside the membrane away from the watery surroundings. Some membrane proteins extend all the way through the phospholipid bilayer and act as gateways for the passage of selected ions and molecules into and out of the cell. Other membrane proteins are used by the cell to detect changes in and signals from the environment outside the cell. Proteins that are not anchored to structures within the cell are free to move sideways within the phospholipid bilayer. This freedom of movement supports what is known as the fluid mosaic model, which describes the plasma membrane as a highly mobile mixture of phospholipids and proteins. This mobility is essential for many cellular functions, including movement of the cell as a whole and the ability to detect external signals.
3. Chloroplasts are found in photosynthetic cells of algae and plants. Each chloroplast has two concentric membranes and an internal network of membranes (thylakoids) that contain the light-absorbing green pigment chlorophyll. Chloroplasts harness light energy to convert carbon dioxide (CO_2) into sugars, splitting water molecules (H_2O) and releasing oxygen gas (O_2) in the process of photosynthesis. Mitochondria are found in nearly all cell types in all eukaryotes, both producers and consumers. Each mitochondrion has two membranes, the inner of which is thrown into many folds (cristae) and contains proteins and other components that enable the organelle to generate the energy carrier ATP, through the oxygen-dependent process of cellular respiration. In this process, the chemical energy of organic molecules (such as sugars) is converted into energy stored within ATP molecules, and CO_2 and H_2O are released as O_2 is consumed.
4. Secreted proteins such as elastin are synthesized by ribosomes attached to the cytosolic side of the rough endoplasmic reticulum (rER). These secreted proteins are released into the lumen of the rER. Transport vesicles carry elastin from the rER to the Golgi apparatus, where the protein may be covalently modified—for example, through the addition of sugar chains on the protein backbone. Transport vesicles containing the modified elastin bud off from the Golgi apparatus and fuse with the plasma membrane in the process of exocytosis, which causes the protein to be released into the extracellular matrix.
5. Multicellularity enables organisms to attain a larger size while still being composed of small cells that collectively present a large surface area for the exchange of materials. Multicellularity also offers the advantage of greater efficiency by enabling division of labor among the highly specialized cell types.
6. The partitioning of the cytoplasm into a variety of highly specialized membrane-enclosed compartments confers speed and efficiency through an intracellular division of labor. The different types of membrane-enclosed organelles serve very specific and unique functions. Unique chemical environments can be maintained within a membrane-enclosed compartment. For example, the reactions that break apart polymers work best under exceptionally acidic conditions, so the organelles that specialize in that task maintain a very low pH, even though the pH of the cytosol outside the organelles is close to neutral. Some chemical reactions produce by-products that could interfere with other vital reactions or even poison the cell. Locking interfering or toxic substances into special compartments avoids such "collateral damage."
7. Parts of a plant cell:

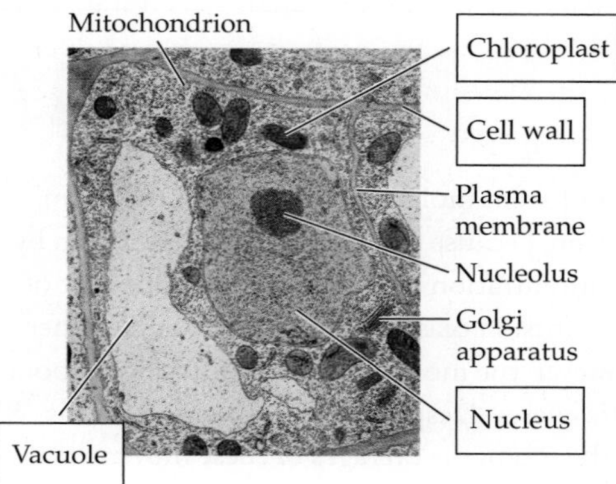

8.

CELLULAR STRUCTURE	MAIN FUNCTION(S)	FOUND IN: PROKARYOTES? EUKARYOTES? PLANTS? ANIMALS?
Plasma membrane	Boundary of cell; controlling movement of substances in and out of cell	Yes, in all
Cytoplasm	Water-based medium; site of thousands of critical chemical reactions	Yes, in all
Nucleus	Repository of DNA, the genetic material; site for replication of DNA and expression of coded information in the form of RNA	Not in prokaryotes, but in all others
Ribosome	Protein-synthesizing unit in the cytoplasm	Yes, in all
Endoplasmic reticulum	Site for synthesis of most lipids and many proteins exported to organelles or the cell exterior	In all eukaryotes, including plants and animals
Golgi apparatus	Site for tagging of lipids and proteins, and for shipping to their final destinations	In all eukaryotes, including plants and animals
Lysosome	Degradation and recycling of molecules and even whole organelles	In animals and some other eukaryotes, such as fungi and some protists

(continued)

CELLULAR STRUCTURE	MAIN FUNCTIONS	FOUND IN: PROKARYOTES? EUKARYOTES? PLANTS? ANIMALS?
Mitochondrion	Energy production in the form of the energy-rich molecule ATP, which is needed by all cells	In all eukaryotes, including plants and animals
Chloroplast	Synthesis of carbohydrate from carbon dioxide and water using light energy	In plants and some protists (algae); not in any other eukaryote
Cytoskeleton	Cell shape; internal organization; intracellular movement of organelles; whole-cell movement	In all eukaryotes, including plants and animals

Chapter 4

1. Yes, the net movement of scent molecules throughout the room is an example of diffusion, because it is a passive process driven by the difference in the concentration of molecules from one part of the room to another. Diffusion ceases at equilibrium because net movement ceases. However, the molecules continue to move about because of their kinetic energy. Despite the continuation of this molecular motion in the room, the averages of these movements are equal in all directions and therefore cancel themselves out.
2. The pond water surrounding the *Paramecium* is hypotonic relative to the cell. The cell has a tendency to gain water through osmosis. It has special organelles, known as contractile vacuoles, that regularly collect excess water and discharge it to the outside of the cell.
3. Phagocytic white blood cells will engulf the invading bacteria. They use a special form of endocytosis, called phagocytosis, to internalize large particles and whole cells. Membranes play a crucial part in the process, which begins when receptor proteins on the outer surface of the plasma membrane recognize surface characteristics of the material to be brought into the cell. Extensions of the membranes called pseudopodia then encircle the invading bacterium to enclose it completely in a vesicle. The vesicle eventually fuses with a lysosome, where the engulfed bacterium is destroyed.
4. Epithelial cells in animals can form leakproof barriers because of tight junctions, which bind cells together with belts of protein embedded in their plasma membranes. The tight junctions prevent molecules from slipping between the cells of the intestinal epithelium to enter the bloodstream on the other side of the cell layer. Molecules and ions can pass to the other side only if they are selectively taken up by the epithelial plasma membrane that faces the space inside the gut.
5. The role of LDL (low-density lipoprotein) particles is to deliver lipids (including cholesterol) from the liver, where these lipids are manufactured, to other cells in the body. A region of the apolipoprotein in the LDL particle is recognized by the LDL receptor. The docking of an LDL particle with an LDL receptor triggers endocytosis of the entire complex. Endocytotic compartments eventually deliver the LDL particle-receptor complex to the lysosome, an organelle that specializes in taking apart biomolecules and releasing the building blocks into the cytosol for reuse.

Chapter 5

1. Hopefully you are not among the 4 in 10 Americans who skip breakfast most days. According to an ABC News poll, 31 percent of Americans who eat breakfast regularly go for cold breakfast cereal, the top choice. The macromolecular carbohydrates, proteins, fats, and nucleic acids in a typical breakfast cereal are broken down to small organic molecules (sugars, amino acids, fatty acids and monoglycerides, nucleotides, and phosphate groups) that are then absorbed into the bloodstream. Some of the chemical energy in these macromolecules is released as heat when they are digested, but beyond warming us slightly, this heat energy is not available for cellular work. In cellular respiration, some of the chemical energy in sugars and fats is turned into the chemical energy of ATP molecules, but a large amount is also released as metabolic heat in accordance with the second law of thermodynamics.

   ATP is used in myriad ways: to contract the heart muscle, thereby pushing blood through the body (chemical energy to kinetic energy); to fire brain cells that help us think (chemical energy to electrical energy); to supply the sodium-potassium pumps that maintain the osmotic balance of every cell (chemical energy being transformed into the potential energy of an electrical gradient); and to synthesize every major biomolecule somewhere in the body (chemical energy being stored as the potential energy of larger molecules). The average cell in the human body uses up about 1–2 billion molecules of ATP per minute, and while some of the chemical energy released in the breaking of a phosphate bond is used to fuel cellular activities, a large proportion is released as metabolic heat. These are just some of the many different kinds of energy transformations that take place in your body every second of your life.
2. The second law of thermodynamics holds that systems tend toward disorder. In a living system such as a cell, the order maintained by chemical reactions is counterbalanced by the release of heat energy (disorder) into the surroundings.
3. "Anabolism" refers to metabolic processes that manufacture larger molecules from smaller units. Anabolic pathways need an input of energy. "Catabolism" refers to metabolic processes that break down macromolecules, releasing energy and small organic molecules. Photosynthesis is an anabolic process because it creates large molecules such as sugars from smaller units such as carbon dioxide and water. It is driven by an input of light energy.
4. Enzymes are biological catalysts that speed up chemical reactions by positioning bound reactant molecules in such a way that they collide more often and more accurately, thereby increasing the reaction rate. Like any other catalyst, an enzyme lowers the activation

energy barrier of a reaction to make a reaction proceed faster, but it does not provide energy to make that reaction happen. In other words, it cannot make a chemical reaction happen that would not proceed without an input of external energy.

5. According to the induced fit model of enzyme action, the binding of the substrate to the active site of an enzyme further molds the active site to create a more stable interaction between an enzyme and its substrate. The bound substrate reshapes the binding site of the enzyme slightly, the way your hand shapes a properly sized glove when you put it on.
6. Ephedrine, the active ingredient in the herbal medicine, is a stimulant that activates the nervous system in ways that increase heart rate and metabolic rate. The increased output of metabolic heat leads to flushing and sweating in most users. The drug was quite effective in weight loss because it increased the metabolism of food stores while also suppressing appetite. However, it also produced serious side effects, and after a number of deaths from heatstroke or fast and irregular heartbeats were linked to the use of ephedrine, sale of the drug was banned by the FDA (Food and Drug Administration).

Chapter 6

1. No, the Calvin cycle cannot be sustained in a plant that is kept in the dark for more than a very short period (a few hours), because these enzymatic reactions depend on the chemical energy delivered by NADPH and ATP that are generated by the light reactions.
2. The transfer of electrons down an electron transport chain (ETC) produces a proton gradient in both chloroplasts and mitochondria. The protons move down that gradient through a membrane channel protein known as ATP synthase. The movement of the protons releases energy, which is used by ATP synthase to phosphorylate ADP to form ATP.
3. The postglycolytic fermentation reactions do not generate any energy. Fermentation swings into action in some cells when the oxygen supply is low, and its main function is to support increased ATP production through increased rates of glycolysis. The only role of the postglycolytic fermentation reactions is to regenerate NAD^+, which is essential for the continued operation of glycolysis.
4. ATP production through oxidative phosphorylation is critically dependent on a proton (H^+) gradient generated by the energy released by electrons as they travel down the electron transport chain (ETC). Electron transport pumps protons from the matrix into the intermembrane space of the mitochondrion, and the potential energy of this gradient powers the phosphorylation of ADP to generate ATP by ATP synthase.

   If the proton gradient collapses before ATP can be generated, all of its potential energy is released as heat, instead of some being transformed into the chemical energy of ATP through oxidative phosphorylation. By collapsing the proton gradient, DNP causes all of its energy to be released as heat. The overall effect is that at low concentrations of DNP, some of the chemical energy in food molecules (the ultimate source of the electrons that travel down the ETC) is released as heat. That means some of the energy in food is "burned off" by the body, so less of it is available as surplus energy to be stashed away as stored fat. This is a dangerous way of tweaking mitochondrial energy output because if the DNP levels become too high in critical tissues, the molecule can kill cells by destroying their ability to generate the life-giving molecule ATP.
5. Mitochondrial respiration is more productive than glycolysis alone. For each molecule of glucose consumed during glycolysis, there is a net yield of 2 ATP molecules and 2 NADH molecules. Much energy remains in the two pyruvate molecules furnished by the glycolytic degradation of one glucose molecule. Mitochondria break up pyruvate through a series of reactions (glycolysis, the Krebs cycle, and oxidative phosphorylation) to yield about 30 ATP molecules per molecule of glucose.
6. Mitochondrial membranes play a crucial role in oxidative phosphorylation, which takes place in the many folds (cristae) of the inner mitochondrial membrane. These folds create a large surface area on which are embedded many electron transport chains (ETCs) and many units of ATP synthase. In contrast, the outer membrane is unfolded because it does not bear any of the components that are responsible for ATP generation. The inner mitochondrial membrane also serves the important function of forming a barrier across which a proton gradient can develop during electron transport. As described in the answer to question 4, the potential energy of this gradient is absolutely necessary for ATP synthesis and hence the energy output from cellular respiration. The thylakoids serve the same function in chloroplasts as the space created by the inner mitochondrial membrane in mitochondria.
7.

	PHOTOSYSTEM I	**PHOTOSYSTEM II**
Located in the stroma of the chloroplast?	no	no
Contains chlorophyll?	yes	yes
Directly involved in splitting of water molecules?	no	yes
Directly involved in generating NADPH?	yes	no
Associated with an electron transport chain?	yes	yes

8. The first product of carbon fixation is a four-carbon molecule in both C_4 and CAM plants, whereas this first product of carbon fixation in C_3 plants is a three-carbon molecule called PGA. C_4 plants open their stomata and produce the four-carbon molecule during the day, whereas CAM plants open their stomata only at night and store the four-carbon molecule they produce at night in large vacuoles. In C_4 plants the light reactions take place in ordinary leaf cells (mesophyll cells), but the Calvin cycle reactions occur in the high-CO_2 environment generated inside specialized cells (bundle-sheath cells) with gas-impermeable walls. In CAM plants, light reactions and Calvin cycle reactions take place in the same cell type in the leaf or stem; the four-carbon molecule stored overnight in vacuoles breaks down to release CO_2 for the Calvin cycle reactions. C_4 plants are some of the most productive plants in the world, but CAM plants—while being able to survive in desert environments—grow more slowly. Examples of C_3 plants: wheat, beans, maples, roses; of C_4 plants: corn, sugarcane, quack grass; of CAM plants: cacti, pineapple, jade plant, stonecrops (members of the Crassulaceae family).
9. Photosynthesis occurs in chloroplasts, uses light energy, synthesizes sugars from carbon dioxide and water, and releases oxygen as a by-product. Cellular respiration occurs in mitochondria, releases energy from organic molecules such as sugars in an oxygen-dependent process, and generates carbon dioxide and water as by-products. Photosynthesis is an anabolic process, while cellular respiration is a catabolic process. Photosynthesis occurs only in producers (algae and plants), while cellular respiration takes place in both producers and consumers. The energy carriers (ATP and NADPH) generated by the light reactions provide the energy and the protons and electrons necessary for converting carbon dioxide into sugar. All three stages of cellular respiration (glycolysis, Krebs cycle, and oxidative phosphorylation) generate ATP and NADH (closely related to NADPH).

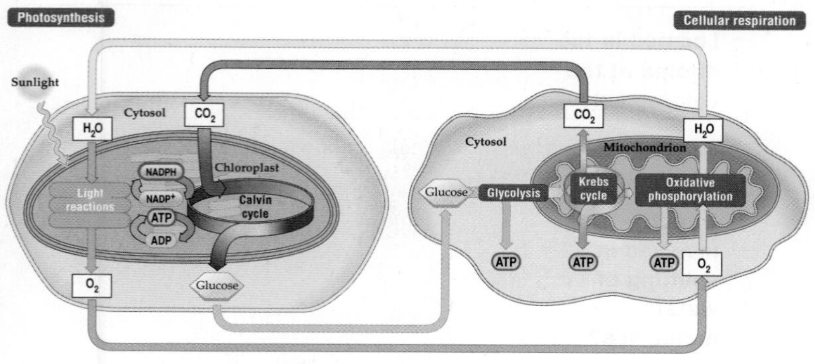

Chapter 7

1. Regular checkpoints ensure that the right cell is dividing at the right time and that cell division occurs with high fidelity. This system of checks and balances is important because cell division is metabolically expensive. Cells with an abnormal chromosome number are also generally nonfunctional and would reduce the fitness of the organism. Finally, in vertebrates, runaway divisions can result in cancer. The cell cycle can progress in only one direction, and because there is no "reverse" button, checks are made to ensure that conditions are optimal before a cell commits to cell division by moving from G_1 to S phase. Cell cycle regulatory proteins can halt the cycle if the cell is too small, the nutrient supply is inadequate, or DNA is damaged. G_2 arrests in the same circumstances, as well as when the DNA duplication that begins in S phase is incomplete for any reason.
2. A horse cell has a total of 128 DNA molecules in the G_2 phase prior to mitosis. At the end of meiosis I, each daughter cell would contain 64 DNA molecules in the form of 32 pairs of linked sister chromatids.
3. Anaphase in mitosis involves (1) the separation of the two chromatids that make up every replicated chromosome and (2) their symmetrical segregation to opposite poles of the cell. The chromatids cannot be segregated evenly unless they are properly aligned at the cell center, which is the main outcome of metaphase. To reduce the risk of mis-segregation of the chromatids, a special anaphase checkpoint called the spindle checkpoint (not described in the chapter) ensures that anaphase does not begin until all replicated chromosomes are properly aligned at the metaphase plate.
4. Homologous chromosome pairs are segregated during meiosis I, whereas sister chromatids are segregated during meiosis II. Meiosis I is therefore the reduction division, resulting in haploid daughter cells (cells with only one chromosome set instead of two). Tetrads are formed and crossing-over occurs during prophase of meiosis I, but there is no equivalent event in prophase of meiosis II. Meiosis II is similar to mitosis in that sister chromatids are segregated into separate daughter cells after cytokinesis is complete. When meiosis begins in a single diploid cell, two cells result at the end of meiosis I, but four daughter cells are seen at the end of meiosis II.
5. If gametes of sexually reproducing organisms were produced by mitosis, the offspring of each succeeding generation would have double the number of chromosomes of the parent generation.
6. Crossing-over occurs during prophase of meiosis I (prophase I). It is the physical exchange of chromosomal segments between non-sister chromatids in paired-off paternal and maternal homologues (tetrads). Crossing-over exchanges alleles between the two homologues. Therefore, chromatids produced by crossing-over are genetic mosaics, bearing new combinations of alleles compared to those originally carried by the paternal and maternal homologues in the diploid parent cell. The mosaic chromatid is said to be recombined, and the creation of new groupings of alleles through the exchange of DNA segments is known as genetic recombination. Crossing-over is biologically significant because it generates genetically diverse daughter cells every time meiosis occurs in the diploid cells of a given individual.

7. Stages of mitosis in HeLa cells:

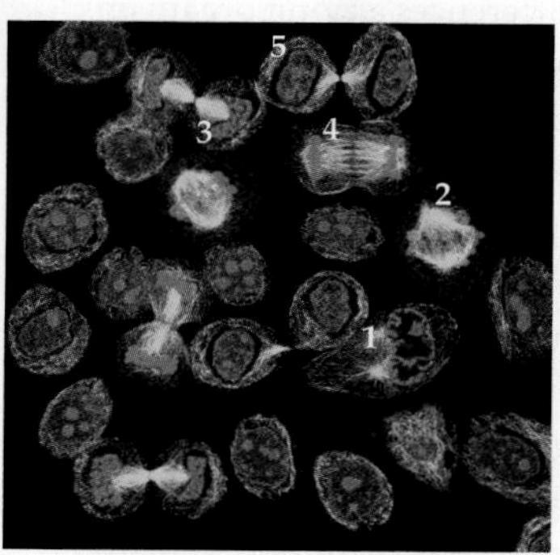

1: Prophase
2: Metaphase
3: Anaphase
4: Telophase
5: Cytokinesis (nearly complete)

8. Comparison of mitosis and meiosis:

	MITOTIC CELL DIVISION	MEIOSIS
1. In humans, the cell undergoing this type of division is diploid.	true	true
2. A total of four daughter cells are produced when one parent cell undergoes this type of division.	false	true
3. Our skin makes more skin cells using this type of cell division.	true	false
4. The daughter cells are genetically identical to the parent cell in this type of division.	true	false
5. This type of division involves two nuclear divisions.	false	true
6. This type of division involves two rounds of cytokinesis.	false	true
7. Maternal and paternal homologues pair up to form tetrads at some point during this type of division.	false	true
8. Sister chromatids separate from each other at some point during this type of cell division.	true	true

Chapter 8

1. Positive growth regulators stimulate cell division and cell survival. Negative growth regulators put the brakes on cell division and can also trigger programmed cell death. Under normal circumstances, these growth regulators are released only when and where they are needed. Runaway cell division results if cell division is excessively stimulated through the pathways controlled by positive growth regulators. But cell division can also be triggered in a cell that ignores the message of negative growth regulators to stop dividing. In either case, excessive cell proliferation sets the stage for the development of cancer.
2. "Angiogenesis" refers to the formation of new blood vessels in a tissue, including in a tumor. Fast-growing tumors often secrete signaling molecules that trigger angiogenesis, thereby recruiting a blood supply to sustain tumor growth. Without the new blood vessels, solid tumors would not be able to grow larger than about 1 or 2 millimeters, because in a larger tumor, the cells buried inside would be starved for nutrients and poisoned by waste products. Angiogenesis, however, enables malignant tumors to grow bigger.
3. The *cdh1* gene is an example of a tumor suppressor gene. It codes for cell adhesion proteins. If the *cdh1* gene becomes nonfunctional, the connections between cells, and between a cell and its extracellular matrix, are lost. The abnormal cell is then able to detach itself from its surroundings, paving the way for it to invade other tissues.
4. Colon cancer begins with a benign growth of cells called a polyp, which is commonly caused by mutations that inactivate both copies of a tumor suppressor gene and/or transform a proto-oncogene into an oncogene. In many patients, loss of a part of chromosome 18 deletes two other tumor suppressor genes, allowing for more aggressive and rapid cell division in the tumor. In most cases of colon cancer, the protective activity of the tumor suppressor gene *p53* is lost, and consequently all remaining cell division controls are removed, enabling the cancer cells of the now malignant tumor to metastasize (spread to other parts of the body).
5. Unavoidable risks include (a) inheritance of oncogenes and other mutated cancer-critical genes, and (b) growing older. There are many avoidable risks, including use of tobacco, excessive exposure to UV light, poor diet, obesity, relatively little physical activity, certain viruses and bacteria, excess alcohol consumption, and exposure to environmental carcinogens.
6. You will have your own opinion on this issue, but all 50 U.S. states place some restrictions on the sale and advertising of tobacco products (for example, a ban on the sale of cigarettes to minors, or a ban on the use of cartoon characters in tobacco advertising). Most states restrict smoking in public places and require warning labels about the harmful effects of tobacco. Public health experts have proposed additional regulations, which include a ban on candy-flavored tobacco products that might be attractive to children; a ban on the labeling of cigarettes as "light" and "mild," because of concerns that such products might be incorrectly perceived as safer; limits on the tar and nicotine content in cigarettes; and larger and more informative warning labels. The FDA (Food and Drug Administration) requires health warnings on all cigarette packaging and advertisements. New warnings implemented in 2012 include very graphic (some say too horrifying)

pictures of the damage caused by tobacco use. Are these justified, in your opinion? Will they have any useful impact? Here's a fact to consider: more than 50 percent of American males smoked in the 1950s; slightly less than 20 percent do so now.

7. Here are some facts to help you formulate an opinion. According to the CDC (Centers for Disease Control and Prevention), the use of tanning beds and sunlamps is linked with skin cancers, including a 59 percent higher risk of melanoma, the deadliest form of skin cancer. Indoor tanning also puts you at risk for premature wrinkling of the skin and eye damage that can result in blindness. Indoor tanning is most common among teenage girls and young women, some of whom may be poorly informed about the health consequences. Indoor tanning by minors is banned in several states, including California, Connecticut, New Jersey, Illinois, Nevada, Oregon, Texas, and Vermont, as well as in countries such as the United Kingdom, Germany, Finland, Italy, and Spain. Despite these restrictions, every year some 30 million Americans use tanning beds, and most of these people are under 30. According to a Minnesota study, in the past 40 years the rates of melanoma have climbed 800 percent among young women and 400 percent among young men. Because teenagers and young people have less spending money than older adults have, the 10 percent tanning tax is expected to deter at least some of them from tanning, and studies in some states have reported a decline. The tanning tax is expected to raise $2.7 billion over 10 years, and these funds are earmarked to offset the cost of the program and help spread sun-safety awareness.
8. It is important to provide sufficient information about cancer risks so that consumers can make informed decisions about their product purchases. Food labels could be simplified to provide more concise information about possible cancer risks, and public health programs could raise awareness of the cancer risks present in certain foods and other products.

Chapter 9

1. Genes are the basic units of inheritance; as such, they carry genetic information for specific traits. Genes are composed of DNA and are located on chromosomes. Most genes govern a genetic trait by influencing the production of specific proteins. There are two copies of every gene in all the diploid cells in a woman's body because she has two of each type of chromosome (a paternal homologue and a maternal homologue), each with one copy of the gene. Like women, men have two copies of all the genes that are located on autosomal chromosomes. However, they have only one copy of the genes that are unique to the X chromosome, since they have only one X chromosome.
2. New alleles arise when genes mutate. A mutation is any change in the DNA that makes up a gene. When a mutation occurs, the new allele that results may contain instructions for a protein with a form different from that of the protein specified by the original allele. By specifying different versions of proteins, the different alleles of a gene cause hereditary differences among organisms.
3. To determine the genotype of the purple-flowered plant (*PP* or *Pp*), you could cross it with a white-flowered plant (*pp*). When either genotype is crossed with the homozygous recessive (the white-flowered plant), the resulting phenotypes of the offspring will indicate whether the purple-flowered parent plant is heterozygous or homozygous. A plant with the genotype *Pp*, when crossed with a plant possessing the *pp* phenotype, will produce offspring that have an equal chance of being either white-flowered or purple-flowered (diagram 1). A dominant homozygous plant (*PP*), when crossed with a recessive homozygous plant (*pp*), will always produce purple-flowered offspring (diagram 2).

(1)

	P	*p*
p	*Pp*	*pp*
p	*Pp*	*pp*

(2)

	P	*P*
p	*Pp*	*PP*
p	*Pp*	*PP*

4. Although identical twins are genetically identical, their phenotypes can differ because environmental factors can alter the effects of genes. One twin who is well nourished in childhood, for example, may grow up to be tall; while the other of the pair, if malnourished in childhood, may grow up to be short. Similarly, exposure to different amounts of sunlight will cause twins' skin color to differ. The phenotypes of twins who are predisposed to a particular genetic disorder can differ radically if one of them is exposed to environmental factors that trigger the onset of that disorder but the other is not.
5. According to the CDC (Centers for Disease Control and Prevention), the four diseases that kill most people in the United States are heart disease, stroke, cancer, and diabetes, with the first three accounting for more than 50 percent of all deaths. Our risk for these diseases is influenced by genes, but lifestyle factors also have a large impact. All four diseases show complex patterns of inheritance because the genetic risk is controlled by numerous, mostly unidentified, genetic loci, and the disease risk is strongly affected by the environment.

   Complex traits are influenced by multiple genes that interact with one another and with the environment. The inheritance of complex traits cannot be predicted by Mendel's laws, because Mendelian traits are controlled by a single gene that is little affected by other genes or by environmental conditions. One way to explain complex traits in simple terms is to say that complex traits are genetic characteristics that are so complicated that we really can't predict how they will be inherited from one generation to the next. The practical value of understanding that many human traits are complex traits is that it can help us avoid the oversimplifications of "pop genetics": claiming to be able to predict a child's height from that of her parents, or setting the odds that someone will get skin cancer because both parents had skin cancer, for example. It also helps us debunk the claims that someone has found, for instance, "a gene for diabetes" or "a gay gene." Finally, complex traits remind us that genes are not destiny; our environment, including our lifestyle, has a large impact on some of the phenotypes that are critical for our health and well-being.

Sample Genetics Problems, Chapter 9

1. a
2. c
3. d
4. a. *A* and *a*
   b. *BC*, *Bc*, *bC*, and *bc*
   c. *Ac*
   d. *ABC*, *ABc*, *AbC*, *Abc*, *aBC*, *aBc*, *abC*, and *abc*
   e. *aBC* and *aBc*
5. a. For *Aa* × *aa*, genotype ratio = 1:1; phenotype ratio = 1:1.

	A	*a*
a	*Aa*	*aa*

   b. For *BB* × *bb*, genotype ratio = 1:0; phenotype ratio = 1:0.

	B
b	*Bb*

   c. For *AABb* × *aabb*, genotype ratio = 1:1; phenotype ratio = 1:1.

	AB	*Ab*
ab	*AaBb*	*Aabb*

   d. For *BbCc* × *BbCC*, genotype ratio = 1 *BBCC*: 1 *BBCc*: 2 *BbCC*: 2 *BbCc*: 1 *bbCC*: 1 *bbCc*; phenotype ratio = 6:2, reduced to 3:1.

	BC	*Bc*	*bC*	*bc*
BC	*BBCC*	*BBCc*	*BbCC*	*BbCc*
bC	*BbCC*	*BbCc*	*bbCC*	*bbCc*

   e. For *AaBbCc* × *AAbbCc*, genotype ratio = 1 *AABbCC*: 2 *AABbCc*: 1 *AABbcc*: 1 *AAbbCC*: 2 *AAbbCc*: 1 *AAbbcc*: 1 *AaBbCC*: 2 *AaBbCc*: 1 *AaBbcc*: 1 *AabbCC*: 2 *AabbCc*: 1 *Aabbcc*; phenotype ratio = 6:2:6:2, reduced to 3:1:3:1.

	ABC	*ABc*	*AbC*	*Abc*	*aBC*	*aBc*	*abC*	*abc*
AbC	*AABbCC*	*AABbCc*	*AAbbCC*	*AAbbCc*	*AaBbCC*	*AaBbCc*	*AabbCC*	*AabbCc*
Abc	*AABbCc*	*AABbcc*	*AAbbCc*	*AAbbcc*	*AaBbCc*	*AaBbcc*	*AabbCc*	*Aabbcc*

6.

	S	*s*
S	*SS*	*Ss*
s	*Ss*	*ss*

   Genotype ratio = 1 *SS*: 2 *Ss*: 1 *ss*: phenotype ratio = 3 healthy: 1 sickle-cell anemia.

   Each time two *Ss* individuals have a child, there is a 25 percent chance that the child will have sickle-cell anemia.
7. 100 percent of the offspring should be chocolate Labs.
8. a. *NN* and *Nn* individuals are normal; *nn* individuals are diseased.
   b.

	N	*n*
N	*NN*	*Nn*
n	*Nn*	*nn*

   Genotype ratio = 1 *NN*: 2 *Nn*: 1 *nn*; phenotype ratio = 3 normal: 1 diseased.

   c.

	N	*n*
N	*NN*	*Nn*

   Genotype ratio = 1:1; phenotype ratio = 2 healthy: 0 diseased.
9. a. *DD* and *Dd* individuals are diseased; *dd* individuals are normal.
   b.

	D	*d*
D	*DD*	*Dd*
d	*Dd*	*dd*

   Genotype ratio = 1 *DD*: 2 *Dd*: 1 *dd*; phenotype ratio = 3 diseased: 1 normal.

   c.

	D	*d*
D	*DD*	*Dd*

   Genotype ratio = 1:1; phenotype ratio = 2 diseased: 0 healthy.
10. The parents are most likely *BB* and *bb*. The white parent must be *bb*. The blue parent could potentially be *BB* or *Bb*, but if it were *Bb*, we would expect about half of the offspring to be white. Therefore, if the cross yields many offspring and all are blue, it is extremely likely that the blue parent's genotype is *BB*.
11. The allele for green fruit pods is dominant. Since each parent breeds true, that means each parent is homozygous. When a homozygous recessive parent is bred to a homozygous dominant parent, the F_1 generation will exhibit only the dominant phenotype. Therefore, the phenotype of the F_1 generation—green fruit pods—is produced by the dominant allele.

Chapter 10

1. A gene is a small region of the DNA molecule in a chromosome. Genes are located on chromosomes.
2. Human females have two X chromosomes, while human males have one X and one Y chromosome; therefore, human males have only one copy of each gene that is unique to either the X or the Y chromosome. As a result, patterns of inheritance for genes located on the X chromosome may differ between males and females. A mother can pass an X-linked allele, such as one for a genetic disorder, to her male or female offspring. A male can pass an X-linked allele only to his female offspring (because his male offspring receive his Y chromosome, not his X chromosome).
3. Yes, a woman can have red-green color blindness if she inherits the allele from both her color-blind father (X^cY) and her carrier or color-blind mother (XX^c or X^cX^c), which would make her a homozygote for this recessive allele (X^cX^c). All the daughters produced by a union between this woman and a man with normal vision will be carriers, and all their sons will be color-blind.

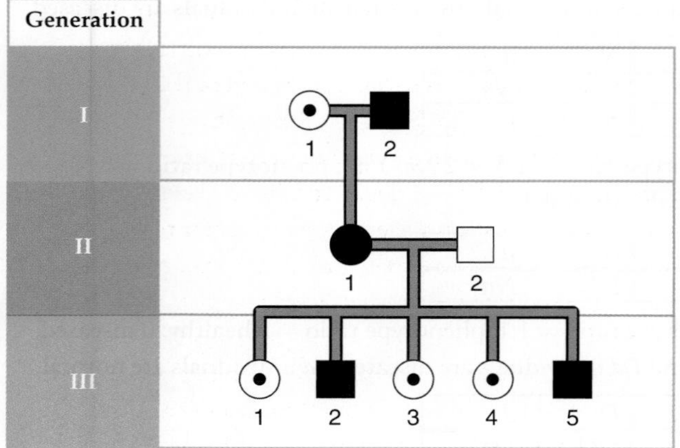

Note: In a pedigree, children are represented by birth order, with the oldest on the left and the youngest on the right. You may imagine any birth order for the three girls and two boys in this pedigree. You can indicate a carrier genotype by filling in half of the relevant circle or square, or by placing a dot in the center of the shape as shown in the example here.

4. Genes located on different chromosomes separate into gametes independently of one another during meiosis; hence, such genes are not linked. If the genes for the traits shown in Figure 10.6 were inherited independently, Morgan would have obtained approximately equal numbers of flies for each of the four genotypes shown in the figure. Since the numbers of the two parental genotypes outnumbered the other two genotypes by a wide margin, Morgan concluded that the genes must be located on the same chromosome. Because they are physically connected, they are inherited together, or linked.
5. Crossing-over occurs when genes are physically exchanged between homologous chromosomes during meiosis. Part of the chromosome inherited from one parent is exchanged with the corresponding region from the other parent. Two genes that are far apart on a chromosome are more likely to be recombined by crossing-over than are two genes that are close to each other. Therefore, one can assume that genes *A* and *C* are more likely to be separated into different gametes because of crossing-over than are genes *A* and *B*.
6. Nonparental genotypes may arise because of crossing-over. The exchange of genes that takes place during crossing-over makes possible the formation of gametes with combinations of alleles that differ from those found in either parent (see Figure 7.16). The independent assortment of homologous chromosome pairs into different gametes is another mechanism that shuffles the allelic combinations found in each parent, generating genetically diverse gametes that each have a random mix of the various paternal and maternal homologues.
7. Relatively few human genetic disorders are caused by inherited chromosomal abnormalities, probably because most large changes in chromosomes kill the developing embryo. Genetic disorders caused by single-gene mutations appear to be more common, because the survival rate of embryos with single-gene mutations is higher.

Sample Genetics Problems, Chapter 10

1. a. A male inherits his X chromosome from his mother, since his Y chromosome must come from his father. His mother does not have a Y chromosome to give him, and he must have one in order to be male.

   b. No, she does not have the disorder. If she has only one copy of the recessive allele, her other X chromosome must then have a copy of the dominant allele. She is a carrier, but she does not have the disorder herself.

   c. Yes, he does have the disorder. The trait is X-linked, he has only one X chromosome, and that X chromosome carries the recessive, disorder-causing allele. His Y chromosome does not carry an allele for this gene, so it cannot contribute to the male's phenotype relative to this trait.

   d. If the female is a carrier of an X-linked recessive disorder, her genotype is X^DX^d, where D is the dominant allele and d is the recessive, disorder-causing allele. This means she can produce two types of gametes relative to this trait: X^D and X^d. Only the X^d gamete carries the disease-causing allele.

   e. None of their children will have the disorder, because the mother will always contribute a dominant, non-disorder-causing allele to each child. However, all of the female children will be carriers, because their second X chromosome comes from their father, who has only one X chromosome to contribute, and it carries the disorder-causing allele.

2. a. For *aa* × *Aa*, 50 percent chance of the cystic fibrosis genotype (*aa*):

	A	*a*
a	*Aa*	*aa*
a	*Aa*	*aa*

   b. For *Aa* × *AA*, 0 percent chance of the cystic fibrosis genotype (*aa*):

	A	*A*
A	*AA*	*AA*
a	*Aa*	*Aa*

   c. For *Aa* × *Aa*, 25 percent chance of the cystic fibrosis genotype (*aa*):

	A	*a*
A	*AA*	*Aa*
a	*Aa*	*aa*

   d. For *aa* × *AA*, 0 percent chance of the cystic fibrosis genotype (*aa*):

	A	*A*
a	*Aa*	*Aa*
a	*Aa*	*Aa*

3. a. For *aa* × *Aa*, 50 percent chance of the Huntington's disease genotype *Aa*:

	A	*a*
a	*Aa*	*aa*
a	*Aa*	*aa*

b. For $Aa \times AA$, 100 percent chance of the Huntington's disease genotype *AA* or *Aa*:

	A	*A*
A	*AA*	*AA*
a	*Aa*	*Aa*

c. For $Aa \times Aa$, 75 percent chance of the Huntington's disease genotype *AA* or *Aa*:

	A	*a*
A	*AA*	*Aa*
a	*Aa*	*aa*

d. For $aa \times AA$, 100 percent chance of the Huntington's disease genotype *Aa*:

	A	*A*
a	*Aa*	*Aa*
a	*Aa*	*Aa*

4. a. For $X^AX^A \times X^aY$, 0 percent chance of a hemophilia genotype:

	X^a	Y
X^A	X^AX^a	X^AY
X^A	X^AX^a	X^AY

b. For $X^AX^a \times X^aY$, 50 percent chance of the hemophilia genotype X^aX^a or X^aY:

	X^a	Y
X^A	X^AX^a	X^AY
X^a	X^aX^a	X^aY

c. For $X^AX^a \times X^AY$, 25 percent chance of the hemophilia genotype X^aY:

	X^A	Y
X^A	X^AX^A	X^AY
X^a	X^AX^a	X^aY

d. For $X^aX^a \times X^AY$, 50 percent chance of the hemophilia genotype X^aY:

	X^A	Y
X^a	X^AX^a	X^aY
X^a	X^AX^a	X^aY

5. No, male and female children do not have the same chance of getting the disease. Male children are more likely to have hemophilia, because they do not possess a second allele for this trait to mask a recessive allele that they may inherit. X-linked hemophilia is extremely rare in females. Because the normal allele (X^H) is dominant over the hemophilia allele (X^h), a girl would have to be homozygous (X^hX^h) to have hemophilia, which would require that she have inherited an X^h allele from her mother and an X^h allele from her father who himself would suffer from hemophilia. Until recently, few males with hemophilia survived into adulthood to father children of their own. Even with the great strides made in recent years in treating hemophilia, the odds are very low that a man with hemophilia would father children with a woman who is a carrier for the X^h allele.
6. The terms "homozygous" and "heterozygous" refer to pairs of alleles for a given gene. Since a male has only one copy of any X-linked gene, it does not make sense to use these pair-related terms when talking about X-linked traits in males.
7. Although neither the mother nor the father expresses the trait in question, some of their children do, so the disease-causing allele (*d*) is recessive and is carried by both parents. The disease-causing allele is located on an autosome. If it were on the X chromosome, the father would express the gene, since we have already determined that he must carry one recessive copy of the gene and he would not have another copy of the gene to mask this recessive allele. Both individuals 1 and 2 of generation I have the genotype *Dd*. Individual 2 of generation I has the disease, so she must have the genotype *dd*, not *Dd*.
8. Designate the dominant, X-linked allele *D* and the recessive normal allele *d*. According to the Punnett squares shown in (a) and (b) below, males are not more likely than females to inherit a dominant, X-linked genetic disorder.
   a. There are two possible Punnett squares, depending on whether the affected female has genotype X^DX^d or genotype X^DX^D:

	X^d	Y
X^D	X^DX^d	X^DY
X^d	X^dX^d	X^dY

or

	X^d	Y
X^D	X^DX^d	X^DY

   b. This cross is $X^dX^d \times X^DY$, which gives the following Punnett square:

	X^D	Y
X^d	X^DX^d	X^dY
X^d	X^DX^d	X^dY

9. The disorder allele is a recessive allele, located on the X chromosome. We know the allele is recessive because individual 2 in generation II carries the allele but does not have the condition. If the allele were located on an autosome, the parents in generation I would be of genotype *AA* (the male) and *aa* (the female). In this case, none of the individuals in generation II could have the condition—yet two of them do have the condition, implying that the allele is on a sex chromosome. Finally, we know that the allele is on the X chromosome, because otherwise only males could have the condition.
10. a. If the two genes were completely linked:

	AB	*ab*
aB	*AaBB*	*aaBb*

b. If the two genes were on different chromosomes:

	AB	*Ab*	*aB*	*ab*
aB	*AaBB*	*AaBb*	*aaBB*	*aaBb*

11. Eight genotypes of gametes are possible: *ADEg, AdEg, ADeg, Adeg, aDEg, adEg, aDeg, adeg.* If *D* and *E* are linked, we expect only four genotypes: *ADEg, Adeg, aDEg, adeg.*

Chapter 11

1. The genetic information of the alleles is contained in the sequence of the nitrogenous bases—adenine (A), cytosine (C), guanine (G), and thymine (T)—found within the segment of DNA that constitutes each allele. At any genetic locus, different alleles differ in the sequence of bases they contain. Therefore, the DNA segments of the two codominant alleles A^1 and A^2 differ in the sequence of bases found in each allele.
2. The double helical structure of DNA and the base-pairing rules theorized by Watson and Crick suggested a simple way that genetic material could be copied. Because A pairs only with T and C pairs only with G, each strand of DNA contains the information needed to produce the complementary strand. In DNA replication, the two strands of the helix separate, and each strand then serves as a template for the construction of its complementary strand, resulting in two identical copies of the original DNA molecule.
3. The sequence of bases in DNA is the basis of inherited variation. A change in the sequence, whether because of an error during replication or because of exposure to a mutagen, is a mutation. Such a change could result in a new allele that would encode a new version of the protein encoded by the gene in which the mutation occurred. If the new allele produced a protein that did not function properly (or at all), serious damage could be done to the cell, and consequently to the organism; such an allele could cause a genetic disorder.
4. DNA is repaired by protein complexes that include enzymes. When DNA is being replicated, enzymes check for and immediately correct mistakes in pair bond formation. Mistakes that escape this process, called mismatch errors, are caught and corrected by repair proteins. DNA repair is essential for cells to function normally because DNA is constantly being damaged by chemical, physical, and biological agents. If none of this damage were repaired, genes that encoded proteins critical to life would eventually cease to function, thereby disabling the production of those proteins and killing the cell, and ultimately the organism.
5. Two features enable cells to pack an enormous amount of DNA into a very small space: the thinness of the DNA molecule, and a highly organized, complex packing system. Each portion of a chromosome, which contains one DNA molecule, is made up of many tightly packed loops. Each loop is composed of a chromatin fiber consisting of many nucleosome spools, which are made of proteins called histones. A segment of DNA winds around each spool, and if that DNA were unwound it would reveal its double helical structure.
6. Prokaryotes have less DNA than eukaryotes have. All DNA in a prokaryote is located on one chromosome; in eukaryotes, by contrast, the DNA is distributed among several chromosomes. Eukaryotes have more genes than prokaryotes do, and genes constitute only a small portion of the eukaryotic genome. Most prokaryotic DNA encodes proteins, and very little of it consists of noncoding DNA and transposons. Functionally related genes in prokaryotes are grouped together on the chromosome; eukaryotic genes with related functions often are not located near one another.
7. The bacterium would begin expressing the gene responsible for encoding the enzyme that breaks down arabinose. Organisms can turn genes on and off in response to short-term changes in food availability or other features of the environment.
8. In multicellular organisms, different cell types express different genes by controlling transcription, along with other methods. By switching specific genes on or off, cells can vary their structure and perform specialized metabolic tasks, even though each has exactly the same genes (and alleles).
9. From gene to protein, here are the steps at which gene expression can be controlled: (1) The expression of tightly packed DNA can be prevented, in part because the proteins necessary for transcription cannot reach them. (2) Transcription can be controlled by regulatory proteins that bind to regulatory DNA, effectively switching a gene on or off. (3) The breakdown of mRNA molecules can be regulated such that mRNA is destroyed hours or weeks after it is made. (4) Translation can be inhibited when proteins bind to mRNA molecules to prevent their translation. (5) Proteins can be regulated after translation, either when the cell modifies or transports them or when they are rendered inactive by repressor molecules. (6) Synthesized proteins can be destroyed.

Chapter 12

1. A gene is a DNA sequence that contains information for the synthesis of one of several types of RNA molecules used to make proteins. A gene stores information in its sequence of nitrogenous bases.
2. Genes control the production of a variety of RNA products, including mRNA, rRNA, and tRNA. Messenger RNA (mRNA) encodes the amino acid sequence of proteins, ribosomal RNA (rRNA) is an essential component of ribosomes (the site of protein synthesis), and transfer RNA (tRNA) carries amino acids to the ribosomes during protein construction. Therefore, each of the RNA products specified by genes functions in the synthesis of proteins. Proteins are essential for many functions that support life. In cells and organisms, proteins provide structural support, transport materials through the body, and defend against disease-causing

organisms. Enzymes are a class of proteins that speed up chemical reactions.

3. Genes commonly contain instructions for the synthesis of proteins. Each gene is composed of a segment of DNA on a chromosome and consists of a sequence of the four bases adenine (A), cytosine (C), guanine (G), and thymine (T). The sequence of bases specifies the amino acid sequence of the gene's protein product. Through transcription and translation, proteins are produced from the information stored in genes. In transcription, mRNA is synthesized directly from the sequence of bases in one DNA strand inside the nucleus of a cell. Translation occurs in the cytoplasm and converts the sequence of bases in an mRNA molecule into the sequence of amino acids in a protein. Proteins, by their many and various functions, influence the phenotype of an individual.
4. For a protein to be made in eukaryotes, the information in a gene must be sent from the gene, which is located in the nucleus, to the site of protein synthesis, on a ribosome. This transfer of information requires an intermediary molecule because DNA does not leave the nucleus but ribosomes are located in the cytoplasm. In eukaryotes, a newly formed mRNA molecule usually must be modified before it can be used to make a protein. The reason is that most eukaryotic genes contain internal sequences of bases (introns) that do not specify part of the protein encoded by the gene. DNA sequences copied from introns must be removed from the initial mRNA product if the protein encoded by the gene is to function properly.
5. RNA splicing is a step in RNA processing in which introns are removed from a newly transcribed mRNA and the remaining exons are joined together to create the mature, export-ready form of the mRNA. RNA splicing is not known to occur in prokaryotes, but it is common among eukaryotes. The great majority of our protein-coding genes produce mRNA that must be spliced before the RNA can exit the nucleus.
6. Messenger RNA is the product of transcription, and is a version of the genetic information stored in a gene. The mRNA moves from the nucleus to the cytoplasm, where it binds with a ribosome to guide the construction of a protein.

   Ribosomal RNA is a major component of ribosomes. Translation occurs at ribosomes, which are molecular machines that make the covalent bonds linking amino acids to form a particular protein.

   Transfer RNA molecules carry the amino acids specified by mRNA to the ribosome. At the ribosome, a three-base sequence (anticodon) on the tRNA binds by means of complementary base pairing with the appropriate codon on the mRNA. Each tRNA molecule carries the amino acid specified by the mRNA codon to which its tRNA anticodon can bind.
7. If a tRNA molecule does not function properly because of a mutation, each protein that it helps to build will be altered in some way. By failing to bind properly with the mRNA codons of many different genes, a mutant tRNA may significantly affect the structure of many different protein products. Because their structure is altered, the function of these protein products may be impaired. Since proteins are key components of many metabolic reactions, changing the function of many different proteins can result in a series of metabolic disorders.
8. A mutation is any alteration in the information coded within an individual's DNA. Sometimes the effects of that change can be detected as a change in the inherited characteristics of that individual (a phenotypic change). But in other cases, the mutation may be "silent," with no outward sign that it has occurred. Most mutations are neutral in their impact on the individual, neither benefiting it nor harming it. Some mutations have harmful effects, and rarely, a mutation might produce a change that enhances the individual's ability to survive and reproduce in a particular environment.

   A very large number of our genes carry information for the construction of specific proteins. That information is carried in the form of a sequence of chemical units, called nitrogenous bases, that in turn specify the sequence in which the amino acids in a protein are strung together (proteins are built from amino acids, and each unique protein has a unique sequence of amino acids). When the base sequence of a protein-coding gene changes, the amino acid sequence of its protein is altered as well. Every protein has special chemical and biological properties that are critical to its function, and most of those properties stem from the precise sequence of amino acids in it. If the amino acid sequence of a protein changes because of a mutation in the gene that codes for it, the biological function of that protein may change as well.

Chapter 13

1. To produce domesticated species, humans have manipulated the reproduction of other organisms, selecting for desirable qualities that, over time, have become standard in domesticated species. Although such selection practices do lead to changes in the DNA of organisms (that is, they lead to an increase in the frequencies of alleles that control the inheritance of the traits we select for), genetic engineering enables us to make much greater changes in a much shorter span of time. Using such methods, we can manipulate the DNA of organisms directly, and we can transfer genes from one species to another. Transfers of DNA from, say, a human to a bacterium (as is done in the production of human insulin) far exceed the scope of DNA transfers that occur in nature or that are possible through conventional breeding of crops and farm animals. We can also selectively change specific DNA sequences—something we could never do before. Overall, we can now manipulate DNA with greater power and precision than we could when we domesticated species such as dogs, corn, and cows.
2. DNA cloning is the introduction of a DNA fragment into a host cell that can generate many copies of the introduced DNA. The

purpose of DNA cloning is to multiply a particular DNA fragment, such as a specific gene, so that a large amount of this DNA is made available for further analysis and manipulation. Bacteria are the most common host cells in DNA cloning. Two of the most common methods of cloning a gene are constructing a DNA library and using the polymerase chain reaction (PCR). To build a DNA library, a vector such as a plasmid is used to transfer DNA fragments from the organism whose gene is to be cloned to a host organism, such as a bacterium. To clone a gene by PCR, primers are synthesized, enabling DNA polymerase to produce billions of copies of the gene in a few hours.

3. The advantage of DNA cloning is that it is easier to study a gene and its function, and to manipulate that function for practical benefits, once you have many copies of it. Once a gene has been cloned, it can be sequenced, transferred to other organisms, or used in various experiments. Today, many lifesaving pharmaceuticals, such as human insulin, human growth hormone, human blood-clotting proteins, and anticancer drugs, are manufactured by bacteria that have been genetically modified by having cloned human genes inserted into them.
4. Genetic engineering is the permanent introduction of one or more genes into a cell, a certain tissue, or a whole organism, leading to a change in at least one genetic characteristic in the recipient. The organism receiving the DNA is said to be genetically modified (GM) or genetically engineered (GE). To create GM organisms, a DNA sequence (often a gene) is isolated, modified, and inserted back into the same species or into a different species.

   Fish such as salmon have been genetically engineered to grow much more rapidly than their unaltered counterparts. Farm-raised, genetically modified fish provide low-cost, high-quality protein with a lower carbon footprint than cattle. Furthermore, consumption of GM fish reduces fishing pressure on wild stocks. However, there are concerns that escaped GM fish could threaten wild fish stocks not only by interbreeding with them and thereby reducing their natural diversity, but also by outcompeting them for resources and thereby driving them toward extinction.

5.–6. These answers depend on your viewpoint, which we hope is at least in part guided by scientific facts.

7. Embryonic stem cells are derived from the inner cell mass in the blastocyst and are pluripotent in that they can give rise to all cell types in the body. Adult stem cells arise in the fetus and persist in small numbers in various tissues and organs in children and adults; they are multipotent or unipotent. Induced pluripotent stem cells (iPSCs) are generated in the lab from ordinary differentiated cells, such as living skin cells from an adult. These iPSCs have the developmental flexibility of embryonic stem cells because they are treated in a way that reprograms their developmental potential, "resetting" it to a stage that resembles the embryonic stage.
8. This answer depends on your viewpoint, which we hope is at least in part guided by scientific facts.

Chapter 14

1. Evolution is change in the genetic characteristics of a population over time, which can occur through mechanisms such as genetic mutation or natural selection. Since the genotypes of individuals do not change, a population can evolve but an individual cannot.
2. In the new habitat, larger lizards will have an advantage over smaller ones. The larger lizards of the species will therefore be more likely to pass the trait of large size on to their offspring, and the average size of lizards in the population will increase over time because of this selective advantage.
3. **Mutation:** A nonlethal mutation of a particular allele can be inherited by offspring, thereby increasing the frequency of the mutant allele in a population over time.

   Gene flow: The exchange of alleles between populations can change the frequencies at which alleles are found in those populations by introducing new alleles. Populations affected by gene flow tend to become more genetically similar to one another.

   Genetic drift: Random events (such as chance events that influence the survival or reproduction of individuals) can cause one allele to become dominant in a small population. By chance alone, drift can lead to the fixation of alleles in small populations; if these alleles are harmful, the population may decrease in size, perhaps to the point of extinction.

   Natural selection: If an inherited trait provides a selective advantage for individuals in a certain population, individuals with that trait will be more likely to reproduce, and the frequency of the allele for that trait will therefore increase in succeeding generations. Likewise, an allele for a disadvantageous trait will be selected against, and will be found with decreasing frequency over time.
4. a. Gene flow is the exchange of alleles between populations. Gene flow makes populations more similar to one another in their genetic makeup.
   b. Genetic drift is a process by which alleles are sampled at random over time. Genetic drift can have a variety of causes, such as chance events that cause some individuals to reproduce and prevent others from reproducing.
   c. Natural selection is a process in which individuals with particular inherited characteristics survive and reproduce at a higher rate than other individuals.
   d. Sexual selection is a form of natural selection in which individuals with certain traits have an advantage in attracting mates, and consequently in passing those traits on to offspring.
5. The chance events that cause genetic drift are much more important in small populations than in large populations. In natural populations, the number of individuals in a population has an effect similar to the number of times a coin is tossed. A small population is analogous to a coin being tossed just a few times. By chance alone, some individuals in a small population may

leave offspring while others do not. One allele may be lost when certain individuals fail to reproduce, and another may become fixed when the individuals bearing it contribute disproportionately to the next generation. When a population has many individuals, the likelihood that each allele will be passed on to the next generation greatly increases. In a large population, it is unlikely that chance events could cause a dramatic change in allele frequencies in a short time (just as a fair coin tossed hundreds of times is unlikely to give any outcome other than approximately 50-50 heads and tails). Genetic drift can occur in large populations, but in these cases its effects are more easily overcome by natural selection and other evolutionary mechanisms. In large populations, genetic drift causes little change in allele frequencies over time.

6. For examples of adaptations in the pronghorn, take a look at Table 1.5 in Chapter 1 (p. 19).
7. Individuals with inherited traits that enable them to survive and reproduce better than other individuals replace those with less favorable traits. This process, by which natural selection improves the match between organisms and their environment over time, is called adaptive evolution. In our efforts to kill or control bacteria that cause infectious diseases, we are creating a new environment in which bacteria that cannot withstand antibiotics are eliminated. Often, some bacteria in the population are not killed by antibiotics; these bacteria reproduce, increasing the frequency of resistant bacteria in the population. These evolutionary changes in disease-causing bacteria are harmful to us because more and more of the diseases we encounter will be resistant to medical treatment.
8. Overwhelming evidence indicates that evolution occurred and continues to occur. Support for evolution comes from five lines of evidence: (1) The fossil record provides clear evidence of the evolution of species over time and documents the evolution of major groups of organisms from previously existing organisms. (2) Organisms contain evidence of their evolutionary history. For example, scientists find that studies of proteins and DNA support the evolutionary relationships determined by anatomical data; that is, the proteins and DNA of closely related organisms are more similar than those of organisms that do not share a recent common ancestor. In this and many similar examples, the extent to which organisms share characteristics other than those used to determine evolutionary relationships is consistent with scientists' understanding of evolution. (3) Scientists' understanding of evolution and continental drift has enabled them to predict the geographic distributions of certain fossils, depending on whether the organisms evolved before or after the breakup of Pangaea. (4) Scientists have gathered direct evidence of small evolutionary changes in thousands of studies by documenting genetic changes in populations over time. (5) Scientists have observed the evolution of new species from previously existing species.

Chapter 15

1. The morphological species concept defines species on the basis of anatomical similarities, while the biological species concept specifies that species are made of individuals that can interbreed. The morphological species concept cannot be applied to organisms with few discernible anatomical characteristics (such as bacteria), species that are physically similar but do not interbreed for other reasons (such as meadowlarks), or species that exhibit a wide variety of anatomical variation within a species (such as *Heliconius* butterflies). The biological species concept cannot be applied to fossil organisms, asexual organisms, or organisms with distinct ranges that prevent scientists from knowing whether they could potentially interbreed.
2. Reproductive isolation prevents gene flow between populations. If two populations exhibit gene flow with one another (that is, exchange alleles), they will become more similar and speciation will be retarded or prevented. As a consequence, the restriction of gene flow through reproductive isolation is a necessary first component to speciation. Once two populations stop sharing alleles, the processes of genetic drift, mutation, and natural selection can impact the two separate gene pools in different ways, causing the populations to diverge genetically.
3. Species that hybridize in nature may still be distinct species, because of a host of alleles that do not affect their ability to interbreed but may cause them to look different or to differ from each other ecologically. For this reason, many people would argue that the rare species is separate from the common species and should remain classified as rare and endangered.
4. Defining species by their inability to reproduce sexually with other species is convenient, but many alleles do not affect reproductive isolation yet could cause the two grasses to be different enough that they could be classified as separate species, even though they would be able to produce hybrids.
5. Because this storm-blown population is now geographically isolated from other populations of its species, there will be little or no gene flow to other populations. As a result, genetic changes due to mutation, genetic drift, and natural selection will accumulate over time. Natural selection is likely to cause genetic change in the population because the new environment is different from the parent population's environment; genetic drift will probably also be important, because the island population is small (making drift more likely). As a by-product of genetic changes due to selection, drift, and mutation, the island population may become reproductively isolated from the parent population. If the island population remains isolated long enough, a sufficient number of genetic changes may accumulate for it to evolve into a new species.
6. Some of the cichlid populations of Lake Victoria may have had so little contact with one another that they can be said to have evolved into separate species in geographic isolation, even though

they live in the same lake. Other populations may have evolved into new species in the absence of geographic isolation.

7. New plant species can form in the absence of geographic isolation as a result of polyploidy, a condition in which an individual has more than two sets of chromosomes. Strong evidence suggests that sympatric speciation occurred in the cichlids that live in Lake Victoria, and evidence is accumulating that apple and hawthorn populations of the apple maggot fly are diverging into two species, despite living in the same area. In apple maggot flies and cichlids, sympatric speciation is promoted by ecological factors (such as selection for specialization on different food items) and sexual selection. Sympatric speciation is thought to be more difficult to achieve because there is a greater potential for gene flow between populations whose geographic ranges overlap than between populations that are geographically isolated from one another. Gene flow tends to cause populations to remain (or become) similar. Therefore, in the absence of geographic isolation, it can be difficult for genetic differences great enough to cause reproductive isolation to accumulate over time. As a result, sympatric speciation occurs less readily than allopatric speciation.

Chapter 16

1. One example of one group of organisms evolving from another is the emergence of mammals from reptiles. The emergence of the mammalian jaw and teeth illustrates the steps in this process. The first step was the development of an opening in the reptilian jaw behind the eye. Then, more powerful jaw muscles and specialized teeth appeared with the therapsids. Finally, a subgroup of these reptiles, the cynodonts, emerged with more specialized teeth and a more forward hinge of the jaws, completing this aspect of mammalian evolution.
2. The emergence of photosynthesis in ancient organisms gradually led to the buildup of O_2 in the atmosphere, which killed many organisms to which oxygen was toxic. However, the oxygen supplied by photosynthetic organisms made possible the evolution of eukaryotes and later multicellular life-forms.
3. The Cambrian explosion was a large increase in the diversity of life-forms over a relatively short time about 530 million years ago. Larger organisms of most phyla emerged during this period, setting the stage for the colonization of land.
4. The colonization of land led to another great increase in the diversity of life-forms. Life on land required different means of mobility and reproduction, adaptations to obtain and retain water, and ways to breathe in air rather than water. Early terrestrial organisms had the opportunity to expand into new types of largely unoccupied habitat, which provided ample resources for organisms that were able to survive the challenges of life on land.
5. A mass extinction event may be associated with rapid environmental changes that have no relation to the conditions that favor a particular adaptation. Organisms with wonderful adaptations can (and have) become extinct during mass extinction events.
6. No. Homeotic genes control the expression of many other genes and in this way determine major developmental features. The homeotic genes from mice were able to initiate eye development in flies because both organisms have eyes. In the case of the pigs, however, introducing the homeotic gene for wing development would not work because pigs lack any genes associated with wings. So, in essence, the homeotic genes would be activating nothing.
7. Gene expression may account for the differences between genetically similar organisms. Although identical twins may have exactly the same genes, those very genes may be activated for different lengths of time or in different environmental contexts so that their physical manifestation (phenotype) may be different.

   At the website maintained by the Integrated Taxonomic Information System (http://www.itis.gov), you can find taxonomic information about more than half a million species belonging to the three domains. You can also try the website of the International Union for Conservation of Nature (http://www.iucn.org) to learn more about the status of your favorite organism, especially if it is endangered. From smallest to largest, the groupings of the Linnaean hierarchy are species, genus, family, order, class, phylum, kingdom. Any two species that belong to the same family must also belong to the same order because order is a more inclusive category than family. The order encompasses not only all members of a family classified within it but also all members of other closely related families.

Chapter 17

1. In both bacteria and eukaryotes, the gel-like cytoplasm that makes up the interior of the cell is bounded by a plasma membrane built mainly from phospholipids. In bacteria, the genetic material, which is DNA, is in direct contact with the fluid part of the cytoplasm (cytosol); in contrast, in eukaryotes the DNA is enclosed in a special organelle, the nucleus. There are many ribosomes—which assist in protein manufacture—in the cytoplasm of both bacteria and eukaryotes. Many bacterial cells have a polysaccharide cell wall, as do some eukaryotes (all plants and fungi, and some protists). Some bacteria also have a slimy protective capsule, numerous projections that they attach themselves with, and one or more whiplike flagella used for swimming in a film of liquid. A eukaryotic cell is larger than a typical bacterial cell and has many complex membrane-enclosed internal compartments that are not found in bacteria.
2. Prokaryotes perform key tasks in ecosystems, including carrying out photosynthesis, providing nitrate to plants, and decomposing dead organisms. Prokaryotes are at the bottom of many food

chains that sustain a great diversity of life. Prokaryotes are useful to humanity in many ways (for example, in cleaning up oil spills and helping with our digestion), but they also cause deadly diseases.

3. Extremophiles (literally, "lovers of the extreme") are organisms that live in extremely harsh habitats that average members of the group could not survive in. Most extremophiles are microbes, including bacteria and archaeans. Some prokaryotic extremophiles live in extreme saltiness and/or acidity, and some can withstand extremely high pressure or very low moisture. Some live at very high or very low temperatures.
4. An endotoxin is a *component* of the bacterial cell wall that triggers illness in a host infected by the bacterium. Endotoxins can produce fever, blood clots, and toxic shock. In contrast, an exotoxin is a poison that an organism *releases into its surroundings*. Exotoxins disrupt cellular processes or damage cell structures such as the plasma membrane. Unlike endotoxins, most exotoxins are destroyed by heat.
5. The flu virus attaches to the plasma membrane of cells in the nose, throat, and lungs of humans and in the intestines of birds. The virus enters the host cell as its fatty envelope merges with the host cell's plasma membrane. Invading viral particles shed their protein coat, releasing the genetic material, which is RNA, into the cytoplasm. Viral RNA enters the host cell's nucleus, where many copies of the viral RNA are produced. The viral genetic material also directs the host cell to produce large amounts of viral coat protein. The viral RNA and coat proteins are assembled into viral particles, which exit the cell enclosed in a "bubble" made by the plasma membrane, without destroying the host cell. Most people start "shedding" the virus into the environment 2–3 days after becoming infected, which is usually a day before symptoms appear, and they remain infective for about 7 days after first catching the virus. The virus can survive on a doorknob for a few days, and on a moist surface for about 2 weeks.

   Within limits, fever can be beneficial because it is a defensive strategy: turning up body temperature slows viral reproduction. But very high fever is an overreaction that can damage the body.
6.

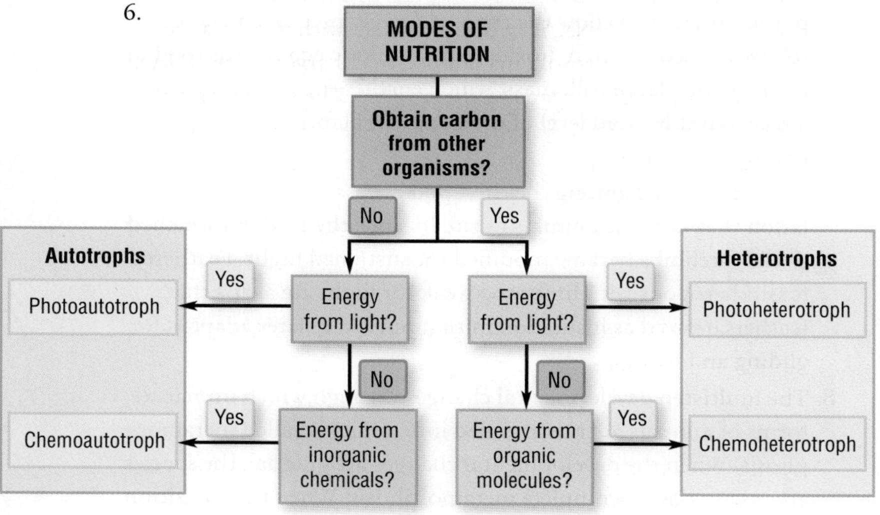

Chapter 18

1. Although prokaryotes are known to acquire genetic information through a process called lateral (or horizontal) gene transfer, they do not exhibit sexual reproduction. Sexual reproduction is common in each of the four kingdoms of the Eukarya. Sexual reproduction combines genetic information from two parents to produce offspring that are genetically different from each other and from both parents. Because it promotes genetic diversity in the offspring, sexual reproduction increases the odds that the population will be varied enough to adapt to environmental changes such as an attack by a new strain of bacterium or virus.
2. Examples of multicellular protists include *Volvox carteri* and other green algae, as well as the red and brown algae that are large enough to be seen with the naked eye and are popularly called seaweed. Multicellularity enabled organisms to grow larger, and large size can be helpful in gathering resources (such as sunshine) and storing them (as surplus food or mineral nutrients), and in escaping potential predators or surviving predation. Multicellularity also enabled more complex and efficient functioning by enabling the division of labor among cell types dedicated to specific functions.
3. A red tide is a population explosion, or bloom, of photosynthetic plankton, usually toxic dinoflagellates. Blooms of these protists sometimes cause the water to turn red because of a reddish pigment concentrated inside their cells. Dinoflagellates and other photosynthetic plankton can produce a variety of toxins, including some that cause nerve and muscle paralysis in humans and other mammals. Some of these toxins accumulate in shellfish and cause paralytic shellfish poisoning in humans and wild animals who eat the shellfish. Scientists do not have a full understanding of what causes the sudden blooms, but the frequency of such blooms has been rising around the world, and pollution from fertilizer runoff and sewage are thought to be among the culprits.
4. Fungi are absorptive heterotrophs, whereas animals (including humans) are ingestive heterotrophs. Fungi are more closely related to animals than they are to plants. Fungi and animals share certain similarities, including these: In the cell walls fungi have chitin, the chemical that also strengthens the exoskeleton of most insects and crustaceans. Fungi and animals both store surplus carbohydrate as a molecule called glycogen. Most telling of all, DNA comparisons show that fungi and animals are closer than fungi and plants.
5. Great piles of unrotted plant material will have accumulated everywhere—in backyards, city parks, and forests. The landfills will be overflowing with undecomposed woody material, and disposing of manure from farms will be even more of a problem. The loss of their mycorrhizal partners will cause nutritional deficiencies in many plants, and some of these plants may become extinct. On the other hand, farmers will raise bumper crops with synthetic fertilizer, since they will lose much less of their crop to the scourge of fungal diseases. There will be no bread leavened with yeast, no

beer and wine as we know them, and sliced mushrooms will not be an option for pizza topping. On the other side of the ledger, we will not have to confront moldy fruit and cheese in the neglected corners of the refrigerator.

6. When plants colonized land, they had to evolve in ways that enabled them to persist in an environment where they were no longer surrounded and supported by water. To deal with the problem of obtaining and retaining water, plants evolved root systems (which enable them to absorb water and nutrients from soil) and the waxy covering over their stems and leaves known as the cuticle (which prevents their tissues from drying out when exposed to sun and air). To combat gravity and grow taller, plants evolved a tough chemical polymer called lignin, which strengthened cell walls, especially those of the water-conducting tubes of xylem.
7. A lichen is a mutualistic association between a photosynthetic microbe and a fungus. The lichen-forming microbe can be a single-celled green alga, a cyanobacterium, or both together. The lichen body is thin and has no protective sheath like the cuticle of a plant or the "armor plating" of some protists. Nor does it have any mechanism for ridding the body of wastes or toxic substances. Therefore, lichens readily absorb and accumulate air- or water-borne pollutants. Lichens are destroyed by acid rain, heavy-metal pollutants, and organic toxins, which is why lichens tend to disappear in heavily industrialized areas with poor pollution controls.

Chapter 19

1. The risk of desiccation (drying out) and the need for mechanical reinforcement of the body were some of the environmental challenges faced by the first insects. The evolution of a chitin-reinforced exoskeleton helped solve both problems. The exoskeleton prevents the soft body tissues from drying out and also strengthens the body. In contrast to the external gills of crustaceans, the gas exchange surfaces for insects are internal, which protects the moist surfaces from desiccation.
2. Sponges lack true tissues and do not have any distinct body symmetry. Jellyfish, in contrast, have true tissues, organs and organ systems, and radial symmetry.
3. A segmented body plan, in which the body consists of repeated units known as segments, was a significant evolutionary innovation because it paved the way for the specialization of body segments, and the appendages that arise from them, for diverse, specialized functions; these in turn enabled animals to adapt to new habitats or to acquire new modes of life. The varied uses of segments and their appendages are exemplified by the body plan of a lobster or crayfish.
4. Muscle and nerve tissues were critical to the development of efficient modes of locomotion, and they are found in all animals except sponges. A fluid-filled body pseudocoelom enabled locomotion through a hydrostatic skeleton. Another evolutionary advance was the evolution of the exoskeleton in many protostomes and the endoskeleton in all chordates; the skeleton provides anchoring for muscles, and collectively the musculoskeletal system enables efficient movement. Bilaterality combined with paired appendages, and cephalization, were adaptations for faster and more efficient modes of locomotion.

   Fish propel themselves through water by moving paired pectoral and pelvic fins and making powerful side-to-side movements with their caudal fins. Tetrapod limbs have to support the animal's weight and move that weight. Accordingly, their limbs are jointed for greater range of motion and strengthened internally by strong bones. Locomotion helps an animal capture prey, eat prey, avoid being captured, attract mates, care for young, and migrate to new habitats.
5. Birds have hollow bones that are light but strong for their weight. They have toothless beaks, and many of their internal organs are reduced (for example, female birds have a single ovary), which lessens body weight. The feathers attached to their forelimbs not only act as insulation to hold body heat, but also enable strong sustained flight, with the tail feathers acting as stabilizers.

   Like dinosaurs and other reptiles, birds lay amniotic eggs and have scaly skin on parts of the body. Unlike reptiles, however, birds have a four-chambered heart that resembles ours: oxygenated blood and nonoxygenated blood are kept in separate chambers, and the blood that is pumped to the rest of the body therefore contains high levels of oxygen, supporting a high metabolic rate. The respiratory system of birds is considered even more efficient than ours. Birds have a one-way flow of air through their respiratory passages, so that incoming air never mixes with outgoing air. In contrast, the air we inhale and exhale takes the same route (although in opposite directions), so there is some mixing of the incoming and outgoing airstreams. Like most mammals, most birds are endotherms and homeotherms.
6. The embryo developing inside a bird's egg is surrounded and protected by layers of extraembryonic membranes, including one called the amnion, that promote gas exchange and waste removal. The calcium-rich protective shell retards moisture loss but is porous enough to allow the entry of life-giving oxygen and the release of waste carbon dioxide. The amniotic egg hoards food in the form of a large yolk mass, which enables the young to achieve a relatively advanced level of development before emerging (hatching) from the shell.
7. Feathers evolved among the theropod dinosaurs as a layer of insulation that kept the animals warm. In time, the feathers attached to the forelimbs became modified for sustained flight. Feathered fossil theropods with limbs too weak for flight are a clue that feathers evolved as insulation initially and were later adapted for gliding and flying.
8. The multistep developmental changes through which immature forms of animals are transformed into adults is called metamorphosis. When the developmental changes are gradual, the species are said to have incomplete metamorphosis. When the transition

from one developmental stage to the next is dramatic, with the immature forms bearing little resemblance to adults, the species are said to have complete metamorphosis.

By having two very different body forms in their life cycle, species with complete metamorphosis can pack very different but highly successful and highly specialized modes of living into the life cycle of one animal. Among lepidopterans, for example, the body plan of the larval stage (the caterpillar) is well suited for voracious eating in a limited area, whereas the winged adult (the butterfly) is best suited for seeking mates and finding the optimal locations for depositing eggs. Together, these very different modes of living acquire a greater variety, and therefore quantity, of available resources than the one body form could on its own.

9. Most mammals are endotherms and homeotherms: they use metabolic energy to generate heat, and they maintain a near-constant body temperature. They can trap body heat with hair, keratin-containing strands on the skin that are effective insulators. Muscles in the skin can raise the hair to trap a thicker layer of air next to the skin, which increases the insulating properties of body hair. Only mammals have sweat glands. Evaporation of sweat cools terrestrial mammals and enables them to maintain a moderate body temperature even in extremely hot and dry environments such as the desert.
10. All mammalian mothers nurse their young with milk, a liquid from their mammary glands that is rich in fat, proteins, salts, and other nutritive substances. Monotremes lay eggs, and the newborns that hatch from them are relatively undeveloped. The mothers secrete milk from their mammary glands directly on the fur, where it is lapped up by the newborns after they hatch from their shells. Marsupial females give birth to somewhat more developed young (joeys) that are further nurtured in a ventral pouch. Marsupial mothers have a nipple in each pouch that the newborn attaches to and feeds from. Eutherians have a longer gestation and give birth to relatively well-developed young. They nurse their newborns from two or more nipples on the ventral side of the body.

Chapter 20

1. The description of the biosphere as an "interconnected web" is apt because all organisms within it are connected by their interactions, as shown by examples ranging from various food chains to symbiotic relationships between species. The global spread of invasive species and their often detrimental effects on native ecosystems, and the change in red kangaroo populations in response to the "dingo fence," are two of the case histories discussed in this chapter that underscore the interrelatedness of organisms throughout the biosphere.
2. Giant convection cells in the atmosphere and ocean currents carry the results of local events (such as a volcanic eruption or an oil spill) to distant areas around the globe. For example, oil spilled into an ocean current next to one continent's shore may be carried by that current and end up coating the shores of other continents. If shorebirds on those continents are killed when they become coated with oil from the spill, they may no longer keep populations of their food organisms under control, and they will no longer be available as a food source for their predators.
3. These are some of Earth's terrestrial biomes: tropical forest, temperate deciduous forest, grassland, chaparral, desert, boreal forest, and tundra. Determine how many of these are located within 100 kilometers (60 miles) of your home. Reread Section 20.3 for a description of the climatic and ecological characteristics of one of these biomes. Your explanation of how those characteristics help determine which life-forms are found in that biome will depend on the biome you choose.
4. The defining feature of a desert is the scarcity of moisture (less than 25 centimeters of precipitation), not high temperature. Antarctica, which receives less than 2 centimeters of precipitation per year, is the largest cold desert in the world. The Sahara desert in northern Africa is the largest hot desert. Because desert air lacks moisture, it cannot retain heat, and therefore it cannot moderate daily temperature fluctuations. As a result, temperatures may be above 45°C (113°F) in the daytime and then plunge to near freezing at night.
5. The leaves of desert plants are small so that they present a smaller surface area for evaporative water loss than do large leaves. Plants native to tropical rainforests do not need special adaptations to reduce water loss, because they live in a habitat that is wet year-round. Some desert plants produce enormously long taproots that are able to reach subsurface water. Desert animals often have light-colored fur, kidneys that help them conserve water, and behavioral adaptations such as hiding in a cool burrow during the hottest parts of the day. Some, like jackrabbits, have large ears that act like a radiator to help dissipate heat.
6. In temperate regions, seasonal changes in temperature cause the oxygen-rich water near the top of a lake to sink in the fall and the spring, bringing oxygen to the bottom of the lake. This seasonal turnover also delivers nutrients from the bottom sediments to the surface layers, where they enhance the growth of photosynthetic organisms. By delivering oxygen to lake-bottom animals and nutrients to producers at the lake surface, lake turnover increases both primary and secondary productivity.
7. An estuary, where a river empties into the sea, is a shallow marine ecosystem. The plentiful light, the abundant supply of nutrients delivered by the river system, and the regular stirring of nutrient-rich sediments by water flow creates a highly productive community of photosynthesizers. The coastal region, which stretches from the shoreline to the continental shelf, is among the most productive marine ecosystems because of the ready availability of nutrients and oxygen. Nutrients delivered by rivers and washed off the surrounding land accumulate in the coastal region. Nutrients that settle to the bottom are stirred up by wave action, tidal movement, and the turbulence produced by storms. The nutrients support the growth of photosynthetic producers, which inhabit

the well-lit upper layers of coastal waters to a depth of about 80 meters (260 feet). The vigorous mixing by wind and waves also adds atmospheric oxygen to the water.

Chapter 21

1. A population may be difficult to define if the boundaries of its range are unclear, if its members move around frequently, or if its members are small and hard to count.
2. If a population is threatened by extinction, possible options for saving it might be to protect it from human disturbances, to treat diseases, to reduce the number of predators, to move the population to an area with greater food supply (limiting death or emigration), to introduce individuals from other populations of the species (increasing immigration), or to institute captive breeding programs (increasing the birth rate).
3.

4. a. Some factors that limit population growth include availability of habitat, availability of food and water, disease, weather, natural disturbances, and predators.
   b. Species new to an area often do not have established predators. In addition, they have not yet reached the carrying capacity of their habitat.
5. Density-dependent factors increase in intensity as the density of the population increases. An example of a density-dependent factor is an infectious disease that spreads more rapidly in densely populated areas. Plants that become overcrowded in a field compete with each other for nutrients, and therefore nutrient availability is another example of a factor that limits population growth in a density-dependent manner. Density-independent factors are not affected by the density of the population. Temperature is a density-independent factor: if the temperature drops below what a certain plant species can tolerate, the plants will die no matter how sparse or dense that species' population is. Natural disturbances like fires and floods also limit the growth of populations in a density-independent way.
6. If the pattern of population growth is understood, managers may be able to manipulate the factors that most directly affect the population's growth rate. If a population of organisms were more successful because, for example, it had adequate access to water, a manager would be sure that nearby rivers were not drained off for agricultural needs. Understanding population growth is especially valuable in pest management. Knowing the population cycles of pests—for example, whether their populations display boom-and-bust growth patterns—can inform growers about when to spray pesticides or release natural predators for biocontrol for maximum effectiveness.
7. Specific actions that could limit the negative impact of humans on our world include the following: (1) Reducing the consumption of unnecessary goods. (2) Reusing and recycling items to promote the sustainable use of resources. (3) Working to develop and follow environmentally friendly policies and activities (for instance, using energy-efficient cars and lightbulbs). (4) Purchasing goods that have a lower impact on the environment, including organically grown clothing and food items. (5) Limiting reproduction to no more than one child per parent (zero population growth). This last factor will affect all the others. Limiting human population growth will reduce all human impacts on the environment.

Chapter 22

1. Fixed behaviors are less obvious in adults, although we are all familiar with some—for example, our tendency to smile when we see friends, or to frown in response to annoying stimuli. "Infectious yawning," the tendency to yawn in response to yawning by another individual, is a fixed behavior seen in humans, as well as in many other mammals. In humans, fixed behaviors are most apparent in newborns and include the grasping reflex, which is the tendency of babies under 6 months of age to hold on tightly to any object placed in their fists.

   A majority of the behaviors that are common in humans are learned behaviors, and many of them are strongly influenced by the time and society we happen to live in. Familiar learned behaviors in Western societies include these common stimulus-response patterns: our mouths water at the sight of a raspberry cheesecake, we stop at a red light and go when it turns green, and we thank a person who does something nice for us.
2. Fear of spiders (arachnophobia) is regularly listed by people as among their greatest phobias, on a par with fear of heights or of enclosed spaces. Behavioral psychologists suggest that a fear of spiders is innate in many people, although there is broad variation in humans and across different cultures. Certainly a person who has suffered a spider bite is likely to develop an aversion, if not a fear, of spiders. Fearful behaviors are difficult to analyze in humans because of the complexity and diversity of human experiences and because we cannot do the necessary controlled

experiments in people. Animal studies show that some insects who are preyed on by spiders also have an innate fear of spiders, while in other species the fear is learned by young observing the reactions of their mothers.

3. The details of your answer will depend on which sport or musical instrument you have chosen to discuss. In general terms, learning to play an instrument or a sport requires learning the techniques and strategies needed to be effective (how to hold the instrument or how to kick a soccer ball, for example) and learning which forms or positions are most effective (which sounds made by the instrument sound nice to the ears or which poses in dance look the most graceful). However, the most talented among us very likely have genetically determined advantages as well. For example, having perfect pitch is a genetic trait, but it becomes evident only in a person with musical training. As an example from sports, the proportion of fast-twitch and slow-twitch muscles we have can determine our prowess in strength versus endurance sports, but once again training plays a large role too.
4. There are many forms of nonverbal communication, some of which tend to be culture-specific. Examples include smiling, frowning, winking, waving goodbye, giving someone an inquiring look, making a "peace" sign, and wearing a uniform (such as a police officer's uniform).
5. Fixed behaviors that might be important to group members could involve responses to stimuli that an organism might never have a chance to learn about. For instance, group-living animals might have fixed behaviors that enable them to identify members of their own group (for example, by recognizing an odor unique to the group). Or a group-living animal might have a fixed behavior that leads it to avoid snakelike objects or to hide at the sight of a silhouette resembling that of a predatory bird. There might not be a chance to learn the appropriate behaviors from others, because one incorrect response could easily be fatal. Behaviors associated with surviving during the first few days following birth or hatching, before the creature has a chance to learn, are likely to be fixed behaviors.
6. The actual rituals of human courtships are societal constructs, but the underlying motivations behind them have their roots in the biological differences between the sexes. Buying an expensive ring, for example, can be seen as a male providing a nuptial gift for the female. Similarly, aggressive behaviors such as those seen in sports or in business can be viewed as demonstrations by males that they are fit as potential mates. As mammals, human females are more invested in the rearing of offspring than males are, so they tend to be the choosier of the sexes.

Chapter 23

1. Mutualism is common because its costs are outweighed by the benefits it provides. Yucca plants, for example, may lose a few seeds to the offspring of their moth pollinators, but they still end up with more seeds than if the moths had not pollinated them to begin with.
2. Organisms eaten by consumers are under selection pressure to develop defenses against those consumers. Likewise, consumers experience selection pressure to overcome the defenses of their food organisms. Adaptations that improve the survival of individuals in either group will therefore be likely to spread throughout the population. An example would be the poison of the rough-skinned newt, which can kill nearly all predators; garter snakes, however, have evolved the capacity to tolerate the toxin and eat the newt.
3. The plant community would have fewer species (a). When the rabbits were removed, the grass they prefer would no longer be eaten, so that grass would assert its dominance as the superior competitor and would probably drive some of the other grass species in the area to extinction. The inferior competitor is still using resources that the superior competitor needs, thereby possibly limiting the superior competitor's distribution or abundance.
4. a. Food webs influence the movement of energy and nutrients through a community. Some species, called keystone species, have a disproportionately large effect, relative to their abundance, on the types and abundances of other species in the community.
   b. Disturbances such as fire occur so often in many ecological communities that the communities are constantly changing and hence may never establish climax communities (relatively stable end points of ecological succession). Depending on the type and severity of the disturbance, a given community may or may not be able to recover.
   c. Climate is a key factor in determining which organisms can live in a given area, so if the climate changes, the community changes.
   d. As continents move to different latitudes, their climates—and therefore their communities—change.
5. The introduction of beard grass to Hawaii has increased the frequency and size of fires on the island. This change is due to the large amount of dry matter that the grass produces, which burns more easily and hotter than does the native vegetation. In this way, the presence of one species in the community has profoundly altered its disturbance pattern, leading to other large changes in the community.
6. The disturbance described in (b) would probably require a longer recovery time, assuming that no other disturbances, such as fire, were to occur. In the disturbance described in (a), the soil and ground cover would be left intact, so new trees would be able to sprout according to natural succession, eventually growing to replace the trees that were removed. In (b), however, the pollutant would have damaged the soil, which would hinder the ability of the trees and ground vegetation to grow. The soil chemistry would need to return to normal before the forest vegetation would be able to grow back and thrive again.

7. Change is a part of all ecological communities. However, human-caused change is unique in that we can consider the impact of our actions, and we can decide whether or not to take actions that cause community change. Whether or not a particular change is viewed as ethically acceptable will depend on the type of change, the reason for it, and the perspective of the person evaluating the change. For example, a person might find it ethically acceptable to alter a region so as to produce a long-term source of food for the growing human population, yet not ethically acceptable to take actions that result in short-term economic benefit but cause long-term economic loss and ecological damage.

Chapter 24

1. An ecosystem consists of a community of organisms together with the physical environment in which those organisms live. The organisms in an ecosystem interact with one another in various ways; organisms can also move from one ecosystem to another. For this reason, determining the boundaries of protection for a particular ecosystem would probably be difficult. Such a plan would require an understanding of the roles that certain organisms play in the overall function of an ecosystem.
2. Energy captured by producers from an external source, such as the sun, is stored in the bodies of producers in chemical forms, such as carbohydrates. At each step in a food chain, a portion of the energy captured by producers is lost from the ecosystem as metabolic heat. This steady loss prevents energy from being recycled.
3. The productivity of marine ecosystems tends to be high close to land but relatively low in the open ocean because nutrients needed by aquatic photosynthetic organisms are in short supply in the open ocean. Nutrients delivered by streams and rivers account for the high productivity of many coastal areas. Nutrients drained off the land stimulate the growth and reproduction of phytoplankton, the small photosynthetic producers that form the foundation of aquatic food webs. Estuaries are some of the most productive habitats because the rich nutrient supply supports large populations of producers, which in turn nourish large populations of consumers. Wetlands such as swamps and marshes can also match the productivity levels of tropical forests and farmland. They trap soil sediments rich in nutrients and organic matter, thereby promoting the growth of flooding-tolerant plants and phytoplankton, which in turn feed a complex community of consumers.

   The relatively high productivity of the Gulf of Alaska is attributed to upwelling. In upwelling, wind and ocean currents drive cold, nutrient-rich layers of water to the surface to replace warm, nutrient-depleted water. Regions that experience upwelling have high NPP because producers that live there are less nutrient-limited than are producers in similar ecosystems that lack upwelling.
4. Decomposers break down the tissues of dead organisms into simple chemical components, thereby returning nutrients to the physical environment so that they can be used again by other organisms.
5. Nutrients can cycle on a global level. When sulfur dioxide pollution, for example, enters the atmosphere in one area of the world, winds can move that pollution around the world, where it can affect other ecosystems.
6. Human economic activity is interwoven with several key ecosystem services. Pollination is essential for the productivity of both commercial crops and backyard gardens. Floodplains act as safety valves for major floods, provided we do not build on them or separate them from the bodies of water they help control. Forests act as water filtration systems. We rely on nutrient cycling to keep us alive. When ecosystem services such as these are damaged, human economic interests are damaged as well.
7. An acre of agricultural land can support many more vegetarians than it can support people who live mainly on animal products such as eggs, poultry, milk, or beef. The reason is the pyramid of energy transfer in food chains. Only about 10 percent of the energy at any trophic level is available to consumers at the next-higher trophic level in a food chain. More of the net primary productivity (NPP) generated by plants is available to primary consumers, less to secondary consumers, and even less to tertiary consumers. Therefore, it takes more land area to grow the same mass of an animal-based food than a plant-based food. According to David Pimentel at Cornell University, all the grain that is currently fed to livestock in the United States would feed nearly 800 million if these people obtained their calories directly from the grain. Raising beef is less efficient in converting plant biomass into human food than is raising chickens. Part of the reason for the reduced efficiency is that less of the larger animal is available as edible food, but the methods used in intensive livestock production are a contributing factor. For example, grass-fed cattle are more efficient in converting plant biomass into meat than are grain-fed cattle.

Chapter 25

1. Major types of global change caused by humans include global warming, land and water transformation, and changes in the chemistry of Earth (for example, changes to nutrient cycles). By altering the conditions under which species live, all of these changes could result in the increased dominance of certain types of species and the disappearance of others from various ecosystems.
2. Human-caused global changes often happen at a much more rapid rate than do changes due to natural causes. Continental drift and natural climate change happen much more slowly than do the measurable increases humans have caused in atmospheric carbon dioxide levels and nitrogen fixation. In addition, humans have a choice about the global changes we cause.
3. The present levels of atmospheric CO_2 are higher than any seen in the previous 420,000 years. Since the middle of the twentieth

century, atmospheric CO_2 levels have been rising at about 2 parts per million (ppm) per year, and they stood at 385 ppm toward the end of 2008. Scientific instruments can directly measure the amount of carbon dioxide in the atmosphere. By measuring CO_2 levels in bubbles of air trapped in ancient ice, scientists can estimate the amount of CO_2 that was present in the atmosphere up to hundreds of thousands of years ago.

4. It would be prudent to take action on global warming sooner rather than later, despite present uncertainties as to its extent. There is already evidence of climate changes that are consistent with the predicted effects of global warming. In addition, the correlation of rising CO_2 levels and worldwide temperature increases suggests that these increases will continue in the future if carbon dioxide emissions are not reduced. If action is delayed too long, it may be too late to undo many of the effects of global warming.
5. According to a New York Times/CBS News poll in 2006, a majority of Americans (55 percent) would support an increase in the federal gasoline tax (currently 18.4 cents per gallon) as long as the money raised is used to reduce global warming or to reduce U.S. dependence on foreign oil. Advocates of the tax increase say it will lower carbon emissions by reducing gasoline consumption as individuals and companies reduce waste and inefficiency in transportation. Many economists argue for a gas tax of up to one dollar a gallon phased in over 5 years, and maintain that most low-income and middle-income citizens will actually stand to benefit, provided some of the tax revenue is invested in improving public transportation.
6. For people to have a sustainable impact on Earth, we must reduce the rate of growth of the human population, and we must reduce the rate of resource use per person. To achieve these goals, many aspects of human society would have to change. For example, we would need to alter our view of nature from looking at it as a limitless source of goods and materials that can be exploited for short-term economic gain to accepting nature's limits and seeking always to take only those actions that can be sustained for long periods of time. Many specific actions would follow from such a change in our view of the world, such as an increase in recycling, the development and use of renewable sources of energy, a decrease in urban sprawl, an increase in the use of technologies with low environmental impact (such as organic farming), and a concerted effort to halt the ongoing extinction of species.

   Examples of actions you could take to make your impact on Earth more sustainable include the following: reducing the quantity of nonfood items purchased; reusing items until they are no longer usable; buying used items rather than getting everything new; recycling paper, plastic, glass, and metal; taking along reusable cloth bags when shopping; rarely using paper cups, plates, or towels; planting trees and other native plants, especially those that help feed native wildlife; reducing water use by not leaving water running when brushing teeth, by adjusting the water level of washing machines to match the size of the load, and by using water-saving fixtures; reducing fossil fuel use by choosing a fuel-efficient car and by using household heating and air-conditioning only as needed; using compact fluorescent lightbulbs and turning off lights that are not in use; supporting organic farmers by purchasing organically grown food.
7. Although ocean acidification by half a pH unit may not seem like much, studies show that the ability of the coral animal (polyp) to calcify is substantially reduced and reef growth slows dramatically. Other reef-building organisms, such as coralline algae, are also debilitated by the acidity. Because coral reefs provide diverse habitats, biodiversity will decline as the reefs decline. A coral reef acts like a breakwater, and biodiversity in sheltered bays and lagoons is likely to be devastated, endangering sea grasses and mangroves, while eroding the shoreline. Tourism, fisheries, and other economic activities supported by coral reefs are valued at $400 billion globally, and all of them are likely to be affected if the current trend in ocean acidification continues.
8. An endocrine disrupter is a chemical that interferes with hormone function to produce negative effects, such as reduced fertility, developmental abnormalities, immune system dysfunction, and increased risk of cancer. DDT and bisphenol A are endocrine disrupters that can be detected in the tissues of most Americans. DDT is a pesticide and also a POP (persistent organic pollutant) that is bioaccumulated and biomagnified along a food chain. Until its use was banned in 1972, DDT was extensively sprayed in the United States to control mosquitoes and protect crops from insect pests. DDT disrupts reproduction in a variety of animals, but predatory birds were especially hard-hit. The chemical interferes with calcium deposition in the developing egg, resulting in thin, fragile eggshells that break easily. Bisphenol A is found in many types of plastics (including those marked with the recycling number 7). In laboratory animals, bisphenol A increases the risk of diabetes, obesity, reproductive problems, and various cancers.

Glossary

abiotic Of or referring to the nonliving environment that surrounds the community of living organisms in an ecosystem; the atmosphere, water, and Earth's crust. Compare *biotic*.

abyssal zone The deeper waters of the open ocean, beyond the continental shelf, at depths greater than 6,000 meters (nearly 20,000 feet).

acid A chemical compound that can give up a hydrogen ion. Compare *base* (definition 1) and *buffer*.

acid rain Precipitation with an unusually low pH, compared to that of unpolluted rain (which has a pH of about 5.2). Acid rain is a consequence of the release of sulfur dioxide and other pollutants into the atmosphere, where they are converted to acids that then fall back to Earth in rain or snow.

activation energy The minimum energy input that enables reactants to overcome the energy barrier, allowing a chemical reaction to proceed at a noticeable rate.

active carrier protein Also called *membrane pump*. A cell membrane protein that, using energy from an energy-rich molecule such as ATP, changes its shape to transport an ion or a molecule across the membrane against a concentration gradient. Compare *passive carrier protein*.

active site The specific region on the surface of an enzyme where substrate molecules bind.

active transport Movement of ions or molecules across a biological membrane that requires an input of energy because it is "uphill" (against a concentration gradient). Compare *passive transport*.

adaptation Also called *adaptive trait*. An inherited characteristic—structural, biochemical, or behavioral—that enables an organism to function well and therefore to survive and reproduce better than competitors lacking the characteristic.

adaptive radiation An evolutionary expansion in which a group of organisms takes on new ecological roles and forms new species and higher taxonomic groups.

adaptive trait See *adaptation*.

adenosine diphosphate See *ADP*.

adenosine triphosphate See *ATP*.

ADP Adenosine diphosphate, the low-energy form of the energy carrier ATP, which constitutes the "energy currency" in all living cells. When ADP becomes phosphorylated (acquires an additional phosphate group), it is transformed into ATP.

adult stem cell Also called *somatic stem cell*. A cell that retains the capacity for self-renewal and persists into adulthood. Compare *embryonic stem cell*.

aerobic Of or referring to a metabolic process or organism that requires oxygen gas. Compare *anaerobic*.

afterbirth See *placenta*.

alga (pl. algae) Any photosynthetic protist. Algae may or may not be motile. Compare *protozoan*.

allele One of two or more alternative versions of a gene that exist in a population of organisms. Each allele has a DNA sequence that is somewhat different from that of all other alleles of the same gene.

allele frequency The proportion (percentage) of a particular allele in a population.

allopatric speciation The formation of new species from populations that are geographically isolated from one another. Compare *sympatric speciation*.

altruism A behavior that benefits another individual but has a cost to the individual performing the behavior.

amino acid A nitrogen-containing, small organic molecule that has an amino group, a carboxyl group, and a variable R group, all covalently attached to a single carbon atom. Proteins are polymers of amino acids.

amniotic egg An egg in which the developing embryo is surrounded and protected by layers of extraembryonic membranes (including the amnion) that promote gas exchange and removal of waste.

anabolism Also called *biosynthesis*. The linked chain of energy-requiring reactions that create complex biomolecules from smaller organic compounds. Compare *catabolism*.

anaerobic Of or referring to a metabolic process or organism that does not require oxygen gas. Compare *aerobic*.

analogous Of or referring to a characteristic shared by two groups of organisms because of convergent evolution, not common descent. Compare *homologous*.

anaphase The stage of mitosis or meiosis during which sister chromatids separate and move to opposite poles of the cell.

ancestral species The original species from which new species descend.

anchorage dependence The inability of a cell to divide if it is detached from its surroundings. Most human cells exhibit this trait.

anchoring junction Also called *desmosome*. A protein complex that acts as a "hook" between two animal cells or between a cell and the extracellular matrix, linking them to brace against rupturing forces. Compare *gap junction* and *tight junction*.

angiosperms Also called *flowering plants*. One of the major groups of plants. Angiosperms have vascular tissues, seeds, flowers, and fruits. They include most plants on Earth today. Compare *bryophytes* and *gymnosperms*.

Animalia The kingdom of Eukarya that is made up of animals—multicellular heterotrophs that have evolved specialized tissues, organs and organ systems, body plans, and behaviors.

Annelida Annelids. The eukaryotic phylum consisting of all worms with true coeloms and segmented bodies. Earthworms are the most familiar example.

antenna complex A disclike grouping of pigment molecules, including chlorophyll, that harvests energy from sunlight in the thylakoid membrane of a chloroplast.

antibiotic A molecule that is secreted by one microorganism to kill or slow the growth of another microorganism.

anticodon A sequence of three nitrogenous bases on a transfer RNA molecule that enables it to form complementary base pairs with a corresponding *codon* on an mRNA molecule.

Archaea One of the three domains of life, encompassing the microscopic, single-celled prokaryotes that arose after the Bacteria. The domain Archaea is equivalent to the kingdom Archaea. Compare *Bacteria* and *Eukarya*.

Arthropoda Arthropods. The largest eukaryotic phylum, consisting of animals characterized by a hard exoskeleton and jointed body parts. Arthropods include millipedes, crustaceans, insects, and spiders.

artificial selection A process in which only individuals that possess certain characteristics are allowed to breed. Artificial selection is used to guide the evolution of crop plants and domestic animals in ways that are advantageous for people. Compare *natural selection*.

ascomycetes Also called *sac fungi*. One of three major groups of fungi. Ascomycetes produce spores in sacs. Compare *basidiomycetes* and *zygomycetes*.

asexual reproduction The production of genetically identical offspring without the exchange of genetic material with another individual. Compare *sexual reproduction*.

atavistic trait A hereditary trait that represents a reversion to an ancestral state.

atmospheric cycle A type of nutrient cycle in which the nutrient enters the atmosphere easily. Compare *sedimentary cycle*.

atom The smallest unit of a chemical element that still has the properties of that element.

atomic mass number The sum of the number of protons and neutrons found in the nucleus of an atom of a particular chemical element.

atomic number The number of protons found in the nucleus of an atom of a particular chemical element.

ATP Adenosine triphosphate, a molecule that is commonly used by cells to store energy and to transfer energy from one chemical reaction to another. ATP fuels a wide variety of cellular activities in every organism on Earth.

ATP synthase A large channel-containing protein complex that spans the thylakoid membrane and catalyzes the conversion of ADP to ATP.

autosome Any chromosome that is not a *sex chromosome*.

autotroph See *producer*. Compare *heterotroph* and *mixotroph*.

Bacteria One of the three domains of life, encompassing the microscopic, single-celled prokaryotes that were the first organisms to arise. The domain Bacteria is equivalent to the kingdom Bacteria. Compare *Archaea* and *Eukarya*.

bacterial conjugation The process in which a bacterium actively trades DNA with another bacterium, usually of the same species.

bacterial culture A lab dish that contains bacteria living off a specially formulated food supply.

base 1. A chemical compound that can accept a hydrogen ion. Compare *acid* and *buffer*. 2. See *nitrogenous base*.

base pair Also called *nucleotide pair*. A pair of complementary nitrogenous bases connected by hydrogen bonds. Base pairs form the "rungs" of the DNA double helical "ladder." In DNA, adenine pairs with thymine (A–T) and cytosine pairs with guanine (C–G); in RNA, uracil replaces thymine.

basidiomycetes Also called *club fungi*. One of three major groups of fungi. Basidiomycetes produce the fruiting bodies we call mushrooms. Compare *ascomycetes* and *zygomycetes*.

behavior Sensing and responding to external cues. More specifically, a coordinated response to a stimulus; particularly, a response that involves movement.

benign tumor A relatively harmless cancerous growth that is confined to a single tumor and does not spread to other tissues in the body.

benthic zone The bottom surface of any body of water.

bilateral symmetry A body arrangement in which only one plane passes vertically from the top to the bottom of the animal, dividing the body into two halves that mirror each other. Bilateral animals have distinct right and left sides, with near-identical body parts on each side. Compare *radial symmetry*.

binary fission A form of asexual reproduction in which a cell divides to form two genetically identical daughter cells that replace the original parent cell.

bioaccumulation An increase in the concentration of a substance within an organism so that it exceeds the concentration in the surrounding abiotic environment. Compare *biomagnification*.

biodiversity The variety of organisms on Earth or in a particular location, ranging from the genetic variation and behavioral diversity of individual organisms or species through the diversity of ecosystems.

biofilm A tough, slime-covered aggregate of bacteria.

biogeochemical cycle See *nutrient cycle*.

biological evolution A change in the overall genetic characteristics of a group of organisms over multiple generations of parents and offspring.

biological hierarchy The nested series in which living things, their building blocks, and their living and nonliving surroundings can be arranged—from atoms at the lowest level to the entire biosphere at the highest level.

biological species concept The idea that a species is defined as a group of populations that can interbreed but are reproductively isolated from other such groups. Compare *morphological species concept*.

biology The study of life.

biomagnification An increase in the tissue concentrations of a bioaccumulated chemical at successively higher trophic levels in a food chain. Compare *bioaccumulation*.

biomass The mass of organisms per unit of area.

biome A large area of the biosphere that is characterized according to its unique climatic and ecological features. Terrestrial biomes are usually classified by their dominant vegetation; aquatic biomes, by their physical and chemical features.

biomolecule Any molecule found within a living cell.

bioremediation The use of organisms to clean up environmental pollution.

biosphere All living organisms on Earth, together with the environments in which they live. Compare *ecosystem*.

biosynthesis See *anabolism*.

biotic Of or referring to the assemblage of interacting organisms—prokaryotes, protists, animals, fungi, and plants—in an ecosystem. Compare *abiotic*.

bivalent See *tetrad*.

blastula A hollow, fluid-filled sphere with a single layer of cells on the surface that forms from repeated division of a zygote. Compare *gastrula*.

boreal forest Also known as *taiga*. A terrestrial biome dominated by coniferous trees that grow in northern or high-altitude regions with cold, dry winters and mild summers.

bryophytes One of the major groups of plants. Bryophytes include mosses, liverworts, and hornworts. They lack true vascular tissues. Compare *angiosperms* and *gymnosperms*.

buffer A chemical compound that can both give up and accept hydrogen ions. Buffers can maintain the pH of water within specific limits. Compare *acid* and *base* (definition 1).

C_3 plant A plant that fixes CO_2 through the Calvin cycle alone. The first stable product of CO_2 fixation in C_3 plants is the three-carbon molecule phosphoglyceric acid. Compare *C_4 plant* and *CAM plant*.

C_4 pathway A photosynthetic pathway that reduces photorespiration by surrounding rubisco in a high-CO_2 environment in special gas-tight cells (bundle-sheath cells). It is most common in plants adapted to hot and sunny climates. Compare *CAM pathway*.

C_4 plant A plant in which the Calvin cycle reactions are carried out in specialized tissues (bundle-sheath cells) that maintain a high CO_2 environment. Carbon dioxide diffusing in through the stomata is converted into four-carbon (C_4) molecules in leaf mesophyll cells before being transported to the bundle sheath. Compare *C_3 plant* and *CAM plant*.

Calvin cycle The series of chemical reactions in photosynthesis that manufacture sugar. The Calvin cycle takes place in the stroma of the chloroplast and synthesizes sugars from carbon dioxide and water. Compare *light reactions*.

CAM See *crassulacean acid metabolism*.

CAM pathway A photosynthetic pathway that enables desert plants to operate the Calvin cycle during the day while keeping stomata closed, to conserve moisture. At night, stomata open and the incoming carbon dioxide is stored for use in the daytime. Compare *C_4 pathway*. See also *crassulacean acid metabolism*.

CAM plant A plant that relies on the CAM (crassulacean acid metabolism) pathway to survive in very dry environments, CAM plants open their stomata only at night, storing the incoming carbon dioxide as four-carbon molecules. The molecules release carbon dioxide for the Calvin cycle reactions in the daytime, when the stomata are shut down to conserve water. Compare *C_3 plant* and *C_4 plant*.

Cambrian explosion A major increase in the diversity of life on Earth that occurred about 530 million years ago, during the Cambrian period. The Cambrian explosion lasted 5–10 million years, during which time large and complex forms of most living animal phyla appeared suddenly in the fossil record.

cancer cell Also called *malignant cell*. A tumor cell that breaks loose from the tumor, invades neighboring tissues, and disrupts the normal function of tissues and organs.

cancer immunotherapy The use of antibodies that stick to cancer cells while leaving healthy cells alone as a strategy to treat cancer.

carbohydrate Any of a class of organic compounds that includes sugars and their polymers, in which each carbon atom is usually linked to two hydrogen atoms and an oxygen atom. Carbohydrates contain carbon, hydrogen, and oxygen in a ratio of 1:2:1.

carbon cycle The transfer of carbon within biotic communities, between living organisms and their physical surroundings, and within the abiotic world.

carbon fixation The process by which carbon dioxide is incorporated into organic molecules. Carbon fixation occurs in the chloroplasts of plants and results in the synthesis of sugars.

carcinogen A physical, chemical, or biological agent that causes cancer.

carrier protein A cell membrane protein that recognizes, binds, and transports a specific cargo molecule across the membrane. See also *active carrier protein* and *passive carrier protein*.

carrying capacity The maximum population size that can be supported indefinitely by the environment in which the population is found.

catabolism The linked chain of reactions that release chemical energy in the process of breaking down complex biomolecules. Compare *anabolism*.

catalyst A substance that speeds up a specific chemical reaction without being permanently altered in the process. Enzymes, which are usually proteins, are an example of biological catalysts.

causation The act of causing something to happen or to exist. Compare *correlation*.

cell The smallest and most basic unit of life, the fundamental building block of all living things.

cell culture A laboratory procedure in which cells are grown in a special nutrient medium in a test tube or petri dish.

cell cycle A series of distinct stages in the life cycle of a cell that is capable of dividing. Cell division is the last stage in the cycle.

cell differentiation The process through which a daughter cell becomes different from its parent cell and acquires specialized functions.

cell division The final stage in the life of an individual cell, in which a parent cell is split up to generate two (mitotic cell division) or four (meiotic cell division) daughter cells.

cell junction A structure that anchors a cell, connects it with a neighbor, or creates communication passageways between two cells.

cell plate A partition, consisting of membrane and cell wall components, that appears during cytokinesis in dividing plant cells. The cell plate matures into a polysaccharide-based cell wall flanked on either side by the plasma membranes of the two daughter cells.

cell theory The theory that every living organism is composed of one or more cells, and that all cells living today came from a preexisting cell.

cellular respiration A metabolic process that extracts chemical energy from organic molecules, such as sugars, to generate the universal energy carrier ATP, consuming oxygen and releasing carbon dioxide and water in the process. Cellular respiration has three phases: glycolysis, the Krebs cycle, and oxidative phosphorylation. Compare *photosynthesis*.

cellulose An extracellular polysaccharide, produced by plants and some other organisms, that strengthens cell walls.

centromere A physical constriction that holds sister chromatids together.

centrosome A cytoskeletal structure in the cytosol that helps organize the mitotic spindle and defines the two poles of a dividing cell.

cephalization In bilateral animals, the evolutionary trend in which structures used for eating are concentrated at the anterior end, thereby allowing the animal to look where it is going.

CFC See *chlorofluorocarbon*.

channel protein Also called *membrane channel*. A cell membrane protein that forms an opening through which specific ions or molecules can pass.

chaparral A terrestrial biome characterized by shrubs and small nonwoody plants that grow in regions with rainy winters and hot, dry summers.

character displacement A process by which intense competition between species causes the forms of the competing species to evolve to become more different over time.

chemical bond The mutual interaction that causes two atoms to associate with each other.

chemical compound A substance in which atoms from two or more different elements are bonded together, each in a precise ratio.

chemical energy Potential energy that is stored in atoms because of their position in relation to other atoms in the system under consideration.

chemical formula The simple shorthand that is used to represent the atomic composition of salts and molecules. A subscript number to the right of the symbol of each element shows how many atoms of that element are contained in the salt or molecule.

chemical reaction The process of breaking or creating chemical bonds.

chemoautotroph An organism that obtains its energy from chemicals and derives its carbon from carbon dioxide in the air. All chemoautotrophs are prokaryotes. Compare *chemoheterotroph* and *photoautotroph*.

chemoheterotroph An organism that obtains its energy from chemicals and derives its carbon from carbon-containing compounds found mainly in other organisms. All fungi and animals, as well as many protists and prokaryotes, are chemoheterotrophs. Compare *chemoautotroph* and *photoheterotroph*.

chemotherapy The administration of high doses of chemical poisons to destroy cancer cells by killing all rapidly dividing cells.

chitin A carbohydrate that serves as an important support material in the cell walls of fungi and in animal exoskeletons.

chlorofluorocarbon (CFC) Any of a class of synthetic chemical compounds whose release into the atmosphere can damage the ozone layer. Until the early 1990s, CFCs were commonly used as refrigerants or propellants.

chlorophyll A green pigment that is used to capture light energy for photosynthesis.

chloroplast An organelle, found in plants and algae, that is the primary site of photosynthesis.

chordate Any of a group of animals that have a notochord, pharyngeal pouches, and a post-anal tail.

chromatin DNA complexed with packaging proteins. Chromosomes are made up of compacted chromatin.

chromosome A threadlike structure composed of a single DNA molecule packaged with proteins. Chromosomes achieve the highest level of compaction when prophase begins during mitosis or meiosis.

chromosome theory of inheritance A theory, supported by much experimental evidence, stating that genes are located on chromosomes.

cilium (pl. cilia) A hairlike structure, found in some eukaryotes, that uses a rowing motion to propel the organism or to move fluid over cells. Compare *flagellum*.

citric acid cycle See *Krebs cycle*.

class In reference to biological classification systems, the level in the Linnaean hierarchy above order and below phylum.

climate The prevailing weather conditions experienced in an area over relatively long periods of time (30 years or more). Compare *weather*.

climate change Significant and long-term change in the average climatic conditions in the biosphere, such as global warming.

club fungi See *basidiomycetes*.

coastal region An ecosystem of the marine biome defined as the area of the ocean stretching from the shoreline to the continental shelf, which is the undersea extension of a continent's edge.

codominance A type of allele interaction in which the effects of both alleles at a given genetic locus are equally visible in the phenotype of a heterozygote. In other words, the influence of each codominant allele is fully displayed in the heterozygote, without being diminished or diluted by the presence of the other allele (as in *incomplete dominance*).

codon A sequence of three nitrogenous bases in an mRNA molecule. Each codon specifies either a particular amino acid or a signal to start or stop the translation of a protein. Compare *anticodon*.

coelom A body cavity that develops within the mesoderm.

coevolution The process by which interactions among species drive evolutionary change in those species.

cohesion The attractive force that holds atoms or molecules of the same kind to one another.

commensalism An interaction between two species in which one partner benefits while the other is neither helped nor harmed. Compare *competition*, *exploitation*, and *mutualism*.

communication A type of behavior that allows one individual to exchange information with another, thereby making it possible for an animal to coordinate its activities with those of other individuals.

community An association of populations of different species that live in the same area.

competition An interaction between two species in which each has a negative effect on the other. Compare *commensalism*, *exploitation*, and *mutualism*.

competitive exclusion Competition in which one species is so superior as a competitor that it ultimately drives the other species extinct.

complex trait A genetic trait whose pattern of inheritance cannot be predicted by Mendel's laws of inheritance.

compound eye An eye, common in insects, that consists of many individual light-receiving units, each with its own lens and small cluster of photoreceptors. Honeybees use their compound eyes to see patterns of color on flowers. Compare *simple eye*.

concentration gradient The difference in concentration that exists between two regions if a substance is more abundant in one region than in the other region.

conditioning A type of behavioral learning in which an organism comes to associate a stimulus with a result.

consumer Also called *heterotroph*. An organism that obtains its energy by eating other organisms or their remains. Consumers include herbivores, carnivores, and decomposers. Compare *producer*.

control group A group of participants in an experiment that are subjected to the same environmental conditions as the *treatment group(s)*, except that the factor or factors being tested in the experiment are omitted.

controlled experiment An experiment in which the researcher measures the value of the dependent variable for two groups of study subjects that are comparable in all respects except that one group is exposed to a systematic change in the independent variable and the other group is not.

convection cell A large and consistent atmospheric circulation pattern in which warm, moist air rises and cool, dry air sinks. Earth has four stable giant convection cells (two in tropical regions and two in polar regions) and two less stable cells (located in temperate regions).

convergent evolution Evolutionary change that occurs when natural selection causes distantly related organisms to evolve similar structures in response to similar environmental challenges.

Coriolis effect The apparent deflection of the path of a moving object that occurs when the start and end points of that path are traveling at different speeds.

correlation A statistical relation indicating that two or more phenomena behave in an interrelated manner. Correlation does not establish *causation*.

covalent bond A strong chemical linkage between two atoms that is based on the sharing of electrons. Compare *ionic bond* and *hydrogen bond*.

crassulacean acid metabolism (CAM) A type of photosynthesis that enables plants adapted to dry environments to conserve water. CAM plants open their stomata to let in CO_2 only at night, convert the incoming CO_2 into four-carbon molecules, and store it overnight. The molecules release carbon dioxide for the Calvin cycle during the day, when stomata are closed.

Cretaceous extinction A mass extinction that occurred 65 million years ago, wiping out many marine invertebrates and terrestrial plants and animals, including the last of the dinosaurs.

crista (pl. cristae) A fold in the inner mitochondrial membrane.

crossing-over A physical exchange of chromosomal segments between paired paternal and maternal members of homologous chromosomes. Crossing-over takes place during prophase of meiosis I.

crustacean Any of a group of aquatic arthropods that are especially diversified in the marine environment. Shrimp, lobster, and crabs are among the most familiar crustaceans.

cryosurgery Surgical procedures that make use of extremely cold temperatures to kill abnormal cells, usually precancerous cells or cancers that are confined to a small region, such as the cervix.

cuticle A waxy layer that covers aboveground plant parts, helping to prevent water loss and to keep enemies, such as fungi, from invading the plant.

cytokinesis The stage of the cell cycle following mitosis, during which the cell physically divides into two daughter cells.

cytoplasm The contents of a cell enclosed by the plasma membrane but, in eukaryotes, excluding the nucleus. Compare *cytosol*.

cytoskeleton A complex network of protein filaments found in the cytosol of eukaryotic cells. The cytoskeleton maintains cell shape and is necessary for cell division and movement.

cytosol The water-based fluid component of the cytoplasm. In eukaryotes, the cytosol consists of all the contents of a cell enclosed by the plasma membrane, but excluding all organelles. Compare *cytoplasm*.

data (s. datum) Information. Typical data answer questions such as where, when, or how much.

decomposer An organism that breaks down dead tissues into simple chemical components, thereby returning nutrients to the physical environment.

dehydration reaction The chemical reaction in which a water molecule is removed as a covalent bond forms. Compare *hydrolytic reaction*.

deletion In genetics, a mutation in which one or more nucleotides are removed from the DNA sequence of a gene, or a piece breaks off from a chromosome and is lost. Compare *insertion*.

denaturation The destruction of a protein's three-dimensional structure, resulting in loss of protein activity.

density-dependent Of or referring to a factor, such as food shortage, that limits the growth of a population more strongly as the density of the population increases. Compare *density-independent*.

density-independent Of or referring to a factor, such as weather, that can limit the size of a population but does not act more strongly as the density of the population increases. Compare *density-dependent*.

deoxyribonucleic acid See *DNA*.

dependent variable Also called *responding variable*. Any variable that responds, or could potentially respond, to changes in the *independent variable*.

desert A terrestrial biome dominated by plants that grow in regions with low precipitation, usually 25 centimeters per year or less.

desertification The conversion of habitats into deserts. Several factors can cause desertification, including deforestation, drought, or improper land management.

desmosome See *anchoring junction*.

deuterostome Any of a group of animals, including sea stars and vertebrates, in which the second opening to develop in the early embryo becomes the mouth. Compare *protostome*.

development The sequence of predictable changes that occur over the life cycle of an organism as it grows and matures to the reproductive stage.

differentiation See *cell differentiation*.

diffusion The passive transport of a substance from a region where it is more concentrated to a region where it is less concentrated. See also *facilitated diffusion* and *simple diffusion*.

dihybrid An individual that is heterozygous for two traits. Compare *monohybrid*.

diploid Of or referring to a cell or organism that has two complete sets ($2n$) of homologous chromosomes. Compare *haploid* and *polyploid*.

directional selection A type of natural selection in which individuals with one extreme of an inherited characteristic have an advantage over other individuals in the population, as when large individuals produce more offspring than do small and medium-sized individuals. Compare *disruptive selection* and *stabilizing selection*.

disaccharide A molecule made up of two linked *monosaccharides*. Some common examples are sucrose, lactose, and maltose. Compare *polysaccharide*.

disruptive selection A type of natural selection in which individuals with either extreme of an inherited characteristic have an advantage over individuals with an intermediate phenotype, as when both small and large individuals produce more offspring than do medium-sized individuals. Compare *directional selection* and *stabilizing selection*.

diversity The composition of an ecological community, which has two components: the number of different species that live in the community and the relative abundances of those species.

DNA Deoxyribonucleic acid, a double-stranded molecule consisting of two spirally wound polymers of nucleotides that store genetic information, including the information needed to synthesize proteins. Each nucleotide in DNA is composed of the sugar deoxyribose, a phosphate group, and one of four nitrogenous bases: adenine, cytosine, guanine, or thymine. Compare *RNA*.

DNA cloning Also called *gene cloning*. The introduction of recombinant DNA into a host cell (usually a bacterium), followed by the copying and propagation of the introduced DNA in the host cells and all its offspring.

DNA fingerprinting The use of DNA analysis to identify individuals and determine the relatedness of individuals.

DNA hybridization An experimental procedure in which DNA from two different sources bind to each other through complementary base pairing.

DNA ligase An enzyme that joins two DNA fragments.

DNA microarray Also called *gene chip*. A solid support about the size of a microscope slide, on which known DNA sequences from a genome have been arranged in an orderly fashion; used to test whether any of the DNA sequences on the microarray are being expressed—and if so, to what degree—in a given biological sample (such as cancer tissue from a patient).

DNA polymerase The key enzyme that cells use to replicate their DNA. In DNA technology, DNA polymerase is used in the polymerase chain reaction to make many copies of a gene or other DNA sequence.

DNA primer A short segment of DNA used in PCR amplification that is designed to pair with one of the two ends of the gene being amplified.

DNA probe A short sequence of DNA (usually tens to hundreds of bases long) that is labeled with a radioactive or fluorescent tag and then allowed to hybridize with a target DNA sample, with the objective of determining whether the two share sequence similarity.

DNA repair A three-step process in which damage to DNA is repaired. Damaged DNA is first recognized, then removed, and then replaced with newly synthesized DNA.

DNA replication The duplication, or copying, of a DNA molecule. DNA replication begins when the hydrogen bonds connecting the two strands of DNA are broken, causing the strands to unwind and separate. Each strand is then used as a template for the construction of a new strand of DNA.

DNA technology The set of techniques that scientists use to manipulate DNA.

domain In reference to biological classification systems, the highest level in the Linnaean hierarchy, immediately above kingdom. The three domains are Bacteria, Archaea, and Eukarya.

dominant allele An allele that controls the phenotype (dominant phenotype) when it is paired with a *recessive allele* in a heterozygous individual.

double-blind experiment An experiment in which neither the study subjects nor the researchers know which participants are receiving the treatment and which are controls. Compare *single-blind experiment*.

duplication In genetics, a type of mutation in which an extra copy of a gene or DNA fragment appears alongside the original, increasing the length of the chromosome.

ecdysis See *molting*.

ECM See *extracellular matrix*.

ecology The scientific study of interactions between organisms and their environment.

ecosystem A community of organisms, together with the physical environment in which the organisms live. Global patterns of air and water circulation link all the world's organisms into one giant ecosystem, the *biosphere*.

ectotherm An organism that relies on environmental heat for most of its heat input. Compare *endotherm*.

electron A negatively charged particle found in atoms. Each atom of a particular element contains a characteristic number of electrons. Compare *proton* and *neutron*.

electron shell One of several defined volumes of space in which electrons move around the nucleus of an atom.

electron transport chain (ETC) A group of membrane-associated proteins and other molecules that can both accept and donate electrons. The transfer of electrons from one ETC component to another releases energy that can be used to drive protons across a membrane and, ultimately, in both chloroplasts and mitochondria, to manufacture ATP.

element In reference to chemicals, a pure substance that has distinctive physical and chemical properties and cannot be broken down to other substances by ordinary chemical methods. The physical world is made up of 92 naturally occurring elements.

elongation The second of three stages of gene transcription (in which RNA polymerase synthesizes the complete RNA from free nucleotides) or translation (in which the ribosomes read the message in the mRNA to produce a complete chain of amino acids). See also *initiation* and *termination*.

embryo The early stage of development in plants and animals, extending from the zygote to the early development of organs and organ systems.

embryonic stem cell A pluripotent stem cell, in or derived from the inner cell mass, that exists only at the blastocyst stage in mammalian embryos. Compare *adult stem cell*.

endocrine disrupter A chemical that interferes with hormone function to produce negative effects, such as reduced fertility, developmental abnormalities, immune system dysfunction, and increased risk of cancer.

endocytosis A process by which a section of a cell's plasma membrane bulges inward as it envelops a substance outside of the cell, eventually breaking free to become a closed vesicle within the cytoplasm. Compare *exocytosis*.

endoplasmic reticulum (ER) An organelle composed of many interconnected membrane sacs and tubes. The ER is the main site for the synthesis of lipids, and certain types of proteins, in eukaryotic cells.

endosymbiosis theory The theory that eukaryotic cells are descended from a predatory ancestral eukaryote that engulfed prokaryotic cells that survived, instead of being digested for food, and evolved into endosymbionts (intracellular symbiotic partners).

endotherm An organism that generates metabolic heat in order to warm itself, instead of depending mainly on heat gain from the environment. Compare *ectotherm*. See also *homeotherm*.

endotoxin A component of bacterial cell walls that triggers illness. Endotoxins can produce fever, blood clots, and toxic shock. Compare *exotoxin*.

energy The capacity of any object to do work, which is the capacity to bring about a change in a defined system.

energy carrier A molecule that can store energy and donate it to another molecule or to a chemical reaction. ATP is the most commonly used energy carrier in living organisms.

energy pyramid A hierarchical representation of the amounts of energy available to organisms in an ecosystem in which each level corresponds to a step in a food chain (a trophic level).

enzyme A macromolecule, usually a protein, that acts as a catalyst, speeding the progress of chemical reactions. Nearly all chemical reactions in living organisms are catalyzed by enzymes.

epistasis (pl. epistases) A gene interaction in which the phenotypic effect of the alleles of one gene depends on which alleles are present for another, independently inherited gene.

ER See *endoplasmic reticulum*.

estuary An ecosystem of the marine biome defined as a region where a river empties into the sea.

ETC See *electron transport chain*.

Eukarya One of the three domains of life, encompassing the eukaryotes. Four kingdoms are included: Animalia, Plantae, Fungi, and Protista. Compare *Archaea* and *Bacteria*.

eukaryote A single-celled or multicellular organism in which each cell has a distinct nucleus and cytoplasm. All organisms other than the Bacteria and the Archaea are eukaryotes. Compare *prokaryote*.

eukaryotic flagellum A whiplike structure, found in eukaryotes, that propels the cell or organism through a lashing action, with waves passing from the base to the tip of the flagellum. Compare *prokaryotic flagellum*.

eutherian Any of a group of mammals whose young are nourished inside the mother's body through the placenta and are therefore born in a relatively well-developed state. Most mammals, including humans, are eutherians. Compare *marsupial* and *monotreme*.

eutrophication A process in which enrichment of water by nutrients (often from sewage or runoff from fertilized agricultural fields) causes bacterial populations to increase and oxygen concentrations to decrease.

evaporation The conversion of a substance from a liquid to a gas. Because this process requires a lot of energy, evaporation is an effective cooling device for animals, transferring heat from water on the body surface to the air in the surrounding environment.

evo-devo Evolutionary developmental biology.

exchange pool In reference to nutrients, a source such as the soil, water, or air where nutrients are available to producers.

exocytosis A process by which a vesicle approaches and fuses with the plasma membrane of a cell, thereby releasing its contents into the cell's surroundings. Compare *endocytosis*.

exon A DNA sequence within a gene that encodes part of a protein. Each exon codes for a stretch of amino acids. Compare *intron*.

exoskeleton An external framework of stiff or hard material that surrounds the soft tissues of an animal.

exotoxin A poison that a bacterium releases into its surroundings. Exotoxins kill tissues, and because the bacteria can spread rapidly, the tissue death can be extensive enough to kill a person in a day or two after the symptoms appear. Compare *endotoxin*.

experiment A repeatable manipulation of one or more aspects of the natural world.

experimental group See *treatment group*.

exploitation An interaction between two species in which one species benefits (the consumer) and the other species is harmed (the food organism). Exploitation includes the killing of prey by predators, the eating of plants by herbivores, and the harming or killing of a host by a parasite or pathogen. Compare *commensalism*, *competition*, and *mutualism*.

exploitative competition A type of competition in which species compete indirectly for shared resources, with each reducing the amount of a resource available to the other. Compare *interference competition*.

exponential growth A type of rapid population growth in which a population increases by a constant proportion from one generation to the next. Exponential growth is represented graphically by a J-shaped curve. Compare *logistic growth*.

extinction The demise of a species or other, larger group of organisms.

extracellular matrix (ECM) A coating of nonliving material, released by the cells of multicellular organisms, that often helps hold those cells together.

extremophile An organism that lives in an extreme environment, such as in a boiling hot geyser or on salted meat. Many archaeans are extremophiles.

F_1 generation The first generation of offspring in a breeding trial involving a series of genetic crosses. Compare *F_2 generation* and *P generation*.

F_2 generation The second generation of offspring in a breeding trial involving a series of genetic crosses. Compare *F_1 generation* and *P generation*.

facilitated diffusion The passive movement of a substance across a membrane with the assistance of membrane transport proteins. Compare *simple diffusion*.

facilitated transport The movement of a substance across a biological membrane with the assistance of membrane proteins.

family In reference to biological classification systems, the level in the Linnaean hierarchy above genus and below order.

fatty acid An organic molecule with a long, strongly hydrophobic hydrocarbon chain and a hydrophilic head group. Fatty acids are found in phospholipids, glycerides such as triglycerides, and waxes.

fermentation A series of catabolic reactions that produce small amounts of ATP through glycolysis and can function without oxygen. In most fermentation pathways, pyruvate from glycolysis is converted to other organic molecules, such as ethanol and carbon dioxide, or lactic acid.

fertilization The fusion of two different haploid gametes (egg cell and sperm) to produce a diploid zygote (the fertilized egg cell).

first law of thermodynamics Also called *law of conservation of energy*. The law stating that energy can be neither created nor destroyed, but only transformed or transferred from one molecule to another.

fixation In genetics, the removal of all alleles within a population at a particular genetic locus except one. The allele that remains has a frequency of 100 percent.

fixed behavior A predictable response to a particular, often simple, stimulus that does not involve learning. Compare *learned behavior*.

flagellum (pl. flagella) A long, whiplike extension from the cell that is lashed (in eukaryotes) or rotated (in prokaryotes) to enable that cell to move. Compare *cilium*. See also *eukaryotic flagellum* and *prokaryotic flagellum*.

flower A specialized reproductive structure that is characteristic of the plant group known as the angiosperms, or flowering plants.

flowering plants See *angiosperms*.

fluid mosaic model The concept of the plasma membrane as a phospholipid bilayer containing a variety of other lipids and embedded proteins, some of which can move laterally in the plane of the membrane.

food chain A single sequence of feeding relationships describing who eats whom in a community. Compare *food web*.

food web A summary of the movement of energy through a community. A food web is formed by connecting all of the *food chains* in the community.

fossil Preserved remains or an impression of a formerly living organism. Fossils document the history of life on Earth, showing that many organisms from the past were unlike living forms, that many organisms have gone extinct, and that life has evolved through time.

founder effect A genetic bottleneck that results when a small group of individuals from a larger source population establishes a new population far from the original population.

frameshift In genetics, the large change in coding information that results when the deletion or insertion in a gene sequence is not a multiple of three base pairs (a codon). The amino acid sequence of the protein that is translated from such a gene is severely altered in most cases, and protein function is typically lost.

fruit A mature ovary surrounding one or more seeds. The ovary wall becomes the fruit wall, which is often juicy and sweet and therefore attractive to animals, which then function as seed dispersal agents by eating the fruit and excreting the seeds.

functional group A specific cluster of covalently bonded atoms, with distinctive chemical properties, that forms a discrete subgroup in a variety of larger molecules.

Fungi The kingdom of Eukarya that is made up of absorptive heterotrophs (consumers that absorb their food after digesting it externally). Fungi include mushroom-producing species, yeasts, and molds, most of which make their living as decomposers.

G_0 phase A resting state during which the cell withdraws from the cell cycle before the S phase. G_0 cells are not competent to divide, unless they receive and respond to signals that direct them to reenter the cell cycle and proceed through the S phase.

G_1 phase The stage of the cell cycle that follows mitosis and precedes the S phase. The cell grows in size during the G_1 phase and makes a commitment to enter the S phase if it receives and responds to cell division signals.

G_2 phase The stage of the cell cycle that follows the S phase and precedes mitosis. The G_2 phase serves as a checkpoint ensuring that mitosis will not be launched under inappropriate conditions (such as inadequate nutrient supply, DNA damage, or incomplete DNA replication).

gamete A haploid sex cell that fuses with another sex cell during fertilization. Egg cells and sperm are gametes. Compare *somatic cell*.

gap junction A cytoplasmic channel, created by a narrow cylinder of proteins, that directly connects two animal cells and allows the passage of ions and small molecules between them. Compare *anchoring junction* and *tight junction*.

gastrula An embryo with at least two cell layers, resulting from the invagination and migration of cells in the ball-like *blastula*.

gel electrophoresis A procedure in which DNA fragments are placed in a gelatin-like substance (a gel) and subjected to an electrical charge, which causes the fragments to move through the gel. Small DNA fragments move farther than large DNA fragments, causing the fragments to separate by size.

gene The smallest unit of DNA that governs a genetic characteristic and contains the code for the synthesis of a protein or an RNA molecule. Genes are located on chromosomes.

gene chip See *DNA microarray*.

gene cloning See *DNA cloning*.

gene expression The creation of a functional product, such as a specific protein or RNA, using the coding information stored in a gene. Gene expression is the means by which a gene influences the cell or organism in which it is found.

gene flow The exchange of alleles between different populations.

gene pool The sum of all the genetic information carried by all the individuals in a population.

gene promoter In genetics, the DNA sequence in a gene that RNA polymerase binds to in order to begin transcription, and that therefore controls gene expression at the transcriptional level.

gene therapy A treatment approach that seeks to correct genetic disorders by repairing the genes responsible for the disorder.

genetic bottleneck A drop in the size of a population that results in low genetic variation or causes harmful alleles to reach a frequency of 100 percent in the population.

genetic carrier A heterozygous individual (*Aa*) that carries the allele for a recessive genetic disorder but, because the allele is recessive, does not get the disorder.

genetic code The code that specifies how information in mRNA is translated to create the specific sequence of amino acids found in the protein encoded by that mRNA. The genetic code consists of all possible three-base combinations (codons) of each of the four nitrogenous bases found in RNA. Of the 64 possible codons, 60 specify a particular amino acid, 3 serve as a "stop translation" signal, and 1 (AUG) acts as a "start translation" signal.

genetic cross A mating experiment, performed to analyze the inheritance of a particular trait, and also to raise crop varieties and particular breeds of animals.

genetic drift The natural process in which chance events cause certain alleles to increase or decrease in a population. The genetic makeup of a population undergoing genetic drift changes at random over time, rather than being shaped in a nonrandom way by natural selection.

genetic engineering The process in which a DNA sequence (often a gene) is isolated, modified, and inserted back into an individual of the same or a different species. Genetic engineering is commonly used to change the performance of the genetically modified organism, as when a crop plant is engineered to resist attack from an insect pest.

genetic heritage A particular set of genes and alleles that is typical of a species but different from that of all other species.

genetic linkage The situation in which different genes that are located close to one another on the same chromosome are inherited together; that is, they do not follow Mendel's law of independent assortment.

genetic recombination The creation of new groupings of alleles through the breaking and re-joining of different DNA segments, as in the crossing-over that takes place between paired homologues during meiosis I.

genetic trait Any inherited characteristic of an organism that can be observed or detected; known examples include body size, coat color, length of fur, and aggressive behavior.

genetic variation The allelic differences among the individuals of a population.

genetically modified organism (GMO) Also called *genetically engineered organism (GEO)* or *transgenic organism*. An individual into which a modified gene or other DNA sequence has been inserted, typically with the intent of improving a particular aspect of the recipient organism's performance.

genetics The scientific study of the inheritance of characteristics encoded by DNA.

genome All the DNA of an organism, including all its genes; in eukaryotes, the term refers to the DNA in a haploid set of chromosomes, such as that found in a sperm or egg.

genotype The allelic makeup of an individual; more specifically, the two alleles of a gene that produce the specific phenotype displayed by an individual. Compare *phenotype*.

genus (pl. genera) In reference to biological classification systems, the level in the Linnaean hierarchy above species and below family.

GEO See *genetically modified organism*.

geographic isolation The physical separation of populations from one another by a barrier such as a mountain chain or a river. Geographic isolation often causes the formation of new species, as when populations of a single species become physically separated from one another and then accumulate so many genetic differences that they become reproductively isolated from one another. Compare *reproductive isolation*.

geologic time A time scale spanning millions of years, relating particular time periods to events in Earth's geologic history.

germ line cell A cell that will develop into a gamete precursor cell. Germ cells remain separate from all other cells of the embryo, and they stay unspecialized until the reproductive organs develop, at which time they migrate to and take up residence inside the egg-producing ovaries or sperm-producing testes.

global change Worldwide change in the environment. There are many causes of global change, including climate change caused by the movement of continents, and changes in land and water use by humans.

global warming A worldwide increase in temperature. Earth appears to be entering a period of global warming caused by human activities—specifically, by the release of large quantities of greenhouse gases such as carbon dioxide into the atmosphere.

glucose A monosaccharide that is the primary metabolic fuel in most cells.

glycogen The main storage polysaccharide in animal cells. Found in humans primarily in the liver and skeletal muscles, glycogen is structurally similar to *starch*, the storage carbohydrate of plants.

glycolysis A series of catabolic reactions that splits glucose to produce pyruvate, which is then used in fermentation in the cytosol or degraded further in the mitochondrion. The glycolytic breakdown of one molecule of glucose yields two molecules of the energy carrier ATP. Glycolysis is the first of the three major phases of cellular respiration, preceding the Krebs cycle and oxidative phosphorylation.

GMO See *genetically modified organism*.

Golgi apparatus An organelle composed of flattened membrane sacs that directs proteins and lipids to various parts of the eukaryotic cell.

grasp reflex A reflex displayed by newborns that enables them to hold on tightly to anything that will fit in their tiny fists.

grassland A terrestrial biome dominated by grasses. Grasslands occur in relatively dry regions, often with cold winters and hot summers.

greenhouse effect The increase in Earth's temperature when absorbed heat that is reemitted by greenhouse gases becomes trapped because it lacks the energy to escape into outer space.

greenhouse gas Any of several gases in Earth's atmosphere that let in sunlight but trap heat. Examples include carbon dioxide, water vapor, methane, and nitrous oxide.

guard cell A specialized cell type found on leaves and stems of plants. Guard cells can inflate and deflate in response to their water content, and through this action they regulate the opening and closing of stomata, thereby controlling the rates at which CO_2 is brought into the plant and O_2 and water are lost from the plant. See also *stoma*.

gymnosperms One of the major groups of plants. Gymnosperms are represented by conifers such as pine or spruce trees. They have vascular tissues and seeds but lack flowers and fruits. Compare *angiosperms* and *bryophytes*.

habitat A characteristic place or type of environment in which an organism lives.

habituation A type of learned behavior in which an animal learns to filter out uninformative stimuli.

haploid Of or referring to a cell or organism that has only one set (n) of homologous chromosomes, such that only one member of each homologous pair (either the paternal or maternal member) is represented in that set. Compare *diploid* and *polyploid*.

harem In polygamous organisms, the group of females to which a single male has sexual access.

heat energy Also called *thermal energy*. Kinetic energy that is inherent in the random motion of particles in a system that can be transferred to other particles in the system; the portion of the total energy of a particle of matter that can flow from one particle of matter to another.

herbivore A consumer that relies on living tissues of producers, such as plants or algae, for nutrients.

heterotroph See *consumer*. Compare *autotroph* and *mixotroph*.

heterozygote An individual that carries one copy of each of two different alleles (for example, an *Aa* individual). Compare *homozygote*.

homeostasis The process of maintaining appropriate and constant conditions inside cells.

homeotherm An endotherm that maintains a near-constant internal temperature.

homeotic gene Any of a class of genes that cause the development of structures in embryological development.

homologous Of or referring to a characteristic shared by two groups of organisms because of their descent from a common ancestor. Compare *analogous*.

homologous chromosomes or **homologous pair** Also called *homologues*. A pair of matched chromosomes in diploid cells, one of which is inherited from the individual's female parent and the other from its male parent.

homozygote An individual that carries two copies of the same allele (for example, an *AA* or an *aa* individual). Compare *heterozygote*.

horizontal gene transfer See *lateral gene transfer*.

hormone A signaling molecule released in very small amounts into the circulatory system of an animal, or into a variety of tissues in a plant, that affects the functioning of target tissues.

hormone therapy Manipulation of the hormone environment in the body to stop or slow cancer cells. This treatment is used in some hormone-responsive cancers, which include some types of breast and prostate cancer.

host The individual, or organism, in which a particular parasite or pathogen lives.

housekeeping gene A gene that has an essential role in the maintenance of cellular activities and is expressed by most cells in the body.

hybrid An offspring that results when two different species, or two different varieties or genotypic lines, are mated.

hydrogen bond A weak electrical attraction between a hydrogen atom that has a slight positive charge and a neighboring atom with a slight negative charge. Compare *covalent bond* and *ionic bond*.

hydrolytic reaction The chemical reaction in which a water molecule is added to break a covalent bond. Compare *dehydration reaction*.

hydrophilic Of or referring to substances, both salts and molecules, that interact freely with water. Hydrophilic molecules dissolve easily in water but not in fats or oils. Compare *hydrophobic*.

hydrophobic Of or referring to molecules or parts of molecules that do not interact freely with water. Hydrophobic molecules dissolve easily in fats and oils but not in water. Compare *hydrophilic*.

hypertonic solution A solution that has a higher solute concentration than the cytosol of a cell has, causing more water to flow out of the cell than into it. Compare *hypotonic solution* and *isotonic solution*.

hypha (pl. hyphae) In fungi, a threadlike absorptive structure that grows through a food source. Mats of hyphae form mycelia, the main bodies of fungi.

hypothesis See *scientific hypothesis*.

hypotonic solution A solution that has a lower solute concentration than the cytosol of a cell has, causing more water to flow into the cell than out of it. Compare *hypertonic solution* and *isotonic solution*.

imprinting A type of learning in which an offspring forms an association or bond with its parent early in its development.

incomplete dominance A type of allelic interaction in which heterozygotes (*Aa* individuals) are intermediate in phenotype between the two homozygotes (*AA* and *aa* individuals) for a particular gene. Compare *codominance*.

independent assortment of chromosomes The random distribution of maternal and paternal chromosomes into gametes during meiosis.

independent variable Also called *manipulated variable*. The single variable that is manipulated in a typical scientific experiment. Compare *dependent variable*.

individual A single organism, usually physically separate and genetically distinct from other individuals.

induced defense A defensive response in plants that is directly stimulated by an attack from herbivores.

induced fit model A model of substrate-enzyme interaction stating that as a substrate enters the active site, the parts of the enzyme shift about slightly to allow the active site to mold itself around the substrate.

induced pluripotent stem cell (iPSC) Any cell, even a highly differentiated cell in the adult body, that has been genetically reprogrammed to mimic the pluripotent behavior of embryonic stem cells.

initiation The first of the three main stages of gene transcription (in which RNA polymerase begins transcribing the template strand of DNA) or translation (in which the ribosome-tRNA complex detects the start codon in an mRNA molecule and protein synthesis begins). See also *elongation* and *termination*.

inner cell mass The cluster of cells inside the blastocyst that eventually develops into the embryo and some of the membranes that surround a mammalian embryo and fetus.

insertion In genetics, a mutation in which one or more nucleotides are inserted into the DNA sequence of a gene. Compare *deletion*.

interference competition A type of competition in which one organism directly excludes another from the use of resources. Compare *exploitative competition*.

intermediate filament One of a diverse class of ropelike protein filaments that serve as structural reinforcements in the cytoskeleton. Compare *microfilament* and *microtubule*.

intermembrane space The space between the inner and outer membranes of a chloroplast or a mitochondrion.

interphase The period of time between two successive mitotic divisions, during which the cell increases in size and prepares for cell division.

intertidal zone The part of the coastal region that is closest to the shore, where the ocean meets the land. It extends from the highest tide mark to the lowest tide mark.

intron A DNA sequence within a gene that does not specify part of the gene's final protein or RNA product. Enzymes in the nucleus must remove introns from mRNA, tRNA, and rRNA molecules for these molecules to function properly. Compare *exon*.

inversion In genetics, a mutation in which a fragment of a chromosome breaks off and returns to the correct place on the original chromosome, but with the genetic loci in reverse order.

ion An atom or group of atoms that has either gained or lost electrons and therefore has a negative or positive charge.

ionic bond A chemical linkage between two atoms that is based on the electrical attraction between positive and negative charges. Compare *covalent bond* and *hydrogen bond*.

iPSC See *induced pluripotent stem cell*.

irregular fluctuation In reference to natural populations, a pattern of population growth in which the number of individuals in the population changes over time in an irregular manner.

isotonic solution A solution that has the same solute concentration as the cytosol of a cell has, resulting in an equal amount of water flowing into the cell and out of it. Compare *hypertonic solution* and *hypotonic solution*.

isotope A variant form of a chemical element that, in its number of neutrons, and therefore in its atomic mass number, differs from the most common form of that element.

J-shaped curve The graphical plot that represents exponential growth of a population. Compare *S-shaped curve*.

jumping gene See *transposon*.

karyotype A display of the specific number and shapes of chromosomes found in the diploid cells of a particular individual, or of a species in general.

keystone species A species that, relative to its own abundance, has a large effect on the presence and abundance of other species in a community.

kinetic energy The energy that a system possesses as a consequence of its state of motion; in simplest terms, energy of motion. Compare *potential energy*.

kingdom In reference to biological classification systems, the level in the Linnaean hierarchy above phylum and below domain. Generally, six kingdoms are recognized: Bacteria, Archaea, Protista, Plantae, Fungi, and Animalia.

Krebs cycle Also called *citric acid cycle*. The second of three major phases of cellular respiration, following glycolysis and preceding oxidative phosphorylation. This series of enzyme-driven oxidation reactions takes place in the mitochondrial matrix and yields many molecules of NADH (and a few of ATP and $FADH_2$).

lake An ecosystem of the freshwater biome defined as a standing body of water of variable size ranging up to thousands of square kilometers.

land transformation Also called *land-use change*. Changes made by humans to the land surface of Earth that alter the physical or biological characteristics of the affected regions. Compare *water transformation*.

language The complex system of human communication, consisting of thousands of words representing everything from objects to actions to abstract ideas.

lateral gene transfer Also called *horizontal gene transfer*. The transfer of genetic material between different species under natural conditions.

law of conservation of energy See *first law of thermodynamics*.

law of independent assortment Mendel's second law, which states that when gametes form, the separation of alleles of one gene is independent of the separation of alleles of other genes. We now know that this law does not apply to genes that are linked.

law of segregation Mendel's first law, which states that the two copies of a gene separate during meiosis and end up in different gametes.

learned behavior A predictable response acquired by trial and error or by watching others. Compare *fixed behavior*.

lichen A mutualistic association between a photosynthetic microbe (alga and/or cyanobacterium, kingdom Protista) and a fungus (kingdom Fungi).

light reactions The series of chemical reactions in photosynthesis that harvest energy from sunlight and use it to produce energy-rich compounds such as ATP and NADPH. The light reactions occur at the thylakoid membranes of chloroplasts and produce O_2 as a by-product. Compare *Calvin cycle*.

lignin A substance that links cellulose fibers in plant cell walls to create a rigid strengthening network. Lignin is one of the strongest materials in nature.

lineage A group of closely related individuals, species, genera, or the like, depicted as a branch on an evolutionary tree.

Linnaean hierarchy The classification scheme used by biologists to organize and name organisms. Its eight levels—from the most inclusive to the least—are domain, kingdom, phylum, class, order, family, genus, and species.

lipid A hydrophobic biomolecule built from chains or rings of hydrocarbons. Lipids are a key component of cell membranes. See also *phospholipid*.

locus (pl. loci) The physical location of a gene on a chromosome.

logistic growth A type of population growth in which a population increases nearly exponentially at first but then stabilizes at the maximum population size that can be supported indefinitely by the environment. Logistic growth is represented graphically by an S-shaped curve. Compare *exponential growth*.

lumen The space enclosed by the membrane of an organelle, or the cavity inside an organ.

lysosome A specialized vesicle with an acidic lumen containing enzymes that break down macromolecules.

macroevolution The rise and fall of major taxonomic groups due to evolutionary radiations that bring new groups to prominence and mass extinctions in which groups are lost; the history of large-scale evolutionary changes over time.

macromolecule A large organic molecule formed by the bonding together of small organic molecules.

malignant cell See *cancer cell*.

mammary gland A modified sweat gland that is the most distinctive feature of mammals. Mammary glands produce a liquid rich in fat, proteins, salts, and other nutritive substances that nourish the newborn.

manipulated variable See *independent variable*.

mantle The layer of Earth's interior that sits just below the crust and extends to the core. It is composed of hot, semisolid rock.

marsupial Any of a group of mammals that protect and feed their newborns with milk in an external pocket or pouch (marsupium). Marsupials include kangaroos and opossums. Compare *eutherian* and *monotreme*.

mass extinction An event during which large numbers of species become extinct throughout most of Earth.

maternal homologue In a homologous pair of chromosomes, the one that comes from the mother. Compare *paternal homologue*.

matrix (pl. matrices) The space interior to the cristae of the inner mitochondrial membrane.

matter Anything that has mass and occupies space.

meiosis A specialized process of cell division in eukaryotes during which diploid cells divide to produce haploid cells. Meiosis has two division cycles, and in animals it occurs exclusively in cells that produce gametes. Compare *mitosis*.

meiosis I The first cycle of cell division in meiosis, in which the members of each homologous chromosome pair are separated into different daughter cells. Meiosis I produces haploid daughter cells, each with half of the chromosome set found in the diploid parent cell.

meiosis II The second cycle of cell division in meiosis, in which the sister chromatids of each duplicated chromosome are separated into different daughter cells. Meiosis II is essentially mitosis, but in a haploid cell.

membrane channel See *channel protein*.

membrane pump See *active carrier protein*.

messenger RNA (mRNA) A type of RNA that specifies the order of amino acids in a protein.

metabolic pathway A series of enzyme-controlled chemical reactions in a cell in which the product of one reaction becomes the substrate for the next.

metabolism The capture, storage, and use of energy by living organisms.

metamorphosis (pl. metamorphoses) A dramatic developmental transformation from a reproductively immature to a reproductively mature form, involving great change in the form and function of the animal.

metaphase The stage of mitosis or meiosis during which chromosomes become aligned at the equator of the cell.

metastasis (pl. metastases) The spread of a disease from one organ to another.

microbe A minute organism visible only with a microscope.

microfilament A protein fiber composed of actin monomers. Microfilaments are part of a cell's cytoskeleton and are important in cell movements. Compare *intermediate filament* and *microtubule*.

microtubule A protein cylinder composed of tubulin monomers. Microtubules are part of the cell's cytoskeleton. Compare *intermediate filament* and *microfilament*.

mimicry A type of adaptation arising from predator-prey interactions in which a species evolves to imitate the appearance of something unappealing to its would-be predator.

mitochondrion (pl. mitochondria) An organelle with a double membrane that is the site of cellular respiration in eukaryotes. Mitochondria break down simple sugars to produce ATP in an oxygen-dependent (aerobic) process.

mitosis The process of cell division in eukaryotes that produces two daughter nuclei, each with the same chromosome number as the parent nucleus. Compare *meiosis*.

mitotic division The process that generates two genetically identical daughter cells from a single parent cell in eukaryotes. A mitotic division consists of mitosis followed by cytokinesis.

mitotic spindle A football-shaped array of microtubules that guides the movement of chromosomes during mitosis.

mixotroph A nutritional opportunist, an organism that can use energy and carbon from a variety of sources (functioning as both *autotroph* and *heterotroph*) to fuel its growth and reproduction.

molecule An association of atoms in which two or more of the atoms are linked through covalent bonds. Compare *salt*.

Mollusca Mollusks. After Arthropoda, the most diverse phylum of animals, characterized by a body that has a muscular foot, a visceral mass, and a mantle. Mollusks include clams and other bivalves (familiar "shellfish"), squid and octopi (cephalopods), and snails (gastropods).

molting Also called *ecdysis*. The process by which juvenile ecdysozoans (a group of protostomes) shed the cuticle that encases them.

monogamous Of or referring to a mating system in which a male and female are bonded to one another solely. Compare *polygamous* and *promiscuous*.

monohybrid An individual that is heterozygous for only one trait. Compare *dihybrid*.

monomer A molecule that can be linked with other related molecules to form a larger *polymer*.

monosaccharide A simple sugar that can be linked to other sugars, forming a *disaccharide* or *polysaccharide*. Glucose is the most common monosaccharide in living organisms.

monotreme Any of a group of mammals that have no placenta and lay eggs. The platypus and echidnas are monotremes. Compare *eutherian* and *marsupial*.

morphological species concept The idea that most species can be identified as a separate and distinct group of organisms by the unique set of morphological characteristics they possess. Compare *biological species concept*.

most recent common ancestor The ancestral organism from which a group of descendants arose.

mRNA See *messenger RNA*.

multicellular organism An organism made up of more than one cell.

multipotent Of or referring to a cell that can differentiate into only a relatively narrow range of cell types. Compare *pluripotent*, *totipotent*, and *unipotent*.

mutagen A substance or energy source that alters DNA.

mutation A change in the sequence of an organism's DNA. Because new alleles arise only by mutation, mutations are the original source of all genetic variation.

mutualism An interaction between two species in which both species benefit. Compare *commensalism*, *competition*, and *exploitation*.

mycelium (pl. mycelia) The main body of a fungus, composed of threadlike hyphae.

mycorrhiza (pl. mycorrhizae) A mutualism between a fungus and a plant, in which the fungus provides the plant with mineral nutrients while receiving organic nutrients from the plant.

NADH The reduced form of nicotinamide adenine dinucleotide (NAD^+), an energy carrier that delivers electrons and protons (H^+) to catabolic reactions (such as cellular respiration, which produces ATP from the breakdown of sugars into water and carbon dioxide). Compare *NADPH*.

NADPH The reduced form of nicotinamide adenine dinucleotide phosphate ($NADP^+$), an energy carrier molecule that delivers electrons and protons (H^+) to anabolic reactions, such as the Calvin cycle reactions of photosynthesis. Compare *NADH*.

natural selection An evolutionary mechanism in which the individuals in a population that possess particular inherited characteristics survive and reproduce at a higher rate than other individuals in the population because those characteristics enable the individuals to function optimally in their particular habitat. Natural selection is the only evolutionary mechanism that consistently improves the survival and reproduction of the organism in its environment. Compare *artificial selection*.

negative growth regulator Any of a variety of external and internal signals and regulatory proteins that control the cell cycle by halting cell division. Compare *positive growth regulator*.

net primary productivity (NPP) The amount of energy that producers capture by photosynthesis, minus the amount lost as metabolic heat. NPP is usually measured as the amount of new biomass produced by photosynthetic organisms per unit of area during a specified period of time.

neutron A particle, found in the nucleus of an atom, that has no electrical charge. Compare *electron* and *proton*.

niche The sum total of the conditions and resources a species or population needs in order to survive and reproduce successfully in its particular habitat.

niche partitioning The differential use of space or resources in a common niche by competing species that enables species to coexist despite their potential for competition.

nicotinamide adenine dinucleotide See *NADH* and *NADPH*.

nitrogen fixation The process by which nitrogen gas (N_2), which is readily available in the atmosphere but cannot be used by plants, is converted to ammonium (NH_4^+), a form of nitrogen that can be used by plants. Nitrogen fixation is accomplished naturally by bacteria and by lightning, as well as by humans in industrial processes such as the production of fertilizer.

nitrogenous base Any of the five nitrogen-rich compounds found in nucleotides. The four nitrogenous bases found in DNA are adenine (A), cytosine (C), guanine (G), and thymine (T); in RNA, uracil (U) replaces thymine.

node The moment in evolution, depicted as a point on an evolutionary tree, at which one lineage split, or diverged, into two separate lineages (such as Archaea and Eukarya).

noncoding DNA A segment of DNA that does not encode proteins or RNA. Introns and spacer DNA are two common types of noncoding DNA.

nonpolar molecule A molecule that has an equal distribution of electrical charge across all its constituent atoms. Nonpolar molecules do not form hydrogen bonds and tend not to dissolve in water. Compare *polar molecule*.

notochord A structure composed of large cells that collectively form a strong but flexible bar running dorsally along the length of an animal, providing support for the rest of the body.

NPP See *net primary productivity*.

nuclear envelope The double membrane that forms the outer boundary of the nucleus, an organelle found only in eukaryotic cells.

nuclear pore One of many openings in the nuclear envelope that allow selected molecules, including specific proteins and RNA, to move into and out of the nucleus.

nucleic acid A polymer made up of nucleotides. There are two kinds of nucleic acids: DNA and RNA.

nucleotide Any of a class of organic molecules that serve as energy carriers and as the chemical building blocks of nucleic acids (DNA and RNA). A nucleotide is made up of a phosphate group, a five-carbon sugar, and one of four *nitrogenous base*). Nucleotides are linked to form a single strand of DNA or RNA.

nucleotide pair See *base pair*.

nucleus (pl. nuclei) The organelle in a eukaryotic cell that contains the genetic blueprint in the form of DNA.

nuptial gift A reward (typically a food item) presented to a female animal by a male for sexual attention.

nutrient In an ecosystem, an inorganic substance (element) required by living organisms.

nutrient cycle Also called *biogeochemical cycle*. The cyclical movement of a nutrient between organisms and the physical environment. There are two main types of nutrient cycles: *atmospheric cycles* and *sedimentary cycles*.

nutrient recycling The breakdown by decomposers of dead organisms or their waste products that releases the chemical elements locked in the biological material and returns them to the environment, where these elements (for example, carbon dioxide, nitrogen, or phosphorus) are used by autotrophs and eventually heterotrophs.

observation A description, measurement, or record of any object or phenomenon. Facts learned in this manner are subsequently used to formulate hypotheses.

oceanic region Also called *open ocean*. An ecosystem of the marine biome defined as the part of the ocean beginning about 40 miles offshore, where the continental shelf, and therefore the coastal region, ends.

oncogene A mutated gene that promotes excessive cell division, leading to cancer.

open ocean See *oceanic region*.

operator A regulatory DNA sequence that controls the transcription of a gene or group of genes.

operon A single promoter sequence that controls the transcription of all the genes in a cluster of genes with related functions.

order In reference to biological classification systems, the level in the Linnaean hierarchy above family and below class.

organ A self-contained collection of different types of tissues, usually of a characteristic size and shape, that is organized for a particular set of functions.

organ system A group of organs of different types that work together to carry out a common set of functions.

organelle A discrete cytoplasmic structure with a specific function. Some cell biologists use the term only for membrane-enclosed cytoplasmic compartments; others include other cytoplasmic structures, such as ribosomes, in the definition.

organic molecule A molecule that contains at least one carbon covalently bonded to one or more hydrogen atoms. Before modern chemistry, organic molecules on Earth were exclusively of biological origin, but now chemists can create many organic molecules artificially (synthetically).

osmoregulation The process of maintaining an internal water and salt balance that supports biological processes.

osmosis The passive movement of water across a selectively permeable membrane.

oxidation The loss of electrons by one atom or molecule to another. Compare *reduction*.

oxidation-reduction reaction See *redox reaction*.

oxidative phosphorylation The third of three major phases of cellular respiration, following glycolysis and the Krebs cycle. In this phase, electrons are shuttled down an electron transport chain in mitochondria that results in the production of ATP.

P generation The parent generation in a breeding trial involving a series of genetic crosses. Compare *F_1 generation* and *F_2 generation*.

Pangaea An ancient supercontinent that contained all of the world's landmasses. Pangaea formed 250 million years ago and began to break apart 200 million years ago, ultimately yielding the continents we know today.

parasite An organism that lives in or on another organism (its host) and obtains nutrients from that organism. Parasites harm and may eventually kill their hosts but do not kill them immediately.

passive carrier protein A cell membrane protein that, without the input of energy, changes its shape to transport a molecule across the membrane from the side of higher concentration to the side of lower concentration. Compare *active carrier protein*.

passive transport Movement of ions or molecules across a biological membrane that requires no input of energy because it is "downhill" (in the same direction as a concentration gradient). Compare *active transport*.

paternal homologue In a pair of homologous chromosomes, the one that comes from the father. Compare *maternal homologue*.

pathogen An organism or virus that infects a host and causes disease, harming and in some cases killing the host.

PCR See *polymerase chain reaction*.

pedigree A chart that shows genetic relationships among family members over two or more generations of a family's history.

peptide bond A covalent bond between the amino group of one amino acid and the carboxyl group of another. Peptide bonds link amino acids together.

permafrost Permanently frozen soil that is found below the surface layers and may be a quarter of a mile deep.

Permian extinction The largest mass extinction in the history of life on Earth; it occurred 250 million years ago, driving up to 95 percent of the species in some groups to extinction.

personalized medicine The practice of tailoring treatments and therapies to a patient's DNA profile.

persistent organic pollutant (POP) Any long-lived organic molecule of synthetic origin that bioaccumulates in organisms and that can have harmful effects. Some of the most damaging and widespread POPs are PCBs (polychlorinated biphenyls) and dioxins.

pH scale A scale that indicates the concentration of hydrogen ions in a solution. The pH scale runs from 1 to 14. A pH of 7 is neutral; values below 7 indicate acids, and values above 7 indicate bases.

phagocytosis A form of endocytosis by which a cell engulfs a large particle, such as another cell; "cell eating." Compare *pinocytosis*.

pharyngeal pouch A structure, found in the early embryo of all chordates, that first appears as a pocket of tissue on either side of the embryonic pharynx and later develops into structures such as gill slits in fishes and larval amphibians, or the larynx and trachea in mammals.

phenotype The particular version of a genetic trait that is displayed by a given individual. For example, black, brown, red, and blond are phenotypes of the hair color trait in humans. Compare *genotype*.

pheromone A chemical signal produced by one individual to communicate that individual's identity and location to another individual.

phloem A tissue composed of living cells through which a plant transports sugars and other organic and inorganic substances. Compare *xylem*.

phosphate group A functional group consisting of a phosphate atom and four oxygen atoms.

phospholipid A lipid consisting of two fatty acids, a glycerol, and a phosphate as part of the hydrophilic head group. Phospholipids are the main component in all biological membranes.

phospholipid bilayer A double layer of phospholipid molecules arranged so that their hydrophobic "tails" lie sandwiched between their hydrophilic "heads." A phospholipid bilayer forms the basic structure of all biological membranes.

photoautotroph An organism that obtains its energy from sunlight and derives its carbon from carbon dioxide in the air. Examples include cyanobacteria, green algae, and plants. Compare *photoheterotroph* and *chemoautotroph*.

photoheterotroph An organism that obtains its energy from sunlight and derives its carbon from carbon-containing compounds found mainly in other organisms. All photoheterotrophs are prokaryotes. Compare *photoautotroph* and *chemoheterotroph*.

photon A massless particle that has wave-like characteristics; the electromagnetic spectrum, which includes visible light, is composed of photons.

photorespiration A series of reactions launched when rubisco links O_2 to RuBP, causing fixed carbon to be lost as carbon dioxide.

photosynthesis A metabolic process by which organisms capture energy from sunlight and use it to synthesize sugars from carbon dioxide and water. Compare *cellular respiration*.

photosystem A large complex of proteins and chlorophyll that captures energy from sunlight. Two distinct photosystems (I and II) are present in the thylakoid membranes of chloroplasts.

photosystem I The photosystem that is primarily responsible for the production of NADPH. Compare *photosystem II*.

photosystem II The photosystem in which light energy is used to initiate an electron flow along the electron transport chain, resulting in ATP synthesis and the release of oxygen gas (O_2) as a by-product. Compare *photosystem I*.

phylogenetics The study of evolutionary relationships between organisms.

phylum (pl. phyla) In reference to biological classification systems, the level in the Linnaean hierarchy above class and below kingdom.

phytoplankton Free-floating, single-celled algae that drift at or near the surface of water bodies. Compare *zooplankton*.

pilus (pl. pili) A short hairlike projection found on the surface of some bacteria.

pinocytosis A form of nonspecific endocytosis by which cells take in fluid; "cell drinking." Compare *phagocytosis*.

placebo A dummy pill or sham treatment that mimics the actual treatment of an experiment.

placebo effect The sense among study participants who are in the control group of an experiment that they are feeling better because they have received a beneficial treatment.

placenta Also called *afterbirth*. A structure found in mammals that transfers nutrients and gases from the blood of the mother to the blood of the fetus developing in her uterus.

plankton Microbes that drift at or near the surface of water bodies. See also *phytoplankton* and *zooplankton*.

Plantae The kingdom of Eukarya that is made up of plants—multicellular autotrophs that live mainly on land and photosynthesize.

plasma membrane The phospholipid bilayer that forms the outer boundary of any cell.

plasmid A small, circular segment of DNA found naturally in bacteria. Plasmids are involved in natural gene transfers among bacteria and can be used as vectors in genetic engineering.

plasmodesma (pl. plasmodesmata) A tunnel-like channel between two plant cells that provides a cytoplasmic connection allowing the flow of small molecules and water between the cells.

plate tectonics A theory suggesting that Earth's crust is composed of several interlocking plates. Several geologic factors can cause these plates to shift position over time, resulting in the movement of continents.

pleiotropy A type of genetic control in which a single gene influences a variety of different traits.

pluripotent Of or referring to a cell that can differentiate into any of the cell types in the adult body. Compare *multipotent*, *totipotent*, and *unipotent*.

point mutation A mutation in which only a single base is altered.

polar molecule A molecule that has an uneven distribution of electrical charge. Polar molecules can easily interact with water molecules and are therefore soluble. Compare *nonpolar molecule*.

pollen The haploid, multicellular, mobile male gametophyte of seed plants, in which sperm are produced.

pollinator An animal that carries pollen grains from the stamens of one flower to the stigmas of other flowers of the same species.

polygamous Of or referring to a mating system in which (usually) a single male mates with multiple females. Compare *monogamous* and *promiscuous*.

polygenic Of or referring to inherited traits that are determined by the action of more than one gene.

polymer A large organic molecule composed of many *monomers* linked together.

polymerase chain reaction (PCR) A method of DNA technology that uses the DNA polymerase enzyme to make multiple copies of a targeted sequence of DNA.

polypeptide A polymer consisting of covalently linked linear chains of amino acids.

polyploid Of or referring to a cell or organism that has three or more complete sets of chromosomes (rather than the usual two complete sets). Populations of polyploid individuals can rapidly form new species without geographic isolation. Compare *diploid* and *haploid*.

polysaccharide A polymer composed of many linked *monosaccharides*. Examples include starch and cellulose. Compare *disaccharide*.

POP See *persistent organic pollutant*.

population A group of interacting individuals of a single species located within a particular area.

population cycle A pattern in which the population sizes of two species increase and decrease together in a tightly linked cycle; this pattern can occur when at least one of the two species involved is very strongly influenced by the other.

population density The number of individuals in a population, divided by the area covered by the population.

population ecology A branch of science concerned with questions that relate to how many organisms live in a particular environment, and why.

population size The total number of individuals in a population.

positive growth regulator Any of a variety of internal signals, including growth factors, hormones, and regulatory proteins, that control the cell cycle by stimulating cell division. Compare *negative growth regulator*.

postzygotic barrier A barrier that prevents zygotes from developing into healthy offspring. Compare *prezygotic barrier*.

potential energy The energy stored in any system as a consequence of its position; in simplest terms, stored energy. Compare *kinetic energy*.

precipitation The falling to Earth of condensed water vapor. Precipitation can take the form of rain, snow, hail, or sleet.

predator An organism that kills other organisms (called *prey*) for food.

prey Animals that *predators* kill and eat.

prezygotic barrier A barrier that prevents a male gamete (like a human sperm cell) and a female gamete (like a human egg cell) from fusing to form a zygote. Compare *postzygotic barrier*.

primary consumer An organism that eats producers and is eaten by *secondary consumers*. Compare also *tertiary consumer*.

primary structure In reference to proteins, the sequence of amino acids in a protein. Compare *secondary structure*, *tertiary structure*, and *quaternary structure*.

primary succession Ecological succession that occurs in newly created habitat, as when an island rises from the sea or a glacier retreats, exposing newly available bare ground. Compare *secondary succession*.

primary tumor The tumor that is located at the initial site of a cancer. Compare *secondary tumor*.

problem solving Figuring out a solution to a challenging issue without the benefit of prior experience.

producer Also called *autotroph*. An organism that uses energy from an external source, such as sunlight, to produce its own food without having to eat other organisms or their remains. Compare *consumer*.

product A substance that is formed by a chemical reaction. Compare *reactant*.

prokaryote A single-celled organism that does not have a nucleus. All prokaryotes are members of the domains Bacteria or Archaea. Compare *eukaryote*.

prokaryotic flagellum A hairlike structure, found in prokaryotes, that propels the cell or organism through a whiplike action. Prokaryotic flagella lack a membrane covering and have a very different internal structure from that of *eukaryotic flagella*. They are believed to have evolved separately from eukaryotic flagella.

promiscuous Of or referring to a mating system in which males and females mate with multiple partners. Compare *monogamous* and *polygamous*.

promoter See *gene promoter*.

prophase The stage of mitosis or meiosis during which chromosomes first become visible under the microscope.

protein A polymer of amino acids that are linked in a specific sequence. Most proteins are folded into complex three-dimensional shapes.

Protista The oldest kingdom of Eukarya, made up of a diverse collection of mostly single-celled but some multicellular organisms. Protista is an artificial grouping, defined only by what members of this group are not: protists are not plants, animals, fungi, bacteria, or archaeans.

proton A positively charged particle found in atoms. Each atom of a particular element contains a characteristic number of protons. Compare *electron* and *neutron*.

proton gradient An imbalance in the concentration of protons across a membrane.

proto-oncogene A gene that promotes cell division in response to growth signals as part of its normal cellular function. Compare *tumor suppressor gene*.

protostome Any of a group of animals, including insects, worms, and snails, in which the first opening to develop in the early embryo becomes the mouth. Compare *deuterostome*.

protozoan Any nonphotosynthetic protist. All protozoans are motile. Compare *alga*.

Punnett square A diagram in which all possible genotypes of male and female gametes are listed on two sides of a grid, providing a graphical way to predict the genotypes of the offspring produced in a genetic cross.

pyruvate A three-carbon molecule produced by glycolysis that is processed in the mitochondria to generate ATP.

quaternary structure In reference to proteins, the three-dimensional arrangement of two or more separate chains of amino acids into a functional protein complex. Compare *primary structure*, *secondary structure*, and *tertiary structure*.

quorum sensing A system of cell-to-cell communication that enables some prokaryotes to sense and respond to bacterial density.

radial symmetry A body arrangement in which an animal could be sliced symmetrically along any number of vertical planes passing through the center of the animal to produce body parts that are nearly identical. Compare *bilateral symmetry*.

radiation The waves of energy, such as light or infrared (heat), that are released by an object and that can be absorbed by another object.

radiation therapy The administration of high-energy radiation to destroy cancer cells by killing all rapidly dividing cells.

radioisotope An unstable, radioactive form of an element that releases energy as it decays to more stable forms at a constant rate over time.

rain shadow An area on the side of a mountain facing away from moist prevailing winds where little rain or snow falls.

random fertilization The idea that any sperm can fertilize any egg, making the number of possible genetic combinations an enormous number.

reactant A substance that undergoes a chemical reaction. Compare *product*.

reaction center A cluster of chlorophyll molecules within an antenna complex whose electrons become excited and are passed to an electron transport chain when the pigment molecules absorb light energy.

receptor A protein in the plasma membrane or cytoplasm of a target cell that binds signaling molecules, allowing those molecules to indirectly affect processes inside the cell.

receptor-mediated endocytosis A form of endocytosis in which receptor proteins embedded in the plasma membrane of a cell recognize certain surface characteristics of materials to be brought into the cell by endocytosis.

receptor protein See *receptor*.

recessive allele An allele that does not affect the phenotype when paired with a *dominant allele* in a heterozygote.

recombinant DNA An artificial assembly of genetic material created by the enzyme-mediated linking of DNA fragments.

redox reaction An oxidation-reduction reaction, a chemical reaction in which electrons are transferred from one molecule or atom to another.

reduction The gain of electrons by one atom or molecule from another. Compare *oxidation*.

regenerative medicine A field of medicine that uses stem cells to repair or replace tissues and organs damaged by injury or disease.

regulatory DNA A DNA sequence that can increase, decrease, turn on, or turn off the expression of a gene or a group of genes. Regulatory DNA sequences interact with regulatory proteins to control gene expression.

regulatory protein Also called *transcription factor*. A protein that signals whether or not a particular gene or group of genes should be expressed. Regulatory proteins interact with regulatory DNA to control gene expression.

relative species abundance The number of individuals of one species in a community, compared to the number of individuals of other species.

releaser In animal behavior, a simple stimulus that triggers a fixed behavior.

repressor A protein that prevents the expression of a particular gene or group of genes.

reproduction The generation of a new individual like oneself.

reproductive cloning A technology used to produce an offspring that is an exact genetic copy (a clone) of another individual. Rather than stem cells being removed from the embryo, as they are in therapeutic cloning, the embryo is transferred to the uterus of a surrogate mother, where, if all goes well, the birth of a healthy offspring ultimately results; this offspring is genetically identical to the individual that provided the donor nucleus.

reproductive isolation A condition in which barriers to reproduction prevent or strongly limit two or more populations from reproducing with one another. Many different kinds of reproductive barriers can result in reproductive isolation, but it always has the same effect: no or few genes are exchanged between the reproductively isolated populations. Compare *geographic isolation*.

respiration See *cellular respiration*.

responding variable See *dependent variable*.

restriction enzyme Any of a class of enzymes that cut DNA molecules at specific target sequences; a key tool of DNA technology.

ribonucleic acid See *RNA*.

ribosomal RNA (rRNA) A type of RNA that is an important component of ribosomes.

ribosome A minute organelle composed of proteins and RNA at which new proteins are synthesized. Ribosomes can be either attached to the endoplasmic reticulum or free in the cytosol.

ring species A species whose populations loop around a geographic barrier (such as a mountain chain) and in which the populations at the two ends of the loop are in contact with one another, yet cannot interbreed.

river An ecosystem of the freshwater biome defined as a body of water that moves continuously in a single direction.

RNA Ribonucleic acid, a single-stranded polymer of nucleotides that is necessary for the synthesis of proteins in living organisms. Each nucleotide in RNA is composed of the sugar ribose, a phosphate group, and one of four nitrogenous bases: adenine, cytosine, guanine, or uracil. Compare *DNA*.

RNA interference (RNAi) A mechanism for selectively blocking the expression of a given gene in which small chunks of RNA silence genes that share nucleotide sequence similarity with them.

RNA polymerase The key enzyme in DNA transcription. RNA polymerase links together the nucleotides of the RNA molecule specified by a gene.

RNA splicing The process by which mRNA introns are snipped out and the remaining pieces of mRNA are re-joined.

RNAi See *RNA interference*.

rough ER A region of the endoplasmic reticulum that has attached ribosomes and specializes in protein synthesis. Compare *smooth ER*.

rRNA See *ribosomal RNA*.

rubisco The enzyme that catalyzes the first reaction of carbon fixation in photosynthesis.

S phase The stage of the cell cycle during which the cell's DNA is replicated.

sac fungi See *ascomycetes*.

salt A compound consisting of ions held together by the mutual attraction between their opposite electrical charge. Compare *molecule*.

saturated fatty acid A fatty acid that has no double bonds between the carbon atoms in its hydrocarbon backbone. Saturated fatty acids are solid or semisolid at room temperature. Compare *unsaturated fatty acid*.

science A method of inquiry that provides a rational way to discover truths about the natural world.

scientific fact A direct and repeatable observation of any aspect of the natural world.

scientific hypothesis (pl. hypotheses) An informed, logical, and plausible explanation for observations of the natural world. An "educated guess." Compare *scientific theory*.

scientific method A series of steps in which the investigator develops a hypothesis, tests its predictions by performing experiments, and then changes or discards the hypothesis if its predictions are not supported by the results of the experiments.

scientific name The unique two-part name given to each species that consists of, first, a Latin name designating the genus and, second, a Latin name designating that species. Scientific names are traditionally italicized.

scientific theory A major explanation about the natural world that has been confirmed through extensive testing in diverse ways by independent researchers. Compare *scientific hypothesis*.

second law of thermodynamics The law stating that all systems, such as a cell or the universe, tend to become more disordered, and that the creation and maintenance of order in a system requires the transfer of disorder to the environment.

secondary consumer An organism that eats *primary consumers* and is eaten by *tertiary consumers*.

secondary structure In reference to proteins, the patterns of local three-dimensional forms in segments of a protein. Alpha helices and pleated beta sheets are common forms in the secondary structure of many proteins. Compare *primary structure*, *tertiary structure*, and *quaternary structure*.

secondary succession Ecological succession that occurs as communities recover from a disturbance, as when a forest grows back after a field ceases to be used for agriculture. Compare *primary succession*.

secondary tumor A tumor that is spawned by a *primary tumor* at a site distant from the primary tumor's location.

sedimentary cycle A type of nutrient cycle in which the nutrient does not enter the atmosphere easily. Compare *atmospheric cycle*.

seed The embryo of a plant, encased in a protective covering.

selective permeability The ability to control which materials can pass through a membrane. Plasma membranes are selectively permeable.

semiconservative replication DNA replication in which one "old" (template) strand is retained, or conserved, in each new helix.

serotype See *viral strain*.

sex chromosome Either of a pair of chromosomes that determines the sex of an individual. Compare *autosome*.

sex-linked Of or referring to genes located on a sex chromosome. See also *X-linked* and *Y-linked*.

sexual dimorphism A distinct difference in appearance between the males and females of a species.

sexual reproduction The combining of genes from two individuals to give rise to a new individual, known as the offspring. Compare *asexual reproduction*.

sexual selection A type of natural selection in which individuals that differ in inherited characteristics differ, as a result of those characteristics, in their ability to get mates.

shared derived trait An evolutionary novelty shared by an ancestor and its descendants but not seen in groups that are not direct descendants of that ancestor.

short tandem repeat (STR) A highly variable region of the human genome. STRs are the DNA segments most commonly used for DNA fingerprinting in North America.

signal transduction pathway A series of cellular events that relay receipt of a signal from protein receptors on the plasma membrane to the cytoplasm.

signaling molecule A molecule produced and released by one cell that affects the activities of another cell (referred to as a *target cell*). Signaling molecules enable the cells of a multicellular organism to communicate with one another and coordinate their activities.

simple diffusion The passive movement of a substance across a membrane without the assistance of any membrane components. Compare *facilitated diffusion*.

simple eye A non-image-forming eye that can distinguish light from dark. Compare *compound eye*.

single-blind experiment An experiment in which the study subjects do not know whether they belong to the control group or the treatment group. Compare *double-blind experiment*.

sister chromatids A pair of identical double helices that are produced by the replication of DNA during the cell cycle.

smooth ER A region of the endoplasmic reticulum that is specialized for lipid synthesis. It is "smooth" because it does not have attached ribosomes. Compare *rough ER*.

social behavior Behavior that involves cooperation among members of a group of animals, usually of the same species.

solute A dissolved substance. Compare *solvent*.

solution Any combination of a solute and a solvent.

solvent A liquid (in biological systems, usually water) into which a *solute* has dissolved.

somatic cell Any cell in a multicellular organism that is not a *gamete* or part of a gamete-making tissue.

somatic mutation A mutation that occurs in a cell other than a sex cell and hence is not passed down to offspring.

somatic stem cell See *adult stem cell*.

spacer DNA A region of noncoding DNA that separates two genes. Spacer DNA is abundant in eukaryotes, but not in prokaryotes.

speciation The splitting of one species to form two or more species that are reproductively isolated from one another.

species (pl. species) 1. All individuals that can interbreed in their natural surroundings to produce fertile offspring. See also *biological species concept* and *morphological species concept*. 2. In reference to biological classification systems, the lowest level in the Linnaean hierarchy, immediately below genus.

species richness The total number of different species that live in a community.

spore A reproductive structure of fungi and plants, usually thick walled, that can survive for long periods of time in a dormant state and will sprout under favorable conditions to produce the body of the organism.

***SRY* gene** A gene, located on the Y chromosome, that functions as a master switch, committing the sex of a developing embryo to "male." *SRY* is short for "sex-determining region of Y."

S-shaped curve The graphical plot that represents logistic growth of a population. Compare *J-shaped curve*.

stabilizing selection A type of natural selection in which individuals with intermediate values of an inherited characteristic have an advantage over other individuals in the population, as when medium-sized individuals produce offspring at a higher rate than do small or large individuals. Compare *directional selection* and *disruptive selection*.

starch A polysaccharide that serves as an energy storage molecule inside plant cells. Compare *glycogen*.

start codon A three-nucleotide sequence on an mRNA molecule (usually the codon AUG) that signals where translation should begin. Compare *stop codon*.

statistics A mathematical science that uses probability theory to estimate the reliability of data.

stem cell An undifferentiated cell that can renew itself through cell division, theoretically indefinitely. Some of the daughter cells generated by stem cells may differentiate into specialized cell types.

sterol Also called *steroid*. Any of a class of lipids whose fundamental structure consists of four hydrocarbon rings fused to each other.

stimulus (pl. stimuli) An environmental cue that elicits a behavioral response.

stoma (pl. stomata) An air pore, found on the leaf or green stem of a plant, that can open and close, thereby controlling the rates at which CO_2 is brought into the plant and O_2 and water are lost from the plant. The flexing of the pair of guard cells that border the stoma determines whether it is open or closed.

stop codon A three-nucleotide sequence on an mRNA molecule that signals where translation should end. Compare *start codon*.

STR See *short tandem repeat*.

stroma (pl. stromata) The space enclosed by the inner membrane of the chloroplast, in which the thylakoid membranes are situated.

substitution In genetics, a mutation in which one nitrogenous base is replaced by another at a single position in the DNA sequence of a gene.

substrate In reference to enzymes, the particular substance on which an enzyme acts. Only the substrate will bind to the active site of the enzyme.

succession A process by which species in a community are replaced over time. For a given location, the order in which species are replaced is fairly predictable. See also *primary succession* and *secondary succession*.

supercontinent A massive landmass formed by the fusing of multiple continents. Several supercontinents have existed throughout Earth's history.

surface tension A force that tends to minimize the surface area of water at an air-water boundary. Surface tension holds the water surface taut, resisting any stretching or breaking of that surface, and it is strong enough to support very light objects.

symbiosis (pl. symbioses) A relationship in which two or more organisms of different species live together in close association.

sympatric speciation The formation of new species from populations that are not geographically isolated from one another. Compare *allopatric speciation*.

taiga See *boreal forest*.

target cell A cell that receives and responds to a *signaling molecule*.

taxon (pl. taxa) Also called *taxonomic group*. A group defined within the Linnaean hierarchy—for example, a species or a kingdom.

taxonomy The branch of biology that deals with the naming of organisms and with their classification in the Linnaean hierarchy.

technology The practical application of scientific techniques and principles.

telomere A unique DNA sequence, associated with special proteins, that caps each end of a chromosome and protects it.

telophase The stage of mitosis or meiosis during which chromosomes arrive at the opposite poles of the cell and new nuclear envelopes begin to form around each set of chromosomes.

temperate deciduous forest A terrestrial biome dominated by trees and shrubs that grow in regions with cold winters and moist, warm summers.

template strand In gene transcription, the strand of DNA (of the two strands in a DNA molecule) that is copied into RNA and is therefore complementary to the RNA synthesized from it.

termination The third of three stages of gene transcription (in which RNA polymerase reaches the terminator sequence of bases) or translation (in which the ribosome reaches the stop codon). See also *initiation* and *elongation*.

terminator In prokaryotic gene transcription, a DNA sequence that, when reached by RNA polymerase, causes transcription to end and the newly formed mRNA molecule to separate from its DNA template.

tertiary consumer An organism that eats *secondary consumers*. Compare also *primary consumer*.

tertiary structure In reference to proteins, the overall three-dimensional form of a protein, created and stabilized by chemical interactions between distantly placed segments of the protein. Compare *primary structure*, *secondary structure*, and *quaternary structure*.

test cross An experimental genetic cross in which one parent is known to be a recessive homozygote, usually conducted to deduce the genotype of the other parent, which could be a dominant homozygote or a heterozygote.

tetrad Also called *bivalent*. The paternal and maternal members of a pair of homologous chromosomes that are aligned parallel to each other during meiosis I.

tetrapod Any terrestrial vertebrate that has four limbs. Amphibians, birds, and mammals are all tetrapods.

thermal energy See *heat energy*.

thylakoid One of a series of flattened, interconnected membrane sacs that lie one on top of another within a chloroplast in stacks called grana.

tight junction A structure made up of rows of proteins associated with the plasma membranes of adjacent animal cells. Neighboring cells are held together tightly by the interlocking of the membrane proteins of adjacent cells. Cells connected by tight junctions form leak-proof sheets that do not permit ions and molecules to pass from one side of the sheet to the other by slipping between the cells. Compare *anchoring junction* and *gap junction*.

tissue A collection of coordinated and specialized cells that together fulfill a particular function for the organism.

totipotent Of or referring to a cell that can differentiate into any cell type. Compare *multipotent*, *pluripotent*, and *unipotent*.

transcription In genetics, the synthesis of an RNA molecule from a DNA template. Transcription is the first of the two major steps in the process by which genes specify proteins; it produces mRNA, tRNA, and rRNA, all of which are essential in the production of proteins. Compare *translation*.

transcription factor See *regulatory protein*.

transfer RNA (tRNA) A type of RNA that transfers the amino acid specified by mRNA to the ribosome during protein synthesis.

transgenic organism See *genetically modified organism*.

translation In genetics, the conversion of a sequence of nitrogenous bases in an mRNA molecule to a sequence of amino acids in a protein. Translation is the second of the two major steps in the process by which genes specify proteins; it occurs at the ribosomes. Compare *transcription*.

translocation In genetics, a mutation in which a segment of a chromosome breaks off and is then attached to a different, nonhomologous chromosome.

transport protein A membrane-spanning protein that provides a pathway by which materials can enter or leave cells.

transport vesicle A small, spherical membrane-enclosed sac that specializes in moving substances from one location to another within the cytoplasm and to and from the exterior of the cell.

transposon Also called *jumping gene*. A DNA sequence that can move from one position on a chromosome to another, or from one chromosome to another.

treatment group Also called *experimental group*. A group of participants in an experiment that are subjected to the same environmental conditions as the *control group* but are exposed to a specific treatment in the form of a change to the independent variable.

triglyceride A lipid in which all three hydroxyl groups in glycerol are bonded to a fatty acid. Animals store most of their surplus energy in the form of triglycerides.

trisomy In diploid organisms, the condition of having three copies of a chromosome (instead of the usual two).

tRNA See *transfer RNA*.

trophic level A level or step in a food chain. Trophic levels begin with producers and end with predators that eat other organisms but are not fed on by other predators.

tropical forest A terrestrial biome dominated by a rich diversity of trees, vines, and shrubs that grow in warm, rainy regions.

tubulin The protein monomer that makes up microtubules.

tumor A solid cell mass formed by the inappropriate proliferation of cells.

tumor suppressor gene A gene that inhibits cell division under normal conditions. Compare *proto-oncogene*.

tundra A terrestrial biome dominated by low-growing shrubs and nonwoody plants that can tolerate extreme cold.

unipotent Of or referring to a stem cell that can differentiate into only one cell type. Compare *multipotent*, *pluripotent*, and *totipotent*.

unsaturated fatty acid A fatty acid that has one or more double bonds between the carbon atoms in its hydrocarbon backbone. Unsaturated fatty acids tend to be liquid at room temperature. Compare *saturated fatty acid*.

upwelling The stirring of bottom sediments in a body of water by air or water currents, or by shifts in water temperature (as in northern lakes during fall and spring).

Vacuole A large, water-filled vesicle found in plant cells. Vacuoles help maintain the shape of plant cells and can also be used to store various molecules, including nutrients and antiherbivory chemicals.

variable A characteristic of any object or individual organism that can change.

vascular system The tissue system in plants that is devoted to internal transport.

vestigial organ A structure or body part that served a purpose in an ancestral species but is currently of little or no use to the organism that has it.

viral strain Also called *serotype*. Any of the variant forms of a particular type of virus.

virus An infectious particle consisting of nucleic acids and proteins. A virus cannot reproduce on its own, and must instead use the cellular machinery of its host to reproduce.

warning coloration Bright colors or striking patterns evolved by many prey organisms to warn potential predators that they are heavily defended.

water transformation Changes made by humans to the waters of Earth that alter their physical or biological characteristics. Compare *land transformation*.

weather Temperature, precipitation, wind speed, humidity, cloud cover, and other physical conditions of the lower atmosphere at a specific place over a short period of time. Compare *climate*.

wetland An ecosystem of the freshwater biome defined as standing water shallow enough that rooted plants emerge above the water surface. Bogs (stagnant, acidic, and oxygen-poor), marshes (grassy), and swamps (dominated by trees and shrubs) are all wetlands.

X-linked Of or referring to sex-linked genes located on an X chromosome. Compare *Y-linked*.

xylem A plant tissue composed of a number of cell types, including tubelike conducting cells that are dead at maturity and that transport water and dissolved minerals from the soil to the leaves. Compare *phloem*.

Y-linked Of or referring to sex-linked genes located on the Y chromosome. Compare *X-linked*.

zooplankton Heterotrophic prokaryotes and protists, together with microscopic animals, that drift at or near the surface of water bodies. Compare *phytoplankton*.

zygomycetes One of three major groups of fungi. Zygomycetes were the first fungal group on land. Compare *ascomycetes* and *basidiomycetes*.

zygote The diploid ($2n$) cell formed by the fusion of two haploid (n) gametes; a fertilized egg.

Credits

Photography Credits

ABOUT THE AUTHORS **Page xix** (both): Courtesy of the authors.

CHAPTER 1 **Page 2**: ZUMA Press/Newscom; **p. 4**: 20th CENTURY FOX/THE KOBAL COLLECTION; **p. 5**: © Trevor Lush/Corbis; **p. 6**: © All Canada Photos/Alamy; **p. 7** (*Science*): From *Science*, 20 December 2013 vol. 342, issue 6165, pages 1405–1544. Reprinted with permission from AAAS; **p. 7** (Nature): Reprinted by permission from Macmillan Publishers Ltd: *Nature*, 497,7451, 30 May 2013; **p. 7** (*Ecology*): The Ecology Society of America; **p. 8**: © Jonathan Larsen/Diadem Images/Alamy; **p. 9**: © Louise Murray/age fotostock; **p. 9**: © Staffan Widstrand/CORBIS; **p. 9**: R. Iegosyn/Shutterstock; **p. 9**: samsonovs/Shutterstock; **p. 10**: ER Degginger/Science Source; **p. 11** (bottom): ElenaGaak/Shutterstock; **p. 11** (top): bikeriderlondon/Shutterstock; **p. 14**: a: Robert Thom (Grand Rapids, MI, 1915–1979, Michigan). *Semmelweis—Defender of Motherhood*. Oil on Canvas. Collection of the University of Michigan Health System. Gift of Pfizer, Inc. UMHS.26; **(b)** Getty Images; **p. 16**: F. Stuart Westmorland/Photo Researchers, Inc.; **p. 17**: Photo by Feng Li/Getty Images; **p. 19**: Roy Morsch/Photolibrary/Getty Images; **p. 20**: © Dave Fleetham/Design Pics/Corbis; **p. 22**: Courtesy of Anu Singh Cundy; **p. 23** (top): ZUMA Press/Newscom; **p. 23** (bottom): Yiming Hu/yiminghu.com; **p. 24**: © Carl Wiens.

CHAPTER 2 **Page 28**: Susan Watts/New York Daily News; **p. 30**: © Alaska Stock/Alamy; **p. 31**: Terrance Klassen/Age Fotostock; **p. 32**: CNRI/Science Photo Library/Photo Researchers; **p. 33**: © Tom Brakefield/CORBIS; **p. 34**: Courtesy of Anu Singh Cundy; **p. 36**: Charles Falco/Photo Researchers; **p. 38**: Eric CHRETIEN/Gamma-Rapho via Getty Images; **p. 38** (bottom): Douglas Steakley Photography; **p. 39**: Nill/Age Fotostock; **p. 40**: Scott Andrews/Science Faction/SuperStock; **p. 45**: woman: Foodfolio/AgeFotostock; glycogen: MedImage/Photo Researchers; cellulose: Biophoto Associates/Photo Researchers; starch: Gary Gaugler/Visuals Unlimited; **p. 47**: Lori Greig/Getty Images/Flickr RF; **p. 53**: Maslowski Wildlife Productions; **p. 55** (top): Skyler Wilder; **p. 55** (bottom): Susan Watts/New York Daily News; **p. 56**: Nitr/Shutterstock; **p. 59**: (rock): Raul Touzon/National Geographic Creative; **p. 59**: (coffee): Cephas Picture Library/Alamy; **p. 59**: (fisherman): Bryan & Cherry Alexander/SeaPics.com.

CHAPTER 3 **Page 60**: Courtesy Matteo Bonazzi, Edith Gouin, and Pascale Cossart/Unité des interactions Bactéries-Cellules, Institut Pasteur; **p. 62**: **(a)** Eye of Science/Photo Researchers, Inc; **(b)** Eric V. Grave/Photo Researchers; **(c)** Wim van Egmond/Visuals Unlimited; **(d)** Biophoto Associates/Photo Researchers; **(e)** Scientifica/Visuals Unlimited; **(f)** Richard Kessel/Visuals Unlimited; **p. 63**: **(a)** Dr. Cecil H. Fox/Photo Researchers; **(b)** Public domain; **p. 64**: Bon Appetit/Alamy; **p. 64**: **(a)** Dr. John D. Cunningham/Visuals Unlimited; **(b)** SPL/Photo Researchers; **(c)** Dr. Dennis Kunkel/Visuals Unlimited; **p. 66**: Courtesy Dr. Ichiro Nishii, Department of Biological Sciences, Nara Women's University, Nara, Japan; **p. 68**: Characterization and functional analysis of osteoblast-derived fibulins in the human hematopoietic stem cell niche, Sonja P. Hergetha, Wilhelm K. Aicherb, Mike Essla, Thomas D. Schreibera, Takako Sasakic, Gerd Klein; **p. 70**: D. Spector/Photolibrary/Getty Images; **p. 71**: Dennis Kunkel Microscopy, Inc./Visuals Unlimited; **p. 71**: Dreamstime; **p. 72**: Dennis Kunkel Microscopy; **p. 73** (top): David M. Philips/Visual Unlimited; **p. 73** (bottom):Biophoto Associates/Science Source/Photo Researchers; Bahnmueller/Alimdi.net; **p. 74**: Universal/The Kobal Collection; **p. 75** (top): Bill Longcore/Science Source/Photo Researchers; **p. 75** (bottom): Dr. George Chapman/Visuals Unlimited; **p. 76**: Torsten Wittmann; **p. 77**: **(a)** Thomas Deerinck/Visuals Unlimited; **(b)** © 1995–2010 by Michael W. Davidson and The Florida State University; **(c)** Image by Sui Huang and Donald E. Ingber, Harvard Medical School; **p. 78**: Louise Cramer; **p. 79**: **(a)** © Dennis Kunkel Microscopy/Visuals Unlimited; **(b)** Eye of Science/Photo Researchers; **(c)** Science/Visuals Unlimited; **(d)** Dr. Gopal Murti/SPL/Science Source/Photo Researchers; **p. 80**: Courtesy Matteo Bonazzi, Edith Gouin, and Pascale Cossart/Unité des interactions Bactéries-Cellules,Institut Pasteur; **p. 81**: Anatomical Travelogue/Science Source; **p. 83**: Biophoto Associates/Science Source.

CHAPTER 4 **Page 84**: © Dr. Dennis Kunkel/Visuals Unlimited; **p. 86**: © Carolina Biological/Visuals Unlimited/Corbis; **p. 86**: Kevin & Betty Collins/Visuals Unlimited/Corbis; **p. 90**: © Steve Lee/age fotostock; **p. 91** (top): Eric V. Grave/Photo Researchers; **p. 91** (bottom): **(a)** Susumu Nishinaga/Photo Researchers; **(b)** David M. Phillips/Photo Researchers; **(c)** David M. Phillips/Photo Researchers; **p. 92**: Ted Kinsman/Science Source; **p. 93**: Image provided courtesy of Abcam Inc. ©2014 Abcam; **p. 97**: Biology Media/Photo Researchers; **p. 98**: Fluorescence Digital Image Gallery/Molecular Expressions; **p. 100**: © Dr. Dennis Kunkel/Visuals Unlimited; **p. 101**: Dancing Fish/Shutterstock.

CHAPTER 5 **Page 104**: Aflo Foto Agency/Photolibrary/Getty Images; **p. 106**: Vladimir Rys Photography/Getty Images; **p. 107**: Glenn Bartley/AgeFotostock; **p. 111**: iStockphoto; **p. 112**: © Brian Bevan/ardea./age fotostock; **p. 113**: **(a)** Jianghongy/Dreamstime; **(b)** Alexstar/Dreamstime; **(c)** The Photo Works; **(d)** Spaxia/Dreamstime; **(e)** The Photo Works; **(f)** Nytumblewe/Dreamstime; **p. 115**: Prof. K. Seddon & Dr. T. Evans, QUB/Photo Researchers; **p. 118**: Aflo Foto Agency/Photolibrary/Getty Images; **p. 119**: Art Wolfe/Science Source; **p. 121**: Scott Camazine/Science Source.

CHAPTER 6 **Page 122**: © Bob Barbour; **p. 125**: Corbis Premium RF/Alamy; **p. 128**: Corbis Premium RF/Alamy; **p. 130**: photka/shutterstock; **p. 132**: © Gary Braasch/CORBIS; **p. 134**: John Elk/Getty Images/Lonely Planet Images; **p. 135**: Wildlife GmbH/Alamy; **p. 136**: **(a)** © MAP/Nathalie Pasquel/age fotostock; **(b)** ChaiyonS021/Shutterstock; **(c)** © Jan Wlodarczyk/age fotostock; **p. 138**: Federico Veronesi/Getty Images; **p. 139**: Patrik Giardino/Corbis; **p. 141**: John Downer/Getty Images/Photolibrary RM; **p. 143** (top): © Bob Barbour; **p. 143** (bottom): Todd Pusser/Nature Picture Library; **p. 144**: fcafotodigital/Getty Images.

CHAPTER 7 **Page 148**: AP Photo; **p. 150**: Peter Skinner/Photo Researchers; **p. 151**: Carolina Biological Supply Co/Visuals Unlimited; **p. 155**: Dr. Gopal Murti/Getty Images; **p. 157** (top): Biophoto Associates/Photo Researchers; **p. 157** (bottom): Photolibrary.com/Getty Images; **pp. 158–59**: (all): Jennifer Waters/Photo Researchers; **p. 160**: Sylvie Corriveau/shutterstock; **p. 161**: Courtesy George von Dassow; **p. 163**: Cezar Serbanescu/Getty Images; **p. 165**: Spike Walker/Wellcome Images; **p. 167**: Paul Kay/Getty Images; **p. 168**: AP Photo; **p. 169**: REUTERS/Pichi Chuang; **p. 171**: Dr. Gopal Murti/Science Source.

CHAPTER 8 **Page 172**: Dr. Gopal Murti/Visuals Unlimited; **p. 172**: New York Public Library/Photo Researchers; **p. 174**: Science Source/Photo Researchers; **p. 175**: © Scott Camazine/Alamy; **p. 177**: Du Cane Medical Imaging Ltd./Photo Researchers; **p. 179**: Jacob Belcher/The Milford Daily News; **p. 182** (bottom): David Scott Smith/Photolibrary/Getty Images; **p. 182** (top): Peter Lansdorp/Visuals Unlimited; **p. 184** (top): Neil McAllister/Alamy; **p. 184** (bottom left): Richard Wheeler/Zephyris/Wikipedia; **p. 184** (bottom center): Voisin/Phanie/SuperStock; **p. 184** (bottom right): JB REED/Landov; **p. 185**: iStockphoto; **p. 186**: Dr. Gopal Murti/Visuals Unlimited; **p. 187**: Jon Kopaloff/FilmMagic/Getty Images.

CHAPTER 9 **Page 190**: Popperfoto/Getty Images; **p. 192** (top): The Mendelianum; **p. 192** (bottom): Eric Isselee/Shutterstock; **p. 193**: Christopher Futcher/Getty Images; **p. 194**: Gerard Lacz Images/SuperStock; **p. 195** (left): Erik Lam/Shutterstock; **p. 195** (right): Jagodka/Shutterstock; **p. 196**: © Mark Conlin/Alamy; **p. 200**: Fred Duval/FilmMagic/Getty Images; **p. 204** (left): Corbis; **p. 204** (center): © Norvia Behling; **p. 204** (right):Courtesy of Jennifer Weske-Monroe; **p. 205**: Zuma Press/Newscom; **p. 206**: imagebroker/Alamy; **p. 207** (top): Ronald C. Modra/Sports Illustrated/Getty Images; **p. 207** (bottom): NCI/Photo Researchers; **p. 208**: Vasiliy Koval/Dreamstime.com; **p. 211** (top): Popperfoto/Getty Images; **p. 211** (bottom left): Romanov Collection, General Collection, Beinecke Library, Yale University; **p. 211** (bottom right): Granger Collection; **p. 212**: jurra8/Shutterstock.

CHAPTER 10 **Page 216**: Courtesy Misty Oto; **p. 218**: Courtesy of Empire Genomics; **p. 219** (top): Alex Wild/Visuals Unlimited; **p. 219** (bottom): Genetix Limited; **p. 220**: WILLIAM H. BOND/NGS Image Collection; **p. 224** (top): Aaron Flaum/Alamy; **p. 224** (bottom): Hubert Boesl/dpa/Corbis; **p. 228**: Byron Purvis/AdMedia/Newscom; **p. 230**: Barcroft/Fame Pictures; **p. 232** (left): Courtesy Dr. George Herman Valentine; **p. 232** (right): Biophoto Associates/Photo Researchers; **p. 233**: Courtesy Misty Oto; **p. 234**: Courtesy Anu Singh Cundy; **p. 237** (left): Ishihara Color Test; **p. 237** (right): Ishihara Color Test.

CHAPTER 11 **Page 240** (right): Vienna Report Agency/Sygma/Corbis; **p. 240** (left): Andrea Solero/AFP/Getty Images; **p. 242** (left): Andreas Keil Image Broker/Newscom; **p. 242** (right): © Kitch Bain/age fotostock; **p. 245**: A. Barrington Brown/Photo Researchers; **p. 249**: Michael J. Daly/Science Source; **p. 250**: Kenneth Greer/Visuals Unlimited; **p. 252** (top): © GYRO PHOTOGRAPHY/a co/age fotostock; **p. 252** (bottom): AP Photo/The Buffalo News, Sharon Cantillon; **p. 253**: D. Hurst/Alamy; **p. 254**: Robert Harding World Imagery/Alamy; **p. 256**: AP Photo; **p. 257**: Peter Menzel/Science Source.

CHAPTER 12 **Page 260**: Courtesy ARS-USDA; **p. 262**: Brunet et al. (1998) Noggin, cartilage morphogenesis, and joint formation in the mammalian skeleton. *Science*, 280, 1455–1457. Courtesy Richard Harland; **p. 264**: AP Photo/The Dominion Post, Jason DeProspero; **p. 265**: George McCarthy/Corbis; **p. 272**: Science Stills/Visuals Unlimited/Getty Images; **p. 275** (left): Manny Millan/Sports Illustrated/Getty Images; **p. 275** (right): Julian Finney/Getty Images; **p. 276**: BOB GIBBONS/SCIENCE PHOTO LIBRARY; **p. 276**: Courtesy ARS-USDA; **p. 277**: Don Pablo/Shutterstock.

CHAPTER 13 **Page 280**: Photo by Rik Sferra, Courtesy of Eduardo Kac; **p. 282** (top): Yann Arthus Bertrand/Corbis; **p. 282** (bottom): Reuters/Corbis; **p. 284**: Steve Granitz/Getty Images; **p. 285**: FEDERICO GAMBARINI/Agentur/Newscom; **p. 286**: AP Photo/Sakchai Lalit; **p. 287** (left): Martin Shields/Science Source; **p. 287** (right): Courtesy of The Innocence Project; **p. 288**: gresei/Shutterstock; **p. 289** (left): aslysun/Shutterstock; **p. 289** (right): Eduardo Kac, 2000. Photo © Chrystelle Fontaine; **p. 289** (bottom): Courtesy of Aqua Bounty Technologies, Inc.; **p. 290** (left): Sarah Cuttle/Getty Images/Dorling Kindersley; **p. 290** (right): © Simko/Visuals Unlimited/Corbis; **p. 291**: Roslin Institute, University of Edinburgh; **p. 292**: RIA Novosti/Science Source; **p. 296**: Ron Sachs/CNP/Corbis; **p. 297** (left): Sandy Huffaker/Corbis; **p. 297** (right): Courtesy J. Pabon M.D./Fertility Center And Applied Genetics of Florida; **p. 298**: Courtesy of Ashanthi De Silva; **p. 299**: DAVID PARRY/PA WIRE HANDOUT/EPA/LANDOV; **p. 300**: Alpha/Landov; **p. 302**: Lawrence Berkeley National Laboratory, Roy Kaltschmidt, Photographer; **p. 304** (top): Photo by Rik Sferra, Courtesy of Eduardo Kac; **p. 304** (bottom): Justin S. Williams and Teh-hui Kao, Biochemistry and Molecular Biology Department, Pennyslvania State University; **p. 305**: Russell Kightley/Science Source.

CHAPTER 14 **Page 308**: © Tui De Roy/Minden Pictures/Corbis; **p. 310** (top left): K Jayaram/Science Source; **p. 310** (top right): © steve bly/Alamy; **p. 310** (bottom left): © Danita Delimont/agefotostock; **p. 310** (bottom right): Peter Vollmert/agefotostock; **p. 311**: Getty Images/The Bridgeman Art Library; **p. 313**: Andres Rodriguez/Alamy; **p. 314**: Bettmann/Corbis; **p. 315** (top): Dr. Jeremy Burgess/Science Source; **p. 315** (bottom): Wayne Bennett/Corbis; (insets) Courtesy of JoGayle Howard, Department of Reproductive Sciences, Smithsonian's National Zoological Park; **p. 316**: Dominique Braud/Wildlife & Nature Photography; **p. 318**: Gerard Lacz/age fotostock/SuperStock; **p. 319**: Reprinted from Hoekstra et al. 2006 Science; **p. 320**: Breck P. Kent/Animals Animals-Earth Scenes; **p. 321** (left): James H. Karales/Peter Arnold/Photolibrary/Getty Images; **p. 321** (right): Courtesy of Tom Smith, UCLA; **p. 322** (top): Premaphotos/Alamy Stock Photo; **p. 322** (bottom both): Erick Greene, University of Montana; **p. 323** (top): John Alcock/Visuals Unlimited; **p. 323** (bottom): © David Denning, BioMedia Associates; **p. 324** (top): Merlin D. Tuttle/Bat Conservation Int'l/Photo Researchers; **p. 324** (bottom): GM Photo Images/Alamy Stock Photo; **p. 325**: Universal Images Group/SuperStock; **p. 326**: Louie Psihoyos/Corbis; **p. 327**: Omikron/Science Source; **p. 330**: Ken Lucas/Visuals Unlimited/Getty Images; **p. 331** (left): © Krishan Lal; **p. 331** (center): Dorling Kindersley/Getty Images; **p. 331** (right): © RM Floral/Alamy; **p. 332** (left): Associated Press; **p. 332** (right): Bill Branson, NIH/Wikimedia Commons; **p. 333** (top): © Tui De Roy/Minden Pictures/Corbis; **p. 333** (middle): Peter Boag; **p. 333** (bottom): Peter Boag; **p. 334**: Miguel Castro/Photo Researchers; **p. 335**: Bill Coster/Alamy.

CHAPTER 15 **Page 338**: Simon Murrell/Alamy; **p. 340**: **(a)** James Urbach/Purestock/Superstock; **(b)** All Canada Photos/Superstock; **p. 341** (all): alslutsky/Shutterstock; **p. 343**: Courtesy Karen Brzys, www.agatelady.com (Gitche Gumee Museum); **p. 343**: Kaibab squirrel: © Tom & Pat Leeson; **p. 343**: Abert's squirrel: Kevin Doxstater/NaturalVisions; **p. 344**: © Peter Schouten. All rights reserved; **p. 345** (top left): © FLPA/David Hosking/age fotostock; **p. 345** (top right): © FLPA/David Hosking/age fotostock;

p. 345 (bottom left): Photo Resource Hawaii/Alamy; **p. 345** (bottom right): Michael Ord/Science Source; **p. 346**: © Maximilian Schönherr/dpa/Corbis; **p. 350**: **(a)** Chris Hellier/Corbis; **(b)** Doug Sokell/Visuals Unlimited; **(c)** Fletcher & Baylis/Photo Researchers; **(d)** George Grall/National Geographic Image Collection; **(e)** Flip Nicklin/Minden Pictures/National Geographic Images; **p. 351**: From "Surgical Treatment of a Patient with Human Tail and Multiple Abnormalities of the Spinal Cord and Column" by Chunquan Cai, Ouyan Shi and Changhong Shen, © 2011 Chunquan Cai et al. This is an open access article distributed under the Creative Commons Attribution License; **p. 352**: Simon Murrell/Alamy; **p. 353**: Philippe Plailly & Atelier Daynes/Science Source.

CHAPTER 16 **Page 356**: Tom Demsey/Photoseek; **p. 360**: **(a)** Ken Lucas/Visuals Unlimited; **(b)** Niles Eldridge; **(c)** B. Miller/Biological Photo Services; **(d)** David Grimaldi, American Museum of Natural History; **(e)** Louie Psihoyos/Corbis; **(f)** Charlie Ott/Photo Researchers; **p. 362**: © Jane Gould/Alamy; **p. 365**: Bernhard Edmaier/Science Source; **p. 368**: **(a)** Rodolfo Coria; **(b)** Jason Edwards/National Geographic Image Collection; **p. 369**: Jose B. Ruiz/naturepl.com; **p. 370** (top left): Eye of Science/Science Source; **p. 370** (top right): Eye of Science/Science Source; **p. 370** (bottom): NHPA/SuperStock; **p. 371** (top): Real World Image; **p. 371** (bottom): Sabena Jane Blackbird/Alamy Stock Photo; **p. 373** (top): The Art Archive/SuperStock; **p. 373** (bottom): Title page of the 10th edition of *Systema naturæ* written by Carl Linnæus, published in 1758 by L. Salvius in Stockholm. Digitized in 2004 from an original copy of the 1758 edition held by Göttingen State and University Library; **p. 374** (top): Tom Demsey/Photoseek; **p. 374** (bottom): © William Stout, Inc.; **p. 375**: AlessandroZocc/Shutterstock.

CHAPTER 17 **Page 378**: Stephanie Schuller/Photo Researchers; **p. 380**: **(a)** Tony Wu/naturepl.com; **(b)** Robert Ziemba/AgeFotostock; **(c)** Satish Arikkath/Alamy; **(d)** John D. Cunningham/Visuals Unlimited; **(e)** National Geographic Image Collection/Alamy; **(f)** AIC/Alamy; **(g)** John Cancalosi/Photolibrary/Getty Images; **(h)** Visuals Unlimited/Kjell Sandved/Getty Images; **(i)** David Hosking/Fran Lane Picture Agency/Corbis; **(j)** Frans Lanting/Corbis; **p. 381**: Courtesy of Anu Singh Cundy; **p. 382**: Wildlife GmbH/Alamy; **p. 384**: Courtesy of Anu Singh Cundy;**p. 386**: Mycoplasma: Phototake/Alamy; *Epulopiscium fishelsoni*: Courtesy Esther Angert, Department of Microbiology, Cornell University; Bacteria: Scimat/Photo Researchers; Glacier: Photo by S. Montross; chimneys: NOAA; lagoons: Aerial Archives/Alamy; **p. 387**: Douglas Faulkner/Science Source; Cyanobacteria: (Dr. Robert Calentine/Visuals Unlimited; cave: Courtesy of Tommy J. Phleps; treatment plant: Jupiterimages/Getty Images; roots: Nigel Cattlin/Alamy; teeth: Rob Byron/Shutterstock; p. 388: Dr. Dennis Kunkel/Visuals Unlimited; **p. 390**: Jerome Wexler/Photo Researchers, Inc.; **p. 391** (top): Krafft/Hoa-qui/Photo Researchers; **p. 391** (bottom): © Darwin Wiggett/All Canada Photos/Corbis; **p. 392**: © Kate Berry; **p. 393** (top): Accent Alaska.com/Alamy; **p. 393** (bottom): AP Photo/Disney-ABC Domestic Television, Ida Mae Astute; **p. 394**: mediaphotos/Getty Images; **p. 396**: Tobacco: Omikron/Photo Researchers; Adenovirus: Dr. Hans Gelderblom/Visuals Unlimited; T-4 phage: Dr. Harold Fisher/Visuals Unlimited; **p. 398**: Design Pics Inc./Alamy; **p. 399**: Nathan Shankar, University of Oklahoma/Stephen Ausmus/USDA/Science Source.

CHAPTER 18 **Page 402**: Frans Lanting/Corbis; **p. 405** (top): John Cancalosi/Photolibrary/Getty Images**p. 405** (bottom): Wim van Egmond/Visuals Unlimited; **p. 406**: Dennis Kunkel/Visuals Unlimited; **p. 408**: Mark Conlin/Alamy; **p. 409**: Peter Falkner/Science Source; **p. 409** (bottom left): Robert Berdan/Science and Art Multimedia; **p. 409** (bottom right): Douvres Photo by Remi Jouan/Wikipedia; **p. 409** (inset): Steve Gschmeissner/Photo Researchers; **p. 410**: salamander: Danita Delimont/Alamy; kayak: ©Frank Borges LLosa/frankly.com; *Noctiluca*: Gerd Guenther/Science Source; *Giardia*: Jerome Paulin/Visuals Unlimited; **p. 411**: foraminifers: Peter Parks/Image Quest 3-D; seaweed: Wim van Egmond/Visuals Unlimited; *Stentor*: Courtesy Antonio Guillén (Proyecto Agua), Spain; photosynthesis: Wim van Egmond/Visuals Unlimited; *Euplotes*: M. I. Walker/Photo Resesearchers; water mold: © Nigel Cattlin/Visuals Unlimited/Corbis; **p. 412**: **(a)** Sankei via Getty Images; **(b)** David Phillips/Visuals Unlimited/Getty Images; **p. 415** (top): © Carson Baldwin Jr./age fotostock; **p. 417**: Ethan Daniels/Alamy; **p. 418**: lily: China Photos/Getty Images; *Wolffia arrhiza* on human fingers. Photo by Christian Fischer/Wikipedia; baobab tree: John Warburton-Lee Photography/Alamy; *Sequoiadendron*: Rob Blakers/WWI/Photolibrary/Getty Images; *Rafflesia*: Compost/Visage/Peter Arnold, Inc./Getty Images; orchid: E. Kajan/AgeFotostock; orchid seeds: © Wayne P. Armstrong; stalk: David M. Dennis/Photolibrary/Getty Images; **p. 420**: AP Photo/Illinois River Biological Station via the Detroit Free Press, Nerissa Michaels; **p. 421**: N. Aubrier/age fotostock; **p. 422** (bottom): NaturePL/SuperStock; **p. 422** (top): FLPA/Gianpiero Ferrari/agefotostock; **p. 423**: *Arthrobotrys*: Courtesy George L. Barron; *Gymnopilus*: Michael P. Gadomski/Photo Researchers; bioluminescent: Ben Nottidge/Alamy; death cap: George McCarthy/Corbis; fairy ring: Wally Eberhart/Getty Images; *Pilobolus*: Carolina Biological Supply Co./Visuals Unlimited; stinkhorn: Sharon Cummings/Dembinsky Photo Associates; **p. 424**: Dr. Edward Ross, California Academy of Sciences; **p. 425**: (top left): Biophoto Associates/Science Source; **p. 425** (top right): **(a)** Courtesy Ganesh Tree and Plant Health Care, Sturbridge, MA. **(b)** Dana Richter/Visuals Unimited; **p. 426**: Frans Lanting/Corbis; **p. 427**: rsmseymour/iStockphoto.

CHAPTER 19 **Page 430**: Mauricio Handler/National Geographic Society/Corbis; **p. 432** (top left): AP Photo; **p. 432** (top right): Reinhard Dirscheri/Visuals Unlimited/Getty Images; **p. 432** (bottom left): David Doubilet; **p. 432** (bottom right): WaterFrame/Alamy; **p. 434** (top): WaterFrame/Alamy; **p. 434** (bottom): © Stefano Piraino, all rights reserved; **p. 440**: Taro Taylor/Wikimedia Commons; **p. 441**: M. I. Walker/Photo Researchers; **p. 443**: **(a)** Paul Kay/Photolibrary/Getty Images; **(b)** © Sea Pics; **(c)** AP Photo**p. 444**: Nancy Nehring/Getty Images; **p. 445** (top): Steve Gschmeissner/Science Source; **p. 445** (center left): Animals Animals/SuperStock; **p. 445** (center right): Francesco Tomasinelli/Science Source; **p. 445** (bottom left): Thomas Shahan/Getty Images; **p. 445** (bottom right): John Mitchell/Getty Images; **p. 448**: Adrian Bicker/Science Source; **p. 449** (top): imagebroker.net/SuperStock; **p. 449** (bottom): Reinhard Dirscheri/Visuals Unlimited/Getty Images; **p. 450**: REX USA/Heritage Auctions; **p. 452** (top left): Blickwinkel/Alamy; **p. 452** (top right): Denis Scott/Corbis; **p. 452** (center left): © Caro/Alamy; **p. 452** (center right): J.Metzner/Photolibrary/Getty Images; **p. 452** (bottom left): Michelle Gilders/Alamy; **p. 453** (top left): © A Hartl/age fotostock; **p. 453** (bottom): Computer Earth/Shutterstock; **p. 456**: marmosets: Asia Images Group Pte Ltd/Alamy; carp: AP Photo; Whale: Denis Scott/Corbis; squid: Reuters/Corbis; sloth: Danita Delimont/Alamy; tern: Malcolm Schuyl/Alamy; marlin: George Holland/AgeFotostock; **p. 457**: clownfish: iStockphoto; elephants: National Geographic Image Collection/Alamy; Komodo dragon: Jason Row/Alamy; mayfly: iStockphoto; turtle: Brandon Cole/Photolibrary/Getty Images; clam: Courtesy Chuck Savall; koala: iStockphoto; catfish: Michael Goulding; anglerfish: Doug Perrine/Photolibrary/Getty Images; **p. 458** (top): Juan Carlos Munoz/Photolibrary/Getty Images; **p. 458** (bottom): Mark Gurney/Smithsonian/Sipa USA/Newscom; **p. 459** (left):

DEA Picture Library/Getty Images; **p. 459** (center): John W Banagan/ Getty Images; **p. 459** (right): Steve Bloom Images/Alamy; **p. 460**: Mauricio Handler/National Geographic Society/Corbis; **p. 461**: © David M. Dennis/ age fotostock.

CHAPTER 20 **Page 464**: Karen Gowlett-Holmes/Getty Images; **p. 464**: AP Photo; **p. 466**: NASA; **p. 468** (top left): Martin Harvey/Corbis; **p. 468** (top right): Frans Lanting/Corbis; **p. 468** (bottom): © John Carnemolla/Corbis; **p. 471**: © akg-images/Alamy; **p. 473** (top): imagebroker/Alamy; **p. 473** (bottom): John E Marriott/Alamy; **p. 474** (top): Michael P. Gadomski/Earth Sciences/Animals Animals; **p. 474** (bottom): AP Photo/HO, the Institute of Cell Biophysics of the Russian Academy of Sciences; **p. 475** (top): Ricardo Reitmeyer/Shutterstock; **p. 475** (bottom): Ken Lucas/Visuals Unlimited; **p. 476** (top): Willard Clay/Dembinsky Photo Associates; **p. 476** (bottom): kavram/Shutterstock; **p. 477** (top): © blickwinkel/Alamy; **p. 477** (bottom): Pixtal Images/Photolibrary/Getty Images; **p. 479**: **(a)** Claire Fackler, NOAA National Marine Sanctuaries; **(b)** © Ambient Images Inc./Alamy; **(c)** © Norbert Wu/Science Faction/Corbis; **(d)** s0ulsurfing—Jason Swain/Getty Images; **p. 480 (a)**: Design Pics Inc/Photolibrary/Getty Images; **p. 480**: **(b)** Anna Gorin/Getty Images; **(c)** James Steinberg/Science Source; **p. 481**: Mark Goodreau/Alamy; **p. 482** (top): Vlad61/Shutterstock; **p. 482** (bottom): Jim Zipp/National Audubon Society Collection/Photo Researchers; **p. 483**: Gregory Ochocki/Science Source; **p. 484**: Karen Gowlett-Holmes/Getty Images; **p. 485**: www.ocean-sole.com.

CHAPTER 21 **Page 488**: Art Wolfe/Stone/Getty Images; **p. 490**: Minden Pictures/SuperStock; **p. 491** (top): Jeremy Burgess/National Audubon Society collection/Photo Researchers Inc; **p. 491** (bottom): Volker Staeger/ SPL/Science Source/Photo Researchers; **p. 492**: Monarch Butterfly Fall Migration Patterns. Base map source: USGS National Atlas; **p. 493**: **(a–b)** Reproduced with permission of the Department of Natural Resources, Queensland, Australia; **p. 493** (moth): Susan Ellis USDA APHIS PPQ Bugwood.org; **p. 494**: STEPHEN KRASEMANN/age fotostock; **p. 494**: Michael & Patricia Fogden/Minden Pictures; p. 495 (top): Laguna Design / Science Picture Library/ Photo Researchers, Inc.; **p. 495** (bottom): Ecoscene/Corbis; **p. 496** (top left): Eye of Science/Science Source; **p. 496** (top right): G A Matthews/SPL/Photo Researchers, Inc.; **p. 496** (bottom): Mark Newman/Tom Stack & Associates; **p. 498**: © Biosphoto/Alain Beignet; **p. 499** (top): © Alex Wild/Visuals Unlimited/Corbis; **p. 501** (top): Art Wolfe/Stone/Getty Images; p. 501 (bottom): Craig Mayhew, Robert Simmon, NASA GSFC; **p. 502**: NASA/Corbis; **p. 503**: AP Photo/Sakchai Lalit.

CHAPTER 22 **Page 506**: Vincent J. Musi/National Geographic Stock; **p. 508**: © Gallo Images/CORBIS; **p. 509**: Joe Raedle/Getty Images; **p. 510**: Larry J. Young, PhD, Center for Translational Social Neuroscience, Yerkes National Primate Research Center; **p. 511** (top): Petit Format/Photo Researchers; **p. 511** (bottom left): Ted Kinsman/Science Source; **p. 511** (bottom right): ER Degginger/Science Source; **p. 512** (top left); imagebroker/Alamy; **p. 512** (top right): Time & Life Pictures/Getty Images; **p. 513** (top): Thomas D. McAvoy/Time & Life Pictures/Getty Images; **p. 513** (bottom): BBC Worldwide Learning; **p. 514** (top left): Thomas Kitchin/Tom Stack & Associates; **p. 514** (top center): Fred Breummer/Peter Arnold/Getty Images; **p. 514** (top right): Ian Tait/ Naturalvisions.co.uk; **p. 514** (bottom left): FLPA/SuperStock; **p. 515** (top left): © Alex Wild/Visuals Unlimited/Corbis; **p. 515** (top right): Danita Delimont/Alamy Stock Photo; **p. 515** (bottom left): Andrey Pavlov/ Shutterstock; **p. 515** (bottom right): Nature's Images/Science Source; **p. 516**: **(a)** Corbis Super RF/Alamy; **(b)** Photolibrary/Getty Images; **(c)** Mauricio Lima/AFP/Getty Images/Newscom; **p. 517**: **(a)** Shutterstock; **(b)** John Cancalosi/Alamy; **p. 518** (top left): Stephen Frink Collection/ Alamy; **p. 518** (top right): Stan Olinsky/Dembinsky Photo Associates; **p. 518** (bottom): James H. Robinson/Science Source; **p. 519** (top left): blickwinkel/Alamy; **p. 519** (top right): Juniors Bildarchiv GmbH/Alamy; **p. 519** (center right): Jane Burton/Getty Images; **p. 519** (center right) Denise Thompson/Shutterstock; **p. 519** (bottom right): DARLYNE A. MURAWSKI/National Geographic Creative; **p. 520**: Vincent J. Musi/ National Geographic Stock;**p. 521**: John Cancalosi/Getty Images.

CHAPTER 23 **Page 524**: Juniors Bildarchiv/Photolibrary/Getty Images; **p. 526**: Vphotographyandart/Getty Images; **p. 527**: MarcusVDT/ Shutterstock; **p. 529** (top): **(a)** Willard Clay/Dembinsky Photo Associates; **(b)** Minden Pictures/Superstock; **p. 529** (bottom): Science Faction/ Superstock; **p. 530**: Maria Sala/University College Dublin; **p. 531** (left): **(a)** Joe McDonald/Visuals Unlimited; **(b)** Gerry Bishop/Visuals Unlimited/ Corbis; (c1) Nancy Nehring/Getty Images; (c2) Stockbyte/Getty Images; **p. 531** (top right): Alan G. Nelson/Dembinsky Photo Associates; **p. 532** (all): Dr. Gerald D. Carr; **p. 535** (top): NHPA/SuperStock; **p. 535** (center): Biosphoto/SuperStock; **p. 535** (bottom): Glenn Price/Shutterstock; **p. 536**: David Wrobel/Getty Images/Visuals Unlimited; **p. 538**: **(a)** Stan Osolinski/Dembinsky Photo Associates; **(b)** Howard Garrett/Dembinsky Photo Associates; **(c)** Walt Anderson/Visuals Unlimited; **pp. 540–41**: Both photos courtesy of Robert Gibbens, Jornada Experimental Range, USDA; **p. 542**: Juniors Bildarchiv/Photolibrary/Getty Images;**p. 543**: Geoff Kidd/ Science Source.

CHAPTER 24 **Page 546**: Sean Gardner/Reuters; **p. 548**: Lester Lefkowitz/ Corbis; **p. 550** (bottom left): Martin Harvey/Corbis; **p. 550** (bottom right): © YAY Media AS/Alamy; **p. 552**: © Sean Tilden/Alamy; **p. 553**: NHPA/ SuperStock; **p. 557** (bottom): AP Photo/Tony Gutierrez; **p. 558**: Hugh Spencer/Photo Researchers; **p. 561** (top): NASA/Goddard Space Flight Center Scientific Visualization Studio; **p. 561** (bottom): Courtesy of the Hubbard Brook Archives; **p. 562** (top): Richard Packwood/Getty Images; **p. 562** (bottom): **(a)** Spencer Platt/Getty Images; **(b)** Spencer Platt/Getty Images; **p. 563** (bottom): Cameron Davidson Stock Connection USA/Newscom; **p. 564**: Sean Gardner/Reuters; **p. 565**: Visuals Unlimited, Inc./Mark Schneider/Getty Images.

CHAPTER 25 **Page 568**: Johnny Johnson/Getty Images; **p. 570** (bottom left): WagTV.com; **p. 570** (bottom right): Popular Science via Getty Images; **p. 571** (top left): US Geological Survey (USGS), NASA/Goddard Space Flight Center; **p. 571** (top right): US Geological Survey (USGS), NASA/Goddard Space Flight Center; **p. 572**: Jacques Jangoux/Photo Researchers; **p. 573**: AP Photo/Andre Penner; **p. 576**: Challenges by Susanne von Caemmerer, W. Paul Quick, Robert T. Furbank, Science 29 June 2012: Vol. 336 no. 6089 pp. 1671–1672; **p. 578** (top left): AP Photo/ Katsumi Kasahara; **p. 578** (top right): NASA/Goddard Space Flight Center Scientific Visualization Studio; **p. 578** (center right): NASA/Goddard Space Flight Center Scientific Visualization Studio; **p. 579** (top left): 1913: W.C. Alden Photo/Courtesy USGS Northern Rocky Mountain Science Center; **p. 579** (center left): © John Scurlock; **p. 579** (bottom right): Patrick Kelley, U.S. Coast Guard; **p. 580** (top left): Harald Pauli; **p. 580** (top right): Visceralim/Dreamstime.com; **p. 580** (center): Harald Pauli; **p. 581**: Qilal Shen/Bloomberg via Getty Images; **p. 582**: AP Photo; **p. 583**: Pascal Goetgheluck/Science Source; **p. 585**: © The AfriCat Foundation, based in the Okonjima Nature Reserve, central Namibia.

Text Credits

ABC SCIENCE STAFF: "Lemur's Long-Buried Secrets Revealed," by ABC Science Staff. ABC Science, May 3, 2013. Reprinted by permission of Agence France-Presse.

AMY BLASZYK: "For a Better, Learner Burger, Get to Know Your Proteins," © Amy Blaszyk courtesy of National Public Radio, Inc. NPR news report titled "For a Better, Leaner Burger, Get to Know Your Proteins" by Amy Blaszyk originally published on NPR.org on August 17, 2012, and is used with the permission of NPR. Any unauthorized duplication is strictly prohibited.

ERYN BROWN: "Gut Microbiome Largely Stable over Time, Feces Study Finds," by Eryn Brown. *Los Angeles Times*, July 5, 2013. Reprinted by permission of the Los Angeles Times.

CBS NEW YORK STAFF: "Seen at 11: New Technology Can Make It So 1 Baby Has 3 Parents," by CBS New York Staff. CBS New York, March 29, 2013. Reprinted by permission of CBS New York.

ERIN ELLIS: "Science of Making Babies Becomes Commonplace," by Erin Ellis. *The Vancouver Sun*, May 18, 2013. Material reprinted with the express permission of: Vancouver Sun, a division of Postmedia Network Inc.

BARRY EVANS: "Prime Number Cicadas," by Barry Evans. Originally printed in *The North Coast Journal*, August 1, 2013. Reprinted by permission of The North Coast Journal.

JAMES GORMAN: "Nectar That Gives Bees a Buzz Lures Them Back for More." From *The New York Times*, March 7, 2013 © 2013 The New York Times. All rights reserved. Used by permission and protected by the Copyright Laws of the United States. The printing, copying, redistribution, or retransmission of this Content without express written permission is prohibited.

JAMES GORMAN: "Study Discovers DNA That Tells Mice How to Construct their Homes." From *The New York Times*, January 16, 2013 © 2013 The New York Times. All rights reserved. Used by permission and protected by the Copyright Laws of the United States. The printing, copying, redistribution, or retransmission of this Content without express written permission is prohibited.

JOSEPH HALL: "The Human Genome Project: How It Changed Biology Forever," by Joseph Hall. *The Toronto Star*, April 28, 2013. Reprinted by permission of Torstar Syndication Services.

THE HUFFINGTON POST STAFF: "BPA Could Affect Brain Development by Impacting Gene Regulation, Study Finds," by The Huffington Post Staff. HuffPost: Healthy Living, February 28, 2013. Reprinted by permission of The Huffington Post.

EDWARD KARDAS: Figure: "Pheromones Can Signal Potential Mates," by Edward Kardas. Southern Arkansas University Lecture, Bombykol. © Edward P. Kardas. Reprinted by permission of the author.

SCOTT KIRSNER: "Gene Therapy Shows New Signs of Promise," by Scott Kirsner. Originally published in *The Boston Globe*, June 2, 2013. Reprinted by permission of the author.

LISA M. KRIEGER: "Stanford University Students Study Their Own DNA," by Lisa M. Krieger. *San Jose Mercury News*, May 24, 2013. Reprinted by permission of The YGS Group.

RENE LYNCH: "Eating Bugs: Would You Dine on Cicadas? Crickets? Buttered Beetles?" by Rene Lynch. *Los Angeles Times*, May 16, 2013. Reprinted by permission of the Los Angeles Times.

DONALD G. MCNEIL, JR.: "Cassava and Mental Deficits." From *The New York Times*, April 22, 2013 © 2013 The New York Times. All rights reserved. Used by permission and protected by the Copyright Laws of the United States. The printing, copying, redistribution, or retransmission of this Content without express written permission is prohibited.

ALANNA MITCHELL: "Beyond the Kiss, Mistletoe Helps Feed Forests, Study Suggests." From *The New York Times*, December 17, 2012 © 2012 The New York Times. All rights reserved. Used by permission and protected by the Copyright Laws of the United States. The printing, copying, redistribution, or retransmission of this Content without express written permission is prohibited.

MONTE MORIN: "Genetically Engineered Tomato Mimics Good Cholesterol," by Monte Morin. *Los Angeles Times*, March 20, 2013. Reprinted by permission of the Los Angeles Times.

JOE MWIHIA: "Kenyan Company Turns Old Sandals into Colorful Array of Toys and Safari Animals," by Joe Mwihia. *The Associated Press*, May 8, 2013. Reprinted by permission of The YGS Group.

BECKY OSKIN: "Too Much Deer Pee Changing Northern Forests," by Becky Oskin. *Live Science: Our Amazing Planet*, June 5, 2013. Reprinted by permission of Wright's Media.

ROXANNE PALMER: "White Tiger Genetic Secret Unveiled: Single Mutation in Single Gene Removes Orange Color." From International Business Times, May 23, 2013 © 2013 IBT Media. All rights reserved. Used by permission and protected by the Copyright Laws of the United States. The printing, copying, redistribution, or retransmission of this Content without express written permission is prohibited.

RONI CARYN RABIN: "Curbing the Enthusiasm on Daily Multivitamins." From *The New York Times*, October 22, 2012 © 2012 The New York Times. All rights reserved. Used by permission and protected by the Copyright Laws of the United States. The printing, copying, redistribution, or retransmission of this Content without express written permission is prohibited.

REX SANTUS: "Cancer Gene Has Led Jolie and Others to Surgery," by Rex Santus. *The Columbus Dispatch*, May 15, 2013. Reprinted by permission of Dispatch Printing Company.

J. WILLIAM SCHOPF: "The Earth's Geographical Timescale Put onto a 24 Hour Clock," by J. W. Schopf, UCLA. http://shellvpower.wordpress.com/2010/02/03/. Reprinted by permission of the author.

JENNIFER VIEGAS: "First Love Child of Human, Neanderthal Found," by Jennifer Viegas. *Discovery News*, March 27, 2013. Reprinted by permission of Discovery Communications, LLC.

NICHOLOAS WADE: "Fish's DNA May Explain How Fins Turned to Feet." From *The New York Times*, April 17, 2013 © 2013 The New York Times. All rights reserved. Used by permission and protected by the Copyright Laws of the United States. The printing, copying, redistribution, or retransmission of this Content without express written permission is prohibited.

ADAM WELZ: "Blind, Starving Cheetahs: the New Symbol of Climate Change?" by Adam Welz. *The Guardian*, June 21, 2013. © Guardian News & Media Ltd. 2013. Reprinted by permission of Guardian News and Media Ltd.

SARAH C. P. WILLIAMS: "Evolution via Roadkill," by Sarah C. P. Williams. Originally published in *Science Now*, March 18, 2013. Reprinted by permission of the author.

Index

Page numbers in *italics* refer to illustrations and tables.

abalones, 443
abiotic environment, 549, 554, 555, *578*
abyssal zone, 483
Acacia trees, 528
acetyl CoA, 140
acidosis, 41
acid rain, 560, 562, *562*
acids, 40
acinar cells, genes in, *244*
acoelmomate animals, 437, *438*
actin, 56, 77
actin filaments, 76
actin ring, 161
activation energy, 114–15
active carrier proteins, 94, *95*
active sites, 113, 114
active transport, 87, *88, 93,* 94, *95*
ACTN3 gene, 275, *275*
Acuña, Jason, *228*
adaptation, *18, 310,* 321–24
  of birds for flight, 454–55
  of cacti, 137
  defensive, 530, *531*
  defined, 19, 318
  dietary, *342,* 343
  of fruits, 417
  of Fungi to absorb nutrients, 421
  Lamarck's explanation for, 311
  limits to, 323–24
  by natural selection, 19, 318, *319,* 321–24
  of Plantae to land, 413
  and reproductive success, 318, 322–23, *323*
  size as, 406
  uses of term, 318
adaptive behaviors, 508
adaptive evolution, 316, 318–21
adaptive fitness, 508
adaptive radiations, *348,* 348–49, 368
adaptive traits, 19
adelgid, *565*
adenine, 53
adenosine deaminase (ADA) deficiency, *225,* 297, *298*
adenosine diphosphate (ADP), 110, *111,* 125
adenosine triphosphate (ATP), *55*
  anaerobic production of, 138–39
  from cellular respiration, 127, 138–41, *139, 142*
  chlorophyl and, 75
  defined, 54, 110
  as energy carrier, 124–26
  from light reactions, *133*
  in metabolism, 110, *111*
  microtubules' use of, 78
  produced by mitochondria, 73, 74, *75*
adhesion proteins, 67
adrenoleukodystrophy (ALD), 74, *225*
adult stem cells, 293, *294,* 295, *295*
aerobic exercise, 140, 143
aerobic heterotrophs, 390
aerobic processes, 139
Afghanistan, *497*
Africa, 144, 288, 474, 476, 477, 585
African seed crackers, 321, *321*
age
  metabolism and, 118
  and risk of cancerous change, 179–80, *180*
Age of Enlightenment, 310
agriculture
  genetically modified organisms in, 288, *288*
  population ecology principles in, 500
  and sea-level rise, 582
  sustainable practices in, 583
Agrobacterium tumefaciens, 304
Alaska, 472, 473
Albanes, Demetrius, 24
albinism, 206, *206*
alcohol abuse, *509,* 509–10, *510*
alcohol dependency (alcoholism), *226*
alcoholic fermentation, 138, *139*
algae, 149, 363, 409
  cell division in, 149
  cell walls of, 68
  chloroplasts in, 74–75, *75*
  defined, 408
  and eutrophication, 559, *560*
  evolution of plants from, 412
  hermaphroditic, 406
  history of, 363, 364
  in hot springs, *409*
  internal compartments in, *406*
  mixotrophic, 412
  photosynthesis in, 108
  as photosynthetic producers, 126
  producers, 16
  reproduction in, 406–7, *407*
algal blooms, *148,* 149, 168
Ali, Muhammad, *296*
alkalosis, 41
alkaptonuria, 262
allele frequencies, 312
alleles, 166
  of *ACTN3,* 275
  codominance, 204, 206
  defined, 194, *196,* 313
  dominant, 195, *196,* 198
  from gene mutation, 195–96
  incomplete dominance, *203,* 203–4, *204*
  and law of independent assortment, 200–202, *201, 202*
  in mutations, 248
  and phenotype, 194, 203
  recessive, 195, *196*
  and risk of disease, 226
allopatric speciation, 344, *344*
Allosaurus, 368
alopecia, 182
alpha helix, 46, *48*
Alpheus shrimp, 528, *528*
alpine ecosystems, 579, *580*
alpine tundra, 472
altruism, 514, 515
Alzheimer's disease, *225, 226,* 272, 295
Amazon lily (*Victoria regia*), *418*
Amazon rainforest, 560, *572,* 573, *573*
Amenmiya, Chris, 375
American Cancer Society, 24
American chestnut, 424, 531, 532
American Heart Association, 29
American Society for Reproductive Medicine, 169
American Southwest, *540,* 541, *541*
amino acids, 46
  carcinogenic compounds from, 185
  codons specifying, 267, 268, *268*
  defined, 46
  diversity of, *47*
  as organic molecules, 43
  structure of, *46*
amino group, *46*
ammonia, 40
amnesia, 29, 100
amniocentesis, 251
amniotic egg, 453, *455*
Amoeba, 76, 405, *405*
amoebocytes, 434
Amorphophallus titanum, 418
AM pathway, 135
amphibians, 119, 327, 365, *365,* 368, 451–52, 580
amylases, 113
amyloidosis, *225*
amyotrophic lateral sclerosis (Lou Gehrig disease), *225,* 295
Anableps anableps, 323, *323*

anabolic steroids, 53
anabolism, 110, *110,* 125, 140
anaerobic exercise, 140
anaerobic heterotrophs, 390
anaerobic processes, 138–39
analogous, 350
anaphase, *159,* 159–61
anaphase I and II, *164,* 165
Anastasia, Princess, of Russia, *190,* 191, 211, *211*
ancestral species, 348
anchorage dependence, 174, 175
anchorage independence, 175
anchoring junctions, 97
Anderson, Anna, 191, *211*
androgen insensitivity syndrome (AIS), 220
Androsace, 580
anemia, 182
angiogenesis, 176, 183
angiosperms, 412, *416,* 416–17
anglerfish, *457, 519*
animal behavior, 506–22
  communication, 515–17
  fixed, 510–11
  genetic basis of, 509–10, 521
  learned, 511–13
  mating, 517–19
  and nature of behavioral responses, 508–10
  social, 513–15
  Toxoplasma gondii's effect on, 525, 542
animal cells
  cell junctions in, 97, 98, *98*
  cytokinesis in, 161, *161*
  microtubules of, 76, *77*
  organization of, *69*
  size of, 64
Animalia, 20, *20, 373, 404,* 430–62
  characteristics of, 433–38
  defined, 20, 432
  evolutionary origins of, 432, *433*
  first invertebrates, 438–40
  in human biome, 398
  protostomes, 440–50
animal products, cancer risk and, 185
animals, *456–57*
  apoptosis in, 155
  asexual reproduction in, 15
  behavior in, 16
  biological hierarchy in, 20, 22
  characteristics of, 433–38
  chocolate and, 71, *71*
  consumers, 16
  cyclopia in, 276
  energy capture by, *16*
  experiments on, 24
  genetically modified, *289*
  hibernation by, 119
  history of, 364, 365
  in marine biome, 481, 483
  mass extinctions of, 366, *367,* 368
  meiosis in, 152
  as multicellular, 63
  originated by polyploidy, 347
  origin of, 431
  plantimals, 281, 304
  pollination by, 416–17
  reproductive cloning of, *291,* 291–92
  seed dispersal by, 417
  sexual reproduction in, 15
  somatic cells of, 152
  sympatric speciation in, 347
  in terrestrial biomes, *see also individual biomes*
  triglycerides in, 50
Annelida, 441
annelids, *441,* 441–42
Antarctica, *356,* 357, 366, 374, *374*
anteaters, 327
antenna complex, 131
anthozoans, 439
anthropogenic climate change, 14, 577, 578
anthropomorphism, 512
antibiotics, *394,* 394–98
antibodies, 183, *183*
anticodon, 269
antioxidants, 185, *185*
Antithamnion plumula, 411
ants, 515, *515,* 528
aphids, 61, 491, *491*
Aphytis chrysomphali, 533, *534*
Aphytis lignanensis, 533, *534*
apolipoprotein A-1, 101
apolipoprotein B, 101
apolipoproteins, 96, *97*
apoptosis, 155
appendages, 438, 445
apple maggot fly (*Rhagoletis pomonella*), 347
applied research, 8
aquaporins, 90, 93
aquatic biomes, *21,* 22, 477–83
  freshwater, *480,* 480–81
  marine, 481–82, *481–83*
  oceanic region, 482, 483
aquatic ecosystems
  and BP oil spill, 547, 564
  carbon acquisition in, 556
  eutrophication in, 559–60, *560*
  fertilizer runoff in, 557
  nutrient loading in, *561*
  sulfur compounds in, 558
aquatic food webs, 552
aquatic protists, movement of, 78, *78*
aquifers, 581
arachnids, *445,* 448
Archaea, 20, *20,* 383, *404*
  defined, 20, 381
  diversity in, *386–87*
  on evolutionary tree, *383*
  gene transfer in, 389
  see also prokaryotes
archaeans
  in abyssal zone, 483
  biosphere role of, 391
  extremism of, 383, *384*
  gas-producing, 391
  of human microbiome, 379
  movement of, 78
  reproduction in, 150
  structure of, 384, 385
Archaeopteryx, 455, *458*
Arctic ecosystems, 579
Arctic ice cap, *568,* 569, 578, *578,* 579
Arctic sea ice, 585
Arctic tern, *456*
Arctic tundra, 22, 425, 472, 473
Argentina, 288, 474
Argentinosaurus, 453
Aristotle, 310
Armillaria ostoyae, 422
arsenic, 23
arsenic bacteria, *2,* 3, 23
artemisinin, 290
arthritis, 226, 295
Arthrobotrys anchonia, 423
arthropods, 444–48
  arachnids, 448
  crustaceans, 445–46
  defined, 444
  of human microbiome, 379
  insects, 446–48, *446–48*
  myriapods, 448
artificial selection, 17, *17, 192,* 329
artiodactyls, 361
asbestosis, 74
ascomycetes, 419
-ase (suffix), 266
asexual reproduction, 150, 151, *151*
  defined, 15, 150
  in eukaryotes, 406–7, *407*
  in fungi, 421–22, *422*
  mitosis, 151–53, *152, 153,* 157, *159,* 159–61, *161, 162*
  in plants, 412
  in prokaryotes, 383, 388
  using binary fission, 151, *151*
Asia, 288, 474, 477, *572*
Asian carp, *420*
Asian elephants, *457*
Aspergillus, 424
asses, 326
asthma, 226, *226,* 398
atavistic traits, 370, *371*
atmosphere
  carbon dioxide in, 556, 575–77, *576*
  CFCs in, 574
  greenhouse gases in, 576
  sulfur in, 560
atmospheric cycle, 555, 557, 558, *558*
atomic mass number, 31–32
atomic number, 31
atomic structure, 30–33, *31*
atoms, 20, *21,* 30, *65*
ATP synthase, 132, 141, *142*
Australia
  Ecological Footprint in, *497*
  Opuntia cacti in, 493, *493,* 494, 526

predator-prey relationships in, 467–68, *468*
Austria, 420
autism spectrum disorders, *226*
autosomal inheritance, *227,* 227–29, *228*
autosomes, 219, *219,* 231, 232
autotrophs, 16, 388, 389, 409
see also producers
Avastin, 183
Avery, Oswald, 242
avian influenza virus, 396
Azolla, 557

bacilli, 384
Bacillus anthracis, 388
Bacteria, 20, *20,* 383, *404*
defined, 20, 381
diversity in, *386–87*
on evolutionary tree, *383*
gene transfer in, 389, *389*
see also prokaryotes
bacteria
anaerobic, 138
and antibiotics, 394
arsenic, *2,* 3, 23
asexual reproduction of, 388, *388*
biofilms of, 385, 386
biosphere role of, 391–92
cancers caused by, 183, 184
cell division in, 151
as cells, 15
direction behavior in, 16
disease-causing, 392–94
fermentation by, *392*
gene expression in, 254–55
in human biome, *378,* 379, 383, 398, 399
and insects, 527
movement of, 78
in nitrogen fixation, 556, 557
organization of, *69*
and parasitism, 61
photoautotrophs, 389
photosynthesis in, 108
photosynthetic, 126, *390*
plasmids in, *283,* 283–84
population growth for, 496
procapsule of, 68
as producers, 16
reproduction in, 150, 151
size of, 64
spores of, 90
structure of, 384, 385
swarming by, 385
bacterial conjugation, 389
bacterial culture, 384
Bacteroides, 398
bald eagle, 499, *499*
ball-and-stick model (molecules), 35, *35*
ballast water (ships), *464,* 465, 484
ballot measures, science-related, *8*
bananas, 254
Banff National Park, Alberta, *473*
baobab tree (*Adansonia rubrostipa*), *418*
bar-headed goose, 141
barnacles, 529, *529,* 533, *533*
basal cell skin cancer, 177
basal metabolic rate (BMR), 118, *118,* 119
base pairs, 245, 246, 248
bases, 40, 41, 245, *246*
basic research, 8
basidiomycetes, 419
Bateson, William, 262
bats, 349, *349*
beach mice, 318, *319*
Beagle, 309, 311, *311*
beard grass, 532
beer brewing, 113
bees, 254, 427, 515, *517*
behavior, 16, 508
see also animal behavior
behavioral adaptations, 322
behavioral mutualism, 528
behavioral responses, 508–10
Belyaev, Dmitry, 507, 520
benign tumors, 174
benthic zone, 482, *483*
benzopyrene, 185
beta-carotene, *130*
beta sheet, 46, 47, *48*
bighorn sheep, 518, *518*
bilateral symmetry, 436
bile salts, 52
Bill and Melinda Gates Foundation, 144
binary fission, 151, *151,* 383, 388, *388*
bioaccumulates (term), 573
bioaccumulation, 573–74
biodiversity
in coastal regions, *482*
defined, 381
loss of, 570
threats to, 420
value of, 382, *382*
see also diversity
biofilms, 385, *388*
biogeochemical cycles, 549, *555*
carbon cycle, 555–56, *556*
defined, 554
in ecosystems, 554–59
humans' impact on, 560, *561, 562*
nitrogen cycle, 556–57, *557, 558*
phosphorus cycle, 558, 559, *559*
sulfur cycle, 557, 558, *558*
biological evolution
defined, 17, 18
by natural selection, *18*
through artificial selection, 17, *17*
and unity/diversity of life, 17–20
biological hierarchy, 20–24, *21, 467*
biological species concept, *341,* 341–42
biology
biological evolution and unity/diversity of life, 17–20
biological hierarchy, 20–24
characteristics of living organisms, *14,* 14–17
defined, 4
and Human Genome Project, 257
biomagnification, 573
biomass, 469, 550, 551
biomes
aquatic, 477–83
in biological hierarchy, 21, *21*
defined, 22, *467,* 471
distribution of, *472*
terrestrial, 471–77
biomolecules
in biological hierarchy, 20
biosynthetic pathways of, 110
carbohydrates, 43–45
defined, 20, 30
lipids, 49–53
nucleic acids, 53–55
organic, 43
proteins, 45–49
see also specific types of biomolecules, e.g.: enzymes
bioremediation, 392, *393*
biosphere
aquatic biomes, 477–83
in biological hierarchy, 21, *21*
climate's effect on, 468–71
defined, 22, 466, *467*
human demand on, *502*
pollutants in, 574
prokaryote roles in, 391–92
terrestrial biomes, 471–77
see also ecology
biosynthetic pathways, 110
biotechnology, 286
see also DNA technology
biotic communities, 549, 555–57
consequences of climate change for, *578,* 579
defined, 549
nitrogen fixation in, 556, 557
birds, 454–55
climate change and, 5, *6,* 580
competition among, 534
courtship behaviors in, 510
as diversity indicator, 543
endangered, 499, *499*
extinct, 454
of Galápagos Islands, 333, *333,* 334, *334*
migratory, *6*
mobbing by, 517
problem solving by, 512–13, *513*
birth weight, 321, *321*
bisphenol A (BPA), 277, 574, 575
bisphenol S (BPS), 277
bivalent homologues, 164
bivalves, 442, 443, *443*
blackbirds, 324, 325
black ghost electric fish, 112, *112*
black marlin, *456*
black swans, *10*
black-tailed prairie dogs, *516*
black widow spiders, *518*
Blanco, Marina, 119
blastocyst, 293

blastopore, 435
blastula, 435
Blaszyk, Amy, 56
blending inheritance, 19
blood
  cholesterol in, 29
  as isotonic solution, 92
  pH of, 41
  types of cells in, 66
blood-brain barrier, 100
blood fluke, 440
blood groups, 195, 204, 205, *206*
blood supply, 140, 176
blood transfusions, 205, *205*
blood vessels, 66, 176
Bluebird Bio, 305
blue jays, 530
blue whale (*Balaenoptera musculus*), *348, 456,* 518
body cavities (animals), 437, *438*
body plan, 433
  annelids, *441*
  chordates, *449*
  insects, *447*
  mollusks, *442*
  sponges, 433–34, *434*
  symmetry in, 436, *436*
body temperature, *17*
Boivin, Michael, 144
Bolivia, 477
bone, 68
bonobos, *242,* 243
bony fishes, 450, 451, *453*
boreal forest, *473,* 473–74
botulism, 393
bovine growth hormone (BGH), 300
bovine spongiform encephalopathy (BSE), 272
box turtles, 453
BP oil spill, *546,* 547, 564
brachydactyly, 224, *224*
Bracken Cave, 494
brain
  and BPA/BPS, 277
  cell cycle in, 155
  energy used by, 105
  nerve cells in, 100
  protein clumps in, 272
  Toxoplasma gondii in, 525, 542
brain cells, *84,* 100
Braudrick, Christian, 426
Brazil, 288, 477, *497*
BRCA1, 179, 187, 193
BRCA2, 179, 187
breast cancer, *175,* 187
  cell of, *174*
  estrogen and, 185
  familial, *225*
  5-year survivability of, 181–82
  inherited, 179
  metastasis of, 177
  physical activity and, 186
  prevalence of, 175
  and working night shifts, 185
breed standards, 197
brittle stars, 449
bromelain, 113, 382
Brontotherium, 348
Brown, Charles, 335
Brown, Eryn, 399
Brown, Mary Bomberger, 335
brown bears (*Ursus arctos*), 341, 351
brown bee orchids (*Ophrys fuscal*), *418*
brown recluse spiders, 580
brown tree snakes, 536
bryophytes, 412–14
buffers, 41
Buffon, George-Louis Leclerc de, 311
bundle-sheath cells, 135, 136
burgers, 56
bush encroachment, 585

cacti, 137, 351, *351, 530*
  photosynthesis in, 134
  prickly pear, 493, *493,* 494, 526
Cactoblastis cactorum, 493, *493,* 494, 526
Caenorhabditiselegans, 444, *444*
caffeine
  in plants, 427
  reaction time and, 13, *13*
calcareous sponges, 439
California, 29, 288, 382, 475, *475,* 476
calories, 106, 118
Calvin cycle, 126, 131, 132, *133,* 134, *134*
Cambrian, *362,* 364
Cambrian explosion, 364, *364*
Cameroon crater lakes, 347
camouflage, *531*
CAM pathway, 135
CAM plants, *136,* 137
campylobacter, 393
Canada
  acid rain in, 562
  biomes in, 472, 473, 475
  Ecological Footprint in, *497*
  GMO crops in, 288
  storms in, 470
  in vitro fertilization in, 169
Canadian lynx, 531, *531*
cancer, 172–88
  avoiding risk of, 183, *184,* 185, 226
  and BPA, 574
  cancer-critical genes, 178–81
  cell division control and, 175–78, *178*
  as complex-trait disease, 226
  conventional therapies for, 182
  cost of diagnosing/treating, 175
  cyclopamine treatment for, 276
  deaths from, 174, *175*
  defined, 155
  development of, 175–76, *176,* 180, *181*
  and gene mutation, 224
  hereditary and nonhereditary, 179
  immunotherapy for, 183
  initiation of, 155, 156, 175, *176*
  and lipid consumption, 51
  metastasis of, 176–77, *177*
  and multivitamin use, 24
  and mutations within cells, 174
  new therapies for, 182
  physiology and diet affecting, 185–86
  stage designations for, 181
  treatment and prevention of, 181–86
  types of, 175, *175*
  viruses causing, 183, 184
  see also specific types of cancer
cancer cells, 156, 176–78, *177*
cancer-critical genes, *178,* 178–81
cancer immunotherapy, 183, *183*
Canidae, 368
Canis lupus familiaris, 329, *329*
canola, 288, *288,* 300
capsule (prokaryotes), 68, 385
carbohydrate-deficient glycoprotein (CDG) syndromes, 74
carbohydrates, 43–45
  defined, 43
  endurance athletes' use of, 123
  and metabolism, 110
  from photosynthesis, 126
  transport of, 71–72
carbon (C), 30, 32, 42, 108, 109, *109*
carbon atoms, 20, *42*
carbon capture, 583
carbon cycle, 555–56, *556,* 575–76
carbon dioxide (CO_2)
  in atmosphere, 556, *575–77, 576*
  in Calvin cycle, *134*
  in ecosystem energy flow, 109, *109*
  isotopes of, 33
  in photorespiration, *135*
  in photosynthesis, 111, 126, 135
  and plant growth, 585
  simple diffusion of, 89
carbon fixation, 132
carbonic anhydrase, 114, *115*
Carboniferous, *362,* 453
carbon monoxide, 144
carboxyl group, *46*
carcinogens, 183–85
cardon, 134
caribou, 579
Carnivora, *373*
carnivores, 550
carotenoids, *130*
carp, *456*
carrier proteins, *94,* 94–95, *95*
carrying capacity, *495,* 496
  defined, 495
  and human population growth, 500–502, *501*
  and insect control, 500
  and quality of life, 503
cartilaginous fishes, 450, *452*
cassava, 144
cata- (prefix), *112*
catabolism, 110, *110,* 112, 125, 140
catalysis, 112
catalysts, 112, *116*

catalytic reactions, 112
catfish, *457*
cats, 511, 520, *524*, 525, 542
Catskill Watershed, 563, *563*
cattle, 534, 541
causation, 9–13
Cedars-Sinai Medical Center, 305
cell(s), *15*, *62*
in biological hierarchy, 20, *21*
defined, 15, 20, 62
fungal, 419
in human body, 15
size of, 64–65, *65*
water content of, 36
see also animal cells; plant cells
cell crawling, 77, *78*
cell culture, 293
cell cycle, 153–56, *156*, 161
cell cycle regulatory proteins, 155, 156
cell death, 155
cell differentiation, 153, 174
cell division, 148–70, *154*
asexual reproduction, 150, 151
in bacteria, 151
binary fission, *151*
biological relevance of, *151*
cancer and, 175–78, *178*
cell cycle, 153–56
and chromosomal organization of genetic material, 156–58
defined, 150, 153
in eukaryotes, 151, 152
and growth of algae, 168
in human body, 153
meiosis, 152, *153*, 161–68, *163–67*
mitosis, 151, 152, *152*, *153*
mitosis and cytokinesis, *159*, 159–61, *161*, *162*
in prokaryotes, *151*
reasons for, 150
sexual reproduction, 150–52, *153*
and test tube babies, 169
"cell eating," 97
cell junctions, 97, 433
cell migration, 77–78
cell plate, 161
cell signaling, 99, *99*
cell specialization, 66, *66*
cell structure, 60–82
cytoskeleton, 76–80
eukaryotic cells, 68, *69*, 70–76
limits on cell size, 64–65
microscopes and study of, 63–64, *64*
multicellularity, 65–66
organelles, *62*, 63
plasma membrane, 66–68
prokaryotic cells, 68, *69*
cell theory, 63
cell transport
diffusion, 87–89, *88*, *89*
endocytosis, *96*, 96–97
exocytosis, 95, *95*
facilitated transport, 92–95, *93*, *94*
osmosis, 89–92, *91*
selective permeability, 86, *87*
transport proteins, 86
cellular connections, 97–98, *98*, 100
cellular respiration, 74, 108, 137–41
ATP production in, 110, 127, 138–41, *139*, *142*
in carbon cycle, 556
defined, 74, 111, 127, 137
in ecosystem energy flow, 109, *109*
fermentation in, 138–39, *139*
glycolysis in, *137*, 137–38
Krebs cycle in, 139, 140, *140*
oxidative phosphorylation in, 140–41, *142*
photosynthesis and, 124
stages of, 127, *128*
cellulases, 113
cellulose, 44, *45*, 413, 414
Cenozoic, *363*
Center for Genome Sciences and Systems Biology, 399
centipedes, 448
central nervous system, 277
centromeres, 157, 160
centrosomes, 160
cephalization, 436
cephalopods, 442
Ceramium pacificum, *62*
Ceratocystis ulmi, 424
cervical cancer, 173, 184
C_4 pathway, 135–36, *136*
C_4 plants, 135
Chain, Ernst, 394
channel proteins, 92–93
chaparral, 475, *475*, 476
character displacement, 534
"characters," 534
Chase, Martha, 242
cheese, 113
cheetahs, 138, 318, *318*, 530, 585, *585*
chemical bonds, 33–36, *34*, *35*, 40
chemical carcinogens, 185
chemical compounds, 33, 37
chemical energy, 106, 107
chemical equations, 39–40
chemical formulas, 33, *35*
chemical reactions, 39–40, 68
see also metabolism
chemistry of life, 28–57
atomic structure, 30–33
building blocks of life, 41–43
carbohydrates, 43–45
chemical bonds, 33–36
chemical reactions, 39–40
elements, 30, 32, 33
lipids, 49–53
matter, 30
nucleotides and nucleic acids, 53–55
pH scale, 40–41
properties of water, 36–39
proteins, 45–49
chemoautotrophs, 389, 390, *391*
chemoheterotrophs, 388, 390–92, *393*
chemolithotrophs, *387*
chemotherapy, 182, *182*
childhood cerebral adrenoleukodystrophy (CCALD), 305
children
grasp reflex in, 510–11, *511*
konzo in, 144
metabolism in, 105
Chile, 475
Chilean sea bass, 499, *499*
chimpanzees, *242*, 243, 371, *371*
China, 474, 497, *497*, *572*, 581
chitin, 419, 444
chlorofluorocarbons (CFCs), 574
chlorophyll, 75, 126, 129, *130*
chloroplast envelope, *75*
chloroplasts
defined, 75, 126
energy capture by, 74–75, *75*
as former prokaryotes, 80
in photosynthesis, 129, *129*, 131
photosystems of, 131, 132
structure of, *129*
choanocytes, 434
choanoflagellates, 431, 460
cholesterol, 29, 52
in brain cells, 100
familial hypercholesterolemia, *225*
genetically engineered tomatoes mimicking, 101
HDL, 101
LDL, 96–97, *97*
statins and, 85, 100
Chordata, 18, *373*, 450
chordates, 432, 449, *450*
see also vertebrates
chorionic villus sampling (CVS), 251
Christmas Bird Count, 5, *6*, 14
Christmas tree worm, *432*, 442
chromatids
in anaphase, 160–61
crossing-over of, 166–67
in meiosis, 163
sister, 157–61, 163–65
in tetrads, 164
chromatin, 156
chromosomes
change in number of, 23–232
defined, 70, 156
in diploid cells, 193, *194*
fluorescent photographs of, 156
in gametes, 161–63, 193
in gametes vs. somatic cells, 161–63, *163*
genes located on, *218*, 218–19, *219*
homologous, 158, 162, 218, 349
in human cells, 158
independent assortment of, 221
inherited abnormalities of, 230–33
karyotype of, *157*, 157–58
linked genes on, 221–22
in meiosis, 161–68
in mitosis, 159–61

chromosomes (*cont.*)
organization of genetic material in, 156–58
recombinant, 166, *166*
sex, *see* sex chromosomes
telomeres of, 182
types of changes in, 230–32, *231*
chromosome theory of inheritance, 218–21
chronic diseases, 226, *226*
Chthamalus stellatus, 533, *533*
Church, Julie, 485
cicadas, 461, *461,* 503, *503*
cichlids, *338,* 339, 347, *347,* 351, 352
cigarettes, 185
cilia, 78, *79,* 408
cis unsaturated fatty acids, 51
citizens, science and, 8
Citizen Science projects, 4–5, *5*
citric acid cycle, 140, *140*
clams, 443
class, 18, 372, *373*
cliff swallows (*Petrochelidon pyrrhonota*), 335, *335*
climate
of Antarctica, 374
biomes and, 471, 472, *472,* 478, *see also individual biomes*
and the biosphere, 468–71
and changes in communities, 538, 539
defined, 468–69
and plate tectonics, 366
climate change, 538, *539,* 576–80, 582–83
alleviating, 583
anthropogenic, 14
and bird migration, 5, *6*
and carbon cycle change, 575, 576
and cheetah population, 585
consequences of, *578,* 578–80, 582, 585
defined, 570
global warming, *see* global warming
from land and water transformation, 572–73
in the media, *570*
as threat to species, 420
clones, 406
cloning
DNA, 283, *283,* 284, *284*
reproductive, *291,* 291–92
Clostridium botulinum, 90, 393
Clostridium difficile, 398
Clostridium tetani, 388
clownfishes, *457,* 543
Cnidaria, 438
cnidarians, 434–36, *435,* 439–40, *440*
cnidocytes, 440, *440*
coastal ecosystems, 571–72
coastal region, 481, 482, *482*
cocci, 384
coccolithophores, 409, *409*
CODIS (Combined DNA Index System), 287
codominance, 204, 206, 207
codons, 267, 268, *268*
coelacanth, 375, *375,* 452
coelom, 437, *438*
coelomate animals, 437, *438*
coenzyme A (CoA), 140
coevolution, 527
cofactors, 114
cohesion, 39
cold-blooded animals, 119
collaborators, *387*
collagen, 56, 67, 113
Collins, Francis, 257, *332*
colon cancer
avoidable risk of, 226, *226*
diet and, 186
familial, *225*
5-year survivability of, 181–82
metastasis of, 177
from multiple mutations, 180, *181*
prevalence of, 175
colonoscopy, 180
color, *128,* 128–29
as adaptation, 318, *319*
by directional selection, 320, *320*
in plants, 130, *130*
colorectal cancer, *175,* 179, 183
colossal squid, *456*
comb jellies, 438
commensalism, 529
communication, 515–17
communities, 524–44
in biological hierarchy, 21, *21*
biotic and abiotic, 549, *555*
change over time in, 536–38, *539*
commensalism in, 529
competition in, 532–34
defined, 22, *467,* 526
in ecosystems, 548
exploitation in, 529–32
human impacts on, 539–41
invasive species in, 535–36
on islands, 532
keystone species in, 535, *536*
mutualism in, 527–29
species diversity factors in, 534–36
species interactions in, *526,* 527–34
competition, 527, 532–34
competitive exclusion, 533
complete metamorphosis, 447
complex-trait diseases, 226, *226*
complex traits, 210
compound eyes, 446
concentration gradient, 88, *89,* 132
concept maps, 5, *7*
Condemi, Silvana, 353
conditioning, 512
conflict avoidance, 516–17
congenital generalized hypertrichosis (CGH), 230, *230*
congenital hypertrichosis, *371*
conifers, 473–74
consumers, *16,* 108, 389, 530–32
cellular respiration and, 126
defined, 16, 549
in defining ecosystems, 550
and ecosystem energy flow, 108, 109, *109,* 549, 553
see also heterotrophs
continental drift, *539*
continents, 330, *330,* 365–66, *366–67*
continuous variation, 210
control group, 12
controlled experiment, 12
convection cell, 469, *470*
convergent evolution, 350, *351*
cooking, 56, 185
Copeland, Amy, 393
Copper Age man, *240,* 241, 256, *256*
Copy Number Variables (CNVs), 257
coral bleaching, 579, 585
coral reefs, 478, 552, 576, 585
corals, 438, 439, 580
Cordyceps, 424
Coriolis effect, 469–71
corn, genetically modified, 288, *288,* 299
corn lilies, 276, *276*
coronary heart disease, 226, *226*
correlation, *11,* 11–12
cortisone, 177
Cotter, Edward, 426
cotton, genetically modified, 288, *288,* 299
covalent bonds, 33–35, *34, 35*
cowbirds, 510, 511, *511*
crassulacean acid metabolism (CAM), 137
Cretaceous, *363, 367*
Cretaceous extinction, 368, *368*
Crick, Francis, *245,* 245–47
cri du chat, 231, *232*
cristae, 73, 141
Crohn disease, 295
crossing-over, 165, 166, *166,* 220–23, *222*
Croton Watershed, 563, *563*
crowdsourcing, 5
crows, 512–13
crustaceans, *445,* 445–46
cryosurgery, 182
Cryphonectria parasitica, 424
Ctenophora, 438
ctenophores, 439–40
C_3 plants, 135–36, *136*
curlews, *535*
cuticle, 413, 444–45
Cuvier, Georges, 311
Cuvier's beaked whales, 143
cyanide, 144
Cyanidinium caldarum, 409
cyanobacteria, 389, *390,* 391
cyclopamine, *276*
cyclopia, *260,* 261, 276
Cymothoa exigua, 530, *530*
cystic fibrosis, 93, *224, 225,* 226
cystic fibrosis transmembrane conductance regulator (CFTR), 93
cyt (term), 96
cytokinesis
in animal cells, 161, *161*
cytoplasm division during, 161

defined, 152, 161
in meiosis, *164,* 165
in mitosis, *159,* 159–61, *161, 162*
in plant cells, 161, *162*
stages of, *158–59*
cytoplasm
chemical components made in, 71
chemical composition of, 86, *87*
chemical reactions in, 66
during cytokinesis, 161
defined, 15, 63
in plants, 73
solutes in, *37*
cytosine, 53
cytoskeleton, *76–79,* 76–80
cytosol, 63, 86, *87,* 127

Damovsky, Marcy, 81
Darwin, Charles, 19, 309, *311,* 311–12, 318, 333, 334, 359, 520
data, 9
daughter cells, 150–53, 161
see also cell division
Davies, Neil, 426
DDT, *31,* 89, 313–14, 499, *499,* 574
dead zones, 559, 560
death cap mushroom (*Amanita phalloides*), 265, *423*
deaths, from cancer, 174, *175,* 177
deciduous trees, 474
decomposers, *387*
in carbon cycle, 556
defined, 391, 549
and ecosystem energy flow, 109, 549, 554
energy for, 550
fungi, 422
in nutrient cycles, 555, *555*
deep-sea vents, 550
Deepwater Horizon oil spill, *546,* 547, 564
deer, 565, *565*
deer antlers, 160
deer mice (*Peromyscus Maniculatus*), 318, *319,* 492, 521, *521*
deforestation, 454, 480, 560
in the Amazon, 573
on Easter Island, 489, 500–502
halting, 583
dehydration reaction, 44
Deinococcus radiodurans, 249
deletion (chromosomes), 230, *231,* 272
deletion mutations, 271
Democratic Republic of Congo, 144
Denali National Park, Alaska, *473*
denaturation, 47, *47,* 49
density-dependent growth, 496
density-independent growth, 496
dental plaque, 384, *387*
dependent variable, 12
desert(s), 476, *476, 550,* 551
desertification, 541
DeSilva, Ashanthi, 297
desmosomes, 97
desmosponges, 439
Desulfotomaculm, 387
deuterostomes, 432, 435, 448–50
development, 15
developmentally regulated genes, 244
Devonian, *362,* 364
diabetes, 226, *226,* 277, 295, 574
diatomaceous earth, 412
diatoms, 409, *409,* 412
diet
cancer risk and, 185–86
and microbiome, 399
dietary lipids, 51
differential reproduction, *18,* 312
diffusion, 87–89, *88, 89, 93*
digestion, enzymes in, 113
digestive system, 399, 528
dihybrids, 200–202, *201*
dingoes, 467, 468, *468*
dinoflagellates, 412, *412*
dinosaurs, 365, 453, *453,* 455
diploid, 162
diploid cells, 193, *194*
diploid set, 152
directional selection, 318, 320, *320*
direct-to-consumer (DTC) genetic testing, 234
disaccharides, 44, *44*
disease(s)
bacterial, 392, 393
complex-trait, 226, *226*
defined, 226
germ theory of, 13–14, *14*
and lipid consumption, 51
mitochondrial, 81
and multivitamin use, 24
organelles and, 74
stem cell therapy for, 295
see also specific diseases
disruptive selection, *320,* 321, *321*
dissociation, 40
diversity, *380,* 380–82, *527*
biological evolution and, 17–20, 349–50
bird, 543
components of, 526
defined, 526
in marine biome, 481, 482
and mass extinctions, 268
species, 349–50, 381, 534–36
see also biodiversity; genetic diversity
DNA (deoxyribonucleic acid), 70, 240–58
of arsenic bacterium, 3
and burrow construction by mice, 521, *521*
and cancer risk, 179
and cellular organization, *15*
of chloroplasts and mitochondria, 80
in chromosomes, 156, *156,* 157
and chromosome theory of inheritance, 218
damage to, 181
defined, 16, 53, 174, 243
in eukaryotes vs. prokaryotes, 250–51, *252*
function of, 246
genetic information carried by, *15,* 16, 243
and genetic traits, 193
genome organization, 250–53
Human Genome Project, 257
identifying descendants by, 211
length of, in human body, 253
mitochondrial, 81
of Ötzi, the Iceman, 256
packing of, 253–54, *254*
and patterns of gene expression, 254–56, *255*
and phenotypes, *243,* 243–44
phosphorus needed for, 23
repairing, 248–50, *249, 250*
replication of, 154, 157, 247, *247,* 248, *248,* 264
reproduction via, 15–16, *see also* reproduction
in resurrecting life-forms, 4, *4*
sequence of bases in, *246*
Stanford students' study of, 234
structure of, *245,* 245–46, 262–64, *263*
DNA capture, 388, 389
DNA cloning, *283,* 284, *284*
DNA fingerprinting, *286,* 286–87, *287*
DNA hybridization, 302
DNA ligase, 301
DNA microarray, 285
DNA polymerase, 247, 248
DNA primers, 302
DNA probes, 302
DNA profile, 284, 286, 287
DNA repair, 248–50, *249, 250*
DNA replication, 154, 157, 247, *247,* 248, *248,* 264
DNA segregation, 159
DNA sequence, 246, *246*
as evidence for evolution, 328
mutations altering, 271, *271*
mutations from change in, 248
DNA sequencing, 302, *302,* 303
DNA synthesis, 302, 303
DNA technology, 280–306
defined, 282–83
DNA fingerprinting, 286–87
ethical and social dimensions of, 299–300
genetic engineering, 287–91
human gene therapy, 297–99, 305
major developments in, 284–85
plantimals, 281, 304
reproductive cloning, 291–92
stem cells, 292–97
strategies and techniques of, 282–84
tools of, 300–303
dogs, 194–95, *195,* 197, 329, *329,* 507, 516, 520
Dolly the sheep, 291, *291*
dolphins, 351, *351*
domains, 19–20, *20,* 372, *373,* 381, *381*
domesticated animals, *506,* 507, 520
dominance, incomplete, *203,* 203–4, *204,* 207
dominant alleles, 195, *196,* 198
dominant genetic disorders, *228,* 228–29
dopamine, 542

double-blind experiment, 13, *13,* 24
double bond, 34, 35
double helix, 16
dowitchers, *535*
down-regulated gene activity, 244
Down syndrome, 231, 232
dragonflies, *448*
Duchenne muscular dystrophy, 229, 230, 264
ducks, *513*
duckweed (*Wolffia globosa*), 417, *418*
duplication, 231, *231*
Dutch elm disease, 424
dyad, 163
dystrophin gene, 264

e- (prefix), 491
Earth
  age of, 359
  anthropogenic climate change on, 14
  average temperatures of, *6,* 13
  carbon in crust, 556
  change in chemistry of, 573–74
  chemical composition of crust, *31*
  elements needed for life on, 23
  water on, 30
earthworms, 441, 442
Easter Island (Rapa Nui), *488,* 489, 500–502, *501*
eastern meadowlark, *340,* 341, 342
Ebola virus, 395
ecdysozoans, 444
echinoderms, 448
eco- (prefix), 466
Ecological Footprint, 497, 502, 581
ecology, 464–86, *466*
  aquatic biomes, 477–83
  biological hierarchy in, *467*
  climate's effect on biosphere, 468–71
  defined, 466
  and interconnected web, 466–68
  invasive species, *464,* 465
  terrestrial biomes, 471–77
ecosystem(s), *22,* 546–66
  biodiversity and, 381, 382
  biogeochemical cycles in, 554–59
  in biological hierarchy, 21, *21*
  defined, 22, *467,* 548
  and DNA technology, 299–300
  energy capture in, 550–52
  energy flow through, 108–9, *109,* 552–54
  in freshwater biome, *480,* 480–81, *481*
  functioning of, 548–50
  fungi' role in, 419, 421
  humans' impacts on, 559–62
  interconnectedness of, 571
  in marine biome, 481–82, *482*
  in overpopulated areas, 497
  plants' significance to, 417
  species interactions in, 534–36
ecosystem processes, 550
ecosystem services, 382, 563
ectotherms, 119, 454
Edunia, *280,* 281, 304
eggs (animal), 64, 65, 163, 168
eggs (female cells), 15, 152, 162, *165,* 169
Einstein, Albert, 10
electrical energy, 106
electromagnetic spectrum, *128,* 129
electronegativity, 36
electrons
  and chemical behavior of atoms, 33
  defined, 30
  in electron transport chains, 131, 132
  ionic bonds, *35,* 35–36
  shared, 33, 34, 37
electron shells, 31, *34,* 34–35, *35*
electron transport, 144
electron transport chain (ETC), 131, 132, *133,* 141, *142*
elements, 30, 32, 33
  atomic mass number of, 31–32
  chemical bonds of, 33–36
  defined, 30
  isotopes of, 32–33
elephant seals, 324, *324*
Ellis, Erin, 169
El Niño, 478, *478*
elongation (transcription stage), 265
elongation (translation stage), 269, 270, *270*
El Yunque National Forest, Puerto Rico, *477*
embryo
  of animals, 435, *435*
  defined, 152
  development of, 153
  evidence of evolution in, 326–27, *327*
  plant, 412
embryonic stem cells, 293, *294,* 296, *297*
emigration, 491, 492
Encephalitozoon cuniculi, 251
endangered species, 292, 345, *345,* 499, *499*
endemism, 345, *345*
endo- (prefix), 96
endocrine disrupter, 574
endocytosis, *96,* 96–97
endoderm, 435
endoplasmic reticulum (ER), 71, *71,* 74
endoskeletons, 77
endosymbiosis theory, 80
endothelial cells, 66
endotherms, 119, 455
endotoxins, 393
energy
  for active transport, 94
  alternative sources of, *583*
  in ATP, 110
  cells' need for, 106
  defined, 106
  flow of, 108–9, 549, *549,* 552–54
  forms of, 106–7
  and laws of thermodynamics, 107–8
  in metabolism, 106–9
  see also light energy
energy barrier, 114–15
energy capture, 16, *16,* 74–75, *75,* 550–52
energy carriers, 124–26, *125*
  defined, 124
  from Krebs cycle, 139, 140, *140*
  from light reactions, 131, *131*
  in light reactions, *133*
energy flow, 108–9, 549, *549,* 552–54
energy pyramid, 553, 554
energy storage, 50
energy transport (nucleotides), 53, 54, *55*
England, 320
Ensatina ring species, 346, *346*
entomophagy, 503
entropy, *108*
environment
  cancer risk factors from, 183, *184,* 185
  chemical carcinogens in, 185
  and complex-trait diseases, 226
  and effects of genes, 208
  energy capture from, 16, *16*
  and gene expression, 254–55, *255*
  responding to, 16
  worldwide change in, *see* global change
enzymes, 71
  in Calvin cycle, 132
  as catalysts, 112, *116*
  defined, 46, 112
  disease and, 74
  in DNA technology, 300–301, *301*
  lysosomal, 72–74
  metabolic pathways and, 110
  in metabolism, *112,* 112–15, *114, 115*
  names of, 266
  properties of, *112*
  restriction, 283
  shapes of, 113–14
  of smooth ER, 71
  telomerase, 182
  in vacuoles, 73
epistasis, *207,* 207–8, *208*
epithelial cells, *244*
Epulopiscium fishelsoni, 386, 406
Equus, 326, 330
Escherichia coli (*E. coli*), 68, 254–55, 272, *378,* 383, *388,* 398
esophageal cancer, 186
estrogen, 52, 53, 177, 185
estuaries, 478, 481, *481,* 552
Euglena, 406, *406,* 408, 412
Eukarya, 20, *20,* 404, *404*
  Animalia, 430–62
  defined, 20, 381
  on evolutionary tree, *383*
  Fungi, 419, 421–26
  origin of, 404–7
  Plantae, 413–19
  Protista, 407–12
eukaryotes
  ATP for, 139
  cell cycle in, 153
  cell division in, 151, 152
  defined, 68
  DNA packing in, 253–54, *254*
  evolution of, 80, *80*

first appearance of, 362
gene expression in, *255,* 255–56, *274*
genome of, 249–51, *250, 252*
movement of, 78, *79*
multicellular, 407
and parasitism, 61
photosynthetic, 126
reproduction in, 150–52, 406–7
size of, 64
termination of transcription in, 266–67, *267*
types of, 63
eukaryotic cells, 68, *69*
cytoskeleton of, 76–78, *76–79*
internal compartments of, 68, 70–76, 406
organization of, *69*
size of, 405–6
eukaryotic flagella, 78, *79*
eumelanin, 212
Euplotes, 411
Europe, 472–74, *572,* 579
European starlings, 514
eutherians, 458, 459, *459*
eutrophication, 559–60, *560*
Evans, Barry, 461
evaporation, 38
evaporative cooling, 38, 39
evo-devo, 369
evolution, 308–36, 531
adaptations, 321–24
adaptive, 316, 318–21
of animals, 431, 432
of eukaryotes, 80, *80*
evidence for, 325–31, *327–31*
of flowers, 416
impact of concept, 331–32
mechanisms of, 312–18
of multicellular organisms, 460
by natural selection, 310–12, 316, 318–21
of plants, 412
and religious beliefs, 331–32, *332*
sexual selection, 324–25
from species interactions, 527
testing populations for, 317
theory of, 14
and types of photosynthesis, 134–37
and unity/diversity of life, 349–50
of vertebrates, 450
evolutionary arms race, 530
evolutionary developmental biology, 369
evolutionary history, 356–76, *362–63*
colonization of land, 364–65
first single-celled organisms, 361–63
fossil record, 359–61
of fungi, 419
macroevolution, 358–59, 368–71
mass extinctions, *269,* 366, 368
multicellular life, 363–67
phylogenetics, 371–73
and plate tectonics, 365–66, *367*
of protists, 408
evolutionary trees, 371, 372, *372*
Animalia, *433*
domains, *383*
Fungi, 419
Plantae, *413*
six kingdoms of life, *404*
vertebrates, *451*
ex- (prefix), 491
exchange pools, 555
excurrent siphon (bivalves), 443
exercise
aerobic, 140, 143
anaerobic, 140
cancer risk and, 185–86
extreme running, 106
lactic acid formed in, 139
metabolism and, 105, 118
strength training, 140
exo- (prefix), 96
exocytosis, 95, *95*
exons, 252, *253,* 266, *267*
exoskeletons, 77, 444, 445
exotoxins, 393
experiments, *11,* 12–13, *13*
exploitation, 527, 530–32
exploitative competition, 533
exponential growth (populations), 492–94, *493,* 500, 501, *501,* 503
extinction
and adaptive radiation, 349
defined, 315
and genetic drift, 315
from human activities, 454
on islands, 345, *345*
and limits of adaptation, 323
and resurrection of life-forms, 4, *4*
and sea-level rise, 582
see also mass extinctions
extracellular matrix (ECM), 67, 68, 433
extraterrestrials, 3, 23
extremophiles, 383, *384*
Exxon Valdez oil spill, 547, 564
eyes, 446
eyespots, 434

facilitated diffusion, 92, *93*
facilitated transport, 92–95, *93, 94*
facts, 13
$FADH_2$, 141, *142*
fairy rings, 423
family, 18, 372, *373*
fast-twitch fibers, 275
fats, 29, 50
fat-tailed lemur, 119
fatty acids, *49,* 49–51, 74
fauna, 433
Felidae, 368, *373*
females
cancer chances for, 174
mating behaviors of, 515–19, *518, 519*
meiosis in, 153, 169
toxoplasmosis in, 542
X-linked dominant traits in, 230
fermentation, 138–39, *139,* 392, *392*
fertilization
defined, 15, 152
genetic variation and, 165–68
of human eggs, *165*
internal, 458
random, 221
in vitro, 169, 251, 296, *297*
fertilizers, 90, 557, 559
fetus, cell death in, 155
F_1 generation, 196, *196*
F_2 generation, 196, *196*
finches, 333–34
first law of thermodynamics, 107
fish
evolution of, 375
extinct, 454
in marine biome, 481, 482
methylmercury from, 573, 574
originated by polyploidy, 347
osmoregulation in, 92
overfishing, 499, *499*
and phytoplankton decline, 569
plastic ingested by, 479
fisheries, 499, 584
fishing, 499, 547, 564
fish oil supplements, 12, *12,* 51
fixation, 315
fixed behaviors, 508, 510–11
flagella, 78, *79,* 385, 408
flatworms, 436–37, *437,* 440–41
flavonoids, *130*
Fleming, Alexander, 394
flies, *370*
flora, 433
Florey, Howard, 394
Florida fish poison tree, 144
Florida panthers, 196, 315, *315*
flowering plants, *see* angiosperms
flowers, 406, 416, 427
fluid mosaic model, 68
flu virus, 396–97, *397*
Fogelman, Alan, 101
food
defined, 108
genetically modified, 282, *282,* 288, *288,* 299–300
insects as, 503
metabolism and, 118
overharvested species, 454
and population growth, 503
Food and Drug Administration (FDA), 29, 234, 574, 575
food chains, *549*
bioaccumulation/biomagnification in, 573–74
defined, 553
energy pyramids in, 553, 554
marine, 584
food poisoning, 61
food webs, 552, 553, *553*
foraminiferans, 412
forest ecosystems, *561,* 580

forests
boreal, *473,* 473–74
deer population damage to, 565, *565*
deforestation, 454, 480, 489, 500–502, 583
increasing, 583
management of, 499, 500, *500*
mistletoe in, 543
reforestation, *572*
second-growth, 540
temperate deciduous, 474, *474*
tropical, 477, *477*
fossil fuels, 452, 556, 560, 581
fossil record, 326, *326,* 359–61, 374
fossils, 326, *327,* 359, *360*
animal, 432
in Antarctica, 357, 374
bird, 455
and climate change, 538
defined, 326
defining species of, 342
eukaryote, 405
as evidence for evolution, 311
geographic locations of, 330
founder effect, 315
four-eyed fish, 323, *323*
Fox, Michael J., *296*
foxes, 507, 520
frameshift mutations, 271, 272
Franklin, Rosalind, 245
free radicals, *130*
freshwater, 502, 571, 581
freshwater biome, *480,* 480–81
frogs, 452, 454, *517,* 585
Froome, Chris, 123
fructose, 43, 44
fruit, 416, 417
fruit flies (*Drosophila*), *342,* 343, 345, *345*
fruiting body, 422
fumigated soils, 426
functional groups, 42, 43, *43*
fungal spike, 422
Fungi, 20, *20, 404,* 419, 421–26
adapted for absorbing nutrients, 421
benefits to humans from, 424
as decomposers, 422
defined, 20, 419
in human biome, 398
parasitic, 42, 424
reproduction in, 421–22, *422*
fungi, 63, 383
cell walls of, 68
of human microbiome, 379
in hypotonic solutions, 92
multicellular, 407, *421*
parasitic, 421
spores of, 90
yeasts, 138

galactose, 113
Galápagos Islands, *308,* 309, *333, 334*
Galápagos turtles, *457*
gametes
alleles in, 198–99
chromosomes in, 161–63, *163,* 193, *see also* meiosis
defined, 152
fertilization, 152
produced in adults, 153
gametocytes, 153
gap junctions, 98
"gap" phases, 154
Gardasil, 184
Garrod, Archibald, 262
gastropods, 442
gastrula, 435
gastrulation, 433, 435
Gaucher disease, *225*
Gause, G. F., 495
geese, 512
gelatin, 113
gel electrophoresis, *301,* 301–2
gender determination, 219–20, *220*
gene(s), 192, 242, 260–72
associated with inherited genetic disorders, *225*
cancer-critical, 178–81
on chromosomes, *218,* 218–19, *219*
Copy Number Variables of, 257
defined, 174, 192, *196,* 252, 262
DNA repair and function of, 248–50
in eukaryotes, 252–53
genetic code, 267–68, *268*
and hereditary cancers, 179
mutations of, 165, 166, 178–80, 195–96, 227–30, 270–72
names of, 193
oncogenes, 179
phenotypes from, *243,* 243–44
proto-oncogenes, 179
sex-linked, 229
somatic mutations of, 179, 180
transcription, 264–67
translation, 269–70, *270*
tumor suppressor, 179
on X and Y chromosomes, 158
see also genetics
gene chip, 285
gene cloning, *see* DNA cloning
gene expression, 244
and BPA/BPS, 277
control of, 272–74, *273, 274*
defined, 174, 243, 254, 272
macroevolution through, 368–71
mutations and, 179
patterns of, 254–56, *255*
process of, 243–44
regulation of, 273–74, *274*
tissue-specific, *244*
gene flow, 314, *314, 315,* 343, 344
gene pool, 312–13, 315
gene promoter, 244, 265
gene therapy, *284,* 297–99, 305, 385
genetically engineered organisms (GEOs), *see* genetically modified organisms (GMOs)
genetically linked, 222
genetically modified animals, 289, *289*
genetically modified foods, 101, 282, *282,* 288, *288*
genetically modified organisms (GMOs)
in agriculture and food production, 288, *288*
defined, 285
as DNA cloning application, *284*
ethical and social issues with, 299–300
opposition to, 300
plants, *280,* 281, 304
genetic bottlenecks, 315, *316*
genetic carriers, 227, 228
genetic characteristics, evolution of, 18
genetic code, 267–68, *268*
genetic condition, 227
genetic crosses, 196, *196,* 202–3
genetic disorders, 74, 226
autosomal inheritance of, *227,* 227–29, *228*
defined, 226, 227
and DNA repair, 249, 250
dominant, *228,* 228–29
inherited, 223–26, *225*
prenatal screening for, 251
recessive, *227,* 227–28
sex-linked inheritance of, *229,* 229–30, *230*
see also individual disorders
genetic diversity, 381
and crossing-over, 220–21
of island species, 345, *345*
meiosis and, 165–68
possibilities of, 167–68
through sexual reproduction, 406–7
genetic drift, *315,* 315–16
genetic engineering, *283,* 285, 287–91
applications of, 289–91
defined, 285
methods of product production, *290*
objectives of, 287, 288
see also genetically modified organisms (GMOs)
genetic heritage, 341
genetic information, *15,* 16, 152
genetic linkages, 221–23, *222, 223*
genetic recombination, 166, *166*
genetics, 216–35
and animal behavior, 507, 508
autosomal inheritance of single-gene mutations, 227–29
chromosomes and inheritance, 218–21
of domesticating animals, 507, 520
genetic linkages, 221–23
human genetic disorders, 223–26
inherited chromosome abnormalities, 230–33
patterns of inheritance, 197–99
principles of, 192–96
as scientific field, 192
sex-linked inheritance of single-gene mutations, 229–30

in the Soviet Union, 507
terminology, *196*
see also inheritance
genetic screening, 251
genetic testing, 187, 217, 233, 234
genetic traits, 192–93
complex, 210
defined, 192, *196*
of domesticated animals, 520
influences on, 192, 193
and patterns of inheritance, 197–99
phenotypes of, *193*
single-gene, 195
see also inheritance
genetic variability, *18*
genetic variation, 313, *313*
for adaptation, 323
continuous, 210
defined, 313
and genetic drift, 315
phenotypes, 193
Genghis Khan, 220
genome(s)
of bonobos and chimps, *242*
defined, 243
as DNA-based information, 243
human, 243, 257, 286
Human Genome Project, 257
of Neandertals, *285*
organization of, 250–53
personal genomics, 275
prokaryotic and eukaryotic, 249–51, *250, 252*
sequencing of, 275
similarities of, 243
size and gene number in, *252*
genotypes
defined, 194, *196*
heterozygous, 219
homozygous, 219
and phenotype, 194, 195, *195*
Punnett square of, *199,* 200
genus, 372, *373*
Genzyme, 305
geographic isolation, *343,* 343–46
geologic time, 359, *359, 362–63,* 368, 369
germ line cells, 153
germ theory of disease, 13–14, *14*
Gey, George, 173, 186
Gey, Margaret, 173, 186
giant clam, *457*
giant kelp, 407, *407*
giant sequoia (*Sequoiadendron giganteum*), *418*
Giardia, 408
Giardia lamblia, 398, *410*
Gibling, Martin, 426
gills, 442, 450
glaciers, 578, *579*
glass sponges, 439
glial cells, *84,* 100, *100*
global change, 568–86
in carbon cycle, 575–76
climate change, 576–80, 582–83
defined, 570
in Earth's chemistry, 573–74
land and water transformation, 570–73
and sustainability, 581
global warming, 13, 14, 576–79, *577*
and atmospheric carbon dioxide, 575
and college attendance, *11*
consequences of, 578–80, 582
defined, 570, 577
effects of, 580
melting of Arctic ice cap, *568,* 569
and phytoplankton decline, 584
and sea-level rise, 582
glucose, 43–44
defined, 43
GLUT protein carriers, 94
in glycolysis, 137–38
from lactose, 113
from photosynthesis, 126
GLUT proteins, 94
glyceraldehyde 3-phosphate (G3P), 134, *135,* 138
glycerides, 50
glycogen, 43, 44, *45*
glycolysis, 127, *137,* 137–38
goats, 520
goby fishes, 528, *528*
Golgi apparatus, 72, *72,* 74
Gomez, Danny, *371*
Gomez, Larry, *371*
Gondwana, 374
Gordon, Jeffrey, 399
Gorman, James, 427, 521
Gould, John, 333
G_0 phase, 154, 156
G_1 phase, 154–56, *156*
G_2 phase, 154, 155, *156*
Grant, Peter, 333
Grant, Rosemary, 333
grasp reflex, 510, 511
grasslands, 474–75, *475*
defined, 474
net primary productivity in, 551
overgrazing of, 534, *540,* 541, *541*
species interactions in, 534
Graveline, Duane, 85, 100
gray whales, *529*
Great Britain, 562
greater prairie chickens, 315, 316, *316*
Great Lakes, 465, 484
Great Pacific Garbage Patch, 479, *479,* 485
Great Plains, 475, *475*
Great Salt Lake, 92
green fluorescent protein (GFP), 288, *289,* 301
greenhouse effect, 576, *577*
see also global warming
greenhouse gases, 576
Griffith, Frederick, 242
grizzly bear (*Ursus arctos*), 341, *348*
ground finch, *308,* 309, 333–34, *334*
groundwater, 581
group living, 513–15, *514*
growth factors, 155, 177, 185
growth regulators, 177–78, *178*
Grypania spiralis, 405, *405*
Guam, 536
guanine, 53
guard cells, 413
Gulf of Mexico, *546,* 547, 560, *561,* 564
Gulf Stream, 470–71, *471*
guppies, *194*
gut inhabitant mutualism, 527–28
Gymnodinium, 412
Gymnopilus spectabilis, 423
gymnosperms, 412, *415,* 415–16

habitat(s), 16
adaptation to, 19
defined, 495
and forest management, 499, 500, *500*
global warming and, 579
hyper-, hypo-, and isotonic environments, 92
loss/deterioration of, 420, 500–502, 570
of mammals, 458
in overpopulated areas, 497
habituation, 512
Hacket, Grant, 143
hair color, 208, *208,* 212, 285, *285*
hairpins, 266
Hall, Joseph, 257
Halobacterium, 386
halophiles, *386*
hantavirus, 492
haploid, 162
haploid set, 152, 165
Hardy, Godfrey, 317
Hardy-Weinberg equilibrium, 317
harem, 519, *519*
Hawaiian Islands, 420, 532
Hawaiian silverswords, *532*
health, 51, 53, 399
health risks, online evaluation of, 179
hearing defects, 295
heart (birds), 455
heart attacks, 12, *12,* 101
heart disease
as complex-trait disease, 226
and fish consumption, 11, *11,* 12
and lipid consumption, 51
measuring risk of, 101
and statin use, 100
and trans fats, 29
heat
defined, 37
metabolic, 107
movement of, 107
stored in water, 38
heat energy, 106, 114, 115
heavy metals, 573
height, nutrition and, 208
HeLa cells, *172,* 173, 184, 186, *186*
Helicobacter pylori, 183, 184
Heliconius butterflies, 341, *341*
Hemichordata, 448
hemlocks, 565

hemoglobin, 47
hemophilia, 230
hepatitis B, 184
hepatitis C, 184
Herbert, Rusty, 81
herbivores, 530, 550
hereditary cancers, 179
hereditary disorders, 113
see also genetic disorders
hermaphrodites, 406, 416
herring gulls, 511, *512*
Hershey, Alfred, 242
heterocyclic amines (HCAs), 185, *185*
heterocyst, 385
heterotrophs, 388
defined, 16, 389
fungi, 419
protists, 409, 412
see also consumers
heterozygotes, 195, *196*
heterozygous genotype, 219
hibernation, 119
high-density lipoprotein (HDL), 101
HIV-1 virus, 396
hoatzin, 455
hobbit people, 344
Hoekstra, Hopi, 318, 521, *521*
homeostasis, 16–17, *17*
homeotherms, 455
homeotic genes, 369–70
Homo florensiensis, 344
homologous chromosomes, 158, 162, 218, 349
homologous pairs, 163–68, 193, *194,* 219, 220, *220*
homologous traits, 349, *349,* 372
homologues, 158, 163, 164
Homo neaderthalensis, 256, 285, *285,* 353
homozygotes, 195, 196, *196*
homozygous genotype, 219
honeycreepers, 345, *345*
Hoogland, John, 335
Hooke, Robert, 63, *63*
hormones, 52, 53
cancer risk and, 185
and cell division, 177
defined, 52, 53, 99
and endocrine disrupters, 574
as signaling molecules, 99
hormone therapy, 182
hornworts, 412
horses, 203, 204, *204,* 326, *327,* 520
hosts, 530
hot springs, 550
housekeeping genes, *255,* 255–56
Hubbard Brook Experimental Forest, *561*
Huffington Post, 3, 277
human activity
aquatic biomes affected by, 478–80
biosphere impact of, 465
changes caused by, *see* global change
communities impacted by, 539–41
ecological effects of, 526, 527
and Ecological Footprint, 497, *497*
ecosystems impacted by, 559–62
extinctions caused by, 454
and nitrogen in atmosphere, 557
plant landscapes modified by, 417, 419
and projected mass extinction, 420
river damage from, 403
and selective environment for wildlife, 335, 339, 352
human behavior, *see* animal behavior
human body
adult stem cells in, *294, 295*
body temperature, *17*
cancer defenses in, 180
cell division in, 153–55
chemical composition of, *31*
fermentation in, 138
fungi on, 421
length of DNA in, 253
microbiome of, 379, 398–99
pH in, 41
types of cells in, 15
human gene therapy, 297–99, 305
human genetic disorders, 223–26
human genome, 243, 257, 286
Human Genome Project (HGP), 257, 284, 302
human growth hormone (hGH), 99, 177
human life cycle, *153*
human papillomavirus (HPV), 184, *184,* 186
human population, 500–503, *501*
humans
characteristics shared with other species, 349, *349*
as deuterostomes, 448
ecological impact of, *502*
flowering plants used by, 417
genes of chimpanzees and, 371, *371*
gut inhabitant mutualism in, 528
notochord in, 449
population growth, 500, 501, *501*
studying, 24
humpback whales, *30*
Huntington's disease, *216,* 217, *225,* 233
Hurricane Katrina, 382
hurricanes, 579
Hutton, James, 359
hybridization, *262,* 352
hybrids, 198
hydras, 438
hydrocarbons, 49
hydrogen (H), 30, 32, 110, 111
hydrogenation, 51
hydrogen bonds, 36–39, *38, 39*
hydrogen gas, *34*
hydrogen ion concentration, 41
hydrolytic reaction, 44
hydrophilic molecules, 37
hydrophobic molecules, 37
hydrothermal vents, 483
hydrozoans, 439
Hyman, Flora Jean, 207, *207*
Hymenoptera, 368
hyper- (prefix), 91
hypercholesterolemia, 97, *225*
hyperthermophiles, *386*
hypertonic solutions, 90, 92
hyphae, 421
hypo- (prefix), 91
hypotheses, 6, *9, 10,* 10–11
hypothesis testing, 6, 7, 11, *11*
hypotonic solution, 91
Hyracotherium, 326

Iberian lynx (*Lynx pardinus*), *369*
ice core samples, 578
Iceland, 265
identical twins, 254
identification, among animals, 516
IGF2 gene, 193
Illinois, 315, 316, *316*
im- (prefix), 491
immigration, 491, 492
immortal jellyfish (*turritopsis nutricula*), 434
immune cells, 150
immune system, 183, *183,* 205, 396, 399
immunoglobulin kappa, 304
immunoglobulins, 292
imprinting, 512, *513*
in- (prefix), 491
-in (suffix), 266
incomplete dominance, *203,* 203–4, *204,* 207
incomplete metamorphosis, 447
incurrent siphon (bivalves), 443
independent assortment
of chromosomes, 166–68, *167,* 221
law of, 200–202, *201, 202*
independent variable, 12
individuals, 21, *21,* 22
induced defenses, 530
induced fit model, 114
induced pluripotent stem cells (iPSCs), 293, 294, 296
Industrial Revolution, 310–11
inert elements, 33
infections, 393–97, 525, 542
influenza virus, 396–97, *397*
infrared radiation, 576
inheritance, 190–213
of chromosomal abnormalities, 230–33
extensions of Mendel's laws, 203–10
and genetic crosses, 196
of genetic disorders, 223–26, *225*
genetic principles, 192–96
Mendel's laws of, 199–203
patterns of, 196–99, 224
role of chromosomes in, 218–21, *see also* genetics
of single-gene mutations, 227–30
initiation (transcription stage), 265
initiation (translation stage), 269, *270*
inner cell mass, 293
insecticides, 499, 500
insects, *445–48,* 446–48
and bacteria, 527

crop-damaging, 500
eating, 503
mass extinction of, 368
pollinators, 427
insertion mutations, 271, 272
in situ hybridization, *262*
insulin, 95, 289, 290
interference competition, 533, *533*
Intergovernmental Panel on Climate Change (IPCC), *577*
intermediate filaments, 76, 77, *77*
intermembrane space, 73
International Union for Conservation of Nature (IUCN), 369
interphase, 153–54, *154, 158*
intertidal zone, 482, *482*
intestinal cancers, 184
introduced species, 420, 532, 535–36, *536*
introns, 252, *252, 253,* 266
Inuits, 13, *38*
invariant traits, 193
invasive species, 466, 535–36
Opuntia cacti, 493, *493,* 494
zebra mussels, *464,* 465, 484
inversion, 231, *231*
invertebrates, 20, 432, 438–40
cnidarians and ctenophores, 439–40
sponges, 439
sterols in, 100
in vitro fertilization (IVF), 169, 251, 296, *297*
ionic bonds, 33, *35,* 35–36, *36*
ions, 33
irregular fluctuations (in population), 498
islands, speciation and extinction on, 345, *345*
iso- (prefix), 91
isotonic solution, 92
isotopes, 32–33

jack jumper ant, 219
Japan, 474, *497*
jaws, 450
jellyfish, 435, *435,* 438, 439
John Paul II (pope), *332*
Johnson, Kristine, 81
Jolie, Angelina, 187
Joseph, Stephen, 29, 55
Joyner, Florence Griffith, *275*
J-shaped curve, 493, 498
Juglans nigra, 62
junk DNA, 252
Jurassic, *363, 366, 453*

Kac, Eduardo, 281, 304
Kaibab squirrel, *343*
Kamilo Beach, Hawaii, 479
Kansas, 325
Karnazes, Dean, 106
karyotype, *157,* 157–58
Kashyap, Sonya, 169
katydids, *531*
Kcc2 gene, 277
Kenya, 485, *497*
keratin, 77, 454, 455
keratinocytes, 155
keystone species, 535, *536,* 537, 543
kidney cancer, 186
kidney disease, polycystic, *225*
kilocalories, 118
Kim, Julian, 187
Kim, Stuart, 234
kinetic energy, 106, 107, *107*
kinetochores, 160
King, Mary-Claire, 187
kingdom(s), *381, 404*
defined, 372
of Eukarya, 20
in Linnaean hierarchy, 372, *373*
in scientific classification, 18
Kirsner, Scott, 305
Kivu, Lake, 339
Klinefelter syndrome, 232
koalas, *457*
Koch, Robert, 13
Komodo dragons, *457*
konzo, 144
Kraft Foods, 29, 55
Krebs cycle (citric acid cycle), 127, 139, 140, *140*
Krieger, Lisa M., 234
krill, 553, *553,* 569
Kuwait, *497*

Laccaria, 426
Lacks, Henrietta, *172,* 173, 184, 186
lac operator, 273
lactase, 113
lactase intolerance, 246
lactic acid fermentation, 139, *139*
lactose, 113
lactose intolerance, 113
lactose synthesis, 272–73, *273*
lakes, 478, *480*
acid rain's effects on, 562
and climate, 471
defined, 480
eutrophic, 559
Lamarck, Jean-Baptiste, 311
lamprey, *452*
land transformation, 570–73, *571*
language, 516
Las Vegas, Nevada, *571*
lateral gene transfer, 389, *389*
latitude, 469
law enforcement, DNA fingerprinting in, 286, 287, *287*
law of conservation of energy, 107
law of independent assortment, 200–202, *201, 202*
law of segregation, 199–200
LCT gene, 243, 244, 246
learned behaviors, 508, 511–13
learning, 459
legless lizards, 370, *370*
lemurs, 119
leopards, 542
Lepidoptera, 368
leukemia, *175,* 298
Levodopa, 305
lichens, 424–25, *425*
Liedtke, Wolfgang, 277
life
atomic ingredients of, 30
chemistry of, *see* chemistry of life
defining, 14
elements required for, 23
origin of, 412
water for, 30
lifestyle choices, 183, *184,* 226
light, *128,* 128–29
light energy, 106
color and, *128,* 128–29
in ecosystem energy flow, 109, *109,* 549, 550
photoautotrophs, 388, 389
photoheterotrophs, 388, 390, 391
as principal energy source, 552–53
see also photosynthesis
light microscope, 63, *63*
lightning, nitrogen from, 557
light reactions, 126, 131, *131, 133*
lignin, 414
limpets, 443
lineages, 371
Linnaean hierarchy, 372–73, *373*
Linnaeus, Carolus, 372
lions, 324, 515, 518, *519,* 542
lipases, 113
lipids, 49–53
made in endoplasmic reticulum, 71, *71*
in meats, 56
and metabolism, 110
trans fats, 51, 55
transport of, 71–72, *72*
Listeria monocytogenes, 60, 61
lithoautotrophs, 390
liver, 105, 154
liver cancer, 183, 184
liver cells, 154
liverworts, 412
living organisms
characteristics of, *14,* 14–17
chemical composition of, 30, *see also* chemistry of life
evolution of, *17,* 17–20
see also specific types of organisms
living systems, energy in, 107–8, *108*
lizards, 347
lobe-finned fishes, 450, *452*
lobsters, *439*
locomotion, 436
locus, 218, 222, 223
locusts, *498*
lodgepole-pine forest, *538*
logistic growth (populations), 494–96, *495*
longitude, 469
longitudinal studies, 24
Lorenzo's Oil (movie), 74
lovebirds, *519*
low-density lipoprotein (LDL), 96–97, *97,* 101

lumen, 71
lung cancer, 175, *175,* 185, 186, 330
lung capacity, 141, 143
lungfish (*Neoceratodus fosteri*), 330, *330*
Luo, Shu-Jin, 212
lutein, *130*
Lyell, Charles, 311, 359
Lynch, Rene, 503
lypha, 421
Lysenko, Trofim, 507
-lysis (suffix), 138
lysosomal storage disorders, 74
lysosomes, 72–73, *73*
-lytic (suffix), 138

macroevolution, 358–59, 368–71
macromolecules, 42, 71–73, *73*
macrophages, 61, 97
macular degeneration, 305
Madagascar dwarf lemurs, 119
Madagascar rosy periwinkle (*Catharanthus roseus*), 382
mad cow disease, 272
Magicicada septendecim, 461, *461*
magnolias, 539
major depressive disorder, *226*
malaria, 290, 412
Maldive Islands, *582*
males
  cancer chances for, 174
  chromosomes of, 219, 220
  mating behaviors of, 515–19, *518, 519*
  meiosis in, 153, 169
  sex-linked disorders in, 229, 230
  toxoplasmosis in, 542
malignant cells, *see* cancer cells
malignant melanoma, *175, 225*
Malthus, Thomas, 312, 318
Mammalia, *373*
mammals, 455, 458–59
  adaptive radiation in, *348*
  with cyclopia, *276*
  evolution of, 365
  hair color in, 208, *208,* 212
  whales, 361
mammary glands, 459
manioc, 144
man-of-war fishes (Nomeidae), 529
mantle, 365
mantle cavity, 442
Mapusarus, 368
Marasmius oreades, 423
Marfan syndrome, 206, 207, *207*
Margulis, Lynn, 80
marijuana cigarettes, 185
marine biome, 479, *479,* 481–82, *481–83*
marine birds, 479
marine ecosystems, 550–52, *551,* 572, 576
marine food chains, 584
marine food web, 553, *553*
marine pollution, *470,* 479, 485
marshes, 382, 481, 552
marsupials, 458, 459, *459*
mass, 30
Massachusetts General Hospital, 118
mass extinctions, 349, 366, *367,* 420
Massospora cicadina, 461
maternal homologue, 163, 193
mating behaviors, 515–19, *516*
matrix, 73
matter, 30
mayflies, *457*
MCIR gene, 193
McNeil, Donald G., Jr., 144
meat marinades, 56
meats, 56, 113, 185
mechanical energy, 106
medical imaging, *33,* 33–34
medicine
  antibiotics, 394–95
  defined, 285
  genetic testing, 187, 217, 233, 234
  and Human Genome Project, 257
  and personal genomics, 275
  personalized, 226, 275, 285
  regenerative, 295
megalodons, *450*
meiosis, *151,* 152, 153, *153,* 161–68, *163–67*
  alleles in gametes, 198–99
  at cellular level, *163*
  chromosomes in gametes, 161–63
  crossing-over in, 165, 166, *166*
  defined, 152
  DNA packing with, 254
  genetic variation and, 165–66
  in human life cycle, *153,* 169
  independent assortment of chromosomes in, 166–68, *167*
  meiosis I, 163–65, *164*
  meiosis II, 164, *164*
meiosis I, 163–65, *164*
meiosis II, 163, 164, *164*
meiotic spindle, 165
melanin, 207, *207,* 208, *208*
melanocytes, 212
melanosomes, 212
melatonin, 185
membrane channels, 92
membrane-enclosed organelles, *80*
membrane proteins, *67,* 67–69
membrane pumps, 94
memory loss, statins and, 85, 100
Mendel, Gregor, 19, 192, *192,* 197–99, 206, 218, 262
Mendel's laws of inheritance, 199–210
mental deficits, from konzo, 144
mercury, 573, 574
"mermaid's purses," 167
Merychippus, 326
mesoderm, 435
Mesohippus, 326
mesophyll cells, 135
Mesozoic, *363,* 453–55
messenger RNA (mRNA), 243, 264, *264*
  defined, 243, 264
  hairpins in, 266
  posttranscriptional processing in, 266
  reading of genetic information in, 267
  and regulation of gene expression, 274
  and in situ hybridization, *262*
  and translation, 269, *270*
metabolic disorders, 74
metabolic heat, 107, 549
metabolic pathways, 110, 115–16, *116, 117*
  see also cellular respiration; photosynthesis
metabolic poisons, 144
metabolic rate, 105, 118, *118*
metabolism, 104–19
  anabolism and catabolism, 110, *110*
  ATP in, 110, *111*
  basal metabolic rate, 118, *118*
  defined, 16, 109
  enzymes in, *112,* 112–15, *114, 115*
  exercise and, 105, 118
  and hibernation, 119
  Krebs cycle in, 140
  and laws of thermodynamics, 111–12
  metabolic pathways, 115–16, *116, 117*
  oxidation-reduction reactions in, 110, 111
  role of energy in, 106–9
  and water in cells, 36
metabolites, 118
metamorphosis, 447, *447*
metaphase, 159, *159,* 160
metaphase I and II, *164,* 165, 167
metaphase plate, *159,* 160
metastasis, 176–77, *177*
metazoans, 433
methionine, 267
methylmercury, 573, 574
Mexican free-tailed bats, 494
Mexico, *497*
Mexico City, 581
mice, 398, 399
Michigan, 540, 565
Micrasterias thomasiana, 411
microbes, 14, 379, 383, *384*
microbiology, 383
microbiome, 379, 398–99
Microceratops, 453
microfilaments, 76–78, *77, 78*
microplastics, 479, *479*
microscopes, *63,* 63–64, *64*
microtubules, 76–77, *77,* 160
migration, 5, *6,* 141, 491, 492
milk, 300
millipedes, 448
mimicry, 530, *531*
Mississippi River, 560, 564
mistletoe, 543, *543*
Mitchell, Alanna, 543
mites, 448
mitochondrial diseases, 81
mitochondrion(-a), 73, 74, *75*
  aerobic exercise and, 140
  cellular respiration in, 139, 141

defined, 63, 73, 127
as former prokaryotes, 80
mitosis, 151, 152, *152, 153*
and cytokinesis, *159,* 159–61, *161, 162*
defined, 152
DNA packing with, 254
in human life cycle, *153*
main phases of, *158–59,* 159
sister chromatids in, 157
in stem cells, 153
mitotic divisions, *151,* 152, 153
mitotic spindle, 160
mixotrophs, 412
Moai, *488,* 489
mockingbirds, 333
Molchanova, Natalia, 123
molds, 419
molecule(s), *21*
in biological hierarchy, 20
chemical formulas for, 33
covalent bonds, 33
defined, 20
hydrophilic, 37
hydrophobic, 37
net movement of, 90, 91
nonpolar, 37
organic, 42
polar, 36–37
size of, *65*
Mollusca, 442
mollusks, *442,* 442–44
bivalves, 442–43, *443*
cephalopods, *443,* 443–44
gastropods, 443, *443*
molting, 444
monarch butterflies, 491, *492,* 530
monogamous (term), 518
monogamous behavior, 510, *510,* 519
monohybrids, 199
Mono Lake, California, *2,* 23, *23*
monomers, 42, *42*
monosaccharides, 43–44, *44*
monotremes, 458, 459, *459*
Moore, Charles, 479
Morgan, Thomas Hunt, 221–23, *222*
Morganucodon, 348
Moriarity, Jay, *122,* 123
Morin, Monte, 101
morphological species concept, 340–42
morula, 293
mosquitoes, 313, 314, 323, 580
mosses, 412
most recent common ancestor, 371, 372
motor proteins, 78
mountains, 471, *471,* 472
Mousterian culture, 353
MRSA, 386
multicellular organisms
animals, 431
biological hierarchy in, 20–22, *21*
body size of, *65,* 65–66
cellular connections in, 97–98, *98*
defined, 15, 66
division of labor in, 65–66, *66*
evolution of, 407, 460
mitotic divisions in, 152
origin of, 363–67
reproduction by, 15
somatic cells in, 152
multiple endocrine neoplasia, type 2, *225*
multiple exostoses, *225*
multiple sclerosis, *226*
multipotent (term), 293
multipotent stem cells, 293
multivitamins, 24
murmurations, 514
Murray, Bryan, 565
muscle cells, 66
muscle fibers, 275
muscular dystrophy, 295
mushrooms, 422, 424
musk oxen, *514*
mussels, 443, 454
mutagens, 248
mutations, 165, 166, 178, 179, 195–96
accumulated with age, 180
beneficial, 271
in cancer-critical genes, *178,* 179, 187
within cells, 174
as change in DNA sequence, 248
defined, 178, 195, 248, 313
from DNA replication errors, *248*
evolution and, *313,* 313–14
and genetic disorders, 223–26
misconceptions about, 195
neutral, 196
and protein synthesis, 270–72, *271*
sex-linked inheritance of single-gene mutations, 229–30
single-gene, autosomal inheritance of, 227–29
somatic, 179, 180, 224
white tigers, 212
mutualism, 80, 424–26, *425,* 527–29
Mwihia, Joe, 485
mycelium, 421
Mycena lampadis, 423
mycoplasmas, 64, *386*
mycorrhizae, 424–26, *425*
myosin, 56
myotonic dystrophy, 226
Myriad Genetics, 187
myriapods, *445,* 448

Nabisco, 29
NAD+, 138
NADH, 125–27, 138, 141, *142*
NADP+, 131, 132
NADPH, 125–26, 131
Namibia, 585
NASA, 3, 23
National Audubon Society, 5
National Institutes of Health (NIH), 8, 257
National Restaurant Association, 29
National Science Foundation (NSF), 8
natural selection, *18,* 19, 310–12, *312*
adaptation by, 321–24
and adaptive evolution, 316, 318–21
and age of Earth, 359
and convergent evolution, 351, *351*
defined, 19, 312
in exploitative relationships, 530
sexual selection, 324–25
and speciation, 343
types of, 319, *320*
Neandertals, *see Homo neaderthalensis*
necrosis, 155
necrotizing fasciitis, 393
nectar, 427
negative growth regulators, 177–78, *178*
Nematoda, 444
Nemoria arizonaria, 322, *322*
Neogene, *363, 367*
nephridia, 442
nerve cells, 155
nerve cords, 441
net movement (term), 90, 91
net primary productivity (NPP), *551, 552*
controlled by humans, 572
defined, 551
and Earth's chemistry, 573
humans' impact on, 559–60
neurofibromatosis, type 2, *225*
neurons, *84,* 100, *100,* 155, *244*
neurotransmitters, 99
neutral mutations, 196
neutrons, 30
neutrophilis, 155
Newton, Isaac, 129
New York City, 562, 563, *563*
New York City Board of Health, *28,* 29
niche partitioning, 533, *535*
niches, 532–33
Nicholas II, czar of Russia, *190,* 191, 211
Nigeria, *497*
night shift work, cancer and, 185
Nile perch, 335, 339, 352, *352*
nitrates, 557, 565
nitrites, 185
nitrogen (N), 30, 32, 565
nitrogen cycle, 556–57, *557, 558,* 565
nitrogen fixation, 392, 556
nitrogen gas, 35
nitrogenous bases, 53, 245
Nixon, Richard, 181
Noctiluca scintillans, 410
node (on evolutionary tree, 371, 372, *372*
Noggin gene, *262*
noncoding DNA, *252,* 252–53, *253*
noncoding genes, 262
nonpolar molecules, 37
nonprotein RNA, 252
North America, 473–75, 538, *539*
North American opossum, 458
North Cascades National Park, Washington, *22*
Nostoc, 385, *387*

notochord, 449
nuclear envelope, 70, *70,* 71, 406
nuclear pores, 70, *70*
nucleic acids, 53–55, *54*
nucleoli, 70–71
nucleotides, 53–55, *54, 55, 245*
  base pairs, 246
  defined, 53
  in DNA vs. RNA, 263, *263*
  sequence of, 243
nucleus (of atoms), 30, 31
nucleus (of cells), *70*
  defined, 16, 63
  eukaryotes, *70,* 70–71, 406
  prokaryotes, 385, *385*
nudibranchs, *432,* 443
nuptial gifts, 518, *519*
nutrient cycles, 549, 554–59, *555*
  defined, 554
  humans' impact on, 560, *561,* 562, 573–76
nutrient flow, 549, *549, 552*
nutrient recycling, 391
nutrients, 549, 553, 554

obesity, 185, 277, 399, 574
objectivity, 7
observation, 9, *9*
observational studies, 11, *11*
ocean currents, 470–71, *471,* 478, *478*
oceanic region, 482, 483
oceans
  acidification of, 578
  carbon stored in, 556
  climate and, 471, 478
  net primary productivity in, *551,* 552
  pH of, 578
  photosynthesis in, 569
  salt content of, 478
  sulfur compounds from, 558
  trash in, 479, *479,* 485
  warming of, 584
Ocean Sole, 485
ochre sea star, *449*
octopi, 442, 443, *443*
offspring, 15, 16, 150
  see also reproduction
Ogallala aquifer, 581
Ohman, Mark, 584
oil production, 581, *581*
oils, 50, 51
oil spills, *546,* 547, 564
oldfield mice (*Peromyscus polionotus*), 318, *319,* 521, *521*
olinguito, *458*
Olympic games, *148,* 149, 168
omega-3 fatty acids, 51
omnivores, 550
oncogenes, 179
On the Origin of Species (Charles Darwin), 312
oocytes, 169
operator, 273
operculum, 450–51
operon, 272, 273
Orange gene, 193
orchids, 323, *418*
order, 18, 372, *373*
Ordovician, *362*
Oregon Health & Science University, 81
Oreo cookies, 29, 55
organelles, *62,* 63, *65,* 68, 74
organic molecules, 42
organisms (term), 42, 466–68, *467*
organs, 20, 21, *21,* 436
organ systems, 21, *21,* 22, 436–37
-osis (suffix), 96
Oskin, Becky, 565
osmoregulation, 92
osmosis, 89–92, *91*
osteoclasts, *244*
ostrich eggs, *64*
Oto, Misty, *216,* 217, 233
Ötzi, the Iceman, *240,* 241, 256, *256*
ovarian cancer, *175*
ovaries, 153
overfishing, 572
overgrazing, 534, *540,* 541, *541*
ovulation, 169
oxidation, 110
oxidation-reduction reactions (redox reactions), 110–12, *111*
oxidative phosphorylation, 127, 140–41, *142*
oxygen (O), 30
  atomic mass number of, 32
  in cellular respiration, 111
  and history of life, 362, 363, *363*
  from light reactions, *133*
  in oxidative phosphorylation, 140–41
  in photosynthesis, 75, 111
  in redox reactions, 110
  simple diffusion of, 89
oxygen gas, *34*
ozone depletion, 579
ozone layer, 574

Pacific Fertility Center, 81
Pacific Ocean garbage patch, 479, *479*
packaging proteins, 253, 254
Paine, Robert, 535
Pakiceptus, 361, *361*
Paleogene, *363, 367*
Paleozoic, *362–63*
Palmer, Roxanne, 212
pampas, 474
pancreatic cancer, 186
pandas, *490,* 553, *553*
Pangaea, 265, 330, *330, 366*
Panthera, 373
Panthera tigris, 373, *373*
papaya, genetically modified, 288, *288*
Pap smear, 184
paralytic shellfish poisoning, 412
Paramecium, 78, 90, *91,* 92
Paramecium caudatum, 62, 66, 76, 495
parasites, 61
  defined, 530
  flatworms, 440, 441
  fungi, 419
  mistletoe, 543, *543*
  Toxoplasma gondii, 524, 525, 542
parasitic fungi, 42, 421, 424
parasitism, 61
parent cells, 150, 152, 157
  see also cell division
Parkinson's disease, *226,* 295, 305
passive carrier proteins, 94, *94*
passive transport, 87, *88, 93, 94*
paternal homologue, 163, 193
pathogens, 393, 412, 530
PAX6, 375
peacocks, *325,* 517
pea plant, *558*
peas, Mendel's work with, 197–98, *198*
pectinases, 113
pedigree, 224
peer-reviewed publications, 7, *7*
Penicillium, 421
Penicillium camembertii, 62
Penicillium notatum, 394, *394*
People's Republic of China, *148,* 149, 168
peppered moth, 320, *320*
peptide bonds, 46, *47*
periwinkles, 443
permafrost, 473
Permian, *363,* 365, 422
Permian extinction, 366
peroxisomes, 74
persistent organic pollutants (POPs), 573–74
personal genomics, 275
personalized medicine, 226, 275, 285
Peru, 477
pesticides, *31,* 313–14
  see also DDT
p53 gene, 180, 181
P generation, 196, *196*
pH
  in human body, 41
  of oceans, 578
  in specialized organelles, 68
phagocytes, 155
phagocytosis, *96,* 97
pharmaceutical products
  from fungi, 424
  genetically engineered, 289, 290
  from plants, 382
  from rainforest species, *477*
  and stem cell technology, 295
pharyngeal pouches, 449, 450
phenotype(s)
  color, *194*
  controlled by more than one gene, 203
  defined, 193, *196*
  and dominant alleles, 195
  genes' effect on, 262
  genetic information for, *243,* 243–44
  of genetic traits, *193*
  genotype and, 194, 195, *195*

and law of independent assortment, 200–202, *201, 202*
Mendel's work with, 197–98, *198*
phenylalanine hydroxylase (PAH), 113
phenylketonuria (PKU), 113, *225*
pheomelanin, 212
pheromones, 254, 515, 516, *516*
phloem, 414, 415
phosphate group, 53
phosphodiester bonds, 53
phospholipid bilayer, 50, *52*, 66–68, 89
phospholipids, 50, *52*
phosphorus, for cell division, 23
phosphorus cycle, 555, 558, 559, *559*
phosphorylation, 141
see also oxidative phosphorylation
photoautotrophs, 388, 389
photodamage, *130*
photoheterotrophs, 388, 390, 391
photolysis, 132
photons, 106, 128, 129
photorespiration, 135, *135*
photosynthesis, *109*, 128–37
ATP production in, 110
in cacti, 134
Calvin cycle reactions in, 132, *133*, 134, *134*
in carbon cycle, 555–56, *556*
cellular respiration and, 124
chloroplasts in, 129, *129*, 131
defined, 16, 75, 111, 126
in ecosystem energy flow, 109, *109*
electron energizing in, 131, 132
energy capture through, 551
evolution of, 391
generation of energy carriers in, 131, *131*
light energy in, 126, *126*
by phytoplankton, 569, 584
in plants, 412
stages of, *127*
types of, 134–37
photosynthetic bacteria, *390*
photosystem, 131, 132
photosystem I, 131, 132
photosystem II, 131, 132
pH scale, 40–41
phthalates, 574
phylogenetics, 371–73
phylum(-a), 18, 364, 372, *373*
physical activity, cancer risk and, 185–86
physiological state, cancer risk and, 185–86
Phytophthora, 411
phytoplankton, 409, *409*, 552, 569, 584
pigments, in plants, 130, *130*
pigs, 520
pikas, *580*
pili, 385
Pilobolus, 423
pinocytosis, 96
pipevine swallowtail caterpillars, 511
Piscidia piscipula, 144
placebo, 13
placebo effect, 13
placenta, 459
plankton, *411*, 431, 552, 553, *553*
defined, 409
in lakes, 480
in oceans, 481
Plantae, 20, *20, 404*, 413–19
adaptation to land, 413
angiosperms, *416*, 416–17
defined, 20, 413
gymnosperms, *415*, 415–16
lignin and height of, 413–14
significance of, 417, 419
vascular systems, 414–15
see also plants
plantain weed (*Plantago major*), *496*
plant cells
cytokinesis in, 161, *162*
in hypotonic solutions, 92
organization of, *69*
plasmodesmata, 98, *98*
size of, 64
walls of, 68
plantimals, 281, 304
plants
adaptive radiation in, 349
asexual reproduction in, 15
and atmospheric carbon dioxide, 575–76, *576*
behavior in, 16
biological hierarchy in, 20, 22
carbon acquisition in, 556
cell division in, *162*
chloroplasts in, 74–75, *75*
of Devonian, 364
energy capture by, *16*, 550, 551
fertilizers for, 90
in freshwater biome, 480, 481
fungal pests of, 424
genetically modified, *280*, 281, 288, *288*, 289, 299–300, 304
in hypertonic environments, 90
induced defenses of, 530
light sources for, 129
in marine biome, 481, 482
as multicellular, 63
nectar of, 427
organization of plant body, *414*
photosynthesis in, 75, *75*, 108, 126, *see also* photosynthesis
pigments in, 130
plantimals, 281, 304
prescription drugs from, 382
producers, 16
and river meanders, 403, 426
somatic cells of, 152
in terrestrial biomes, 471, *see also individual biomes*
vacuoles in, 73, *73*
see also Plantae
plant sterols, 100
plasma (blood), 92
plasma membrane, 66–68, 86–97
cell junctions in, 97–98, *98*
defined, 15, 50, 63
function of, 86–89, *86–89*
osmosis, 89–92, *91*
phospholipids in, 50, *52*
transport across, 92–97, *93, 94*, *see also* cell transport
plasmids, 283–84, *289*, 389
plasmodesmata, 98, *98*
Plasmodium falciparum, 61, 412
plasmolysis, 90
plate tectonics, 330, 365–66, *366–67*
platypus, 458
pleiotropy, *206*, 206–7, *207*
pluripotent, 293
pneumatocysts, 407, *407*
Pneumocystis carinii, 424
Pocono Mountains, Pennsylvania, *474*
point mutations, 271
poison dart frog, *531*
poison ivy, 580
poisons, respiratory (metabolic), 144
polar bears (*Ursus maritimus*), 351, *568*, 569, 585
polar compounds, 37
polar molecules, 36–37
pollen, 415–17
pollination, 416–17, 427
pollinator mutualism, 528
pollinators, 528
pollution
biosphere changes from, 574
and ecosystem services, 563
and human population growth, 502
large-scale, 570
in overpopulated areas, 497
watershed, 563
water transformation by, 572
poly-A tail, 266
polychaet marine annelids, 442
polychlorinated biphenyls (PCBs), 573–74
polycistronic mRNA, 273
polycyclic aromatic hydrocarbons (PAHs), 185
polycystic kidney disease, *225*
polygamous (term), 518
polygamous mating systems, 519
polygenic traits, 208–10, *209, 210*
polymerase chain reaction (PCR), 283, 287, 302–3, *303*
polymers, 42, *42*, 43, 68
polynucleotides, 53, 243, 245
polypeptides, 46–47, *48*
polyploidy, 346, 347
polyposis of the colon, *225*
polyps (cniderians), 439
polyps, colon, 180, 181, *181*
polysaccharides, 44, *45*, 68
population(s)
in biological hierarchy, 21, *21*
defined, 18, 22, *467*, 490
evolution of, 18, 19
genetic variation within, 165, 166
of introduced species, 532
testing for evolution in, 317

population cycle, 531, *531*
population density, 490, 491, 496, 498
population ecology, 488–504
applications of, 498–500
changes in population size, 491–92
defined, 490
exponential growth, 492–94, *493*
growth-limiting factors, 496, 498
logistic growth, 494–96, *495*
population growth
on Easter Island, 501, 502
exponential, 492–94, *493*
human, 500–503, *501*
irregular fluctuations in, *498,* 498–99
limiting factors for, 496, 498
logistic, 494–96, *495*
as threat to biodiversity, 420
population size
changes in, 491–92, *492*
defined, 490
determining, 491
irregular fluctuations in, *498,* 498–99
limits on, 494–98
Porifera, 438, 439
positive growth regulators, 177, 178, *178*
posttranscriptional processing, 266
postzygotic barriers, 341, *341*
potato blight, *411,* 412
potential energy, 106, 107, *107*
prairies, 474, 475, *475*
Precambrian, *362,* 363
precipitation, 469, 471, *471,* 472
see also individual biomes
predators, 530
predictions, *9,* 10
preimplantation genetic diagnosis (PGD), 251
prenatal genetic screening, 251
prey, 514, 530–32
prezygotic barriers, 341, *341*
prickly pear cactus (*Opuntia*), 493, *493,* 494, 526
primary consumers, 553
primary structure (polypeptides), 46, *48*
primary succession, *537, 537*
primary tumor, 176
Primula floribunda, 331
Primula kewensis, 331, *331*
Primula verticillata, 331
Principles of Geology (Charles Lyell), 311
probability, 202–3
problem solving, 512
processed meats, 185
process of science, 5, 9–13
see also scientific method
producers, *16*
ATP production in, 110
cellular respiration and, 126
defined, 16, 549
in defining ecosystems, 550
and ecosystem energy flow, 108, 109, *109,* 549–51, *550,* 553
energy for, 108
productivity of, 469
protists, 409
see also autotrophs
productivity, 469
see also net primary productivity (NPP)
products (of chemical reactions), 39, 112
programmed cell death (PCD), 155, 177
prokaryotes, 383–95, *386–87*
and antibiotic use, *394,* 394–98
asexual reproduction of, 383, 388
biological success of, 383–84
biosphere roles of, 391–92
cell division in, *151*
chloroplasts and mitochondria from, 80
defined, 20, 68, 381
diseases caused by, 392–93
DNA capture by, 388, 389
economic value of, 392
extinction of, 363
as first life forms, 362
genome of, 249–51, *250, 252*
in human biome, 398
metabolic diversity of, 389–91, *390*
movement of, 78, *79*
nutrition modes among, *390*
organization of, *69*
reproduction in, 151, *151*
as single-celled organisms, 63
size of, 64
social behavior of, 385, 388, *388*
structural plan of, 384–85, *385–87*
termination of transcription in, 265, 266
see also archaeans; bacteria
prokaryotic cells, 68, *69, 385*
prokaryotic flagella, 78, *79*
prokaryotics, 68, *69*
prometaphase, 160
promiscuous (term), 518
promiscuous mating systems, 519
pronghorn, 18, 19, *19,* 531
prophase, *158,* 159, 160
prophase I and II, 164, *164,* 165
prostate cancer, 175, *175,* 185–87
proteases, 113
protection mutualism, 528
protein(s), 45–49
adhesion, 67
carrier, *94,* 94–95, *95*
cell cycle regulatory, 155, 156
channel, 92–93
damaged, 272
defined, 45
DNA as, 242
enzymes as, 112
from gene mutations, 178, 179
genes' manufacture of, 192
growth factors, 177
insects as source of, 503
as macromolecules, 43
made in endoplasmic reticulum, 71, *71*
in meats, 56
names of, 266
in names of genes, 193
in phenotype generation, 243
in plasma membrane, *67,* 67–69
receptor, 67, 96, 99, *99*
regulation of, 274
regulatory, 272
structural levels of, 46–47, *48*
transport, 67, 71–72, *72,* 86, 92
tumor markers, 175
protein-coding genes, 262
protein synthesis, 270–72, *271*
Protista, 20, *20,* 383, 398, *404,* 407–12
protists, 63, *410–11*
autotrophic, 409
cell walls of, 68
evolutionary history of, 408
heterotrophic, 409, 412
in hypotonic environments, 92
movement of, 78
pathogens, 412
photosynthesis in, *see* photosynthesis
single-cell, 408–9
size of, 408–9
see also algae; protozoans
proton gradient, 132, 141
protons, 30
proto-oncogenes, 179
protostomes, 440–50
annelids, *441,* 441–42
arthropods, 444–48
chordates, *450,* 450–51
defined, 440
deuterostomes, 448–50
ecdysozoans, 444
embryo development in, 435
mollusks, 442–44
phyla of, 432
rotifers and flatworms, 440–41
vertebrates, 450–59
protozoans, 379, 408, 410
pseudocoelomate animals, 437, *438*
pseudopodia, 408
pseudoscience, 10
Psilocybe, 423
Psychrobacter, 386
psychrophiles, *386*
public-funded research, 8
puffball mushrooms, 494, *494*
Punnett square, 199, *199,* 200, 204
pygmy marosets, *456*
Pyrodictium abyssi, 386
pyruvate, 127, 138, 139
pythons, *349*

Quaternary, *363, 367*
quaternary structure (polypeptides), 47, *48*
quinine, 382
quorum sensing, 385, *388*

rabbits, *468*
Rabin, Roni Caryn, 24
Radcliffe, Daniel, *200*

Radcliffe, Paula, *275*
radi- (prefix), 360
radial symmetry, 436
radiation poisoning, 249
radiation therapy, 182
radioisotopes, 32–34, *33,* 359
radula (gastropods), 443
Rafflesia arnoldii, 418
rain, acid, 560, 562, *562*
rainfall patterns, 579
  see also precipitation
rainforest destruction, 454, 560
rainforests, 473, 474, 527
  see also tropical rainforests
rain shadow, 471, *471*
random fertilization, 221
randomization, 12
Rapanui people, *488,* 501–2
rats, 511, 512, *512, 524,* 525, 542
ray-finned fishes, 450, *452*
reactants, 39, 112
reaction center, 131
reading frame, 268
receptor-mediated endocytosis, 96, *96*
receptor proteins, 67, 99, 177
receptors, 96, 99, *99*
recessive (term), 195
recessive alleles, 195, *196*
recessive genetic disorders, *227,* 227–28
recombinant chromosomes, 166, *166*
recombinant DNA, *283,* 283–84, 289
recombinant proteins, *284*
recycling, 485
red blood cells, 66, 155
Redfield, Rosie, 23
red kangaroos, 468
Red List, 369, 585
redox reactions, 110
red-pine forests, 540
red tides, *412*
reduction, 110
reduction division, 165
red-winged blackbirds, 516–17, *517*
redwoods, 414
reforestation, *572*
regenerative medicine, 295
regulatory DNA, *253,* 272
regulatory proteins, 272
reindeer, 496, *496,* 579
relative species abundance, 526
releaser, 511
Relman, David, 399
renewable energy, 583
rennin, 113
repressor proteins, 273, *273*
reproduction
  of angiosperms, *416*
  asexual, *see* asexual reproduction
  defined, 15
  differential, *18,* 312
  and endocrine disrupters, 574
  in fungi, 421–22, *422*
  in mammals, 458–259
  and natural selection, 324
  in plants, 412
  seasonal, 461
  sexual, *see* sexual reproduction
  via DNA, 15, 16
reproductive cloning, *291,* 291–92
reproductive isolation, 342, *342,* 343
reptiles, 119, 365, 452–54
research, 8
reservoir, 554
respiration, 108
  in carbon cycle, 555–56, *556*
  and static apnea, 123, 143
  whole-body vs. cellular, 127, *see also* cellular respiration
respiratory poisons, 144
respiratory system, 455
resting metabolic rate (RMR), 118
restriction enzymes, 283, 300–301, *301*
resurrection fern (*Pleopeltis polypodioides*), 415
retinitis pigmentosa, *225*
retinoblastoma, *225*
retroviruses, 396
Rhizobium, 387, 392
ribosomal RNA (rRNA), 264, *264*
ribosomes, 63, 243
Rim Fire, *23*
ring species, 346, *346*
ring-tailed lemur, *508*
Riparo di Mezzena, 353
rivers, 478, *480*
  defined, 480
  meandering, *402,* 403, 426, *426*
  nutrient flow in, 552
RNA (ribonucleic acid), 70, 243
  defined, 53, 243
  as enzyme, 112
  genetic information carried by, *262,* 262–64, *263*
  messenger, *see* messenger RNA (mRNA)
  nonprotein, 252
  ribosomal, 264, *264*
  structure of, 262–64, *263*
  transfer, 264, *264, 269,* 269–70
  types of, 264
RNA interference (RNAi), 298–99
RNA-only genes, 262
RNA polymerase, 265, *266*
RNA splicing, 267
rockweed, 406, *407*
rod and coccal bacteria, *386*
Rodhocetus, 361, *361*
Roos, Andrew, 234
Roos, Thomas, 234
roots, plant, 415
Ross Ice Shelf, Antarctica, *356,* 357
rotenone, 144
rotifers, 440–41, *441*
rough ER, 71
rubisco, 132, 135
Russia, 472–74
rusts, 424

Saccharomyces cerevisiae, 424
salamanders, 346, *346*
salmon, *288,* 289, 299
Salmonella, 393
Salmonella typhimurium, 62
salsify, 331
salts, 33, *36, 37,* 114
Santus, Rex, 187
saturated fat, 185
saturated fatty acids, 49, *49*
savanna, 474, 585
scales, fish, 450
Scandinavia, 473, 562
scanning electron microscope (SEM), 64
Schanzkowska, Franziska, 191, 211
Scherer, Stephen, 257
science
  defined, 5
  facts and theories in, 13–14
  nature of, 4–7
  process of, 5–13
  statewide ballot measures on issues of, *8*
science journals, *7*
scientific fact, 13
scientific hypotheses, 6, *9, 10,* 10–11
  see also hypothesis testing
scientific method, 5, 7, *7,* 8, *9*
scientific name, 372
scientific theory, 13
scientific thinking, 4
scorpions, 448
Scotland, 533, *533*
scrub jays, 514, *514*
scyphozoans, 439
sea anemones, 436, 439, 543
sea lettuces, 168
sea-level rise, 582, *582*
sea slugs, *432,* 443
sea squirts, *432,* 448, 449
sea star (*Pisaster ochraceus*), 448, 449, 535, *536*
seawater, microorganisms in, 431
seaweeds, 409
secondary consumers, 553
secondary oocytes, 169
secondary structure (polypeptides), 46, *48*
secondary succession, 537, *538*
secondary tumors, 176
second law of thermodynamics, 107–8, *108,* 112
sedimentary cycle, 555, 558, 559, *559*
seed coat, 416
seed dispersal mutualism, 528
seed fern, *360*
seed masting, 461
seeds, 415–17, *417*
segments (animals), 438, *439*
segregation, law of, 199–200
selective permeability, 67, 86, *87*
Semibalanus balonoides, 533, *533*
semiconservative replication, 247, *247*
sensory perception, *20*

septum (fungi), 421
sex chromosomes, 162
autosomes vs., 219, *219*
changes in number of, 232
changes in structure of, 231
defined, 158, 219
X and Y, 158, 162, 219–20, *220,* 229
sex hormones, 52, 53
sex-linked genes, 229
sex-linked inheritance, *229,* 229–30, *230*
sexual behaviors, genetic basis of, 510, *510*
sexual dimorphism, 324
sexual reproduction, *15,* 150–52, *151, 153*
defined, 15, 150
in eukaryotes, 406, *407*
genetic diversity from, 150–51
mating behaviors in animals, 515–19, *517, 518, 519*
meiosis, *151,* 152, 153, *153,* 161–67, *163–67*
in plants, 412
sexual selection, 324–25, 517
shared derived traits, 372
sharks, *20,* 167, 351, *351*
Shar-Peis, *192,* 193
Shaw, Rosie, *216,* 217
shellfish, 442–43, *443*
Shepard Glacier, *579*
shh gene, 276
short tandem repeats (STRs), 286, 287
shrimp, 528, *528*
Shubin, Neil, 375
Siamese cats, 208, *208*
sickle-cell anemia, *225,* 226
Sierra Nevada (North America), 471
Siesta, Tom, 123
signaling molecules, 99, *99*
signal transduction pathways, 99, 177
Silene steophylla, 474
silicon (Si), 30
silkworm moths, 515
Silurian, *362*
silver foxes, *506,* 507, 520
simple diffusion, 89, *93*
simple eyes, 446
single-blind experiments, 12–13
single-celled organisms, 361–63, 408–9, 421
see also specific organisms
single-gene disorders, *225*
single-gene mutations
autosomal inheritance of, *227,* 227–29, *228*
sex-linked inheritance of, *229,* 229–30, *230*
sister chromatids, 157–61
defined, 157
in meiosis II stage, 164
in meiosis I stage, 163, 164
in mitosis, 157
separating, 165
skeletal muscles, 140, 275
skeleton (term), 77
skin, cell division in, *150*
skin cancer, 177, 249, 250
skin color, 207–9, *209*
skunks, *33, 34*
SLC45A3, 212
slime molds, 409
slow-twitch fibers, 275
slugs, 442, 443
Smith, Alan, 305
smoking, cancer and, 185
Smoothened protein receptor, 276
smooth ER, 71
smuts, 424
snails, 442, 443
Snake River valley, *548*
snowy owls, *519*
social behavior
in animals, 513–15
defined, 513
of prokaryotes, 385, 388, *388*
sodium-potassium pump, 95
solar radiation, 469, *469,* 551, 576
see also light energy
solutes, 37, 91
solution, 37
solvent, 37
somatic cells
chromosomes in, 157–58, 161–63
defined, 152
genes in, *194*
somatic mutations, 179, 180, 224
somatic stem cells, 293
see also adult stem cells
-some (suffix), 220
-somy (suffix), 231
sonic hedgehog *gene,* 276
Sonoran Desert, *476*
soredia, 425
sound energy, 107
South Africa, 475
South America, 474, 477, 540
Southeast Asia, 540
soybeans, genetically modified, 288, *288*
space-filling model (molecules), 35, *35*
spacer DNA, 252, *252, 253*
space shuttle, *40*
spadefoot toad, *452*
specialization, cell, 66, *66,* 153, 407, *408,* 433
speciation, 342–47, *345,* 351–52
species, 338–54
adaptive radiations, 348–49
ages of, 360
ancestral, 348
climate change effects on, 579–80
defined, 18, 372
in domains of life, 19–20, *20*
ecosystem health and number of, 382
endangered, 292, 345, *345,* 499, *499*
evolution of, 17, 18, 349–50
genetic changes within, 328–30, 329
interactions in communities, *526,* 527–34
introduced, 420
invasive, *464,* 465, 466, 484, 493, *493,* 494, 535–36
on islands, 532
keystone, 535, *536,* 537, 543
in Linnaean hierarchy, 372, *373*
morphological distinctions among, 340–42
new, 19, 331, *331, see also* speciation
number of, 358
in rainforests, 527
reproductive isolation of, 341–42
ring, 346, *346*
speciation, 342–47, 351–52
threatened, 369, 585, *585*
unity/diversity of, 349–50
species diversity, 349–50, 381, 534–36
species interactions, 527–34
species richness, 526
sperm, 15, 152, 162, *165*
S phase, 154
spicules, 434
spiders, 448
spinocerebellar ataxia, *225*
spirilla, 384
spleen, 155
sponges, *430,* 433–34, *434,* 438, 439, 460, *460*
spores, 90, 168, 388, 421–22, *422*
sports, *ACTN3* alleles and, 275, *275*
spotted owl, 499, 500, *500*
spotted salamander, *410*
spurious correlations, 11
squamous cell skin cancer, 177
squid, 442
squirting cucumber (*Ecballium elaterium*), 417
SRY gene, 220
S-shaped curve, 495, 498
St. Paul Island, 496, *496*
stabilizing selection, *320,* 321, *321*
Stalin, Joseph, 507
Stanford University, 234
staphylococci, 398
Staphylococcus aureus, 398
starches, 43, 44, *45*
start codon, 267
static apnea, *122,* 123, 143
statins, 85, 100
statistics, 11
stem cells, 153, 292–97
adult, 293, *294,* 295, *295*
controversy over, 296, *297*
defined, 292
embryonic, 293, *294,* 296, *297*
induced pluripotent, 293, 294, 296
mitosis in, 153
as source of new cells, 292–93
stem cell technologies, 295, *296*
Stentor coeruleus, 411
steppe, 474
sterols (steroids), 52–53, *53*
stimuli, 508
stinkhorn mushroom (*Phallus impudicus*), 423
stomach, pH in, 41
stomach cancer, *179,* 184
stomata, 129, 413
stop codons, 267
strawberries, 346

streams, 478, 552
Streptococcus mutans, 398
Streptococcus pyogenes, 393
stroke, 226, *226,* 295
stroma, 129
stromatolites, 361, 362, *362*
structural DNA, *253*
structural formulas, *35*
sturgeon, 163
subalpine ecosystem, 22, *22*
substitution mutation, 271
substrates, 113–16
succession, 537, *537, 538*
sucrose, 44, 134
sugar beets, genetically modified, 288, *288*
sugarcane, 136
sugars, 43–44, *44*
  in ATP production, 137
  from Calvin cycle, 132, 134, *134*
  for cell attachment, 67
  as organic molecules, 43
  source of, *135*
  synthesized in plants, 75
Sulfolobus, 391
sulfur cycle, 557, 558, *558*
sunlight, *see* light energy; solar radiation
supercontinents, 265, 365, 374
surface tension, 39, *39*
survival, group living for, 513–15, *514*
survival of the fittest, 312
sustainable resource use, 581
  in agriculture, 583
  and Easter Island, 489
  and Ecological Footprint, 497, *497*
  and human population growth, 502
swamps, 481, 552
sweat glands, 458
Sweden, *497*
swim bladders, 451
sycamore trees, 351
symbiosis, 61, 80, 379, 424, 527, 528
sympatric speciation, 346, *347*
synaptic cleft, 99
synchronized reproduction, 461
system (term), 106

taiga, 473
tapeworms, 440
tapioca, 144
tardigrades, 444
target cell, 99
Taxol, 382
taxon (taxa), 373
taxonomy, 371, 373
Tay-Sachs disease, 74, *225,* 233
T-DNA, 304
technology, defined, 4
teeth, 450
telomerase, 182, 186
telomeres, 182, 186
telophase, 159, *159,* 161
telophase I and II, *164*
temperate deciduous forests, 474, *474*
temperatures
  coldest and warmest, 476
  defined, 37
  global rise in, *see* global warming
  in human body, *17*
  and migratory birds' winter range, *6*
  moderated by water, 38
template strand, 265
termination (transcription stage), 265, 266
termination (translation stage), 270, *270*
terminator, 266
terminator genes, in GM plants, 300
termites, *360,* 515, 528
terrestrial biomes, 22, 471–77
  aquatic biomes influenced by, 478
  boreal forest, *473,* 473–74
  chaparral, 475, *475,* 476
  desert, 476, *476*
  grasslands, 474–75, *475*
  temperate deciduous forests, 474, *474*
  tropical forests, 477, *477*
  tundra, 472–73, *473*
terrestrial ecosystems
  carbon acquisition in, 556
  energy capture in, 550–52
  excess nitrogen in, 557
  net primary productivity in, 551, *551*
  sulfur compounds in, 558
territory marking, 508, *508,* 517
tertiary consumers, 553
tertiary structure (polypeptides), 47, *48*
test cross, 221
testes, 153
testosterone, 52, 53, 185
test tube babies, 169
tetrad, 164, *165*
tetrapods, 451–53
Texas, 581
theories, scientific, 13–14
theory (term), 13
thermal energy, 106
thermodynamics, 107, 111–12
Thibodeaux, Damon, *287*
third variable, 11
Thomson's gazelle, 530
threatened species, 369, 585, *585*
three-toed sloth, *456*
thylakoid membrane, 131
thylakoids, 75, 131
thylakoid space, 131
thymine, 53
thyroid gland, 32, *32,* 33
ticks, 448
tigers, 212
tight junctions, 97, *98*
Tinbergen, Niko, 511
tissues
  in animals, 434–35, *435*
  in biological hierarchy, 20, *21*
  defined, 20
  from stem cells, 295
tobacco, cancer and, 185
tobacco smoke, carcinogens in, 184
tomatoes, 101, 300, *300*
tortoises, 453
totipotent, 293
totipotent cells, 293
Toxoplasma gondii, 398, *524,* 525, 542
toxoplasmosis, 525, 542
transcription, *264–66,* 264–67
  defined, 243, 264
  and gene control, 272–73
  regulation of, 274
  stages of, 265
trans fats, 51, 55
transfer RNA (tRNA), 264, *264, 269,* 269–70
transformation, 389
transgene, 287
transgenic organisms, 287
  see also genetically modified organisms (GMOs)
translation, 243, 264, *264, 265,* 269–70, *270*
translocation, 231, *231*
transmission electron microscope (TEM), 64
transport proteins, 67, 86, 92
transport vesicles, 71–72, *72*
  defined, 71
  in endocytosis, *96,* 96–97
  in exocytosis, 95, *95*
transposons, 249, *253*
trastuzumab, 183, 186
treatment group, 12
"tree of life," 19–20, *20*
trees, 415, 461, 585
Triassic, *363*
Triceratops, *326*
Trichomonas vaginalis, 412
trichomoniasis, 412
triglycerides, 50, *50,* 110, 140
trilobite, *360*
triple bonds, 35
trisomy, 231
trisomy 21 (Down syndrome), 231, 232
Triulus cistiodes, 333
-troph (suffix), 388
trophic levels, 553, 554, *554*
tropical ecosystems, 579
tropical forests, 477, *477,* 550, 551
tropical rainforests, *550,* 560, *572,* 573, *573*
truffles, 424
Trut, Lyudmila N., 520
tube feet, *449*
tubulin, 76
tumor cells, 175–78, *176*
tumor markers, 175
tumors
  benign, 175
  defined, 156, 174
  origin of, 175
  primary, 176
  secondary, 176, *177*
  stage designations for, 181
tumor suppressor genes, 179–81

tundra, 472–73, *473, 474*
túngara frogs, 324, *324,* 517
turgor pressure, 73, *73,* 92
Turner syndrome, 232
turtles, 453
23andMe, 234
twins, 254
Type 1 diabetes, 295
Type 2 diabetes, 226, *226*
Tyrannosaurus rex, 453
tyrosinase, 207, 208, 212

Ulva prolifera, 168
Union for International Cancer Control, 226
unipotent (term), 293
unipotent stem cells, 293
United Arab Emirates, *497*
United Kingdom, 81, 497, *497*
United Nations Food and Agriculture Organization, 503
United States
acid rain from, 562
biomes in, 475
Ecological Footprint in, 497, *497,* 502
GMO crops in, 288, 299
human cancers in, 174, *175*
invasive species in, 466
nonnative species in, 466
public-funded research in, 8
science-related ballot measures in, 8, *8*
storms in, 470
wetlands in, 571
unity of life, 17–20, 349–50
University of Texas, 277
unsaturated fatty acids, *49,* 50, 51
up-regulated gene activity, 244
upwelling, 552
uracil, 53
urbanization, species diversity and, 535
Ursidae, 368
U.S. Department of Agriculture (USDA), 8
U.S. Department of Energy (DOE), 8, 257
U.S. Geological Survey, 23
U.S. Navy SEALs, 13
Ussher, James, 310
Utahraptor, 453
uterine cancer, 186
uterus, 459
UV radiation, DNA damage from, *250*

vaccines, genetically engineered, 290, 291
vacuoles, 73, *73*
valence shell, 34
vampire finch, *308,* 309, 333–34, *334*
variables, 11, 12
vascular system (plants), 414–15
vegetables, 416
Velociraptor, 360, 453
Veratrum californicum, 276, *276*
vertebrae, 450
vertebral column, 450
vertebrates, 20, 365, 432, 450–59, *456–57*
amphibians, 451–52
birds, 454–55
cell junctions in, 97–98
cholesterol in, 100
evolution of, 450–51
mammals, 455, 458–59
reptiles, 452–54
very long chain fatty acids (VLFAs), 74
vestigial organs, 349–51, *349, 351*
vestigial traits, 370
Vibrio cholerae, 393
viceroy butterfly, *531*
Victoria, Lake, 339, *347,* 351, 352
Viegas, Jennifer, 353
viral strains (serotypes), 395
Virginian opossum, 458
viruses, *252, 395,* 395–97, *396*
visible light, 129
vision defects, stem cell therapy for, 295
vitamin D, 52
vitrification, 169
Volvox, 76
Volvox carteri, 66, *66*

Wade, Nicholas, 375
Wallace, Alfred, 19, 312, 318
warm-blooded animals, 119
warning coloration, 530
wasps, 323, *323,* 533, *534*
water
adding salt to, 36
aquatic biomes, *21, 22,* 477–83
in cells, 30, 89
chemical formula for, 33
dissociation in, 40
on Earth, 30
in ecosystem energy flow, *109*
electron sharing in, 34–35, *35*
need for, 30
and net primary productivity, 551
osmosis, 89–92, *91*
in photosynthesis, 111, 126
properties of, 36–39
water currents, climate and, 470–71, *471*
water cycle, *572*
water lilies, 417, *418*
water mold, *411,* 412
water pollution, *470,* 479, 485, 563
water transformation, 570–73
Watson, David, 543
Watson, James, *245,* 245–47
wavelength (light), 128, *128,* 129
weather, 468, 469, 579
weaver ants, 322, *322*
weevil, *424*
weight
cancer risk and, 185, 186
and gut bacteria, 399
maintaining, 118
Weinberg, Wilhelm, 317
Weismann, August, 218
Welz, Adam, 585
western meadowlark, *340,* 341, 342
wetlands, 478, *480,* 481
defined, 480
destruction of, 478
humans' impact on, 571
net primary productivity in, *552*
whales, 327, 349, *349, 358,* 358–59, *361,* 458, 529, *529*
whelks, 443
whimbrels, *535*
white blood cells, 66, 97, *244*
white-pine forests, 540
white tigers, 212
wild mustard, *17*
Wilkins, Maurice, 245
Williams, Sarah, 335
Williams, Vanessa, *284*
Wilson, Edward O., 454
wind, climate and, 469–70, *470*
wind pollination, 416
Wolfe-Simon, Felisa, *2,* 23, *23*
wolves, *514, 516*
wood pigeons, 514, *514*
woody plants, 585
woolly mammoths, *4, 292, 348*
World Wildlife Fund, 485
worms, 440–42
Wright, Geraldine, 427

X chromosomes, 158, 162, 219–20, *220,* 229
xeroderma pigmentosum (XP), 249, 250, *250*
X-linked genes, 229, 230
X-SCID, 297, 298, 305
xylem, 415

Y chromosomes, 158, 162, 219–20, *220,* 229
yeasts, 138, 419, 421
Yellowstone National Park, *538*
Yeti crab, *432*
Y-linked genes, 229
Yosemite National Park, *23*
Yssichromis pyrrhocephalus, 352
yucca moth, 528–29, *529*
yucca plant, 528–29, *529*
Yurovsky, Yakov, 191

zebra mussels, *464,* 465, 484
zebras, 326
zoion, 433
zooplankton, 409, 412
zygomycetes, 419
zygotes, 15, 153
in animals, 435
defined, 152, 293
number of chromosomes in, 162
plant, 412
sets of chromosomes in, 163
of sharks, 167